10th Anniversary of Technologies—Recent Advances and Perspectives

10th Anniversary of Technologies—Recent Advances and Perspectives

Guest Editors

Manoj Gupta
Eugene Wong
Gwanggil Jeon

MDPI

Basel • Beijing • Wuhan • Barcelona • Belgrade • Novi Sad • Cluj • Manchester

Guest Editors
Manoj Gupta
Department of Mechanical Engineering
NUS, Singapore
Singapore

Eugene Wong
Singapore Institute of Technology
Singapore
Singapore

Gwanggil Jeon
Department of Embedded Systems Engineering
Incheon National University
Incheon
Korea, South

Editorial Office
MDPI AG
Grosspeteranlage 5
4052 Basel, Switzerland

This is a reprint of the Special Issue, published open access by the journal *Technologies* (ISSN 2227-7080), freely accessible at: www.mdpi.com/journal/technologies/special_issues/10th_Anniversary_Technologies.

For citation purposes, cite each article independently as indicated on the article page online and using the guide below:

Lastname, A.A.; Lastname, B.B. Article Title. *Journal Name* **Year**, *Volume Number*, Page Range.

ISBN 978-3-7258-3084-8 (Hbk)
ISBN 978-3-7258-3083-1 (PDF)
https://doi.org/10.3390/books978-3-7258-3083-1

Contents

About the Editors

Manoj Gupta

Professor Manoj Gupta was a former Head of the Materials Division of the Mechanical Engineering Department and Director designate of Materials Science and Engineering Initiative at NUS, Singapore. He received his Ph.D. at the University of California, Irvine, USA (1992), and postdoctoral research at the University of Alberta, Canada (1992). He is currently among the top 0.6% researchers as per Stanford's List, among the top 1% of scientists of the World Position by The Universal Scientific Education and Research Network, and among the top 1% of scientists as per ResearchGate. Some of his accredited works include the following: (i) disintegrated melt deposition technique; (ii) hybrid microwave sintering technique, an energy-efficient solid-state processing method; and (iii) turning-induced deformation technique to synthesize alloys/micro/nanocomposites. He has published over 700 peer-reviewed journal papers and owns two US patents and two Trade Secrets. His current h-index is 87, with citations greater than 28000 and reads greater than 160,000 (Research Gate). He has also co-authored eight books, published by John Wiley, Springer, and MRF, USA. As a multiple-award winner, he actively collaborates/visits Japan, France, Saudi Arabia, Qatar, China, the USA, and India as a visiting researcher, professor, and chair professor.

Eugene Wong

Associate Professor Eugene Wong is currently the Director of Programmes for the Engineering Cluster at the Singapore Institute of Technology (SIT) where he oversees the management of the engineering programs offered by SIT. Prior to joining SIT, Eugene was with Newcastle University International Singapore from July 2011 to February 2023, progressing from Lecturer to Associate Professor and Director of Undergraduate Studies. He obtained his Ph.D. and B.Eng in Mechanical Engineering with a specialization in Materials Engineering in Design from the National University of Singapore in 2008 and 2003 respectively.

In addition to his role at SIT, Eugene is the Deputy Head of the National Additive Manufacturing Innovation Cluster (NAMIC) Hub at SIT. His research expertise is in the area of additive manufacturing, lightweight composite materials, and microwave processing of materials.

He was a recipient of the NUS President Graduate Fellowship (2006–2007), The Institute of Engineers Singapore (IES) Outstanding Volunteer and Best Committee (2014), and Singapore Armed Forces National Serviceman of the Year (2015). Eugene is also a member of the Engineering Advisory Committee for Ngee Ann Polytechnic.

Gwanggil Jeon

Gwanggil Jeon received his B.S., M.S., and Ph.D. (summa cum laude) degrees from the Department of Electronics and Computer Engineering, Hanyang University, Seoul, Korea, in 2003, 2005, and 2008, respectively. From 2009.09 to 2011.08, he was with the School of Information Technology and Engineering, University of Ottawa, Ottawa, ON, Canada, as a Post-Doctoral Fellow. From 2011.09 to 2012.02, he was with the Graduate School of Science and Technology, Niigata University, Niigata, Japan, as an Assistant Professor. From 2014.12 to 2015.02 and 2015.06 to 2015.07, he was a Visiting Scholar at Centre de Mathématiques et Leurs Applications (CMLA), École Normale Supérieure Paris-Saclay (ENS-Cachan), France.

From 2019 to 2020, he was a Prestigious Visiting Professor at Dipartimento di Informatica, Università degli Studi di Milano Statale, Italy. From 2019 to 2020 and 2023 to 2024, he was a Visiting Professor at Faculdade de Ciência da Computação, Universidade Federal de Uberlândia, Brazil. He is currently a professor at Incheon National University, Incheon. He was a general chair of IEEE SITIS 2023 and served as a workshop chair in numerous conferences.

Dr. Jeon is an Associate Editor of IEEE Transactions on Circuits and Systems for Video Technology (TCSVT), Elsevier Sustainable Cities and Society, IEEE Access, Springer Real-Time Image Processing, Journal of System Architecture, and Wiley Expert Systems.

Dr. Jeon was a recipient of the IEEE Chester Sall Award in 2007, ACM's Distinguished Speaker in 2022, the ETRI Journal Paper Award in 2008, and the Industry-Academic Merit Award by the Ministry of SMEs and Startups of Korea Minister in 2020.

Preface

In 2022, *Technologies* celebrated a landmark achievement: its 10th anniversary as an influential international, peer-reviewed, open-access journal published by MDPI. Since its first issue in 2013, the journal has made significant strides in promoting interdisciplinary research across diverse technological fields, becoming a respected platform for high-impact studies. Indexed in prestigious databases such as ESCI, Inspec, and INSPIRE, Technologies has established itself as a valuable resource for researchers worldwide. Ranked 46th among 170 journals in the "Engineering and Multidisciplinary" category, and holding a Q2 rating in the Journal Citation Indicator (2021), Technologies received its first Impact Factor in 2021, marking a testament to the journal's dedication to quality and relevance in scientific publishing.

To commemorate a decade of contributions to the field, Technologies launched a Special Issue titled "10th Anniversary of Technologies—Recent Advances and Perspectives". This collection welcomed submissions of original research articles and in-depth reviews across a spectrum of emerging and impactful topics, including quantum technologies, innovations in materials processing, construction technologies, environmental and medical technologies, biotechnologies, and advancements in computer and information technologies. The response from the global research community was outstanding, culminating in the publication of 36 exceptional papers that collectively showcase the journal's mission to foster innovation and knowledge sharing in rapidly evolving technological fields.

This reprint brings together the articles featured in the Special Issue, offering readers an insightful overview of current advancements and future perspectives in technology research. Each chapter provides a window into critical research areas, reflecting the expertise and forward-thinking approaches of the authors. We extend our sincere gratitude to the contributors and reviewers whose efforts made this commemorative edition possible. Their dedication has not only highlighted the accomplishments of *Technologies* over the past decade but has also set the stage for continued innovation in the years to come.

As *Technologies* embarks on its second decade, we are excited to continue serving the global scientific community, sharing insights and discoveries that will shape the future of technological progress.

Manoj Gupta, Eugene Wong, and Gwanggil Jeon
Guest Editors

technologies

MDPI

Editorial

Guest Editorial on 10th Anniversary of *Technologies*—Recent Advances and Perspectives

Manoj Gupta [1], Eugene Wong [2] and Gwanggil Jeon [3,*]

[1] Department of Mechanical Engineering, National University of Singapore, Singapore 117576, Singapore; mpegm@nus.edu.sg
[2] Engineering Cluster, Singapore Institute of Technology, 10 Dover Drive, Singapore 138683, Singapore; eugene.wong@singaporetech.edu.sg
[3] Department of Embedded Systems Engineering, Incheon National University, 119 Academy-ro, Yeonsu-gu, Incheon 22012, Republic of Korea
* Correspondence: gjeon@inu.ac.kr

1. Introduction

In 2022, *Technologies* (ISSN: 2227-7080) celebrated its 10th anniversary. This international, peer-reviewed, open access journal is published by MDPI (Basel, Switzerland) and indexed by ESCI, Inspec, and INSPIRE, among others. It ranks 46/170 (Q2) in "Engineering and Multidisciplinary" in the Journal Citation Indicator (2021) and has received its first Impact Factor. The journal released its inaugural issue in 2013 and published its 500th paper in 2021.

To commemorate this milestone, a Special Issue titled "10th Anniversary of Technologies—Recent Advances and Perspectives" was launched. The Special Issue welcomed high-quality original research articles and reviews on topics like quantum technologies, innovations in materials processing, construction technologies, environmental technologies, biotechnologies, medical technologies, and computer and information technologies. Contributors were invited to submit papers on trendy or emerging topics for peer review and possible publication. A total of 36 papers were published in this Special Issue.

check for updates

Citation: Gupta, M.; Wong, E.; Jeon, G. Guest Editorial on 10th Anniversary of *Technologies*—Recent Advances and Perspectives. *Technologies* **2024**, *12*, 177. https://doi.org/10.3390/technologies12100177

Received: 19 September 2024
Accepted: 25 September 2024
Published: 29 September 2024

2. Overview of Contributions

In the contribution by Yeh and Zhu, titled "Forecasting by Combining Chaotic PSO and Automated LSSVR", a novel automatic least square support vector regression (LSSVR) optimization method, using mixed kernel chaotic particle swarm optimization (CPSO), was introduced to tackle regression problems [item 1 in the List of Contributions]. The LSSVR model consisted of the following three steps: chaotic sequence positioning for randomness and ergodicity, binary particle swarm optimization (PSO) for selecting potential input feature combinations, and a chaotic search to refine the input features. These steps were combined to form the CP-LSSVR model. The method was evaluated using datasets from UCI, showing a strong predictive capability and efficient model building with a limited number of features.

The contribution by Brischetto et al., titled "A Layer-Wise Coupled Thermo-Elastic Shell Model for Three-Dimensional Stress Analysis of Functionally Graded Material Structures", presented a coupled 3D thermo-elastic shell model for analyzing thermal stress in one-layered and sandwich plates and shells with functionally graded material (FGM) layers [item 2 in the List of Contributions]. The model combined three-dimensional (3D) equilibrium equations and the Fourier heat conduction equation for spherical shells into four coupled equations. Solved using the exponential matrix method, the model assumed simply supported boundary conditions. Static responses were evaluated in terms of displacements and stresses. The model's accuracy, showing less than 0.5% difference from uncoupled models, was validated for various thickness ratios, geometries, and temperatures. The FGM layers were metallic at the bottom and ceramic at the top.

The contribution by Rehman et al., titled "FogTrust: Fog-Integrated Multi-Leveled Trust Management Mechanism for Internet of Things", introduced FogTrust, which is a lightweight trust management mechanism designed to enhance security in the Internet of Things (IoT) [item 3 in the List of Contributions]. With a multi-layer architecture, it includes edge nodes, a trust agent, and a fog layer. The trust agent acts as an intermediary, calculating trust degrees and transmitting encrypted values to the fog layer for computation, reducing node burden and maintaining a trustworthy environment. FogTrust was tested against various attacks, such as on–off, good mouthing, and bad mouthing. The simulation results showed its effectiveness in assigning low trust degrees to malicious nodes, even with varying percentages of malicious actors in the network.

The contribution by Manjarrez et al., titled "Estimation of Energy Consumption and Flight Time Margin for a UAV Mission Based on Fuzzy Systems", presented a method for estimating the energy required for UAV missions to ensure safety and efficient operation [item 4 in the List of Contributions]. A fuzzy Takagi–Sugeno system, optimized using fuzzy C-means and particle swarm optimization, was implemented to estimate power requirements during mission stages. Additionally, a fuzzy model of a battery's equivalent circuit was used to determine the state of charge, combined with an extended Kalman filter. A methodology was developed to calculate the minimum allowable battery charge and the available flight time margin. A physical experiment with a hexarotor UAV showed a maximum prediction error of 7 s, or 2% of the total mission time.

The contribution by Bozorgpanah et al., titled "Privacy and Explainability: The Effects of Data Protection on Shapley Values", explored the impact of privacy methods on explainability techniques, based on Shapley values in machine learning models [item 5 in the List of Contributions]. Explainability is crucial for understanding model behavior, while privacy is essential for protecting sensitive data. The study examined how privacy-preserving methods influenced Shapley values across four machine learning models. The results suggested that, while some degree of protection could result in the maintenance of valuable Shapley information, linear models were the most affected by privacy measures. The paper highlighted the balance between ensuring data privacy and maintaining effective model explainability, particularly when using Shapley-based methods.

The contribution by Carneiro et al., titled "Simulation Analysis of Signal Conditioning Circuits for Plants' Electrical Signals", discussed plant electrophysiology and presented low-cost signal conditioning circuits for acquiring electrical signals generated by plants in response to environmental stimuli like touch, light, and heat [item 6 in the List of Contributions]. These signals informed the entire plant structure almost instantly. Two specific signal conditioning circuits, depending on the signal type, were detailed, with electrical simulations performed using OrCAD Capture Software. Monte Carlo simulations were also conducted to assess the impact of component variations on circuit accuracy. The results showed that, despite variations, the filters' cut-off frequencies deviated by no more than 4% from the mean, indicating reliable performance.

In the contribution by Tarasov et al., titled "Friction Stir Welding of Ti-6Al-4V Using a Liquid-Cooled Nickel Superalloy Tool", the authors introduced friction stir welding (FSW) of titanium alloy, which was performed using a heat-resistant nickel superalloy tool cooled by circulating water [item 7 in the List of Contributions]. The FSW joints were analyzed for microstructures and mechanical strength. The results showed that the mechanical strength of the welded joints exceeded that of the base metal, demonstrating the effectiveness of liquid cooling in improving the quality and strength of FSW joints in titanium alloys.

The contribution by Lin et al., titled "Modelling the Trust Value for Human Agents Based on Real-Time Human States in Human-Autonomous Teaming Systems", proposed a multi-evidence human trust model to address the challenges of calibrating human trust in human–autonomous teaming (HAT) systems [item 8 in the List of Contributions]. Human trust was influenced by dynamic cognitive states, making it harder to estimate than robotic trust. The model used real-time data from eye trackers, heart rate monitors, and human awareness to assess attention, stress, and perception abilities. Fuzzy reinforcement learning fused these data and handled uncertainty in physiological signals. Simulations showed that the model improved human trust estimation and boosted HAT system efficiency by over 50%. These findings suggested that the model could enhance future HAT systems through real-time adaptation based on human states.

In a competitive global market, construction companies can enhance their competitiveness by selecting qualified personnel for construction engineering manager roles. Traditional selection methods often rely on qualitative techniques, leading to suboptimal decisions. The contribution by Phan and Nguyen, titled "Evaluation Based on the Distance from the Average Solution Approach: A Derivative Model for Evaluating and Selecting a Construction Manager", introduced a new model using the Evaluation Based on the Distance from the Average Solution Approach (EDASA) for selecting construction managers [item 9 in the List of Contributions]. EDASA effectively addresses personnel evaluation by incorporating quantitative criteria, improving decision-making. The research findings demonstrated that EDASA was efficient, particularly when the number of evaluation criteria or alternatives increased, offering a faster and more reliable selection process for construction managers.

The contribution by Yurova et al., titled "Design and Implementation of an Anthropomorphic Robotic Arm Prosthesis", discussed the development of a low-cost, anthropomorphic prosthetic arm with twenty-one degrees of freedom (DOFs), for use in robotic research and education [item 10 in the List of Contributions]. This robotic hand replicated human hand functions, with four degrees of freedom per finger, three for the thumb, and two for hand positioning. It was designed using open-source mechanical components, closely mimicking human hand dimensions and motor parameters. The prosthesis can operate autonomously via battery power and supports various control systems, including computer interfaces, electroencephalographs, and touch gloves. The study highlighted the practical implementation of this artificial hand and its control system.

The contribution by Saeidi Aminabadi et al., titled "An Automatic, Contactless, High-Precision, High-Speed Measurement System to Provide In-Line, As-Molded Three-Dimensional Measurements of a Curved-Shape Injection-Molded Part", presented a high-precision, high-speed, contactless 3D measurement system for inspecting piano-black injection-molded parts [item 11 in the List of Contributions]. The system, capable of ± 5 μm precision and measuring a part in 24 s, operated in real time to enable closed-loop and predictive quality control. A multicolor confocal sensor, along with a linear and cylindrical moving axis, performed measurements on the part's glossy, curved surface. A six DOF robot handled part transfer, while communication was managed via OPC UA protocol. Repeatability tests confirmed an accuracy within ± 5 μm at speeds under 60 mm/s, with increased error (up to ± 10 μm) from fixture and suction effects.

The contribution by Almeida et al., titled "Extraction and Characterization of β-Viginin Protein Hydrolysates from Cowpea Flour as a New Manufacturing Active Ingredient", investigated the antimicrobial potential of cowpea (*Vigna unguiculata* L.) vicilin (7S) protein against antibiotic-resistant pathogens [item 12 in the List of Contributions]. Due to genetic similarities between vicilins from soybean and vicilins from adzuki beans, cowpea was chosen for its high protein content. The beta viginin protein from cowpea was isolated, characterized, and hydrolyzed, both in silico and in vitro, using pepsin and chymotrypsin. The resulting hydrolysate fractions were tested for antimicrobial activity against *Staphylococcus aureus* and *Pseudomonas aeruginosa*, showing promising inhibitory effects. These findings suggested that cowpea-derived peptides could be used as potential innovative agents for combating antibiotic resistance.

The contribution by Algredo-Badillo et al., titled "Analysis and Hardware Architecture on FPGA of a Robust Audio Fingerprinting Method Using SSM", addressed the rise in the unauthorized use of digital media, particularly in audio applications, due to increased digital sharing during the pandemic [item 13 in the List of Contributions]. To secure audio content, acoustic fingerprint technology was employed to identify the unique properties of audio files. The paper presented two hardware architectures for audio fingerprinting, utilizing spectrogram saliency maps (SSM) and a brute-force search. The first system processed 33 maps of 32 × 32 pixels. A second, optimized architecture reduced the map size to 27 × 25 pixels, cutting hardware usage by 75.67%, power consumption by 64.58%, and improving efficiency by 3.19 times through a 22.29% throughput reduction.

The contribution by Hadi et al., titled "Efficient Supervised Machine Learning Network for Non-Intrusive Load Monitoring", addressed the challenge of energy disaggregation, or non-intrusive load monitoring (NILM), which estimated individual appliance energy consumption from a home's overall electrical usage [item 14 in the List of Contributions]. While AI-based models were effective for NILM, they often required significant computational resources, making them impractical for devices with limited capabilities. The study proposed an efficient non-parametric supervised machine learning (ENSML) architecture, designed to reduce size and computational costs while maintaining high performance. The ENSML model allowed for fast inference and accurately predicted appliance-level consumption. The results demonstrated that the model improved energy prediction accuracy in 99% of cases, offering a resource-efficient solution for NILM.

The contribution by Stopka et al., titled "Optimization of the Pick-Up and Delivery Technology in a Selected Company: A Case Study" examined pick-up and delivery processes in a company distributing gastronomic products and suggested improvements for efficiency [item 15 in the List of Contributions]. It began by defining key logistics optimization concepts, followed by an analysis of current delivery routes. The article then applied operations research methods, including the Hungarian method, Vogel approximation method, nearest neighbor method, and the Routin route planner (based on the Greedy algorithm), to minimize the total distance traveled. The findings were technically and economically evaluated, comparing the results of each method. Ultimately, optimized delivery routes were selected, aiming to streamline the company's distribution activities and reduce costs.

The contribution by Gabbar et al., titled "Demonstration of Resilient Microgrid with Real-Time Co-Simulation and Programmable Loads", presented a real-time simulation of a micro energy grid (MEG) system designed for resilience and sustainability, aimed at reducing fossil fuel dependence and enhancing grid stability [item 16 in the List of Contributions]. The system ensured reliable energy flow by backing up renewable energy sources, mitigating peak demand effects, and providing fail-safe operation through redundant control. It integrated real hardware components like inverters, battery chargers, and controllers with emulated components, via OPAL-RT OP4510, for real-time testing. The setup supported modular, expandable, and flexible scenarios, including fault imitations, using various energy sources like solar panels, wind turbines, and energy storage systems to optimize energy management and grid operation.

The multivehicle routing problem (MVRP) is a variation of the vehicle routing problem (VRP), focusing on finding optimal routes for multiple vehicles to serve multiple customers at minimal cost, while tolerating traffic delays. This NP problem is typically solved using metaheuristics like evolutionary algorithms. The contribution by Li et al., titled "Distribution Path Optimization by an Improved Genetic Algorithm Combined with a Divide-and-Conquer Strategy", proposed an optimal distribution path optimization method using a divide-and-conquer strategy inspired by dynamic programming [item 17 in the List of Contributions]. An improved genetic algorithm (GA) was employed, incorporating preprocessing, elitist strategy, two-point crossover, and reversion mutation operators. The improved GA outperformed the simple GA in cost, route feasibility, and efficiency, benefiting logistics, transportation, and manufacturing enterprises for flow-shop scheduling.

The contribution by Gradov, titled "Exciting of Strong Electrostatic Fields and Electromagnetic Resonators at the Plasma Boundary by a Power Electromagnetic Beam", explored the interaction of an electromagnetic beam with the sharp boundary of a dense cold semi-limited plasma under normal wave incidence [item 18 in the List of Contributions]. It revealed the possibility of an electrostatic field forming outside the plasma, with its intensity diminishing, according to a power law with distance from the plasma and beam center. The study also identified the potential to form cavities with reduced electron density, which act as electromagnetic resonators that penetrate deep into the plasma. These cavities can exist in a stable state for extended periods, offering insights into plasma behavior and electromagnetic interactions.

The contribution by Yeh et al., titled "Solving Dual-Channel Supply Chain Pricing Strategy Problem with Multi-Level Programming Based on Improved Simplified Swarm Optimization", addressed the pricing strategy in capital-constrained dual-channel supply chains, where companies sell through both traditional and online third-party platforms [item 19 in the List of Contributions]. Using game theory, specifically Stackelberg game theory, they modeled the pricing negotiations between manufacturers and other parties. The study proposed a multi-level improved simplified swarm optimization (MLiSSO) method to solve the multi-level programming problem (MLPP) associated with supply chain pricing strategies. The method was tested on three MLPPs from previous studies, demonstrating its effectiveness, stability, and applicability to other multi-level optimization problems. The results confirmed MLiSSO's capability in solving complex supply chain decision problems.

The contribution by Zimeras, titled "Patterns Simulations Using Gibbs/MRF Auto-Poisson Models", focused on pattern analysis in big data, particularly in image recognition, using spatial models like Markov random fields (MRFs) [item 20 in the List of Contributions]. It highlighted auto-Poisson models, which leveraged local characteristics of images to improve pattern recognition. By employing advanced statistical techniques such as Monte Carlo Markov Chains (MCMC), specifically the Gibbs sampler, the study aimed to define an MRF model under Poisson distribution and demonstrate its effectiveness through simulations. The results illustrated the model's performance on both simulated and real pattern data, showcasing its ability to accurately capture and explain underlying data structures.

The contribution by Jarfors et al., titled "An a Priori Discussion of the Fill Front Stability in Semisolid Casting", reviewed the filling front behavior in metal casting, particularly focusing on semisolid casting processes, which offer design flexibility, productivity, and cost-effectiveness while addressing filling-related defects [item 21 in the List of Contributions]. It emphasized the importance of solid fraction and gate design, providing a fresh perspective on gate configurations in semisolid processing compared to conventional high-pressure die-casting. The study highlighted that optimizing gate widths and managing solid fractions were crucial to preventing instability and issues like spraying during the casting process, ultimately enhancing the quality and reliability of cast products.

The contribution by Bauer et al., titled "Palachandran, M.; Wadehn, F.O.; Wolfschmidt, C.; Grothe, T.; Güth, U.; Ehrmann, A. Electrospinning for the Modification of 3D Objects for the Potential Use in Tissue Engineering", explored the use of electrospinning in biotechnological applications, particularly for tissue engineering and cell growth, by investigating the influence of 3D-printed substrates on the orientation and diameter of electrospun nanofiber mats [item 22 in the List of Contributions]. It examined how conductive and insulating 3D-printed objects affected fiber characteristics, using 3D-printed ear models as a case study. The research highlighted the impact of shadowing on fiber formation and demonstrated the potential of integrating electrospun nanofibers with 3D-printed scaffolds to create tissue structures in desired shapes, advancing applications in tissue engineering.

The contribution by Khan et al., titled "Study of Joint Symmetry in Gait Evolution for Quadrupedal Robots Using a Neural Network", investigated the impact of joint symmetry on the gait of bio-inspired legged robots, focusing on their ability to navigate uneven terrains efficiently [item 23 in the List of Contributions]. Using a spider-like robot morphology simulated in PyroSim, the study tested various joint symmetries, including diagonal, adjacent, and random configurations. Each robot, equipped with eight joints and controlled by an artificial neural network optimized through a genetic algorithm, underwent simulations on a flat surface. The results indicated that joint symmetry enhanced gait optimization, producing stable and effective movements reminiscent of natural gaits. Certain symmetries demonstrated superior performance in stability, speed, and distance traveled.

The contribution by Papachristou and Anastassiu, titled "Application of 3D Virtual Prototyping Technology to the Integration of Wearable Antennas into Fashion Garments", addressed the integration of wearable antennas into everyday clothing, highlighting a gap in the existing literature, which focuses primarily on antenna efficiency without considering garment design [item 24 in the List of Contributions]. Utilizing two-dimensional pattern and 3D virtual prototyping technology, the study developed market-available clothing with embedded antennas, ensuring that the garment's elegance and comfort were maintained. The paper detailed the functionality of various commercial software modules used in this automated design process and presented specific design examples that demonstrated the effectiveness of the approach. This work paved the way for creating more complex configurations of wearable antennas within garments.

The contribution by Haque et al., titled "Comparative Analysis on Suicidal Ideation Detection Using NLP, Machine, and Deep Learning", focused on detecting suicidal ideation through social media analysis, specifically using Twitter data [item 25 in the List of Contributions]. It addressed the challenges of identifying early symptoms of suicidal thoughts by comparing various machine learning and deep learning models. The research aimed to improve model performance by enhancing its accuracy when recognizing suicidal indicators, in order to potentially save lives. Using a dataset of 49,178 instances derived from live tweets, the study employed text preprocessing and feature extraction techniques. The results showed that the random forest (RF) model achieved a 93% accuracy, while the BiLSTM deep learning model, enhanced by word embedding, reached 93.6% accuracy and an F1 score of 0.93.

The contribution by Kodakkal et al., titled "An Optimized Enhanced Phase Locked Loop Controller for a Hybrid System", addressed the need for efficient control algorithms in hybrid power systems that integrate renewable energy sources, specifically wind and solar energy [item 26 in the List of Contributions]. It proposed a controller, based on the enhanced phase locked loop (EPLL) algorithm, to maintain power quality and manage load fluctuations without affecting source current. EPLL overcomes the double-frequency error common in standard phase locked loops and offers simplicity, precision, and stability. The paper employed optimization techniques to tune the proportional-integral (PI) controller gains, with the salp swarm algorithm yielding the best results. Additionally, maximum power point tracking (MPPT) was implemented using the perturb and observe method to enhance solar power efficiency.

The contribution by Ma et al., titled "Aging Mechanism and Models of Supercapacitors: A Review", explored electrochemical supercapacitors as a promising energy storage technology with diverse applications [item 27 in the List of Contributions]. It outlined their fundamental working principles and applications, while analyzing aging mechanisms that affect performance. The study reviewed existing supercapacitor models, evaluating their characteristics and application scopes. By assessing the current state and limitations of supercapacitor modeling research, the paper highlighted the need for more accurate models to enhance rational utilization, performance optimization, and system simulation. It also identified future development trends and key research focuses in the field of supercapacitor modeling, emphasizing the significance of improving these models for better energy storage solutions.

The Internet of Medical Things (IoMT) has significantly transformed healthcare by enabling efficient patient monitoring and data management. However, security and privacy concerns arise from the increased connectivity and potential cyber threats to sensitive data. The contribution by Pritika et al., titled "Risk Assessment of Heterogeneous IoMT Devices: A Review", reviewed existing IoT and IoMT applications, risks, and common attacks, emphasizing the inadequacy of current risk assessment frameworks for heterogeneous IoMT devices [item 28 in the List of Contributions]. It analyzed established frameworks like NIST, ISO 27001, and TARA, highlighting the need for new methodologies to address diverse risks. The proposed framework aimed to enhance risk assessment for IoMT devices, ensuring better security and privacy for users in healthcare settings.

The contribution by Barbosa et al., titled "Production Technologies, Regulatory Parameters, and Quality Control of Vaccine Vectors for Veterinary Use", examined the impact of the Internet of Medical Things (IoMT) on healthcare, emphasizing its ability to facilitate remote patient monitoring and streamline hospital data management [item 29 in the List of Contributions]. However, it also addressed significant security and privacy concerns arising from the increasing number of cyber threats targeting sensitive user information. The study reviewed existing risk assessment frameworks for IoMT devices, including NIST, ISO 27001, TARA, and IEEE213-2019, noting their limitations in addressing the heterogeneous risks associated with IoMT. It advocated for new methodologies to improve risk assessment and proposed a comprehensive framework based on NIST and ISO 27001 to enhance security for IoMT users.

The contribution by Uddin et al., titled "Thermal Inkjet Printing: Prospects and Applications in the Development of Medicine", discussed the significant advancements in inkjet printing technologies over the past decade, particularly their applications in the pharmaceutical and biomedical sectors [item 30 in the List of Contributions]. Thermal inkjet printing was highlighted for its versatility in developing bioinks for cell printing and biosensors, as well as its potential for fabricating personalized medications, including films and tablets. The paper provided an overview of the principles underlying inkjet printing, detailing its advantages and limitations. Additionally, it presented a variety of case studies showcasing the use of inkjet printing in precision medicine, emphasizing its growing relevance in tailored healthcare solutions.

The contribution by Guzmán and Maestro, titled "Synthetic Micro/Nanomotors for Drug Delivery", focused on synthetic micro/nanomotors (MNMs), which are self-propelled devices that convert chemical energy into motion, making them promising tools for biomedical applications, particularly in drug delivery [item 31 in the List of Contributions]. MNMs offer advantages over conventional drug carriers by enhancing drug transport to specific targets, thereby improving bioavailability in tissues. However, to ensure safe in vivo applications, further research is needed to address biocompatibility and biodegradability of these systems. The review provided an updated perspective on the potential of synthetic MNMs in drug delivery, while discussing key performance factors and biosafety considerations necessary for their clinical use.

The contribution by Rizzoli et al., titled "Multimodal Semantic Segmentation in Autonomous Driving: A Review of Current Approaches and Future Perspectives", addressed the challenges of achieving accurate semantic scene representation for autonomous driving systems using only RGB information [item 32 in the List of Contributions]. The lack of geometric details and sensitivity to weather and lighting conditions necessitates the use of multiple sensors, such as color, depth, thermal cameras, LiDARs, and RADARs. The paper presented commonly employed acquisition setups and datasets, followed by a review of various deep learning architectures for multimodal semantic segmentation. It discussed techniques for integrating color, depth, and LiDAR data at different stages of learning architectures, highlighting how effective fusion strategies can enhance performance compared to relying on a single data source.

The contribution by Madanu et al., titled "Explainable AI (XAI) Applied in Machine Learning for Pain Modeling: A Review", reviewed the role of artificial intelligence (AI) in pain assessment, highlighting its potential to enhance understanding of patient discomfort through physiological and behavioral changes [item 33 in the List of Contributions]. Pain, which varies in intensity and can arise from injuries, illnesses, or medical procedures, is often reflected in facial expressions, providing valuable information for clinicians. Recent advancements in machine learning and deep learning have improved the automatic assessment of pain. The review focused on explainable AI (XAI) and its applications for evaluating different types of pain, emphasizing the growing importance of AI in biomedical and healthcare settings for better patient outcomes.

The contribution by Ali et al., titled "Advanced Security Framework for Internet of Things (IoT)", aimed to propose a secure framework for the Internet of Things (IoT) in response to the vulnerabilities posed by the widespread interconnectivity of IoT devices [item 34 in the List of Contributions]. Utilizing a systematic literature review (SLR) approach, the study analyzed around 568 articles, ultimately focusing on 260 articles and 54 reports to identify key constructs and themes related to data security, confidentiality, and integrity. The analysis was conducted using MAXQDA (MAXQDA11), leading to the development of a qualitative model. This model, grounded in existing literature, was designed to assist IT managers, developers, and users in enhancing IoT security.

The contribution by Pearce, titled "Strategic Investment in Open Hardware for National Security", explored the potential of free and open-source hardware (FOSH) development to enhance national security by undermining imports and exports from targeted countries posing threats [item 35 in the List of Contributions]. A formal methodology was proposed for selecting strategic national investments in FOSH, which included identifying the threatening country, quantifying key imports, and identifying hardware that could reduce reliance on these imports. The methodology was illustrated through a case study of a current military aggressor and fossil-fuel exporter, revealing opportunities for FOSH development in energy conservation and renewable energy. The widespread adoption of FOSH could mitigate pollution and decrease financing for military activities.

The contribution by Yu et al., titled "Developments and Applications of Artificial Intelligence in Music Education", explored the integration of artificial intelligence (AI) in music education, highlighting its advantages and applications [item 36 in the List of Contributions]. With advancements in information technology, AI introduces innovative elements that enhance traditional teaching methods. By addressing the lack of personalization in conventional music education, AI facilitates a more individualized learning experience, fostering greater student engagement and interest. The paper systematically analyzed various AI applications in music education and discussed future development prospects, emphasizing the potential of intelligent technology to revolutionize the educational landscape in music. Overall, AI serves as a valuable tool for improving teaching effectiveness and student outcomes in music learning.

3. Conclusions

This Special Issue presents 36 groundbreaking research findings on Recent Advances and Perspectives in *Technologies*. It is expected that the insights shared here will help in further development and research in future technologies.

Funding: This research received no external funding.

Acknowledgments: I thank the authors who published their research results in this Special Issue and the reviewers who reviewed their papers. I also thank the editors for their hard work and perseverance in making this Special Issue a success.

Conflicts of Interest: The author declares no conflicts of interest.

List of Contributions:

1. Yeh, W.-C.; Zhu, W. Forecasting by Combining Chaotic PSO and Automated LSSVR. *Technologies* **2023**, *11*, 50. https://doi.org/10.3390/technologies11020050.
2. Brischetto, S.; Cesare, D.; Torre, R. A Layer-Wise Coupled Thermo-Elastic Shell Model for Three-Dimensional Stress Analysis of Functionally Graded Material Structures. *Technologies* **2023**, *11*, 35. https://doi.org/10.3390/technologies11020035.
3. Rehman, A.; Awan, K.A.; Ud Din, I.; Almogren, A.; Alabdulkareem, M. FogTrust: Fog-Integrated Multi-Leveled Trust Man-agement Mechanism for Internet of Things. *Technologies* **2023**, *11*, 27. https://doi.org/10.3390/technologies11010027.

4. Manjarrez, L.H.; Ramos-Fernández, J.C.; Espinoza, E.S.; Lozano, R. Estimation of Energy Consumption and Flight Time Margin for a UAV Mission Based on Fuzzy Systems. *Technologies* **2023**, *11*, 12. https://doi.org/10.3390/technologies11010012.
5. Bozorgpanah, A.; Torra, V.; Aliahmadipour, L. Privacy and Explainability: The Effects of Data Protection on Shapley Values. *Technologies* **2022**, *10*, 125. https://doi.org/10.3390/technologies10060125.
6. Carneiro, M.; Oliveira, V.; Oliveira, F.; Teixeira, M.; Pinto, M. Simulation Analysis of Signal Conditioning Circuits for Plants' Electrical Signals. *Technologies* **2022**, *10*, 121. https://doi.org/10.3390/technologies10060121.
7. Tarasov, S.; Amirov, A.; Chumaevskiy, A.; Savchenko, N.; Rubtsov, V.E.; Ivanov, A.; Moskvichev, E.; Kolubaev, E. Friction Stir Welding of Ti-6Al-4V Using a Liquid-Cooled Nickel Superalloy Tool. *Technologies* **2022**, *10*, 118. https://doi.org/10.3390/technologies10060118.
8. Lin, C.-T.; Fan, H.-Y.; Chang, Y.-C.; Ou, L.; Liu, J.; Wang, Y.-K.; Jung, T.-P. Modelling the Trust Value for Human Agents Based on Real-Time Human States in Human-Autonomous Teaming Systems. *Technologies* **2022**, *10*, 115. https://doi.org/10.3390/technologies10060115.
9. Phan, P.T.; Nguyen, P.T. Evaluation Based on the Distance from the Average Solution Approach: A Derivative Model for Evaluating and Selecting a Construction Manager. *Technologies* **2022**, *10*, 107. https://doi.org/10.3390/technologies10050107.
10. Yurova, V.A.; Velikoborets, G.; Vladyko, A. Design and Implementation of an Anthropomorphic Robotic Arm Prosthesis. *Technologies* **2022**, *10*, 103. https://doi.org/10.3390/technologies10050103.
11. Saeidi Aminabadi, S.; Jafari-Tabrizi, A.; Gruber, D.P.; Berger-Weber, G.; Friesenbichler, W. An Automatic, Contactless, High-Precision, High-Speed Measurement System to Provide In-Line, As-Molded Three-Dimensional Measurements of a Curved-Shape Injection-Molded Part. *Technologies* **2022**, *10*, 95. https://doi.org/10.3390/technologies10040095.
12. Almeida, T.S.; da Cruz Souza, C.A.; de Cerqueira e Silva, M.B.; Batista, F.P.R.; Ferreira, E.S.; Santos, A.L.S.; Silva, L.N.; Melo, C.R.; Bani, C.; Bianconi, M.L.; et al. Extraction and Characterization of β-Viginin Protein Hydrolysates from Cowpea Flour as a New Manufacturing Active Ingredient. *Technologies* **2022**, *10*, 89. https://doi.org/10.3390/technologies10040089.
13. Algredo-Badillo, I.; Sánchez-Juárez, B.; Ramírez-Gutiérrez, K.A.; Feregrino-Uribe, C.; López-Huerta, F.; Estrada-López, J.J. Analysis and Hardware Architecture on FPGA of a Robust Audio Fingerprinting Method Using SSM. *Technologies* **2022**, *10*, 86. https://doi.org/10.3390/technologies10040086.
14. Hadi, M.U.; Suhaimi, N.H.N.; Basit, A. Efficient Supervised Machine Learning Network for Non-Intrusive Load Monitoring. *Technologies* **2022**, *10*, 85. https://doi.org/10.3390/technologies10040085.
15. Stopka, O.; Gross, P.; Pečman, J.; Hanzl, J.; Stopková, M.; Jurkovič, M. Optimization of the Pick-Up and Delivery Technology in a Selected Company: A Case Study. *Technologies* **2022**, *10*, 84. https://doi.org/10.3390/technologies10040084.
16. Gabbar, H.A.; Elsayed, Y.; Isham, M.; Elshora, A.; Siddique, A.B.; Esteves, O.L.A. Demonstration of Resilient Microgrid with Real-Time Co-Simulation and Programmable Loads. *Technologies* **2022**, *10*, 83. https://doi.org/10.3390/technologies10040083.
17. Li, J.; Wang, Y.; Du, K.-L. Distribution Path Optimization by an Improved Genetic Algorithm Combined with a Divide-and-Conquer Strategy. *Technologies* **2022**, *10*, 81. https://doi.org/10.3390/technologies10040081.
18. Gradov, O.M. Exciting of Strong Electrostatic Fields and Electromagnetic Resonators at the Plasma Boundary by a Power Electromagnetic Beam. *Technologies* **2022**, *10*, 78. https://doi.org/10.3390/technologies10040078.
19. Yeh, W.-C.; Liu, Z.; Yang, Y.-C.; Tan, S.-Y. Solving Dual-Channel Supply Chain Pricing Strategy Problem with Multi-Level Programming Based on Improved Simplified Swarm Optimization. *Technologies* **2022**, *10*, 73. https://doi.org/10.3390/technologies10030073.
20. Zimeras, S. Patterns Simulations Using Gibbs/MRF Auto-Poisson Models. *Technologies* **2022**, *10*, 69. https://doi.org/10.3390/technologies10030069.
21. Jarfors, A.E.W.; Zhang, Q.; Jonsson, S. An a Priori Discussion of the Fill Front Stability in Semisolid Casting. *Technologies* **2022**, *10*, 67. https://doi.org/10.3390/technologies10030067.
22. Bauer, L.; Brandstäter, L.; Letmate, M.; Palachandran, M.; Wadehn, F.O.; Wolfschmidt, C.; Grothe, T.; Güth, U.; Ehrmann, A. Electrospinning for the Modification of 3D Objects for the Potential Use in Tissue Engineering. *Technologies* **2022**, *10*, 66. https://doi.org/10.3390/technologies10030066.

23. Khan, Z.; Naseer, F.; Khan, Y.; Bilal, M.; Butt, M.A. Study of Joint Symmetry in Gait Evolution for Quadrupedal Robots Using a Neural Network. *Technologies* **2022**, *10*, 64. https://doi.org/10.3390/technologies10030064.
24. Papachristou, E.; Anastassiu, H.T. Application of 3D Virtual Prototyping Technology to the Integration of Wearable Antennas into Fashion Garments. *Technologies* **2022**, *10*, 62. https://doi.org/10.3390/technologies10030062.
25. Haque, R.; Islam, N.; Islam, M.; Ahsan, M.M. A Comparative Analysis on Suicidal Ideation Detection Using NLP, Machine, and Deep Learning. *Technologies* **2022**, *10*, 57. https://doi.org/10.3390/technologies10030057.
26. Kodakkal, A.; Veramalla, R.; Kuthuri, N.R.; Salkuti, S.R. An Optimized Enhanced Phase Locked Loop Controller for a Hybrid System. *Technologies* **2022**, *10*, 40. https://doi.org/10.3390/technologies10020040.
27. Ma, N.; Yang, D.; Riaz, S.; Wang, L.; Wang, K. Aging Mechanism and Models of Supercapacitors: A Review. *Technologies* **2023**, *11*, 38. https://doi.org/10.3390/technologies11020038.
28. Pritika; Shanmugam, B.; Azam, S. Risk Assessment of Heterogeneous IoMT Devices: A Review. *Technologies* **2023**, *11*, 31. https://doi.org/10.3390/technologies11010031.
29. Barbosa, R.d.M.; Silva, A.M.; Silva, C.F.d.; Cardoso, J.C.; Severino, P.; Meirelles, L.M.A.; Silva-Junior, A.A.d.; Viseras, C.; Fonseca, J.; Souto, E.B. Production Technologies, Regulatory Parameters, and Quality Control of Vaccine Vectors for Veterinary Use. *Technologies* **2022**, *10*, 109. https://doi.org/10.3390/technologies10050109.
30. Uddin, M.J.; Hassan, J.; Douroumis, D. Thermal Inkjet Printing: Prospects and Applications in the Development of Medicine. *Technologies* **2022**, *10*, 108. https://doi.org/10.3390/technologies10050108.
31. Guzmán, E.; Maestro, A. Synthetic Micro/Nanomotors for Drug Delivery. *Technologies* **2022**, *10*, 96. https://doi.org/10.3390/technologies10040096.
32. Rizzoli, G.; Barbato, F.; Zanuttigh, P. Multimodal Semantic Segmentation in Autonomous Driving: A Review of Current Approaches and Future Perspectives. *Technologies* **2022**, *10*, 90. https://doi.org/10.3390/technologies10040090.
33. Madanu, R.; Abbod, M.F.; Hsiao, F.-J.; Chen, W.-T.; Shieh, J.-S. Explainable AI (XAI) Applied in Machine Learning for Pain Modeling: A Review. *Technologies* **2022**, *10*, 74. https://doi.org/10.3390/technologies10030074.
34. Ali, A.; Mateen, A.; Hanan, A.; Amin, F. Advanced Security Framework for Internet of Things (IoT). *Technologies* **2022**, *10*, 60. https://doi.org/10.3390/technologies10030060.
35. Pearce, J.M. Strategic Investment in Open Hardware for National Security. *Technologies* **2022**, *10*, 53. https://doi.org/10.3390/technologies10020053.
36. Yu, X.; Ma, N.; Zheng, L.; Wang, L.; Wang, K. Developments and Applications of Artificial Intelligence in Music Education. *Technologies* **2023**, *11*, 42. https://doi.org/10.3390/technologies11020042.

technologies

MDPI

Article

Forecasting by Combining Chaotic PSO and Automated LSSVR

Wei-Chang Yeh [1,*] and Wenbo Zhu [2]

1 Integration and Collaboration Laboratory, Department of Industrial Engineering and Engineering Management, National Tsing Hua University, Hsinchu 300, Taiwan
2 School of Mechatronical Engineering and Automation, Foshan University, Foshan 528000, China
* Correspondence: yeh@ieee.org

Abstract: An automatic least square support vector regression (LSSVR) optimization method that uses mixed kernel chaotic particle swarm optimization (CPSO) to handle regression issues has been provided. The LSSVR model is composed of three components. The position of the particles (solution) in a chaotic sequence with good randomness and ergodicity of the initial characteristics is taken into consideration in the first section. The binary particle swarm optimization (PSO) used to choose potential input characteristic combinations makes up the second section. The final step involves using a chaotic search to narrow down the set of potential input characteristics before combining the PSO-optimized parameters to create CP-LSSVR. The CP-LSSVR is used to forecast the impressive datasets testing targets obtained from the UCI dataset for purposes of illustration and evaluation. The results suggest CP-LSSVR has a good predictive capability discussed in this paper and can build a projected model utilizing a limited number of characteristics.

Keywords: mixed kernel; particle swarm optimization; support vector regression (SVR); least squares SVR

Citation: Yeh, W.-C.; Zhu, W. Forecasting by Combining Chaotic PSO and Automated LSSVR. *Technologies* **2023**, *11*, 50. https://doi.org/10.3390/technologies11020050

Academic Editors: Manoj Gupta, Eugene Wong and Gwanggil Jeon

Received: 23 February 2023
Revised: 24 March 2023
Accepted: 27 March 2023
Published: 30 March 2023

1. Introduction

In addition to using the sample distributions that are provided, traditional statistical methods also base their estimation of the parameter's value on the assumption that samples are infinite. The application of some outstanding statistics methods for the real-world issue is severely constrained. The major popular approach for nonlinear modeling, the artificial neural network (ANN), overcomes the limitations of conventional approaches for parameter estimation and may be built entirely from historical input-output data. The empirical risk minimization (ERM) principle-based ANN, however, there are several significant drawbacks, including the need for more training data, a lack of a consistent theoretical mathematical framework, the failure to find the fractional answers, overtraining, and dimension fatal event.

Support vector machine (SVM), a cutting-edge, potent machine learning technique created within the body of statistical learning theory (SLT), carries out the structural risk minimization (SRM) principle rather than the equivalent risk minimization (ERM) principle, giving it excellent generalization abilities in the case of small samples. SVM can effectively minimize modeling complexity, establish network structure automatically, and dimension disaster-free without local minima. Support vector regression (SVR) that had been proven its outstanding strength in many areas such as identification of patterns, regression analysis, forecasting in time-series, and optimization in numerous systems is expanded for resolving non-linear regression analysis.

A novel technique recently presented is termed the least squares SVR (LSSVR) [1], which uses equality constraints similar to that of a traditional artificial neural network (ANN). As the problem's solution can be discovered using linearization, it is substantially simplified. It can be used to create a classification and prediction model, as seen in [2,3]. Yet, the right LSSVR meta parameter configuration determines how well LSSVR models perform. As a result, only "professional" users with a solid understanding of the SVR approach can handle the LSSVR program.

The proper three LSSVR settings determine the quality of LSSVR models. Secondly, in big-size cases, the LSSVR function is quite slow since a quadratic programming (QP) issue needs to be solved. Second, several crucial parameters that endure and impair LSSVR's recapitulation capabilities, are not optimal in the LSSVR modeling. Finally, a predictive LSSVR model also has a hard time choosing some crucial properties. Furthermore, when the model interpretability is crucial, the issue could become more difficult to solve. How to choose these variables to guarantee outstanding recapitulation expression is a fundamental challenge in utilizing LSSVR for nonlinear systems. Evolutionary algorithms (EAs), particularly genetic algorithms (GAs), are the most popular method for determining parameters and characteristics, and they have already been used to choose characteristics and optimize parameters for the LSSVR model [4,5]. While particle swarm optimization (PSO), which is inspired by the swarm action from creatures, is extremely simple for installation as well as has fewer parameters for the tune, GA is challenging and is short of computational power. The simulation findings demonstrate that GA and PSO readily trap into local optimal solution, despite the fact that PSO can effectively employ in handling optimal problems of more dimensional [6–11].

Thus, this study must address the following three problems:

1. Initialization of the parameters: This is due to the possibility that it is unaware of the location of the global minimum at which the prior optimization problem was resolved.
2. Characteristic extract or the characteristic evolution component can typically accomplish the decrease in data dimensionality. Often, this has been accomplished using characteristic extract methods like principal component analysis (PCA). The PCA is ineffective for this study's objectives since it also wants to produce highly precise predictive models, not just reduce the dimensionality of the data. Nevertheless, the PCA doesn't take into account the link for variables of input and variables of reply throughout the data reduction process, making it challenging to create a model that is extremely precise. Furthermore, if the input variables' dimensionality is really high, it might be challenging to interpret the main components that are produced by the PCA. On the other hand, for data sets with high dimensionality, the PSO has been shown to perform better than other methods [11]. A simplified LSSVR model with improved generalization can be created by selecting more information for any data set provided when employing the fewest characteristics possible throughout the characteristic evolution phase.
3. Another PSO is employed in the parameter evolution component to optimize the LSSVR's parameters. Generally, LSSVR generalization ability is governed by the type of kernel, parameters' kernel, and parameter's upper bound. Every form of the kernel has benefits and drawbacks, hence a mixed kernel makes sense [12–14]. Additionally, computational time and complexity in the training of the algorithm equals the total execution generations multiplied by the number of total solutions and multiplied by the time complexity of the update for each solution.

In this study, the chaotic particle swarm optimization (CPSO) approach has been used to tackle optimization issues. This novel PSO is based on chaos investigating, the logical model, and the tent model [15,16]. The advantage and innovation of the new regression method CP-LSSVR presented in this work is demonstrated as follows:

1. The CP-LSSVR is used to initialize the parameters for the parameters initialization issue of LSSVR applications.
2. A binary PSO is utilized for feature selection in the input data to improve the model's interpretability for the issue of requiring LSSVR to preprocess the input characteristics if the dimensions of input space or input characteristics are quite vast.
3. A third PSO is applied to optimize its parameters to boost the LSSVR's capacity for normalization.

SVR, LSSVR, kernel function, Particle Swarm Optimization (PSO) algorithm, and chaotic sequences are all described in Section 2. In Section 3, the CP-LSSVR learning paradigm is thoroughly explained. Benchmark datasets, comparative methods, and the reported experimental results are described in Section 4. Conclusion and additional research have been analyzed in Section 5.

2. SVR and LSSVR

By applying a nonlinear function (m > d) to transfer an input of d-dimensionality onto an m-dimensional characteristic space, SVM regression (SVR) creates a linear model in the characteristic space. The process is depicted in Figure 1.

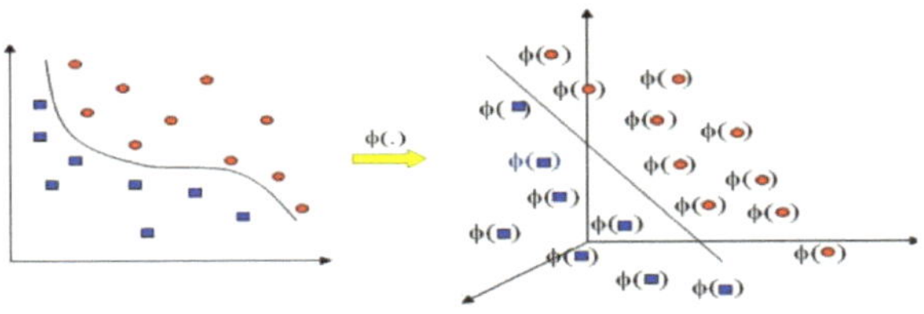

Figure 1. Example of the nonlinear transformation used to get from the input space to the characteristic space (d = 2; m = 3), where colors represent different characteristic and symbol $\phi(\cdot)$ signifies a sequence of nonlinear transformations.

Let's assume a training data set $\{x_k, y_k\}_{k=1}^{N}$, where $x_k \in \mathbb{R}^n$ as well as $y_k \in \mathbb{R}$ for $k = 1, \ldots, N$ and represent characteristics' input space and objective value, respectively. N denotes the training data's size. In order to translate the input data to an advanced dimensional characteristic space, the SVR must identify a nonlinear map from input space to output space. Then, Equation (1) shows the linear regression using the following estimate function [17]:

$$l(x) = a\phi(x) + e \tag{1}$$

where a is the coefficients, $\phi(x)$ signifies a sequence of nonlinear transformations that translates the input space into the characteristic space, and e means a real number. To reduce the risk is the goal is shown in Equation (2) [1]:

$$\begin{array}{ll} \min\limits_{a,e,E_k^u,E_k^l} & F(a,e,E_k^u E_k^l) = {||a||^2}/{2} + C\sum\limits_{k=1}^{N}(E_k^u + E_k^l) \\ s.t & y_k - (a\phi(x_k)) + e \le \sigma + E_k^l \\ & (a\phi(x_k)) + e - y_k \le \sigma + E_k^u \\ & E_k^u, E_k^l \ge 0 \qquad k = 1,\ldots,N \end{array} \tag{2}$$

The characteristic map ϕ vector has transferred the dataset of k-sample's vector to a advanced-dimensional space, and E_k^u is training error of upper bound and E_k^l presents training error of lower bound for the σ-insensitive tube.

$$|y - (a\phi(x) + e)| \le \sigma \tag{3}$$

The dual formulations solution may be sparse due to the σ-insensitive loss model that also precludes the complete training set from fulfilling the boundary requirements. A balance achieved between the flatness of F and its accurateness in catching the training data is determined by the item ${||a||^2}/{2}$, which is known as the normalization item, and C, which presents the normalization constant.

Equation (3) suggests that the tube σ contains the majority of the data α_k. An error E_k^l or E_k^l has been typically tried to reduce in the objective function if α_k is outside the tube. This is depicted in Figure 2. By minimizing the normalization term $\|a\|^2/2$ as well as the training error $C\sum_{k=1}^{N}(E_k^u + E_k^l)$, SVR prevents the training data to be underfitted and overfitted.

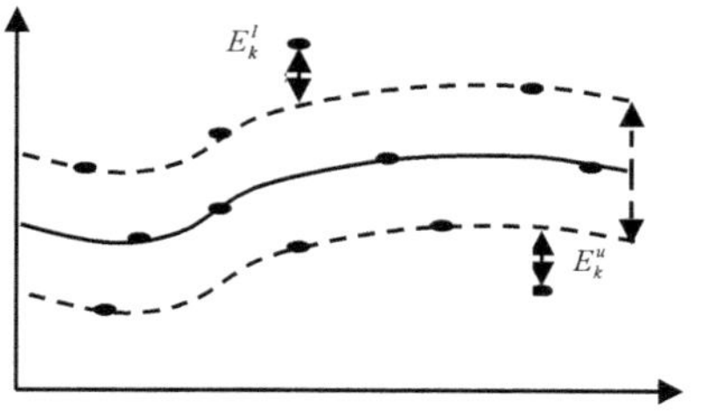

Figure 2. For SVR, a σ-insensitive tube.

The Karush-Kuhn-Tucker methods [18] require introduction of Lagrange multipliers γ_k, γ_k^*, and the SVR approach aggregates to solving the convex quadratic model given in Equation (4)

$$\begin{array}{ll} \min\limits_{\gamma,\gamma^*} & \frac{1}{2}\sum\limits_{k,s=1}^{N}(\gamma_k - \gamma_k^*)(\gamma_s - \gamma_s^*)K(x_k, x_s) + \sigma\sum\limits_{k=1}^{N}(\gamma_k + \gamma_k^*) - \sum\limits_{k=1}^{N} y(\gamma_k - \gamma_k^*) \\ s.t & \sum\limits_{k=1}^{N}(\gamma_k - \gamma_k^*) = 0 \\ & 0 \le \gamma_k, \gamma_k^* \le C \end{array} \tag{4}$$

SVMs are superior to other regression techniques because they solve the quadratic programming (QP) problem without hitting the local minima that depend on the statistical learning theory and the structural risk minimization concept [18]. The non-linear SVR function in this work is LSSVR. LSSVR employed was selected as the estimation approach due to its superior normalization power and ability to produce an almost global answer in a reasonable amount of training time [19]. The optimization problem's basic formulation of an LSSVR regression model in characteristic space is Equation (5) [19]:

$$\begin{array}{lll} \min\limits_{a,e,\pi} & F(a,e,\pi) = \|a\|^2/2 + C\sum\limits_{k=1}^{N}\pi_k^2 \Big/ 2 & \\ s.t & y_k - (a^T \cdot \phi(x_k) + e) = \pi_k & k = 1,2\ldots,N \end{array} \tag{5}$$

Due to the weighting vector's extremely high dimension, the calculation of Equation (5) is very challenging. This issue can be resolved by computing the model through a Lagrangian stated in Equation (6) in a dual space as opposed to the primal space.

$$L(a,e,\pi,\gamma) = F(a,e,\pi) - C\sum_{k=1}^{N}\gamma_k\{a^T \cdot \phi(x_k) + e + \pi_k - y_k\} \tag{6}$$

where γ_k is the support vector that belongs to the real number, often known as the Lagrange multiplier. In light of this, Equation (7) lists the prerequisites for optimality.

$$\begin{array}{l} \frac{\partial L}{\partial a} = 0 \Rightarrow a = \sum_{k=1}^{N} \gamma_k \phi(x_k) \\ \frac{\partial L}{\partial e} = 0 \Rightarrow -\sum_{k=1}^{N} \gamma_k = 0 \\ \frac{\partial L}{\partial \pi_k} = 0 \Rightarrow \gamma_k = C\pi_k \\ \frac{\partial L}{\partial \gamma_k} = 0 \Rightarrow a^T \phi(x_k) + e + \pi_k - y_k = 0 \\ k = 1, \ldots, N \end{array} \quad (7)$$

After removing π and a, the answer is obtained as Equation (8).

$$\begin{bmatrix} 0 & 1^T \\ 1 & \Omega + \tau^{-1} I \end{bmatrix} \begin{bmatrix} e \\ \gamma \end{bmatrix} = \begin{bmatrix} 0 \\ y \end{bmatrix} \quad (8)$$

where $y = (y_1, \ldots, y_n)^T$, $I = (1, \ldots 1)^T$, $\gamma = (\gamma_1, \ldots, \gamma_n)^T$, $\Omega_{ks} = (\phi(x_k))^T \phi(x_s)$ for k,s = 1, ..., N. There is a mapping and an expression that may be written as $k(x, y) = \sum_k \phi_k(x)^T \phi_k(y)$ based on Mercer's condition. Hence, the kernel $k(.,.)$ is constructed such that Equation (9).

$$k(x_k, x_s) = \phi(x_k)^T \phi(x_s) \quad (9)$$

where ϕ represents the formula that simulates the real non-linear mapping formula and x_k as well as x_s denoted as both goals for the data set.

In Equation (10), the outcome LS-SVR model for function estimation is found.

$$l(x) = \sum_{k=1}^{N} \gamma_k \cdot v(x_k, x_s) + e \quad (10)$$

where x_k and e are the answers to Equation (10).

Complex non-linear data can be mapped using kernels into an advanced-dimensional characteristic space that linear modeling is feasible. Due to the difficulty in determining the mapping model and the overall lack of prior knowledge, the characteristic space is completely created by calling a common kernel model. A kernel model (K) works on two input vectors as shown in Equation (9).

To build a linear function in the characteristic space, one does not need to be aware of the actual underlying feature map when using a kernel function. In literature, a number of kernel functions are frequently utilized.

The width of the tube, the mapping function, and the error cost C are the three parameters in this study that define the LSSVR quality. The nonlinear mapping is performed by the Mercer kernel's approximate characteristic map. Equations (11)–(13) show the common kernel functions in machine learning theories [20].

Linear kernel:

$$\kappa(x_k, x_s) = x_k^T \cdot x_s \quad (11)$$

Kernel of a polynomial:

$$\kappa(x_k, x_s) = (x_k^T \cdot x_s + h)^g \quad (12)$$

Gaussian (RBF) kernel:

$$\kappa(x_k, x_s) = \exp\left(-\frac{\|x_k - x_s\|^2}{2\sigma^2}\right) \quad (13)$$

The parameters x_k and x_s are vectors in the input space; **g** signifies the polynomial's level and T presents the item of intercept constant in Equation (12) and σ^2 indicates the Gaussian kernel's width in Equation (13).

The polynomial kernel (a global kernel), among the three standard kernels, stated earlier, exhibits stronger extrapolation capabilities at lower orders of levels but needs higher orders of levels for effective inserted values. The RBF kernel, a local kernel, excels in inserted values but falls short in extrapolating across longer distances.

It is difficult to declare which kernel is the greatest across the board, though, because each has pros and cons. According to the studies [13,21], combining or hybridizing various kernel functions can enhance SVM's generalization capabilities. In this paper, the LSSVR model using a mixed kernel is trained. The three kernels mentioned above are combined to form the mixed kernel. It is possible to write the convex combination kernel by Equation (14).

$$\begin{aligned} &\kappa = \lambda_1 \kappa_{linear} + \lambda_2 \kappa_{poly} + \lambda_3 \kappa_{rbf} \\ &\textit{where } \lambda_1 + \lambda_2 + \lambda_3 = 1,\ 0 \leq \lambda_1, \lambda_2, \lambda_3 \leq 1 \end{aligned} \tag{14}$$

Since all kernel functions in the proposed mixed kernel fulfill Mercer's theory, a convex combination of them fulfills the theory, too.

3. LSSVR Based on Chaotic Particle Swarm Optimization (CPSO) Algorithm

The suggested automatic LSSVR learning paradigm is further detailed in this section. The automatic LSSVR learning paradigm is first introduced in its common structure. Then, every step of the chaotic map and PSO-based LSSVR parameter initialization, characteristic selection, and parameter optimization is discussed.

3.1. *Automatic LSSVR Learning Paradigm*

Several practical studies have shown that the LSSVM is a successful learning technique for regression issues [21–23]. For LSSVR applications, there are still three key issues.

1. Parameters initialization.
2. It is required for LSSVR to preprocess the input characteristics if the dimensions of input space or input characteristics are quite vast, in order to improve the interpretability of the LSSVR-based forecasting model.
3. This work adopts a mixed kernel model to get beyond the effect of kernel types because LSSVR normalization ability is frequently governed via (a) kernel type. The next two things, however, heavily rely on the researchers' artistic ability. (b) kernel parameters: convex combination coefficients (λ_1, λ_2, λ_3), and kernel parameters ($\mathbf{g}$, σ). C is the upper bound parameter.

Software algorithms are employed to solve these three issues. A chaotic map is used to initialize the parameters for the first issue. In order to improve the model's interpretability for the second issue, a binary PSO is utilized for feature selection in the input data. To boost the LSSVR's capacity for normalization, a third PSO is applied to optimize its parameters. The automatic LSSVR learning paradigm, shown in Figure 3, is developed based on the three techniques.

It is simple to see that the automatic LSSVR learning paradigm has three basic components that address the aforementioned three major issues.

Use the chaotic map in the first section to defeat the PSO algorithm's initialization's randomly produced solutions.

The binary PSO searches a subset of characteristic variable subsets in the exponential space in the second section before sending that subset of characteristics to an LSSVR model. From each subgroup, the LSSVR extracts forecasting data and gains knowledge of the schemes. A trained LSSVR is tested on a holdout data set that was not utilized for training after learning the schemes in the data, and it then sends the calculation rule as a fitness function for PSO. The PSO biases its search direction according to fitness values to maximize the assessment aim. It is important to note that LSSVM just adopts the chosen characteristic variables during the training and evaluation processes.

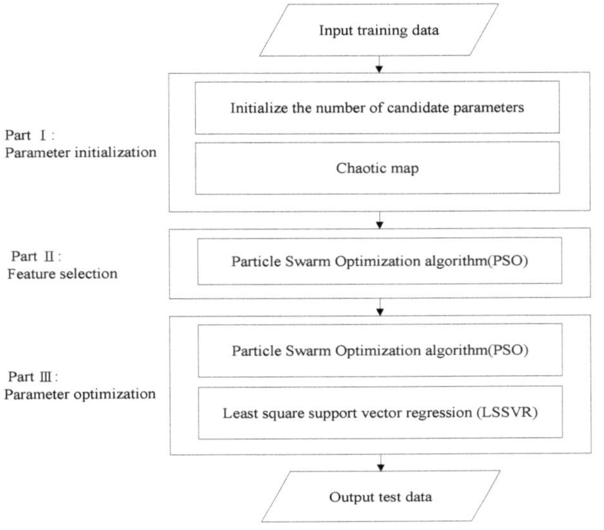

Figure 3. General framework of the automatic LSSVR learning paradigm.

The six unknown parameters (λ_1, λ_2, λ_3, g, σ, C) are optimized in the third section using PSO. Sections 3.2–3.4 define in full the contents of each section.

To combine the two elements of evolution is feasible. It is possible to simultaneously optimize the characteristic evolution and the parameter evolution. It means the parameter optimization technique can be carried out prior to the characteristic selection procedure in the autonomous LSSVR learning paradigm. Large input characteristic dimensions are not recommended in reality due to the excessive computational workload. In this regard, it makes it more logical to undertake characteristic selection before parameter evolution.

3.2. Chaotic Sequences-Based Parameters Initialization

It might not know the position of the global minimum before an optimization problem is solved [13]. The PSO-generated solutions, however, use a random mechanism in the beginning stages, making it simple to reach the local optimal. This paper makes an effort to use chaotic sequences to tackle this issue.

Step 0. Generated by Logistic map chaotic sequence by Equation (15), following:

$$L(k+1) = n{\cdot}L(k){\cdot}(1-L(k)),\ L(k) \in [0,1];\ n \in [3.56,4] \tag{15}$$

Step 1. For the m particles in the D-dimensional space, the first generates a random initial value m:

$$L_1(1), L_2(1), \ldots, L_m(1)$$

Step 2. Chaotic sequence to the initial value of m-Equation (15). At that point, m will be the trajectory after Z iterations.

Step 3. Substituting the chaotic trajectory of the article from m in the selected Z iteration value into the Equation (16). One can compute $x_{v,k}$

$$x_{v,k} = L_v(k)(\max_k - \min_k)v/m + \min_k,\ v = 1,2,\ldots,m;\ k = 1,2,\ldots,Z \tag{16}$$

where $x_{v,k}$ denotes the position of the v particles in the k-dimensional space. $L_v(k)$ is for the first v particles in the randomly generated initial value of Equation (15) after k multiplying the value by the number of iterations.

Using Equation (16) calculated for all $x_{v,k}$ components m row column Z Matrix as following Equation (17):

$$\begin{bmatrix} x_{1,1} & x_{1,2} & \cdots & x_{1,Z} \\ x_{2,1} & x_{2,2} & & x_{2,Z} \\ \vdots & & \ddots & \\ x_{m,1} & x_{m,2} & \cdots & x_{m,Z} \end{bmatrix} \quad (17)$$

where each row vector represents the initial position of a particle.

3.3. PSO-Based Input Features Evolution

The level of potential input variables might be rather big for many practical issues. It's possible that some of these input variables are redundant. A significant input variables' level will also raise LSSVR's size, necessitating more training data and longer training periods to achieve a respectable level of normalization [22,23]. As a result, characteristic selection should be used to reduce input characteristics. Typically, the procedure of selecting a subset of the original characteristics by eliminating any duplicate or poorly-informed characteristics is referred to as characteristic selection [24].

The second challenge in the addressed automatic LSSVR learning paradigm is to choose crucial features for LSSVR learning. Two things are the major goals of characteristic selection:

1. To eliminate some less-important characteristics for decreasing the input characteristics' size and enhance forecasting capability.
2. Additionally, it is to pinpoint several crucial characteristics that influence model performance, hence bringing down model complexity.

PSO, the most prevalent kind of software algorithm to date, has developed into a significant stochastic optimization technique, in contrast to most conventional optimization algorithms, because it frequently finds the optimal optimum. In this study, the input characteristic subset for LSSVR modeling is extracted using PSO.

The required particle number is initially set using the principles of particle swarm optimization, and the starting coding alphabetic string for every particle has been then generated at random. In this work, every particle is coded to mimic a chromosome using a common method; every particle is converted to a binary alphabetic string $S = A_1, A_2, \ldots, A_m$, $m = 1, 2, \ldots, N$, where bit value {1} presents a characteristic that has been chosen and bit value {0} denotes a characteristic that has not been chosen.

When utilizing particle swarm optimization to examine a 10-dimensional data set ($m = 10$), for instance, any characteristics' level can be chosen fewer than m, i.e., it can randomly select six characteristics, as shown in the accompanying Figure 4.

Origin S_{10}	A_1	A_2	A_3	A_4	A_5	A_6	A_7	A_8	A_9	A_{10}
Encode	1	0	1	0	1	0	1	0	0	1
After S_6	A_1	A_3	A_5	A_7	A_{10}					

Figure 4. Choosing characteristics using PSO.

The six characteristics in each data set serve as a representation of the data dimension and have been assessed via LSSVR while calculating the fitness score. Adaptive value serves as a foundation for every particle renewal. The best adaptive value within a group of p-best is g-best, while the best fitness value for every particle renewal is p-best. After obtaining p-best and g-best, it may monitor the characteristics of p-best and g-best particles in terms of their location and speed. The binary version of PSO is utilized in this investigation [25]. Each particle's location is specified as a binary string that corresponds to a characteristic selection scenario [26].

The fitness function is the last evaluation criterion and is used to assess each string's quality. The following Equation (18) can be used to build the fitness function for the PSO variable selection.

$$F_i = \frac{1}{f_i + mic} \tag{18}$$

where "mic" is a decimal that is used to avoid the denominator being zero, f_i representing the ith solution's objective.

In this case, the relevancy of the input variables and the response variable is modeled using the LSSVR, as you may have noticed. The LSSVR models are then trained using training data, tested using holdout data, and the suggested model is assessed using the reduction LSSVR-error quadratic sum of the solution.

The aforementioned formulae determine how often each particle is updated. The PSO procedure's pseudo code is shown below Algorithm 1.

Algorithm 1: PSO—Based Input Features Evolution

Goal: Reducing (1-hit ratio)
Input: training data set.
Output: The PSO-LSSVR's characteristics set.
BEGIN
Establish the population
While (number of generations, or the halting requirement is not fulfilled)
For $i = 1$ (particles' number)
When one's fitness level X_k exceeds another's p-best,
next update p-best$_k$ = X_k
For υ belongs to neighborhood of X_k
    If fitness X_υ is higher than fitness of g-best,
Next update g-best = X_υ
    then υ
Every dimension g
$V_{k,g}(h+1) = aV_{k,g}(h) + \tau_1 c_1(P_{k,g} - x_{k,g}(h)) + \tau_2 c_2(P_{j,g} - x_{k,g}(h))$
$S(V_{k,g}(h+1)) = \frac{1}{1+\pi^{-V_{k,g}(h+1)}}$
when rand() < $S(V_{k,g}(h+1))$
then $X_{k,g}(h+1) = 1$
else $X_{k,g}(h+1) = 0$
    Next g
Next k
Next generation till the ending criteria
END

The function in Equation (19) determines the characteristic following renewal.

$$S(V_{k,g}(h+1)) = \frac{1}{1+\pi^{-V_{k,g}(h+1)}} \tag{19}$$

If $S(V_{k,g}(h+1))$ is greater than a disorder number generated at random and falling within the range of (0, 1), then its position value A_m ($m = 1, 2, \ldots, N$) denotes this characteristic has been chosen as a claimed characteristic for the next renewal as {1} otherwise, denotes this characteristic has not been chosen as a claimed characteristic for the next renewal as {0}.

3.4. PSO-Based Parameters Optimization

The current GA algorithm has some speed and accuracy restrictions for the convergence of high-dimensional problems, in addition to being complex in terms of selection, crossover, and mutation. The PSO algorithm, in contrast, is a parallel global search based on population strategy, the idea of which is straightforward and simple to implement. It also has a faster convergence speed, is able to handle high-dimensional problems while still having some advantages, and is a population-based stochastic optimization technique.

Theoretical benefits of the PSO algorithm, which was used in this study to perform a solution search for the six parameters of LSSVR.

Up until the termination condition is met, the process is repeated. The undetermined parameter, containing controlled parameters and Lagrange multipliers, constitutes the objects of evolution after the up-construction of the LSSVR model with adjustment [13,19,21]. This is due to the selection of these parameters significantly based on the theories of researchers. The parameters in this study are set up in common and appropriate ranges ($g : [0,6]$, $\sigma^2 : (0,1]$ and $C : (0,20]$) to reduce computing costs.

As a result, a vector is described as a common notation of the parameters employed in the PSO-LSSVR training procedure as shown in Equation (20):

$$\Psi = (\lambda_1, \lambda_2, \lambda_3, g, \sigma, C, \gamma_k, e) \tag{20}$$

It can now reformulate using the forecasting parameter constraints and the following model Equation (21).

$$\begin{array}{ll} \min & F(\Psi) = {||a||^2}/{2} + {C \sum_{k=1}^{N} \pi_k^2}/{2} \\ s.t & y_k - (a^T \cdot \phi(x_k) + e) = \pi_k \qquad k = 1, \ldots N \\ & \lambda_1 + \lambda_2 + \lambda_3 = 1 \\ & \lambda_1, \lambda_2, \lambda_3 \geq 0 \\ & 1 \leq n \leq \#attributions, n \in N+ \\ & 0 \leq g \leq 6, \quad 0 \leq \sigma \leq 1, \quad 0 \leq C \leq 20 \end{array} \tag{21}$$

The automatic LSSVR learning paradigm with a mixed kernel, the best input characteristics, and the optimized parameters are created by the aforementioned evolutionary processes. Six outstanding datasets from the UCI dataset can be used as testing targets in Section 4 for the developed automatic LSSVR learning paradigm as an example and evaluation tool.

4. Experiment Findings

The benchmark datasets and similar approaches are initially introduced in this section. Then, it shows how to identify the essential features that influence the outcomes of the prediction and describe the optimization procedure. Lastly, it evaluates the effectiveness of the suggested automatic LSSVR in comparison to a few other predicting models.

4.1. Benchmark Datasets and Compared Approaches

It tested six datasets referred to the UCI Machine Learning Repository [27] to confirm the robustness of our method, as shown in Table 1 and in more detail in Table 2.

Individual CP-LSSVR models with polynomial, RBF, and tangent kernels, collectively referred to as CP-LSSVRlinear, CP-LSSVRpoly, and CP-LSSVRRBF, were trained using the PSO algorithm (PSO-LSSVR) for further comparison. The data used for training and testing was for the six datasets is listed in Table 1. For example, among the total observations of Boston Housing Data is 506, 304 data are used for training data and the remaining 202 data are used for testing data.

Table 1. Data sets from the UCI.

No.	Data Sets	Observations	Training (100%)	Testing	Attributions
1	Boston Housing Data	506	304	202	13
2	Auto-Mpg	398	239	159	8
3	machine CPU	209	125	84	6
4	Servo	167	100	67	4
5	Concrete Compressive Strength	1030	618	412	8
6	Auto Price	159	95	64	14

Table 2. Detail data sets from the UCI.

Attribution	Boston Housing Data	Auto-Mpg	Machine CPU	Servo	Concrete Compressive Strength	Auto Price
1	CRIM	cylinders	MYCT	motor	Cement	normalized-losses
2	ZN	displacement	MMIN	screw	Blast Furnace Slag	wheel-base
3	INDUS	horsepower	MMAX	pgain	Fly Ash	length
4	CHAS	weight	CACH	vgain	Water	width
5	NOX	acceleration	CHMIN		Superplasticizer	height
6	RM	model year	CHMAX		Coarse Aggregate	curb-weight
7	AGE	origin			Fine Aggregate	engine-size
8	DIS	car name			Age	bore
9	RAD					stroke
10	TAX					compression-ratio
11	PTRATIO					horsepower
12	B					peak-rpm
13	LSTAT					city-mpg
14						highway-mpg
15						price

4.2. *Characteristic Selection Using PSO*

This study conducted six benchmarks to approve the PSO-based characteristic selection algorithm's resilience.

To create a characteristic selection to avoid overlapping, the training data is randomly picked. As the training data's size may be smaller than the test data or it might be too small to be regarded as typical training data, it reasoned that CP-LSSVR training with less than 50% training data is insufficient.

Three steps were included in each experiment.

Step 1: Use the initial value that the chaotic map process created in step 1 (for instance, the dataset Auto-Mpg Figure 5).

origin	A1	A2	A3	A4	A5	A6	A7	A8
encode	0	1	1	1	0	1	1	0
after	A2	A3	A4	A6	A7			

Figure 5. PSO to choose characteristics.

Step 2: Assess the fitness function.
Step 3: Before MCN.set $w = 0.9 - 0.5 \cdot j/MCN$, $j =$ iteration, choose the number of input characteristics using Binary PSO.

Step 4: Selected training data $k\%$ ($k = 50, 60, \ldots, 100$) are used to train the LSSVR with PSO.
Step 5: The trained LSSVR was tested for a set size.

Also, an AMD Turion (tm) 642 Mobile Technology TL-64 computer running at 2.2 GHz with 3 GB of memory was used to construct the PSO algorithm in the C++ programming language. The ideal values for these parameters are established in Table 3 following a test approach.

Table 3. The best values of these parameters.

No.	Data Sets	Number of Particles	Iteration	c_1	c_2	w_0	u
1	BostonHousing Data	50	200	2	2	0.9	4
2	Auto-Mpg	50	200	2	2	0.9	4
3	machine CPU	50	200	2	2	0.9	4
4	Servo	50	200	2	2	0.9	4
5	Concrete Compressive Strength	50	200	2	2	0.9	4
6	Auto Price	50	200	2	2	0.9	4

Tables 4–9 display the chosen characteristics. It should be noted that the major characteristics chosen indicate the best characteristic sets for all tests and were chosen through six separate experiments using various data sets created via data partition approach.

Table 4. Selected features for Boston Housing Data.

Training (%)	Selected Feature ID	#Features
50	1, 2, 4, 5, 8, 9, 11, 12	8
60	1, 2, 4, 6, 8, 13	6
70	1, 2, 3, 9, 11	5
80	1, 2, 4, 8, 9	5
90	1, 2, 4, 6, 9, 11	6
100	1, 2, 4, 8, 9, 11	6
Average		6.0000

Table 5. Selected features for Auto-Mpg.

Training (%)	Selected Feature ID	#Features
50	1, 2, 3, 4, 5, 8	6
60	1, 3, 4, 6, 8	5
70	1, 4, 7, 8	4
80	2, 3, 4, 6	4
90	1, 2, 4, 5, 7	5
100	1, 4, 6, 8	4
Average		4.6667

Table 6. Selected features for MACHINE CPU.

Training (%)	Selected Feature ID	#Features
50	1, 2, 3, 5, 6	5
60	1, 2, 4, 6	4
70	2, 3, 4, 6	4
80	1, 2, 4	3
90	1, 3, 4, 6	4
100	1, 2, 3, 4	4
Average		4.0000

Table 7. Selected features for Servo.

Training (%)	Selected Feature ID	#Features
50	1, 2, 3, 4	4
60	1, 2, 3	3
70	1, 2, 3, 4	4
80	1, 2, 3, 4	4
90	1, 2, 3, 4	4
100	1, 2, 3, 4	4
Average		3.8333

Table 8. Selected features for Concrete Compressive Strength.

Training (%)	Selected Feature ID	#Features
50	2, 3, 4, 5, 6, 8	6
60	1, 2, 3, 6, 7, 8	6
70	1, 2, 3, 7	4
80	1, 2, 4, 7	4
90	1, 2, 4, 6, 7	5
100	1, 2, 4, 7	4
Average		4.8333

Table 9. Selected features for Auto Price.

Training (%)	Selected Feature ID	#Features
50	1, 2, 3, 4, 5, 7, 9, 10, 12, 13, 14	11
60	2, 3, 4, 6, 7, 9, 10, 12, 13	9
70	1, 3, 4, 510, 12, 13	7
80	1, 2, 4, 5, 7, 10, 12, 13	8
90	1, 2, 4, 5, 7, 10, 12, 13	8
100	1, 2, 4, 5, 10, 12, 13	7
Average		8.3333

4.3. PSO-Based Parameter Optimization for CP-LSSVR

The mixed kernel's use in this paper also results in more unknown parameters. As a result, the chaotic initialization strategies are used, and the uncertain parameters comprise one feature election number, six controlled parameters, and N Lagrange multipliers.

Before providing the experimental findings, the evaluation criteria are specified for assessing the effects of the suggested algorithms. Present m is the quantity of the testing samples, $\hat{y}_i$ denotes the forecast value of $\hat{y}$, and $\overline{y} = \sum_i y_i / m$ is the mean of $y_1, \ldots, y_m$ without losing generality. Then, for algorithm evaluation, the following criteria are employed.

SSE: Sum squared error of testing, $SSE = \sum_{i=1}^{m} (y_i - \hat{y}_i)^2$. SSE stands for fitting accuracy; the lower the SSE, the more accurately the estimate fits the data. If noises have been employed as testing samples, a low SSE likely indicates that the regressor is overfitted.

SST: $SST = \sum_{i=1}^{m} (y_i - \overline{y})^2$ stands for the sum squared deviation of testing samples and represents the underlying variation of the testing samples, which often includes noise- and input-related volatility.

SSR: The sum squared deviation that the estimator can account for is referred to as SSR, $SSR = \sum_{i=1}^{m} (\hat{y}_i - \overline{y})^2$. The SSR reveals the regressor's capacity for explanation. SSR gathers more statistical data from test samples as it grows in size.

SSE/SST: Also known as the ratio of the sum squared error to the sum squared deviation of testing samples, $SSE/SST = \sum_{i=1}^{m} (y_i - \hat{y}_i)^2 / \sum_{i=1}^{m} (y_i - \overline{y})^2$. SSR/SSE is the ratio of the real sum squared deviation of testing samples to the interpretable sum squared deviation, $SSR/SST = \sum_{i=1}^{m} (\hat{y}_i - \overline{y})^2 / \sum_{i=1}^{m} (y_i - \overline{y})^2$ as defined. Little SSE/SST typically presents a strong compact between estimates and true data, and getting smaller SSE/SST typically requires raising SSR/SST [28,29].

The extraordinarily low value of SSE/SST is actually a bad thing because it suggests that the regressor is definitely overfitted. Due to this, an effective estimate should balance SSR/SST and SSE/SST.

The big Lagrange multiplier samples won't be shown here, but the forecast decision outcomes in Table 10 might be used to demonstrate the effectiveness of ideal Lagrange multipliers.

Table 10. Optimal Solution of Different Parameters and Prediction Performance.

Training (100%)	Weights of Mixed Kernel			Kernel Parameters		Upper Bound
	lada_1	**lada_2**	**lada_3**	**d**	**sigma**	**C**
Boston Housing Data	0.2140	0.3711	0.4148	2.8264	0.4331	3.0288
Auto-Mpg	0.2970	0.3417	0.3613	2.8863	0.3420	2.2992
machine CPU	0.1661	0.2654	0.5684	2.5853	0.4018	3.4588
Servo	0.3223	0.2743	0.4034	2.6001	0.5901	2.4512
Concrete Compressive Strength	0.2827	0.3197	0.3976	2.7553	0.6408	3.2270
Auto Price	0.3466	0.3828	0.2705	2.3582	0.3772	2.4982

The performance of the predictions is explained in detail. Secondly, it can be seen from the kernel mixed coefficients that the scales for three kernels are chosen with data characters for all test instances, even when several partition training data tests favor one or two kernels. Then, the kernel parameters have values that can be adjusted for various data sets. The upper bound parameter C reacts to how difficult it is to forecast data. By way of illustration, the big value C results in a narrow margin due to the high likelihood of misclassification. In conclusion, the PSO-based feature selection method used in the growing CP-LSSVR learning paradigm is quite reliable.

4.4. Comparisons and Discussion

Every similar regression model mentioned in the previous section is calculated using the training data in accordance with the experiment design. An empirical analysis depends on the testing data was then conducted after the model estimation selection procedure.

At this point, SSE/SST and SSR/SST were used to gauge how well the models predicted the future. Table 11 presents the results the comparable best results are marked in bold for each dataset.

The results are displayed in Table 11 and discussed as follows.

Initially, it is possible to see the differences between the models. For instance, the SSE/SST and SSR/SST for the CP-LSSVR for "Boston Housing Data" are 1.0437 and 0.1563, respectively.

1. The proposed CP-LSSVR performs best among the comparable methodologies for Servo in SSR/SST = 1.7869,
2. SVR performs best among the comparable methodologies for Boston Housing Data in both SSE/SST = 0.1274 and SSR/SST = 0.9032, Servo in SSE/SST = 0.1315, Concrete Compressive Strength in SSR/SST = 0.9425, Auto Price in SSE/SST = 0.1278.
3. LSSVR performs best among the comparable methodologies for Auto-Mpg in both SSE/SST = 0.1064 and SSR/SST = 0.9897, machine CPU in both SSE/SST = 0.1017 and

SSR/SST = 0.9877, Concrete Compressive Strength in SSE/SST = 0.1226, Auto Price in SSR/SST = 0.9952.

Table 11. Prediction Performance Percentages.

Data Sets	Regressor	SSE/SST	SSR/SST
Boston Housing Data	SVR	0.1274	0.9032
	LSSVR	0.1293	0.8964
	PSO-LSSVR	0.9091	0.1720
	CP-LSSVR	0.9900	0.1484
	CP-LSSVR	1.0000	0.1276
	CP-LSSVR	1.0030	0.1302
	CP-LSSVR	1.0437	0.1563
Auto-Mpg	SVR	0.1134	0.9873
	LSSVR	0.1064	0.9897
	PSO-LSSVR	0.9941	0.4608
	CP-LSSVR	0.9560	0.5095
	CP-LSSVR	1.0409	0.4609
	CP-LSSVR	1.0071	0.4688
	CP-LSSVR	1.0010	0.4884
machine CPU	SVR	0.1048	0.9813
	LSSVR	0.1017	0.9877
	PSO-LSSVR	0.9585	0.0064
	CP-LSSVR	0.9652	0.0103
	CP-LSSVR	0.9552	0.0104
	CP-LSSVR	0.9700	0.0055
	CP-LSSVR	0.9547	0.0058
Servo	SVR	0.1315	0.9774
	LSSVR	0.1331	0.9756
	PSO-LSSVR	0.9713	1.7185
	CP-LSSVR	1.0034	1.7251
	CP-LSSVR	1.0044	1.7869
	CP-LSSVR	1.0043	1.6734
	CP-LSSVR	1.0234	1.7577
Concrete Compressive Strength	SVR	0.1237	0.9425
	LSSVR	0.1226	0.9338
	PSO-LSSVR	0.9395	0.2030
	CP-LSSVR	0.9604	0.1944
	CP-LSSVR	0.9802	0.1836
	CP-LSSVR	0.9700	0.1942
	CP-LSSVR	0.9692	0.1936
Auto Price	SVR	0.1278	0.9821
	LSSVR	0.1288	0.9952
	PSO-LSSVR	0.9913	0.1858
	CP-LSSVR	0.9843	0.1803
	CP-LSSVR	0.9950	0.1982
	CP-LSSVR	1.0515	0.1639
	CP-LSSVR	1.0562	0.1862

The findings suggest that for mining and investigating prediction data, the proposed CP-LSSVR learning paradigm significantly outperforms the SVR model for the Servo dataset in SSR/SST. However, the SVR and LSSVR significantly outperform the compared methods including the proposed CP-LSSVR for the six datasets in both SSE/SST and SSR/SST.

Second, it, according to a mixed kernel model among the four CP-LSSVRs with various kernel functions, shows its expected performance in comparison to the other three single kernel models. It is primarily due to the mixed kernel's ability to absorb advantages and outweighs the negatives in each individual kernel function, as each has pros and cons of its own. Moreover, the chaotic PSO-based input characteristic selection method significantly lowers the function input variable, improving the function's capacity to be understood and effective. This is the main factor behind how the PSO-LSSVR performs less well than individual CP-LSSVR models.

Last but not least, the suggested CP-LSSVR learning paradigm exhibits comparative advantages over standalone CP-LSSVR models and contemporary techniques reported in the literature.

1. The CP-LSSVR has an SVR feature that can get beyond some of the BP-neural network's drawbacks, like overfitting and local minima.
2. Because it employs a mixed kernel, the CP-LSSVR offers better generalization capabilities for prediction. Intriguingly, Table 10 data that favors a higher percentage of particular kernels also revealed an outperforming result in Table 11 for that specific CP-LSSVR.
3. The chaotic PSO parameter optimization method can help improve the normalization effect. Fourth, the character development in the CP-LSSVR can quickly identify important factors that influence model performance, improving the LSSVR's interpretability.

5. Conclusions

This paper provides a least square support vector regression (LSSVR) algorithm that is automatically optimized using CPSO with a mixed kernel to address data forecasting issues. The CP-LSSVR model is composed of three components. In the first step, parameters were initialized using a chaotic map. In the second and third steps, PSO was adopted to choose the input characteristic combinations and optimize the LSSVR's parameters. Finally, the CP-LSSVR was used to forecast the six outstanding datasets that were acquired from the UCI dataset.

The CP-LSSVR model has two distinctly strong points. One is that the lesser number of features employed makes it easier to create an understandable forecasting model. Another is that all of its model parameters have been optimized, making it possible to construct the best forecasting model. It demonstrates the proposed CP-LSSVR model's ability to not only minimize forecasting error but also choose the most affordable model with the most crucial characteristics through a series of experiments. They approve the suggested approach can be utilized as a workable substitute for projected data mining and exploration.

However, it is important to keep in mind that the suggested CP-LSSVR learning paradigm might one day be enhanced in ways like ensemble learning and ensemble evolution with LSSVR. Additionally, this suggested approach can be used to solve practice issues in addition to regression problems, and it may even employ novel algorithms that enhance computation speed and solution quality. For instance, a number of other studies have attempted to offer effective strategies for parameter selection [30–51], and our method would benefit from incorporating these methods. Future research will examine these crucial concerns.

In future works, the other statistical figures of merits (like R2, RMSE, or MSE ...) can be incorporated. Additionally, the regression results in a plot (prediction vs. true) can be shown in future studies. Additionally, results can be provided considering different holdout % and holdout validation approaches [52]. More examples will be also considered to be included to further verify CP-LSSVR with more datasets in future work.

Author Contributions: Conceptualization, W.-C.Y. and W.Z.; methodology, W.-C.Y.; software, W.-C.Y.; validation, W.-C.Y.; formal analysis, W.-C.Y. and W.Z.; investigation, W.-C.Y. and W.Z.; resources, W.-C.Y. and W.Z.; data curation, W.-C.Y. and W.Z.; writing—original draft preparation, W.-C.Y. and W.Z.; writing—review and editing, W.-C.Y.; visualization, W.-C.Y.; supervision, W.-C.Y.; project administration, W.-C.Y.; funding acquisition, W.-C.Y. All authors have read and agreed to the published version of the manuscript.

Funding: This research was supported in part by National Natural Science Foundation of China (Grant No. 621060482), Research and Development Projects in Key Areas of Guangdong Province (Grant No. 2021B0101410002) and National Science and Technology Council, R.O.C (MOST 107-2221-E-007-072-MY3, MOST 110-2221-E-007-107-MY3, MOST 109-2221-E-424-002 and MOST 110-2511-H-130-002).

Data Availability Statement: The datasets referred to UCI Machine Learning Repository [27].

Acknowledgments: This article was once submitted to arXiv as a temporary submission that was just for reference and did not provide the copyright.

Conflicts of Interest: The authors declare no conflict of interest.

References

1. Vapnik, V.N. *Statistical Learning Theory*; Wiley: New York, NY, USA, 1998.
2. Yeh, W.C.; Jiang, Y.; Tan, S.Y.; Yeh, C.Y. A New Support Vector Machine Based on Convolution Product. *Complexity* **2021**, *2021*, 9932292. [CrossRef]
3. Ma, J.; Xia, D.; Guo, H.; Wang, Y.; Niu, X.; Liu, Z.; Jiang, S. Metaheuristic-based support vector regression for landslide displacement prediction: A comparative study. *Landslides* **2022**, *19*, 2489–2511. [CrossRef]
4. Samantaray, S.; Das, S.S.; Sahoo, A.; Satapathy, D.P. Monthly runoff prediction at Baitarani river basin by support vector machine based on Salp swarm algorithm. *Ain Shams Eng. J.* **2022**, *15*, 101732. [CrossRef]
5. Bansal, M.; Goyal, A.; Choudhary, A. A comparative analysis of K-Nearest Neighbour, Genetic, Support Vector Machine, Decision Tree, and Long Short Term Memory algorithms in machine learning. *Decis. Anal. J.* **2022**, *3*, 100071. [CrossRef]
6. Song, X.F.; Zhang, Y.; Gong, D.W.; Gao, X.Z. A Fast Hybrid Feature Selection Based on Correlation-Guided Clustering and Particle Swarm Optimization for High-Dimensional Data. *IEEE Trans. Cybern.* **2021**, *52*, 9573–9586. [CrossRef]
7. Hu, Y.; Zhang, Y.; Gao, X.; Gong, D.; Song, X.; Guo, Y.; Wang, J. A federated feature selection algorithm based on particle swarm optimization under privacy protection. *Knowl. Based Syst.* **2022**, *260*, 110122. [CrossRef]
8. Liu, X.; Wang, G.G.; Wang, L. LSFQPSO: Quantum particle swarm optimization with optimal guided Lévy flight and straight flight for solving optimization problems. *Eng. Comput.* **2022**, *38*, 4651–4682. [CrossRef]
9. Wei, C.L.; Wang, G.G. Hybrid Annealing Krill Herd and Quantum-Behaved Particle Swarm Optimization. *Mathematics* **2020**, *8*, 1403. [CrossRef]
10. You, G.R.; Shiue, Y.R.; Yeh, W.C.; Chen, X.L.; Chen, C.M. A weighted ensemble learning algorithm based on diversity using a novel particle swarm optimization approach. *Algorithms* **2020**, *13*, 255. [CrossRef]
11. Kennedy, J.; Eberhart, R. Particle swarm optimization. In Proceedings of the IEEE International Conference on Neural Networks (IJCNN), Perth, WA, Australia, 27 November–1 December 1995; Volume 4, pp. 1942–1948.
12. Hsieh, T.J.; Hsiao, H.F.; Yeh, W.C. Mining financial distress trend data using penalty guided support vector machines based on hybrid of particle swarm optimization and artificial bee colony algorithm. *Neurocomputing* **2012**, *82*, 196–206. [CrossRef]
13. Hsieh, T.J.; Yeh, W.C. Knowledge discovery employing grid scheme least squares support vector machines based on orthogonal design bee colony algorithm. *IEEE Trans. Syst. Man Cybern. Part B (Cybern.)* **2011**, *41*, 1198–1212. [CrossRef] [PubMed]
14. Mao, Y.; Wang, T.; Duan, M.; Men, H. Multi-objective optimization of semi-submersible platforms based on a support vector machine with grid search optimized mixed kernels surrogate model. *Ocean Eng.* **2022**, *260*, 112077. [CrossRef]
15. Li, M.; Yang, S.; Zhang, M. Power supply system scheduling and clean energy application based on adaptive chaotic particle swarm optimization. *Alex. Eng. J.* **2022**, *61*, 2074–2087. [CrossRef]
16. Silva-Juarez, A.; Rodriguez-Gomez, G.; Fraga, L.G.d.l.; Guillen-Fernandez, O.; Tlelo-Cuautle, E. Optimizing the Kaplan–Yorke Dimension of Chaotic Oscillators Applying DE and PSO. *Technologies* **2019**, *7*, 38. [CrossRef]
17. Smola, A.J.; Scholkopf, B. *A Tutorial on Support Vector Regression*; NeuroCOLT Tech. Rep. NC-TR-98-030; Royal Holloway College, Univ.: London, UK, 1998.
18. Jiao, L.; Bo, L.; Wang, L. Fast sparse approximation for least squares support vector machine. *IEEE Trans. Neural Netw.* **2007**, *18*, 685–697. [CrossRef]
19. Suykens, J.A.K.; Gestel, T.V.; Brabanter, J.D.; Moor, B.D.; Vandewalle, J. *Least Squares Support Vector Machines*; World Scientific: Singapore, 2002.
20. Schölkopf, B. Support Vector Learning. Ph.D. Thesis, Technische Universität, Berlin, Germany, 1997.
21. Yu, L.; Chen, H.; Wang, S.; Lai, K.K. Evolving Least Squares Support Vector Machines for Stock Market Trend Mining. *IEEE Trans. Evol. Comput.* **2009**, *13*, 87–102.
22. Yeh, W.C.; Yeh, Y.M.; Chang, P.C.; Ke, Y.C.; Chung, V. Forecasting wind power in the Mai Liao Wind Farm based on the multi-layer perceptron artificial neural network model with improved simplified swarm optimization. *Int. J. Electr. Power Energy Syst.* **2014**, *55*, 741–748. [CrossRef]
23. Yeh, W.C. A squeezed artificial neural network for the symbolic network reliability functions of binary-state networks. *IEEE Trans. Neural Netw. Learn. Syst.* **2017**, *28*, 2822–2825. [CrossRef]
24. Yeh, W.C.; Yeh, Y.M.; Chiu, C.W.; Chung, Y.Y. A wrapper-based combined recursive orthogonal array and support vector machine for classific ation and feature selection. *Mod. Appl. Sci.* **2014**, *8*, 11.

25. Chapelle, O.; Vapnik, V. Model selection for support vector machines. In Proceedings of the 13th Annual Conference on Neural Information Processing Systems (NIPS), Cambridge, MA, USA, 1 January 2000; pp. 230–236.
26. Tu, C.J.; Chuang, L.Y.; Chang, J.Y.; Yang, C.H. Feature Selection using PSO-SVM. *IAENG Int. J. Comput. Sci.* **2007**, *33*, IJCS_33_1_18.
27. UCI Machine Learning Repository. Available online: http://archive.ics.uci.edu/ml/index.html (accessed on 11 January 2023).
28. Staudte, R.G.; Sheather, S.J. *Robust Estimation and Testing: Wiley Series in Probability and Mathematical Statistics*; Wiley: New York, NY, USA, 1990.
29. Bates, D.M.; Watts, D.G. *Nonlinear Regression Analysis and its Applications*; Wiley: New York, NY, USA, 1988.
30. Zhou, J.; Zheng, W.; Wang, D.; Coit, D.W. A resilient network recovery framework against cascading failures with deep graph learning. *Proc. Inst. Mech. Eng. Part O J. Risk Reliab.* **2022**. [CrossRef]
31. Yousefi, N.; Tsianikas, S.; Coit, D.W. Dynamic maintenance model for a repairable multi-component system using deep reinforcement learning. *Qual. Eng.* **2022**, *34*, 16–35. [CrossRef]
32. Yeh, W.C. Novel Recursive Inclusion-Exclusion Technology Based on BAT and MPs for Heterogeneous-Arc Binary-State Networ k Reliability Problems. *Reliab. Eng. Syst. Saf.* **2023**, *231*, 108994. [CrossRef]
33. Liu, T.; Ramirez-Marquez, J.E.; Jagupilla, S.C.; Prigiobbe, V. Combining a statistical model with machine learning to predict groundwater flooding (or infiltration) into sewer networks. *J. Hydrol.* **2021**, *603*, 126916. [CrossRef]
34. Borrelli, D.; Iandoli, L.; Ramirez-Marquez, J.E.; Lipizzi, C. A Quantitative and Content-Based Approach for Evaluating the Impact of Counter Narratives on Affective Polarization in Online Discussions. *IEEE Trans. Comput. Soc. Syst.* **2021**, *9*, 914–925. [CrossRef]
35. Su, P.C.; Tan, S.Y.; Liu, Z.Y.; Yeh, W.C. A Mixed-Heuristic Quantum-Inspired Simplified Swarm Optimization Algorithm for scheduling of real-time tasks in the multiprocessor system. *Appl. Soft Comput.* **2022**, *1131*, 109807. [CrossRef]
36. Yeh, W.C.; Liu, Z.; Yang, Y.C.; Tan, S.Y. Solving Dual-Channel Supply Chain Pricing Strategy Problem with Multi-Level Programming Based on Improved Simplified Swarm Optimization. *Technologies* **2022**, *10*, 10030073. [CrossRef]
37. Yeh, W.C.; Tan, S.Y. Simplified Swarm Optimization for the Heterogeneous Fleet Vehicle Routing Problem with Time-Varying Continuous Speed Function. *Electronics* **2021**, *10*, 10151775. [CrossRef]
38. Bajaj, N.S.; Patange, A.D.; Jegadeeshwaran, R.; Pardeshi, S.S.; Kulkarni, K.A.; Ghatpande, R.S. Application of metaheuristic optimization based support vector machine for milling cutter health monitoring. *Intell. Syst. Appl.* **2023**, *18*, 200196. [CrossRef]
39. Patange, A.D.; Pardeshi, S.S.; Jegadeeshwaran, R.; Zarkar, A.; Verma, K. Augmentation of Decision Tree Model Through Hyper-Parameters Tuning for Monitoring of Cutting Tool Faults Based on Vibration Signatures. *J. Vib. Eng. Technol.* **2022**. [CrossRef]
40. Yeh, W.C. A new branch-and-bound approach for the n/2/flowshop/αF+ βCmax flowshop scheduling problem. *Comput. Oper. Res.* **1999**, *26*, 1293–1310. [CrossRef]
41. Yeh, W.C. Search for MC in modified networks. *Comput. Oper. Res.* **2001**, *28*, 177–184. [CrossRef]
42. Yeh, W.C.; Wei, S.C. Economic-based resource allocation for reliable Grid-computing service based on Grid Bank. *Future Gener. Comput. Syst.* **2012**, *28*, 989–1002. [CrossRef]
43. Hao, Z.; Yeh, W.C.; Wang, J.; Wang, G.G.; Sun, B. A quick inclusion-exclusion technique. *Inf. Sci.* **2019**, *486*, 20–30. [CrossRef]
44. Yeh, W.C. Novel binary-addition tree algorithm (BAT) for binary-state network reliability problem. *Reliab. Eng. Syst. Saf.* **2021**, *208*, 107448. [CrossRef]
45. Corley, H.W.; Rosenberger, J.; Yeh, W.C.; Sung, T.K. The cosine simplex algorithm. *Int. J. Adv. Manuf. Technol.* **2006**, *27*, 1047–1050. [CrossRef]
46. Yeh, W.C. A new algorithm for generating minimal cut sets in k-out-of-n networks. *Reliab. Eng. Syst. Saf.* **2006**, *91*, 36–43. [CrossRef]
47. Yeh, W.C.; He, M.F.; Huang, C.L.; Tan, S.Y.; Zhang, X.; Huang, Y.; Li, L. New genetic algorithm for economic dispatch of stand-alone three-modular microgrid in DongAo Island. *Appl. Energy* **2020**, *263*, 114508. [CrossRef]
48. Yeh, W.C.; Chuang, M.C.; Lee, W.C. Uniform parallel machine scheduling with resource consumption constraint. *Appl. Math. Model.* **2015**, *39*, 2131–2138. [CrossRef]
49. Lee, W.C.; Yeh, W.C.; Chung, Y.H. Total tardiness minimization in permutation flowshop with deterioration consideration. *Appl. Math. Model.* **2014**, *38*, 3081–3092. [CrossRef]
50. Lee, W.C.; Chuang, M.C.; Yeh, W.C. Uniform parallel-machine scheduling to minimize makespan with position-based learning curves. *Comput. Ind. Eng.* **2014**, *63*, 813–818. [CrossRef]
51. Bae, C.; Yeh, W.C.; Wahid, N.; Chung, Y.Y.; Liu, Y. A New Simplified Swarm Optimization (SSO) using Exchange Local Search Scheme. *Int. J. Innov. Comput. Inf. Control* **2012**, *8*, 4391–4406.
52. Bajaj, N.S.; Patange, A.D.; Jegadeeshwaran, R.; Kulkarni, K.A.; Ghatpande, R.S.; Kapadnis, A.M. A Bayesian Optimized Discriminant Analysis Model for Condition Monitoring of Face Milling Cutter Using Vibration Datasets. *J. Nondestruct. Eval.* **2022**, *5*, 021002. [CrossRef]

technologies

Article

A Layer-Wise Coupled Thermo-Elastic Shell Model for Three-Dimensional Stress Analysis of Functionally Graded Material Structures

Salvatore Brischetto *, Domenico Cesare and Roberto Torre

Department of Mechanical and Aerospace Engineering, Politecnico di Torino, Corso Duca degli Abruzzi 24, 10129 Torino, Italy
* Correspondence: salvatore.brischetto@polito.it

Abstract: In this work, a coupled 3D thermo-elastic shell model is presented. The primary variables are the scalar sovra-temperature and the displacement vector. This model allows for the thermal stress analysis of one-layered and sandwich plates and shells embedding Functionally Graded Material (FGM) layers. The 3D equilibrium equations and the 3D Fourier heat conduction equation for spherical shells are put together into a set of four coupled equations. They automatically degenerate in those for simpler geometries thanks to proper considerations about the radii of curvature and the use of orthogonal mixed curvilinear coordinates α, β, and z. The obtained partial differential governing the equations along the thickness direction are solved using the exponential matrix method. The closed form solution is possible assuming simply supported boundary conditions and proper harmonic forms for all the unknowns. The sovra-temperature amplitudes are directly imposed at the outer surfaces for each geometry in steady-state conditions. The effects of the thermal environment are related to the sovra-temperature profiles through the thickness. The static responses are evaluated in terms of displacements and stresses. After a proper and global preliminary validation, new cases are presented for different thickness ratios, geometries, and temperature values at the external surfaces. The considered FGM is metallic at the bottom and ceramic at the top. This FGM layer can be embedded in a sandwich configuration or in a one-layered configuration. This new fully coupled thermo-elastic model provides results that are coincident with the results proposed by the uncoupled thermo-elastic model that separately solves the 3D Fourier heat conduction equation. The differences are always less than 0.5% for each investigated displacement, temperature, and stress component. The differences between the present 3D full coupled model and the the advantages of this new model are clearly shown. Both the thickness layer and material layer effects are directly included in all the conducted coupled thermal stress analyses.

Keywords: 3D coupled thermo-elastic shell model; 3D Fourier heat conduction equation; 3D elastic equilibrium equations; functionally graded materials; plates; shells

Citation: Brischetto, S.; Cesare, D.; Torre, R. A Layer-Wise Coupled Thermo-Elastic Shell Model for Three-Dimensional Stress Analysis of Functionally Graded Material Structures. *Technologies* **2023**, *11*, 35. https://doi.org/10.3390/technologies11020035

Academic Editor: Manoj Gupta

Received: 25 January 2023
Revised: 21 February 2023
Accepted: 21 February 2023
Published: 24 February 2023

1. Introduction

The thermal stress analysis, for innovative and modern aerospace industries, is fundamental for new projects about space vehicles and aircraft where high temperature gradients and fast changes in temperature are classical operational conditions [1–4]. The study of the temperature gradient influence on strains and stresses is mandatory for the correct analysis of the performances of avant-garde structures. The main topic is related to the heat flux loads acting on the external structures of the airplanes and launch vehicles. These thermal loads and the mechanical loads can cause serious damage to the structure. For this reason, a proper mathematical formulation to analyse all these load effects must be developed. In addition, thanks to the technological improvement, the Functionally Graded Materials (FGMs), capable of continuously varying their mechanical and thermal properties along a given direction thanks to two or more constituent phases that vary over a defined volume, permit enhancing performances of the classical composite materials. The FGMs can vary their properties in each direction, but, in the aerospace sector, the thickness direction is the most common one. FGMs are even used for the thermal applications because they can better stand high temperatures at the same strength-to-weight ratio and there are not any layer interfaces [5]. In this paper, a full coupled thermo-elastic model is presented where the 3D equilibrium equations and the 3D Fourier heat conduction equation for spherical shells are employed. In this way, the sovra-temperature and the three displacement components are unknown variables of the problem. The most important characteristics of full coupled thermo-elastic models were shown in [6–12] where divergence and gradient equations, constitutive equations, boundary conditions, variational principles, field equations, proportionality equations between the heat flux and gradient of the thermal variable, energy balance equations, and the initial conditions were deeply discussed.

Several papers about the thermal stress analysis of FGM structures, both numerical and analytical models, are present in the open literature. In order to better remark the new characteristics of this new formulation, thermo-mechanical models for plates and shells embedding FGM layers are grouped as those related to 1D exact solutions, 1D numerical models, 2D exact solutions, 2D numerical models, 3D exact solutions, and 3D numerical models.

For what concerns the literature about the 1D exact solutions, Kapuria et al. [13] presented a third-order zigzag theory, together with the modified rule of mixtures for the effective modulus of elasticity for layered FGM beams. This model was evaluated for static and free vibration analyses. Ghiasian et al. [14] presented the buckling of FGM beams under different thermal loads. The beam was resting over a three-parameter elastic foundation with hardening/softening cubic nonlinearity, which acted in tension as well as in compression. Kiani and Eslami [15] showed the static and dynamic buckling of an FGM beam under uniform temperature rise and uniform compression. The material properties in [14,15] varied along the thickness direction. The Timoshenko beam model resting on a two-parameter non-linear elastic foundation was implemented in [16] in the case of the thermal buckling of FGM beams subjected to a temperature rise. Ma and Lee [17] presented a closed-form solution for the nonlinear static responses of FGM beams under uniform in-plane thermal loads. The three governing equations for the axial and transverse deformations of beams were reduced to a single nonlinear fourth-order integral–differential equation. The geometrically non-linear post-buckling load–deflection behaviour of FGM Timoshenko beams under in-plane thermal loadings was discussed in [18]. The thermal loads were applied by providing a non-uniform temperature rise across the beam thickness in steady-state conditions. Zhang et al. [19] proposed a thermal buckling of ceramic–metal FGM beams subjected to a transversely non-uniform temperature rise. The investigation method was the symplectic theory in the Hamiltonian system.

In the framework of 1D numerical models, Chakraborty et al. [20] presented a beam element for thermoelastic analysis based on the first-order shear deformation theory where elastic and thermal properties changed along the thickness direction. The interpolating polynomials were created thanks to the exact solution of the same beam model. In [21], a thermo-elastic vibration analysis of FGM beams with general boundary conditions was presented. A higher-order shear beam deformation theory with material properties dependent on the temperature was used. An improved Finite Element Method (FEM) was shown in [22]. Thanks to this new FE model, the transverse and axial vibrations of FGM beams under thermal fields and exposed to a moving mass were analysed. Esfahani et al. [23] proposed a thermal buckling and post-buckling analysis of FGM structures using the Timoshenko beam model resting on a non-linear elastic foundation. Both thermal and mechanical properties were functions of the temperature and position. Tang and Li [24] proposed FGM slender beam analyses. The numerical model was created via the principle of the minimum for the total potential energy in order to derive the non-linear governing equations of the structures. The buckling of FGM beams with different boundary conditions was presented in [25]. The numerical model was based on the Timoshenko beam theory and the critical buckling load of the structures was evaluated using the Ritz method. Ziane et al. [26] presented a numerical model, using the Galerkin method, to analyse the thermal buckling of simply supported and clamped–clamped FGM box beams.

In the case of 2D exact solutions, Jahaveri and Eslami [27] presented stability and equilibrium equations for rectangular plates constituted of FGM layers under thermal loads. The equations were derived from the high-order shear deformation theory. Akbaş [28] proposed free vibration and static analyses of simply supported FGM plates; the porosity effect was included. Saad and Hadji [29] showed a thermal buckling problem for porous thick rectangular plates constituted of FGM layers using a high-order shear deformation theory. Sangeetha et al. [30] presented a closed form solution for FGM plates under thermal loads using a refined model based on the first-order shear deformation theory. The effects of thermal stresses were studied for several temperature variations across the thickness direction of the plate. In [31], Zenkour and Mashat presented thermal buckling responses using a Sinusoidal shear deformation Plate Theory (SPT). The governing equations were derived using SPT and solved in a closed form. Yaghoobi and Ghannad [32] proposed a thermal analysis for FGM cylinders under non-uniform heat fluxes. The governing equations were based on the first-order temperature theory and the energy method. Zeighami and Jafari [33] proposed a solution for the thermo-mechanical analysis of functionally graded carbon nanotube-reinforced composite plates with a central hole. This solution was possible thanks to the Lekhnitskii complex potential approach and the proper conformal mapping functions.

In the area of 2D numerical models, Praveen and Reddy [34] proposed a finite element able to take into account transverse shear strains, rotary inertia, and large rotations for FGM ceramic–metal plates. Static and dynamic analyses were conducted. Thai et al. [35] presented a four-unknown shear and normal deformation theory for static, dynamic and buckling analyses of FGM plates where the 3D material matrix was used. The system of equations was derived using the Galerkin weak form and the isogeometric analysis. In [36], a non-linear finite element model was presented to study the dynamic response of FGM structures under the thermal and mechanical harmonic loads. In this case, the FGM properties depended on the temperature and they varied in the thickness direction via a power law distribution. The effects of the material variation through the thickness and the size of the FGM were studied using the finite element method in [37] using the Crank–Nicolson–Galerkin scheme. Alibeigloo [38] showed the bending analysis of the FGM sandwich circular plates under thermo-mechanical loads. The used method was the Generalized Differential Quadrature method (GDQ) and the temperature distribution in the 3D form was computed by solving the heat conduction governing the equation in closed form. Hong [39] proposed a GDQ third-order shear deformation plate theory for FGM structures under thermal vibrations. Karakoti et al. [40] showed an eight-nodes isoparametric finite element in order to obtain a nonlinear transient response of porous FGM sandwich plates and shells. The FGM structure was subjected to blast loadings and thermal loads. Jooybar et al. [41] presented a numerical model where the equations of motion (and related boundary conditions), derived thanks to the Hamilton principle, were solved with the use of the differential quadrature method. The embedded FGM was temperature dependent. In [42], a thermal buckling analysis of the FGM sandwich plates, using an improved mesh-free Radial Point Interpolation Method (RPIM), was presented. This buckling formulation for plates was derived from an improvement of RPIM employing a new radial basis function. Therefore, the shape functions were built without any supporting fixing parameters based on the higher-order shear deformation plate theory. Qi et al. [43] showed a dynamic analysis for stiffened doubly curved sandwich composite panels with an FGM core and two isotropic layers under thermal loads. This analysis was based on von Kármán non-linear strain–displacement relationships and classical plate theory. The mathematical problem was solved by adopting the finite difference model and the Newmark method. Taj et al. [44] presented the static analysis of FGM plates using the HSDT (High Shear Deformation Theory) where the transverse shear stress was represented as quadratic along the thickness direction; the material properties of the FGM varied along the same direction.

For what concerns the 3D analytical solutions, they can be used to validate all the previously discussed models because they can consider all the peculiarities for geometry and lamination. Reddy and Cheng [45] proposed a theory for the bending of rectangular plates embedding FGM layers and piezoelectric actuators. In this way, when the FGM surface was subjected to thermal loads, displacements and stresses can be controlled. In [46], an analytical solution was presented for 3D steady and transient heat conduction problems of double-layer plates (including a coating layer and an FGM layer) with a local heat source. For this solution, the Poisson method and the layerwise approach were employed. Chen et al. [47] proposed a method based on state–space formulations with laminate approximations. The employed FGM was temperature dependent. Ootao e Tanigawa [48] showed a theoretical method for transient thermoelastic problems involving orthotropic FGM rectangular plates and non-uniform heat supply. The transient 3D temperature was analysed with an exponential law in the thickness direction. The same authors also [49] proposed a theoretical analysis of a 3D thermal stress problem for FGM structures subjected to a partial heat supply in a transient state. Jabbari et al. [50] discussed an exact solution for the steady state thermo-elastic problem of 3D simply supported circular FGM plates. Thermal and mechanical loads were axisymmetrically applied at the outer surfaces. Vel and Batra [51] presented an exact solution for 3D deformations of simply supported FGM rectangular plates subjected to mechanical and thermal loads at the external surfaces. Proper temperature and displacement functions were used in order to satisfy the boundary conditions at the edges and to reduce the system of partial differential equations for the thermo-elastic problem. Liu [52] discussed a 3D axisymmetric FGM circular plate under thermal loads at outer surfaces. A proper temperature function for the thermal boundary conditions at the edges was used and the variable separation method was employed to reduce the order of the set of governing equations in steady-state heat conduction. Alibeigloo [53] showed a 3D thermo-elastic model for FGM rectangular plates with simply supported edges under thermo-mechanical loads. The analytical solutions for temperature, stress, and displacement fields were proposed by using the Fourier series and the state–space method.

In the framework of 3D numerical (FEM, meshless methods, and GDQ) models, a study about the thermal elastic residual stresses occurring in Ni–Al_2O_3, Ni–TiO_2, and Ti–SiC FG plates, due to different temperature fields through the plate thickness, was presented in [54]. A 3D eight-nodes isoparametric-layered finite element with three degrees of freedom per node was implemented here. In [55], Hajlaoui et al. proposed a modified first-order enhanced solid-shell element formulation for the thermal buckling of functionally graded shells. The material properties varied in the thickness direction via a power law. In [56], a 3D free vibration analysis of shells constituted of laminated FGMs was shown thanks to the use of the quadrature element method. The shell geometry was analysed both in thermal and non-thermal configurations. Burlayenko et al. [57] presented a 3D analysis for free vibrations of thermally FGM sandwich plates. The material properties varied along the thickness direction and the analysis was conducted using the ABAQUS FE code. A 3D analysis of FGM cylinders containing semi-elliptical circumferential surface crack and thermo-mechanical loading was presented in [58]. The variation law for the Young modulus was exponential through the thickness. Naghdabadi and Kordkheili [59] proposed an FE formulation for the thermoelastic analysis of FGM plates and shells. The power law distribution for the composition of the constituent phase varied in the thickness direction. Qian and Batra [60] proposed a transient heat conduction analysis for thick FGM plates by using a higher-order plate theory and a meshless local Petrov–Galerkin method. Mian and Spencer [61] proposed models for both laminated and FGM structures in a Cartesian coordinate system. The comparisons between the two materials were discussed.

This new 3D coupled thermo-elastic shell model is valid for several geometries: spherical shells and cylindrical panels, cylinders, and plates. A closed-form solution is proposed using the simply supported hypotheses for the edges and the harmonic forms for the primary variables. The materials embedded in the proposed structures could be classical or functionally graded along the thickness direction. The 3D equilibrium equations coupled with the 3D heat conduction equation for shells are solved thanks to the exponential matrix method along the thickness direction. A layer-wise approach is employed. The present 3D exact coupled thermo-elastic shell model can be seen as the general case of the pure mechanical model already developed by Brischetto in [62,63] in the case of free vibration and bending analyses, respectively. The addition of the thermal-related equation in the orthogonal mixed curvilinear coordinates (see [64–67]) to 3D equilibrium relations creates a homogeneous differential equation set that can be solved via the procedure shown in [68,69]. The new results are presented in terms of displacements, stresses, and temperature profiles. They can be used as benchmarks for the development and testing of new 3D, 2D, and 1D numerical models for thermal stress analyses of FGM structures.

The paper is organized as follows. Section 2 is about the coupled thermo-elastic governing equations for spherical shells and the related solution methodology. Section 3 covers the results and is split in a first subsection for preliminary assessments and a second subsection for new benchmarks. Section 4 provides the main conclusions.

2. 3D Exact and Coupled Thermo-Elastic Governing Equations for Spherical Shells

This section shows the development of the 3D coupled exact thermo-elastic shell model for FGM structures. Plates, cylinders, cylindrical, and spherical shells (see the differences between the four geometries in Figure 1) are analysed using the same model thanks to appropriate considerations about the radii of the curvature R_α and R_β.

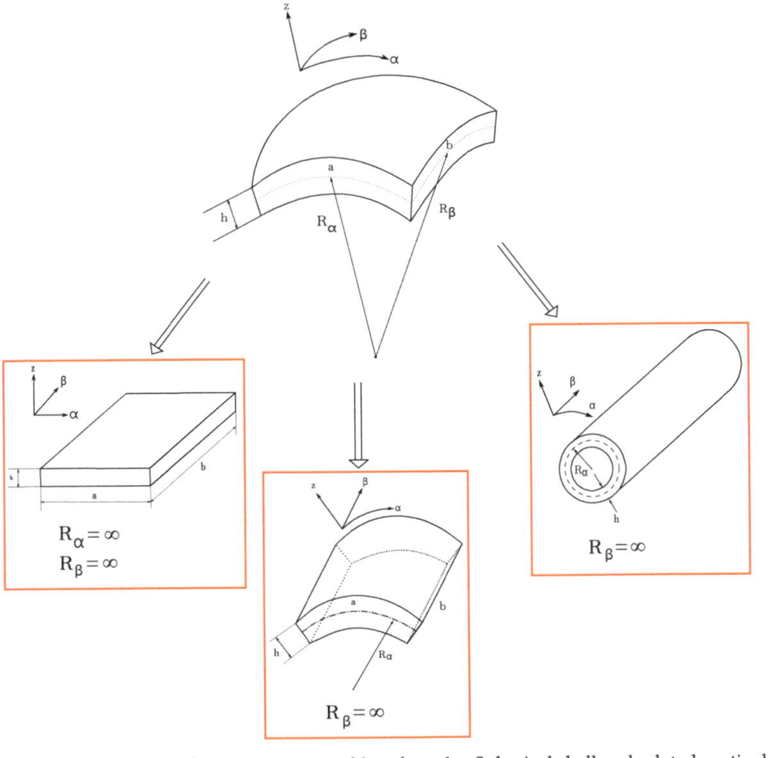

Figure 1. Geometries for assessments and benchmarks. Spherical shell and related particular cases as plate, cylindrical shell, and cylinder.

In Figure 1, the orthogonal mixed curvilinear reference system (α, β, z) has its origin in a corner. The in-plane directions α and β are parallel to the lateral curved sides and they lie on the middle surface Ω_0. The Ω_0 surface is the reference surface for the computation of all the geometrical parameters. The thickness direction z is normal to the Ω_0 surface and directed from the bottom to the top surface. This model has, as unknown variables, the three displacement components u, v, and w in the three directions of the reference system and the scalar sovra-temperature θ. The 3D elastic equilibrium equations and the 3D Fourier heat conduction equation are put together in the same system that is solved as a system of second-order partial differential equations where the unknowns are displacements, temperature, and related derivatives created with respect to z.

2.1. 3D Equilibrium and Heat Conduction Equations for Spherical Shells

The starting point for the mathematical formulation of the present 3D full coupled exact thermo-elastic shell model is the writing of the 3D equilibrium equations and the 3D Fourier heat conduction relation for spherical shells:

$$H_\beta(z)\frac{\partial\sigma^k_{\alpha\alpha}}{\partial\alpha} + H_\alpha(z)\frac{\partial\sigma^k_{\alpha\beta}}{\partial\beta} + H_\alpha(z)H_\beta(z)\frac{\partial\sigma^k_{\alpha z}}{\partial z} + \Big(\frac{2H_\beta(z)}{R_\alpha} + \frac{H_\alpha(z)}{R_\beta}\Big)\sigma^k_{\alpha z} = 0, \quad (1)$$

$$H_\beta(z)\frac{\partial\sigma^k_{\alpha\beta}}{\partial\alpha} + H_\alpha(z)\frac{\partial\sigma^k_{\beta\beta}}{\partial\beta} + H_\alpha(z)H_\beta(z)\frac{\partial\sigma^k_{\beta z}}{\partial z} + \Big(\frac{2H_\alpha(z)}{R_\beta} + \frac{H_\beta(z)}{R_\alpha}\Big)\sigma^k_{\beta z} = 0, \quad (2)$$

$$H_\beta(z)\frac{\partial\sigma^k_{\alpha z}}{\partial\alpha} + H_\alpha(z)\frac{\partial\sigma^k_{\beta z}}{\partial\beta} + H_\alpha(z)H_\beta(z)\frac{\partial\sigma^k_{zz}}{\partial z} - \frac{H_\beta(z)}{R_\alpha}\sigma^k_{\alpha\alpha} - \frac{H_\alpha(z)}{R_\beta}\sigma^k_{\beta\beta} + \\ + \Big(\frac{H_\beta(z)}{R_\alpha} + \frac{H_\alpha(z)}{R_\beta}\Big)\sigma^k_{zz} = 0, \quad (3)$$

$$\kappa_1^{*k}(z)\frac{\partial^2\theta}{\partial\alpha^2} + \kappa_2^{*k}(z)\frac{\partial^2\theta}{\partial\beta^2} + \kappa_3^{*k}(z)\frac{\partial^2\theta}{\partial z^2} = 0. \quad (4)$$

Equations (1)–(3) are the 3D equilibrium equations, and they are linked with the 3D Fourier heat conduction equation in steady-state condition (Equation (4)) in order to couple the mechanical field with the thermal one. The 3D Fourier heat conduction relation in Equation (4) is proposed in a steady-state form. Therefore, the dependence on the time is discarded. R_α and R_β are the radii of curvature in the α and β directions, respectively, and they are constant values. The $H_\alpha(z)$ and $H_\beta(z)$ coefficients presented in Equations (1)–(4) are linear functions of the thickness coordinate z or $\tilde{z}$. They introduced the curvature effects of the shells and they are defined, for each direction, as:

$$H_\alpha(z) = \Big(1 + \frac{z}{R_\alpha}\Big) = \Big(1 + \frac{\tilde{z} - h/2}{R_\alpha}\Big), \quad (5)$$

$$H_\beta(z) = \Big(1 + \frac{z}{R_\beta}\Big) = \Big(1 + \frac{\tilde{z} - h/2}{R_\beta}\Big), \quad (6)$$

$$H_z = 1. \quad (7)$$

The term h shown in Equations (5)–(7) represents the total thickness of the structure. h is always considered as constant. The variable z is in the range between $-h/2$ and $h/2$ and $\tilde{z}$ is in the range between 0 and h. In Equation (4), the conduction coefficients $\kappa_1^{*k}(z)$, $\kappa_2^{*k}(z)$, and $\kappa_3^{*k}(z)$ are defined as:

$$\kappa_1^{*k}(z) = \frac{\kappa_1^k(z)}{H_\alpha^2(z)}, \quad \kappa_2^{*k}(z) = \frac{\kappa_2^k(z)}{H_\beta^2(z)}, \quad \kappa_3^{*k}(z) = \kappa_3^k(z)\,. \quad (8)$$

where $\kappa_1^k(z)$, $\kappa_2^k(z)$ and $\kappa_3^k(z)$ are the conduction coefficients in the three directions of the mixed orthogonal reference system. If the k layer is an FGM, they depend on z. $\kappa_1^{*k}(z)$, $\kappa_2^{*k}(z)$, and $\kappa_3^{*k}(z)$ also depend on the curvature via the use of $H_\alpha(z)$ and $H_\beta(z)$. It must be noted that, in the case of a plate, the $\kappa_i^{*k}(z)$ conduction coefficients degenerate in $\kappa_i^k(z)$ because there are no curvatures involved ($H_\alpha = H_\beta = 1$).

2.2. 3D Geometrical and Constitutive Relations

The geometrical equations used for this thermo-mechanical model consider both the strains linked to the displacements $\boldsymbol{u}^k$ and the strains linked to the sovra-temperature θ^k. The equations can be written in matrix form as:

$$\boldsymbol{\epsilon}^k = \Big(\boldsymbol{\Delta}(z) + \boldsymbol{G}(z)\Big)\boldsymbol{u}^k - \boldsymbol{\mu}^k(z)\theta^k \quad (9)$$

where $\boldsymbol{\epsilon}^k$ is the 6×1 strain vector, $\boldsymbol{\Delta}(z)$ is a 6×3 matrix containing the differential terms for the shell configuration, $\boldsymbol{G}(z)$ is a 6×3 matrix that includes the pure geometrical curvature terms, $\boldsymbol{u}^k$ is the 3×1 displacement vector, and $\boldsymbol{\mu}(z)^k$ is the 6×1 vector containing the thermal expansion coefficients evaluated in α, β, and z directions. The explicit form of these matrices and vectors are here presented:

$$\boldsymbol{\epsilon}^k = \begin{bmatrix} \epsilon^k_{\alpha\alpha} \\ \epsilon^k_{\beta\beta} \\ \epsilon^k_{zz} \\ \gamma^k_{\beta z} \\ \gamma^k_{\alpha z} \\ \gamma^k_{\alpha\beta} \end{bmatrix}, \quad \boldsymbol{\Delta}(z) = \begin{bmatrix} \frac{\partial}{\partial\alpha}\frac{1}{H_\alpha(z)} & 0 & 0 \\ 0 & \frac{\partial}{\partial\beta}\frac{1}{H_\beta(z)} & 0 \\ 0 & 0 & \frac{\partial}{\partial z} \\ 0 & \frac{\partial}{\partial z} & \frac{\partial}{\partial\beta}\frac{1}{H_\beta(z)} \\ \frac{\partial}{\partial z} & 0 & \frac{\partial}{\partial\alpha}\frac{1}{H_\alpha(z)} \\ \frac{\partial}{\partial\beta}\frac{1}{H_\beta(z)} & \frac{\partial}{\partial\alpha}\frac{1}{H_\alpha(z)} & 0 \end{bmatrix},$$

$$\boldsymbol{G}(z) = \begin{bmatrix} 0 & 0 & \frac{1}{H_\alpha(z)R_\alpha} \\ 0 & 0 & \frac{1}{H_\beta(z)R_\beta} \\ 0 & 0 & 0 \\ 0 & -\frac{1}{H_\beta(z)R_\beta} & 0 \\ -\frac{1}{H_\alpha(z)R_\alpha} & 0 & 0 \\ 0 & 0 & 0 \end{bmatrix},$$

$$\boldsymbol{u}^k = \begin{bmatrix} u^k \\ v^k \\ w^k \end{bmatrix}, \quad \boldsymbol{\mu}^k(z) = \begin{bmatrix} \mu^k_\alpha(z) \\ \mu^k_\beta(z) \\ \mu^k_z(z) \\ 0 \\ 0 \\ 0 \end{bmatrix}. \quad (10)$$

The superscript k indicates that the matrix and the vector are valid for each physical layer. In order to obtain these coefficients, a rotation from the material reference system (1,2,3) to the structural reference system (α, β, and z) has been employed.

Thanks to the constitutive equations, the strain components can be related to the six stress components $\boldsymbol{\sigma}^k = [\sigma^k_{\alpha\alpha}\ \sigma^k_{\beta\beta}\ \sigma^k_{zz}\ \sigma^k_{\beta z}\ \sigma^k_{\alpha z}\ \sigma^k_{\alpha\beta}]^T$. The constitutive relation is the well-known Hooke law:

$$\boldsymbol{\sigma}^k = \boldsymbol{C}^k(z)\boldsymbol{\epsilon}^k \quad (11)$$

where the vector of mechanical strains $\boldsymbol{\epsilon}^k$ is the algebraic summation of geometrical strains and thermal strains. The elastic coefficient matrix $\boldsymbol{C}^k(z)$ has 6×6 dimension and it depends on z in the case of a k layer constituted of an FGM. The elastic coefficient matrix used in this formulation has $C^k_{16}(z) = C^k_{26}(z) = C^k_{36}(z) = C^k_{45}(z) = 0$ (it implies only $0°/90°$ orthotropic angle) to solve, in closed form, the mathematical problem with respect to the unknown displacements and temperature amplitudes by means of the Navier methodology in the α and β plane directions and the exponential matrix method in the z direction.

Using the constitutive relation expressed in Equation (11) and substituting Equation (9) in it:

$$\begin{aligned}\boldsymbol{\sigma}^k &= \boldsymbol{C}^k(z)\boldsymbol{\epsilon}^k = \boldsymbol{C}^k(z)\Big[\Big(\boldsymbol{\Delta}(z) + \boldsymbol{G}(z)\Big)\boldsymbol{u}^k - \boldsymbol{\mu}^k(z)\theta^k\Big] = \\ &= \boldsymbol{C}^k(z)\Big(\boldsymbol{\Delta}(z) + \boldsymbol{G}(z)\Big)\boldsymbol{u}^k - \boldsymbol{C}^k(z)\boldsymbol{\mu}^k(z)\theta^k = \mathscr{M}^k(z)\boldsymbol{u}^k - \boldsymbol{\lambda}^k(z)\theta^k\end{aligned} \tag{12}$$

where $\mathscr{M}^k(z)$ has 6×3 dimension and it indicates the pure mechanical coefficients; $\boldsymbol{\lambda}^k(z)$ denotes the presence of the thermo-mechanical coupling coefficients in the structural reference system. Vector $\boldsymbol{\lambda}^k(z)$ is defined as:

$$\boldsymbol{\lambda}^k(z) = \begin{bmatrix}\lambda^k_\alpha(z)\\ \lambda^k_\beta(z)\\ \lambda^k_z(z)\\ 0\\ 0\\ 0\end{bmatrix} = \begin{bmatrix}C^k_{11}(z) & C^k_{12}(z) & C^k_{13}(z) & 0 & 0 & 0\\ C^k_{12}(z) & C^k_{22}(z) & C^k_{23}(z) & 0 & 0 & 0\\ C^k_{13}(z) & C^k_{23}(z) & C^k_{33}(z) & 0 & 0 & 0\\ 0 & 0 & 0 & C^k_{44}(z) & 0 & 0\\ 0 & 0 & 0 & 0 & C^k_{55}(z) & 0\\ 0 & 0 & 0 & 0 & 0 & C^k_{66}(z)\end{bmatrix}\begin{bmatrix}\mu^k_\alpha(z)\\ \mu^k_\beta(z)\\ \mu^k_z(z)\\ 0\\ 0\\ 0\end{bmatrix}. \tag{13}$$

According to the previously described substituting steps, the stresses are linked with u, v, w, and θ, which are the unknown variables of the formulation.

2.3. *Exponential Matrix Methodology and Layer Wise Approach*

In order to solve the problem in an exact and closed form, there is the necessity of the harmonic form for the primary unknowns u, v, w, and θ:

$$u^k(\alpha,\beta,z) = U^k(z)cos(\bar{\alpha}\alpha)sin(\bar{\beta}\beta)\,, \tag{14}$$

$$v^k(\alpha,\beta,z) = V^k(z)sin(\bar{\alpha}\alpha)cos(\bar{\beta}\beta)\,, \tag{15}$$

$$w^k(\alpha,\beta,z) = W^k(z)sin(\bar{\alpha}\alpha)sin(\bar{\beta}\beta)\,, \tag{16}$$

$$\theta^k(\alpha,\beta,z) = \Theta^k(z)sin(\bar{\alpha}\alpha)sin(\bar{\beta}\beta)\,. \tag{17}$$

Thanks to these impositions, the boundary conditions for all the structures are the simply supported sides. The amplitudes of the unknowns are identified with the appropriate capital letter (e.g., the amplitude of the displacement u is written as U). The $\bar{\alpha}$ and $\bar{\beta}$ coefficients depend on the in-plane dimensions a and b and on the half-wave numbers m and n in the in-plane directions as follows:

$$\bar{\alpha} = \frac{m\pi}{a}, \quad \bar{\beta} = \frac{n\pi}{b}. \tag{18}$$

Substituting the unknowns written in harmonic form (Equations (14)–(17)), the geometrical relations (Equation (9)) and the constitutive relations (Equation (11)) into the equilibrium relations (Equations (1)–(4)), four second-order differential equations are written as follows:

$$A_1^k(z)U^k(z) + A_2^k(z)V^k(z) + A_3^k(z)W^k(z) + J_1^k(z)\Theta^k(z) + A_4^k(z)U_{,z}^k(z) + \\ + A_5^k(z)W_{,z}^k(z) + A_6^k(z)U_{,zz}^k(z) = 0\,, \tag{19}$$

$$A_7^k(z)U^k(z) + A_8^k(z)V^k(z) + A_9^k(z)W^k(z) + J_2^k(z)\Theta^k(z) + A_{10}^k(z)V_{,z}^k(z) + \\ + A_{11}^k(z)W_{,z}^k(z) + A_{12}^k(z)V_{,zz}^k(z) = 0\,, \tag{20}$$

$$A_{13}^k(z)U^k(z) + A_{14}^k(z)V^k(z) + A_{15}^k(z)W^k(z) + J_3^k(z)\Theta^k(z) + A_{16}^k(z)U_{,z}^k(z) + \\ + A_{17}^k(z)V_{,z}^k(z) + A_{18}^k(z)W_{,z}^k(z) + J_4^k(z)\Theta_{,z}^k(z) + A_{19}^k(z)W_{,zz}^k(z) = 0\,, \tag{21}$$

$$\left(J_5^k(z) + J_6^k(z)\right)\Theta^k(z) + J_7^k(z)\Theta_{,zz}^k(z) = 0\,, \tag{22}$$

where coefficients $A_s^k(z)$ (s rises from 1 to 19) and $J_r^k(z)$ (r varies from 1 to 7) are not constant (they depend on z). Note that Equations (19)–(22) are written in a generic k-th physical layer.

The unknowns, from this moment forward, are the amplitudes of the displacements and sovra-temperature and the related first-order derivatives (only created along the z direction because the ones created in in-plane directions are exactly evaluated and they become constant values). All four equations have coefficients that depend on z because of the curvature terms $H_\alpha(z)$ and $H_\beta(z)$ (see Equations (5)–(7)) and some FGM layers. In order to transform this set of non-constant coefficient second-order differential equations into a constant coefficients set, each k physical layer is split up into several fictitious layers. So, a number of M fictitious layers are employed along the thickness direction of the structure to discretise curvature terms and FGM properties. To define these fictitious layers, a new index has to be introduced: j is the index that counts these layers and it varies from 1 to M (total number of fictitious layers employed in the thickness direction). M can be easily calculated with the relation $M = s \cdot k$ where s indicates the number of subdivisions of each k physical layer. In these fictitious layers, the coefficients $H_\alpha(z)$ and $H_\beta(z)$ and FGM properties can be exactly calculated in their middle points and the obtained values are constant within each fictitious layer. From this point forward, all the equations proposed are written in the j fictitious layer and they have constant coefficients. After the introduction of the fictitious layers, the system expressed in Equations (19)–(22) can be transformed into a first-order system by redoubling the number of unknowns (as stated in [68,69]). After the redoubling of the unknowns, the set of first-order differential equations can be reported in compact form as:

$$\begin{bmatrix} A_6^j & 0 & 0 & 0 & 0 & 0 & 0 & 0 \\ 0 & A_{12}^j & 0 & 0 & 0 & 0 & 0 & 0 \\ 0 & 0 & A_{19}^j & 0 & 0 & 0 & 0 & 0 \\ 0 & 0 & 0 & J_7^j & 0 & 0 & 0 & 0 \\ 0 & 0 & 0 & 0 & A_6^j & 0 & 0 & 0 \\ 0 & 0 & 0 & 0 & 0 & A_{12}^j & 0 & 0 \\ 0 & 0 & 0 & 0 & 0 & 0 & A_{19}^j & 0 \\ 0 & 0 & 0 & 0 & 0 & 0 & 0 & J_7^j \end{bmatrix} \begin{bmatrix} U^j \\ V^j \\ W^j \\ \Theta^j \\ U^{j'} \\ V^{j'} \\ W^{j'} \\ \Theta^{j'} \end{bmatrix}' =$$

$$\begin{bmatrix} 0 & 0 & 0 & 0 & A_6^j & 0 & 0 & 0 \\ 0 & 0 & 0 & 0 & 0 & A_{12}^j & 0 & 0 \\ 0 & 0 & 0 & 0 & 0 & 0 & A_{19}^j & 0 \\ 0 & 0 & 0 & 0 & 0 & 0 & 0 & J_7^j \\ -A_1^j & -A_2^j & -A_3^j & -J_1^j & -A_4^j & 0 & -A_5^j & 0 \\ -A_7^j & -A_8^j & -A_9^j & -J_2^j & 0 & -A_{10}^j & -A_{11}^j & 0 \\ -A_{13}^j & -A_{14}^j & -A_{15}^j & -J_3^j & -A_{16}^j & -A_{17}^j & -A_{18}^j & -J_4^j \\ 0 & 0 & 0 & -(J_5^j + J_6^j) & 0 & 0 & 0 & 0 \end{bmatrix} \begin{bmatrix} U^j \\ V^j \\ W^j \\ \Theta^j \\ U^{j'} \\ V^{j'} \\ W^{j'} \\ \Theta^{j'} \end{bmatrix}, \tag{23}$$

where $[U^j, V^j, W^j, \Theta^j, U^{j'}, V^{j'}, W^{j'}$, and $\Theta^{j'}]^T$ is the new unknown vector containing both the displacement and sovra-temperature amplitudes and the related first-order derivatives along the z direction (indicated with the apex $'$). Different from the model proposed by Brischetto and Torre in [70], the sovra-temperature profile along the z direction is computed without using an external tool. It is calculated as a primary variable by means of Equation (4) combined with the equilibrium equations. Equation (23) can be written in a compact form as:

$$\boldsymbol{D}^j \boldsymbol{X}^{j'} = \boldsymbol{A}^j \boldsymbol{X}^j, \tag{24}$$

A further rewriting of Equation (24) is:

$$\boldsymbol{X}^{j'} = \boldsymbol{A}^{*j} \boldsymbol{X}^j, \tag{25}$$

with $\boldsymbol{A}^{*j} = \boldsymbol{D}^{j^{-1}} \boldsymbol{A}^j$. $\boldsymbol{X}^j$ is the 8×1 vector containing the unknowns and $\boldsymbol{X}^{j'}$ includes the derivatives along z of these unknowns. To solve the problem written in Equation (25), it is possible to use the method of the exponential matrix proposed in [68,69]. The solution of the problem in Equation (25) can be written as:

$$\boldsymbol{X}^j_t = e^{(\boldsymbol{A}^{*j} \tilde{z}^j)} \boldsymbol{X}^j_b = \boldsymbol{A}^{**j} \boldsymbol{X}^j_b . \tag{26}$$

Knowing the exponential matrix term $e^{(\boldsymbol{A}^{*j} \tilde{z}^j)}$, it is possible to obtain the unknown vector $\boldsymbol{X}^j_t$ (corresponding to the unknown vector at the top of the j-th fictitious layer) related with the unknown vector $\boldsymbol{X}^j_b$ (bottom of the j-th fictitious layer). The $e^{(\boldsymbol{A}^{*j} \tilde{z}^j)}$ term must be calculated introducing the thickness value h^j of each j layer. The exponential matrix can be expanded in a power series and it must be computed for each fictitious layer for the values h^j:

$$\boldsymbol{A}^{**j} = e^{(\boldsymbol{A}^{*j} h^j)} = \boldsymbol{I} + \boldsymbol{A}^{*j} h^j + \frac{\boldsymbol{A}^{*j^2}}{2!} h^{j^2} + \frac{\boldsymbol{A}^{*j^3}}{3!} h^{j^3} + \dots + \frac{\boldsymbol{A}^{*j^N}}{N!} h^{j^N}, \tag{27}$$

where $\boldsymbol{I}$ is the 8×8 identity matrix. Equation (26) links the top with the bottom within each j fictitious layer. To link the top of the j fictitious layer with the bottom of the $j+1$ fictitious layer, proper interlaminar continuity conditions must be imposed. These interlaminar continuity conditions must be imposed to u, v, and w displacements, sovra-temperature θ, transverse shear, and transverse normal stresses $\sigma_{\alpha z}$, $\sigma_{\beta z}$, σ_{zz}, and transverse normal heat flux q_z at each fictitious layers' interface. The continuity of all these terms can be written in matrix form as:

$$\boldsymbol{x}^{j+1}_b = \begin{bmatrix} u^{j+1}_b \\ v^{j+1}_b \\ w^{j+1}_b \\ \theta^{j+1}_b \end{bmatrix} = \boldsymbol{x}^j_t = \begin{bmatrix} u^j_t \\ v^j_t \\ w^j_t \\ \theta^j_t \end{bmatrix}, \tag{28}$$

$$\boldsymbol{\sigma_n}^{j+1}_b = \begin{bmatrix} \sigma^{j+1}_{\beta z_b} \\ \sigma^{j+1}_{\alpha z_b} \\ \sigma^{j+1}_{zz_b} \end{bmatrix} = \boldsymbol{\sigma_n}^j_t = \begin{bmatrix} \sigma^j_{\beta z_t} \\ \sigma^j_{\alpha z_t} \\ \sigma^j_{zz_t} \end{bmatrix}, \tag{29}$$

$$q^{j+1}_{z_b} = q^j_{z_t}. \tag{30}$$

where Equation (28) is related to the displacements and sovra-temperature, Equation (29) is related to stresses and Equation (30) to the heat flux along the z direction. The conditions expressed in Equation (28) can be easily imposed for the amplitudes U^j, V^j, W^j, and Θ^j. It is possible to derive an amplitude form of Equations (28)–(30) using the constitutive

Equation (11) and the harmonic forms in Equations (14)–(17). These conditions can be rewritten in matrix form as:

$$\begin{bmatrix} U \\ V \\ W \\ \Theta \\ U' \\ V' \\ W' \\ \Theta' \end{bmatrix}_b^{j+1} = \begin{bmatrix} 1 & 0 & 0 & 0 & 0 & 0 & 0 & 0 \\ 0 & 1 & 0 & 0 & 0 & 0 & 0 & 0 \\ 0 & 0 & 1 & 0 & 0 & 0 & 0 & 0 \\ 0 & 0 & 0 & 1 & 0 & 0 & 0 & 0 \\ T_1 & 0 & T_2 & 0 & T_3 & 0 & 0 & 0 \\ 0 & T_4 & T_5 & 0 & 0 & T_6 & 0 & 0 \\ T_7 & T_8 & T_9 & \tau_1 & 0 & 0 & T_{10} & 0 \\ 0 & 0 & 0 & 0 & 0 & 0 & 0 & \tau_2 \end{bmatrix}^{j+1,j} \begin{bmatrix} U \\ V \\ W \\ \Theta \\ U' \\ V' \\ W' \\ \Theta' \end{bmatrix}_t^{j}, \tag{31}$$

where the diagonal submatrix including 1 is Equation (28) written in matrix form. The other coefficients T_i and τ_i are Equations (29) and (30) in matrix form. A compact form of Equation (31) is:

$$\boldsymbol{X}_b^{j+1} = \boldsymbol{T}^{j+1,j}\boldsymbol{X}_t^{j}, \tag{32}$$

where $\boldsymbol{T}^{j+1,j}$ is the transfer matrix. All the analyses can be conducted using this mathematical formulation considering the simply supported boundary conditions. This constraint configuration is automatically assured by Equations (14)–(17). It is possible to write:

$$\theta = 0,\ w = v = 0,\ \sigma_{\alpha\alpha} = 0 \quad \text{for} \quad \alpha = 0, a\,, \tag{33}$$

$$\theta = 0,\ w = u = 0,\ \sigma_{\beta\beta} = 0 \quad \text{for} \quad \beta = 0, b\,. \tag{34}$$

The load boundary conditions must be enforced at the outer faces of the structure; it is possible to write them as:

$$\sigma_{zz} = 0,\ \sigma_{\alpha z} = 0,\ \sigma_{\beta z} = 0,\ \Theta = T - T_0 \quad \text{for} \quad z = \pm h/2. \tag{35}$$

Thanks to Equation (35), the sovra-temperature is directly imposed at the outer faces of the structure. Equation (35) can be rewritten in a compact way as:

$$\begin{bmatrix} -\frac{C_{13}^M}{H_{\alpha_t}^M}\bar{\alpha} & -\frac{C_{23}^M}{H_{\beta_t}^M}\bar{\beta} & \frac{C_{13}^M}{H_{\alpha_t}^M R_\alpha} + \frac{C_{23}^M}{H_{\beta_t}^M R_\beta} & -\lambda_z^M & 0 & 0 & C_{33}^M & 0 \\ 0 & -\frac{C_{44}^M}{H_{\beta_t}^M R_\beta} & \frac{C_{44}^M}{H_{\beta_t}^M}\bar{\beta} & 0 & 0 & C_{44}^M & 0 & 0 \\ -\frac{C_{55}^M}{H_{\alpha_t}^M R_\alpha} & 0 & \frac{C_{55}^M}{H_{\alpha_t}^M}\bar{\alpha} & 0 & C_{55}^M & 0 & 0 & 0 \\ 0 & 0 & 0 & 1 & 0 & 0 & 0 & 0 \end{bmatrix} \begin{bmatrix} U \\ V \\ W \\ \Theta \\ U' \\ V' \\ W' \\ \Theta' \end{bmatrix}_t^M = \begin{bmatrix} 0 \\ 0 \\ 0 \\ \Theta_t \end{bmatrix}, \tag{36}$$

$$\begin{bmatrix} -\frac{C_{13}^1}{H_{\alpha_b}^1}\bar{\alpha} & -\frac{C_{23}^1}{H_{\beta_b}^1}\bar{\beta} & \frac{C_{13}^1}{H_{\alpha_b}^1 R_\alpha} + \frac{C_{23}^1}{H_{\beta_b}^1 R_\beta} & -\lambda_z^1 & 0 & 0 & C_{33}^1 & 0 \\ 0 & -\frac{C_{44}^1}{H_{\beta_b}^1 R_\beta} & \frac{C_{44}^1}{H_{\beta_b}^1}\bar{\beta} & 0 & 0 & C_{44}^1 & 0 & 0 \\ -\frac{C_{55}^1}{H_{\alpha_b}^1 R_\alpha} & 0 & \frac{C_{55}^1}{H_{\alpha_b}^1}\bar{\alpha} & 0 & C_{55}^1 & 0 & 0 & 0 \\ 0 & 0 & 0 & 1 & 0 & 0 & 0 & 0 \end{bmatrix} \begin{bmatrix} U \\ V \\ W \\ \Theta \\ U' \\ V' \\ W' \\ \Theta' \end{bmatrix}_b^1 = \begin{bmatrix} 0 \\ 0 \\ 0 \\ \Theta_b \end{bmatrix}. \tag{37}$$

Equations (36) and (37) can be further compacted as:

$$\boldsymbol{B}_t^M \boldsymbol{X}_t^M = \mathscr{P}_t, \tag{38}$$

$$\boldsymbol{B}_b^1 \boldsymbol{X}_b^1 = \mathscr{P}_b, \tag{39}$$

where vectors $\mathscr{P}_t$ and $\mathscr{P}_b$ contain the impositions related to the mechanical load conditions and the sovra-temperature. For the present thermal stress analysis, the mechanical load conditions are enforced to zero.

In order to include Equations (38) and (39) into an algebraic system written in compact form, it is possible to write $\boldsymbol{X}_t^M$ in terms of $\boldsymbol{X}_b^1$ to link the top of the last fictitious layer with the bottom of the first fictitious layer. This operation can be achieved by recursively introducing Equation (32) into Equation (26) as follows:

$$\boldsymbol{X}_t^M = \left(\boldsymbol{A}^{**M} \boldsymbol{T}^{M,M-1} \boldsymbol{A}^{**M-1} \boldsymbol{T}^{M-1,M-2} \ldots\ldots \boldsymbol{A}^{**2} \boldsymbol{T}^{2,1} \boldsymbol{A}^{**1} \right) \boldsymbol{X}_b^1 = \boldsymbol{H}_m \boldsymbol{X}_b^1. \tag{40}$$

Equation (40) defines the 8×8 matrix $\boldsymbol{H}_m$ for structures embedding FGM layers. This matrix has a different size with respect to the $\boldsymbol{H}_m$ matrix proposed in [63] for the pure mechanical case developed by Brischetto. The matrix size difference is due to the coupling of the 3D Fourier heat conduction equation directly with the elastic equilibrium relations of the problem. Introducing Equation (40) in Equation (38), this last one can be rewritten in terms of $\boldsymbol{X}_b^1$ as:

$$\boldsymbol{B}_t^M \boldsymbol{H}_m \boldsymbol{X}_b^1 = \mathscr{P}_t, \tag{41}$$

Equations (39) and (41) can be now compacted as:

$$\begin{bmatrix} \boldsymbol{B}_t^M \boldsymbol{H}_m \\ \boldsymbol{B}_b^1 \end{bmatrix} \boldsymbol{X}_b^1 = \boldsymbol{E} \boldsymbol{X}_b^1 = \begin{bmatrix} \mathscr{P}_t \\ \mathscr{P}_b \end{bmatrix} = \mathscr{P} \Rightarrow \boldsymbol{E} \boldsymbol{X}_b^1 = \mathscr{P}. \tag{42}$$

The main characteristic of the final system written in Equation (42) is the fact that matrix $\boldsymbol{E}$ independently has an 8×8 dimension by the number of fictitious layers employed, even if the method uses a layer-wise approach. This new formulation can be considered as the generalization of the pure mechanical model proposed in [63] by Brischetto. Vector $\mathscr{P}$ now contains all the load impositions, both mechanical and thermal ones. The system in Equation (42) is formally the same as shown in [63,70], and in [71], but the addition of the 3D Fourier heat conduction equation to the 3D equilibrium relations is now considered.

This formulation can be implemented in a Matlab code where stresses, strains, and displacements can be evaluated along the thickness direction z of several structures embedding different FGM configurations. Once the displacements at the bottom of the first fictitious layer have been calculated, Equations (26) and (32) can be progressively used to compute the displacements and sovra-temperature (and related derivatives with respect to z) through all points in the z direction of the structure.

3. Results

This section is related to the comparison of results between the presented coupled thermo-elastic model and the past uncoupled thermo-elastic model proposed by Brischetto and Torre [70–73]. The section is divided into two different parts: in the first one, two preliminary assessments are presented to validate the proposed general 3D exact coupled thermo-elastic shell model and to clearly understand the proper choice of N (order of expansion for the exponential matrix in Equation (27)) and the appropriate number of M (mathematical layers) for the calculation of FGM properties and constant curvature terms related to shell geometries. In the second part, four new benchmarks are proposed; after the opportune choice of N and M parameters, the present 3D coupled thermo-elastic model is considered validated and it allows for the discussion of effects due to the geometry of structures, thickness ratios, FGM laws, and temperature impositions for several cases. All the presented assessments and benchmarks consider an FGM layer with two constituent phases: a metallic one (Monel 70Ni-30Cu) and a ceramic one (Zirconia). These two constituent phases have the following elastic and thermal properties:

$$K_m = 227.24\,\text{GPa}, \quad G_m = 65.55\,\text{GPa}, \quad \mu_m = 15 \cdot 10^{-6}\,\frac{1}{K}, \quad k_m = 25\,\frac{W}{mK} \tag{43}$$

$$K_c = 125.83\,\text{GPa}, \quad G_c = 58.077\,\text{GPa}, \quad \mu_c = 10 \cdot 10^{-6}\,\frac{1}{K}, \quad k_c = 2.09\,\frac{W}{mK} \tag{44}$$

where K_m and K_c indicate the bulk modulus of the metallic and ceramic phase, G_m and G_c indicate the shear modulus of the metallic and ceramic phase, μ_m and μ_c indicate the thermal expansion coefficient of the metallic and ceramic phase, and k_m and k_c indicate the conductivity coefficient of the metallic and ceramic phase. For each presented case (both assessments and benchmarks), the volume fraction of the ceramic phase V_c follows the indicated law:

$$V_c = \left(\frac{\tilde{z}_{FGM}}{h_{FGM}}\right)^p \tag{45}$$

where $\tilde{z}_{FGM}$ is the local thickness coordinate for the FGM layer (it goes from 0 at the bottom to h_{FGM} at the top of the FGM layer), h_{FGM} is the thickness of the FGM layer, and p is the related exponential coefficient. The FGM layer used for the proposed results is a full metallic at the bottom and full ceramic at the top. This consideration is the same for both sandwich and one-layered configurations. The bulk and shear moduli along the thickness direction depend on the volume fraction of the ceramic phase V_c. These two material properties are estimated using the Mori–Tanaka model as follows:

$$\frac{K - K_m}{K_c - K_m} = \frac{V_c}{1 + (1 - V_c)\frac{K_c - K_m}{K_m + \frac{4}{3}G_m}}, \quad \frac{G - G_m}{G_c - G_m} = \frac{V_c}{1 + (1 - V_c)\frac{G_c - G_m}{G_m + f_m}} \tag{46}$$

where $f_m = \frac{G_m(9K_m + 8G_m)}{6(K_m + 2G_m)}$. The heat conduction coefficient k is a function of the volume fraction of the ceramic phase V_c as:

$$\frac{k - k_m}{k_c - k_m} = \frac{V_c}{1 + (1 - V_c)\frac{k_c - k_m}{3k_m}}, \tag{47}$$

and the thermal expansion coefficient μ can be computed using:

$$\frac{\mu - \mu_m}{\mu_c - \mu_m} = \frac{\frac{1}{K} - \frac{1}{K_m}}{\frac{1}{K_c} - \frac{1}{K_m}}. \tag{48}$$

In Equation (48), the dependence on V_c is delegated to the bulk modulus K. The material data employed in this section were proposed by Reddy and Cheng in [72].

3.1. Preliminary Assessments

The new proposed 3D coupled thermo-elastic exact solution for shells embedding FGM layers, here indicated as 3D-u-θ, is validated using two preliminary assessments. A square one-layered FGM plate and a one-layered FGM cylindrical shell are investigated considering different thickness ratios. After these assessments, the 3D full coupled shell model will be defined as validated: the validation occurs for $N = 3$ (order of expansion for the exponential matrix) and $M = 300$ mathematical layers. For the cases presented in this section, the convergence of the 3D-u-θ model is obtained even for fewer mathematical layers than those used for the 3D uncoupled thermo-elastic model presented in [70] and called 3D(θ_c, 3D) in Tables 1 and 2. However, higher values for N and M are chosen in a conservative sense. The results for these preliminary assessments are provided in non-dimensional forms as:

$$\{\bar{u}, \bar{v}, \bar{w}\} = \frac{\{u, v, w\}}{aC}, \quad \{\bar{\sigma}_{\alpha\alpha}, \bar{\sigma}_{\beta\beta}, \bar{\sigma}_{\alpha\beta}, \bar{\sigma}_{zz}, \bar{\sigma}_{\alpha z}, \bar{\sigma}_{\beta z}\} = \frac{\{\sigma_{\alpha\alpha}, \sigma_{\beta\beta}, \sigma_{\alpha\beta}, \sigma_{zz}, \sigma_{\alpha z}, \sigma_{\beta z}\}}{CK^*} \tag{49}$$

where $C = 10^{-6}$ is a constant value, a is the in-plane dimension of the structure in the α direction, and $K^* = 10^9\ Pa$.

Table 1. First preliminary assessment, one-layered FGM (p = 2) square plate with external sovra-temperature amplitudes $\Theta_t = +1\,K$ and $\Theta_b = 0\,K$ for $m = n = 1$. Reference solution is the 3D uncoupled thermoelastic model by Brischetto and Torre [70] with the temperature profile calculated using the 3D Fourier heat conduction Equation (3D(θ_c,3D)). The new 3D coupled thermoelastic solution is 3D-u-θ.

	3D(θ_c,3D) [70]	**3D-u-θ**
	a/h = 4	
$\bar{w}$ (a/2, b/2, h)	3.042	3.042
$\bar{w}$ (a/2, b/2, h/2)	2.142	2.142
$\bar{w}$ (a/2, b/2, 0)	1.900	1.900
$\bar{u}$ (0, b/2, h)	−1.680	−1.680
$\bar{u}$ (0, b/2, h/2)	−0.6819	−0.6819
$\bar{u}$ (0, b/2, 0)	0.08245	0.08245
	a/h = 10	
$\bar{\sigma}_{\alpha z}$ (0, b/2, h/2)	1.584	1584
$\bar{\sigma}_{zz}$ (a/2, b/2, h/2)	1.015	1.015
	a/h = 50	
$\bar{\sigma}_{\alpha\alpha}$ (a/2, b/2, h)	−1009	−1009
$\bar{\sigma}_{\alpha\alpha}$ (a/2, b/2, h/2)	−251.7	−250.5
$\bar{\sigma}_{\alpha\alpha}$ (a/2, b/2, 0)	−76.15	−76.15

Table 2. Second preliminary assessment, one-layered FGM (p = 2) cylindrical shell with external sovra-temperature amplitudes $\Theta_t = +1\,K$ and $\Theta_b = 0\,K$ for $m = n = 1$. Reference solution is the 3D uncoupled thermoelastic model by Brischetto and Torre [70] with the temperature profile calculated using the 3D Fourier heat conduction Equation (3D(θ_c,3D)). The new 3D coupled thermoelastic solution is 3D-u-θ.

	3D(θ_c,3D) [70]	**3D-u-θ**
	$R_\beta/h = 50$	
$\bar{w}$ (a/2, b/2, h)	7.1325	7.1325
$\bar{w}$ (a/2, b/2, h/2)	6.4120	6.4120
$\bar{w}$ (a/2, b/2, 0)	6.1931	6.1931
$\bar{u}$ (0, b/2, h)	−3.5461	−3.5461
$\bar{u}$ (0, b/2, h/2)	−1.4530	−1.4530
$\bar{u}$ (0, b/2, 0)	0.4832	0.4832
	$R_\beta/h = 1000$	
$\bar{\sigma}_{\beta\beta}$ (a/2, b/2, h)	−1164.9	−1164.9
$\bar{\sigma}_{\beta\beta}$ (a/2, b/2, h/2)	159.05	159.92
$\bar{\sigma}_{\beta\beta}$ (a/2, b/2, 0)	990.89	990.89
$\bar{\sigma}_{\alpha z}$ (0, b/2, h/2)	−5.2234	−5.2234
$\bar{\sigma}_{zz}$ (a/2, b/2, h/2)	0.2392	0.2392

The first preliminary assessment shows a simply supported square plate ($a = b$) with thickness $h = 1\,m$ and bi-sinusoidal ($m = n = 1$) sovra-temperature imposed at the top and bottom surfaces ($\Theta_t = +1\,K$ and $\Theta_b = -1\,K$). The structure is composed of a single FGM layer with the top part constituted of a ceramic phase and the bottom a metallic phase, as it is defined in Equations (43)–(48). The volume fraction of the ceramic phase V_c is a function of the thickness coordinate, with the exponential coefficient $p = 2$ in Equation (45). The reference results are based on the 3D uncoupled thermo-elastic solution proposed by Brischetto and Torre [70] where the temperature is calculated by separately solving the 3D Fourier heat conduction Equation (3D(θ_c, 3D)) by means of hyperbolic functions (see the mathematical formulation proposed by the authors in [70]). The present new solution uses an order of expansion $N = 3$ for the calculation of the exponential matrix and $M = 300$ mathematical layers for the calculation of curvature terms and FGM properties. The novelty of the 3D-u-θ model is the inclusion of the 3D Fourier heat conduction equation directly in the system including the 3D elastic equilibrium equations. This feature permits to take into account both the material layer and the thickness layer effects without the calculating the temperature profile using an external tool. Table 1 shows no-dimensional transverse and in-plane displacements and no-dimensional in-plane normal and transverse shear/normal stresses for different thickness ratios a/h. The present 3D coupled model provides the same results as the reference solution [70] for each thickness ratio a/h because, in both models, the material and thickness layer effects are properly evaluated.

The second preliminary assessment shows a simply supported cylindrical shell with radii of curvature $R_\alpha = \infty$ and $R_\beta = 10\,m$, bi-sinusoidal ($m = n = 1$) sovra-temperature imposed at the top and bottom surfaces ($\Theta_t = +1\,K$ and $\Theta_b = 0\,K$), and in-plane dimensions $a = 1\,m$ and $b = \frac{\pi}{3}R_\beta$. The cylindrical shell is composed of a single FGM layer whose top part is constituted of a ceramic phase and the bottom of a metallic phase, as it is defined in Equations (43)–(48). The power law for the volume fraction of the ceramic phase V_c is quadratic ($p = 2$) and it depends on the thickness coordinate $\tilde{z}_{FGM}$. The results used as a reference are based on the previously discussed uncoupled 3D thermo-elastic solution proposed by Brischetto and Torre in [70]. This second assessment uses the same values of N and M seen in the first one. The new proposed 3D coupled shell model (3D-u-θ) includes the 3D Fourier heat conduction equation in the same way seen in the first assessment. Table 2 shows no-dimensional transverse and in-plane displacements and no-dimensional in-plane and transverse stresses for two different thickness ratios ($R_\beta/h = 50$ and $R_\beta/h = 1000$). The 3D-u-θ model is always coincident with the reference solution [70] because both use the 3D Fourier heat conduction equation: coupled with the 3D elastic equilibrium Equations (3D-u-θ model) or separately solved using an external tool (3D(θ_c,3D) model). The results confirm all the previous considerations.

The benchmarks proposed in the next section consider the cases where different geometries, thickness ratios, FGM configurations, and temperature profiles are deeply evaluated. For this purpose, $M = 300$ mathematical layers combined with an $N = 3$ order of expansion will be employed in these new benchmarks, as suggested by the validation here conducted using the preliminary assessments.

3.2. New Benchmarks

Here, four new benchmarks are proposed considering plates, cylinders, cylindrical shells, and spherical shells. The involved geometries can be seen in Figure 1. Several sovra-temperature impositions, different m and n half-wave numbers and different FGM laws are considered for each case. The variation of the FGM law occurs with the variation of the parameter p in the exponential law of Equation (45). For each possible geometry, a FGM layer is involved (both single-layer and sandwich configurations). In all the proposed 3D coupled results, the $N = 3$ order of expansion for the exponential matrix and $M = 300$ mathematical layers are used. The new 3D coupled exact thermo-elastic model (3D-u-θ) will be used for these benchmarks and it will be compared with previously uncoupled 3D results where the 3D Fourier heat conduction equation was separately solved (3D(θ_c,3D)), the 1D Fourier heat conduction equation was separately solved (3D(θ_c,1D)), and the assumed linear temperature profiles were considered a priori (3D(θ_a)). The results presented in this subsection can be useful for those scientists involved in the development of 2D and 3D analytical and numerical models for the thermal stress analysis of plates and shells embedding FGM layers.

The first benchmark proposes a simply supported sandwich square plate ($a = b = 10\,m$) with an FGM core (see Figure 1). The thickness of the entire structure is h and the analysed thickness ratios are $a/h = 2, 5, 10, 20, 50, 100$. The thickness of the top and bottom skins is $h_1 = h_3 = 0.2\,h$ and the FGM core has $h_2 = 0.6\,h$. The sovra-temperature has an amplitude value at the top of $\Theta_t = +0.5\,K$ and at the bottom $\Theta_b = -0.5\,K$ with bi-sinusoidal form (half-wave numbers $m = 1$ and $n = 1$). The volume fraction law of the ceramic phase V_c for the FGM core has exponential $p = 2$; all the mechanical and thermal property variations of the FGM core through the thickness are described by means of Equations (46)–(48). The bottom skin is full metallic and the top skin is full ceramic, while the FGM core has a continuous variation (see Figure 2). Figure 3 shows the temperature profile through the thickness of a moderately thick ($a/h = 5$) and a moderately thin ($a/h = 20$) plate. In the case of the moderately thin configuration ($a/h = 20$), the calculated temperature profiles using 3D Fourier heat conduction Equation (θ_c,3D), using 1D Fourier heat conduction Equation (θ_c,1D), and using the coupled model (u-θ) are perfectly coincident. The assumed linear temperature profile (θ_a) is inappropriate for this case. For the moderately thick plate ($a/h = 5$), the temperature profile calculated using the full coupled model (u-θ) is coincident with the uncoupled model that computes the temperature profile via the 3D Fourier heat conduction Equation (θ_c,3D). These two models introduce both the thickness and material layer effects. The temperature profile using the 1D Fourier heat conduction Equation (θ_c,1D) does not consider the thickness layer effect. The uncoupled model that considers the temperature profile using an a priori linear assumption (θ_a) does not take into account both the effects. Table 3 shows the in-plane and transverse displacement components and the in-plane normal, in-plane shear, transverse shear, and transverse normal stress components in different positions through the thickness for different a/h ratios. In the case of very thin plates ($a/h = 100$), the last 3D models presented in Table 1 are coincident. For thick or moderately thick plates, the 3D(θ_c,3D) and the 3D-u-θ model provide the same results because they properly consider the thickness layer and material layer effects thanks to the implementation of the 3D Fourier heat conduction equation. The 3D(θ_a) model provides different results because it did not properly consider any material layer and thickness layer effect. Figure 4 shows the in-plane and transverse displacement components and two stress components (σ_{zz} and $\sigma_{\alpha z}$) through the thickness of a moderately thick plate ($a/h = 10$). The in-plane displacement and transverse displacement are continuous through the thickness direction because the mechanical properties of the FGM layer vary with continuity along the thickness direction and the compatibility conditions are correctly introduced in the shell model. The transverse normal stress and the transverse shear stress satisfies the boundary load conditions imposed at the top and at the bottom external surfaces of the structures ($\sigma^t_{zz} = \sigma^b_{zz} = P_z = 0$ and $\sigma^t_{\beta z} = \sigma^b_{\beta z} = P_\beta = 0$) and they are continuous because of the correct impositions of equilibrium conditions.

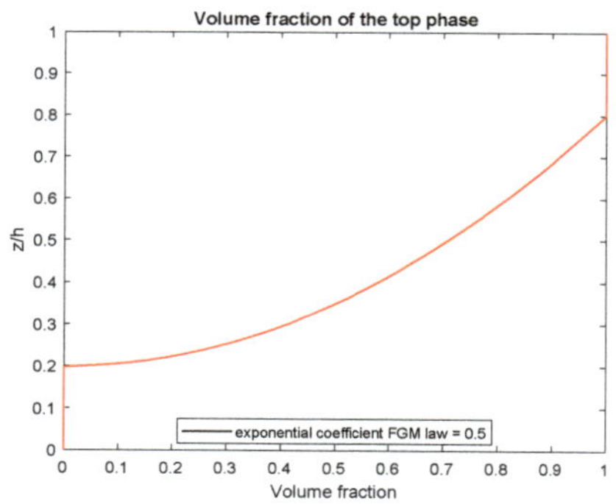

Figure 2. First benchmark, volume fraction of the ceramic phase for a simply supported sandwich square plate with FGM (p = 0.5) core.

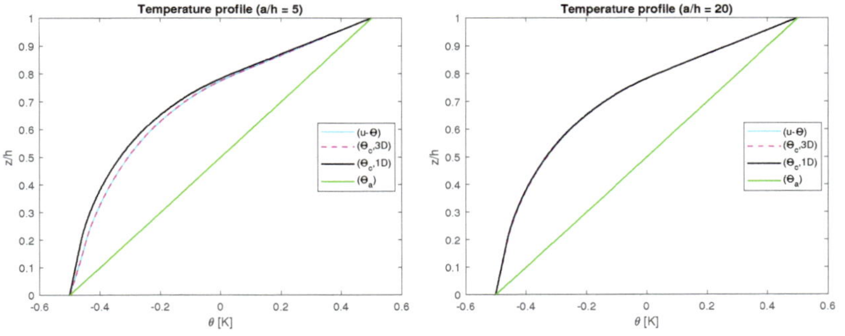

Figure 3. First benchmark, temperature profiles for thick and moderately thick simply supported sandwich square plates with FGM (p = 0.5) core. The maximum amplitude of the temperature $\theta(\alpha, \beta, z)$ is evaluated at the centre of the plate ($a/2$,$b/2$).

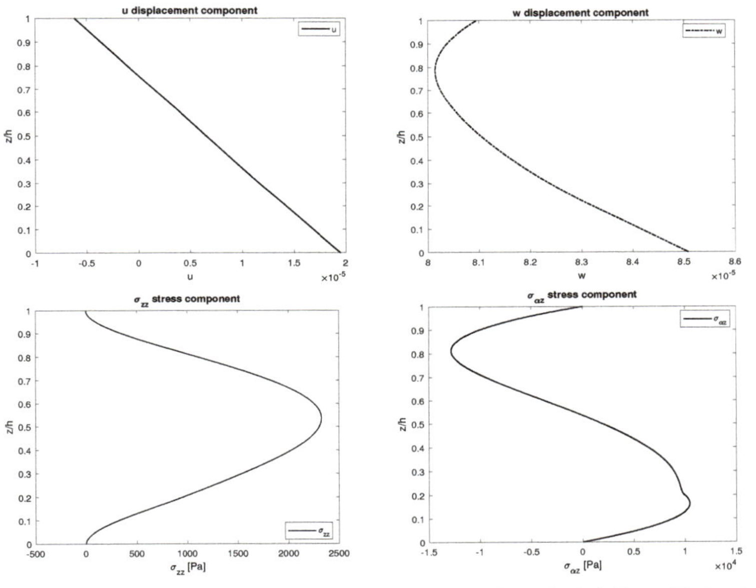

Figure 4. First benchmark, displacements and stresses for a thick (a/h = 10) simply supported sandwich square plate with FGM (p = 0.5) core obtained via the 3D-u-θ model. Maximum amplitudes: w and σ_{zz} at ($a/2$,$b/2$); u and $\sigma_{\alpha z}$ at (0,$b/2$).

Table 3. First benchmark, simply supported sandwich square plate with FGM (p = 0.5) core. Sovra-temperature imposed as $\Theta_t = +0.5\,K$ and $\Theta_b = -0.5\,K$ for $m = 1$ and $n = 1$. 3D uncoupled thermoelastic models from [70]. The new 3D coupled thermoelastic solution is 3D-u-θ.

a/h	2	5	10	20	50	100
			u[10^{-5} m] at ($\alpha = 0, \beta = b/2, \bar{z} = 0$)			
3D(θ_a) [70]	1.6223	1.4866	1.4659	1.4606	1.4592	1.4589
3D(θ_c,1D) [70]	2.2351	2.0009	1.9613	1.9511	1.9482	1.9478
3D(θ_c,3D) [70]	1.9180	1.9463	1.9475	1.9477	1.9477	1.9477
3D-u-θ	1.9180	1.9463	1.9475	1.9477	1.9477	1.9477

Table 3. *Cont.*

a/h	2	5	10	20	50	100
			w$[10^{-5}$ m] at ($\alpha = a/2, \beta = b/2, \tilde{z} = h/2$)			
3D(θ_a) [70]	1.4865	4.1617	8.4399	16.937	42.382	84.776
3D(θ_c,1D) [70]	1.4159	3.9997	8.1190	16.297	40.783	81.577
3D(θ_c,3D) [70]	1.3743	3.9724	8.1047	16.289	40.780	81.575
3D-u-θ	1.3743	3.9724	8.1047	16.289	40.780	81.575
			$\sigma_{\beta\beta}[10^3$ Pa] at ($\alpha = a/2, \beta = b/2, \tilde{z} = h$)			
3D(θ_a) [70]	−239.78	−258.49	−261.30	−262.01	−262.21	−262.24
3D(θ_c,1D) [70]	−731.19	−676.95	−665.61	−662.63	−661.78	−661.66
3D(θ_c,3D) [70]	−603.01	−652.62	−659.41	−661.07	−661.53	−661.60
3D-u-θ	−603.01	−652.62	−659.41	−661.07	−661.53	−661.60
			$\sigma_{\alpha\beta}[10^3$ Pa] at ($\alpha = 0, \beta = 0, \tilde{z} = h/4$)			
3D(θ_a) [70]	292.33	311.02	314.19	315.00	315.23	315.26
3D(θ_c,1D) [70]	498.38	515.05	518.41	519.29	519.54	519.58
3D(θ_c,3D) [70]	404.32	496.08	513.46	518.04	519.34	519.53
3D-u-θ	404.32	496.08	513.46	518.04	519.34	519.53
			$\sigma_{\alpha z}[10^3$ Pa] at ($\alpha = 0, \beta = b/2, \tilde{z} = h/4$)			
3D(θ_a) [70]	−30.682	−17.013	−8.8995	−4.4999	−1.8056	−0.9032
3D(θ_c,1D) [70]	77.609	21.539	9.8156	4.7825	1.8988	0.9484
3D(θ_c,3D) [70]	37.694	18.520	9.4317	4.7343	1.8957	0.9480
3D-u-θ	37.693	18.521	9.4317	4.7343	1.8957	0.9480
			$\sigma_{\beta z}[10^3$ Pa] at ($\alpha = a/2, \beta = 0, \tilde{z} = 3h/4$)			
3D(θ_a) [70]	28.092	14.681	7.6209	3.8461	1.5424	0.7715
3D(θ_c,1D) [70]	−84.322	−25.702	−12.028	−5.9059	−2.3501	−1.1742
3D(θ_c,3D) [70]	−44.805	−22.672	−11.642	−5.8573	−2.3470	−1.1738
3D-u-θ	−44.805	−22.672	−11.642	−5.8573	−2.3470	−1.1738
			$\sigma_{zz}[10^3$ Pa] at ($\alpha = a/2, \beta = b/2, \tilde{z} = 3h/4$)			
3D(θ_a) [70]	−17.007	−3.0277	−0.7695	−0.1932	−0.0310	−0.0077
3D(θ_c,1D) [70]	49.914	6.4569	1.5345	0.3784	0.0603	0.0151
3D(θ_c,3D) [70]	28.983	5.8135	1.4934	0.3758	0.0602	0.0151
3D-u-θ	28.983	5.8135	1.4934	0.3758	0.0602	0.0151

For the case related to the second benchmark, a simply supported one-layered FGM cylinder (see Figure 1) is analysed. The radii of curvature of the structure are $R_\alpha = 10\,m$ and $R_\beta = \infty$. The global thickness of the structure is h. The in-plane dimensions are $a = 2\pi R_\alpha$ and $b = 30\,m$. The applied sovra-temperatures at the external surfaces are the same as those already proposed for the first benchmark, but half-wave numbers $m = 2$ and $n = 1$ are now imposed. The volume fraction of the ceramic phase V_c is linear ($p = 1$), the bottom of the structure is full metallic and the top is full ceramic as can be seen in Figure 5. The reference equations for all the material characteristics are proposed in Equations (46)–(48). The temperature profiles through the thickness proposed in Figure 6 do not change when the thickness ratio varies. This feature is due to the symmetry and rigidity of the cylinder. Therefore, the effect of the thickness is negligible even if the structure is really thick ($R_\alpha/h = 5$ case). Table 4 shows the results in terms of displacements and stresses for different R_α/h ratios. For thicker cylinders, the 3D(θ_c,3D) and 3D-u-θ models are in accordance because they take into account both the material layer and the thickness layer effects. The 3D(θ_c,1D) only considers the material layer effect and 3D(θ_a) only considers a linear assumed temperature amplitude: for this reason, the results are not correct. For thin cylinders, only the 3D(θ_a) model shows important differences with respect to the other three models (3D(θ_c,1D), 3D(θ_c,3D), and 3D-u-θ. Figure 7 provides the displacements and stresses for a moderately thick ($a/h = 10$) cylinder with one FGM layer. In-plane and transverse displacements are continuous because the mechanical and thermal properties of the FGM layer vary continuously in the thickness direction and the compatibility conditions are correctly imposed for each mathematical interface. The same consideration is also valid for the stresses $\sigma_{\alpha\alpha}$ and $\sigma_{\beta z}$ shown in Figure 7. The transverse shear stress satisfies the free mechanical load conditions at the external surfaces ($\sigma^t_{\beta z} = \sigma^b_{\beta z} = P_\beta = 0$).

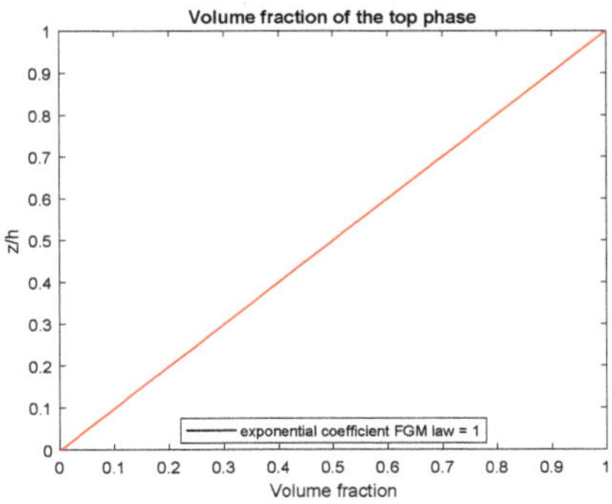

Figure 5. Second benchmark, volume fraction of the ceramic phase for a simply supported cylinder with one FGM (p = 1) layer.

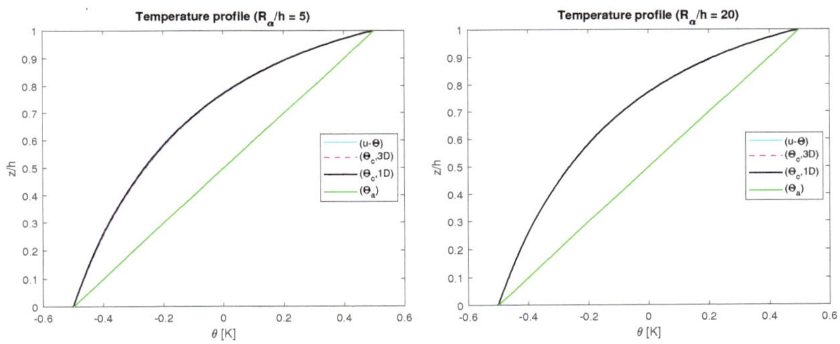

Figure 6. Second benchmark, temperature profiles for thick and moderately thick simply supported cylinders with one FGM (p = 1) layer. The maximum amplitude of the temperature $\theta(\alpha, \beta, z)$ is evaluated at the centre of the cylinder ($a/2$,$b/2$).

Table 4. Second benchmark, simply supported cylinder with one FGM (p = 1) layer. Sovra-temperature imposed as $\Theta_t = +0.5\,K$ and $\Theta_b = -0.5\,K$ for $m = 2$ and $n = 1$. 3D uncoupled thermoelastic models from [70]. The new 3D coupled thermoelastic solution is 3D-u-θ.

R_α/h	2	5	10	20	50	100
			v[10^{-6} m] at ($\alpha = a/2, \beta = 0, \tilde{z} = h$)			
3D(θ_a) [70]	−8.6733	2.8714	5.3717	6.2023	6.5527	6.6441
3D(θ_c,1D) [70]	24.724	30.279	30.248	29.774	29.340	29.170
3D(θ_c,3D) [70]	22.657	29.994	30.184	29.758	29.337	29.170
3D-u-θ	22.657	29.994	30.184	29.758	29.337	29.170
			w[10^{-5} m] at ($\alpha = a/2, \beta = b/2, \tilde{z} = h/2$)			
3D(θ_a) [70]	1.2362	−0.1105	−0.7159	−1.0306	−1.2201	−1.2831
3D(θ_c,1D) [70]	−2.9978	−4.6191	−5.2275	−5.5225	−5.6934	−5.7490
3D(θ_c,3D) [70]	−2.7332	−4.5726	−5.2158	−5.5196	−5.6929	−5.7489
3D-u-θ	−2.7332	−4.5726	−5.2158	−5.5196	−5.6929	−5.7489

Table 4. *Cont.*

R_α/h	2	5	10	20	50	100
			$\sigma_{\alpha\alpha}[10^3$ Pa] at ($\alpha = a/2, \beta = b/2, \bar{z} = 0$)			
3D(θ_a) [70]	2245.9	2104.6	2023.8	1977.7	1948.6	1938.6
3D(θ_c,1D) [70]	1802.3	1517.9	1396.1	1331.9	1292.8	1279.7
3D(θ_c,3D) [70]	1829.7	1523.9	1397.7	1332.4	1292.9	1279.7
3D-u-θ	1829.7	1523.9	1397.7	1332.4	1292.9	1279.7
			$\sigma_{\alpha\beta}[10^3$ Pa] at ($\alpha = 0, \beta = 0, \bar{z} = h/4$)			
3D(θ_a) [70]	165.97	62.049	28.841	13.745	5.3213	2.6301
3D(θ_c,1D) [70]	143.90	44.761	19.011	8.5799	3.2023	1.5624
3D(θ_c,3D) [70]	143.38	44.833	19.024	8.5818	3.2025	1.5624
3D-u-θ	143.38	44.833	19.024	8.5818	3.2025	1.5624
			$\sigma_{\alpha z}[10^3$ Pa] at ($\alpha = 0, \beta = b/2, \bar{z} = h/4$)			
3D(θ_a) [70]	−177.52	−70.179	−34.135	−16.746	−6.6109	−3.2901
3D(θ_c,1D) [70]	−163.27	−57.830	−26.873	−12.859	−4.9976	−2.4739
3D(θ_c,3D) [70]	−162.57	−57.869	−26.881	−12.860	−4.9977	−2.4739
3D-u-θ	−162.57	−57.869	−26.881	−12.860	−4.9977	−2.4739
			$\sigma_{\beta z}[10^3$ Pa] at ($\alpha = a/2, \beta = 0, \bar{z} = 3h/4$)			
3D(θ_a) [70]	−104.50	−52.635	−27.225	−13.710	−5.4896	−2.7442
3D(θ_c,1D) [70]	−142.28	−61.257	−30.160	−14.825	−5.8512	−2.9110
3D(θ_c,3D) [70]	−138.41	−61.063	−30.139	−14.823	−5.8511	−2.9110
3D-u-θ	−138.41	−61.063	−30.139	−14.823	−5.8511	−2.9110
			$\sigma_{zz}[10^3$ Pa] at ($\alpha = a/2, \beta = b/2, \bar{z} = 3h/4$)			
3D(θ_a) [70]	92.991	46.640	24.829	12.766	5.1869	2.6064
3D(θ_c,1D) [70]	118.96	53.212	27.251	13.742	5.5187	2.7624
3D(θ_c,3D) [70]	116.03	53.050	27.232	13.740	5.5186	2.7624
3D-u-θ	116.03	53.050	27.232	13.740	5.5186	2.7624

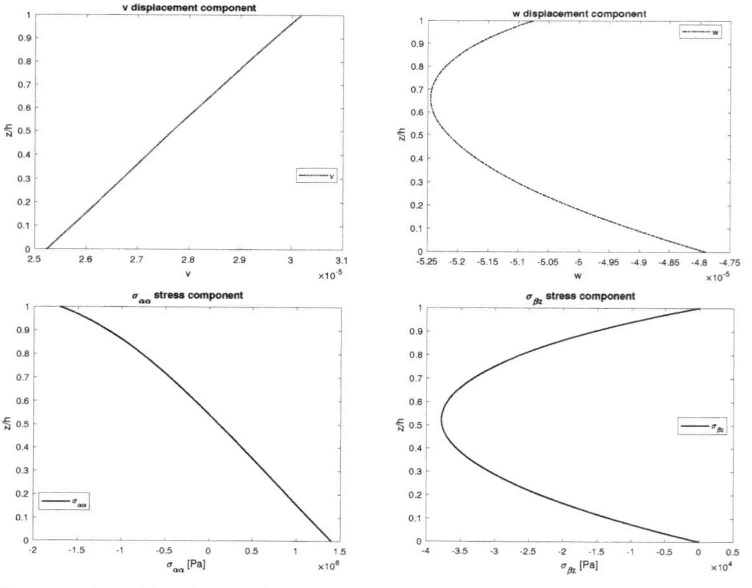

Figure 7. Second benchmark, displacements and stresses for a thick ($R_\alpha/h = 10$) simply supported cylinder with one FGM (p = 1) layer obtained using the 3D-u-θ model. Maximum amplitudes: w and $\sigma_{\alpha\alpha}$ at ($a/2$,$b/2$); v and $\sigma_{\beta z}$ at ($a/2$,0).

The third benchmark presents a simply supported sandwich cylindrical shell with an FGM core (see Figure 1). The radii of curvature are $R_\alpha = 10\,m$ and $R_\beta = \infty$ and the in-plane dimensions are $a = \frac{\pi}{3}R_\alpha$ and $b = 30\,m$. The FGM core has thickness $h_2 = 0.6\,h$ and the two skins have $h_1 = h_3 = 0.2\,h$; h is the global thickness of the structure. The material properties of this configuration are based on Equations (46)–(48) with a quadratic exponential coefficient ($p = 2$). The bottom skin is full metallic and the top skin is full ceramic with thermal and mechanical properties expressed as in Equations (43) and (44) (as is visible in Figure 8). The investigated thickness ratios are $R_\alpha/h = 2, 5, 10, 20, 50, 100$. The sovra-temperature in harmonic form (half-wave numbers $m = 1$ and $n = 1$) has amplitude values at the top $\Theta_t = -1\,K$ and at the bottom $\Theta_b = 0\,K$. The temperature profiles are provided in Figure 9 for a moderately thick ($R_\alpha/h = 5$) and a moderately thin ($R_\alpha/h = 20$) cylindrical shell. For thicker configurations, the thickness layer effect is a little bit more evident than the second benchmark because of the rigidity of the previous closed and symmetric cylinder. For thinner configurations, 3D(θ_c,1D), 3D(θ_c,3D), and 3D-u-θ models have no differences in terms of their temperature profiles. The 3D(θ_a) model always proposes an inadequate temperature profile because it did not consider any material layer and thickness layer effect. The same considerations are still valid for the discussion of displacements and stresses in Table 5 where the 3D(θ_c,1D), 3D(θ_c,3D), and 3D-u-θ models are quite coincident for all the R_α/h ratios proposed. The great difference involves only the 3D(θ_a) model that is always inadequate for this configuration. Figure 10 shows displacement and stress evaluations through the thickness of the simply supported sandwich cylindrical shell with an FGM core. The analysed cylindrical shell is moderately thick and both proposed displacements and stresses are continuous because the FGM properties continuously vary along the thickness direction.

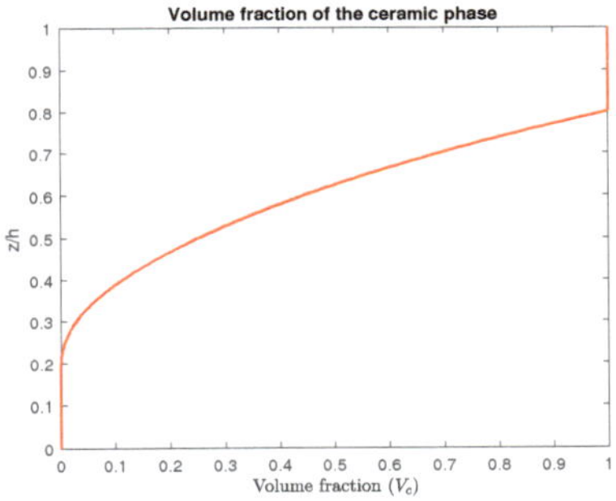

Figure 8. Third benchmark, volume fraction of the ceramic phase for a simply supported sandwich cylindrical shell with FGM (p = 2) core.

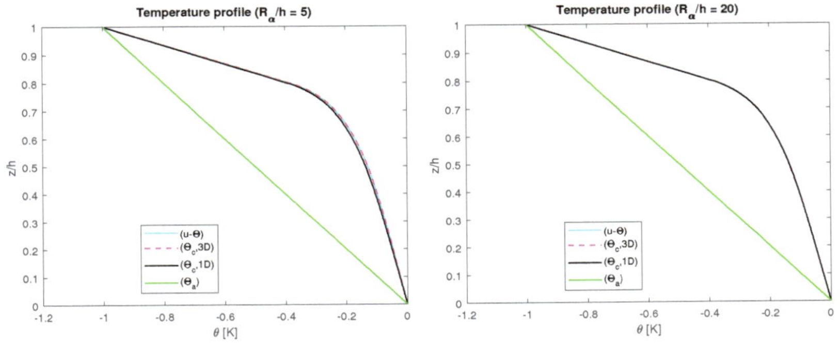

Figure 9. Third benchmark, temperature profiles for thick and moderately thick simply supported sandwich cylindrical shells with FGM (p = 2) core. The maximum amplitude of the temperature $\theta(\alpha, \beta, z)$ is evaluated at the centre of the cylindrical shell at $(a/2, b/2)$.

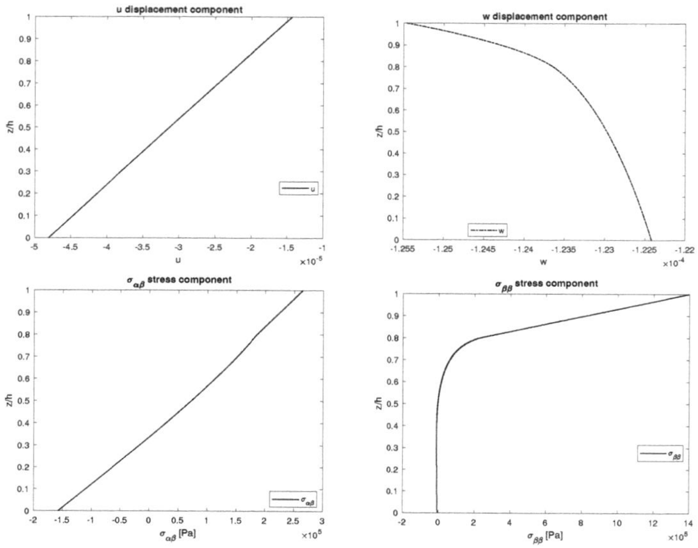

Figure 10. Third benchmark, displacements and stresses for a thick ($R_\alpha/h = 10$) simply supported sandwich cylindrical shell with FGM (p = 2) core obtained using the 3D-u-θ model. Maximum amplitudes: w and $\sigma_{\beta\beta}$ at $(a/2,b/2)$; u at $(0,b/2)$; $\sigma_{\alpha\beta}$ at (0,0).

Table 5. Third benchmark, simply supported sandwich cylindrical shell with FGM (p = 2) core. Sovra-temperature imposed as $\Theta_t = -1\,K$ and $\Theta_b = 0\,K$ for $m = 1$ and $n = 1$. 3D uncoupled thermoelastic models from [70]. The new 3D coupled thermoelastic solution is 3D-u-θ.

R_α/h	2	5	10	20	50	100
		u[10^{-5} m] at ($\alpha = 0, \beta = b/2, \bar{z} = 0$)				
3D(θ_a) [70]	−0.1032	−2.5775	−6.5549	−13.028	−18.648	−18.423
3D(θ_c,1D) [70]	−1.0043	−2.5320	−4.8150	−8.0331	−9.7702	−8.8619
3D(θ_c,3D) [70]	−1.1076	−2.5335	−4.8060	−8.0263	−9.7682	−8.8613
3D-u-θ	−1.1076	−2.5335	−4.8060	−8.0263	−9.7682	−8.8613
		w[10^{-5} m] at ($\alpha = a/2, \beta = b/2, \bar{z} = h/2$)				
3D(θ_a) [70]	−2.5005	−8.0358	−18.792	−37.833	−57.181	−58.624
3D(θ_c,1D) [70]	−2.0740	−5.8451	−12.332	−22.256	−29.339	−27.831
3D(θ_c,3D) [70]	−2.0327	−5.8020	−12.297	−22.235	−29.333	−27.829
3D-u-θ	−2.0327	−5.8020	−12.297	−22.235	−29.333	−27.829
		$\sigma_{\beta\beta}$[10^3 Pa] at ($\alpha = a/2, \beta = b/2, \bar{z} = h$)				
3D(θ_a) [70]	1100.6	1034.2	823.07	553.81	506.00	655.96
3D(θ_c,1D) [70]	1521.1	1495.3	1395.3	1292.9	1352.4	1468.5
3D(θ_c,3D) [70]	1564.2	1504.9	1398.5	1293.9	1352.5	1468.6
3D-u-θ	1564.2	1504.9	1398.5	1293.9	1352.5	1468.6
		$\sigma_{\alpha\beta}$[10^3 Pa] at ($\alpha = 0, \beta = 0, \bar{z} = h/4$)				
3D(θ_a) [70]	211.18	46.212	50.865	−66.576	−120.91	−83.255
3D(θ_c,1D) [70]	46.213	7.5615	−39.847	−90.780	−89.957	−55.402
3D(θ_c,3D) [70]	28.473	4.5755	−40.378	−90.820	−89.951	−55.400
3D-u-θ	28.473	4.5755	−40.378	−90.820	−89.951	−55.400
		$\sigma_{\alpha z}$[10^3 Pa] at ($\alpha = 0, \beta = b/2, \bar{z} = h/4$)				
3D(θ_a) [70]	23.813	−15.719	−17.016	−7.0833	5.0945	5.3470
3D(θ_c,1D) [70]	−79.295	−40.804	−21.620	−7.0251	2.5966	2.8247
3D(θ_c,3D) [70]	−89.122	−41.263	−21.634	−7.0233	2.5961	2.8246
3D-u-θ	−89.122	−41.263	−21.634	−7.0233	2.5961	2.8246

Table 5. *Cont.*

R_α/h	2	5	10	20	50	100
			$\sigma_{\beta z}[10^3$ Pa] at ($\alpha = a/2, \beta = 0, \bar{z} = 3h/4$)			
3D(θ_a) [70]	−13.098	−15.405	−12.417	−6.3749	−0.1390	0.8218
3D(θ_c,1D) [70]	29.138	7.4634	2.3569	2.0687	2.5041	1.7405
3D(θ_c,3D) [70]	32.873	7.8962	2.4322	2.0798	2.5046	1.7406
3D-u-θ	32.873	7.8962	2.4322	2.0798	2.5046	1.7406
			$\sigma_{zz}[10^3$ Pa] at ($\alpha = a/2, \beta = b/2, \bar{z} = 3h/4$)			
3D(θ_a) [70]	42.679	16.454	10.492	5.0412	−0.1093	−0.8396
3D(θ_c,1D) [70]	−40.599	−11.996	−4.5234	−2.8784	−2.5498	−1.6952
3D(θ_c,3D) [70]	−48.215	−12.534	−4.5992	−2.8887	−2.5503	−1.6952
3D-u-θ	−48.215	−12.534	−4.5992	−2.8887	−2.5503	−1.6952

The last benchmark takes into account a simply supported spherical shell with one FGM layer (see Figure 1). The global thickness is h, the radii of curvature are $R_\alpha = R_\beta = 10\,m$, the imposed sovra-temperatures are $\Theta_t = 0\,K$ and $\Theta_b = -1\,K$, the in-plane dimensions are $a = \frac{\pi}{3}R_\alpha = b = \frac{\pi}{3}R_\beta$, and the half-wave numbers are $m = 2$ and $n = 1$. The configuration and material characteristics are the same as seen in the second benchmark (see Equations (46)–(48) for the bulk modulus, shear modulus, thermal expansion coefficient, and conductivity coefficient and see Equation (45) and Figure 11 for the volume fraction of the ceramic phase V_c with $p = 2$). In Figure 12, the temperature profile for the four models are shown. The 3D(θ_c,3D) and 3D-u-θ models are mandatory for the correct analysis of the thick shells. The 3D(θ_c,1D) model is a good approximation only for thin spherical shells, but it always denotes some differences with respect to the previous two models. The 3D(θ_a) model is always inappropriate for this benchmark for each R_α/h ratio considered. For the spherical shell, the thick configuration shows an important thickness layer effect. As in the previous benchmark, the 3D-u-θ model provides the same results obtained with the 3D(θ_c,3D) model because they consider both the thickness layer and the material layer effects. The displacements and stresses for several R_α/h are proposed in Table 6 where it is evident how the results obtained using the 3D(θ_a) model are inadequate for each thickness ratio because the actual temperature profile is never linear along the thickness direction. In Figure 13, the displacement and stress evaluations through the thickness of a spherical shell with one FGM layer are shown. In-plane and transverse displacements, transverse shear, and transverse normal stress are continuous because the compatibility and equilibrium conditions have been correctly imposed and also because the FGM layer continuously varies its own mechanical and thermal properties along the thickness direction. Transverse normal stress σ_{zz} satisfies the load boundary conditions ($\sigma^t_{zz} = \sigma^b_{zz} = P_z = 0$) as can be seen in Figure 13.

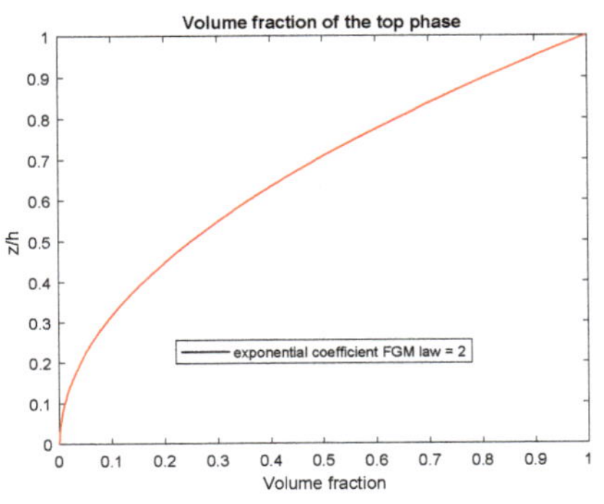

Figure 11. Fourth benchmark, volume fraction of the ceramic phase for a simply supported spherical shells with one FGM (p = 2) layer.

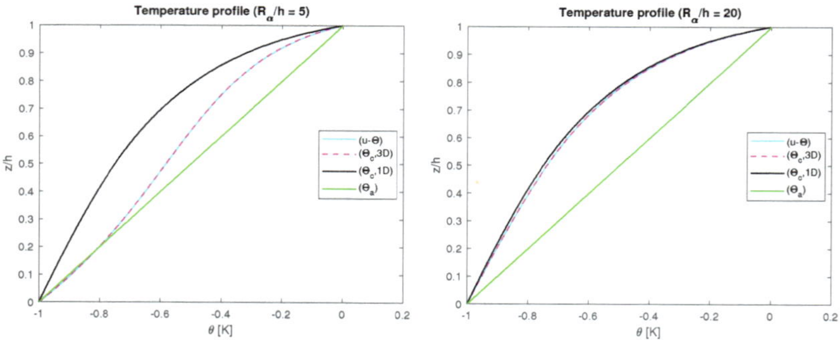

Figure 12. Fourth benchmark, temperature profiles for thick and moderately thick simply supported spherical shell with one FGM (p = 2) layer.

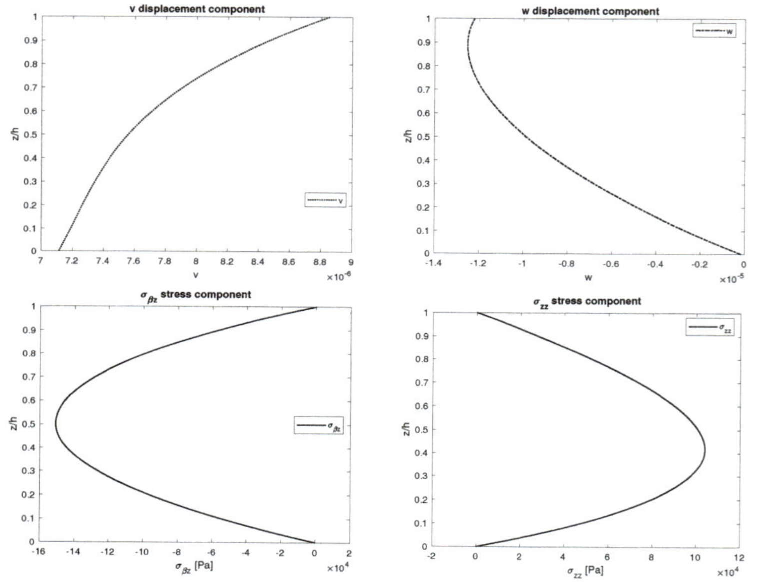

Figure 13. Fourth benchmark, displacements and stresses for a thick ($R_\alpha/h = 10$) simply supported spherical shell with one FGM (p = 2) layer obtained using the 3D-u-θ model. Maximum amplitudes: w and σ_{zz} at ($a/2$,$b/2$); v and $\sigma_{\beta z}$ at ($a/2$,0).

Table 6. Fourth benchmark, simply supported spherical shell with one FGM (p = 2) layer. Sovra-temperature imposed as $\Theta_t = 0\,K$ and $\Theta_b = -1\,K$ for $m = 2$ and $n = 1$. 3D uncoupled thermoelastic models from [70]. The new 3D coupled thermoelastic solution is 3D-u-θ.

R_α/h	2	5	10	20	50	100
		v[10^{-6} m] at ($\alpha = a/2$, $\beta = 0$, $\bar{z} = h$)				
3D(θ_a) [70]	3.8133	4.3442	6.2723	6.3118	3.4996	1.8525
3D(θ_c,1D) [70]	8.1671	8.0198	9.3872	8.2942	4.2336	2.1785
3D(θ_c,3D) [70]	1.5635	6.0279	8.8526	8.1968	4.2271	2.1778
3D-u-θ	1.5635	6.0279	8.8526	8.1968	4.2271	2.1778

Table 6. *Cont.*

R_α/h	2	5	10	20	50	100
			$w[10^{-5}\ \mathrm{m}]$ at $(\alpha = a/2, \beta = b/2, \bar{z} = h/2)$			
3D(θ_a) [70]	−0.0045	0.7679	0.1533	−2.3272	−5.4440	−6.5416
3D(θ_c,1D) [70]	−0.2283	0.1565	−1.1720	−4.4728	−8.0033	−9.1116
3D(θ_c,3D) [70]	0.1726	0.4099	−0.9875	−4.3848	−7.9847	−9.1067
3D-u-θ	0.1726	0.4099	−0.9875	−4.3848	−7.9847	−9.1067
			$\sigma_{\alpha\alpha}[10^3\ \mathrm{Pa}]$ at $(\alpha = a/2, \beta = b/2, \bar{z} = 0)$			
3D(θ_a) [70]	−627.94	1354.3	2306.7	2822.4	2666.3	2442.3
3D(θ_c,1D) [70]	−1140.7	1156.2	2169.3	2555.0	2157.7	1832.7
3D(θ_c,3D) [70]	797.05	1425.5	2219.4	2570.5	2161.7	1833.9
3D-u-θ	797.11	1425.5	2219.4	2570.5	2161.7	1833.9
			$\sigma_{\alpha\beta}[10^3\ \mathrm{Pa}]$ at $(\alpha = 0, \beta = 0, \bar{z} = h/4)$			
3D(θ_a) [70]	827.55	754.23	583.00	310.60	85.349	30.478
3D(θ_c,1D) [70]	981.89	833.16	595.79	276.15	52.204	10.609
3D(θ_c,3D) [70]	500.42	747.63	584.62	276.18	52.348	10.634
3D-u-θ	500.41	747.63	584.62	276.18	52.348	10.634
			$\sigma_{\alpha z}[10^3\ \mathrm{Pa}]$ at $(\alpha = 0, \beta = b/2, \bar{z} = h/4)$			
3D(θ_a) [70]	21.383	−227.08	−223.11	−149.47	−58.821	−27.092
3D(θ_c,1D) [70]	149.49	−219.90	−228.25	−145.56	−51.490	−22.143
3D(θ_c,3D) [70]	−125.09	−225.85	−226.36	−145.42	−51.517	−22.149
3D-u-θ	−125.09	−225.84	−226.36	−145.42	−51.517	−22.149
			$\sigma_{\beta z}[10^3\ \mathrm{Pa}]$ at $(\alpha = a/2, \beta = 0, \bar{z} = 3h/4)$			
3D(θ_a) [70]	−184.45	−104.69	−91.933	−63.362	−25.499	−11.800
3D(θ_c,1D) [70]	−296.96	−158.20	−121.36	−75.344	−27.707	−12.309
3D(θ_c,3D) [70]	−60.985	−120.11	−114.83	−74.530	−27.670	−12.306
3D-u-θ	−60.984	−120.11	−114.83	−74.530	−27.670	−12.306
			$\sigma_{zz}[10^3\ \mathrm{Pa}]$ at $(\alpha = a/2, \beta = b/2, \bar{z} = 3h/4)$			
3D(θ_a) [70]	237.90	70.115	51.648	35.425	15.234	7.3766
3D(θ_c,1D) [70]	391.90	107.16	67.497	41.558	16.381	7.6416
3D(θ_c,3D) [70]	73.490	79.064	63.665	41.100	16.360	7.6394
3D-u-θ	73.490	79.064	63.665	41.100	16.360	7.6394

4. Conclusions

A full coupled thermo-elastic 3D exact shell solution for thermal stress analysis of shells and plates embedding FGM layers has been proposed. The sovra-temperature amplitudes have been directly imposed at the external surfaces in steady-state conditions and the sovra-temperature profile was evaluated along the z direction. The sovra-temperature profile is a primary unknown variable similar to the displacements; this is possible because the 3D Fourier heat conduction equation and the 3D equilibrium equations for shells are put together into a set of four second-order differential equations. This temperature profile considers both the thickness layer and material layer effects for each possible geometry, without separately solving the related 3D Fourier heat conduction equation. The set of four second-order differential equations for shells is solved in a closed-form thanks to the Navier solution and the exponential matrix method. Different analyses, in terms of displacements, temperature profiles, in-plane, and out-of-plane stresses have been shown for several thickness ratios, geometries, FGM configurations, and temperature impositions. The proposed results showed a complete match between the model that separately solves the 3D Fourier heat conduction equation and the present 3D full coupled thermo-elastic model. For each investigated variable (temperatures, displacements and stresses) and for each thickness ratio, geometry, FGM configuration, and load imposition, the results proposed differences that were always less than 0.5%. This new coupled method permits taking into account both the material and the thickness layer effects using a simpler and more consistent mathematical formulation. Moreover, a reduced number of fictitious layers

M is requested in comparison with the uncoupled 3D model; this feature permits having the same accurate results for a less complicated fictitious layer discretization.

Author Contributions: Methodology, S.B. and R.T.; Software, D.C. and R.T.; Validation, D.C.; Formal analysis, R.T.; Investigation, D.C.; Data curation, S.B.; Writing—original draft, D.C.; Writing—review & editing, S.B.; Supervision, S.B. All authors have read and agreed to the published version of the manuscript.

Funding: This research received no external funding.

Institutional Review Board Statement: Not applicable.

Informed Consent Statement: Not applicable.

Data Availability Statement: Not applicable.

Conflicts of Interest: The authors declare no conflict of interest.

References

1. Librescu, L.; Marzocca, P. *Thermal Stresses '03, Vol. 1*; Virginia Polytechnic Institute and State University: Blacksburg, VA, USA, 2003.
2. Librescu, L.; Marzocca, P. *Thermal Stresses '03, Vol. 2*; Virginia Polytechnic Institute and State University: Blacksburg, VA, USA, 2003.
3. Nowinski, J.L. *Theory of Thermoelasticity with Applications*; Sijthoff & Noordhoff: Alphen aan den Rijn, The Netherlands, 1978.
4. Noor, A.K.; Burton, W.S. Computational models for high-temperature multilayered composite plates and shells. *Appl. Mech. Rev.* **1992**, *45*, 419–446. [CrossRef]
5. Swaminathan, K.; Sangeetha, D.M. Thermal analysis of FGM plates—A critical review of various modeling techniques and solution methods. *Compos. Struct.* **2017**, *160*, 43–60. [CrossRef]
6. Altay, G.A.; Dökmeci, M.C. Fundamental variational equations of discontinuous thermopiezoelectric fields. *Int. J. Eng. Sci.* **1996**, *34*, 769–782. [CrossRef]
7. Altay, G.A.; Dökmeci, M.C. Some variational principles for linear coupled thermoelasticity. *Int. J. Solids Struct.* **1996**, *33*, 3937–3948. [CrossRef]
8. Altay, G.A.; Dökmeci, M.C. Coupled thermoelastic shell equations with second sound for high-frequency vibrations of temperature-dependent materials. *Int. J. Solids Struct.* **2001**, *38*, 2737–2768. [CrossRef]
9. Cannarozzi, A.A.; Ubertini, F. A mixed variational method for linear coupled thermoelastic analysis. *Int. J. Solids Struct.* **2001**, *38*, 717–739. [CrossRef]
10. Das, N.C.; Das S.N.; Das, B. Eigenvalue approach to thermoelasticity. *J. Therm. Stress.* **1983**, *6*, 35–43. [CrossRef]
11. Kosìnski, W.; Frischmuth, K. Thermomechanical coupled waves in a nonlinear medium. *Wave Motion* **2001**, *34*, 131–141. [CrossRef]
12. Wauer, J. Free and forced magneto-thermo-elastic vibrations in a conducting plate layer. *J. Therm. Stress.* **1996**, *19*, 671–691. [CrossRef]
13. Kapuria, S.; Bhattacharyya, M.; Kumar, A.N. Bending and free vibration response of layered functionally graded beams: A theoretical model and its experimental validation. *Compos. Struct.* **2007**, *82*, 390–402. [CrossRef]
14. Kiani, Y.; Eslami, M.R. Thermal buckling analysis of functionally graded material beams. *Int. J. Mech. Mater. Des.* **2010**, *6*, 229–238. [CrossRef]
15. Ghiasian, S.E.; Kiani, Y.; Eslami, M.R. Dynamic buckling of suddenly heated or compressed FGM beams resting on nonlinear elastic foundation. *Compos. Struct.* **2013**, *106*, 225–234. [CrossRef]
16. Sun, Y.; Li, S.-R.; Batra, R.C. Thermal buckling and post-buckling of FGM Timoshenko beams on nonlinear elastic foundation. *J. Therm. Stress.* **2016**, *39*, 11–26. [CrossRef]
17. Ma, L.S.; Lee, D.W. Exact solutions for nonlinear static responses of a shear deformable FGM beam under an in-plane thermal loading. *Eur. J. Mech. A/Solids* **2012**, *31*, 13–20. [CrossRef]
18. Paul, A.; Das, D. Non-linear thermal post-buckling analysis of FGM Timoshenko beam under non-uniform temperature rise across thickness. *Eng. Sci. Technol. Int. J.* **2016**, *19*, 1608–1625. [CrossRef]
19. Zhang, J.; Chen L.; Lv, Y. Elastoplastic thermal buckling of functionally graded material beams. *Compos. Struct.* **2019**, *224*, 111014. [CrossRef]
20. Chakraborty, A.; Gopalakrishnana, S.; Reddy, J.N. A new beam finite element for the analysis of functionally graded materials. *Int. J. Mech. Sci.* **2003**, *45*, 519–539. [CrossRef]
21. Chen, Y.; Jin, G.; Zhang, C.; Ye, T.; Xue, Y. Thermal vibration of FGM beams with general boundary conditions using a higher-order shear deformation theory. *Compos. Part B* **2018**, *153*, 376–386. [CrossRef]
22. Esen, I. Dynamic response of functional graded Timoshenko beams in a thermal environment subjected to an accelerating load. *Eur. J. Mech. A/Solids* **2019**, *78*, 103841. [CrossRef]

23. Esfahani, S.E.; Kiani, Y.; Eslami M.R. Non-linear thermal stability analysis of temperature dependent FGM beams supported on non-linear hardening elastic foundations. *Int. J. Mech. Sci.* **2013**, *69*, 10–20. [CrossRef]
24. Li, Y.; Tang, Y. Application of Galerkin iterative technique to nonlinear bending response of three-directional functionally graded slender beams subjected to hygro-thermal loads. *Compos. Struct.* **2022**, *290*, 115481. [CrossRef]
25. Şimşek, M. Buckling of Timoshenko beams composed of two-dimensional functionally graded material (2D-FGM) having different boundary conditions. *Compos. Struct.* **2016**, *149*, 304–314. [CrossRef]
26. Ziane, N.; Meftah, S.A.; Ruta, G.; Tounsi, A. Thermal effects on the instabilities of porous FGM box beams. *Eng. Struct.* **2017**, *134*, 150–158. [CrossRef]
27. Javaheri, R.; Eslami, M.R. Thermal buckling of functionally graded plates based on higher order theory. *J. Therm. Stress.* **2002**, *25*, 603–625. [CrossRef]
28. Akbaş, S.D. Vibration and static analysis of functionally graded porous plates. *J. Appl. Comput. Mech.* **2017**, *3*, 199–207.
29. Saad, M.; Hadji, L. Thermal buckling analysis of porous FGM plates. *Mater. Today Proc.* **2022**, *53*, 196–201. [CrossRef]
30. Sangeetha, D.M.; Naveenkumar, D.T.; Vınaykuma, V.; Prakash, K.E. Temperature stresses in Functionally graded (FGM) material plates using deformation theory—Analytical approach. *Mater. Today Proc.* **2022**, *49*, 1936–1941. [CrossRef]
31. Zenkour, A.M.; Mashat, D.S. Thermal buckling analysis of ceramic-metal functionally graded plates. *Nat. Sci.* **2010**, *2*, 968–978. [CrossRef]
32. Yaghoobi, M.P.; Ghannad, M. An analytical solution for heat conduction of FGM cylinders with varying thickness subjected to non-uniform heat flux using a first-order temperature theory and perturbation technique. *Int. Commun. Heat Mass Transf.* **2020**, *116*, 104684. [CrossRef]
33. Zeighami, V.; Jafari, M. A closed-form solution for thermoelastic stress analysis of perforated asymmetric functionally graded nanocomposite plates. *Theor. Appl. Fract. Mech.* **2022**, *118*, 103251. [CrossRef]
34. Praveen, G.N.; Reddy, J.N. Nonlinear transient thermoelastic analysis of functionally graded ceramic-metal plates. *Int. J. Solid Strcut.* **1998**, *35*, 4457–4476. [CrossRef]
35. Thai, C.H.; Zenkour, A.M.; Wahab, M.A.; Nguyen-Xuan, H. A simple four-unknown shear and normal deformations theory for functionally graded isotropic and sandwich plates based on isogeometric analysis. *Compos. Struct.* **2016**, *139*, 77–95. [CrossRef]
36. Parandvar, H.; Farid, M. Large amplitude vibration of FGM plates in thermal environment subjected to simultaneously static pressure and harmonic force using multimodal FEM. *Compos. Struct.* **2016**, *141*, 163–171. [CrossRef]
37. Cho, J.R.; Oden, J.T. Functionally graded material: A parametric study on thermal-stress characteristics using the Crank-Nicolson-Galerkin scheme. *Comput. Methods Appl. Mech. Eng.* **2000**, *188*, 17–38. [CrossRef]
38. Alibeigloo, A. Thermo elasticity solution of sandwich circular plate with functionally graded core using generalized differential quadrature method. *Compos. Struct.* **2016**, *136*, 229–240. [CrossRef]
39. Hong, C.C. GDQ computation for thermal vibration of thick FGM plates by using fully homogeneous equation and TSDT. *Thin-Walled Struct.* **2019**, *135*, 78–88. [CrossRef]
40. Karakoti, A.; Pandey, S.; Kar, V.R. Nonlinear transient analysis of porous P-FGM and S-FGM sandwich plates and shell panels under blast loading and thermal environment. *Thin-Walled Struct.* **2022**, *173*, 108985. [CrossRef]
41. Jooybar, N.; Malekzadeh, P.; Fiouz, A.; Vaghefi, M. Thermal effect on free vibration of functionally graded truncated conical shell panels. *Thin-Walled Struct.* **2016**, *103*, 45–61. [CrossRef]
42. Tao, C.; Dai, T. Analyses of thermal buckling and secondary instability of post-buckled S-FGM plates with porosities based on a meshfree method. *Appl. Math. Model.* **2021**, *89*, 268–284. [CrossRef]
43. Qi, Y.-N.; Dai, H.-L.; Deng, S.-T. Thermoelastic analysis of stiffened sandwich doubly curved plate with FGM core under low velocity impact. *Compos. Struct.* **2020**, *253*, 112826. [CrossRef]
44. Gulshan Taj, M.N.A.; Chakrabarti, A.; Sheikh, A.H. Analysis of functionally graded plates using higher order shear deformation theory. *Appl. Math. Model.* **2013**, *37*, 8484–8494. [CrossRef]
45. Reddy, J.N.; Cheng, Z.-Q. Three-dimensional solutions of smart functionally graded plates. *J. Appl. Mech.* **2001**, *68*, 234–241. [CrossRef]
46. Jiang, H.-J.; Dai, H.-L. Analytical solutions for three-dimensional steady and transient heat conduction problems of a double-layer plate with a local heat source. *Int. J. Heat Mass Transf.* **2015**, *89*, 652–666. [CrossRef]
47. Chen, W.-Q.; Bian, Z.-G.; Ding, H.-J. Three-dimensional analysis of a thick FGM rectangular plate in thermal environment. *Int. Zhejiang Univ. Sci. A* **2003**, *4*, 1–7. [CrossRef]
48. Ootao, Y.; Tanigawa, Y. Three-dimensional solution for transient thermal stresses of an orthotropic functionally graded rectangular plate. *Compos. Struct.* **2007**, *80*, 10–20. [CrossRef]
49. Ootao, Y.; Tanigawa, Y. Three-dimensional transient thermal stresses of functionally graded rectangular plate due to partial heating. *J. Therm. Stress.* **2010**, *22*, 35–55.
50. Jabbari, M.; Shahryari, E.; Haghighat, H.; Eslami, M.R. An analytical solution for steady state three dimensional thermoelasticity of functionally graded circular plates due to axisymmetric loads. *Eur. J. Mech. A/Solids* **2014**, *47*, 124–142. [CrossRef]
51. Vel, S.S.; Batra, R.C. Exact solution for thermoelastic deformations of functionally graded thick rectangular plates. *AIAA J.* **2002**, *40*, 1421–1433. [CrossRef]
52. Liu, W.-X. Analysis of steady heat conduction for 3D axisymmetric functionally graded circular plate. *J. Cent. South Univ.* **2013**, *20*, 1616–1622. [CrossRef]

53. Alibeigloo, A. Exact solution for thermo-elastic response of functionally graded rectangular plates. *Compos. Struct.* **2010**, *92*, 113–121. [CrossRef]
54. Apalak, M.K.; Gunes, R. Thermal residual stress analysis of Ni–Al_2O_3, Ni–TiO_2, and Ti–SiC functionally graded composite plates subjected to various thermal fields. *J. Thermoplast. Compos. Mater.* **2005**, *18*, 119–152. [CrossRef]
55. Hajlaoui, A.; Chebbi, E.; Dammak, F. Three-dimensional thermal buckling analysis of functionally graded material structures using a modified FSDT-based solid-shell element. *Int. J. Press. Vessel. Pip.* **2021**, *194*, 104547–104568. [CrossRef]
56. Liu, B.; Shi, T.; Xing, Y. Three-dimensional free vibration analyses of functionally graded laminated shells under thermal environment by a hierarchical quadrature element method. *Compos. Struct.* **2020**, *252*, 112733–112746. [CrossRef]
57. Burlayenko, V.N.; Sadowski, T.; Dimitrova, S. Three-dimensional free vibration analysis of thermally loaded FGM sandwich plates. *Materials* **2019**, *12*, 2377. [CrossRef]
58. Nami, M.R.; Eskandari, H. Three-dimensional investigations of stress intensity factors in a thermo-mechanically loaded cracked FGM hollow cylinder. *Int. J. Press. Vessel. Pip.* **2012**, *89*, 222–229. [CrossRef]
59. Naghdabadi, R.; Kordkheili, S.A.H. A finite element formulation for analysis of functionally graded plates and shells. *Arch. Appl. Mech.* **2005**, *74*, 375–386. [CrossRef]
60. Qian, L.F.; Batra, R.C. Three-dimensional transient heat conduction in a functionally graded thick plate with a higher-order plate theory and a meshless local Petrov-Galerkin method. *Comput. Mech.* **2005**, *35*, 214–226. [CrossRef]
61. Mian, M.A.; Spencer, J.M. Exact solutions for functionally graded and laminated elastic materials. *J. Mech. Phys. Solids* **1998**, *46*, 2283–2295. [CrossRef]
62. Brischetto, S. Exact elasticity solution for natural frequencies of functionally graded simply-supported structures. *CMES-Comput. Model. Eng. Sci.* **2013**, *95*, 391–430.
63. Brischetto, S. A general exact elastic shell solution for bending analysis of funcionally graded structures. *Compos. Struct.* **2017**, *175*, 70–85. [CrossRef]
64. Özişik, M.N. *Heat Conduction*; John Wiley & Sons, Inc.: New York, NY, USA, 1993.
65. Povstenko, Y. *Fractional Thermoelasticity*; Springer International Publishing: Cham, Switzerland, 2015.
66. Moon, P.; Spencer, D.E. *Field Theory Handbook Including Coordinate Systems, Differential Equations and Their Solutions*; Springer: Berlin, Germany, 1988.
67. Mikhailov, M.D.; Özişik, M.N. *Unified Analysis and Solutions of Heat and Mass Diffusion*; Dover Publications Inc.: New York, NY, USA, 1984.
68. Boyce, W.E.; DiPrima, R.C. *Elementary Differential Equations and Boundary Value Problems*; John Wiley & Sons, Ltd.: New York, NY, USA, 2001.
69. Systems of Differential Equations. Available online: http://www.math.utah.edu/gustafso/ (accessed on 30 May 2013).
70. Brischetto, S.; Torre, R. 3D shell model for the thermo-mechanical analysis of FGM structures via imposed and calculated temperature profiles. *Aerosp. Sci. Technol.* **2019**, *85*, 125–149. [CrossRef]
71. Brischetto, S. A 3D layer-wise model for the correct imposition of transverse shear/normal load conditions in FGM shells. *Int. J. Mech. Sci.* **2018**, *136*, 50–66. [CrossRef]
72. Reddy, J.N.; Cheng, Z.-Q. Three-dimensional thermomechanical deformations of functionally graded rectangular plates. *Eur. J. Mech.-A/Solids* **2001**, *20*, 841–855. [CrossRef]
73. Brischetto, S.; Torre, R. Thermo-elastic analysis of multilayered plates and shells based on 1D and 3D heat conduction problems. *Compos. Struct.* **2018**, *206*, 326–353. [CrossRef]

technologies

MDPI

Article

FogTrust: Fog-Integrated Multi-Leveled Trust Management Mechanism for Internet of Things

Abdul Rehman [1], Kamran Ahmad Awan [1], Ikram Ud Din [1,*], Ahmad Almogren [2,*] and Mohammed Alabdulkareem [2]

1 Department of Information Technology, The University of Haripur, Haripur 22620, Pakistan
2 Chair of Cyber Security, Department of Computer Science, College of Computer and Information Sciences, King Saud University, Riyadh 11633, Saudi Arabiaa
* Correspondence: ikramuddin205@yahoo.com (I.U.D.); ahalmogren@ksu.edu.sa (A.A.)

Abstract: The Internet of Things (IoT) is widely used to reduce human dependence. It is a network of interconnected smart devices with internet connectivity that can send and receive data. However, the rapid growth of IoT devices has raised security and privacy concerns, with the identification and removal of compromised and malicious nodes being a major challenge. To overcome this, a lightweight trust management mechanism called FogTrust is proposed. It has a multi-layer architecture that includes edge nodes, a trusted agent, and a fog layer. The trust agent acts as an intermediary authority, communicating with both IoT nodes and the fog layer for computation. This reduces the burden on nodes and ensures a trustworthy environment. The trust agent calculates the trust degree and transmits it to the fog layer, which uses encryption to maintain integrity. The encrypted value is shared with the trust agent for aggregation to improve the trust degree's accuracy. The performance of the FogTrust approach was evaluated against various potential attacks, including On-off, Good-mouthing, and Bad-mouthing. The simulation results demonstrate that it effectively assigns low trust degrees to malicious nodes in different scenarios, even with varying percentages of malicious nodes in the network.

Keywords: Internet of Thing; fog-computing; trust management; security; privacy preservation; trustworthiness

Citation: Rehman, A.; Awan, K.A.; Ud Din, I.; Almogren, A.; Alabdulkareem, M. FogTrust: Fog-Integrated Multi-Leveled Trust Management Mechanism for Internet of Things. *Technologies* **2023**, *11*, 27. https://doi.org/10.3390/technologies11010027

Academic Editors: Manoj Gupta, Eugene Wong and Gwanggil Jeon

Received: 21 December 2022
Revised: 30 January 2023
Accepted: 2 February 2023
Published: 7 February 2023

1. Introduction

The concept of the Internet of Things (IoT) [1] has become more prevalent, though the idea of connected devices dates back to the 1970s. The term "Internet of Things" was introduced in 1999 by Kevin Ashton [2]. IoT is a technology that connects devices and machines to communicate with each other [3]. It is used in various fields, such as smart homes [4], wearable technology [5], and smart agriculture [6]. Despite its wide range of applications, IoT faces several challenges, including security, connectivity, privacy, interoperability, and energy consumption [7–9]. In 2018, over 23 billion devices were connected, which is twice the population [10]. The future projection is that the number of IoT devices will increase to a minimum of 80 billion [11]. The main goal of IoT is to make all devices autonomous through the power of the internet. However, the vast number of devices presents major privacy and security challenges [12]. These are critical issues that IoT companies must address for a promising future. The major security challenges include authentication [13], access control [14], policy enforcement [15], mobile security [16], secure middleware, confidentiality, and latency [8]. Security is a crucial concern for organizations, governments, and individuals, as they become increasingly digital-centric [17]. With the growing complexity of IoT attacks, it is important to detect, defend against, and respond to these threats. Hackers now have additional access points that can affect the real world [18].

Fog computing promotes IoT innovation through an open architecture [19]. It is a decentralized form of computing, where applications and data storage are located between the data source and cloud [20]. Fog computing performs computation, storage, and communication from edge devices, which control the flow of data between two networks such as routers, switches, access devices, gateways, hubs, etc. Fog operates in a DIST network environment [21] that is closely connected to the cloud and IoT/edge devices. It processes selected data locally before sending it to the cloud, reducing bandwidth [22] and latency [23] needs. An important benefit of fog computing is improved security, as it provides computing security locally rather than remotely.

In this article, a mechanism named FogTrust is proposed to detect and eliminate compromised and malicious nodes. The proposed system uses the fog to provide data integrity, which helps to prevent potential IoT attacks such as on-off, good-mouthing, and bad-mouthing attacks. To ensure security and integrity, a lightweight mechanism is implemented to maintain the integrity and aggregate the computed trustworthiness data (TD) for aggregation purposes. The TD of IoT nodes will reduce the impact of malicious and compromised nodes in good and bad-mouthing attacks. The use of a trust agent as an intermediary between the fog and IoT nodes performs a trust evaluation, reducing the computational burden on less capable nodes to improve security and reduce vulnerabilities caused by such nodes. The proposed approach can be summarized as:

1. The proposed mechanism, FogTrust, uses a multi-layer trust management (TM) architecture with central authorities to maintain a secure environment by detecting and eliminating malicious and compromised nodes with low trust.
2. The fog is integrated into the architecture to encrypt and maintain the integrity of the trust degree (TD) computed by trust agents.
3. The proposed system aggregates the current trust (CT) with previous trust (PT) to form the aggregated TD of a node, providing robustness against potential IoT attacks.

The structure of the rest of the paper is as follows:

Section 2 summarizes the existing literature and provides a comparative analysis to highlight the limitations. Section 3 explains the working of the PM, including the proposed architecture, trust parameters and computations, direct trust computation, indirect trust computation, trust development, and decision-making. Section 4 presents simulation results and discusses the performance of FogTrust compared to existing literature. Section 5 concludes the paper.

2. Related Work

To manage the IoT trust, various TM mechanisms have been proposed, including DIST, and CENT. DIST relies on nodes to manage trust between nodes, while IoT nodes in CENT depend on a CA for trust management. Despite several techniques for addressing trust management, the privacy and security challenges in fog computing remain significant, as sensitive information is transmitted between IoT devices or the edge layer and fog layer. Identifying malicious nodes and protecting data from attacks in fog computing is a major issue, but no notable solution has been proposed to address these security challenges in data sharing in fog computing.

A novel context-based trust management model is proposed for the Social IoT [24]. The proposed "ConTrust" approach uses a novel combination of parameters (satisfaction, commitment, and capability) to increase system efficiency. ConTrust measures job characteristics, honesty, job capability, and behavior of malicious nodes. Its architecture includes three components: job requester, trust management, and prospective provider. When a job requester requests a service, the trust evaluation process starts by computing trust parameters. If a node is trustworthy, the prospective provider will provide services to the requester. ConTrust is limited to covering IoT-related potential attacks. A multi-dimensional trust management model based on SLA is also proposed for Fog computing [25]. It contributes to applications, peers, and fog editors for fog service providers and measures their trustworthiness. The architecture comprises five components: smart application client, fog auditors, SLA agent, service providers, and smart applications. The model works when a service provider advertises their services and the application interacts with them for the first time.

In [26], a lightweight trust mechanism is presented that uses trust agents to manage communication certificates, which identify the trustworthiness of nodes using parameters. The mechanism employs a statistical probabilistic model to compute the degree of trust with high precision and adaptability. Its main role is to provide a solid and reliable mechanism for edge device information exchange [27]. The system can be enhanced with a hybrid approach to detect malicious nodes and network attacks. Ref. [28] presents research focusing on privacy and security in fog computing with IoT applications. It uses a subjective logic-based (SL-B) trust approach to improve IoT security and address challenges related to data transmission protection and protection against compromised attacks. The proposed system maintains the trustworthiness of each node in the network, calculates and updates their trust values, and stores them in a local list with a node ID.

In the field of cloud computing, industry TM is a recurring research trend [29]. It is expected that similar problems will arise in the emerging fog realm. Although the fog and cloud are similar, evaluating the trust in fog is more challenging than evaluating the trust in the cloud due to its mobility, distributed nature, and proximity to the end-user [18]. Unlike clouds, fog has little to no human involvement and is not redundant, meaning that disruptions may occur at any time, making it difficult to trust. These unique characteristics can be used as metrics to assess fog trust, along with existing features. In [30], a fuzzy logic approach was proposed to evaluate trust in fog and identify configurations that can alter its trust value. A campus scenario was presented as an example application, where various fog resources (FRs) were evaluated for reliability using the proposed metrics. The scenario discussed the FRs and attributes used to assess their trustworthiness, and an adopted fuzzy-logic approach was used to handle the complex trust values. The approach follows the steps of a fuzzy inference system, first evaluating the attributes of distance, latency, and reliability, then using the AND operator rather than the OR operator in the second step.

In [31], a TM framework is proposed that uses the MAPE-K feedback control loop to evaluate the trust levels. The framework includes trust agents and a consumer layer of TMS nodes that interact with clients. The cloud filters trust parameters into an adaptive trust parameters pool and assists with trust evaluation via the MAPE-K loop. The input framework takes into account the previous history to standardize the effect of anomalies. False decisions caused by malicious information decrease the effectiveness of the MAPE-K loop.

In [32], a TM scheme called COMITMENT is presented for fog computing. It uses the fog node reputation to construct a global reputation language and provides secure and trusted environments for information exchange. The DIST fog topology is considered, with nodes connected through communication protocols and a unique identity. Each node computes the trust evaluation of its nearest nodes to create a list of trusted nodes. The COMITMENT is a set of protocols installed on fog nodes that select trusted nodes for information sharing and provide a secure environment for resource sharing and information exchange. The goal is to build trust between parties to facilitate sensitive information exchange. The approach requires a central trust authority for trust level evaluation.

In [33], a two-way trust management system (TMS) is presented that allows both the service requester (SR) and service provider (SP) to evaluate each other's trustworthiness. The TMS aggregates trust using subjective logic theory, which is useful when uncertainty and proposition are involved. Clients request services from fog servers and the fog server evaluates the trustworthiness of the clients through direct observation and consultation with the nearest fog server. The clients also consult the nearest server to determine the trust level of the fog server. Both clients and servers in the fog share information about other clients and servers. The system must simultaneously calculate the trustworthiness of both the SR and SP. In a recent study, Trust2Vec [34], a trust management system for large-scale IoT systems, is proposed. The system has the ability to manage trust relationships in large IoT systems and mitigate attacks from malicious devices. It uses a network structure to build trust relationships among devices and has a key phase to detect malicious nodes through determining device communities, generating random walk algorithms, and leveraging trust relationships in clusters. The proposed system has an overall detection rate of 94% for malicious devices or nodes, and its key contribution is the use of a random-walk algorithm for navigating trust relationships and a parallelization method for attack detection.

In [35], a TM model is proposed to improve security, social relationships, and services in fog computing. The model evaluates trust through direct trust, recommendations, and reputation, and uses fuzzy logic to aggregate trust and handle uncertainty in mobile fog computing. The detection and mitigation rate is approximately 71%, with 70% of clients and fogs being malicious and 74% of attacks detected. However, like other TMS in fog computing, it assumes fog nodes are static, making it challenging to handle dynamic nodes. The contributions and limitations of the existing approaches are provided in Table 1.

Table 1. Comparative Analysis of Existing Literature.

Ref.	Contribution	Limitation
[24]	A trust management model is presented in social IoT that is context-dependent to compute the trust.	Need to check the proposed system against potential attacks that are related to trust.
[25]	A multi-dimensional trust management system is presented to check the trustworthiness of FSP.	Need to evaluate the malicious behavior of applications that enter in fog environment.
[26]	Utilizes a lightweight mechanism that manages trust in IIoT-Edge nodes.	Requires improved prediction capabilities to increase performance.
[28]	Establishes a secure environment for fog applications by using a TM.	A hybrid technique is required to ensure robust network security.
[30]	Utilizes a fuzzy approach to evaluate trust in fog computing.	Requires a broker that acts as a fog TM.
[31]	Utilizes a MAPE-K feedback control loop for evaluation of trust level.	Requires trust to be calculated before the fog layer and data to be protected in the fog layer.
[32]	Utilizes the COMITMENT approach for security in fog computing.	Requires a CA that evaluates trust before the fog layer.
[33]	TW-TMS evaluates the trust level of SP and then checks the TD of SD.	Requires the trustworthiness of the SP and SR to be calculated at the same time.
[34]	Utilizes a random-walk algorithm for the navigation of trust relationships and parallelization method for attack detection.	The work can be extended by including the TM of data entities.
[35]	Utilizes fuzzy logic for trust aggregation to handle uncertainty in fog computing.	Static nodes handling is difficult.

3. Proposed FogTrust Mechanism

The proposed model will use the fog computing to ensure data integrity, which will reduce the possibility of various IoT attacks, including on-off attacks, good-mouthing attacks, and bad-mouthing attacks. The system proposes a lightweight encryption method to protect the TD and aggregate its evaluation to maintain integrity. The encrypted TD from IoT nodes reduces the impact of malicious and compromised nodes in good and bad-mouthing attacks. A trust agent, acting as an intermediary between the fog and IoT nodes, performs trust evaluation, reducing the computational burden on less capable nodes, thereby improving security and reducing vulnerabilities posed by such nodes.

3.1. Proposed Architecture of FogTrust

The proposed architecture of FogTrust consists of three layers: community layer, trust agent layer, and fog layer. The working of the proposed architecture is shown in Figure 1; the FogTrust includes communities separated into different domains, each of which has nodes that can connect with one another to complete specific tasks. The IoT nodes have a unique identity, and the message file includes their identification, community, and domain information. When a node (TE) requests communication from another node (TR), TR provides material to the trust agents for trust evaluation. The community layer, edge layer, or IoT node layer consists of IoT devices such as smart cameras, smartwatches, smartphones, smart laptops, sensors, and other IoT-related devices that can generate and transmit data or information autonomously. Before the data are transmitted to the fog, the trust agent in the trust agent layer evaluates its trustworthiness, determining whether the information is trustworthy or not.

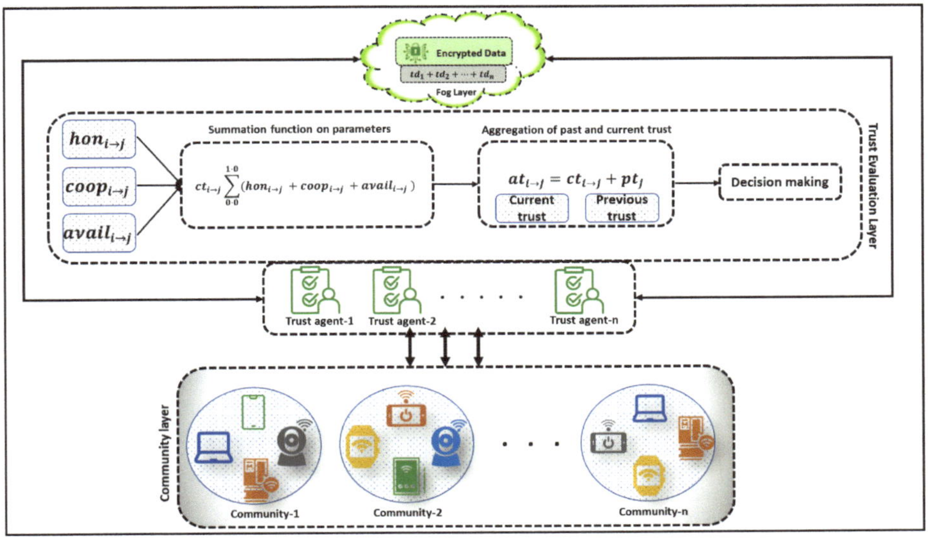

Figure 1. The proposed FogTrust Architecture.

The proposed system in FogTrust performs trust evaluation using a combination of three trust parameters: honesty, cooperativeness, and availability. The evaluation combines current trust and previous trust values to compute the aggregated trust values, which are used to make trust decisions. The trustworthiness of a device is determined by comparing its trust data (TD) with a threshold value that ranges from 0.0 to 1.0, with 0.0 being the minimum trust and 1.0 being the maximum trust. Newly joined nodes are assigned a default trust value of 0.5.

3.2. Trust Parameters and Computation

The trust evaluation combines three parameters: availability, honesty, and cooperativeness to enhance the reliability and security in the IoT network. Availability refers to the accessibility of resources to end-users, while cooperativeness reflects a node's ability to collaborate with others. Honesty is determined based on the observations of one node (i) towards another node (j). The cooperativeness is measured by analyzing response time and calculated as the ratio of prompt responses to the total number of responses. The evaluation considers the previous and current trust values to make the final trust decision. The threshold for trust ranges from 0.0 to 1.0, with 0.0 being the minimum and 1.0 being the maximum trust. New nodes are assigned a default trust value of 0.5.

3.3. Direct Trust Computations

The evaluation of the direct trust procedure begins with the *TE* being identified using their unique ID. The Algorithm 1 represents a direct trust observation procedure that takes place when a TR needs to evaluate the *TD*. TE requests services from TR during the joining of the network.

Algorithm 1 DOB-Trust Computation

procedure TRUST EVALUATION($i \to j$)
    j_{i_d} ▷ Identification of TE
    $j_{req} \to i$ ▷ Request TE towards TR
    $pt_{o_b}: I \to J\ hon_{i \to j}, coop_{i \to j}, avail_{i \to j}$
    if ($pt_{ob}: i \to j == Yes$) **then**
        $GotoStep - 9;$
    else
        $GotoAlgorithm - 2;$
    $\mathcal{E}va_{Trust}: i \to j[hon_{i \to j}, coop_{i \to j}, avail_{i \to j}]$
    $\sum_{0.0}^{1.0} ct_{i \to j}^{direct} = \sum(hon_{i \to j} + coop_{i \to j} + avail_{i \to j})$
    $at_{i \to j} = ct_{i \to j} + pt_j$ ▷ Aggregated Trust Formulation
    if ($at_{i \to j} \geq threshold$) **then**
        $ProvideServices;$
    else
        $Decline;$
    Exit.

The j_{id} shows the identification of the TE that requested to gather the services from the *TR*. Where j represents the TE and *id* is the identification, it is initialized when a *TR* receives a request from the *TE* for services. In j_{req}, j represents the *TE*, *req* is the request, and i demonstrates *TR*. This is where the *TE* requests services from the *TR*.

$$pt_{ob}: I \to J[hon_{i \to j}, coop_{i \to j}, avail_{i \to j}] \tag{1}$$

The evaluation process starts with determining the trust level of the TE node using the trust parameters of honesty, cooperativeness, and availability, as described in Equation (1). The TE is identified and initialized into the network. In Equation (1), *pt* represents past trust, *ob* represents observation, *I* and *J* represent TR and TE, respectively, *hon* represents honesty, *coop* represents cooperativeness, and *avail* represents availability.

$$If(pt_{ob}: \ \to j == Yes) \tag{2}$$

The Equation (2) represents the observations that the TE (j) must gather for the TR (i) before service can be provided. TR (i) will evaluate TE (j) only if the required observations are equal to "yes".

$$\mathcal{E}va_{Trust}: i \to j[hon_{i \to j}, coop_{i \to j}, avail_{i \to j}] \tag{3}$$

In Equation (3), the evaluation process of TE j through TR i. Here, *εva* represents the trust evaluation.

$$\sum_{0.0}^{1.0} ct_{i \to j}^{direct} = \sum(hon_{i \to j} + coop_{i \to j} + avail_{i \to j}) \tag{4}$$

In Equation (4), the current trust (*ct*) is evaluated by combining the direct trust evaluation (*direct*) between the TR (i) and TE (j).

$$at_{i \to j} = ct_{i \to j} + pt_j \tag{5}$$

In Equation (5), *at* represents the aggregated trust which is calculated as the mean of the current trust (*ct*) and previous trust (*pt*) of node j and i and j, which are, respectively, the TE and the TR. The $i \to j$ symbol represents the trust of TE towards TR. The final TD is formulated by aggregating the past and current trust values.

$$If(at_{i \to j} \geq threshold) \tag{6}$$

In Equation (6), *at* represents the aggregated trust, and *threshold* is the predetermined threshold value, which is compared to the aggregated trust value. If the aggregated trust value is greater than or equal to the threshold value, then the TR (i) starts providing services to the TE (j) and communication starts.

3.4. Absolute TD Formulation

The evaluation process of the absolute TD formulation starts with identifying the TE using its unique ID. The algorithm referred to as Algorithm 2 represents the procedure of absolute observations that take place when a TR needs to evaluate the degree of trust. The TE requests services from the TR after joining the network, and the Algorithm 2 thoroughly explains the TD formulation procedure.

Algorithm 2 ATD-Formulation

procedure TRUST EVALUATION($i \rightarrow j$)
  j_{id} ▷ Identification of TE
  if (j==new) **then** ▷ Newly joined node check
    $GotoStep - 7;$
  else
    $GotoAlgorithm - 3;$
  $\mathcal{E}_{Trust}: i \rightarrow j[hon_{i \rightarrow j}, coop_{i \rightarrow j}, avail_{i \rightarrow j}]$
  $\sum_{0.0}^{1.0} ct_{i \rightarrow j}^{form} = \sum(hon_{i \rightarrow j} + coop_{i \rightarrow j} + avail_{i \rightarrow j})$ ▷ Direct Trust formulation
  $at_{i \rightarrow j} = ct_{i \rightarrow j}^{form} + pt_j$ ▷ Aggregated Trust Formulation
  if ($at_{i \rightarrow j} \geq threshold$) **then**
    $ProvideServices;$
  else
    $Decline;$
  Exit.

$$If(j == new) \tag{7}$$

The Equation (7) is used to determine whether the TE is new to the network or not. If the TE j is determined to be new, then its trust is evaluated. Otherwise, trust is measured using the Algorithm 3.

$$\mathcal{E}_{Trust}: i \rightarrow j[hon_{i \rightarrow j}, coop_{i \rightarrow j}, avail_{i \rightarrow j}] \tag{8}$$

In Equation (8), the process of the trust value evaluation is illustrated. The variable $\rightarrow$ represents the evaluation, and *hon*, *coop*, and *avail* represent the honesty, cooperativeness, and availability of the TE, respectively.

$$\sum_{0.0}^{1.0} ct_{i \rightarrow j}^{form} = \sum(hon_{i \rightarrow j} + coop_{i \rightarrow j} + avail_{i \rightarrow j}) \tag{9}$$

In Equation (9), *ct* represents the current trust, while i and j are the TE and the TR, respectively.

$$at_{i \rightarrow j} = ct_{i \rightarrow j}^{form} + pt_j \tag{10}$$

In Equation (10), *at* represents the aggregated trust which is formulated as the mean of *ct* current trust, where *form* is the formulation of trust and *pt* is the previous trust of node i and j, respectively.

After formulation of the Algorithm 2, it will further evaluate the honesty, cooperativeness, and availability as described in Algorithm 1 and as elaborated earlier in Equations (1) and (3). The algorithm then formulates the direct overall degree of trust by aggregating the current and previous trust evaluations and checking the final aggregated TD against the threshold value to determine if it can provide the services. The function and description of these Equations (4) and (10) have been explained earlier.

3.5. Recommendations-Based Indirect Trust Evaluation

When direct observation of the TE is not available, the TR must rely on recommendations to evaluate the trust level. The indirect trust evaluation will be conducted by gathering recommendations from nearby nodes based on their knowledge of the TE. If the available observations are insufficient, these algorithms pass a request to Algorithm 3 to compute the indirect trust. If the information shows that the TE is not from the same network, Algorithm 3 will perform an indirect evaluation.

Algorithm 3 RB-Indirect Trust Evaluation

procedure TRUST EVALUATION($i \rightarrow j$)
  Generating Request to gather Recommendations $\rightarrow r_j \rightarrow k_{th}$
  j_{id} ▷ TE Identification
  $rec^{check}[i \rightarrow j]$
  $rec^{eva}_{j \rightarrow k_{th}} : [r_{k_1 \rightarrow j} + r_{k_2 \rightarrow j} + \ldots\ldots\ldots + r_{k_n \rightarrow j}]$
  $\sum_{0.0}^{1.0} r^{re}_{j \rightarrow k_{th}} = r^{je_1}_{j \rightarrow k_{th}} + r^{je_2}_{j \rightarrow k_{th}} + \ldots + r^{je_n}_{j \rightarrow k_{th}}$
  $\sum_{0.0}^{1.0} r^{indirect}_{j \rightarrow t} = \sum_{r=0.0}^{r=1.0} r^{re}_{j \rightarrow k_{th}}$
  $pt_j = r^{indirect}_{j \rightarrow t}$
  $at_{i \rightarrow j} = ct_i \rightarrow j + pt_j$
  if $(at_{i \rightarrow j} > Yes)$ **then**
    $ProvideServices;$
  else
    $Decline;$
  Exit.

The Algorithm 3 begins by sending requests for recommendations to nearby nodes. Equation (11) represents the generation of these requests to gather the necessary information for evaluating the TD of a TE.

$$GeneratingRequesttogatherRecommendations \rightarrow r_j \rightarrow k_{th} \tag{11}$$

In Equation (11), k_{th} represents the nearest nodes (k) and th represents the number of nodes to which a system sends requests for recommendations for a TE evaluation.

$$\sum_{0.0}^{1.0} r^{re}_{j \rightarrow k_{th}} = r^{je_1}_{j \rightarrow k_{th}} + r^{je_2}_{j \rightarrow k_{th}} + \ldots + r^{je_n}_{j \rightarrow k_{th}} \tag{12}$$

After gathering the recommendations, they are arranged correctly. In Equation (12), r represents the recommendations, re represents the number of received recommendations, k represents the neighboring node, and th represents the number of generated requests.

$$\sum_{0.0}^{1.0} r^{indirect}_{j \rightarrow t} = \sum_{r=0.0}^{r=1.0} r^{re}_{j \rightarrow k_{th}} \tag{13}$$

The algorithm evaluates trust by calculating the total degree of trust after gathering the recommendations. The mean value of the recommendations is used to compute the overall degree of trust, which results in a final degree of trust with a value between 0.0 and 1.0.

$$pt_j = r^{indirect}_{j \rightarrow t} \tag{14}$$

In Equation (14), the algorithm calculates the indirect trust value by aggregating it with the previous trust (PT) value. pt represents previous trust, r represents the recommendation, *indirect* represents the indirect trust evaluation, and $j \rightarrow t$ indicates that the j TE generates a request to gather recommendations from $k_t h$ nodes.

The rest of the Algorithm 3 operates similarly to what was described earlier. It combines the current trust value with the previous one and compares the aggregated trust value to the threshold value, as outlined in Algorithm 1. If the TE's trust value surpasses the threshold value, the TR offers services. Otherwise, the TR declines and ceases further communication.

3.6. Trust Development

Trust agents can calculate the whole trust value through trust development. They evaluate three separate parameters and use the standard function sigma to obtain the aggregated trust value from the trust parameters' output. The final TD is then formulated and shared with the fog layer. Nodes with low TD are not allowed to share information or communicate. However, nodes with supreme trust or TD higher than the threshold are allowed to communicate further. To determine a node's trustworthiness, the trust evaluation layer computes its trust value and compares it to a predefined threshold. Trust agents can evaluate the aggregated trust value, allowing trust development. They calculate three different parameters and use the sigma function to obtain the aggregated trust value from the trust parameters output.

3.7. Decision Making

The IoT network uses the absolute trust value to make quick decisions for improving system efficiency. The TD of nodes is calculated by evaluating parameters with a comparison to a threshold, with a range of 0.0 to 1.0 and a default trust level of 0.5. A trust value of 0.0 to 0.49 is untrustworthy, 0.51 to 0.79 is moderately trustworthy, and 0.8 to 1.0 is supremely trustworthy. Nodes with trust values above 0.5 are allowed to communicate in the network. The TM must have an effective and reliable technique for determining the absolute trust value.

4. Experimental Simulation and Outcomes

This section presents the simulation results of FogTrust with the existing TM mechanism. The authors have evaluated the trustworthiness of the system in terms of good and bad-mouthing attacks, as well as various on-off attack scenarios. They also compare their proposed approach to existing TM mechanisms such as ConTrust and SLA-Trust. The criteria used for evaluating their work include Aggregation Impact, Good and Bad-mouthing attacks, and On-Off attacks. The simulation results were generated using MATLAB, a multi-paradigm programming language and computing development framework developed by MathWorks. MATLAB is mainly used for matrix operations, data visualization, algorithm implementation, user interface creation, and interfacing with other programming languages. Although symbolic computation is not a primary function of MATLAB, it can be performed through an optional toolbox that uses the MuPAD symbolic engine. Additionally, the Simulink tool provides visual simulation capabilities for dynamic and integrated systems. The data used in the simulation analysis is experimental and is generated when an IoT node joins the network. The proposed approach assigns a pre-defined default trust degree to each node, allowing for communication between nodes.

The simulation setup for the proposed FogTrust mechanism is shown in Table 2. The simulation uses data from the table, which includes the "area" parameter set at 200 square meters and "number of devices" set at 600 randomly distributed. The simulation runs for 100 s, with a data transmission rate of 6 to 8 Mbps. The malicious node detection rate during the simulation is between 50% and 75%.

Table 2. Simulation Environment Implementation Setup.

Parameters	Value
Network area	200 m^2
Number of devices	600
Simulation duration	100 (s)
Degree of trust	0.0∼1.0
Default trust	0.5
Node distribution	Random
Transmission rate	6∼8 Mbps
Malicious nodes percentage	50∼75%

4.1. Analysis of the Trust Aggregation

This section presents the impact of using the aggregation process on the trust degree computation. The comparison is made between using the previous trust with the present computed trust degree and the computation performed without the aggregation process. The use of the aggregation process has a significant impact on the trust degree computation, resulting in more consistent values, as shown in Figure 2.

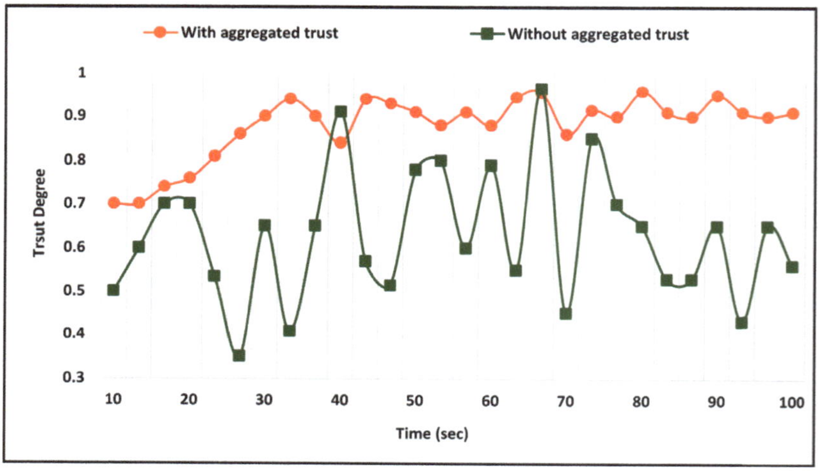

Figure 2. Previous Trust Aggregation Impact on Direct Trust Evaluation.

The comparison shows that the use of aggregation in the trust calculation process results in more consistent trust values and improved reliability compared to the scenario where aggregation is not used. This highlights the importance of considering past trust data in determining the current trust level, which helps to reduce errors and improve the security of the network by accurately identifying malicious nodes.

4.2. Analysis of Detection Rate

The detection rate is a crucial metric for evaluating the performance of any trust management system, as it reflects the system's ability to accurately identify trustworthy entities. Our proposed approach in this article enhances the detection rate by aggregating previous trust degrees with the current computed trust, resulting in more accurate and reliable trust decisions. In this simulation setup, each node has several close neighbors that offer various services over time, while the percentage of malicious and compromised nodes is 70%.

Figure 3 presents the simulation results of the proposed approach in terms of the number of interactions and detection percentage. The results demonstrate that the proposed mechanism has an initial detection rate of 70% and steadily increases over time, reaching over 80% after 25 interactions and exceeding 90% after 45 interactions. This indicates that the proposed approach outperforms other existing mechanisms, such as SLA-Trust, which has a continuous improvement in detection rate, reaching a peak of 81%. While ConTrust has a higher initial detection rate of 80%, it decreases to 66.5% after 20 interactions. Its highest detection rate is 89%, but is still lower than that of FogTrust. The average detection rate of FogTrust is 84.32%, which is higher than that of SLA-Trust (67.89%) and ConTrust (79.66%).

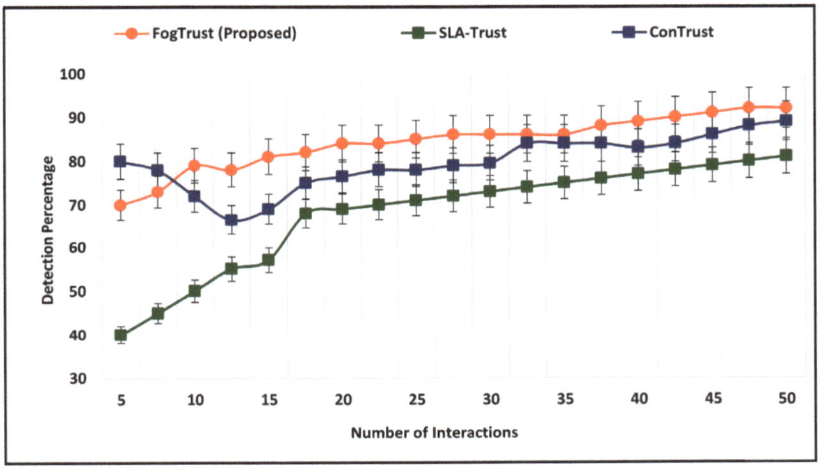

Figure 3. The Detection Rate Comparison of FogTrust with Exiting Approaches.

4.3. On-Off Attack

The simulation results demonstrate that in the occurrence of an on-off attack, the TD of compromised nodes decreases dramatically from 0.5 to 0.2 within seconds. This highlights the effectiveness of the proposed mechanism in detecting and mitigating the impact of such attacks. However, it should be noted that in the case of ConTrust [24], the malicious node may regain its trust after a certain period, which suggests the need for continuous monitoring and updating of trust values.

Figure 4 shows the malicious nodes' level of trust, which decreases and is still unable to regain the highest trust level. In comparison to SLA-Trust [25], the proposed mechanism successfully detects the on-off attack and the malicious node's degree of trust. Furthermore, in the case of ConTrust, the trust value of the malicious node goes down. The malicious node regains its trust to 0.35 in 70 s, but after 70 s it again increases. Similarly, the SLA-Trust value of the malicious node also decreases, which shows that FogTrust can detect the malicious node at a low level of trust.

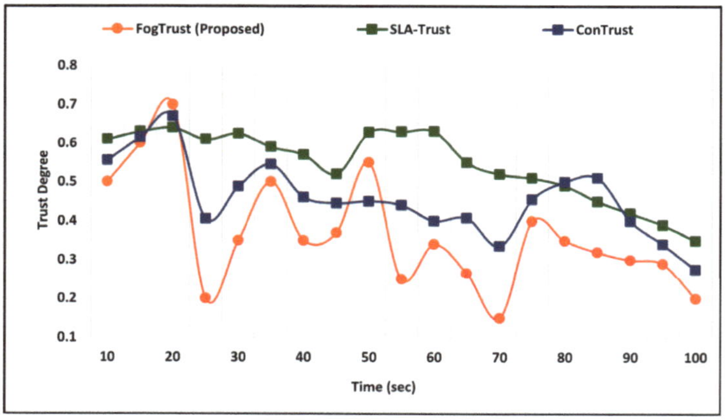

Figure 4. Comparative Analysis of FogTrust Against On-off Attacks.

4.4. Good and Bad Mouthing Attack

This section discusses the comparative simulation outcomes against good and bad-mouthing attacks. The trust value is a predefined threshold value ranging from 0.0 to 1.0. Time (s) is 100 and trust defaults to 0.5. To test the efficiency of the proposed approach against attacks involving goodmouth, we put three trust management models into practice. When the number of negative recommendations grows over time, the level of trust is shown to be declining as shown in Figure 5.

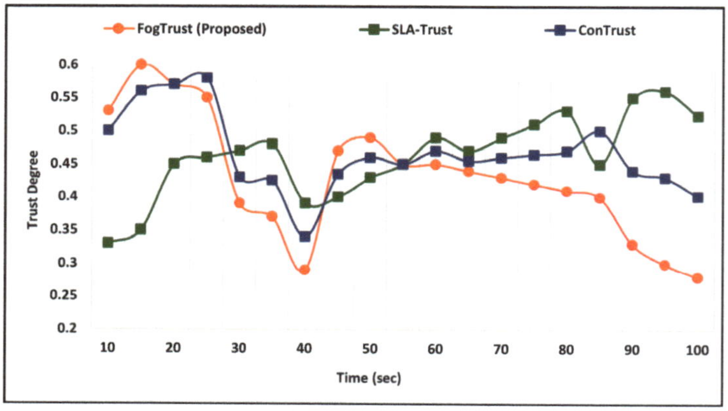

Figure 5. Comparative Analysis Against Good-Mouthing Attacks.

The effectiveness of the PM against bad-mouthing attacks has also been evaluated. The results indicate that the PM is effective in preventing such attacks. Three models were implemented, each with different trust and threshold values. As depicted in Figure 6, as the trust value increases, the detection rate also increases, but if the trust value increases too much, then the detection rate decreases. This indicates that the PM can detect bad-mouthing attacks even when they are at an increasing rate.

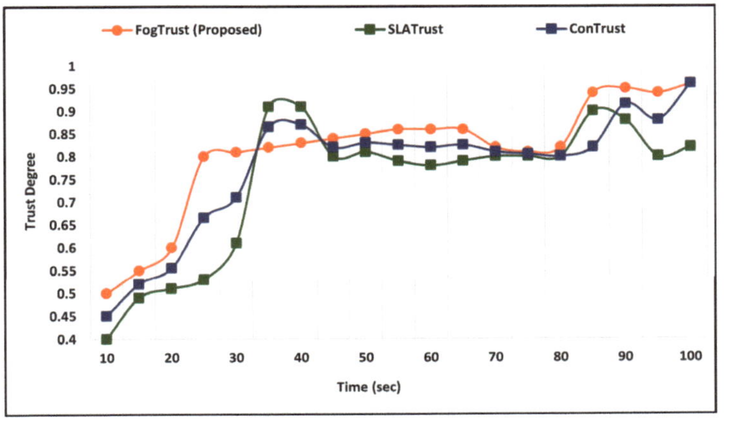

Figure 6. Comparative Analysis Against Bad-Mouthing Attacks.

5. Conclusions

The Internet of Things (IoT) is widely used in various industries, however, IoT nodes often struggle to maintain security on their own, making them susceptible to various attacks. To mitigate these risks, many mechanisms based on privacy and trust management have been proposed. However, current approaches neglect some features of central trust authority communications and the importance of central authority trust management, such as trust agents. The proposed FogTrust is effective in managing trust in the communication of fog computing with IoT devices. Other trust management mechanisms have been proposed, but they ignore the deployment of a centralized trust authority before the fog layer. To enhance the accuracy and reliability of FogTrust, a central authority, i.e., trust agents, is deployed. This central trust authority improves accuracy while reducing the computational weight on IoT nodes, which enhances resistance against attacks, reduces vulnerability, and provides standard security. The overall detection of malicious nodes in the proposed FogTrust mechanism ranges between 50% to 75% when compared with existing approaches. The *PM* can be further enhanced by identity, naming, and certificate allocation, and the security can be increased by encrypting the shared trust degree with the fog.

Author Contributions: Conceptualization, K.A.A. and I.U.D.; methodology, A.A.; software, A.R. and K.A.A.; validation, K.A.A., I.U.D. and A.A.; formal analysis, K.A.A. and M.A.; investigation, I.U.D., A.A. and M.A.; resources, A.A.; data curation, M.A. and A.A.; writing—original draft preparation, A.R.; writing—review and editing, K.A.A., I.U.D. and A.A.; visualization, A.A.; supervision, I.U.D.; project administration, A.A.; funding acquisition, A.A. All authors have read and agreed to the published version of the manuscript.

Funding: This work was supported by the Deanship of Scientific Research at King Saud University, Riyadh, Saudi Arabia, through the Vice Deanship of Scientific Research Chairs: Chair of Cyber Security.

Institutional Review Board Statement: Not applicable.

Informed Consent Statement: Not applicable.

Data Availability Statement: Not applicable.

Conflicts of Interest: The authors declare no conflict of interest.

Abbreviations

CA	Central Authority
CENT	Centralized
CT	Current Trust
DIST	Distributed
DO	Direct Observation
FS	Fog Server
IO	Indirect Observation
PM	Proposed Mechanism
PT	Previous Trust
SP	Service Provider
SR	Service Requester
TD	Trust Degree
TE	Trustee
TM	Trust Management
TMS	Trust Management System
TR	Trustor

References

1. Koohang, A.; Sargent, C.S.; Nord, J.H.; Paliszkiewicz, J. Internet of Things (IoT): From awareness to continued use. *Int. J. Inf. Manag.* **2022**, *62*, 102442. [CrossRef]
2. Ashton, K. That 'internet of things' thing. *RFID J.* **2009**, *22*, 97–114.
3. Abid, M.A.; Afaqui, N.; Khan, M.A.; Akhtar, M.W.; Malik, A.W.; Munir, A.; Ahmad, J.; Shabir, B. Evolution towards smart and software-defined internet of things. *AI* **2022**, *3*, 100–123. [CrossRef]
4. Babangida, L.; Perumal, T.; Mustapha, N.; Yaakob, R. Internet of Things (IoT) Based Activity Recognition Strategies in Smart Homes: A Review. *IEEE Sens. J.* **2022**, *22*, 8327–8336. [CrossRef]
5. Trovato, V.; Sfameni, S.; Rando, G.; Rosace, G.; Libertino, S.; Ferri, A.; Plutino, M.R. A Review of Stimuli-Responsive Smart Materials for Wearable Technology in Healthcare: Retrospective, Perspective, and Prospective. *Molecules* **2022**, *27*, 5709. [CrossRef]
6. Awan, K.A.; Ud Din, I.; Almogren, A.; Almajed, H. AgriTrust—A trust management approach for smart agriculture in cloud-based internet of agriculture things. *Sensors* **2020**, *20*, 6174. [CrossRef]
7. Mishra, V.K.; Tripathi, R.; Tiwari, R.G.; Misra, A.; Yadav, S.K. Issues, Challenges, and Possibilities in IoT and Cloud Computing. In *Proceedings of the International Conference on Computational Intelligence in Pattern Recognition*; Springer: Singapore, 2022; pp. 326–334.
8. George, A.; Ravindran, A.; Mendieta, M.; Tabkhi, H. Mez: An adaptive messaging system for latency-sensitive multi-camera machine vision at the iot edge. *IEEE Access* **2021**, *9*, 21457–21473. [CrossRef]
9. Bhat, S.A.; Huang, N.F.; Sofi, I.B.; Sultan, M. Agriculture-Food Supply Chain Management Based on Blockchain and IoT: A Narrative on Enterprise Blockchain Interoperability. *Agriculture* **2021**, *12*, 40. [CrossRef]
10. Farhan, L.; Kharel, R.; Kaiwartya, O.; Quiroz-Castellanos, M.; Alissa, A.; Abdulsalam, M. A concise review on Internet of Things (IoT)-problems, challenges and opportunities. In Proceedings of the 2018 11th International Symposium on Communication Systems, Networks & Digital Signal Processing (CSNDSP), Budapest, Hungary, 18–20 July 2018; pp. 1–6.
11. Abiodun, O.I.; Abiodun, E.O.; Alawida, M.; Alkhawaldeh, R.S.; Arshad, H. A review on the security of the internet of things: Challenges and solutions. *Wirel. Pers. Commun.* **2021**, *119*, 2603–2637. [CrossRef]
12. Din, I.U.; Guizani, M.; Hassan, S.; Kim, B.S.; Khan, M.K.; Atiquzzaman, M.; Ahmed, S.H. The Internet of Things: A review of enabled technologies and future challenges. *IEEE Access* **2018**, *7*, 7606–7640. [CrossRef]
13. Zhang, J.; Shen, C.; Su, H.; Arafin, M.T.; Qu, G. Voltage over-scaling-based lightweight authentication for IoT security. *IEEE Trans. Comput.* **2021**, *71*, 323–336. [CrossRef]
14. Ouaddah, A.; Mousannif, H.; Abou Elkalam, A.; Ouahman, A.A. Access control in the Internet of Things: Big challenges and new opportunities. *Comput. Netw.* **2017**, *112*, 237–262. [CrossRef]
15. Sahay, R.; Meng, W.; Estay, D.S.; Jensen, C.D.; Barfod, M.B. CyberShip-IoT: A dynamic and adaptive SDN-based security policy enforcement framework for ships. *Future Gener. Comput. Syst.* **2019**, *100*, 736–750. [CrossRef]
16. Garg, S.; Kaur, K.; Kaddoum, G.; Garigipati, P.; Aujla, G.S. Security in IoT-driven mobile edge computing: New paradigms, challenges, and opportunities. *IEEE Netw.* **2021**, *35*, 298–305. [CrossRef]
17. Tanwar, S.; Gupta, N.; Iwendi, C.; Kumar, K.; Alenezi, M. Next Generation IoT and Blockchain Integration. *J. Sens.* **2022**, *2022*, 9077348. [CrossRef]
18. Mendieta, M.; Neff, C.; Lingerfelt, D.; Beam, C.; George, A.; Rogers, S.; Ravindran, A.; Tabkhi, H. A Novel Application/Infrastructure Co-design Approach for Real-time Edge Video Analytics. In Proceedings of the 2019 SoutheastCon, Huntsville, AL, USA, 11–14 April 2019; pp. 1–7.
19. Haseeb, K.; Alzahrani, F.A.; Siraj, M.; Ullah, Z.; Lloret, J. Energy-Aware Next-Generation Mobile Routing Chains with Fog Computing for Emerging Applications. *Electronics* **2023**, *12*, 574. [CrossRef]

20. Saad, Z.M.; Mhmood, M.R. Fog computing system for internet of things: Survey. *Tex. J. Eng. Technol.* **2023**, *16*, 1–10.
21. Ruan, H.; Gao, H.; Qiu, H.; Gooi, H.B.; Liu, J. Distributed operation optimization of active distribution network with P2P electricity trading in blockchain environment. *Appl. Energy* **2023**, *331*, 120405. [CrossRef]
22. Gupta, P.; Saini, D.K. Introduction to Optimization in Fog Computing. In *Bio-Inspired Optimization in Fog and Edge Computing Environments;* Auerbach Publications: New York, NY, USA, 2023; pp. 1–24.
23. Kar, B.; Yahya, W.; Lin, Y.D.; Ali, A. Offloading using Traditional Optimization and Machine Learning in Federated Cloud-Edge-Fog Systems: A Survey. *IEEE Commun. Surv. Tutor.* **2023**. [CrossRef]
24. Latif, R. ConTrust: A novel context-dependent trust management model in social Internet of Things. *IEEE Access* **2022**, *10*, 46526–46537. [CrossRef]
25. Chang, V.; Sidhu, J.; Singh, S.; Sandhu, R. SLA-based Multi-dimensional Trust Model for Fog Computing Environments. *J. Grid Comput.* **2023**, *21*, 1–19. [CrossRef]
26. Din, I.U.; Bano, A.; Awan, K.A.; Almogren, A.; Altameem, A.; Guizani, M. LightTrust: Lightweight trust management for edge devices in industrial internet of things. *IEEE Internet Things J.* **2021**. [CrossRef]
27. George, A.; Ravindran, A. Scalable approximate computing techniques for latency and bandwidth constrained IoT edge. In *Proceedings of the International Summit Smart City 360°*; Springer: Cham, Switzerland, 2021; pp. 274–292.
28. Al Muhtadi, J.; Alamri, R.A.; Khan, F.A.; Saleem, K. Subjective logic-based trust model for fog computing. *Comput. Commun.* **2021**, *178*, 221–233. [CrossRef]
29. Baghalzadeh Shishehgarkhaneh, M.; Keivani, A.; Moehler, R.C.; Jelodari, N.; Roshdi Laleh, S. Internet of Things (IoT), Building Information Modeling (BIM), and Digital Twin (DT) in Construction Industry: A Review, Bibliometric, and Network Analysis. *Buildings* **2022**, *12*, 1503. [CrossRef]
30. Rahman, F.H.; Au, T.W.; Newaz, S.S.; Suhaili, W.S. Trustworthiness in fog: A fuzzy approach. In Proceedings of the 2017 VI International Conference on Network, Communication and Computing, Kunming, China, 8–10 December 2017; pp. 207–211.
31. Namal, S.; Gamaarachchi, H.; MyoungLee, G.; Um, T.W. Autonomic trust management in cloud-based and highly dynamic IoT applications. In Proceedings of the 2015 ITU Kaleidoscope: Trust in the Information Society (K-2015), Barcelona, Spain, 9–11 December 2015; pp. 1–8.
32. Al-Khafajiy, M.; Baker, T.; Asim, M.; Guo, Z.; Ranjan, R.; Longo, A.; Puthal, D.; Taylor, M. COMITMENT: A fog computing trust management approach. *J. Parallel Distrib. Comput.* **2020**, *137*, 1–16. [CrossRef]
33. Alemneh, E.; Senouci, S.M.; Brunet, P.; Tegegne, T. A two-way trust management system for fog computing. *Future Gener. Comput. Syst.* **2020**, *106*, 206–220. [CrossRef]
34. Dhelim, S.; Kechadi, T.; Aung, N.; Ning, H.; Chen, L.; Lakas, A. Trust2Vec: Large-Scale IoT Trust Management System based on Signed Network Embeddings. *arXiv* **2022**, arXiv:2204.06988.
35. Ogundoyin, S.O.; Kamil, I.A. A trust management system for fog computing services. *Internet Things* **2021**, *14*, 100382. [CrossRef]

technologies

MDPI

Article

Estimation of Energy Consumption and Flight Time Margin for a UAV Mission Based on Fuzzy Systems

Luis H. Manjarrez [1], Julio C. Ramos-Fernández [2], Eduardo S. Espinoza [1,3,*] and Rogelio Lozano [1,4]

1 Center for Research and Advanced Studies of the National Polytechnic Institute, Mexico City 07360, Mexico
2 Department of Mechatronics Engineering, Polytechnic University of Pachuca, Hidalgo 43830, Mexico
3 National Council for Science and Technology, Mexico City 07360, Mexico
4 HEUDIASYC CNRS, Université de Technologie de Compiègne, 60319 Compiegne, France
* Correspondence: eduardo.espinoza@cinvestav.mx or eespinoza@conacyt.mx; Tel.: +52-5557-47-3800 (ext. 4263)

Abstract: An essential aspect to achieving safety with a UAV is that it operates within the limits of its capabilities, the available flight time being a key aspect when planning and executing a mission. The flight time will depend on the relationship between the available energy and the energy required by the UAV to complete the mission. This paper addresses the problem of estimating the energy required to perform a mission, for which a fuzzy Takagi–Sugeno system was implemented, whose premises were developed using fuzzy C-means to estimate the power required in the different stages of the mission. The parameters used in the fuzzy C-means algorithm were optimized using particle swarm optimization. On the other hand, an equivalent circuit model of a battery was used, for which fuzzy modeling was employed to determine the relationship between the open-circuit voltage and the state of charge of the battery, which in conjunction with an extended Kalman filter allows determining the battery charge. In addition, we developed a methodology to determine the minimum allowable battery charge level. From this, it is possible to determine the available flight time at the end of a mission defined as the flight time margin. In order to evaluate the developed methodology, a physical experiment was performed using an hexarotor UAV obtaining a maximum prediction error equivalent to the energy required to operate for 7 s, which corresponds to 2% of the total mission time.

Keywords: SoC estimation; fuzzy clustering; multirotor UAV

check for updates

Citation: Manjarrez, L.H.; Ramos-Fernández, J.C.; Espinoza, E.S.; Lozano, R. Estimation of Energy Consumption and Flight Time Margin for a UAV Mission Based on Fuzzy Systems. *Technologies* **2023**, *11*, 12. https://doi.org/10.3390/technologies11010012

Academic Editors: Manoj Gupta, Eugene Wong and Gwanggil Jeon

Received: 20 December 2022
Revised: 6 January 2023
Accepted: 9 January 2023
Published: 12 January 2023

1. Introduction

UAVs are a booming technology, since they represent a versatile platform for a wide range of applications. This technology has found wide acceptance in the energy, construction, and agriculture industries, where it is mainly used for mapping, inspection, photography, and filming. This situation has led to the global drone market being valued at USD 30.6 billion in 2022, and it is estimated that this could reach USD 55.8 billion by 2030 [1]. However, these platforms are susceptible to emerging risks due to technical and operational issues, such as environmental factors, tampering, technical failures, and even cyber-attacks [2,3].

If we combine the continuing growth in the use of these platforms with the risks associated with their operation, it is evident that establishing safety measures in their operations is critical. This is reflected in the rules and regulations adopted worldwide, which limit the types of vehicles, along with the allowed flying zones and operating conditions [4,5].

One of the key factors to guarantee integrity during an operation is that the assigned tasks are aligned with the capabilities of the vehicle. The maximum reachable flight time is an essential parameter since, in order to establish a safe mission profile, in addition to the determination of the time required to complete every mission stage, a safe energy margin must be considered to allow operating for an additional amount of time to successfully complete it. This represents a safety measure against variations in energy consumption or situations not contemplated that could compromise the aircraft's integrity, people's security, and the environment in which the mission is carried out.

Multirotor UAVs are mostly battery-powered, and therefore, their maximum flight time depends on the available battery energy and the discharge rate. In turn, as stated in [6], the discharge rate will depend on many factors, such as:

- Vehicle design: aerodynamic design, weight, number of actuators, avionics, and energy efficiency.
- Operating environment: air density, wind speed, relative wind direction.
- Dynamics: speed, acceleration, and direction of motion.
- Mission: payload and area of operation.

Furthermore, there are other factors that can affect the energy consumption of the system, such as rotor and hardware failures. In this scenario, the remaining faultless rotors are forced to operate in a region of lower energy efficiency [7], reducing the available energy of the battery due to saturation phenomena in the actuators. Therefore, the information provided by the manufacturers, or that obtained from a performance test under specific conditions, should only be considered as a reference when a mission profile is established.

In this sense, predicting the behavior of the discharge rate and the available energy in a battery makes it possible to know whether the planned mission can be completed successfully, and even to anticipate whether or not it can be completed under conditions that cause unforeseen changes in consumption.

There are two main ways of estimating the energy required to complete the trajectory: (i) using mathematical models that employ the physical characteristics of the vehicle, and its operating speed [8,9]; or (ii) using empirical models employing regressions for a predefined data set [10]. However, such techniques do not consider possible fluctuations in consumption that could affect the capacity to complete a mission successfully.

Fuzzy systems have been shown to be suitable for managing energy-related aspects of UAVs, as can be seen in [11], where a fuzzy system in conjunction with the PSO algorithm was employed to manage the power supply of a hybrid-powered system, showing favorable results in fuel economy while maintaining robustness to variations in power consumption variation. In addition, fuzzy systems have been employed in other UAV-related tasks, such as in control [12] and decision making during the mission [13]; however, these systems have not been used for the calculation of required energy during a mission.

In order to provide a solution for energy estimation in a multirotor UAV, so that it can operate under persistent changes in the energy requirement, we developed a fuzzy-based methodology to determine the total energy required to complete a specific mission based on the vertical and horizontal velocities, the period during which it travels at those velocities, and the power-estimation error for a given state of the UAV.

The proposed methodology is based on fuzzy systems, which are some of the empirical methods. The use of Takagi–Sugeno fuzzy systems was due to their ability to recreate with adequate accuracy the existing functions among the parameters affecting energy consumption. In addition, the structure of the method allows it to be extended to include other factors that affect the energy required without major modifications to the structure. Unlike to the works presented in the literature, it has been conceived for use during the execution of the mission, and not only as a way of estimating the energy required a priori. This provides an important advantage for the safe operation of UAVs, since it not only allows one to know in advance if the mission to be performed is feasible, but also, once it is in progress, it allows one to anticipate variations in consumption that could jeopardize the operation.

In addition, a methodology was developed to determine the minimum charge level to which the battery can be brought considering the relationship between the thrust control signal and the battery voltage. This allows knowing the available flight time, and moreover, if we combine this with the estimate of the energy needed to execute a mission, we can determine the flight time during which the UAV will be able to operate with the expected remaining energy.

The main contributions of this research work are summarized as follows:

1. We developed a required-energy estimation system capable of adapting to persistent variations in energy consumption based on fuzzy C-means.
2. We propose a new methodology to determine the aircraft flight time, which to the best of our knowledge, is the first to consider the effect of the battery's state of charge on the control signals.

The rest of the paper is organized as follows. Section 2 presents the works related to the estimation of energy required during a flight and the estimation of flight time. Section 3 presents the proposed methodology divided into energy estimation (Section 3.2), state-of-charge estimation (Section 3.3) and flight-time-margin estimation (Section 3.4). Section 4 shows the application of the proposed methodology in a hexarotor UAV. Finally, Section 5 presents the conclusions and future improvements that can be applied to the proposed methodology.

2. Related Work

In the process of estimating the energy required to conduct a specific mission, three different approaches can be distinguished: (i) methods based on aerodynamic models, (ii) methods using regressions, and (iii) those based on intelligent systems. Some of the principal solutions that have been developed in the field of energy estimation are discussed below.

One of the most widely used models is presented in [14]. This model provides a simple way to approximate the required power based on the total weight of the vehicle, its displacement speed, the efficiency in the transfer of energy from the motor to the propeller, and the drag–lift ratio of the vehicle, in addition to a term corresponding to the power consumed by the vehicle's electronics. While this methodology provides an easy way to estimate the consumed energy, it neglects significant factors such as the wind and the air density, which could affect the vehicle's energy consumption. Therefore, this methodology should be used with caution when determining the energy required for a specific mission.

In [15], the authors proposed a power-estimation method wherein the vehicle motion is decomposed into its horizontal and vertical components. For each component, the required power is evaluated considering the acceleration and velocity at which it moves. The aerodynamic effect is considered assuming that the reference surface will have the characteristics of a flat plate. Although the presented model showed favorable performance in numerical simulations, it is complex to find the area affected by the airflow, which will depend on the direction of flight, and the wind speed and direction. In addition, parameters such as propeller tip speed are not available for most UAVs.

A simulation model was presented in [16] where aerodynamic, motor, and battery models are considered. For the estimation of the torque required by each rotor, the blade element moment theory is used, from which the consumed power is determined considering the efficiency of the motor. An equivalent circuit model is used for the battery, considering that the effective capacity is determined using a correction factor as a function of the required power. Finally, the flight time is calculated by dividing the effective battery energy by the required power. This method, despite the positive relation between the measured results and those obtained by a simulation, was not conceived as a method for online energy estimation during a mission.

Regression-based estimation methods, such as the one presented in [17], estimate the required energy based on the vehicle's operation. The authors divided the mission phases into: the idle mode, armed, takeoff, vertical and horizontal flight, and the effect of the payload. For each of these stages, a polynomial regression was performed based on data obtained from experimental tests. Although this method provides an easy way to estimate the energy required to complete a mission, it is not able to adapt to conditions different from those of the flights in which the modeling data were obtained.

The method presented in [18] uses a set of regressors using elastic net regression. The regressors were set up in two stages for ascent, descent, and horizontal movement. The first stage determines the time during which a maneuver will be performed, and the second stage determines the energy required based on the determined time. In the second stage, the required energy for the moments when the vehicle is in hovering flight is added. Although this method showed high accuracy in energy estimation, the method does not include a way to adapt its estimation depending on the mission performance.

The problem of the lack of adaptability to variations in consumption can also be observed in [19], where the authors employed a deep neural network that used as inputs the velocity, acceleration, altitude, wind speed, weight, and surface of the load. Although the proposed method considers multiple variables, which increases its accuracy in energy estimation, it is also unable to adapt to changes not considered in the model's training.

As can be seen in the literature review, data-driven energy-estimation methods, the ones using regression analysis and the ones that employ neural networks, have shown good performance for applications of energy estimation during a mission. However, it is necessary to solve the lack of adaptability of the presented techniques to conditions not considered during their training.

With respect to the flight time, it will depend on the energy required and the available energy. Some of the principal methodologies developed to determine the flight time are discussed below.

In [20], a method is presented to estimate the flight time using regressions and deep learning, considering factors such as known flight time without load, payload, battery capacity, and the onboard computer. However, this method only allows an a priori estimation, since it does not provide a way to update the estimation during the execution of the mission.

In [21,22], the flight time was obtained from the division of the battery capacity by the discharge rate, where it was assumed that the available energy is known. However, there are factors that can modify the amount of usable battery energy that must be considered to provide an accurate estimation of the flight time.

Based on the above methodologies for flight time estimation, it can be observed that it is necessary to have solutions to dynamically adapt the flight time estimations considering that the usable energy may change in situations such as increasing of the payload weight, adverse weather conditions, or system failures.

3. Methods

Energy estimation for a mission is a complex problem due to the multiple factors and variables involved in the process. Nevertheless, to estimate the energy requirements is crucial to guaranteeing the feasibility and safety of a mission. Moreover, given the dynamism of the environment in which a multirotor UAV can fly, and the variations to which it may be subject either by lowered energy efficiency due to wear of its components or malfunctioning of one of its parts, it is necessary to continually reevaluate the energy required to complete the mission. In addition, the knowledge of the flight time during which the multirotor UAV can continue operating can be used in decision-making process by an autonomous system or by a human operator.

The architecture of our proposed system is depicted in Figure 1. The system works with a methodology consisting of three parts, which are described as follows:

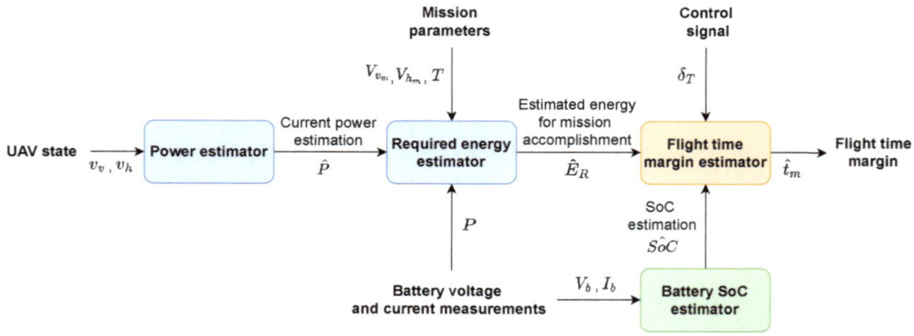

Figure 1. Overview of the available energy and flight time estimation process. The system consists of three sections, energy estimation, battery-charge estimation, and flight-time-margin estimation.

(i) The first part of the methodology corresponds to the estimation of the energy required in the mission once the vehicle is in flight, based on the knowledge of the horizontal and vertical velocities at which it will move along the different stages of the mission, and the time during which it will move at that velocity. To do this, we propose the use of a cascaded Takagi–Sugeno fuzzy system, using the C-means algorithm for the premise of the rules, to estimate the power required to move at a given speed. From the knowledge of the power required to move at a given velocity and the time during which it performs such action, it is possible to know the energy required for the mission. It is also proposed to use the PSO algorithm for the optimization of the parameters used in the fuzzy C-means algorithm to find a balance between the execution time of the system, which is affected by the number of clusters, and the accuracy of the system, which is affected by both the number of clusters and the weighting exponent.

(ii) The second part of the methodology consists of determining the state of charge of the battery, for which it is proposed to use an extended Kalman filter based on the equivalent circuit model of the battery, for which it is proposed to use fuzzy modeling to define the relationship between the open-circuit voltage and the state of charge.

Finally, (iii) the third part of the methodology consists of determining the flight-time margin considering the effect of the battery's voltage change during discharge on the thrust control signal. For this stage, it is proposed to use recursive least-squares to determine the relationship between the battery voltage and the thrust control signal in order to determine the minimum voltage at which the vehicle can operate considering a maximum value for the average value of the thrust control signal. From this voltage, we determine the associated battery charge level considering also the constraints given by the operator. Finally, based on the knowledge of the energy required to complete the mission, the battery charge, and the minimum allowable charge level, we calculate the flight time margin.

The methods used in each of the stages are detailed below.

3.1. Preliminaries

A brief introduction to Takagi–Sugeno fuzzy systems with premises given by trapezoidal membership functions and using the C-means method is presented, and a way to determine the values of the consequent parameters of the fuzzy rules based on the membership values and output values of a training data set. Additionally, we present the operation of the particle swarm optimization algorithm with a constraint factor on the particle velocity.

3.1.1. Takagi–Sugeno Fuzzy Systems with Fuzzy C-Means

Takagi–Sugeno (T-S) fuzzy systems have been adopted in several applications, since they are able to approximate a function with adequate accuracy in a closed set, maintain a structure that is easy to interpret thanks to its high transparency, and maintain low computational complexity [23]. These systems are defined by a set of M rules in the form:

$$\textbf{If} \quad x \quad \text{is} \quad A_i, \quad \textbf{Then} \quad y_i = a_i x + b_i \tag{1}$$

where the premise of the rule is formed by the input of the system x and the fuzzy set A_i, and the consequent of the rule is defined as y_i, a_i and b_i being design parameters.

The output of the T-S fuzzy system is given by

$$y = \frac{\sum_{i=1}^{M} u_i y_i}{\sum_{i=1}^{M} u_i} \tag{2}$$

where $u_i \in [0, 1]$ indicates the degree of membership of the input x to each of the fuzzy sets A_i.

The form of determining the degree of membership will depend on the way in which the fuzzy set is defined. One of the most common used ways to define the fuzzy set is with a trapezoidal function [24], such as the one shown in Figure 2.

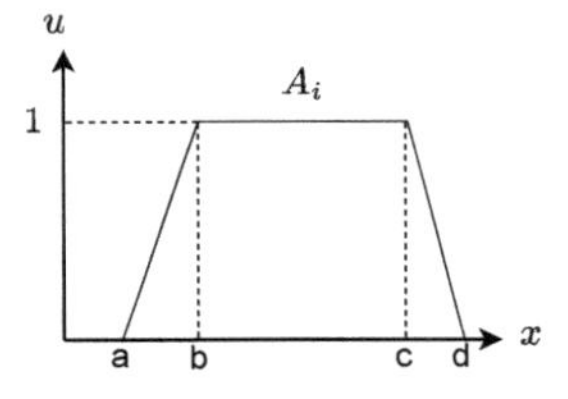

Figure 2. Trapezoidal membership function. This function is defined by four points and will have a value of $0 < u_i \leq 1$ for $a < x < d$.

In these sets, the level of membership is calculated as:

$$u_i = max\left(min\left(\frac{x-a}{b-a}, 1, \frac{d-x}{d-c}\right), 0\right) \tag{3}$$

Another way to determine the fuzzy sets is through the fuzzy C-means method, which determines the membership value of $x \in \mathbb{R}^n$ to a fuzzy set, from the closeness to the cluster center, defined by a vector $v \in \mathbb{R}^n$ [25]. For a set of N data, grouped in M fuzzy sets, we must minimize the function

$$J_m = \sum_{k=1}^{N} \sum_{i=1}^{M} (u_{ik})^m \|x_k - v_i\|^2 \tag{4}$$

where $m > 1$ is the weighting exponent, which is a design parameter that modifies the performance of the fuzzy C-means method.

The membership value used in (4) is given by

$$u_i = \frac{1}{\sum_{j=1}^{M} \left(\frac{\|x_i - v_i\|}{\|x_i - v_j\|}\right)^{\frac{2}{m-1}}} \tag{5}$$

where the optimal values of the centers v for a given number of clusters M and exponent m are obtained through the iterative process presented in [25].

3.1.2. Computation of Parameters for the Consequents in T-S Systems

Consider a set of training input data $X \in \mathbb{R}^{N \times n}$, for which the membership value $U_i \in \mathbb{R}^N$ corresponding to each of the M fuzzy sets A_i is known, and the expected output of the fuzzy system is $Y \in \mathbb{R}^N$. It is possible to employ the least-squares method to determine the design parameters a_i and b_i of each of the fuzzy rules as follows [26].

Let us define the extended matrix X_e as:

$$X_e = \begin{bmatrix} X & \mathbf{1} \end{bmatrix} \tag{6}$$

where $\mathbf{1} \in \mathbb{R}^N$ is a vector of ones. Then, using the matrix X_e and the membership values U_i, the following matrix is formed:

$$X' = \begin{bmatrix} \text{diag}(U_1)X_e & \cdots & \text{diag}(U_M)X_e \end{bmatrix} \tag{7}$$

Finally, with a global approach using the X' matrix, the vector of parameters for the consequents will be calculated as follows.

$$\theta = [(X')^T X']^{-1}(X')^T Y \tag{8}$$

where the values obtained in the parameter vector have the following structure:

$$\theta = \begin{bmatrix} a_1 & b_1 & \cdots & a_M & b_M \end{bmatrix} \tag{9}$$

which correspond to the design parameters required in the consequent of each of the rules.

3.1.3. Particle Swarm Optimization

The particle swarm optimization (PSO) method is a bio-inspired optimization technique that was presented in [27]. In this method, a set of particles $p_i \in \mathbb{R}^n$ is proposed, which are moved through a search space to find the minimum (or maximum) of a function $J(p_i)$. The displacement is performed based on the best solutions found by each of the particles, which are initialized with random values within a search space.

In [28], a variant of the original PSO method was developed to improve the convergence capabilities. In this variant, the velocity of each particle v_i at instant k is calculated as:

$$v_i[k] = \chi(v_i[k-1] + c_1\text{rand}()(p_{b_i} - p_i[k-1]) + c_2\text{rand}()(p_{b_g} - p_i[k-1])) \tag{10}$$

where χ is a velocity constraint factor, c_1 and c_2 are constants that weigh the effect of the cognitive and social components, p_{b_i} is the value of the particle that generated the minimum of that particle, and p_{b_g} is the value of the particle that generated the global minimum among the particles. The factor χ is calculated as a function of c_1 and c_2 as:

$$\chi = \frac{2}{|2 - \phi - \sqrt{\phi^2 - 4\phi}|} \quad , \quad \phi = c_1 + c_2 \quad , \quad \phi > 4 \tag{11}$$

Given the velocity of each particle, the position of each particle is updated as:

$$p_i[k] = v_i[k] + p_i[k-1] \tag{12}$$

In order to find the minimum of $J(p_i)$, an iterative process of velocity calculation and position update of each particle is performed, in which the values of p_{b_i} and p_{b_g} are acquired. The updating of p_{b_g} can be performed synchronously—i.e., this value is updated when the cost function for all particles has been evaluated; or it can be asynchronous, where the value of p_{b_g} is updated each time a new global minimum is encountered, regardless of whether the iteration has not been completed. The performance of the synchronous and asynchronous methods was evaluated in [29], where it was shown that the asynchronous method improves the convergence speed of the method.

3.2. Estimation of the Required Energy in Flight

As discussed above, the energy consumption of a multirotor UAV depends on a wide variety of factors, and since the fuzzy C-means technique performs well for systems up to about 20 dimensions [30], it is used in the development of the energy estimation system.

3.2.1. System for the Estimation of the Required Energy

To estimate the required energy during a mission, it is assumed that during the development of a mission with automatic navigation, the vehicle follows a series of waypoints, which consist of coordinates related to information of the flight altitude, speed, and hover time. To cover the points defined in a desired trajectory, the vehicle will move following velocity profiles, which can be decomposed into their vertical and horizontal components. Based on the behavior of the navigation algorithm used, it is possible to anticipate the velocity profile that will be present along the trajectory, and therefore, the proposed energy estimation system will use this information. It consists of two cascaded subsystems which use fuzzy clustering, as shown in Figure 3.

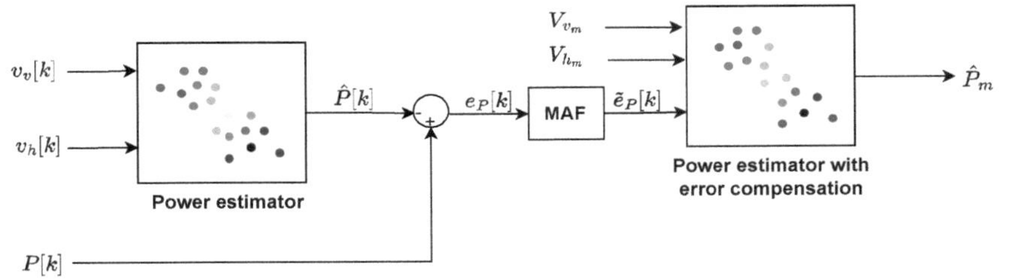

Figure 3. Proposed system for the estimation of the required power during a flight. It is composed of two cascaded subsystems based on fuzzy clustering. The first subsystem must determine the required power for the current system velocity, and the second subsystem determines the required power for subsequent mission stages based on the expected velocities and the determined power-estimation error.

The first subsystem evaluates the current state of the system, having as input the vertical velocity v_v, positive in the downward direction; and the horizontal velocity, $v_h \geq 0$. The output corresponds to the power estimation ($\hat{P}$) for the present state of the vehicle. Such power estimation is performed continuously as the flight develops. The value of $\hat{P}$ is determined from a set of M_1 fuzzy rules of the form

$$\textbf{If} \quad [v_v \quad v_h] \quad \text{is} \quad A_i, \quad \textbf{Then} \quad y_i = a_{1i}v_v + a_{2i}v_v + b_i \tag{13}$$

From the defined rules, the value of $\hat{P} = y$ is calculated using Equation (2). The output of the first subsystem is used in combination with the power measurement P, obtained from the UAV sensors, to calculate the power-estimation error as:

$$e_p = P - \hat{P} \tag{14}$$

The estimation error e_p will be smoothed by a moving average filter (MAF) given by:

$$\tilde{e}_p = \frac{1}{j} \sum_{i=k-j}^{k} e_p \tag{15}$$

where j corresponds to the number of data samples used in the filter.

The second subsystem utilizes as input the sets of vertical velocities $V_{v_m} = \{v_{v_1}, \cdots, v_{v_s}\}$ and horizontal velocities $V_{h_m} = \{v_{h_1}, \cdots, v_{h_s}\}$ of the s segments that constitute the expected velocity profile for the rest of the mission, and $\tilde{e}_p$. This subsystem is formed by a set of M_2 fuzzy rules of the form:

$$\textbf{If} \quad [v_{v_i} \quad v_{h_i} \quad e_p] \quad \text{is} \quad A_i, \quad \textbf{Then} \quad y_i = a_{1i}v_v + a_{2i}v_h + a_{3i}e_p + b_i \tag{16}$$

The output of this subsystem, given by Equation (2), corresponds to the required power to fly at the velocities defined in the different stages of the mission $\hat{P}_m = \{\hat{P}_{m_1}, \cdots, \hat{P}_{m_s}\}$.

Finally, the estimation of the energy required to complete a mission is given by

$$\hat{E}_R = \sum_{i=1}^{s} \hat{P}_{m_i} T_i \tag{17}$$

where T_i corresponds to the period during the multirotor UAV moves at velocities v_{v_i} and v_{h_i}.

3.2.2. Computation of the Parameters of the Power-Estimation System

The performance of the proposed subsystems for power estimation depends on the quality of the training set, the number of clusters, and the chosen fuzzy exponent (m). There is also a trade-off between the number of clusters and the accuracy of the power estimation, since a larger number of clusters can lead to more accurate results, but, in a resource-constrained application, such as in embedded systems, it is necessary to employ algorithms with low computational cost that can be executed in a short period of time. In order to find a balance between power-estimation accuracy and speed of execution, we propose the use of the PSO method presented in Section 3.1.3.

In this sense, the structure of the particles to be used during the optimization process is defined as follows:

$$p_i = \begin{bmatrix} M_i' & m_i \end{bmatrix} \tag{18}$$

where M_i' is an auxiliary variable that allows one to define the number of clusters, and m_i is the weighting exponent associated with the particle. In order to carry out the PSO algorithm, the following cost function is proposed:

$$J = w_1 \frac{1}{N} \sum_{i=1}^{N} |e_p| + w_2 \lfloor M_i' \rceil \tag{19}$$

where N is the number of samples in the validation set, and w_1 and w_2 are constants that represent the trade-off between system accuracy and execution speed. The function $\lfloor \cdot \rceil$ indicates the value rounded to the nearest integer. Although the above function does not explicitly include the m_i parameter of the particle, it is reflected through e_p.

For the first subsystem, the training data set will be composed of the vertical velocities $V_v = \{v_{v_1}, \cdots, v_{v_N}\}$, horizontal velocities $V_h = \{v_{h_1}, \cdots, v_{h_N}\}$, and measured power $P = \{P_1, \cdots, P_N\}$. To apply the PSO method to the subsystem, the clustering process [25] is performed with $x = [v_{v_i} \quad v_{h_i}]$ and $y = P_i$. The number of clusters is given by $M_1 = \lfloor M_i' \rceil$ and the weighting exponent $m_1 = m_i$.

Using the obtained cluster centers v_i, the membership values are calculated with Equation (5) for the training set, obtaining a set $U_i = \{u_{i1}, \cdots, u_{iN}\}$ for each cluster. The obtained values are used in the least-squares method presented in Section 3.1.2 with $X = [V_v \quad V_h]$ and $Y = P$, to obtain the parameters of the consequent of the rule shown in Equation (13). Once the parameters of the M_1 rules have been determined, using Equation (2), the values of $\hat{P} = \{\hat{P}_1, \cdots, \hat{P}_i\}$ are calculated for the validation set. Finally, the function (19) is calculated with the values of e_p, as given in (14) between the validation set measurements and the power estimations.

The process described above is conducted for each particle until the defined maximum number of iterations is reached. The values of M_1 y m_1 will correspond to the last value obtained for p_{b_g}.

For the training of the second subsystem, we calculate the set $\tilde{E}_p = \{\tilde{e}_{p_1}, \cdots, \tilde{e}_{p_N}\}$ using Equation (15) for the values of e_p obtained from the power-estimation process of the training data set used in the first subsystem.

The training of the second subsystem is performed using the training data set of the first subsystem, extended with E_p, $x = [v_{v_i} \quad v_{h_i} \quad e_{p_i}]$ and $y = P_i$. The training process is performed using the procedure indicated for subsystem one to obtain M_2 and m_2, and the parameters in the consequent of each rule (16).

3.3. Estimation of the Battery's State of Charge

This section presents the mathematical model of the equivalent circuit used for batteries using a single time constant, proposing the use of fuzzy modeling for the relationship between the battery's open-circuit voltage and the state of charge. After that, the estimation of the state of charge using an extended Kalman filter is addressed.

3.3.1. Equivalent Circuit Model for Batteries

An electric battery is an element that stores energy in chemical form, which can be released in a controlled way. A model that has been widely used to determine its behavior is the equivalent circuit model [31]. Figure 4 presents the model of a single time constant, which allows studying its behavior with a suitable degree of accuracy when the objective is the estimation of the state of charge (SoC) for practical applications [32].

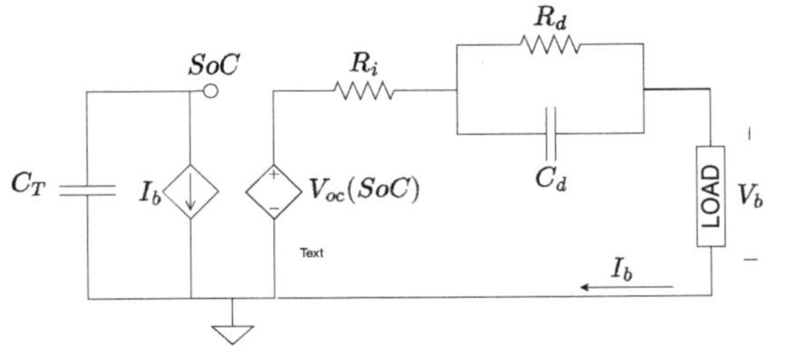

Figure 4. Equivalent circuit model with a single time constant for batteries.

On the left-hand side, a capacitor C_T models the battery charge capacity, and a current source whose flow is equal to that flowing through the load at the battery terminals I_b. The capacitance of C_T is given by:

$$C_T = 3600 Q_B \eta \tag{20}$$

where Q_B is the capacity of the battery expressed in Ah and η is a factor that will depend on the temperature and health of the battery. The voltage at C_T, whose value is between zero and one, represents the SoC of the battery and is calculated by:

$$SoC(t) = SoC(t_0) - \frac{1}{C_T} \int_{t_0}^{t} I_b(\tau) d\tau \tag{21}$$

On the right-hand side, a voltage source models the open-circuit voltage V_{oc} as a function of the SoC. R_i is the internal resistance, and the pair $R_d C_d$ represents the transient behavior of the battery. The dynamics of the voltage $\dot{V}_d$ on $R_d C_d$, and the battery terminal voltage V_b, are given by:

$$\dot{V}_d = \frac{I_b}{C_d} - \frac{V_d}{R_d C_d} \tag{22}$$

$$V_b = V_{oc}(SoC) - I_b R_i - V_d \tag{23}$$

The function between the V_{oc} and the SoC has been approximated from a set of linear functions or polynomial functions [33,34]. In the present work, it is proposed to approximate such a function with a Takagi–Sugeno fuzzy model with rules of the form given in Equation (1), as shown in Figure 5. It can be observed that the function is segmented using trapezoidal fuzzy sets. This approximation allows having the simplicity given by a set of linear functions while maintaining the smoothness of the transitions between regions of the curve, as can be observed in polynomial approximations.

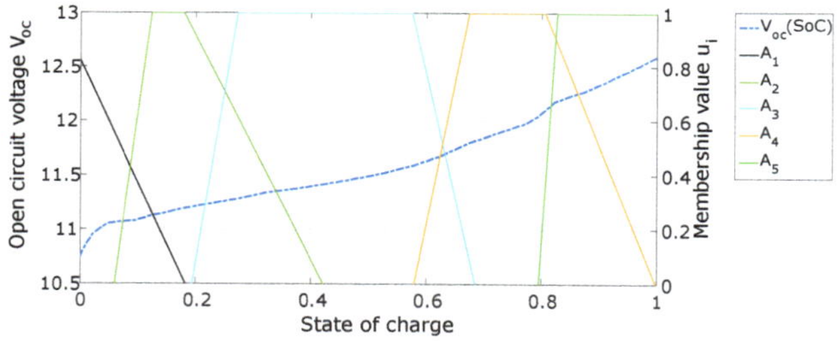

Figure 5. Modeling the relationship between the V_{oc} and SoC using a set of trapezoidal fuzzy sets.

Each membership function of the set of rules defining the functions $V_{oc}(SoC)$ or $SoC(V_{oc})$ is given by four points as seen in Figure 2, whose membership value will be given by Equation (3), and the output of the model is given by Equation (2).

3.3.2. Estimation of the SoC

A widely used methodology for the estimation of the SoC of the battery is the use of the Kalman filter, which, unlike the Coulomb count, allows compensating the differences between the estimated initial SoC and the real one; moreover, it presents lower vulnerability to noise in the current measurements [35].

To implement the extended Kalman filter (EKF) [36], Equations (21) and (22) are discretized using Euler's method and grouped as follows:

$$\underbrace{\begin{bmatrix} SoC[k] \\ V_d[k] \end{bmatrix}}_{x[k]} = \underbrace{\begin{bmatrix} 1 & 0 \\ 0 & \left(1 - \frac{\Delta t}{R_d C_d}\right) \end{bmatrix}}_{A} \underbrace{\begin{bmatrix} SoC[k-1] \\ V_d[k-1] \end{bmatrix}}_{x[k-1]} + \underbrace{\begin{bmatrix} -\frac{\Delta t}{C_T} \\ \frac{\Delta t}{C_d} \end{bmatrix}}_{B} \underbrace{I_b[k-1]}_{u[k-1]} \tag{24}$$

where Δt is the sampling time interval. In Equation (24), the terms of the battery state-space model x, u, A, and B are obtained. In addition, from the linearization of Equation (23), we obtain:

$$\underbrace{V_b[k]}_{y[k]} = \underbrace{\begin{bmatrix} \frac{\partial V_{oc}}{\partial SoC}[k] & -1 \end{bmatrix}}_{C[k]} \underbrace{\begin{bmatrix} SoC[k] \\ Vd[k] \end{bmatrix}}_{x[k]} + V_{oc}(SoC[k]) - \frac{\partial V_{oc}}{\partial SoC}[k] SoC[k] - \underbrace{R_i}_{D} \underbrace{I_b[k]}_{u[k]} \tag{25}$$

Using the terms obtained in (24) and (25), the SoC of the battery is calculated by employing the EKF given in Algorithm 1, where P is the covariance error matrix, K is the estimator gain, Q is the process noise covariance, and R is the measurement noise covariance. The value of $\frac{\partial V_{oc}}{\partial SoC}$ can be obtained from the fuzzy model for the function $V_{oc}(SoC)$ using the partial derivative of its consequent.

3.4. Estimation of Flight Time Margin

This section presents an analysis of the relationship between the battery charge and the thrust control signal, from which a methodology is developed to determine the minimum charge level at which it is possible to operate the multirotor UAV and the time for which it can operate after the end of its mission with the remaining energy.

Algorithm 1 Estimation of the SoC.

Require: $P[k-1] = P_0$, $\hat{x}[k-1] = x_0, Q, R$

1: **loop**

2: Perform measurement of $u[k] = I_b[k-1]$ and $y = V_b[k]$

3: Obtain the value of $\frac{\partial V_{oc}}{\partial SoC}$

4: Predict the estimated state

$$\hat{x}^-[k] = A\hat{x}[k-1] + Bu[k-1] \tag{26}$$

5: Prediction of the covariance error

$$\hat{P}^-[k] = AP[k-1]A^T + BQB^T \tag{27}$$

6: Calculate the Kalman gain

$$K = \hat{P}^-[k]C^T(C\hat{P}^-[k]C[k]^T + DRD^T)^{-1} \tag{28}$$

7: Update the estimated state estimate with the measurement of $y[k]$

$$\hat{x}[k] = \hat{x}^-[k-1] + K(y[k] - y(\hat{x}^-[k], u[k-1])) \tag{29}$$

8: Update covariance error

$$P[k+1] = (I - K[k]C)P^-[k] \tag{30}$$

9: **end loop**

3.4.1. Relationship between Control Signals and the SoC

In the controller field of multirotor UAVs, a widely used architecture is the one that uses a set of cascaded controllers, as shown in Figure 6 [37,38]. This architecture begins with a position controller for a reference ξ_d, from which an attitude reference η_d is obtained, along with a thrust control signal δ_T. From the previous reference, an attitude controller generates an angular velocity reference ω_d. The angular velocity reference is sent to an angular velocity controller that will generate a set of control signals for the angular velocity δ_p, δ_q, and δ_r. The four control signals obtained are used by a control allocation system, also known as mixer, which will generate a set of control signals U for each of the rotors of the multicopter.

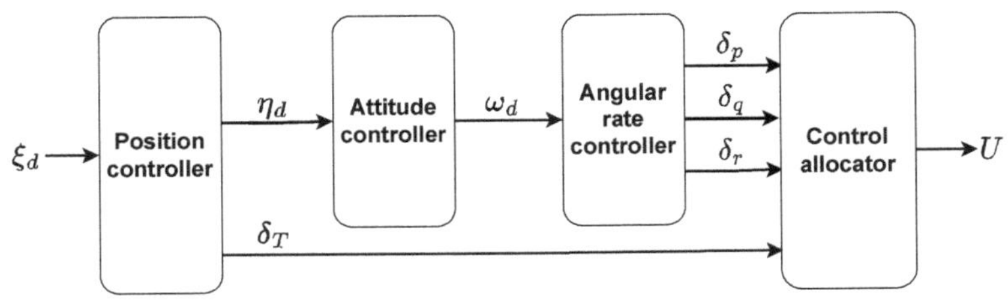

Figure 6. Control architecture employed in widely used autopilots such as PX4 or Arducopter.

For the operation of a multirotor UAV at a constant altitude, the thrust control signal is of a magnitude to provide the force required to compensate for the weight of the multirotor UAV, for which there is a control signal for each of the rotors. The angular velocity control signals will generate variations in the control signals of each rotor around the point required to generate the thrust force. The force generated by each rotor f_r is calculated as:

$$f_r = k_f(k_\omega V_m)^2 \tag{31}$$

where k_f is the propeller force constant, k_ω is the motor angular velocity constant, and V_m is the average voltage applied to the motor.

As can be seen in Equation (31), for a given rotor, the force generated will depend on the average voltage applied. In turn, the average voltage applied to the rotor will be proportional to the duty cycle of the applied signal and is computed as [39]:

$$V_m = V_b u_d \tag{32}$$

where u_d is the duty cycle with $0 \leq u_d \leq 1$.

As can be seen in Equation (32), there is a linear relationship between the duty cycle and the battery voltage. Therefore, for a given flight condition, the duty cycle will increase as the battery voltage decreases. For a normalized thrust control signal (δ_T) we have the behavior shown in Figure 7, where it is observed that the δ_T control signal increases linearly as the voltage (V_b) decreases.

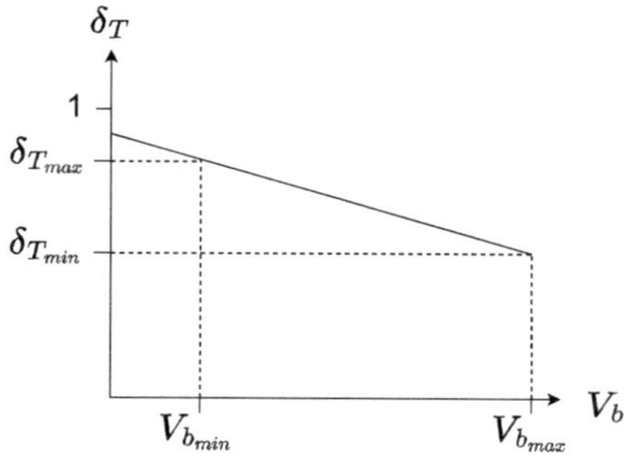

Figure 7. Relationship between battery voltage and thrust control signal. For a constant flight condition, a linear relationship will be maintained between both values.

The δ_T value is minimal when the battery presents its maximum voltage, and should not exceed a maximum level, which allows one to maintain a safe operating margin for variations due to angular velocity controls.

Since the applied voltage will depend on the *SoC* of the battery, as seen in Figure 8, the minimum allowable battery voltage will determine the minimum *SoC* at which it is safe to operate.

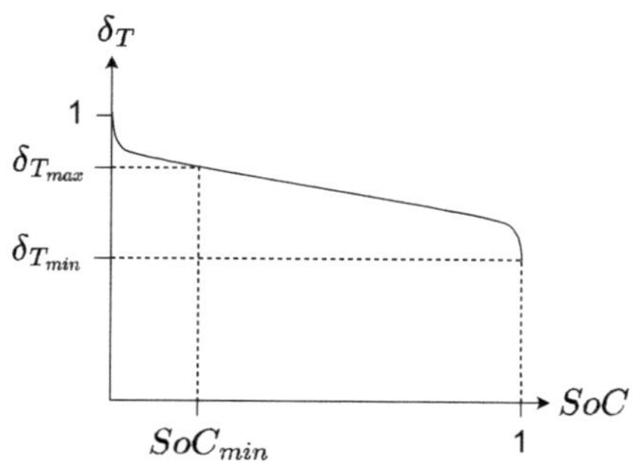

Figure 8. Relationship between *SoC* and thrust control signal. For a constant flight condition, the required δ_T increases depending on the battery discharge curve.

3.4.2. Estimation of the Flight Time Margin

The remaining flight time is approximated from the average power consumed P_m and the usable energy in the battery. To determine the usable energy in the battery, the minimum charge level at which it is safe to operate is determined based on the expected performance of the thrust control signal.

Using a recursive least squares (RLS) [40] process, the following relationship is determined:

$$V_{b_\delta} = \alpha_1 \delta_T + \alpha_2 \tag{33}$$

where V_{b_δ} is the expected battery voltage for a given δ_T, and α_1 and α_2 are parameters obtained from the RSL process.

Then, using the maximum admissible value of the thrust control signal $\delta_{T_{max}}$, the minimum admissible voltage at the battery terminals $V_{b_{min}}$ is calculated as:

$$V_{b_{min}} = \max\{V_{b_s}, V_{b_\delta}(\delta_{T_{max}})\} \tag{34}$$

where V_{b_s} is the minimum voltage at which the battery can be safely operated.

Based on the equivalent circuit model (Figure 4), it can be seen that in the worst-case scenario, the open-circuit voltage when the terminal voltage is $V_{b_{min}}$, for a value of P_m, is given by:

$$V_{oc_{min}} = V_{b_{min}} + \frac{P_m}{V_{b_{min}}}(R_i + R_d) \tag{35}$$

Therefore, the minimum charge level at which it is possible to operate the multirotor UAV under the constraints of $\delta_{T_{max}}$ is given by:

$$\hat{SoC}_{min} = \max\{SoC_s, SoC(V_{oc_{min}})\} \tag{36}$$

where $SoC_s \geq 0$ is the minimum charge at which it is desired to operate the battery.

For a defined mission whose required energy is given by $\hat{E}_R$ when the battery has a charge SoC_0, the estimated charge at the end of the mission $\hat{SoC}_f$ is calculated as:

$$\hat{SoC}_f = SoC_0 - \frac{\hat{E}_R}{E_T} \tag{37}$$

where E_T is the energy stored in the battery when the battery is fully charged and is calculated by:

$$E_T = Q_B V_{b_{nom}} \tag{38}$$

where $V_{b_{nom}}$ is the nominal voltage of the battery.

A mission may be considered as an achievable mission if $\hat{SoC}_f \geq SoC_{min}$. If the mission is achievable, then the flight time margin $\hat{t}_m$ is estimated as:

$$\hat{t}_m = \frac{(\hat{SoC}_f - SoC_{min})E_T}{P_m} \tag{39}$$

In the case of a manually controlled flight, this method may be used to estimate the remaining flight time using $\hat{E}_R = 0$.

4. Physical Experiments and Results

In order to evaluate the performance of the proposed methods, a series of physical experiments were carried out using a hexarotor UAV as the case of study. The experiments consisted of (i) performing discharge tests of the battery used on the hexarotor UAV in order to obtain the parameters of its equivalent circuit and the relationship between the open-circuit voltage and the state of charge; (ii) executing a series of flights to obtain information for the training process of the power-estimation system, including the optimization of its parameters; and finally, (iii) evaluating the performance of the proposed methods through a validation flight.

The hexarotor used to conduct the experiments, shown in Figure 9, has the specifications given in Table 1.

Figure 9. Hexarotor UAV used as the case study for the estimation of the required energy.

Table 1. Hexarotor specifications.

Parameter	Values
Flight controller	Pixhawk 2.1 Cube Black
Motor	T-motor Air 2213 920KV
ESC	T-motor Air 20A
Propeller	T-motor 9.45 × 4.5 in
Power module	Mauch HS-050-LV
Battery	Turnigy graphene 4Ah 3S 45C
Weight	1.8 kg
Dimensions	55 cm × 55 cm × 23 cm

The power module uses an Allegro ACS758LCB-050U Hall-effect linear current sensor with a maximum capacity of 50A that is able to provide current measurements with an accuracy of 2%. The development of the aforementioned experiments is detailed below.

4.1. Experiment 1: Battery Characterization

The battery used in this experiment was characterized at room temperature using the electronic load model KP-184 shown in Figure 10. The characterization process consisted of performing a series of discharges at constant current. After each discharge, a relaxation period of 15 min was maintained since, after this time, each cell was meant to be within 3 mV of the V_{oc} [41]. This process was performed until an amount of energy equivalent to the nominal capacity of the battery was reached. The discharge process performed to characterize the battery is shown in Figure 11.

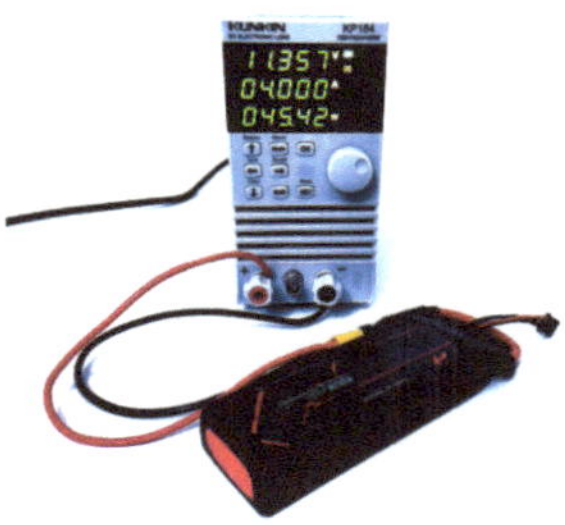

Figure 10. KP-184 programmable electronic load used to characterize the battery employed in the experimental platform.

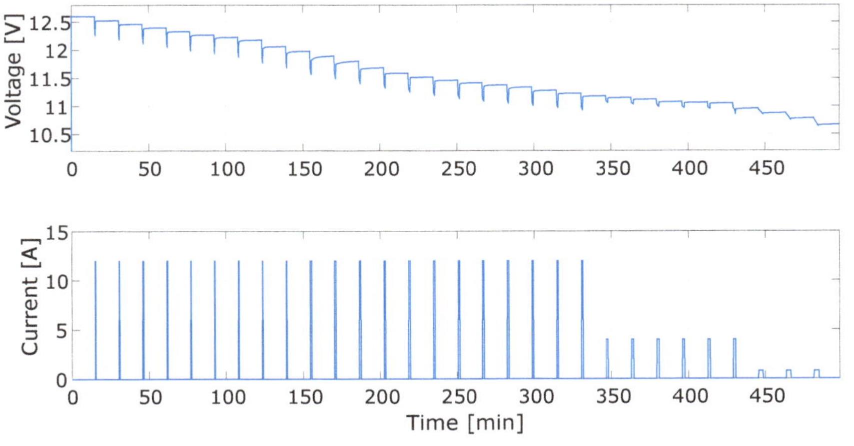

Figure 11. Voltage and current graphs obtained from the battery characterization process.

The information obtained from the battery discharge test was analyzed with the process presented in [42], which performed an optimization process using nonlinear least squares to determine the values of the equivalent circuit elements shown in Table 2.

Table 2. Equivalent circuit parameters of the used battery.

Parameter	Value
R_i	41.6 mΩ
R_d	9.6 mΩ
C_d	1016 F
C_T	14,400 F

Based on the obtained open-circuit voltage values, two T-S fuzzy models with five rules were generated using trapezoidal membership functions. The first model represents the function $V_{oc}(SoC)$ using the parameters shown in Table 3. The second model represents the inverse function $SoC(V_{oc})$ using the parameters shown in Table 4.

Table 3. Parameters of the membership functions and consequents of the rules used for the model that determines the V_{oc} for a given SoC. The points correspond to trapezoidal membership functions, as shown in Figure 2.

Rule	Membership Function Points				Consequent
	a	b	c	d	
1	−0.2073	−0.1045	−0.2485	0.1568	$V_{oc} = 14.14SoC + 10.76$
2	0.0033	0.0887	0.2246	0.4007	$V_{oc} = 2.565SoC + 10.71$
3	0.1731	0.3055	0.5641	0.6686	$V_{oc} = 1.325SoC + 10.83$
4	0.5565	0.6668	0.7931	1	$V_{oc} = 1.827SoC + 10.57$
5	0.7793	0.8354	1.077	1.18	$V_{oc} = 1.912SoC + 10.69$

Table 4. Parameters of the membership functions and consequents of the rules used for the model that determines the *SoC* for a given V_{oc}. The points correspond to trapezoidal membership functions, as shown in Figure 2.

Regla	Membership Function Points a	b	c	d	Consecuente
1	9.948	10.17	10.53	10.77	$SoC = 0.1089V_{oc} - 1.155$
2	10.46	10.75	10.92	11.32	$SoC = 0.1716V_{oc} - 1.1852$
3	11.05	11.07	11.59	11.78	$SoC = 0.8967V_{oc} - 9.82$
4	11.57	11.83	12.21	12.43	$SoC = 0.4059V_{oc} - 4.109$
5	12.23	12.43	12.77	12.99	$SoC = 0.3735V_{oc} - 3.706$

4.2. Experiment 2: Training of the Energy Estimation System

To conduct the training of the energy estimation system, we used data from a series of flights, performing various patterns of horizontal and vertical movements outdoors. The flights were performed at an altitude of 2245 m in diverse weather conditions, including winds between 3 and 8 km/h, with gusts of up to 22 km/h. The goal of this experiment was that the energy estimation system would provide an energy estimate as an average of the energy requirements of the different conditions in which it can operate.

The optimization process for obtaining the parameters was performed as indicated in Section 3.2.2, involving a series of 100 iterations for each subsystem. According to the results presented in [29], a set of 30 particles was used for each subsystem. Considering that the values of the exponent m must be greater than one, and its upper limit for practical applications is given in [43], the search space is defined by $1.01 \leq m_i \leq 3.5$. Similarly, the search space for the number of clusters was defined to be $2 \leq M_i' \leq 20$ based on the criterion for the maximum number of clusters $M_{max} \leq 2 \ln N$ presented in [44].

Figure 12 presents the value of the cost function (19) evaluated at the value of p_{b_g} for each of the subsystems. Table 5 presents the parameters obtained for the implementation of the first subsystem, and Table 6 shows the parameters obtained for the implementation of the second subsystem.

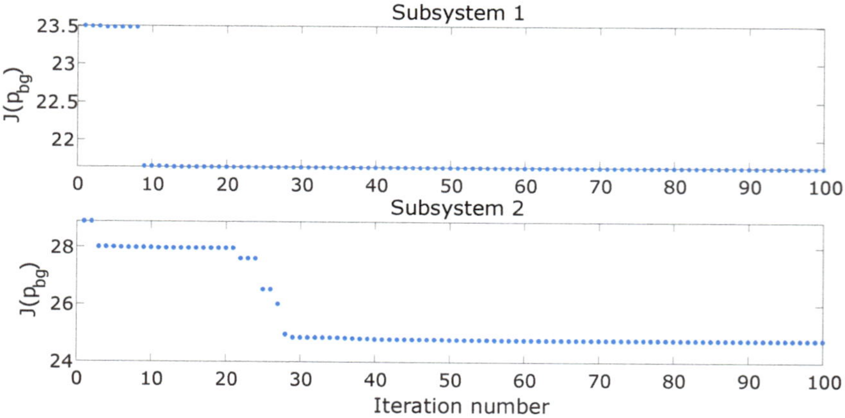

Figure 12. Value of the cost function used in the optimization of the power-estimation system, evaluated according to the value of the particle with the best cost of each subsystem.

Table 5. Parameters of the first power-estimation subsystem.

Number of Clusters M_1			Weighting Exponent m_1		
2			1.4628		
	Cluster Centers		**Consequent Parameters**		
Cluster	v_v	v_h	a_{1i}	a_{2i}	b_i
1	0.0069	4.0424	−19.011	−3.5064	262.2819
2	−0.0022	1.2231	−17.4258	−1.5127	258.4393

Table 6. Parameters of the second power-estimation subsystem.

Number of Clusters M_2				Weighting Exponent m_2			
2				1.0338			
	Cluster Centers			**Consequent Parameters**			
Cluster	v_v	v_h	e_p	a_{1i}	a_{2i}	a_{3i}	b_i
1	−0.0014	2.0883	−12.6536	−15.6601	−3.4644	0.9661	260.9404
2	−0.0072	1.4581	50.9937	−16.694	−3.3685	0.684	277.0185

4.3. Experiment 3: System Evaluation

To validate the performance of the energy estimation system, we conducted a validation flight. The mission profile consisted of a series of climb and descent maneuvers while performing increments in translation speed, as shown in Figure 13. The flight was executed by using the PX4 system for the vehicle control and navigation.

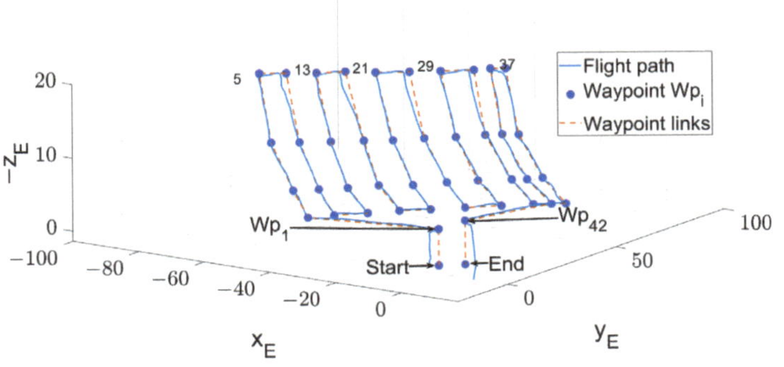

Figure 13. Mission profile conducted by the hexarotor UAV to validate the performance of the developed energy estimation system. The blue dots represent the waypoints that defined the multirotor UAV mission during the flight. The dotted orange line is the trajectory linking 42 waypoints, and the solid blue line corresponds to the multirotor UAV trajectory.

The energy required to complete the mission was estimated from each of the waypoints that defined the trajectory. Figure 14 shows the comparison between the required energy prediction $\hat{E}_R$ and the ones measured by the Mauch HS-05-LV sensor E_R. Calculating the required energy estimation error $e_{E_R} = E_R - \hat{E}_R$, and dividing this by the average measured power value P_m, produces the graph shown in Figure 15, which is interpreted in terms of flight time.

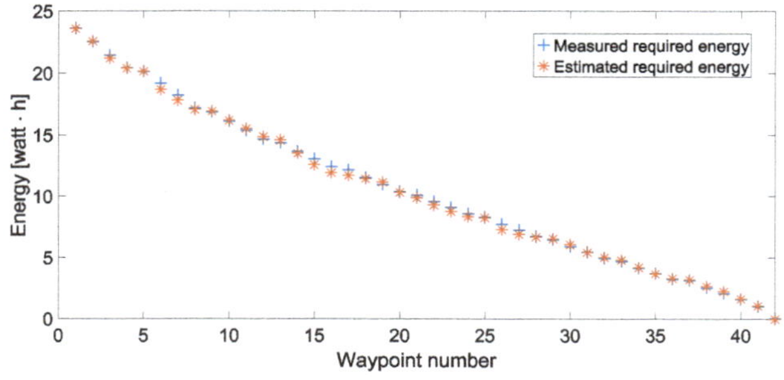

Figure 14. Comparison between the measurement of the energy required to complete the mission from a waypoint and the estimation obtained with the developed system.

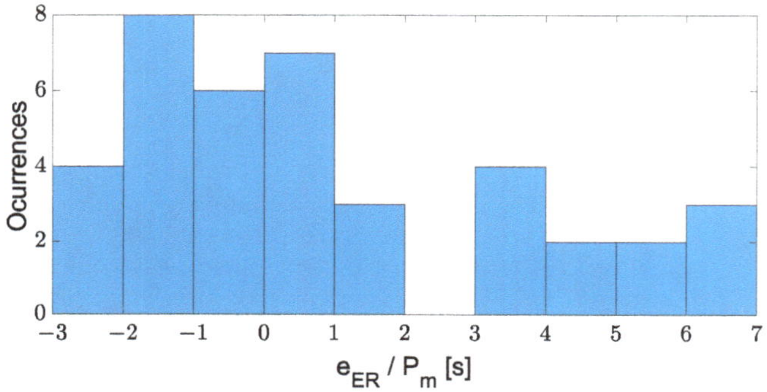

Figure 15. Number of occurrences for the values obtained from dividing the energy estimation error by the average required power in the mission. This value can be interpreted in terms of flight time.

From the comparison between the required power estimation and the power use measured by the system, the maximum estimation error was found to correspond to the power required to fly for 7 s at the average power.

As part of the flight time margin estimation process, the RLS method was applied to obtain the relationship given in Equation (33) to determine the expected voltage at the maximum allowable δ_T. In Figure 16, the comparison between the recorded δ_T and the function $\delta_T(V_b)$ derived from Equation (33) can be observed. As can be seen, the function obtained by the RLS process averages the behavior of δ_T, so it is possible to obtain the V_b for a given average δ_T.

Figure 17 shows the obtained values related to the horizontal and vertical velocities, the attitude, and the estimations of the $S\hat{o}C$, $S\hat{o}C_{min}$, and $\hat{t}_m$ from the validation flight considering a $V_{b_s} = 9.6v$, $SoC_s = 10\%$ and $\delta_{T_{max}} = 0.78$. It is possible to see four intervals where the estimation of the $S\hat{o}C_{min}$ exceeds the established SoC_s. These intervals correspond to the instants after changes in the vehicle attitude in the roll, pitch, and yaw angles simultaneously, which increase the control signal δ_T, and consequently, it is estimated that the $\delta_{T_{max}}$ will be reached at a higher V_b, so the $S\hat{o}C_{min}$ is higher. The value of $\hat{t}_m$ was computed by considering an $\hat{E}_R = 0$, in which it is possible to observe intervals where the flight time estimation decreases due to the increase in $S\hat{o}C_{min}$.

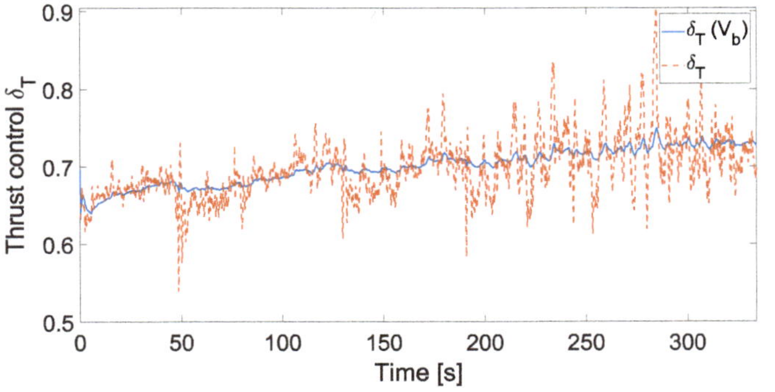

Figure 16. Comparison between the control signal δ_T recorded during the validation flight and the one obtained from the inverse function of Equation (33).

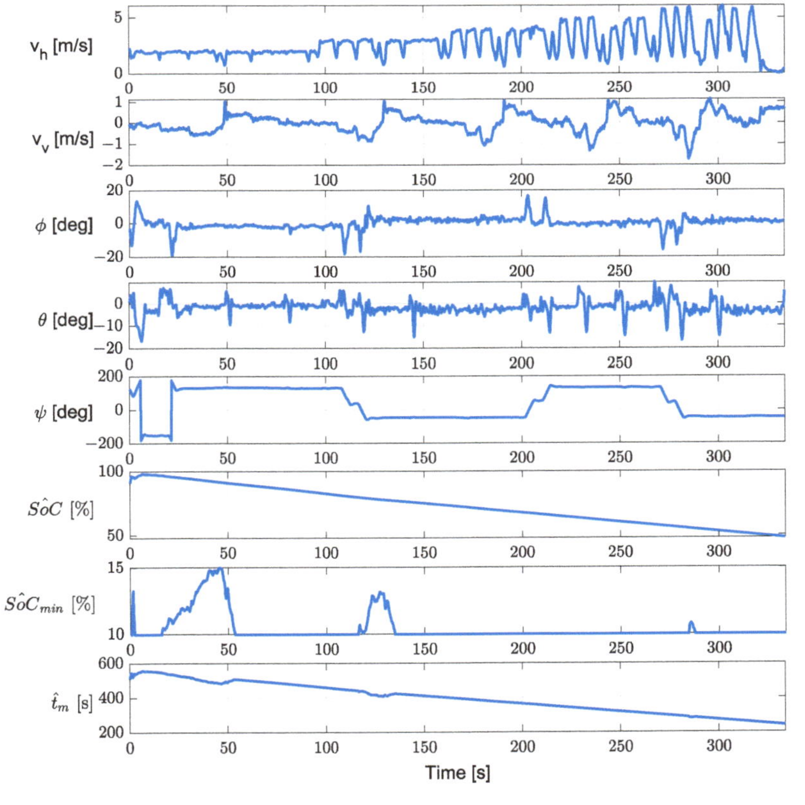

Figure 17. Parameters obtained from the validation flight. From top to bottom, horizontal velocity (v_v), vertical velocity (v_h); roll (ϕ), pitch (θ), yaw (ψ) angles; estimated state of charge, $S\hat{o}C$; minimum admissible state of charge, $S\hat{o}C_{min}$; and flight time margin, $\hat{t}_m$.

5. Conclusions and Future Work

In this research work, a new configuration for the estimation of the energy required to perform a multirotor UAV mission using fuzzy c-means was presented. This estimator was able to make predictions of the required energy with a maximum error equivalent to the energy required to fly for 7 s.

The optimization process of the clustering parameters using PSO allowed us to determine the optimal value for the weighting exponent m and the number of clusters for each subsystem. Although the proposed methodology only considers the multirotor UAV's velocities, for which the use of two clusters per sub-system was determined, this architecture can be extended to include other parameters that affect the energy consumption of the vehicle without modifying the overall system structure, and we employed the proposed PSO-based methodology to determine the optimal system parameters.

The relationship between the state of charge of the battery and the thrust control signal was analyzed, and a methodology was presented to determine the minimum admissible charge level to operate the multirotor UAV safely. This method can be especially useful when the vehicle is operated outside the design conditions, as in the case of a failure during a mission or in environments different from those considered in its design.

With respect to the way to determine the state of charge, although the EKF-based methodology has been widely used, its use was proposed in conjunction with T-S fuzzy models for the relationship between V_{oc} and the SoC, which allows combining the simplicity of linear functions with the smoothness of transitions between function segments.

Since energy estimation is a complex multiparameter-dependent problem, it is required to extend the number of input parameters to include factors such as the payload weight, operating altitude, wind speed, and relative wind direction. To generate the training set, a combination of experimental flight data and high-precision simulations were proposed so that the system could determine the energy required in situations where the vehicle has not been exposed.

It is proposed to evaluate new meta-heuristic algorithms for the optimization of the parameters of the energy estimation system, which can offer more effective alternatives in terms of execution time, which is essential if a greater number of parameters for energy estimation are to be added.

Regarding the battery parameters, it is proposed to incorporate the effect of battery health and temperature into the model to increase the accuracy of the SoC estimation, and consequently, of the flight time margin.

Author Contributions: Conceptualization, R.L. and E.S.E.; methodology, L.H.M. and E.S.E.; software, L.H.M. and J.C.R.-F.; validation, L.H.M. and J.C.R.-F.; formal analysis, L.H.M. and E.S.E.; investigation, L.H.M. and E.S.E.; resources, R.L. and E.S.E.; writing—original draft preparation, L.H.M. and E.S.E.; writing—review and editing, L.H.M., E.S.E., and J.C.R.-F.; visualization, L.H.M. and J.C.R.-F.; supervision, E.S.E. and R.L.; project administration, E.S.E. and R.L.; funding acquisition, E.S.E. and R.L. All authors have read and agreed to the published version of the manuscript.

Funding: This work was partially supported by the Mexican National Council for Science and Technology through Project 321224: National Laboratory of Autonomous Vehicles and Exoskeletons

Institutional Review Board Statement: Not applicable.

Informed Consent Statement: Not applicable.

Data Availability Statement: Not applicable.

Conflicts of Interest: The authors declare no conflict of interest.

Abbreviations

The following abbreviations are used in this manuscript:

Acronyms	Description
EKF	Extended Kalman Filter.
MAF	Moving Average Filter.
PSO	Particle Swarm Optimization.
RLS	Recursive Least Squares.
SoC	State of Charge.
T-S	Takagi–Sugeno.
UAV	Unmanned Aerial Vehicle.
Fuzzy system variables	Description
x, x_k	Input to the fuzzy system.
A_i	Fuzzy set.
y_i	Output of the fuzzy rule.
a_i, b_i	Design parameters of the consequent fuzzy rule.
M	Number of rules of the fuzzy system.
u_i	Membership degree of the input x to the rule.
a, b, c, d	Trapezoidal membership function characteristic points.
N	Number of elements of the data set.
J_m	Cost function of the fuzzy C-means algorithm.
v	Center vector of the cluster.
m	Weighting exponent for fuzzy C-means.
X	Set of training input data.
U_i	Set of membership values of the i-th rule for the set of training data set.
X_e, X'	Auxiliary matrices for calculating of T-S consequent. parameters.
θ	Vector of parameters of the T-S rules.
PSO algorithm variables	Description
J	Function to minimize by the PSO algorithm.
p_i	Position of the i-th particle.
p_{b_i}	Best local of the i-th particle.
p_{b_g}	Best global among all particles.
v_i	Velocity of the i-th particle.
χ	Velocity constrain factor of the particle.
c_1	Weighting factor of the cognitive component.
c_2	Weighting factor of the social component.
ϕ	Auxiliary variable for the calculation of χ.
Energy estimation system variables	Description
P_i	Measured power consumption at current state.
$\hat{P}$	Estimated required power at current state.
e_P	Estimated required power error.
v_v	Vertical velocity.
v_h	Horizontal velocity.
V_{v_m}	Set of vertical velocities of the mission profile.
V_{h_m}	Set of horizontal velocities of the mission profile.
T_i	Period during the UAV moves at a given velocity.
M_i'	Auxiliary variable to determine the number of clusters.
$\tilde{e}_p$	Output of the MAF with input e_p.
Battery equivalent circuit variables	Description
C_T	Capacitance to model the battery capacity.
C_d	Capacitance to model battery transient.
R_i	Internal battery resistance.
R_d	Resistance to model battery transient.
Q_B	Energy stored in the battery.

η	Efficiency factor of the battery.
I_b	Current through battery terminals.
V_b	Voltage on battery terminals.
V_d	Battery transient voltage.
V_{oc}	Open circuit battery voltage.
Flight time margin variables	Description
ξ_d	Position reference.
ω_d	Angular velocity reference.
δ_T	Thrust control signal.
$\delta_{T_{max}}$	Maximum admissible value of δ_T.
$\delta_p, \delta_q, \delta_r$	Control signals for angular velocity .
k_f	Force constant of the rotor.
k_ω	Angular velocity constant of the rotor.
u_d	Duty cycle of the control signal.
V_m	Average voltage applied to the motor.
$V_{b_{nom}}$	Nominal voltage of the battery.
V_{b_δ}	Battery voltage for a given δ_t.
$V_{oc_{min}}$	Open circuit voltage at minimum admissible SoC.
V_{b_s}	Minimum admissible voltage at the battery terminals defined by the operator.
$V_{b_{min}}$	Minimum voltage at the battery terminals at which it is possible to operate the UAV.
P_m	Average required power.
SoC_s	Minimum charge of the battery defined by the operator.
$S\hat{o}C_{min}$	Minimum battery charge at which it is possible to operate the UAV.
$S\hat{o}C_0$	Estimated SoC at the time of flight time evaluation.
$S\hat{o}C_f$	Expected SoC at the end of the mission.
$\hat{E}_R$	Estimated required energy to mission accomplishment.
E_T	Energy stored in the battery when fully charged.
e_{E_R}	Required energy estimation error.
$\hat{t}_m$	Estimated flight time margin.

References

1. Alvarado, E. Drone Blog—UAV Market Insights. 2022. Available online: https://droneii.com/drone-publications (accessed on 25 October 2022).
2. Yaacoub, J.P.; Noura, H.; Salman, O.; Chehab, A. Security analysis of drones systems: Attacks, limitations, and recommendations. *Internet Things* **2020**, *11*, 100218. [CrossRef]
3. Ghasri, M.; Maghrebi, M. Factors affecting unmanned aerial vehicles' safety: A post-occurrence exploratory data analysis of drones' accidents and incidents in Australia. *Saf. Sci.* **2021**, *139*, 105273. [CrossRef]
4. Henderson, I.L. Aviation safety regulations for unmanned aircraft operations: Perspectives from users. *Transp. Policy* **2022**, *125*, 192–206. [CrossRef]
5. Lee, D.; Hess, D.J.; Heldeweg, M.A. Safety and privacy regulations for unmanned aerial vehicles: A multiple comparative analysis. *Technol. Soc.* **2022**, *71*, 102079. [CrossRef]
6. Zhang, J.; Campbell, J.F.; Sweeney, D.C., II; Hupman, A.C. Energy consumption models for delivery drones: A comparison and assessment. *Transp. Res. Part D Transp. Environ.* **2021**, *90*, 102668. [CrossRef]
7. Schacht-Rodrígue, R.; Ponsart, J.C.; García-Beltrán, C.D.; Astorga-Zaragoza, C.M. Analysis of energy consumption in multirotor UAV under actuator fault effects. In Proceedings of the 2019 4th Conference on Control and Fault Tolerant Systems (SysTol), Casablanca, Morocco, 18–20 September 2019; pp. 104–109.
8. Sajid, M.; Mittal, H.; Pare, S.; Prasad, M. Routing and scheduling optimization for UAV assisted delivery system: A hybrid approach. *Appl. Soft Comput.* **2022**, *126*, 109225. [CrossRef]
9. Ghorbel, M.B.; Rodríguez-Duarte, D.; Ghazzai, H.; Hossain, M.J.; Menouar, H. Joint position and travel path optimization for energy efficient wireless data gathering using unmanned aerial vehicles. *IEEE Trans. Veh. Technol.* **2019**, *68*, 2165–2175. [CrossRef]
10. Singh, S.P.; Sharma, S. Genetic-algorithm-based energy-efficient clustering (GAEEC) for homogenous wireless sensor networks. *IETE J. Res.* **2018**, *64*, 648–659. [CrossRef]

11. Lei, T.; Wang, Y.; Jin, X.; Min, Z.; Zhang, X.; Zhang, X. An Optimal Fuzzy Logic-Based Energy Management Strategy for a Fuel Cell/Battery Hybrid Power Unmanned Aerial Vehicle. *Aerospace* **2022**, *9*, 115. [CrossRef]
12. Ferdaus, M.M.; Anavatti, S.G.; Pratama, M.; Garratt, M.A. Towards the use of fuzzy logic systems in rotary wing unmanned aerial vehicle: A review. *Artif. Intell. Rev.* **2020**, *53*, 257–290. [CrossRef]
13. Ragab, M.; Ashary, E.B.; Aljedaibi, W.H.; Alzahrani, I.R.; Kumar, A.; Gupta, D.; Mansour, R.F. A novel metaheuristics with adaptive neuro-fuzzy inference system for decision making on autonomous unmanned aerial vehicle systems. *ISA Trans.* **2022**, *in press*. [CrossRef]
14. D'Andrea, R. Guest editorial can drones deliver? *IEEE Trans. Autom. Sci. Eng.* **2014**, *11*, 647–648. [CrossRef]
15. Yan, H.; Chen, Y.; Yang, S.H. New energy consumption model for rotary-wing uav propulsion. *IEEE Wirel. Commun. Lett.* **2021**, *10*, 2009–2012. [CrossRef]
16. Bauersfeld, L.; Scaramuzza, D. Range, Endurance, and Optimal Speed Estimates for Multicopters. *IEEE Robot. Autom. Lett.* **2022**, *7*, 2953–2960. [CrossRef]
17. Abeywickrama, H.V.; Jayawickrama, B.A.; He, Y.; Dutkiewicz, E. Comprehensive energy consumption model for unmanned aerial vehicles, based on empirical studies of battery performance. *IEEE Access* **2018**, *6*, 58383–58394. [CrossRef]
18. Prasetia, A.S.; Wai, R.J.; Wen, Y.L.; Wang, Y.K. Mission-based energy consumption prediction of multirotor uav. *IEEE Access* **2019**, *7*, 33055–33063. [CrossRef]
19. Hong, D.; Lee, S.; Cho, Y.H.; Baek, D.; Kim, J.; Chang, N. Least-energy path planning with building accurate power consumption model of rotary unmanned aerial vehicle. *IEEE Trans. Veh. Technol.* **2020**, *69*, 14803–14817. [CrossRef]
20. Sarkar, S.; Totaro, M.W.; Kumar, A. An intelligent framework for prediction of a uav's flight time. In Proceedings of the 2020 16th International Conference on Distributed Computing in Sensor Systems (DCOSS), Marina del Rey, CA, USA, 25–27 May 2020; pp. 328–332.
21. Jung, S.; Jo, Y.; Kim, Y.J. Flight time estimation for continuous surveillance missions using a multirotor UAV. *Energies* **2019**, *12*, 867. [CrossRef]
22. Bershadsky, D.; Haviland, S.; Johnson, E.N. Electric multirotor UAV propulsion system sizing for performance prediction and design optimization. In Proceedings of the 57th AIAA/ASCE/AHS/ASC Structures, Structural Dynamics, and Materials Conference, San Diego, CA, USA, 4–8 January 2016; p. 0581.
23. Klement, E.P.; Koczy, L.T.; Moser, B. Are fuzzy systems universal approximators? *Int. J. Gen. Syst.* **1999**, *28*, 259–282. [CrossRef]
24. Barua, A.; Mudunuri, L.S.; Kosheleva, O. Why trapezoidal and triangular membership functions work so well: Towards a theoretical explanation. *J. Uncertain Syst.* **2013**, *8*, 1–6.
25. Bezdek, J.C.; Ehrlich, R.; Full, W. FCM: The fuzzy c-means clustering algorithm. *Comput. Geosci.* **1984**, *10*, 191–203. [CrossRef]
26. Babuška, R. Fuzzy systems, modeling and identification. *Delft Univ. Technol. Dep. Electr. Eng. Control Lab. Mekelweg* **1996**, *4*, 1–31.
27. Kennedy, J.; Eberhart, R. Particle swarm optimization. In Proceedings of the ICNN'95-International Conference on Neural Networks, Perth, WA, Australia, 27 November 27–1 December 1995; Volume 4, pp. 1942–1948.
28. Eberhart, R.C.; Shi, Y. Comparing inertia weights and constriction factors in particle swarm optimization. In Proceedings of the 2000 Congress on Evolutionary Computation, CEC00 (Cat. No. 00TH8512), La Jolla Marriott Hotel, La Jolla, CA, USA, 16–19 July 2000; Volume 1, pp. 84–88.
29. Anthony Carlisle, G.D. An Off-The-Shelf PSO. In *Proceedings of the Workshop on Particle Swarm Optimization*; Purdue School of Engineering and Technology: Indianapolis, IN, USA, 2001.
30. Winkler, R.; Klawonn, F.; Kruse, R. Fuzzy c-means in high dimensional spaces. *Int. J. Fuzzy Syst. Appl. (IJFSA)* **2011**, *1*, 1–16. [CrossRef]
31. Chen, M.; Rincon-Mora, G.A. Accurate electrical battery model capable of predicting runtime and IV performance. *IEEE Trans. Energy Convers.* **2006**, *21*, 504–511. [CrossRef]
32. Huria, T.; Ceraolo, M.; Gazzarri, J.; Jackey, R. High fidelity electrical model with thermal dependence for characterization and simulation of high power lithium battery cells. In Proceedings of the 2012 IEEE International Electric Vehicle Conference, Greenville, SC, USA, 4–8 March 2012; pp. 1–8.
33. Rahimi-Eichi, H.; Baronti, F.; Chow, M.Y. Modeling and online parameter identification of Li-Polymer battery cells for SOC estimation. In Proceedings of the 2012 IEEE International Symposium on Industrial Electronics, Hangzhou, China, 28–31 May 2012; pp. 1336–1341.
34. Yu, Q.Q.; Xiong, R.; Wang, L.Y.; Lin, C. A comparative study on open circuit voltage models for lithium-ion batteries. *Chin. J. Mech. Eng.* **2018**, *31*, 1–8. [CrossRef]
35. Shrivastava, P.; Soon, T.K.; Idris, M.Y.I.B.; Mekhilef, S. Overview of model-based online state-of-charge estimation using Kalman filter family for lithium-ion batteries. *Renew. Sustain. Energy Rev.* **2019**, *113*, 109233. [CrossRef]
36. Terejanu, G.A. *Extended Kalman Filter Tutorial*; University at Buffalo: Buffalo, NY, USA, 2008.
37. ArduPilot Dev Team.. Copter Attitude Control. Available online: https://ardupilot.org/dev/docs/apmcopter-programming-attitude-control-2.html (accessed on 15 November 2022).
38. PX4 User Guide. Controller Diagram. Available online: https://docs.px4.io/main/en/flight_stack/controller_diagrams.html (accessed on 15 November 2022).
39. Krykowski, K.; Hetmańczyk, J. Constant current models of brushless DC motor. *Sci. J. Riga Tech. Univ.-Electr. Control Commun. Eng.* **2013**, *3*, 19–24. [CrossRef]

40. Islam, S.A.U.; Bernstein, D.S. Recursive least squares for real-time implementation [lecture notes]. *IEEE Control Syst. Mag.* **2019**, *39*, 82–85. [CrossRef]
41. Pei, L.; Lu, R.; Zhu, C. Relaxation model of the open-circuit voltage for state-of-charge estimation in lithium-ion batteries. *IET Electr. Syst. Transp.* **2013**, *3*, 112–117. [CrossRef]
42. The MathWorks Inc. Generate Parameter Data for Equivalent Circuit Battery Block. Available online: https://la.mathworks.com/help/autoblks/ug/generate-parameter-data-for-estimations-circuit-battery-block.html (accessed on 28 November 2022).
43. Pei, J.; Yang, X.; Gao, X.; Xie, W. Weighting exponent m in fuzzy C-means (FCM) clustering algorithm. In Proceedings of the Object Detection, Classification, and Tracking Technologies, Wuhan, China, 24 September 2001 ; Volume 4554, pp. 246–251.
44. Yu, J.; Cheng, Q. The upper bound of the optimal number of clusters in fuzzy clustering. *Sci. China Ser. Inf. Sci.* **2001**, *44*, 119–125. [CrossRef]

technologies

MDPI

Article

Simulation Analysis of Signal Conditioning Circuits for Plants' Electrical Signals

Mirella Carneiro [1], Victor Oliveira [1], Fernanda Oliveira [1], Marco Teixeira [2,*] and Milena Pinto [3]

1 Electrical Engineering Program, COPPE, Federal University of Rio de Janeiro, Rio de Janeiro 21941-972, Brazil
2 Department of Software Engineering, Federal University of Technology-Parana (UTFPR), Curitiba 80230-901, Brazil
3 Department of Electronics Engineering, Federal Center for Technological Education of Rio de Janeiro, Rio de Janeiro 20271-110, Brazil
* Correspondence: marcoteixeira@utfpr.edu.br

Abstract: Electrical signals are generated and transmitted through plants in response to stimuli caused by external environment factors, such as touching, luminosity, and leaf burning. By analyzing a specific plant's electrical responses, it is possible to interpret the impact of external aspects in the plasma membrane potential and, thus, determine the cause of the electrical signal. Moreover, these signals permit the whole plant structure to be informed almost instantaneously. This work presents a brief discussion of plants electrophysiology theory and low-cost signal conditioning circuits, which are necessary for the acquisition of plants' electrical signals. Two signal conditioning circuits, which must be chosen depending on the signal to be measured, are explained in detail and electrical simulation results, performed in OrCAD Capture Software are presented. Furthermore, Monte Carlo simulations were performed to evaluate the impact of components variations on the accuracy and efficiency of the signal conditioning circuits. Those simulations showed that, even after possible component variations, the filters' cut-off frequencies had at most 4% variation from the mean.

Keywords: plant electrophysiology; electrical signals; information acquisition; simulation software; electronic instrumentation

Citation: Carneiro, M.; Oliveira, V.; Oliveira, F.; Teixeira, M.; Pinto, M. Simulation Analysis of Signal Conditioning Circuits for Plants' Electrical Signals. *Technologies* **2022**, *10*, 121. https://doi.org/10.3390/technologies10060121

Academic Editors: Gwanggil Jeon, Manoj Gupta and Eugene Wong

Received: 10 November 2022
Accepted: 21 November 2022
Published: 25 November 2022

1. Introduction

Plants are organisms aware of diverse factors in the habitat they are placed. Furthermore, they continuously adapt their metabolism and growth in response to environmental changes. Due to this adaptation mechanism, plants have developed techniques to react right after they detect habitat modification aspects and external stimuli. They respond to these factors by transmitting electrical responses through their structure. Plants' electrical activities are related to transient modifications in the plasma membrane potential [1]. The flow of ions and the activation of ion channels induces a transient and local change in the potential of the cell membrane. Taking into account the main reason for this change is that all cells (mainly root cells associated with ions uptake) hold the whole time ions essentially crossing the plasma membrane [2,3].

Distinct sorts of disturbances, like an abrupt light variation, soil moisture content, and leaf burning, can generate these specific electrical signals in living plant cells, according to [1,4,5]. In contrast to chemical signals, electrical responses originated by these stimuli can conduct information quickly over long distances, from the top of the stem to the roots, in either direction [6]. Besides, once initiated, these responses spread to adjoining excitable cells. The coordination of internal processes and their balance with the environment is connected to plant cells' excitability [7].

Plants have four different types of electrical signals, which are: *(i)* action potentials (APs); *(ii)* variation potentials (VPs); *(iii)* local electrical potentials (LEPs); and *(iv)* system potentials (SPs).

LEP is a local electrical signal generated from natural changes related to the environment, such as luminosity, soil nutrients, and air humidity. These changes cause a sub-threshold electrical response in plants [8]. SP was detected in the plant leaves after caterpillar feedings. Besides, it is a long-distance signal with duration and amplitude dependent on the stimulus [9]. AP is induced by a non-damaging disturbance to the plant (electrical, mechanical stimulus, or thermal shock [6]) and is characterized by transmitting information over long distances along the plant in a short amount of time. VP is caused by harmful stimuli to the plant, such as burning and cutting. The plant type and the disturbance's intensity have influence on the VP signal shape and magnitude [10].

The way some characteristics of the electrical signal, like amplitude, duration and speed of the electrical signal behaves while propagating through the plant structure depends primarily on the type of the signal, i.e., if it is an AP, VP, LEP ou SP. Since each signal has got their own peculiarities, which will be explained further in Section 2.

Furthermore, two methods to measure electrical potential in plants can be employed: extracellular and intracellular [11].

According to [12], real-time monitoring of these electrical signals enable the user to be informed about what happens in the habitat where the plants are placed. With this information, it is possible to identify the presence of landslides [13], acid rain [14], an increase in air pollution, whether the plant receives too much light or if pests are attacking a certain plant in the plantation [9]. Figure 1 shows the necessary steps to acquire plants' electrical responses. This work addresses the fifth step: Signal Conditioning Circuit.

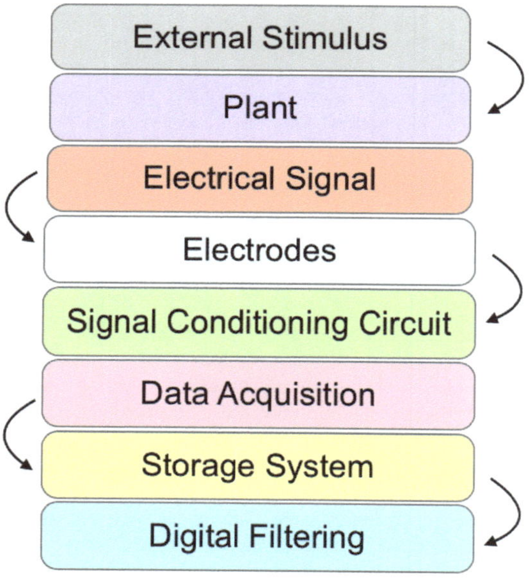

Figure 1. Flowchart of the steps used to acquire plants electrical signals. Adapted from [3,12].

Usually, when a sensor is used to measure a signal, the sensor reads not only the desired stimulus, but also noise. Furthermore, the measured electrical signal amplitude from the plant may not be large enough for proper data acquisition. Most signal conditioning circuits purpose is to filter, to reduce noise, and/or amplify the original signal in order to enable data acquisition.

When measuring plants' electrical responses, an analog-to-digital converter (ADC) is used to convert the waveform of the electrical signal into digital data. Furthermore, it is necessary to carefully choose an ADC with appropriate sampling frequency according to the measured signal. In addition, the greater the input impedance of the ADC, the closer the ADC input signal value is to the signal conditioning circuit output. Therefore, the ADC input impedance has to be at least 100 times greater than the output impedance of the signal conditioning circuit. Besides, since environmental factors influence the plants' signals, it is important to employ an environmental parameters-acquiring system to measure such factors. Furthermore, the plant, along with the unshielded components of the measuring system, must be placed inside a Faraday cage, in order to improve the signal-to-noise ratio (SNR) of the measured signal [12]. The complete acquisition system is shown in Figure 2.

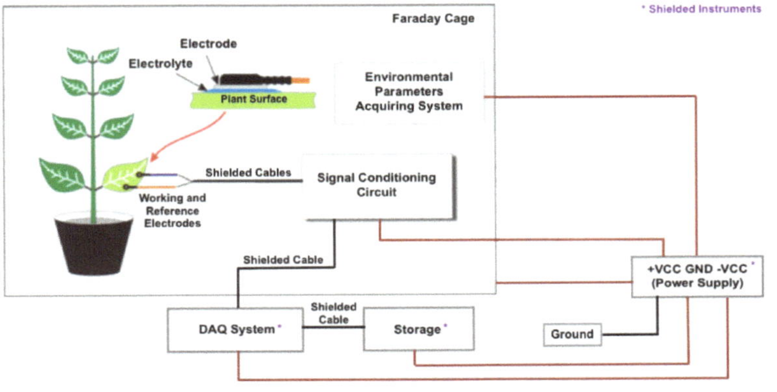

Figure 2. The complete acquisition system for measuring plants electrical responses [12].

More details about plants acquisition system can be found in [3,12].

1.1. Related Works and Main Contributions

In most of the works carried out in the area of measuring electrical signals in plants, the equipment applied in the process is expensive, as shown in [6,7,14–16]. Besides, there are only a few authors that design their own signal conditioning circuits. Most authors employ ready-for-use instruments to perform this task, which contributes to the increased cost of capturing electrical responses emitted by plants. A requirement of the equipment (or the developed signal conditioning circuit) that reads the plant signal measured by the sensor (the electrode) is an input impedance in the order of GΩ [16]. Equipment with an input impedance in this order of magnitude usually cost thousands of dollars.

In articles where the authors develop their own circuits, they generally do not explain the circuits thoroughly [17,18]. In other words, the circuits and their functionality are not explained in detail, and their efficiency is not authenticated with consistent results. Besides, those circuits are usually not robust [19–21], i.e., they just amplify the signal and do not have filtering steps.

When analysing the plants' electrical response, the shape of the response depends on the type stimulus. Consequently, by analyzing the format of the excitation, along with other aspects, like propagation velocity and amplitude, we can infer which stimulus caused the response. The proposed system, i.e., the sensing step, the digital filters and the signal conditioning circuit, can be used for developing a low-cost equipment that has the purpose of monitoring and informing environmental changes, such as the ones mentioned in Section 1, in the habit of a plant.

The main contributions of this work are: to present fundamental knowledge about plants electrophysiology, focusing on the types of plants' electrical responses; the instructions to develop two types of robust signal conditioning circuits that must be selected based on the electrical signal measured. In this sense, even a user who has an intermediary familiarity with these matters, can comprehend plants electrophysiology and implement a complete signal conditioning circuit to make the electrical response clearer before it goes to the ADC.

1.2. Organization

This research work is divided as follows. Section 2 presents essential information to understand the types of electrical signals that might be emitted by plants. Section 3 explains in detail the methodology employed in order to develop the entire signal conditioning circuit. Section 4 addresses the results and discussion considering Monte Carlo simulations. Finally, Section 5 presents the conclusions about the research established in the work and future works taking into account the field studied in the article.

2. Types of Electrical Signals

The action potential is an electrical response characterized by quickly transmitting and disseminating the disturbances along the phloem, which is one of the tissues of vascular plants, over long distances [22]. AP was the first plant's electrical response recorded and it is provoked by non-invasive excitation (e.g., electrical stimulation, thermal stress, mechanical stimulus) [11,23]. When comparing AP to Variation potential, an expressing AP attribute is that an increase in the magnitude of the excitation above a certain threshold does not modify the electrical response's shape and amplitude, as stated in [8]. One of the most important aspects of the AP is that it follows the all-or-nothing principle. To put it in other words, the tentative to cause a stimulus weaker than a certain threshold cannot trigger an action potential. Additionally, the cell membrane enters a refractory period after the period AP is triggered, in which another action potential cannot be generated or transmitted [23]. Furthermore, action potentials are able to spread through the plant structure without loss of amplitude and with constant speed, unlike VPs [1]. APs transmission speed of most plants studied previously range from 0.5 cm/s to 20 cm/s, according to [22].

Variation potential, also known as slow-wave potential, is an electrical signal generated by plants caused by damaging disturbances such as wounding, herbivore attack, and burning [23]. This signal consists of a local variation in the plasma membrane due to the transit of some other signal (chemical, hydraulic, or both combined), as stated in [24]. The xylem, which is one of the tissues of vascular plants, is the main pathway through which the VP spreads [8]. VP, unlike the AP, is defined by a decrease in the amplitude and speed of the response's propagation as it moves away from the local, which has suffered an excitation [13,24]. Besides, the plant chosen and the intensity of the disturbance influence the shape and magnitude of the VP. Variation potentials detain a great variety of changes in their shape, according to [1]. This signal can penetrate into poisoned or dead regions of the plant. Furthermore, the VP can be suppressed by a scenario of prolonged darkness and high humidity since the tension of xylem tissue becomes irrelevant, and the generation of a VP is linked to the pressure difference between the intact interior of the plant and the external environment [25]. VPs propagation speed range is from 0.1 cm/s to 1 cm/s [22].

Local electrical potential is generated at the stimuli site, which causes a sub-threshold electrical response in plants as a consequence of natural modifications in aspects connected with the external environment, such as phytohormones, fertility, and air temperature. This signal type significantly influences the plant's physiological status. The local electrical potential has a limited location, not being transmitted to other parts of the plant's body. Additionally, the intensity and duration of the excitation influence its amplitude. In addition, it can be generated using changes in the ion channel and by the transient inactivation of H^+-ATPase, according to [11].

System potential was first noticed by [26], being detected in leaves dozens of centimeters distant from the local that suffered the stimulus after caterpillars feeding. This signal is a self-propagate systemic signal with duration and magnitude that depends on the nature of the disturbance caused. System potential initialization is associated with the activation of H^+-ATPase, which induces the hyperpolarisation of the plasma membrane [11]. According to [27], this signal is strongly dependent on the conditions and treatments of the experiments. Besides, different from AP, SP does not follow the all-or-none rule. Weak stimuli that are not enough to initiate APs, since they do not reach the critical intensity, can trigger system potentials. Additionally, SP is triggered by a hyperpolarization of the plasma membrane. This is unlike AP and VP, which begin with a depolarization. System potential has a propagation speed that ranges from 5 cm/min to 10 cm/min [11].

3. Proposed Methodology

Figure 3 illustrates the steps of the proposed methodology. This diagram shows two signal conditioning circuits options, each for a different frequency range. Since plants' electrical responses have frequency components that range from very low frequencies [28,29] to several hundreds of Hertz, according to [30], it is necessary to design the conditioning circuit taking into account the measured signal range. Plants' signal frequency depends on their species, growth stage, measured tissue, and the excitation source.

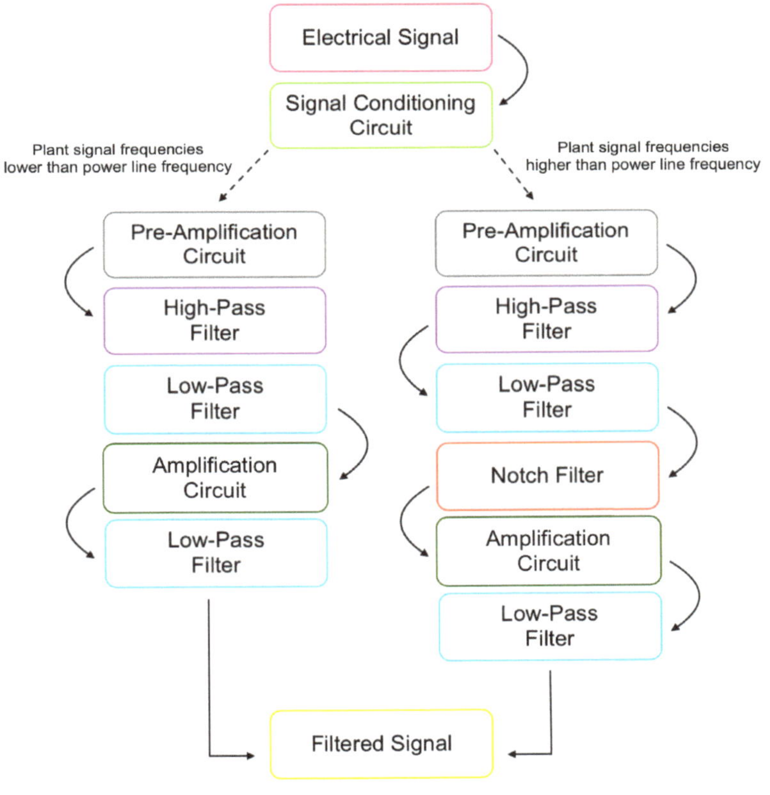

Figure 3. Flowchart of the proposed methodology.

Electrical signals generated by plants have low amplitude, in the order of tens of μV to tens of V [28]. So, a signal conditioning circuit is crucial to improve the SNR of the electrical response. SNR compares the level of a desired signal with respect to the level of background noise.

Moreover, the electrical response amplitude must fit within the ADC dynamic range. An ADC's dynamic range is the range of signal amplitudes which the analog-to-digital converter is able to resolve. To use the full resolution provided by the ADC, the input signal range must be the same as the ADC operation range.

The electrical signal conditioning circuit structures presented in this work are shown in Figures 4 and 5.

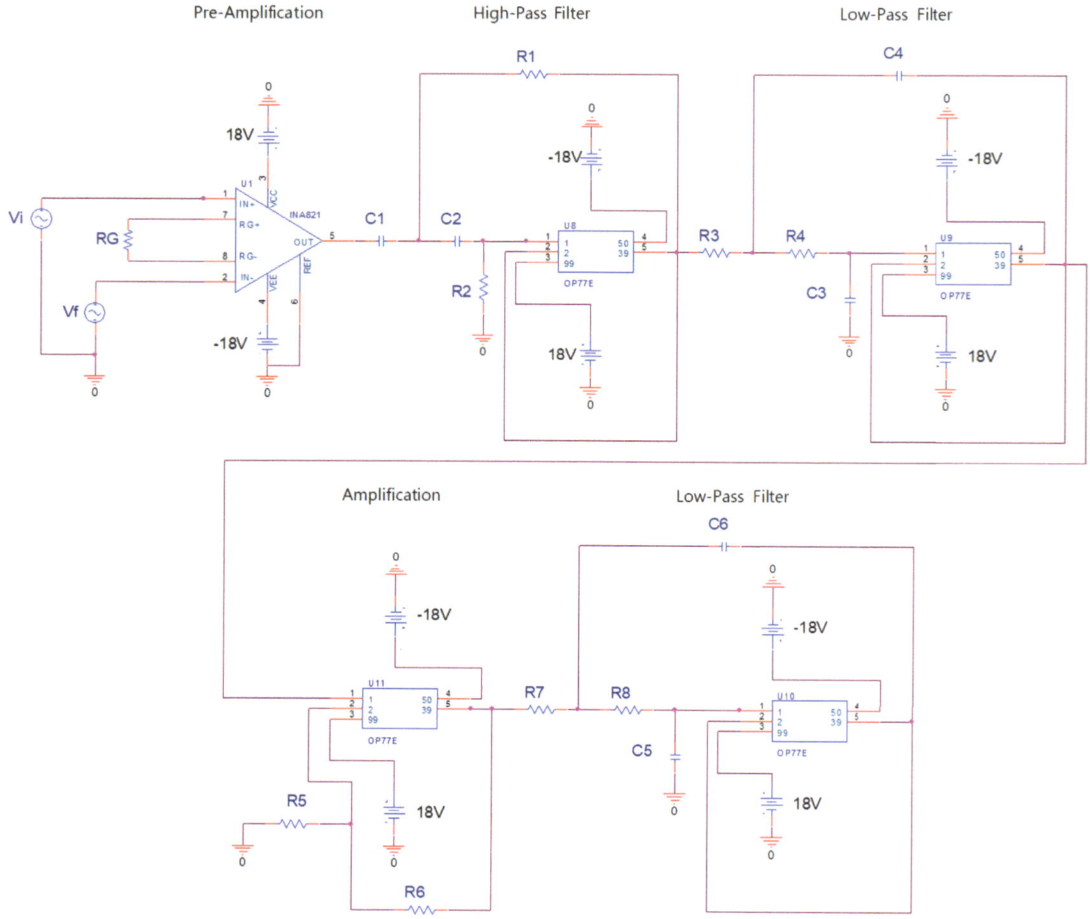

Figure 4. Schematic diagram of the first developed signal conditioning circuit. In the pre-amplification stage, an INA821 instrumentation amplifier is applied. In the high-pass and low-pass filters steps, and in the amplification stage, an OP77 op-amp is employed as well.

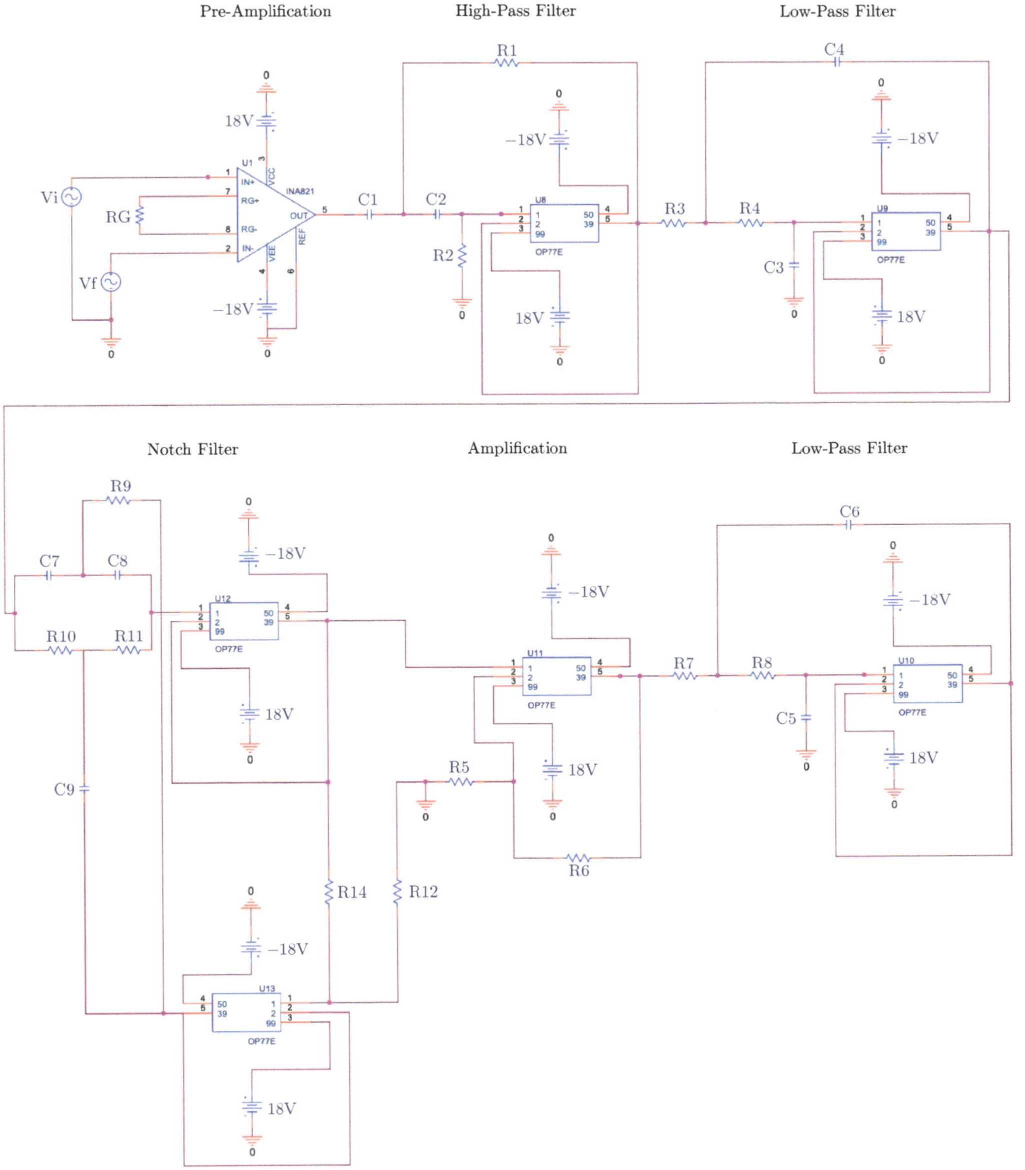

Figure 5. Schematic diagram of the second developed signal conditioning circuit. In the pre-amplification stage, an INA821 instrumentation amplifier is employed. In the high-pass, low-pass, and notch filters steps, and in the amplification stage, an OP77 op-amp is employed as well.

3.1. Pre-Amplification Circuit

The pre-amplification stage is the most crucial of the entire signal conditioning circuit because, if it is adequately built with a differential amplifier, an appreciable part of the common mode noise that interferes with the plant's electrical response can be minimized. The input impedance of the pre-amplification circuit must be in the order of GΩ. The reason for this input impedance order is that the impedance value of $Ag/AgCl$ electrodes, which are the most commonly used electrodes in this application, is in the order of a few kΩ, and the source impedance (plant) often has a value in the order of hundreds or thousands of kΩ [31]. As a consequence, the value of the electrical response that appears at the input of the pre-amplification step is approximately equal to the plant signal if the circuit input impedance is as large as possible.

The input offset voltage temperature coefficient of the op-amp employed in the pre-amplification circuit has to be less than 10 μV/°C, as in [17]. Usually, the gain value applied to the signal in this stage range from 10 to 50.

Additionally, as stated in [18], the common-mode rejection ratio has to be at least 100 dB, so the power line frequency interference, which is present on the non-inverting and inverting op-amp inputs, can be attenuated adequately [32]. This parameter indicates how much an undesired common-mode signal influences the measurements, which is a crucial criterion.

In the pre-amplification stage, it is recommended to employ the classic instrumentation amplifier structure using three op-amps, as used in [32]. In this classic architecture there are two amplifiers in the voltage follower configuration with a third op-amp, as in [17], or an instrumentation amplifier integrated circuit. The differential amplifier configuration employing only one op-amp cannot be used at this step for the sake of does not offer the necessary input impedance. Some instrumentation amplifier integrated circuits that can be applied in the first step are AD8221, INA821 and INA128, as stated in [12]. INA821 was chosen to be used in the simulation, and the expression related to the gain of this stage can be seen in Equation (1).

$$G = 1 + \frac{49.4\ \text{k}\Omega}{R_G} \tag{1}$$

The output of the pre-amplification circuit is given by Equation (2).

$$V_{OUT} = G(V_{IN+} - V_{IN-}) + V_{REF} \tag{2}$$

3.2. Low-Pass and High-Pass Filters

Sallen-Key configuration, which is non-inverting, is applied to the second and third stages. Sallen-Key filter topology was selected because it is a low-complexity second-order filter, and it is the least dependent on the frequency response of the chosen op-amp, according to [33]. The second stage consists of a high-pass filter, and the third one is a low-pass filter, both of which have unit gain. OP07 and OP77 are op-amps that can be employed in these steps. Moreover, it is feasible to use some possible filter approximations, like Butterworth, Chebyshev, and Bessel, depending on the adjustment of the quality factor Q. It is important to cite that these approximations dictate the format of the frequency response.

A bandpass filter was made by cascading a high-pass filter with a low-pass filter opposite to applying only one op-amp. The advantage of customizing the filter to have an asymmetrical response is the motivation for this procedure. A Sallen-Key bandpass filter using only one op-amp has got cut-off frequencies symmetrically apart from the center frequency f_0. Note that OP77 was selected to be applied in the simulation. The Sallen-Key equations for the high-pass filter are given by Equations (3)–(5), which represent the transfer function respectively, f_c, and Q [12].

$$\frac{V_{39}}{V_1} = \frac{s^2(R_1R_2C_1C_2)}{s^2(R_1R_2C_1C_2) + sR_1(C_1 + C_2) + 1} \tag{3}$$

$$f_c = \frac{1}{2\pi\sqrt{R_1R_2C_1C_2}} \tag{4}$$

$$Q = \frac{\sqrt{R_1R_2C_1C_2}}{R_1(C_1 + C_2)} = \frac{1}{2\pi f_c R_1(C_1 + C_2)} \tag{5}$$

Sallen-Key equations for low-pass filter are given by Equations (6)–(8), which represent respectively the transfer function, f_c, and Q [12].

$$\frac{V_{39}}{V_1} = \frac{1}{s^2(R_3R_4C_4C_3) + sC_3(R_3 + R_4) + 1} \tag{6}$$

$$f_c = \frac{1}{2\pi\sqrt{R_3R_4C_4C_3}} \tag{7}$$

$$Q = \frac{\sqrt{R_3R_4C_4C_3}}{C_3(R_3 + R_4)} = \frac{1}{2\pi f_c C_3(R_3 + R_4)} \tag{8}$$

3.3. Notch Filter

The notch filter step is only used in the signal conditioning circuit shown in Figure 5, which is employed for plants' signals with frequency components higher than the power line frequency. Notch filters are part of a special class of band-stop filters capable of rejecting a very narrow range of frequencies. It acts almost exclusively on the selected frequency, in this case, the power line frequency.

The one applied in this project was Twin-T Notch Active Filter, and it was chosen instead of this filter's passive implementation because the latter has a significant shortcoming of a Q fixed at 0.25 [33]. The active configuration holds a variable Q, allowing the user to set its value in a way that can achieve the best compromise between rejection at the notch frequency f_n and bandwidth BW since these two variables are related by Equation (9).

$$Q = \frac{f_n}{BW} \tag{9}$$

The amount of the signal feedback determines the value of Q of the circuit, which, in turn, defines the notch depth. This parameter is set by R_{14}/R_{12} ratio. The design equations for the Twin-T notch filter are shown in Equations (10) and (11) [33]. $V_{39'}$ is the input and refers to the output of the low-pass filter.

$$\frac{V_{39}}{V_{39'}} = \frac{s^2 + \omega_o^2}{s^2 + \frac{s\omega_o}{Q} + \omega_o^2} = \frac{s^2 + (\frac{1}{RC})^2}{s^2 + s(\frac{1}{RC})\left(\frac{4}{1+\frac{R_{12}}{R_{14}}}\right) + (\frac{1}{RC})^2} \tag{10}$$

$$f_n = \frac{1}{2\pi RC} \tag{11}$$

OP77 was chosen to be employed in the simulation but OP07 and TLV2252ID [34] are op-amps that can be applied in this step.

3.4. Amplification Circuit

The last but one stage of the signal conditioning circuit includes a non-inverting configuration in that the gain is determined from the selected resistor values. Normally, the gain value applied to the signal in this stage range from 10 to 1000. It is needful to point out that the higher the gain value, the narrower the bandwidth op-amp will work without the signal being attenuated. Therefore, it is important to guarantee the bandwidth the op-amp is working with a certain gain covers all frequencies of the plant's signal selected to perform measurements without suffering attenuation. OP07 and OP77 are op-amps that can be used in this amplification stage [12]. The mathematical statements related to this stage can be seen in Equations (12)–(16).

$$V_1 = V_2 = V_{in} \tag{12}$$

Resulting in:

$$V_{39} = V_{out} \tag{13}$$

$$\frac{V_{in} - 0}{R_5} + \frac{V_{in} - V_{out}}{R_6} = 0 \tag{14}$$

$$V_{in}R_6 + V_{in}R_5 - V_{out}R_5 = 0 \tag{15}$$

The gain is defined by:

$$A_v = 1 + \frac{R_6}{R_5} \tag{16}$$

3.5. Anti-Aliasing Filter

In conclusion, the last step of the circuit is the anti-aliasing filter, which is a low-pass filter with the cut-off frequency set to the Nyquist frequency. Sallen-Key low-pass topology is used in this step too. Besides, OP77 and OP07 are op-amps that can be employed in this stage.

At the end of the whole process, the electrical signal shows up clearer at the signal conditioning circuit output, stronger and with undesired frequencies attenuated, ready for the ADC and digital filtering step.

4. Results and Discussion

The signal conditioning circuits, shown in Figures 4 and 5, were simulated in OrCAD Capture 16.6 software to testify the functionality for which they were proposed. The values chosen for the components are shown in Tables 1 and 2.

Table 1. Components values of the first signal conditioning circuit.

Components	Values	
Resistors	$R_G = 2.4$ kΩ $R_1 = 39$ kΩ $R_2 = 82$ kΩ $R_3 = 10$ kΩ $R_4 = 10$ kΩ	$R_5 = 1$ kΩ $R_6 = 100$ kΩ $R_7 = 5$ kΩ $R_8 = 5$ kΩ
Capacitors	$C_1 = 5.6$ µF $C_2 = 5.6$ µF $C_3 = 0.27$ µF	$C_4 = 560$ nF $C_5 = 0.28$ µF $C_6 = 0.56$ µF

Table 2. Components values of the second signal conditioning circuit.

Components	Values	
Resistors	$R_G = 2.4$ kΩ $R_1 = 39$ kΩ $R_2 = 82$ kΩ $R_3 = 3$ kΩ $R_4 = 3$ kΩ $R_5 = 1$ kΩ $R_6 = 100$ kΩ	$R_7 = 3$ kΩ $R_8 = 3$ kΩ $R_9 = 13$ kΩ $R_{10} = 27$ kΩ $R_{11} = 27$ kΩ $R_{12} = 80$ kΩ $R_{14} = 20$ kΩ
Capacitors	$C_1 = 5.6$ µF $C_2 = 5.6$ µF $C_3 = 0.37$ µF $C_4 = 0.75$ µF $C_5 = 0.18$ µF	$C_6 = 0.37$ nF $C_7 = 0.1$ µF $C_8 = 0.1$ µF $C_9 = 0.2$ µF

Furthermore, Monte Carlo simulations were carried out employing the same software with the intention of checking the behavior of the circuits, taking into account possible variations in the nominal components value.

4.1. Filters Frequency Response

In the second stages of Figures 4 and 5, values of capacitors and resistors of the high-pass filters were selected so that they could have $Q = 0.707$ and $f_c = 0.5$ Hz. In the third stages of Figures 4 and 5, which are low-pass filters, values of the capacitors and resistors were chosen so they could have $Q = 0.707/f_c = 40$ Hz and $Q = 0.707/f_c = 100$ Hz, respectively. In the last stages, which are anti-aliasing filters, values of resistors and capacitors were selected so that they could have $f_c = 100$ Hz/$Q = 0.707$ and $f_c = 200$ Hz/$Q = 0.707$, respectively. Taking into account the notch filter, the f_n chosen was 60 Hz, because this is the power line frequency employed in Brazil. Besides, the Q value is 2.5.

The cut-off frequencies of Figure 4 were chosen to take into account a plant electrical signal with frequency components between 5 Hz and 25 Hz [18,35,36]. Additionally, the cut-off frequencies of Figure 5 were set considering a plant electrical signal with frequency components between 5 Hz and 85 Hz [30]. It is important to highlight that it is not suggested to choose the f_c exactly equal to the minimum and maximum frequencies components of the signal to be measured. Note that the user commonly does not know the minimum/maximum frequency components of a determined electrical response of a specific plant. If the low-pass filter f_c set with the slack is higher than the power line frequency, it will be necessary to apply the circuit of Figure 5. Even if the signal to be measured is supposed to have frequency components lower than the power line frequency. Figures 6–8 show the magnitude responses for the high-pass, notch and low-pass ($f_c = 100$ Hz) filters, respectively.

Taking into account Figure 6 high-pass filter, it is needful to cite that the cut-off frequency found in the simulation was $f_c = 0.517$ Hz. The notch frequency achieved for the notch filter shown in Figure 7 was $f_n = 63.096$ Hz. Considering Figure 8 low-pass filter, the cut-off frequency obtained in the simulation was $f_c = 101.141$ Hz. For the low-pass filters with cut-off frequencies equal to 200 Hz, 80 Hz, and 40 Hz, the simulation results provided $f_c = 203.667$ Hz, 80.157 Hz, and 41.517 Hz, respectively.

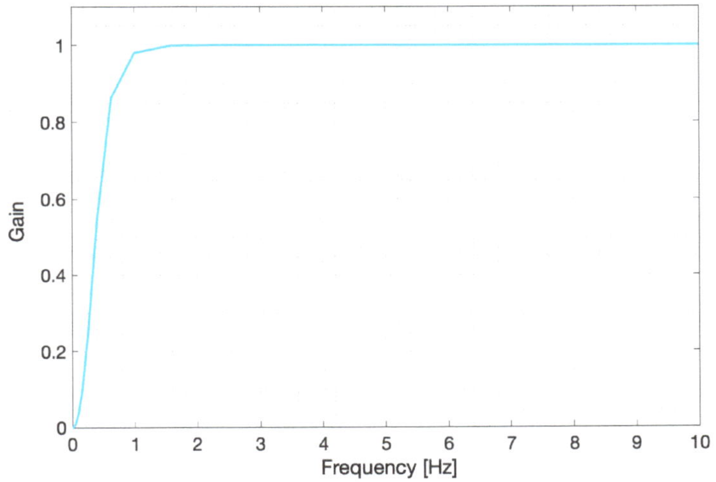

Figure 6. Magnitude response for the high-pass filter. The plot shows the filter's gain for different sinusoidal inputs frequencies. The filter is configured to reject frequencies lower than $f_c = 0.5$ Hz.

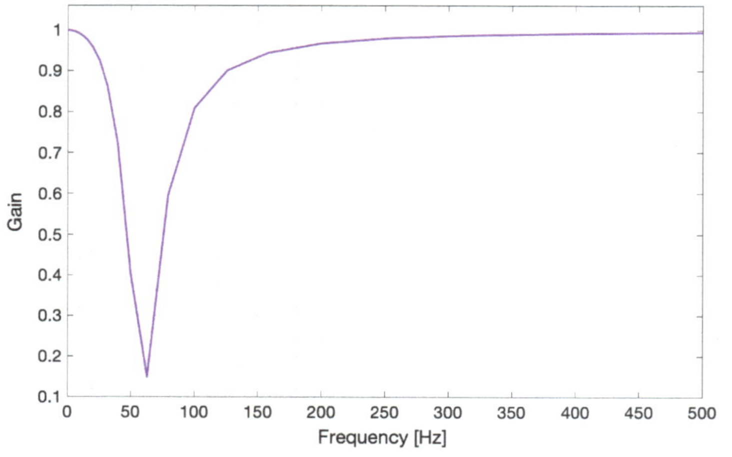

Figure 7. Magnitude response for the notch filter. The plot shows the filter's gain for different sinusoidal inputs frequencies. The filter is configured to reject the power line frequency of $f_n = 60$ Hz.

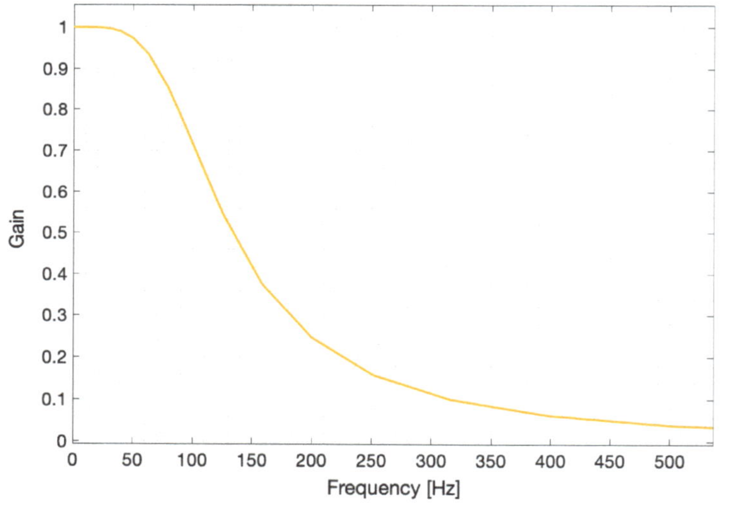

Figure 8. Magnitude response for the low-pass filter. The plot shows the filter's gain for different sinusoidal inputs frequencies. The filter is configured to reject frequencies higher than $f_c = 100$ Hz.

4.2. *Signal Conditioning Circuit Simulation*

In the pre-amplification stage, a gain of 21.58× (26.68 dB) was set, and $V_{REF} = 0$. In the stage in which the electrical signal is amplified, values of the resistors were selected employing Equation 16 so that the gain setting could be 101× (40.09 dB). Figures 9 and 10 show the gain of each stage of circuits 1 and 2, respectively, when they are submitted to sinusoidal inputs of varying frequencies. These figures show the filters cascade response, meaning that each stage refers to the output of that stage and all previous stages combined. At the first stage, pre-amplification, it is possible to see that the gain value of 26.69 dB given to the differential signal at the input (V_{IN+}-V_{IN-}) is constant for the range of tested frequencies and is very close to what was expected. Then, at the high-pass filter stage, which is the cascade of the pre-amplification circuit and a high-pass filter, both Figures 9 and 10 show that the gain stays the same, but the high-pass filter introduces a cut-off frequency at 0.49 Hz. Following the cascade, at the low-pass filter stage, the gain and lower cut-off frequencies remain unaltered. Figures 9 and 10 show that the low-pass filter introduces high cut-off frequency at 41.56 Hz and 101.1 Hz, respectively. At the end of the low-pass filter stage, we can see that, due to the filter cascade, both circuits work as band-pass filters with a gain defined by the pre-amplification step and cut-off frequencies defined by the low-pass and high-pass filters stages.

As can be seen in Figure 9, for circuit 1, the step after the low-pass filter stage is the amplification step, which adds a gain of 40.09 dB. Due to the cascade of filters, the total gain at the amplification stage is 66.78 dB (the sum of the pre-amplification and amplification steps). Moreover, for circuit 1 (Figure 9), the last stage, anti-aliasing filter, is a low-pass filter which has a cut-off frequency (theoretically 100 Hz) higher than the previous low-pass filter in the cascade (41.56 Hz). Consequently, the result of the cascade has a high cut-off frequency, smaller than both low-pass filters, at 40.3 Hz.

For circuit 2, as can be seen in Figure 10, the stage that follows the low-pass filter step is the notch filter. At this stage, the notch filter introduces a rejection peak in which the minimum is at 59.5 Hz. Similarly to circuit 1, at the amplification stage of circuit 2, the total gain is 66.78 dB due to the cascade of amplifications. Finally, in the last stage, the anti-aliasing filter has a similar frequency to the low-pass filter introduced earlier in the cascade (100 Hz versus 101.1 Hz). Therefore, the result of the cascade has a similar shape before and after the anti-aliasing filter, but the gain at frequencies higher than 100 Hz decreases faster with respect to the increase in frequency.

In summary, Figures 9 and 10 show that when input frequencies are within the passband of the high-pass and low-pass filters, the electrical signal is amplified throughout the stages, and is attenuated otherwise. Taking into account the notch filter of Figure 10, frequencies around the 60 Hz notch frequency are also rejected. As stated in [15] for plants' electrical signals, voltage values employed in the circuits are in the range from tens of µV to tens of V.

In Section 4.1, the filters' frequency response was presented when each filter was simulated individually. For both signal conditioning circuit designs (with or without the notch filter), the filters are connected in cascade, as shown in Figures 4 and 5. When two filters are connected, the first output serves as the second's input. Because of this, the input of the second filter is already modified by the first filter. As a result, when two low-pass filters are attached, as in circuit 1, the cut-off frequency of the entire cascade is not the same as one of the filters' cut-off frequencies.

The tolerance of the components may influence the gain given in the first and last stages. Moreover, this parameter can influence Q and f_c of the filters because they are dependent on component values.

Due to their maximally flat magnitude response in the passband, Butterworth filters were chosen to be applied. However, this filter type introduces a customarily undesired phase shift into the filtered data, as shown in Figures 11 and 12. The delay length elevates with increasing filter order and decreasing f_c.

In Figure 11, the amplitude of the first, second, and third stages are similar to each other. Moreover, their amplitude is much lower than in the following stages since the gain given to the electrical signal in the fourth stage is 40 dB. As a result, the first, second, and third stages graphs are not clearly shown in the figure. The same situation occurs in Figure 12 because the first four stages have lower amplitude than the fifth and sixth stages.

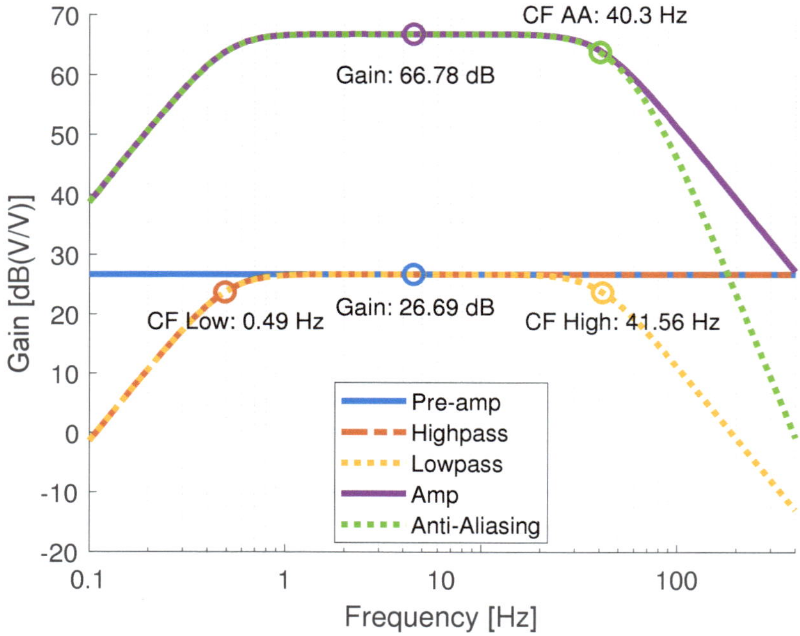

Figure 9. Magnitude response of all stages of circuit 1. "CF" stands for cut-off frequency and "AA" stands for anti-aliasing.

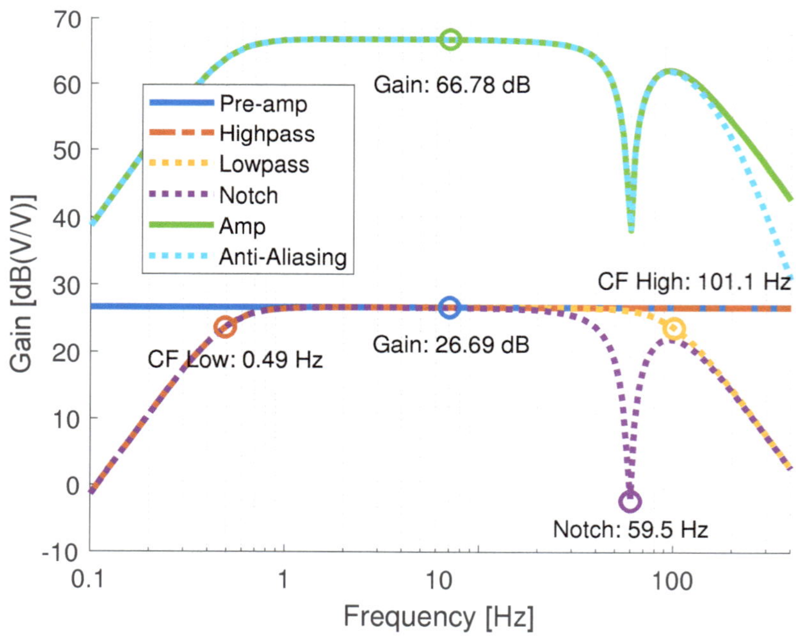

Figure 10. Magnitude response of all stages of circuit 2. "CF" stands for cut-off frequency and "Notch" refers to notch frequency.

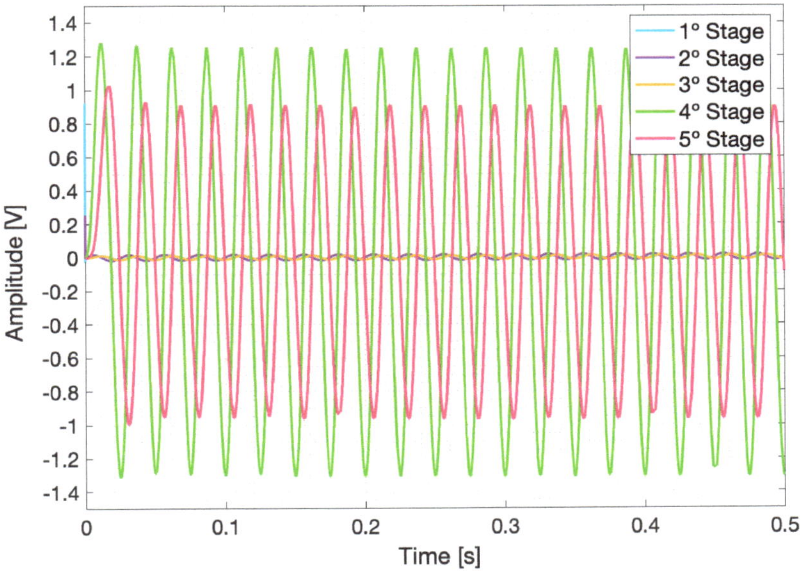

Figure 11. Graphs of each conditioning circuit 1 stage when f = 40 Hz.

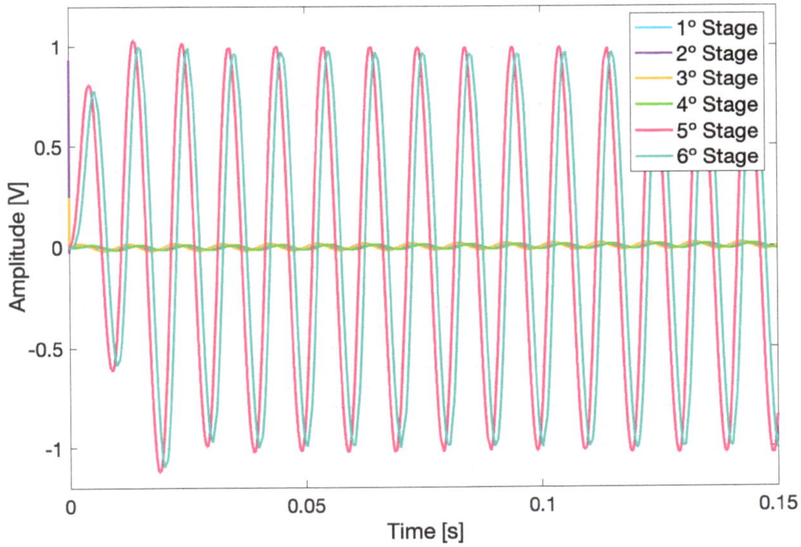

Figure 12. Graphs of each conditioning circuit 2 stage when f = 100 Hz.

4.3. Components Variation Simulation

Electronic components have a nominal, as labeled by the manufacturer, and a real value. There are imperfections in manufacturing these components; therefore, their real value is not always the same as the nominal value. Simulations can be performed to verify the behavior of a circuit, taking into consideration possible variations in the nominal value of components.

To further characterize circuits 1 and 2, and validate their performance for real-world cases, Monte Carlo simulations are performed. These simulations use a given statistical distribution to slightly alter the value of each component within the specified tolerance range. Each Monte Carlo sample refers to a possible set of random component values. To produce statistically relevant results, usually, hundreds of samples are simulated.

In this work, the model used for the Monte Carlo simulations is the default model provided by OrCAD, and is defined as follows. Resistors and capacitors have their values independently randomized following a Gaussian distribution for each circuit. The distribution is adjusted so that the resulting values (after being randomized) fall within the components' 1% tolerance. A different Gaussian distribution is generated for each component, which has a mean equal to the respective nominal component value and with three standard deviations being the nominal value after 1% variation. A hundred Monte Carlo samples are simulated for each circuit.

Figure 13 shows the magnitude response for the last stage of circuit 1. Each Monte Carlo sample represents a different set of component values and, thus, generates a different magnitude response. For circuit 1, the required (without component value variation) output behavior is a band-pass filter. As can be seen in Figure 13, all Monte Carlo variations have the demanded output behavior.

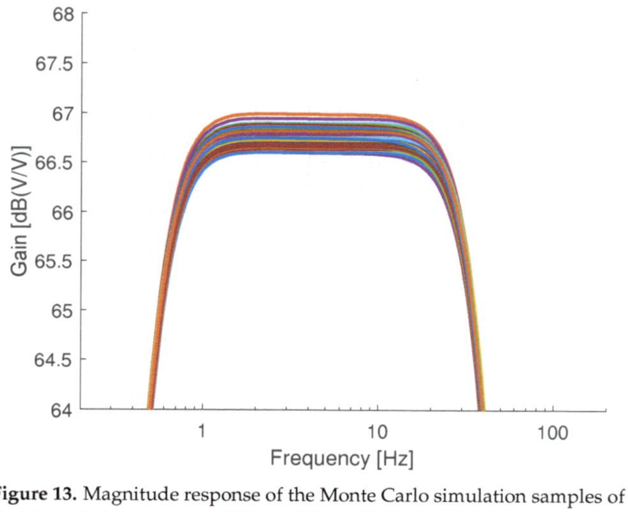

Figure 13. Magnitude response of the Monte Carlo simulation samples of the last stage of circuit 1. Each line plot represents a different Monte Carlo sample.

To closely examine Monte Carlo samples, for each sample of circuit 1, the lower and higher cut-off frequencies are computed. When observing all Monte Carlo samples of circuit 1's output, the lower cut-off frequency holds an average value of 0.49 Hz and variance of 1.1×10^{-5}, and the higher cut-off frequency has an average value of 40.39 Hz and variance of 1.8×10^{-1}.

Figure 14 presents the distribution of both cut-off frequencies. The computed frequencies are normalized regarding their respective mean to show the relative variation between samples better. As can be seen in Figure 14, for the simulations performed, the lower and higher cut-off frequencies have got at most 2% and 4% variation from the mean, respectively. However, most Monte Carlo samples resulted in frequencies within less than 1% variation.

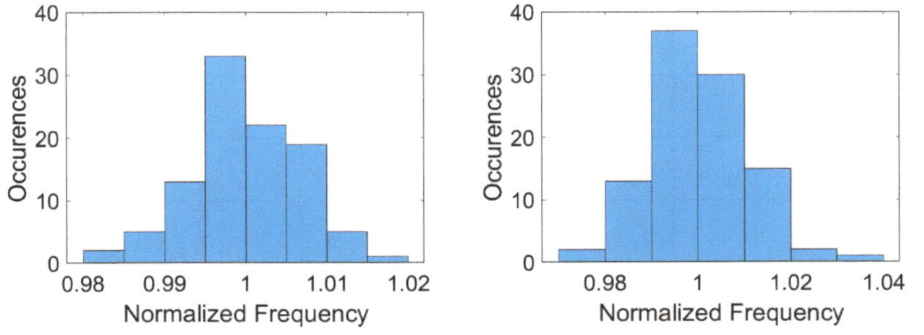

(a) (b)

Figure 14. Histogram of the (**a**) lower and (**b**) higher cut-off frequency of the last stage of circuit 1. Both frequencies are normalized with respect to the mean of all 100 Monte Carlo samples of each respective frequency.

Figure 15 shows the magnitude response for the last stage of circuit 2. The Monte Carlo samples appear more similar to each other when compared to circuit 1, because of the plot scale. For circuit 2, the desired output behavior is also a band-pass filter, but with a notch filter with f_n = 60 Hz. As shown in Figure 15, all Monte Carlo variations have the desired output behavior.

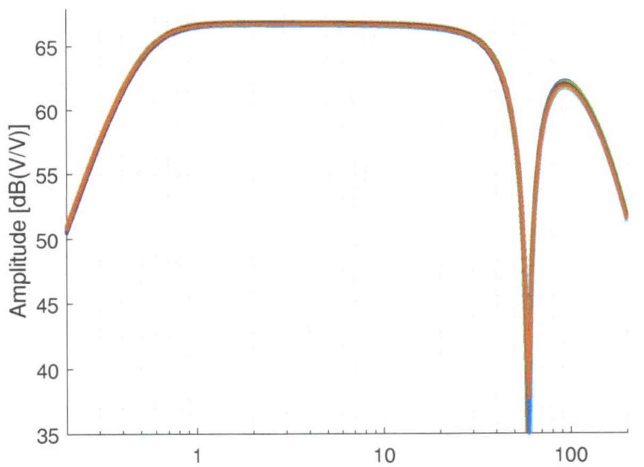

Figure 15. Magnitude response of the Monte Carlo simulation samples of the last stage of circuit 2. Each line plot represents a different Monte Carlo sample.

When observing all Monte Carlo Samples of circuit 2, the lower cut-off frequency has a mean value of 0.49 Hz and variance of 1.2×10^{-5} and the center frequency of the notch stopband has a mean value of 59.46 Hz and variance of 1.2×10^{-1}.

Figure 16 shows the distribution of the notch filter's lower cut-off frequency and center frequency for circuit 2's output. As can be seen in Figure 16, for the simulations performed, the lower cut-off and notch frequencies have at most 2% and 1.5% variation from the mean, respectively. However, most Monte Carlo samples resulted in frequencies within less than 1% variation.

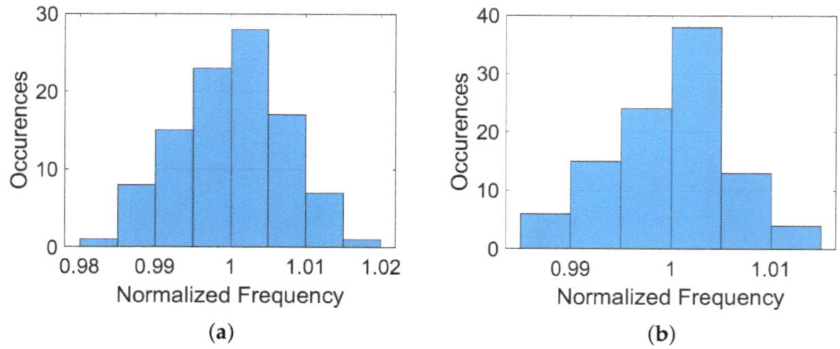

Figure 16. Histogram of the (**a**) lower cut-off frequency and (**b**) center frequency of the notch band-stop filter of the last stage of circuit 2. Both frequencies are normalized with respect to the mean of all 100 Monte Carlo samples of each respective frequency.

5. Conclusions and Future Work

This work has presented valuable signal conditioning circuits that operate efficiently. Computer simulations allowed the authors to validate the circuits' behavior via software without carrying out bench tests. The results obtained with this project are similar to the ones expected by the theory. The methodology presented can be followed and adjusted according to the type of plant and its electrical signals. In addition, through Monte Carlo simulations, OrCAD Capture software is able to generate hundreds of possible variations in the circuit's parameters. With this approach, results that more closely resemble real-world performance were also showed. This information makes it possible to determine which circuits are suitable for the required application.

Therefore, this work opens the possibility of several improvements in terms of implementation. In future works, the authors intend to include experiments of these signal conditioning circuits using different species of plants, employing the project developed in this work.

Author Contributions: Conceptualization, M.C. and V.O.; methodology, M.C., V.O. and M.P.; software, M.C. and V.O.; validation, M.C. and V.O.; formal analysis, M.C. and V.O.; investigation, M.C. and V.O.; resources, F.O., M.C., V.O. and M.T.; data curation, M.C. and V.O.; writing—original draft preparation, M.C., V.O. and M.P.; writing—review and editing, M.C., V.O., F.O. and M.P.; visualization, M.P. and M.T.; supervision, M.P., F.O. and M.T.; project administration, F.O. and M.P.; funding acquisition, M.T. All authors have read and agreed to the published version of the manuscript.

Funding: This research received no external funding.

Data Availability Statement: Not applicable.

Acknowledgments: The authors would like to thank UFRJ, CEFET/RJ, UTFPR and the brazilian research agencies CAPES, CNPq, and FAPERJ.

Conflicts of Interest: The authors declare no conflict of interest.

Abbreviations

The following abbreviations are used in this manuscript:

ADC	Analog-to-Digital Converter
APs	Action Potentials
LEPs	Local Electrical Potentials
SNR	Signal-to-Noise Ratio
SPs	Systems Potentials
VPs	Variation Potentials

References

1. Fromm, J.; Lautner, S. Electrical signals and their physiological significance in plants. *Plant Cell Environ.* **2007**, *30*, 249–257. [CrossRef]
2. Davies, E. Electrical signals in plants: Facts and hypotheses. In *Plant Electrophysiology*; Springer: Berlin/Heidelberg, Germany, 2006; pp. 407–422.
3. Mirella, M.D.O.; Pinto, M.F.; Manhães, A.G.; da Graça, U.D.F. Performance Analysis of Digital Filters Employed to Plants Electrical Signals. In Proceedings of the Simpósio Brasileiro de Automação Inteligente-SBAI; Rio Grande, Brazil, 17–20 October 2021; Volume 1.
4. Trebacz, K.; Dziubinska, H.; Krol, E. Electrical signals in long-distance communication in plants. In *Communication in Plants*; Springer: Berlin/Heidelberg, Germany, 2006; pp. 277–290.
5. Hagihara, T.; Toyota, M. Mechanical Signaling in the Sensitive Plant *Mimosa pudica* L. *Plants* **2020**, *9*, 587. [CrossRef]
6. Volkov, A.G.; Lang, R.D.; Volkova-Gugeshashvili, M.I. Electrical signaling in Aloe vera induced by localized thermal stress. *Bioelectrochemistry* **2007**, *71*, 192–197. [CrossRef]
7. Labady, A., Jr.; Shvetsova, T.; Volkov, A.G. Plant bioelectrochemistry: Effects of CCCP on electrical signaling in soybean. *Bioelectrochemistry* **2002**, *57*, 47–53. [CrossRef]
8. Yan, X.; Wang, Z.; Huang, L.; Wang, C.; Hou, R.; Xu, Z.; Qiao, X. Research progress on electrical signals in higher plants. *Prog. Nat. Sci.* **2009**, *19*, 531–541. [CrossRef]
9. Zimmermann, M.R.; Maischak, H.; Mithöfer, A.; Boland, W.; Felle, H.H. System potentials, a novel electrical long-distance apoplastic signal in plants, induced by wounding. *Plant Physiol.* **2009**, *149*, 1593–1600. [CrossRef]
10. Gurovich, L. *Electrophysiology of Woody Plants*; IntechOpen: London, UK, 2012. [CrossRef]
11. de Toledo, G.R.; Parise, A.G.; Simmi, F.Z.; Costa, A.V.; Senko, L.G.; Debono, M.W.; Souza, G.M. Plant electrome: The electrical dimension of plant life. *Theor. Exp. Plant Physiol.* **2019**, *31*, 21–46. [CrossRef]
12. Mirella, M.D.O.; Pinto, M.F.; Manhães, A.G.; dos Reis, M.S. Development of a Low Complexity System to Measure Electrical Signals in Plants. In Proceedings of the Simpósio Brasileiro de Automação Inteligente-SBAI, Rio Grande, Brazil, 17–20 October 2021; Volume 1.
13. Aditya, K.; Freeman, J.D.; Udupa, G. An Intelligent Plant EMG Sensor System for Pre-detection of Environmental Hazards. *Int. J. Sci. Environ. Technol.* **2013**, *2*, 28–37.
14. Shvetsova, T.; Mwesigwa, J.; Labady, A.; Kelly, S.; Lewis, K.; Volkov, A.G. Soybean electrophysiology: Effects of acid rain. *Plant Sci.* **2002**, *162*, 723–731. [CrossRef]
15. Volkov, A.G. Signaling in electrical networks of the Venus flytrap (Dionaea muscipula Ellis). *Bioelectrochemistry* **2019**, *125*, 25–32. [CrossRef]
16. Macedo, F.D.C.O.; Daneluzzi, G.S.; Capelin, D.; da Silva Barbosa, F.; da Silva, A.R.; de Oliveira, R.F. Equipment and protocol for measurement of extracellular electrical signals, gas exchange and turgor pressure in plants. *MethodsX* **2021**, *8*, 101214. [CrossRef] [PubMed]
17. Wang, Z.Y.; Leng, Q.; Huang, L.; Zhao, L.L.; Xu, Z.L.; Hou, R.F.; Wang, C. Monitoring system for electrical signals in plants in the greenhouse and its applications. *Biosyst. Eng.* **2009**, *103*, 1–11. [CrossRef]
18. Wu, H.; Tian, L.G.; Hu, S.; Chen, Z.L.; Li, M. Detection system on weak electrical signal in plants. In *Applied Mechanics and Materials*; Trans Tech Publications Ltd.: Zürich, Switzerland, 2013; Volume 427, pp. 2037–2040.
19. Pranav, S.B.N.; Ganesan, M. Plant signal extraction and Analysis with the influence of sound waves. In Proceedings of the 2020 5th International Conference on Communication and Electronics Systems (ICCES), Coimbatore, India, 10–12 June 2020; pp. 542–547.
20. Cai, W.; Qi, Q. Study on electrophysiological signal monitoring of plant under stress based on integrated Op-Amps and patch electrode. *J. Electr. Comput. Eng.* **2017**, *2017*, 4182546. [CrossRef]
21. Ochiai, T.; Tago, S.; Hayashi, M.; Fujishima, A. Highly sensitive measurement of bio-electric potentials by boron-doped diamond (BDD) electrodes for plant monitoring. *Sensors* **2015**, *15*, 26921–26928. [CrossRef]
22. Gallé, A.; Lautner, S.; Flexas, J.; Fromm, J. Environmental stimuli and physiological responses: The current view on electrical signalling. *Environ. Exp. Bot.* **2015**, *114*, 15–21. [CrossRef]
23. Szechyńska-Hebda, M.; Lewandowska, M.; Karpiński, S. Electrical signaling, photosynthesis and systemic acquired acclimation. *Front. Physiol.* **2017**, *8*, 684. [CrossRef] [PubMed]
24. Vodeneev, V.; Akinchits, E.; Sukhov, V. Variation potential in higher plants: Mechanisms of generation and propagation. *Plant Signal. Behav.* **2015**, *10*, e1057365. [CrossRef] [PubMed]
25. Stahlberg, R.; Cleland, R.E.; Volkenburgh, E.V. Slow wave potentials—A propagating electrical signal unique to higher plants. In *Communication in Plants*; Springer: Berlin/Heidelberg, Germany, 2006; pp. 291–308.
26. Zimmermann, M.R.; Mithöfer, A.; Will, T.; Felle, H.H.; Furch, A.C. Herbivore-triggered electrophysiological reactions: Candidates for systemic signals in higher plants and the challenge of their identification. *Plant Physiol.* **2016**, *170*, 2407–2419. [CrossRef]
27. Choi, W.G.; Hilleary, R.; Swanson, S.J.; Kim, S.H.; Gilroy, S. Rapid, long-distance electrical and calcium signaling in plants. *Annu. Rev. Plant Biol.* **2016**, *67*, 287–307. [CrossRef] [PubMed]
28. Tian, L.; Meng, Q.; Wang, L.; Dong, J.; Wu, H. Research on the effect of electrical signals on growth of Sansevieria under Light-Emitting Diode (LED) lighting environment. *PLoS ONE* **2015**, *10*, e0131838. [CrossRef] [PubMed]

29. Mousavi, S.A.; Chauvin, A.; Pascaud, F.; Kellenberger, S.; Farmer, E.E. Glutamate receptor-like genes mediate leaf-to-leaf wound signalling. *Nature* **2013**, *500*, 422–426. [CrossRef] [PubMed]
30. Zhao, D.J.; Chen, Y.; Wang, Z.Y.; Xue, L.; Mao, T.L.; Liu, Y.M.; Wang, Z.Y.; Huang, L. High-resolution non-contact measurement of the electrical activity of plants in situ using optical recording. *Sci. Rep.* **2015**, *5*, 13425. [CrossRef] [PubMed]
31. Jovanov, E.; Volkov, A.G. Plant electrostimulation and data acquisition. In *Plant Electrophysiology*; Springer: Berlin/Heidelberg, Germany, 2012; pp. 45–67.
32. Joachin, H. Biopotential amplifiers. In *The Biomedical Engineering Handbook*, 2nd ed.; Bronzino, J., Ed.; Taylor & Francis: Oxfordshire, UK; 2000.
33. Zumbahlen, H. *Linear Circuit Design Handbook*; Newnes/Elsevier: Amsterdam, The Netherlands 2011.
34. Ji, N.; Jiang, Y.; Yang, Z.; Jing, X.; Wang, H.; Zheng, Y.; Xia, Z.; Chen, S.; Xu, L.; Li, G. An active electrode design for weak biosignal measurements. In Proceedings of the 2016 IEEE 13th International Conference on Signal Processing (ICSP), Chengdu, China, 6–10 November 2016; pp. 502–507.
35. Cabral, E.F.; Pecora, P.C.; Arce, A.I.C.; Tech, A.R.B.; Costa, E.J.X. The oscillatory bioelectrical signal from plants explained by a simulated electrical model and tested using Lempel–Ziv complexity. *Comput. Electron. Agric.* **2011**, *76*, 1–5. [CrossRef]
36. Lu, J.; Ding, W. Study and evaluation of plant electrical signal processing method. In Proceedings of the 2011 4th International Congress on Image and Signal Processing, Shanghai, China, 15–17 October 2011; Volume 5, pp. 2788–2791.

technologies

Article

Privacy and Explainability: The Effects of Data Protection on Shapley Values

Aso Bozorgpanah [1], Vicenç Torra [1,*] and Laya Aliahmadipour [2]

[1] Department of Computing Science, Umeå University, SE-90185 Umeå, Sweden
[2] Department of Computer Science, Faculty of Mathematics and Computer, Shahid Bahonar University of Kerman, Kerman 7616913439, Iran
* Correspondence: vtorra@cs.umu.se

Abstract: There is an increasing need to provide explainability for machine learning models. There are different alternatives to provide explainability, for example, local and global methods. One of the approaches is based on Shapley values. Privacy is another critical requirement when dealing with sensitive data. Data-driven machine learning models may lead to disclosure. Data privacy provides several methods for ensuring privacy. In this paper, we study how methods for explainability based on Shapley values are affected by privacy methods. We show that some degree of protection still permits to maintain the information of Shapley values for the four machine learning models studied. Experiments seem to indicate that among the four models, Shapley values of linear models are the most affected ones.

Keywords: data protection; masking; anonymization; explainability; machine learning; Shapley values

check for updates

Citation: Bozorgpanah, A.; Torra, V.; Aliahmadipour, L. Privacy and Explainability: The Effects of Data Protection on Shapley Values. *Technologies* **2022**, *10*, 125. https://doi.org/10.3390/technologies10060125

Academic Editors: Manoj Gupta, Eugene Wong and Gwanggil Jeon

Received: 25 October 2022
Accepted: 21 November 2022
Published: 1 December 2022

1. Introduction

The importance of data privacy has increased in recent years. Data are being gathered and stored in huge quantities and then extensively used for profiling and recommendations. This is a threat for individual privacy. People's concern has increased in parallel with this increasing storage and use of data. Legislation has been adapted to take into account new threats. European data protection regulation (GDPR) is one of the initiatives to support individual rights. GDPR not only supports data protection and privacy but also requirements on how decision making affecting people should be done. One of them is the requirement that automated decisions should be explainable and that individuals affected by these decisions can request explanations of these decisions.

Data privacy [1,2] provides tools for data anonymization. These tools typically perturb a data set in a way that the modified data do not lead to disclosure. At the same time, perturbation needs to be performed so that the data are still useful [3–5]. There are different ways to understand disclosure; this has led to different definitions of privacy. Formal definitions of privacy are known as privacy models. Then, a plethora of data protection mechanisms exists providing solutions according to the different privacy models. These methods can be compared in terms of their privacy guarantees but also with respect to the quality of the resulting data. That is, given a data set and a privacy level, some methods behave better than others for a particular data use. A very simple example is the following: if our goal is to compute a mean of the data set, then microaggregation is better than noise addition. This is so because microaggregation will not change the mean of the data, and noise addition can. For a more complex data analysis, similar studies have been performed. This usually corresponds to study how a data protection mechanism, a masking method or anonymization technique is able to produce a machine learning model of good quality.

The need for explainability [6,7] adds a new element in the machine learning process. A machine learning model needs to be good enough with respect to accuracy or prediction error, in terms of selected performance measures. Nevertheless, this is not enough. We need to provide tools to understand the predictions. Several tools have been developed for this purpose.

To interpret the prediction of machine learning models, there are different methods. They are categorized into main categories. We can distinguish between model-specific and model-agnostic and between global and local methods. For example, global methods focus on the average behavior of the model. They are especially helpful when the user wants to comprehend the general mechanism behind the data. In contrast, models' individual predictions are explained through local interpretation techniques. In this paper, we focus on explaining individual prediction. For this, we use local models. There are different local model-agnostic methods. They include the Individual Conditional Expectation (ICE), Local Interpretable Model-agnostic Explanations (LIME), counterfactual explanation, Scoped Rules (Anchors), and Shapley values (e.g., SHapley Additive exPlanations). In this work, we use Shapley values [8,9]. Shapley values were introduced by Shapley in 1953 [10] in game theory. We selected this approach because, in the context of explainability, we build a game for a machine learning model that takes into account the interaction of all features. Then, the Shapley value distributes these interactions among the features in a fair way. The theoretical properties of Shapley values have been extensively studied [10–12]. So, in short, they provide a summary of interactions between features. In addition, Shapley values have been extensively used in the literature on explainability, and it is easy to compare Shapley values corresponding to different models based on data described in the same features.

Explainability poses a threat to privacy. In short, the more we explain in a model and the less opaque it is, the more information we give in the training data set. Similarly, when data are protected, and models are learned from the data, are explanations still valid? Are the explanations going to change? This is an open problem. Note that there are researchers that state that, from a legal perspective, it is impossible to have both privacy and explainability (see Grant and Wischik [13]). This paper tries to provide some initial results about this research question from a technical perspective. In a previous paper [14], some effects of two anonymization methods (microaggregation and noise addition) on importance features were studied. TreeSHAP [9] was used, which is based on tree-based machine learning models. In this paper, we further study this process with extensive experimentation.

The objective of this paper is to better understand how masking methods affect explanations when these explanations are based on Shapley values. We have conducted extensive experiments with a variety of alternatives. For example, we used three different data sets, four different machine learning algorithms, seven masking strategies, each with a large number of parameters, and different analyses of the results based on the Shapley values. Masking methods include well-established anonymization techniques but also a recently introduced method based on non-negative matrix factorization. The paper does not focus on disclosure risk or utility (from a more classical machine learning perspective). These topics have been studied in several papers, as reported in the literature [1,2,15].

Our results show that

- Data protection, through masking, does permit explainability using Shapley values, as they are not significantly affected under moderate protection;
- The use of different machine learning models causes different behaviors in Shapley values. For example, we see that among the methods, linear models are the ones in which Shapley values change the most.

The structure of this paper is as follows. In Section 2, we describe some masking methods we use in this paper. In Section 3, we describe the methodology. In Section 4, we describe the experiments and results. The paper concludes with a summary of our results and some new research directions.

2. Preliminaries

In this section, we review some masking methods for data protection and anonymization and discuss Shapley values as a tool for explainability.

2.1. Masking Methods

Data privacy [1,2,15] provides several methods for data anonymization. They protect a data set by means of modifying it so that sensitive information cannot be disclosed. Masking methods are useful for data publishing, that is, when we need to share data with third parties (e.g., researchers, software engineers, decision makers, etc.) and, particularly, when the data usage is ill-defined or not defined at all. Privacy models for this type of release are k-anonymity [16,17], privacy for re-identification [18,19], and local differential privacy [20,21].

There are three main families of masking methods. Perturbative methods, non-perturbative, and synthetic data generators. Perturbative methods modify the data introducing some kind of error. Noise addition, where a value is replaced by a noisy one, is an example. Rank swapping is another example, in which values are swapped between individuals in order to protect them. In contrast, non-perturbative modifies the data, changing the level of detail but without making it erroneous. For example, replacing a numerical value by an interval, or a town by a county or sets of towns. The interval is more general than the numerical value and, thus, less informative, but there is no error in the information supplied (i.e., the interval). Synthetic data are about replacing the original data by artificial data generated by a model. That is, a machine learning or statistical model is trained with the data, and then the model is used to create artificial data.

In this paper, we used perturbative methods. These methods are preferred to non-perturbative ones because the latter make data processing more complex (e.g., having mixtures of numerical data and intervals, data at different levels of generalization, and sets of values). Synthetic methods are increasingly being used, but we leave them for future work. We discuss below the methods we used in this work.

We use X to denote the original file to be protected, ρ_p to denote a masking method with parameter p, and $X' = \rho_p(X)$, the protected version of X using masking method ρ with parameter p. The following methods are considered in our work.

Microaggregation. This method consists of building small clusters of the original data and then replacing each original record by the cluster center. Protection is achieved by means of controlling the minimum number of records in a cluster. This corresponds to the parameter k. The larger the k, the larger the protection and the larger the distortion. Microaggregation has been proven to provide a good trade-off between privacy and utility. We used two methods of microaggregation: MDAV [22,23] and Mondrian [24]. That is, two different ways of building the clusters.

Noise addition. This method replaces each numerical value x by $x + \epsilon$, where ϵ follows a given distribution. We use two types of distributions: a normal distribution with mean zero and standard deviation $\sqrt{(variance * k)}$ and a Laplace distribution with mean zero and standard deviation as above. Here, k is the parameter. The larger the k, the larger the protection and the larger the distortion.

SVD. We apply a singular value decomposition to the file, and then rebuild the matrix but only with some of the components. The number of components is a parameter of the system. We use k to denote this parameter. The smaller the number of components, the larger the distortion and larger the privacy.

PCA. This is similar to the previous method using principal components. We use k to denote the number of components. Therefore, the smaller the k and the number of components, the larger the protection and distortion.

NMF. This approach corresponds to non-negative matrix factorization [25]. The first use of NMF in data privacy seems to be by Wang et al. [26]. Our approach follows

Algorithm 1, and it is based on the implementation of one of the authors [27]. Again, the smaller the number of components k, the larger the privacy. NMF needs the data to be positive, thus, data are scaled into [0,1] before the application of NMF.

Algorithm 1: Algorithm for masking data using NMF. Here, X is the original file with N records and $|V|$ attributes. Protected files $X'^1, \ldots, X'^K$ are produced.

Input: $X = [x_1, \ldots, x_N] \in \mathbb{R}^{|V| \times N}$; K: maximum rank to consider

Output: $\mathcal{A} = \{X'_k | k \in 1, \ldots, K\}$, a family of masked data sets

Step 1. For all ranks $k \in 1, \ldots, K$
apply $NMF(X, k)$ and find matrices W^k and H^k

Step 2. For all ranks $k \in 1, \ldots, K$, do

Step 2.1. For each record $j = 1, \ldots, N$
construct masked data vectors a_j^k as follows:

$$a_j^k := \sum_{l=1}^{k} H_{lj}^k W_l^k \in \mathbb{R}^{|V|},$$

Step 2.2. Define the masked matrix X'^k as:

$$X'^k = [a_j^k]_{j=1,\ldots,N}.$$

We mentioned above three privacy models related to data sharing. We briefly review these methods and discuss the relationship of the above methods with the privacy models.

Privacy for re-identification is about avoiding identity disclosure. That is, avoiding intruders finding records in the published database. If intruders have information on a particular person (e.g., a record x), then they will try to find x in the protected file X'. As data are protected, x will not appear as such in X'. So, intruders will try to guess which record x' in X' corresponds to x. For example, selecting the most similar record $x' = \arg\max_{x' \in X'} d(x', x)$. All masking methods are defined to provide privacy for re-identification. Different parameters provide different guarantees. i.e., the larger the distortion, the stronger the guarantee.

Another privacy model is k-anonymity. The goal of this privacy model is to hide a record (or individual) in a set of indistinguishable records (or individuals). A file X' satisfies k-anonymity (for a given set of features) when, for each combination of values of the features, we have at least k indistinguishable records. Microaggregation is one of the tools to provide k-anonymity. When we force clusters to have at least k-records, and we replace each record by the cluster centers, we will have that there will be for each combination k indistinguishable records.

Differential privacy is a privacy model focusing on computations. Given a function f and a database X, the goal is to produce a value $f(X)$ that does not depend on particular records in X. More formally, a function K_f satisfies differential privacy when the result of $K_f(X)$ is very similar to $K_f(X_1)$, where X and X_1 differ on a single record. The definition presumes that the function K_f is a randomized version of f, and then very similar is understood in terms of the similarity between the distribution functions on the space of possible outputs. Local differential privacy is a variation of differential privacy that is appropriate for databases. In this case, individual records are protected independently, with each feature also protected independently. There are different mechanisms to provide differential privacy. The use of Laplacian noise is usual for numerical data. Randomized response (which is equivalent to PRAM) is usual for categorical data. Among the methods discussed above, noise addition with a Laplace distribution is the one that can provide differential privacy. The larger the noise, the larger the privacy guarantees in differential privacy.

2.2. Shapley-Value-Based Explainability

The use of Shapley values as a tool for explainability was introduced by Lundberg and Lee [8]. The motivation is to use game theory machinery [28] as the basis of explanation. A game is a set function defined on a reference set.

In our context, explanations are values for the features expressing their relevance to the outcome of instances (i.e., the columns or attributes of our records). Let us consider some notation. Let x be a record in a data set X and a model ML built from our training data set X. Then, $ML(x)$ is the prediction of our model. We consider that X is defined in terms of the features, attributes or variables V.

Then, the game is a set function on sets of features. That is, we consider a subset of features $A \subset V$ and define for $x \in X$ a function $\mu_x(A)$. To compute the $\mu_x(A)$, we consider the output of our model ML if we only knew the attributes in A; for the others, we just have "don't know", or e.g., the mean value of the database. Then, $\mu_X(A)$ is the difference between this output and the mean output.

Game theory provides a tool to determine the importance of each feature for a given game. This is known as the Shapley value. In short, given a game μ on the reference set V, its Shapley value is a function that assigns to each feature in V a value in [0,1]. In addition, the addition of all Shapley values is equal to one. These properties hold when the game is positive and normalized. This is not the case here. We may have negative values because $\mu_x(A)$ is a difference that can be negative (the output of a prediction can be smaller than the mean output), and, naturally, is not normalized. Nevertheless, the Shapley values are still useful because they gives a magnitude of the importance of each feature. We have features with positive Shapley values and features with negative Shapley values. The former mean that the feature has a positive influence in the outcome of the model, and the latter represent a negative influence. Then, larger values (in absolute terms) represent larger influence in the outcome. In this way, we know the relevance of features on computing the outcome of a model for a given instance x.

3. Methodology

We implemented the process described in more detail below. It mainly consists of producing different alternative protected files. For a given protected file, we computed a machine learning model, and then for the pair (protected data and machine learning model), we used some records to compute its explanation in terms of the Shapley value. Shapley values obtained through the masked file and through the original file are compared. Different ways of comparison were used. In this way, we can analyze the effects of masking on the Shapley values.

We detail now the methodology for an original data file X. We describe in Section 4 the three actual data sets used in our experiments. The process is described for a particular machine learning algorithm. We use $ML := A(X)$ to denote that ML is the machine learning model trained from data X using algorithm A and use $ML(x)$ to denote the outcome of the model when applied to record x (and all features in x are used). We use $ML^S(x)$ to denote the outcome of the model when applied to record x, and only the features in S are used. The actual 4 machine learning algorithms used in our experiments are also described in Section 4. A summary of the notation used in this section is given in Table 1.

Table 1. Notation used.

Notation	Explanation
X	Data file
X^{te}	Test data set
X^{tr}	Training data set
ρ_p	Masking method ρ with parameter p
A	Machine learning algorithm
ML_o	Machine learning model from original data
ML_{ρ_p}	Machine learning model from masked data using ρ_p
ML^S	Machine learning model that uses as input only attributes in S
$\phi_{ML}(x)$	Shapley value of a machine learning model ML for an instance/ record x
$\bar{\phi}_{ML,X}$	Mean Shapley value of a machine learning model ML for all instances/ records x in X

The methodology is described below. We consider different masking methods ρ with parameters p_ρ. We use the notation p_ρ because parameters depend on the method. When clear, we just use p for the parameters for the sake of conciseness.

- Split the data set X in training X^{tr} and testing X^{te}.
- Define $ML_o := A(X^{tr})$ as the machine learning model learned from the original data.
- For each $x \in X^{te}$, define its game $\mu_{ML_o,x}$ according to the existing literature. Formally, for a set of features S, we define $\mu_{ML_o,x}(S) = ML_o^S(x) - ML_o^{\emptyset}(x)$. Then, compute the Shapley value $\phi_{ML_o}(x)$ of this game. Use all records in X^{te} to compute the mean Shapley value. We obtain a mean Shapley value for each masking method and parameter. That is, $\bar{\phi}_{ML_o,X^{te}}$.
- Produce $X_{\rho_p} = \rho_p(X^{tr})$ for each pair masking method ρ and parameter p_ρ.
- Produce the corresponding machine learning model $ML_{\rho_p} := A(X_{\rho_p})$.
- For each $x \in X^{te}$, compute the games and the corresponding Shapley values associated to models ML_{ρ_p}. We denote them by $\mu_{ML_{\rho_p},x}$ and $\phi_{ML_{\rho_p}}(x)$ for each $x \in X^{te}$. Use all records in X^{te} to compute the mean Shapley value $\bar{\phi}_{ML_{\rho_p},X^{te}}$.
- The following comparisons are considered:
  - Compare the mean Shapley of the original and masked files using the Euclidean distance. That is, $||\bar{\phi}_{ML_o,X^{te}} - \bar{\phi}_{ML_{\rho_p},X^{te}}||$.
  - Compare the mean Shapley of the original and masked files using Spearman's rank correlation.
  - Compare the Shapley values for each x using the Euclidean distance, and then compute the average distance. Formally, this corresponds to:

$$\frac{\sum_{x \in X^{te}} ||\phi_{ML_o}(x) - \phi_{ML_{\rho_p}}(x)||}{|X^{te}|}$$

  - Compare the Shapley values for each x using Spearman's rank coefficient.

We considered four different comparisons, because we consider that they provide different types of information. The use of mean Shapley values gives information on a global level. Mean Shapley values permit us to know which are the most relevant features in general terms. So, we can observe if these important features are changed because of data protection. Nevertheless, important features in general terms do not need to coincide with the relevant features for a particular example. When the machine learning models are non-linear, this is not necessarily the case. That is why it is also relevant to see if masking data causes changes at the local level. This can be observed with a direct comparison of the Shapley values for $x \in X^{te}$ and then averaging these comparisons.

We used the Euclidean distance to compare the Shapley values but also the Spearman rank coefficient. The Shapley values are numerical values, but from the point of view of relevant attributes, the relative order is what matters. We used the Spearman rank coefficient because it only takes into account the relative position and not the values themselves.

4. Experiments and Analysis

In this section, we detail the experiments we have conducted and discuss the results.

4.1. Implementation

Our experiments were conducted in Python. We have our own implementation of the masking methods. We used the `sklearn` package for machine learning. That is, to find machine learning models from training data. We have our own implementation for computing games and for computing the Shapley value of these games. The Spearman rank correlation coefficient is from the `scipy` package. Code is available here: [29].

4.2. Parameters

We considered the following parameters for the masking methods described above. In practice, parameter selection depends on the privacy requirements and data utility requirements. For microaggregation, a value of k around 5 is used. Noise addition requires values that depend on the available data and their sensitivity (when implementing differential privacy). PCA and SVD parameters close to the number of features may imply low levels of privacy.

- Microaggregation. As explained above, we considered two different microaggregation algorithms: MDAV and Mondrian. The difference in the algorithms is in how clusters are built. For both algorithms, the cluster centers are defined in terms of the means of the associated records. The following values of the parameter k were used: $k = \{2,3,4,5,6,7,8,9,10,12,14,15,16,18,20\}$.
- Noise addition. We considered Normal and Laplacian distribution. The following values of k were used: $k = \{0.2, 0.3, 0.4, 0.5, 0.6, 0.7, 0.8, 0.9, 1.0, 1.2, 1.4, 1.5\}$.
- SVD. We considered the singular value decomposition and the reconstruction of the matrix using different number of values. In our experiments, we considered $k = \{2,3,4,5,6,7,8\}$.
- PCA. As in the case of SVD, we considered $k = \{2,3,4,5,6,7,8\}$.
- NMF. The selection of the parameter that approximates well the matrix is a difficult problem [30]. We considered here a different number of components in the factorization. We used $k = \{2,3,4,5,6,7,8\}$.

4.3. Data Sets and Machine Learning Algorithms

We applied our method to the following data sets. They were selected because they are well-known in the literature and used before in both machine learning as well as data privacy [31] research. Only numerical data were considered. Data are available in the UCI repository [32] and in the `sklearn` Python library. We leave non-numerical data for future work.

- Tarragona. This data set contains 834 records described in terms of 13 attributes. We used the first 12 attributes as the independent ones and the 13th attribute (last column in the file) as the dependent one.
- Diabetes. This data set contains 442 records with information on 10 attributes. An additional numerical attribute is also included in the data set, for prediction.
- Iris. This data set contains 150 records described in terms of 4 attributes and a class (which corresponds to a fifth attribute). We used the 4 attributes as the independent variables, used the class as a numerical value, and used one as a numerical dependent.

4.4. Machine Learning Algorithms

We considered different machine learning algorithms, supplied by `sklearn`. In particular, we considered the methods (used in all cases with default parameters)

- `linear_model.LinearRegression` (linear regression);
- `sklearn.linear_model.SGDRegressor` (linear model implemented with stochastic gradient descent);
- `sklearn.kernel_ridge.KernelRidge` (linear least squares with l2-norm regularization, with the kernel trick);
- `sklearn.svm.SVR` (Epsilon-Support Vector Regression).

These algorithms were applied using dependent and independent attributes, as described in the previous section. The standard versions of these algorithms were used.

4.5. Results

An analysis of the results leads to the following conclusions.

The first observation is that both the mean distance between Shapley values and the distance between Shapley values can be very large. Note that when the game is defined for a particular machine learning algorithm, the game is unbounded and depends on the values of the prediction. That is, the value of the game for a set may be very large if the prediction is large. Because of this, the Shapley values can be large and, thus, the distance between two Shapley values can also be large. This makes comparisons cumbersome. This is illustrated in Figure 1, which shows (left) the distances for the Tarragona data set and (right) the distances for the Diabetes data set. It is not so easy to compare the scales of the two figures. Moreover, considering 11 or 12 independent inputs (left and middle figures) changes the scale. In contrast, the rank correlation is always in the $[-1,1]$ interval, which makes comparisons easier. This is illustrated in Figure 2.

These figures also show that larger distances do not mean larger rank correlation. That is, the the distances between Shapley values do not mean that the order of these values are changed so much. Observe that, for the set Diabetes, in Figure 1, Mondrian give larger distances than MDAV (i.e., curves lo_dm and lo_md have larger values than curves ld_dm and ld_md). That is, MDAV seems to behave better with larger amounts of noise. In contrast, in Figure 2, it is MDAV which shows a worse performance, as Mondrian has a rank correlation near to 1 for larger parameters. The set Tarragona seems to have a more erratic behavior on the distances and rank correlations with respect to the parameters but is more consistent if we compare Figures 1 and 2. It can also be seen that when considering more input attributes, the curves seem to have a better shape. Compare left and middle curves in these figures, where the distances are smaller and correlations are larger.

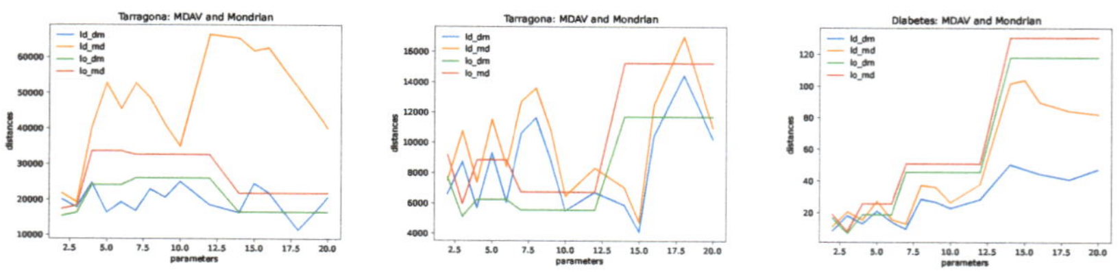

Figure 1. Distance of mean Shapley values (_dm) and mean distance of Shapley values (_md) for MDAV and Mondrian (letters d and o) using linear regression as the machine learning algorithm. Experiments with the Tarragona file were performed considered only the first 11 inputs (**left**), all 12 independent inputs (**middle**), and the Diabetes file (**right**).

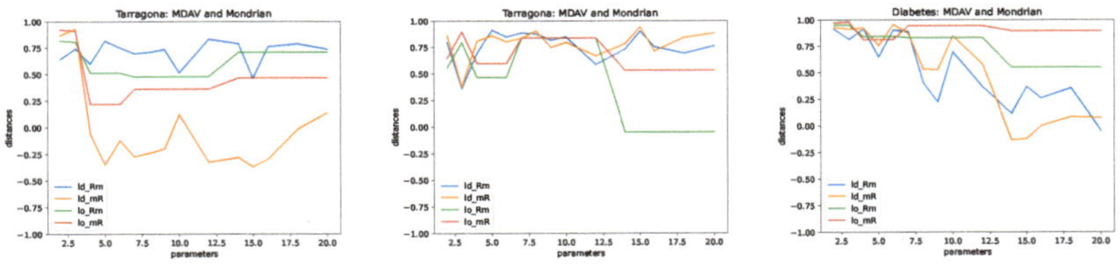

Figure 2. Rank correlation of mean Shapley values (_Rm) and mean correlation of Shapley values (_mR) for MDAV and Mondrian (letters d and o) using linear regression as the machine learning algorithm. Experiments with the Tarragona file were performed considered only the first 11 inputs (**left**), all 12 independent inputs (**middle**), and the Diabetes file (**right**).

Now, we show that we obtain similar changes on the rank correlation independently of the machine learning method used. Figure 3 includes the results for MDAV (left) and Mondrian (right). We compare the mean rank correlation of all Shapley values computed using the four different machine learning algorithms considered in the paper. We can see that the results are quite similar, except for the case of linear regression and MDAV. The scale of the figure was set to [0.75,1] to better visualize the results.

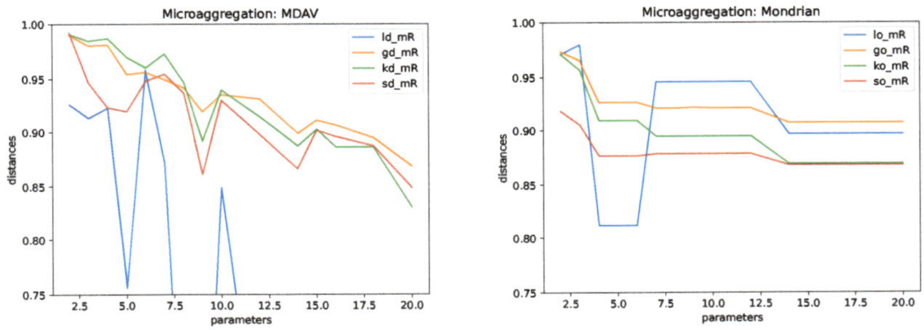

Figure 3. Rank correlation of mean Shapley values (_Rm) for MDAV (**left**) and Mondrian (**right**) (letters d and o) using linear regression (letter l), SGD Regressor (letter g), Kernel Ridge (letter k), and SVM (letter s). That is, ld_mR reads for linear regression as the machine learning method for data protected using MDAV and the curve corresponding to mean rank correlation. Computations for the Diabetes file.

This similar behavior appears also with other masking methods. In Figure 4, we have the case of noise addition, with both types of noise (Gaussian noise and Laplacian noise). It is interesting to underline that the linear model is the one that has a larger effect on the rank correlation, and as it can be seen in the figure for microaggregation, it also happens in MDAV. In fact, the same behavior is also reproduced for protection with SVD, PCA, and NMF. Figure 5 includes the curves for PCA and NMF. The one for SVD is not included, but the resulting figure is almost the same as the one for PCA. It is relevant to underline that the parameters of SVD, PCA, and NMF are a kind of reversal to the ones of microaggregation and noise. That is, the smaller the parameter k, the larger the protection. That is why the curves in Figure 5 are increasing instead of decreasing.

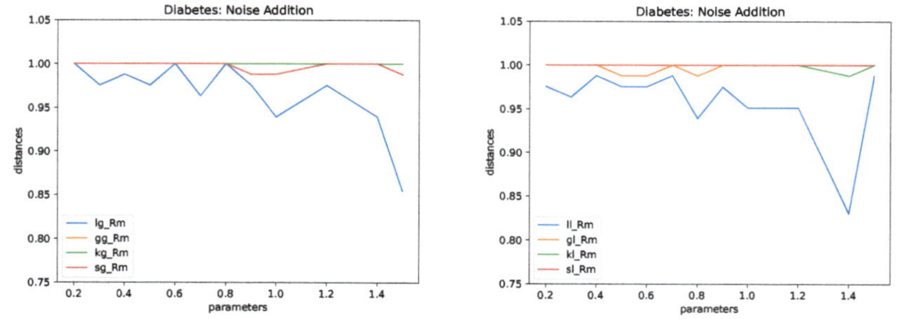

Figure 4. Rank correlation of mean Shapley values (_Rm) for noise addition for Gaussian noise (**left**) and Laplacian noise (**right**) considering the four types of machine learning models: linear regression (letter l), SGD Regressor (letter g), Kernel Ridge (letter k), and SVM (letter s). Computations for the Diabetes file.

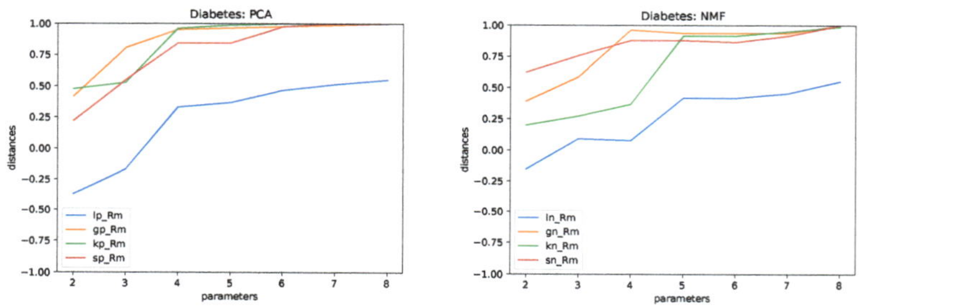

Figure 5. Rank correlation of mean Shapley values (_Rm) for data protected using PCA (**left**) and NMF (**right**) considering the four types of machine learning models: linear regression (letter l), SGD Regressor (letter g), Kernel Ridge (letter k), and SVM (letter s). Computations for the Diabetes file.

The figures discussed so far correspond to the Tarragona and Diabetes files. The results for the Iris data set are consistent with the findings of these two files, although the curves have additional noise. We consider that this is due to the fact that the data file is smaller, and the effects of the same amount of masking on the machine learning models are larger. This affects the rank correlation of the Shapley value of the variables. Compare Figure 6 with the results of masking with Mondrian and PCA for the Iris data set and Figure 3 (left, Mondrian for Diabetes) and Figure 5 (right, PCA for Diabetes).

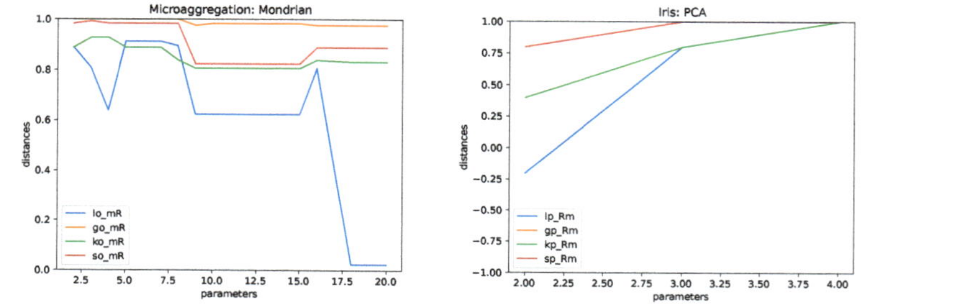

Figure 6. Rank correlation of mean Shapley values (_Rm) for Microaggregation (Mondrian) and PCA. The four types of machine learning models considered are linear regression (letter l), SGD Regressor (letter g), Kernel Ridge (letter k), and SVM (letter s). Computations for the diabetes file.

5. Conclusions

There is an increasing need for explainability in the context of machine learning models and automated decisions. Nevertheless, machine learning models and automated decisions need to be compliant with privacy requirements. At present, there is no clear understanding of how explainability and privacy are incompatible, or if some levels of explainability are possible when privacy guarantees are ensured. There are claims [13] that having both is impossible. This work studied this problem in a particular scenario.

More particularly, we studied the effect of machine learning algorithms on explainability, when the latter are implemented in terms of the Shapley value. That is, we studied how masking affects Shapley values. Different analyses were performed: one based on differences in the Shapley values and another based on rank correlation of these Shapley values.

These results seem to indicate that protection does not prevent explainability when this is implemented using Shapley values. That is, that under some assumptions, explainability and privacy are not incompatible. We saw that the results based on rank correlation have a sounder behavior (they change more smoothly with respect to protection) and have a similar behavior for different machine learning models than the results based on the difference of the values (difference computed in terms of the norm). In this case, the fact that rank correlation is better than the norm means that what seems to be relevant is the order of the variables with respect to the Shapley values and not the values themselves.

The analysis has also shown that among the four machine learning models, the linear model is the one that has the worst performance with respect to the Shapley value. That is, the relevance of the features changes the most. This seems to be a constant independent of the masking method applied to the data.

It is important to note that tools for explainability [6,7] are to be used by humans when decisions are being automated. Then, the study of explainability is incomplete without the user perspective. This also applies here. We considered and compared the results of the Shapley values, but we did not perform any user study on what users can consider relevant in this setting. In our analysis, we considered all Shapley values; future work may consider the most significant Shapley values. Note that in our context, the most significant seem to be the larger ones in absolute value, as the game can take negative values.

In this study, we focused on numerical data files of a relatively small size. The computational requirements of the analysis become challenging for larger files. The results seem to indicate that the larger the file, the more robust the results of Shapley. We plan to study if this is the case. In addition, we plan to further analyze local effects. We considered Shapley because it is good as a way to evaluate local explainability. For large data sets, it is difficult to analyze and compare these local results. We need to study these local effects in large data sets along with other criteria.

In this paper, we studied the effects of masking into explainability when the latter is expressed in terms of Shapley values. We showed that explainability is not incompatible with privacy for this limited scenario. We plan to extend this work considering other tools related to explainability as, for example, logic-based explanations.

Author Contributions: Conceptualization, A.B. and V.T.; software, V.T. (NMF-based masking by L.A.); validation, A.B. and L.A.; writing—original draft preparation, V.T., with contributions by A.B. and L.A.; writing—review and editing, A.B., L. A. and V.T.; funding acquisition, V.T. All authors have read and agreed to the published version of the manuscript.

Funding: This study was partially funded by the Wallenberg AI, Autonomous Systems and Software Program (WASP) funded by the Knut and Alice Wallenberg Foundation.

Institutional Review Board Statement: Not applicable.

Informed Consent Statement: Not applicable.

Data Availability Statement: We used publicly available data. References were supplied.

Conflicts of Interest: The authors declare no conflict of interest.

References

1. Hundepool, A.; Domingo-Ferrer, J.; Franconi, L.; Giessing, S.; Nordholt, E.S.; Spicer, K.; de Wolf, P.-P. *Statistical Disclosure Control*; Wiley: Hoboken, NJ, USA, 2012.
2. Torra, V. *Data Privacy: Foundations, New Developments and the Big Data Challenge*; Springer: Berlin/Heidelberg, Germany, 2017.
3. Abowd, J.; Ashmead, R.; Cumings-Menon, R.; Garfinkel, S.; Kifer, D.; Leclerc, P.; Sexton, W.; Simpson, A.; Task, C.; Zhuravlev, P. An Uncertainty Principle Is a Price of Privacy-Preserving Microdata. In Proceedings of the 35th Conference on Neural Information Processing Systems, Virtual, 6–14 December 2021.
4. Pastore, A.; Gastpar, M.C. Locally differentially-private randomized response for discrete distribution learning. *J. Mach. Learn. Res.* **2021**, *22*, 1–56.
5. Reimherr, M.; Awan, J. Elliptical Perturbations for Differential Privacy. In Proceedings of the NeurIPS 2019, Vancouver, BC, Canada, 8–14 December 2019.
6. Riveiro, M.; Thill, S. The challenges of providing explanations of AI systems when they do not behave like users expect. In Proceedings of the 30th ACM Conference on User Modeling, Adaptation and Personalization, Barcelona, Spain, 4–7 July 2022; pp. 110–120.
7. Riveiro, M.; Thill, S. "That's (not) the output I expected!" On the role of end user expectations in creating explanations of AI systems. *Artif. Intell.* **2021**, **298**, 103507. [CrossRef]
8. Lundberg, S.M.; Lee, S.I. A unified approach to interpreting model predictions. In Proceedings of the NeurIPS 30, Long Beach, CA, USA, 4–9 December 2017.
9. Lundberg, S.M.; Erion, G.; Chen, H.; DeGrave, A.; Prutkin, O.M.; Nair, B.; Katz, R.; Himmelfarb, J.; Bansal, N.; Lee, S.-I. Explainable AI for Trees: From Local Explanations to Global Understanding. *arXiv* **2019**, arXiv:1905.04610.
10. Shapley, L. A value for *n*-person games. *Ann. Math. Stud.* **1953**, *28*, 307–317.
11. Dubey, P. On the uniqueness of the Shapley value. *Int. J. Game Theory* **1975**, *4*, 131–140. [CrossRef]
12. Roth, A.E. (Ed.) *The Shapley Value*; Cambridge University Press: Cambridge, MA, USA, 1988.
13. Grant, T.D.; Wischik, D.J. Show Us the Data: Privacy, Explainability, and Why the Law Can't Have Both. *Geo. Wash. L. Rev.* **2020**, *88*, 1350.
14. Bozorgpanah, A.; Torra, V. Explainable machine learning models with privacy. 2021, manuscript.
15. Torra, V. *A Guide to Data Privacy*; Springer: Berlin/Heidelberg, Germany, 2022.
16. Samarati, P. Protecting Respondents' Identities in Microdata Release. *IEEE Trans. Knowl. Data Eng.* **2001**, *13*, 1010–1027. [CrossRef]
17. Samarati, P.; Sweeney, L. *Protecting Privacy When Disclosing Information: k-Anonymity and Its Enforcement through Generalization and Suppression*; SRI International Technical Report; 1998. Available online: https://epic.org/wp-content/uploads/privacy/reidentification/Samarati_Sweeney_paper.pdf (accessed on 23 September 2022).
18. Jaro, M.A. Advances in record-linkage methodology as applied to matching the 1985 Census of Tampa, Florida. *J. Am. Stat. Assoc.* **1989**, *84*, 414–420. [CrossRef]
19. Winkler, W.E. Re-identification methods for masked microdata, PSD 2004. *Lect. Notes Comput. Sci.* **2004**, *3050*, 216–230.
20. Evfimievski, A.; Gehrke, J.; Srikant, R. Limiting privacy breaches in privacy preserving data mining. In Proceedings of the Twenty-Second ACM SIGMOD-SIGACT-SIGART Symposium on Principles of Database Systems, San Diego, CA, USA, 9–12 June 2003.
21. Kasiviswanathan, S.P.; Lee, H.K.; Nissim, K.; Raskhodnikova, S.; Smith, A. What can we learn privately? In Proceedings of the Annual Symposium on Foundations of Computer Science, Washington, DC, USA, 25–28 October 2008.
22. Domingo-Ferrer, J.; Mateo-Sanz, J.M. Practical data-oriented microaggregation for statistical disclosure control. *IEEE Trans. Knowl. Data Eng.* **2002**, *14*, 189–201. [CrossRef] [PubMed]
23. Domingo-Ferrer, J.; Martinez-Balleste, A.; Mateo-Sanz, J.M.; Sebe, F. Efficient Multivariate Data-Oriented Microaggregation. *Int. J. Very Large Databases* **2006**, *15*, 355–369. [CrossRef]
24. LeFevre, K.; DeWitt, D.J.; Ramakrishnan, R. *Multidimensional k-Anonymity*; Technical Report 1521; University of Wisconsin: Madison, WI, USA, 2005.
25. Lee, D.D.; Seung, H.S. Learning the parts of objects by non-negative matrix factorization. *Nat. Vol.* **1999**, *401*, 788–791. [CrossRef] [PubMed]
26. Wang, J.; Zhang, J. NNMF-based factorization techniques for high-accuracy privacy protection on non-negative-valued datasets. In Proceedings of the Sixth IEEE International Conference on Data Mining—Workshops (ICDMW'06), Hong Kong, China, 18–22 December 2006.
27. Aliahmadipour, L.; Valipour, E. A new method for preserving data privacy based on the non-negative matrix factorization clustering. *Fuzzy Syst. Its Appl.* **2022**. (In Persian) [CrossRef]
28. Myerson, R.B. *Game Theory*; Harvard University Press: Cambridge, MA, USA, 1991.
29. Code Python of Our Software Available online: www.mdai.cat/code (accessed on 23 September 2022).
30. Berry, M.W.; Browne, M.; Langville, A.M.; Pauca, V.P.; Plemmons, R.J. Algorithms and applications for approximate nonnegative matrix factorization. *Comput. Stat. Data Anal.* **2007**, *52*, 155–173. [CrossRef]

31. Brand, R.; Domingo-Ferrer, J.; Mateo-Sanz, J.M. *Reference Datasets to Test and Compare SDC Methods for Protection of Numerical Microdata*; Technical Report; European Project IST-2000-25069 CASC; 2002. Available online: Available online: https://research.cbs.nl/casc/CASCrefmicrodata.pdf (accessed on 23 September 2022).
32. Dua, D.; Graff, C. *UCI Machine Learning Repository*; University of California, School of Information and Computer Science: Irvine, CA, USA, 2019. Available online: http://archive.ics.uci.edu/ml (accessed on 23 September 2022).

technologies

MDPI

Article

Modelling the Trust Value for Human Agents Based on Real-Time Human States in Human-Autonomous Teaming Systems

Chin-Teng Lin [1,2,*], Hsiu-Yu Fan [2], Yu-Cheng Chang [1], Liang Ou [1], Jia Liu [1], Yu-Kai Wang [1] and Tzyy-Ping Jung [3]

1 CIBCI Lab, Australian Artificial Intelligence Institute, University of Technology Sydney, Ultimo, NSW 2007, Australia
2 Institute of Imaging and Biomedical Photonics, National Yang Ming Chiao Tung University, Hsinchu City 30010, Taiwan
3 Institute of Engineering in Medicine and Institute for Neural Computation, University of California San Diego, La Jolla, CA 92093, USA
* Correspondence: chin-teng.lin@uts.edu.au; Tel.: +61-2-95142000

Abstract: The modelling of trust values on agents is broadly considered fundamental for decision-making in human-autonomous teaming (HAT) systems. Compared to the evaluation of trust values for robotic agents, estimating human trust is more challenging due to trust miscalibration issues, including undertrust and overtrust problems. From a subjective perception, human trust could be altered along with dynamic human cognitive states, which makes trust values hard to calibrate properly. Thus, in an attempt to capture the dynamics of human trust, the present study evaluated the dynamic nature of trust for human agents through real-time multievidence measures, including human states of attention, stress and perception abilities. The proposed multievidence human trust model applied an adaptive fusion method based on fuzzy reinforcement learning to fuse multievidence from eye trackers, heart rate monitors and human awareness. In addition, fuzzy reinforcement learning was applied to generate rewards via a fuzzy logic inference process that has tolerance for uncertainty in human physiological signals. The results of robot simulation suggest that the proposed trust model can generate reliable human trust values based on real-time cognitive states in the process of ongoing tasks. Moreover, the human-autonomous team with the proposed trust model improved the system efficiency by over 50% compared to the team with only autonomous agents. These results may demonstrate that the proposed model could provide insight into the real-time adaptation of HAT systems based on human states and, thus, might help develop new ways to enhance future HAT systems better.

Keywords: trust modelling; information fusion; human-autonomous teaming

check for updates

Citation: Lin, C.-T.; Fan, H.-Y.; Chang, Y.-C.; Ou, L.; Liu, J.; Wang, Y.-K.; Jung, T.-P. Modelling the Trust Value for Human Agents Based on Real-Time Human States in Human-Autonomous Teaming Systems. *Technologies* **2022**, *10*, 115. https://doi.org/10.3390/technologies10060115

Academic Editor: Mohammed Mahmoud

Received: 16 September 2022
Accepted: 30 October 2022
Published: 8 November 2022

1. Introduction

The emerging cooperation of artificial intelligence and advanced automation systems provides an opportunity to ease the requirements of human labor and minimise risk in various tasks. In many instances, human and autonomous agents are coupled in a human-autonomous teaming (HAT) system to address complex problems where the tasks could be either unreachable or dangerous for humans or not suitable for autonomous agents with conventional automation [1–5]. Such problems often contain a series of factors that can easily cause mistakes and result in a high cost, including, but not limited to, navigation, patrolling, medical health insurance, rescue and scientific research [6–9].

As a critical factor in coordinating agents or allocating tasks, the evaluation of trust values for human agents becomes an essential issue in the cooperation of human and autonomous agents [10]. Previous studies proposed trust-based approaches to explore either human or teammate trust for the optimization of interactions among agents in specified tasks [5,11–20]. The trust in autonomous agents can be well-modelled based on their previous experience, states, and actions, where humans can judge the trustworthiness of autonomous agents by observing their actions, e.g., whether they can act as expected. Additionally, measuring human cognitive states may benefit the identification of under what circumstances and contexts autonomous agents' performance can be higher or lower than expected [21,22]. However, it is challenging to fairly measure an individual's states, as cognitive states, such as mental stress and attention, are easily affected by human behaviours, which could cause human cognitive states to change from time to time[5,23]. Therefore, in an attempt to properly evaluate the trustworthiness of human agents, this study captured the human state in real-time and investigated the distributed human trust dynamics in an HAT system. We considered human trust to be affected by human psychological state and situational awareness as factors that indicated individuals' bias when making decisions during human-autonomous interactions. This definition aligns with the concept proposed by Guo et al. [24] and Azevedo-Sa et al. [25].

In this study, we introduced a fusion mechanism in the proposed trust model to estimate human trust values by fusing multiple pieces of information from human agents. To obtain an adaptive fusion mechanism for the human trust model, we leveraged a reinforcement learning (RL) algorithm to learn fusion weights from an external reward via a simulation-based training process. One advantage of using RL is that it can learn without prior knowledge, which avoids bias based on forepassed data [26,27]. However, a mathematical equation to describe reward values is still difficult to define for a system with multiple sources with RL. Moreover, uncertainty and noise are additional issues, as external or ineffective information could confuse rewards with reinforcement learning. To overcome these issues, we applied the fuzzy inference system (FIS) in our model. FIS is well known as an effective method for generating rewards for complex scenarios and deal with uncertainty from the environment. Several recent studies present the implementation of FIS-based reward structures for different complex scenarios [28–30]. Evidence shows that with the aid of its member functions and If-Then-Rule structure, FIS has an inherent capability to overcome uncertainty and noise from the environment [23,31–33]. Therefore, we used FIS in this study to generate rewards for the proposed trust model to overcome the above issues.

To verify the effectiveness of the proposed trust model, we use a robot simulator to design a ball collection task scenario that includes an HAT team working together. The HAT team involves a human agent who has to cooperate with one or two robot agents to collect balls with collision-free movements while performing the task. The human agent's sight is restricted; the environment can only be observed through a fixed monitoring camera in the simulator. Robot agents can determine whether they follow human commands or not, based on the human trust values evaluated by the proposed trust model. We used a training scenario to learn the fusion method with the Q-learning algorithm and tested it in three test scenarios with different settings. We further compare the performance of the HAT, only human agents, and only robot agents. The comparison results demonstrate that the proposed trust model can improve coordination in the HAT teams with different human participants in all test scenarios, which also suggests that the proposed model can adapt to various levels of human performance and generate reliable trust values via the Q-learning algorithm.

The main contributions of this research are three-fold:

- This paper proposes a trust model to estimate human trust value in real-time. The proposed trust model was applied to a ball collection task with robot agents, which presents uses of the proposed trust model in the human-autonomous teaming framework.
- The proposed trust model considers multiple pieces of information from a human agent, e.g., attention level, stress index and situational awareness, by leveraging a fuzzy fusion model. In this research, the attention level and the stress index are

evaluated based on pupil response and heart rate variability, respectively; situational awareness is measured from the environment through visual perception.

- We further use a Q-learning algorithm with a fuzzy reward to adaptively learn the fusion weight of the fusion model. The fuzzy reward is generated by a TSK-type fuzzy inference system, which facilitates the defending reward for complex scenarios and is able to handle the uncertainty of human information.

This paper is organised as follows. Section 2 introduces related work on human trust modelling. Section 3 describes the proposed Multi-Human-Evidence Based Trust Evaluation Model. This section first describes the details of trust evaluation metrics, followed by the details of the Trust Metric Fusion Model and the Reinforcement Learning Algorithm. Section 4 presents the experimental methods, including scenario design, human agent setup and recording and experimental procedures. Section 5 presents the experimental results. Section 6 shows the discussion based on the experimental results. Finally, Section 6 presents conclusions.

2. Related Works

Several studies have modelled human trust by analysing human physiological signals and behaviours. Sadrfaridpour et al. proposed a mutual trust model between human and autonomous agents to coordinate collaboration [14]. The authors defined human performance based on muscle fatigue and the recovery dynamics of the human body when performing repetitive tasks, and the performance of autonomous agents was evaluated using the difference between human and autonomous agents' behaviours. Mahani, Jiang and Wang applied a Bayesian mechanism to predict human-to-autonomy trust based on human trust feedback to each individual robot and human intervention [16]. With the aid of a data-driven approach, Hu et al. proposed a human trust model that classified an individual's trust and untrust with electroencephalography (EEG) signals and galvanic skin response (GSR) data [15]. Similar work is also presented in [34]; the researchers exploited human cognitive states extracted from EEG data to model human workers' trust in Collaborative Construction Robots. In addition, the pupillary response is observed to be an effective index for human trust estimation. Lu and Sarter exploited eye tracking metrics to infer human trust in real time [17]. Alves et al. [18] consider kinesic courtesy cues from human to machine as an important factor in establishing human trust in HAT collaboration. Apart from pure human factors, some work considers human behaviour and machine performance when modelling the mutual trust between human users and machines. Inspired by human social behaviour, Jacovi et al. [19] proposed a formalisation method to model mutual trust between a human user and a machine. Furthermore, some researchers proposed computational trust models for HAT systems. In [11], the model of pupil dilation is extended as a computational trust model to facilitate the interaction between humans and robots. The computational model of human-robot mutual trust presented in [20] considers multiple pieces of information in physical human-robot collaboration, such as robot motion, robot safety, robot singularity, and human performance. Although all of the above studies provide valuable perspectives on human trust modelling, some trust models consider subjective feedback or the historical behaviour of humans, which is not reliable enough and may lead to bias in the evaluation of present status and condition. Other approaches that generate trust based on a single human cognitive state might also result in a miscalibration of trust in complex scenarios. Thus, this study attempted to remedy the lack of multievidence in current trust models by modelling information from multiple sources that can be optimised without prior knowledge and provides a comprehensive evaluation of human trust.

3. Multi-Human-Evidence-Based Trust Evaluation Model

This section introduces the proposed trust evaluation framework used to generate human trust values based on real-time human cognition signals. The objective of the proposed framework is to enable autonomous agents to be aware of the real-time human states and, therefore, to make a decision of the proper action based on the current human conditions. By generating a single trust value without historical data, the proposed model could reduce the complexity of the cooperation task and surely eliminate the bias from previous behaviour. Figure 1 shows the structure of our model. The proposed model combines multiple human evidence to estimate a single human trust value. The evidence contains three human states, including attention level, stress index and human perception. The information fusion block is responsible for combining the three pieces of evidence with sorting and weight learning via fuzzy Q-learning. The final output of the framework is the human trust value. Note that the human trust value is produced in real time, although the learning of the fusion weights requires offline training. Details of the components of the framework are presented in the following subsections.

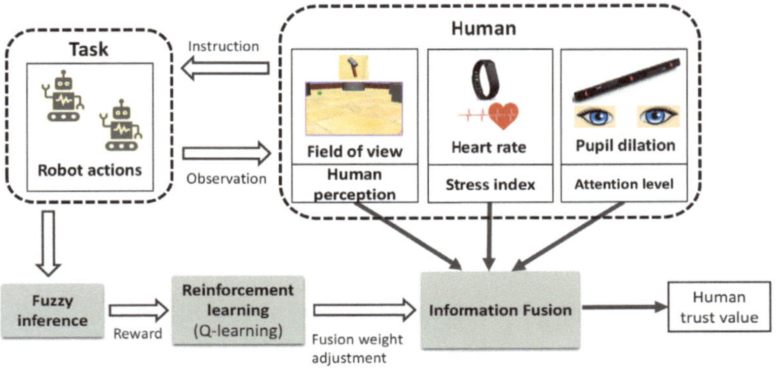

Figure 1. Structure of the proposed model.

3.1. Trust Evaluation Metrics

3.1.1. Attention Level

As one of our three evaluation metrics for human performance, the attention level is calculated based on pupil response. Research evidence has shown that the dynamics of pupil response are an effective characteristic for estimating the human state of concentration or distraction [35]. The attention level is computed based on Equation (1) proposed by Hoeks and Levelt [35]:

$$y(t) = h(t) * x(t), \tag{1}$$

where $y(t)$ is the pupillary response, $h(t)$ is a system constant called the impulse response, $x(t)$ is the attention level and $*$ is the convolution operator. The variables y, h, and x indicate the functions for the independent variable, time t.

The impulse response $h(t)$, derived from the approach introduced by Hoeks and Levelt [35], is set to represent the relation between attention and the pupillary response. The computing equation of impulse response $h(t)$ is presented in (2).

$$h(t) = s \times (t^n) \times e^{\left(\frac{-n \times t}{t_{\max}}\right)}, \tag{2}$$

where n is the number of layers that is set to 10.1, $t_{\max} = 5000$ ms is the maximum response time of participants, $s = \frac{1}{10^{33}}$ is a constant used to scale the impulse response, and t is the response time, which are the same settings as in the cited equation [35].

3.1.2. Stress Index

It has been well accepted that the stress level of individuals can affect the corresponding performance in a task. Thus, we also calculated human stress as one metric used for human trust evaluation. In this paper, we used heart rate variability (HRV) as the stress index of the participants. HRV is computed based on the measurement of the duration from a series of continuous heart cycles, known as the interbeat interval (IBI), which is used to evaluate the human body's autonomic regulation. A normal heart rate is in the range of 60–70 beats/min, which is controlled by the parasympathetic nervous system. During a cognitive state with stress, human sympathetic nervous system activity increases, which affects the duration of IBI and heart rate. To quantitatively evaluate the stress level, we applied geometric methods to analyze the distribution and shapes of the IBI. The stress index (SI) was computed with the IBI data by means of Baevsky's equation in (3) [36].

$$SI = \frac{AMo}{2 \times Mo \times MxDMn}, \tag{3}$$

where AMo is the pattern amplitude expressed as a percentage, Mo is the mode that represents the most frequently occurring RR interval (the interval between successive heartbeats), and $MxDMn$ is the variation range, which reflects the variability degree of the RR interval, as shown in Figure 2. The mode of Mo is simply taken as the median of the RR intervals, and AMo is the height of the histogram of the normalized RR interval (the width is 50 ms). $MxDMn$ represents the difference between the shortest and longest RR intervals for each participant.

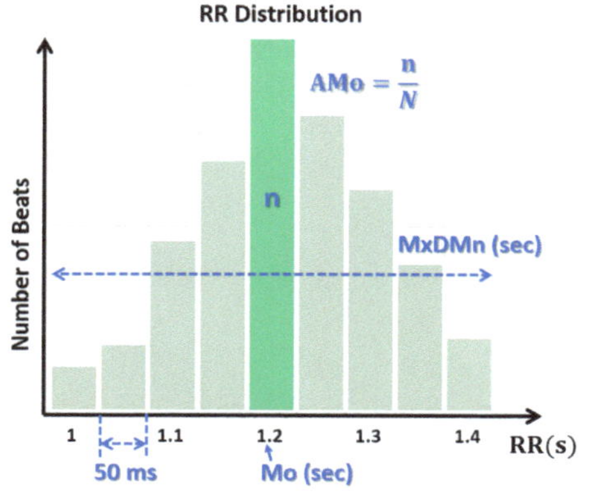

Figure 2. Histogram of the RR distribution for Baevsky's stress index, in which n is the number of beats, N is successive beat intervals, AMo is the height of the histogram of the normalised RR interval, $MxDMn$ represents the difference between the shortest and longest RR intervals, and Mo is the median of RR intervals.

3.1.3. Human Perception

Human perception measures confidence in the decision-making of the human agent based on the situational awareness of the human in a HAT environment through visual perception. The simulation task applied in this study is a target (ball)-collection task. As shown in Table 1, based on human perceptions of the autonomous agent and target, four situations could occur during the task. The first situation is that the human can see both the autonomous agent and target; the second and third are that the human only observes either the target or the agent, respectively; and last, the human can see neither the autonomous agents nor the target. We use four factors, including the indexes to indicate position ($S1$), orientation ($S2$), distance ($S3$) and view angle ($S4$), to estimate human perception ability, where the human perception evaluation level equals $f(S1) + f(S2) + f(S3) + f(S4)$. More details of our experiment and indexes are elaborated in Section 4.

Table 1. Four situations of human perception.

	Human Perception
First situation	Agent + Target
Second situation	No Agent + Target
Third situation	Agent + No Target
Fourth situation	No Agent + No Target

3.2. Trust Metric Fusion Model

The fusion model combines three pieces of evidence from human states in real time to produce a trust value. The function is defined as $F: [0,1]^n \rightarrow [0,1]$, through which multiple values located in the interval $[0, 1]$ are assigned to a single final value. We use the Hamacher product [37] to implement the fusion for the proposed trust model. The Hamacher product is a nonlinear transformation operation that uses confidence values from detectors to produce the final confidence for the fusion. If the evaluated single input values improve, the final trust value will increase such that $F(0, 0, \ldots, 0) = 0$ and $F(1, 1, \ldots, 1) = 1$. When all single input values evaluated are zero, the final trust value falls to the minimum; in other words, the human is completely untrustworthy. However, if all the values of the evaluated input are 1, the upper limit of the trust value is 1. Assuming that the result of the fusion $F(\boldsymbol{E})$ satisfies the constraint $\min\left(E^1, E^2, \ldots, E^n\right) \leq F(\boldsymbol{E}) \leq \max\left(E^1, E^2, \ldots, E^n\right)$, we can define an aggregation function as follows:

$$F(\boldsymbol{E}) = \sum_{i=1}^{n} f(E_i - E_{i-1}, w_i), \tag{4}$$

where $\boldsymbol{E} = (E_1, E_2, \ldots, E_n) \in [0, 1]^n$ is an increasing permutation of evaluations such that $0 \leq E_1 \leq E_2 \leq \ldots \leq E_n$, $\boldsymbol{w} = [w_1, w_2, \ldots, w_n]$ is the fusion weight vector, and $w_1 + \ldots + w_n = 1$.

We use the Hamacher product to fuse each pair of evidence. The Hamacher t-norm involves the use of a fuzzy measure [35]. Therefore, the fusion model that produces the trust value based on the three pieces of evidence is defined as follows:

$$F_h(\boldsymbol{E}) = \frac{g(\boldsymbol{E}) \times w_i}{g(\boldsymbol{E}) + w_i - g(\boldsymbol{E}) \times w_i}, \tag{5}$$

where $F_h(\boldsymbol{E})$ represents the human trust value, and $g(\boldsymbol{E}) = \sum_{i=1}^{n} (E_i - E_{i-1})$. The fusion weights $\boldsymbol{w} = [w_1, w_2, w_3]$ are learnt by Q-learning based on collective human state data, including the pupil, HRV and human perception signals. In addition, we used the min-max normalisation for the pupil and HRV data to normalise the value in the range of $[0,1]$.

3.2.1. Reinforcement Learning

This section discusses the Q-learning method used to update the fusion weights. As a model-free, off-policy reinforcement learning method, Q-learning tracks what has been learnt and finds the best course of action for the agent to gain the greatest reward [38,39]. As discussed above, the final trust value is calculated by multiplying the evaluated values of three human states by the corresponding weights and then summing the results. Weights represent the relative importance of each individual evaluation, and the vector of weights is initialised randomly. Thus, since Q-learning is capable of transferring functions or reward functions with random factors [40], we applied its algorithm to determine which weight vector is used to fuse the estimated trust values from a random perspective. The equation to update the Q-value with action i and state s in each step is shown below:

$$Q(s, i) \leftarrow Q(s, i) + \alpha \times \left(w(s, i) \times \nabla + \gamma \times \sum_{j=1}^{n} (w(s', j) \times Q(s', j)) - Q(s, i) \right) \tag{6}$$

where α is a fixed value used as the learning rate that satisfies the condition $0 < \alpha \leq 1$, $w(s,\ i)$ represents the value of the weight in state s and action i, the parameter γ is a temporal discount factor that satisfies the condition $0 < \gamma \leq 1$, s' is the state after performing the action under state s, and ∇ is the reward. Here, we set α to 0.1 and γ to 0.2. Additionally, we update the weight vector based on the Q-values after updating the Q-tables. Two conditions are used when applying the value of weight $w(s,\ i)$: (1) The summation of $w(s,\ i)$ is normalized to 1. (2) Parameter δ is within the range $(0,\ 1]$.

Formula (7) shows the weight updating rule:

$$w'(s,i) \leftarrow w(s,i) + \begin{cases} (1 - w(s,i)) \times \delta \times \left(\frac{1}{1+e^{-a\times Q(S,i)+b}}\right), if\ i = \arg\max_j Q(s,j) \\ (0 - w(s,\ i)) \times \delta \times \left(\frac{1}{1+e^{-a\times Q(S,i)+b}}\right), otherwise. \end{cases} \tag{7}$$

Then, we normalize the weight value in (8) so that $\sum_{i=1}^{n} w(s,\ i) = 1$:

$$w(s,\ i) \leftarrow \frac{w'(s,\ i)}{\sum_{j=1}^{n} w'(s,\ j)}, \tag{8}$$

3.2.2. Fuzzy Reward

This section presents the FIS used to produce fuzzy rewards for RL to learn fusion weights and adjust the Q-learning reward. The FIS is composed of a zero-order Takagi–Sugeno–Kang (TSK) fuzzy system [41–43], which can be defined as

$$\begin{aligned} R_I : If\ x_1(k)\ is\ A_{i1}\ And\ \ldots\ And\ x_n(k)\ is\ A_{in} \\ Then\ y_1(k)\ is\ a_i, \end{aligned} \tag{9}$$

where $x_1(k), \ldots, x_n(k)$ represents the input variables at time k, $A_{i1}, \ldots, A_{in}$ are the fuzzy sets, and a_i represents the singleton consequence. Moreover, $\mu_{A_{ij}}$ is the membership value of A_{ij}, and Φ_i is the firing strength of rule R_i. We use algebraic multiplication to implement the fuzzy AND operation. Then, Φ_i with input data set $\vec{x}(k) = [x_1(k), \ldots, x_n(k)]$ can be described as

$$\Phi_i(\vec{x}(k)) = \prod_{j=1}^{n} \mu_{A_{ij}}(x_j(k)) \tag{10}$$

Supposing the FIS consists of r rules, the output of the FIS $y(k)$ can be calculated by the weighted average defuzzification method in (11).

$$y(k) = \frac{\sum_{i=1}^{r} \Phi_i(\vec{x}(k))a_i}{\sum_{i=1}^{r} \Phi_i(\vec{x}(k))} \tag{11}$$

To properly score the relationship between the generated trust value and human performance, we used the fuzzy reward to feed back the score to the proposed trust model to tune the fusion weights. We defined four rules for reward evaluation based on human performance for the Q-learning algorithm. There are two input variables for each fuzzy rule: human reaction time τ_h and human trust value F_h. Here, the reaction times are divided into fast and slow camps, and the trust values also contain high and low levels. Thus, four combinations exist. The rules are defined as follows.

R_1: If $\tau_h(k)$ is A_{fast} and $F_h(k)$ is B_{high}, then $r(k) = 1$.
R_2: If $\tau_h(k)$ is A_{slow} and $F_h(k)$ is B_{low}, then $r(k) = 1$.
R_3: If $\tau_h(k)$ is A_{fast} and $F_h(k)$ is B_{low}, then $r(k) = -1$.
R_4: If $\tau_h(k)$ is A_{slow} and $F_h(k)$ is B_{high}, then $r(k) = -1$.

where A_{fast} and A_{slow} are fuzzy sets describing fast and slow human reaction times, respectively. Specifically, under R_1, humans make decisions faster and are trustworthy, the trust value of the fusion result is high, which represents a positive result, and the reward value of R_1 is 1. Under R_2, humans are slow to make decisions and, thus, untrustworthy. In addition, the fusion result of the trust value is low. The two input variables of R_2 both show a consistently negative result, so the reward value of R_2 is 1. However, if the trend is inconsistent, as in R_3 and R_4, the reward is -1.

The membership value of A_{fast} and A_{slow} is computed by the membership function, as follows

$$\mu_{A_{fast}} = \begin{cases} f\left(\tau_h \middle| m_{fast}\,,\,\sigma_{fast}\right), & \tau_h > m_{fast}, \\ 1, & otherwise \end{cases} \tag{12}$$

and

$$\mu_{A_{slow}} = \begin{cases} f(\tau_h | m_{slow}\,,\,\sigma_{slow}), & \tau_h > m_{slow}, \\ 1, & otherwise \end{cases} \tag{13}$$

where $f(x|m,\sigma) = exp\left[-\frac{(m-x)^2}{\sigma^2}\right]$,$B_{high}$ and B_{low} are the fuzzy sets describing high and low human trust values, respectively. The membership value of B_{high} and B_{low} is computed by the membership function as follows:

$$\mu_{B_{high}} = \begin{cases} f\left(F_h \middle| m_{high}\,,\,\sigma_{high}\right), & F_h > m_{high}, \\ 1, & otherwise \end{cases} \tag{14}$$

and

$$\mu_{B_{low}} = \begin{cases} f(F_h | m_{low}\,,\,\sigma_{low}), & F_h > m_{low}, \\ 1, & otherwise \end{cases} \tag{15}$$

where m_{high} and m_{low} of the reaction time are calculated using the average reaction time and standard deviation of reaction times from all participants. The slow reaction time is defined as the time values that are twice the standard deviation more than the average reaction time, and the fast reaction time is twice the standard deviation less than the average reaction time. Figure 3 shows a schematic diagram of the four rules.

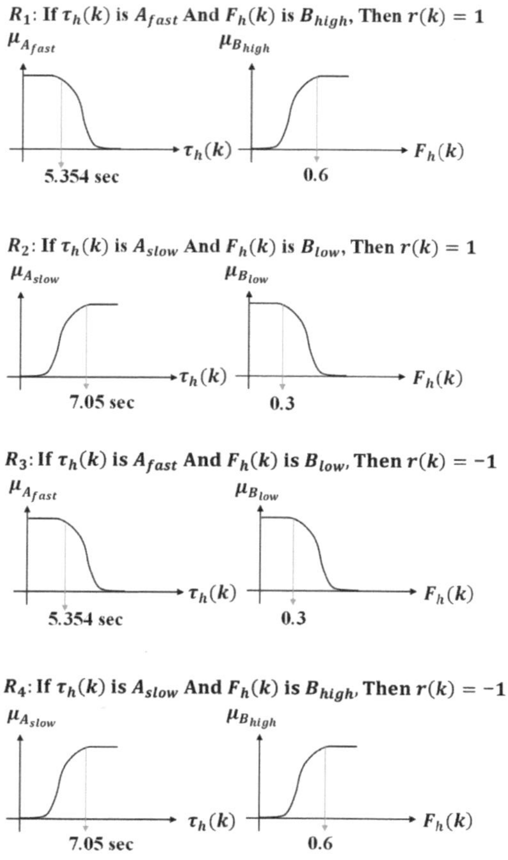

Figure 3. Four rules of the fuzzy neural network.

4. Methods

4.1. Participants

Six healthy male participants aged 21 to 24 years participated in this study. Following an explanation of the experimental procedure, all participants received an informed consent form and signed before participating in the study. This study received the approval of the Institute's Human Research Ethics Committee of National Chiao Tung University, Hsinchu, Taiwan. None of the participants reported a history of psychological disorder, which could have affected the experimental results.

4.2. Scenario Design

The built simulation scenario of ball collection is designed by a professional robot simulator, Webots 8.6.2 (Cyberbotics Ltd., Lausanne, Switzerland). As shown in Figure 4, the environment is fenced with several balls and obstacles inside. A human agent and a robot agent are expected to work together to collect the balls without collision between the robot and the wall or obstacles. Humans can only observe the entire environment with restricted sight through a monitoring camera located in the top left corner of the scenario. On the basis of the observation, the human can instruct the robot to search for the ball, and the robots are also allowed to explore the environment by themselves when there is no instruction from the human or the human trust values are not high enough to be trusted.

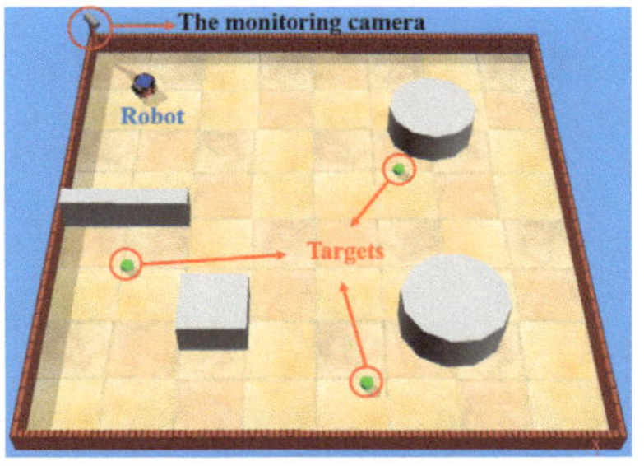

(a) Scenario with one robot and three balls.

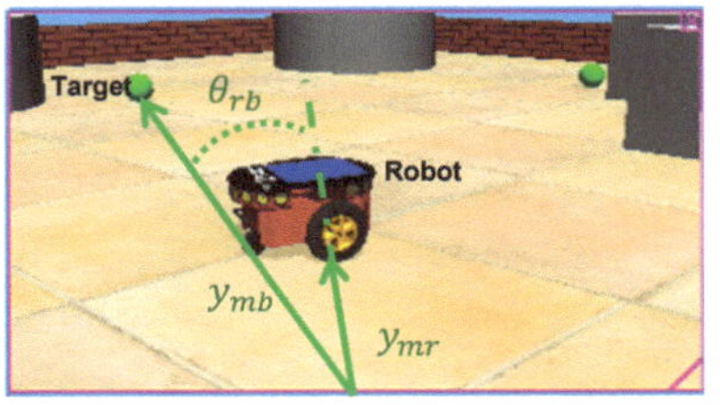

(b) Participant's view via the monitoring camera during the experiment.

Figure 4. Scenario for training data collection (Scenario_1).

4.3. Human-Agent Setup and Recording

The design of the whole scenario is affected by not only the autonomous agents' actions but also human physiological factors. Here, we use two instruments, eye-tracking and a heart rate monitoring watch, to measure the human physiological state in real time, as shown in Figure 1. Eye tracking data were recorded using the Tobii Pro X2-30 screen-based eye tracker (Tobii AB Corp., Stockholm, Sweden). We corrected the pupil data and gaze location of each participant to monitor their concentration level and fixation pathways. Heart rate data was recorded using the Empatica-E4 wristband (Empatica Inc., Cambridge, MA, USA). We used the real-time heart rate of each participant to estimate the current stress level of the human agent while performing the task.

Human perception ability was identified by the monitoring camera used to provide sight of the scenario situation for the human agent, as presented in Figure 4a. Following our definition of human perception in Section Trust Evaluation Metrics, we categorised human perception into four classes (see Table 1). Real-time perception ability is calculated according to the following formula:

$$E^a = k_1 \times \left(1 - \frac{y_{mr}}{y}\right) + k_2 \times \left(1 - \frac{y_{mb}}{y}\right) + k_3 \times \left(1 - \frac{\theta_{rb}}{\pi}\right), \tag{16}$$

where E^a is the value of current perception ability, k_1, k_2, k_3 are predefined weights, y_{mr} is the distance between the monitoring camera and the robot, y_{mb} is the distance between the monitor and ball, θ_{rb} is the deviation value between the robot and ball, and y is the distance the monitor can measure, as shown in Figure 4b.

We set k_3 as the largest weight because it represents the situation in which both the robot and ball can be perceived, in which humans have the best chance of completing the task successfully. k_2 is given the second highest weight that represents the situation in which the human knows the exact position of the ball, although the robot is not visible. Finally, k_1 has the smallest weight, which indicates the situation in which the human does not know the location of the ball, although the robot is visible. Every situation is transformed into a corresponding evaluation value, which ranges in value between 0 and 1. Here, if the human cannot see the ball or robot, the corresponding terms are set to zero in the equation. Then, if neither the robot nor ball can be seen, the third term in the equation is set to 0. Furthermore, if an object does not exist in some conditions, the value of that item is also set to 0.

4.4. Experimental Procedures

During the experiment, participants sat in front of the computer screen while performing the task. Each participant first performed a calibration for the eye tracker was performed by each participant first. Next, we introduced the operation and process of the whole experiment, including how the robot is controlled and various other considerations in the experiment. While conducting the ball-collection task, participants used two keys on the keyboard to control the clockwise or anticlockwise rotation of the robot, following our instructions. The balls were scattered around the environment, including invisible or blind areas. The participant can only monitor the scenario from a fixed perspective, as mentioned above, and an example of the participant's view via the monitoring camera is shown in Figure 4b. The task would be completed once all the balls have been found by the robot.

For the cooperative work between human and robot agents, we define the process of their interaction in each trial. First, the human has eight seconds to set the facing direction for the robot through rotation control. Then, the robot moves in the direction the human agent has selected for 15 s. To discard the direction setting for the robot, the human agent is allowed to select robot self-exploration. The robot may move for more than fifteen seconds if it detects a target by itself during self-exploration. A schematic diagram of each trial is shown in Figure 5. Here, t_p represents the time that the human controls the direction of the robot, $t_{p_{\max}}$ represents the maximum time for the human to control the robot, and t_r represents the time that the robot acts and follows the command from the human.

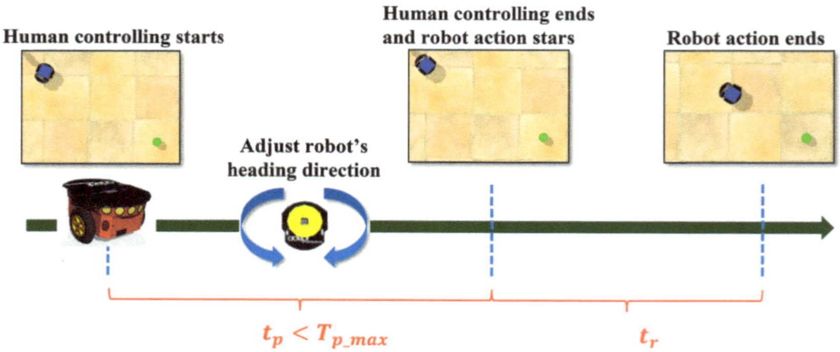

Figure 5. Robot and human interaction in each trial.

5. Results

This section describes the results of our simulation experiment with our proposed trust model in the human-autonomous cooperation task scenario. We first discuss the training results by presenting the convergence of the fuzzy reward and its standard deviation. Then, we provided our testing results based on three different scenarios with our trained trust model.

5.1. Training Results

The training process inputs the data collected in Figure 4 into the Q-learning. In (6), we set the learning rate, α, to 0.1 and the discount factor, γ, to 0.2. These settings are commonly applied to various scenarios with the fuzzy neural network to obtain the reward value [43–45]. We implemented 100 episodes to train our model to eliminate the impact of the unstable reward in the first ten episodes. The visualized convergence result of the reward values from each episode is shown in Figure 6. The data recorded in Figure 5 were used to train the reinforcement learning method, including three pieces of human evidence signals and the reaction time of humans t_p. The training results combined cross-subject data. One of the best-performing weights with the best reward values is $[w_1, w_2, w_3] = [0.2440, 0.3688, 0.3872]$.

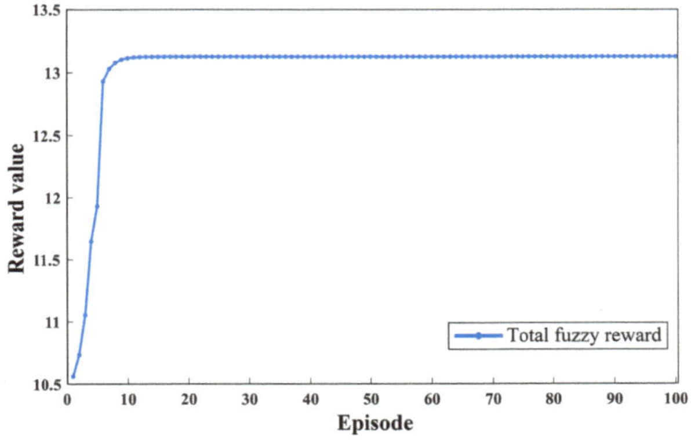

Figure 6. Convergence of the fuzzy reward during the training process of the fusion mechanism.

5.2. Testing Results

We used the trained weights to fuse the three human states. The fusion results from all six participants are used to assess the human trust value, which ranges from 0 to 1. Then, we conducted the tests in one training (Scenario_1) and three test scenarios (Scenario_2–4). Figure 7 presents the three test scenarios, which we refer to as Scenario_2, Scenario_3 and Scenario_4. Tables 2 and 3 provide all the test results. We statistically analyzed the execution time of three task-performing modes, including human instruction (robot follows human instruction only to find the targets), a collaboration of human and robot agents (HAT) and robot random search in Table 2. The results contain both the completion time and decision time. Additionally, the number of operations in the collaboration and human instruction modes are presented in Table 3.

Table 2. Completion time under Scenario_1–4, HAT represents experiments with control switching between the human and robot.

Evaluation of Completion Time		Participant					
		1	2	3	4	5	6
Scenario	**Setting**	**Time (s)**					
Scenario_1	human instruction	223	175	194	196	177	216
	HAT	194	138	171	140	131	143
	random search	287					
Scenario_2	human instruction	165	181	168	121	132	129
	HAT	117	119	123	98	100	90
	random search	185					
Scenario_3	human instruction	408	428	407	469	427	450
	HAT	372	343	371	403	380	359
	random search	573					
Scenario_4	human instruction	209	237	248	229	222	242
	HAT	178	208	195	201	197	213
	random search	330					

Table 3. Number of decisions made in Scenarios_1–4.

Evaluation of Number of Decisions Made		Participant					
		1	*2*	*3*	*4*	*5*	*6*
Scenario	**Setting**	**Number of Decisions**					
Scenario_1	human instruction	6	6	6	6	6	6
	human/robot	4 / 2	4 / 2	4 / 2	4 / 2	5 / 1	4 / 2
Scenario_2	human instruction	6 / 6	5 / 6	6 / 6	4 / 5	4 / 6	4 / 5
	human/blue	4 / 2	4 / 1	5 / 1	3 / 1	3 / 1	2 / 2
	human/pink	2 / 4	3 / 3	4 / 2	3 / 2	5 / 1	2 / 3
Scenario_3	human instruction	12	12	13	12	12	11
	human/robot	9 / 3	9 / 3	7 / 6	7 / 5	8 / 4	6 / 5
Scenario_4	human instruction	5 / 7	6 / 7	6 / 8	6 / 7	6 / 7	8 / 7
	human/blue	3 / 2	3 / 3	2 / 4	2 / 4	4 / 2	4 / 2
	human/pink	7 / 0	7 / 0	5 / 3	5 / 2	4 / 3	4 / 3

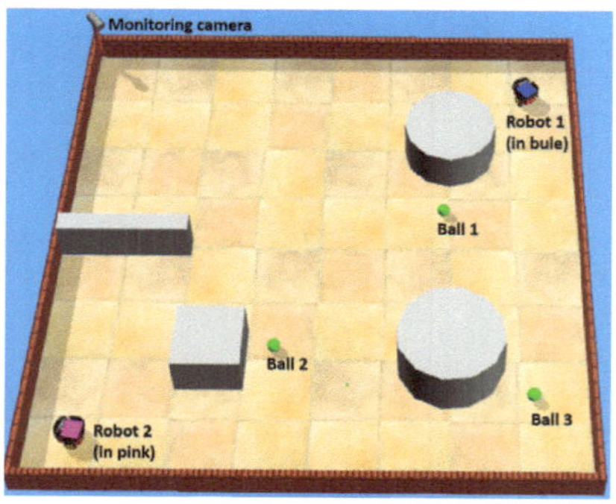

(**a**) Small scenario with two robots and three balls. (Scenario_2)

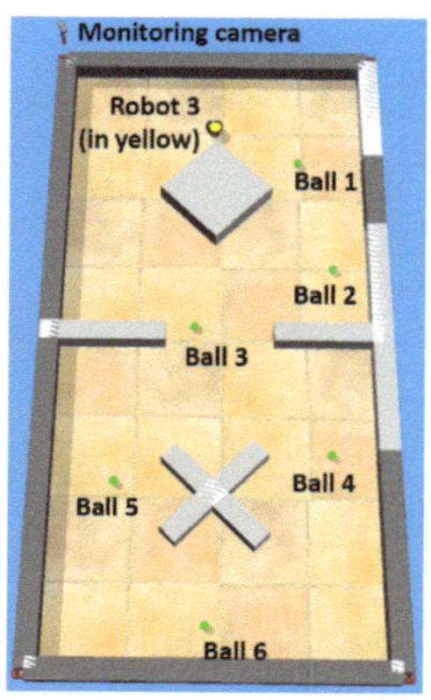

(**b**) Large scenario with one robot and six balls. (Scenario 3)

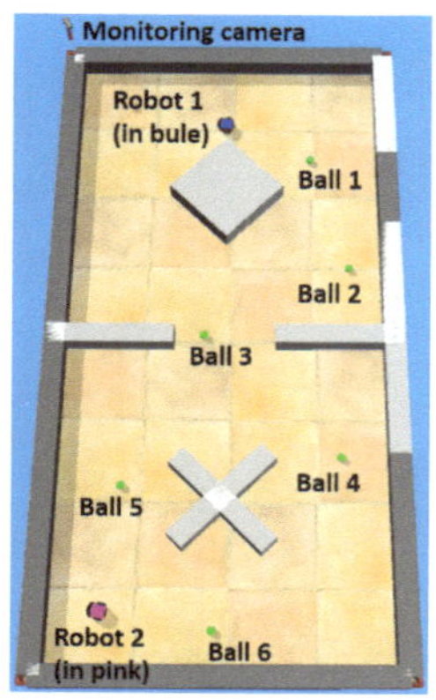

(**c**) Large scenario with two robots and six balls. (Scenario_4)

Figure 7. Scenarios for testing.

Overall, the completion time in the collaboration mode is always the shortest compared to the other two modes for all participants. Specifically, in Scenario_1 shown in Table 2, Participant 5 spent the shortest time completing the task in collaboration mode, and the proportion of manipulation by humans was also the highest, indicating that a high level of trust was evaluated for this participant while performing the ball collection task.

In Scenario_2, Participant 3 took the longest time to complete the task in collaboration mode. However, the proportion of manipulations by humans was also the highest in this case. This may be because Participant 3 maintained a high level of trust, but did not control the robot well, which led to a longer time required to complete the task. In Scenario_3, the second participant took the shortest time to achieve the task in collaboration mode, and the robot was involved in the least amount of autonomous control. This result may suggest that the second participant was trustworthy and able to identify the shortest route to save a considerable amount of time during the experiment. In Scenario_4, the result of the decisions indicates that Participants 1 and 2 controlled the robot all by themselves without robot intervention, and the completion time varied greatly. This may suggest that both participants were trustworthy, but the first participant could identify a better path than the second.

6. Discussion

This section discusses the improvement of efficiency among the three modes (robot random search, human instruction and collaboration/HAT) and explores the reasons for the improved efficiency. The magnitude of the improvement for each scenario is shown in Table 4.

Table 4. A Comparison of improvement rate in Scenarios_1–4. HAT represents experiments with control switching between the human and robot, H represents experiments with only human instructions, and RS represents experiments with only robot random search

Scenario	Setting	Participant						
		1	2	3	4	5	6	Avg
		Improvement Rates						
Scenario_1	H vs. RS	22.29	39.02	32.4	31.71	38.33	24.74	31.42
	HAT vs. RS	32.4	51.92	40.42	51.22	54.36	50.17	46.75
	HAT vs. H	13.01	21.14	11.86	28.57	25.99	33.79	22.39
Scenario_2	H vs. RS	10.81	2.16	9.19	34.59	28.65	30.27	19.28
	HAT vs. RS	36.76	35.68	33.51	47.02	45.95	51.35	41.71
	HAT vs. H	29.09	34.25	26.79	19.01	24.24	30.23	27.27
Scenario_3	H vs. RS	28.8	25.31	28.97	18.15	25.48	21.47	24.69
	HAT vs. RS	35.08	40.14	35.25	29.67	33.68	37.35	35.19
	HAT vs. H	8.82	19.86	8.85	14.07	11.01	20.22	13.81
Scenario_4	H vs. RS	36.67	28.18	24.85	30.61	32.73	26.67	29.95
	HAT vs. RS	46.06	36.97	40.91	39.09	40.3	35.45	39.79
	HAT vs. H	14.83	12.24	21.37	12.22	11.26	11.98	13.98

In Scenario_2, Participant 3 conducted the task with the largest number of decisions made by humans and the longest completion time, and, on the contrary, Participant 6 had the largest number of decisions made by two robots and the shortest completion time. The two robots follow Participant 3's instructions nine times in 12 trials, which is 75% of decisions made by the human, and follow Participant 2's instructions seven times in 11 trials, which is 63.6% of decisions made by the human. Whereas, Participant 6 made decisions four times for the two robots in nine trials, which is 44.4% of decisions made by the human. To visualise how these two participants conducted the tasks, we present the robot paths and control decisions in Figures 8 and 9 for each participant, respectively. As revealed in Figure 8, Participant 3 controlled the pink robot in the third trial, but did not adjust it in the right direction, causing the pink robot to take a long detour to find the ball and wasted a substantial amount of time. In contrast, the pathways of robots in Figure 9 indicate that Participant 6 could well control both robots and guide them on a shorter route to find the balls. Furthermore, as the greatest improvement achieved between human instruction and HAT, we also visualised the route of the task of Participant 2, as shown in Figure 10. In the human instruction condition, Participant 2 failed to adjust the robot in the right direction in the fifth trial, which resulted in a miss-out of the targets for the robot. However, in the collaboration condition, due to the lower trust value in the fifth trial for Participant 2 than the threshold value, the pink robot did not receive human instructions and proceeded forward to collect the ball. In other words, along with the successful awareness of human states, the robot made the decision itself and achieved better performance through an efficient evaluation of human states by our model.

In addition to Scenario_2, the collaboration controlled by our proposed model also greatly improves in the more complicated scenarios. The performance of Participant 2 in Scenario_3 indicates that the human decision was estimated to be trustworthy to make the shortest choice of path through the aid of our evaluation model on real-time human states, which achieved an optimal decision on the route to save a lot of time to complete the task. Similarly, in Scenario_4, Participant 1 and Participant 2 were successfully evaluated as trustworthy agents, although Participant 1 could choose the better path. The test results shown in Table 2 suggest that the fusion weights trained with the Q-learning algorithm in Scenario_1 can be directly applied to more complicated scenarios without retraining. The fusion weights were trained with collected cross-subject human data and can be used in real-time scenarios, implying that the FIS can overcome the subject difference in human data and compute appropriate rewards for the Q-learning algorithm to tune the fusion weights.

In summary, the proposed multievidence-based trust evaluation model could generate a trust-considering value for human agents that reflects the dynamics of human states in real-time. The comparison among robot random search, human instruction only, and collaboration modes demonstrates that the collaboration between human and autonomous robots controlled by the proposed trust model has adaptability and robustness for the ball-collection task under different levels, which greatly improves the task completion time compared to the other two modes.

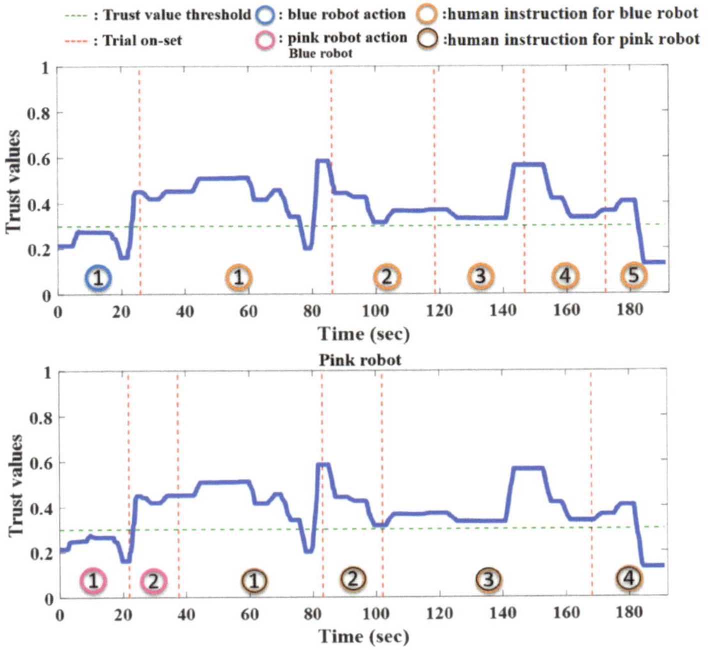

(**a**) Control switch between human and robot.

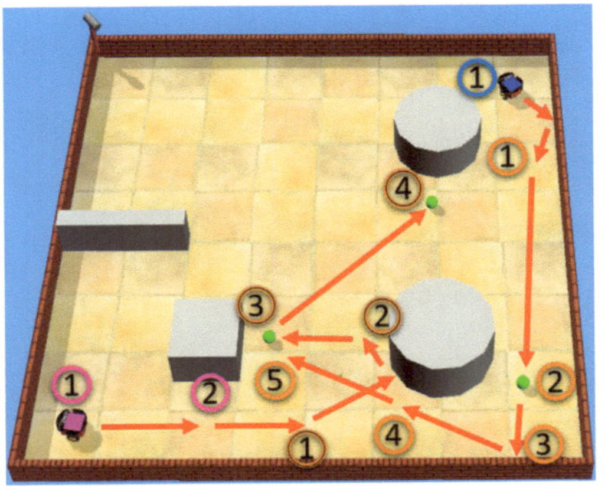

(**b**) Robot path trajectory.

Figure 8. Experimental results made by Participant 3.

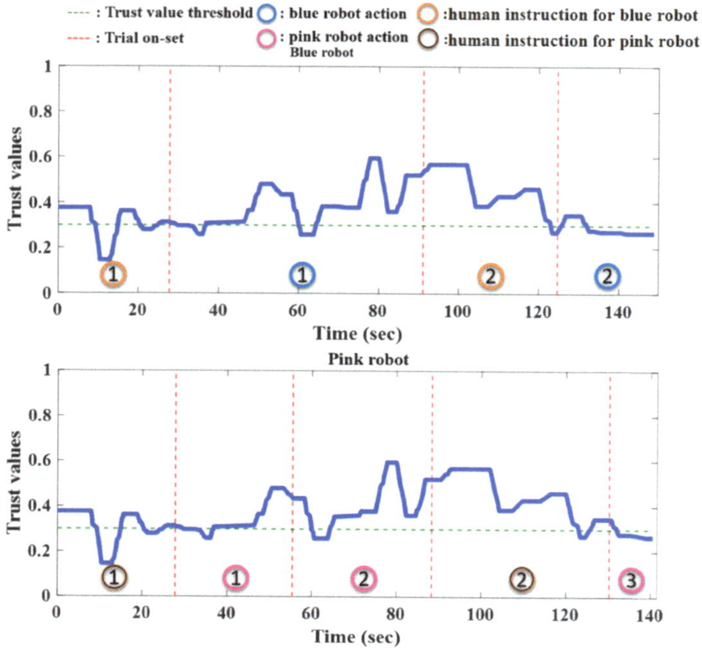

(**a**) Control switch between human and robot.

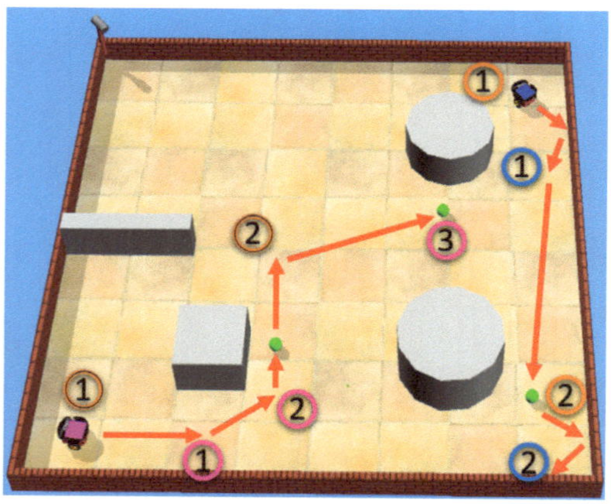

(**b**) Robot path trajectory.

Figure 9. Experimental results made by Participant 6.

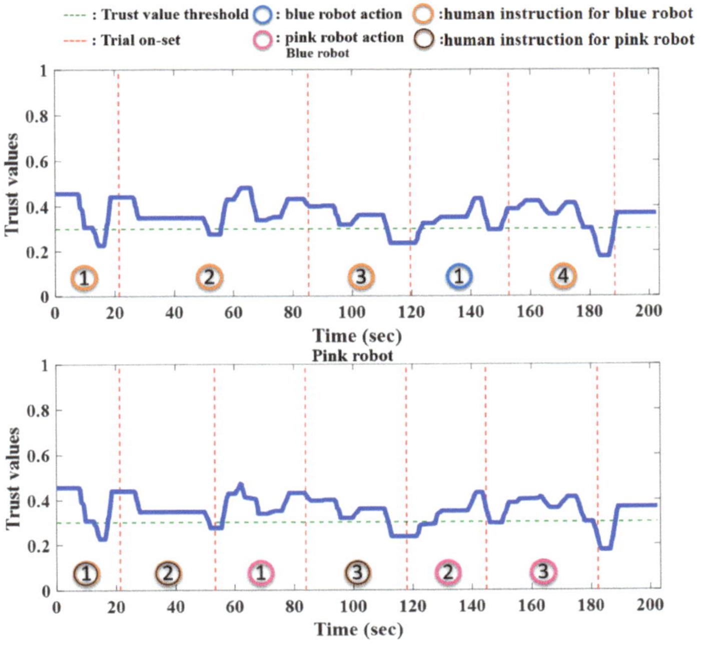

(a) Control switch between human and robot.

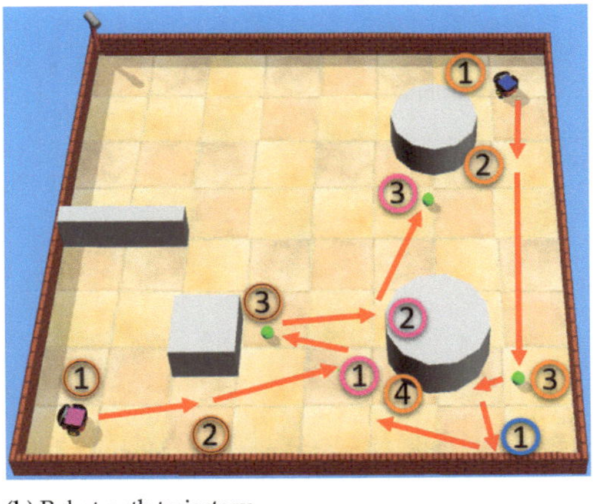

(b) Robot path trajectory.

Figure 10. Experimental results of Participant 2.

7. Conclusions

This study proposed an adaptive trust model considering multiple real-time human cognitive states. The proposed trust model uses a fusion mechanism to combine various types of information, namely human attention level, stress index, and human perception. To verify the performance of the proposed trust model, we implemented four environmental settings, including different types of obstacles and different numbers of robot agents. We compare the performance of the HAT with those of pure human agents and those of robot agents. The results of the comparison show that the HAT team with the proposed trust model can improve the efficiency of the given task by at least 13% in different scenario settings; The HAT team coordinated by the proposed trust model can complete the given task faster than others. Our results also suggest that the trust value generated based on these three pieces of evidence can reflect the performance of a human agent more accurately, which contributed to an improvement in efficiency for the cooperation between human and autonomous robot agents in all test scenarios. These results demonstrate that the proposed model can adapt to various levels of human performance and generate reliable trust values via the reinforcement learning algorithm. The main limitation of this study is our participant pool; only male participants were involved in our experiments. For future works, we will enlarge the participant pool and consider gender balance to conduct more comprehensive research. Furthermore, we will develop trust modelling to assess the trust of robot agents and then create a mutual trust model to provide more informatic reasoning for interaction in the HAT systems.

Author Contributions: Conceptualization, C.-T.L. and T.-P.J.; methodology, C.-T.L., H.-Y.F. and Y.-C.C.; software, H.-Y.F., Y.-C.C. and L.O.; validation, Y.-C.C. and Y.-K.W.; formal analysis, H.-Y.F.; investigation, H.-Y.F. and L.O.; resources, C.-T.L. and Y.-K.W.; data curation, H.-Y.F. and L.O.; writing—original draft preparation, H.-Y.F.; writing—review and editing, Y.-C.C., L.O. and J.L.; visualization, H.-Y.F., Y.-C.C., L.O. and J.L.; supervision, C.-T.L.; project administration, C.-T.L. and Y.-K.W.; funding acquisition, C.-T.L. All authors have read and agreed to the published version of the manuscript.

Funding: This work was supported in part by the Australian Research Council (ARC) under discovery grants DP180100670, DP180100656 and DP210101093. The research was also sponsored in part by the Australia Defence Innovation Hub under Contract No. P18-650825, US Office of Naval Research Global under Cooperative Agreement Number ONRG - NICOP - N62909-19-1-2058, and AFOSR – DST Australian Autonomy Initiative agreement ID10134, and AFOSR Grant No. FA2386-22-1-0042. We also thank the NSW Defence Innovation Network and NSW State Government of Australia for financial support in part of this research through grant DINPP2019 S1-03/09.

Institutional Review Board Statement: The study was conducted in accordance with the Declaration of Helsinki and approved by the Institutional Human Research Ethics Committee of National Chiao Tung University, Hsinchu, Taiwan.

Informed Consent Statement: Informed consent was obtained from all subjects involved in the study.

Data Availability Statement: Not applicable.

Conflicts of Interest: The authors declare no conflict of interest. The funders had no role in the design of the study; in the collection, analysis, or interpretation of data; in the writing of the manuscript; or in the decision to publish the results.

References

1. Shneiderman, B. Human-centered artificial intelligence: Reliable, safe & trustworthy. *Int. J. Hum. Comput. Interact.* **2020**, *36*, 495–504.
2. Doroodgar, B.; Ficocelli, M.; Mobedi, B.; Nejat, G. The search for survivors: Cooperative human-robot interaction in search and rescue environments using semi-autonomous robots. In Proceedings of the 2010 IEEE International Conference on Robotics and Automation, Anchorage, AK, USA, 3–7 May 2010; pp. 2858–2863.
3. Söffker, D. From human–machine-interaction modeling to new concepts constructing autonomous systems: A phenomenological engineering-oriented approach. *J. Intell. Robot. Syst.* **2001**, *32*, 191–205. [CrossRef]
4. Benderius, O.; Berger, C.; Lundgren, V.M. The best rated human–machine interface design for autonomous vehicles in the 2016 grand cooperative driving challenge. *IEEE Trans. Intell. Transp. Syst.* **2017**, *19*, 1302–1307. [CrossRef]
5. Demir, M.; McNeese, N.J.; Gorman, J.C.; Cooke, N.J.; Myers, C.W.; Grimm, D.A. Exploration of teammate trust and interaction dynamics in human-autonomy teaming. *IEEE Trans. Hum. Mach. Syst.* **2021**, *51*, 696–705. [CrossRef]
6. Dorronzoro Zubiete, E.; Nakahata, K.; Imamoglu, N.; Sekine, M.; Sun, G.; Gomez, I.; Yu, W. Evaluation of a home biomonitoring autonomous Mobile Robot. *Comput. Intell. Neurosci.* **2016**, *2016*, 9845816. [CrossRef]
7. Robinette, P.; Howard, A.M.; Wagner, A.R. Effect of robot performance on human–robot trust in time-critical situations. *IEEE Trans. Hum. Mach. Syst.* **2017**, *47*, 425–436. [CrossRef]
8. Pippin, C.; Christensen, H. Trust modeling in multi-robot patrolling. In Proceedings of the 2014 IEEE International Conference on Robotics and Automation (ICRA), Hong Kong, China, 31 May–7 June 2014; pp. 59–66.

9. Holbrook, J.; Prinzel, L.J.; Chancey, E.T.; Shively, R.J.; Feary, M.; Dao, Q.; Ballin, M.G.; Teubert, C. Enabling urban air mobility: Human-autonomy teaming research challenges and recommendations. In Proceedings of the AIAA AVIATION 2020 FORUM, Virtual, 15–19 June 2020; p. 3250.
10. Huang, L.; Cooke, N.J.; Gutzwiller, R.S.; Berman, S.; Chiou, E.K.; Demir, M.; Zhang, W. Distributed dynamic team trust in human, artificial intelligence, and robot teaming. In *Trust in Human-Robot Interaction*; Elsevier: Amsterdam, The Netherlands, 2021; pp. 301–319.
11. Tjøstheim, T.A.; Johansson, B.; Balkenius, C. A computational model of trust-, pupil-, and motivation dynamics. In Proceedings of the 7th International Conference on Human-Agent Interaction, Kyoto, Japan, 6–10 October 2019; pp. 179–185.
12. Pavlidis, M.; Mouratidis, H.; Islam, S.; Kearney, P. Dealing with trust and control: A meta-model for trustworthy information systems development. In Proceedings of the 2012 Sixth International Conference on Research Challenges in Information Science (RCIS), Valencia, Spain, 16–18 May 2012; pp. 1–9.
13. Kaniarasu, P.; Steinfeld, A.M. Effects of blame on trust in human robot interaction. In Proceedings of the The 23rd IEEE International Symposium on Robot and Human Interactive Communication, Edinburgh, UK, 25–29 August 2014; pp. 850–855.
14. Sadrfaridpour, B.; Saeidi, H.; Burke, J.; Madathil, K.; Wang, Y. Modeling and control of trust in human-robot collaborative manufacturing. In *Robust Intelligence and Trust in Autonomous Systems*; Springer: Berlin/Heidelberg, Germany, 2016; pp. 115–141.
15. Hu, W.L.; Akash, K.; Jain, N.; Reid, T. Real-time sensing of trust in human-machine interactions. *IFAC-PapersOnLine* **2016**, *49*, 48–53. [CrossRef]
16. Mahani, M.F.; Jiang, L.; Wang, Y. A Bayesian Trust Inference Model for Human-Multi-Robot Teams. *Int. J. Soc. Robot.* **2020**, *13*, 1951–1965.
17. Lu, Y.; Sarter, N. Modeling and inferring human trust in automation based on real-time eye tracking data. In *Proceedings of the Human Factors and Ergonomics Society Annual Meeting*; SAGE Publications Sage CA: Los Angeles, CA, USA, 2020; Volume 64, pp. 344–348.
18. Alves, C.; Cardoso, A.; Colim, A.; Bicho, E.; Braga, A.C.; Cunha, J.; Faria, C.; Rocha, L.A. Human–Robot Interaction in Industrial Settings: Perception of Multiple Participants at a Crossroad Intersection Scenario with Different Courtesy Cues. *Robotics* **2022**, *11*, 59. [CrossRef]
19. Jacovi, A.; Marasović, A.; Miller, T.; Goldberg, Y. Formalizing trust in artificial intelligence: Prerequisites, causes and goals of human trust in AI. In Proceedings of the 2021 ACM Conference on Fairness, Accountability, and Transparency, Virtual, 3–10 March 2021; pp. 624–635.
20. Wang, Q.; Liu, D.; Carmichael, M.G.; Aldini, S.; Lin, C.T. Computational Model of Robot Trust in Human Co-Worker for Physical Human-Robot Collaboration. *IEEE Robot. Autom. Lett.* **2022**, *7*, 3146–3153. [CrossRef]
21. Xing, Y.; Lv, C.; Cao, D.; Hang, P. Toward human-vehicle collaboration: Review and perspectives on human-centered collaborative automated driving. *Transp. Res. Part C Emerg. Technol.* **2021**, *128*, 103199. [CrossRef]
22. Liu, Y.; Habibnezhad, M.; Jebelli, H. Brainwave-driven human-robot collaboration in construction. *Autom. Constr.* **2021**, *124*, 103556. [CrossRef]
23. Chang, Y.C.; Wang, Y.K.; Pal, N.R.; Lin, C.T. Exploring Covert States of Brain Dynamics via Fuzzy Inference Encoding. *IEEE Trans. Neural Syst. Rehabil. Eng.* **2021**, *29*, 2464–2473. [CrossRef] [PubMed]
24. Guo, Y.; Yang, X.J. Modeling and Predicting Trust Dynamics in Human–Robot Teaming: A Bayesian Inference Approach. *Int. J. Soc. Robot.* **2021**, *13*, 1899–1909. [CrossRef]
25. Azevedo-Sa, H.; Jayaraman, S.K.; Esterwood, C.T.; Yang, X.J.; Robert, L.P.; Tilbury, D.M. Real-time estimation of drivers' trust in automated driving systems. *Int. J. Soc. Robot.* **2021**, *13*, 1911–1927. [CrossRef]
26. Nian, R.; Liu, J.; Huang, B. A review on reinforcement learning: Introduction and applications in industrial process control. *Comput. Chem. Eng.* **2020**, *139*, 106886. [CrossRef]
27. Joo, T.; Jun, H.; Shin, D. Task Allocation in Human–Machine Manufacturing Systems Using Deep Reinforcement Learning. *Sustainability* **2022**, *14*, 2245. [CrossRef]
28. Yang, Y.; Li, Z.; He, L.; Zhao, R. A systematic study of reward for reinforcement learning based continuous integration testing. *J. Syst. Softw.* **2020**, *170*, 110787. [CrossRef]
29. Chen, M.; Lam, H.K.; Shi, Q.; Xiao, B. Reinforcement learning-based control of nonlinear systems using Lyapunov stability concept and fuzzy reward scheme. *IEEE Trans. Circuits Syst. II Express Briefs* **2019**, *67*, 2059–2063. [CrossRef]
30. Kofinas, P.; Vouros, G.; Dounis, A.I. Energy management in solar microgrid via reinforcement learning using fuzzy reward. *Adv. Build. Energy Res.* **2018**, *12*, 97–115. [CrossRef]
31. Jafarifarmand, A.; Badamchizadeh, M.A.; Khanmohammadi, S.; Nazari, M.A.; Tazehkand, B.M. A new self-regulated neuro-fuzzy framework for classification of EEG signals in motor imagery BCI. *IEEE Trans. Fuzzy Syst.* **2017**, *26*, 1485–1497. [CrossRef]
32. Lin, F.C.; Ko, L.W.; Chuang, C.H.; Su, T.P.; Lin, C.T. Generalized EEG-based drowsiness prediction system by using a self-organizing neural fuzzy system. *IEEE Trans. Circuits Syst. I Regul. Pap.* **2012**, *59*, 2044–2055. [CrossRef]
33. Zhang, L.; Shi, Y.; Chang, Y.C.; Lin, C.T. Hierarchical Fuzzy Neural Networks With Privacy Preservation for Heterogeneous Big Data. *IEEE Trans. Fuzzy Syst.* **2020**, *29*, 46–58. [CrossRef]
34. Shayesteh, S.; Ojha, A.; Jebelli, H. Workers' Trust in Collaborative Construction Robots: EEG-Based Trust Recognition in an Immersive Environment. In *Automation and Robotics in the Architecture, Engineering, and Construction Industry*; Springer: Berlin/Heidelberg, Germany, 2022; pp. 201–215.

35. Hoeks, B.; Ellenbroek, B.A. A neural basis for a quantitative pupillary model. *J. Psychophysiol.* **1993**, *7*, 315–315.
36. Baevsky, R.M.; Chernikova, A.G. Heart rate variability analysis: Physiological foundations and main methods. *Cardiometry* **2017**, 66–76. [CrossRef]
37. Silambarasan, I.; Sriram, S. Hamacher sum and Hamacher product of fuzzy matrices. *Intern. J. Fuzzy Math. Arch.* **2017**, *13*, 191–198.
38. Watkins, C.J.; Dayan, P. Q-learning. *Mach. Learn.* **1992**, *8*, 279–292. [CrossRef]
39. Clifton, J.; Laber, E. Q-learning: Theory and applications. *Annu. Rev. Stat. Its Appl.* **2020**, *7*, 279–301. [CrossRef]
40. Lin, J.L.; Hwang, K.S.; Wang, Y.L. A simple scheme for formation control based on weighted behavior learning. *IEEE Trans. Neural Netw. Learn. Syst.* **2013**, *25*, 1033–1044.
41. Qin, B.; Chung, F.L.; Wang, S. KAT: A Knowledge Adversarial Training Method for Zero-Order Takagi-Sugeno-Kang Fuzzy Classifiers. *IEEE Trans. Cybern.* **2020**, *52*, 6857–6871. [CrossRef]
42. Tkachenko, R.; Izonin, I.; Tkachenko, P. Neuro-Fuzzy Diagnostics Systems Based on SGTM Neural-Like Structure and T-Controller. In *Proceedings of the International Scientific Conference "Intellectual Systems of Decision Making and Problem of Computational Intelligence"*; Springer: Berlin/Heidelberg, Germany, 2021; pp. 685–695.
43. Lin, C.J.; Lin, C.T. Reinforcement learning for an ART-based fuzzy adaptive learning control network. *IEEE Trans. Neural Netw.* **1996**, *7*, 709–731. [PubMed]
44. Xie, J.; Xu, X.; Wang, F.; Liu, Z.; Chen, L. Coordination Control Strategy for Human-Machine Cooperative Steering of Intelligent Vehicles: A Reinforcement Learning Approach. *IEEE Trans. Intell. Transp. Syst.* **2022**, 1–15. [CrossRef]
45. Lin, C.T.; Kan, M.C. Adaptive fuzzy command acquisition with reinforcement learning. *IEEE Trans. Fuzzy Syst.* **1998**, *6*, 102–121.

technologies

Article

Evaluation Based on the Distance from the Average Solution Approach: A Derivative Model for Evaluating and Selecting a Construction Manager

Phuong Thanh Phan [1,2] and Phong Thanh Nguyen [1,2,*]

[1] Department of Project Management, Faculty of Civil Engineering, Ho Chi Minh City Open University, Ho Chi Minh City 700000, Vietnam
[2] Professional Knowledge & Project Management Research Team (K2P), Ho Chi Minh City Open University, Ho Chi Minh City 700000, Vietnam
* Correspondence: phong.nt@ou.edu.vn

Abstract: In the current market of integration and globalization, the competition between engineering and construction companies is increasing. Construction contractors can improve their competitiveness by evaluating and selecting qualified personnel for the construction engineering manager position for their company's civil engineering projects. However, most personnel evaluation and selection models in the construction industry rely on qualitative techniques, which leads to unsuitable decisions. To overcome this problem, this paper presents evaluation criteria and proposes a new model for selecting construction managers based on the evaluation based on the distance from the average solution approach (EDASA). The research results showed that EDASA has many strengths, such as solving the problem faster when the number of evaluation criteria or the number of alternatives is increased.

Keywords: construction manager; construction project; engineering management; EDASA; resource management; personnel selection; project management

Citation: Phan, P.T.; Nguyen, P.T. Evaluation Based on the Distance from the Average Solution Approach: A Derivative Model for Evaluating and Selecting a Construction Manager. *Technologies* **2022**, *10*, 107. https://doi.org/10.3390/technologies10050107

Academic Editors: Manoj Gupta, Eugene Wong and Gwanggil Jeon

Received: 15 September 2022
Accepted: 12 October 2022
Published: 14 October 2022

1. Introduction

Resource management is just as critical as challenging engineering project management themes such as schedule management, time management, cost management, quality management, and risk management [1–4]. In resource management, evaluating and recruiting personnel for engineering projects are always given top priority. Any project's success may be attributed to the fundamental human principle of selecting the appropriate personnel, delivering the correct product, and delivering the product at the right time [5–10]. Therefore, appropriate candidate evaluation criteria are needed for evaluating and selecting personnel for the position of construction manager in civil engineering projects [11–18]. A new scientific and objective selection method is needed for the company to select a qualified candidate. However, a portion of the currently used models for personnel selection relies on qualitative methods, often resulting in inappropriate decisions [19,20]. The goal of this study is to present evaluation criteria and propose a new method for choosing a construction manager using an EDASA to address this issue.

Next, this paper presents a literature review on personnel competence in construction projects to provide the foundation for identifying basic criteria for selecting a construction manager for civil engineering projects.

Competence is the ability to use skills, knowledge, and personal characteristics to improve efficiency in work performance, increasing the likelihood of project success [21]. According to the Project Management Institute (PMI), there are three types of project management competencies: knowledge, performance, and personal competence [22]. When a project manager applies methods, tools, and techniques to project activities, they are said to have knowledge competency. The project manager's ability to implement their project management expertise to complete the project's needs is performance competence. Finally, personal competencies, in addition to attitudes and fundamental personality qualities, describe how project managers perform when engaging in activities within the context of a project. The capacity framework identifies ten management implementation capacities, including managing project (1) integration, (2) scope, (3) time and schedule, (4) cost, (5) quality, (6) resource, (7) risk, (8) procurement, (9) communication, and (10) stakeholders. The six personal competencies include (1) communication, (2) leadership, (3) management, (4) cognitive ability, (5) efficiency, and (6) professionalism.

Construction managers have an important role in projects. Knowledge and skills are two core factors for construction managers [23]. The development and implementation of personnel training methods in the enterprise will help the management apparatus be flexible in assigning personnel, permitting maximum project efficiency. This benefits the construction manager and helps the company, which has a key human resource for long-term development. El-Sabaa [24] identifies the characteristics and skills of an effective construction manager. The author considers communication skills as the top criterion of project managers, while technical skills were less influential. In addition, the authors also highlight the difference between a project manager and a construction company executive. While both require resourcefulness, a construction manager requires extensive, broad knowledge to make the best use of resources. In addition, construction managers must have soft skills, accept change, and be proactive in their work. The construction manager should be the leader throughout the project lifecycle. In that role, the construction manager must be the individual who knows how to plan and monitor the entire project for the best efficiency.

Gharehbaghi and McManus [17] explore the necessary leadership qualities for successful construction projects. They depend on the task, team, work environment, resources, schedule, and budget. The author also suggests four important criteria that construction management engineers need, including (1) knowing other people, (2) knowing yourself well, (3) being able to communicate, and (4) decisiveness. A good leader must know and understand the wishes of their subordinates and demonstrate concern for their lives. In other words, understand personnel at the construction site, share experiences, and unite to accomplish individual goals. Construction managers must understand themselves and continue to learn and develop. A good leader must communicate well and be decisive in all situations. In addition, a construction manager must possess good general knowledge and skills and thoroughly understand the company culture and the construction site. These conditions require construction companies to equip themselves with the necessary additional knowledge through training, including short-term training courses.

Dainty, et al. [25] identify the core competencies related to the construction manager's role and deploy a predictive model to make selection decisions and train personnel for construction managers for large construction companies. The authors reveal that many project manager candidates participate in surveys in which their employees are asked to recount problems and solutions. This practice allows managers to understand their capabilities. The authors provide a logistic regression model for assessing candidate competence, and their results show that self-control and team leadership are the dominant factors determining a construction manager's competence. In addition to 12 performance-related abilities important for project managers, the study identified 10 additional competency characteristics: accomplishment orientation, initiative, information seeking, attention, impact, and efficacy in meeting client needs, direction, teamwork and collaboration, analytical and conceptual thinking, and agile execution.

Based on interviews with 13 project leaders, civil engineers, and construction managers, as well as 7 team leaders, in 13 construction projects in Sweden, Styhre and Josephson [26] find the importance of specific roles in project success. The authors also show that, although they are required to manage a substantial amount of work in their projects, most construction management engineers are satisfied with their work. The authors have shown that the position of construction engineers is indispensable to ensuring the project's success.

Construction enterprises should establish training courses for construction engineers and consider core skills for advanced training according to job characteristics. Technical skills alone are insufficient to create a successful project manager. Fisher [27] suggests six soft skills necessary for human resource management and corresponding behaviors for an effective construction manager, including (i) understanding employee behavioral characteristics, (ii) the ability to lead the team, (iii) the ability to influence, (iv) committing clear and honest actions, (v) the ability to resolve conflicts, and (vi) perceiving personality differences of project team members.

Zulch [28] recognizes essential characteristics that a construction manager must possess for successful communication. The managers should know that all leadership styles will have varying degrees of influence on the success of a project. Knowledge of leadership will help managers flexibly solve work problems according to specific situations, permitting project success. Evaluation of the capacity of the construction manager cannot be complete without assessing their experience because, without experience, competence cannot be demonstrated or improved [29]. Moreover, experience is considered an important factor for successful personal growth. To successfully fulfill their assigned role, individuals need to accumulate the necessary experience and thus complement their potential.

According to the APM Competence Framework, project managers' competencies include 20 technical competencies, 15 behavioral competencies, and 11 contextual competencies [30]. Construction project managers must have both technical knowledge and proficiency and abilities to coordinate and communicate effectively with various stakeholders. To ensure project success, construction managers must possess technical expertise, people skills, and a work ethic. Nuwan, et al. [11] discover management development approaches. The authors use the Delphi method, including 12 experts and 44 respondents, to develop 20 factors of specialized knowledge, soft skills, and working attitude that are meaningful for construction engineers. The most important of these are planning and managing progress. The most important soft skills regarding working attitude are time management and leadership.

Based on the list of capacity assessment criteria surveyed above, construction experts in Vietnam have selected the 15 most important criteria (within three groups) to select construction managers in Table 1.

Table 1. Criteria for the evaluation and selection of a construction manager.

	Code	Criteria for the Evaluation and Selection of a Construction Manager
CE		Construction Expertise
	CE1	Construction technical knowledge
	CE2	Knowledge of construction organization and management
	CE3	Knowledge of the construction schedule
	CE4	Knowledge of occupational safety and environmental sanitation
	CE5	Understanding of construction quality and volume management
SS		Soft Skills
	SS1	Communication and presentation skills
	S2	Construction problem-solving skills
	S3	Ability to lead and guide construction workers
	S4	Information management skills (documents, construction records)
	S5	Creative innovation ability
WE		Work Experience
	WE1	Similar projects and works completed
	WE2	Experience working with owner, project management unit, and supervisory unit
	WE3	Experience working with contractors, project teams, and construction suppliers
	WE4	Professional degrees and certificates in construction
	WE5	Ability to use construction specialized software

The rest of the paper is organized as follows. Section 2 provides the EDASA research method employed in Section 3. This section describes the empirical results and discusses the EDASA application. The final section concludes the study.

2. Methodology

Keshavarz et al. invented the distance from the average solution approach EDASA method in 2015 [31,32]. The best alternative is selected using EDASA by measuring the distance of each choice from the ideal value. This method is especially useful in situations with contradicting attributes or conflicting criteria. EDASA has been applied in the evaluation of airline services [33], solving air traffic problems [34], personnel selection [35], green supplier selection [36], material selection [37], and hospital site selection [38]. Using this method, suppose there are n construction manager candidates and m evaluation and selection criteria. The steps for using the proposed method are presented as follows [31–33,35–60]:

Step 1: Calculate the weight of each criterion.

Step 2: Create a decision-making matrix, shown as follows:

$$X = \begin{bmatrix} x_{11} & x_{12} & \cdots & x_{1n} \\ x_{21} & x_{21} & \cdots & x_{2n} \\ \vdots & \vdots & & \vdots \\ x_{1n} & x_{2n} & \cdots & x_{mn} \end{bmatrix}; \ i = 1,\ 2,\ \ldots,\ m;\ j = 1,\ 2,\ \ldots,\ n \tag{1}$$

where

x_{ij} denotes the performance value of the ith alternative on the jth criterion. Moreover, the assessor weight of the criteria $w = [w_1, w_2, \ldots, w_n]$.

Step 3: Identify the average solution based on each of the following criteria:

$$\overline{x}_j = (x_1,\ x_2, \ldots, x_n), \tag{2}$$

where

$$\overline{x}_j = \frac{\sum_{i=1}^{m} x_{ij}}{m};\ j = 1,\ 2, \ldots,\ n.$$

Step 4: Determine the positive and negative distances from the average solution.

The positive distances from the average (PDA) and the negative distances from the average (NDA) are dependent on the type of criteria (benefit and cost), calculated as follows:

$$d_{ij}^{+} = \begin{cases} \frac{\max(0,(x_{ij}-\overline{x}_j))}{\overline{x}_j}, j \in \Omega_{\max} \\ \frac{\max(0,(\overline{x}_j-x_{ij}))}{\overline{x}_j}, j \in \Omega_{\min} \end{cases}. \tag{3}$$

and

$$d_{ij}^{-} = \begin{cases} \frac{\max(0,(\overline{x}_j-x_{ij}))}{\overline{x}_j}, j \in \Omega_{\max} \\ \frac{\max(0,(x_{ij}-\overline{x}_j))}{\overline{x}_j}, j \in \Omega_{\min} \end{cases}. \tag{4}$$

where

d_{ij}^{+} and d_{ij}^{-} denote the positive and negative distance of ith candidates from the average solution of jth factors, respectively;

$\Omega_{\max}$ and $\Omega_{\min}$ are positive real numbers that represent the set of benefit criteria and the cost criteria, respectively.

Step 5: Determine the weighted sum of PDA, and the weighted sum of NDA, for all alternatives, shown as follows:

$$Q_i^{+} = \sum_{j=1}^{n} w_j d_{ij}^{+};\ i = 1,\ 2,\ \ldots,\ m \tag{5}$$

$$Q_i^{-} = \sum_{j=1}^{n} w_j d_{ij}^{-};\ i = 1,\ 2,\ \ldots,\ m \tag{6}$$

where

w_j denotes the nonnegative weight of the criterion j.

Step 6: Normalize the values of the weighted sums of PDA and NDA for each of the candidates, as shown below:

$$S_i^+ = \frac{Q_i^+}{\max\limits_k Q_k^+}. \tag{7}$$

$$S_i^- = 1 - \frac{Q_i^-}{\max\limits_k Q_k^-}. \tag{8}$$

where

S_i^+ and S_i^- denotes the normalized weighted sum of the PDA and the NDA, respectively.

Step 7: The appraisal scores S_i for all project managers are computed as follows:

$$S_i = \frac{S_i^+ + S_i^-}{2}. \tag{9}$$

where

$0 \le S_i \le 1; i = 1, 2, \ldots, m$

The appraisal scores for construction manager candidates are listed in descending order. Among the applicants, the one with the highest S_i is the best option.

3. Results

We applied the EDASA through a case study in one construction project in Vietnam. The recruitment committee consists of five professionals who must evaluate and select one of three candidates (A1, A2, A3) for the construction manager position. First, construction experts used Saaty's scale of 1–9 to make a pairwise comparison of evaluation and selection criteria for construction managers. The results of the weight calculation of these criteria are presented in Table 2.

Table 2. The weight of criteria for the evaluation and selection of a construction manager.

	Code	Criteria for the Evaluation and Selection of a Construction Manager	Weight
CE		Construction Expertise	
	CE1	Construction technical knowledge	0.1760
	CE2	Knowledge of construction organization and management	0.0920
	CE3	Knowledge of the construction schedule	0.0630
	CE4	Knowledge of occupational safety and environmental sanitation	0.2900
	CE5	Understanding of construction quality and volume management	0.0380
SS		Soft Skills	
	SS1	Communication and presentation skills	0.0070
	SS2	Construction problem-solving skills	0.0500
	SS3	Ability to lead and guide construction workers	0.0300
	SS4	Information management skills (documents, construction records)	0.0110
	SS5	Creative innovation ability	0.0170
WE		Work Experience	
	WE1	Similar projects and works completed	0.0270
	WE2	Experience working with owner, project management unit, and supervisory unit	0.1040
	WE3	Experience working with contractors, project teams, and construction suppliers	0.0580
	WE4	Professional degrees and certificates in construction	0.0230
	WE5	Ability to use construction specialized software	0.0140

Second, five construction experts created the decision-making matrix and calculated the average solution using Equation (2) according to all selection criteria, as shown in Table 3.

Table 3. The average solution of criteria for the evaluation and selection of a construction manager.

Code	Criteria for Evaluation and Selection of Construction Manager	A1	A2	A3	$\bar{x}_j$
CE	Construction Expertise	75	60	82	72.3333
CE1	Construction technical knowledge	83	62	74	73.0000
CE2	Knowledge of construction organization and management	84	71	64	73.0000
CE3	Knowledge of the construction schedule	72	62	82	72.0000
CE4	Knowledge of occupational safety and environmental sanitation	62	84	71	72.3333
CE5	Understanding of construction quality and volume management	71	85	63	73.0000
SS	Soft Skills	73	62	82	72.3333
SS1	Communication and presentation skills	82	73	63	72.6667
SS2	Construction problem-solving skills	74	81	61	72.0000
SS3	Ability to lead and guide construction workers	62	83	71	72.0000
SS4	Information management skills (documents, construction records)	84	60	74	72.6667
SS5	Creative innovation ability	72	63	81	72.0000
WE	Work Experience	63	73	80	72.0000
WE1	Similar projects and works completed	83	62	74	73.0000
WE2	Experience working with owner, project management unit, and supervisory unit	64	81	71	72.0000
WE3	Experience working with contractors, project teams, and construction suppliers	75	60	82	72.3333
WE4	Professional degrees and certificates in construction	83	62	74	73.0000
WE5	Ability to use construction specialized software	84	71	64	73.0000

The positive and negative distances from the average solution are calculated using Equations (3) and (4), as shown in Tables 4 and 5.

Table 4. Values of the positive distances from the average (PDA).

Code	Criteria for the Evaluation and Selection of a Construction Manager	A1	A2	A3
CE1	Construction technical knowledge	0.0369	0.0000	0.1336
CE2	Knowledge of construction organization and management	0.1370	0.0000	0.0137
CE3	Knowledge of the construction schedule	0.1507	0.0000	0.0000
CE4	Knowledge of occupational safety and environmental sanitation	0.0000	0.0000	0.1389
CE5	Understanding of construction quality and volume management	0.0000	0.1613	0.0000
SS1	Communication and presentation skills	0.0000	0.1644	0.0000
SS2	Construction problem-solving skills	0.0092	0.0000	0.1336
SS3	Ability to lead and guide construction workers	0.1284	0.0046	0.0000
SS4	Information management skills (documents, construction records)	0.0278	0.1250	0.0000
SS5	Creative innovation ability	0.0000	0.1528	0.0000
WE1	Work experience	0.1560	0.0000	0.0183
WE2	Similar projects and works completed	0.0000	0.0000	0.1250
WE3	Experience working with owner, project management unit, and supervisory unit	0.0000	0.0139	0.1111
WE4	Experience working with contractors, project teams, and construction suppliers	0.1370	0.0000	0.0137
WE5	Professional degrees and certificates in construction	0.0000	0.0000	0.0000

The weighted sum and the weighted normalized sum of PDA and NDA for the candidates are calculated using Equations (5)–(8). Finally, the appraisal score of each construction manager candidate is calculated using Equation (9). All results are shown in Table 6.

Table 5. Values of the negative distances from the average (NDA).

Code	Criteria for the Evaluation and Selection of a Construction Manager	A1	A2	A3
CE1	Construction technical knowledge	0.0000	0.1705	0.0000
CE2	Knowledge of construction organization and management	0.0000	0.1507	0.0000
CE3	Knowledge of the construction schedule	0.0000	0.0274	0.1233
CE4	Knowledge of occupational safety and environmental sanitation	0.0000	0.1389	0.0000
CE5	Understanding of construction quality and volume management	0.1429	0.0000	0.0184
SS1	Communication and presentation skills	0.0274	0.0000	0.1370
SS2	Construction problem-solving skills	0.0000	0.1429	0.0000
SS3	Ability to lead and guide construction workers	0.0000	0.0000	0.1330
SS4	Information management skills (documents, construction records)	0.0000	0.0000	0.1528
SS5	Creative innovation ability	0.1389	0.0000	0.0139
WE1	Work experience	0.0000	0.1743	0.0000
WE2	Similar projects and works completed	0.0000	0.1250	0.0000
WE3	Experience working with owner, project management unit, and supervisory unit	0.1250	0.0000	0.0000
WE4	Experience working with contractors, project teams, and construction suppliers	0.0000	0.1507	0.0000
WE5	Professional degrees and certificates in construction	0.0000	0.0000	0.0000

Table 6. The weighted normalized sum of PDA and NDA and the appraisal score.

	A1	A2	A3
Q_i^+	0.0406	0.0122	0.0920
Q_i^-	0.0152	0.1142	0.0153
S_i^+	0.4410	0.1326	1.0000
S_i^-	0.8666	0.0000	0.8657
S_i	0.6538	0.0663	0.9329

The calculation results in Table 6 show that candidate A3 has the highest appraisal score (0.9329). Therefore, this person is prioritized to be selected as the construction manager. The research results showed that EDASA has many strengths. First, some qualitative attributes could be converted into quantitative attributes. Second, compared with traditional assessment methods (e.g., AHP), EDASA can consider conflicting criteria in the same problem. Third, the time to apply EDASA to solve the problem was faster when the number of evaluation criteria or the number of alternatives increased. Finally, this method can be combined with other theories such as fuzzy logic or grey system theory to reflect the complexity or uncertainty of the real world because it has a solid mathematical basis [39,43,61].

4. Conclusions

The fundamental human principle of choosing the right personnel, delivering the right product, and delivering the product on time is necessary for the success of any engineering and construction project. This paper presents fifteen evaluation criteria for selecting a construction manager and proposes a new quantitative methodology for this selection utilizing EDASA. This method is practically applied through a case study of the evaluation and selection of construction managers, demonstrating its effectiveness, especially in the event of the evaluation of many construction manager candidates. In addition, in some situations where the selection problem is complex or has more selection criteria, the EDASA deterministic approach should be combined with another method or theory (such as fuzzy logic theory or grey system theory) to reflect the uncertainty in the judgment of the decision maker.

Author Contributions: Conceptualization, P.T.P.; Data curation, P.T.P.; Formal analysis, P.T.N.; Funding acquisition, P.T.P.; Investigation, P.T.N.; Methodology, P.T.N.; Project administration, P.T.P.; Resources, P.T.N.; Writing—original draft, P.T.P.; Writing—review and editing, P.T.N. All authors have read and agreed to the published version of the manuscript.

Funding: This research is funded by Ho Chi Minh City Open University under the grant number E2019.11.3.

Data Availability Statement: Not applicable.

Conflicts of Interest: The authors declare no conflict of interest.

References

1. Huemann, M.; Keegan, A.; Turner, J.R. Human resource management in the project-oriented company: A review. *Int. J. Proj. Manag.* **2007**, *25*, 315–323. [CrossRef]
2. Samimi, E.; Sydow, J. Human resource management in project-based organizations: Revisiting the permanency assumption. *Int. J. Hum. Resour. Manag.* **2020**, *32*, 49–83. [CrossRef]
3. Ling, F.Y.Y.; Ning, Y.; Chang, Y.H.; Zhang, Z. Human resource management practices to improve project managers' job satisfaction. *Eng. Constr. Arch. Manag.* **2018**, *25*, 654–669. [CrossRef]
4. Apollo, M.; Miszewska-Urbańska, E. Analysis of the Increase of Construction Costs in Urban Regeneration Projects. *Adv. Sci. Technol. Res. J.* **2015**, *9*, 68–74. [CrossRef]
5. Sulieman, H.A.; Alfaraidy, F.A.; Sulieman, H.A.; Alfaraidy, F.A. Influences of Project Management Capabilities on the Organizational Performance of the Saudi Construction Industry. *Eng. Technol. Appl. Sci. Res.* **2019**, *9*, 4144–4147. [CrossRef]
6. Belout, A.; Gauvreau, C. Factors influencing project success: The impact of human resource management. *Int. J. Proj. Manag.* **2004**, *22*, 1–11. [CrossRef]
7. Loosemore, M.; Dainty, A.; Lingard, H. *Human Resource Management in Construction Projects: Strategic and Operational Approaches*; Routledge: New York, NY, USA, 2003.
8. Yurova, V.A.; Velikoborets, G.; Vladyko, A. Design and Implementation of an Anthropomorphic Robotic Arm Prosthesis. *Technologies* **2022**, *10*, 103. [CrossRef]
9. Gemünden, H.G. Success Factors of Global New Product Development Programs, the Definition of Project Success, Knowledge Sharing, and Special Issues of Project Management Journal®. *Proj. Manag. J.* **2015**, *46*, 2–11. [CrossRef]
10. Zhou, Z.; Zhang, J.; Gong, C. Automatic detection method of tunnel lining multi-defects via an enhanced You Only Look Once network. *Comput. Civ. Infrastruct. Eng.* **2022**, *37*, 762–780. [CrossRef]
11. Nuwan, P.M.M.C.; Perera, B.A.K.S.; Dewagoda, K.G. Development of Core Competencies of Construction Managers: The Effect of Training and Education. *Technol. Knowl. Learn.* **2020**, *26*, 945–984. [CrossRef]
12. Abdullah, A.H.; Yaman, S.K.; Mohammad, H.; Hassan, P.F. Construction manager's technical competencies in Malaysian construction projects. *Eng. Constr. Arch. Manag.* **2018**, *25*, 153–177. [CrossRef]
13. Mohammad, H.; Tun, U.; Onn, H.; Hassan, P.F.; Khalijah, Y.S.; Tun, U.; Onn, H. Quantitative Significance Analysis for Technical Competency of Malaysian Construction Managers. *Issues Built Environ.* **2018**, 77–107.
14. Liu, H.; Zhang, H.; Zhang, R.; Jiang, H.; Ju, Q. Competence Model of Construction Project Manager in the Digital Era—The Case from China. *Buildings* **2022**, *12*, 1385. [CrossRef]
15. Mohammad, H.; Hassan, F.; Abd Rashid, R.; Yaman, S.K. Dimensionality analysis of technical competency for Malaysian construction managers. In Proceedings of the International UNIMAS STEM Engineering Conference, Kuching, Malaysia, 24–27 October 2016.
16. Chai, A.H.R. Competencies of Construction Manager. Ph.D. Thesis, Universiti Tunku Abdul Rahman, UTAR, Kampar, Malaysia, 2016.
17. Arditi, D.; Balci, G. Managerial Competencies of Female and Male Construction Managers. *J. Constr. Eng. Manag.* **2009**, *135*, 1275–1278. [CrossRef]
18. Gharehbaghi, K.; McManus, K. The construction manager as a leader. *Leadersh. Manag. Eng.* **2003**, *3*, 56–58. [CrossRef]
19. Sajjad, A.; Ahmad, W.; Hussain, S. Decision-Making Process Development for Industry 4.0 Transformation. *Adv. Sci. Technol. Res. J.* **2022**, *16*, 1–11. [CrossRef]
20. Zavadskas, E.K.; Turskis, Z.; Tamošaitienė, J. Multicriteria Selection of Project Managers by Applying Grey Criteria/Projektų Valdytojo Parinkimo Daugiatikslio Vertinimo Modelis. *Technol. Econ. Dev. Econ.* **2008**, *14*, 462–477. [CrossRef]
21. Moradi, S.; Kähkönen, K.; Aaltonen, K. Comparison of research and industry views on project managers' competencies. *Int. J. Manag. Proj. Bus.* **2019**, *13*, 543–572. [CrossRef]
22. Cartwright, C.; Yinger, M. Project management competency development framework. In Proceedings of the PMI Global Congress, Budapest, Hungary, 14–16 May 2007.
23. Edum-Fotwe, F.; McCaffer, R. Developing project management competency: Perspectives from the construction industry. *Int. J. Proj. Manag.* **2000**, *18*, 111–124. [CrossRef]
24. El-Sabaa, S. The skills and career path of an effective project manager. *Int. J. Proj. Manag.* **2001**, *19*, 1–7. [CrossRef]

25. Dainty, A.R.J.; Cheng, M.-I.; Moore, D.R. Competency-Based Model for Predicting Construction Project Managers' Performance. *J. Manag. Eng.* **2005**, *21*, 2–9. [CrossRef]
26. Styhre, A.; Josephson, P.E. Revisiting site manager work: Stuck in the middle? *Constr. Manag. Econ.* **2006**, *24*, 521–528. [CrossRef]
27. Fisher, E. What practitioners consider to be the skills and behaviours of an effective people project manager. *Int. J. Proj. Manag.* **2011**, *29*, 994–1002. [CrossRef]
28. Zulch, B. Leadership Communication in Project Management. *Procedia Soc. Behav. Sci.* **2014**, *119*, 172–181. [CrossRef]
29. IPMA. *Individual Competence Baseline*; International Project Management Association: Nijkerk, The Netherlands, 2015; p. 432.
30. De Rezende, L.B.; Blackwell, P. Project management competency framework. *Iberoam. J. Proj. Manag.* **2019**, *10*, 34–59.
31. Turskis, Z.; Morkunaite, Z.; Kutut, V. A Hybrid Multiple Criteria Evaluation Method of Ranking of Cultural Heritage Structures for Renovation Projects. *Int. J. Strat. Prop. Manag.* **2017**, *21*, 318–329. [CrossRef]
32. Ghorabaee, M.K.; Zavadskas, E.K.; Olfat, L.; Turskis, Z. Multi-Criteria Inventory Classification Using a New Method of Evaluation Based on Distance from Average Solution (EDAS). *Informatica* **2015**, *26*, 435–451. [CrossRef]
33. Ghorabaee, M.K.; Amiri, M.; Zavadskas, E.K.; Turskis, Z.; Antucheviciene, J. A new hybrid simulation-based assignment approach for evaluating airlines with multiple service quality criteria. *J. Air Transp. Manag.* **2017**, *63*, 45–60. [CrossRef]
34. Kikomba, M.; Mabela, R.; Ntantu, D. Applying EDAS method to solve air traffic problems. *Int. J. Sci. Innov. Math. Res. (IJSIMR)* **2016**, *4*, 15–23.
35. Stanujkic, D.; Popovic, G.; Brzakovic, M. An approach to personnel selection in the IT industry based on the EDAS method. *Transform. Bus. Econ.* **2018**, *17*, 44.
36. Wei, G.; Wei, C.; Guo, Y. EDAS method for probabilistic linguistic multiple attribute group decision making and their application to green supplier selection. *Soft Comput.* **2021**, *25*, 9045–9053. [CrossRef]
37. Kumar, R.; Dubey, R.; Singh, S.; Singh, S.; Prakash, C.; Nirsanametla, Y.; Królczyk, G.; Chudy, R. Multiple-Criteria Decision-Making and Sensitivity Analysis for Selection of Materials for Knee Implant Femoral Component. *Materials* **2021**, *14*, 2084. [CrossRef] [PubMed]
38. Adalı, E.A.; Tuş, A. Hospital site selection with distance-based multi-criteria decision-making methods. *Int. J. Health Manag.* **2019**, *14*, 534–544. [CrossRef]
39. Stanujkic, D.; Zavadskas, E.K.; Ghorabaee, M.K.; Turskis, Z. An Extension of the EDAS Method Based on the Use of Interval Grey Numbers. *Stud. Inform. Control* **2017**, *26*. [CrossRef]
40. Stević, Z.; Vasiljević, M.; Zavadskas, E.K.; Sremac, S.; Turskis, Z. Selection of carpenter manufacturer using fuzzy EDAS method. *Eng. Econ.* **2018**, *29*, 281–290. [CrossRef]
41. Maksimović, M.; Brzakovic, M.; Grahovac, M.; Jovanovic, I. An approach for evaluation the safety and quality of transport at the open pit mines, based on the EDAS method. *Min. Met. Eng. Bor* **2017**, 139–144. [CrossRef]
42. Zavadskas, E.K.; Kaklauskas, A.; Turskis, Z.; Tamošaitienė, J. Multi-Attribute Decision-Making Model by Applying Grey Numbers. *Informatica* **2009**, *20*, 305–320. [CrossRef]
43. Vesković, S.; Stević, Z.; Karabašević, D.; Rajilić, S.; Milinković, S.; Stojić, G. A New Integrated Fuzzy Approach to Selecting the Best Solution for Business Balance of Passenger Rail Operator: Fuzzy PIPRECIA-Fuzzy EDAS Model. *Symmetry* **2020**, *12*, 743. [CrossRef]
44. Keshavarz-Ghorabaee, M.; Amiri, M.; Zavadskas, E.K.; Turskis, Z.; Antucheviciene, J. A Dynamic Fuzzy Approach Based on the EDAS Method for Multi-Criteria Subcontractor Evaluation. *Information* **2018**, *9*, 68. [CrossRef]
45. Ghorabaee, M.K.; Zavadskas, E.K.; Amiri, M.; Turskis, Z. Extended EDAS Method for Fuzzy Multi-criteria Decision-making: An Application to Supplier Selection. *Int. J. Comput. Commun. CONTROL* **2016**, *11*, 358–371. [CrossRef]
46. Naik, G.; Kishore, R.; Dehmourdi, S.A.M. Modeling a Multi-Criteria Decision Support System for Prequalification Assessment of Construction Contractors using CRITIC and EDAS Models. *Oper. Res. Eng. Sci. Theory Appl.* **2021**, *4*, 79–101. [CrossRef]
47. Das, P.P.; Chakraborty, S. Multi-response Optimization of Hybrid Machining Processes Using Evaluation Based on Distance from Average Solution Method in Intuitionistic Fuzzy Environment. *Process Integr. Optim. Sustain.* **2020**, *4*, 481–495. [CrossRef]
48. Zhang, S.; Wei, G.; Gao, H.; Wei, C.; Wei, Y. Edas Method for Multiple Criteria Group Decision Making with Picture Fuzzy Information and Its Application to Green Suppliers Selections. *Technol. Econ. Dev. Econ.* **2019**, *25*, 1123–1138. [CrossRef]
49. Schitea, D.; Deveci, M.; Iordache, M.; Bilgili, K.; Akyurt, I.Z.; Iordache, I. Hydrogen mobility roll-up site selection using intuitionistic fuzzy sets based WASPAS, COPRAS and EDAS. *Int. J. Hydrog. Energy* **2019**, *44*, 8585–8600. [CrossRef]
50. Karaşan, A.; Kahraman, C. A novel interval-valued neutrosophic EDAS method: Prioritization of the United Nations national sustainable development goals. *Soft Comput.* **2018**, *22*, 4891–4906. [CrossRef]
51. Ecer, F. Third-Party Logistics (3pls) Provider Selection Via Fuzzy Ahp and Edas Integrated Model. *Technol. Econ. Dev. Econ.* **2017**, *24*, 615–634. [CrossRef]
52. Peng, X.; Liu, C. Algorithms for neutrosophic soft decision making based on EDAS, new similarity measure and level soft set. *J. Intell. Fuzzy Syst.* **2017**, *32*, 955–968. [CrossRef]
53. Peng, X.; Dai, J.; Yuan, H. Interval-valued Fuzzy Soft Decision Making Methods Based on MABAC, Similarity Measure and EDAS. *Fundam. Inform.* **2017**, *152*, 373–396. [CrossRef]
54. Ghorabaee, M.K.; Amiri, M.; Zavadskas, E.K.; Turskis, Z.; Antucheviciene, J. Stochastic EDAS method for multi-criteria decision-making with normally distributed data. *J. Intell. Fuzzy Syst.* **2017**, *33*, 1627–1638. [CrossRef]

55. Keshavarz-Ghorabaee, M.; Amiri, M.; Zavadskas, E.K.; Turskis, Z.; Antucheviciene, J. A new multi-criteria model based on interval type-2 fuzzy sets and EDAS method for supplier evaluation and order allocation with environmental considerations. *Comput. Ind. Eng.* **2017**, *112*, 156–174. [CrossRef]
56. Keshavarz-Ghorabaee, M.; Amiri, M.; Zavadskas, E.K.; Turskis, Z. Multi-criteria group decision-making using an extended edas method with interval type-2 fuzzy sets. *E+M Èkon. A Manag.* **2017**, *20*, 48–68. [CrossRef]
57. Kahraman, C.; Keshavarz-Ghorabaee, M.; Zavadskas, E.K.; Onar, S.C.; Yazdani, M.; Oztaysi, B. Intuitionistic Fuzzy Edas Method: An Application to Solid Waste Disposal Site Selection. *J. Environ. Eng. Landsc. Manag.* **2017**, *25*, 1–12. [CrossRef]
58. Ahn, C.W.; An, J.; Yoo, J.-C. Estimation of particle swarm distribution algorithms: Combining the benefits of PSO and EDAs. *Inf. Sci.* **2012**, *192*, 109–119. [CrossRef]
59. Zhang, X. Criteria for Selecting the Private-Sector Partner in Public–Private Partnerships. *J. Constr. Eng. Manag.* **2005**, *131*, 631–644. [CrossRef]
60. Sun, H.; Wei, G.-W.; Chen, X.-D.; Mo, Z.-W. Extended EDAS method for multiple attribute decision making in mixture z-number environment based on CRITIC method. *J. Intell. Fuzzy Syst.* **2022**, *43*, 2777–2788. [CrossRef]
61. Stević, Z.; Vasiljević, M.; Puška, A.; Tanackov, I.; Junevičius, R.; Vesković, S. Evaluation of Suppliers under Uncertainty: A Multiphase Approach Based on Fuzzy Ahp and Fuzzy Edas. *Transport* **2019**, *34*, 52–66. [CrossRef]

technologies

MDPI

Article

Friction Stir Welding of Ti-6Al-4V Using a Liquid-Cooled Nickel Superalloy Tool

Sergei Tarasov, Alihan Amirov, Andrey Chumaevskiy *, Nikolay Savchenko, Valery E. Rubtsov, Aleksey Ivanov, Evgeniy Moskvichev and Evgeny Kolubaev

Institute of Strength Physics and Materials Science, Siberian Branch of Russian Academy of Sciences, 634055 Tomsk, Russia
* Correspondence: tch7av@ispms.com; Tel.: +7-(3822)-28-68-63

Abstract: Friction stir welding (FSW) of titanium alloy was carried out using liquid cooling of the FSW tool made of heat-resistant nickel superalloy. Cooling of the nickel superalloy tool was performed by means of circulating water inside the tool. The FSW joints were characterized by microstructures and mechanical strength. The mechanical strength of the joints was higher than that of the base metal.

Keywords: friction stir welding; titanium alloys; weld strength; microhardness; X-ray structure analysis; fractography; tool wear

check for updates

Citation: Tarasov, S.; Amirov, A.; Chumaevskiy, A.; Savchenko, N.; Rubtsov, V.E.; Ivanov, A.; Moskvichev, E.; Kolubaev, E. Friction Stir Welding of Ti-6Al-4V Using a Liquid-Cooled Nickel Superalloy Tool. *Technologies* **2022**, *10*, 118. https://doi.org/10.3390/technologies10060118

Academic Editors: Manoj Gupta, Eugene Wong and Gwanggil Jeon

Received: 31 August 2022
Accepted: 17 November 2022
Published: 18 November 2022

1. Introduction

Titanium alloys are known for having a number of important functional characteristics that include, among others, high corrosion resistance, high specific strength, and biological compatibility [1–3]. Therefore, these alloys are widely used in aerospace, power, and chemical industries, as well as in shipbuilding and surgery [4–6]. Depending on the dominating phase state, titanium alloys can be classified into: α-Ti alloys, which include technical purity grade titanium and titanium alloyed with α-Ti supporting elements; (α + β)-Ti alloys, containing both α-Ti and β-Ti supporting elements; and β-Ti alloys containing up to 30 wt.% of the β-Ti supporting elements [7,8]. The phase composition of the alloy determines its mechanical characteristics so that ductile alloys contain more β-Ti, while α-Ti attains more strength [9,10].

When it comes to obtaining fusion-welded joints on the titanium alloys, some problems may occur such as overheating, excess grain growth, and high residual stresses due to low heat conductivity [11]. Solutions to these problems may be found when applying thermal post-treatment or/and surface impact treatment [12–14]. An alternative solution may be using friction stir welding (FSW), which has been widely and successfully used for building welded structures from aluminum alloys in aerospace, transportation, and power industries, and the advantage of which is that the joining is by intermixing the solid plasticized and refined metal [15–20].

Earlier experiments with the FSW on titanium alloys have revealed a number of problems, among which the fast wear of the FSW tool and intermixing the wear particles into the weld joint were the most prominent. The most popular materials for fabricating the FSW tools intended for titanium alloys are those with refractory metals, such as tungsten. However, cemented tungsten carbide tools demonstrated high wear despite their good heat resistance [21]. The most stable against wear in FSP (friction stir processing) were the tools made of tungsten-rhenium alloys, though production costs were too high [22]. Polycrystalline cubic boron carbide showed promising results in FSW on steels [23], but was unstable on titanium alloys, as it reacted with titanium, and formed brittle titanium borides and nitrides that were detrimental for the welded joint strength [24,25].

High heat resistance is the main characteristic to provide acceptable wear resistance of the FSW tool, and therefore, one of the possible solutions may be to use nickel superalloys, which usually work at 900–1000 °C as turbine blades [26]. Such an approach has already been used for friction stir processing on titanium alloys [27]. For example, acceptable results were obtained on technically pure titanium and α′-Ti alloys [28,29]. The wear rate of nickel superalloy tools was also quite high [30] and required some measures to reduce it. Poor heat conductivity may lead to overheating of the tool, loss of strength, as well as enhancing the reaction-diffusion between the tool and the stirred alloy.The natural remedy may be cooling the tool by means of fluid flow.

This work was focused on studying the effect of the FSW nickel superalloy tool cooling on the FSW tool wear, FSW joint characteristics and microstructures of the FSW joint zones.

2. Materials and Methods

2.1. Experimental Set-Up and Materials

The FSW on a Ti-6Al-4V was carried out with the use of argon blow shielding against oxidizing the seam metal. Argon was supplied via an inlet directly to the welding zone, as shown in Figure 1. Titanium alloy sheets with 2.5 mm thickness were secured on an AISI 304 stainless steel substrate using special clamps. On plunging the tool into the metal, the plunging force was maintained at the constant level. The tool inclination with respect to the horizontal plane was 1.5°. A liquid flow cooling system was used to supply a coolant via an axial hole, and thus, limit the tool's heating (Figure 2).

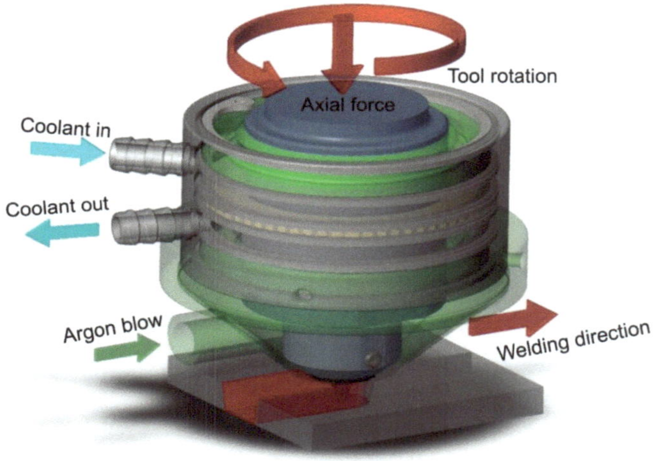

Figure 1. Schemes of friction stir welding titanium alloys with argon gas blanket of the welding joint and liquid tool cooling. Blue and green arrows show the coolant fluid and gas flow directions, respectively. Red arrows show the tool rotation and transverse motion and axial plunging force. Directions.

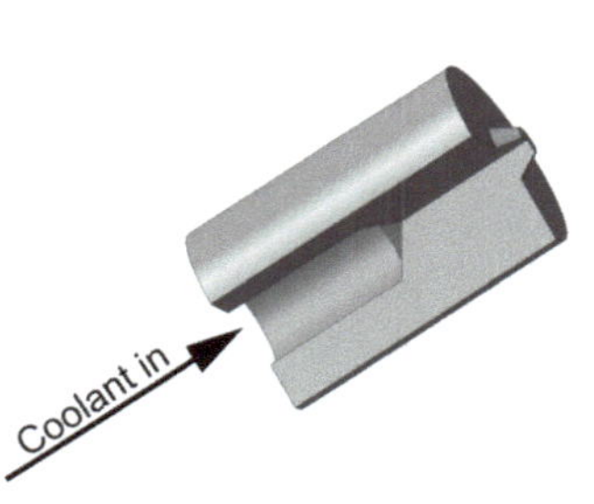

Figure 2. Scheme of working tool from ZhS6U alloy for FSW titanium alloys. The arrow shows the coolant inflow direction.

The tool's shoulder and pin diameters were 20 mm and 3 mm, respectively, with the pin's height at 2.3 mm. The compositions of nickel superalloy and titanium alloys are shown in Tables 1 and 2.

Table 1. Element composition of ZhS6U superalloy (wt%).

Fe	Nb	Ti	Cr	Co	W	Ni	Al	Mo	C
≤1	0.8–1.2	2–2.9	8–9.5	9–10.5	9.5–11	54.3–62.7	5.1–6	1.2–2.4	0.13–0.2
Ce	**Si**	**Mn**	**P**	**S**	**Zr**	**Bi**	**B**	**Y**	**Pb**
≤0.02	≤0.4	≤0.4	≤0.015	≤0.01	≤0.04	≤0.0005	≤0.035	≤0.01	≤0.01

Table 2. Element composition of Ti-6Al-4V alloy (wt%).

Fe	C	Si	V	N	Ti	Al	Zr	O	H	Other
≤0.6	≤0.1	≤0.1	3.5–5.3	≤0.05	86.45–90.9	5.3–6.8	≤0.3	≤0.2	≤0.015	0.3

The FSW was initially performed with parameters as reported elsewhere [31], but later, their values were corrected for a new tool design and were as follows: axial force in plunging and welding were Fp = 43 kN and Fw = 45 kN, respectively; welding speed V = 86 mm/min; rotation rates n1 = 340, n2 = 360, and n3 = 380 RPM for samples 1, 2, and 3, respectively.

2.2. Equipment and Sample Preparation

The metallographic views were prepared by cutting the joint in a plane perpendicular to the welding direction, then grinding and polishing the section on abrasive papers with grain sizes P180 to P2000 and diamond paste ACM 1/0. Etching was carried out in a reagent composed of $C_3H_8O_3$—30 mL, HNO_3—10 mL, and HF—10 mL, for 30–35 s.

Microstructural examination was performed using an optical microscope «Altami Met 1S». An SEM field emission cathode instrument FEG SEM Apreo 2 S (Thermo Fisher Scientific, Waltham, MA, USA) attached to an EDS Octane Elect Super (EDAX) analyzer was used for fractography.

An XRD diffractometer DRON 7 operated at 36 kV, 22 mA, CoKα radiation with wavelength 1.7902 Å, diffraction angle 2θ interval 15–102°, with 0.05° step, and exposition of 40 s was used to characterize phases formed. A symmetrical Bragg–Brentano XRD configuration (θ/2θ) was applied for identifying phases formed in basic Ti-6Al-4V alloy, basic nickel superalloy, and in the welded joints. Grazing-incidence X-ray diffraction was used to detect phases on the worn surfaces of the nickel superalloy at a beam incidence angle of 13°.

The XRD peak identification was performed using Crystal Impact's "Match!" software version 3.9 (Crystal Impact, Bonn, Germany). The relative peak intensities of α-Ti and β-Ti phases *R*(*x*), were calculated from Formula (1) as follows [32]:

$$R(x) = (I(x)/\sum I(A)) \times 100 \quad (1)$$

where *I*(*x*) is the intensity of the '*x*' phase reflection; $\sum I(A)$ is the sum of all reflection intensities.

A microhardness tester «Affairs DM8» at 100 g load and a dwell time 10 s allowed for the obtaining of microhardness numbers at 1 mm steps along the lines 11 mm away from the centerline. Tensile tests were carried out using a tensile machine UTS 110M-100 at room temperatures on samples cut off the welded joints, as shown in Figure 3.

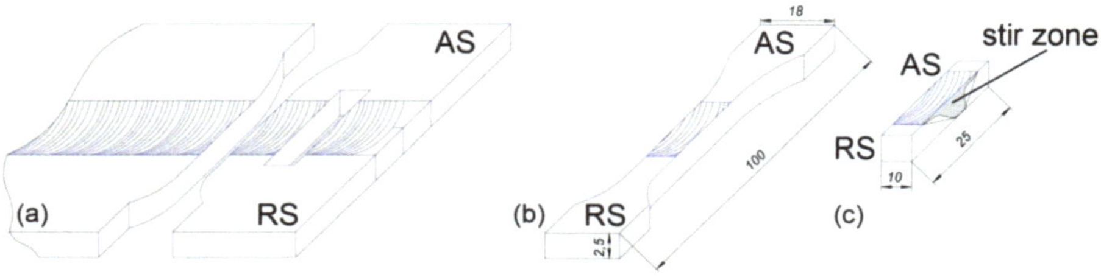

Figure 3. Scheme of the FSW seam sectioning (**a**) for cutting off tensile (**b**) and metallographic (**c**) specimens. AS and RS are the advancing and retreating sides of the seam, respectively.

3. Results

The as-received Ti-6Al-4V alloy and nickel superalloy are represented by 73% vol. of α-Ti + 27 vol.% of β-Ti (Figure 4a) and γ(γ′) + MC carbides (Figure 4b), respectively.

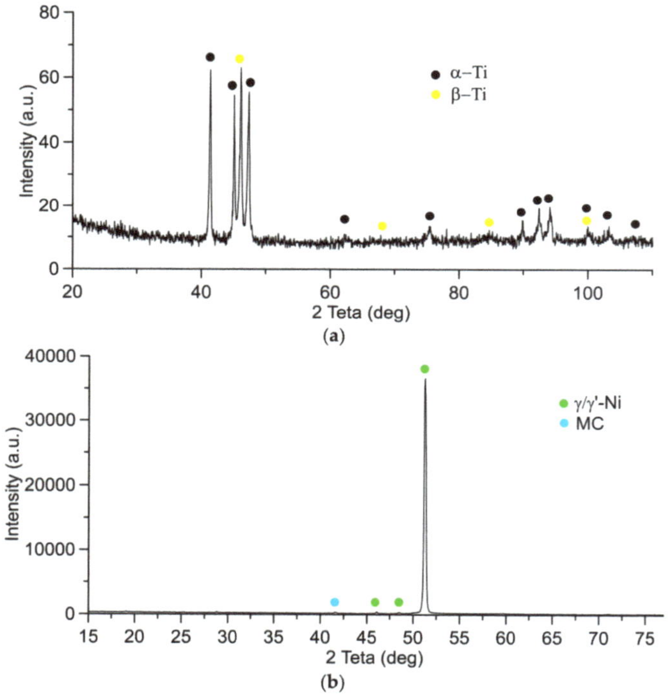

Figure 4. The X-ray diffraction patterns of the basic Ti-6Al-4V alloy (**a**) and nickel superalloy (**b**).

According to the XRD pattern in Figure 5a, the welded joint metal contains both these phases, but this time the β-Ti content decreased to 16 vol.% because of the phase transformations in heating and cooling [33]. More dramatic changes occurred to the FSW tool surface, which has been covered by a tribological layer where intermetallic compound (IMC) Ti_2Ni and carbides, such as MC, M_2C, and M_3C_2, are detected using the grazing-incidence X-ray diffraction; here, M stands for W, Cr, Nb (Figure 5b).

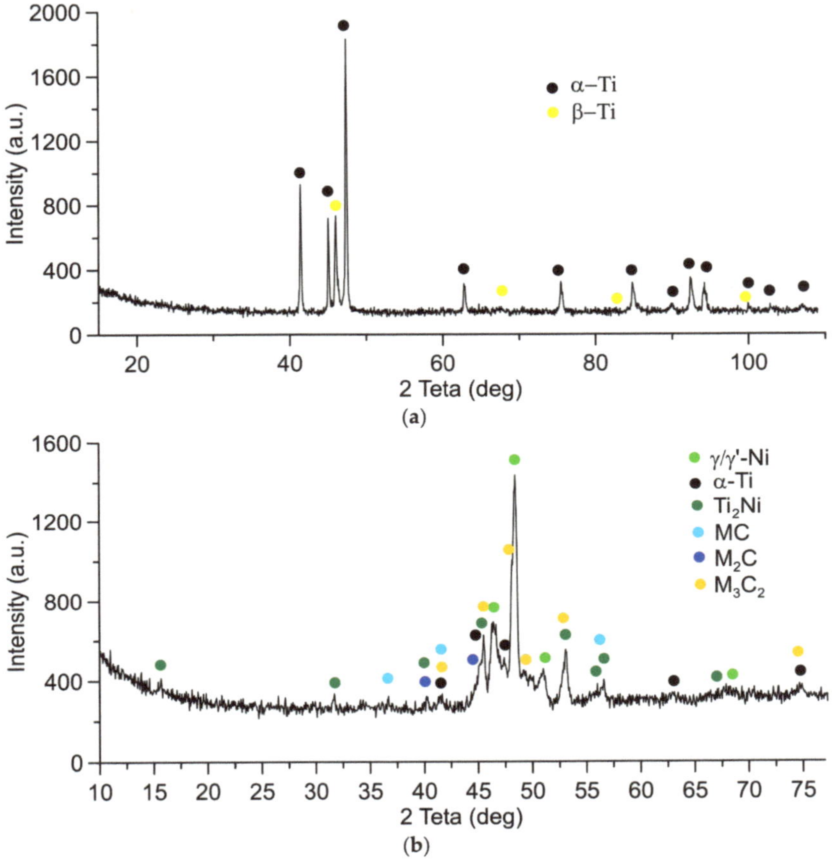

Figure 5. The X-ray diffraction patterns from the stir zone on the welded titanium alloy (**a**) and tribological layer on the corresponding FSW tool shoulder surface (**b**).

In general, the macrostructures typical of the FSW joints [34] can be observed on the optical cross-section views in Figures 6–8 and with some specificity stemming from the poor heat conductivity and high strength of the titanium alloy. Such a specificity mainly relates to forming a narrow heat-affected zone (HAZ) [35,36] and absence of a thermomechanically affected zone (TMAZ) [36–39]. A wormhole defect can be observed in a joint obtained according to regime 1 with the lowest rotation rate (Figure 8). No wormholes were detected in the samples welded according to regimes 2 and 3 with the increased rotation rates.

All samples demonstrate some specific branching structures located closer to the bottom part of the stir zone. It has been shown previously [25,27] that these structures are formed by intermixing the tool's wear particles with the SZ metal. The higher the rotation rate, the higher the temperature and more titanium alloy is transferred on the tool's surface, where reaction diffusion between titanium and nickel occurs with ensuing formation of intermetallic compounds (IMC) and wear particles, which then intermix with the metal welded. This coarse wear particle can be observed in sample 3, with the SZ obtained at the highest rotation rate (Figure 8).

The microstructure (Figure 9) of the as-received base metal is characterized by α-Ti and β-Ti grains of mean sizes 4.1 ± 1.5 μm and 1.2 ± 0.3 μm, respectively. The SZ metal is represented by recrystallized α-Ti 0.53 ± 0.2 μm grains; i.e. at least 87% grain refining effect was achieved that resulted in the increased ultimate tensile strength and reduced plasticity of the SZ metal by grain boundary hardening mechanism.

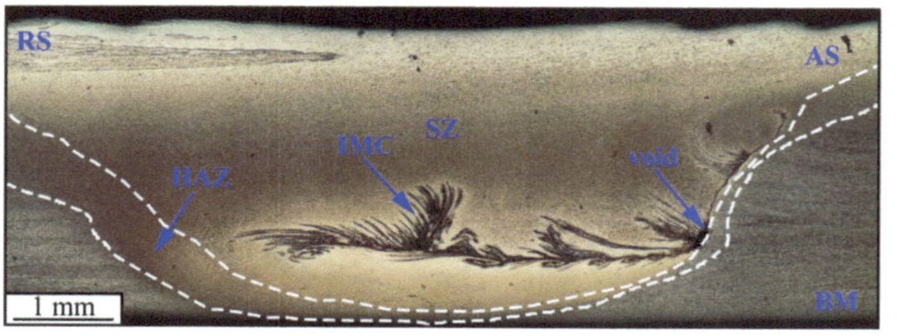

Figure 6. Cross-sectional metallographic image of specimen No. 1.

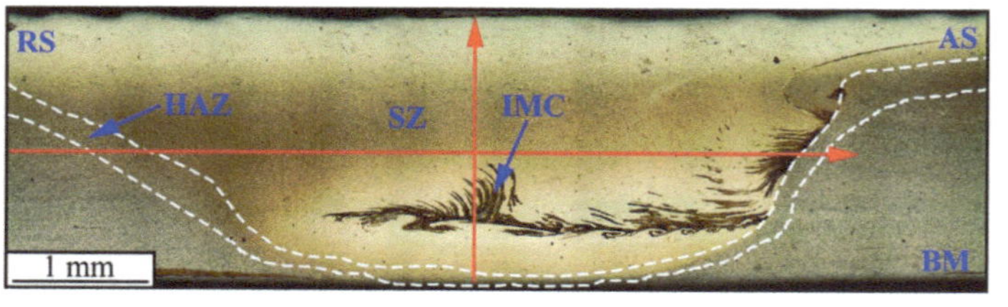

Figure 7. Cross-sectional metallographic image of specimen 2. Red lines identify the lines of microhardness profiles.

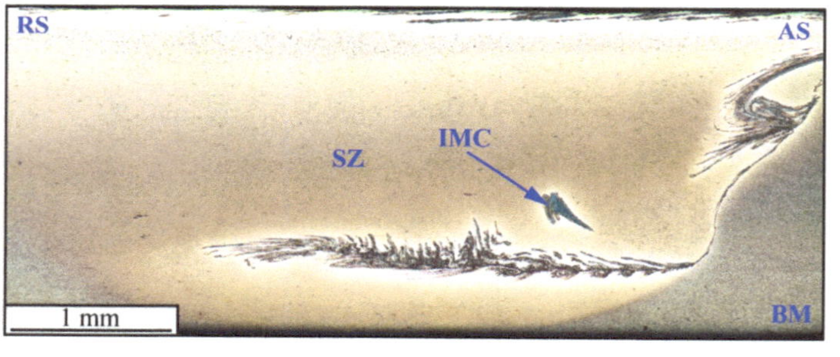

Figure 8. The stir zone image of specimen 3.

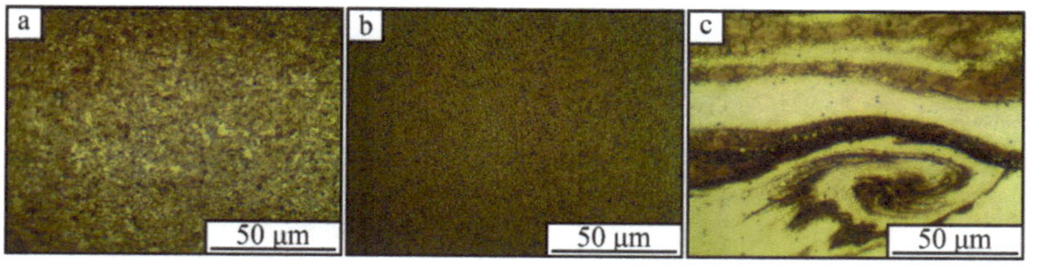

Figure 9. Microstructure of Ti-6Al-4V alloy in the base metal (**a**), SZ (**b**), and IMC in SZ (**c**).

The IMC zones intermixed with the IMC-free areas are represented in more detail in Figure 10a,b. Figure 10b denotes the EDS probe zones with compositions shown below in Table 3. The dark areas in Figure 10 (Table 3, pos. 8–13) contain elements inherent to the

Ti-6Al-4V alloy, while areas with the light-gray particles additionally contain Cr, Co, Ni, and W; i.e., elements that initially belonged to the tool alloy.

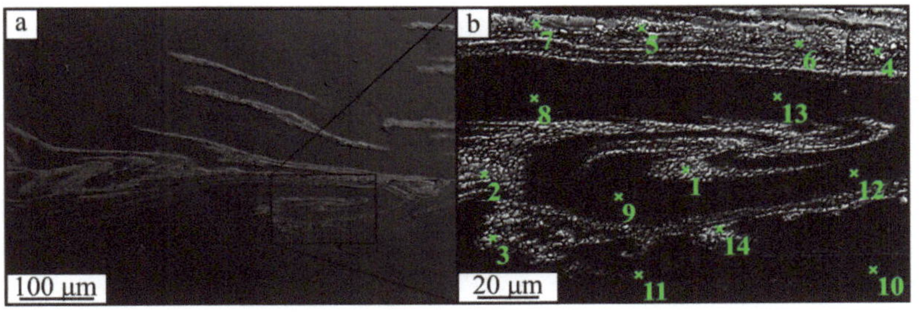

Figure 10. SEM image the IMCs intermixed with the stir zone (**a**) and enlarged view of these IMC structures with the EDS probe zones numbered from 1 to 14 (**b**), whose EDS spectra are identified in Table 3 below.

Table 3. The EDS element compositions of the FSW TiAl64V stir zone in points as shown in Figure 10.

Spectrum	Chemical Element Content, Atomic/Weight %						
	Al	Ti	V	Cr	Co	Ni	W
1	9.91/5.68	79.35/80.71	3.48/3.76	1.13/1.24	0.92/1.16	4.85/6.05	0.36/1.40
2	9.55/5.50	78.41/80.16	3.56/3.87	1.09/1.21	1.10/1.39	6.28/7.86	-
3	9.93/5.70	80.22/81.79	3.65/3.96	1.00/1.10	0.68/0.85	4.18/5.22	0.35/1.37
4	9.65/5.34	70.04/68.82	3.11/3.25	1.94/2.07	1.75/2.11	12.69/15.28	0.83/3.12
5	8.17/4.55	74.06/73.23	3.07/3.22	1.80/1.93	1.83/2.22	10.53/12.76	0.55/2.08
6	9.97/5.63	74.24/74.40	3.29/3.51	1.31/1.42	1.31/1.62	9.39/11.54	0.49/1.88
7	10.09/5.79	79.23/80.69	3.44/3.72	0.82/0.91	1.04/1.31	5.06/6.31	0.32/1.27
8	9.86/5.79	86.42/90.09	3.72/4.12	-	-	-	-
9	9.72/5.70	86.59/90.20	3.70/4.10	-	-	-	-
10	9.99/5.87	86.14/89.84	3.87/4.29	-	-	-	-
11	10.25/6.03	86.28/90.12	3.47/3.85	-	-	-	-
12	9.77/5.73	86.19/89.79	4.04/4.48	-	-	-	-
13	10.13/5.96	85.95/89.69	3.92/4.35	-	-	-	-
14	10.08/5.79	80.18/81.69	3.43/3.71	0.84/0.93	0.67/0.84	4.41/5.51	0.39/1.53

These areas were characterized by microhardness numbers at the level of 4.40–5.30 GPa, while microhardness of the stirring zone was in the range 3.75–4.05 GPa; i.e., about 20% higher than that of the base metal (Figure 11).

Tensile testing showed that, despite there were regions with intermixed tool wear particles, the joint strength values of all samples were higher than that of the base metal (Figure 12), while the maximum strength was achieved on samples obtained according to regime 3, with fracture occurred outside of the stir zone at a ~45° angle and with respect to the tensile axis (Figure 13). Samples 1 and 2 demonstrated fracture localization inside their stir zones closer to the retreating side (RS) and in a normal direction to the tensile axis. These samples had higher strength values than that of the as-received Ti6Al4V, but the lowest strain-to-fracture values. It seems that neither the presence of a void in sample 1 nor the large amount of IMCs formed on the advanced sides (AS) of all samples had any detrimental effect on the sample's tensile strength.

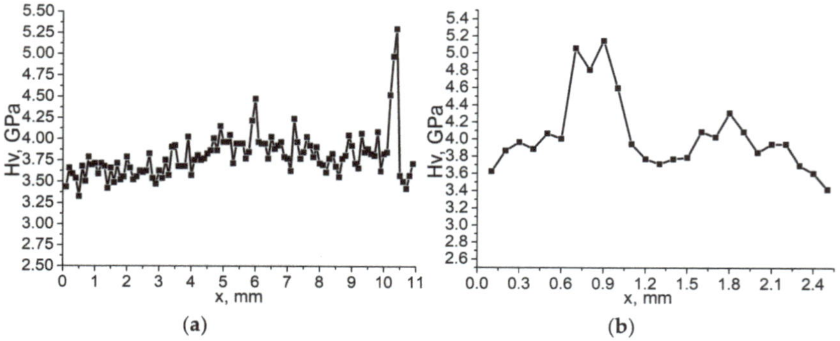

Figure 11. Microhardness of the welded joint Ti-6Al-4V alloy obtained by friction stir welding with nickel-base heat-resistant tool along the green lines shown in Figure 7: (**a**) vertical profile; (**b**) horizontal profile.

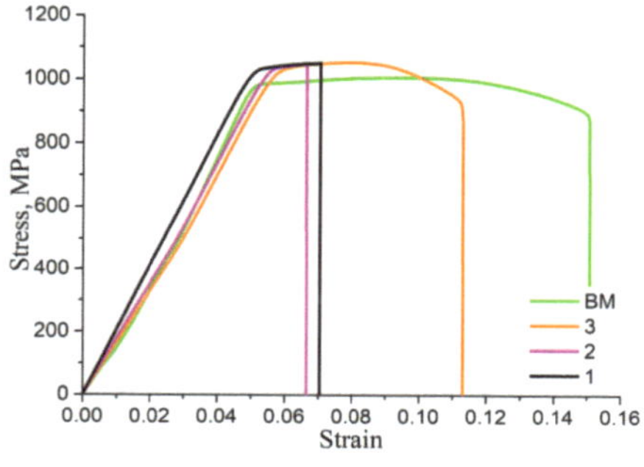

Figure 12. The tensile stress-strain curves obtained on samples 1, 2 and 3 produced with 340, 360 and 380 RPM rotation rates.

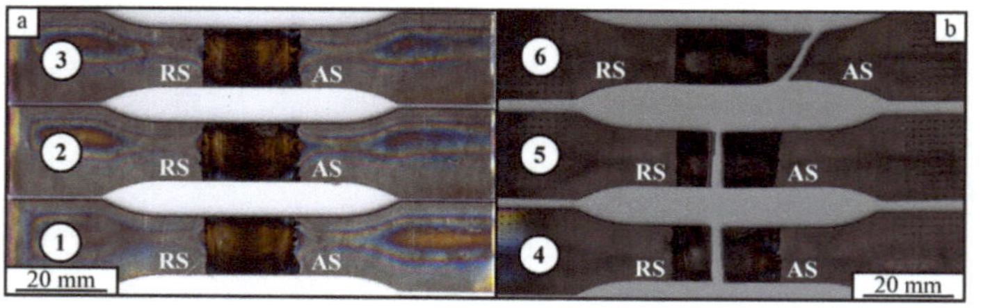

Figure 13. The tensile samples cut off the FSW welds obtained at tool rotation rates 340 (1 and 4), 360 (2 and 5) and 380 (3 and 6) RPM before (**a**) and after (**b**) the tensile tests.

The same conclusion may be obtained from the tensile curve of sample 3. Moreover, this time the fracture was localized on the RS outside of the stir zone with intermetallic compounds. This can be interpreted at least as the lack of tensile strength sensitivity to the IMC structures, as well as defects.

The XRD diffractogram obtained from the fracture surface of sample 2 shows the presence of both α-Ti and β-Ti (Figure 14) without any reflections from Ti_2Ni IMCs that were formed on the surface of the superalloy tool and then intermixed with the titanium alloy, thus forming those IMC branched structures, as shown in Figures 6–8.

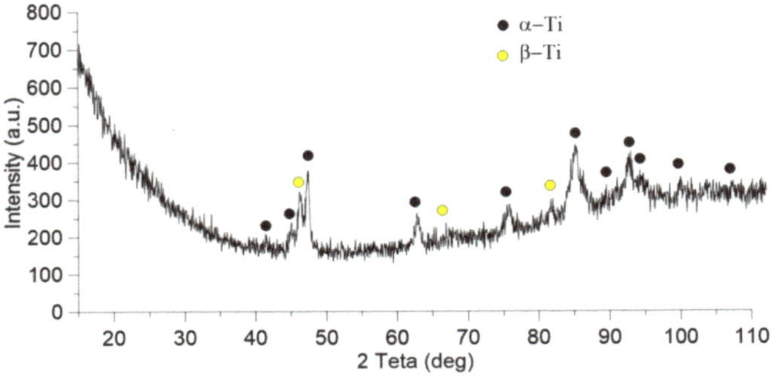

Figure 14. The X-ray diffraction pattern of the weld fracture surface after mechanical testing on the sample 2.

The fracture surfaces SEM SE images obtained from samples 1 and 2 present large bulges, ledges, and dimples, testifying on the development of a crack in inhomogeneous structures (Figure 15a–e). The SEM BSE images (Figure 15b–f), however do not reveal any regions with the BSE contrast, other than that of the titanium alloy, and therefore, are interpreted as nickel-rich IMCs. This result, being combined with the XRD pattern in Figure 14, confirms the absence of IMCs on the fracture surface and at least 15 μm below it, which is the maximum X-ray penetration in a titanium alloy, at $2\Theta = 160°$. Nevertheless, the small scale surface images suggest a viscous type of fracture in the fine-crystalline stir zones. The fracture surface of sample 3 is located outside of the stir zone with some necking (Figure 15b, sample 3), which could also be interpreted in terms of the viscous type of fracture developing through the SZ away from the IMC structures.

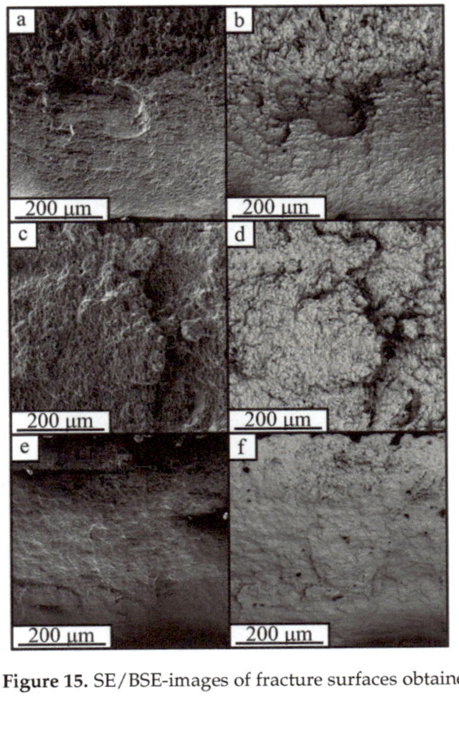

Figure 15. SE/BSE-images of fracture surfaces obtained from sample 1 (**a**,**b**), 2 (**c**,**d**), and 3 (**e**,**f**).

The FSW tool resistance against wear during FSW on the titanium alloys can be evaluated from a comparison between the new and worn tools in Figure 16a,b. The worn tool had a smaller height pin and shoulder surface coated by transferred titanium alloy (Figure 16b). The most intense wear, however, occurred on the pin root surface where an annular groove had formed. The use of water cooling reduced the FSW tool pin wear after welding at least a 2 m long weld seam, while partially retaining the pin shape (Figure 16b,c).

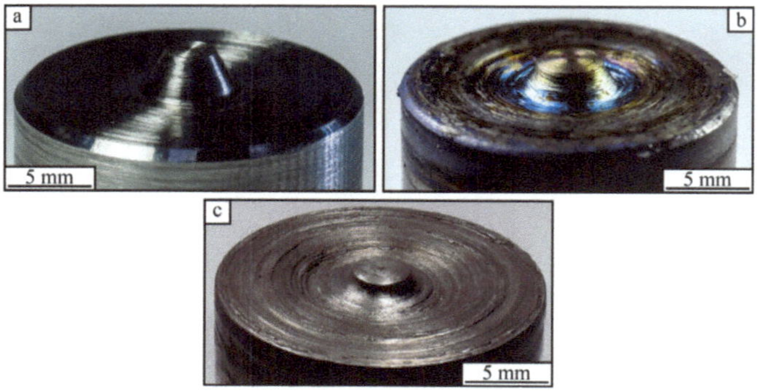

Figure 16. FSW tool produced from alloy ZhS6U before welding (**a**), after ≈2 m of welding (**b**), and after ≈2 m of welding without water cooling (**c**).

As shown above, the use of a grazing incidence angle XRD allowed identifying IMCs and carbides in the tribological layer that covered the FSW tool shoulder worn surface. However, there were zones on the tool surface that differed morphologically (Figure 17a–c). It has been shown, when studying the FSW tool wear in FSP on Ti-Cu system [27], that the shoulder worn surface revealed the lowest wear rate, generating a thick and anti-wear tribological layer. The same type of layer was generated on the FSW tool's shoulder surface in the present work (Figure 17a). This layer also contained gray Ti_2Ni IMC regions (Figure 17a, pos. 1) and bright carbide particles (Figure 17a, pos. 2). The cross-section view of the FSW tool subsurface allowed observing a rather thick tribological layer that could be structurally divided into an upper transfer layer, almost fully consisting of adhesion transferred Ti (Figure 18a), and a transition layer, composed of Ti_2Ni with a tungsten-based carbide network (Figure 18a, pos 5, 6). The EDS element profiles across the tribological layer confirmed the above suggestions (Figure 19).

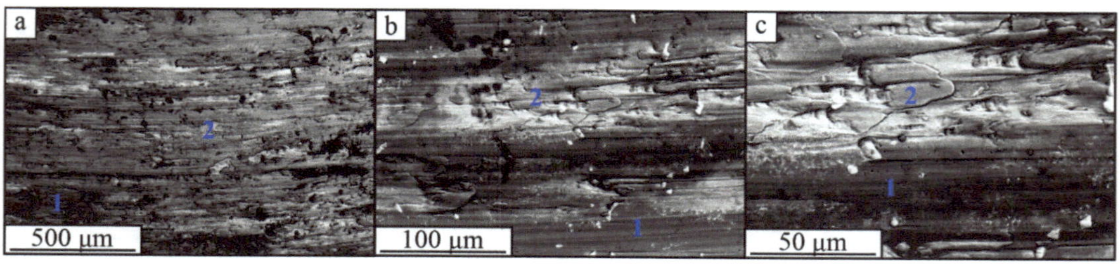

Figure 17. SEM BSE images of the worn surface zones on the FSW tool shoulder (**a–c**): 1—intermetallic regions, 2—carbides.

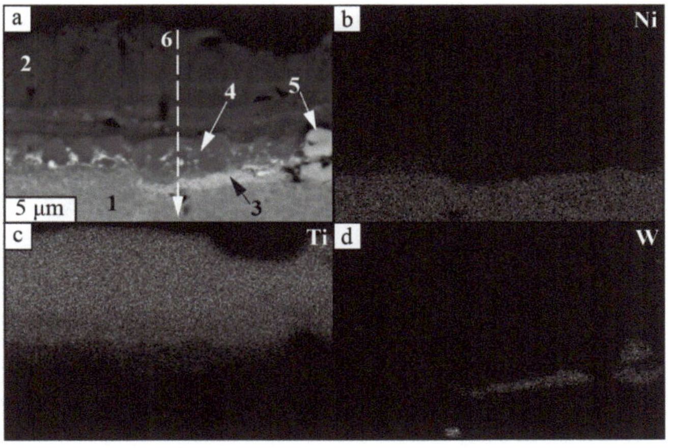

Figure 18. The SEM BSE image of tribological layer structures on the FSW tool worn surface (**a**) and corresponding EDS element distribution maps (**b**–**d**): 1—base superalloy metal; 2—transfer layer, intermetallic compound; 3—fine tungsten carbides; 4—carbide network in the transition layer; 5—large carbide particles; 6 is the EDS probe trajectory.

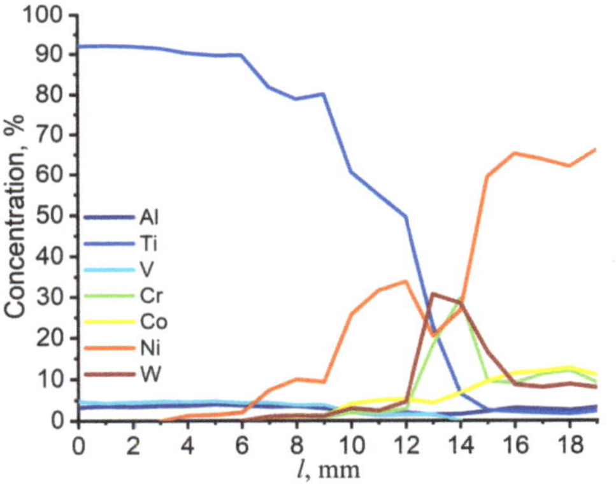

Figure 19. EDS element profiles obtained along trajectory 6 in Figure 18a.

The most intensive wear occurred on the pin/shoulder fillet surface, so that corresponding worn surfaces demonstrated areas deprived of any tribological layer (Figure 20a,c, pos. 1, 2) or tribological layer fragments (Figure 20, pos. 3) with primary carbides (Figure 20b, pos. 4). The cross-section views of pin/shoulder fillet area demonstrated either full absence of the tribological layer (Figure 20d) or the presence of its fragments (Figure 20e, pos. 5) structurally consisting of Ti_2Ni in the Ti matrix. This meant that this area experienced intensive wear with the removal of the tool material. The subsurface superalloy structure carbides could be seen on the worn surface, while no carbide network could be found in the tribological layer fragments (Figure 20e).

The worn surface of the pin end could be characterized by the absence of any tribological layer (Figure 21a) with transfer layer areas (Figure 21b, pos. 2) and concentric wear grooves (Figure 21a,c) with the detached fragments (Figure 21c, pos. 3).

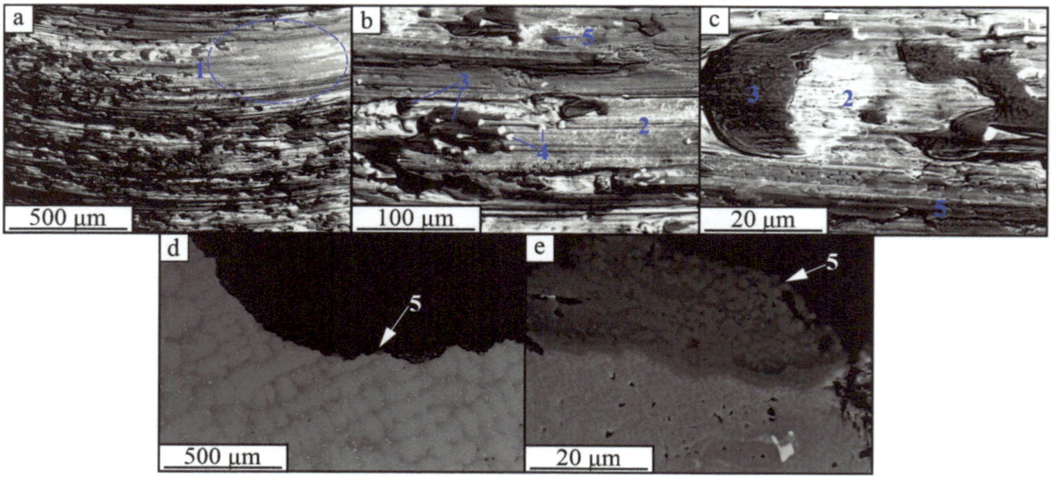

Figure 20. SEM BSE images of subsurface FSW tool structures on the worn surface (**a**–**c**) and below the pin/shoulder fillet worn surface without (**d**) and with tribological layer fragments (**e**): 1, 2—tribological layer-free regions; 3—transfer layer fragments; 4—primary carbides; 5—intermetallic layer fragments.

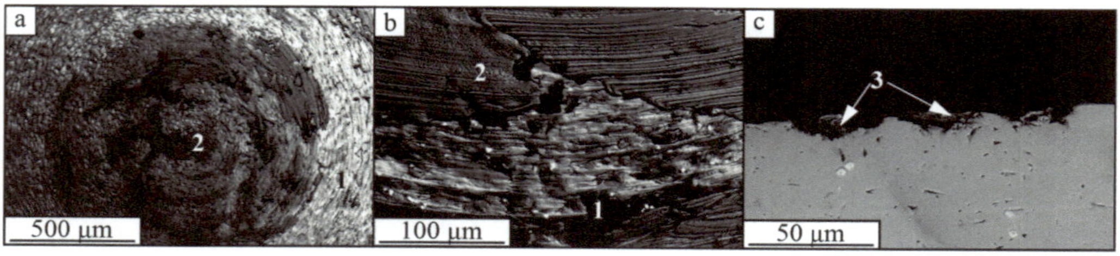

Figure 21. SEM BSE images of the pin end worn surface with transfer layer areas ((**a**,**b**), pos. 2) and detached wear debris ((**c**), pos. 3).

4. Discussion

It was observed that the tensile characteristics of all samples were not sensitive to the IMC structures formed in them by intermixing with wear particles detached from practically consumable FSW tool. Considering the data obtained, the improved strength and loss of ductility of the welded joint, as compared to those of the as-received Ti-6Al-4V (Figure 12), could be caused by at least two reasons. The first and well-known reason could be the SZ strengthening by grain refinement according to the Hall-Petch mechanism. The second reason could be the phase transformation occurred in SZ during cooling. It is known that the dual-phase titanium alloys were developed especially for combining high strength with acceptable ductility, so that increasing the content of β-Ti attains more ductility and less hardening and vice versa; increasing the content of α-Ti corresponds to higher ultimate strength and less ductility [2,12]. Therefore, the higher strength of the welded joints in samples 1 and 2 may be due to reduced content of β-Ti and higher content of α-Ti, as well as extra α′-Ti formed in the welded joint (Figure 5). Heating and stirring in FSW was accompanied by $\alpha + \beta \rightarrow \beta$ transformation, while during cooling there occurred $\beta \rightarrow \alpha' + \alpha + \beta$ [40]. The higher plasticity of sample 3 FSWed at the highest tool rotation rate and, therefore, at higher temperature, could have resulted from slow cooling and forming a higher amount of residual β-Ti as compared to those in samples 1 and 2.

The third reason could be that fine intermetallic compounds, such as Ti_2Ni, may be distributed in the SZ metal and actually serve as reinforcing particles (Figure 11). These types of structures have been observed in SZ of TiAl6V4 friction stir processed and intermixed with copper powder [27]. The nanosized Ti_2Ni precipitates formed on the TiAl6V4 grains and caused a dislocation of the barriers (Figure 22). The same IMC structures were formed in this FSW work. The IMC refining can be an additional factor in the SZ hardening.

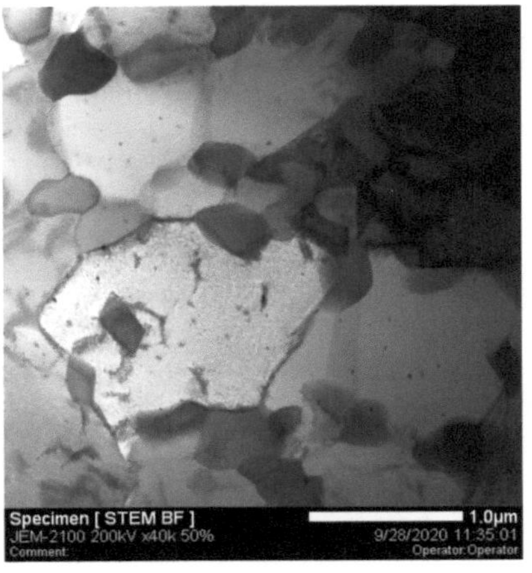

Figure 22. Ti_2Ni grain boundary precipitates in TiAl6V4 + Cu powder after friction stir processing.

The presence of a clearly observable tribological layer on the shoulder surface (Figure 16b) can be taken as a good indication from the point of view of the actual mechanism of degradation of the tool material during welding. It is known that the formation of so-called mechanically mixed layers (MML) is one of the most important aspects of the interaction between the FSW tool and the material being welded [27]. For FSW, these MMLs formed on the tool surface due to the high adhesion of the heated and plasticized Ti6Al4V titanium alloy material. Despite the water-cooling system being applied to avoid overheating of the tool, due to the high temperature reactivity and low thermal conductivity of Ti6Al4V, the diffusion-reaction occurred between the MML and the tool material to form the Ti_2Ni intermetallic compound (Figure 12). At the same time, such a diffusion reaction led to the formation of a continuous protective film, and did not form brittle intermetallic protrusions, outgrowths, spikes, and other similar formations that might contribute to the appearance of stress concentrators, due to which large wear particles would be formed by brittle fracture, and then, accordingly, caused the undesirable abrasive wear. In other words, the improved resistance to thermal degradation was observed in case of using the liquid-cooled nickel superalloy tool for FSW on Ti6Al4V.

Generally, the worn surfaces of the FSW tool used in this welding experiment with cooling can be characterized by the less wear intensity as compared to those reported previously [27]. Such a conclusion mostly relates to the pin end and pin root worn surfaces which demonstrated less deep grooves and adhesion wear traces as those obtained by FSP intermixing of copper powder in TiAl64V [27], where intense exothermic diffusion reaction between Ti and Cu occurred that additionally increased the temperature during contact between the tool and alloy.

5. Conclusions

The FSW experiments were conducted using a heat-resistant nickel superalloy tool on a titanium (α + β)-alloy under conditions of water cooling. Such an approach allowed obtaining a weld joint with a tensile strength higher than that of the base metal (sample 3), despite the formation of the Ti_2Ni intermetallic compounds on the tool surface and intermixing them with the stirring zone metal in the form of IMC-branched structures. As shown by fractography, these IMC structures were not present on the fracture surfaces of samples after tensile testing, and therefore, did not embrittle the stir zone. The use of a FSW tool water cooling reduced the wear of the FSW tool made of nickel superalloy.

Author Contributions: Conceptualization, A.C. and S.T.; methodology, A.A., A.C. and S.T.; validation, A.A., V.E.R. and A.C.; formal analysis, A.I. and A.A.; investigation, A.A., A.C., N.S. and E.M.; resources, E.K. and V.E.R.; writing—original draft preparation, A.A. and S.T.; writing—review and editing, S.T. and A.C.; visualization, A.I. and A.A.; supervision, S.T.; project administration, S.T. and A.C.; funding acquisition, A.C. and S.T. All authors have read and agreed to the published version of the manuscript.

Funding: This work was performed under the Russian Science Foundation grant No. 22-29-01621.

Institutional Review Board Statement: Not applicable.

Informed Consent Statement: Not applicable.

Data Availability Statement: All data included in the main text.

Acknowledgments: The investigations have been carried out using the equipment of Share Use Centre "Nanotech" of the ISPMS SB RAS.

Conflicts of Interest: The authors declare no conflict of interest.

References

1. Pilchak, A.L.; Juhas, M.C.; Williams, J.C. Microstructural Changes Due to Friction Stir Processing of Investment-Cast Ti-6Al-4V. *Met. Mater. Trans. A* **2007**, *38*, 401–408. [CrossRef]
2. Balasundar, I.; Raghu, T.; Kashyap, B.P. Hot working and geometric dynamic recrystallisation behaviour of a near-α titanium alloy with acicular microstructure. *Mater. Sci. Eng. A.* **2014**, *600*, 135–144. [CrossRef]
3. Zhu, Y.Y.; Bo, C.H.E.N.; Tang, H.B.; Cheng, X.; Wang, H.M.; Jia, L.I. Influence of heat treatments on microstructure and mechanical properties of laser additive manufacturing Ti-5Al-2Sn-2Zr-4Mo-4Cr titanium alloy. *Trans. Nonferr. Met. Soc. China* **2018**, *28*, 36–46. [CrossRef]
4. Steele, M.C.; Hein, R.A.; He, T.; Chen, W.; Wang, W.; Ren, F.; Stock, H.R. Superconductivity of Titanium. *J. Phys. Rev.* **1953**, *92*, 243–247. [CrossRef]
5. Gorsse, S.; Miracle, D.B. Mechanical properties of Ti-6Al-4V/TiB composites with randomly oriented and aligned TiB reinforcements. *Acta Mater.* **2003**, *51*, 2427–2442. [CrossRef]
6. Qian, T.T.; Dong, L.I.U.; Tian, X.J.; Liu, C.M.; Wang, H.M. Microstructure of TA2/TA15 graded structural material by laser additive manufacturing process. *Trans. Nonferr. Met. Soc. China* **2014**, *24*, 2729–2736. [CrossRef]
7. Jiang, L.; Huang, W.; Liu, C.; Chai, L.; Yang, X.; Xu, Q. Microstructure, texture evolution and mechanical properties of pure Ti by friction stir processing with slow rotation speed. *Mater. Charact.* **2019**, *148*, 1–8. [CrossRef]
8. Mironov, S.; Sato, Y.S.; Kokawa, H. Development of grain structure during friction stir welding of pure titanium. *Acta Mater.* **2009**, *57*, 4519–4528. [CrossRef]
9. Liu, F.C.; Liao, J.; Gao, Y.; Nakata, K. Influence of texture on strain localization in stir zone of friction stir welded titanium. *J. Alloys Compd.* **2015**, *626*, 304–308. [CrossRef]
10. Zhang, W.; Ding, H.; Cai, M.; Yang, W.; Li, J. Ultra-grain refinement and enhanced low-temperature superplasticity in a friction stir-processed Ti—6Al—4V alloy. *Mater. Sci. Eng. A.* **2018**, *727*, 90–96. [CrossRef]
11. Junaid, M.; Rahman, K.; Khan, F.; Bakhsh, N.; Baig, M. Comparison of microstructure, mechanical properties, and residual stresses in tungsten inert gas, laser, and electron beam welding of Ti–5Al–2.5 Sn titanium alloy. *Proc. Inst. Mech. Eng. Part L-J. Mater.-Des. Appl.* **2019**, *233*, 1336–1351. [CrossRef]
12. Wang, B.; Yao, X.; Liu, L.; Zhang, X.; Ding, X. Mechanical properties and microstructure in a fine grained Ti–5Al–5Mo–5V–1Cr–1Fe titanium alloy deformed at a high strain rate. *Mater. Sci. Eng. A* **2018**, *736*, 202–208. [CrossRef]
13. Liu, H.; Fujii, H. Microstructural and mechanical properties of a beta-type titanium alloy joint fabricated by friction stir welding. *Mater. Sci. Eng. A* **2018**, *711*, 140–148. [CrossRef]
14. Ashton, P.J.; Jun, T.; Zhang, Z.; Britton, T.B.; Harte, A.M.; Leen, S.B.; Dunne, F.P.E. The effect of the beta phase on the micromechanical response of dual-phase titanium alloys. *Int. J. Fatigue* **2017**, *100*, 377–387. [CrossRef]
15. Kolubaev, E.A.; Luo, A.A. A Distinctions of structure forming of welded joints produced by friction stir welding. *Prog. Mater. Sci.* **2008**, *53*, 980–1023. [CrossRef]
16. Threadgill, P.L. Terminology in friction stir welding. *Sci. Technol. Weld. Join.* **2007**, *12*, 357–360. [CrossRef]

17. Sato, Y.S.; Park, S.H.C.; Matsunaga, A.; Honda, A.; Kokawa, H. Novel production for highly formable Mg alloy plate. *J. Mater. Sci.* **2005**, *40*, 637–642. [CrossRef]
18. Li, G.; Zhou, L.; Luo, S.; Dong, F.; Guo, N. Microstructure and mechanical properties of bobbin tool friction stir welded ZK60 magnesium alloy. *Mater. Sci. Eng. A* **2020**, *776*, 138953. [CrossRef]
19. Kulkarni, S.S.; Truster, T.; Das, H.; Gupta, V.; Soulami, A.; Upadhyay, P.; Herling, D. Microstructure-Based Modeling of Friction Stir Welded Joint of Dissimilar Metals Using Crystal Plasticity. *J. Manuf. Sci. Eng.* **2021**, *143*, 121008. [CrossRef]
20. Kulkarni, S.S.; Gupta, V.; Ortiz, A.; Das, H.; Upadhyay, P.; Barker, E.; Herling, D. Determining cohesive parameters for modeling interfacial fracture in dissimilar-metal friction stir welded joints. *Int. J. Solids Struct.* **2021**, *216*, 200–210. [CrossRef]
21. Rai, R.; De, A.; Bhadeshia, H.K.D.H.; DebRoy, T. Review: Friction stir welding tools. *Sci. Technol. Weld. Join.* **2011**, *16*, 325–342. [CrossRef]
22. Farias, A.; Batalha, G.F.; Prados, E.F.; Magnabosco, R.; Delijaicov, S. Tool wear evaluations in friction stir processing of commercial titanium Ti–6Al–4V. *Wear* **2013**, *302*, 1327–1333. [CrossRef]
23. Zhang, Y.; Sato, Y.S.; Kokawa, H.; Park, S.H.C.; Hirano, S. Stir zone microstructure of commercial purity titanium friction stir welded using pcBN tool. *Mater. Sci. Eng. A* **2008**, *488*, 25–30. [CrossRef]
24. Wu, L.H.; Wang, D.; Xiao, B.L.; Ma, Z.Y. Tool wear and its effect on microstructure and properties of friction stir processed Ti–6Al–4V. *Mater. Chem. Phys.* **2014**, *146*, 512–522. [CrossRef]
25. Mironov, S.; Sato, Y.S.; Kokawa, H. Friction-stir welding and processing of Ti-6Al-4V titanium alloy: A review. *J. Mater. Sci. Technol.* **2018**, *34*, 58–72. [CrossRef]
26. Gurianov, D.A.; Fortuna, S.V.; Nikonov, S.Y.; Moskvichev, E.N.; Kolubaev, E.A. Heat Input Effect on the Structure of ZhS6U Alloy. *Russ. Phys. J.* **2021**, *64*, 1415–1421. [CrossRef]
27. Zykova, A.; Vorontsov, A.; Chumaevskii, A.; Gurianov, D.; Gusarova, A.; Kolubaev, E.; Tarasov, S. Structural evolution of contact parts of the friction stir processing heat-resistant nickel alloy tool used for multi-pass processing of Ti6Al4V/ (Cu+Al) system. *Wear* **2022**, *488*, 204138. [CrossRef]
28. Amirov, A.I.; Eliseev, A.A.; Rubtsov, V.E.; Utyaganova, V.R. Butt friction stir welding of commercially pure titanium by the tool from a heat-resistant nickel alloy. *AIP Conf. Proc.* **2019**, *2167*, 020016. [CrossRef]
29. Amirov, A.I.; Eliseev, A.A.; Beloborodov, V.A.; Chumaevskii, A.V.; Gurianov, D.A. Formation of α'titanium welds by friction stir welding. *J. Phys. Conf. Ser.* **2020**, *1611*, 012001. [CrossRef]
30. Amirov, A.; Eliseev, A.; Kolubaev, E.; Filippov, A.; Rubtsov, V. Wear of ZhS6U nickel superalloy tool in friction stir processing on commercially pure titanium. *Metals* **2020**, *10*, 799. [CrossRef]
31. Amirov, A.I.; Chumaevskii, A.V.; Vorontsov, A.V. Formation of (α+ β) titanium welds by friction stir welding using heat-resistant alloy tool. *AIP Conf. Proc.* **2020**, *2310*, 020017. [CrossRef]
32. Balla, V.K.; Soderlind, J.; Bose, S.; Bandyopadhyay, A. Microstructure, mechanical and wear properties of laser surface melted Ti6Al4V alloy. *J. Mech. Behav. Biomed. Mater.* **2014**, *32*, 335–344. [CrossRef]
33. Mironov, S.; Sato, Y.S.; Kokawa, H. Grain structure evolution during friction-stir welding. *Phys. Mesomech.* **2020**, *23*, 21–31. [CrossRef]
34. Liu, H.J.; Zhou, L. Microstructural zones and tensile characteristics of friction stir welded joint of TC4 titanium alloy. *Trans. Nonferr. Met. Soc. China* **2010**, *20*, 1873–1878. [CrossRef]
35. Lippold, J.C.; Livingston, J.J. Microstructure evolution during friction stir processing and hot torsion simulation of Ti—6Al—4V. *Metall. Mater. Trans. A* **2013**, *44*, 3815–3825. [CrossRef]
36. Edwards, P.; Ramulu, M. Fracture toughness and fatigue crack growth in Ti—6Al—4V friction stir welds. *J. Mater. Eng. Perform.* **2015**, *24*, 3263–3270. [CrossRef]
37. Zhou, L.; Liu, H.J.; Liu, Q.W. Effect of rotation speed on microstructure and mechanical properties of Ti—6Al—4V friction stir welded joints. *Mater. Des.* **2010**, *31*, 2631–2636. [CrossRef]
38. Ji, S.; Li, Z.; Wang, Y.; Ma, L. Joint formation and mechanical properties of back heating assisted friction stir welded Ti—6Al—4V alloy. *Mater. Des.* **2017**, *113*, 37–46. [CrossRef]
39. Fall, A.; Fesharaki, M.H.; Khodabandeh, A.R.; Jahazi, M. Tool wear characteristics and effect on microstructure in Ti—6Al—4V friction stir welded joints. *Metals* **2016**, *6*, 275. [CrossRef]
40. Zykova, A.P.; Vorontsov, A.V.; Chumaevskii, A.V.; Gurianov, D.A.; Gusarova, A.V.; Savchenko, N.L.; Kolubaev, E.A. The Influence of Multipass Friction Stir Processing on Formation of Microstructure and Mechanical Properties of VT6 Alloy. *Russ. J. Non-Ferr. Met.* **2022**, *63*, 167–176. [CrossRef]

technologies

MDPI

Article

Design and Implementation of an Anthropomorphic Robotic Arm Prosthesis

Valentina A. Yurova [1,2,*], Gleb Velikoborets [1,3] and Andrei Vladyko [1]

1 Faculty of Fundamental Training, The Bonch-Bruevich Saint Petersburg State University of Telecommunications, 193232 Saint Petersburg, Russia
2 Department of Medical Informatics and Physics, North-Western State Medical University named after I.I. Mechnikov (NWSMU), 195067 Saint Petersburg, Russia
3 Design Department, Federal Scientific Center of Rehabilitation of the Disabled named after G.A. Albrecht, 195067 Saint Petersburg, Russia
* Correspondence: v.a.yurova@gmail.com

Abstract: The development and manufacture of prosthetic limbs is one of the important tendencies of the development of medical techniques. Taking into account the development of modern electronic technology and automated systems and its mobility and compactness, the actual task is to create a prosthesis that will be close to a fully functioning human limb in its anthropomorphic properties and will be capable of reproducing its basic actions with a high accuracy. The paper analyzes the main directions in the development of a control system for electronic limb prostheses. The description and results of the practical implementation of a prototype of an anthropomorphic prosthetic arm and its control system are presented in the paper. We developed an anthropomorphic multi-finger artificial hand for utilization in robotic research and teaching applications. The designed robotic hand is a low-cost alternative to other known 3D printed robotic hands and has 21 degrees of freedom—4 degrees of freedom for each finger, 3 degrees for the thumb and 2 degrees responsible for the position of the robotic hand in space. The open-source mechanical design of the presented robotic arm has mass-dimensional and motor parameters close to the human hand, with the possibility of autonomous battery operation, the ability to connect different control systems, such as from a computer, an electroencephalograph, a touch glove.

Keywords: robotics; anthropomorphic prosthetic arm; medical electronics; microcontrollers

Citation: Yurova, V.A.; Velikoborets, G.; Vladyko, A. Design and Implementation of an Anthropomorphic Robotic Arm Prosthesis. *Technologies* **2022**, *10*, 103. https://doi.org/10.3390/technologies10050103

Academic Editors: Manoj Gupta, Eugene Wong and Gwanggil Jeon

Received: 30 July 2022
Accepted: 15 September 2022
Published: 21 September 2022

1. Introduction

Robotic arms are widely used in different fields: remote manipulation of objects in unsafe conditions [1,2], industry [3,4], and surgery [5]. One of the broad areas of application is prosthesis. Robotic prostheses give people back the ability to interact with the world around them. However, most of the prostheses that are used in modern rehabilitation do not fully provide users with these abilities. They allow people without arms to perform simple actions, such as grabbing and holding objects, and some other simple operations.

The major criteria used for the design of the anthropomorphic robotic arm were the number of joints, degrees of freedom, number and type of actuators, weight, and the mobility ranges of each of the finger joints and the hand. This paper presents the possibility of the practical implementation of a low-cost design of an anthropomorphic arm, approximated or improved in properties to available analogues, with the ability to work from various power sources, and with simple controls and connections to various control systems. This will; make the prototype universal for different utilizations in robotics research, design of arm–brain control systems and teaching applications and simplify and make it more ergonomic to connect it to control systems.

The purpose of the work was to design a low-cost, easy-to-maintain anthropomorphic robotic arm prosthesis as close in functionality and mass-dimensional dimensions as possible to a real human hand. The control system of the arm–hand prototype presented in the work was designed so that it could be connected to any control system (such as from a computer, an electroencephalograph, a touch glove, etc.) to expand the possibilities of application in medicine, robotic research and teaching.

The designs of robotic arms that can provide smooth and precise manipulations are called anthropomorphic. Anthropomorphic design implies the presence of a mechanism similar in structure to a human hand (joints and connection tissues). Most of the existing prototypes are under actuated, that is, the number of actuators is less than the number of degrees of freedom (DoF). In previous research, [6–8] it was determined that only 1–6 activators are usually contained in constructions of underactuated robotic arms (compared to 34 muscles controlling the human hand).

What problems appear when designing fully actuated mechanisms? The human hand has 27 degrees of freedom: 4 on each finger, 3 for extension and flexion and 1 for extension and compression. The biomechanics of the thumb and wrist are more intricate. The thumb has five degrees of freedom, and for the rotation and movement of the wrist there are six degrees of freedom. Therefore, fully actuated robotic arms imply the presence of a drive mechanism for each of the degrees of freedom. This significantly increases the weight and size characteristics of the structure and its automated control system due to the large number of drives and the choice of their effective placement in the housing.

The results of the investigation of this issue [9] showed that the weight of the prosthetic hand should be no more than 500 g, based on patient feedback. This corresponds to the data in [10] about the average weight of the body segment, where the average arm weight (men and women) was 580 g. Based on [10] data about the average weight of the body segment, the average weight of the forearm is about 3 times the average weight of the hand. Therefore, using the Chappel limit of 500 g as an ideal reference point for limiting the weight of the prosthesis, the forearm should weigh less than 1.5 kg (if it is included in the design at all) in the design of anthropomorphic constructions.

In the work in [11], the modern existing designs of upper limb prostheses were investigated. From their data, it can be concluded that most of robotic arm constructions are insufficiently equipped (on average four actuators per 10 DoF). Such a DoF/actuators ratio does not allow a user with a lost limb to make up for the entire range of movement that was previously available to him.

Taking into account the growing need for electronically controlled robotics systems in various fields of technology and rehabilitation medicine, the aim of the work was to design and create an experimental prototype of a multi-functional anthropomorphic robotic arm with the possibility of using it as a manipulator and prosthesis of a lost limb.

The main stages of our research are presented in the form of a flowchart in Figure 1.

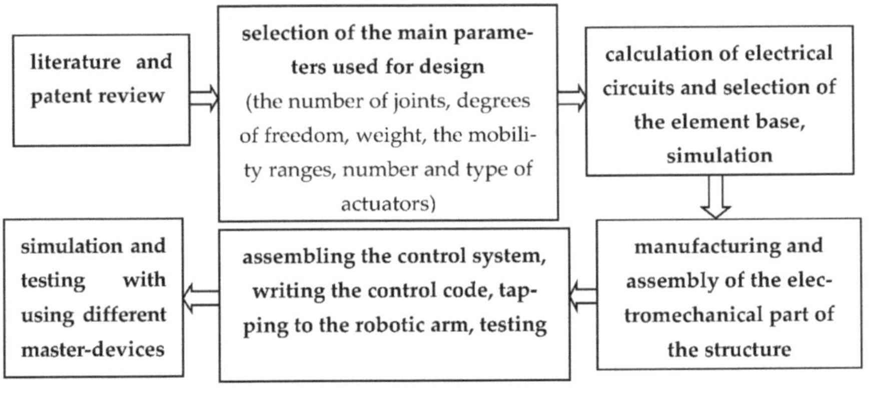

Figure 1. The flowchart of the main stages of the research.

Motivated by the described state-of-art, this work presents the results of the analysis, calculation, design and the practical realization of a working prototype of a robotic anthropomorphic arm–hand with justifications of engineering solutions. The completely independent creation of a working prototype was carried out, which is characterized by a high degree of freedom of movement of individual components in comparison with existing analogues, simplicity and accuracy of the control system and mass and weight dimensions close in parameters to the human hand. Our novel contributions can be summarized as follows:

1. Analyzing the basic requirements for the parameters of an anthropomorphic robotic arm;
2. Designing and developing the anthropomorphic robotic arm and manufacturing and assembling the mechanical parts for construction;
3. The calculation, design and development of the power supply circuits and the control system of the anthropomorphic robotic arm were carried out to ensure the simultaneous control of all drives;
4. Testing the possibilities of connecting to different control systems.

The rest of the article is structured as follows. Section 2 presents the brief analysis of the main directions in the development of robotic arm designs. Section 3 presents the results of the development, manufacture and assembly of the design of the anthropomorphic robotic arm and the data on the mass-dimensional parameters and properties of the mechanical design of the developed prototype of an anthropomorphic robotic arm. Moreover, Section 3 presents the structure of automated and control systems, analysis and calculations of the power supply and control circuits of the robotic anthropomorphic arm. The obtained results and links to video files demonstrating the work of the anthropomorphic robotic arm are presented in Section 4. Finally, the conclusions and future work are presented in Section 5.

2. Related Works

There are many developments of robotic arms, but not all of them fully correspond to anthropomorphic design. In [12,13], the authors developed robotic arms based on the kinematics of the human hand, that is, the most satisfying signs of anthropomorphism, but each of them was underactuated, similar to most other developments of robotic prostheses.

A number of developments, such as in [14–17], implemented only some parts of the joints of the human hand, with others staying strictly fixed in one position, in particular the MCPjoint, removing its adduction and abduction, leaving only compression and stretching. Each of the developments was also underactuated.

However, most of the developments deviate from anthropomorphic design in order to provide the prosthesis with only the basic movements that are necessary in everyday life. For example, in the studies in [8,18,19], two types of joints were combined, and thus the finger had two phalanges, instead of three.

In other developments [7,20], it was structurally made so that the joints were kinematically connected to each other, which excludes a separate contraction of the phalanges.

The number of DoF and actors of some modern designing robotic arms are shown in Table 1.

Most of the robotic arm–hand designs can be divided into the following groups: finger, hand; arm forearm; and full arm construction up to the shoulder. In the first group, generally, a good reproducibility of movements has been worked out, as close as possible to the real part of the limb. This group is more often used in research to design touch sensors and research systems in relation to the reproducibility of tactile sensations. However, the wrist or the base of the finger (in the case of a design in the form of 1–2 fingers) remains stationary. Less detail and accuracy of reproducing movements, avoiding anthropomorphism, are characteristic of arm–shoulder prosthesis designs. This group uses more complex electrical circuits and electromechanical systems to control and recreate hand movements.

Table 1. The comparison of the main parameters of some modern designs of robotic arms.

Project	DoF	Number of Actuators
Anthropomorphic robotic arm	22	17
Robotic arm [12]	20	6
Robotic arm [13]	20	5
Robotic arm [14]	18	5
Robotic arm [15]	15	3
Galileo Hand [16]	15	6
Robotic arm [8]	11	6
Hannes hand [18]	9	1
MyoAdapt Hand [19]	10	6
ALARIS hand [7]	6	6
Robotic arm [20]	10	7

3. Methods

3.1. Prototype Design and Manufacture

Most of the hand parts were 3D printed. Such a solution is increasingly used in many devices, because of a significant reduction in the weight of structures, the simplicity, fast and low cost of manufacture, the possibility of manufacturing variety of shapes, and the high repeatability of the parameters of details, taking into account individual characteristics [14,21–24]. The latter property is especially relevant for devices used in medicine and rehabilitation. The plastic used in three-dimensional printing is currently characterized by high strength and wear resistance, which explains the possibility of its use in medical devices.

In our prototype of an anthropomorphic robotic arm our goal was to closely copy the properties of the hand rather than its intrinsic structure. Based on the results of the analysis of the biomechanics of the hand, the design of the finger of the robotic arm was developed (Figure 2), in which each phalanx is connected to the next one with springs of twisting.

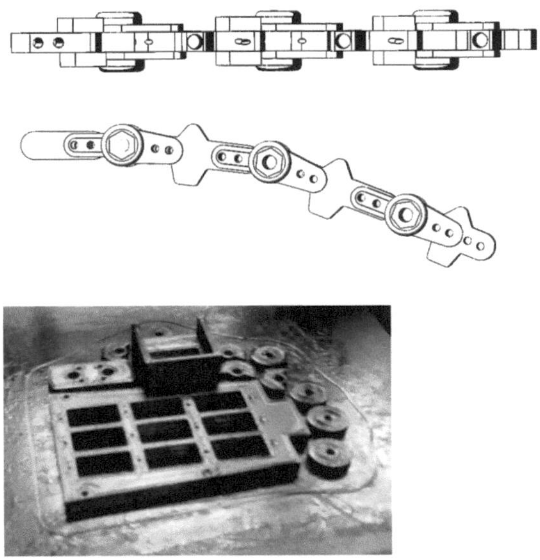

Figure 2. Designed and assembled finger construction made using three-dimensional printing technology; the process of manufacturing components of the mechanism.

A multiturn winder transfers the rotational gear motion to the tendons. The phalanges are driven by threads attached to servos, as shown in Figure 3a. The servos are located in the forearm. It contains the motor, the position sensors, the wave generator of the harmonic drive gear and the electronics. To reduce the distal phalanges, mg90s servos were used, which provide a compression force of 1.8 kg, and DS–939MG servos were used to reduce the proximal phalanges, which provide a compression force of 2.5 kg. With the help of a disk mounted at the output of the drive gearbox, the rotational motion is converted into linear. In the future, it is planned to use linear actuators to reduce the proximal phalanges, in order to reduce the volume occupied by it.

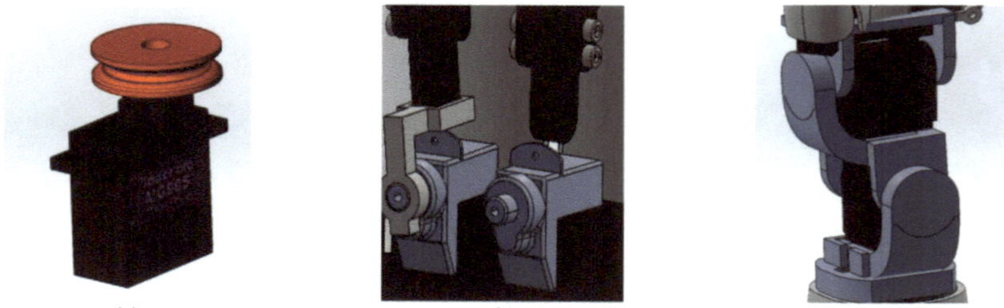

(a) (b) (c)

Figure 3. The main mechanical blocks of the projected robotic arm: (**a**) a mechanism for converting the rotational movement of the drive into linear; (**b**) a mechanism for finger extension and flexion; (**c**) a wrist mechanism.

The movement of the fingers is carried out directly by fixing the base of the finger on the mg995 servo gearboxes installed in the wrist of the robotic arm, as shown in Figure 3b. The wrist part of the robotic arm was made using the mechanism shown in Figure 3c. This inclusion of the servos allows the greatest effort to be developed; the design can withstand loads up to 15 kg.

The wrist was made on the basis of the spherical antiparallelogram mechanism, which provides rotation of the whole hand, combining reliability and simplicity, and allows for sideways motion, flexion and extension of the robotic wrist as a real human wrist [25,26]. The main difference from the designs taken as a basis is that the actuators are placed directly in the wrist joints, which provides maximum reliability and the smallest shoulder of force action, and, as a result, the maximum possible force exerted.

The range of movement of the mechanism for the fingers is shown in Figure 4. The number of servos and the design of the fingers provide controlled flexion and extension of the entire finger and each phalanx separately. The position can be fixed at any angle within the operating limits.

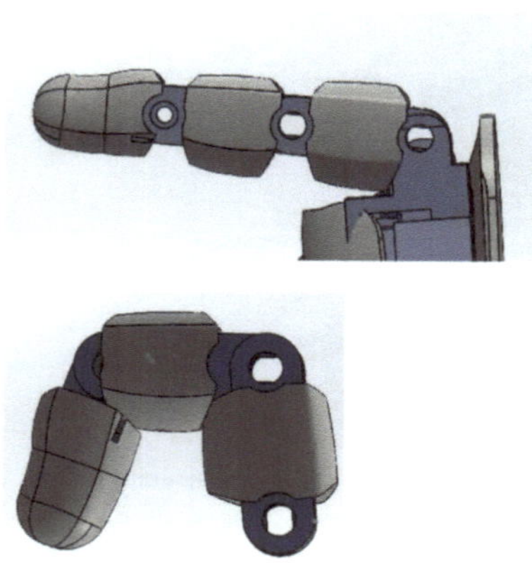

Figure 4. Simulation of the angular displacement on the example of the index finger.

The final view of the developed and assembled forearm and hand of the anthropomorphic robotic prototype is shown in Figure 5.

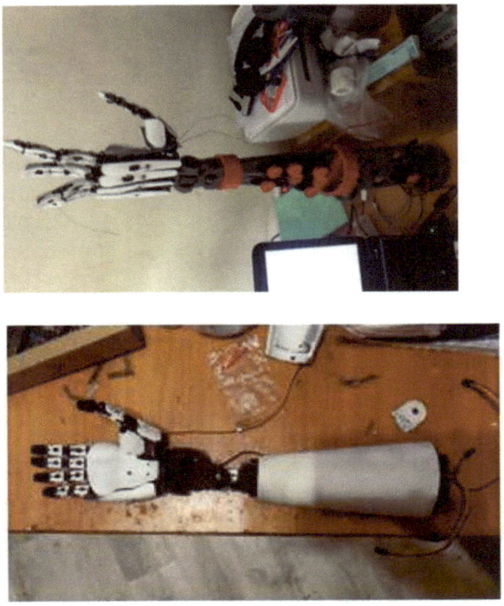

Figure 5. Engineered construction of an anthropomorphic robotic arm.

Taking into account the nominal angular velocity of micromotors, the maximum displacement of the phalanx of fingers and kinematic characteristics, we obtained that the maximum rotation speed was almost 5 rad/s, the response time to the execution of the command of flexion and extension of the phalanx of 90 degrees for this type of movement was less than 400 ms. The range of the finger and wrist movements were comparable with the movement of a human hand [27].

3.2. Control System

When designing the structures of an anthropomorphic robotic arm, the automated control system is crucial to the smooth regulation of actions and the accuracy of their reproduction. The lack of an analytical approach, difficult-to-solve dynamics and inconsistency with standard robotic manipulators are the reasons that reflect the importance of research on the ways of designing and automating an anthropomorphic arm for the next stage in the development of robotic manipulators and anthropomorphic prosthetic arms. To work with it, developments are underway to create control interfaces that differ in the types of signals used (speech, electromagnetic, etc.) and the methods of their processing and transmission for reproducing actions with robotic manipulators or a prosthesis. The typical structure of a control system consists of setting devices and regulators performing algorithms and control methods.

The methods of prosthesis control are based on fundamental principles: open-loop control (without feedback), feedback and compensation. Technical feedback is usually implemented either by the force or by the position. To increase the efficiency of the operations performed, an operator needs to receive more information about the state of the object (for example, about the temperature of the object, the humidity of the surface), that requires extended feedback, which in the vast majority of modern systems has not yet been implemented. The choice of the upper limb prosthesis control system is largely determined by of the type of the master signal. For example, the mechanical movement of arm segments, bioelectric signals of contracting muscles, varying impedance (total resistance) to the alternating current of a contracting muscle can be used as signals.

There is a transition to electromechanical prostheses of lost limbs in rehabilitation medicine. In spite of a great variety of modern technological achievements, the widely used electromechanical prostheses are controlled by means of electrical signals associated with the contraction and relaxation of the forearm muscles [28,29]. The signals are taken by surface electromyography electrodes from two groups of stump muscles (flexors and extensors) and fed through amplifiers to the electromechanical hand control system. The information from the sensors is transmitted to the microprocessor of the robotic hand and through computer algorithms is converted into motor commands and the prosthesis performs a certain gesture or grip. This control system is based on the electromyography and can be used only in the cases of the amputation of a part of a limb (hand and/or forearm), when electrical activity is preserved in intact muscle fibers associated with the control of the missing part.

Another large group of prosthetic control methods are neurocomputer interfaces. A neurocomputer interface is understood as a system that allows neural signals of the brain related to some part of the body, for example, to an arm or leg, to be decoded. There are several competing approaches to the creation of neurocomputer interfaces, which differ in the way electrical signals are transmitted from the brain to the computer. Invasive systems are based on implanting a matrix of ultrathin electrodes into certain areas of the brain [30]. However, the implantation of the electrode matrix requires unsafe surgery. Moreover, the questions remain open about the long-term biocompatibility of the electrode material and brain tissue and the change in work efficiency over time. Currently, the record of life with an invasive brain–computer interface is 7 years and 3 months [31].

Non-invasive systems are based on capturing electrical signals of the brain from the surface of the scalp. They use an electroencephalogram. Electroencephalography (EEG) electrodes that record the brain activity of the operator are used as the method that registers the master signals from a biological object (the operator of the prosthesis) [8]. The control of the prosthesis by this method consists of registering a signal of electrical activity of the brain using surface electrodes, then the signal is processed using the input circuits of the amplifier and converted into a digital code, the digital code is analyzed by the microcontroller of the control unit and converted into a command for the actuator of the prosthesis. To conduct this, EEG electrodes are fixed on the head, which are connected by wires to the control system of the prosthesis. Non-invasive interfaces are inferior to invasive ones in the accuracy of executing commands, because the electrical signals of the brain are significantly loosened and distorted when passing through the bones of the skull and skin. Accordingly, patients using non-invasive interfaces need longer training. However, these disadvantages are compensated by the security of non-invasive interfaces.

The analysis showed that the main part of the developments uses a certain control device and its corresponding algorithms. This limits the possibilities of using a robotic arm. For example, in rehabilitation medicine, it is first possible to connect to a sensory glove or augmented reality means in the process of teaching the patient. In the future, it will be possible to connect to the brain–computer interface using an electroencephalograph.

Moreover, the control system can be connected with another master device or software application that is used in different teaching applications, automated systems and rehabilitation systems with augmented reality simulators. However, all these control systems have the same universal functional block of connection between the master device and the prosthesis. Our task was also to create a control system that could be connected via the anthropomorphic robotic arm to various master-devices.

Figure 6 shows the block diagram of the hand drive control system. The structure is generally similar to that used in [25]. Control commands are sent to the hand controller via Bluetooth protocol. The controller regulates the current consumed by the servos so that its value does not exceed if allowed value.

Figure 6. Block diagram of the control system.

A current shunt is used to measure the current flowing through each servo. The current is measured by measuring the voltage drop on the shunt. For example, with a given value of the maximum allowable current in I_{max} = 1 A and a shunt resistance of 0.02 ohms, the voltage drop on the shunt will be:

$$U = I \cdot R = 1\ A \cdot 0.02\ \Omega = 20\ mV. \quad (1)$$

The resulting voltage value must be compared with a value that cannot be exceeded. After the measurement, the signal enters the ADC input of the AtTiny24 microcontrollers. It has a bit depth of 10 bits. However, a signal with an amplitude of 20 mV cannot be compared with the set value. The ADC has a measurement range U_{max} = 3.3 V, that is, at 10 bits of bit depth, we obtain a quantization step:

$$U_q = U_{max}/2^{10} = 3.3\ V/1024 = 3\ mV. \quad (2)$$

The number of steps for a range of 0 . . . 20 mV, which corresponds to a current change from 0 to 1 A, is:

$$n = U/U_q = 20\ mV/3\ mV = 7. \quad (3)$$

In this case, the bit depth of the current measurement will be:

$$k = I_{max}/n = 1\ A/7 = 143\ mA. \quad (4)$$

Equations (3) and (4) show that to increase the accuracy, it is necessary to amplify the measured signal before feeding it to the ADC input. To perform this, we used an operational amplifier (OP amp). The LM358ADR chip, which has two op-amps, was used as an op-amp in the designed electrical control circuit. Calculation of the output voltage was made with Equation (5) of the non-inverting scheme of the OP amp.

$$U_{out} = U_{in} \cdot (1 + R1/R2) = 20\ mV \cdot (1 + 220\ k\Omega/1.2\ k\Omega) = 3.68\ V. \quad (5)$$

Based on the results of the calculations, a printed circuit board was designed (Figure 7a), including AtTiny24 microcontrollers and two dual op-amp chips, which allows the current from all four drives of each finger to be measured. In total, five boards will be used for the entire anthropomorphic robotic arm.

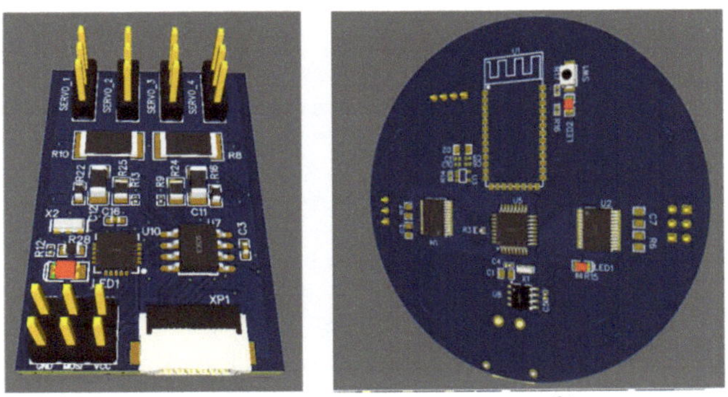

(a) (b)

Figure 7. Appearance of the designed board: (**a**) current measurements via servos; (**b**) control boards.

After processing the signals on the microcontroller, a signal is sent to the main control board (Figure 7b), on which the ATmega328-based main processor opens the key on the field-effect transistor if it is log. 0, and closes the key if it is log. 1, thereby providing protection of the drives from over current.

Twenty-one servos were assembled and used, one for each degree of freedom, with a peak current consumption of up to 1000 mA to ensure a high degree of similarity and the ability of the prototype to work as real human arm. Moreover, a battery was developed. The following requirements were drawn up for it: the ability to provide the prototype with power for a long period of time, being compact in mass and size in order to create a mobile and compact prosthesis that is close to a real human hand. The battery consists of four series-connected VTC 6 18650 lithium-ion batteries with a capacity of 3000 mAh and a peak current of 30 A (Figure 8). A balancing board was soldered to the batteries, the output was connected to a step-down DC-DC converter, which outputs 5.18 V, that was necessary and sufficient to power the servos of the prototype created.

Figure 8. Designed and assembled 16.7 V battery with a peak current of 30 A for autonomous power supply of the prototype.

4. Results and Discussion

Tests of individual components [12] and the operation of the control system [13] were carried out after the manufacture and assembly of a prototype of an anthropomorphic robotic arm. The developed experimental prototype of an anthropomorphic robotic arm with a large number of degrees of freedom provided a sufficient range of movements of the prosthesis, with the required degree of similarity with the movements of the human hand.

The mass-dimensional parameters of the created prototype of the anthropomorphic robotic arm were ergonomic and similar in properties to the human hand. The development had the smallest weight in comparison to the mass-dimension parameters of the currently used analogues (Table 2).

Table 2. The comparison of the mass and size parameters of anthropomorphic robotic arms [32–35].

Project	Anthropomorphic Robotic Arm	Bebionic 3 (England)	Motorica Manifesto Hand (RF)
Mass, of the arm part kg	0.520	0.698	0.482
maximum static weight, kg	12	45	20
battery options, mAh	3000	2200	1200 ... 3500
degree of freedom (DoF)	25	6	6

For an experimental characterization of the force of compression and gripping, weighing scales with a precision of 0.1 mg were placed on each fingertip with different durations of exposure. Holding a load weighing less than 1 kg was stable for 15 min for the fingers and more than 4 h for the arm mechanism. In the future, it will be possible to improve these parameters by using a material of the connecting elements of the phalanges with larger stiffness.

Simulations in CAD software and the prototype tests were carried out in which the anthropomorphic robotic arm emulated the different types of fingers and wrist movements, such as shown, for example, on Figure 3. The results obtained from the software and real simulation of different types of movements are summarized in Table 3.

Table 3. The experimental results for the angular displacements and the required time of each finger and wrist of anthropomorphic robotic arm.

Parameter	The Required Time to Perform the Flexion/Extension, ms	Finger Angles of Full Compression	Finger Angles of Full Extension	Proximal Phalanx	Distal Phalanx
Little finger	0.1	89°	90°	90°	83°
Ring finger	0.1	89°	90°	90°	82°
Middle finger	0.1	89°	90°	90°	82°
Index finger	0.1	89°	90°	90°	82°
Wrist rotation	0.12	-	-	-	-

Signals sufficient in magnitude to control the prototype of the anthropomorphic robotic arms were received, when connecting and sending a signal from a sensor glove and an electroencephalograph. It was experimentally established that the proposed control system is universal for connecting various types of control signal sources. The example of the measured transient response a test signal from a servo control for flexion/extension of a prosthetic finger is shown in Figure 9.

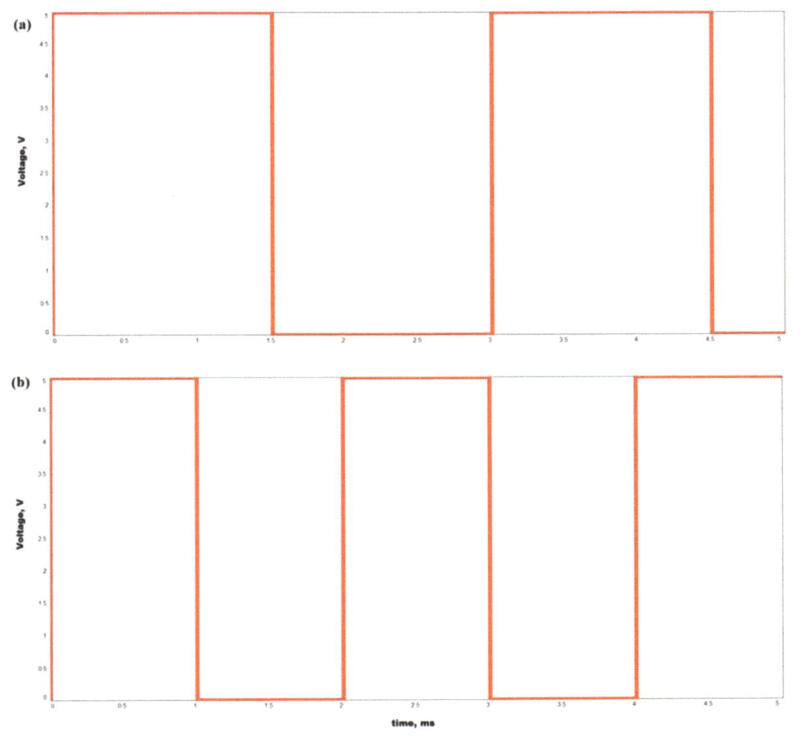

Figure 9. The sample of test impulse response of the servo of the index finger flexion (**a**) and extension (**b**).

The next stage of the research was to increase the accuracy of reproducing the actions of the sensory glove, to refine the software for reproducing the movements of an anthropomorphic robotic arm in an augmented reality and to assemble and debug the brain–hand interface by means of an electroencephalograph.

The designed and assembled prototype of the robotic arm during testing showed a fairly good accuracy of repeating the movement of individual finger joints, combining the simplicity and reliability of the design in real time. The developed prototype differs from its analogues in a large number of degrees of freedom, which allows for more precise operations, ensuring the movement of each individual phalanx.

The novelty of the project lies in the design of an anthropomorphic robotic arm that simulates real human movements more accurately than existing analogues of bionic hands.

5. Conclusions

The robotic arm was developed in conjunction with a glove that will detect small movements in the user's hand to control the prosthesis as a manipulator. Thus, the projected anthropomorphic robotic arm will be able to quickly and accurately simulate the movement of the user's hand.

The designed and assembled prototype of the hand during testing showed a fairly good accuracy of reproducing the movement of individual finger joints, combining the simplicity and reliability of the design in real time. The developed prototype differs from its analogues in a large number of degrees of freedom, which allows for more precise manipulation operations, the movement of each individual phalanx is ensured.

The development can be used as a prosthesis for the purpose of the rehabilitation of people, or as a high-precision manipulator in cases where it is necessary to replace a person with a sufficiently accurate reproduction of actions by hand.

Author Contributions: Conceptualization, V.A.Y., G.V. and A.V.; methodology, V.A.Y., G.V. and A.V.; software, V.A.Y. and G.V.; validation, V.A.Y. and A.V.; investigation, V.A.Y. and G.V.; writing—original draft preparation, V.A.Y. and G.V.; writing—review and editing, V.A.Y. and A.V.; visualization, G.V. and A.V.; supervision, A.V.; project administration, A.V. All authors have read and agreed to the published version of the manuscript.

Funding: This research received no external funding.

Institutional Review Board Statement: Not applicable.

Informed Consent Statement: Not applicable.

Data Availability Statement: Not applicable.

Conflicts of Interest: The authors declare no conflict of interest.

References

1. Lee, S.U.; Choi, Y.S.; Jeong, K.M.; Jung, S. Development of an underwater manipulator for maintaining nuclear power reactor. In Proceedings of the 2007 International Conference on Control, Automation and Systems, Seoul, Korea, 17–20 October 2007; pp. 1006–1010. [CrossRef]
2. Marinceu, D.; Murchison, A.; Hatton, C. Use of robotic equipment in a Canadian Used Nuclear Fuel Packing Plant. In Proceedings of the 2012 2nd International Conference on Applied Robotics for the Power Industry (CARPI), Zurich, Switzerland, 11–13 September 2012; pp. 139–144. [CrossRef]
3. Bae, H.Y.; Jae-Paeng, I.; Kim, S.; Park, S.Y.; Shin, H.S.; Kim, D.B.; Han, S.H. A Robust Control of Robotic Hand with Nine Axis for Assembly and Handling of Parts in Forging Manufacturing Process. In Proceedings of the 2018 International Conference on Information and Communication Technology Robotics (ICT-ROBOT), Busan, Korea, 6–8 September 2018; pp. 1–3. [CrossRef]
4. Park, C.; Kyung, J.H.; Choi, T.Y.; Do, H.M.; Kim, B.I.; Lee, S.H. Design of an industrial dual arm robot manipulator for a Human-Robot hybrid manufacturing. In Proceedings of the 2012 9th International Conference on Ubiquitous Robots and Ambient Intelligence (URAI), Daejeon, Korea, 26–28 November 2012; pp. 616–618. [CrossRef]
5. Zhang, J.; Wang, W.; Cai, Y.; Li, J.; Zeng, Y.; Chen, L.; Yuan, F.; Ji, Z.; Wang, Y.; Wyrwa, J. A Novel Single-Arm Stapling Robot for Oral and Maxillofacial Surgery—Design and Verification. *IEEE Robot. Autom. Lett.* **2022**, *7*, 1348–1355. [CrossRef]
6. Tavakoli, M.; Enes, B.; Santos, J.; Marques, L.; de Almeida, A.T. Underactuated anthropomorphic hands: Actuation strategies for a better functionality. *Robot. Auton. Syst.* **2015**, *74*, 267–282. [CrossRef]

7. Nurpeissova, A.; Tursynbekov, T.; Shintemirov, A. An Open-Source Mechanical Design of ALARIS Hand: A 6-DOF Anthropomorphic Robotic Hand. In Proceedings of the 2021 IEEE International Conference on Robotics and Automation (ICRA), Xi'an, China, 30 May–5 June 2021; pp. 1177–1183. [CrossRef]
8. Sanchez-Velasco, L.E.; Arias-Montiel, M.; Enrique Guzmán-Ramírez, E.; Lugo-González, E. A Low-Cost EMG-Controlled Anthropomorphic Robotic Hand for Power and Precision Grasp. *Biocybern. Biomed. Eng.* **2020**, *40*, 221–237. [CrossRef]
9. Light, C.M.; Chappell, P.H. Development of a lightweight and adaptable multiple-axis hand prosthesis. *Med. Eng. Phys.* **2000**, *22*, 679–684. [CrossRef]
10. Plagenhoef, S.; Evans, F.G.; Abdelnour, T. Anatomical Data for Analyzing Human Motion. *Res. Q. Exerc. Sport* **1983**, *54*, 169–178. [CrossRef]
11. ten Kate, J.; Smit, G.; Breedveld, P. 3D-printed upper limb prostheses: A review. *Disabil. Rehabil. Assist. Technol.* **2017**, *12*, 300–314. [CrossRef] [PubMed]
12. He, Z.; Yurievich, R.R.; Shimizu, S.; Fukuda, M.; Kang, Y.; Shin, D. A Design of Anthropomorphic Hand based on Human Finger Anatomy. In Proceedings of the 2020 International Symposium on Community-centric Systems (CcS), Tokyo, Japan, 23–26 September 2020; pp. 1–5. [CrossRef]
13. Atasoy, A.; Toptaş, E.; Kuchimov, S.; Gulfize, S.; Turpçu, M.; Kaplanoglu, E.; Güçlü, B.; Özkan, M. Biomechanical Design of an Anthropomorphic Prosthetic Hand. In Proceedings of the 2018 7th IEEE International Conference on Biomedical Robotics and Biomechatronics (Biorob), Enschede, The Netherlands, 26–29 August 2018; pp. 732–736. [CrossRef]
14. Devaraja, R.R.; Maskeliūnas, R.; Damaševičius, R. Design and Evaluation of Anthropomorphic Robotic Hand for Object Grasping and Shape Recognition. *Computers* **2021**, *10*, 1. [CrossRef]
15. Estay, D.; Basoalto, A.; Ardila-Rey, J.; Cerda, M.; Barraza, R. Development and Implementation of an Anthropomorphic Underactuated Prosthesis with Adaptive Grip. *Machines* **2021**, *9*, 209. [CrossRef]
16. Fajardo, J.; Ferman, V.; Cardona, D.; Maldonado, G.; Lemus, A.; Rohmer, E. Galileo Hand: An Anthropomorphic and Affordable Upper-Limb Prosthesis. *IEEE Access* **2020**, *8*, 81365–81377. [CrossRef]
17. Kashef, S.R.; Amini, S.; Akbarzadeh, A. Robotic hand: A review on linkage-driven finger mechanisms of prosthetic hands and evaluation of the performance criteria. *Mech. Mach. Theory* **2020**, *145*, 103677. [CrossRef]
18. Laffranchi, M.; Boccardo, N.; Traverso, S.; Lombardi, L.; Canepa, M.; Lince, A.; Semprini, M.; Saglia, J.A.; Naceri, A.; Sacchetti, R.; et al. The Hannes hand prosthesis replicates the key biological properties of the human hand. *Sci. Robot.* **2020**, *5*, 46. [CrossRef] [PubMed]
19. Leddy, M.T.; Dollar, A.M. Preliminary Design and Evaluation of a Single-Actuator Anthropomorphic Prosthetic Hand with Multiple Distinct Grasp Types. In Proceedings of the 2018 7th IEEE International Conference on Biomedical Robotics and Biomechatronics (Biorob), Enschede, The Netherlands, 26–29 August 2018; pp. 1062–1069. [CrossRef]
20. Owen, M.; Au, C.; Fowke, A. Development of a Dexterous Prosthetic Hand. *J. Comput. Inf. Sci. Eng.* **2018**, *18*, 010801. [CrossRef]
21. Baspinar, U.; Barol, H.S.; Senyurek, V.Y. Performance Comparison of Artificial Neural Network and Gaussian Mixture Model in Classifying Hand Motions by Using sEMG Signals. *Biocybern. Biomed. Eng.* **2013**, *33*, 33–45. [CrossRef]
22. Barabulut, D.; Ortes, F.; Arslan, Y.Z.; Adli, M.A. Comparative evaluation of EMG signal features for myoelectric controlled human arm prosthetics. *Biocybern. Biomed. Eng.* **2017**, *37*, 326–335. [CrossRef]
23. Hakonen, M.; Piitulainen, H.; Visala, A. Current state of digital signal processing in myoelectric interfaces and related applications. *Biomed. Signal Processing Control* **2015**, *18*, 334–359. [CrossRef]
24. Mott, R.L.; Vavrek, E.M.; Wang, J. *Machine Elements in Mechanical Design*; Pearson: New York, NY, USA, 2018.
25. Grebenstein, M.; Albu-Schäffer, A.; Bahls, T.; Chalon, M.; Eiberger, O.; Friedl, W.; Gruber, R.; Haddadin, S.; Hagn, U.; Haslinger, R.; et al. The DLR Hand Arm System. In Proceedings of the 2011 IEEE International Conference on Robotics and Automation (ICRA), Shanghai, China, 9–13 May 2011; pp. 3175–3182. [CrossRef]
26. Grebenstein, M.; van der Smagt, P. Antagonism for a Highly Anthropomorphic Hand–Arm System. *Adv. Robot.* **2008**, *22*, 39–55. [CrossRef]
27. Ben-Tzvi, P.; Zhou, M.A. Sensing and Force-Feedback Exoskeleton (SAFE) Robotic Glove. *IEEE Trans. Neural Syst. Rehabil. Eng.* **2015**, *23*, 992–1002. [CrossRef] [PubMed]
28. Abougarair, A.J.; Shashoa, N.A.A.; Elmelhi, A.M.; Gnan, H.M. Real Time Classification for Robotic Arm Control Based Electromyographic Signal. In Proceedings of the 2022 IEEE 2nd International Maghreb Meeting of the Conference on Sciences and Techniques of Automatic Control and Computer Engineering (MI-STA), Sabratha, Libya, 23–25 May 2022; pp. 155–160. [CrossRef]
29. Controzzi, M.; Clemente, F.; Barone, D.; Ghionzoli, A.; Cipriani, C. The SSSA-MyHand: A dexterous lightweight myoelectric hand prosthesis. *IEEE Trans. Neural Syst. Rehabil. Eng.* **2017**, *25*, 459–468. [CrossRef] [PubMed]
30. Young, M.J.; Lin, D.J.; Hochberg, L.R. Brain-Computer Interfaces in Neurorecovery and Neurorehabilitation. *Semin. Neurol.* **2021**, *41*, 206–216. [CrossRef] [PubMed]
31. Mullin, E.; This Man Set the Record for Wearing a Brain-Computer Interface. Wired 2022. Available online: https://www.wired.com/story/this-man-set-the-record-for-wearing-a-brain-computer-interface/ (accessed on 1 September 2022).
32. Robohand. Hand and Forearm Prototype. YouTube. Available online: https://www.youtube.com/watch?v=PXw_nXA1t0c (accessed on 30 July 2022).
33. Motion Capture Glove Frame. YouTube. Available online: https://www.youtube.com/watch?v=tT_MFeVPESs (accessed on 30 July 2022).

34. Motorica Manifesto Hand. BionicsForEveryone.Com. Available online: https://bionicsforeveryone.com/motorica-manifesto-hand/#grip-patterns-control (accessed on 30 July 2022).
35. 8E70—bebionic Hand EQD. Ottobock. Available online: https://www.ottobock.com/en-us/product/8E70 (accessed on 30 July 2022).

technologies

MDPI

Article

An Automatic, Contactless, High-Precision, High-Speed Measurement System to Provide In-Line, As-Molded Three-Dimensional Measurements of a Curved-Shape Injection-Molded Part

Saeid Saeidi Aminabadi [1,*], Atae Jafari-Tabrizi [2], Dieter Paul Gruber [2], Gerald Berger-Weber [3] and Walter Friesenbichler [1,*]

1 Department of Polymer Engineering and Science, Montanuniversitaet Leoben, Otto Gloeckel Str. 2, 8700 Leoben, Austria
2 Polymer Competence Center Leoben GmbH, Roseggerstrasse 12, 8700 Leoben, Austria
3 Institute of Polymer Processing and Digital Transformation, Johannes Kepler University Linz, Altenberger Street 69, 4040 Linz, Austria
* Correspondence: s.saeidi-aminabadi@stud.unileoben.ac.at (S.S.A.); walter.friesenbichler@unileoben.ac.at (W.F.)

Abstract: In the manufacturing of injection-molded plastic parts, it is essential to perform a non-destructive (and, in some applications, contactless) three-dimensional measurement and surface inspection of the injection-molded part to monitor the part quality. The measurement method depends strongly on the shape and the optical properties of the part. In this study, a high-precision (±5 µm) and high-speed system (total of 24 s for a complete part dimensional measurement) was developed to measure the dimensions of a piano-black injection-molded part. This measurement should be done in real time and close to the part's production time to evaluate the quality of the produced parts for future online, closed-loop, and predictive quality control. Therefore, a novel contactless, three-dimensional measurement system using a multicolor confocal sensor was designed and manufactured, taking into account the nominal curved shape and the glossy black surface properties of the part. This system includes one linear and one cylindrical moving axis, as well as one confocal optical sensor for radial R-direction measurements. A 6 DOF (degrees of freedom) robot handles the part between the injection molding machine and the measurement system. An IPC coordinates the communications and system movements over the OPC UA communication network protocol. For validation, several repeatability tests were performed at various speeds and directions. The results were compared using signal similarity methods, such as MSE, SSID, and RMS difference. The repeatability of the system in all directions was found to be in the range of ±5 µm for the desired speed range (less than 60 mm/s–60 degrees/s). However, the error increases up to ±10 µm due to the fixture and the suction force effect.

Keywords: automatic in-line measurement; cylindrical three-dimensional measurement; confocal sensor; shrinkage and warpage measurement; piano-black surface part; OPC UA communication protocol

check for updates

Citation: Saeidi Aminabadi, S.; Jafari-Tabrizi, A.; Gruber, D.P.; Berger-Weber, G.; Friesenbichler, W. An Automatic, Contactless, High-Precision, High-Speed Measurement System to Provide In-Line, As-Molded Three-Dimensional Measurements of a Curved-Shape Injection-Molded Part. *Technologies* **2022**, *10*, 95. https://doi.org/10.3390/technologies10040095

Academic Editors: Manoj Gupta, Eugene Wong and Gwanggil Jeon

Received: 29 June 2022
Accepted: 12 August 2022
Published: 17 August 2022

1. Introduction

Injection molding of plastics is a non-linear and considerably complex process with several dependent process parameters that drive the quality of the produced parts [1,2]. There is an increasing demand for in-line and real-time inspection of the quality of injection-molded parts, which requires in-line quality feature measurements, machine learning for quality feature prediction, and smart adaptive control of the injection molding process. The quality of the produced parts depends on various production parameters, such as plastic material, part geometry, required surface quality (Gruber et al. [3–6]), mold design, and process parameters [2,7–14]. The quality requirements vary based on the application of the produced part. Generally, quality disturbances of an injection-molded part include sink marks, weld lines, diesel effect, matt points, jetting, grooves, streaks, flashing, blister, underfilling, flaking, cold slug, voids, shrinkage, or warpage. Usually, the weight and the dimensional properties of the part are considered as optimization goals [1,2,7–13,15], as they are easy to evaluate. In industry, the dimensional properties of the produced part caused by shrinkage and warpage during cooling from liquid to solid state are the most important features conforming to the application of the part, typically in an assembly group.

Since 1998, researchers have attempted to bring these quality features under control [13]. However, due to the complexity and expenditure ofa quality feedback system for injection molding, most researchers have attempted to build and apply an offline quality feedback system [11–13] to achieve this optimization goal. Shrinkage and warpage, as the most quantitate variables in injection-molded part production, are directly related to the three-dimensional (3D) measurement of the part [7,8]. Regarding the complete 3D measurement time consumption and the part properties, some simplified measurements have been approved by researchers for shrinkage and warpage calculations; therefore, only selected dimensional properties are measured and compared. The applied method/instrument for measurement is relevant in terms of the part geometric properties and the required precision.

In their research on as-molded and post-molded shrinkage measurement, Jansen et al. [16] used a Strasmann traveling microscope to measure the length and width of specimens. Pomerleau et al. [17] reviewed a range of research applying optical, profile projector, coordinate measuring machines (CMM) methods or mechanical tools, such as calipers and micrometers, to measure dimensions. They applied a profilograph to measure the distance between several points of the samples. Régnier et al. [18] developed a special camera tool for dimensional measurement of engravings on polymer plates. Liao et al. [15] employed a Cyclone scanner to create a cloud point file of cell phone cover parts. They used PolyCAD and PolyWORKS software to calculate the shrinkage and warpage in groups of width, length, and thickness of the part. Use of a caliper or micrometer is common for dimensional measurement [1,8], even in recent research. Given that using a caliper or micrometer for high-precision measurement could intervene in the applied force, a coordinate measuring machine (CMM [9]) has also been employed in many studies. Other measurement methods, such as computed tomography (X-ray μ-CT measurement) [7], can achieve high-precision measurement, although typically in a time longer than common manufacturing cycle times. Most of the applied measurement methods for injection-molded parts lack speed and automatability to achieve high-precision, automatic, in-line, and as-molded dimensional measurement with the ability to analyze numerous complex dimensional part features within seconds, which is crucial for real-time, closed-looped quality control.

Two types of non-destructive measurement solutions can be applied for glossy-surface parts: 3D camera-aided methods and optical sensor methods. Camera-aided methods for high-speed and high-precision measurements have been used widely in previous works [19–21]. Whereas camera methods are dependent on light and reflection, the glossy surface of our sample encouraged us to use optical laser distance measurement methods. Two studies [22,23] compared the measurement accuracy and capabilities of triangulation laser (TL) sensors and confocal laser (CL) sensors, showing that the CL sensors are among the most efficient, reliable, and accurate sensors for profile measurement applications. Boltryk [23] studied CL sensors on a cylindrical surface. Confocal optical sensors were previously used for surface characterization and in surface profilometer [24,25] using surface confocal microscopy (SCM), which inspired us to further study surface features of the injection-molded parts using CL sensors. Yang et al. [26] reviewed a range of 3D surface measurement methods and determined that confocal laser measurement (CLM) is a suitable method for microscale surface characterization with high axial resolution and high signal-to-noise ratio. Nouira et al. [27] studied confocal sensor behavior with respect to surface material and errors, which ensures the reliability of these types of sensors.

Based on the molded part properties and required precision of the measurements, a CL-3000 series Keyence confocal displacement sensor [28] which is a multicolor confocal sensor, was selected as the measurement tool for our research. Detailed information on this sensor is provided in Section 3.1. The temperature of the confocal head influences the output accuracy, as proven by Berkovic et al. [29], who tested a Micro-Epsilon confocal sensor in a temperature range of −5 °C to 55 °C, in which the error increased to 100 μm. However, our selected laser has a heat-eliminated head, and its temperature nonlinearity error is 0.005% of full scale (F.S.) per °C [18]. The full scale (F.S.) range of the measurement is ±3 mm for a more accurate output of the sensor, and the ambient temperature is always between 20 and 25 °C (as measured). Therefore, the error for a maximum 5 Kelvin temperature deviation is approximately 1.5 μm. The accuracy of cylindrical surface measurements is associated with the roundness measurement. Although cylindricality and roundness have been defined in the ISO 1101-2012 and 12181 standards [30,31], respectively, researchers have proposed other methods to test the profile measurement model versus systematic errors [32,33]. Calculating the accurate profile of a cylindrical surface is advantageous in comparing the measured real part with the reference.

The aim of our research is to study the dimensional variations of the measured parts with respect to the changes in the injection molding machine parameters in an in-line, fast, and high-precision manner. Therefore, the precision target is the significance of measurement repeatability. Shrinkage (and, consequently, warpage) of injection-molded parts is divided into three groups depending on the measurement time: in-mold shrinkage (shrinkage during the injection molding process), as-molded shrinkage (just after demolding of the part), and post-molded shrinkage (shrinkage over the time after demolding, measured 16 h later earliest) [16]. In most cases, researchers measure the post-molded shrinkage to relate the final shrinkage and warpage of the part to the process parameters. Nevertheless, the as-molded shrinkage measurement type is indispensable with respect to in-line dimensional measurement, real-time process parameter optimization, and closed-loop control.

In the current research, a novel measurement system was created for a particular piano-black curved part as a "hard example" for a contactless and high-speed measurement problem. Therefore, experiences from previous projects with high- and ultra-high-gloss surfaces were applied [5,34]. Measurements of injection-molded parts should be provided as a feedback system for a closed-loop, real-time optimization problem. This proposed measurement system provides three quality properties: linear length, arc length, and sink marks of/on the injection-molded part (sink marks measurement is a combination of the cylindrical measurement system and a camera system, which is beyond the scope of this paper). In particular, speed, as well as the repeatability of the measurement system, is important for the sustainability of the measurement and optimization systems. Chiaroitti et al. [35] presented a high-accuracy dimensional measurement system for cylindrical components based on a confocal chromatic sensor. The presented system includes an X-Y micrometric stage to move the specimen to correct the center of the measuring cylinder, a 90° tilted confocal sensor with a linearity of 13 μm, and two moving stages (one on the Z axis and one rotational) to move the confocal sensor for the measurements. The authors also suggested methods for thermal self-compensation and self-calibration checks.

In our research, we conducted a cylindrical surface measurement to reach the required measurement repeatability precision correlating with the production parameters and dimensional variations, excluding systematic errors. In this paper, we first present the properties of the part and requirements of the measurements. Then, we introduce the manufactured measurement system. Subsequently, we present the results of measurements conducted on a ground gauge, as well as the results of a repeatability test on the actual part, to validate the precision of the measurements under varied process conditions.

2. Part and Measurement Properties

2.1. Part Properties

The inspected part is a partially cylindrically shaped sample with mold dimensions of length (L) = 120.2 mm, outer radius (r) = 123 mm, curve (bow) length (C) = 125.36 mm, and arc angle (α) = 58.39°, as illustrated in Figure 1a. The dimensional properties after ejection of the part from the mold vary depending on its shrinkage and warpage originating from the employed material, process parameters, and geometric features of the part. The post-molded volumetric shrinkage ratio for ABS (acrylonitrile butadiene styrene, i.e., the selected material) is approximately 0.4 to 0.7% [36]. Therefore, the part dimensions, after complete shrinkage for more than 24 h, are expected to be L = 119.4–119.7 mm, r = 122.1–122.5 mm, and C = 124.5–124.9 mm. Warpage is measured as the deviation from centroid shrinkage, i.e., as deviation from the mean radius.

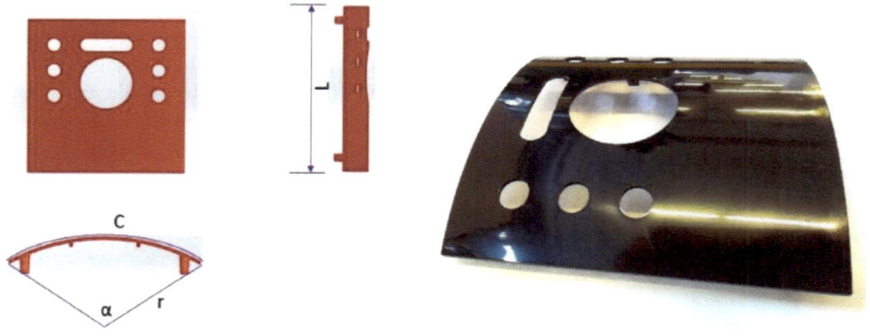

(**a**) (**b**)

Figure 1. Illustration of the measurement sample. (**a**) Isometric view of the sample. (**b**) The highly reflective surface of the sample, called a piano-black surface.

The highly reflective surface of the part, as illustrated in Figure 1b, makes optical measurement challenging with respect to the optical dimensional measurement and surface defects. The combination of cylindricality, high reflectivity, and black color makes the measurement problem a difficult case for conventional optical measurement methods.

2.2. Measurement Requirements

Based on initial simulations of the injection molding process for this part, the minimum required resolution of the measurement should be ±5 μm to detect the effects of all studied injection molding process parameters on dimensions of the produced part. Accordingly, the goal of our research is to achieve such precision for maximum traceability of the effects of the process parameters.

2.3. Part Manipulation: Gripper

A Kuka® KR 5 arc (KUKA CEE GmbH, Steyregg, Austria) robot is responsible for delivering the produced parts from the opened injection mold to the measurement system, completing the in-line measurement, and assuring time consistency between production and measurement. The handling system for the part should avoid any dimensional distortion before shrinkage and warpage measurements, as well as any surface intervention, to avoid disturbing sink marks and weld line detections. Therefore, the central circular hole of the part is used for handling between the mold and the fixture in the measurement system. The handling method is demonstrated in Figure 2a. The gripper is designed with a conical shape to provide maximal surface contact with the part and reduce the force and, consequently, the strain on the part (Figure 2b).

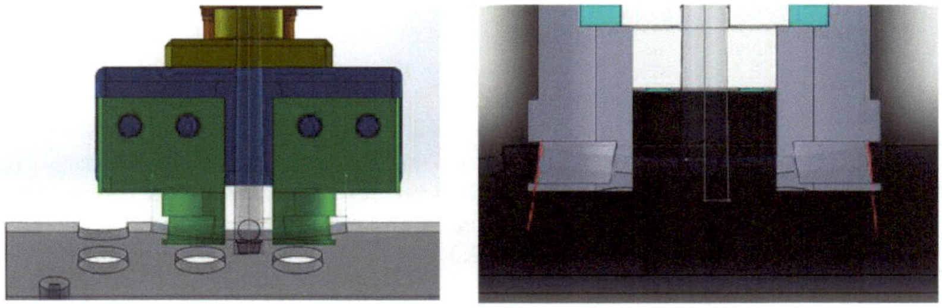

(a) (b)

Figure 2. The robot handling gripper. (**a**) Picking up the part through its center hole. (**b**) Matching the gripper slope with the hole slope for surface contact.

2.4. Fixture on the Measurement Stations

A fixture (Figure 3) mounted on the dimensional measurement system protects the part against unwanted movements during the measurements. The fixture must avoid any surface or dimensional distortion. Therefore, a fixture with four support pins and one vacuum suction cup was designed. The functions of the fixture parts (see Figure 3a) are as follows:

Pins 1 and 2 Fix the part through the lowest small eccentric holes in its surface against rotations and surface movements, limiting the downward movement of the part.

Pin 3 limits the downward movement and rotation of the part.

Pin 5 prevents the part from falling down when the vacuum suction is turned off.

The needle pin (highlighted in Figure 3b) is used to prevent positioning of the part in the incorrect direction on the fixture during troubleshooting by manual operation.

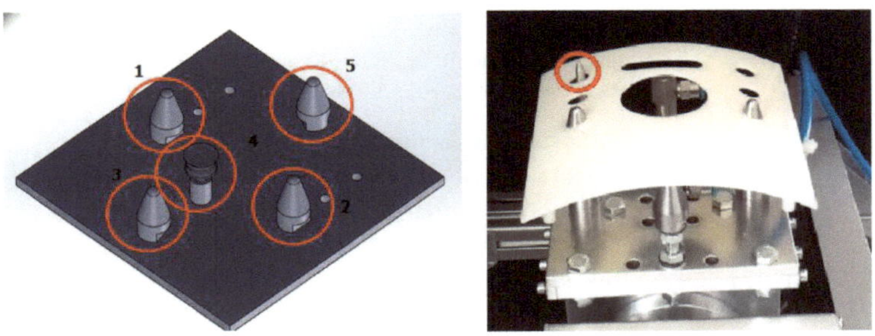

(a) (b)

Figure 3. Illustration of the measurement station fixture. (**a**) Positions of Support pins 1, 2, 3, 5 and vacuum suction 4. (**b**) A part positioned on the fixture.

3. Measurement System and Experiments

3.1. Feasibility Test

The confocal principle uses focused light emanating from an aperture onto the object to detect the light reflected from the object back into the aperture. The source light is transmitted using an optical fiber to the measurement head lens, which emits light at varying focal distances for each wavelength. The light reflected from the target surface passes through the lens into the spectrometer. The received light is split by wavelength and focused onto a high-resolution image sensor. Finally, a processor calculates the distance to the target based on the received light signals on the image sensor using processing techniques [22,24,25]. There are various types of optical measurement sensors; we initially studied those produced by KEYENCE company. In the interest of brevity, the results are not presented in this paper.

Based on the results of the initial experiments, a CL-3000 series confocal displacement sensor with a CL-P070 head [28] (KEYENCE INTERNATIONAL, Salzburg, Austria) was selected because of its appropriate response on black convex mirrored surfaces, its wide response range, high accuracy, and small laser spot diameter. Additionally, the selected CL sensor head type eliminates excess heat generation at the sensor head, which is important for high-precision measurement with confocal sensors [29]. The spot diameter of the laser head is 50 μm, and the output resolution is 0.25 μm; a resolution of 1.0 μm was selected in the controller for the output declaration. The linearity of the output within a range of ±3.0 mm is ±2.0 μm, and the maximum sampling rate of the laser controller is 10 kHz [28].

We initially expected that a two-dimensional linear system should be able to scan the part surface while moving the part under the confocal laser sensor. However, during the experiments, we observed that the confocal laser sensor was not able to read the reflection of the mirror surface at angle of more than 20 degrees. Therefore, about 17 mm of each side of the part (Figure 4) could not be measured by the confocal laser sensor with linear movements only, resulting in the conclusion that a fully Cartesian measurement system is not feasible.

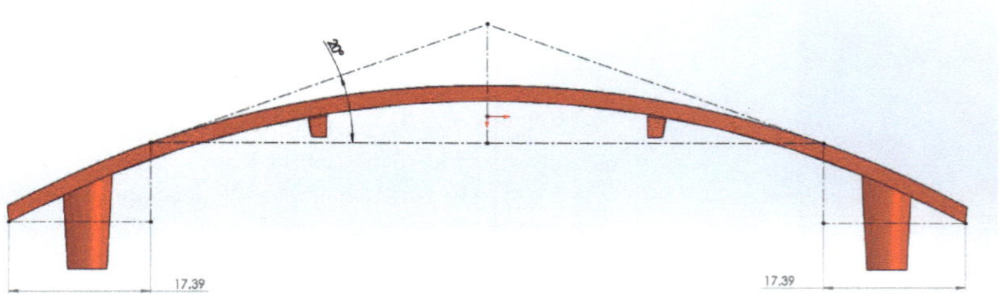

Figure 4. A local surface slope of more than 20° hindered the confocal sensor from achieving proper measurement, i.e., about 17 mm from either edge inward was unmeasurable with pure linear movements of the part under the sensor.

3.2. Measurement Methodology

The adopted solution for non-contact measurement of the part involves rotating the part around its cylindrical center, i.e., the measured surface section is always perpendicular to the sensor axis. Linear and rotary scans should be performed for 3D measurement of the surface of the part (shown in Figure 5c,d). Three-dimensional measurement consists of the relative rotated angle (θ), the relative position of linear movement (Z), and the measured radius of the surface element (R) (Figure 5a,b). For synchronization of the R, θ, and Z measurements, one linear encoder and one rotary encoder are used to provide pulse commands to the confocal sensor for sampling. The confocal controller takes one sample as soon as it receives a pulse command on the encoder input card. The stored values in the confocal controller for points that are beyond the part surface are represented by maximum values, whereas points on the part surface are represented by values in the range of ±3000 μm. The linear length of the measurement is determined by the number of scan points on the sample surface during the linear scan and the resolution value of the linear encoder in one pulse. Similarly, the arc length of the sample depends on the number of rotary encoder pulses and the measured radius (R) at the scan point.

The linear encoder is a FAGOR 'MX-420' with 1μm resolution and 5μm accuracy. The rotary encoder is an RLS magnetic encoder with an 'LM13IC2D0AA50F00' head with a resolution of 1 μm and a max scan frequency of 8 MHz, in addition to an 'MR100S071A152B00' disk with a 100 mm diameter and 152 poles, with a pole length of 2 mm. The rotary encoder provides 304,000 pulses per revolution (PPR) in A-B mode, and the linear encoder nominally provides 1000 pulses per millimeter (PPM).

The sampling speed of the confocal sensor is 10 kHz, and a divider is applied to reach the cycle time limit with a 5 μm resolution in the Z (divided by 5) and arc direction (divided by 2). The arc length of the sample in the design is 125.36 mm, which was expected to be in the range of 124.5 to 124.9 mm and is equivalent to a maximum of 24,980 steps for a resolution of 5 μm. The angle of the sample arc is approximately 60° (58.36°), which comprises 149,880 steps for a complete revolution (PPR). The closest value to the required PPR is half of the encoder output, i.e., 152,000 PPR, which is provided by using a pulse divider to be divided by two.

Based on the given descriptions and Figure 5b, the following applies:

$$Z = \sum_{m} \Delta Z_m = m \times \Delta Z \tag{1}$$

$$C = \sum_{n} R_n \times \Delta\theta_n = \sum_{n} R_n \times \Delta\theta \tag{2}$$

where ΔZ_m is the length value for each encoder pulse to the confocal sensor in the Z direction through linear movements, which nominally equals ΔZ = 0.005 mm per pulse for all of the m points of sampling in the Z direction. Subsequently, $\Delta\theta_n$ is the rotation angle value for each encoder pulse to the confocal sensor during the rotary movements, which nominally equals $\Delta\theta$ = 0.00236842° per pulse for all n sampling points. The nominal curve length (ΔC) for each rotation element is:

$$\Delta C = 122.56 \times 0.00236842 \times \frac{pi}{180} = 0.005066 \text{ mm} \tag{3}$$

where the radius of the designed part is 123 mm, which is expected to be 122.1 to 122.5 mm after shrinkage. During the rotation, the actual radius (R_n) differs depending on the local (shrinkage- and warpage-induced) radial deviation of the surface element. R_n includes a constant distance from the rotation center of the rotary axis to the zero position of the confocal sensor, i.e., R_0 = 122.56, and the distance measured by the confocal sensor ($R_{CL,n}$) from the zero position of the confocal sensor.

$$R_n = R_0 + R_{CL,n} \tag{4}$$

Therefore, the arc length of the part can be measured as (keeping in mind that $\Delta\theta_n = \Delta\theta$):

$$C = \sum_{n} R_n \times \Delta\theta = \sum_{n} (R_0 + R_{CL_n}) \times \Delta\theta = \sum_{n} R_{CL_n} \times \Delta\theta + n \times R_0 \times \Delta\theta \tag{5}$$

Because the total measurement duration should be less than the molding cycle time, it would be impossible to perform a complete scan of the surface of the part. Hence, three linear and three rotary lines on the samples are measured as illustrated in Figure 6. The four close-to-corner lines are positioned approximately above the center of the backside screw pins, allowing for a future study on sink marks measurements in these areas. The other two lines are approximately positioned at the center of the part and the center of the large circular hole, respectively. Later, the dimension of each line will be compared with similar lines of the other samples (or the golden sample) to observe shrinkage alterations. Warpage is defined based on the shrinkage balance between scanned profiles of the samples.

We used linear and rotary units from SMC company (SMC Austria GmbH, Korneuburg, Austria). The linear drive is an "LEFSH40B-300-R5CE17" series, and the rotary drive is an "LERH50K" series. The applied vacuum suction actuator exhibits a minimum load/displacement error in the design and fixture load. The nominal positioning accuracy of the actuators is ±0.01 mm for the linear and ±0.03° for the rotary axis. Although the accuracies of the actuators do not match with the target precision of the measurements, the precision of the measurements is obtained by the encoders and the confocal sensor sampling. Precise positioning of the axes is required to position the scan lines and home position for coworking with the pick–place robot. The scan line movements start from an absolute position (based on the axes coordinates and accuracies) outside the injection-molded part. The rotary unit moves to the absolute positions of 17.25°, 40.00° (center), and 65.30° angles for linear scans, starting from the absolute linear position of 30.00 mm and running to the absolute linear position of 160.00 mm. Subsequently, the linear axis moves to the 49.05 mm, 88.00 mm, and 147.50 mm positions for a rotated scan starting from 5.00° to 75.00° angles.

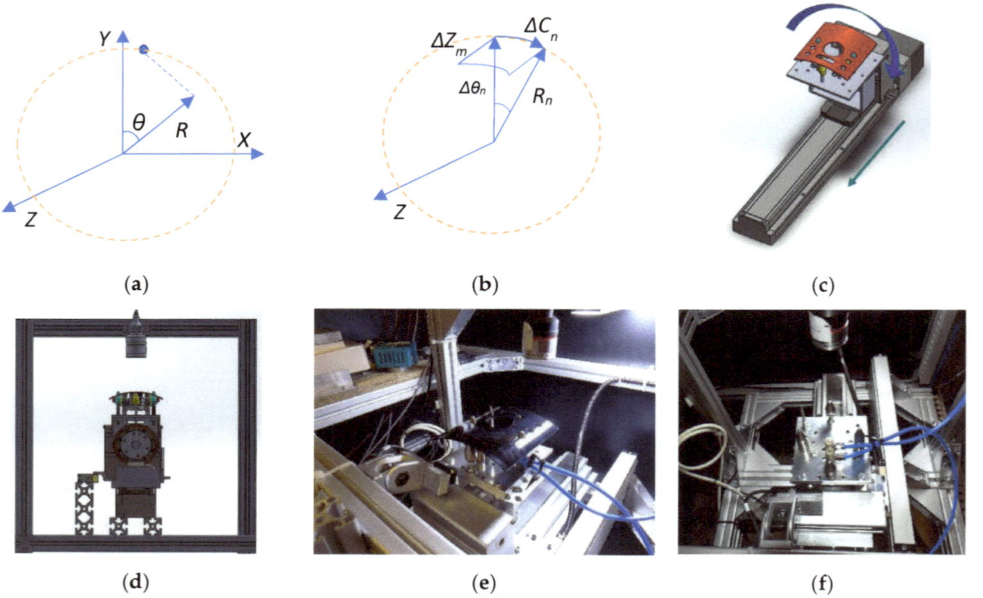

Figure 5. The final measurement system design. (**a**) Cylindrical coordinate system; (**b**) movement details during measurement; (**c**) two-dimensional movements; (**d**) side view of the rotary unit and confocal head; (**e**) isometric view of the installed dimensional measurement system; (**f**) top view of the installed dimensional measurement system.

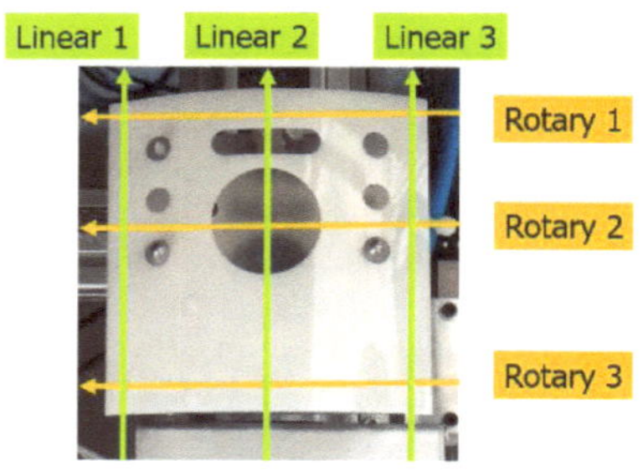

Figure 6. The position of the six scanning lines.

A BECKHOFF® IPC modelCP6600 (BECKHOFF Automation GmbH, Buers, Germany) was programmed to control axes movements and communicate with the robot and software (running on a separate computer for data acquisition) to read and store the data from the confocal sensor controller. The robot takes the part from the opened mold, checks the orientation of the part (for a possible rotation during the ejection and picking process) in a fork sensor, places the part on the balance, and finally, places the part on the fixture of the presented dimensional measurement system (see also Figure 7). Moreover, the time difference between the production and measurement is recorded (30 to 32 s); parts with excess time difference are excluded from the final shrinkage/warpage comparison due to time-dependent shrinkage.

The communication between the injection molding machine, IPC, and the data acquisition computer (DAQ) takes place on the OPC UA platform, as shown in Figure 8. The actual part number of the molded part, along with the production time, is extracted from the injection molding machine over OPC UA. Simultaneously, the time of measurement is read from the IPC and stored. The raw measurement data of the confocal sensor is stored on the DAQ computer for every point, including the pulse count number (received from the linear or rotary encoder) and the height distance from the confocal zero position of the sensor. A program written in Python code calculates the length of the scanned lines for each molded part after each measurement cycle.

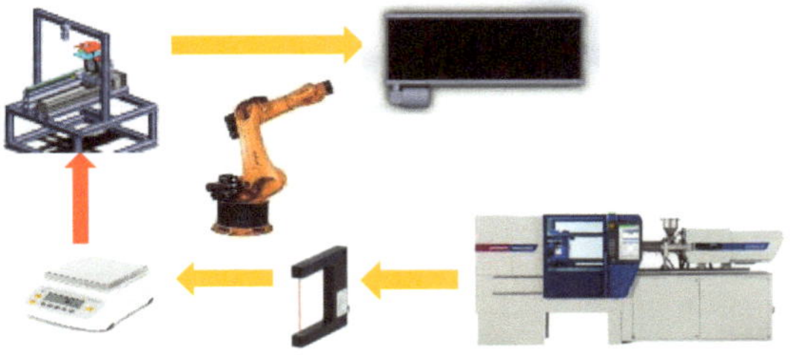

Figure 7. The handling and measurement procedure of the injection-molded part.

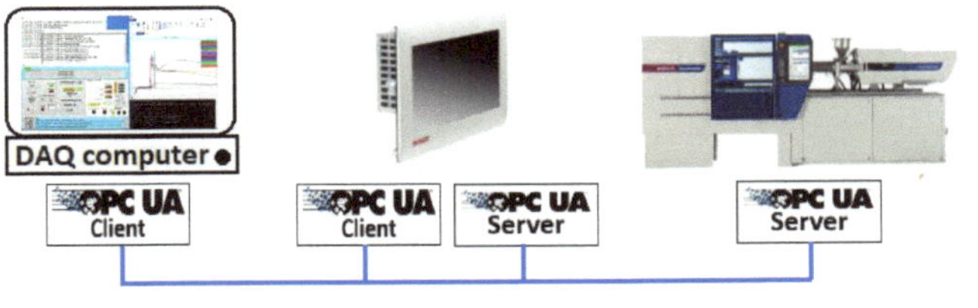

Figure 8. OPC UA communication platform between the IPC, injection molding machine, and data acquisition computer.

3.3. Experiment 1: Precision Validation Using a Self-Built Gauge

A special gauge (Figure 9) was built to test the repeatability and precision of the built system. The test measurements were conducted without encoders; therefore, a different scan strategy was adopted. Three linear scans at various angles and three rotary scans at different longitudinal positions were performed on the gauge (shown in Figure 9). The axes speeds were selected to as 30, 60, and 120 mm/s for linear scans and 30 and 60°/s for rotary scans (shown in Table 1). The linear scans have a length of 100 mm, and the rotary scans have a bow length of 200°. Each scan position was measured five times in both directions with similar test conditions, such as ambient temperature and environmental noise. The confocal sensor was set to a sampling frequency of 1kHz (to cover all speeds). We compared the measurements obtained with each measurement position separately to exclude systematic errors in the measurements [32].

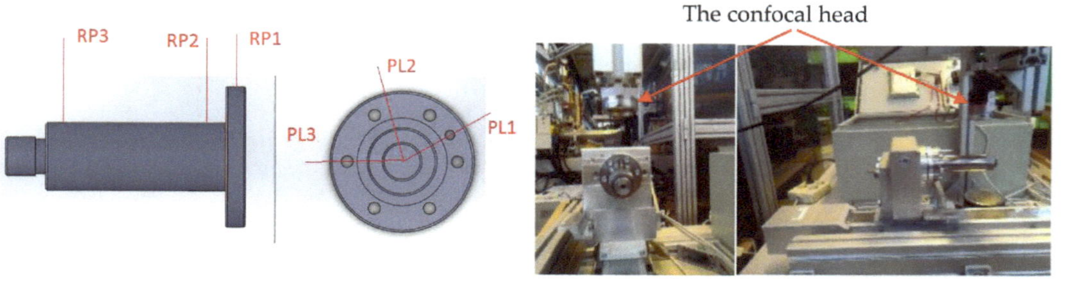

(a) (b)

Figure 9. (a) The designed gauge for initial precision validation and (b) the gauge installed on the axis.

Table 1. Experiments performed for repeatability evaluation.

Exp.	Position	Type	Speed	Unit
1-1	PL1	Linear	30	mm/s
1-2	PL1	Linear	60	mm/s
1-3	PL1	Linear	120	mm/s
1-4	PL2	Linear	30	mm/s
1-5	PL2	Linear	60	mm/s
1-6	PL2	Linear	120	mm/s
1-7	PL3	Linear	30	mm/s
1-8	PL3	Linear	60	mm/s
1-9	PL3	Linear	120	mm/s
1-10	RP1	Rotary	30	°/s
1-11	RP1	Rotary	60	°/s
1-12	RP2	Rotary	30	°/s
1-13	RP2	Rotary	60	°/s
1-14	RP3	Rotary	30	°/s
1-15	RP3	Rotary	60	°/s

The raw results of experiments 1–4 (linear scan for PL1 at a speed of 30 mm/s) are presented in Figure 10. Each scan took about 3.5 s at this speed, and there was a delay of 2 to 5 s due to manual movement commands. All the scans were stored continuously, and later, an algorithm based on the line slope was applied to identify the beginning of the scan lines (blue vertical lines in Figure 10). Results showed that the gauge, including the axes and production error, was tilted to a depth of about 0.03 mm with a length of 100 mm (about 0.017°). The scans in the reverse movement direction were rotated, and the ends of all scans were cut, resulting in the same length after extraction. To evaluate the repeatability of the measurements, the measurement values were subtracted from the average of the measurements, once considering only unidirectional movement and once considering only bidirectional movements, in order to compare the deviations of the scans and therefore the repeatability of the measurements. The comparison results and a histogram of the error are presented in Figure 11.

The comparison of errors in the R direction shows a satisfactory unidirectional result of ± 2 µm with a Gaussian distribution around the center, indicating that the error is a random type. However, a phase-shift problem occurred in the bidirectional comparison between the forward and backward directions, as shown in Figure 11b. Therefore, the measurement approach should be unidirectional for higher measurement repeatability.

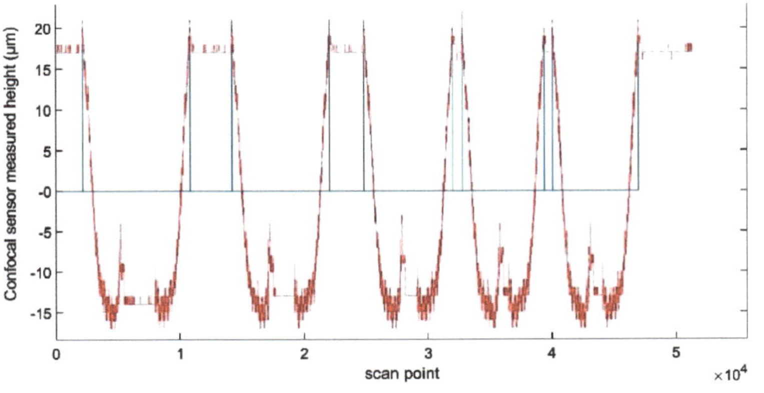

Figure 10. Raw result of confocal scan at the PL1 position with a speed of 30 mm/s.

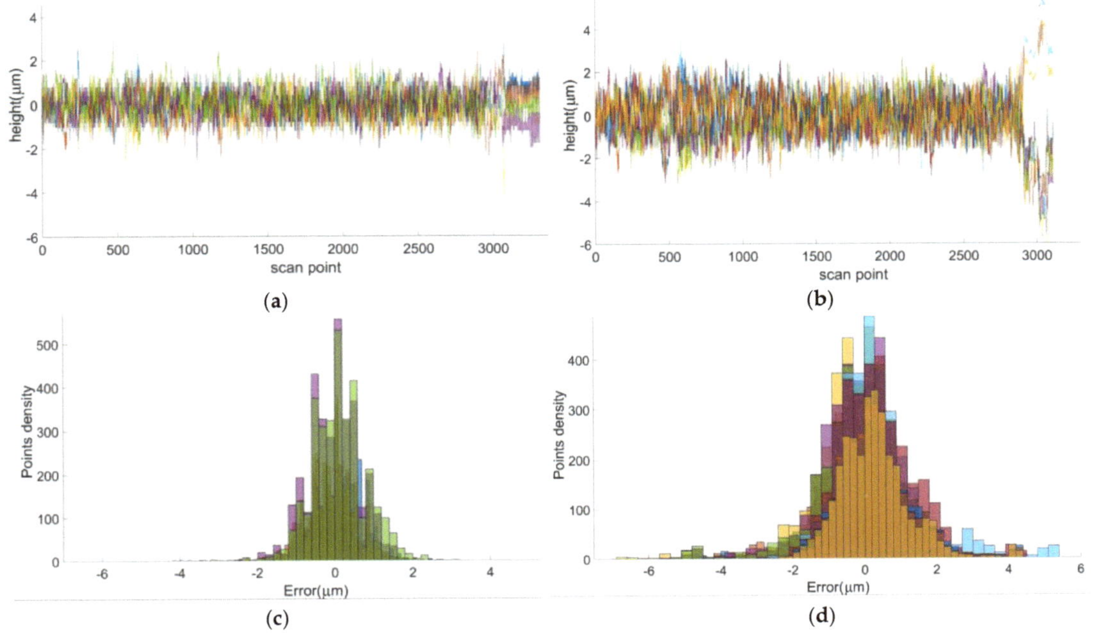

Figure 11. Five PL1 scans with a speed of 30 mm/s are subtracted from the average value: (**a**) unidirectional movements; (**b**) bidirectional movements; (**c**) histogram of error distribution for unidirectional movements; (**d**) histogram of error distribution for bidirectional movements.

The precision of the measurement system was examined using a precision gauge in this experiment. The unidirectional movements were found to be more accurate (less than 5 μm) in the confocal sensor direction, whereas the error distribution was normal. The standard deviation of the error (illustrated in Figure 12) proves the repeatability of the measurements in the sensor direction, with a maximum standard deviation of 1.15 μm for the unidirectional and 1.9 μm for bidirectional movements. Additionally, a group of similarity tests were derived to test the similarity of the measurement signals given in Figure 11a for all five unidirectional scans; the results are presented in Figure 13. The measurements similarity results show a very low RMS (root mean square) difference, and the cross-correlations between the mean of the measurements and each measurement match perfectly, which shows the best matching signal at zero lag.

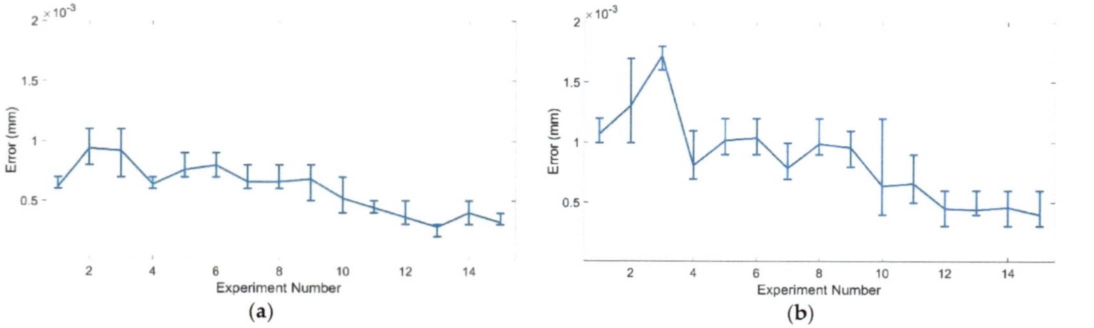

Figure 12. Standard deviation of errors for experiment 1 tests. (**a**) Standard deviation of errors for unidirectional movements; (**b**) standard deviation of errors for bidirectional movements.

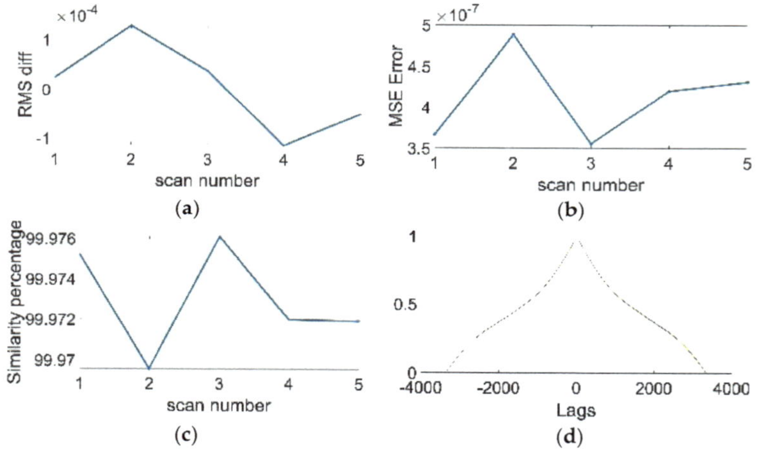

Figure 13. Similarity measures, (**a**) RMS level, (**b**) MSE error, (**c**) SSIM in percentage, and (**d**) cross correlation between the mean of measurements and each measurement.

3.4. Experiment 2: Precision Validation on the Molded Part

After the rotary encoder installation, 14 molded parts with different production settings were produced and sampled under stable production conditions to quantify the precision of the measurements. The produced parts were stored close to the measurement system at a temperature of about 23 °C for at least 48 h to ensure dimensional stability of the injection-molded parts. For the experiment, the parts were manually positioned on a fixture on the scale. Afterwards, the Kuka robot picked up the part from the fixture on the scale and positioned it on the fixture of the measurement system. Each part was measured 11 times continuously with a time difference of about 50 s to duplicate the in-line measurement conditions.

The measurements were compared with respect to the length and the stability in the confocal sensor direction (R). The first measurement was used as the reference, and other measurements were subtracted from the reference to observe the possible effect of the suction force. The results, illustrated for a sample in Figure 14, show that the difference from the first measurement increases with successive measurements. This effect appears strongest at the center of the part for line R1 and the corners of the part for R3 due to the positions of the support pins on the fixture, whereas the center of R3 is fixed under the support pin 3 (Figure 3) and shows no movement during the measurements. The linear line (L2) in the center rotates downward, whereas lines L1 and L3 exhibit a rotation effect combined with the deformation effect of R1 and R3. Based on the results shown in Figure 14, it can be concluded that the suction has a deformation effect of up to 10 μm in a time duration of about 9 min. The duration of the measurement for each part is about 50 s, and the total suction effect on the dimension in the first 50 s is less than 3 μm on the cooled part. However, the temperature of the as-molded part is higher; therefore, a higher deformation effect in 50 s can be expected, rather than the cooled part.

The length of the measured lines for the same sample is provided in Table 2. Despite the suction deformation effect during the 9 min of measurement, the maximum error for the length of linear lines fits in ±5 μm. However, the maximum error for the length of rotary lines is ±6 μm. The number of encoder pulses from the starting absolute position of the dimensional measurement system to the detected edge of the part for each scan line is given in Table 3.

Table 2. The calculated length of the lines (in mm) and the error range for lines 1 to 6.

Line	Mean Val.	Min	Max	Error
R1	122.742	122.736	122.747	0.011
R2	122.943	122.940	122.948	0.008
R3	122.831	122.822	122.834	0.012
L1	119.704	119.700	119.710	0.010
L2	119.785	119.780	119.790	0.010
L3	119.657	119.660	119.655	0.005

Table 3. The distance (number of encoder pulses) from the starting absolute point to the part edge for each line.

Distance to Line	Min	Max	Difference
R1	2488	2490	2
R2	2492	2494	2
R3	2474	2477	3
L1	425	426	1
L2	414	416	2
L3	383	384	1

The experiment was conducted on 14 parts produced under different machine settings. The results show a difference of a maximum of ±10 μm for the length of the lines (Figure 15) and a maximum pulse difference of ±2 for the distance from the absolute starting point (Figure 15a). However, this error includes the deformation resulting from the suction force, which was applied for 9 min duration of the successive measurement. The result of this experiment shows that the measurement repeatability error of the presented measurement system is better than ±10 μm for the presented measuring part, despite the repeatability error for most of the measurements being limited to ±5 μm. Results for the distance from absolute starting point to the edge of the part is shown in Figure 16.

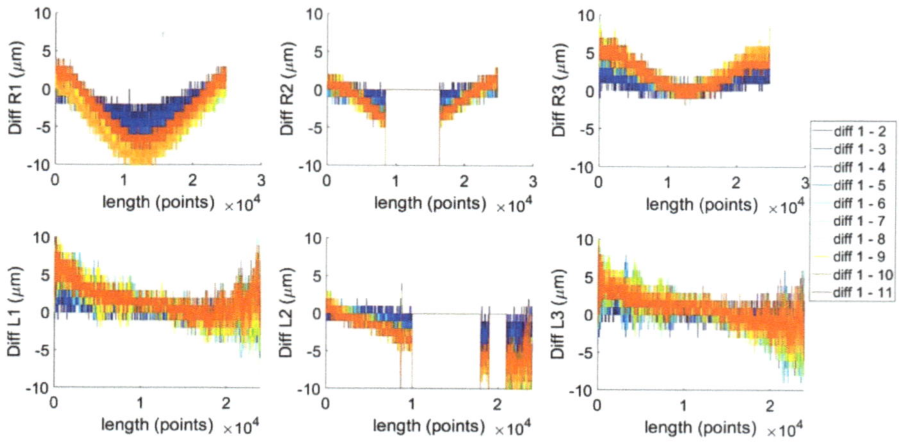

Figure 14. The difference of the scan lines between the first and the next 10 lines indicates the suction force effect during the successive measurements.

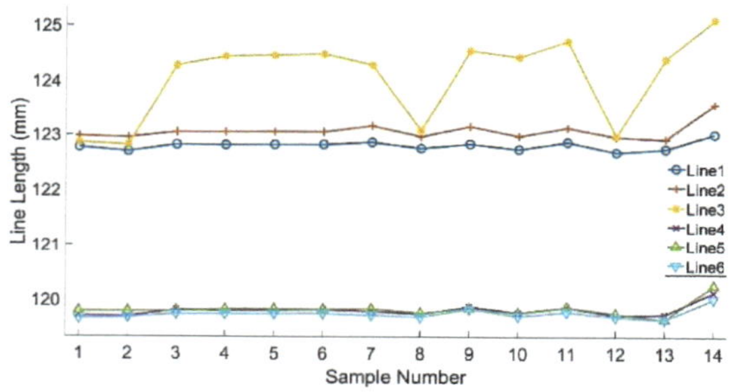

(**a**)

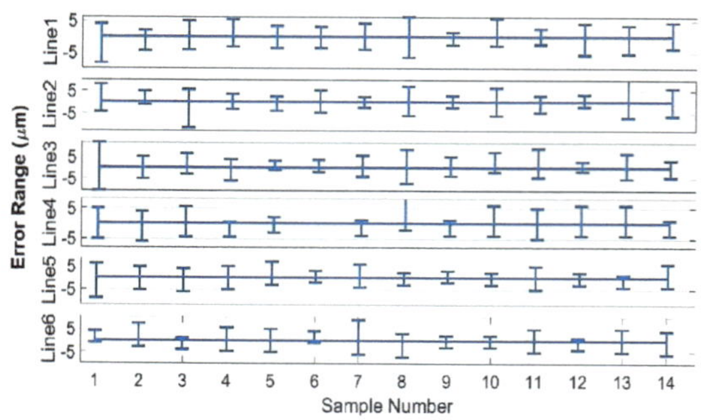

(**b**)

Figure 15. (**a**) Average measured line length for 14 samples. (**b**) Repeatability error range for the 14 samples.

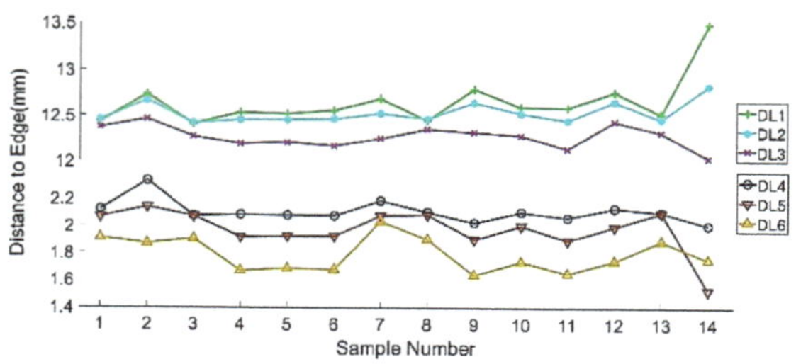

(**a**)

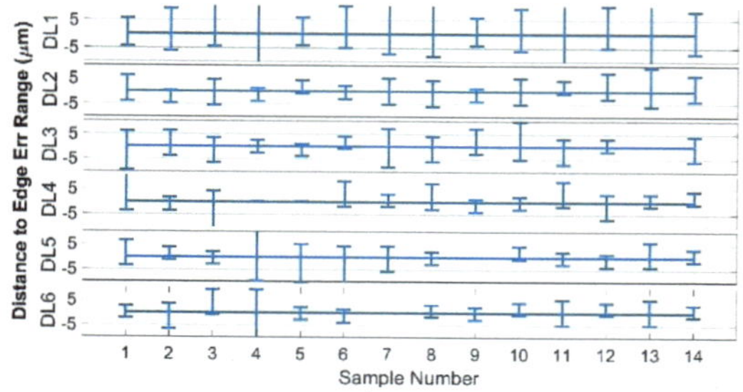

(**b**)

Figure 16. (**a**) Distance from the absolute starting point for the 14 samples. (**b**) Repeatability error range for the distance to the starting point.

3.5. Experiment 3: Positioning Error on the Fixture

After the robot positions the part on the fixture and releases it, the suction forces the part down onto the fixture, and the part settles through the fixture pins (Figure 3, pins 2 and 4). Several factors, such as the shrinkage of the holes of the pin positions and friction between the part surface and the pins, influence the part settlement. Further study of the errors of part positioning on the fixture is needed to perceive the error sources and possible required modifications.

Six types of minor part movements can occur on the fixture, as shown in Figure 17. Positioning errors 'a' and 'f' do not affect the results of the measurement. Errors 'b' and 'e' affect the length of linear and rotary lines. These errors can be calculated based on the distance from the absolute starting point to the edge of the part. Errors 'c' and 'd', in particular, affect the length of the rotary lines. Therefore, further study is required with respect to the level of these errors and the effect on the line length.

A set of 10 parts is selected with different machine settings after the stability of the production is verified. The robot places each part on the measurement system 13 times. It picks up the part after finishing the measurement each time and replaces the part on the measurement system to conclude the experiment. The first scan is considered the reference, and other scans are subtracted from the first scan to observe deviations during the successive replacements and scans.

The results of the replacement of the part for one sample are illustrated in Figure 18. The selected sample has moved between 15 and 30 μm in all directions (based on the distance from the absolute starting point). The R direction shows the movements of the sample included with the deformation resulting from the suction force. The effects of replacement are as large as ±40 μm (although mostly ±20 μm) in some scans in the R direction. The error of the distance to the edge of the part and the error of the line length for all parts remain within ±10 μm tolerance.

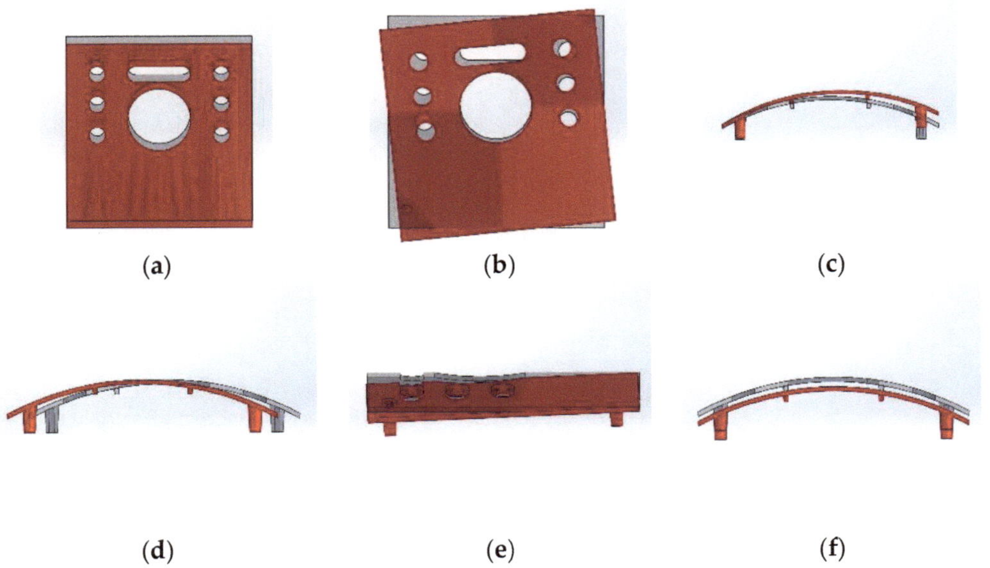

Figure 17. The six types of possible positioning errors of the part on the fixture.

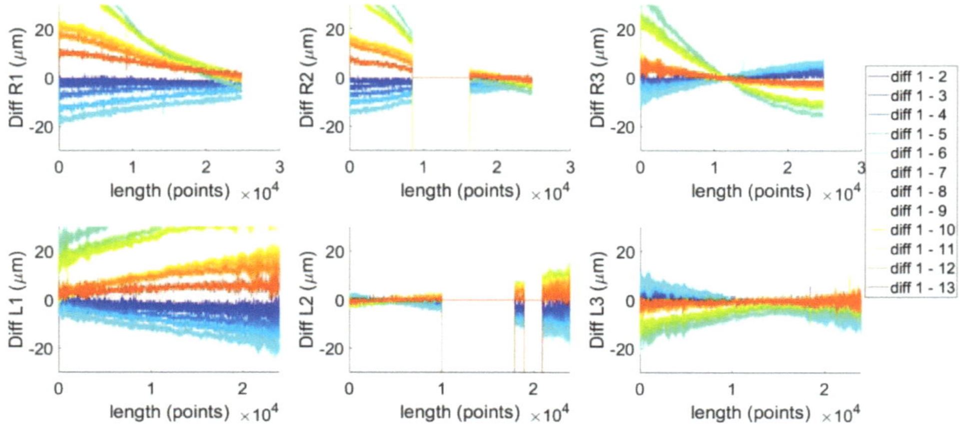

Figure 18. Repositioning of the part on the fixture in the R direction. The next 12 scans are subtracted from the first scan for comparison.

4. Results

Three experiments were carried out for precision validation of the presented cylindrical dimensional measurement system. Each experiment validated some properties of the measurement system. Experiment 1 validated the R direction (confocal sensor direction) during linear and rotary movements. A list of experiments is given in Table 1. The variation range of the standard deviation during each experiment is illustrated in Figure 12. The standard deviation of the errors (Figure 12) shows that the error of the linear movement increases with the increment of the speed for experiments 1 to 9 (linear movements), whereas the error decreases with the increment of the speed for experiments 10 to 15 (rotary movements). Moreover, Figure 12 shows that the error for bidirectional movements is higher than the error for unidirectional movements. Although the error is in the boundary of ±5 μm, the bidirectional movements are rejected, and unidirectional movements are accepted for higher precision. To test the results of scans with respect to additional aspects, the mean square error (MSE), structural similarity index (SSIM), root mean square (RMS) level, and cross correlation between the mean of the measurements and each measurement were calculated as the measures of signal similarity. The results for experiments 1–4 are presented in Figure 13. The high similarity index, very low MSE error, and very low RMS difference indicate that measurements are close to each other. The matching cross correlations between the mean of the signals and each signal with a central peak at zero and a value of 0.9998 is another measure to validate the similarity of the measurements. These measures were repeated for the other experiments, with similar results for both unidirectional movements.

Experiment 2 showed the actual repeatability error of the dimensional measurement system with the in-line conditions, despite the effect of the suction force in the R direction. The results are given in Figures 15 and 16. This experiment was conducted on 14 samples with different machine settings. Figure 15 shows the variations in the line length and error distribution versus the variation in machine settings. We expected that all lines would approximately follow a similar gradient. Because Line 3 did not follow the rule, we investigated and found a flashing problem on the end side of the line. The error tolerance for most of the samples is below ±5 μm. However, for some samples, the error increases up to ±10 μm. Regarding the deformation caused by suction force, the large error could be a result of increased deformation (lower stiffness) during the 9 min of measurement.

The distance from the absolute starting point of each line to the part edge is given in Figure 16. This distance is useful for reshaping the three-dimensional part for shrinkage and warpage calculations. The distance to Line 1 (DL1) has a larger error tolerance than the other lines, as the position of Line1 on the part has the largest distance to the support pins on the fixture. The error tolerance of the distance to the edge of the part for most of the lines fits in the range of ±5 µm; however, the error tolerance for all the measurements and lines fits in the range of ±10 µm.

With experiment 3, we studied the error produced by the fixture during placement of the part by the robot. The results show a displacement of ±20 µm in the R direction during the replacement of the part on the fixture. The linear lines are less sensitive to the replacement, whereas the rotary lines seem to be very sensitive. The middle of Line 3 (R3), which is fixed on support pin 3, is a very repeatable point of measurement. Despite the displacement error and the suction force effect, the tolerance of the line length fits within ±10 µm for all the measurements.

5. Conclusions

The final result shows that the repeatability of the scans in the R direction is dependent on the speed of movement, although the signal similarity error is always less than ±5 µm. The repeatability experiments with the injection-molded part show that the total precision of the measurement system is around ±5 µm for a short period of measurement (under 60 s). However, the suction force and slight movement of the part increase the error to a value of a maximum of ±10 µm.

In addition to the suction force, positioning the part on the fixture includes multiple errors, which have a minor influence on the part dimensional measurement and a major influence on the absolute positioning of the part in the cylindrical coordinate system.

In this study, the zero-position radius of the confocal sensor was calculated experimentally. For a future system generalization, we plan to develop an algorithm that automatically determines the zero-position radius. Moreover, we will build a new fixture to reduce the suction effect on the part and increase the precision of the measurements.

Author Contributions: Conceptualization, S.S.A.; data curation, G.B.-W.; funding acquisition, D.P.G., G.B.-W. and W.F.; investigation, S.S.A., A.J.-T. and G.B.-W.; methodology, S.S.A.; project administration, W.F.; resources, D.P.G. and W.F.; software, S.S.A. and A.J.-T.; supervision, D.P.G. and G.B.-W.; validation, S.S.A., A.J.-T., D.P.G. and G.B.-W.; visualization, S.S.A.; writing—original draft, S.S.A.; writing—review and editing, A.J.-T., D.P.G., G.B.-W. and W.F. All authors have read and agreed to the published version of the manuscript.

Funding: This research was funded by FFG research promotion agency in Austria as a part of the project INQCIM.

Institutional Review Board Statement: Not applicable.

Informed Consent Statement: Not applicable.

Data Availability Statement: The data presented in this study are available upon request from the corresponding authors.

Acknowledgments: This research was performed at Montanuniversitaet Leoben, Department of Polymer Engineering and Science, Injection Molding of Polymers, and in partnership with Polymer Competence Center Leoben (PCCL) as a part of the project INQCIM, which was supported by the WITTMAN BATTENFELD GmbH, MAHLE Filtersysteme Austria GmbH, EKB Elektro-u. Kunststofftechnik GmbH, Miraplast Kunststoffverarbeitungs GmbH, Julius Blum GmbH, and the Institute of Production Engineering and Photonic Technologies at TU WIEN University.

Conflicts of Interest: The authors declare no conflict of interest.

References

1. Ahmed, T.; Sharma, P.; Karmaker, C.L.; Nasir, S. Warpage Prediction of Injection-Molded PVC Part Using Ensemble Machine Learning Algorithm. *Mater. Today Proc.* **2020**. [CrossRef]
2. Singh, G.; Pradhan, M.K.; Verma, A. Multi Response Optimization of Injection Moulding Process Parameters to Reduce Cycle Time and Warpage. *Mater. Today Proc.* **2018**, *5*, 8398–8405. [CrossRef]

3. Gruber, D.P.; Buder-Stroisznigg, M.; Wallner, G.; Strauss, B.; Jandel, L.; Lang, R.W. A Novel Methodology for the Evaluation of Distinctness of Image of Glossy Surfaces. *Prog. Org. Coat.* **2008**, *63*, 377–381. [CrossRef]
4. Gruber, D.P. Method for Automatically Detecting a Defect on a Surface of a Molded Part. WO2010102319, 16 September 2010.
5. Gruber, D.P.; Buder-Stroisznigg, M.; Wallner, G.; Strauß, B.; Jandel, L.; Lang, R.W. Characterization of Gloss Properties of Differently Treated Polymer Coating Surfaces by Surface Clarity Measurement Methodology. *Appl. Opt.* **2012**, *51*, 4833–4840. [CrossRef]
6. Gruber, D.P. Method and Device for the Optical Analysis of the Surface of an Object. EP14186013, 1 April 2015.
7. Masato, D.; Rathore, J.; Sorgato, M.; Carmignato, S.; Lucchetta, G. Analysis of the Shrinkage of Injection-Molded Fiber-Reinforced Thin-Wall Parts. *Mater. Des.* **2017**, *132*, 496–504. [CrossRef]
8. Sreedharan, J.; Jeevanantham, A.K. Analysis of Shrinkages in ABS Injection Molding Parts for Automobile Applications. *Mater. Today Proc.* **2018**, *5*, 12744–12749. [CrossRef]
9. Azad, R.; Shahrajabian, H. Experimental Study of Warpage and Shrinkage in Injection Molding of HDPE/RPET/Wood Composites with Multiobjective Optimization. *Mater. Manuf. Processes* **2019**, *34*, 274–282. [CrossRef]
10. Barghash, M.A.; Alkaabneh, F.A. Shrinkage and Warpage Detailed Analysis and Optimization for the Injection Molding Process Using Multistage Experimental Design. *Qual. Eng.* **2014**, *26*, 319–334. [CrossRef]
11. Chen, W.-C.; Fu, G.-L.; Tai, P.-H.; Deng, W.-J.; Fan, Y.-C. ANN and GA-Based Process Parameter Optimization for MIMO Plastic Injection Molding. In Proceedings of the 2007 International Conference on Machine Learning and Cybernetics, Hong Kong, China, 19–22 August 2007; Volume 4, pp. 1909–1917. [CrossRef]
12. Ozcelik, B.; Erzurumlu, T. Comparison of the Warpage Optimization in the Plastic Injection Molding Using ANOVA, Neural Network Model and Genetic Algorithm. *J. Mater. Process. Technol.* **2006**, *171*, 437–445. [CrossRef]
13. Petrova, T.; Kazmer, D. Hybrid Neural Models for Pressure Control in Injection Molding. *Adv. Polym. Technol.* **1999**, *18*, 19–31. [CrossRef]
14. Kazmer, D.O. Injection Mold Design Engineering. In *Injection Mold Design Engineering*, 2nd ed.; Kazmer, D.O., Ed.; Hanser Publishers: Munich, Germany, 2016; pp. I–XXIV. ISBN 978-1-56990-570-8.
15. Liao, S.J.; Chang, D.Y.; Chen, H.J.; Tsou, L.S.; Ho, J.R.; Yau, H.T.; Hsieh, W.H.; Wang, J.T.; Su, Y.C. Optimal Process Conditions of Shrinkage and Warpage of Thin-Wall Parts. *Polym. Eng. Sci.* **2004**, *44*, 917–928. [CrossRef]
16. Jansen, K.M.B.; Van Dijk, D.J.; Husselman, M.H. Effect of Processing Conditions on Shrinkage in Injection Molding. *Polym. Eng. Sci.* **1998**, *38*, 838–846. [CrossRef]
17. Pomerleau, J.; Sanschagrin, B. Injection Molding Shrinkage of PP: Experimental Progress. *Polym. Eng. Sci.* **2006**, *46*, 1275–1283. [CrossRef]
18. Régnier, G.; Trotignon, J.P. Local Orthotropic Shrinkage Determination in Injected Moulded Polymer Plates. *Polym. Test.* **1993**, *12*, 383–392. [CrossRef]
19. Gao, J.; Gindy, N.; Chen, X. An Automated GD&T Inspection System Based on Non-Contact 3D Digitization. *Int. J. Prod. Res.* **2006**, *44*, 117–134. [CrossRef]
20. Liu, Y.; Zhang, Q.; Liu, Y.; Yu, X.; Hou, Y.; Chen, W. High-Speed 3D Shape Measurement Using a Rotary Mechanical Projector. *Opt. Express* **2021**, *29*, 7885–7903. [CrossRef]
21. Li, W.; Zhou, L.; Yan, S.-J. A Case Study of Blade Inspection Based on Optical Scanning Method. *Int. J. Prod. Res.* **2015**, *53*, 2165–2178. [CrossRef]
22. Alkmal, J.S. Investigation of Optical Distance Sensors for Applications in Tool Industry: Optical Distance Sensors. Master's Thesis, Saimaa University of Applied Science, South Kariland, Finland, 2013.
23. Boltryk, P.J.; Hill, M.; McBride, J.W.; Nascè, A. A Comparison of Precision Optical Displacement Sensors for the 3D Measurement of Complex Surface Profiles. *Sens. Actuators A Phys.* **2008**, *142*, 2–11. [CrossRef]
24. Jordan, H.-J.; Wegner, M.; Tiziani, H. Highly Accurate Non-Contact Characterization of Engineering Surfaces Using Confocal Microscopy. *Meas. Sci. Technol.* **1998**, *9*, 1142–1151. [CrossRef]
25. Yang, L.; Wang, G.; Wang, J.; Xu, Z. Surface Profilometry with a Fibre Optical Confocal Scanning Microscope. *Meas. Sci. Technol.* **2000**, *11*, 1786–1791. [CrossRef]
26. Yang, Y.; Dong, Z.; Meng, Y.; Shao, C. Data-Driven Intelligent 3D Surface Measurement in Smart Manufacturing: Review and Outlook. *Machines* **2021**, *9*, 13. [CrossRef]
27. Nouira, H.; El-Hayek, N.; Yuan, X.; Anwer, N.; Salgado, J. Metrological Characterization of Optical Confocal Sensors Measurements (20 and 350 Travel Ranges). *J. Phys. Conf. Ser.* **2014**, *483*, 012015. [CrossRef]
28. Keyence Confocal Displacement Sensors CL-3000. 2019. Available online: www.keyence.com (accessed on 24 March 2019).
29. Berkovic, G.; Zilberman, S.; Shafir, E. Temperature Effects in Chromatic Confocal Distance Sensors. In Proceedings of the SENSORS, 2013 IEEE, Baltimore, MD, USA, 3–6 November 2013; pp. 1–3. [CrossRef]
30. Geometrical Product Specifications (GPS)—Roundness 2011, no. ISO12181-2. Available online: https://www.iso.org/standard/53621.html (accessed on 20 March 2021).
31. Geometrical Product Specifications (GPS)—Geometrical Tolerancing—Tolerances of Form, Orientation, Location and Run-Out, 2017, no. ISO1101. Available online: https://www.iso.org/obp/ui/#iso:std:iso:1101:ed-4:v1:en (accessed on 20 March 2021).
32. Sun, C.; Wang, H.; Liu, Y.; Wang, X.; Wang, B.; Li, C.; Tan, J. A Cylindrical Profile Measurement Method for Cylindricity and Coaxiality of Stepped Shaft. *Int. J. Adv. Manuf. Technol.* **2020**, *111*, 2845–2856. [CrossRef]

33. Zeng, W.; Jiang, X.; Scott, P.J. Roundness Filtration by Using a Robust Regression Filter. *Meas. Sci. Technol.* **2011**, *22*, 035108. [CrossRef]
34. Gosar, Z.; Gruber, D.P. IN-LINE Quality Inspection of Freeform Plastic High Gloss Surfaces Aided by Multi-Axial Robotic Systems. In Proceedings of the International Electrotechnical and Computer Science Conference (ERK'2017), Portotoz, Slovenia, 26 September 2017; pp. 445–448.
35. Chiariotti, P.; Fitti, M.; Castellini, P.; Zitti, S.; Zannini, M.; Paone, N. High-Accuracy Dimensional Measurement of Cylindrical Components by an Automated Test Station Based on Confocal Chromatic Sensor. In Proceedings of the 2018 Workshop on Metrology for Industry 4.0 and IoT, Brescia, Italy, 16–18 April 2018; pp. 58–62. [CrossRef]
36. Brinkmann, O.B.; Schmachtenberg, O. *International Plastics Handbook*; HANSER: Munich, Germany, 2012; pp. 547–553.

technologies

MDPI

Article

Optimization of the Pick-Up and Delivery Technology in a Selected Company: A Case Study

Ondrej Stopka [1,*], Patrik Gross [1], Jan Pečman [1], Jiří Hanzl [1], Mária Stopková [1] and Martin Jurkovič [2]

[1] Department of Transport and Logistics, Faculty of Technology, Institute of Technology and Business in Ceske Budejovice, Okruzni 517/10, 37001 Ceske Budejovice, Czech Republic; gross@mail.vstecb.cz (P.G.); pecman@mail.vstecb.cz (J.P.); hanzl@mail.vstecb.cz (J.H.); stopkova@mail.vstecb.cz (M.S.)

[2] The Faculty of Operation and Economics of Transport and Communications, University of Zilina, Univerzitna 1, 01026 Zilina, Slovakia; martin.jurkovic@fpedas.uniza.sk

* Correspondence: stopka@mail.vstecb.cz

Abstract: This article deals with pick-up and delivery activities in a selected company that focuses on the distribution of products in the gastronomic sector of the market and suggests how to make the present approach more efficient. The introductory part of the article clarifies the meanings of basic concepts related to the issue of optimizing the logistics processes in the company. The crucial goal is to analyze the existing pick-up and delivery technology and then, in the application part of the article, to propose adequate measures in the context of streamlining these activities with their technical and economic evaluation. An analysis of current delivery routes, which are used for the distribution of gastronomic products, is first performed. Thereafter, the routes are optimized with the aim of minimizing the total distance traveled by using the Operations Research methods, namely: the Hungarian method, Vogel approximation method, nearest neighbor method and the Routin route planner which is based on a principle of the Greedy algorithm. At the end of the article, a technical and economical evaluation of the findings is discussed, wherein the individual results of optimization through selected methods are first compared and then, new optimized routes are selected.

Keywords: logistics center; Operations Research; distribution problem; vehicle routing problem; Hungarian method; Vogel approximation method; nearest neighbor method

check for updates

Citation: Stopka, O.; Gross, P.; Pečman, J.; Hanzl, J.; Stopková, M.; Jurkovič, M. Optimization of the Pick-Up and Delivery Technology in a Selected Company: A Case Study. *Technologies* **2022**, *10*, 84. https://doi.org/10.3390/technologies10040084

Academic Editors: Manoj Gupta, Eugene Wong and Gwanggil Jeon

Received: 15 June 2022
Accepted: 12 July 2022
Published: 14 July 2022

Publisher's Note: MDPI stays neutral with regard to jurisdictional claims in published maps and institutional affiliations.

1. Introduction

By optimizing pick-up and delivery routes of various transport-logistics companies, when using the methods of Operations Research, it is possible to reduce transport costs and other attributes for operating such routes. The savings in transport performance or fuel consumption may not be significant in the short term (one transport or one day), but, e.g., in a one-year period, the value of savings could be remarkable. The saved values could be used, for example, for other transport performance, reinvesting in company development, marketing and so forth.

This article deals with the analysis of the current pick-up and delivery activity in a selected company that deals with the distribution of gastronomic equipment. After this analysis, the optimization of current pick-up routes is performed by methods of operation research. There are three selected and applied methods: the Hungarian method, Vogel approximation method, the nearest neighbor method and the Routin route planner. The individual results after the optimization of the initial routes are compared with each other and, on the basis of this comparison, new pick-up routes with an adjusted order of unloading points are recommended to the selected company. These recommendations are supplemented by a technical and economical evaluation of the results, where the potential saving in transport performance, which represents an optimization criterion, and saving in fuel consumption costs are calculated after the application of the proposed route modification. Thus, the objective of this research is to define optimal delivery routes in terms of supplying predetermined customers when minimizing the total distance traveled.

The optimization of specific pick-up and delivery routes of the distribution enterprise when using multiple Operations Research methods as well as a selected web application along with a technical and economical evaluation of the final outcomes is where the novelty of our research lies. Based on a literature review in the following section of the manuscript, it was found that no similar scientific work when using identical vehicle routing problem methods to those applied in this manuscript has been published yet. Hence, our research clearly contributes to the gaps in the existing literature.

2. Literature Review

The optimization of transport networks, among others, includes addressing the distribution tasks—pick-up and delivery problem—using mathematical (Operations Research) methods. According to Cheng [1], Hamiltonian circuits have a crucial role in terms of optimization tasks, especially in addressing distribution tasks where each vertex (customer, supplier, logistics center and so forth) needs to be visited just once. To address this problem, a number of conditions have been defined that the graph must adhere to including the Hamiltonian circuit [2]. These days, optimization in transport networks has many options to use and can save considerable resources. This is the case in both the corporate sphere, where the goal is to optimize the distribution cost and transport network operation, and in the personal sphere, where the emphasis is placed on searching for the shortest possible route from point A to point B.

As described by numerous authors in a variety of publications, e.g., [3–9], in practice, many different distribution systems are utilized. All of the below methods were taken into account as potential tools for the optimization purpose, however, due to specific input conditions set in this research, only some of them can be deemed as adequate (see Section 3.1):

- Gradual distribution (intermediate warehousing)—each stage represents placement of a product in a warehouse. It is a system in which warehouses are used to a maximum extent. Distribution centers completing sales requirements are typical examples of such a distribution.
- Direct delivery system—products are delivered to the point of consumption directly from one or more storage locations, or directly from the production factory. The supplier has at his disposal one central warehouse to which he collects individual consignments, and from which he also handles them. This system includes cross-dock operations as well, which are mainly applied to high-volume product flows towards the retail network. The distribution center is integrated directly into the chain segment between a larger number of suppliers on one side and a retail network on another. Deliveries from all suppliers are collected to this center, stored in appropriate warehouse departments, and completed (assembled) according to retail network requirements. Consequently, the delivery itself is usually carried out at an exact time.
- Combined systems—the combination of the previous two systems is most commonly used. It determines which products will be distributed directly and which through intermediate warehousing. These systems also make it possible to deliver supplies in an alternative way.
- Postponement strategies for final operations—modern distribution systems do not only wait for the final order, but are also based on forecasting. This is also related to the risk that actual orders will differ from those anticipated. If some production distribution operations can be postponed until a specific order arrives, this risk can be substantially reduced. The basis is to keep the products in the production process in the unfinished state for as long as possible and to make the final adjustment up to confirming the customer order. The main effect of this process is to reduce the product range in stock, minimize the risk of poor inventory location and make better utilization of storage capacity for completing operations [10].

- Coupling methods—these are carried out due to an effort to cut down shipping cost. The larger the shipment, the lower the shipping cost per unit. Coupling also improves shipping cost control.

The optimal distribution concept consists in the optimal number of locations, combination of own and contracted warehouses, appropriate ratio of in-house and external transport—outsourcing, including a method of planning and management—and all of these while complying with capacity and customer requirements, and minimum costs. The distribution efficiency is affected by the geographical distribution of the partners (stakeholders) involved in the distribution process. It considerably affects the level of customer service and distribution cost.

The concept of distribution and distribution systems is undoubtedly related to pick-up and delivery tasks. In most models, the pick-up and delivery of shipments from the logistics center (LC)—often referred to as the "first and last mile" of the entire transport chain—in terms of the issue of city logistics are provided by road carriers with their own vehicles. The only exception is those shipments that are delivered directly to the recipient's own railway siding or to public reloading tracks at the destination station by the system of preferential or ordinary train formation [11,12].

Pick-up and delivery technology should be managed according to the following principles, as described in [13]:

- consistent operational management according to the current needs of the network and contracted transport volumes; i.e., exact transport requirements will be assigned to a road carrier in a short-term period, within a long-term contracted capacity;
- the principle of maximum utilization of road vehicles, which leads to maximum profitability of transport;
- providing transport services directly from home to home by the relevant regional road carrier, direct contact with the customer, delivery of shipment and shipping documents are highly desirable;
- effort to minimize handling cost to a maximum extent; i.e., using appropriate transshipment mechanisms in an LC, prompt cargo transshipment to the customer with respect to an option of using a vehicle for further carriage;
- distribution by railways only when delivering (or dispatching) to the recipient's own siding or to public reloading tracks at the destination station, i.e., without reloading and other logistics operations in relevant LC.

As presented by Karakikes et al., the implementation of appropriate mechanisms for the management of technological processes and quality of services is a key element to operate the LC on its own in the context of pick-up and delivery tasks, which means [14]:

- tracking shipments on international and domestic routes;
- monitoring of technological processes in the LC;
- monitoring of road vehicles during collection and distribution;
- checking the collection of load and transition between routes;
- operational planning of capacities, means of transport, routes and operation of LCs and the network as a whole;
- evidence of vehicles, wagons, containers and other means, tracking the movement of means of transport in LCs, in the network, to customers;
- addressing deviations from the plan and extraordinary traffic;
- service quality management, i.e., just-in-time delivery, timeliness delivery, accuracy, flexibility and reliability;
- providing transport and logistics services;
- dispatching management of the LC and the network as a whole.

Planning the pick-up and delivery of shipments as well as the movement of means of transport to the customer or to intermediate warehouses must be optimized in real time depending on various criteria (such as cost, distance traveled, empty journeys, etc.), as stated by the authors Graf and Stadlmann [15], and Masuda et al. [16]. This service should be designed so that the final customer does not have to be equipped with appropriate handling equipment for reloading the shipment.

As an extension to our research topic, Musollino et al. in [17] present the integration of the Mansky paradigm principle as path choice problem and general vehicle routing problem tools when applying methodological and experimentation approaches; i.e., an analysis of similarity of criteria to create various alternatives for distribution routes and creating a choice route model regarding freight vehicles. Similarly, even in [18], Croce et al. deal with a path choice problem and vehicle routing problem specifically in the Calabria region (southern Italy) in order to compare specific delivery routes with simulated and optimized routes of commercial vehicles with an aim to assess the similarity and coverage levels.

In addition to the vehicle routing problem, specific approaches based on network theory and game theory in regard to a distribution problem could also be considered. For instance, Arena et al. discuss the Parrondo paradox concerning the role of chaos when proving that two separate losing games can be combined following a random or periodic strategy to have a resulting winning game [19]. In analogy, Guanhui focuses on analyzing various game theory models, particularly in the supply chain. The author suggests a multi-enterprise output game theory approach under the circumstances of information asymmetry from the point of view of function, hypothesis parameters and modeling basis, and evaluates the impact of producer's output adjustment speed attributes on the entire supply chain [20].

3. Methodology of the Addressed Problem

Operations Research is a multidisciplinary subject that combines knowledge, experience and skills from different industries. The advantage of the Operations Research methods lies in their wide use to address problems of varying complexity. Operations Research can be viewed as a scientific discipline that includes a wider range of scientific subdisciplines focused on analyzing and managing activities in terms of addressing decision-making problems. These areas of Operations Research can be used in decision-making problems themselves, but also as a combination of several of them [21].

The aim of Operations Research is to set up operations and their interconnections so that the examined system is as effective as possible. Effectiveness must be assessed on the basis of objective or subjective criteria. A mathematical or physical model of a system is often created in order to perform tests of its functionality [21]. Section 3.1 describes the selected research methods in more detail.

3.1. Research Methods

Considering the issue addressed, its scale, transport territory, all the input conditions and other possible aspects and intricacies, it was decided by the authors as well as a panel of experts dealing with the issue of vehicle routing problems that three mathematical instruments will be applied to optimize the distribution problem.

The Hungarian method is a combinatorial optimization algorithm, which falls into special methods for addressing assignment transport problems. It was invented by the Hungarian author Egervary and belongs to the most effective methods of addressing transport problems. It is also referred to as the Kuhn–Munkres algorithm. The advantage of the Hungarian method is mainly its universality due to its use in assignment problems or vehicle routing problems, and it is also universally applicable for several types of matrices. The disadvantage of this method is relatively long computational time [21].

Several suppliers operate in the system and import various goods at different points of consumption. When suppliers accomplish their task, they return their vehicle back to the origin point. A supplier only visits the site once. The goal of this problem is to minimize the total distance traveled as much as possible. Due to these conditions, the optimal order of vertices to be operated by suppliers is compiled. In order to solve the task by the Hungarian method, the following conditions must be met [22]:

- equality of the number of rows and columns (symmetric distance matrix); if this condition is not met, it is necessary to add a fictitious row with prohibitive rates (when minimizing, values are to be higher than the highest value of the distance matrix, when maximizing, we assign the value of 0);
- the distance matrix must be quantifiable;
- suppliers' capacities and customers' requirements must be homogeneous (any customer can be served by any supplier).

The procedure is given as follows [21]:

- Step 1. Listing distances—compilation of the distance matrix.
- Step 2. Row reduction—select the lowest value in each row; this value is then subtracted in individual rows, and this step gives us the required zeros in each row. This step is not repeated in the calculation process.
- Step 3. Column reduction—this step is similar to the second step, except that the lowest number is now selected in each column and subtracted from the given values in a particular column.
- Step 4. Placement of cross rows—in this step, the independent zeros that are individually in a column or row are identified. They are marked (crossed out) by either vertical or horizontal rows to use as few crossed rows as possible.
- Step 5. Modifying a matrix and selection of a minimum value—in the matrix, non-crossed numbers are searched and the number with the lowest value is identified. This number is designated as, for example, the letter n. Values of numbers that are crossed out once do not change. Numbers that are crossed out twice are increased by a value of n. From numbers that are not crossed out, the value of n is subtracted.
- Step 6. Finding a path—in the matrix, zero-value cells are nodes. A path can pass through this place provided that the shortest path is met. Here, it applies that it is possible to pass through each site only once. The aim is to pass through all the sites so that the circuit distance is as short as possible. As a result, the route starts at site number 1 and ends at site number 1.
- Step 7. Final procedure—repeat Steps 4 and 5 until the final solution is reached. The end of the calculation process occurs at the moment of closing the entire circuit, where the route leads through all the sites. Steps 4 and 5 are carried out together, this is called iteration. After modifying the matrix by Step 5, we obtain the first iteration.
- Step 8. Route distance calculation—the calculation is based on the first unmodified matrix in which we write the distances of each route. Here, we indicate the individual values that result from the assigned route. The values are summed and the final circuit distance is calculated.

Following the abovementioned steps, it can be stated that even the Hungarian method is suitable to be applied for addressing the objective of this publication, due to its appropriateness for scenarios where one or multiple suppliers serve more customers or are operated by several other suppliers to travel over short or longer distances among each other as well as the necessity to have balanced distribution tasks.

The Vogel approximation method (hereinafter VAM) is an approximation method used to deal with transport problems and is a member of the distribution tasks that fall into the tasks of linear programming. This method is one of the most widely used approximate methods by which transport problems are usually solved. Its main advantage lies in the fact that even for large-scale tasks, its results are very close to the optimum and its procedure is not time-consuming [22].

According to [23], the VAM is an improved version of the least cost method and the northwest corner method. In its general procedure, better initial basic feasible solutions, which are understood as basic feasible solutions that report a smaller value in the objective (minimization) function of a balanced transport problem (sum of the supply = sum of the demand), are obtained. The Vogel approximation method is also called the penalty method because the difference costs chosen are nothing but the penalties of not choosing the least cost routes. It consists in an iterative approach that can be used to address a single-circuit transport problem. The procedure of using the Vogel approximation method is as follows [22]:

- Step 1. The basic element of this method is to compile a default symmetric (balanced) distance table among individual locations of one circuit route.
- Step 2. For each row and column of a default distance table, it is necessary to calculate the difference between the two lowest values. The difference value is written on the table's right side for rows and at the bottom for columns.
- Step 3. The highest possible value of all the difference values is then selected. For the row or column with the highest difference value, the lowest value in the distance table is identified. This value represents the first segment of the circuit and presents the order in which the circuit will be operated.
- Step 4. Both the row and the column for the selected value must be removed (crossed out). Furthermore, it is imperative to remove a value which, with the value currently occupied, could close the circuit route without operating all the necessary locations.
- Step 5. The next step is to recalculate the differences for the remaining rows and columns, followed by the same procedure as for Step 3.
- Step 6. We repeat the above procedure until all the necessary locations are ranked in one circuit route.

This method is suitable for models where one or multiple operators distribute cargo to multiple customers (i.e., delivery tasks) or there are several other operators (i.e., pick-up tasks) over shorter or longer distances traveled. Furthermore, each transport problem dealing with determination of the optimal transport plan by this method needs to be balanced, i.e., requirements of the destination sites must be equal to source capacities and, besides that, all the capacities and requirements must be depleted. On the basis of the aforesaid reasons, it is appropriate to apply the Vogel approximation method for the purpose of this work.

The nearest neighbor method is considered one of the simplest heuristic methods for addressing routing transport problems. It was decided to apply this algorithm due to the fact that it is suitable for types of tasks where only one supplier collects or delivers products to predetermined locations even in urban or suburban territory. After passing through all the planned stops (vertices), the vehicle returns to the point of origin. Each vertex can only be visited once. The aim of this method is to help find a solution specifying the optimal operation order of individual locations while minimizing the distance traveled or total shipping cost. This heuristic method is a simple technique and does not need more complicated calculations. The data source consists of a distance matrix among individual vertices, which is searched sequentially [22,24].

According to formulations written in a research study [25], this algorithm is one of the effective methods used to address a vehicle routing problem. The principle of the nearest neighbor algorithm starts by choosing an origin point from which the most advantageous connection to another point is to be found, and this procedure is applied until all the defined vertices are visited (operated). Once we connect all the vertices, we will return to the origin point. This method's algorithm is summarized in the following steps [25]:

- Step 1. Identify a point of origin and, in the distance matrix, the column corresponding to the given location is marked (crossed).
- Step 2. Seek a row corresponding to the given location and, in that row, find the field with a minimum value, and thereby another place to visit is determined.
- Step 3. Find a column with this new location and cross it. Search for a row corresponding to the given location and, in that row, find the field with the minimum value; thus, apply Steps 2 and 3 until all the columns are crossed out.
- Step 4. In the last row, occupy the field in a column corresponding to an origin point, so the whole circuit is actually closed.
- Step 5. Select another location as an origin point and, applying Steps 2–5, define the circuit route for this origin point.

As stated in [24], in the distance matrix with n vertices, we come to a situation where we have n circuit routes and, from these routes, the best one needs to be determined, i.e., the one with the lowest sum of values. If the task has an asymmetric distance matrix, it is also necessary to find a "backward" route for each location, either by crossing (marking) the rows, and then searching for the minimum values in the relevant columns, or by converting the original matrix to transposed type, and then applying the original procedure to it.

Following the previously mentioned statements, the nearest neighbor method appears to be perfectly suitable in terms of its application for the objective of this research work, i.e., to specify optimal delivery routes to operate defined unloading points when minimizing distance traveled.

3.2. *Presentation of the Addressed Problem*

The issue addressed is based on the need to optimize the already existing delivery routes of the presented company at the branch in the city of České Budějovice, Czech Republic. This branch distributes gastronomic equipment throughout the year on optimized routes with the full utilization capacity of service vehicles [26]. However, the problem arises during the main season, when the seasonal demand of the operators of camps and restaurant facilities increases for the regular served customers, mainly due to the increased tourist traffic. This demand lasts only for a certain part of the year, from March to November. Due to the increased demand, the company does not have enough standard delivery routes to operate given locations, so it gains brigade strength for this period and introduces special seasonal delivery to customers with whom the company has a collective agreement during the main season. To cover the mentioned seasonal demand, a collective agreement with customers is used, which guarantees them delivery of the same amount of ordered goods three times a week. Thus, delivery days are set to Monday, Tuesday, Thursday and Friday, when the company delivers goods to customers in the main season. In this way, 3 routes A, B, C are operated including a total of 32 unloading points, which have not been optimized yet using adequate methods. The initial operation order of the unloading points on the selected routes is based only on the experience of the company's employees.

Traffic in České Budějovice (congestions, lower travel speed, etc.) has a significant effect on the driving time of the delivery vehicle(s), however, given that all the defined unloading points are located near small towns outside the agglomeration of larger cities, it was not necessary to take into account the urban traffic. The journey of a delivery vehicle traveling along a part of the route leading in the extra-urban area (i.e., extra-urban part of the route) exceeds the journey of such a vehicle along the part of the same route leading in the urban area (i.e., urban part of the route) by several times in terms of kilometers traveled and time consumed. Any delay of the delivery vehicle in city traffic is therefore negligible for the purposes of this case study and was not further included in the application of single mathematical methods.

3.2.1. Default State: Route A

Route A serves 12 unloading points and is focused on serving the area northeast of České Budějovice. The following Figure 1 shows the default route A before optimization.

Table 1 shows basic data about route A.

Table 1. Route A—default state.

Number of Unloading Points	12
Length of the route	166.6 km
Average speed	47.2 km/h
Driving time	3 h 32 min
Average time spent at a stop	8 min
Preparation and loading of goods	50 min
Total route time	5 h 58 min

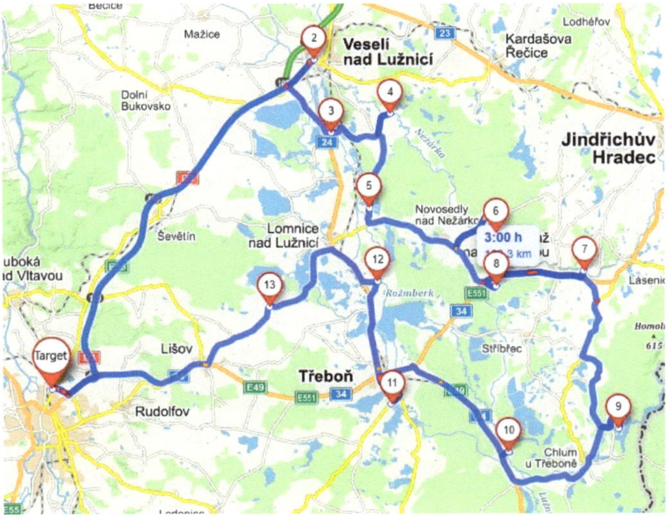

Figure 1. Route A—default state.

3.2.2. Default State: Route B

Route B serves 10 unloading points and is focused on serving the area south of České Budějovice. The following Figure 2 shows the default route B before optimization.

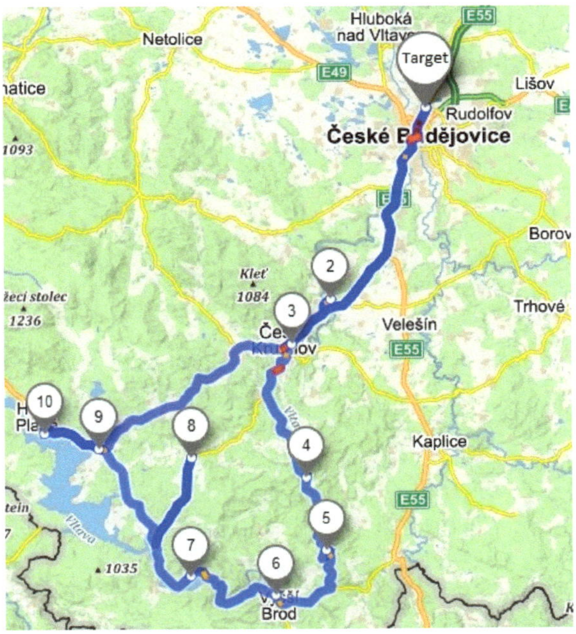

Figure 2. Route B—default state.

Table 2 presents basic information about this route.

Table 2. Route B—default state.

Number of Unloading Points	10
Length of the route	181.2 km
Average speed	49.2 km/h
Driving time	3 h 41 min
Average time spent at a stop	12 min
Preparation and loading of goods	42 min
Total route time	6 h 10 min

3.2.3. Default State: Route C

Route C serves 10 unloading points and is focused on serving the area northwest of České Budějovice. The following Figure 3 shows the default route C before optimization.

Figure 3. Route C—default state.

Table 3 summarizes basic information about this route.

Table 3. Route C—default state.

Number of Unloading Points	10
Length of the route	166.2 km
Average speed	46.2 km/h
Driving time	3 h 36 min
Average time spent at a stop	8 min
Preparation and loading of goods	42 min
Total route time	5 h 38 min

4. Optimization of the Pick-Up Technology

All the abovementioned methods are gradually used to optimize pick-up routes. In addition, to compare the quality of the results, the individual routes are optimized using the mobile application Routin: Smart Route Planner (hereinafter referred to as Routin), which is freely available on Google Play.

Creating default matrices

For each path individually, first, it is necessary to build the default matrices. This matrix is formed so that for each individual unloading point on each route it is necessary to separately measure the distance and travel time from that point to each other point on the same route [27].

Default matrix: Route A

For route A, it is necessary to create a default matrix for a total of 13 unloading points, including the company's headquarters. In total, it is necessary to make 78 separate measurements of distances between two points to create the following Table 4. This obtained matrix will be used to optimize route A.

Table 4. Route A—default matrix of times and distances.

Unloading Points		A1	A2	A3	A4	A5	A6	A7	A8	A9	A10	A11	A12	A13
Sving	A1		29.7	31.6	36.6	27.1	39.8	38	31.9	44.3	33.1	25	29.3	17.5
Penzion Veselí nad Lužnicí	A2	19		6.7	7.2	17.4	27.6	31.7	25.6	44.9	33.7	25.9	18.2	21.2
Kemp Vlkov	A3	20	7		5.6	11.3	21.6	25.7	19.6	38.9	27.6	19.9	12.2	15.2
Kemp Hamr	A4	28	9	11		7.1	19.7	36.6	17.7	44.4	33.2	25.5	17.8	20.7
Penzion Klec	A5	30	19	13	12		12.6	16.6	10.6	30.2	23.9	16.1	8.4	11.4
Kemp Jemčina	A6	38	30	24	31	19		14.9	8.8	28.5	23.7	20.2	18.7	21.6
Autokemp Dolní Lhota	A7	32	31	25	36	19	17		6.1	14.4	22	18.5	22.4	25.7
Kemp Mláka	A8	27	26	20	26	14	12	5		19.6	15.9	12.4	16.3	19.6
Autokemp Staňkov	A9	46	47	41	51	39	37	21	25		12.3	24.7	28.6	38.1
Kemp Majdalena	A10	31	30	25	35	24	25	19	14	19		13.5	17.4	26.9
Autokemp Třeboň	A11	23	24	18	29	19	23	16	12	31	14		9.6	19.1
Kemp Lužnice	A12	27	17	11	22	12	23	20	15	34	18	11		11.4
Kemp Dolní Slovětice	A13	18	22	16	27	17	28	29	24	44	28	21	14	

In the same way, time and distance matrices were created for routes B and C.

Speed difference coefficient

The company's vehicles do not reach the same average speed as in the case of application measurements, which was identified according to 25 investigations of speed during standard deliveries. This difference must be taken into account when creating matrices or interpreting the results. The simplest variant is to modify the resulting numbers when interpreting these data, so it was necessary to measure the average speed of vehicles on existing routes directly in practice and compare with the speed measured using the Mapy.cz application. The share of the obtained values expresses the difference coefficient calculated in Table 5, by which it will be necessary to multiply the final data appearing as results from individual methods.

Table 5. Calculation of the velocity difference coefficient.

		Route A	Route B	Route C
Default values from Mapy.cz	Distance (km)	166.6	181.2	166.2
	Time (min)	191	197	192
	Speed (km/h)	52.3	55.2	51.9
The resulting velocity difference coefficient *		1.110	1.122	1.125
Real values in the company	Speed (km/h)	47.2	49.2	46.2
	Time (min)	212	221	216
	Distance (km)	166.6	181.2	166.2

* The obtained coefficient expresses the ratio of the speed obtained from Mapy.cz and real vehicles' speed in practice.

4.1. Optimization of Default Routes by the Hungarian Method

Route A

The first step of the Hungarian method is to compile a default distance matrix. It has already been created, so it is possible to proceed to the next step, the so-called row reduction, where the lowest value (see Table 6, column "Min") in a given row is subtracted from all values in each row, see Table 7 [27].

Table 6. Route A—row reduction.

	A1	A2	A3	A4	A5	A6	A7	A8	A9	A10	A11	A12	A13	Min
A1		19	20	28	30	38	32	27	46	31	23	27	18	18
A2	19		7	9	19	30	31	26	47	30	24	17	22	7
A3	20	7		11	13	24	25	20	41	25	18	11	16	7
A4	28	9	11		12	31	36	26	51	35	29	22	27	9
A5	30	19	13	12		19	19	14	39	24	19	12	17	12
A6	38	30	24	31	19		17	12	37	25	23	23	28	12
A7	32	31	25	36	19	17		5	21	19	16	20	29	5
A8	27	26	20	26	14	12	5		25	14	12	15	24	5
A9	46	47	41	51	39	37	21	25		19	31	34	44	19
A10	31	30	25	35	24	25	19	14	19		14	18	28	14
A11	23	24	18	29	19	23	16	12	31	14		11	21	11
A12	27	17	11	22	12	23	20	15	34	18	11		14	11
A13	18	22	16	27	17	28	29	24	44	28	21	14		14

Table 7. Route A—column reduction.

	A1	A2	A3	A4	A5	A6	A7	A8	A9	A10	A11	A12	A13
A1		1	2	10	12	20	14	9	28	13	5	9	0
A2	12		0	2	12	23	24	19	40	23	17	10	15
A3	13	0		4	6	17	18	13	34	18	11	4	9
A4	19	0	2		3	22	27	17	42	26	20	13	18
A5	18	7	1	0		7	7	2	27	12	7	0	5
A6	26	18	12	19	7		5	0	25	13	11	11	16
A7	27	26	20	31	14	12		0	16	14	11	15	24
A8	22	21	15	21	9	7	0		20	9	7	10	19
A9	27	28	22	32	20	18	2	6		0	12	15	25
A10	17	16	11	21	10	11	5	0	5		0	4	14
A11	12	13	7	18	8	12	5	1	20	3		0	10
A12	16	6	0	11	1	12	9	4	23	7	0		3
A13	4	8	2	13	3	14	15	10	30	14	7	0	
Min	4	0	0	0	1	7	0	0	5	0	0	0	0

There is now a zero value in each row of the matrix that will be needed to find the optimal path [28]. The state after the row reduction is shown in Table 7. In this table, it is now necessary to search for columns in which there is no zero, if there are such columns. In the found columns, it is necessary to find the lowest value in each such column (see Table 7, row "Min") and subtract it from each value in the selected column. This step of the procedure is called column reduction and its initial state together with the state after row reduction is shown in Table 7.

In the following Table 8, the selection of independent zeros and the location of cover rows are already in progress. In this step, it is necessary to make sure that there is a maximum of one selected independent zero in each row or column. Independent zeros are highlighted in bold and the cover rows are highlighted in gray.

Table 8. Route A—selection of independent zeros and construction of cover rows.

	A1	A2	A3	A4	A5	A6	A7	A8	A9	A10	A11	A12	A13
A1		1	2	10	11	13	14	9	23	13	5	9	**0**
A2	8		**0**	2	11	16	24	19	35	23	17	10	15
A3	9	0		4	5	10	18	13	29	18	11	4	9
A4	15	0	2		2	15	27	17	37	26	20	13	18
A5	14	7	1	**0**		0	7	2	22	12	7	0	5
A6	22	18	12	19	6		5	0	20	13	11	11	16
A7	23	26	20	31	13	5		**0**	11	14	11	15	24
A8	18	21	15	21	8	0	**0**		15	9	7	10	19
A9	23	28	22	32	19	11	2	6		**0**	12	15	25
A10	13	16	11	21	9	4	5	0	**0**		0	4	14
A11	8	13	7	18	7	5	5	1	15	3		**0**	10
A12	12	6	0	11	**0**	5	9	4	18	7	0		3
A13	**0**	8	2	13	2	7	15	10	25	14	7	0	

Now all the elements in the matrix not covered by the cover rows are reduced by the lowest uncovered value of the element α. At the point where the cover rows intersect, the value of these elements is increased by α. In this case, the lowest uncovered value is α = 2 [24]. In the following Table 9, the value of α is again subtracted from the uncovered values and added to the values where the cover rows intersect.

Table 9. Route A—adjusting the matrix to the lowest uncovered value.

	A1	A2	A3	A4	A5	A6	A7	A8	A9	A10	A11	A12	A13
A1		1	0	8	9	11	12	9	21	13	3	7	**0**
A2	8		**0**	2	11	16	24	21	35	25	17	10	17
A3	7	**0**		2	3	8	16	13	27	18	9	4	9
A4	13	0	0		**0**	13	25	17	35	26	18	11	18
A5	14	9	1	**0**		0	7	4	22	14	7	0	7
A6	20	18	10	17	4		3	0	18	13	9	9	16
A7	21	26	18	29	11	3		**0**	9	14	9	13	24
A8	18	23	15	21	8	**0**	0		15	11	7	10	21
A9	21	28	20	30	17	9	0	6		**0**	10	13	25
A10	13	18	11	21	9	4	5	2	**0**		0	4	16
A11	8	15	7	18	7	5	5	3	15	5		**0**	12
A12	12	8	0	11	0	5	9	6	18	9	**0**		5
A13	**0**	10	2	13	2	7	15	12	25	16	7	0	

It is still necessary to monitor the matrix to prevent premature closing of the circle route. Accordingly, it is necessary to choose the independent zeros that make up circuit path [25]. Table 10 shows the final optimized version of this method.

Table 10. Route A—optimized matrix.

	A1	A2	A3	A4	A5	A6	A7	A8	A9	A10	A11	A12	A13
A1		1	**0**	8	9	11	12	12	15	13	3	7	0
A2	6		0	**0**	9	14	22	22	27	23	15	8	10
A3	7	**0**		2	3	8	16	16	21	18	9	4	4
A4	13	0	0		**0**	13	25	20	29	26	18	11	13
A5	14	9	1	0		**0**	7	7	16	14	7	0	2
A6	17	15	7	14	1		0	**0**	9	10	6	6	8
A7	21	23	15	26	8	0		0	**0**	11	6	10	16
A8	18	23	15	21	8	0	**0**		9	11	7	10	16
A9	21	28	20	30	17	9	0	9		**0**	10	13	20
A10	13	18	11	21	9	4	5	5	0		**0**	4	11
A11	8	15	7	18	7	5	5	6	9	5		**0**	7
A12	12	8	0	11	0	5	9	9	12	9	0		**0**
A13	**0**	10	2	13	2	7	15	15	19	16	7	0	

The final optimized route according to Table 10 will lead through the points in the following order: A1 ≥ A3 ≥ A2 ≥ A4 ≥ A5 ≥ A6 ≥ A8 ≥ A7 ≥ A9 ≥ A10 ≥ A11 ≥ A12 ≥ A13 ≥ A1.

It is now necessary to calculate the time and distance value of the route thus optimized in the default matrix for this route. The results are shown in the following Table 11.

Table 11. Route A—final table of values.

Length of the Route (km)	**158.8**
Operating time (min)	181

Routes B and C were optimized in the same way.

4.2. Optimization of Default Routes by the Vogel Approximation Method

In this part, the optimization of circle routes using the VAM for each separate default route A, B and C is described.

Route A—route optimization by Vogel approximation method

To calculate the optimal route by the VAM, the same default matrix is needed, which has already been used in the case of the Hungarian method. In the first step, it is necessary to specify the two lowest values in each row (row and column). The difference between these values is called the "difference" and is listed for each row on the right and bottom edge of the table, see Table 12 [29,30].

Table 12. Route A—determining the differences from the default table.

	A1	A2	A3	A4	A5	A6	A7	A8	A9	A10	A11	A12	A13	Min	Dif
A1		19	20	28	30	38	32	27	46	31	23	27	18	18	1
A2	19		7	9	19	30	31	26	47	30	24	17	22	7	2
A3	20	7		11	13	24	25	20	41	25	18	11	16	7	4
A4	28	9	11		12	31	36	26	51	35	29	22	27	9	2
A5	30	19	13	12		19	19	14	39	24	19	12	17	12	0
A6	38	30	24	31	19		17	12	37	25	23	23	28	12	5
A7	32	31	25	36	19	17		5	21	19	16	20	29	5	11
A8	27	26	20	26	14	12	5		25	14	12	15	24	5	7
A9	46	47	41	51	39	37	21	25		19	31	34	44	19	2
A10	31	30	25	35	24	25	19	14	19		14	18	28	14	5
A11	23	24	18	29	19	23	16	12	31	14		11	21	11	1
A12	27	17	11	22	12	23	20	15	34	18	11		14	11	1
A13	18	22	16	27	17	28	29	24	44	28	21	14		14	2
Min	18	7	7	9	12	12	5	5	19	14	11	11	14		
Dif	1	2	4	2	0	5	11	7	2	5	1	0	2		

There are two values with a difference of 11 in Table 12. Now, it is necessary to select the lowest value in the rows with the largest difference and specify it as the starting point of the route. In this case, it does not matter so much which of the smallest values in the row and column with the largest difference will be chosen, as both are the same for the same route and only determine the direction in which the optimization will take place [29].

Therefore, the value connecting the route from point A8 to point A7 with the value 5 is selected. For clarity, the whole row and column in which the selected value is located, as well as its symmetrical counterpart connecting the route from point A7 to point A8, are excluded from the matrix [27]. The differences are recalculated and the lowest value in the row with the largest difference is selected again as the next value to be included in the circle route. This step is illustrated in Table 13 below.

Table 13. Route A—selection of the first and second section of the route.

	A1	A2	A3	A4	A5	A6	A7	A8	A9	A10	A11	A12	A13	Min	Dif
A1		19	20	28	30	38		27	46	31	23	27	18	18	1
A2	19		7	9	19	30		26	47	30	24	17	22	7	2
A3	20	7		11	13	24		20	41	25	18	11	16	7	4
A4	28	9	11		12	31		26	51	35	29	22	27	9	2
A5	30	19	13	12		19		14	39	24	19	12	17	12	0
A6	38	30	24	31	19			12	37	25	23	23	28	12	7
A7	32	31	25	36	19	17			21	19	16	20	29	16	1
A8							5								
A9	46	47	41	51	39	37		25		19	31	34	44	19	6
A10	31	30	25	35	24	25		14	19		14	18	28	14	5
A11	23	24	18	29	19	23		12	31	14		11	21	11	1
A12	27	17	11	22	12	23		15	34	18	11		14	11	1
A13	18	22	16	27	17	28		24	44	28	21	14		14	2
Min	18	7	7	9	12	17		12	19	14	11	11	14		
Dif	1	2	4	2	0	2		2	2	5	3	0	2		

Now, in Table 14, there is a row with the largest difference 7, which contains the lowest value 12, which connects the route from point A8 to point A7. The previous procedure is repeated, when the whole row and the column in which the selected value is located, as well as its symmetrical counterpart connecting the route from point A6 to point A8, are excluded from the matrix [31]. The differences are recalculated and the lowest value in the row with the largest difference is selected again as the next value to be included in the roundabout. This selection is shown in Table 14 below.

Table 14. Route A—selection of the third section of the route.

	A1	A2	A3	A4	A5	A6	A7	A8	A9	A10	A11	A12	A13	Min	Dif
A1		19	20	28	30	38			46	31	23	27	18	18	1
A2	19		7	9	19	30			47	30	24	17	22	7	2
A3	20	7		11	13	24			41	25	18	11	16	7	4
A4	28	9	11		12	31			51	35	29	22	27	9	2
A5	30	19	13	12		19			39	24	19	12	17	12	0
A6								12							
A7	32	31	25	36	19	17			21	19	16	20	29	16	1
A8							5								
A9	46	47	41	51	39	37				19	31	34	44	19	12
A10	31	30	25	35	24	25			19		14	18	28	14	4
A11	23	24	18	29	19	23			31	14		11	21	11	3
A12	27	17	11	22	12	23			34	18	11		14	11	0
A13	18	22	16	27	17	28			44	28	21	14		14	2
Min	18	7	7	9	12	17			19	14	11	11	14		
Dif	1	2	4	2	0	2			2	5	3	0	2		

In Table 14, the rows related to the selection of the second section of the route have been removed. This is followed by the recalculation of the differences after removing these rows and searching for the highest difference in the row and selecting the lowest value in it [29]. In this case, it is the difference 12 in the row with point A9, which is connected to point A10. In the next step, the values from the rows that belong to this selected value and its symmetrical counter-value showing the route from point A10 to point A9 will be removed again for this value. They will then be recalculated throughout the table and the process outlined in these steps will be repeated. The following steps of the method will be skipped and in Table 15 the penultimate step of the VAM is presented [30,31].

Table 15. Route A—penultimate step.

	A1	A2	A3	A4	A5	A6	A7	A8	A9	A10	A11	A12	A13	Min	Dif
A1			20		30									20	10
A2				9											
A3		7													
A4			11		12									11	1
A5						19									
A6								12							
A7									21						
A8							5								
A9										19					
A10											14				
A11												11			
A12													14	14	1
A13	18														
Min			11		12								14		
Dif			9		18								13		

Table 15 shows the penultimate phase of the method calculation. In the previous step, the value connecting the route from point A12 to point A13 was selected. In this step, the differences in the rows and the selected minimum connecting the route from point A4 to A6 are recalculated in the highest row [32]. After removing the last rows belonging to the selected point, it is no longer necessary to calculate the differences, because the last unconnected route remains in the table and that is the connection from point A1 to point A3 [24]. The resulting matrix is shown in Table 16.

Table 16. Route A—final optimization.

	A1	A2	A3	A4	A5	A6	A7	A8	A9	A10	A11	A12	A13
A1			20										
A2				9									
A3		7											
A4					12								
A5						19							
A6								12					
A7									21				
A8							5						
A9										19			
A10											14		
A11												11	
A12													14
A13	18												

The final optimized route according to the VAM for route A based on Table 16 will be: A1 ≥ A3 ≥ A2 ≥ A4 ≥ A5 ≥ A6 ≥ A8 ≥ A7 ≥ A9 ≥ A10 ≥ A11 ≥ A12 ≥ A13 ≥ A1.

The last step is the calculation of the time this route takes and the total length of this route. The result is shown in the following Table 17.

Table 17. Route A—final table of values.

Length of the Route (km)	**158.8**
Operating time (min)	181

Routes B and C were optimized in the same way.

4.3. Optimization of Initial Routes by the Nearest Neighbor Method

In this part, the optimization of the initial circle routes using the nearest neighbor method is described. Gradually, the routes from the initial routes A, B, C are optimized here [33].

Route A—route optimization by the nearest neighbor method

In the nearest neighbor method, a circular path is created from the starting point gradually to the next nearest point. However, it is necessary to calculate the value of the purpose function so that each of the possible points of the original route is gradually selected for the beginning of the circular route. In the case of the route A optimization, it is necessary to calculate the optimization for each of the 13 points separately and then select the best from the offered solutions. The following Table 18 shows the results of the nearest neighbor method for each individual point, including other alternative circular paths that have been calculated for some points [34].

Table 18. Route A—circle routes from points.

Route from the Point	Value of the Purpose Function
A1	202
A1–variant 2	214
A2	221
A2—variant 2	236
A3	211
A4	227
A5	220
A6	205
A7	206
A7—variant 2	213
A7—variant 3	226
A8	205
A9	211
A10	209
A10—variant 2	222
A10—variant 3	211
A11	215
A12	204
A12—variant 2	210
A13	220
A13—variant 2	214

The smallest value of the purpose function was reached by the circular route leading from point A1 [35]. How this table came out is shown in the following Table 19. Remaining tables for calculating the path from other points are not part of this text due to the allowed length of the article. All these tables were created in the same way as Table 19, which came out as the shortest.

Table 19. Route A—the shortest selected route.

	A1	A2	A3	A4	A5	A6	A7	A8	A9	A10	A11	A12	A13
A1		19	20	28	30	38	32	27	46	31	23	27	18
A2	19		7	9	19	30	31	26	47	30	24	17	22
A3	20	7		11	13	24	25	20	41	25	18	11	16
A4	28	9	11		12	31	36	26	51	35	29	22	27
A5	30	19	13	12		19	19	14	39	24	19	12	17
A6	38	30	24	31	19		17	12	37	25	23	23	28
A7	32	31	25	36	19	17		5	21	19	16	20	29
A8	27	26	20	26	14	12	5		25	14	12	15	24
A9	46	47	41	51	39	37	21	25		19	31	34	44
A10	31	30	25	35	24	25	19	14	19		14	18	28
A11	23	24	18	29	19	23	16	12	31	14		11	21
A12	27	17	11	22	12	23	20	15	34	18	11		14
A13	18	22	16	27	17	28	29	24	44	28	21	14	

Table 19 shows the most advantageous circular route that can be achieved by the nearest neighbor method in this case. The starting point for this route is A1. To create a route starting at this point, the smallest value in the row is selected, which is the value in column A13. The lowest value in row A13 is now searched for. The route also includes the minimum value in this row, which is located in column A12. Next, the lowest value in the row that belongs to the top A12 is searched again. From this step, care must be taken not to select a value in a column that has already been included in the solution, and at the same time the route must not return to the first column A1 until it has passed all other points. In this way, a circuit route is gradually created, which includes all points [36].

This optimal route passes through the following points: A1 ≥ A13 ≥ A12 ≥ A3 ≥ A2 ≥ A4 ≥ A5 ≥ A6 ≥ A8 ≥ A7 ≥ A11 ≥ A10 ≥ A9 ≥ A1.

The length of the resulting optimized route using the nearest neighbor method is shown in the following Table 20.

Table 20. Route A—the resulting table of values.

Length of the Route (km)	178.2
Operating time (min)	202

Routes B and C were optimized in the same way.

4.4. *Optimization of Initial Routes Using the Routin Application*

In order to compare the success of the solution of Operations Research methods used to address routing problems and modern route planner applications, the optimization of individual routes using the Routin application is performed in this section [37]. In the application web interface, first of all, it is necessary to search for and assign all vertices from each route. Thereafter, it is possible to optimize each route individually. The advantage of this application lies in the fact that searching and assignment of the vertices is the most time-consuming optimization, and then the application very quickly suggests the final routes that can be used to operate the defined transport network. However, it is not specified which principles and which optimization method the given application uses.

In the application, it is first necessary to search for and place all points from each route. Then, each route can be optimized individually [38]. The advantage of this application is that the most time-consuming task to optimize is the search and location of points, then the application very quickly designs its own routes, which can operate the selected network. Essentially, this application works on the principle of the Greedy algorithm, which is described, for instance, in [39]. For route A, the application proposed the order of the points which is shown in the following Figure 4.

It is now necessary to compare the route thus obtained with the initial route and to determine the order of points as shown in a previous study [39]. From the initial matrix, it is then necessary to find the length and operating time of the selected route [40]. This optimized route leads through the points: A1 ≥ A11 ≥ A10 ≥ A9 ≥ A7 ≥ A8 ≥ A6 ≥ A5 ≥ A4 ≥ A2 ≥ A3 ≥ A12 ≥ A13 ≥ A1.

For route B, the Routin application designed the following order of points: B1 ≥ B2 ≥ B10 ≥ B9 ≥ B7 ≥ B6 ≥ B5 ≥ B4 ≥ B8 ≥ B3 ≥ B1.

For route C, the Routin application designed the following order of points: C1 ≥ C2 ≥ C3 ≥ C4 ≥ C11 ≥ C10 ≥ C9 ≥ C8 ≥ C5 ≥ C6 ≥ C7 ≥ C1.

Final length and the operating time for all individual routes are summarized in the following Table 21.

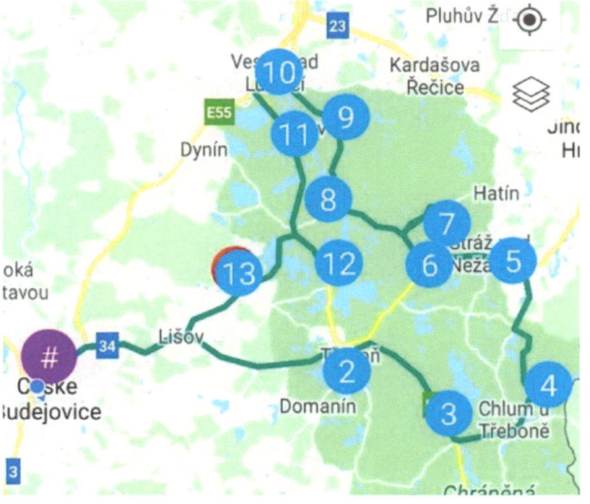

Figure 4. Route A—optimization of the route by the Routin application.

Table 21. Optimization with the Routin application—final table for all routes.

Route Name	Operating Time (min)	Length of the Route (km)
Route A	184	154.8
Route B	198	177.9
Route C	187	154.3

5. Discussion

In this section, the individual proposed routes are assessed in terms of route length (km) or operating time (min). In the case of both of these indicators, the percentage savings compared to the original route are given. For each route, the operating time is multiplied by the speed difference coefficient, which takes into account real speed measurements in practice [41].

Technical evaluation of route A

The following Table 22 contains a comprehensive summary of optimized routes using the methods of Operations Research, including the original values for the initial route A.

Table 22. Route A—final table of the optimized routes.

Route A	Length of the Route (km)	Percentage Saving Compared to the Length of the Initial Route	Operating Time (min)	Speed Difference Coefficient	Final Operating Time (min)	Percentage Saving Compared to the Operating Time of the Initial Route
Initial route	166.6				212	
Hungarian method	158.8	4.68%	181	1.11	201	5.23%
VAM	158.8	4.68%	181	1.11	201	5.23%
Nearest neighbor method	178.2	−6.96%	202	1.11	224	−5.76%
Routin application	154.8	7.08%	184	1.11	204	3.66%

Optimization using the nearest neighbor method proves to be the least advantageous for route A. The Hungarian and VAM methods bring identical results for this route [42]. The Routin application shortens the initial route the most. It therefore depends on the required aspect whether the shortest route or the fastest route is searched. In the case of route A, the shortest route is chosen, as there are greater fuel savings [43]. The following route is selected as the optimal solution, which is created via the Routin application: A1 ≥ A11 ≥ A10 ≥ A9 ≥ A7 ≥ A8 ≥ A6 ≥ A5 ≥ A4 ≥ A2 ≥ A3 ≥ A12 ≥ A13 ≥ A1.

Through this route, the traffic performance of vehicles on route A will be reduced by 7.08% and the time required to operate the route will be reduced by 3.66%.

Technical evaluation of route B

The following Table 23 shows the length of optimized routes and their final operating time for route B.

Table 23. Route B—final table of the optimized routes.

Route B	Length of the Route (km)	Percentage Saving Compared to the Length of the Initial Route	Operating Time (min)	Speed Difference Coefficient	Final Operating Time (min)	Percentage Saving Compared to the Operating Time of the Initial Route
Initial route	181.2				221	
Hungarian method	178.3	1.60%	199	1.122	223	−1.03%
VAM	178.3	1.60%	199	1.122	223	−1.03%
Nearest neighbor method	177.9	1.82%	198	1.122	222	−0.52%
Routin application	177.9	1.82%	198	1.122	222	−0.52%

To find the optimized route, the nearest neighbor method and the Routin application can be used with final length of 177.9 km and an operating time of 222 min. Both of these routes will reduce traffic performance of vehicles on route B by 1.82%, while the time needed to operate the route will increase by 0.52%.

The nearest neighbor method chooses the following order of service points: B1 ≥ B9 ≥ B10 ≥ B7 ≥ B6 ≥B5 ≥ B4 ≥ B8 ≥ B3 ≥ B2 ≥ B1.

The Routin application chooses the following order of service points: B1 ≥ B2 ≥ B10 ≥ B9 ≥ B7 ≥ B6 ≥ B5 ≥ B4 ≥ B8 ≥ B3 ≥ B1.

Technical evaluation of route C

The following Table 24 shows the length of optimized routes and their final operating time for route C.

Table 24. Route C—final table of the optimized routes.

Route C	Length of the Route (km)	Percentage Saving Compared to the Length of the Initial Route	Operating Time (min)	Speed Difference Coefficient	Final Operating Time (min)	Percentage Saving Compared to the Operating Time of the Initial Route
Initial route	166.2				216	
Hungarian method	155.5	6.44%	183	1.125	206	4.69%
VAM	170.1	−2.35%	191	1.125	215	0.52%
Nearest neighbor method	155	6.74%	184	1.125	207	4.17%
Routin application	154.3	7.16%	187	1.125	210	2.60%

The route obtained by optimization through the Routin application is based on the shortest route and passes through the following points: C1 ≥ C2 ≥ C3 ≥ C4 ≥ C11 ≥ C10 ≥ C9 ≥ C8 ≥ C5 ≥ C6 ≥ C7 ≥ C1.

In the following Table 25, a summary calculation of savings by using all three methods is presented.

The economic evaluation listed in the above table complements the technical evaluation and focuses on the calculation of operating costs associated with fuel consumption [44]. The total expected financial savings after optimization reached 5.27%.

Table 25. Summary calculation—calculation of savings by used methods.

Route	Total Fuel Costs	Total Fuel Costs for Each Route	Total Fuel Cost Savings for Each Route	Percentage Fuel Cost Savings on Each Route
A	EUR 3280.21	EUR 3047.88	EUR 232.33	7.08%
B	EUR 3548.45	EUR 3483.83	EUR 64.62	1.82%
C	EUR 3314.65	EUR 3077.32	EUR 237.33	7.16%
	Total fuel costs in 2019		Total fuel costs in 2020	Total expected financial savings after optimization
	EUR 10,143.31	EUR 9609.03	EUR 534.28	5.27%

6. Conclusions

This paper was devoted to the optimization of pick-up and delivery activities and to the technical and economical evaluation of such an optimization. For the application part, i.e., optimization of individual routes, first of all, it was necessary to compile the professional context of the problem, which forms the theoretical part of the work. The application part of the manuscript includes the introduction of the addressed problem as well as the methodological section.

The main part of this study then deals with the very optimization of the pick-up routes and the technical and economic evaluation of this optimization. To this end, in order to address the vehicle routing problem, the Hungarian method, the Vogel approximation method and the nearest neighbor method were determined to be the adequate methods of Operations Research. The Hungarian method is based on a uniform distance matrix and its application is universal. The Vogel approximation method and the nearest neighbor method were used since they use the same input matrix as the Hungarian method and are thus suitable for mutual comparison. To complement these methods, the Routin route planner was applied, which is a publicly available intuitive application that optimizes distribution routes.

For each distribution route separately, input matrices were generated, which contain all the operated unloading points of the given route and their mutual distance value. These matrices are necessary for optimization using the defined techniques being applied to the discussed distribution problem.

This was followed by the technical and economic evaluation of the work results, which assesses the results of the optimization in terms of saving time and transport performance, as well as the economic aspect. As for route A, the newly designed route managed to reduce transport performance by 7.08% and the time required to operate this route also decreased by 3.66%. In regard to route B, transport performance decreased by 1.82%, whereby the time required to operate the route increased by 0.52%. As far as route C is concerned, transport performance decreased by 7.16% and the time required to operate the route was reduced by 2.6%.

Regarding the used methods, it can be stated that the Hungarian method and optimization using the Routin application brought the most efficient results. However, in general, we must state that it is not possible to choose the best possible method of optimization, because each can bring different results in terms of route length and in terms of operation speed, and it depends on which of these variables is preferred for optimization. In our case, the shortest route was sought, and in all the cases, the Routin application found it. Nevertheless, it is appropriate to use a wider range of methods, because then the ability to compare the results obtained increases and thus approaches the optimal solution.

The economic evaluation provided in the Discussion section focuses on the calculation of fuel costs valid in the case that the selected company decides to start these new optimized routes and change its current distribution routes. The economic evaluation compares the fuel costs on the original routes with the newly optimized routes. Assuming the same fuel costs, a saving of EUR 534.28 per year is calculated here, which means a reduction in fuel costs by 5.27% for all the routes together.

As for the further research, the introduction of some specific telematics tools should represent an option in terms of searching for optimal distribution routes. Currently, telematics devices are important both when providing logistics services and when executing transport operations, and their interconnection with the surroundings is inevitable. To maintain an efficient distribution system (i.e., delivery routes), it is imperative to design the concept of telematics interconnection of on-line information related to several transport modes and kinds of logistics services—their optimal deployment, utilization of their capacities with regard to transport infrastructure capacity, fuel prices, tolls, charges for infrastructure with respect to the environment, etc. The basic idea is to create a platform, by corresponding HW and SW, for the telematics flow of processes inside logistics objects and among individual parties involved. To this end, it is important to know the development outlook directions of the transport and logistics market, the participants and requirements of customers in terms of services provided.

Author Contributions: Conceptualization, O.S. and P.G.; methodology, O.S., P.G. and J.P.; software, P.G. and J.P.; validation, O.S., J.H. and M.S.; formal analysis, J.P., J.H. and M.J.; investigation, J.P., M.S. and M.J.; resources, O.S. and J.H.; data curation, O.S. and P.G.; writing—original draft preparation, P.G. and J.P.; writing—review and editing, O.S., J.H. and M.S.; visualization, O.S., J.P. and M.J.; supervision, O.S. and M.J.; project administration, O.S.; funding acquisition, O.S. All authors have read and agreed to the published version of the manuscript.

Funding: No external funding.

Institutional Review Board Statement: Not applicable.

Informed Consent Statement: Not applicable.

Data Availability Statement: Not applicable.

Acknowledgments: Project VEGA No. 1/0128/20: Research on the Economic Efficiency of Variant Transport Modes in the Car Transport in the Slovak Republic with Emphasis on Sustainability and Environmental Impact, Faculty of Operation and Economics of Transport and Communications: University of Zilina, 2020–2022.

Conflicts of Interest: The authors declare no conflict of interest.

References

1. Cheng, D. Hamiltonian paths and cycles pass through prescribed edges in the balanced hypercubes. *Discret. Appl. Math.* **2019**, *262*, 56–71. [CrossRef]
2. Onete, C.E.; Onete, M.C.C. Building Hamiltonian networks using the cycles Laplacian of the underlying graph. In Proceedings of the IEEE International Symposium on Circuits and Systems 2015, Lisbon, Portugal, 24–27 May 2015; Article no. 7168591. pp. 145–148. [CrossRef]
3. Arab, R.; Ghaderi, S.F.; Tavakkoli-Moghaddam, R. Solving a new multi-objective inventory-routing problem by a non-dominated sorting genetic algorithm. *Int. J. Eng. Trans. B Appl.* **2018**, *31*, 588–596.
4. Björklund, M.; Johansson, H. Urban consolidation centre—A literature review, categorisation, and a future research agenda. *Int. J. Phys. Distrib. Logist. Manag.* **2018**, *48*, 745–764. [CrossRef]
5. Cempírek, V. Basic conditions of the logistics center establishment. *Logistika* **2009**, *9*, 58–59, ISSN 1211-0957.
6. Hiohi, L.; Burciu, S.; Popa, M. Collaborative systems in urban logistics. *UPB Sci. Bull. Ser. D Mech. Eng.* **2015**, *77*, 71–84.
7. Scavarda, M.; Seok, H.; Nof, S.Y. The constrained–collaboration algorithm for intelligent resource distribution in supply networks. *Comput. Ind. Eng.* **2017**, *113*, 803–818. [CrossRef]
8. Van Heeswijk, W.J.A.; Mes, M.R.K.; Schutten, J.M.J. The Delivery Dispatching Problem with Time Windows for Urban Consolidation Centers. *Transp. Sci.* **2019**, *53*, 203–221. [CrossRef]
9. Xu, L.; Zhai, W. Stochastic model used for temporal-spatial analysis of vehicle-track coupled systems. *J. China Railw. Soc.* **2018**, *40*, 74–79. [CrossRef]
10. Oluwaseyi, J.A.; Onifade, M.K.; Odeyinka, O.F. Evaluation of the Role of Inventory Management in Logistics Chain of an Organisation. *LOGI Sci. J. Transp. Logist.* **2017**, *8*, 1–11. [CrossRef]
11. Bin Othman, M.S.; Shurbevski, A.; Karuno, Y.; Nagamochi, H. Routing of carrier-vehicle systems with dedicated last-stretch delivery vehicle and fixed carrier route. *J. Inf. Process.* **2017**, *25*, 655–666. [CrossRef]
12. Ferdinand, F.N.; Ferdinand, F.V. A study on network design for the shortest path in expedition company. *J. Telecommun. Electron. Comput. Eng.* **2018**, *10*, 1–4.

13. Verlinde, S.; Macharis, C.; Milan, L.; Kin, B. Does a mobile depot make urban deliveries faster, more sustainable and more economically viable: Results of a pilot test in Brussels. *Transp. Res. Procedia* **2014**, *4*, 361–373. [CrossRef]
14. Karakikes, I.; Nathanail, E.; Savrasovs, M. Techniques for smart urban logistics solutions' simulation: A systematic review (Book Chapter). *Lect. Notes Netw. Syst.* **2019**, *68*, 551–561. [CrossRef]
15. Graf, H.; Stadlmann, B. Automated internet-shopping terminals for self-service pick-ups. In Proceedings of the International Conference on Industrial Logistics 2014, ICIL 2014, Bol on Island Brac, Croatia, 11–13 June 2014; Code 106754. pp. 82–88.
16. Sarkar, B.; Ullah, M.; Kim, N. Environmental and economic assessment of closed-loop supply chain with remanufacturing and returnable transport items. *Comput. Ind. Eng.* **2017**, *111*, 148–163. [CrossRef]
17. Musolino, G.; Rindone, C.; Vitetta, A. A modelling framework to simulate paths and routes choices of freight vehicles in sub-urban areas. In Proceedings of the 7th International Conference on Models and Technologies for Intelligent Transportation Systems (MT-ITS), Heraklion, Greece, 16–17 June 2021; pp. 1–6. [CrossRef]
18. Croce, A.I.; Musolino, G.; Rindone, C.; Vitetta, A. Route and Path Choices of Freight Vehicles: A Case Study with Floating Car Data. *Sustainability* **2020**, *12*, 8557. [CrossRef]
19. Arena, P.; Fazzino, S.; Fortuna, L.; Maniscalco, P. Game theory and non-linear dynamics: The Parrondo Paradox case study. *Chaos Solitons Fractals* **2003**, *17*, 545–555. [CrossRef]
20. Guanhui, W. Chaos analysis of the output game between multi-role enterprises in supply chain. In Proceedings of the 2nd International Conference on Electronics and Communication, Network and Computer Technology, ECNCT 2020, Chengdu, China, 23–25 October 2020; p. 012126. [CrossRef]
21. Jablonský, J. *Operační výzkum*, 3rd ed.; University of Economics in Prague: Prague, Czech Republic, 2001; ISBN 80-245-0162-7.
22. Friebelová, J. *Vybrané Metody z Operační Analýzy*; University of South Bohemia in České Budějovice, Faculty of Economics: České Budějovice, Czech Republic, 2008; ISBN 978-80-7394-124-6.
23. Vaněčková, E. *Ekonomicko—Matematické Metody—Lineární Programování. Sít'ová Analýza*; University of South Bohemia in České Budějovice, Faculty of Agriculture: České Budějovice, Czech Republic, 1996; ISBN 80-7040-187-7.
24. Kučera, P. Metodologie Řešení Okružního Dopravního Problému. Ph.D. Thesis, Czech University of Life Sciences Prague, Prague, Czech Republic, 2009.
25. Volek, J.; Linda, B. *Teorie Grafů: Aplikace v Dopravě a Veřejné Správě*, 1st ed.; University of Pardubice: Pardubice, Czech Republic, 2012; ISBN 978-80-7395-225-9.
26. Nam, D.; Park, M. Improving the operational efficiency of parcel delivery network with a bi-level decision making model. *Sustainability* **2020**, *12*, 8042. [CrossRef]
27. Tagorda, I.P.; Elwyn Calata, L.; Limjoco, W.J.R.; Dizon, C.C. Development of a vehicle routing system for delivery services. In Proceedings of the 2020 IEEE Region 10 Conference (TENCON), Osaka, Japan, 16–19 November 2020; IEEE: New York City, NY, USA, 2020; pp. 1187–1191. [CrossRef]
28. Sumathi, P.; Sathiabama, C.V. Minimizing the cost using an inventive proposal in transportation problems. In Proceedings of the IOP Conference Series: Materials Science and Engineering, Kolaghat, India, 15 July 2020; pp. 62–65. [CrossRef]
29. Hussein, H.A.; Shiker, M.A.K. A Modification to Vogel's approximation method to solve transportation problems. *J. Phys. Conf. Ser.* **2020**, *912*, 062065. [CrossRef]
30. Hussein, H.A.; Shiker, M.A.K.; Zabiba, M.S.M. A New Revised Efficient of VAM to Find the Initial Solution for the Transportation Problem. *J. Phys. Conf. Ser.* **2020**, *1591*, 012032. [CrossRef]
31. Ezekiel, I.D.; Edeki, S.O. Modified Vogel approximation method for balanced transportation models towards optimal option settings. *Int. J. Civil. Eng. Technol.* **2018**, *9*, 358–366, ISSN 0976-6308.
32. Agarana, M.C.; Omogbadegun, Z.O.; Makinde, S.O. VAM–MODI mathematical modelling method for minimizing cost of transporting perishables from markets to cafeterias in covenant university. In Proceedings of the International Conference on Industrial Engineering and Operations Management, Washington, DC, USA, 27–29 September 2018; pp. 2088–2103, ISSN 2169-8767.
33. Pečený, L.; Meško, P.; Kampf, R.; Gašparík, J. Optimisation in transport and logistic processes. *Transp. Res. Procedia* **2020**, *44*, 15–22. [CrossRef]
34. Ziólkowski, J.; LęGas, A. Problem of modelling road transport. *J. Konbin* **2019**, *49*, 159–193. [CrossRef]
35. Gupta, R.; Gulati, N. Survey of transportation problems. In Proceedings of the International Conference on Machine Learning, Big Data, Cloud and Parallel Computing: Trends, Prespectives and Prospects, COMITCon 2019, Faridabad, India, 14–16 February 2019; pp. 417–422. [CrossRef]
36. Bansal, S.; Goel, R.; Maini, R. Ground vehicle and UAV collaborative routing and scheduling for humanitarian logistics using random walk based ant colony optimization. *Sci. Iran.* **2022**, *29*, 632–644. [CrossRef]
37. Dudziak, A.; Droździel, P.; Stoma, M.; Caban, J. Market electrification for BEV and PHEV in relation to the level of vehicle autonomy. *Energies* **2022**, *15*, 3120. [CrossRef]
38. Fedorko, G.; Neradilova, H.; Sutak, M.; Molnar, V. Application of simulation model in terms of city logistics. In Proceedings of the 20th International Scientific Conference of Transport Means 2016, Juodkrante, Lithuania, 5–7 October 2016; pp. 169–174.

39. Thinakaran, N.; Jayaprakash, J.; Elanchezhian, C. Greedy algorithm for inventory routing problem in a supply chain—A review. In *Materials Today: Proceedings, Proceedings of the 2017 International Conference on Advances in Materials, Manufacturing and Applied Sciences, ICAMMAS 2017, Tamil Nadu, India, 30–31 March 2017*; Elsevier: Amsterdam, The Netherlands, 2019; Volume 16, pp. 1055–1060. [CrossRef]
40. Šedivý, J.; Čejka, J.; Guchenko, M. Possible application of solver optimization module for solving single-circuit transport problems. *LOGI–Sci. J. Transp. Logist.* **2020**, *11*, 78–87. [CrossRef]
41. Droździel, P.; Komsta, H.; Krzywonos, L. An analysis of unit repair costs as a function of mileage of vehicles in a selected transport company. *Transp. Probl.* **2014**, *9*, 73–81.
42. Šego, D.; Hinić, M.; Poljičak, A. Methods of Goods Delivery to the Historic Core of the City of Šibenik during the Tourist Season. *LOGI–Sci. J. Transp. Logist.* **2020**, *11*, 88–98. [CrossRef]
43. Trotta, M.; Archetti, C.; Feillet, D.; Quilliot, A. Pickup and delivery problems with autonomous vehicles on rings. *Eur. J. Oper. Res.* **2022**, *300*, 221–236. [CrossRef]
44. Siragusa, C.; Tumino, A.; Mangiaracina, R.; Perego, A. Electric vehicles performing last-mile delivery in B2C e-commerce: An economic and environmental assessment. *Int. J. Sustain. Transp.* **2022**, *16*, 22–33. [CrossRef]

technologies

MDPI

Article

Distribution Path Optimization by an Improved Genetic Algorithm Combined with a Divide-and-Conquer Strategy

Jiaqi Li [1], Yun Wang [2] and Ke-Lin Du [3,*]

[1] College of Shipbuilding Engineering, Harbin Engineering University, Harbin 150001, China; jiaqili1999@outlook.com
[2] Faculty of Mechanical Engineering and Automation, Zhejiang Sci-Tech University, Hangzhou 310018, China; yunwang8789@outlook.com
[3] Department of Electrical and Computer Engineering, Concordia University, Montreal, QC H3G 2W1, Canada
* Correspondence: kldu@ece.concordia.ca

Abstract: The multivehicle routing problem (MVRP) is a variation of the classical vehicle routing problem (VRP). The MVRP is to find a set of routes by multiple vehicles that serve multiple customers at a minimal total cost while the travelling-time delay due to traffic congestion is tolerated. It is an NP problem and is conventionally solved by metaheuristics such as evolutionary algorithms. For the MVRP in a distribution network, we propose an optimal distribution path optimization method that is composed of a distribution sequence search stage and a distribution path search stage that exploits a divide-and-conquer strategy, inspired by the idea of dynamic programming. Several optimization objectives subject to constraints are defined. The search for the optimal solution of the number of distribution vehicles, distribution sequence, and path is implemented by using an improved genetic algorithm (GA), which is characterized by an operation for preprocessing infeasible solutions, an elitist's strategy, a sequence-related two-point crossover operator, and a reversion mutation operator. The improved GA outperforms the simple GA in terms of total cost, route topology, and route feasibility. The proposed method can help to reduce costs and increase efficiency for logistics and transportation enterprises and can also be used for flow-shop scheduling by manufacturing enterprises.

Keywords: vehicle routing problem; multivehicle routing problem; improved genetic algorithm; divide-and-conquer strategy; dynamic programming

check for updates

Citation: Li, J.; Wang, Y.; Du, K.-L. Distribution Path Optimization by an Improved Genetic Algorithm Combined with a Divide-and-Conquer Strategy. *Technologies* **2022**, *10*, 81. https://doi.org/10.3390/technologies10040081

Academic Editors: Manoj Gupta, Eugene Wong and Gwanggil Jeon

Received: 9 June 2022
Accepted: 4 July 2022
Published: 5 July 2022

1. Introduction

The issues of reducing logistics costs, ensuring timeliness of cargo distribution, and optimizing the path of delivery vehicles are crucial to the competitivity of logistics and transportation enterprises [1]. The vehicle routing problem (VRP) is a major problem in distribution, logistics, and transportation. The VRP was first described in 1959 by Dantizg and Ramser [2], as the truck dispatch problem. It is a combinatorial integer programming problem, which is NP-hard. In [3], the VRP was solved using a genetic algorithm (GA) with the idea of implementing different genetic operators, modified for the VRP. In [4], these issues were transformed into a mathematical model, which was solved by using an adaptive evolution algorithm.

The multivehicle routing problem (MVRP) is more common where multiple vehicles of a carrier are used for logistics or transportation. In [5], the problem of mixed fleet vehicles with time windows was solved by self-adaptive neighbor search algorithms. In [6], a mathematical model that minimized the total cost was proposed for a green hybrid fleet with time window and the charging strategy, and it was solved by heuristic algorithms.

In [7], in order to minimize the total cost of a hybrid team composed of traditional fuel, plug-in hybrid, and electric vehicles, a mathematical model was defined and then solved by a metaheuristic algorithm based on the GA and neighborhood search. The metaheuristic was further hybridized with an integer programming solver over a set-partitioning formulation, so as to recombine high-quality routes from the search history for better solutions.

The optimization of the cross-docking distribution network and the internal scheduling of the cross-docking center were studied in [8,9]. The VRP with cross-docking consists in finding a set of routes to distribute products from a set of suppliers to a set of customers through a cross-docking facility at minimal costs, without violating the vehicle capacity and time horizon constraints. In [10], a two-phase metaheuristic based on column generation was proposed for the VRP with cross-docking. A set of destroy and repair operators were used in order to explore a large neighborhood space.

In a production environment of the re-entrant flow-shop (RFS), all jobs have the same routing over the machines of the shop, and the same sequence is traversed several times to complete the jobs. In [11], a GA was used to minimize the makespan for RFS scheduling problems. Hybrid GAs were proposed to enhance the performance of the simple GA.

Cold chain distribution route optimization for fresh agricultural products is formulated as the minimization of the operator's total expenditure that includes emission cost due to carbon tax and comprehensive distribution cost. In [12], this problem was implemented by using the bacterial foraging optimization algorithm. In [13], a model was integrated to determine the delivery time for each order in the multitemperature distribution logistics by minimizing a carrier's total spending.

In [14], a simplified physical road network model was defined by representing the path with the shortest distance, the shortest time, or the lowest cost between two points as arcs, and an exact solution algorithm was proposed. In [15], the VRP was formulated as an integer linear programming model that minimized the distribution cost, the emission cost, or the sum of the distribution and emission costs. It was solved by an ant colony optimization (ACO)-based metaheuristic.

In [16], the shared customer collaboration VRP was introduced and formulated as a mathematical programming problem, and then solved by using a branch-cut-set algorithm. The shared customer collaboration VRP aims at reducing the overall operational cost in a collaboration framework, where several carriers operate and some of their customers have demand of service from more than one carrier.

For the VRP with stochastic demands, a stochastic programming model, composed of a route-planning stage and an execution stage, was introduced in [17]. If a vehicle cannot meet a customer's random demand requested during the execution process, it needs to return to the distribution center for replenishment and resume its planned route at the point of failure. The objective is to minimize the sum of the planned route cost and the expected recourse cost. A local branching metaheuristic was implemented for the MVRP with stochastic demands in [17].

The VRP with hard time windows under demand and travel time uncertainty was studied in [18]. A robust optimization model was built based on route-dependent uncertainty sets. By using a modified adaptive variable neighborhood search heuristic, the designed two-stage algorithm first minimized the total number of vehicle routes, and then minimized the total travel distance.

The fleet size and mix VRP with synchronized visits (FSM-VRPS) is an extension of the VRP with synchronization, where a mixed fleet composed of electric and conventional bikes, and passenger cars having different acquisition costs are considered. Multipath routing can use the resources of multiple networks to transport at the same time, and the transport ability of multiple networks are aggregated [19]. A multistart adaptive large neighborhood search heuristic with threshold accepting has been proposed.

Dozens of prominent VRP variants as well as their respective mixed integer linear-programming formulations were surveyed in [20].

In this paper, we define a variant MVRP and then solve it using an improved GA. We introduce a new divide-and-conquer strategy for calculating the cost of the vehicles during driving, in consideration of the starting cost of the vehicles. In order to reduce the overall cost, a manufacturer provides a reasonable number of vehicles with a limited carrying capacity to provide customers with fast and convenient distribution services. We conduct a simulation and prove our improved GA combined with the divide-and-conquer strategy is effective.

This paper is organized as follows. In Section 2, we give a distribution path optimization model of the MVRP. A divide-and-conquer strategy for MVRP is introduced in Section 3. In Section 4, the distribution path optimization is implemented based on the improved GA. Simulation results are given and analyzed in Section 5. Section 6 concludes this paper.

2. Distribution Path Optimization Model of the MVRP

The VRP may be considered a generalized variation of the traveling salesman problem (TSP) [21]. The TSP consists in finding the shortest path between n cities, which passes all the cities and returns to the starting point, given the distances between the cities. There are $n!$ Feasible route solutions for a visit of n cities. Thus, it is difficult to find the optimal solution. Both the VRP and the TSP can be modelled as combinatorial integer programming problems and are NP-hard.

Consider the route optimization of a distribution service consisting of N_d local retailers and a manufacturer. Therefore, we need to consider how to distribute goods from a manufacturer's distribution node to retailers' demand nodes at the least cost. To this end, we need to find the optimal number of vehicles and the shortest path of the vehicles for the minimum distribution cost, given the quantities of products demanded by the nodes.

A vehicle route is represented by a weighted diagraph $\boldsymbol{G} = (\boldsymbol{R}, \boldsymbol{E}, \boldsymbol{D})$, where $\boldsymbol{R} = \{R_0, R_1, R_2, \ldots, R_n\}$ is the set of nodes, with R_0 being the distribution node and $R_1, R_2, \ldots, R_n$ being demand nodes, $\boldsymbol{E} = \{(R_i, R_j) | R_i, R_j \in \boldsymbol{R}, i \neq j\}$ is the set of edges, and $\boldsymbol{D} = \{d_{ij}, i \neq j\}$ is the set of distances between nodes.

The MVRP is a variant of the classical VRP. The MVRP consists in finding a set of routes by multiple vehicles that serve multiple customers at the least total cost, while the travelling-time delay due to traffic congestion is tolerated. The MVRP is illustrated in Figure 1, where three vehicles undertake the distribution jobs.

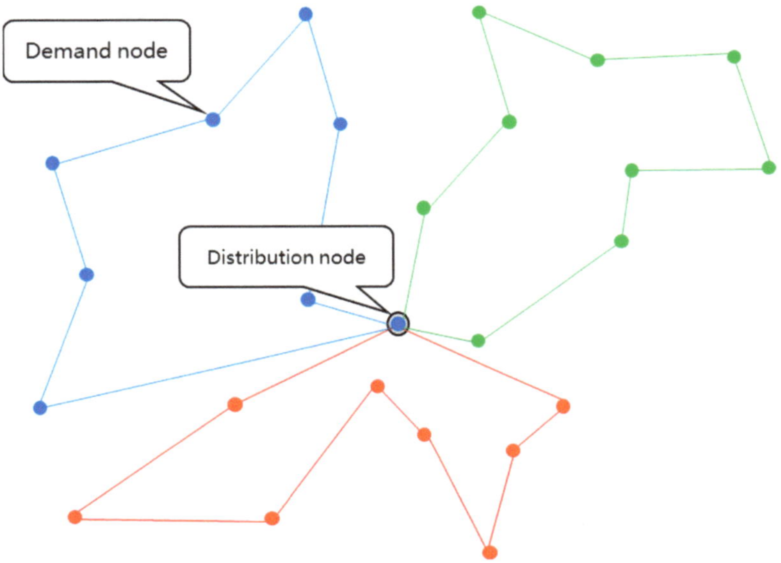

Figure 1. MVRP with a valid solution.

A distribution path optimization model for the MVRP is presented in the following. Our objective is to find a path with the number of vehicles as small as possible such that the vehicles' total itinerary is the shortest, for the sake of the least cost. We implement the search process in two steps. We first find an optimal distribution sequence, and then search for a specific path between two neighboring nodes of the sequence.

Objective function 1. The total cost is minimized,

$$\min C = \sum_{i=0}^{N_d} \sum_{j=0}^{N_d} \sum_{k=0}^{N_v} \left(c_{ijk} x_{ijk} d_{ijk} + p_k \right), \tag{1}$$

where N_d is the number of demand nodes, N_v is the number of vehicles used, c_{ijk} is the unit transport cost from node i to node j for vehicle k, d_{ijk} is the distance between nodes i and j for vehicle k, p_k is the starting cost for vehicle k, and x_{ijk} is the state of vehicle k from node i to node j,

$$x_{ijk} = \begin{cases} 1, & \text{vehicle } k \text{ from node } i \text{ to node } j \\ 0, & \text{otherwise} \end{cases}. \tag{2}$$

Objective function 2. The total length of the route is minimized,

$$\min L = \sum_{i=0}^{N_d} \sum_{j=0}^{N_d} \sum_{k=1}^{N_v} x_{ijk} d_{ijk}. \tag{3}$$

Notice that the calculation of d_{ijk} may consider the real road sections on a map, and it is not the Euclidean distance between the nodes.

When the c_{ijk}s are the same, $p_k = 0$, and the two objective functions, namely, the total cost and the total route length, are equivalent.

Constraint 1. The delivered goods shall not exceed the maximum load capacity of a vehicle,

$$\sum_{i=1}^{N_d} y_{ik} q_i \le Q,\ k = 1, 2, \ldots, N_v, \tag{4}$$

where q_i is a quantity demanded by node i, $q_0 = 0$, Q is the load capacity, and

$$y_{ik} = \begin{cases} 1, & \text{node i's job is completed by vehicle } k \\ 0, & \text{otherwise} \end{cases}, i = 1, 2, \ldots, N_d. \tag{5}$$

Constraint 2. The distribution job of each node is completed by only one vehicle, and all the distribution jobs are completed by N_v vehicles,

$$\sum_{k=1}^{N_v} y_{ik} = \begin{cases} 1, & i = 1, 2, \ldots, N_d \\ N_v, & i = 0 \end{cases}. \tag{6}$$

Constraint 3. There is only one vehicle that reaches or leaves a demand node,

$$\sum_{i=1}^{N_d} x_{ijk} = y_{ik},\ j = 1, 2, \ldots N_d;\ k = 1, 2, \ldots, N_v, \tag{7}$$

$$\sum_{j=0}^{N_d} x_{ijk} = y_{ik},\ i = 1, 2, \ldots N_d;\ k = 1, 2, \ldots, N_v. \tag{8}$$

Constraint 4. There is no duplicate or loop route section on the optimized path,

$$\sum_{i=1,\, i \ne j}^{N_d} z_{ij} \le 1,\ j = 2, 3, \ldots, N_d, \tag{9}$$

$$\sum_{j=1,j\neq i}^{N_{\mathrm{d}}} z_{ij} \leq 1,\ i = 2,\ 3, \ldots, N_{\mathrm{d}}, \tag{10}$$

where z_{ij} is the state of the path between nodes i and j being on the optimal path,

$$z_{ij} = \begin{cases} 1, & \text{on the optimized path} \\ 0, & \text{otherwise} \end{cases}. \tag{11}$$

In a practical case, there are road sections between any two nodes, and we can treat all the road intersections as nodes and search for an optimal path between the two nodes.

Objective function 3. For any two demand/distribution nodes A and B, assume that there is $n-2$ road intersection nodes between them. A and B are treated as nodes 1 and n, respectively, and all the other $n-2$ nodes are permutated and then renamed as 2, 3, ..., $n-1$. Then, the path length of any sequence between the two nodes is minimized,

$$\min S = \sum_{i=1,j=i+1}^{i=n-1,j=n} c_{ij} z_{ij} d_{ij}, \tag{12}$$

where c_{ij} is the connectivity between nodes i and j,

$$c_{ij} = \begin{cases} 1, & \text{direct access between nodes } i \text{ and } j \\ \infty, & \text{otherwise} \end{cases}, \tag{13}$$

and d_{ij} is the Euclidean distance between nodes i and j.

Through the above model, the distribution path is searched. With minimum L or C as the goal subject to the constraints, a distribution sequence is obtained.

When using multiple vehicles to distribute products to multiple demand nodes, we need to determine the number of vehicles and the route of each vehicle. There is a limitation on the load capacity of each vehicle, and the loads and demand nodes are relatively balanced unless there is a supercustomer, whose demand does not exceed 60% of the load capacity.

3. A Divide-and-Conquer Strategy for the MVRP

We propose a divide-and-conquer strategy for the MVRP, which employs the idea of dynamic programming to decompose the problem into multiple steps. For the route search, when the number of demand nodes and road intersections within the range of delivery is very large, it is very difficult to find an optimal route since it is an NP-hard problem. Inspired by the idea of dynamic programming, we find an optimal sequence of the demand nodes at first, and then further find an optimal route between any two nodes.

For the search of a suboptimal sequence, we assume that all the demand nodes and the distribution node are directly accessible. We solve this problem by using an improved GA algorithm. Once an optimal sequence for all the demand nodes is found, we search for an optimal route between any two nodes, and the complete route is then obtained step by step.

In Figure 2, the two nodes are defined as the starting point and ending point. To find a route between two nodes, the center of the two nodes is selected as an auxiliary point. The intersection that is closest to the auxiliary point is named point 1. The distances from point 1 to the two demand nodes are calculated and the larger value is selected as the search radius. The number of intersections may be excessive if the radius is very large.

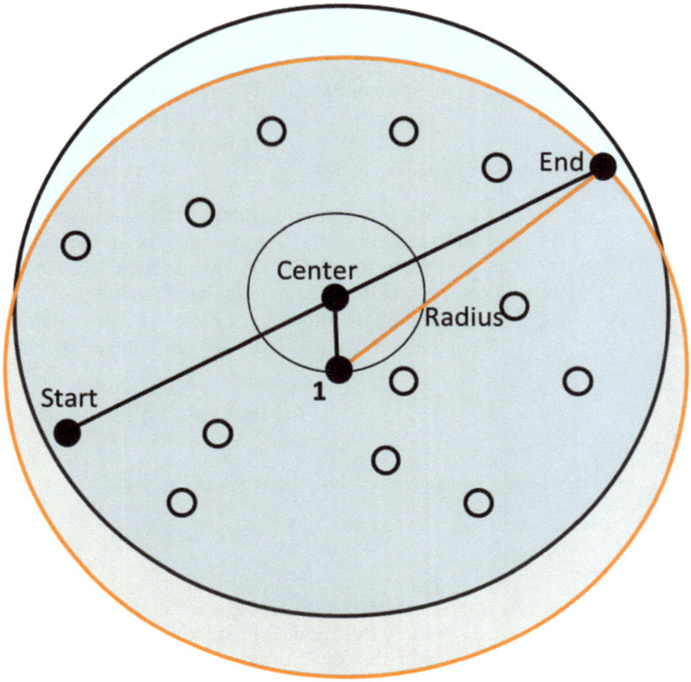

Figure 2. Determination of the search center and search radius.

In Figure 3a, given a search radius, in order to deliver from node 1 to node 3, the distance between the two nodes cannot be covered by a circle. The circle shown is the largest search region, and node 3 is out of the region. We select the crossing point 2 as a reference point, and then select point 4, which is the node closest to point 2 in the search region, as the endpoint of the search. We then connect node 1 and point 4, and get the center of the line segment, point 5. Treating point 5 as the center and setting a search radius, Figure 3b is obtained. Once the optimal route from node 1 to point 4 is found, point 4 is further treated as a new starting point, and still treating point 3 as the end point to find the optimal route. This operation is implemented iteratively until node 3 is reached.

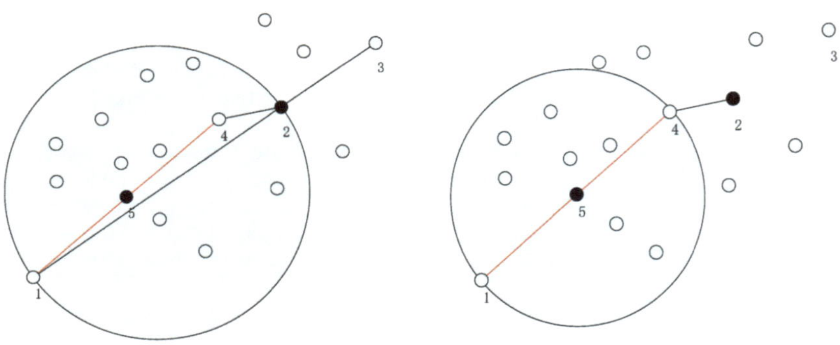

Figure 3. Determination of the search radius. (**a**) When node 3 is out of search range, and (**b**) The new search range.

4. An Improved GA for Distribution Path Optimization of the MVR

In a GA, each solution called an individual is coded as a chromosome. In this paper, we implemented a GA as follows. By selecting multiple chromosomes to form a population, genetic operations are applied. Individuals in the population are selected using the roulette wheel selection and an elitist's strategy, and the selected individuals are then used for recombination and mutation according to certain probabilities, and finally a new population is formed by selection. The process is repeated until a termination criterion is met.

The presented model is solved by the proposed GA. The individuals are encoded as integers. The demand nodes are numbered in an increasing order of distances from the demand nodes to the distribution node. In case of a tie between distances, two continuous integers are randomly assigned. In order to search for a distribution sequence, we start with vehicle 1, from the distribution node 0 to the first demand node, and then to the next demand node, and so on, until it returns to the distribution node 0. We need to determine a search center and a radius. The nodes within the search region are numbered according to the distance to the search center. Then, all the nodes in its region are coded and the optimal sequence is solved by the GA. This process is repeated for all the N_v vehicles.

The initial coding string of an individual is formed by a random combination of all the N_d nodes, and then $N_v - 1$ random but different integers between 2 and $N_d - 1$ are generated and ordered in an increasing order to serve as the breakpoints for assigning jobs to the vehicles.

As an example, assume there are a distribution node and $N_d = 40$ demand nodes. The distribution node is numbered 0, and the demand nodes are numbered 1 to 40 successively. An example individual is a string of the 40 nodes: [32-40-22-34-35-6-3-16-11-30- 33-7-38-28-17-14-8-36-29-21-25-37-31-27-26-19-15-1-36-23-2-4-18-24-39-13-9-20-10-12]. For $N_v = 5$, four breakpoints at positions [7 | 12 | 20 | 32] are generated. The distribution sequences for the vehicles are given by

vehicle 1: [0-32-40-22-34-35-6-3-0];
vehicle 2: [0-16-11-30-33-7-0];
vehicle 3: [0-38-28-17-14-8-36-29-21-0];
vehicle 4: [0-25-37-31-27-26-19-15-1-36-23-2-4-0];
vehicle 5: [0-18-24-39-13-9-20-10-12-0].

The N_v subchromosomes constitute an individual [32-40-22-34-35-6-3 | 16-11-30-33-7 | 38-28-17-14-8-36-29-21 | 25-37-31-27-26-19-15-1-36-23-2-4 | 18-24-39-13-9-20-10-12].

For each of the N_v subchromosomes, we evaluate the load capacity constraint; if a subchromosome cannot satisfy the load constraint, we discard the solution and regenerate a new one, or repair only those infeasible subchromosomes by recombination, until a feasible individual is generated. By adjusting the breakpoints of some infeasible solutions in the population, some infeasible solutions are made feasible. In the case when the loading capacity of the vehicles cannot meet the constraint after a certain number of tries, it is necessary to increase the number of vehicles.

This procedure is repeated until N_P feasible individuals are generated to form a population. The fitness of all the individuals in the population is calculated. The individual with the best fitness is maintained in the population by the elitist's strategy [22]. Through the roulette wheel selection, we randomly select two individuals for recombination and then by mutation and repeat this process until a new generation of N_P individuals are generated. For recombination, we use a two-point crossover, and for mutation we randomly select two points on the chromosome and reverse the genes between the two points.

For two individuals, we select two crossover positions and exchange the segments between the two positions. For the first individual, we remove the same genes acquired from the second individual and obtain a shortened chromosome, and then we insert the acquired segment into the first crossover position, and a new individual is obtained. This operator is known as the Syswerda crossover operator [23].

For example, given two parents p_1 = [2-6-4 | 7-3-5-8 | 9-1] and p_2 = [4-5-2 | 1-8-6-7 | 9-3], if the two crossover points are positions 3 and 7, the segments [7-3-5-8] and [1-8-7-6] will be exchanged. To start with, we delete [1-8-7-6] from p_1 and q_1 = [2-4-3-5-9] remains, then we fill in q_1 the acquired segment [1-8-7-6] in the position of the original p_1, and we get the offspring q_1 = [2-4-3 | 1-8-7-6 | 5-9]. Likewise, we get q_2 = [4-2-1 | 7-3-5-8 | 6-9].

For mutation, we randomly select two mutation points on the chromosome, and then reversed the in-between segment. As an example, an individual q_3 = [2-4 | 3-1-8-6-7 | 5-9] is obtained by mutating individual p_3 = [2-4 | 7-6-8-1-3 | 5-9].

For a complete path of the distribution path optimization, an appropriate distribution sequence is found at first, and the path between any two nodes is then searched. The implementation of the distribution path search includes Algorithm 1 for the search of the distribution sequence and Algorithm 2 for the search of the distribution route between two nodes.

For both algorithms, the elitist's method is used to retain the best individual, and subsequently $N_{\mathrm{P}} - 1$ individuals are selected by the roulette wheel selection to form a new population. A two-point crossover operator is used with crossover probability p_{c}. The mutation operator modifies an individual with mutation probability p_{m}. The mutation operator is the reversal operator. The algorithms stop when they converge or run for T generations.

Algorithm 1 [Distribution Sequence Search]

```
Input: N_P, p_c, p_m, T; N_d, N_v, Q, q_i, node positions, ...
Output: Distribution sequence of demand nodes.
Begin:
  Load node positions;
  Code demand nodes into continual integers;
  Generate initial population;
  while (TRUE):
    Generate a population;
    Preprocess infeasible individuals;
    Calculate fitness defined by (1) or (3);
    Perform selection, combination, and mutation;
    Apply elitist's strategy;
  until Termination criterion is met.
End
```

Algorithm 2 [Route Search Between Two Nodes]

```
Input: N_P, p_c, p_m, T; node positions, intersection positions, ...
Output: Optimal route between the two nodes.
Begin:
  while (TRUE):
    Apply the divide-and-conquer strategy:
      Find a center between two nodes;
      Draw a search region using a radius;
      Identify the road intersections within the range as additional nodes;
    Generate a population using nodes in the region;
    Preprocess infeasible individuals;
    Calculate fitness defined by (12);
    Perform selection, combination, and mutation;
    Apply elitist's strategy;
  until Termination criterion is met.
End
```

For example, the distribution sequence of vehicle 1 is [0-32-40-22-34-35-6-3-0]. If there are 7 demand nodes that need to be delivered, the distribution route of the search is extracted in order. The road paths between nodes are numbered as

$$[0 - 32 - 40 - 22 - 34 - 35 - 6 - 3 - 0]$$
$$1 \quad 2 \quad 3 \quad 4 \quad 5 \quad 6 \quad 7 \quad 8,$$

where the road path 1 denotes the path from the distribution node 0 to the demand node 32, and so on. A search procedure for road path 1 is illustrated in Section 3.

Figure 4 gives an illustration of Algorithm 2 for the road path 1. We renumbered the road intersections covered in the range, the distribution node, and the demand node 32. The fitness is defined by (12), where the connectivity of each road section within the search range is available from the map. A path is an individual encoded as an integer string, such as $E = [1\text{-}0\text{-}4\text{-}7\text{-}3\text{-}0\text{-}2\text{-}8]$. Infeasible individuals are made feasible by preprocessing. Once a population is generated, it is subject to selection, recombination, and mutation. This procedure continues for all the 8 road paths between the nodes for a complete route of a vehicle, and then for the distribution paths of all the vehicles.

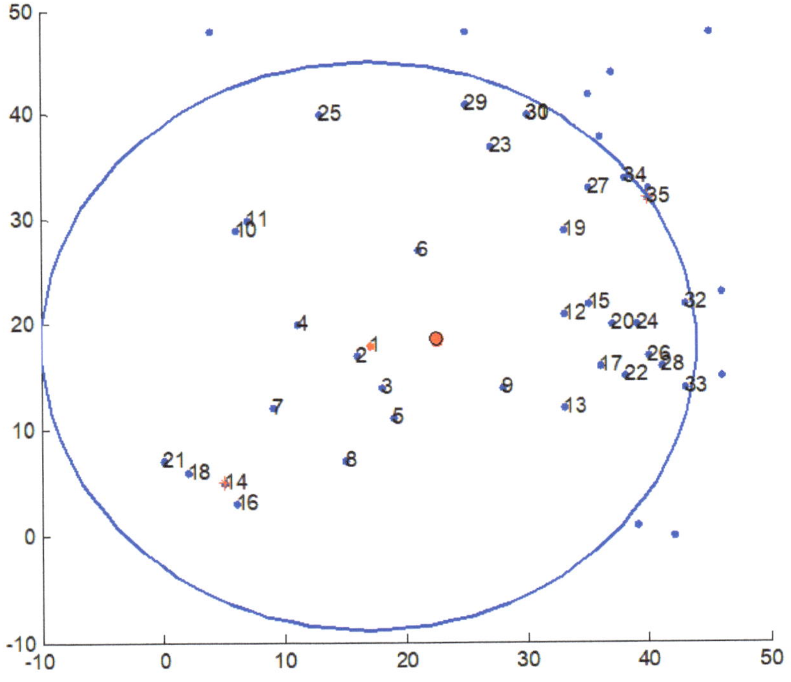

Figure 4. Search range for a distribution path.

5. Simulation

As an example, a manufacturing enterprise in Hangzhou delivers products to $N_d = 40$ customers, and the load capacity of a vehicle is $Q = 250,000$ products. Let the coordinate of the distribution node be (50, 50). We position on a map the coordinates of the $N_d = 40$ demand nodes by using ARCGIS software, and the coordinates and quantities of the demand nodes are listed in Table 1. The coordinates of the demand nodes are plotted in Figure 5.

Table 1. The coordinates and demand quantities of the demand nodes.

No.	x (km)	y (km)	Demand ($\times 10^3$ pcs)	No.	x (km)	y (km)	Demand ($\times 10^3$ pcs)
0	50	50	no	21	88	51	0.3
1	47	42	8.1	22	65	29	8.4
2	79	26	9	23	44	64	9.3
3	99	91	1.2	24	91	64	6.7
4	30	33	9.1	25	74	92	7.5
5	65	20	6.3	26	1	68	7.4
6	53	10	0.9	27	53	35	3.9
7	40	83	2.7	28	44	35	6.5
8	83	95	5.4	29	58	51	1.7
9	53	7	9.5	30	88	10	7.0
10	35	17	9.6	31	49	36	0.3
11	72	90	1.5	32	71	13	2.7
12	35	80	9.7	33	39	62	0.4
13	57	69	9.5	34	84	57	0.9
14	43	22	4.8	35	30	30	8.2
15	44	14	8.0	36	17	49	6.9
16	46	46	1.4	37	10	25	3.1
17	94	40	4.2	38	18	74	9.5
18	42	19	9.1	39	18	13	0.3
19	66	91	7.9	40	8	55	9.5
20	70	47	6.5	Total			224.9

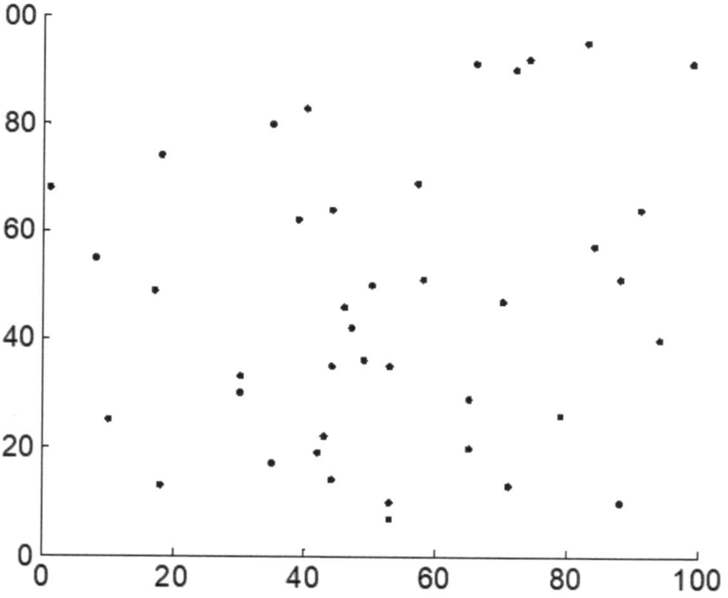

Figure 5. The coordinates of the demand nodes. The x- and y-coordinates correspond to a position on a map.

5.1. *Sequence Search Using Algorithm 1*

In order to simplify the simulation, we optimized the total route length (objective function 2). We also assumed there were a direct connection between any two nodes. Thus, we did not need to use Algorithm 2 and objective function 3 for the route search.

For the GA parameters, we set the number of generations $T = 5000$, population size $N_P = 50$, crossover probability $p_c = 0.8$, and mutation probability $p_m = 0.1$. A population was selected from 200 individuals produced at random. The best result was obtained from 20 random runs. Figure 6 shows the distribution routes generated by our improved GA.

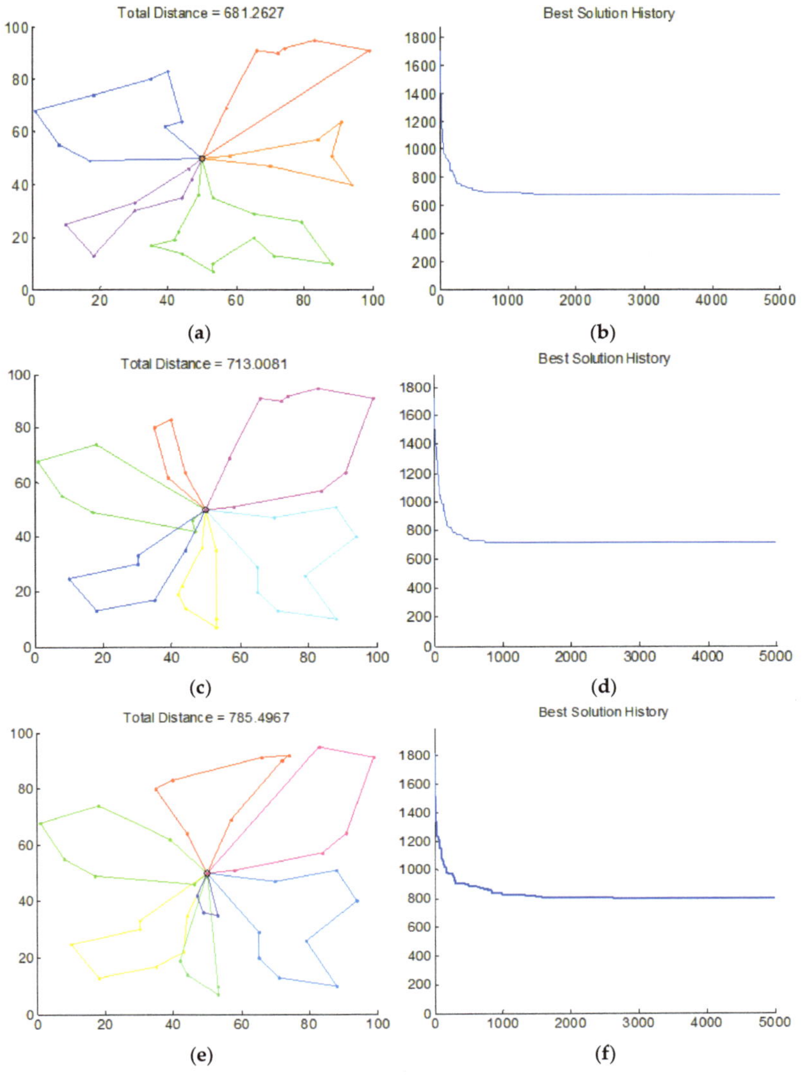

Figure 6. The distribution sequence, using the improved GA, (**a**,**b**) for 5 vehicles, (**c**,**d**) for 6 vehicles, and (**e**,**f**) for 7 vehicles. For (**a**,**c**,**e**), the x- and y-coordinates correspond to a position on a map. For (**b**,**d**,**f**), the x-coordinate corresponds to the number of generation, and the y-coordinate corresponds to the total distance in km.

The best results for the total length of the distribution route are 681.26 km for $N_v = 5$, 713.01 km for $N_v = 6$, and 785.50 km for $N_v = 7$, respectively. It can be seen there are few intersecting routes. The paths are reasonable. When the number of vehicles is five, the total length is 681.26 km, which is the best result. The route of each vehicle does not intersect, while satisfying all the constraints.

Due to the elitist strategy, the evolution is stable. It can be seen that the distribution sequence rarely overlaps and intersects. The result is reasonable and desirable.

Figure 7 gives the results when the simple GA with an elitist's strategy was used. For 20 random runs, the generated total path length can be shorter, but the result was unpredictable. For five vehicles, the shortest path is 662.00 km, but it is an infeasible solution. One of the vehicles only distributes to two demand nodes, while another vehicle distributes to 12 demand nodes, exceeding the maximum load limit of the vehicle. For the case of six and seven vehicles, the solutions are also infeasible due to violation of the constraints. There are many intersecting routes and infeasible solutions are often produced.

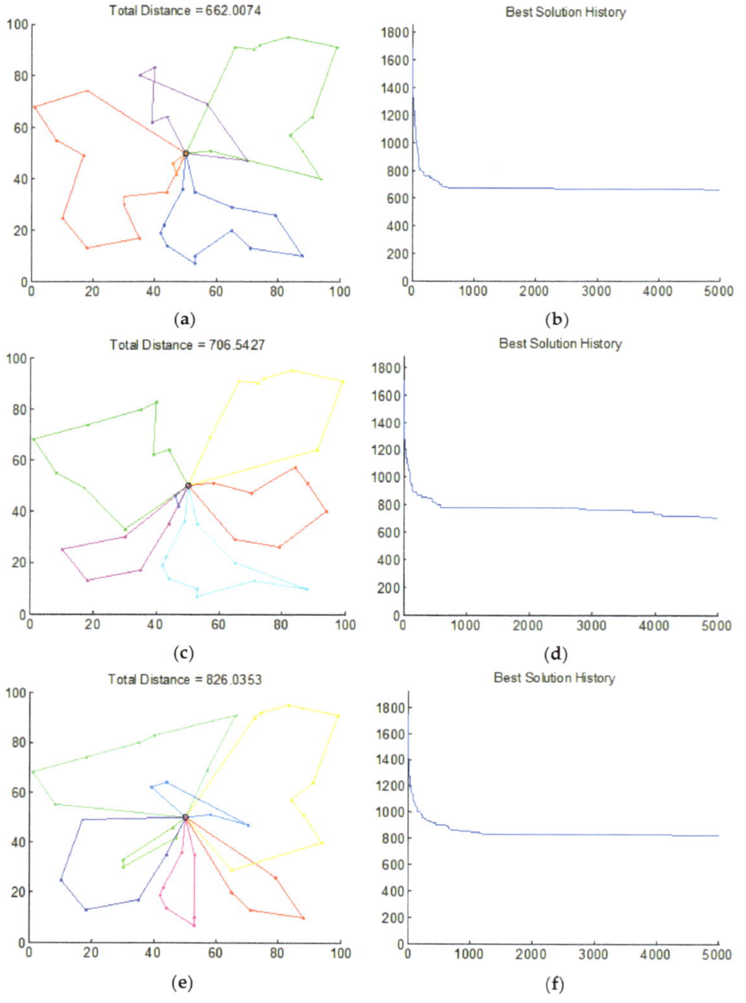

Figure 7. Distribution sequence, using the simple GA, (**a**,**b**) for 5 vehicles, (**c**,**d**) for 6 vehicles, and (**e**,**f**) for 7 vehicles. For (**a**,**c**,**e**), the x- and y-coordinates correspond to a position on a map. For (**b**,**d**,**f**), the x-coordinate corresponds to the number of generation, and the y-coordinate corresponds to the total distance in km.

The experiments show that the improved GA outperforms the simple GA for the MVRP, the proposed MVRP mathematical model is practical, and the solution found using the improved GA is more reasonable.

5.2. Route Search Using Algorithm 2

Finally, after searching for the optimal distribution sequence, we searched for a shortest route between two demand nodes or between the distribution node and a demand. As an illustration of Algorithm 2, we give a simple illustration. Objective function 3 was used for the route search.

We specified the number of generations $T = 200$, population size $N_P = 100$, crossover probability $p_c = 0.9$, and mutation probability $p_m = 0.1$. The optimal path is shown in Figure 8a, where nodes 23 and 27 are demand nodes, the coordinate of node 23 is (5.00, 5.00), and the optimal path is 23-21-14-5-2-1-6-19-27; Figure 8b shows the optimal path is obtained at the 62nd generation.

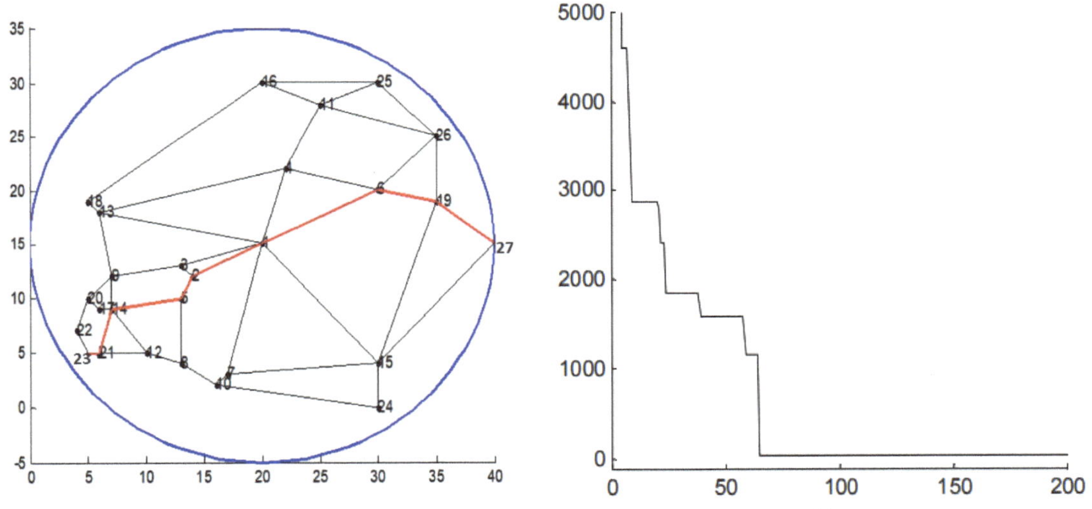

Figure 8. Distribution path (**a**) and the optimization curve (**b**). For (**a**), the x- and y-coordinates correspond to a position on a map. For (**b**), the x-coordinate corresponds to the number of generations, and the y-coordinate corresponds to the total route length in km.

Algorithm 2 is shown to be very effective at searching for an optimal route between two nodes.

6. Conclusions

The MVRP was investigated in this paper. We proposed a mathematical model for the MVRP, based on which the optimal distribution route for multiple vehicles was searched. Using the proposed improved GA combined with our unique divide-and-conquer strategy, we verified the feasibility and rationality of the model in search of the optimal solution to the MVRP. Applying a roulette wheel selection and elitist's strategy, preprocessing of the infeasible individuals, and well-designed crossover and mutation operators, the improved GA accelerated the convergence to the optimal solution, when compared with the simple GA. The performance of the improved GA combined with the divide-and-conquer strategy was validated when solving the distribution path optimization of a manufacturer and 40 demand nodes. In the future, we will compare our method with some state-of-the-art algorithms.

Author Contributions: Conceptualization, J.L. and Y.W.; methodology, J.L., Y.W. and K.-L.D.; software, J.L. and Y.W.; writing—original draft preparation, J.L.; writing—review and editing, K.-L.D.; supervision, K.-L.D.; project administration, K.-L.D.; funding acquisition, K.-L.D. All authors have read and agreed to the published version of the manuscript.

Funding: This work was funded by the National Natural Science Foundation of China under Grant no. 51475434.

Data Availability Statement: All data included in the main text.

Acknowledgments: The authors would like to thank the anonymous reviewers for their constructive comments, which improve the manuscript.

Conflicts of Interest: The authors declare no conflict of interest.

References

1. Fisher, M.L.; Jaikumar, R. A Generalized Assignment Heuristic for Vehicle Routing. *Networks* **1981**, *11*, 109–124. [CrossRef]
2. Dantzig, G.B.; Ramser, R.H. The Truck Dispatching Problem. *Manag. Sci.* **1959**, *6*, 80–91. [CrossRef]
3. Ochelska-Mierzejewska, J.; Poniszewska-Maranda, A.; Maranda, W. Selected Genetic Algorithms for Vehicle Routing Problem Solving. *Electronics* **2021**, *10*, 3147. [CrossRef]
4. Sabar, N.R.; Bhaskar, A.; Chung, E.; Turky, A.; Song, A. A Self-adaptive Evolutionary Algorithm for Dynamic Vehicle Routing Problems with Traffic Congestion. *Swarm Evol. Comput.* **2019**, *44*, 1018–1027. [CrossRef]
5. Goeked, D.; Schneider, M. Routing a Mixed Fleet of Electric and Conventional Vehicles. *Eur. J. Oper. Res.* **2015**, *245*, 81–99. [CrossRef]
6. Macrina, G.; Pugliese, L.D.P.; Guerriero, F.; Laporte, G. The Green Mixed Fleet Vehicle Routing Problem with Partial Battery Recharging and Time Windows. *Comput. Oper. Res.* **2018**, *101*, 183–199. [CrossRef]
7. Hiermann, G.; Hartl, R.F.; Puchinger, J.; Vidal, T. Routing a Mix of Conventional, Plug-in Hybrid, and Electric Vehicles. *Eur. J. Oper. Res.* **2019**, *272*, 235–248. [CrossRef]
8. Goodarzi, A.H.; Zegordi, S.H.; Alpan, G.; Kamalabadi, I.N.; Kashan, A.H. Reliable Cross-docking Location Problem Under the Risk of Disruptions. *Oper. Res.* **2021**, *21*, 1569–1612. [CrossRef]
9. Gelareh, S.; Glover, F.; Guemri, O.; Hanafi, S.; Nduwayo, P.; Todosijevic, R. A Comparative Study of Formulations for a Cross-dock Door Assignment Problem. *Omega* **2020**, *91*, 102015. [CrossRef]
10. Gunawan, A.; Widjaja, A.T.; Vansteenwegen, P.; Yu, V.F. A Matheuristic Algorithm for the Vehicle Routing Problem with Cross-docking. *Appl. Soft Comput.* **2021**, *103*, 107163. [CrossRef]
11. Chen, J.-S.; Pan, J.C.-H.; Lin, C.-M. A Hybrid Genetic Algorithm for the Re-entrant Flow-shop Scheduling Problem. *Expert Syst. Appl.* **2008**, *34*, 570–577. [CrossRef]
12. Ning, T.; An, L.; Duan, X. Optimization of Cold Chain Distribution Path of Fresh Agricultural Products Under Carbon Tax Mechanism: A Case Study in China. *J. Intell. Fuzzy Syst.* **2021**, *40*, 10549–10558. [CrossRef]
13. Chen, W.T.; Hsu, C.-I. Optimal Scheduling for Multi-temperature Joint Distribution Under Carbon Tax. *Int. J. Oper. Res.* **2019**, *16*, 45–62. [CrossRef]
14. Ticha, H.B.; Absi, N.; Feillet, D.; Quilliot, A.; Woensel, T.V. A Branch-and-Price Algorithm for the Vehicle Routing Problem with Time Windows on a Road Network. *Networks* **2019**, *73*, 401–417. [CrossRef]
15. Jabir, E.; Panicker, V.V.; Sridharan, R. Design and Development of a Hybrid Ant Colony-variable Neighbourhood Search Algorithm for a Multi-depot Green Vehicle Routing Problem. *Transp. Res. Part D* **2017**, *57*, 422–457. [CrossRef]
16. Fernández, E.; Mireia, R.-R.; Speranza, M.G. The Shared Customer Collaboration Vehicle Routing Problem. *Eur. J. Oper. Res.* **2018**, *265*, 1078–1093. [CrossRef]
17. Hernandez, F.; Gendreau, M.; Jabali, O.; Rey, W. A Local Branching Metaheuristic for the Multi-vehicle Routing Problem with Stochastic Demands. *J. Heuristics* **2019**, *25*, 215–245. [CrossRef]
18. Hu, C.; Lu, J.; Liu, X.; Zhang, G. Robust Vehicle Routing Problem with Hard Time Windows Under Demand and Travel Time Uncertainty. *Comput. Oper. Res.* **2018**, *94*, 139–153. [CrossRef]
19. Masmoudi, M.A.; Hosny, M.; Koc, C. The Fleet Size and Mix Vehicle Routing Problem with Synchronized Visits. *Transp. Lett.* **2021**, *14*, 427–445. [CrossRef]
20. Marinakis, Y.; Marinaki, M.; Migdalas, A. Variants and Formulations of the Vehicle Routing Problem. In *Open Problems in Optimization and Data Analysis*; Pardalos, P., Migdalas, A., Eds.; Springer Optimization and Its Applications; Springer: Cham, Switzerland, 2018; Volume 141, pp. 91–127. [CrossRef]
21. Du, K.-L.; Swamy, M.N.S. *Neural Networks in a Softcomputing Framework*; Springer: London, UK, 2006. [CrossRef]
22. Du, K.-L.; Swamy, M.N.S. *Search and Optimization by Metaheuristics*; Springer: New York, NY, USA, 2016. [CrossRef]
23. Barbulescu, L.; Howe, A.E.; Whitley, L.D.; Roberts, M. Understanding Algorithm Performance on an Oversubscribed Scheduling Application. *J. Artif. Intell. Res.* **2006**, *27*, 577–615. [CrossRef]

technologies

MDPI

Article

Analysis and Hardware Architecture on FPGA of a Robust Audio Fingerprinting Method Using SSM

Ignacio Algredo-Badillo [1,†], Brenda Sánchez-Juárez [2,†], Kelsey A. Ramírez-Gutiérrez [1], Claudia Feregrino-Uribe [3], Francisco López-Huerta [4,*] and Johan J. Estrada-López [5,*]

1 Consejo Nacional de Ciencia y Tecnología-Instituto Nacional de Astrofísica, Óptica y Electrónica (CONACYT-INAOE), Luis Enrique Erro 1, Santa María Tonanzintla, Puebla 72840, Mexico; algredobadillo@inaoep.mx (I.A.-B.); kramirez@inaoe.mx (K.A.R.-G.)
2 Universidad Politécnica de Tlaxcala, Av. Universidad Politécnica No.1, San Pedro Xalcaltzinco, Zacatelco 90180, Mexico; brenda.sanchez.j7@gmail.com
3 Instituto Nacional de Astrofísica, Óptica y Electrónica, Luis Enrique Erro 1, Santa María Tonanzintla, Puebla 72840, Mexico; cferegrino@inaoep.mx
4 Facultad de Ingeniería Eléctrica y Electrónica, Universidad Veracruzana, Boca del Río 94294, Mexico
5 Physics and Engineering Department, Westmont College, Santa Barbara, CA 93108, USA
* Correspondence: frlopez@uv.mx (F.L.-H.); jestradalopez@westmont.edu (J.J.E.-L.)
† These authors contributed equally to this work.

Abstract: The significant volume of sharing of digital media has recently increased due to the pandemic, raising the number of unauthorized uses of these media, such as emerging unauthorized copies, forgery, the lack of copyright, and electronic fraud, among others. In particular, several applications integrate services or products such as music distribution, content management, audiobooks, streaming, and so on, which require users to demonstrate and guarantee their audio ownership. The use of acoustic fingerprint technology has emerged as a solution that is widely used to secure audio applications. This technique extracts and analyzes certain information that identifies the inherent properties of a partial or complete audio file. In this paper, we introduce two audio fingerprinting hardware architectures with a feature extraction system based on spectrogram saliency maps (SSM) and a brute-force search. The first of these conducts a search in 33 saliency maps of 32×32 pixels in size. After analyzing the first algorithm, a second architecture is proposed, in which the saliency map is reduced to 27×25 pixels, requiring 75.67% fewer hardware resources, lowering the power consumption by 64.58%, and improving the efficiency by 3.19 times via a throughput reduction of 22.29%.

Keywords: FPGA; audio fingerprinting; hardware architecture; SSM

Citation: Algredo-Badillo, I.; Sánchez-Juárez, B.; Ramírez-Gutiérrez, K.A.; Feregrino-Uribe, C.; López-Huerta, F.; Estrada-López, J.J. Analysis and Hardware Architecture on FPGA of a Robust Audio Fingerprinting Method Using SSM. *Technologies* **2022**, *10*, 86. https://doi.org/10.3390/technologies10040086

Academic Editors: Manoj Gupta, Eugene Wong and Gwanggil Jeon

Received: 3 June 2022
Accepted: 28 June 2022
Published: 19 July 2022

1. Introduction

The aim of digital technology consists of transforming analog information to digital information that can be interpreted by electronic devices. Thanks to this technology, thousands of books, songs, and images can be stored, shared, and consulted. However, digital technology also entails negative aspects that can affect the author's interests regarding the exploitation of their rights and works since unauthorized uses of their work can occur. In addition to being economical and relatively simple, digital works retain their quality in the generation of subsequent copies, which is attractive to the final consumer [1].

Nowadays, piracy is a problem that lurks in the digital world, affecting: (1) creators, including the authors and holders of their related rights, because their primary source of income is reduced; (2) workers from all industries due to the loss of their jobs; (3) industries, which are not able to commercialize as many original products and thus suffer decreased profits; and (4) the government, which receives fewer taxes, because piracy is carried out outside the law. One of the technological methods used for copyright protection is fingerprinting, which is intended to identify and collect information about works; this information can be used later to recognize pirates responsible for illegal copies [2].

Currently, users share audiovisual material that is not always protected, making illegal traffic a common problem. However, there are two possible ways of preventing this: (1) a priori protection with watermarking policing, in which a digital watermark is embedded in the content, and (2) posterior forensics using fingerprinting that identifies an authorized user or purchaser of the material. In addition, fingerprinting suffers from fewer technical problems than watermarking since it does not require widespread key distribution [3].

Over the years, a significant amount of studies have proposed watermarking and fingerprinting solutions to protect audiovisual material. For example, the authors in [4] proposed a blind adaptive audio watermarking algorithm based on singular value decomposition (SVD) in the discrete wavelet transform domain; this scheme showed low error probability rates. The authors in [5] developed a perceptual audio hashing algorithm based on the Zernike moment and maximum-likelihood watermark detection to enable content-oriented searches in a database. The algorithm obtained smaller samples than those in the conventional broadcast monitoring system based on a comparison of the whole sample set.

Several algorithms can be used to obtain an acoustic fingerprint; however, not all resist the compression and cropping attacks required by specific audio applications. Moreover, they are usually composed of elementary keys, also known as sub-fingerprints, based on small parts of the signal. Acoustic fingerprints are often composed of consecutive keys used to identify any part of a signal [6].

The design and development of these security systems require highly complex operations, and they are usually implemented on non-specialized machines or general-purpose processors. This situation occurs when the SSM algorithm is used, and it conducts searches in large databases. Additionally, many computational resources are wasted because they may not be used or are used in other processes. Today, some lines of research focus on providing hardware architectures or new algorithms for improving performance results, depending on the system's requirements, the application, or the user. For example, one user may need a system to display graphics quickly, whereas another may require the system to search efficiently in a database, or they may require low power consumption.

Audio fingerprinting algorithms are commonly resource-consuming tasks, and they are time-consuming when implemented in software running on non-specialized machines, which can be executing other tasks necessary for operating systems or other applications. For this reason, demanding operations such as the SSM algorithm or the requirements of an extensive database search require high computational resources. In this way, different hardware architectures are necessary due to the specialized needs of different systems and users. For example, a collusion-resistant fingerprinting system was implemented in [7] and was found to be suitable for a massive online music distribution applications.The authors in [8] proposed a security technique (MixLock) based on logic locking of the digital section of a mixed-signal circuit, which could be employed to mitigate reverse engineering and counterfeiting. They proposeda device identification protocol that leverages the frequency response of a speaker and a microphone from two wireless devices as an acoustic hardware fingerprint. A device identification protocol uses an acoustic hardware fingerprint extracted from the frequency response of a speaker and a microphone from two wireless devices, as proposed in [9]. Furthermore, an audio fingerprint algorithm that balances the ideal amount of data embedded to enable a comparison, while keeping the fingerprints lightweight for manageable access, indexing, searches, and storage, was embedded on an ARM 7-LPC2148 device [10]. In [11], the design and testing of a music information retrieval algorithm was conducted based on fingerprinting techniques implemented in a low-cost, embedded, and reconfigurable platform. Different hardware can thus be implemented to satisfy different requirements, such as a fast graphics display system, an efficient database search, or low power consumption requirements.

In this study, a robust method based on the saliency maps of the audio signal spectrogram [12] was implemented, proposing two hardware architectures on FPGA. After analyzing the window size, the use of hardware resources decreases, memory requirements are reduced, efficiency is increased, and power consumption is improved at the cost of a slight loss in performance. In addition, the search of the acoustic fingerprint in a database through correlation is presented. The architectures allow a significant parallelization of the computations, which results in a higher efficiency by 3.19 orders of magnitude.

The paper is organized as follows: Section 2 presents the state of art in Fingerprinting, Section 3 introduces two versions of the fingerprinting algorithm, while Section 4 presents their hardware implementation. Section 5 describes analysis and implementation results of the fingerprinting algorithms. Section 6 presents a discussion of the proposed architectures and the main highlights of the analysis, whereas Section 7 compares the obtained results with related works. Finally, conclusions and future work are discussed in Section 8.

2. Fingerprinting and SSM

Some technological measures of protecting digital documents require an effective transmission and processing of the information contained in a protected work. In general, there are two types of technological protection measures in the digital domain: (1) those that manage the access, processing, and transmission of the work and (2) those that only protect its integrity and transmission. The first type involves extending the control of the digital use of files in an inter-operative manner. The second preserves the integrity and protects copyright, preventing any non-authorized modification, alteration, or distribution of the work.

Protection measures, such as anti-copying systems, encryption, and watermarking are not entirely secure. For example, digital watermarks have been proposed as an efficient solution to protect copyright and ownership of multimedia files (image, audio, or video), by making it possible to identify their source. However, digital watermarks are based on the code's identification inserted directly into the content of the file, and it is possible to detect them only by using a specific algorithm and a key. On the other hand, acoustic fingerprints are used to identify audio, for search, navigation, monitoring, and other monetary purposes, such as music recovery and video identification. Acoustic fingerprints are extracted from audio, video, or images. However, they are not embedded in the file. Thus, the signal is not altered before its transmission.

The most popular audio transformations are resampling, compression, noise addition, recording, and temporal resynchronization [13]. Audio compression reduces the size of an audio file, requiring smaller storage capacity. However, compressing an audio file many times results in a low fidelity of sound. Noise is all unwanted sound, and recording consists of D/A and A/D conversion or re-recording. Temporary desynchronization occurs when audio is delayed or advancing in time. Therefore, an audio fingerprint algorithm used to detect copies should be robust to these attacks.

An acoustic fingerprint is an identifier for audio files based on their content. With them, it is possible to identify a pattern or signature in audio files which can then be recognized from an audio database. In [6], it is mentioned that a fingerprinting system usually consists of two components: (1) a mathematical process that calculates the fingerprint (i.e., fingerprint extraction) and (2) a search algorithm to scan a database of previously derived acoustic fingerprints in search of similarities (i.e., fingerprint search).

2.1. Fingerprint Extraction

Several extraction algorithms can obtain an acoustic fingerprint from an audio file, but not all of them resist compression and cropping. A few seconds of audio are needed to extract an acoustic fingerprint. A common technique is to divide the piece of audio into small segments and extract their characteristics. There are numerous strategies for this division process. The most common are the use of Fourier Coefficients, Cepstral Coefficients in Mel Frequencies (CCMF), Linear Predictive Coding, and Mean and Variance of characteristics.

The next step is to map the extracted characteristics into a more compact representation, using Hidden Markov Models, Quantization, or other methods [14]. In addition, fingerprints are usually composed of elementary keys (called sub-fingerprints) based on small parts of the signal. Often, the acoustic fingerprints are composed of consecutive keys used to identify any part of the signal. It is also possible to use the Spectrogram Saliency Maps (SSM) algorithm as a fingerprint extraction method, representing an audio signal through a spectrogram combined with a fingerprint extraction. The extraction of fingerprints is based on the saliency maps of the audio signal spectrogram.

Fingerprints are not exclusive to human fingers. They also exist in documents, but they must be extracted. For that extraction, there are many algorithms, such as Winnowing, Karp–Rabin, All-to-all matching, and Shazam.

In [15], the Karp–Rabin algorithm for matching sub-strings is the first fingerprint version based on k-grams. It consists of finding the matches of a particular string s of length k within a longer string. On the other hand, Winnowing [16] presents an efficient local fingerprinting algorithm that selects the minimum value of a hash window. If there is more than one hash with the minimum value, the algorithm selects the rightmost occurrence. Then, all selected hashes are saved as the fingerprints of the document. The authors in [17] propose that by using an unknown audio's acoustic fingerprint, a query can be made in a fingerprint database (from an extensive library of songs) to identify the audio. This system requires a robust method of fingerprint extraction, and a very efficient search strategy capable of working with limited computer resources. A copy detection algorithm should have three properties: (1) blankness insensitivity, (2) noise suppression, and (3) independent position. For the search of traces, previous works describe some methods such as a hierarchical search, reduction in candidates, and a search based on the tree.

On the one hand, Ref. [18] describes a system of acoustic fingerprints consisting of a generation algorithm and the searching algorithm to find the matches of the fingerprints in a database. In addition, the fingerprint extraction includes a front-end where audio is divided into frames, and a series of robust discriminating features are extracted in each frame. Subsequently, these features are transformed into a fingerprint by a modeling unit that compacts the representation of fingerprints. On the other hand, Ref. [19] presents an audio detection system robust to various attacks, such as pitch and tempo change. In that work, a two-dimensional representation is proposed for audio signals called chroma time images. A pitch change in the audio signal appears as a circular shift along the chroma axis of that image, and a change in tempo in the audio signal appears as a scale change along the time axis of that image. In [20], the authors consider chromatic characteristics and compare the performance of the systems based on them, with the use of the timbral characteristics in the same experimental frame. When making system classification based on the equal error rate, they conclude that the best audio segmentation uses detectors grouped by octaves and sub-bands for music and noise. For the voice, the timbral characteristics use the CCMF-SDC (Cepstral Coefficients in Mel Frequencies–Shifted Delta Coefficients).

In [21], the Shazam algorithm is described, which is based on local acoustic fingerprints, and uses the peaks observed in the spectrogram of the audio signal as the points of local characteristics. This algorithm is resistant to noise and distortion, is efficient and scalable, and it is able to quickly identify a segment of music captured through the phone's microphone from a base of more than one million songs. Furthermore, the algorithm uses a combinatorial method of time-frequency analysis for the audio constellation, producing unusual properties such as transparency, in which several mixed tracks can be identified.

2.2. Fingerprint Search

An essential point for the usability of a fingerprint system is how to make comparisons between unknown audio and, possibly, millions of fingerprints. In general, the methods depend on the representation of the fingerprint [2].

In this regard, the search is an important operation since it allows recovering previously stored data. The search result is successful if the information is found, or unsuccessful if it is not found. The search can be applied to ordered or unordered elements. Additionally, Ref. [22] describes that search methods can be classified as follows: (1) *Sequential.* This method consists of reviewing the data structure element by element, until the data that are being looked for are reached. This method works for ordered or unordered data. (2) *Binary*, where the method divides the total of the elements in two, comparing the searched element with the central one. This method only works if the data have been previously sorted. (3) *Hash.* The key transformation method increases the search speed without requiring the elements to be previously ordered. This method allows access to the data by a key that directly indicates where the searched data are stored.

There are also searching systems based on the Index of Inverted File. For example, Ref. [23] describes a system that uses a look-up table (LUT) of possible entries of sub-fingerprints, with pointers to fingerprints in the base of data. The applicability depends on the alphabet and the size of the sub-fingerprint. It may be feasible to generate a list containing all possible entries and corresponding pointers. The index uses code words extracted from binary sequences representing the audio. In addition, in [17], an index of possible track pieces that point to positions in the songs is proposed. Since a piece of the candidate track is free of errors (exact match), a list of candidate songs can be efficiently obtained to exhaustively search in.

On the other hand, the filtering of unlikely candidates is proposed, where [24] describes an effective way to reduce the search space using a simple similarity measure to quickly eliminate many candidates, ensuring that false rejections do not occur. During the comparison process, candidates with the worst score can be excluded.

There are methods based on a hierarchical search; for example, Ref. [25] presents a hierarchical search using an abstract version of the problem to dynamically generate heuristic values. In addition, there is a regressive switchover, which reduces the number of expansions and, therefore, the execution time.

Finally, some methods use a tree-based search since, in essence, the search for a fingerprint is similar to the search for the nearest neighbor. Often, trees are used to locate the nearest neighbor. Authors in [26] propose an algorithm in which, every 5 s, a binary fingerprint block (8192 bits) is considered as a point in the fingerprint space. The fingerprint block is divided into 1024 8-bit patterns. The value of each consecutive 8-bit pattern determines which of the 256 possible children descends. A path from the root node to a leaf defines a block of fingerprints. When a query fingerprint is consulted in the database, each 8-bit pattern is compared with the tree elements; the error between the query fingerprint and the best sheet is estimated at each tree level. As soon as the error is estimated, the best result is found, and the search stops.

3. Algorithm and Analysis

This section describes two key elements: (1) the audio fingerprinting algorithm, which has two different versions of the searching module, depending on their storage: the *brute-force search* and the *optimized brute-force search*; (2) the analysis necessary to reduce the window size, optimize the storage, and, consequently, the search process. The two versions of the algorithm are implemented on hardware in the next section.

3.1. Audio Fingerprinting Algorithm

The audio fingerprinting algorithm uses fingerprint extraction and the searching process in the database. The searching procedure has two versions depending on the storage. However, the algorithm simultaneously exposes the SSM algorithm as a fingerprint extraction method, and the correlation as a searching process. The first version is based on brute-force searching. The other is based on an optimized search, with a reduced window size, which results in an improvement in several performance characteristics.

Figure 1 shows the general diagram of the algorithm proposed by [12], where the audio signal is represented by a spectrogram, combined with a fingerprint extraction method based on the saliency maps of the audio signal's spectrogram. On the other hand, the correlation function is used to search the acoustic fingerprint in the audio database.

Figure 1. General diagram of the SSM algorithm.

3.1.1. Fingerprint Extraction

Fingerprint extraction is based on the SSM algorithm [12] and consists of three fundamental steps:

- Decreasing the resolution of the audio signal.
  Subsampling the signal means keeping each $N - th$ sample and eliminating the remaining samples. Compact Discs (CD), most FM radio stations, TV channels, and satellite TV all transmit stereo audio signals. The purpose of recording the sound in stereo is to recreate a more natural experience when listening to it. Although the term commonly refers to two-channel systems (left and right channels), it can also be applied to any system that uses more than one channel. On the other hand, the mono-aural sound is the one that is defined by a single channel. A mono-aural file requires half the space occupied by a stereo file, since it only contains one track, while a stereo file contains two (one for the signal on the left and one for the signal on the right). That is why the conversion from stereo to mono-aural sound is realized, as shown in Figure 2.

Figure 2. Conversion from stereo to mono-aural sound.

  As far as down-sampling is concerned, it refers to decreasing the frequency by the factor of an entire number, as shown in Figure 3.

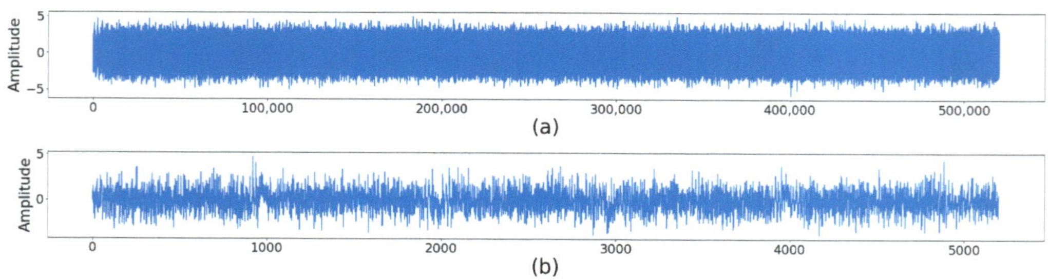

Figure 3. Downsampling: (**a**) Original Signal and (**b**) decreased frequency.

- Change the signal to the frequency domain.
  When the signal is changed to the frequency domain, a spectrogram is created, see Figure 4. A spectrogram consists of the graphic representation of the frequency spectrum or amplitude modulations and their variation over time. Usually, a spectrogram represents time on the horizontal axis, frequency on the vertical axis, and the amplitude is represented by gray-scale or colors. In this sense, a saliency map is a kind of

global feature that represents the most prominent visual regions of an image; that is, this mechanism filters the interesting information and ignores the irrelevant [27]. The spectrogram's creation consists of two fundamental steps: (1) *frame analysis* and (2) *selecting a window to choose the limited number of samples to process*. This window is a compromise between the size of the spectrogram, the process, and the signal analysis.

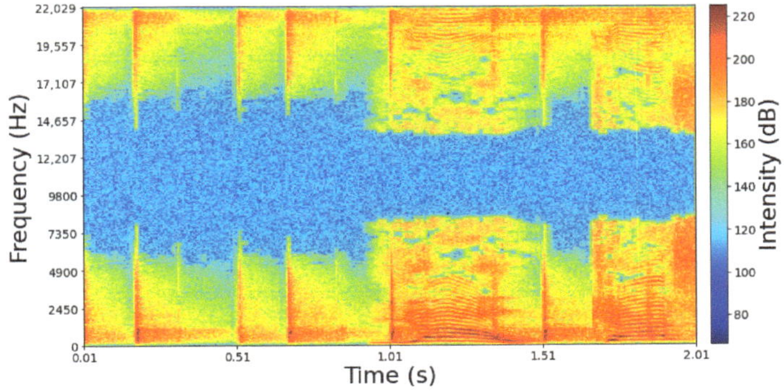

Figure 4. Example of a spectrogram.

- Extracting fingerprints.
  The saliency maps are used to extract the fingerprints. Figure 5 illustrates how they are generated. First, the image is decomposed into different channels (color, intensity, and orientation). Then, the main characteristics of each channel are extracted, and at the end, the features are added into a single image (saliency map).

Figure 5. Creation of saliency maps.

Figure 6 shows the fingerprint storage. First, the audio signal is fixed by from stereo to mono-aural, down-sampling, and dividing it into segments. Next, the signal is changed to the frequency domain by converting it into a spectrogram. Then, the spectrogram is treated as an image, and the saliency map is obtained, which will finally be saved in a database.

Figure 6. Storing of the fingerprints based on saliency maps.

3.1.2. Search Process

The search process consists of two sub-processes: (1) *extracting the fingerprint* and (2) *matching*. The extraction has been described previously, but instead of performing it from the complete audio, it is performed from segments of the query audio. Later, a matching process based on correlation is applied to compare the query track and the tracks stored on the database. See Figure 7.

Figure 7. Search or matching system.

The brute force search presented in [12] consists of comparing between unknown audio and possibly millions of fingerprints using correlation. Algorithm 1 presents the pseudo-code for this process. Its main operation is the correlation function named $corr()$, which is described by Algorithm 2.

Algorithm 1 Brute Force Search.

Require: I binary edge image
Input M_{DB}, M_{TT}
Output $M_{DB}[index], Corr_{max}$
  for each track $M_{DB}(i)$ **do**
    $C = corr(M_{DB}, M_{TT})$
    if $C > Corr_{max}$ **then**
      $Corr_{max} = C$
      $index = i$
      return $M_{DB}[index]$
    else
    end if
  end for

Algorithm 2 Correlation Function $corr()$.

Require: I binary edge image
Input $M_{DB}[][], M_{TT}[][]$
Output $M_{DB}[index], Corr_{max}$
  for $i < 32j < 32$ **do**
    $A_{temp} = A_{temp} + A[i][j]$
    $B_{temp} = B_{temp} + B[i][j]$
    $count = count + 1$
  end for
  for $i < 32j < 32$ **do**
    $A[i][j] = A[i][j] - A_{mean}$
    $B[i][j] = B[i][j] - B_{mean}$
  end for
  if $i < 32j < 32$ **then**
    $AB[i][j] = A[i][j] * B[i][j]$
    $A[i][j] = A[i][j] * A[i][j]$
    $B[i][j] = A[i][j] * B[i][j]$
  end if
  for $i < 32j < 32$ **do**
    $countAB = countAB + AB[i][j]$
    $countB = countB + A[i][j]$
    $countB = countB + B[i][j]$
  end for
  $CORR = countAB/sqrt(countA * countB)$
  return $CORR$

3.2. Sample Size Analysis

The analysis focuses on identifying the correlation between the song and its sample, including the amount of data in each data set register. Each saliency map is of a 32 × 32 size, and each datum is 8 bits long. The data set is made up of 33 songs of 10 seconds each (tracks $T1$, $T2$, ..., $T33$), where each song is divided into three parts: (1) the first 5 s (segment Tn_1), (2) the last 5 s (segment Tn_2) and (3) from second 2.5 to second 7.5 (segment Tn 3). For each segment, the saliency map is obtained, so the data set has 99 records or saliency maps, 5 s each segment. The 32 × 32 salience map can be observed as an array with indices (1:32, 1:32), the first index for rows, and columns. More details about the algorithm design and parameters are found in [12,13].

The hypothesis used to perform this analysis is that not all data from the 32 × 32 matrix are necessary. Therefore, the number of operations in the correlation process can be reduced, benefiting the search process and the architecture's performance. Table 1 shows the correlation between saliency maps of different sizes and one of three segments of the saliency map stored in the dataset. For example, $T1/T1\_3$ refers to Track 1 ($T1$) and segment 3 of Track 1 ($T1\_3$). On the other hand, the first column (16:32, 8:24) means that the saliency map sample is extracted from rows 16 to 32 and from columns 8 to 24 of the original to reduce the 32 × 32 original size map. In this way, the highest (cells in green) and lowest (cells in yellow) correlation indices are identified.

Figure 8 shows the regions where the three sample maps with the best success averages intersect. These are (8:32, 1:24) in green, (2:28, 4:28) in white, and (8:28, 4:24) in gray; all of them obtained from the complete map colored in the common region: (2:28,4:28).

After computing the average accuracy, it is determined that the slice of the main map (2:28, 4:28) obtains the same accuracy when comparing it with the complete map, 95.27% with the same data set, while other settings have lower results. For example, Figure 9 shows the complete map (32 × 32) on the left, and the smallest map (27 × 25) on the right is framed in yellow. Therefore, the accuracy is not affected while significantly reducing the number of required calculations, both temporal and spatial, as demonstrated with the implementation in the next section.

Table 1. Correlation between samples.

Window	T1/T_1_1	T1/T_1_2	T1/T_1_3	T2/T_2_1	T2/T_2_2	T2/T_2_3
(1:32, 1:32)	0.98630806	0.98500662	0.98745433	0.87798847	0.93625615	0.98640428
(1:16, 1:16)	0.98079073	0.98635301	0.98646834	0.87286297	0.89285985	0.98660328
(8:24, 8:24)	0.98538054	0.95933728	0.98853678	0.85405802	0.91543184	0.98890805
(16:32, 16:32)	0.9944561	0.98415072	0.99835338	0.98424574	0.9127887	0.99766129
(1:16, 16:32)	0.98226984	0.98417877	0.97647128	0.96208663	0.93775528	0.98172735
(16:32, 1:16)	0.9952343	0.99794383	0.99853883	0.97165405	0.97768967	0.99959274
(12:20, 12:20)	0.9798889	0.95821871	0.96536814	0.90881263	0.96118717	0.99199828
(16:32, 1:32)	0.99479573	0.98556999	0.99836617	0.97555804	0.93672673	0.99830267
(16:32, 8:24)	0.99396493	0.98424148	0.99830195	0.97834078	0.96649479	0.99798179
(1:16, 1:32)	0.98126411	0.97477707	0.98091944	0.79509093	0.90298747	0.97811535
(1:32, 1:16)	0.98589129	0.99090808	0.99086538	0.91667433	0.93281902	0.99136852
(8:32, 1:24)	0.99100237	0.98236547	0.99451482	0.90180133	0.94759758	0.99398419

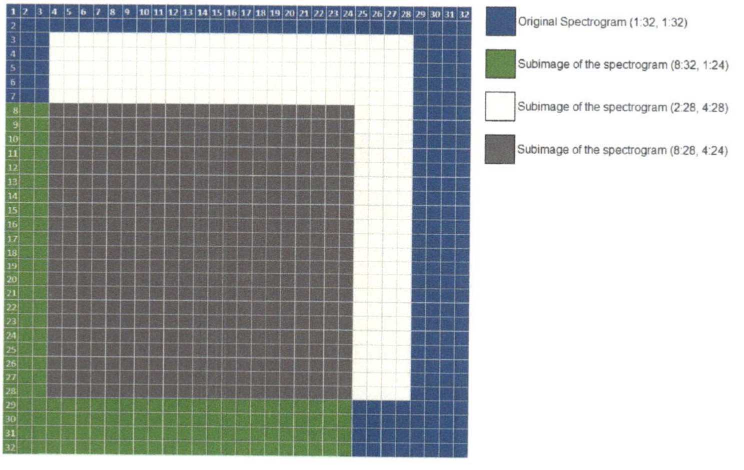

Figure 8. Intersection of selected saliency maps.

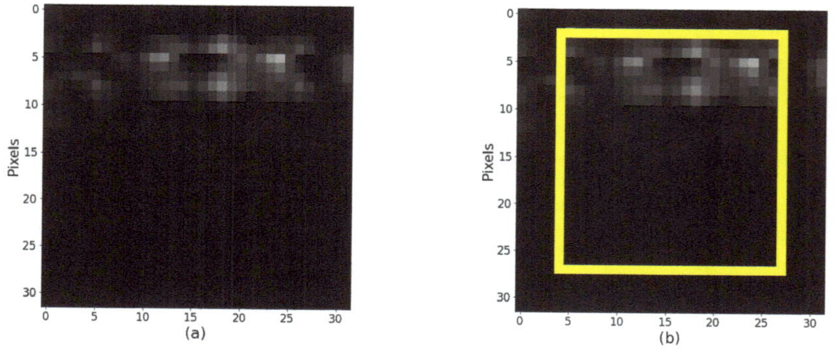

Figure 9. Results on maps according to the analysis: (**a**) original size of 32 × 32 pixels and (**b**) subimage (yellow box) with a smaller size of 27 × 25 pixels. Note: Saliency maps have few pixels, that is, they have low resolution.

A contribution of this work is demonstrating that most of the samples obtained from the SSM algorithm have relevant information in the center of the fingerprint, obtaining the same accuracy and reducing the searching time, with less computational complexity in software and hardware implementations.

4. Hardware Implementations

In this section, the design of hardware architectures for the *search* and the *correlation modules* is presented. The hardware designs of the architecture were implemented in the System Generator. It is essential to highlight that the two hardware proposals use brute-force search, although the second one utilizes an optimized process in the search based on its generation and storage.

4.1. Search by Brute Force

The implementation of the hardware architectures for the search consists of two deterministic finite automata. The first one is used to control the search and the second to perform the correlation. In addition, it uses a module that acts as a voting system under the condition of keeping the record of which saliency map has the highest correlation.

Figure 10 shows the complete architecture of the *Search System*: the *Signal Builder* that generates a signal, a block called *MCode* that performs the function of automaton, then two *counters* and two *ROMs*, where memory *A0* symbolizes the audio or unknown saliency map and *B0* the database. Finally, there are two modules. The first one was named *Correlation Function*, and the second one was named *Comparison*.

This proposed hardware architecture can be used for map sizes presented on the analysis, 32×32 and 27×25, with 1024 and 675 elements, respectively.

Figure 10. Proposed hardware architecture.

4.1.1. Correlation Function

The *Correlation Function* is one of the most critical modules; its output is essential for the system's functionality. In this case, the correlation function is the search criteria; its hardware architecture design is presented in Figure 11. The module of this function contains three inputs and one output. The first input corresponds to the start signal sent from the *MCode1* block, and the other two correspond to the used saliency maps. This function is composed of an automaton that coordinates and synchronizes the data and several subsystems (*Data Input, Mean, Multiplication, Accumulator*, and *Correlator)*, and blocks with different functionalities, such as store, accumulate, or multiply data. Additionally, two multiplexers reset the RAM values used in the first subsystem called *Data Input*, the output of which will be the saliency map.

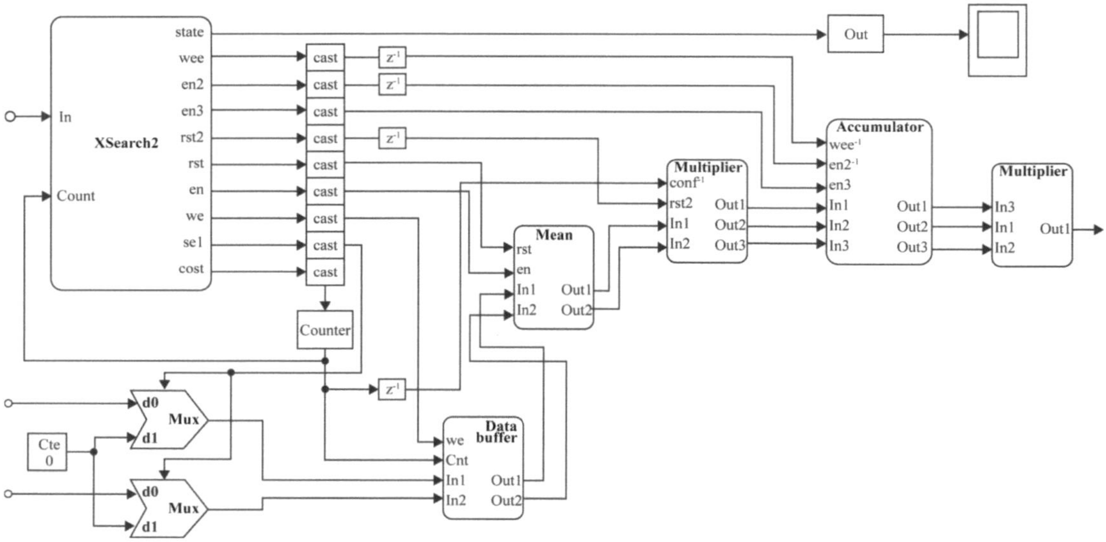

Figure 11. Block diagram for the *Correlation Function*.

DataInput is a data buffer, which stores data inputs according to the counter controlled by the Finite State Machine (FSM) and outputs the saliency maps. Then, in the *Mean* module, saliency maps are added and divided by their number of elements, obtaining the mean, subtracting them from each of the saliency map data (see Figure 12). This result will be the subsystem output and the entrance to the *Multiplication* subsystem, which performs the multiplication of the maps among themselves and with each other, having three outputs, that the *Accumulator* subsystem will return. Finally, the *Correlation Function* delivers one output datum that corresponds to the correlation value.

Figure 12. Block diagram for the *Mean* submodule.

4.1.2. Comparison Module

The second important module in the *Search System* is called *Comparison* (see Figure 11), which records the highest value that determines the corresponding audio on the database. The *Comparison* module comprises a relational block that decides whether the input datum is higher than the one stored in the record. If that is the case, it sends a signal indicating in which register the new datum should be saved.

To obtain the correlation value requires comparing two saliency maps according to (a) the audio that we are trying to determine if we have it stored and (b) each audio of the data set so that only a correlation value is generated in a given time. In this way, it is only necessary to store the audio data where the correlation is the highest. There are two possibilities in the *Comparison* module while comparing the new correlation value with the stored one. First, if the new value is higher than the stored one, the index is updated with the latter. Second, if the new correlation value is smaller or equal to the stored one, then it is unnecessary to update the data, and the previous index is maintained.

Both proposed architectures are constructed from the modules previously described, where the analysis of the saliency map size demonstrates that there are improvements in the optimized hardware architecture while maintaining the accuracy, which will be explained in the next section.

4.2. Optimized Brute-Force Search

It is important to highlight that both proposed architectures use brute-force search, and the analysis enables reducing computational complexity through modifying the map size. In this way, the optimized brute-force search uses the same design as the brute-force search previously described. The variant occurs when processing the saliency maps of size 27×25, i.e., processing 675 elements, only 65% of the original map. For example, Figure 9 shows two bars: the left one exemplifies the 1024 data contained in the 32×32 saliency map, representing 100% of the map, while the bar on the right represents only 65% of the data, delimited by the yellow box in the saliency map on the right.

The architecture is optimized in computational complexity since the analysis shows a similar performance when working with 27×25 maps than with 32×32 maps. Furthermore, this reduction in the map size decreases the number of floating-point operations, where 675 elements are now evaluated instead of 1024 elements. This improves the whole process because the algorithm must evaluate and analyze each fingerprint track within the database to find if it is registered. There is also an improvement concerning track storage, which is reflected when processing the entire set of tracks. More details of the advantages are found in the next section about the area, efficiency, and throughput.

5. Analysis and Results

In this section, the obtained results are presented, with a comparison between the brute-force and the optimized methods. The data obtained by each implemented module are reported separately, describing the number of required LUTs, FFs, and BRAMs, among other blocks. The designs are implemented using the Xilinx Vivado v2015.2 software in a xc7v2000tflg1925-1 FPGA. The operation of each design is observed, including the state of the state-machines, the percentage of correlation that exists between two maps, and which track has a higher correlation compared to the other tracks.

Table 2 shows the results of the hardware implementation for individual modules. It is observed that the *ROM* module uses approximately 77% of the total LUTs in the entire system because it is the input of all data used by the system. On the other hand, the block that uses most of the FFs is the Accumulator, a sub-block of the *Correlation* module, with 40% of the total FFs. In addition, the only segments that use RAM blocks are *Correlation_DataEntry* and *Correlation_Multiplication*, where all processed information is stored. Finally, the segment that consumes the most power is the *Correlation_DataInput*, followed by the *Counters*.

These results show that most of the resources are used for the storage of the track samples used to evaluate the system, see Table 2. In the future, it is necessary to examine RAM memories or external memories, where the first ones occupied specialized resources and the second ones increased the critical path. In the reported case, excluding the use of the ROMs allows for evaluating the hardware resource requirements of the other modules.

Table 2. Hardware implementations of the individual modules.

Module Name	LUT	FF	BRAM	DSP	Power (W)	Minimum Period (ns)
State Machine 1	22	2	0	0	0.636	1.335
Counters	2	26	0	0	0.712	1.408
ROM Memories	17,059	59	0	0	0.636	-
Correlation_Cast_Counter	2	25	0	0	0.66	1.300
Correlation_StateMachines	8	1	0	0	0.636	1.088
Correlation_DataEntry	0	0	2	0	0.756	2.183
Correlation_Mean	2798	128	0	0	0.636	40.244
Correlation_Multiplication	0	0	2	3	0.636	3.229
Correlation_Accumulator	998	192	0	0	0.636	11.131
Correlation_Correlation	1385	80	0	1	0.636	42.917
Comparison	8	16	0	0	0.642	2.195
Total	**22,282**	**529**	**4**	**4**	**7.222**	

Table 2 shows results of the implementation provided by the Vivado tool, but it is important to have metrics of the behavior of the complete architecture, which are described next. In order to compare the obtained results some equations are applied:

$$Throughput = bits\_of\_data\_block/(latency \times minimum\_period), \tag{1}$$

$$Processing\_time\_per\_track = latency \times minimum\_period, \tag{2}$$

$$Maximum\_frequency = 1/minimum\_period, \tag{3}$$

Table 3 shows a 4x difference in terms of LUTs between the search by brute force and the optimized search. Furthermore, the BRAMs are reduced by 50% in the optimized search; the number DSP and FFs are maintained, the period increases by 48%, and the power decreases by 65%. New architectures for the optimized search must be designed, which could increase the performance of this compact architecture.

Table 3. Results comparison of searching method.

Search	LUTs	FFs	BRAMs	DSP	Minimum Period (ns)	Power (W)	Performance (Mbps)	Efficiency (Kbps/LUT)
Brute Force	22,538	720	8	12	44.703	1.796	361.34	16.03
Optimized	5484	717	4	12	61.209	0.636	280.80	51.20

LUT: Look-up table, FF: flip flop, BRAM: Block random access memory, DSP: Digital signal processor, W: Watt.

Additionally, a software implementation was carried out in Matlab (running on Intel Core i7-7500U at 2.7 GHz, two cores, 16 GB SDRAM, and Windows 10) using the same algorithm used in the hardware implementations of the brute-force and the optimized searches, using saliency maps of 32 × 32 and 25 × 27, respectively, and a database with 33 tracks. As a result, the brute-force search took approximately 11.53 ms, while the optimized search was 9 ms, as shown in Table 4.

Table 4. Results comparison software–hardware.

	Hardware	Software
Brute-Force Search	2.99 ms	11.53 ms
Optimized Brute-Force Search	2.54 ms	9.00 ms

It can be observed that the hardware implementation is 3.85 times and 3.54 times faster than the software implementation for the brute-force and optimized search algorithms, respectively. This will change, however, when evaluating with hardware platforms of different specifications. Nevertheless, these results still represent valid reference values.

Until now, the designs, an optimization analysis, the implementations, and a comparison between the proposed architectures have been reviewed. Comparisons with related work are presented below.

6. Comparisons

In this section, a comparison with related works is presented in Table 5. The comparison is not equivalent, because different algorithms, models, processes, and FPGA technologies were used, but it still provides essential elements of evaluation about hardware architectures. Due to the diversity of the used platforms, the analysis is based on the works implemented in FPGAs.

Table 5. State-of-the-art comparison.

Work—Design Technique		Hardware Resources	Technique	Platform
[7]	Without Pipeline	2056 LUTs, 549 FF, 549 Slice Registers, 80 DSP48	MCLT	FPGA XC7VX330T-1FFG1157
	Pipeline	2056 LUTs, 1227 FF, 1672 Slice Registers, 80 DSP48		
[10]	Without Pipeline	7949 LUTs, 24800 FF, 11 BRAM, 25 DSP	FFT, 48 Filter Banks, Square Root	FPGA XC7A35T-1CPG236C
[28]	Embedded system (software)	32-bit finger-print, 60 MHz Laptop, Board	Random LSB coding	Device LPC2148 with ARM7 core on MCB2140 board
[29]	Software	Not provided	DWT, locally linear embedding	Not provided
This work—Without Pipeline		22538 LUTs, 720 FF, 8 BRAM, 12 DSP	Brute Force Search	FPGA XC7V2000T-FLG1925
		5484 LUTs, 717 FF, 4 BRAM, 12 DSP	Optimized Search	

FPGA: Field programmable gate array, LUT: Look-up table, FF: flip flop, BRAM: Block random access memory, DSP: Digital signal processor, FFT: Fast Fourier transform, MCLT: Modulated complex lapped transform.

Table 5 shows that some hardware implementations for audio fingerprinting have been proposed already. For example, Ref. [7] presents a fingerprinting system resistant to collision, based on a spread spectrum algorithm in the modulated complex lapped transform domain. In addition, authors in [28] present the implementation of the windowing, FFT, filter banks, and square root functions as parts of the feature extraction. Next, three points are described for comparison.

First, in the included works, three iterative architectures and one pipeline architecture are presented. In general, a pipeline architecture allows multiple blocks to be processed simultaneously, which should increase performance while at the same time increasing power consumption and operating frequency. On the other hand, non-pipelined architecture work iteratively across tracks, using fewer hardware resources, but reducing the throughput.

Second, considering the brute force search, the architecture proposed in this work requires considerable hardware resources. The proposed architecture stores different fingerprints (it stores the database of the fingerprints). The proposed optimized brute force search reduces the required amount of hardware resources because fewer blocks are required for storage and processing. The resource reduction is approximately 75.66%, an additional advantage to that mentioned in Section 4.2. According to Table 2, the database requires 17,059 LUTs, and the rest of the hardware architecture requires 22,282 − 17,059 = 5223 LUTs, which is similar to the related works. Related works [7,28] report that they had to create specialized modules for their architectures and operations, which is the same situation in our case: several different specialized modules were designed for the architecture proposed in this article. Additionally, the DSP and FF amounts are similar but show the consumption of state-of-the-art hardware resources.

Third, the throughput seems to affect the proposed optimized search technique. However, in the optimized case, the data amount of the map has to be 25 × 27 32-bit, single-precision floating-point numbers, that is, 25 × 27 × 32 = 21,600 bits, while the brute force implementation requires 32 × 32 × 32 bits = 32,768 bits. Additionally, if the difference of 2.54 ms (optimized search) versus 2.99 ms (brute-force search) is considered, then the amount of bits processed per unit of time is reflected in a lower throughput in the optimized search. It is important to highlight that the optimized search processes a small amount of data and requires a short processing time, reducing it from 2.99 ms to 2.54 ms and reporting an improvement in the processing time of 15.05%, which is reflected in the output for the user. Then, the optimized architecture is faster than the brute-force search architecture. However, it processes less data (the key point of the proposal), leading to faster processing of large sets of tracks to determine if the track in evaluation has been found.

7. Discussion

The analysis, design, and implementation of the hardware architectures gave a set of results and discussion elements, which will be described below. Three types of results and contributions can be highlighted:

- Proposals for both the non-optimized and optimized hardware architectures, in which specialized modules are designed to carry out the brute-force search and to correlate the saliency maps of the track sample with each saliency map of the stored map set. The correlation factor allows identifying, locating, and pointing to the index (address) with the highest correlation value (pointing to the ROM address) between the saliency maps of the query input with some saliency map in the set of maps in the ROMs.
- An analysis focused on the sample size, where the saliency maps are reduced, requiring fewer pixels and decreasing the computational complexity. That is, fewer operations are performed in the correlation, only 65% of the pixels are stored, and fewer hardware resources are required. This was achieved without affecting the average accuracy of 95.27%. Therefore, the analysis leads to the design and implementation of an optimized hardware architecture that improves various parameters. However, future research is still necessary to reduce the critical path and, consequently, to improve performance and efficiency.
- Finally, two comparative analyses were conducted. The first one focused on the optimized and non-optimized architectures, whose results are adequate according to their design, platform, and architecture. The second comparative analysis focused on examining architectures proposed in related works. Although different platforms, devices, and architectures were used in previous works, this comparison still enables reporting reference values. On the one hand, the first comparative analysis shows a reduction in hardware resources, such as LUTs, FFs, and RAMs. This is because fewer data in the database or set of saliency maps must be stored, requiring only 76.55% of the LUTs. On the other hand, the second analysis shows that there are various algorithms that can be implemented for audio fingerprinting, and that our proposed architecture has a competitive consumption of hardware resources according to the optimized version.

Two limitations of the proposed architectures are described next. The first limitation focuses on the critical path time, where the throughput and efficiency of the optimized hardware architecture are affected because they depend on the minimum period or critical path time (this determines maximum frequency). Related works show that it is necessary to explore other design techniques such as pipelining for improving throughput (efficiency will be consequently improved because it is related to the throughput). This technique generally reduces the critical path and increases the processing capacity, thus improving the throughput which is one of the metrics that must be increased in our optimized architecture. The second limitation is found in the growth of the dataset. Since the resources of the FPGA are limited, large size tracks cannot be stored in this device, so other alternatives such as embedded RAM memory and external RAM memories must be explored.

8. Conclusions

Three main contributions are presented in this paper. The first one is the hardware implementation of a fingerprint extraction algorithm, which has two searching versions depending on their storage: (1) the *brute-force search* and (2) the *optimized brute-force search*. This means that the searching module (generation and storing of the fingerprinting) is different for both versions. Second, this paper presents the analysis that allows reducing the window size and optimizing the storage and, consequently, the search module. Third, two comparative analyses for reference are described, using our hardware architectures and related works for evaluating different metrics and showing advantages (fewer hardware resources and operations to execute) and disadvantages (larger critical path and fewer throughput).

It is demonstrated that the results of a 32×32 map and a reduced map of 27×25 have similar accuracy, errors, and success rates. Furthermore, with the reduction in the saliency map, both the number of operations and the storage space are decreased. In the hardware implementation, ROM blocks are reduced to 50%, and the number of clock cycles decreases. In general, fewer resources are used, less power is consumed, and there is a decrease in processing times. Furthermore, the hardware implementation is approximately 3.85 times faster than the software implementation, where the software can hardly be improved. In contrast, there are still several techniques that can be used to improve the hardware implementation, such as increasing parallelization and improving the individual modules' design, among others.

Future work will focus on using other hardware design techniques such as pipelining, which it is expected to improve the throughput, by reducing the critical path time. Pipelining accomplishes this time requirement, in addition to increasing the processing capacity.

Author Contributions: Conceptualization, I.A.-B., B.S.-J., and C.F.-U.; methodology, I.A.-B. and B.S.-J.; validation, B.S.-J., F.L.-H., and J.J.E.-L.; formal analysis, I.A.-B. and K.A.R.-G.; investigation, I.A.-B., B.S.-J., and C.F.-U.; resources, C.F.-U.; writing—original draft preparation, I.A.-B., B.S.-J., K.A.R.-G., and C.F.-U.; writing—review and editing, I.A.-B., K.A.R.-G., F.L.-H., and J.J.E.-L.; visualization, K.A.R.-G. and C.F.-U.; supervision, I.A.-B.; funding acquisition, F.L.-H. and J.J.E.-L. All authors have read and agreed to the published version of the manuscript.

Funding: This research received no external funding.

Institutional Review Board Statement: Not applicable.

Informed Consent Statement: Not applicable.

Data Availability Statement: Not applicable.

Acknowledgments: Authors express their gratitude to the Mexican National Council for Science and Technology (CONACYT) for enabling this work through the Research Project 882.

Conflicts of Interest: The authors declare no conflict of interest.

Abbreviations

The following abbreviations are used in this manuscript:

A/D	Analog to Digital
bps	Bits per second
BRAM	Block RAM
BRMA	Block-Recursive Matching Algorithm
CCMF	Cepstral Coefficients in Mel Frequencies
CD	Compact Disc
D/A	Digital to Analog
DSP	Digital Signal Processor
DWT	Discrete Wavelet Transform
FF	Flip Flop
FFT	Fast Fourier Transform
FM	Frequency Modulation
FPGA	Field Programmable Gate Array
FSM	Finite State Machine
GTCC	GammaTone Cepstral Coefficients
LUT	Look-Up Table
MCLT	Modulated Complex Lapped Transform
ms	Millisecond
ns	Nanosecond
RAM	Random-Access Memory
ROM	Read-Only Memory
SDC	Shifted Delta Coefficients
SSM	Spectrogram Saliency Maps
SVD	Singular Value Decomposition
TV	Television
W	Watt

References

1. Li, H.; Jain, S.; Kannan, P.K. Optimal Design of Free Samples for Digital Products and Services. *J. Mark. Res.* **2019**, *56*, 419–438. [CrossRef]
2. Megías, D.; Kuribayashi, M.; Qureshi, A. Survey on Decentralized Fingerprinting Solutions: Copyright Protection through Piracy Tracing. *Computers* **2020**, *9*, 26. [CrossRef]
3. Becker, E.; Buhse, W.; Günnewig, D.; Rump, N. *Digital Rights Management: Technological, Economic, Legal and Political Aspects Lecture Notes in Computer Science 2770*; Springer: Cham, Switzerland, 2004; pp. 93–100.
4. Bhat, V.; Sengupta, I.; Das, A. An adaptive audio watermarking based on the singular value decomposition in the wavelet domain. *Digit. Signal Process.* **2010**, *20*, 1547–1558. [CrossRef]
5. Chen, N.; Xiao, H.-D. Perceptual audio hashing algorithm based on Zernike moment and maximum-likelihood watermark detection. *Digit. Signal Process.* **2013**, *23*, 1216–1227. [CrossRef]
6. Lebossé, J.; Brun, L.; Pailles, J.C. A robust audio fingerprint extraction algorithm. In Proceedings of the Fourth Conference on IASTED International Conference: Signal Processing, Pattern Recognition, and Applications (SPPR'07), Innsbruck, Austria, 14–16 February 2007; pp. 269–274.
7. Garcia-Hernandez, J.J.; Gomez-Ricardez, J.J. Hardware architecture for an audio fingerprinting system. *Comput. Electr. Eng.* **2019**, *74*, 210–222. [CrossRef]
8. Leonhard, J.; Louërat, M.-M.; Aboushady, H.; Sinanoglu, O.; Stratigopoulos, H.-G. Mixed-Signal Hardware Security Using MixLock: Demonstration in an Audio Application. In Proceedings of the International Conference on Synthesis, Modeling, Analysis and Simulation Methods and Applications to Circuit Design (SMACD), Laussane, Switzerland, 15–18 July 2019; pp. 185–188.
9. Chen, D.; Mao, X.; Qin, Z.; Wang, W.; Li, X.-Y. Wireless Device Authentication Using Acoustic Hardware Fingerprints. In Proceedings of the Big Data Computing and Communications, BigCom 2015, Lecture Notes in Computer Science, Taiyuan, China, 1–3 August 2015; Volume 9196, pp. 193–204._16. [CrossRef]
10. Janakiraman, S.; Thenmozhi, K.; Rayappan, J.B.B.; Amirtharajan, R. Audio Fingerprint Indicator in Embedded Platform: A Way for Hardware Steganography. *J. Artif. Intell.* **2014**, *7*, 82–93. [CrossRef]
11. Martínez, J.I.; Vitola, J.; Sanabria, A.; Pedraza, C. Fast parallel audio fingerprinting implementation in reconfigurable hardware and GPUs. In Proceedings of the 2011 VII Southern Conference on Programmable Logic (SPL), Cordoba, Argentina, 13–15 April 2011; pp. 245–250.

12. Guzman-Zavaleta, Z.J.; Feregrino-Uribe, C.; Menendez-Ortiz, A.; Garcia-Hernandez, J.J. A robust audio fingerprinting method using spectrograms saliency maps. In Proceedings of the 9th International Conference for Internet Technology and Secured Transactions (ICITST–2014), London, UK, 8–10 December 2014; pp. 47–52.
13. Guzman-Zavaleta, Z.J. An Effective and Efficient Fingerprinting Method for Video Copy Detection. Ph.D. Thesis, Instituto Nacional de Astrofísica, Óptica y Electrónica, Puebla, Mexico, 2017.
14. Nimo, E.S. *Detection and Identification of Radio and Television Ads in Real Time*; Technical Report; Universidad de Sevilla: Sevilla, Spain, 2007.
15. Iffath, F.; Kayes, A.S.M.; Rahman, M.T.; Ferdows, J.; Arefin, M.S.; Hossain, M.S. Online Judging Platform Utilizing Dynamic Plagiarism Detection Facilities. *Computers* **2021**, *10*, 47. [CrossRef]
16. Saul, S.; Daniel, W.; Alex, A. Winnowing: Local Algorithms for Document Fingerprinting. In Proceedings of the ACM SIGMOD International Conference on Management of Data, San Diego, CA, USA, 9–12 June 2003.
17. Jaap, H.; Ton, K. A Highly Robust Audio Fingerprinting System. In Proceedings of the 4th International Conference on Music Information Retrieval, Baltimore, MD, USA, 26–30 October 2003.
18. Cano, P.; Batlle, E.; Kalker, T.; Haitsma, J. A review of audio fingerprinting. *J. VLSI Signal Process. Syst.* **2005**, *41*, 271–284. [CrossRef]
19. Malekesmaeili, M.; Ward, R.K. A local fingerprinting approach for audio copy detection. *Signal Process.* **2014**, *98*, 308–321. [CrossRef]
20. Rincón, E.G. Audio Segmentation through Chromatic Features in News Files. Ph.D. Thesis, Universidad Autónoma de Madrid, Madrid, Spain, 2015.
21. Wang, A.L.-C. An industrial-strength audio search algorithm. In Proceedings of the 4th International Conference on Music Information Retrieval, Baltimore, MD, USA, 26–30 October 2003.
22. Patil, V.H. *Data Structures Using C++*; Oxford University Press: Oxford, UK, 2012; pp. 420–527.
23. Kurth, F. A ranking technique for fast audio identification. In Proceedings of the IEEE Workshop on Multimedia Signal Processing, St. Thomas, VI, USA, 9–11 December 2002; pp. 186–189.
24. Doets, P.J.O. Modeling Audio Fingerprints: Structure, Distortion, Capacity. Ph.D. Thesis, Electrical Engineering, Mathematics and Computer Science, Technische Universiteit Delf, Delft, The Netherlands, 2010.
25. Leighton, M.J.; Ruml, W.; Holte, R.C. Faster Optimal and Suboptimal Hierarchical Search. In Proceedings of the 4th Annual Symposium on Combinatorial Search (SoCS 2011), Catalonia, Spain, 15–16 July 2011.
26. Miller, M.L.; Rodriguez, M.A.; Cox, I.J. Audio fingerprinting: Nearest neighbor search in high dimensional binary spaces. *J. VLSI Signal Process. Syst.* **2005**, *41*, 285–291. [CrossRef]
27. Borji, A.; Cheng, M.; Jiang, H.; Li, J. Salient object detection: A benchmark. *IEEE Trans. Image Process.* **2015**, *24*, 5706–5722. [CrossRef] [PubMed]
28. Hervás, M.; Alsina-Pagès, R.M. An FPGA Platform Proposal for Real-Time Acoustic Event Detection: Optimum Platform Implementation for Audio Recognition with Time Restrictions. *Proceedings* **2017**, *1*, 2. [CrossRef]
29. Jia, M.; Li, T.; Wang, J. Audio Fingerprint Extraction Based on Locally Linear Embedding for Audio Retrieval System. *Electronics* **2020**, *9*, 1483. [CrossRef]

technologies

MDPI

Article

Extraction and Characterization of β-Viginin Protein Hydrolysates from Cowpea Flour as a New Manufacturing Active Ingredient

Taline S. Almeida [1,2], Caio A. da Cruz Souza [3], Mariana B. de Cerqueira e Silva [3], Fabiana P. R. Batista [3], Ederlan S. Ferreira [3], André L. S. Santos [4], Laura N. Silva [4], Carlisson R. Melo [1,2], Cristiane Bani [5], M. Lucia Bianconi [6], Juliana C. Cardoso [1,2], Ricardo L. C. de Albuquerque-Júnior [1,2], Raquel de Melo Barbosa [7], Matheus M. Pereira [8], Eliana B. Souto [9,10,*], Cleide M. F. Soares [1,2] and Patrícia Severino [1,2,*]

1 Institute of Technology and Research (ITP), Av. Murilo Dantas, 300, Aracaju 49010-390, Brazil; talinealmeida2009@hotmail.com (T.S.A.); carlisson_melo@hotmail.com (C.R.M.); juliana.cordeiro@souunit.com.br (J.C.C.); ricardo_albuquerque@unit.br (R.L.C.d.A.-J.); cleide18@yahoo.com.br (C.M.F.S.)
2 University of Tiradentes (Unit), Av. Murilo Dantas, 300, Aracaju 49010-390, Brazil
3 School of Pharmacy, Federal University of Bahia, Salvador 40170-115, Brazil; caioacs1@gmail.com (C.A.d.C.S.); marianabarros.cs@gmail.com (M.B.d.C.e.S.); fabianaprb@gmail.com (F.P.R.B.); ederlan.ferreira@ufba.br (E.S.F.)
4 Instituto de Microbiologia Paulo de Góes (IMPG), Universidade Federal do Rio de Janeiro (UFRJ), Rio de Janeiro 21941-902, Brazil; andre@micro.ufrj.br (A.L.S.S.); lauransilva@gmail.com (L.N.S.)
5 Department of Morphology, Federal University of Sergipe, Aracaju 49100-000, Brazil; crisbani@gmail.com
6 Instituto de Bioquímica Médica Leopoldo de Meis, Universidade Federal do Rio de Janeiro, Rio de Janeiro 21941-902, Brazil; bianconi@bioqmed.ufrj.br
7 Department of Pharmacy, Federal University of Rio Grande do Norte, R. Gen. Gustavo Cordeiro de Faria, S/N-Petrópolis, Natal 59012-570, Brazil; m.g.barbosafernandes@gmail.com
8 CICECO-Aveiro Institute of Materials, Departament of Chemistry, University of Aveiro, 3810-193 Aveiro, Portugal; matheus.pereira@ua.pt
9 Department of Pharmaceutical Technology, Faculty of Pharmacy, University of Porto, Rua de Jorge Viterbo Ferreira, 228, 4050-313 Porto, Portugal
10 REQUIMTE/UCIBIO, Faculty of Pharmacy, University of Porto, Rua de Jorge Viterbo Ferreira, 228, 4050-313 Porto, Portugal
* Correspondence: ebsouto@ff.up.pt (E.B.S.); patricia_severino@itp.org.br (P.S.)

Citation: Almeida, T.S.; da Cruz Souza, C.A.; de Cerqueira e Silva, M.B.; Batista, F.P.R.; Ferreira, E.S.; Santos, A.L.S.; Silva, L.N.; Melo, C.R.; Bani, C.; Bianconi, M.L.; et al. Extraction and Characterization of β-Viginin Protein Hydrolysates from Cowpea Flour as a New Manufacturing Active Ingredient. *Technologies* **2022**, *10*, 89. https://doi.org/10.3390/technologies10040089

Academic Editors: Manoj Gupta, Eugene Wong and Gwanggil Jeon

Received: 28 May 2022
Accepted: 12 July 2022
Published: 21 July 2022

Abstract: The increased mortality rates associated with antibiotic resistance has become a significant public health problem worldwide. Living beings produce a variety of endogenous compounds to defend themselves against exogenous pathogens. The knowledge of these endogenous compounds may contribute to the development of improved bioactive ingredients with antimicrobial properties, useful against conventional antibiotic resistance. Cowpea is an herbaceous legume of great interest due to its high protein content and high productivity rates. The study of genetic homology of vicillin (7S) from cowpea (*Vigna unguiculata* L.) with vicilins from soybean and other beans, such as adzuki, in addition to the need for further studies about potential biological activities of this vegetable, led us to seek the isolation of the vicilin fraction from cowpea and to evaluate the potential in vitro inhibitory action of pathogenic microorganisms. The cowpea beta viginin protein was isolated, characterized, and hydrolyzed in silico and in vitro by two enzymes, namely, pepsin and chymotrypsin. The antimicrobial activity of the protein hydrolysate fractions of cowpea flour was evaluated against *Staphylococcus aureus* and *Pseudomonas aeruginosa*, confirming the potential use of the peptides as innovative antimicrobial agents.

Keywords: *Vigna unguiculata* L.; vicilin; cowpea bean; antimicrobial peptides; fibroblasts cell line

1. Introduction

The average mortality caused by bacterial resistance to antibiotics is expected to reach about 10 million people by 2050 globally, but with a higher rate in sub-Saharan African countries due to limited access to viable drugs [1]. In countries of higher income, antibiotic resistance results from the intensive and/or inappropriate use of antimicrobial drugs, triggering prominent multiresistant microorganisms. In addition, the report of new cases of resistance has been higher than the number of new drug substances with antibiotic activity that are being launched on the market [2].

Living beings produce a variety of substances against invasive pathogens. Improving the bioactive compounds from these organisms is therefore of great interest in searching for promising alternatives over conventional antibiotic drugs. In recent years, legume proteins have gained prominence, mainly due to the rapid expansion of knowledge about their bioactive peptides [3].

Peptides derived from legume seed proteins have been described to show various biological activities in vitro and in vivo, namely, with effects on the control of hunger [4], on the cardiovascular system [5,6], on inflammatory processes [7], cancer [8], and also with antimicrobial activity [9,10]. Antimicrobial peptides (AMPs) are a new class of biopharmaceuticals widely studied as important therapeutic alternatives [11].

AMPs are small molecules, playing an essential part of the defense system from various plant species [12]. These peptides are considered multifunctional since they show antibacterial, antifungal, antiparasitic, and antiviral properties, in addition to some antitumoral activity, capacity for insulin release, and immunomodulatory response mediated by cytokines [13]. Recently, several studies describe the action of peptides obtained from the digestion of legume proteins with antimicrobial properties [9,10]. There is growing interest in the production of food protein hydrolysates for potential therapeutic applications. Therefore, the ability of proteins to generate peptides during their gastrointestinal digestion, with favorable characteristics to reach the bloodstream, has been considered a fundamental condition [14]. Furthermore, the knowledge about the sequence of peptides is instrumental to understand the correlation between the composition of hydrolysates and their biological activity [14].

Peptides from the leguminous species *Vigna unguiculata* (L.) Walp, popularly known as cowpea, can be a natural and abundant source for the commercial production of antimicrobial peptides [15], since their dried seeds are a valuable source of proteins. The protein content of the seeds varies from 20 to 35%, with 7S globulins being the primary reserve proteins of cowpea [16,17]. Cowpea also has a genetic similarity with soybeans (64%) and adzuki beans (81%), and studies have shown that these legumes have significant antimicrobial, anticancer, antidiabetic, antioxidant, and hypocholesterolemic activities. In addition, *V. unguiculata* stands out for its low production cost, high nutritional value, and is abundant in the northeast region of Brazil [18]. The present study aimed to identify and characterize the beta vignin protein and its protein hydrolysates from cowpea with antimicrobial potential for use in pharmaceutics.

2. Materials and Methods

2.1. Materials

Cowpea (*Vigna unguiculata*, L. Walp) seeds were kindly providedby Dr. RogérioFaria Vieira (Agricultural Research Company from Minas Gerais at the Federal University of Viçosa, Minas Gerais, Brazil). All others reagents were bought fromSigma-Aldrich (San Luis, MO, USA).

2.2. Preparation of Cowpea Flour

The seeds were selected and soaked in distilled water at 4 °C for 12 h. Afterward, the seeds were dried in an oven at 50 °C for 12 h and powdered to 60 mesh size. The flour was stored at 4 °C and used for protein extraction.

2.3. Isolation of Cowpea β-Vignin

The β-vignin was isolated according to the methods described by Ferreira et al. (2015) [19]. Aliquots of the isolated protein (80 mg of β-vignin) were solubilized in potassium phosphate solution (0.05 M) at pH 7.5, NaCl (0.5 M), and sodium azide (0.01%), for a Sepharose CL-6B column (1.0 cm × 100 cm) with filtration. The flow rate was 0.45 mL/min, and the protein elution was monitored by measuring the absorbance at 280 nm. The major fraction (peak tube) was dialyzed, precipitated, and lyophilized. The protein concentration was determined as described by Lowry et al. (1951) [20].

2.4. Gel Electrophoresis

Samples of total protein extract and β-vignin isolated by chromatography were analyzed by one-dimensional sodium dodecyl sulfate–polyacrylamide gel electrophoresis (SDS–PAGE), as described by Laemlli (1970) [21], using an electrophoresis system Hoefer-MiniVE (Amersham Biosciences®, Hercules, CA, EUA). Aliquots of 10 µg of the sample were applied to the gel, and low molecular weight proteins (between 97 and 14.4 kDa) were used as molecular markers (GE Healthcare®, Little Chalfont, UK). The gel images were analyzed by the software AlphaEase® (Alpha Innotech, San Leandro, CA, USA).

2.5. Enzymatic Hydrolysis and Fractionation

β-vignin was hydrolyzed in vitro according to the procedure described by Akeson and Stahman (1964) [22]. Enzymatic hydrolysis was performed with pepsin and chymotrypsin. To perform pepsin (EC 34231) hydrolysis, a 1:66 enzyme/substrate ratio was used at 37 °C for 3 h (pH = 2.0); for chymotrypsin (EC3421), a 1:25 enzyme/substrate ratio was used at 37 °C for 3 h (pH 7.0). The total hydrolyzed extract was ultrafiltered from >30 kDa to >3 kDa (MWCO) using ultrafiltration membrane filters (Merck® Millipore, Darmstadt, Germany).

2.6. High-Performance Liquid Chromatography

The chromatographic profiles of the hydrolyzate of the fractions containing 30–10 kDa and 10–3 kDa peptides were determined by high-performance liquid chromatography (HPLC) using a PerkinElmer system with a reversed-phase column (C18 × 0.45 cm × 25 cm) and a UV/VIS detector (HPLC, PerkinElmer system, Waltham, MA, USA). The gradient was used for 10 min at 95% A and 50 min to reach 25% B. The solvent system comprised 0.045% trifluoroacetic acid in ultrapure water (A) and 0.036% trifluoroacetic acid in acetonitrile (B), with a flow rate of 1.0 mL/min at temperature of 30 °C. Readings were recorded at 220 nm.

2.7. In Silico Screening of Peptides with Antimicrobial Properties

The primary sequences of β-vignin (NCBI/GenBank Blast: AM905848 and UniProtKB: A8YQH5_VIGUN), adzuki bean 7S globulin (NCBI/GenBank Blouse: AB292246.1; UniProtKB: A4PI98_PHAAN), and α subunit of soybean β-conglycinin (NCBI)/GenBank Blast: AY221105.1; UniProtKB: UniProtKB: GLCAP_SOYBN) were compared with protein modeling software [23]. Then, the β-vignin primary sequence was virtually hydrolyzed by the sequential action of the enzymes pepsin (EC 3.4.23.1) and chymotrypsin (EC 3.4.21.1), as available on the BIOPEP server [24]. Subsequently, the probability of bioactivity of β-vignin-derived peptides was analyzed according to the physicochemical characteristics that were presented.

2.8. Minimum Inhibitory Concentration

Minimum inhibitory concentration (MIC) assays for the peptides were performed using the microdilution technique following the National Committee for Clinical Laboratory Standards guidelines [25]. Strains of the *Staphylococcus aureus* (ATCC 25923) and *Pseudomonas aeruginosa* (ATCC27853) were used. Colonies were harvested and resuspended to 1.5×10^8 CFU/mL (turbidity equivalent to 0.5 McFarland standard scale). Samples of hydrolyzed proteins and total proteins were diluted in dimethyl sulfoxide at concentrations ranging from 1.0 through 0.0019531 mg/mL, and added to Mueller–Hinton broth (Merck, Darmstadt, Germany). The negative control was 0.1 mL of Mueller–Hinton broth, and the positive control was ciprofloxacin (Merck, Darmstadt, Germany). Plates were incubated at

37 °C for 24 h. At the end of the incubation time, MIC was visually identified as the lowest concentration of the test compound that inhibits visible growth.

2.9. Agar Disk Diffusion Method

The agar disk diffusion method was carried out according to the National Committee for Clinical Laboratory Standards guidelines [26]. Bacterial suspensions were cultured in Mueller–Hinton broth for 24 h at 35 °C, standardized in sterile saline solution (0.9%) at a concentration of 10^8 CFU/mL, a 0.5 McFarland standard. The strains were sown with the sterile swab. After 10 min, three holes were made on the surface of the inoculated medium using light pressure, and the peptides were inserted according to their minimum inhibitory concentration values. The plates were incubated in a bacteriological oven at 37 °C for 24 h. The antimicrobial activity results of the tested sample were expressed through the diameter size of the inhibition halo.

2.10. Cell Viability in L929 Cell-Line

The human fibroblast line-L929 was used for the colorimetric method using methylthiazolyl-tetrazolium (MTT assay), following the ISO 10993-5 (2009) guidelines [27]. L929 cells were seeded in 96-well culture plates (2×10^4 cells/well). A solution of MTT was placed in contact with the cells, and then incubated at 37 °C for 3 h. After removal of the MTT, dimethyl sulfoxide was placed for solubilization of the tetrazole salt crystals. Then, the optical density reading was performed on an automated plate reader at 570 nm wavelength. The tests were conducted in quadruplicate and then normalized [28]. The results are expressed as a relative percentage of cell viability compared to the control, calculated by applying the following equation:

$$CV\ (\%) = \frac{Ab_{sample}}{Ab_{control}} \times 100$$

where CV (%) is the percentage of cell viability, Ab_{sample} is the absorbance recorded for the sample, and $Ab_{control}$ is the absorbance recorded for the control.

3. Results

The spectrochromatographic profile of the cowpea 7S protein isolate on the Sepharose CL 6B column showed a single absorbance peak in tube 78 (Figure 1). Electrophoresis indicated that 7S globulins represent the major proteins of cowpea, being composed of bands corresponding to two significant polypeptides (55 to 60 kDa). Other smaller bands were registered by polyacrylamide gel electrophoresis of the protein fractions (Figure 2). The recorded data ensured the identification and quantification the β-vignin protein, resulting in a 50% yield of the 7S fraction of cowpea extract.

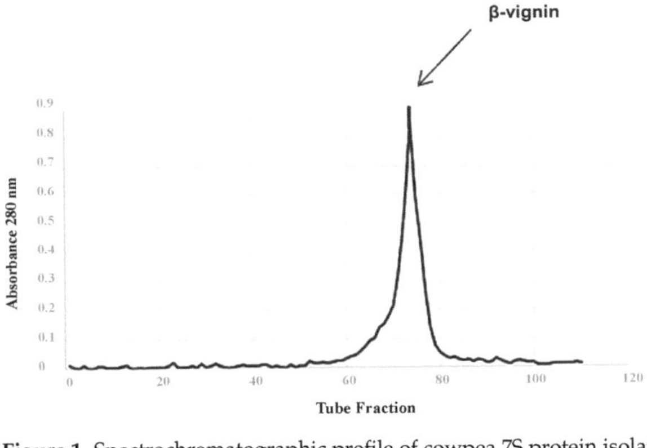

Figure 1. Spectrochromatographic profile of cowpea 7S protein isolate on a Sepharose CL 6B column.

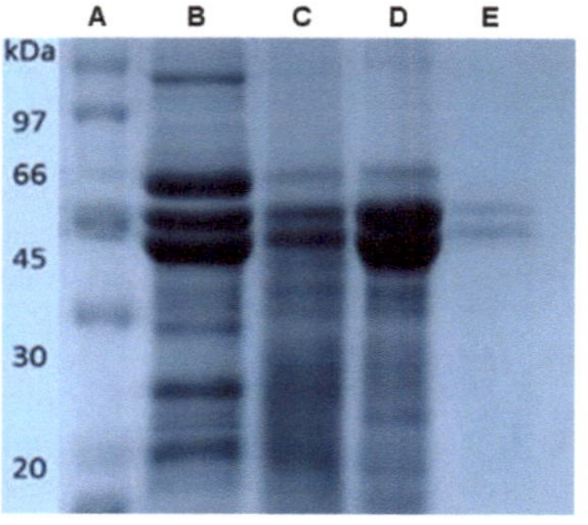

Figure 2. Polyacrylamide gel electrophoresis of protein fractions from cowpea. The columns represent the protein marker (**A**), the total protein extract (**B**), the defatted and purified beta vignin (**C**), and finally, the enzymatic hydrolysis (**D**,**E**).

The chromatographic profiles of the protein hydrolyzate of the 30–10 kDa and 10–3 kDa fractions of chymotrypsin (A and B) and in vitro pepsin hydrolyzate (C and D) of beta vignin are shown in Figure 3. The registered peaks are the indication of peptides present in samples under study. Although many peptides were produced by this protocol (Table 1), few significant peaks were observed in the 30–10 kDa hydrolysates for both enzymes. There was higher peak intensity in this case, but the retention times were the same. For hydrolyzates < 30 kDa, the test was not performed since the column would not support this molecular weight, causing the clogging of the column.

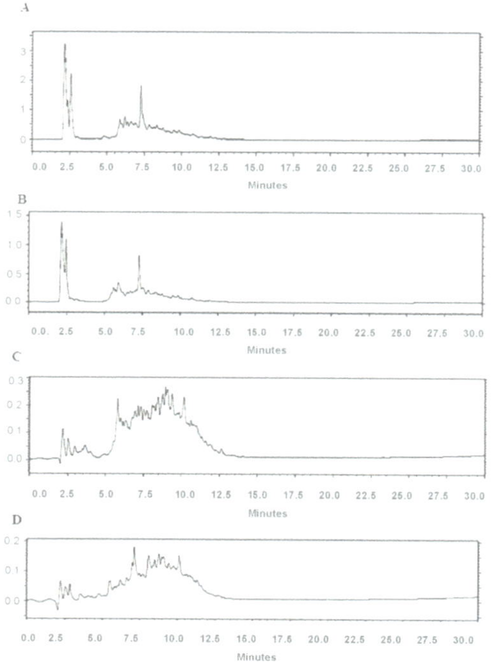

Figure 3. Chromatogram of RP-HPLC chymotrypsin hydrolyzate of the fraction containing 30–10 kDa peptides (**A**); hydrolyzate of the fraction containing 10–3 kDa smaller peptides (**B**); pepsin hydrolyzate fraction containing 30–10 kDa smaller peptides (**C**), and hydrolysate fraction containing smaller peptides 10–3 kDa (**D**).

Table 1. Quantification of the fraction consisting of peptides larger than 30, 30–10, and 10–3 kDa.

Peptides Fractions	Chymotrypsin (mg/mL)	Pepsin (mg/mL)
>30 kDa	33.7	45.0
30–10 kDa	2.48	3.9.0
10–3 kDa	2.48	5.39

The computer simulation of enzymatic hydrolysis performed on cowpea beta vignin produced hundreds of peptide fragments. Most predicted bioactive peptides (Tables 2 and 3) have between 10 and 30 amino acid residues, molecular mass from 1 to 5 kDa, with hydrophobicity values ranging from 0.9 to −0.1 kcal, positive charge ranging from +2 to 4.1, and predominance of the hydrophobic residues, namely, phenylalanine, tyrosine, or leucine. According to the simulation, the peptides that presented the most favorable characteristics and prediction for antimicrobial activity were those numbered 7 and 21 (Table 2) for protein hydrolyzates with chymotrypsin, and peptides numbered 19 and 21 (Table 3) for protein hydrolyzates with pepsin.

Table 2. Screening prediction of β-vignin-derived peptides from cowpea hydrolyzed with chymotrypsin in silico and their characteristics of isoelectric point, molecular mass, and high-performance chromatography time.

Peptide	Localization	Molecular Mass	Charge	Hydrophobicity	Ip	Sequence
1	1–10	1083.72	0.0	−1.5	5.70	VPLLLLGVLF
2	11–18	823.46	0.0	−0.9	5.70	LASLSVSF
3	19–41	2632.22	−1.8	0.6	5.52	GIVHRGHQESQEESEPRGQNNPF
4	44–48	678.28	−1.0	1.2	3.71	DSDRW
7	**58–66**	**1125.66**	**2.1**	**−0.2**	**12.50**	**GHLRVLQRF**
8	67–79	1635.81	0.0	0.6	6.57	DQRSKQIQNLENY
9	80–84	649.37	0.0	0.1	6.36	RVVEF
10	85–101	1903.97	−0.8	0.0	6.18	QSKPNTLLLPHHADADF
11	102–123	2398.35	0.0	−0.3	6.05	LLVVLNGRAILTLVNPDGRDSY
12	124–139	1698.88	0.0	−0.3	7.20	ILEQGHAQKTPAGTTF
13	141–165	2923.56	0.2	0.1	7.37	LVNHDDNENLRIVKLAVPVNNPHRF
14	170–179	1113.51	−1.0	−0.1	3.85	LSSTEAQQSY
15	180–183	464.25	0.0	−1.0	5.70	LQGF
16	184–192	1008.54	0.0	0.0	5.91	SKNILEASF
17	193–196	483.17	−2.0	1.0	−0.01	DSDF
18	197–204	1018.60	1.0	0.2	9.00	KEINRVLF
19	205–252	5612.82	−1.9	0.9	5.35	GEEEQKQQDEESQQEGVIVQLKREQ IRELMKHAKSTSKKSLSTQNEPF
20	**253–261**	**1118.63**	**2.0**	**0.1**	**10.30**	**NLRSQKPIY**
22	266–285	2377.26	−0.9	0.5	5.55	GRLHEITPEKNPQLRDLDVF
23	286–301	1749.91	−1.0	−0.2	4.13	LTSVDIKEGGLLMPNY
24	302–335	3893.02	−2.0	0.2	4.69	NSKAIVILVVNKGEANIELVGQREQQQQ QQEESW
25	336–340	694.35	0.0	0.5	6.36	EVQRY
26	341–350	1152.52	−3.0	0.9	2.92	RAEVSDDDVF
27	351–368	1878.00	0.0	−0.8	5.69	VIPASYPVAITATSNLNF
28	372–382	1276.60	0.0	0.2	6.36	GINAENNQRNF
29	383–422	4403.16	−5.9	0.4	4.10	LAGEEDNVMSEIPTEVLDVTFPASGE KVEKLINKQSDSHF
30	423–433	1343.67	0.1	1.5	7.21	TDHSSKREERV

Table 3. Screening prediction of β-vignin-derived peptides from cowpea hydrolyzed with pepsin in silico and their characteristics of isoelectric point, molecular mass, and high-performance chromatography time.

Peptide	Localization	Molecular Mass	Charge	Hydrophobicity	Ip	Sequence
1	19–41	2632.75	−1.8	0.6	5.40	GIVHRGHQESQEESEPRGQNNPF
2	44–49	824.85	−1.0	0.6	4.21	DSDRWF
3	54–60	886.97	1.1	−0.2	8.75	RNQYGHL
4	67–76	1229.36	1.0	0.6	8.75	DQRSKQIQNL
5	77–84	1055.16	−1.0	0.2	4.53	ENYRVVEF -
6	85–91	786.88	1.0	0.2	8.75	QSKPNTL
7	94–101	908.93	−1.8	0.2	5.05	PHHADADF
8	107–112	642.76	1.0	−0.1	9.75	NGRAIL
9	115–125	1248.36	−1.0	0.2	4.21	VNPDGRDSYIL
10	126–139	1472.58	0.1	0.1	6.85	EQGHAQKTPAGTTF
11	142–150	1069.05	−2.9	0.6	4.02	VNHDDNENL
12	151–155	627.83	2.0	0.2	11.00	RIVKL
13	156–165	1150.31	1.1	−0.3	9.80	AVPVNNPHRF
14	171–180	1113.15	−1.0	−0.1	4.00	SSTEAQQSYL
16	197–203	871.05	1.0	0.6	8.75	KEINRVL
17	205–225	2430.52	−6.0	0.9	3.77	GEEEQKQQDEESQQEGVIVQL
18	226–233	1071.24	1.0	1.4	8.75	KREQIREL
19	**234–245**	**1345.62**	**4.1**	**0.7**	**10.48**	**MKHAKSTSKKSL**
20	246–252	821.84	−1.0	0.1	4.00	STQNEPF
21	**255–265**	**1367.57**	**3.0**	**0.3**	**10.29**	**RSQKPIYSNKF**
22	269–279	1305.45	−0.9	0.4	5.40	HEITPEKNPQL
23	287–296	1018.13	−1.0	0.4	4.37	TSVDIKEGGL
24	298–309	1362.65	1.0	−0.6	8.34	MPNYNSKAIVIL
25	310–320	1185.34	−1.0	0.2	4.53	VVNKGEANIEL
26	321–350	3697.85	−5.0	0.7	4.12	VGQREQQQQQEESWEVQRYRAEVSDDDVF
27	351–366	1616.87	0.0	−0.8	5.49	VIPASYPVAITATSNL
28	372–382	1276.33	0.0	0.2	6.00	GINAENNQRNF
29	384–399	1732.88	−5.0	0.4	3.45	AGEEDNVMSEIPTEVL
30	404–413	1057.21	0.0	0.9	6.56	PASGEKVEKL
31	414–422	1075.15	0.1	0.2	6.74	INKQSDSHF
32	423–437	1.343.42	0.1	1.5	6.43	TDHSSKREERV

These MIC values were recorded against two main strains, selected on the basis of commonly reported clinical diagnostic infections, as examples of Gram-positive (*Staphylococcus aureus*) and Gram-negative (*Pseudomonas aeruginosa*) bacteria. *Staphylococcus aureus* is one of the most common human pathogens and *Pseudomonas aeruginosa* is a common cause of nosocomial pneumonia, urinary tract infection, and surgical site infection. It has become a less frequent cause of bacteremia in patients with neutropenia in most parts of the world, but remains the most important pathogen in patients with cystic fibrosis. The MIC values for all peptide hydrolyzates were 512 µg/mL against *Pseudomonas aeruginosa* and *Staphylococcus aureus* (Table 4). These results shown in Table 2 do not depict the fraction of peptides with the highest antimicrobial activity since the hydrolyzates presented the same MIC. However, when evaluated by the disk diffusion test, the protein hydrolyzates of both enzymes showed bacterial inhibition and dose-dependent effect against the tested strains (Table 5). In general, larger inhibition halos were observed for pepsin hydrolyzates for *Staphylococcus aureus* (with a diameter of 11.11 mm at a concentration of 250 µg/mL and 12.1 mm at a concentration of 500 µg/mL) and *Pseudomonas aeruginosa* at a concentration of 500 µg/mL (10.9 mm in diameter).

Table 4. MIC of fractionated protein hydrolysates of cowpea 7S globulin against *Pseudomonas aeruginosa* and *Staphylococcus aureus*. Results are expressed as (µg/mL) the mean ± standard deviation ($n = 4$).

Strains	Samples	Chymotrypsin	Pepsin
Staphylococcus aureus	>30 kDa	512	512.00
	30–10 kDa	512	512
	10–3 kDa	512	512
	Ciprofloxacin	0.125	0.125
Pseudomonas aeruginosa	>30 kDa	512	512
	30–10 kDa	512	512
	10–3 kDa	512	512
	Ciprofloxacin	0.125	0.125

Table 5. Antimicrobial activity of cowpea beta vignin total protein hydrolyzate against Gram-positive and Gram-negative bacteria indicated by halo inhibition (mm) at two concentrations.

Bacteria	Chymotrypsin	Pepsin
		250 µg/mL
Staphylococcus aureus	9.00	11.11
Pseudomonas aeruginosa	10.80	9.99
		500 µg/mL
Staphylococcus aureus	11.00	12.10
Pseudomonas aeruginosa	10.90	13.66

The inhibitory effect of the derivatives of the hydrolysis of beta vignin with pepsin on the proliferation of mammalian cells at a concentration of 512 µg/mL was greater than 80%, indicating biocompatibility (Figure 4A). However, for the chymotrypsin hydrolysates (Figure 4B) at a concentration of 512 µg/mL, the sample showed cytotoxicity (48% cell viability). For the other concentrations of chymotrypsin hydrolysates, cytotoxicity can be considered insignificant for mammalian cells.

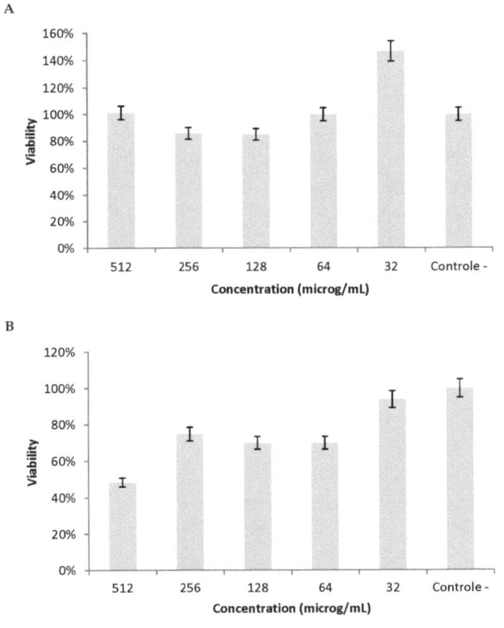

Figure 4. Evaluation of the viability of L929 human fibroblasts determined by the MTT assay after 24 h of incubation in pepsin (**A**) and chymotrypsin (**B**) hydrolysates.

4. Discussion

Globins constitute about 80% of the total protein in cowpea [29]. Studies indicate that this protein fraction is formed mainly by α-, β-, and γ-vignin proteins. However, in our study we found 7S globulins (β-vignina), the main proteins in cowpea, with a yield similar to that reported in the literature [6,19].

Our study demonstrated that the total globulin fraction consists of eight polypeptide chains (Figure 2B). However, the β-vignin protein obtained by chromatography had three polypeptide chains (Figure 2C), comprising two main glycosylated polypeptide chains with 50 and 55 kDa molecular weights. According to the literature, β-vignin is composed of two main chains of glycosylated polypeptides with molecular weights of 60 and 55 kDa and other smaller chains [30]. However, under denaturing and reducing conditions, cowpea vicilins are a heterogeneous mixture of polypeptides of various sizes [19,29,31]. Molecular mass analysis by electrophoresis showed a difference between the values calculated from the amino acid sequences. These differences are attributed to the glycosylation that these polypeptides undergo in the post-translational processing of their precursors [17].

The computer simulation of enzymatic hydrolysis performed on beta vignin from cowpea produced hundreds of peptide fragments, most of them with a molecular mass of 1 to 5 kDa. Although expected, this same pattern was not observed when this enzymatic proteolysis was performed in vitro. There was a higher amount of hydrolyzates with molecular mass greater than 30 kDa when compared to the other fractions. However, some peaks in the chromatographic samples were seen, showing many peptides within the samples under study. Several review studies do highlight the therapeutic potential of food-derived bioactive peptides, which have an antimicrobial function [32]. In our work, we used bacterial strains of *Pseudomonas aeruginosa* and *Staphylococcus aureus* to determine the MIC. The MIC value for all hydrolyzates was the same (512 μg/mL) when evaluated in both bacterial strains. There is no consensus on the acceptable standard for hydrolyzed protein isolates of legume proteins compared to conventional antibiotics. Some authors consider results only similar to known antibiotics, provided they work with a fraction already determined. In the present study, we did not work with the predetermined peptide fraction of beta vignin of cowpea. We followed the criteria suggested by Holetz et al. (2002) [33] and Carvalho et al. (2014) [34], who consider that a MIC below 100 mg/mL has appropriate antimicrobial activity and while concentrations above 500 mg/mL have poor activity and are difficult to use in the treatment of bacterial infections.

The antimicrobial activity of peptides is primarily based on the interaction between the peptide structure and the microorganism's cell membrane [35]. Because of their cationic or amphiphilic character, peptides bind to lipid membranes because of the attraction of the arginine and lysine residues of the peptide structure to the phospholipids present in the bilayer and through the interaction between the hydrophobic amino acid residues of the peptide and the membrane. These interactions allow the peptides to cross the lipid bilayer and reach the inner side of the cell, causing membrane dysfunction through the formation of pores with extravasation of ions and metabolites, depolarization, loss of membrane coupled respiration, and, ultimately, cell death [35–37].

Previous studies have highlighted the difficulty in identifying bioactive peptides derived from enzymatic proteolysis due to the significant variability of the primary sequences that were found [38,39]. Therefore, an in silico simulation was performed, during which it was possible to identify four hydrolyzates that could show more favorable characteristics and prediction for antimicrobial activity. These hydrolyzates are small sequences with a mass of less than 3 kDa. However, it is worth noting that a pool of peptides that were not yet isolated was used in our study, and they may be modulating the antimicrobial activity evaluated against *S. aureus* and *P. aeruginosa*. The dissociation of these peptides into even smaller fractions is necessary to identify and isolate the peptide with the best antimicrobial activity [40].

In addition, it is worth considering that *Pseudomonas aeruginosa* species becomes increasingly difficult to control due to a diversity of intrinsic and acquired drug-resistance mechanisms [41]. There are even records in the literature of resistance to some antimicrobial peptides against Gram-negative bacteria [42].

For the cytotoxicity evaluation, the MTT test was used against fibroblast cell line-L929 as recommended in the ISO 10993/2009 guideline [27]. Additionally, the L929 is highly proliferative and is widely used in cytotoxicity testing, mainly to check toxicity toward cellular viability and proliferation. Fibroblast cells are the most common cells of all types of connective tissues, being actively engaged in the synthesis and upkeep of the collagenous extracellular matrix, and also modulating adjacent cell behavior, including migration, proliferation, and differentiation. In this way, biological evaluation with fibroblast cell cultures might be regarded as a general bioassay, providing reliable information concerning basal cytotoxicity. The results showed that the hydrolyzed fractions of beta vignin from cowpea generally have low cytotoxicity (Figure 4). Thus, the peptide fractions used in the present work are safe and show antimicrobial potential. The unfolding is exactly in the sequencing and identification of the major peptides of this fraction, bearing in mind that the MIC of beta vignin protein hydrolyzates showed satisfactory concentrations.

5. Conclusions

Cowpea protein hydrolyzatesare shownto be safe and can be considered a potential alternative for developing innovative antimicrobials. Further studies of the peptide composition on amino acid sequences of beta vignin are needed to understand the structure–activity relationships of these peptides to elucidate their antimicrobial mechanisms of action and possible applicability in the market as new biotechnological drugs for human health or in the construction of transgenic plants with resistance to pathogens.

Author Contributions: T.S.A., C.A.d.C.S., M.B.d.C.e.S., F.P.R.B., E.S.F., A.L.S.S., L.N.S. and C.R.M. contributed in the conceptualization, methodology, validation, formal analysis, investigation, and writing—original draft preparation. C.B., M.L.B., J.C.C., R.L.C.d.A.-J., R.d.M.B., M.M.P., E.B.S., C.M.F.S. and P.S. contributed for the methodology, supervision, writing—review and editing, project administration, resources, and funding acquisition. All authors have contributed substantially to the work. All authors have read and agreed to the published version of the manuscript.

Funding: This research was funded by CNPq, FAPERJ and CAPES (financial code-001), and by the Fundação Carolina (*Movilidad de profesoradoBrasil-España, Movilidad. Estancias de Investigación*, C.2020) granted to P.S.

Institutional Review Board Statement: Not applicable.

Informed Consent Statement: Not applicable.

Data Availability Statement: Not applicable.

Conflicts of Interest: The authors declare no conflict of interest.

References

1. Murray, C.J.L.; Ikuta, K.S.; Sharara, F.; Swetschinski, L.; Robles Aguilar, G.; Gray, A.; Han, C.; Bisignano, C.; Rao, P.; Wool, E.; et al. Global burden of bacterial antimicrobial resistance in 2019: A systematic analysis. *Lancet* **2022**, *399*, 629–655. [CrossRef]
2. Brown, D. Antibiotic resistance breakers: Can repurposed drugs fill the antibiotic discovery void? *Nat. Rev. Drug Discov.* **2015**, *14*, 821–832. [CrossRef] [PubMed]
3. Okoye, C.O.; Ezeorba, T.P.C.; Okeke, E.S.; Okagu, I.U. Recent Findings on the Isolation, Identification and Quantification of Bioactive Peptides. *Appl. Food Res.* **2022**, *2*, 100065. [CrossRef]
4. Martinez-Villaluenga, C.; Rupasinghe, S.G.; Schuler, M.A.; de Mejia, E.G. Peptides from purified soybean beta-conglycinin inhibit fatty acid synthase by interaction with the thioesterase catalytic domain. *FEBS J.* **2010**, *277*, 1481–1493. [CrossRef] [PubMed]
5. Lammi, C.; Zanoni, C.; Arnoldi, A.; Vistoli, G. Two Peptides from Soy β-Conglycinin Induce a Hypocholesterolemic Effect in HepG2 Cells by a Statin-Like Mechanism: Comparative in Vitro and in Silico Modeling Studies. *J. Agric. Food Chem.* **2015**, *63*, 7945–7951. [CrossRef] [PubMed]
6. Marques, M.R.; Fontanari, G.G.; Pimenta, D.C.; Soares-Freitas, R.M.; Arêas, J.A.G. Proteolytic hydrolysis of cowpea proteins is able to release peptides with hypocholesterolemic activity. *Food Res. Int.* **2015**, *77*, 43–48. [CrossRef]
7. Guha, S.; Majumder, K. Structural-features of food-derived bioactive peptides with anti-inflammatory activity: A brief review. *J. Food Biochem.* **2019**, *43*, e12531. [CrossRef]

8. Moreno, C.; Mojica, L.; González de Mejía, E.; Camacho Ruiz, R.M.; Luna-Vital, D.A. Combinations of Legume Protein Hydrolysates Synergistically Inhibit Biological Markers Associated with Adipogenesis. *Food* **2020**, *9*, 1678. [CrossRef]
9. Farkas, A.; Maróti, G.; Kereszt, A.; Kondorosi, É. Comparative Analysis of the Bacterial Membrane Disruption Effect of Two Natural Plant Antimicrobial Peptides. *Front. Microbiol.* **2017**, *8*, 51. [CrossRef]
10. Ageitos, J.M.; Sánchez-Pérez, A.; Calo-Mata, P.; Villa, T.G. Antimicrobial peptides (AMPs): Ancient compounds that represent novel weapons in the fight against bacteria. *Biochem. Pharmacol.* **2017**, *133*, 117–138. [CrossRef]
11. Luong, H.X.; Thanh, T.T.; Tran, T.H. Antimicrobial peptides—Advances in development of therapeutic applications. *Life Sci.* **2020**, *260*, 118407. [CrossRef] [PubMed]
12. Benko-Iseppon, A.M.; Galdino, S.L.; Calsa, T., Jr.; Kido, E.A.; Tossi, A.; Belarmino, L.C.; Crovella, S. Overview on plant antimicrobial peptides. *Curr. Protein Pept. Sci.* **2010**, *11*, 181–188. [CrossRef] [PubMed]
13. Conlon, J.M.; Mechkarska, M.; Lukic, M.L.; Flatt, P.R. Potential therapeutic applications of multifunctional host-defense peptides from frog skin as anti-cancer, anti-viral, immunomodulatory, and anti-diabetic agents. *Peptides* **2014**, *57*, 67–77. [CrossRef]
14. Aiello, G.; Lammi, C.; Boschin, G.; Zanoni, C.; Arnoldi, A. Exploration of Potentially Bioactive Peptides Generated from the Enzymatic Hydrolysis of Hempseed Proteins. *J. Agric. Food Chem.* **2017**, *65*, 10174–10184. [CrossRef]
15. Osman, A.; Enan, G.; Al-Mohammadi, A.-R.; Abdel-Shafi, S.; Abdel-Hameid, S.; Sitohy, M.Z.; El-Gazzar, N. Antibacterial Peptides Produced by Alcalase from Cowpea Seed Proteins. *Antibiotics* **2021**, *10*, 870. [CrossRef] [PubMed]
16. Duan, Y.; Guan, N.; Li, P.; Li, J.; Luo, J. Monitoring and dietary exposure assessment of pesticide residues incowpea (*Vigna unguiculata* L. Walp) in Hainan, China. *Food Control* **2016**, *59*, 250–255. [CrossRef]
17. Rocha, A.J.; Sousa, B.L.; Girão, M.S.; Barroso-Neto, I.L.; Monteiro-Júnior, J.E.; Oliveira, J.T.A.; Nagano, C.S.; Carneiro, R.F.; Monteiro-Moreira, A.C.O.; Rocha, B.A.M.; et al. Cloning of cDNA sequences encoding cowpea (*Vigna unguiculata*) vicilins: Computational simulations suggest a binding mode of cowpea vicilins to chitin oligomers. *Int. J. Biol. Macromol.* **2018**, *117*, 565–573. [CrossRef]
18. Pina-Pérez, M.C.; Ferrús Pérez, M.A. Antimicrobial potential of legume extracts against foodborne pathogens: A review. *Trends Food Sci. Technol.* **2018**, *72*, 114–124. [CrossRef]
19. Ferreira, E.S.; Amaral, A.L.S.; Demonte, A.; Zanelli, C.F.; Capraro, J.; Duranti, M.; Neves, V.A. Hypocholesterolaemic effect of rat-administered oral doses of the isolated 7S globulins from cowpeas and adzuki beans. *J. Nutr. Sci.* **2015**, *4*, e7. [CrossRef]
20. Lowry, O.H.; Rosebrough, N.J.; Farr, A.L.; Randall, R.J. Protein measurement with the Folin phenol reagent. *J. Biol. Chem.* **1951**, *193*, 265–275. [CrossRef]
21. Laemmli, U.K. Cleavage of structural proteins during the assembly of the head of bacteriophage T4. *Nature* **1970**, *227*, 680–685. [CrossRef] [PubMed]
22. Akeson, W.R.; Stahmann, M.A. A pepsin pancreatin digest index of protein quality evaluation. *J. Nutr.* **1964**, *83*, 257–261. [CrossRef] [PubMed]
23. Tamura, K.; Stecher, G.; Kumar, S. MEGA11: Molecular Evolutionary Genetics Analysis Version 11. *Mol. Biol. Evol.* **2022**, *38*, 3022–3027. [CrossRef] [PubMed]
24. Minkiewicz, P.; Iwaniak, A.; Darewicz, M. BIOPEP-UWM Database of Bioactive Peptides: Current Opportunities. *Int. J. Mol. Sci* **2019**, *20*, 5978. [CrossRef]
25. Clinical and Laboratory Standards Institute (CLSI). *Methods for Antimicrobial Broth Dilution and Disk Diffusion Susceptibility Testing of Bacteria Isolated From Aquatic Animals*, 2nd ed.; CLSI Guideline VET03; Clinical and Laboratory Standards Institute: Wayne, PA, USA, 2020.
26. Clinical and Laboratory Standards Institute (CLSI). *Methods for Dilution Antimicrobial Susceptibility Tests for Bacteria That Grow Aerobically*, 11th ed.; CLSI Standard M07; Clinical and Laboratory Standards Institute: Wayne, PA, USA, 2018.
27. *ISO 10993-1:2009*; Biological Evaluation of Medical Devices—Part 5: Tests for In Vitro Cytotoxicity. British Standards: London, UK, 2009.
28. Severino, P.; Chaud, M.V.; Shimojo, A.; Antonini, D.; Lancelloti, M.; Santana, M.H.A.; Souto, E.B. Sodium alginate-cross-linked polymyxin B sulphate-loaded solid lipid nanoparticles: Antibiotic resistance tests and HaCat and NIH/3T3 cell viability studies. *Colloids Surf. B Biointerfaces* **2015**, *129*, 191–197. [CrossRef]
29. Aluko, R.E.; Girgih, A.T.; He, R.; Malomo, S.; Li, H.; Offengenden, M.; Wu, J. Structural and functional characterization of yellow field pea seed (*Pisum sativum* L.) protein-derived antihypertensive peptides. *Food Res. Int.* **2015**, *77*, 10–16. [CrossRef]
30. Freitas, R.L.; Teixeira, A.R.; Ferreira, R.B. Characterization of the proteins from *Vigna unguiculata* seeds. *J. Agric. Food Chem.* **2004**, *52*, 1682–1687. [CrossRef]
31. Gonçalves, A.; Goufo, P.; Barros, A.; Domínguez-Perles, R.; Trindade, H.; Rosa, E.A.; Ferreira, L.; Rodrigues, M. Cowpea (*Vigna unguiculata* L. Walp), a renewed multipurpose crop for a more sustainable agri-food system: Nutritional advantages and constraints. *J. Sci. Food Agric.* **2016**, *96*, 2941–2951. [CrossRef]
32. Awika, J.M.; Duodu, K.G. Bioactive polyphenols and peptides in cowpea (*Vigna unguiculata*) and their health promoting properties: A review. *J. Funct. Foods* **2017**, *38*, 686–697. [CrossRef]
33. Holetz, F.B.; Pessini, G.L.; Sanches, N.R.; Cortez, D.A.G.; Nakamura, C.V.; Dias Filho, B.P. Screening of some plants used in the Brazilian folk medicine for the treatment of infectious diseases. *Memórias Instituo Oswaldo Cruz* **2002**, *97*, 1027–1031. [CrossRef]
34. Carvalho, A.F.; Silva, D.M.I.; Silva, T.R.C.; Scarcelli, E.; Manhani, M.R. Evaluation of the antibacterial activity of ethanolic and cyclohexane extracts of chamomile flowers (*Matricaria chamomilla* L.). *Rev. Bras. Plantas Med.* **2014**, *16*, 521–526. [CrossRef]

35. Andersson, D.I.; Hughes, D.; Kubicek-Sutherland, J.Z. Mechanisms and consequences of bacterial resistance to antimicrobial peptides. *Drug Resist. Updates* **2016**, *26*, 43–57. [CrossRef] [PubMed]
36. Bahar, A.A.; Ren, D. Antimicrobial Peptides. *Pharmaceuticals* **2013**, *6*, 1543–1575. [CrossRef] [PubMed]
37. Salas, C.E.; Badillo-Corona, J.A.; Ramírez-Sotelo, G.; Oliver-Salvador, C. Biologically active and antimicrobial peptides from plants. *BioMed Res. Int.* **2015**, *2015*, 102129. [CrossRef] [PubMed]
38. Silva, M.; Souza, C.; Philadelpho, B.O.; Cunha, M.; Batista, F.P.R.; Silva, J.R.D.; Druzian, J.I.; Castilho, M.S.; Cilli, E.M.; Ferreira, E.S. In vitro and in silico studies of 3-hydroxy-3-methyl-glutaryl coenzyme A reductase inhibitory activity of the cowpea Gln-Asp-Phe peptide. *Food Chem.* **2018**, *259*, 270–277. [CrossRef]
39. Fassini, P.G.; Noda, R.W.; Ferreira, E.S.; Silva, M.A.; Neves, V.A.; Demonte, A. Soybean glycinin improves HDL-C and suppresses the effects of rosuvastatin on hypercholesterolemic rats. *Lipids Health Dis.* **2011**, *10*, 165. [CrossRef]
40. Xiang, N.; Lyu, Y.; Zhu, X.; Bhunia, A.K.; Narsimhan, G. Methodology for identification of pore forming antimicrobial peptides from soy protein subunits β-conglycinin and glycinin. *Peptides* **2016**, *85*, 27–40. [CrossRef]
41. Pang, Z.; Raudonis, R.; Glick, B.R.; Lin, T.J.; Cheng, Z. Antibiotic resistance in Pseudomonas aeruginosa: Mechanisms and alternative therapeutic strategies. *Biotechnol. Adv.* **2019**, *37*, 177–192. [CrossRef]
42. Mangoni, M.L.; Shai, Y. Short native antimicrobial peptides and engineered ultrashort lipopeptides: Similarities and differences in cell specificities and modes of action. *Cell. Mol. Life Sci.* **2011**, *68*, 2267–2280. [CrossRef]

technologies MDPI

Article

Demonstration of Resilient Microgrid with Real-Time Co-Simulation and Programmable Loads

Hossam A. Gabbar [1,2,*], Yasser Elsayed [1], Manir Isham [1], Abdalrahman Elshora [2], Abu Bakar Siddique [1] and Otavio Lopes Alves Esteves [1]

[1] Faculty of Energy Systems and Nuclear Science, Ontario Tech University, 2000 Simcoe St. North, Oshawa, ON L1G0C5, Canada; yasser.elsayed@ontariotechu.ca (Y.E.); manir1@gmail.com (M.I.); abubakar.siddique@ontariotechu.net (A.B.S.); otavio.lopesalvesesteves@ontariotechu.net (O.L.A.E.)

[2] Faculty of Engineering and Applied Science, Ontario Tech University, 2000 Simcoe St. North, Oshawa, ON L1G0C5, Canada; abdalrahman.elshora@ontariotechu.net

* Correspondence: hossam.gaber@ontariotechu.ca; Tel.: +1-9057218668

Abstract: In recent years, the foment for sustainable and reliable micro energy grid (MEG) systems has increased significantly, aiming mainly to reduce the dependency on fossil fuels, provide low-cost clean energy, lighten the burden, and increase the stability and reliability of the regional electrical grid by having interconnected and centralized clean energy sources, and ensure energy resilience for the population. A resilient energy system typically consists of a system able to control the energy flow effectively by backing up the intermittent output of renewable sources, reducing the effects of the peak demand on the grid side, considering the impact on dispatch and reliability, and providing resilient features to ensure minimum operation interruptions. This paper aims to demonstrate a real-time simulation of a microgrid capable of predicting and ensuring energy lines run correctly to prevent or shorten outages on the grid when it is subject to different disturbances by using energy management with a fail-safe operation and redundant control. In addition, it presents optimized energy solutions to enhance the situational awareness of energy grid operators based on a graphical and interactive user interface. To expand the MEG's capability, the setup integrates real implemented hardware components with the emulated components based on real-time simulation using OPAL-RT OP4510. Most hardware components are implemented in the lab to be modular, expandable, and flexible for various test scenarios, including fault imitation. They include but are not limited to the power converter, inverter, battery charger controller, relay drivers, programmable AC and DC loads, PLC, and microcontroller-based controller. In addition, the real-time simulation offers a great variety of power sources and energy storage such as wind turbine emulators and flywheels in addition to the physical sources such as solar panels, supercapacitors, and battery packs.

Keywords: resiliency design; microgrid; fault-tolerant control; real-time co-simulation

check for updates

Citation: Gabbar, H.A.; Elsayed, Y.; Isham, M.; Elshora, A.; Siddique, A.B.; Esteves, O.L.A. Demonstration of Resilient Microgrid with Real-Time Co-Simulation and Programmable Loads. *Technologies* **2022**, *10*, 83. https://doi.org/10.3390/technologies10040083

Academic Editors: Manoj Gupta, Eugene Wong and Gwanggil Jeon

Received: 22 May 2022
Accepted: 9 July 2022
Published: 12 July 2022

1. Introduction

Increasingly, the world is becoming more dependent on electricity, directly and indirectly. Almost every sector in the globalized world has its process linked to the electric industry. In addition to the conventional use of electricity, an unreliable energy system negatively impacts sectors such as transportation, security, food, health, etc. [1]. For this reason, outages are more than an inconvenience; they seriously threaten community safety, economic stability, and national security [2]. Some factors can significantly impact an energy grid reliability, such as more frequent natural disasters, the current aged energy networks, the asynchrony between energy demand and supply, and the vulnerable energy control system [3].

The rise in the severity and frequency of power outages due to extreme weather conditions has made studying grid resilience crucial. Furthermore, weather events are expected to become more severe due to climate change and global warming in the following years [4]. Nowadays, several reliability indices are used to evaluate the power system performance, such as the System Average Interruption Duration Index (SAIDI), the System Average Interruption Frequency Index (SAIFI), and the Customer Average Interruption Index (CAIDI), among others [5,6]. However, these indices are not long enough to evaluate the whole system because they are focused only on high probability low impact disturbances, and the grid operates passively. They do not account for the primary power outages caused by extreme weather events; consequently, the recovery mechanism is limited to a specific point where the outage occurs [7]. The entire grid reliability is only analyzed when a considerable and broad disruption happens. For that reason, new studies are needed to encompass the grid reliability to deal with both the impact caused by extreme weather and the fact that the current energy grid is aging and overdue for an upgrade [8].

Additionally, with technological advances, smart grids are expected to spread in the following years, consequently resulting in intelligent homes. This scenario means that basically, everything in a house will need electricity, such as turning on or off the heating, cooling, doors, windows, everything, resulting in a society more connected and more dependent on electricity than ever [9]. The same happens with businesses and other buildings that have their daily activity linked directly to electricity use. This interconnectedness means that every sector is negatively affected when part of the grid needs to be repaired [10]. That is why it is important to design resilient energy systems; to prevent power disruptions or restore electricity quickly if an outage occurs [11]. Additionally, the increase in advanced energy loads, such as smart appliances and electric vehicles, leads to another concern in the electric sector: the volatility and the asynchrony between demand and load, resulting in a struggle to provide a stable energy flow. In most cases, the utility grid needs to be oversized to supply peak demand, even for a short period [12].

Research has shown that the way to improve the resiliency of the energy grid is to invest in microgrids integrated with renewable sources, especially by using a distributed system. Distributed generation has grown considerably all over the world, and this is basically due to the incentives for the use of renewable sources as the key to fighting climate change, as well as the advances in technologies and market expansion, which have increased the viability of the installation of small generators connected directly to the distribution network, as well as the advancement of new technologies and the need for the small power grid in remote places [12]. To be considered a microgrid, a system needs to consist of an energy load and generators with a control capability, which means it can disconnect from the traditional grid and operate autonomously [13]. In other words, a microgrid is generally connected to a utility grid; however, in the case of extreme weather and power outages, it can break off and operate on its own by using a local energy generation and, consequently, keep the load supplied.

According to the U.S. Department of Energy Microgrid Exchange Group, a microgrid can be defined as a group of interconnected loads and distributed energy resources within clearly defined electrical boundaries that act as a single controllable entity concerning the grid, which can be controlled as a grid-connected system or works in an off-grid mode. Different designs of microgrids are being developed to provide reliable energy using emissions-free sources integrated with energy storage systems, with a tendency to increase over time due to the falling cost of renewable energy sources associated with the efforts to replace fossil fuels and the frequent transmission line failures [14].

A lot of research has been conducted regarding the possible scenarios and microgrid benefits. In [15], the author states an overview of the integration of MEG with the existing utility grid and its main problems and points out the most relevant research, including distributed generation, applications with energy converters, management and control, protection, and communications. However, not much is found about the many challenges that must be dealt with. In [16], the author summarizes different approaches and technologies that have been studied to address the complexity and challenges of microgrids, mainly regarding the power quality, which includes energy flow balancing, real-time management, frequency control, efficiency, and economical operation, as well as highlights the importance of focusing on fail-safe control and fault tolerance systems to improve the resiliency in advanced microgrids. In [17], the paper identifies possible controller designs used in existing MEG, including the control system's challenges, and proposes research systems with fault-tolerant control applied to a hierarchical architecture to enhance the smartness of control systems. More recently, many studies have focused on research and development in the area of fail-safe and resilience techniques for MG [18–20]. Due to its multiple functions and many possible solutions proposed in the literature, the design of control systems for MEG is a complex engagement [21].

Since microgrids are typically composed of different technologies and energy physics, power electronics converters are essential components that must be included in the MEG systems to integrate various energy networks, which consist of electronic devices able to convert electric energy from one form to another, such as alternating (AC) and direct current (DC), and also allowing the adjustment in energy voltage, current, and frequency [22]. Additionally, power converters can be used to increase maneuver abilities in hybrid systems, especially with renewable sources, by controlling the extracted power in each source and stabilizing the energy flows between components [23]. For these reasons, researchers are constantly exploring power converters technologies, and topologies to meet the more complex microgrid components integration [24]; however, most of them have some limitations in adapting different energy flows and technologies.

In [25] is presented a multi-input convert to be used in renewable energy applications; however, by using only a unidirectional power flow with limited components and a non-dispatchable current source. In [26], the load can be supplied by different voltage levels prevenient from two sources without circulating current, using the multi-input converter, but it lacks modularity. The proposed converter in [27] has fewer conduction losses, and the load supply can be performed individually or simultaneously from two different sources, which have different voltage–current characteristics with three power switches only; however, the proposed converter can be used only in DC microgrid applications, and the sources cannot transfer the power between them. For this paper, the proposed multi-input converter aims to work in bidirectional power flow capability and exchange the energy between the DC sources with minimum components parts. In addition, it has the modularity feature, so it can easily be adapted to different energy sources and consequently increase the microgrid reliability.

To develop reliable microgrids, it is crucial to focus on research and lab experiments and ensure that existing infrastructure adapts to new technologies. Lab experiments must simulate threats and responses and validate new technologies to help grid operators against outages. Performing experiments also help visualize the impact of climate change and the increasing of sophisticated cyber attackers to enhance the electric grid. One way to do this is by performing real-time simulations, which consists of testing computer modeling and small-scale energy system with real-life input and output. OPAL-RT Technologies is the leading developer of open real-time digital simulators and hardware in the loop testing equipment in the energy sector. OPAL-RT can be used to design, test, and optimize control and protection systems for power grids, power electronics, etc. In addition, the RT-LAB, core OPAL-RT software, enables users to develop models suitable for real-time simulation. RT-LAB models are fully integrated with MATLAB/Simulink [28]. Simulink, developed by MathWorks, is data-flow graphical programming software for modeling, simulating, and analyzing dynamic systems. It supports simulation, automatic code generation, continuous tests, and verification of embedded systems. Integrated with MATLAB, it can incorporate MATLAB algorithms into models and export simulation results to MATLAB for further analysis [29].

This paper aims to introduce an experimental platform for a micro energy grid with unique merits such as having sizable and extensible AC and DC loads, hybrid power and energy storage sources through real-time co-simulation, and a redundant control system for enabling the fail-safe and resilient control to the MEG. A sizeable load is considered the first merit and is developed through an array of relays to dynamically change the arrangement of series and parallel connections between loads to determine the designated load value, which is applicable for AC and DC loads, including capacitive, inductive, and resistive load types. The second merit is combining a real-time simulation to emulate power and energy sources that are impracticable to be considered in lab sizes, e.g., wind turbines as a power source and flywheels as energy storage. The real-time simulation is designed based on OPAL-RT simulator OP4510 to connect the real-time hardware signals with the Simulink models. The signals that could be manipulated and connected are input signals from sensors and switches, while output signals like relaying simulation output signals are used to activate and run actuators, e.g., motors. The third merit is the redundant control system which utilizes two controllers: the first is based on a microcontroller array, and the second is based on a PLC controller aiming to back up each other and maintain the system availability.

The paper is outlined starting with the introduction section, followed by Section 2 presenting the design and co-simulation of the microgrid, and introduces the proposed microgrid's design and structure, highlighting the proposed system innovation. Section 3 demonstrates the testing and validation. Section 4 shows the resiliency, fail-safe control, and fault-tolerant capacity of the proposed microgrid. Finally, Section 5 concludes the work.

2. Design and Co-Simulation of Microgrid

The MEG consists of three main components: power sources, controllers, and energy loads. Therefore, it is possible to develop hybrid energy systems by combing different components technologies with hardware and emulators subsystems. The main features of the microgrid proposed in this paper consist of:

- Resiliency technique based on redundant control between microcontroller and PLCs;
- Hybrid energy sources, i.e., PV, utility grid, and energy storage system (battery bank);
- Programmable DC load and AC load based on relay control and load management;
- Hybrid power bus topologies, i.e., DC and AC networks;
- Master–slave networking topology between master and slave PLCs (1211C and 1214C CPUs).

The system schematic proposed in this paper consists of a microgrid with an interconnected power system, including physical sources, such as the utility grid and a solar panel, and emulated power sources, such as a wind turbine, using a real-time simulator. In addition, the system includes energy storage systems, physically and simulated, such as battery bank and flywheel, respectively. Furthermore, the system has a redundant control system based on microcontrollers and PLCs, as well as a dynamic load system, which consists of an AC and DC relay controlled sizable load. The following subsections elaborate on the system structure, including system components and parameters and the co-simulator based on Opal-RT OP4510.

2.1. Microgrid Structure and Resources

The structure of the proposed microgrid is shown in Figure 1. The system consists of AC and DC load, controllers, and power sources, including PV panels, emulated wind turbines, and flywheels based on real-time simulation, battery bank, and supercapacitors. Due to the complexity of using actual wind turbine and flywheel systems on a small scale for laboratory purposes, both approaches are emulated by using real-time simulations, which mimic the actual behavior of these technologies. The wind turbine and flywheel emulators using MATLAB/Simulink and testing using an OPAL-RT real-time simulator.

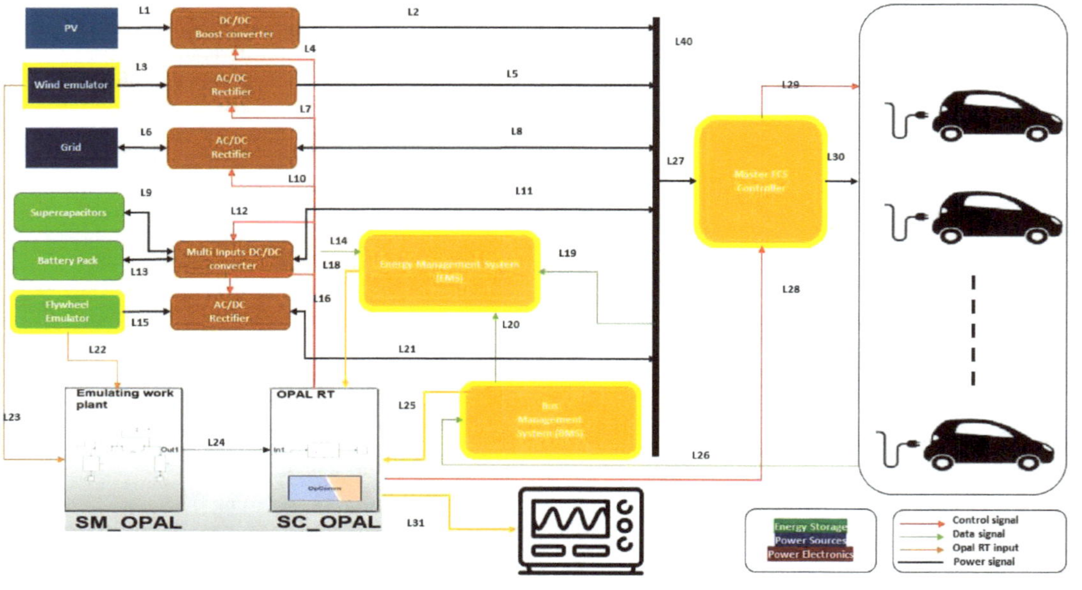

Figure 1. Microgrid structure.

Regarding the energy loads, the system includes AC and DC programmable loads by changing the series and parallel arrangement of the loads using a bank of relays controlled by microcontroller-based relay drivers. In addition, the controller can insert or withdraw some loads based on the system capacity limit or according to the user's request. Figure 1 shows two types of power sources: physical power sources such as PV and the utility grid are highlighted in dark solid blue and emulated power sources such as wind emulators are highlighted in dark blue with a yellow border. Additionally, it shows two types of energy sources: physical energy sources such as supercapacitor and battery pack are highlighted in green and emulated energy sources such as emulated flywheel are highlighted in green and yellow border. The figure shows the bus and energy management controllers in the middle. In addition, the co-simulation based on OPAL-RT (OP4510) is connected between the hardware and simulation Simulink for cooperating between the emulated components, called emulated plant, e.g., flywheel or wind turbine, and the hardware components of the systems.

Table 1 lists the system components describing their types and parameters, stating the physical and emulated power sources, energy storage, AC and DC loads, converter, inverter, and real-time simulator. It shows each component's capacity and limits, voltage, input, and output parameters.

Table 1. MEG system components and parameters.

Components	Type	Capacity (kW)
Utility Grid	physical power source	unlimited
PV	physical power source	1
Wind Turbine	emulated power source	1
Battery Pack	physical energy storage	0.576
Supercapacitor	physical energy storage	0.5
Flywheel	Emulated energy storage	1

Table 1. *Cont.*

Components	Type	Capacity (kW)
Converter	DC-DC multi-input	3
Inverter	pure sine wave	1
DC Load	resistive load	up to 3
AC Load	inductive and capacitive	up to 1
Real-Time Co-Simulator	Opal-RT4510	
DC Bus	12VDC	
AC Bus	110 AC/60 HZ	
Relay Array	50 SPDT relays 12VDC 10A	
Microcontrollers	3 Arduino uno R3	
PLC controllers	2 PLCs S7-1200 CPU 1211C and 1214C	
Converter	DC-DC multi-input converter 12VDC-300VDC	1
Inverter	pure sine wave inverter	1

2.2. Circuit Design

Figure 2 depicts the schematic of the MEG circuitry, including relay array and microcontrollers, PLCs, inverter, converter interfaces, and dynamic loads. Since the system has been designed in the lab and consists of modular sectors, i.e., controlling units, signal acquisition boards, inverter, and power converter circuits, it makes easy maintenance and upgrading flexibility possible. Regarding the schematic details, the system consists of relay driver boards for driving the variable DC and AC loads, a variable capacitor for power factor correction, a charging controller of the battery pack, sensor boards for adapting the sensors' readings with the acceptable limits of microcontrollers, Arduino boards, communication devices for TCP ethernet connections, Modbus connectivity, pure sine wave inverter, bidirectional and multi-input power converter, and the PLC controllers in master-to-slave topology. Additionally, the system provides hybrid AC and DC bus systems that can feed different loads.

2.3. Fail-Safe and Resiliency Design of Microgrid

The concept of a resilient, intelligent, and effective microgrid brings forth promising enhancements for the current energy scenario without the necessity of redesigning the existed transmission lines, increasing the ability to integrate different types of energy demand with advanced energy storage systems, and with a high share of distributed power generation from renewable energy sources. However, as the diversity of distributed generators and energy load increases, the complexity of operating, controlling, and monitoring all the nodes within the system to maintain the power flow stability increases. Therefore, in addition to a myriad of benefits that a theoretical MEG can provide, there are some challenges that an effective energy management system needs to bear to become reliable [30].

2.3.1. Fail-Safe Algorithm

An essential factor that impacts the power quality is the ability to smoothly switch between energy sources and controllable loads that become undetectable for emergency loads. The necessity to change between the energy source surge for different reasons, mainly due to problems on the transmission side, such as climate destruction or utility grid maintenance, resulting in the necessity of switching between the grid-tied and islanded mode. This transition needs to be smooth enough to avoid or cause a less possible interruption in the energy supply. When it is necessary to change to an island mode, the energy management needs to identify a better scenario to meet the energy demands, activating the energy storage system and auxiliary generators.

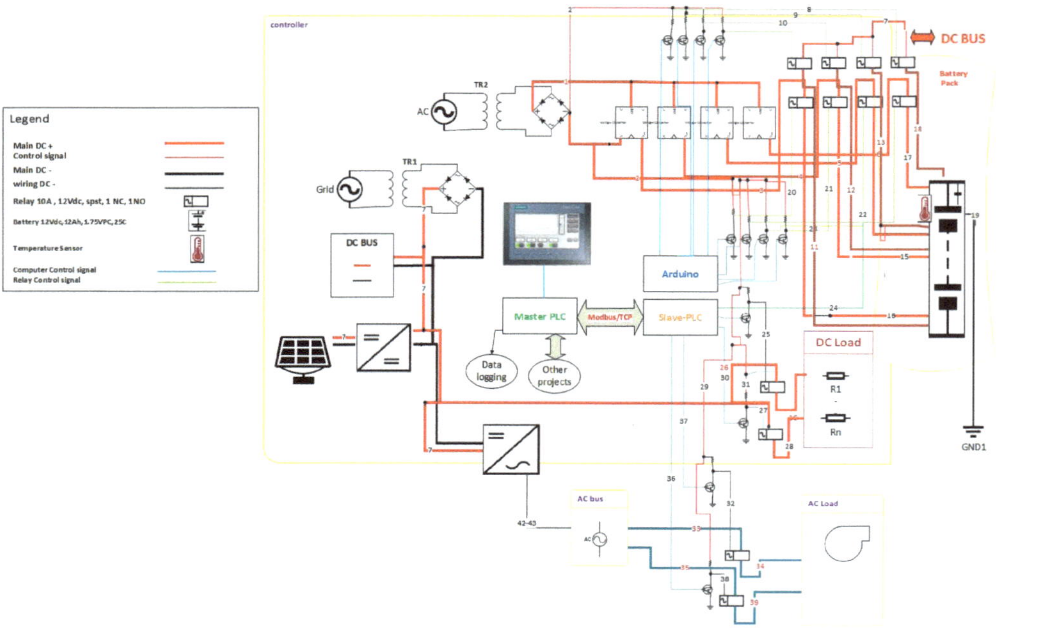

Figure 2. Circuit design.

This mechanism is commonly performed by using electric actuates, which require a constant energy source to operate and that can identify power loss in the system and drive actuators to a predetermined safe position to maintain the energy supply for the system [31,32]. Figure 3 charts the logic flow of the fail-safe algorithm that guarantees the balance between load demand and available power source capacity by monitoring the status of loads and sources. It calls for a fail-safe routine to detect a change in any contact's position by calculating the load demand and source capacity smartly based on current statuses and checking the power availability to fulfil the request load demand. If there is plenty of power to satisfy the loads, it updates the system accordingly. However, if there is a power shortage, it calculates for emergence loads. Then, if the amount of available power sources can satisfy the emergency loads, it activates the emergency loads only. Or, it loads the previous safe settings of loads and sources.

2.3.2. The Resiliency of the Microgrid

Figure 4 demonstrates the resilient design of the microgrid by showing the remote access ability to monitor, control, and perform maintenance to the microgrid by using the modbus-based communication between two PLC controllers. This way, the microgrid can be accessed from anywhere without having operators on site. In addition, the redundant control topology between PLC and microcontrollers maintains the microgrid secured when one controller goes down. Thanks to the firm and safe control algorithm, the microgrid can simultaneously be commanded from both sides.

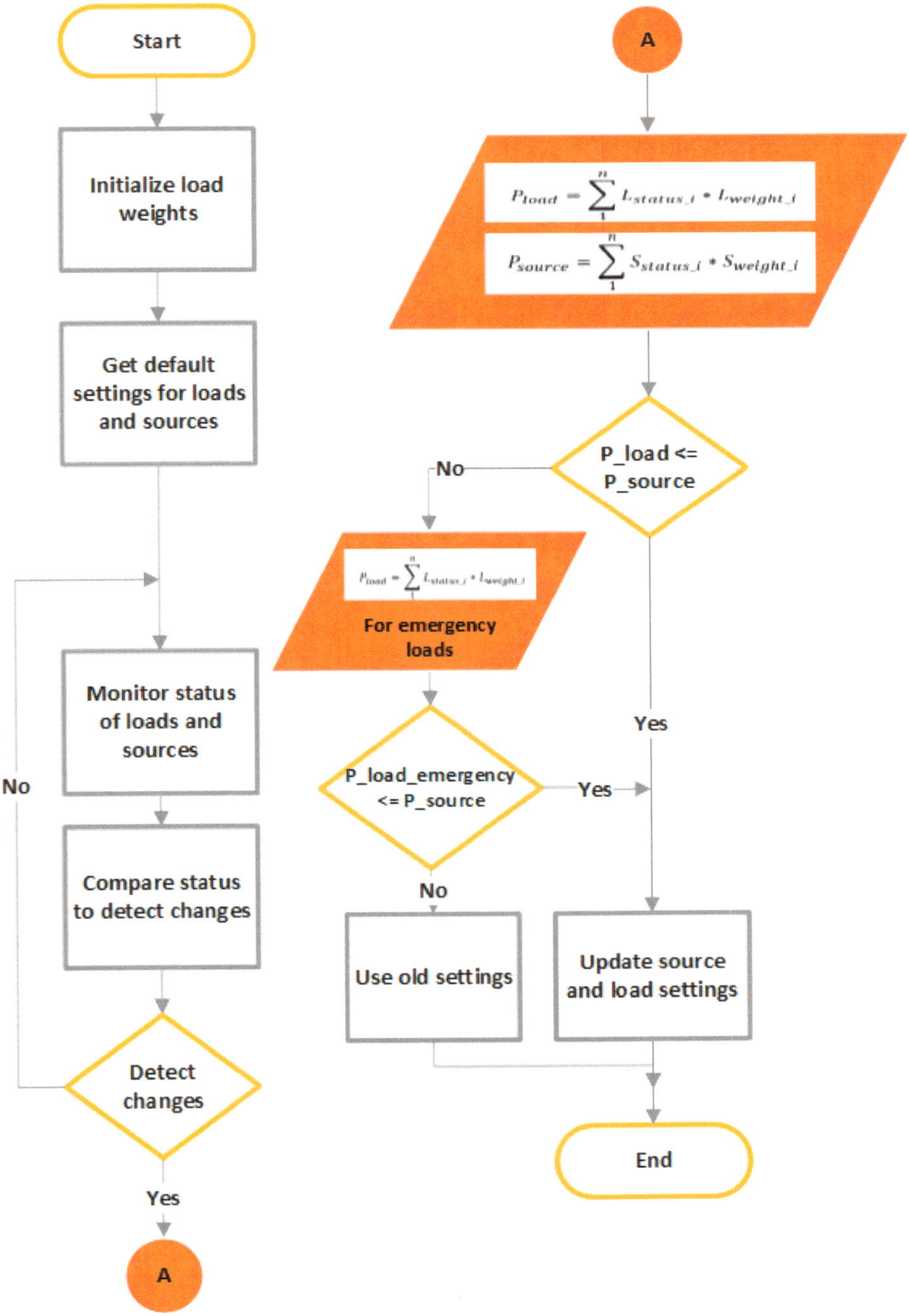

Figure 3. Fail-safe algorithm of the microgrid.

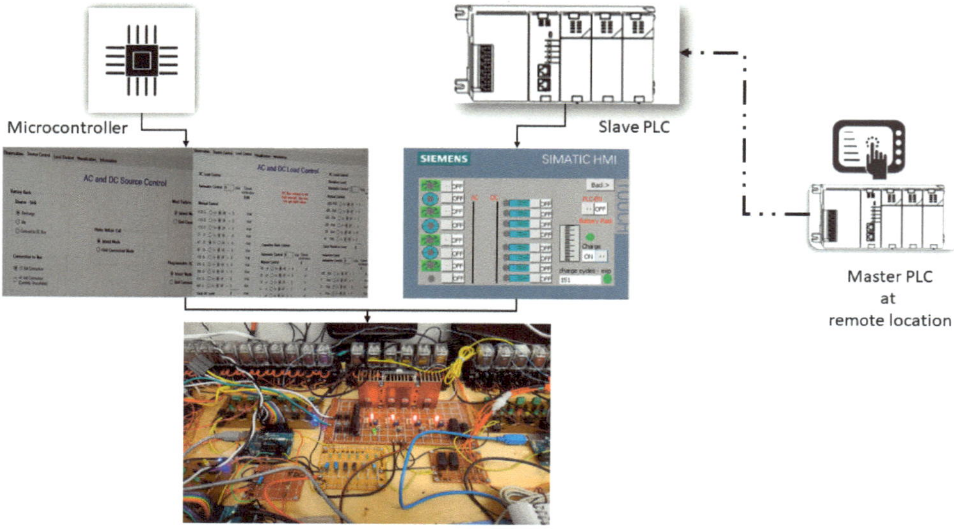

Figure 4. The resilient design of the microgrid.

2.4. Fault-Tolerant Control of Microgrid

The concept of fault-tolerant power systems consists of a technique in which the energy system can ensure continuous and uninterrupted functionality. One of the requirements for a resilient system is to avoid outages. The system must be immune to single or multiple failures in the primary energy supply, which is made by developing architectures with backup and redundancy power sources.

Fault-tolerant control can be classified into active management and passive control. Passive control is widely used due to its robustness. Prior knowledge of possible faults is required for such control that may alter the system's stability. Due to those faults, the elements installed, such as consumers and subsystem components, might affect their fundamental performance. Therefore, fault tolerance is required to provide the continuity of the microgrid functionality, and the following techniques can achieve it:

- Fault detection: it consists of the appearance and detection of possible faults within the microgrid system.
- Recovery: it means the replacement of the fault states with a particular state. It consists of two mechanisms:
  - Fault management: fault state needs to be eliminated to clear the erroneous state of the microgrid.
  - Fault handling: diagnosis, isolation, reconfiguration, and reset are required to resolve the fault.

This experimental work, stated in Figure 5, uses a fault tolerance algorithm based on PLC and relays logic control. PLC scans the loads, power sources, and energy storage statuses by reading the relays contacts statuses. Then, it checks the type of faults; if it is related to load, it isolates the faulty load(s), but if it is related to source(s), it isolates that source. Then, it calculates balanced load demands and power source generation. Ultimately, it activates and deactivates load(s) accordingly. In other words, it carries out a corrective action based on balancing the demanded load with the available power after isolating the faulty source(s) and or load. The main features of the proposed control are:

- Regulating microgrid voltages considering faults within the supplies and loads.

- The control strategy is robust against those faults, disturbances, and harmonics as changing the controller structure is not required like the existing controller. Therefore, even when some part of the system fails due to fault conditions, the proposed controller enables the continuity of the safe operation of the microgrid and thus increases the system's reliability.
- Finally, two case scenarios have been designed, the controller has been implemented in hardware to demonstrate its performance, and the results are described.

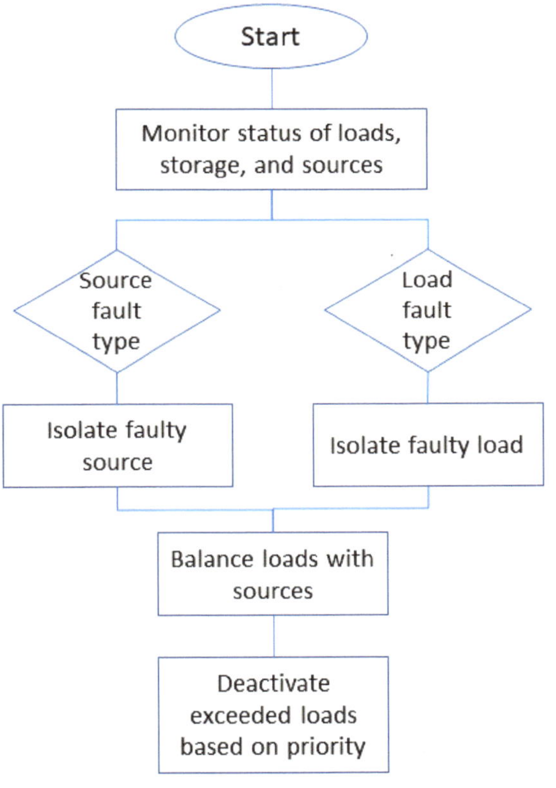

Figure 5. Fault tolerance algorithm based on PLC and relay control.

2.5. Microgrid Control

The redundant control technique has been designed based on three microcontrollers and two PLC controllers connected based on master-to-slave topology. Figure 6 shows the design of the microgrid control system that has been driven redundantly by microcontrollers and PLC, meaning when one controller is down, the other takes over the control to increase the reliability of the MEG. In addition, the system may receive commands from any of them to maintain a fail-safe system. Additionally, by applying master-to-slave control, one PLC works as a master controller (remote) that can send commands to the slave controller (local) to enable remote access to the system from anywhere. In addition to the remote access, the system can monitor all metrics, faults, and measurements of the system that can be used in analysis work for concluding and forecasting purposes. The figure also shows the interconnected system components, including the connection between the master PLC and slave PLC and the HMI screen connected to the master controller as a user interface to the control system. In addition, it shows the ethernet communication based on eight ports ethernet switch and how the two ways controllers, Arduino-based and PLC-based controllers can share the responsibility of controlling loads.

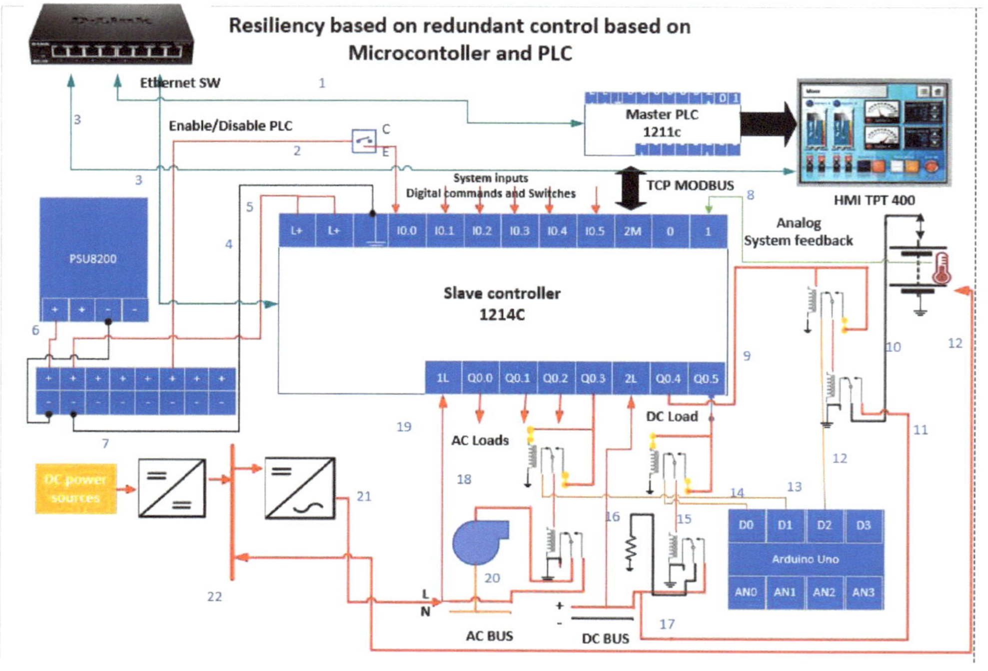

Figure 6. Microgrid control.

2.6. Co-Simulation Based on OPAL-RT

Real-time simulation is a technique used to mimic and analyze the system performance at the same rate as the actual physical system using computer modelling. Real-time simulations are used widely in several engineering fields and can be configured in hardware-in-the-loop simulations by testing controllers using real-life conditions.

OPAL-RT Technologies is the leading developer of open real-time digital simulators and hardware in the loop testing equipment in the energy sector. Figure 7 depicts the outstanding features of OPAL-RT (Model OP4510), which include 32 channels of digital inputs (DIs), 32 channels of digital outputs (DOs), and 16 channels of analog inputs (AIs), and 16 channels of analog outputs (AOs). These features allow the MEG system to emulate complex subsystems, such as flywheels, wind turbines, and nuclear reactors, which are unlikely to be hosted indoor, e.g., on a lab scale. In addition, OPAL-RT enables interfacing with external world devices, including switches, relays, motors, and sensors. So, these capabilities of OPAL-RT OP4510 have been utilized to process all the digital inputs from the user, such as the switches and feedback signals that relays' contacts have relaid. Furthermore, analog inputs from measurement devices and sensors such as voltages and currents, and battery temperature, are transferred from hardware to the Simulink model by using OPAL-RT to be processed. Similarly, the generating outputs from the Simulink model are converted into physical signals to either be displayed on scopes or be used to switch actuators on and off or control signals that drive the power switches in controller and converter devices.

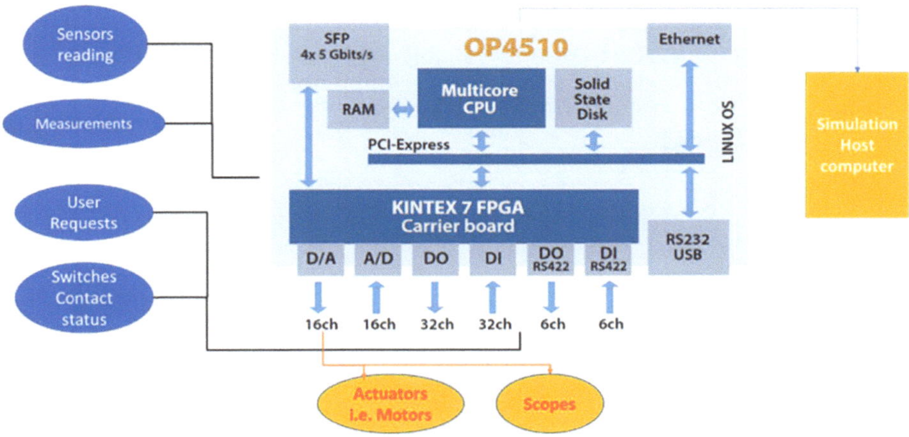

Figure 7. Real-time co-simulation for microgrid applications.

2.7. Experimental Setup of MEG

Based on the schematic design of the MEG circuitry described above, the experimental setup of the microgrid is pictured in Figure 8. It shows the integration of multiple power sources, including the grid, PV panels, battery pack, and supercapacitors in addition to the other emulated power and energy sources by OPAL-RT co-simulation. The figure also shows the redundant control system consists of a redundancy system including three microcontroller boards and two PLCs.

Figure 8. Microgrid experimental setup.

The setup provides two common power buses, AC and DC buses, for supplying different types of loads, which are sizable programmatically based on relay control logic. The load can be decided manually by switching on and off any load according to user demands. Additionally, it can be set automatically by selecting based on the total amount of load the user requested. It is important to highlight one crucial contribution that, for any load mode control, restrictions of balancing between the available power source and the requested loads can be selected by considering the priority of emergency loads according to a predefined plan.

2.8. Design of the Multi-Input Converter

The proposed multi-input converter is shown in Figure 9. The system is projected to charge the battery bank from DC-link or PV panel and discharge with the PV panel in the DC link to supply the AC and DC loads. The battery is used as an energy storage device with high energy density, and the PV is used as a renewable energy source that will reduce the supply from the grid. The converter consists of six MOSFETs, two inductors, and one capacitor, as shown in Figure 9. The design can be modular by adding two power switches and one inductor to combine the switching leg to add more renewable energy sources or energy storage devices with high power density such as supercapacitors. Table 2 shows the equations used to calculate the minimum values of inductors L_1 and L_2, and Table 3 shows the design specifications of the proposed converter.

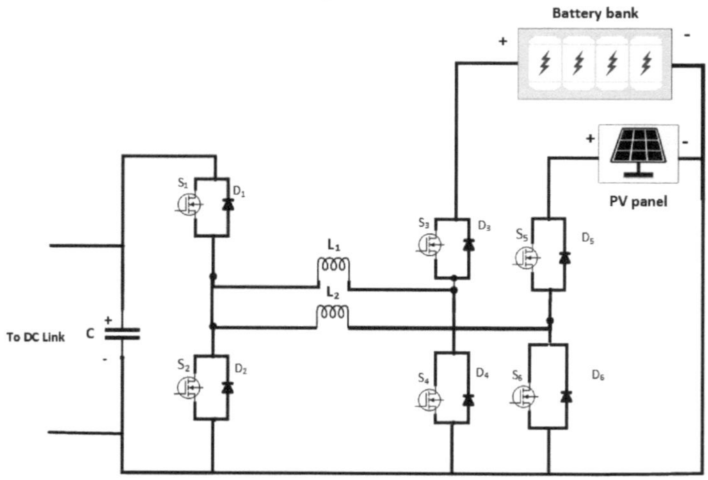

Figure 9. The configuration of the multi-input converter with MEG DC sources.

Table 2. Inductance equations for different operation modes.

Mode	1	2	3	4
L_1, min	$\frac{(1-D)*T_s}{\Delta I_{L_1}} * V_{DC}$	-	$\frac{D*T_s}{2\Delta I_L} * V_{Bt}$	$\frac{V_{DC}*(1-d_2)*T_s}{\Delta I_L}$
L_2, min	-	$\frac{(1-D)*T_s}{\Delta I_{L_2}} * V_{DC}$	$\frac{D*T_s}{2\Delta I_L} * V_{Bt}$	$\frac{V_{DC}*(1-d_2)*T_s}{\Delta I_L}$

Table 3. Design specifications of the proposed converter.

Specification	Battery Bank Voltage (V_{BT})	PV Panel Voltage (V_{PV})	DC Link Voltage (V_{DC})	Switching Frequency (f_s)	Inductors (L_1 and L_2)	Capacitor (C)	Power
Values	48 V	12 V	30 V	30 kHz	2 mH and 2 mH	100 μF	1 kW

2.8.1. Converter Control

The control strategy was implemented using the proportional–integral (PI) controller; therefore, the system's parameters can be adjusted easily by adjusting the PI gains. Flexibility and easy implementation are the main advantages of the PI controller. The error between the output voltage and the reference voltage is minimized using the PI controller, as shown in Figure 10, to generate the desired duty ratio for the pulses of the MOSFETs in the different operation modes. To make the system underdamped, the integral coefficient is regulated to be KI = 0.3, and the steady-state error and overshoot are minimized by controlling the proportional coefficient to be Kp = 0.001.

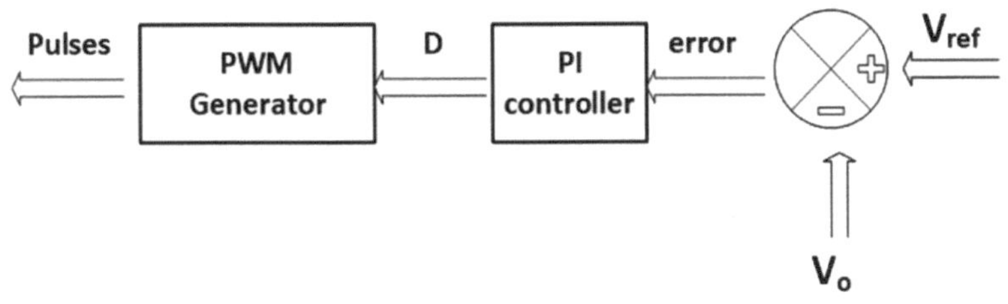

Figure 10. Converter controller.

2.8.2. Operation Modes of the Converter

The proposed converter, presented in Figure 9, works in four different operation modes, as we will discuss later to transfer the power between the DC link and the battery bank, discharge from the PV panel and battery bank simultaneously, and exchange the energy between the PV panel and battery bank. The four operating modes of the converter are:

Mode 1: Battery Bank to DC link

The inductor L_1 is charged and discharged during this operation mode in three time intervals T_1, T_2, and T_3 to discharge the battery bank only in the DC link to charge the DC and AC loads. The switches S_2 and S_3 are turned in the first time interval to charge the L_1 and hence the inductor voltage V_{L_1} is described by the following equation:

$$V_{L_1} = L\frac{di}{dt} = V_{Bt} \tag{1}$$

In the second interval, L_1 discharge through diodes D_1 and D_4 while switches S_2 and S_3 are turned off to supply the DC link. To maintain the continuous discharging of energy from L_1 in the DC link, the switches S_1 and S_4 are turned during the third time interval instead of the diodes in T_2. The switches S_1 and S_4 operate as synchronous rectifiers to reduce the voltage drop to a level of about 0.2 V; hence, the system efficiency improved. The voltage across L_1 in T_2, and T_3 is described by the following equation:

$$V_{L_1} = L\frac{di}{dt} = -V_{DC} \tag{2}$$

By simplifying Equation (3), and by applying the principle of volt-second balance for inductor L_1 using Equations (1) and (2), the following equation can be deduced:

$$V_{L_1} = D * V_{Bt} + (1 - D)(-V_{DC}) = 0 \tag{3}$$

Under the steady-state condition, the relation between DC-link voltage as an output and battery bank voltage as input is expressed using Equation (4).

$$V_{DC} = \frac{T_1}{T_2 + T_3} * V_{BT} = \frac{D}{1 - D} * V_{BT} \tag{4}$$

where D is the duty cycle ratio defined by $\frac{T_1}{T_S}$ where T_S is the total period of the switching cycle, and T_1 can be expressed as:

$$T_1 = L\frac{\Delta I_{L1}}{V_{BT}} \tag{5}$$

Additionally, T_S can be expressed as:

$$T_s = \frac{1}{f_s} \tag{6}$$

The battery bank voltage is boosted to the DC link with working in duty cycle equal $D > 0.5$. The DC link charges the battery bank by reversing the current in L_1. Under the steady-state condition and taking into consideration D to be T_1/T_S, the relation between DC-link voltage and battery bank voltage can be expressed using Equation (7).

$$V_{BT} = \frac{T_1}{T_2 + T_3} \times V_{DC} = \frac{D}{1-D} \times V_{DC} \quad (7)$$

Mode 2: PV Panel to DC Link

In this operating mode, the inductor L_2 is discharged in the DC link in three time intervals T_1, T_2, and T_3, similar to the previous operation mode to supply the DC link from the PV only. The switches S_2 and S_5 are turned in the first time interval to charge the L_2. The following equation describes the voltage across the inductor L_2.

$$V_{L_2} = L\frac{di_{L_2}}{dt} = V_{PV} \quad (8)$$

In the second interval, the L_2 discharge through diodes D_1 and D_6 while switches S_2 and S_5 are turned off to supply the DC link. In the third interval, the switches S_1 and S_6 are turned on for continued discharging of L_2 in the DC link and to reduce the voltage drop of the diodes. During these intervals, V_{L_2} can be described by the following equations:

$$V_{L_2} = L\frac{di_{L_2}}{dt} = -V_{DC} \quad (9)$$

By applying the principle of volt-second balance for inductor L_2 using Equations (8) and (9), the following equation can be deduced:

$$V_{L_2} = D \times V_{SC} + (1-D)(-V_{DC}) = 0 \quad (10)$$

Equation (11) shows the voltage of the DC link V_{DC} as an output voltage as a function of the PV panel V_{PV} input voltage.

$$V_{DC} = \frac{T_1}{T_2 + T_3} \times V_{PV} = \frac{D}{1-D} \times V_{PV} \quad (11)$$

where D is the duty cycle ratio equal $\frac{T_1}{T_S}$, and T_S is the total period of the switching cycle. The following equation expresses the relation between input and output voltages, and T_1 can be expressed as:

$$T_1 = L\frac{\Delta I_{L2}}{V_{DC}} \quad (12)$$

Mode 3: Battery Bank and PV Panel

In mode 3, the battery bank can be charged from the PV panel to extend the lifetime of the batteries during the peak hours and in sunny conditions. The PV panel work in buck operation mode to charge the battery bank, as shown in Equation (13), using the switches S_3, S_5, and S_6.

$$V_{BT} = \frac{T_1}{T_s} \times V_{PV} = D \times V_{PV} \quad (13)$$

$$\text{Where } T_1 = L\frac{2\Delta I_L}{V_{Bt}}, \text{ and } D = \frac{T_1}{T_s} \quad (14)$$

The switching sequence of the power switches in the three time intervals of the previous operation modes is summarized in Table 4.

Table 4. Modes (1), (2), and (3) switching states.

	Mode 1 (a)	Mode 1 (b)	Mode 2	Mode 3
T_1	S_2 , S_3	S_1 , S_4	S_2 , S_5	S_3 , S_6
T_2	D_4 , D_1	D_2 , D_3	D_6 , D_1	S_3, D_5
T_3	S_1 , S_4	S_2 , S_3	S_6 , S_1	S_3 , S_5

Mode 4: Battery Bank and PV Panel to DC Link

In this operation mode, the load from the grid can be mitigated during peak hours by simultaneously charging the loads from the battery bank and the PV panel. Inductors L_1 and L_2 are charged and discharged in five time intervals. Table 4 shows the DC link simultaneously from the battery bank and the PV panel using Equations (15) and (16).

$$V_{BT} = \frac{T_4 + T_5}{T_1} * V_{DC} = \frac{T_s - d_2 T_s}{d_1 T_s} * V_{DC} = \frac{1 - d_2}{d_1} * V_{DC} \tag{15}$$

$$V_{PV} = \frac{T_4 + T_5}{T_1 + T_2 + T_3} * V_{DC} = \frac{T_s - d_2 T_s}{d_2 T_s} * V_{DC} = \frac{1 - d_2}{d_2} * V_{DC} \tag{16}$$

where d_1 is the ratio of the on-time of switch S_3 to total switching period T_S and, similarly, d_2 corresponds to switch S_2. The switching sequence of the power switches in the five time intervals of the fourth operation mode is summarized in Table 5.

Table 5. Switching states in modes (4).

	T_1	T_2	T_3	T_4	T_5
Mode 4	S_2, S_3, S_5	S_2, D_4 , S_5	S_2, S_4, S_5	D_1, S_4, D_6	S_1, S_4, S_6

3. Testing and Validation

Table 6 summarizes the system's preliminary test and validation scenarios, which examine the proposed Microgrid's capabilities, function, and features. These tests and validations have been certified in the lab, including grid control, and switching between standalone and grid-connected systems considering islanding mode. In addition, battery charging and discharging control have been tested, showing the safe and accurate battery disconnection at threshold voltage values for a full charge and discharge checkpoints.

The variable AC and DC load based on relay-based logic control has been tested for manual and automatic modes. Bus management control for AC and DC buses has been evaluated at normal and fault conditions. Additionally, protection functions for load, battery pack, and power sources have been validated successfully. The functions of the co-simulation have been validated, showing how OPAL-RT interfaces the hardware components with the Simulink components of the MEG. In addition, the remote access control and data logging for monitoring purposes have been tested. The listed functions below can be considered the whole set of tasks that can be carried out with the proposed MEG setup. However, only some key functions and potential works that have been listed as the merits of the proposed MEG are selected to be presented in this section.

Because the proposed MEG has a redundant control system based on PLCs and microcontrollers, two graphical user interfaces (GUI) have been approached to facilitate the system's operation. Figure 11 shows the microcontroller-based GUI that includes a power source, battery charging, and load center control. The interface lets the user control the load, power sources, and energy storage. Figure 12 shows the HMI interface of PLC-based control that can carry out the load, power source, energy storage control, and battery charging control.

Table 6. Testing and validation parameters.

Grid side control	Grid control
	Islanding control
Load management	AC load control
	DC Load control
Power factor correction	PF capacitor control
Battery management	Battery charging
	Battery discharging
Bus management	AC bus control
	DC bus control
Collecting feedback signals and monitoring	Reading DC bus data
	Reading AC bus data
	Reading battery data
PV power source control	Maximum power tracking (PV)
	Connection mode
Protection	Battery protection
	DC side Faults
	AC side faults
	Battery protection
Remote access and monitoring	Remote access
	Monitoring and data logging

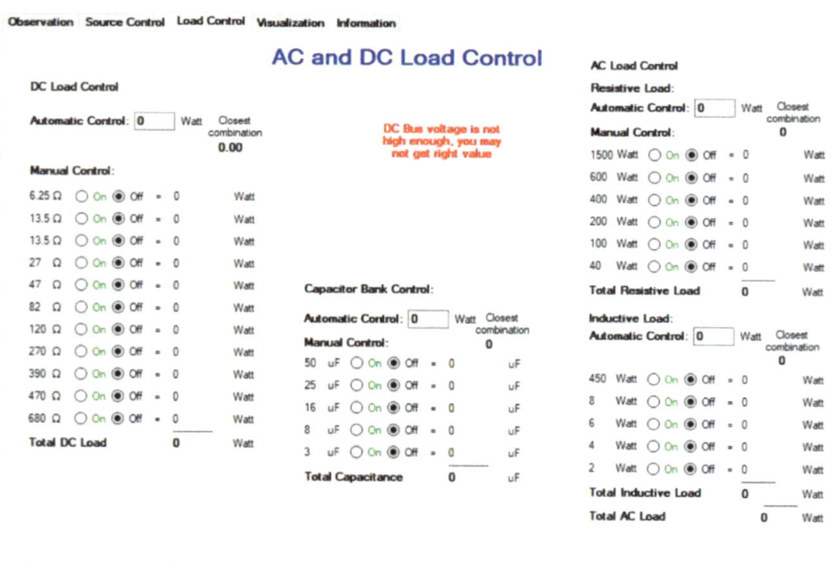

Figure 11. Microcontroller control user interface.

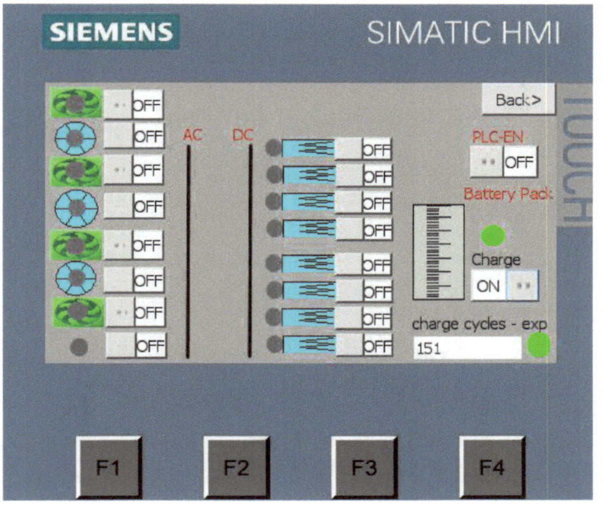

Figure 12. HMI of PLC controller interface.

4. Co-Simulation Based on OPAL-RT

As mentioned above, the main objective of co-simulation is to interface the hardware with the simulation software, the Simulink. To enable a Simulink model, flywheel and wind turbine, to act out like a physical power source or energy storage, the study utilizes OPAL-RT OP4510 with the input mentioned above and output resources to link between the physical signals in the experimental world with the input and output (IOs) data in a Simulink model. In this test scenario, Figure 13 shows Simulink's and OPAL-RT's analog inputs and outputs (IOs) interface. It shows how the IOs are connected to tags and variables in the Simulink model. This test scenario shows how the OPAL-RT IOs tags are connected to the simulation.

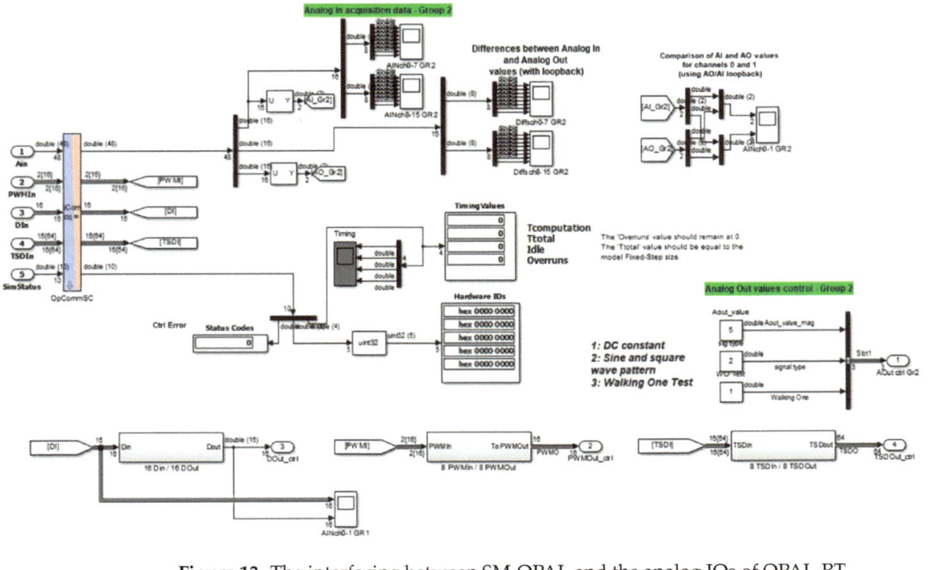

Figure 13. The interfacing between SM-OPAL and the analog IOs of OPAL-RT.

Figure 14 shows the analogous signals received by OPAL-RT from sensors and measurement devices and generates switching pulses for driving the power switch elements, i.e., MOSFETS in power converters accordingly.

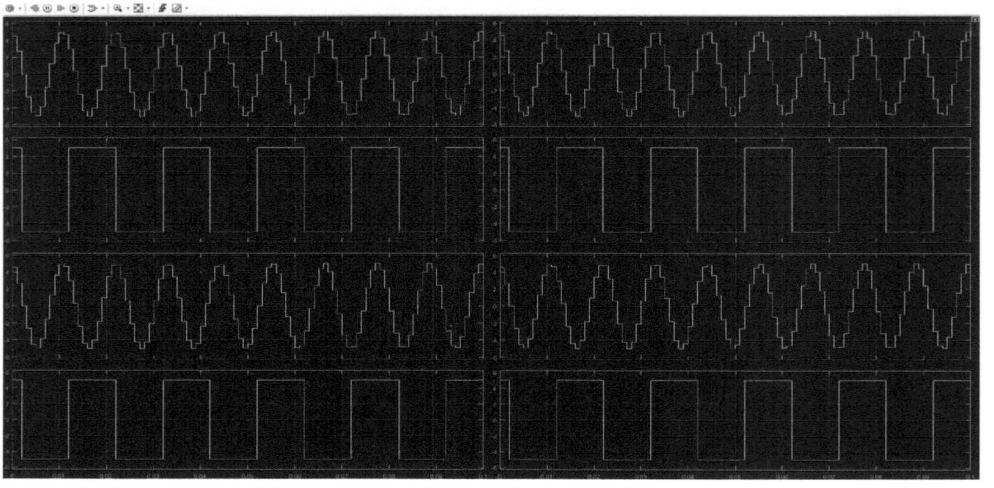

Figure 14. Input and output analog channels signals in OPAL-RT.

4.1. Testing Fail-Safe Logic

In this test case scenario, according to data listed in Table 7 that tabulates the collected statues of loads and power sources and their weights, the previous status of the loads and sources (On/Off), the total power of loads that were in ON status was 350 W (nodes 4, 5 and 6). Additionally, the full source power was 364 W (nodes 1, 2, and 3).

Table 7. Collected status of load and power sources.

Node ID	Type	Description	Status 1: Connected, 0: Disconnected	New Status Requested	Weight
1	DC source	Battery pack	1	1	144
2	DC source	Supercapacitor	1	0	100
3	DC source	Converter from grid	1	1	120
4	Resistive load	Bank of resistors	1	1	150
5	AC load	AC lamp 1	1	1	100
6	AC load	AC lamp 2	1	1	100
7	AC load	AC inductive load	0	1	450

The fifth column, which represents the new requested statues, shows a total load of 800 w (nodes 4 to 7) when the available power of sources was reported from the previous status to be 264 W (nodes 1 and 3). Therefore, emergency loads are the only ones that be allowed to be energized. As a result, nodes number 4 and 5 that are highlighted in red will be expected to be energized, and node 6 will be turned off to keep the balance between load demand and available power. Figure 15 shows the microgrid setup before and after requesting changes. The supper capacitor has been disconnected due to either fault-tolerant or user request as in Table 7, as shown by the meter reading in Figure 15. Consequently, the intelligent fail-safe algorithms determine that the system can continue running with the emergency loads only that is why AC lamp 1 keeps switched on as an emergency load. However, AC lamp 2 has been switched off as it is a normal load. Finally, after having the supercapacitor removed out of the system, the remaining available power will be sufficient for the demanded power.

Figure 15. Fail-safe test scenario results on microgrid setup.

4.2. *Fail-Safe Testing*

4.2.1. Source out of Service

In this scenario, depicted in Figure 16, a fault condition has been considered within the power source to make an overload condition. In this condition, the controller will detect the faulty power source, calculate the total power generation, and compare it with the load demand. If the demand is high, the controller will disconnect some of the loads from the bus connector and only support the emergency load; thus, the system can achieve a continuation.

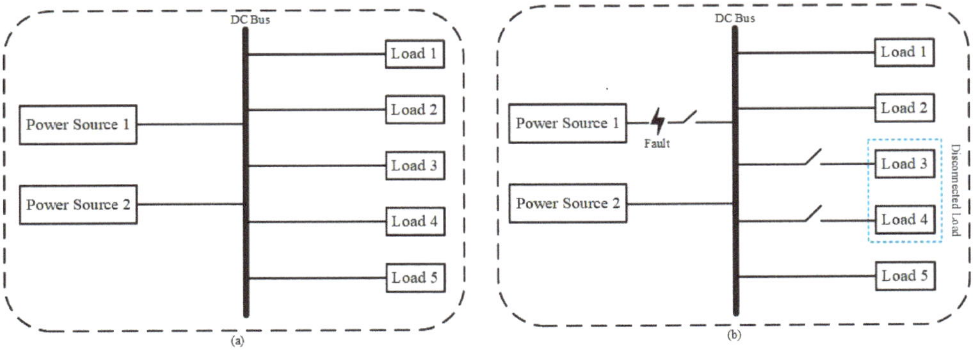

Figure 16. (**a**) Normal operation; (**b**) operation during fault condition.

4.2.2. Short Circuit Condition

In this scenario, depicted in Figure 17, a short circuit condition has been applied within the system to make the system stop by force. A circuit breaker is used within the components and controlled by the PLC to bypass this fault. Whenever a component gets short-circuited, the controller will detect and activate the breaker to bypass this component, and the system will continue its safe operation.

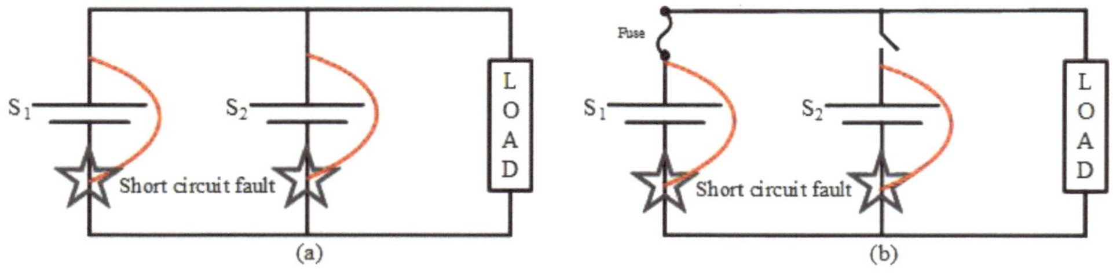

Figure 17. (**a**) Short circuit fault without FTC; (**b**) short circuit fault with FTC.

Figure 18 shows the results of the fault-tolerant test on the microgrid experimental setup. The test scenario is that one of the power sources has gotten out of the system due to a breaker's reaction to a fault. Without fault-tolerant considerations, the system will be totally down. However, the fault-tolerant feature enables the system to disconnect the faulty power source and maintain the system operation. Because the remaining amount of power decreased by losing one of the power sources, the fault-tolerant combines with a fail-safe algorithm to calculate the emergence load. Therefore, the emergency loads are only allowed to continue energized, as shown before and after the fault.

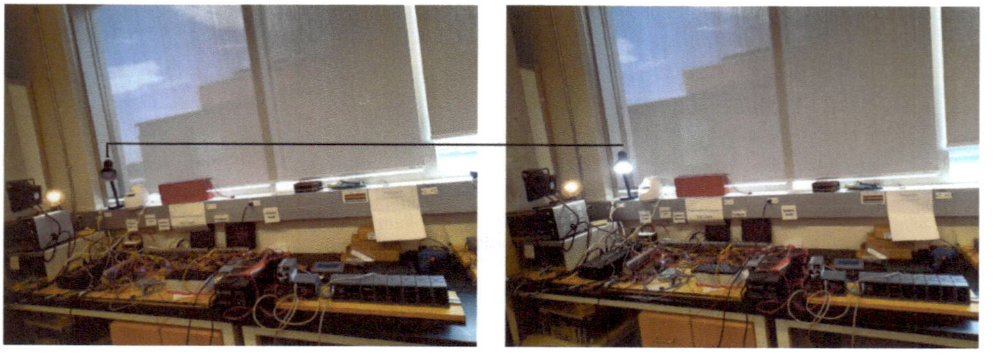

Figure 18. Fault tolerance test results of the microgrid.

5. Conclusions

This study introduces an experimental platform for a microgrid with distinct features, such as enabling extensible and sizable AC and DC load and combining physical and emulated power sources and storage systems, aiming to increase the system flexibility by utilizing real-time simulation OPAL-RT OP4515. In addition, the design includes fail-safe and resiliency features by developing a redundant control system based on microcontrollers and PLCs. Furthermore, PLC and relay logic control enable the sizing control of the load demand.

The system combines physical and emulated power sources and energy storage, aiming to increase flexibility and diminish the limitations of including different types and sizes of power and energy sources. For example, the wind turbine as a power source and flywheel as energy storages are impractical to incorporate within an indoor experimentation setup. Using co-simulation based on interfacing hardware components with Simulink based on OPAL-RT real-time simulation opened the door to include simulated components to act on the system the way the physical sources acts. Utilizing PLC combined with relay logic control successfully shows the ability to take the system from total failure and shutdown to fail-safe with balanced load demands and power generation. Applying the redundant control by utilizing PLC and microcontrollers increased the system reliability because controllers can cover each other. It only reinforces the benefits that microgrids integrate physical and emulated energy sources with energy management and storage systems to pursue resiliency, fail-safe, and fault-tolerant capabilities.

Furthermore, a complex and long-term transition is necessary to achieve a resilient and reliable energy system, primarily based on renewable energy and high energy efficiency, aiming for sustainability, robustness, and adaptiveness. The paper presents the design of a proposed microgrid that is enriched with various power sources and energy storage along with variable AC and DC loads. This paper demonstrates a complete scenario of real-time simulation based on emulated energy sources to be integrated with the physical sources to complement the ecosystem of the proposed microgrid. Additionally, a novel fault-tolerant and fail-safe algorithm have been implemented and tested. In addition, a redundant control system based on a microcontroller array and PLC has been implemented and employed to provide the proposed microgrid's resiliency. Testing and validation of the claimed key features and potential merits of the implemented MEG platform yield results that prove the capabilities of reliability based on the redundant control system, fail-safe operation, fault tolerance, and resiliency of the proposed system.

Author Contributions: Conceptualization, All; methodology, H.A.G., Y.E. and M.I.; software, All; validation, All; formal analysis, All; investigation, H.A.G.; resources, All; data curation, All; writing—original draft preparation, All; writing—review and editing, All; visualization, All; supervision, H.A.G.; funding acquisition, H.A.G. All authors have read and agreed to the published version of the manuscript.

Funding: This research was funded by NSERC grant number [210320].

Institutional Review Board Statement: Not applicable.

Informed Consent Statement: Not applicable.

Data Availability Statement: The study did not report any data.

Acknowledgments: This research is supported by NSERC Discovery Grant. The authors would like to thank members of the Smart Energy Systems Lab (SESL) at Ontario Tech University, Canada for their support.

Conflicts of Interest: The authors declare no conflict of interest.

References

1. Bostan, I.; Gheorghe, A.; Dulgheru, V.; Sobor, I.; Bostan, V.; Sochirean, A. *Resilient Energy Systems: Wind, Solar, Hydro*; Springer: Berlin/Heidelberg, Germany, 2012.
2. Jesse, B.-J.; Heinrichs, H.U.; Kuckshinrichs, W. Adapting the theory of resilience to energy. *Energy Sustain. Socie* **2019**, *9*, 27. [CrossRef]
3. Energy Weekly News. EEI Statement on the U.S. Department of Energy's Energy Grid Reliability Study. *Gale Acad. OneFile* **2017**, 87. Available online: https://www.lelezard.com/en/news-14489102.html (accessed on 15 April 2022).
4. Abi-Samra, N.C. One Year Later: Superstorm Sandy Underscores Need for a Resilient Grid. IEEE Spectrum, 4 November 2013. Available online: https://spectrum.ieee.org/one-year-later-superstorm-sandy-underscores-need-for-a-resilient-grid (accessed on 2 April 2022).
5. *IEEE Std 1366–2003*; 1366–2012—IEEE Guide for Electric Power Distribution Reliability Indices. IEEE: Piscataway Township, NJ, USA, 2012.
6. Roege, P.E.; Collier, Z.A.; Mancillas, J.; McDonagh, J.A.; Linkov, I. Metrics for energy resilience. *Energy Policy* **2014**, *72*, 249–256. [CrossRef]
7. Fox, J. *Sustainable Electricity—Case Study from Electric Power Companies in North America*; Springer: Berlin/Heidelberg, Germany, 2016.
8. Oh, S.; Heo, K.; Jufri, F.H.; Choi, M.; Jung, J. Storm-Induced Power Grid Damage Forecasting Method for Solving Low Probability Event Data. *IEEE Access* **2021**, *9*, 20521–20530. [CrossRef]
9. Hatti, M. *Smart Energy Empowerment in Smart and Resilient Cities*; ICAIRES: Taghit-Bechar, Algeria, 2019.
10. Jasiūnas, P.D.J.M.J. Energy system resilience—A review. *Renew. Sustain. Energy Rev.* **2021**, *150*, 111476. [CrossRef]

11. Egert, R.; Daubert, J.; Marsh, S.; Muhlhauser, M. Exploring energy grid resilience: The impact of data, prosumer awareness, and action. *Patterns* **2021**, *2*, 100258. [CrossRef] [PubMed]
12. Borlase, S. *Smart Grids—Advanced Technologies and Solutions*; CRC Press: Boca Raton, FL, USA, 2017.
13. Lantero, A. Department of Energy—Energy.gov. Available online: https://energy.gov/articles/how-microgrids-work (accessed on 15 April 2022).
14. Berkeley Lab. Optimizing Energy Resources on the Grid. Grid Integration Group. 2019. Available online: https://building-microgrid.lbl.gov/about-microgrids (accessed on 17 June 2014).
15. Parhizi, S.; Lotfi, H.; Khodaei, A.; Bahramirad, S. State of the Art in Research on Microgrids: A review. *IEEE Access* **2015**, *3*, 890–925. [CrossRef]
16. Ortiz, L.; Gonzalez, J.W.; Gutierrez, L.B.; Llanes-Santiago, O. A review on control and fault-tolerant control systems of AC/DC microgrids. *Heliyon* **2020**, *6*, e04799. [CrossRef] [PubMed]
17. Estefania, P.; Asier, G.-d.-M.; Andreu, J.; Kortabarria, I.; de Alegria, I.M. General aspects, hierarchical controls and droop methods in microgrids: A review. *Renew. Sustain. Energy Rev.* **2013**, *17*, 147–159.
18. Hare, J.; Shi, X.; Gupta, S.; Bazzi, A. Fault diagnostics in smart micro-grids: A survey. *Renew. Sustain. Energy Rev.* **2016**, *60*, 1114–1124. [CrossRef]
19. Simani, S.; Castaldi, P. Active actuator fault-tolerant control of a wind turbine. *Int. J. Robust Nonlinear Control* **2013**, *24*, 1283–1303. [CrossRef]
20. Aldeen, M.; Saha, S.; Gholami, S. Fault tolerant control of electronically coupled distributed energy. *Electr. Power Energy Syst.* **2018**, *95*, 327–340.
21. Bevrani, H.; François, B.; Ise, T. *Microgrid Dynamics and Control*; Wiley: Hoboken, NJ, USA, 2017.
22. Alvarez-Diazcomas, A.; Lopez, H.; Carrillo-Serrano, R.V.; Rodríguez-Reséndiz, J.; Vazquez, N.; Herrera-Ruiz, G. ANovelIntegrated Topology to Interface Electric Vehicles and Renewable Energies with the Grid. *Energies* **2019**, *12*, 4091. [CrossRef]
23. Kumar, G.V.B.; Sarojini, R.K.; Palanisamy, K.; Padmanaban, S.; Holm-Nielsen, J.B. Large Scale Renewable Energy Integration: Issues and Solutions. *Energies* **2019**, *12*, 1996. [CrossRef]
24. Ming, W. *Power Electronic Converters for Microgrids*; IntechOpen: Sankt Veit am Flaum, Croatia, 2021.
25. Dhananjaya, M.; Pattnaik, S. Design and implementation of a multi-input single-output DC-DC covnerter. In Proceedings of the 2019 IEEE International Conference on Sustainable Energy Technologies and Systems, Bhubaneswar, India, 26 February–1 March 2019; pp. 194–199.
26. Athikkal, A.; Sundaramoorthy, K.; Sankar, A. Development and performance analysis of dual-input DC-DC converters for DC microgrid application. *IEEJ Trans. Electr. Electron. Eng.* **2018**, *13*, 1034–1043. [CrossRef]
27. Jeong, Y.; Park, J.; Rorrer, R.; Kim, K.; Lee, B. A Novel multi-input and single-output DC/DC converter for small unmanned aerial vehicle. In Proceedings of the IEEE Applied Power Electronics Conference and Exposition (APEC), New Orleans, LA, USA, 15–19 March 2020; pp. 1302–1308.
28. OPAL-RT Technologies. Professional Real Time Digital Simulation Software. OPAL-RT. Available online: https://www.opal-rt.com/product/software-rt-lab/-professional-real-time-digital-simulation-%20software (accessed on 5 May 2022).
29. Sailer, R.; Oliver, H.; Budzisz, Ł. NMLab: A Co-Simulation Framework for Matlab. In Proceedings of the 2010 Second International Conference on Advances in System Simulation (SIMUL), Nice, France, 22–27 August 2010; pp. 152–157.
30. Sabzehgar, R. Overview of Technical Challenges, Available Technologies and Ongoing Developments of AC/DC Microgrids. In *Development and Integration of Microgrids*; IntechOpen: Sankt Veit am Flaum, Croatia, 2016.
31. Wang, J.; Feng, T.; Wang, B.Q. Fault Tolerant Control of Sensor Faults in Microgrid Inverter Control System. *J. Phys. Conf. Ser.* **2021**, *1993*, 012003. [CrossRef]
32. Fields, S. What Are Microgrids and How do They Work? Energysage—Smart Enegy Decidions. Available online: https://news.energysage.com/what-are-microgrids/ (accessed on 17 January 2019).

technologies

MDPI

Article

Solving Dual-Channel Supply Chain Pricing Strategy Problem with Multi-Level Programming Based on Improved Simplified Swarm Optimization

Wei-Chang Yeh *, Zhenyao Liu, Yu-Cheng Yang and Shi-Yi Tan

Integration and Collaboration Laboratory, Department of Industrial Engineering and Engineering Management, National Tsing Hua University, Hsinchu 30013, Taiwan; liuzhenyao49@gmail.com (Z.L.); catherineyang@gapp.nthu.edu.tw (Y.-C.Y.); s108034871@m108.nthu.edu.tw (S.-Y.T.)

* Correspondence: yeh@ieee.org

Abstract: With the evolution of the Internet and the introduction of third-party platforms, a diversified supply chain has gradually emerged. In contrast to the traditional single sales channel, companies can also increase their revenue by selling through multiple channels, such as dual-channel sales: adding a sales channel for direct sales through online third-party platforms. However, due to the complexity of the supply chain structure, previous studies have rarely discussed and analyzed the capital-constrained dual-channel supply chain model, which is more relevant to the actual situation. To solve more complex and realistic supply chain decision problems, this paper uses the concept of game theory to describe the pricing negotiation procedures among the capital-constrained manufacturers and other parties in the dual-channel supply chain by applying the Stackelberg game theory to describe the supply chain structure as a hierarchical multi-level mathematical model to solve the optimal pricing strategy for different financing options to achieve the common benefit of the supply chain. In this study, we propose a Multi-level Improved Simplified Swarm Optimization (MLiSSO) method, which uses the improved, simplified swarm optimization (iSSO) for the Multi-level Programming Problem (MLPP). It is applied to this pricing strategy model of the supply chain and experiments with three related MLPPs in the past studies to verify the effectiveness of the method. The results show that the MLiSSO algorithm is effective, qualitative, and stable and can be used to solve the pricing strategy problem for supply chain models; furthermore, the algorithm can also be applied to other MLPPs.

Keywords: dual-channel supply chain; pricing strategy; Stackelberg game; multi-level programming; improved simplified swarm optimization

Citation: Yeh, W.-C.; Liu, Z.; Yang, Y.-C.; Tan, S.-Y. Solving Dual-Channel Supply Chain Pricing Strategy Problem with Multi-Level Programming Based on Improved Simplified Swarm Optimization. *Technologies* **2022**, *10*, 73. https://doi.org/10.3390/technologies10030073

Academic Editors: Manoj Gupta, Eugene Wong and Gwanggil Jeon

Received: 20 May 2022
Accepted: 9 June 2022
Published: 11 June 2022

1. Introduction

Supply chain systems have been progressively diversifying besides conventional retailing manners. Nowadays, the increased competition and globalization of the market have become necessary for different individuals in the supply chain to cooperate to achieve mutual benefits. The competitions within the supply chain have catching researchers' attention [1,2].

Supply chain management (SCM) handles the entire production flow of a good or service, which is a network that moves the product along from the suppliers of raw materials to those organizations that deal directly with users. In addition, due to the invention and growth of the internet, the prosperity of third-party platforms—online retailing, has gradually increased; companies can engage an additional sales channel in direct selling their products to customers and create various ways of sales to increase income. The appearance of multi-channel supply chain management issues is due to the rapid growth of e-commerce, which has led some manufacturers to sell their products online and increase their sales channels to remain competitive and increase the accessibility of their products. The increase in sales channels represents the complexity of their competition as well as coordination. The market demand is sensitive to the selling price set by the seller, so in a supply chain system, pricing strategy is a complex and tedious decision with numerous factors that affect it.

With all these buying and selling behaviors in business activities, many variables are considered by either the seller or the buyer. Each party is expected to achieve its desired benefit or goal in conducting the activity. Under this premise, the parties coordinate to reach a consensus and gain equilibrium through repeated communication for possible requests to achieve compromises. Therefore, replacing traditional corporate goals with overall value maximization through a holistic approach to the supply chain is a key issue for companies to consider nowadays. Decentralized decision-making occurs when there is a conflict between decision-makers. A hierarchical structure of decentralized decision-making should be carried out according to organizational departments. The objectives of the decision-makers are independent, and they are aimed at maximizing their own profits.

In reality, however, the situation is actually not that simple. There may be some conflict of interest because the asymmetry of the market position causes an asymmetry in the order of decision making for this German economist Heinrich Freiherr von Stackelberg proposed the Stackelberg model in 1934 to describe this situation of priority order decision making with leaders and followers [3]. In addition, the equilibrium point of this problem is determined through the solution of the Stackelberg game. The Nash equilibrium does not guarantee the best resolution for all decision-makers, the result may not be the most favorable situation, but it is an acceptable outcome for all parties. As a result, some studies use the method of multi-level programming (MLP), a mathematical model for solving decentralized decision-making problems, as an extension of the Stackelberg game to find the solution [4–6]. The key feature of this model is that the decision-makers have independent objective functions at each level of the hierarchy and control over the selection of decision variables.

The earliest proofs that a multi-level programming problem (MLPP) is an NP-hard problem are Ben & Blair (1990) [7] and Bard (1991) [8], and the bi-level programming problem (BLPP) they solved and proposed is derived from the MLPP problem, so the more complex MLPP also belongs to the NP-hard problem. Because of its limitations and complexity, it is more difficult to solve large-scale problems by mathematical planning methods. In recent years, researchers have adopted the more efficient meta-heuristic algorithm to obtain approximate solutions [6,9–12], which may not always lead to the best solution, but can handle more complex MLPP problems.

Nowadays, with increasingly complex supply chain relationships, companies need to be equipped with better decision models to manage their own goals. These problems can be solved by proposing suitable algorithms that can solve MLPP in a reasonable time and, at the same time, obtain an acceptable quality of the solution. The relationship between the supply chain network and the logistics distribution scheduling as regards applying swarm optimization algorithms proposed by some scholars, they harnessed the machine learning method, algorithms in the retailing environment in dynamic assessment to determine the users' trends and patterns and grasp customer attitudes and feelings [13–16].The improved Simplified Swarm Optimization (iSSO) designed by Yeh [17] in 2015 is one of the evolutionary algorithms and stochastic optimization algorithms. It is characterized by the simplicity and efficiency of the iterative method. The algorithm demonstrates its excellent efficiency and generates high-quality solutions in solving most of the continuity problems.

Based on the above-mentioned excellent features, we propose an approach that uses iSSO to further optimize the pricing strategy by constructing an MLPP model that can effectively maximize the profit among all parties.

The purpose of this study is to investigate the use of an MLPP to solve the pricing problem of the supply chain, considering the financing decision options under dual sales channels where different options generate different interactions among the parties in the supply chain that will affect the pricing strategy. Therefore, this paper first uses the study of Zhen (2020) [18] as the basis of the mathematical model of a dual-channel supply chain system to analyze the profit formula of each party under different financing strategies.

This study aims to develop a method to solve the NP-hard problem of MLPP and construct and investigate the multi-level supply chain system by exploring how the decision-makers in the supply chain system should decide on the best pricing strategy to maximize the profit. By considering various competing influences, the benefits of the supply chain system are maximized by making pricing decisions that satisfy all parties. Thus, the ideal situation is to prioritize the manufacturer's best interests and minimize the costs of all parties while achieving the best interests of the other parties.

The study objectives can be summarized as follows,

1. Build an MLPP model to obtain the equilibrium solution of pricing strategy in the dual-channel supply chain system.
2. Study and analyze the best decision for the manufacturer on finance strategy.
3. Apply the improved, simplified swarm optimization algorithm to multi-level programming problems.

The rest of this article is organized as follows. Section 2 lists the theoretical basis of the research. Section 3 describes the supply chain model, including proposed symbols, assumptions, and mathematical models. In Section 4, we introduce the research method, including the concept of MLPP and iSSO, and discuss the novelty and steps of the proposed MLiSSO. In Section 5, we analyzed the effects of the proposed method and described in detail the results of the above-mentioned supply chain pricing strategy. Finally, our conclusions are given in Section 6.

2. Literature Review

2.1. Dual-Channel Supply Chain

Nowadays, due to the rapid development of technology, various sales models have emerged in our society. In response to the purchasing habits of the new generation, the development of online retailing has become more and more prevalent.

In addition to traditional sales channels, upstream manufacturers in the supply chain are gradually developing channels to sell their products directly online. In this way, sales can be managed through a third-party platform without expanding your physical store or website and only require the costs associated with the platform, such as profit sharing and rent; the structure of the dual-supply chain is shown in Figure 1. Various studies on the supply chain phenomenon are also available in the market with service competition [19], channel selection [20], pricing strategies [21], and dual-channel supply chains [22].

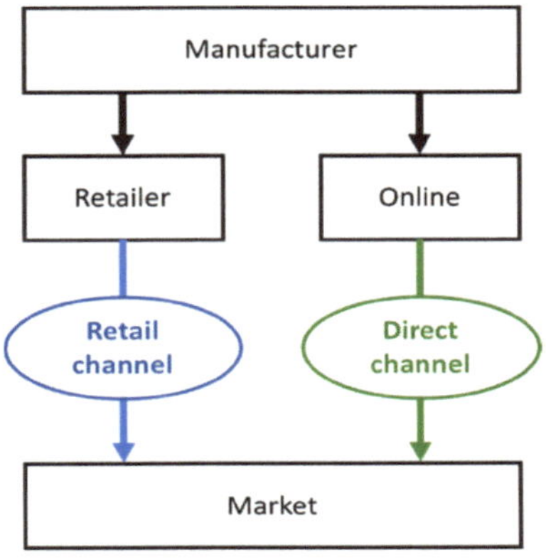

Figure 1. Structure of the dual-channel supply chain.

Channel competition holds an important role in dual-channel supply chain management. For example, Bernstein et al. (2009) address how competition between both retail and direct channels affects decisions made by manufacturers on supply chain structure [23]. Ryan et al. (2012) discussed the price competition and coordination in a dual-channel model [24]. Saha (2016) compared the performance of the manufacturer, the distributor, the retailer, and the entire supply chain in three different supply chain structures to prove that under some conditions that a dual-channel can outperform a single retail channel [25].

However, the studies on dual-channel supply chains mostly do not assume that firms are capital constrained; therefore, this study uses the financing strategy preferences of capital-constrained firms in the dual-channel supply chain proposed by Zhen in 2020 [18] while considering the financing strategies of third-party platforms in SCM. As a result, this study considers the operational management and financing strategy preferences of supply chain systems in the above-mentioned points.

Therefore, based on the model proposed by Zhen [18], this study examines the two aforementioned financing approaches for dual-channel competition and consumer considerations and presents the decision relationships between manufacturers and retailers with three different financing strategies. In addition, we compare the impact of cost and revenue on manufacturers, retailers, and lenders in the supply chain, maximizing profit and minimizing each cost to obtain the best pricing decision for the entire supply chain.

2.2. Supply Chain Finance

As the members of the supply chain gain benefits by selling their products while the market demand is sensitive to the selling price of the products, therefore, the pricing decision plays an important role in the profit optimization of the supply chain [6]. In considering changes in the correlation between product prices and market demand, companies can make profit analysis and pricing strategies efforts [26].

Lack of funding may be a hindrance to business development. There are two types of financing discussed in the literature on supply chain financing. One type of financing is external financing, which is defined as loans from institutions outside the supply chain, such as banks, third-party logistics, or other financial institutions. The other is internal financing, defined as loans from companies in the supply chain to their upstream or downstream companies, such as trade credit and buyer's credit [27].

Most research on internal financing has examined trade credit financing, with the majority of studies focusing on contract coordination and operational decisions under credit risk [28,29]. For external financing, the emphasis is on how the financing affects inventory or operations management and supply chain coordination [30,31]. Unlike the previous studies, Zhen (2020) focuses on the capital constraints of upstream firms under channel competition. This study is significant in examining how the capital constraints of upstream manufacturers affect the operation of dual channels [18].

When sales are not limited to traditional retail channels, to maximize the overall revenue is to develop a multi-channel pricing strategy, and it is cooperation and negotiation between each member in the supply chain system which can be considered as a game. For example, Matsui (2017) proposed that it would be appropriate for the manufacturer to release the direct selling price before the wholesale price is set. A sub-game perfect Nash equilibrium with the non-cooperative game of channel members is reached, and the manufacturer's profit is maximized [1]. The subgame perfect Nash equilibrium of the non-cooperative game with channel members is reached, and the manufacturer's profit is maximized.

2.3. Game Theory

Game theory is considered to be one of the most effective tools for dealing with these management problems. The well-known Prisoner's Dilemma and the Nash Equilibrium of modern noncooperation have become important concepts in game theory.

The strategic interactions between players are what game theory studies as the real-life dilemma that we often encounter. A strategic interaction means that the optimal choices of one player depend on other players' optimal choices and vice versa. Assume that each player is aware of the equilibrium strategies of the other players. In addition, none of the players gains any benefit by unilaterally changing its own strategy.

Increasingly, research papers are applying game theory to supply chain management [2,32,33]. Cachon and Zipkin [34] addressed the Nash equilibrium in a non-cooperative supply chain with one supplier and multiple retailers. Hennet and Arda [35] evaluated the efficiency of different types of contracts among industrial parties in a supply chain. Tian et al. [36] proposed a dynamic system model for green supply chain management based on evolutionary game theory, which applied game-theoretic methods to decision-making purposes.

2.4. Stackelberg Game

Several researchers have studied through game theory about coordination between manufacturers and retailers [37,38]. Each member attempts to maximize their own profit, a situation known as a non-cooperative game.

Since the market position asymmetry leads to the asymmetry of the decision sequence, there may be some conflict of interest. For this reason, the German economist Heinrich Freiherr von Stackelberg proposed the Stackelberg model in 1934 [3]. The Stackelberg model emphasizes the sequential relationship of decisions. In a game, the player who decides on a decision firstly is called the first player, while the other player is called the follower. When the first player decides his own strategy, he has already taken into account the possible decisions made by the followers in response to the first player's decision. After the first player's decision, the follower observes the first player's decision and thinks about the effect of the strategy on itself, and then makes the best response decision. The whole process means that both sides in the game make decisions based on the pursuit of their own best goals while considering the possible best response of the other side.

In the Stackelberg non-cooperative game, the dominant (leader) member controls the other members who act after the leader (followers). After estimating the reactions of other members, the leader will take the first decision [39]. The aforementioned hierarchical structure and the sequential nature of decision-making are consistent with the context set by Stackelberg's theory. Therefore, the main modeling framework in this paper applies the Stackelberg model in the tournament.

According to the number of participants, the Stackelberg game can be divided into four main structures, as shown in Figure 2.

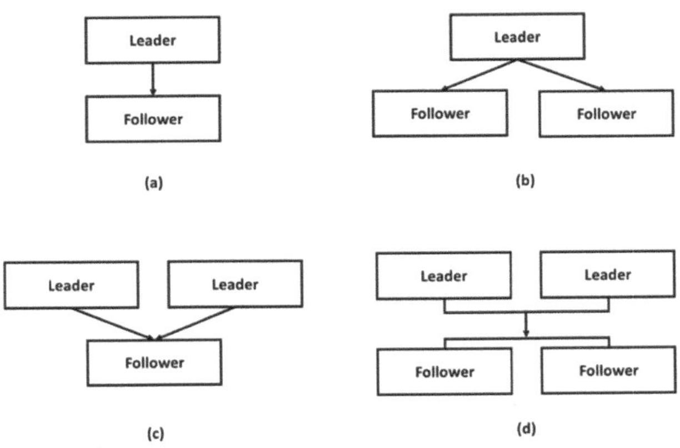

Figure 2. Relationship between participants in the Stackelberg game. (**a**) represents a single leader and follower, (**b**) represents a single leader and multiple followers, (**c**) represents multiple leaders and single followers, and (**d**) represents multiple leaders with multiple followers.

2.5. Multi-Level Programming Problem

In this section, we first review the development of techniques for solving the Stackelberg game problem, then addresses the general formulation of the bi-level programming problem model and multi-level programming problem model.

The multilevel programming problem (MLPP) is an extension of the Stackelberg game [39]. It aims to solve decentralized planning involving multiple decision-makers, where each member seeks to maximize its own interests in a hierarchical organization. This mathematical model has been widely used in practical problems such as resource allocation [4], transportation network design [5], and pricing and lot-sizing [6].

When decision-makers conflict with each other, a decentralized decision-making problem arises. The decentralized decision-making should be by organization departments and form a kind of hierarchical structure. The decision makers' objectives are independent and may have some conflict of profit. Every decision-maker always wants to achieve a win-win situation called "dominant strategic equilibrium." However, in reality, the situation is actually not that simple. Nash equilibrium does not guarantee the best solution for every decision-maker, but it can get the best solution under the consideration of the entire group; therefore, multilevel programming (MLP) would be needed to find a solution. Zhou (2012) used game theory to determine the optimal pricing strategy to maximize the multilevel remanufacturing reverse supply chain [40]. Sadigh et al. (2012) found the optimal equilibrium of price, advertising spending, and production strategy in a bi-level programming approach [41].

The multilevel programming model has more advantages compared to the traditional single-level programming model. Its main benefits are (1) multilevel planning can be applied to analyze both different or even conflicting objectives in the decision process; (2) The multi-criteria approach of bi-level planning for decision-making can better reflect the actual problem; (3) The multi-level planning approach can denote the interactions between decision-makers.

In the current development of multi-level programming, several challenges emerge (1) Large scale—due to high dimensional decision variables for multi-level decision problems which become complex; (2) Uncertainty—with the uncertain information causing imprecise or unclear decision parameters and conditions for the decision subjects concerned; (3) Variety—with the possibility of the existence of multiple decision subjects with various relationships among them in each decision level. Yet, existing decision models or solution methods cannot fully and effectively handle these large-scale, uncertain and diverse multilevel decision problems [42].

There are two fundamental problems in solving MLPPs from a practical point of view. The first is the way to construct a multilevel decision model that describes the hierarchical decision process. Depending on the number of objectives involved, including dual objectives or multiple objectives; the number of members involved, including single leaders and followers or multiple leaders and followers; and the number of layers in the structure, including the bi-level programming problem (BLPP) or the MLPP. The BLPP is a special type of MLPP, and most of the research has been devoted to the BLPP study [9–11,41,43]. In addition, MLPP making that it is more complex than BLPP has been studied in depth in model building.

A further problem is how to identify methods for optimizing decisions. Several solving methods have been developed to solve these problems, broadly classified as exact algorithms and intelligent optimization algorithms. On the basis of the complexity of solving MLPP solutions, Ben & Blair (1990) proved through the well-known knapsack problem that BLPP is an NP-hard problem [7], and Bard (1991) even proved that BLPP is also an NP-hard problem through the search for locally optimal solutions [8]. This leads to exact algorithms that are time-consuming in solving nonlinear, discrete, and multi-optimal versions of large-scale problems that rely heavily on target function differentiability, which is not universally applicable [42].

At present, to obtain the optimal solution of MLPP, metaheuristic algorithms or innovative computations have been designed and widely used to solve BLPP and MLPP, i.e., Liu (1998) proposed a genetic algorithm for solving the Stackelberg-Nash equilibrium problem for generic MLPP with multiple followers [12], and Ma et al. (2013) using Particle Swarm Optimization (PSO) to solve BLPP on supply chain model [6]. Moreover, extending these algorithms to solve MLPPs is difficult and sometimes almost impossible. The main reason why solving MLPPs remains difficult is the lack of efficient algorithms; this is the biggest obstacle to the MLP problem [35,37].

Consequently, a more efficient algorithm has to be developed to solve large-scale BLPP and these algorithms can also be extended to solve MLPP. Thus, in this paper, we propose a multi-level improved, simplified swarm optimization (MLiSSO) method to solve the complex pricing strategy problem of a dual-channel supply chain involving multi-decision-makers, which are applied with a multi-level structure.

2.5.1. Bi-Level Programming Problem

A special case of a multi-level programming problem(MLPP) with a two levels structure is the bi-level programming problem (BLPP) [44]. The general form of the BLPP structure is shown in Figure 3.

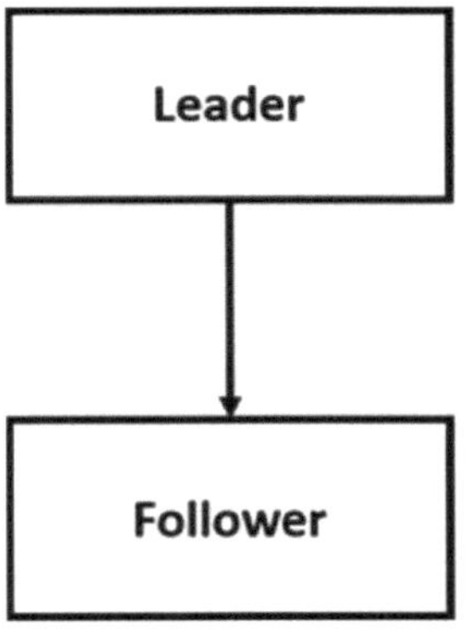

Figure 3. The general form of BLPP structure.

Assume that upper-level decision-makers are given control over X, and lower-level decision-makers are given control over Y. Thus, we have $x \in X \subset R^P$, $y \in Y \subset R^q$, and $F,\ f : R^P \times R^q \rightarrow R^1$. The general BLPP can be formulated as follows:

$$\underset{x \epsilon X}{Min} F(x,y) \text{ (Leader)} \tag{1}$$

$$s.t. \quad G(x,y) \leq 0 \tag{2}$$

where y, for each x fixed, solves the problems Equations (3) and (4).

$$\underset{y \epsilon Y}{Min} f(x,y) \text{ (Follower)} \tag{3}$$

$$s.t. \quad g(x,y) \leq 0 \tag{4}$$

The leader is the upper-level decision-maker Equation (1), and the follower is the lower-level decision-maker Equation (3). Depending on the demands of the model, x and y may have some additional restrictions, such as integer restrictions or limits on upper and lower bounds.

Based on these, we have the following definitions [45]:

Definition 1.1.

1. *The problem constraint region,*

$$S = \{(x,\ y) \in X \times Y : G(x,y) \leq 0,\ g(x,\ y) \leq 0\} \tag{5}$$

2. *The follower feasible set for each fixed x,*

$$S(x) = \{y \in Y : g(x,\ y) \leq 0\} \tag{6}$$

3. *The follower rational reaction set,*

$$P(x) = \{y \in Y : y \in arg\ min[f(x,y) : y \in S(x)]\}. \tag{7}$$

4. *The problem inducible region (IR),*

$$IR = \{(x,y) : (x,y) \in S, y \in P(x)\}. \tag{8}$$

5. *The problem optimal solution set,*

$$OS = \{(x,y) : (x,y) \in argmin[F(x,y) : (x,y) \in IR]\} \tag{9}$$

Definition 1.2. *This section may be divided into subheadings. It should provide a concise and precise description of the experimental results, their interpretation, as well as the experimental conclusions that can be drawn.*

Definition 1.3. *For* $\forall(x,y) \in IR,\ if\ \exists_{(x^*,y^*)} \in IR,\ F(x^*,y^*) \leq F(x,y)$, *then* (x^*,y^*) *is an optimal solution of problem.*

2.5.2. Multi-Level Programming Problem

In many applications, the problem of decentralized decision-making within a hierarchical system tends to include more than two levels, which are known as multi-level programming problems (MLPP). The general form of MLPP—tri-level structure is shown in Figure 4.

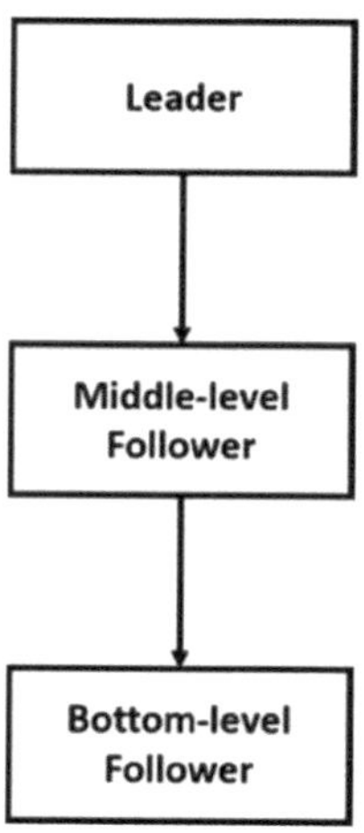

Figure 4. The general form of MLPP—tri-level structure.

For $x \in X \subset R^P$, $y \in Y \subset R^q$, $z \in Z \subset R^r$, The general tri-level decision problem presented by Faísca [41] is defined as follows:

$$\underset{x\epsilon X}{\text{Min}} f_1(x,y,z) \text{ (Leader)} \tag{10}$$

$$\text{s.t.} g_1(x,y,z) \leq 0 \tag{11}$$

where (y, z), for each x fixed, solves the problems Equations (12)–(15)

$$\underset{y\epsilon Y}{\text{Min}} f_2(x,y,z) \text{ (Middle − level follower)} \tag{12}$$

$$\text{s.t.} g_2(x,y,z) \leq 0 \tag{13}$$

where z, for each (x, y) fixed, solves the problems Equations (14) and (15)

$$\underset{z\epsilon Z}{\text{Min}} f_3(x,y,z) \text{ (Bottom − level follower)} \tag{14}$$

$$\text{s.t. } g_3(x,y,z) \leq 0 \tag{15}$$

where x, y, z are the decision variables of the leader, the middle-level follower, and the bottom-level follower, respectively; $f_1, f_2, f_3 : R^P \times R^q \times R^r \to R$ are the objective functions of the three decision entities, respectively; $g_i : R^p \times R^q \times R^r \to R^{k_i}$, $i = 1,2,3$ are the constraint conditions of the three decision entities respectively.

Based on these, we have the following definitions [46]:

Definition 2.1.

6. *The problem constraint region,*

$$S = \{(x,\ y,z) \in X \times Y \times Z : g_i(x,y,z) \leq 0,\ i = 1,2,3\} \tag{16}$$

7. *The middle-level follower feasible set for each fixed x,*

$$S(x) = \{(y,z) \in Y \times Z : z \in g_2(x,y,z) \leq 0, g_3(x,y,z) \leq 0\} \tag{17}$$

8. *The bottom-level follower feasible set for each fixed (x, y),*

$$S(x,y) = \{z \in Z : g_3(x,y,z) \leq 0\} \tag{18}$$

9. *The bottom-level follower rational reaction set,*

$$P(x,y) = \{z \in Z : z \in argmin[f_3(x,y,z) : z \in S(x,y)]\} \tag{19}$$

10. *The middle-level follower rational reaction set,*

$$P(x) = \{y,z \in Y \times Z : (y,z) \in argmin[f_2(x,y,z) : (y,z) \in S(x), z \in P(x,y)]\} \tag{20}$$

11. *The problem inducible region,*

$$IR = \{(x,y,z) : (x,y,z) \in S, (y,z) \in P(x)\} \tag{21}$$

12. *The problem optimal solution set,*

$$OS = \{(x,y,z) : (x,y,z) \in argmin[f_1(x,y,z) : (x,y,z) \in IR]\} \tag{22}$$

To develop an efficient algorithm to solve a 3 levels decision problem, it is necessary to explore the geometry of the solution space and the associated theoretical properties. The following assumptions are usually made to ensure that the problem is well formulated in terms of the existence of a solution.

Assumption 2.1. *$f_1, f_2, f_3, g_1, g_2, g_3$ are continuous functions, whereas f_2, f_3, g_1, g_2, g_3 are continuously differentiable.*

Assumption 2.2. *f_3 is strictly convex in z for $z \in S(x,y)$ where $S(x,y)$ is a compact convex set, while f_2 is strictly convex in (y,z) for $(y,z) \in S(x)$ where $S(x)$ is a compact convex set.*

Assumption 2.3. *f_1 is continuous convex in x, y, and z.*

Under Assumptions 2.1 and 2.2, the rational reaction sets of the bottom-level follower and the middle-level follower $P(x,y)$ and $P(x)$ are point-to-point maps and closed, which implies that IR is compact. Thus, under Assumption 2.3, solving the tri-level decision problem is equivalent to optimizing the leader's continuous function f_1 over the compact set IR. It is well known that the solution to such a problem is guaranteed to exist.

It is noticeable that if the bottom-level follower's problem is a convex parametric programming problem that satisfies the Karush–Kuhn–Tucker Conditions (KKT) for each fixed (x,y) [45,47], the bottom-level follower's problem is equivalent to the following KKT Equations (23)–(26):

$$\nabla_z L(x,y,z,u) = \nabla_z f_3(x,y,z) + u\nabla_z g_3(x,y,z), \tag{23}$$

$$u g_3(x,y,z) = 0, \tag{24}$$

$$g_3(x,y,z) \leq 0, \tag{25}$$

$$u \geq 0 \tag{26}$$

where $\nabla_z f_3(x,y,z) + u\nabla_z g_3(x,y,z)$ is the Lagrangian function of the bottom-level follower, $\nabla_z L(x,y,z,u)$ denotes the gradient of the function, for z and u is the vector of Lagrangian multipliers. A necessary and sufficient condition that $(y,z) \in P_{(x)}$ is that the row vector u exists such that $(x,\ y,\ z,\ u)$ satisfies the KKT Equations (23)–(26).

On this basis, by replacing the bottom-level follower problem with the KKT Equations (23)–(26), the tri-level programming problem can be transformed into a bi-level programming problem. The converted equation is shown below:

$$\underset{x}{Min} f_1(x,y,z) \text{ (Leader) } (11) \tag{27}$$

where (y, z), for each x fixed, solves the problems Equations (22)–(25)

$$\underset{y,z,u}{Min} f_2(x,y,z) \text{ (Follower) } (12) \tag{28}$$

$$\nabla_z f_3(x,y,z) + u\nabla_z g_3(x,y,z) = 0 \text{ } (24)\text{–}(26) \tag{29}$$

In this research, the proposed MLiSSO algorithm is extended to solve a multi-level supply chain pricing problem to find a solution (x,y,z) based on Equations (11), (12) and (27)–(29).

2.6. Improved Simplified Swarm Optimization (iSSO)

In this study, because of the NP-hard nature of the multi-level model, we propose a solution procedure based on a novel, convenient and efficient heuristic algorithm called improved Simplified Swarm Optimization (iSSO) [17], which is based on the Simplified Swarm Optimization (SSO) [48] that can perform a full domain search over a large feasible solution space and enhance the solution quality of the algorithm during the search process.

In 2009, Yeh designed the Simplified Swarm Optimization (SSO) [43] to overcome the shortcomings of PSO proposed by Kennedy and Eberhart [49], which was developed based on human observation of birds foraging behavior and a little weak for discrete problems. The targeting principle was used to update variables quickly, which only uses one random number, two multiplications, and one comparison after c_w, c_p, and c_g are given in SSO. According to the results of Yeh [50,51], SSO is more efficient in converging to high-quality solution spaces in some problems.

The update mechanism of SSO is very simple, efficient and flexible [48,50–56], and can be presented as a stepwise-function update:

$$x_{ij}^{t+1} = \begin{cases} g_j, & if\rho_{[0,1]} \in [0, C_g) \\ p_{ij}, & if\rho_{[0,1]} \in [C_g, C_p) \\ x_{ij}^t, & if\rho_{[0,1]} \in [C_p, C_w) \\ x, & if\rho_{[0,1]} \in [C_w, 1) \end{cases} \tag{30}$$

All variables need to be updated in traditional SSO (called all-variable update), $i = 1, 2, \ldots, Nsol$, $j = 1, 2, \ldots, Nvar$, $t = 0, 1, 2, \ldots, Ngen - 1$. Let $X_i^t = \{x_{i1}^t, x_{i2}^t, \ldots, x_{iNvar}^t\}$ represent the ith solution in the t generation, and in the formula of Equation (30), x_{ij}^t is expressed as the jth variable in X_i^t; N_{var} represents the number of variables; c_w, c_p, and c_g are a preset constant; p_{ij}^t is the best solution in its evolutionary history; g_j is the jth variable of the best solution ever, and x is a random number between the lower bound and the upper bound of the jth variable.

Then to further improve the ability of SSO to solve continuous type problems, Yeh introduced the improved Simplified Swarm Optimization (iSSO) in 2015 [17]. A continuous version of SSO with a new update mechanism is proposed in this work to enhance the ability to solve continuous problems with traditional SSO. To date, iSSO has been successfully applied to many sequential problems, as shown in Yeh [57,58], with experimental results demonstrating its effectiveness in solving sequential problems and its ability to produce high-quality solutions. The update mechanism of iSSO is much simpler than the major soft computing technique-PSO (which must calculate both the velocity and position functions) [18,48,54–56].

The update mechanism of iSSO can be presented as follows:

$$x_{ij}^{t+1} = \begin{cases} x_{ij}^t + r_{ij[-0.5,0.5]}^t \cdot u_j & if\ x_{ij}^t = g_i\ or\ \rho_{ij[0,1]}^t \in [0, C_r = c_r) \\ g_j + r_{ij[-0.5,0.5]}^t \cdot u_j & if\ x_{ij}^t \neq g_i\ and\ \rho_{ij[0,1]}^t \in [C_r, C_g = c_r + c_g) \\ x_{ij}^t + r_{ij[-0.5,0.5]}^t \cdot (x_{ij}^t - g_j) & if\ x_{ij}^t \neq g_i\ and\ \rho_{ij[0,1]}^t \in [C_g, 1 = c_r + c_g + c_w) \end{cases} \tag{31}$$

$$u_j = \frac{x_j^{min} - x_j^{max}}{2 \cdot Nvar} \tag{32}$$

As defined in Equation (36), $C_r = c_r, C_g = c_r + c_g$. In addition, in Equation (32), u_j is calculated with the variable's lower-bound x_j^{min}, the upper-bound x_j^{max}, and the numbers of variables. For each update, a random number ρ_{ij}^t that is uniformly distributed between [0, 1] is randomly generated first, and r_{ij}^t is a random number that is uniformly distributed between [−0.5, 0.5]. To compare ρ_{ij}^t with the three constants C_r, C_g, and C_w, if $0 < \rho_{ij}^t < C_r$, the variable is updated according to the first term of Equation (31); if t, the variable is updated according to the second term of Equation (31) to find the adjacent values of g. If $C_r < \rho_{ij}^t < C_g$, the variable will be updated according to the third term of Equation (36) to find a value between the interval from itself to g_j.

If the variable does not meet the upper and lower bound restrictions, the variable will be set to the nearest boundary value. If X_i^{t+1} does not outperform X_i^t in the target function, then $X_i^{t+1} = X_i^t$ and will not be updated.

So far, only a few papers have studied dual-channel supply chains under capital constraints, which can be regarded as an MLPP. To solve these problems, we apply a continuous-type algorithm iSSO on MLPP to deal with these pricing strategy problems. The detailed algorithmic procedure will be presented and explained in Section 4.

3. Statement

3.1. Model Description

To solve the optimal pricing strategy for the overall supply chain and to further illustrate the hierarchical and interactive relationships among the supply chain decisions, we use a multi-level programming problem to describe the master-slave decision structure of the proposed capital constraint dual-channel supply chain model by Zhen [18]. The Stackelberg game is applied to the model due to the aforementioned level structure of the supply chain system and the sequential relation of decision-making is consistent with the context set. As a result and according to the different financing strategies, the supply chain structure can be divided into two types: bi-level and tri-level planning models to present the decisions made by all members of the supply chain in pursuit of their own optimal goals while considering the optimal responses of each other. This chapter introduces the assumptions, notations, and mathematical models of the problem.

3.2. Assumptions

All assumptions regarding the study are described below.

1. This study constructs a dual-channel supply chain model with three levels of the supply chain (manufacturer → retailer → customer) to profit maximization.
2. The manufacturer's initial capital is zero and must repay the entire capital liability.
3. The basic principle of profitability is that the price must be designed to meet the conditions of profitability for all parties.
4. In the model, neither the upstream manufacturer nor the downstream manufacturer considers the inventory problem. The upstream manufacturer ships as much product as it makes to the downstream retailer. The downstream manufacturer buys as much as it can and sells it all to the market.

3.3. Notations

According to the article published by Zhen [18], the notation in Table 1 is used in the capital-constrained dual-channel supply chain model.

Table 1. Notations of the dual-channel supply chain model.

Type	Symbol	Description
parameter	a	The total potential market size.
	λ	The underlying market share of the retailer for the manufacturer is $(1-\lambda)$. $0 \le \lambda \le 1$.
	b	Demand sensitivity to its selling/retail price. $0 < b \le 1$.
	d	The coefficient of cross − price sensitivity. $0 < d \le 1$.
	c	Product production cost.
	η	Revenue sharing of 3rd party platform. $0 < \eta \le 1$.
	i	Finance strategy for: $i = \begin{cases} B, \text{bank finance strategy.} \\ T, \text{3rd platform finance strategy.} \\ R, \text{retailer finance strategy.} \end{cases}$

Table 1. *Cont.*

Type	Symbol	Description
variables	w^i	Wholesale price, for $i = \begin{cases} B \\ T \\ R \\ None \end{cases}$, $w^i \geq 0$.
	P_R^i	Retailer′s retail channel retail price, with finance strategy for $i = B,\ T\ or\ R$. $P_R^i \geq 0$.
	P_M^i	Manufacturer′s direct channel selling price through 3rd party platform, with finance strategy for $i = B,\ T\ or\ R$. $P_M^i \geq 0$.
	q_R^i	Retail channel demand, with finance strategy for $i = B,\ T\ or\ R$.
	q_M^i	Direct channel demand, with finance strategy for $i = B,\ T\ or\ R$.
	r^i	Revenue sharing rate, with finance strategy for $i = B,\ T\ or\ R$. $0 < r^i \leq 1$.

In general, in each channel, the demand is mainly influenced by its own price; therefore, it is assumed that $b > d$. In addition, we also assume $a > (b-d)(P_M + P_R)$ as the demand is not negative.

3.4. The Mathematical Model Description

A dual-channel supply chain with the aim of profit maximization is considered in this article, with a manufacturer to produce a unit product at cost c. The manufacturer has two sales channels in this market. One is the retail channel, where the manufacturer sells the product at wholesale price w to the retailer, which sells it at retail price P_R to the consumer, and this channel is also known as the traditional channel. The other channel is the direct sales channel. The manufacturer sells the products directly to the consumers at the selling price P_M through a third-party platform, also called the third-party platform channel. The structure of the supply chain model is shown in Figure 5; the solid black line indicates that the products are sold to retailers through wholesale; the dotted black line indicates that the products are sold directly through a third-party online platform, and the platform fee η is paid for the cooperation.

In addition, the demands in this market are variables, defined as manufacturer demand q_M and retailer demand q_R, respectively. It is assumed that the demand structure of this supply chain is a linear price dependence, which is widely used in the literature [59]. The demand functions are as follows,

$$q_R = \lambda a - bP_R + dP_M \tag{33}$$

$$q_M = (1-\lambda)a - bP_M + dP_R \tag{34}$$

where b means that a unit of price reduction increases the demand by b, corresponding to marginal and switching customers. A large value of d corresponds to switching customers who are sensitive to differences between the selling price and the retail price. In other words, the degree of differentiation between direct and retail channels decreases as d increases. Thus, d captures the degree of competition between the two channels [60,61].

To find the optimal financing decision and related pricing outcome for this model, we try to maximize the profits for manufacturers, retailers, third-party platforms, and the bank. The objective function and constraints of the model are described in the next part.

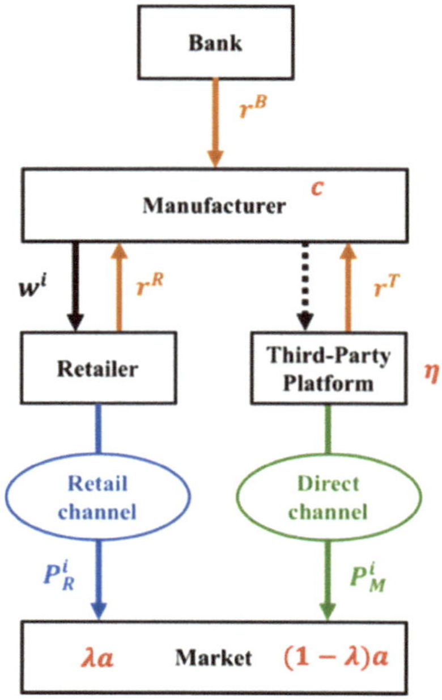

Figure 5. Supply chain model.

3.5. Model Construction

Based on the proposed dual-channel supply chain model proposed in [18] and due to the hierarchical decision relationship of the model, we formulate the supply chain model as an MLPP according to the literature review as follows:

(a) Retailer

$$Max = w^R q_R^R + (1-\eta)P_M^R q_M^R - c\left(1+r^R\right)\left(q_M^R + q_R^R\right) \tag{35}$$

$$Max = \left(P_R^R - w^R\right)q_R^R + r^R c\left(q_M^R + q_R^R\right) \tag{36}$$

(b) Bank

$$Max\ f_1 = w^B q_R^B + (1-\eta)P_M^B q_M^B - c\left(1+r^B\right)\left(q_M^B + q_R^B\right) \tag{37}$$

$$Max\ f_2 = \left(P_R^B - w^B\right)q_R^B \tag{38}$$

$$Max\ f_3 = r^B c\left(q_M^B + q_R^B\right) \tag{39}$$

(c) 3rd Party Platform

$$Max\ f_1 = w^T q_R^T + (1-\eta)P_M^T q_M^T - c\left(1+r^T\right)\left(q_M^T + q_R^T\right) \tag{40}$$

$$Max\ f_2 = \left(P_R^T - w^T\right)q_R^T \tag{41}$$

$$Max\ f_3 = \eta P_M^T q_M^T + r^T c\left(q_M^T + q_R^T\right). \tag{42}$$

(d) Constraints of all

$$\text{s.t. } a > (b-d)\left(P_M^i + P_R^i\right) \tag{43}$$

$$w^i q_R + (1-\eta)P_M^i q_M^i > c\left(q_R^i + q_M^i\right)\left(1 + r^i\right) \tag{44}$$

$$P_R^i \geq w^i \geq c\left(1 + r^i\right) \tag{45}$$

$$P_M^i \geq c\left(1 + r^i\right) \tag{46}$$

$$P_M^i \geq w^i \tag{47}$$

Equation (35) means the manufacturer's maximum profit with retailer finance strategy, and Equation (36) means the retailer's maximum profit with retailer finance strategy.

Equation (37) means the manufacturer's maximum profit with bank finance strategy, Equation (43) means the retailer's maximum profit with bank finance strategy, and Equation (38) means the bank's maximum profit with bank finance strategy.

Equation (39) means the manufacturer's maximum profit with a third-party platform finance strategy, Equation (40) means the retailer's maximum profit with a third-party platform finance strategy, and Equation (41) means the third-party platform's maximum profit with a third-party platform finance strategy.

These three problems all share the same constraints for each layer of Equations (43)–(47). For Equation (43) means demand should not be negative. To fulfill Equation (44), manufacturers must set a selling price that ensures their revenue is greater than its cost. For Equations (45) and (46), we assume that channel prices must exceed marginal costs. For Equation (47), the retailer cannot purchase from the direct channel, so the selling price of the direct channel must not be lower than the wholesale price.

4. Methodology

4.1. Multi-Level Improved Simplified Swarm Optimization (MLiSSO)

In this study, we propose an MLiSSO approach to apply iSSO to MLPP, including the following functional concepts, which are described in detail in the subsequent subsections.

The basic idea of MLPP can be explained as follows, for a strategy given by the leader, followers are assumed to react rationally. The resulting decisions of the leader and the followers can be considered as the "outcome" of the problem. If the leader chooses a different strategy, the outcome will change accordingly. The inducible region is then defined as the set of these outcomes for all the leader's strategies. Thus, the best outcomes that the leader can induce are the best results in the inducible region.

Therefore, in this paper, based on the main constraints of MLPP solving, we use the following methods to ensure the implementation of the above ideas and to avoid the problem of the solving process falling into the best solution of the region, as well as to ensure that it presents the characteristics of the MLPP model and the generation of feasible solutions.

4.1.1. Improved Simplified Swarm Optimization (iSSO)

For the update mechanism iSSO we used in this paper, which is mentioned in the literature review [17], to maintain the diversity of the solution, we add the condition of partial best p_{ij} back into the formula, set the random range for turbulence to $[-0.001, 0.001]$, and make the value of u_j decrease as the number of generations increases. Equations (48) and (49) are the core formulas when iSSO evolves. Thus, there are four evolution scenarios for x_{ij}^{t+1}: oscillating near g_j, oscillating near P_{ij}, oscillating near the original value, or evolving to the point between the original value and g_j. The evolution decision is determined by ρ_{ij}^t.

The modified formula is as follows:

$$x_{ij}^{t+1} = \begin{cases} g_j + \rho_{[-0.001,0.001]} \cdot u_j & \text{if } x_{ij}^t \neq g_j \text{ and } \rho_{[0,1]} \in [0, C_g) \\ p_{ij} + \rho_{[-0.001,0.001]} \cdot u_j & \text{if } x_{ij}^t \neq g_j \text{ and } \rho_{[0,1]} \in [C_g, C_p) \\ x_{ij}^t + \rho_{[-0.001,0.001]} \cdot u_j & \text{if} x_{ij}^t = g_j \text{ or } \rho_{[0,1]} \in [C_p, C_w) \\ x_{ij}^t + \rho_{[-0.001,0.001]} \cdot \left(x_{ij}^t - g_j\right) & \text{if } x_{ij}^t \neq g_j \text{ and } \rho_{[0,1]} \in [C_w, 1] \end{cases} \tag{48}$$

$$u_j = \frac{x_j^{min} - x_j^{max}}{2 \cdot N_{gen} N_{var}} \tag{49}$$

If any variable violates the boundary condition, it is set to the nearest boundary after using Equation (53). The steps are shown in detail in the following Section 4.1.7

4.1.2. Fixed-Variables Local Search

Based on the hierarchical properties of the MLP problem, we introduce a local search method with fixed variables in MLiSSO. In traditional SSO, the initial solution is generated randomly between the lower and upper bounds at the same time. When SSO updates the position, the solutions of all dimensions are changed simultaneously. In this study, only the solution of the decision variable of that level is changed when it is updated and then the local search is executed. The solutions of the remaining levels keep the original results. The calculation process is explained in detail in the following Section 4.1.7.

4.1.3. Fitness Function

For the BLPP structure mentioned in the literature review Equations (1)–(4), the upper and lower levels of the programming problem are both standard constraints optimization problems that do not consider the information interaction between the leader and the follower. We treat the lower-level programming problem as a separate constraint optimization problem without losing the general approach to describe constraint processing techniques. In this case, the fitness of all particle updates can be calculated according to Equation (50); the fitness of the best solution is calculated and evaluated according to Equation (51):

$$fitness(x,y) = \begin{cases} f(x,y), & ify \in S(x) \\ F(x,y), & ify \in S(x)/S \end{cases} \tag{50}$$

where $S(x)$ denotes the lower-level programming problem feasible set and S denotes the constraint region. The fitness value is calculated differently according to the updated level. For the upper-level update, we generate the value of $F(x,y)$, and for the lower-level update, we generate the value of $f(x,y)$, due to the level having different objective functions to obtain their optimal value.

$$fitness(x,y) = F(x,y), \ ify \in S(x)/S \tag{51}$$

The MLiSSO targets the leader's priority first. Therefore we use the higher-level objective functions for the best solution evaluation to ensure that we always put the leader's interest first in a multilevel programming situation. The calculation of all processes is explained in detail in the following Section 4.1.7.

4.1.4. Constraint Handling

In this study, to make the solutions obtained by MLiSSO conform to the problem constraints, we propose a simple but effective constraint method that ensures that the solutions generated during its operation are all conforming to the various constraints of the problem.

We use conditional constraints to enforce domain integrity by restricting the solutions generated after iterative updates to acceptable values that match the domain restrictions. A Boolean operator is used to establish the constraints, and when a solution is generated, it is determined whether it satisfies the constraints, and the result is returned. If the result meets the constraint, it is accepted as True and proceeds to the next step of the process; if it violates any of the constraints, it is rejected as False and generates a random set of solutions that meet the variable limitations (upper bound and lower bound), then redo the Boolean evaluation, repeating this step until it is accepted.

4.1.5. Stopping Criteria

There are two major stopping criteria used:

1. The generation number.
2. The maximum iteration.

It will terminate the MLiSSO algorithm after it has reached the maximum number of iterations or generations.

4.1.6. Level Conversion

Based on the literature review that we referred to above, the problem of tri-level supply chains required to be solved in this study can be transformed into a bi-level programming problem through the use of Kuhn–Tucker conditions Equations (23)–(26) to convert the problem to the term as Equations (27)–(29) and Equations (11) and (12). The transformed supply chain equation is shown in Equations (52)–(57), Equations (11), Equations (13) and Equation (13) below:

$$\underset{x}{Max}F = w^i q_R^i + (1-\eta)P_M^i q_M^i - c\left(1+r^i\right)\left(q_M^i + q_R^i\right)\ (11) \tag{52}$$

where (y, z, u), for each x fixed, solves the problems Equations (59)–(66)

$$\underset{y,z,u}{Max}f = \left(P_R^i - w^i\right)q_R^i\ (13) \tag{53}$$

$$0.4\left(q_M^i + q_R^i\right) - u_1\left[0.4\left(q_M^i + q_R^i\right)\right] - 0.4u_2 - 0.4u_3 = 0 \tag{54}$$

$$u_1\left[-w^i q_R^i - (1-\eta)P_M^i q_M^i + 0.4\left(1+r^i\right)\left(q_M^i + q_R^i\right)\right] = 0, \tag{55}$$

$$u_2\left[-w^i + 0.4\left(1+r^i\right)\right] = 0, \tag{56}$$

$$u_3\left[-P_M^i + 0.4\left(1+r^i\right)\right] = 0,\ (15),\ (26) \tag{57}$$

4.1.7. Steps of MLiSSO for Solving MLPP

The steps of MLiSSO to solve MLPP are described in this section. With one main program and two subprograms are included, which are based on iSSO algorithms. The details are explained as follows.

Main Program: The best solution to solving

STEP 1-1	Maximum iteration T_{max}.
STEP 1-2	Set T_{max} $T = 0$.
STEP 1-3	Call Subprogram1 and generate the initial solution $\left(X_i^T, Y_i^T\right)$.
STEP 1-4	Evaluate $F\left(X_i^T, Y_i^T\right)$and let $(X^*, Y^*) = \left(X_i^T, Y_i^T\right)$.
STEP 1-5	Fixed Y_i^T to the upper-level programming model.
STEP 1-6	Let $T = T + 1$.
STEP 1-7	Call **Subprogram2** to generate X_i^T.
STEP 1-8	Fixed the solution X_i^T.
STEP 1-9	Call **Subprogram2** to generate Y_i^T.
STEP 1-10	Fixed $\left(X_i^T, Y_i^T\right)$ into the objective function to evaluate the value of the objective function. If $F\left(X_i^T, Y_i^T\right) > F(X^*, Y^*)$, $\left(X_i^T, Y_i^T\right)$ it is recorded as (X^*, Y^*).
STEP 1-11	Stopping criterion : if $T \geq Tmax$ go to **STEP 1-12**; otherwise, go to **STEP 1-6**.
STEP 1-12	Output (X^*, Y^*) and the objective function value of the upper − level $F(X^*, Y^*)$ and the lower − level $f(X^*, Y^*)$.

Subprogram 1: Solution initialization

STEP 2-1	Initiate N_{sol}, N_{gen}, N_{var}, C_g, C_p, and C_w, and the upper and lower bounds of each variable.
STEP 2-2	Set N_{gen} $t = 0$ and $i = 1$, where $i = 1, 2, \ldots, N_{sol}$.
STEP 2-3	Generate $\left(X_i^{Tt}, Y_i^{Tt}\right)$. Let $P_{fi}^T = \left(X_i^{Tt}, Y_i^{Tt}\right)$, and calculate $f\left(P_{fi}^T\right) = f\left(X_i^{Tt}, Y_i^{Tt}\right)$ for $i = 1, 2, \ldots, N_{sol}$. And find Gbest such that $f\left(P_{fG}^T\right)$ is the best, and then let $t = 1$ and $i = 1$.
STEP 2-4	Generate ρ and calculate u_j.
STEP 2-5	Generate r to update the X_i^{Tt} and Y_i^{Tt}, and calculate $f\left(X_i^{Tt}, Y_i^{Tt}\right)$.
STEP 2-6	If $f\left(X_i^{Tt}, Y_i^{Tt}\right) > f\left(P_{fi}^T\right)$, then $P_{fi}^T = \left(X_i^{Tt}, Y_i^{Tt}\right)$; Otherwise, go to **STEP 2-8**.
STEP 2-7	If $f\left(P_{fi}^T\right) > f\left(P_{fG}^T\right)$, then $P_{fG}^T = P_{fi}^T$.
STEP 2-8	If $i \leq N_{sol}$ then $i = i + 1$ and return to **STEP 2-4**.
STEP 2-9	If $t < N_{gen}$ then $t = t + 1$ and $i = 1$, and return to **STEP 2-4**. Otherwise, go to **STEP 2-10**.
STEP 2-10	Output $P_{fG}^T = \left(X_i^T, Y_i^T\right)$.

Subprogram 2: Level updating solving

STEP 3-1	Initiate N_{sol} for both levels, N_{genl} (if updating with upper-level $l = 1$; otherwise, $l = 2$), N_{var}, C_g,C_p, and C_w, and the upper and lower bounds of each variable.
STEP 3-2	Set N_{genl} $t = 0$ and $i = 1$, where $i = 1, 2, \ldots, N_{sol}$.
STEP 3-3	Generate X_i^{Tt} or Y_i^{Tt}. Let $P_{Fi}^T = \left(X_i^{Tt}, Y_i^T\right)$, $P_{fi}^T = \left(X_i^T, Y_i^{Tt}\right)$, and calculate $F\left(P_{Fi}^T\right) = F\left(X_i^{Tt}, Y_i^T\right)$, $f\left(P_{fi}^T\right) = f\left(X_i^T, Y_i^{Tt}\right)$ for $i = 1, 2, \ldots, N_{sol}$. And find Gbest such that $F\left(P_{FG}^T\right)$ or $f\left(P_{fG}^T\right)$ is the best, and then let $t = 1$ and $i = 1$.
STEP 3-4	Generate ρ and calculate u_j.
STEP 3-5	Generate r to update X_i^{Tt} and Y_i^{Tt} and calculate $F\left(X_i^{Tt}\right)$ and $f\left(Y_i^{Tt}\right)$.
STEP 3-6	For upper − level update, If $F\left(X_i^{Tt}, Y_i^T\right) > F\left(P_{Fi}^T\right)$, then $P_{Fi}^T = \left(X_i^{Tt}, Y_i^T\right)$; for lower − level update, if $f\left(X_i^T, Y_i^{Tt}\right) > f\left(P_{fi}^T\right)$, then $P_{fi}^T = \left(X_i^T, Y_i^{Tt}\right)$; Otherwise, go to **STEP 3-8**.

STEP 3-7	For upper − level update, if $F(P^T_{Fi}) > F(P^T_{FG})$, then $P^T_{FG} = P^T_{Fi}$; for lower − level update, if $f\left(P^T_{fi}\right) > f\left(P^T_{fG}\right)$, then $P^T_{fG} = P^T_{fi}$.
STEP 3-8	If $i \le N_{sol}$, then $i = i + 1$ and return to **STEP 3-4**.
STEP 3-9	If $t < N_{genl}$ then $t = t + 1$ and $i = 1$, and return to **STEP 3-4**. Otherwise, stop.
STEP 3-10	Output P^T_{FG} *or* P^T_{fG}.

5. Data Analysis and Results

Section 5 is divided into two subsections. The first subsection presents a comparative analysis of the differences between the performance of the proposed algorithms in this thesis and other algorithms based on other papers. In the second section, the proposed methodology is applied to the actual supply chain problem, and the pricing decision results are analyzed.

5.1. Numerical Experiments

To test and demonstrate the above concept, three different types of numerical examples taken from the literature are presented. For comparison, in this study, 20 runs were performed (for problem 1 is 30 runs according to the compared algorithm results) for each problem. The standard deviation was calculated with the formula listed below in Table 2, where the standard deviation is based on the upper-level objective function.

Table 2. Comparison formula.

	Formula	Description
Standard deviation (SD)	$\sqrt{\frac{\sum_{i=1}^{R}\left(F^*_{Mi}-F^*_{MA}\right)^2}{R}}$ where $i = 1, 2, \ldots, R$. $R = 30$ in this paper.	F^*_{Mi}= The optimal solution for MLiSSO in ith run. F^*_{MA}= The average of R optimal solutions for MLiSSO.

5.1.1. Experimental Datasets

In this study, we used four questions used in previous literature as the datasets for the validation tests; the functions are as shown in Tables 3–5. The dataset parameters were set according to the parameters used in the reference source data.

Table 3. Functions for problem 1.

No.	Problem Functions
Problem 1 [62]	$Max\ F = 8x_1 + 4x_2 - 4y_1 + 40y_2 + 4y_3$, where (y_1, y_2, y_3) solves, $Max\ f = -x_1 - 2x_2 - y_1 - y_2 - 2y_3$ s.t. $y_1 - y_2 - y_3 \ge -1$ $-2x_1 + y_1 - 2y_2 + 0.5y_3 \ge -1$ $-2x_2 - 2y_1 + y_2 + 0.5y_3 \ge -1$ $x_1, x_2, y_1, y_2, y_3 \ge 0$

Table 4. Functions for problem 2.

No.	Problem Functions
Problem 2 [63]	$Min\ F = -x_1{}^2 - 3x_2{}^2 - 4y_1 + y_2{}^2$, where (y_1, y_2) solves, s.t. $x_1{}^2 + 2x_2 \leq 4$ $x_1, x_2 \geq 0$ $Min\ f = 2x_1{}^2 + y_1{}^2 - 5y_2$ s.t. $x_1{}^2 - 2x_1 + x_2{}^2 - 2y_1 + y_2 \geq -3$ $4x_2 + 3y_1 - 4y_2 \geq 4$ $y_1, y_2 \geq 0$

Table 5. Functions for problem 3.

No.	Problem Functions
Problem 3 [6]	$Min\ F = x^2 + (y-10)^2$, where y solves, s.t. $x + 2y - 6 \leq 0,$ $-x \leq 0$ $Min\ f = x^3 - 2y^3 + x - 2y - x^2$ s.t. $-x + 2y - 3 \leq 0,$ $-y \leq 0$

5.1.2. Experiments with Orthogonal Arrays

The experimental design of the MLiSSO setup was carried out using a two-factor, two-level full factorial design with four experimental combinations. Including the parameter pbest, and the modification of the u value according to the above mentioned in Section 4.

Each of the above-mentioned three experimental datasets was used to perform independent configuration experiments to identify the most suitable configurations, and Tables 6 and 7 show the configuration combinations.

Table 6. Factor level table.

Level\Factor	Parameter *Cp*	*u* Value Setting
1	Without	Constant
2	Add-in	Dynamic

Table 7. Full factorial design table.

Setting\Factor	Parameter *Cp*	*u* Value Setting
1	Without	Constant
2	Without	Dynamic
3	Add-in	Constant
4	Add-in	Dynamic

The following experiments were compiled using python 3.8 with the same basic parameters, $Cg = 0.2$, $Cp = 0.3$, $Cw = 0.5$, number of particles = 20, number of generations = 200 (for subprogram), and iterations= 500 (for main program).

Each experiment was run 20 times, and the results were evaluated and analyzed by using the leader's target function value results. Assuming that the samples conformed to the norm, an analysis of variance (ANOVA) with $\alpha = 0.05$ was conducted to select the most suitable configuration.

Dataset: Problem 1

Table 8 shows that the p-values of factors A is smaller than $\alpha = 0.05$, so the factors did cause significant differences, and the p-value of factor B were greater than $\alpha = 0.05$, so the factors did not cause significant differences. However, factor B was more likely to cause differences than factor A, as shown in Table 9, which shows the mean value of 20 experiments for each of the four groups of experiments

Table 8. ANOVA table of Dataset problem 1.

Source	DF	SS	MS	*F*-Value	*p*-Value
A	1	23.401	23.4015	4.53	0.036
B	1	0.413	0.4126	0.08	0.778
Error	76	392.193	5.1604		
Total	79	427.444	0.000		
S = 2.27166	R − Sq = 8.25%		R − Sq (adj) = 4.63%		

Table 9. Response table of Dataset problem 1.

Level	A	B
1	27.5	27.0
2	28.0	26.0
Delta	0.5	1.0
Rank	2	1

From Figure 6, it can be concluded that the A factor has better performance at level 2 than level 1, and the B factors have better performance at level 1 than at level 2. But, according to the interaction plot, as shown in Figure 7, it indicates the existence of interaction, and we cannot tell if the configuration settings will have better performance by all set to level 2.

Dataset: Problem 2

Table 10 shows that the p-values of both factors A and B are smaller than $\alpha = 0.05$, so the factors did cause significant differences; furthermore, the factor B was more likely to cause differences than the factor A, as shown in Table 11, which is the mean values of 20 experiments for each of the four groups of experiments.

Table 10. ANOVA table of Dataset problem 2.

Source	DF	SS	MS	*F*-Value	*p*-Value
A	1	0.02050	0.020505	5.13	0.026
B	1	0.04019	0.040187	10.06	0.002
Error	76	0.30370	0.003996		
Total	79	0.45029	0.000000		
S = 0.0632139	R − Sq = 32.56%		R − Sq (adj) = 29.89%		

From Figure 8, it can be concluded that both A and B factors have better performance at level 2 than at level 1, which also has a significant difference in the performance. However, the interaction plot, as shown in Figure 9, it indicates the existence of interaction, and we cannot tell if they could have better performance by all set to level 2.

Table 11. Response table of Dataset problem 2.

Level	A	B
1	18.59	18.63
2	18.625	18.585
Delta	0.035	0.045
Rank	2	1

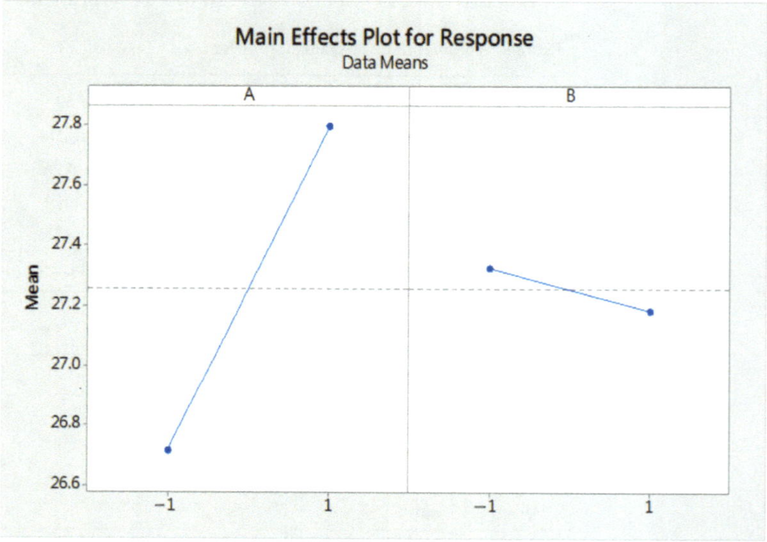

Figure 6. Main effects plot of problem 1. A represents factor A in Table 8, B represents factor B in Table 8.

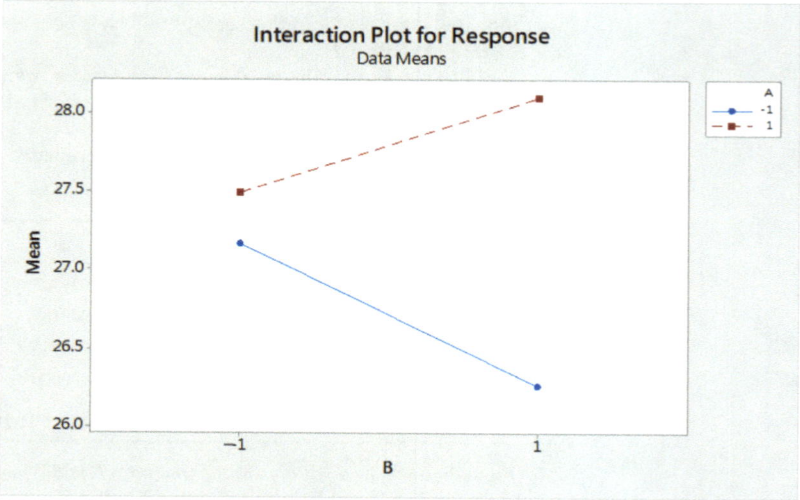

Figure 7. Interaction plot of problem 1. A represents factor A in Table 8, B represents factor B in Table 8.

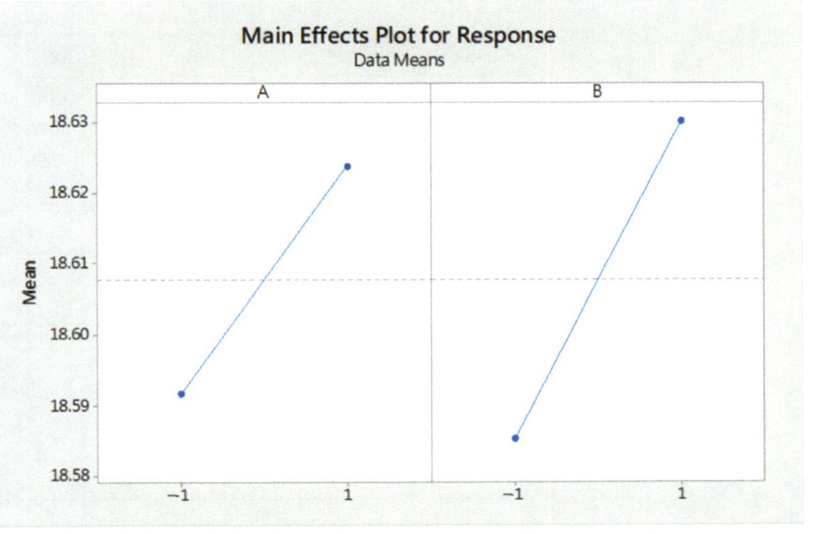

Figure 8. Main effects plot of problem 2. A represents factor A in Table 8, B represents factor B in Table 8.

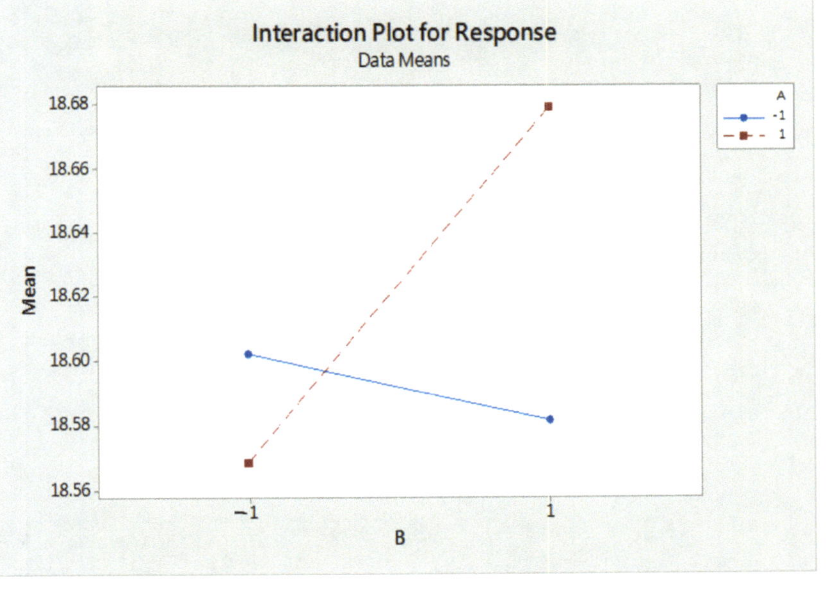

Figure 9. Interaction plot of problem 2. A represents factor A in Table 8, B represents factor B in Table 9.

Dataset: Problem 3

Table 12 shows that the *p*-values of factors A and B were greater than $\alpha = 0.05$, so the factors did not cause significant differences; however, factor A was more likely to cause differences than factor B, as shown in Table 13, which is the mean values of 20 experiments for each of the four groups of experiments.

Table 12. ANOVA table of Dataset problem 3.

Source	DF	SS	MS	*F*-Value	*p*-Value
A	1	3.1344	3.1344	3.24	0.076
B	1	0.6697	0.6697	0.69	0.408
Error	76	73.4531	0.9665		
Total	79	78.8582	0.0000		
S = 0.983102	R−Sq = 6.85%		R−Sq (adj) = 3.18%		

Table 13. Response table of Dataset problem 3.

Level	A	B
1	−67.5	−67.2
2	−67.1	−67.4
Delta	0.4	0.2
Rank	1	2

From Figure 10, it can be concluded that the A factor has better performance at level 2 than level 1, and the B factors have better performance at level 1 than at level 2. However, the interaction plot, as shown in Figure 11, it indicates the existence of interaction, and we cannot tell if the configuration settings will have better performance by all set to level 2.

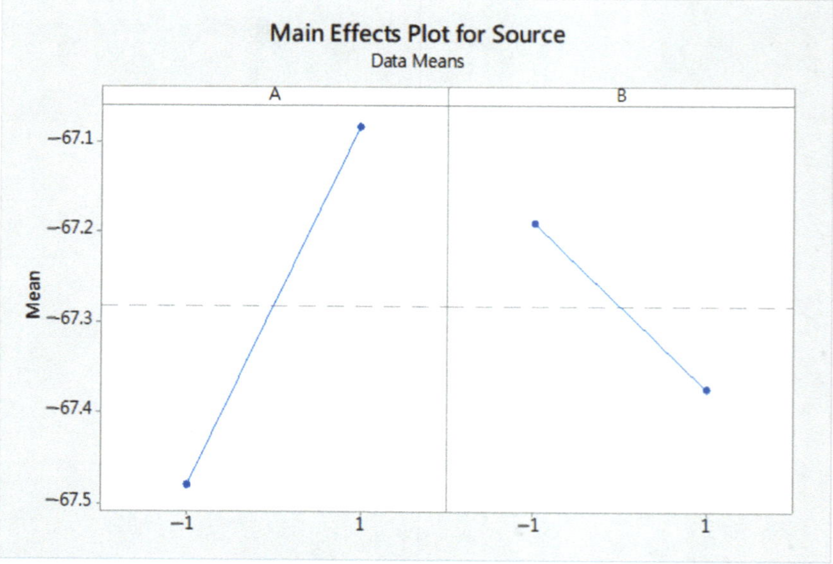

Figure 10. Main effects plot of problem 3. A represents factor A in Table 8, B represents factor B in Table 8.

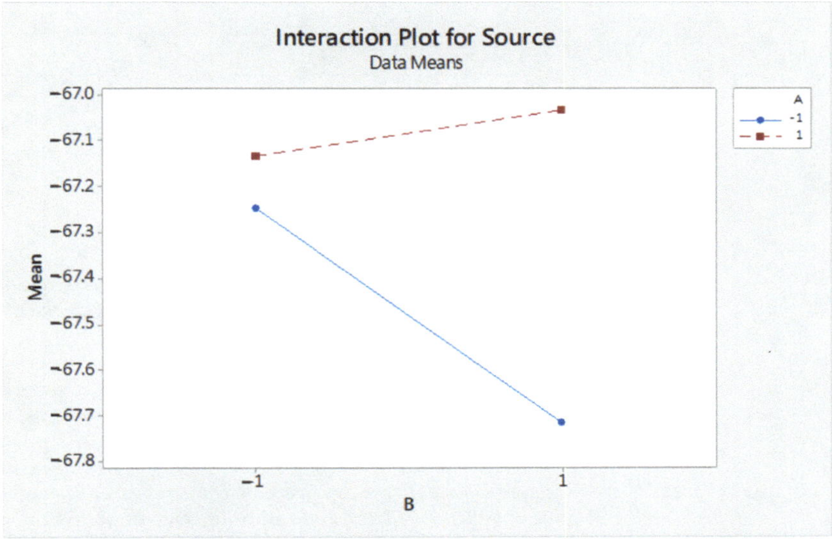

Figure 11. Interaction plot of problem 3. A represents factor A in Table 8, B represents factor B in Table 8.

Result Summary

The results of the experiments for the above four configurations are listed and discussed. As shown in Tables 14–16, it can be concluded that when both factors A and B are set to level 2 (setting 4), the results obtained for this configuration are superior to those in the other configurations in all three experiments. This setting 4 is also the MLiSSO configuration proposed in this study. Therefore, based on this result, the proposed MLiSSO will be used for other experiments and analyses in the following.

Table 14. Results of dataset problem 1.

Setting	F_{avg}	F_{stdev}	F_{stdev}	F_{min}
1	27.1654863	2.7327978	2.7327978	17.3713556
2	26.2656556	3.0503393	3.0503393	20.0050241
3	27.4909892	1.7662205	1.7662205	21.7917118
4	28.1035510	0.8657080	0.8657080	26.4103750

Table 15. Results of dataset problem 2.

Setting	F_{avg}	F_{stdev}	F_{stdev}	F_{min}
1	18.6020324	0.0710975	18.7231808	18.4959183
2	18.5813208	0.0616991	18.7136139	18.4883674
3	18.5685148	0.0636811	18.7202904	18.4420441
4	18.6788774	0.0553813	18.8334935	18.6205643

Table 16. Results of dataset problem 3.

Setting	F_{avg}	F_{stdev}	F_{stdev}	F_{min}
1	67.24701323	0.977282509	68.93931587	65.6916659
2	67.71292941	1.031981155	69.28061716	65.54469136
3	67.13405382	0.969031058	69.17334034	65.33748003
4	67.03412608	0.952295648	67.03412608	65.23260349

5.1.3. Comparison Experiment Results

In this study, we solved three sets of MLPP problems with different levels of complexity by using MLiSSO and compared the results with those of algorithms proposed in other related literature.

Dataset: Problem 1

We constructed a linear BLPP with multiple leaders and followers from [62] as a numerical example to analyze more complex problems; the functions of problem 1 are listed in Table 3.

In this problem, the parameters setting of two different algorithms, GA [64] and PSO, are given in [11], then we used the trial-and-error method for the setting of MLiSSO, and summarized in Table 17. In addition, as mentioned above, the number of iterations of MLiSSO is indicated by generation(update) and iteration(main), while other algorithms used for comparison are indicated by iteration if not specified. The best optimal solution, mean, and standard deviation values of the solutions for 30 runs are presented in Tables 18 and 19. Figure 12 shows the convergence of the optimal solution target function value $F(X, Y)$ in MLiSSO.

Table 17. Parameters setup for problem 1.

GA [64]	PSO [11]	MLiSSO
opulation: 20, Crossover rate: 0.9, Mutation rate: 0.1, Iterations: N/A	Population: 20, Vmax: 10, Inertial weight: 1.2–0.2, Iterations: 150	Population: 20, Cg: 0.3, Cp: 0.6, Cw: 0.8, Generations:100/150, Iterations: 500/150

Table 18. Best results of problem 1.

	GA	PSO	MLiSSO (500)	MLiSSO (As Literature/150)
x_1	0.000	0.0004	0.0002	0.0266
x_2	0.898	0.8996	0.8991	0.0205
y_1	0.000	0.0000	0.0000	0.7969
y_2	0.599	0.5995	0.5993	0.7944
y_3	0.399	0.3993	0.3986	0.1503
F	29.1480	29.1788	29.6631	29.4853
f	−3.193	−3.1977	−3.1948	−1.9594
Runtime(s)	N/A	N/A	35	32

Table 19. Average results & SD of problem 1.

	GA	PSO	MLiSSO (500)	MLiSSO (As Literature/150)
x_1	0.15705	0.02192	0.00078	0.01579
x_2	0.86495	0.86693	0.89607	0.18669
y_1	0.00000	0.00000	0.00000	0.41225
y_2	0.47192	0.56335	0.59701	0.66371
y_3	0.51592	0.34108	0.39351	0.19149
F	21.52948	24.81256	29.04494	26.53842
f	−3.39072	−3.1977	−3.17696	−1.84811
F *stdev*	3.14432	1.55374	0.10689	2.23245
Runtime(s)	N/A	N/A	45	35

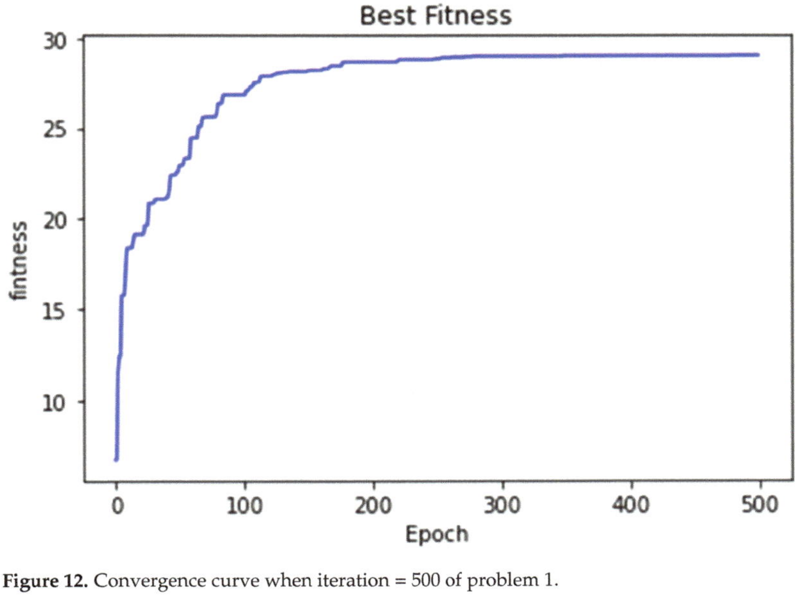

Figure 12. Convergence curve when iteration = 500 of problem 1.

Tables 18 and 19 indicate that MLiSSO has the smallest standard deviation according to the objective value priority of the leader in the case of linear bi-level decision-making with multiple leaders and multiple followers. It returns better results than the GA, just after the results of PSO with a difference of 0.0157 in terms of the best result. In addition, the average solutions of MLiSSO return significantly better than the solutions of other algorithms in the average result, and the standard deviation of the MLiSSO method is lower than that of other algorithms. This indicates that MLiSSO has higher stability and provides better solution quality for solving complex problems.

In the study of Kuo & Huang [11], for the initial solution, they adopt the float coding method to generate the random numbers for the upper-level variables and program for variables in the lower level. Then, update the velocity and position for every particle at once. To compare with the results, we use the same structure and iteration = 150 to generate the results, which are listed in the right column (as literature) in Tables 18 and 19.

The results show that if we only use the proposed modified iSSO with the same kind of structure, the best result and the average result are both superior to the other two methods, and the average result is just after the original MLiSSO. However, the purpose of MLiSSO is for general use on other types of MLPP when this method is only used on linear programming problems mentioned in the literature, so we use the following non-linear problems to emphasize the commonality of MLiSSO.

Dataset: Problem 2

For the example of nonlinear BLPP, which was constructed from [63], the functions of problem 2 are listed in Table 4. In this problem, the two different algorithms, evolutionary algorithm (EA) and PSO-CST are given in [65,66], and the setup of the parameters is listed in Table 20. In addition, as mentioned above, the number of iterations of MLiSSO is indicated by generation (update) and iteration (main), while other algorithms used for comparison are indicated by iteration if not specified. The best optimal solution, mean, and standard deviation values of the solutions for 20 runs are shown in Table 21, and because the average results are not given in the literature [65,66], we only list the result of MLiSSO. Figure 13 shows the convergence of the optimal solution target function value $F(X, Y)$ in MLiSSO.

Table 20. Parameters setup for problem 2.

EA [65]	PSO-CST [66]	MLiSSO
Population: 30, Crossover rate: 0.8, Mutation rate: 0.2, Iterations: 100	Population : 45, Numbers of particles : $m = 40$ (first update), $n = 5$ (CST particles), $V_{max} = 2$, $c_1 = c_2 = 2$, Iterations: 8	Population: 20, Cg: 0.2, Cp: 0.3, Cw: 0.5, Generations: 100, Iterations: 100

Table 21. Best results of problem 2.

	EA [65]	PSO-CST [66]	MLiSSO
x_1	0.00000044	0.3844	0.0115
x_2	2	1.6124	1.9765
y_1	1.875	1.8690	1.8466
y_2	0.9063	0.8041	0.7988
F	−12.68	−14.7772	−18.4633
f	−1.016	−0.2316	−6.1174
F $stdev$	N/A	N/A	0.1396
F avg	N/A	N/A	−18.4566
Runtime(s)	N/A	N/A	5

In the case of this problem, we conclude from Table 21 that MLiSSO outperforms the other algorithms in terms of the objective priority of the leaders and the standard deviation of the solutions obtained is only 0.1396, which means that its solutions remain fairly stable. Thus, MLiSSO shows better performance than these two algorithms for the nonlinear BLPP. This result also implies that the proposed MLiSSO is suitable for solving nonlinear multi-player BLPP.

Dataset: Problem 3

We constructed a nonlinear BLPP with a single leader and follower from [6] as a numerical example to analyze more complex problems(cube); the functions of problem 1 are listed in Table 5.

In this problem, the two different algorithms, HPSOBLP and IBPSO, are given in [6,67], with parameter settings summarized in Table 22. In addition, as mentioned above, the number of iterations of MLiSSO is indicated by generation (update) and iteration (main), while other algorithms used for comparison are indicated by iteration if not specified. The best optimal solution, mean, and standard deviation values of the solutions for 20 runs are presented in Tables 23 and 24. Figure 14 shows the convergence of the optimal solution target function value F(X,Y) in MLiSSO.

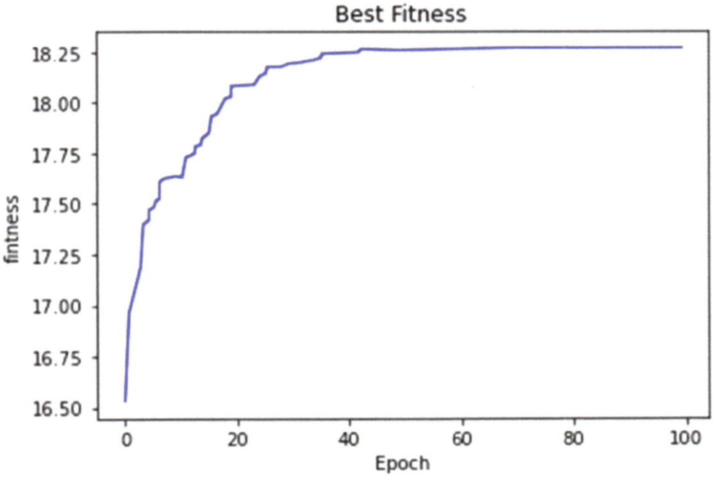

Figure 13. Convergence curve when iteration = 100 of problem 2.

Table 22. Parameters setup for problem 3.

HPSOBLP [67]	IBPSO [6]	MLiSSO
Population : $N_{max} = 20,\ 40$, $c_1 = c_2 = 2$, V_{max} = bounds, w = decrease linearly from 1.2 to 0.1, Iterations: 120, 30	Population : $N_1 = N_2 = 20$, $V_{max} = 10$, $c_1 = c_2 = 2.5$, Iteration : $T_1 = T_2 = 100$	Population: 20, Cg: 0.2, Cp: 0.3, Cw: 0.5, Generations: 100, Iterations: 100

Table 23. Best results of problem 3.

	HPSOBLP	IBPSO	MLiSSO
x	N/A	0.4960	1.0186
y	N/A	1.7356	1.9753
F	88.77571	68.5459	65.8663
f	−0.7698	−13.5561	−18.9133
Runtime(s)	N/A	N/A	5

Table 24. Average results & SD of problem 3.

	HPSOBLP	IBPSO	MLiSSO
x	N/A	1.1985	1.1036
y	N/A	1.7791	1.8756
F	88.7835	69.0192	67.4949
f	N/A	−13.3375	−15.8649
F *stdev*	0.0016	N/A	1.0366
Runtime(s)	N/A	N/A	5

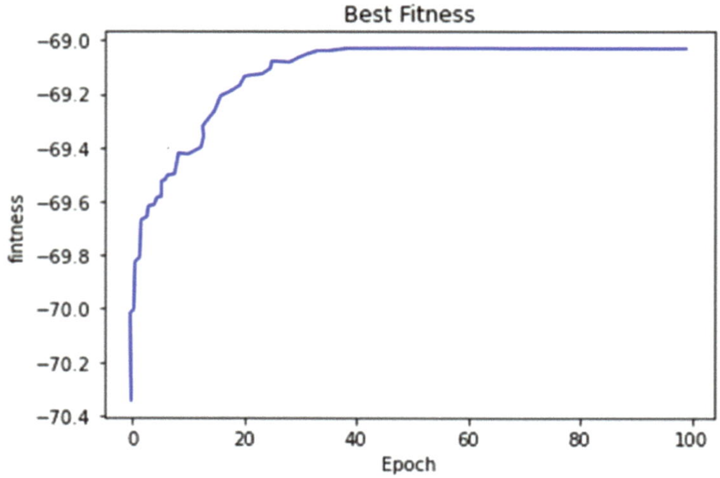

Figure 14. Convergence curve when iteration = 100 for problem 3.

Tables 23 and 24 indicate that MLiSSO outperforms the other algorithms in terms of the objective priority of the leaders, which means that MLiSSO has a better performance than these two algorithms. In addition, also the average results show that the solution can be obtained with better quality in several independent experiments.

5.2. Model Evaluation

Based on the aforementioned experimental results, it can be stated that MLiSSO can be used to solve the MLPP with a higher stability quality of optimal solution results. This section further verifies the practicality of the MLiSSO on supply chain problems by using the supply chain model with three different financing strategies Equations (40)–(47) and constraints of them Equations (48)–(52). The parameters setup of the supply chain model is, according to Zhen [18], listed in Table 25, and the parameter setup of MLiSSO is listed in Table 26. The corresponding solutions of the three financing strategy models are shown in Tables 27 and 28. Figures 15 and 16 show the convergence of the optimal solution target function value $F(X,Y)$ in MLiSSO

Table 25. Model parameters setup.

Parameter	a	λ	b	d	c	η
Setup	1	0.4	1	0.5	0.4	0.15

Table 26. Parameters setup for supply chain model.

MLiSSO
Population: 20 Cg: 0.2 Cp: 0.3 Cw: 0.5 Generations:100 Iterations: 500

Table 27. The best result of the supply chain model.

	RF	BF	3PF
w	0.64952	0.40506	0.40506
p_M	0.83284	0.70765	0.70765
p_R	0.71113	0.62951	0.62951
q_M	0.10529	0.12432	0.12432
q_R	0.12272	0.20710	0.20710
r	0.34810	0.01265	0.01265
f_1	0.03231	0.04068	0.04068
f_2	0.00756	0.00284	0.00284
f_3	N/A	0.00168	0.01487

Table 28. Average results & SD of supply chain model.

	RF	BF	3PF
w	0.73053	0.40495	0.40495
p_M	0.88513	0.71293	0.71293
p_R	0.76135	0.61588	0.61588
q_M	0.08122	0.14058	0.14058
q_R	0.09554	0.19501	0.19501
r	0.53832	0.01237	0.01237
f_1	0.02197	0.03898	0.03898
f_2	0.00355	0.00271	0.00271
f_3	N/A	0.00161	0.01664
f_1 *stdev*	0.00524	0.00068	0.00068
f_2 *stdev*	0.00434	0.00265	0.00265
f_3 *stdev*	N/A	0.00153	0.00106

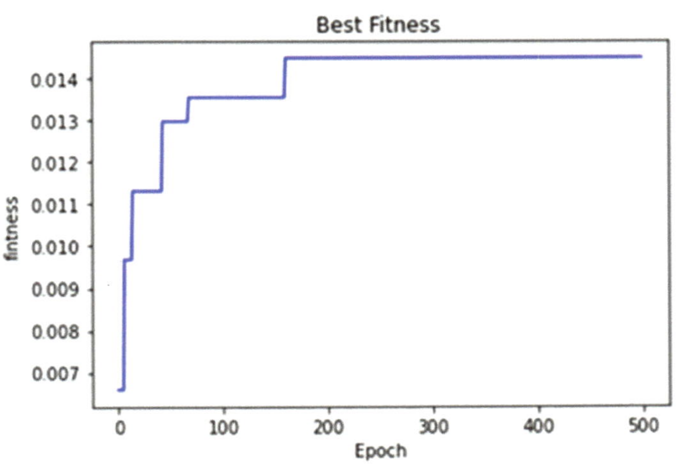

Figure 15. Convergence curve of RF.

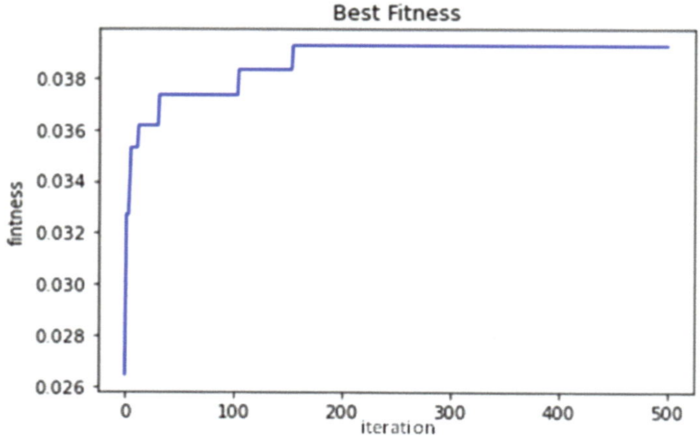

Figure 16. Convergence curve of 3PF & BF.

As shown in Table 27 (1) for the retailer-financed case, the objective value of the leader has converged to $f_1(x, y, z) = 0.03231$, while the objective value of layers 2 have converged to $f_2(x, y, z) = 0.00756$; (2) for the bank-financed case, the objective value of the leader has converged to $f_1(x, y, z) = 0.04068$ while the objective values for layers 2 and 3 converge to $f_2(x, y, z) = 0.00284$ and $f_3(x, y, z) = 0.00168$; (3) the third-party platform-financed case, the objective value of the leader has converged to $f_1(x, y, z) = 0.04068$, while the objective values for layers 2 and 3 converge to $f_2(x, y, z) = 0.00284$ and $f_3(x, y, z) = 0.01487$. We also list the average and standard deviation of the solutions we obtained in 20 runs in Table 28.

It can be noted that under this group of market conditions of parameters and after conversion calculation, as shown in Table 27, we learn that among all financing options, the financing strategy with third-party platforms and banks has absolutely favorable conditions for manufacturers. Thus, it can be concluded that this approach can be applied to complex and practical decision problems to solve MLPP.

6. Conclusions

First, we review this paper; our proposed method uses hierarchical updates of fixed variables, trivial problem transformations, computation of objective functions, and iSSO algorithms. Although it does not outperform the best current algorithms for the related small-scale problems, it surpasses the performance of other algorithms for large-scale problems. In conclusion, due to the average and standard deviation of the results, it provides a relatively stable, feasible, and effective solution to the MLPP problem and can be applied to the relevant decision-making process. On the other hand, this paper uses a fixed-variable approach to search, which can express the concept of hierarchical decision-making more effectively and can be implemented on higher-level MLPs that introduce multiple leaders and multiple followers to achieve a more realistic large-scale goal problem. It is also easier to extend to solve a complex problem. If further exploration and experimentation can be done, it may further enhance the ability of this solution to solve problems.

In recent years, many researchers have been studying hybrid algorithms for solving MLPP problems, and as the complexity of the problems increases, mathematical research will become more practical. Therefore, it is expected that more ways and improvements will be developed to solve related problems to meet the industry's current needs.

With the results of this study, the necessity of investigating many of these issues is highlighted, especially to improve the methodology of MLPP. Among the many topics to be explored in future research, there are several major extensions that we intend to focus on.

(1) Hybridization of other heuristic mechanisms to improve MLiSSO solving
(2) Consider the dynamical mechanism for adjusting the upper and lower terms in terms of the turbulence of the update mechanism to improve the generated solutions towards the desired optimal solution to improve the efficiency and quality of the solutions.

Author Contributions: Conceptualization, Z.L. and W.-C.Y.; methodology, Z.L., Y.-C.Y. and W.-C.Y.; writing—original draft preparation, Z.L.; writing—review and editing, Z.L. and S.-Y.T. All authors have read and agreed to the published version of the manuscript.

Funding: This research received no external funding.

Institutional Review Board Statement: Not applicable.

Informed Consent Statement: Not applicable.

Data Availability Statement: All data included in the main text.

Conflicts of Interest: The authors declare no conflict of interest.

References

1. Matsui, K. When should a manufacturer set its direct price and wholesale price in dual-channel supply chains? *Eur. J. Oper. Res.* **2017**, *258*, 501–511. [CrossRef]
2. Ma, J.; Wang, H. Complexity analysis of dynamic noncooperative game models for closed-loop supply chain with product recovery. *Appl. Math. Model.* **2014**, *38*, 5562–5572. [CrossRef]
3. Von Stackelberg, H. *Marktform Und Gleichgewicht*; Springer: Berlin/Heidelberg, Germany, 1934.
4. Grand, S.; Von Krogh, G.; Leonard, D.; Swap, W. Resource allocation beyond firm boundaries: A multi-level model for open source innovation. *Long Range Plan.* **2004**, *37*, 591–610. [CrossRef]
5. Dotoli, M.; Fanti, M.P.; Meloni, C.; Zhou, M.C. A multi-level approach for network design of integrated supply chains. *Int. J. Prod. Res.* **2005**, *43*, 4267–4287. [CrossRef]
6. Ma, W.; Wang, M.; Zhu, X. Improved particle swarm optimization based approach for bilevel programming problem-an application on supply chain model. *Int. J. Mach. Learn. Cybern.* **2014**, *5*, 281–292. [CrossRef]
7. Ben-Ayed, O.; Blair, C.E. Computational difficulties of bilevel linear programming. *Oper. Res.* **1990**, *38*, 556–560. [CrossRef]
8. Bard, J.F. Some properties of the bilevel programming problem. *J. Optim. Theory Appl.* **1991**, *68*, 371–378. [CrossRef]
9. Segall, R. Using branch-and-bound to solve bi-level geometric programming problems: A new optimization model. *Appl. Math. Model.* **1990**, *14*, 271–274. [CrossRef]
10. Kasemset, C.; Kachitvichyanukul, V. A PSO-based procedure for a bi-level multi-objective TOC-based job-shop scheduling problem. *Int. J. Oper. Res.* **2012**, *14*, 50–69. [CrossRef]
11. Kuo, R.; Huang, C. Application of particle swarm optimization algorithm for solving bi-level linear programming problem. *Comput. Math. Appl.* **2009**, *58*, 678–685. [CrossRef]
12. Liu, B. Stackelberg-Nash equilibrium for multilevel programming with multiple followers using genetic algorithms. *Comput. Math. Appl.* **1998**, *36*, 79–89. [CrossRef]
13. Kliestik, T.; Zvarikova, K.; Lăzăroiu, G. Data-driven machine learning and neural network algorithms in the retailing environment: Consumer engagement, experience, and purchase behaviors. *Econ. Manag. Financ. Mark.* **2022**, *17*, 57–69.
14. Hopkins, E. Machine Learning Tools, Algorithms, and Techniques. *J. Self-Gov. Manag. Econ.* **2022**, *10*, 43–55.
15. Nica, E.; Sabie, O.M.; Mascu, S.; Luţan, A.G. Artificial Intelligence Decision-Making in Shopping Patterns: Consumer Values, Cognition, and Attitudes. *Econ. Manag. Financ. Mark.* **2022**, *17*, 31–43.
16. Kliestik, T.; Kovalova, E.; Lăzăroiu, G. Cognitive decision-making algorithms in data-driven retail intelligence: Consumer sentiments, choices, and shopping behaviors. *J. Self-Gov. Manag. Econ.* **2022**, *10*, 30–42.
17. Yeh, W.-C. An improved simplified swarm optimization. *Knowledge-Based Syst.* **2015**, *82*, 60–69. [CrossRef]
18. Zhen, X.; Shi, D.; Li, Y.; Zhang, C. Manufacturer's financing strategy in a dual-channel supply chain: Third-party platform, bank, and retailer credit financing. *Transp. Res. Part E: Logist. Transp. Rev.* **2020**, *133*, 101820. [CrossRef]
19. Tsay, A.A.; Agrawal, N. Channel Conflict and Coordination in the E-Commerce Age. *Prod. Oper. Manag.* **2009**, *13*, 93–110. [CrossRef]
20. Cai, G.G. Channel selection and coordination in dual-channel supply chains. *J. Retail.* **2010**, *86*, 22–36. [CrossRef]
21. Yan, R.; Pei, Z. Information asymmetry, pricing strategy and firm's performance in the retailer-multi-channel manufacturer supply chain. *J. Bus. Res.* **2011**, *64*, 377–384. [CrossRef]
22. Chiang, W.-Y.K.; Chhajed, D.; Hess, J.D. Direct Marketing, Indirect Profits: A Strategic Analysis of Dual-Channel Supply-Chain Design. *Manag. Sci.* **2003**, *49*, 1–20. [CrossRef]
23. Bernstein, F.; Song, J.-S.; Zheng, X. Free riding in a multi-channel supply chain. *Nav. Res. Logist. (NRL)* **2009**, *56*, 745–765. [CrossRef]

24. Ryan, J.; Sun, D.; Zhao, X. Coordinating a Supply Chain with a Manufacturer-Owned Online Channel: A Dual Channel Model Under Price Competition. *IEEE Trans. Eng. Manag.* **2012**, *60*, 247–259. [CrossRef]
25. Saha, S. Channel characteristics and coordination in three-echelon dual-channel supply chain. *Int. J. Syst. Sci.* **2014**, *47*, 740–754. [CrossRef]
26. Huang, W.; Swaminathan, J.M. Introduction of a second channel: Implications for pricing and profits. *Eur. J. Oper. Res.* **2009**, *194*, 258–279. [CrossRef]
27. Tang, C.S.; Yang, S.A.; Wu, J. Sourcing from Suppliers with Financial Constraints and Performance Risk. *Manuf. Serv. Oper. Manag.* **2018**, *20*, 70–84. [CrossRef]
28. Lee, C.H.; Rhee, B.-D. Trade credit for supply chain coordination. *Eur. J. Oper. Res.* **2011**, *214*, 136–146. [CrossRef]
29. Kouvelis, P.; Zhao, W. Who Should Finance the Supply Chain? Impact of Credit Ratings on Supply Chain Decisions. *Manuf. Serv. Oper. Manag.* **2018**, *20*, 19–35. [CrossRef]
30. Kouvelis, P.; Zhao, W. Supply Chain Contract Design Under Financial Constraints and Bankruptcy Costs. *Manag. Sci.* **2016**, *62*, 2341–2357. [CrossRef]
31. Caldentey, R.; Haugh, M.B. Supply Contracts with Financial Hedging. *Oper. Res.* **2009**, *57*, 47–65. [CrossRef]
32. Aydin, R.; Kwong, C.; Ji, P. Coordination of the closed-loop supply chain for product line design with consideration of remanufactured products. *J. Clean. Prod.* **2016**, *114*, 286–298. [CrossRef]
33. Yang, D.; Jiao, J.; Ji, Y.; Du, G.; Helo, P.; Valente, A. Joint optimization for coordinated configuration of product families and supply chains by a leader-follower Stackelberg game. *Eur. J. Oper. Res.* **2015**, *246*, 263–280. [CrossRef]
34. Cachon, G.P.; Zipkin, P.H. Competitive and cooperative inventory policies in a two-stage supply chain. *Management science* **1999**, *45*, 936–953. [CrossRef]
35. Hennet, J.-C.; Arda, Y. Supply chain coordination: A game-theory approach. *Eng. Appl. Artif. Intell.* **2008**, *21*, 399–405. [CrossRef]
36. Tian, Y.; Govindan, K.; Zhu, Q. A system dynamics model based on evolutionary game theory for green supply chain management diffusion among Chinese manufacturers. *J. Clean. Prod.* **2014**, *80*, 96–105. [CrossRef]
37. Cachon, G.P.; Netessine, S. *Game Theory in Supply Chain Analysis, in Models, Methods, and Applications for Innovative Decision Making*; INFORMS: Catonsville, MD, USA, 2006; pp. 200–233.
38. Leng, M.; Parlar, M. Game Theoretic Applications in Supply Chain Management: A Review. *INFOR Inf. Syst. Oper. Res.* **2005**, *43*, 187–220. [CrossRef]
39. Stackelberg, H.V.; Von, S.H. *The Theory of the Market Economy*; Oxford University Press: Oxford, UK, 1952.
40. Zhou, Z.F. Research on Pricing Decision of Multi-Level Remanufacturing Reverse Supply Chain Based on Stackelberg Game. *Appl. Mech. Mater.* **2012**, *220–223*, 290–293. [CrossRef]
41. Sadigh, A.N.; Mozafari, M.; Karimi, B. Manufacturer–retailer supply chain coordination: A bi-level programming approach. *Adv. Eng. Softw.* **2012**, *45*, 144–152. [CrossRef]
42. Lu, J.; Han, J.; Hu, Y.; Zhang, G. Multilevel decision-making: A survey. *Inf. Sci.* **2016**, *346–347*, 463–487. [CrossRef]
43. Luo, H.; Liu, L.; Yang, X. Bi-level programming problem in the supply chain and its solution algorithm. *Soft Comput.* **2019**, *24*, 2703–2714. [CrossRef]
44. Colson, B.; Marcotte, P.; Savard, G. Bilevel programming: A survey. *4OR* **2005**, *3*, 87–107. [CrossRef]
45. Bard, J. *Practical Bilevel Optimization: Applications and Algorithms, in Series: Nonconvex Optimization and Its Applications*; Springer: Berlin/Heidelberg, Germany, 1998.
46. Faísca, N.P.; Saraiva, P.M.; Rustem, B.; Pistikopoulos, E.N. A multi-parametric programming approach for multilevel hierarchical and decentralised optimisation problems. *Comput. Manag. Sci.* **2007**, *6*, 377–397. [CrossRef]
47. Dempe, S. *Foundations of Bilevel Programming*; Springer Science & Business Media: Berlin, Germany, 2002.
48. Yeh, W.-C. A two-stage discrete particle swarm optimization for the problem of multiple multi-level redundancy allocation in series systems. *Expert Syst. Appl.* **2009**, *36*, 9192–9200. [CrossRef]
49. Eberhart, R.; Kennedy, J. Particle Swarm Optimization. In Proceedings of the IEEE International Conference on Neural Networks, Perth, Australia, 27 November–1 December 1995.
50. Yeh, W.-C. Orthogonal simplified swarm optimization for the series–parallel redundancy allocation problem with a mix of components. *Knowl.-Based Syst.* **2014**, *64*, 1–12. [CrossRef]
51. Yeh, W.-C. Optimization of the Disassembly Sequencing Problem on the Basis of Self-Adaptive Simplified Swarm Optimization. *IEEE Trans. Syst. Man, Cybern.-Part A Syst. Hum.* **2011**, *42*, 250–261. [CrossRef]
52. Huang, C.-L.; Jiang, Y.-Z.; Yin, Y.; Yeh, W.-C.; Chung, V.Y.Y.; Lai, C.-M. Multi Objective Scheduling in Cloud Computing Using MOSSO. *IEEE Congr. Evol. Comput.* **2018**, *12*, 1–8.
53. Yeh, W.-C. A new exact solution algorithm for a novel generalized redundancy allocation problem. *Inf. Sci.* **2017**, *408*, 182–197. [CrossRef]
54. Yeh, W.-C. New parameter-free simplified swarm optimization for artificial neural network training and its application in the prediction of time series. *IEEE Trans. Neural Netw. Learn. Syst.* **2013**, *24*, 661–665.
55. Yeh, W.-C. Novel swarm optimization for mining classification rules on thyroid gland data. *Inf. Sci.* **2012**, *197*, 65–76. [CrossRef]
56. Yeh, W.-C. Simplified swarm optimization in disassembly sequencing problems with learning effects. *Comput. Oper. Res.* **2012**, *39*, 2168–2177. [CrossRef]

57. Yeh, W.-C. A novel boundary swarm optimization method for reliability redundancy allocation problems. *Reliab. Eng. Syst. Saf.* **2019**, *192*, 106060. [CrossRef]
58. Lin, P.; Cheng, S.; Yeh, W.; Chen, Z.; Wu, L. Parameters extraction of solar cell models using a modified simplified swarm optimization algorithm. *Sol. Energy* **2017**, *144*, 594–603. [CrossRef]
59. Shen, Y.; Willems, S.P.; Dai, Y. Channel Selection and Contracting in the Presence of a Retail Platform. *Prod. Oper. Manag.* **2018**, *28*, 1173–1185. [CrossRef]
60. Abhishek, V.; Jerath, K.; Zhang, Z.J. Agency Selling or Reselling? Channel Structures in Electronic Retailing. *Manag. Sci.* **2016**, *62*, 2259–2280. [CrossRef]
61. Liu, Y.-H.; Hart, S.M. Characterizing an optimal solution to the linear bilevel programming problem. *Eur. J. Oper. Res.* **1994**, *73*, 164–166. [CrossRef]
62. Bard, J.F.; Falk, J.E. An explicit solution to the multi-level programming problem. *Comput. Oper. Res.* **1982**, *9*, 77–100. [CrossRef]
63. Amouzegar, M.A.; Cybernetics, P.B. A global optimization method for nonlinear bilevel programming problems. *IEEE Trans. Syst. Man Cybern. Part B (Cybern)* **1999**, *29*, 771–777. [CrossRef]
64. Guang-Min, W.; Zhong-Ping, W.; Xian-Jia, W.; Ya-Lin, C. Genetic Algorithms for Solving Linear Bilevel Programming. In Proceedings of the IEEE Sixth International Conference on Parallel and Distributed Computing Applications and Technologies (PDCAT'05), Dalian, China, 5–8 December 2005; pp. 920–924.
65. Wang, Y.; Jiao, Y.-C.; Li, H. An Evolutionary Algorithm for Solving Nonlinear Bilevel Programming Based on a New Constraint-Handling Scheme. *IEEE Trans. Syst. Man Cybern. Part C Appl. Rev.* **2005**, *35*, 221–232. [CrossRef]
66. Wan, Z.; Wang, G.; Sun, B. A hybrid intelligent algorithm by combining particle swarm optimization with chaos searching technique for solving nonlinear bilevel programming problems. *Swarm Evol. Comput.* **2013**, *8*, 26–32. [CrossRef]
67. Li, X.; Tian, P.; Min, X. A Hierarchical Particle Swarm Optimization for Solving Bilevel Programming Problems. In *International Conference on Artificial Intelligence and Soft Computing*; Springer: Berlin/Heidelberg, Germany, 2006; pp. 1169–1178.

technologies

MDPI

Article

Electrospinning for the Modification of 3D Objects for the Potential Use in Tissue Engineering

Laura Bauer [1], Lisa Brandstäter [1], Mika Letmate [1], Manasi Palachandran [1], Fynn Ole Wadehn [1], Carlotta Wolfschmidt [1], Timo Grothe [1], Uwe Güth [2] and Andrea Ehrmann [1,*]

1 Faculty of Engineering and Mathematics, Bielefeld University of Applied Sciences, 33619 Bielefeld, Germany; laura.bauer@fh-bielefeld.de (L.B.); lisa.brandstaeter@fh-bielefeld.de (L.B.); mika.letmate@fh-bielefeld.de (M.L.); manasi.palachandran@fh-bielefeld.de (M.P.); fynn_ole.wadehn@fh-bielefeld.de (F.O.W.); carlotta.wolfschmidt@fh-bielefeld.de (C.W.); timo.grothe@fh-bielefeld.de (T.G.)

2 Department of Physical and Biophysical Chemistry (PC III), Faculty of Chemistry, Bielefeld University, 33615 Bielefeld, Germany; uwe.gueth@uni-bielefeld.de

* Correspondence: andrea.ehrmann@fh-bielefeld.de

Abstract: Electrospinning is often investigated for biotechnological applications, such as tissue engineering and cell growth in general. In many cases, three-dimensional scaffolds would be advantageous to prepare tissues in a desired shape. Some studies thus investigated 3D-printed scaffolds decorated with electrospun nanofibers. Here, we report on the influence of 3D-printed substrates on fiber orientation and diameter of a nanofiber mat, directly electrospun on conductive and isolating 3D-printed objects, and show the effect of shadowing, taking 3D-printed ears with electrospun nanofiber mats as an example for potential and direct application in tissue engineering in general.

Keywords: needleless electrospinning; poly(lactic acid) (PLA); poly(acrylonitrile) (PAN); nanospider; cell adhesion; cell proliferation; 3D printing

Citation: Bauer, L.; Brandstäter, L.; Letmate, M.; Palachandran, M.; Wadehn, F.O.; Wolfschmidt, C.; Grothe, T.; Güth, U.; Ehrmann, A. Electrospinning for the Modification of 3D Objects for the Potential Use in Tissue Engineering. *Technologies* **2022**, *10*, 66. https://doi.org/10.3390/technologies10030066

Academic Editors: Manoj Gupta, Eugene Wong and Gwanggil Jeon

Received: 29 April 2022
Accepted: 26 May 2022
Published: 29 May 2022

1. Introduction

Electrospinning enables the production of nanofibers in a relatively fast and simple way. Generally, a polymer solution or melt is inserted into a strong electric field between two electrodes, one of which is typically the needle through which the spinning solution is pressed, or a wire coated with the spinning solution [1,2]. The field leads to the formation of Taylor cones from which a polymer jet is extruded towards the counter-electrode. The spiraling shape of this jet results in strong elongation while the solvent is evaporated, until ultrathin nanofibers are deposited on the counter-electrode or a substrate that shields the counter-electrode [3–5].

The fiber orientation on the substrate depends on the collector. A static collector, as is mostly used in wire-based electrospinning, usually leads to arbitrary fiber orientations [6,7]. For several applications, it can be supportive to use roughly parallel oriented fibers. This can be reached, e.g., with a fast-rotating cylinder as collector [8,9]. Another possibility to prepare mats of aligned nanofibers is given by adding dielectric or conductive areas to the substrate, which deform the electric field and in this way allow for the tailoring of the position of the deposited nanofibers, as well as their orientation to a certain amount [10–12].

Such oriented nanofibers are often supportive for oriented cell growth and increased cell proliferation, both of which are important factors in tissue engineering [13,14]. Another important factor is the material of the electrospun nanofibers. Many biomaterials, such as gelatin, are water-soluble and thus have the disadvantage that they need an additional crosslinking step after spinning before they can be used in a fluid medium [15,16]. Other polymers need toxic solvents, which makes a sophisticated post-treatment necessary to avoid reducing the biocompatibility of the nanofiber mats [17,18]. Only few water-stable polymers can be electrospun from the low-toxic solvent dimethyl sulfoxide (DMSO) [19], amongst them poly(acrylonitrile) (PAN) [20,21]. While pure PAN does not serve as an ideal substrate for cell adhesion and proliferation, water-stable blends of PAN with gelatin, maltodextrin, casein, etc. can be used to support cell growth [22,23].

Here we report on electrospinning PAN nanofiber mats on different 3D printed shapes, prepared from various polymers, some of which have conductive properties. Generally, the combination of 3D-printed shapes with an electrospun nanostructure was reported to be an interesting method to combine the desired morphology, mimicking the extracellular matrix, with a desired macroscopic shape [24–26].

Opposite to a previous study in which nanofibers were electrospun on a flat 3D-printed structure [27], here higher and partly irregular shapes are investigated, especially regarding shadowing effects, taking 3D-printed ears with nanofiber mats as an example. Optical investigations reveal strongly different fiber orientations, depending on the shape and the material of the 3D-printed substrates.

2. Materials and Methods

Electrospinning was performed on the wire-based electrospinning machine Nanospider Lab (Elmarco, Liberec, Czech Republic) applying the following unchanged spinning parameters during the experiments: nozzle diameter 0.9 mm; distance between electrode and substrate 240 mm; carriage speed 100 mm/s; the substrate was not moved. The temperature in the spinning chamber was 22–23 °C, the relative humidity 32–33%. The varying spinning parameters are given in Table 1.

Table 1. Assignment of the sample description, the associated 3D printed parts and its spinning parameters. Due to the overview, only the altered parameters are shown.

Description	3D Printed Part	Spinning Solution	Voltage	Current	Duration
V1	None	16% PAN	80 kV	0.116 mA	30 min
V2	None	16% PAN	80 kV	0.08 mA	30 min
V3	None	16% PAN	80 kV	0.08 mA	30 min
V4	None	16% PAN + 5% dextran	80 kV	0.04 mA	31 min
V5	None	14% PAN	80 kV	0.04 mA	45 min
V6	Various 3D parts	14% PAN	81 kV	0.032 mA	30 min
V6-1	Various 3D parts	14% PAN	81 kV	0.032 mA	30 min
V6-2	Various 3D parts	14% PAN	81 kV	0.032 mA	30 min
V7	Aluminum foil	14% PAN	81 kV	0.032 mA	30 min
V9-2	3D filaments	14% PAN	80 kV	0.03 mA	30 min
V10	3D printed ear	12% PAN + 2% dextran	80 kV	0.03 mA	17 min
V11	3D printed ears from different filaments	13% PAN	80 kV	0.03 mA	25 min
V12-1	3D printed funnel in profile	13% PAN	82 kV	0.03 mA	16 min
V12-2	3D printed ears (partly grounded)	13% PAN	50 kV	0.016 mA	30 min

The spinning solutions were prepared from 13–16% PAN (X-PAN, Dralon, Dormagen, Germany) and 2–5% Dextran 500 (for biochemistry, 500 kDa, purchased from Carl Roth GmbH & Co. KG, Karlsruhe, Germany) dissolved in DMSO (min 99.9%, S3 chemicals, Bad Oeynhausen, Germany) by stirring for 24 h under ambient conditions.

The following 3D-printing materials were investigated:

- Filaflex 82A (Recreus, Elda, Spain)
- Conductive PLA (Proto-pasta, Vancouver, Canada)
- XT-CF20 (Colorfabb, Belfeld, The Netherlands)
- Bronzefill (Colorfabb, Belfeld, The Netherlands)
- Growlay brown (Lay-Filaments, Cologne, Germany)
- Carbon X2–85 (3DXTech, Grand Rapids, MI, USA)
- CarbonFil (Formfutura, Nijmegen, The Netherlands)
- Poly(lactic acid) (PLA) (Filamentworld, Neu-Ulm, Germany)

Only the conductive PLA shows a measurable conductivity (R~200 Ω/cm); the others partly include conductive carbon fibers, etc., but apparently without forming sufficient percolation paths.

3D printing was performed using an Orcabot XXL (Prodim, The Netherlands) with a nozzle diameter of 0.4 mm, nozzle temperature of 210 °C, printing bed temperature of 60 °C, layer thickness of 0.2 mm and 100% infill (linear).

The ear model was taken from Thingiverse (https://www.thingiverse.com/thing:304657, created by addamay123, published under a CC-BY-SA license, accessed on 7 March 2022).

All 3D-printed specimens and 3D-printing filaments were mounted below the standard polypropylene substrate, as depicted in Figure 1.

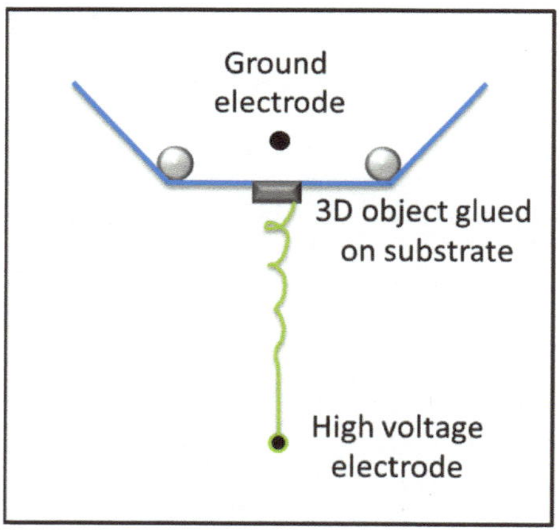

Figure 1. Sketch of the nanofiber setup: The high-voltage electrode wire (**black**) is coated by the polymer solution (**green**). The latter is dragged by the strong electric field towards the ground electrode, before which it is deposited on a substrate (**blue**) or on objects glued onto the substrate (**grey**).

The following samples were prepared, varying spinning parameters and 3D printed parts:

The morphology of the samples was investigated by a confocal laser scanning microscopy (CLSM) VK-8710 (Keyence, Neu-Isenburg, Germany). Exemplary images were taken by a scanning electron microscope (SEM) FEI XL30 ESEM (Philips, Amsterdam, The Netherlands), after sputtering the samples with palladium. Macroscopic images were taken by a Sony Cybershot DSC-RX100 IV camera.

3. Results and Discussion

The first nanofiber mats (V1-V5) were used to investigate the reproducibility of the gained nanofibers mats as well as the influence of additional dextran in the solution, which was shown to result in relatively thick, straight fibers [28]. As expected, the CLSM images showed similar PAN nanofiber mats on an intermediate scale, while sample V4 with a PAN/dextran blend had significantly thicker fibers (Figure 2). Some areas of some of the nanofiber mats contained nonfibrous areas, as visible in Figure 2c. This happens especially in case of slightly increased relative humidity or not-completely exhausted solvent vapor in the chamber after long spinning durations (45 min in case of sample V5).

(**a**) (**b**) (**c**)

Figure 2. CLSM images of nanofiber mats: (**a**) V3 containing pure PAN; (**b**) V4 containing a PAN/dextran blend; (**c**) V5 containing pure PAN. Scale bars indicate 20 µm.

Another possible difference between nominally identical nanofiber mats is depicted more in Figure 3, using the example of samples V1 and V2. Here, the time dependence of the electrospinning solution was investigated. While the solution for V1 (Figure 3a) was left in the lab for two weeks, the solution used for V2 (Figure 3b) was directly electrospun after stirring for 24 h. It is clearly visible that although no macroscopic differences between the spinning solutions could be recognized, the results differ strongly, with V1 showing relatively thick, straight fibers, while V2 has significantly thinner fibers with beads. These beads typically occur when the spinning solution does not contain a sufficient solid content [29], while thicker fibers are typical for spinning solutions with a higher amount of PAN [30]. This comparison indicates that usual stirring by a magnetic stirrer for some hours does not fully dissolve PAN in DMSO, so that the duration between preparation of the solution and electrospinning should be taken into account as an additional parameter. Here, all other nanofiber mats were electrospun approx. 1–2 days after preparation of the solution.

(**a**) (**b**)

Figure 3. SEM images of (**a**) sample V1; (**b**) sample V2. Scale bars indicate 7 µm.

Next, the influence of nonconductive 3D objects glued on the substrate was tested. Figure 4 depicts a comparison of different areas of sample V6-1, electrospun on (Figure 4a) or next to (Figure 4b) a 3D printed object from PLA with a ratchet-like surface. Interestingly, a clear fiber orientation parallel to the maxima of the ratchet is found on the 3D-printed object, as it was also recognized in an earlier study [12], while no such orientation is visible next to the 3D object (Figure 4b).

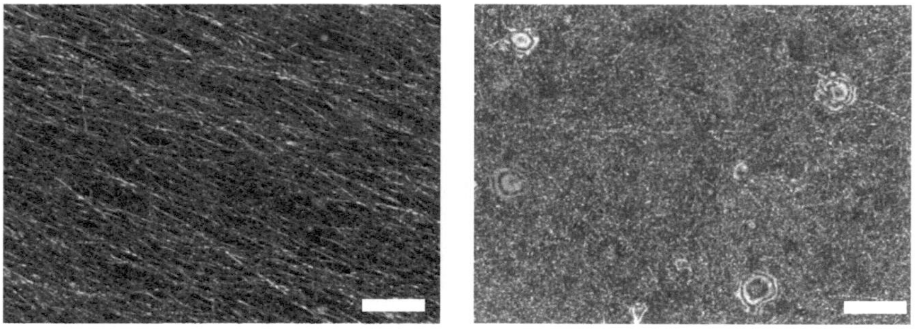

(a) (b)

Figure 4. CLSM images of sample V6-1 (**a**) on a 3D-printed object; (**b**) next to a 3D-printed object. Scale bars indicate 20 μm.

Investigating the nanofiber mat on top of the 3D printed object by SEM reveals a similar finding, as visible in Figure 5. The nanofiber mat on top of the object (on the right-hand side of the dotted line) shows a clear fiber orientation, which directly changes when the nanofiber mat is examined on the common polypropylene (PP) nonwoven substrate (left of the dotted line). This underlines the influence of a substrate variation on the nanofiber mat morphology.

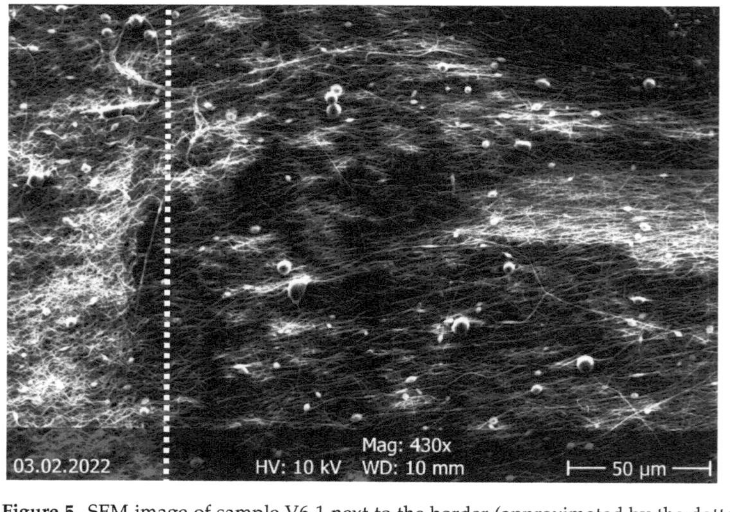

Figure 5. SEM image of sample V6-1 next to the border (approximated by the dotted line) of the 3D-printed object.

To investigate the effect of different materials as substrate modifications further, seven 3D-printing filaments with partly conductive filling (cf. Section 2) were glued on the PP substrate before electrospinning with standard parameters (V9-2) was performed. Figure 6 depicts CLSM images of the surfaces. Most of them look very similar, partly with visible beads or nonfibrous areas (visible as bright, round spots). Only the conductive filament "Conductive PLA" (Figure 6d) shows a clear fiber orientation. It should be mentioned that the optical properties of such PAN nanofiber mats, independent of the fiber orientation, generally show a total transmission around 40–70% (depending on the nanofiber mat thickness) and a specular transmission near 0% throughout the whole visible spectrum without any maxima or minima, corresponding to the typical white color of such nanofiber mats [31,32].

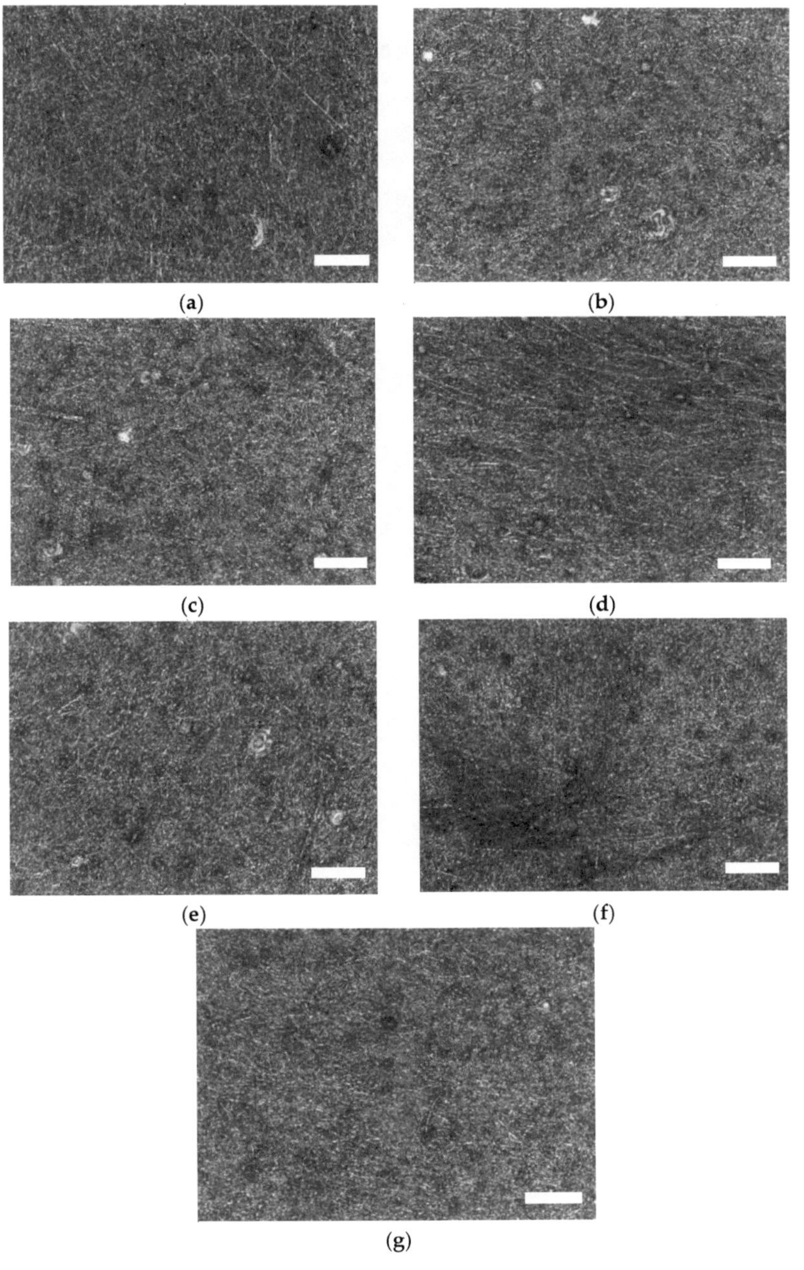

Figure 6. CLSM images of PAN nanofiber mats on 3D-printing filaments: (**a**) Bronzefill; (**b**) CarbonFil; (**c**) Carbon X2-85; (**d**) Conductive PLA; (**e**) Filaflex; (**f**) Growlay; (**g**) XT-CF20. Scale bars indicate 20 μm.

Since CLSM images can only show the fiber near the sample surface and do not allow the depiction of the thickness of the nanofiber mat, the same samples are depicted by macroscopic photographs in Figure 7. Here, it becomes clear that the filaments Carbon X2-85 (Figure 7c), Conductive PLA (Figure 7d) and XT-CF20 (Figure 7g) attract the highest quantity of nanofibers and thus show the thickest nanofiber mats, while the other filaments seem to repel the nanofiber mats. Such an effect has already been recognized in a previous study [12]. The next tests in which nanofiber mats were grown on different 3D-printed shapes were thus performed with Conductive PLA as the most conductive filament, here showing the most regular nanofiber mat on top of it.

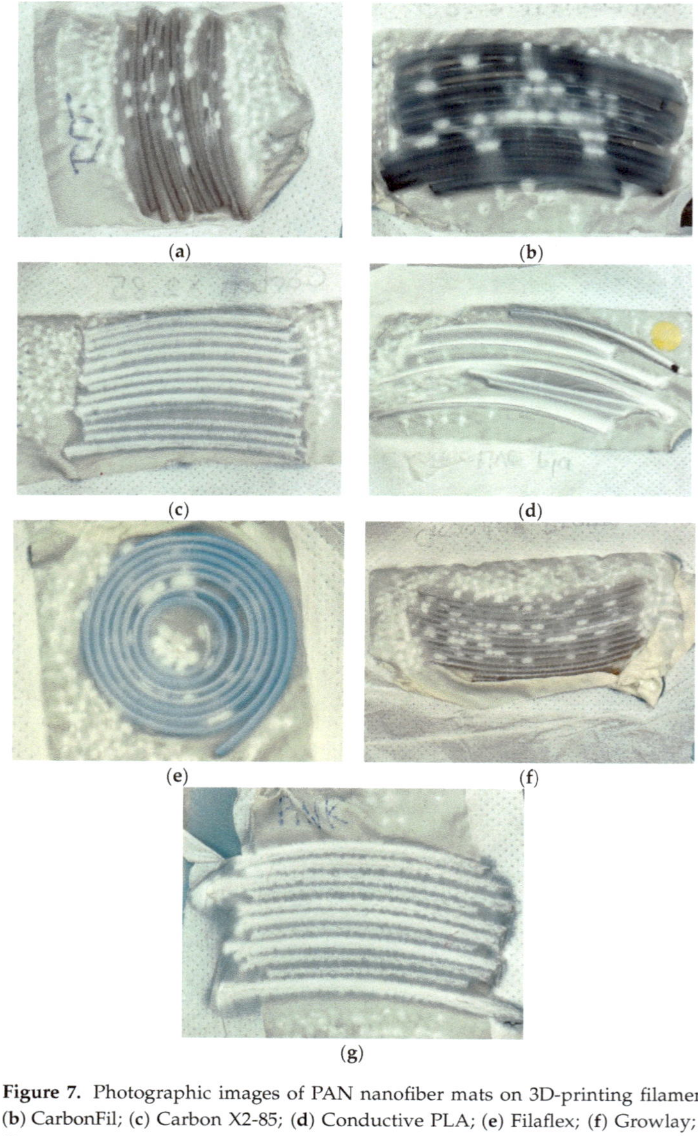

Figure 7. Photographic images of PAN nanofiber mats on 3D-printing filaments: (**a**) Bronzefill; (**b**) CarbonFil; (**c**) Carbon X2-85; (**d**) Conductive PLA; (**e**) Filaflex; (**f**) Growlay; (**g**) XT-CF20. All filament diameters are 1.75 mm.

Next, 3D-printed funnels were 3D printed to investigate possible shadowing effects at the rounded edges (experiment V12-1, Figure 8). Generally, the surface of both funnels is completely covered with nanofiber mat, with the nanofiber mats following the surface steps of the funnels due to the layer-wise printing, thus causing steps of 0.15 mm height. However, a deeper look at both samples reveals that a thicker nanofiber mat is placed on the funnel with additional copper foil, i.e., a system which modifies the electric field of the electrospinning apparatus more than the pure conductive 3D-printed object. Moreover, nearly no nanofibers are visible on the PP substrate around the conductive print with copper foil below. Comparing both surfaces shows that the nanofiber mat on the pure Conductive PLA object has lower irregularities. The black holes in the nanofiber mat on the funnel on copper foil (Figure 8b) were burnt by small flash-arcs, which can occur in areas with a highly concentrated electric field if the relative humidity is high enough or the spinning solution has a sufficient conductivity.

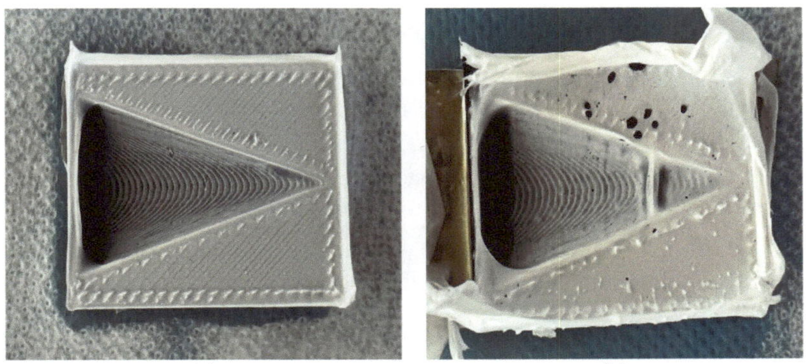

(a) (b)

Figure 8. PAN electrospun on funnels printed with Conductive PLA: (**a**) glued on the PP substrate; (**b**) glued on a copper foil on the PP substrate. The funnels have a length of 40 mm and width of 31 mm.

CLSM images of the apex of the funnel in Figure 8a are depicted in Figure 9 at different magnifications. On both scales, there are no fiber orientations visible, which may be attributed to the small height gradient inside the funnel. The height plot in Figure 9b shows two of the steps due to the printing process (from the orange plateau to the green layer below and the blue layer below the green one). These steps are no longer visible at higher magnification (Figure 9d).

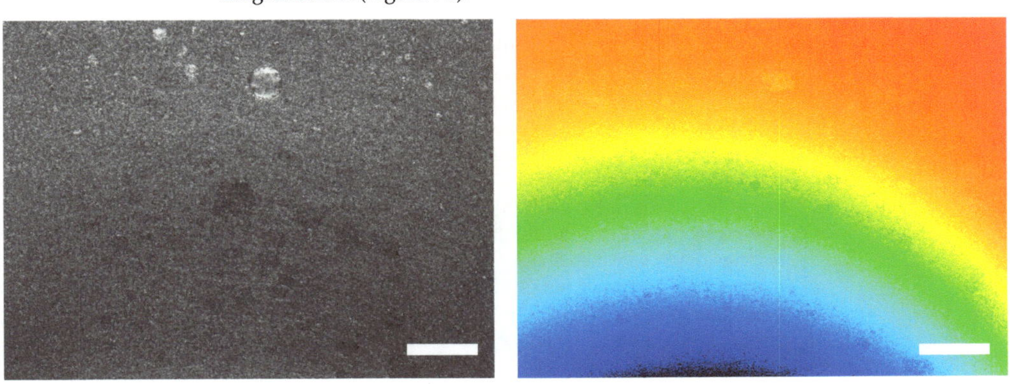

(a) (b)

Figure 9. *Cont.*

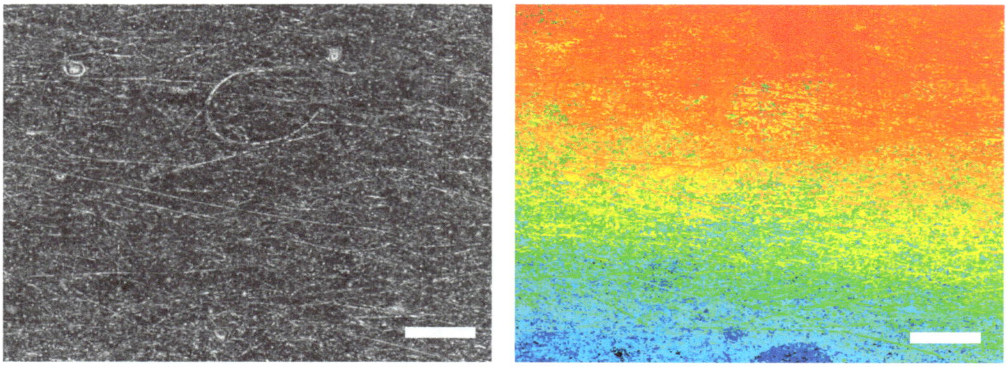

Figure 9. CLSM images of the apex of the funnel in Figure 8a: (**a**) morphology with 200× magnification; (**b**) height plot with 200× magnification; (**c**) morphology with 2000× magnification; (**d**) height plot with 2000× magnification. Scale bars correspond to 200 µm (**a**,**b**) and 20 µm (**c**,**d**), respectively.

As a stronger 3D-shaped object, 3D-printed ears were tested as substrates (V10, V11 and V12-2). Tests were performed comparing PLA and Conductive PLA as printing materials; the ears were partly placed on additional conductive copper foils, and they were partly additionally grounded. PAN/dextran and pure PAN nanofiber mats were electrospun on them. Figure 10 depicts some of the results of these tests.

As already expected, the pure PLA ears strongly repelled the nanofibers, while ears printed from Conductive PLA showed a nanofiber mat similar to the surrounding PP nonwoven (Figure 10a). No macroscopic differences are visible comparing PAN/dextran (Figure 10b) and PAN nanofiber mats. The inner areas of the ear, however, are not covered by nanofibers in these tests.

This is why subsequent tests were performed with reduced voltage to examine the influence of this parameter on the covering of the 3D-printed objects (Figure 10c,d). However, the shadowing effect became even stronger, as compared to Figure 10a,b. This can be explained by the nanofibers impinging on the substrate at a smaller speed if the voltage is lower, in this way being stronger directed towards the highest conductive areas and thus leaving more lower areas inside the ear uncovered. Furthermore, the influence of an additional highly conductive copper foil below the ear from Conductive PLA (Figure 10d) is clearly visible, as it was already recognized in Figure 8. Apparently, Conductive PLA has a well-suitable conductivity to avoid repelling a nanofiber mat without deforming the electric field so strongly that highly irregular nanofiber mats are formed, as visible in Figure 9d. This shows that material and shape have to be tailored carefully to enable the covering of the whole surface, possibly even by combining different 3D-printing polymers in one object, which is possible with several recent 3D printers.

In order to prepare 3D substrates for tissue engineering with a nanostructured surface, it is nevertheless necessary to enable coating the whole surface by a nanofiber mat. One possible approach to reach this aim is by using so-called 4D printing, i.e., 3D-printing a plane object that can be deformed afterwards by heat or other stimuli [33]. Since PLA belongs to the so-called shape-memory polymers, which enable 4D printing [34], the shape-memory properties of the Conductive PLA under investigation in the recent study will be investigated in a future study.

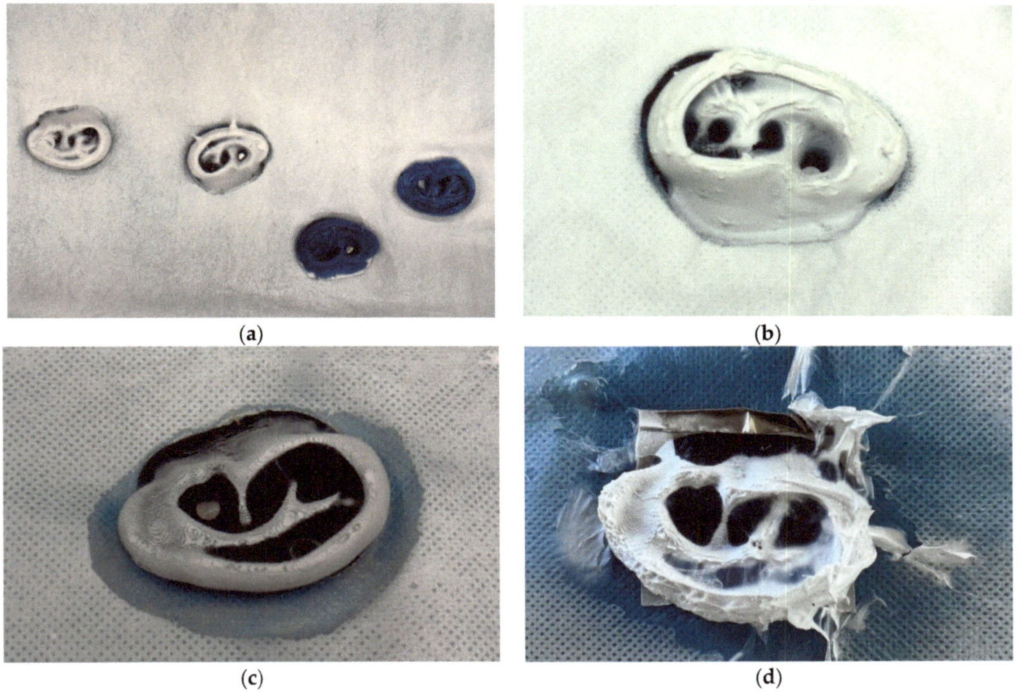

(a) (b) (c) (d)

Figure 10. 3D-printed ears decorated with electrospun nanofiber mats: (**a**) from left to right: PAN on Conductive PLA on copper foil, Conductive PLA on aluminum foil, PLA on copper foil, PLA on PP substrate (V11); (**b**) PAN/dextran on Conductive PLA; (**c**) PAN on Conductive PLA (50 kV); (**d**) PAN on Conductive PLA on copper foil (50 kV). All ears have a length of 59 mm (longest side).

4. Conclusions

Electrospinning PAN and PAN/dextran nanofiber mats was performed on diverse 3D-printing polymers. Depending on their shape, thickness and conductivity, the nanofibers were repelled or strongly attracted. 3D-printed ears from conductive PLA were covered along the higher parts, while varying spinning and solution parameters did not enable covering the whole surface of the structure. Oppositely, 3D-printed funnels with lower slope could be completely covered, with the electrospun nanofiber mat following the surface structure given by the 3D-printing process. Along the borders of some 3D-printed materials, a clear fiber orientation was found, which can be used for oriented cell growth.

As a possible solution, 4D-printing of conductive shape-memory polymers will be investigated in a future study. Moreover, biocompatibility in general, as well as mammalian cell adhesion and proliferation, will be tested for different conductive PLA materials. Finally, degradation of PLA in cell-culture medium has to be evaluated, especially related to the potential influence of the nanofiber mat grown on it.

Author Contributions: Conceptualization, T.G.; methodology, T.G. and A.E.; validation, T.G. and A.E.; formal analysis, T.G. and A.E.; investigation, all authors; writing—original draft preparation, T.G. and A.E.; writing—review and editing, all authors.; visualization, U.G., T.G. and A.E. All authors have read and agreed to the published version of the manuscript.

Funding: This research was partly funded by the German Federal Ministry for Economic Affairs and Energy as part of the Central Innovation Program for SMEs (ZIM) via the AiF, based on a resolution of the German Bundestag, grant number KK5129703CR0.

Institutional Review Board Statement: Not applicable.

Informed Consent Statement: Not applicable.

Data Availability Statement: All data are shown in this paper.

Conflicts of Interest: The authors declare no conflict of interest. The funders had no role in the design of the study; in the collection, analyses, or interpretation of data; in the writing of the manuscript; or in the decision to publish the results.

References

1. Arkoun, M.; Daigle, F.; Heuzey, M.-C.; Ajji, A. Antibacterial electrospun chitosan-based nanofibers: A bacterial membrane perforator. *Food Sci. Nutr.* **2017**, *5*, 865–874. [CrossRef] [PubMed]
2. Yalcinkaya, F. A review on advanced nanofiber technology for membrane distillation. *J. Eng. Fibers Fabr.* **2019**, *14*, 1558925018824901. [CrossRef]
3. Agarwal, S.; Wendorff, J.H.; Greiner, A. Use of electrospinning technique for biomedical applications. *Polymer* **2008**, *49*, 5603–5621. [CrossRef]
4. Cengiz, F.; Krucinska, I.; Gliscinska, E.; Chrzanowski, M.; Göktepe, F. Comparative analysis of various electrospinning methods of nanofibre formation. *Fibres Text. East. Eur.* **2009**, *72*, 13–19.
5. Bhardwaj, N.; Kundu, S.C. Electrospinning: A fascinating fiber fabrication technique. *Biotechnol. Adv.* **2010**, *28*, 325–347. [CrossRef]
6. Greiner, A.; Wendorff, J.H. Electrospinning: A fascinating method for the preparation of ultrathin fibers. *Angew. Chem. Int. Ed.* **2007**, *46*, 5670–5703. [CrossRef]
7. Li, D.; Xia, Y. Electrospinning of nanofibers: Reinventing the wheel? *Adv. Mater.* **2004**, *16*, 1151–1170. [CrossRef]
8. Bertocchi, M.J.; Simbana, R.A.; Wynne, L.H.; Lundin, J.G. Electrospinning of tough and elastic liquid crystalline polymer-polyurethane composite fibers: Mechanical properties and fiber alignment. *Macromol. Mater. Eng.* **2019**, *304*, 1900186. [CrossRef]
9. Bazrafshan, Z.; Stylios, G.K. A novel approach to enhance the spinnability of collagen fibers by graft polymerization. *Mater. Sci. Eng.* **2019**, *94*, 108–116. [CrossRef]
10. Storck, J.L.; Grothe, T.; Mamun, A.; Sabantina, L.; Klöcker, M.; Blachowicz, T.; Ehrmann, A. Orientation of electrospun magnetic nanofibers near conductive areas. *Materials* **2020**, *13*, 47. [CrossRef]
11. Nguyen, D.N.; Hwang, Y.; Moon, W. Electrospinning of well-aligned fiber bundles using an end-point control assembly method. *Eur. Polym. J.* **2016**, *77*, 54–64. [CrossRef]
12. Hellert, C.; Storck, J.L.; Grothe, T.; Kaltschmidt, B.; Hütten, A.; Ehrmann, A. Positioning and aligning electrospun PAN fibers by conductive and dielectric substrate patterns. *Macromol. Symp.* **2021**, *395*, 2000213. [CrossRef]
13. Guarino, V.; Iannotti, V.; Ausanio, G.; Ambrosio, L.; Lanotte, L. Elastomagnetic nanofiber wires by magnetic field assisted electrospinning. *Express Polym. Lett.* **2019**, *13*, 419–428. [CrossRef]
14. Johnson, C.D.L.; Ganguly, D.; Zuidema, J.M.; Cardina, T.J.; Ziemba, A.M.; Kearns, K.R.; McCarthy, S.M.; Thompson, D.M.; Ramanath, G.; Borca-Tasciuc, D.A.; et al. Injectable, magnetically orienting electrospun fiber conduits for neuron guidance. *ACS Appl. Mater. Interfaces* **2019**, *11*, 356–372. [CrossRef] [PubMed]
15. Cooper, A.; Oldinski, R.; Ma, H.; Bryers, J.D.; Zhang, M. Chitosan-based nanofibrous membranes for antibacterial filter applications. *Carbohydr. Polym.* **2013**, *92*, 254–259. [CrossRef]
16. Ehrmann, A. Non-toxic crosslinking of electrospun gelatin nanofibers for tissue engineering and biomedicine—A review. *Polymers* **2021**, *13*, 1973. [CrossRef]
17. Nam, J.; Huang, Y.; Agarwal, S.; Lannutti, J. Materials selection and residual solvent retention in biodegradable electrospun fibers. *J. Appl. Polym. Sci.* **2008**, *107*, 1547–1554. [CrossRef]
18. Fatih Canbolat, M.; Tang, C.; Bernacki, S.H.; Pourdeyhimi, B.; Khan, S. Mammalian cell viability in electrospun composite nanofiber structures. *Macromol. Biosci.* **2011**, *11*, 1346–1356. [CrossRef]
19. Wortmann, M.; Frese, N.; Sabantina, L.; Petkau, R.; Kinzel, F.; Gölzhäuser, A.; Ehrmann, A. New polymers for needleless electrospinning from low-toxic solvents. *Nanomaterials* **2019**, *9*, 52. [CrossRef]
20. Plamus, T.; Savest, N.; Viirsalu, M.; Harz, P.; Tarasova, E.; Krasnou, I.; Krumme, A. The effect of ionic liquids on the mechanical properties of electrospun polyacrylonitrile membranes. *Polym. Test.* **2018**, *71*, 335–343. [CrossRef]
21. Krasonu, I.; Tarassova, E.; Malmberg, S.; Vassiljeva, V.; Krumme, A. Preparation of fibrous electrospun membranes with activated carbon filler. *IOP Conf. Ser. Mater. Sci. Eng.* **2019**, *500*, 012022. [CrossRef]
22. Wehlage, D.; Blattner, H.; Mamun, A.; Kutzli, I.; Diestelhorst, E.; Rattenholl, A.; Gudermann, F.; Lütkemeyer, D.; Ehrmann, A. Cell growth on electrospun nanofiber mats from polyacrylonitrile (PAN) blends. *AIMS Bioeng.* **2020**, *7*, 43–54. [CrossRef]
23. Wehlage, D.; Blattner, H.; Sabantina, L.; Böttjer, R.; Grothe, T.; Rattenholl, A.; Gudermann, F.; Lütkemeyer, D.; Ehrmann, A. Sterilization of PAN/gelatin nanofibrous mats for cell growth. *Tekstilec* **2019**, *62*, 78–88. [CrossRef]
24. Yu, Y.X.; Hua, S.; Yang, M.K.; Fu, Z.Z.; Teng, S.S.; Niu, K.; Zhao, Q.H.; Yi, C.Q. Fabrication and characterization of electrospinning/3D printing bone tissue engineering scaffold. *RSC Adv.* **2016**, *6*, 110557–110565. [CrossRef]
25. Muerza-Cascante, M.L.; Shokoohmand, A.; Khosrotehrani, K.; Haylock, D.; Dalton, P.D.; Hutmacher, D.W.; Loessner, D. Endosteal-like extracellular matrix expression on melt electrospun written scaffolds. *Acta Biomater.* **2017**, *52*, 145–158. [CrossRef]

26. De Mori, A.; Pena Fernández, M.; Blunn, G.; Tozzi, G.; Roldo, M. 3D Printing and Electrospinning of Composite Hydrogels for Cartilage and Bone Tissue Engineering. *Polymers* **2018**, *10*, 285. [CrossRef]
27. Trabelsi, M.; Mamun, A.; Klöcker, M.; Sabantina, L.; Großerhode, C.; Blachowicz, T.; Ehrmann, A. Increased mechanical properties of carbon nanofiber mats for possible medical applications. *Fibers* **2019**, *7*, 98. [CrossRef]
28. Böttjer, R.; Grothe, T.; Wehlage, D.; Ehrmann, A. Electrospraying poloxamer/(bio-) polymer blends using a needleless electrospinning machine. *J. Text. Fibrous Mater.* **2018**, *1*, 2515221117743079. [CrossRef]
29. Sabantina, L.; Rodríguez Mirasol, J.; Cordero, T.; Finsterbusch, K.; Ehrmann, A. Investigation of needleless electrospun PAN nanofiber mats. *AIP Conf. Proc.* **2018**, *1952*, 020085.
30. Grothe, T.; Storck, J.L.; Dotter, M.; Ehrmann, A. Impact of solid content in the electrospinning solution on the physical and chemical properties of polyacrylonitrile (PAN) nanofibrous mats. *Tekstilec* **2020**, *63*, 225–232. [CrossRef]
31. Kerker, E.; Steinhäußer, D.; Mamun, A.; Trabelsi, M.; Fiedler, J.; Sabantina, L.; Juhász Junger, I.; Schiek, M.; Ehrmann, A.; Kaschuba, R. Spectroscopic investigation of highly-scattering nanofiber mats during drying and film formation. *Optik* **2020**, *208*, 164081. [CrossRef]
32. Grothe, T.; Böhm, T.; Habashy, K.; Abdullaeva, O.S.; Zablocki, J.; Lützen, A.; Dedek, K.; Schiek, M.; Ehrmann, A. Optical Index Matching, Flexible Electrospun Substrates for Seamless Organic Photocapacitive Sensors. *Phys. Status Solidi B* **2021**, *258*, 2000543. [CrossRef]
33. Koch, H.C.; Schmelzeisen, D.; Gries, T. 4D textiles made by additive manufacturing on pre-stressed textiles—An overview. *Actuators* **2021**, *10*, 31. [CrossRef]
34. Ehrmann, G.; Ehrmann, A. 3D printing of shape memory polymers. *J. Appl. Polym. Sci.* **2021**, *138*, 50847. [CrossRef]

technologies

MDPI

Article

Study of Joint Symmetry in Gait Evolution for Quadrupedal Robots Using a Neural Network

Zainullah Khan [1], Farhat Naseer [1], Yousuf Khan [1,*], Muhammad Bilal [2] and Muhammad A. Butt [3,4]

[1] Embedded Systems Research Group, Department of Electronic Engineering, Balochistan University of Information Technology, Engineering and Management Sciences, Quetta 87300, Pakistan; zain.9496@gmail.com (Z.K.); farhaty16@gmail.com (F.N.)
[2] Department of Telecommunication Engineering, Balochistan University of IT, Engineering and Management Sciences, Quetta 87300, Pakistan; muhammad.bilal4@buitms.edu.pk
[3] Institute of Microelectronics and Optoelectronics, Warsaw University of Technology, Koszykowa 75, 00-662 Warszawa, Poland; ali.butt@pw.edu.pl
[4] Samara National Research University, 443086 Samara, Russia
* Correspondence: yousuf.khan@buitms.edu.pk

Abstract: Bio-inspired legged robots have the potential to traverse uneven terrains in a very efficient way. The effectiveness of the robot gait depends on the joint symmetry of the robot; variations in joint symmetries can result in different types of gaits suitable for different scenarios. In the literature, symmetric and asymmetric gaits have been synthesized for legged robots; however, no relation between the gait effectiveness and joint symmetry has been studied. In this research work, the effect of joint symmetry on the robot gait is studied. To test the suggested algorithm, spider-like robot morphology was created in a simulator. The simulation environment was set to a flat surface where the robots could be tested. The simulations were performed on the PyroSim software platform, a physics engine built on top of the Open Dynamics Engine. The quadrupedal robot was created with eight joints, and it is controlled using an artificial neural network. The artificial neural network was optimized using a genetic algorithm. Different robot symmetries were tested, i.e., diagonal joint symmetry, diagonal joint reverse symmetry, adjacent joint symmetry, adjacent joint reverse symmetry and random joint symmetry or joint asymmetry. The robot controllers for each joint symmetry were evolved for a set number of generations and the robot controllers were evaluated using a fitness function that we designed. Our results showed that symmetry in joint movement could help in generating optimal gaits for our test terrain, and joint symmetry produced gaits that were already present in nature. Moreover, our results also showed that certain joint symmetries tended to perform better than others in terms of stability, speed, and distance traveled.

Keywords: quadrupedal robot; genetic algorithm; gait evolution; neural networks; robot morphology; robot generations

check for updates

Citation: Khan, Z.; Naseer, F.; Khan, Y.; Bilal, M.; Butt, M.A. Study of Joint Symmetry in Gait Evolution for Quadrupedal Robots Using a Neural Network. *Technologies* **2022**, *10*, 64. https://doi.org/10.3390/technologies10030064

Academic Editors: Manoj Gupta, Eugene Wong and Gwanggil Jeon

Received: 26 April 2022
Accepted: 18 May 2022
Published: 22 May 2022

Publisher's Note: MDPI stays neutral with regard to jurisdictional claims in published maps and institutional affiliations.

1. Introduction

Legged robots (that have been inspired by the morphology of real animals) can traverse uneven terrains in a more efficient way than their wheeled counterparts [1,2]. Among legged robots, quadrupedal robots are the most agile and the most stable [3–5].This motivates to mimic nature and thereby adapting robots and vehicles to suit uneven terrains rather than adapting the environment to cater to the needs of wheeled vehicle. However, such legged robot morphologies can prove to be quite a challenge to control manually [6]; therefore, it is very common to automate the process of gait generation for such robots. A robot gait is the movement of actuators that allows the robot to traverse an area [7]. The gaits for quadrupedal robots are categorized into two types according to [8,9]: the first type has joint symmetry and the second type has joint asymmetry. Joint symmetry occurs when the movement of one joint is replicated by another joint in the robot, and joint asymmetry occurs when the movement of two is completely different from one another, i.e., there is no correlation between the movement of one joint with the movement of another joint.

Quadrupedal animals in nature possess both aforementioned gait varieties. The three most common quadrupedal robot gaits are walking, trotting and galloping [10]. Out of these, the walking gait is asymmetric, and the movement of all the legs is different from one another [11]. On the other hand, the trot and gallop gaits are symmetric. In the trot gait, the diagonal legs of a quadruped are in symmetry [12], while in the gallop gait, the front and rear leg pairs are in symmetry [13]. It should also be noted that the trot and gallop gaits are used for running and, hence, among these mentioned gaits, running gaits are symmetric while walking gaits are asymmetric [14].

Inspired by nature, multiple approaches have been taken to synthesize gaits for quadrupedal robots. In [15,16], a static gait for a quadrupedal robot is developed that allows it to traverse uneven terrains; the gait is asymmetric and the robot is controlled using a path planning algorithm. The robot can successfully traverse a rough terrain. Contrary to this, refs. [17–19] develop gaits for quadrupedal robots with joint symmetry. Trotting gaits are developed in simulation, and then later the same gaits are tested on a physical robot. The results show that the robot can traverse plain areas. In [20–22], running and turning gaits are synthesized for an under-actuated quadrupedal robot; the gallop gait is chosen for running, making the running gait symmetric, and the turning gait is also symmetric.

All the methods stated above allow the robots to successfully traverse different terrains; however, the gaits are all fixed and cannot change if the terrain is slightly altered. The robots have no means of adjusting their gaits to match the environment that they are in. Therefore, to adapt well to the environment, the gaits must be synthesized using an optimization process. Some of the most common gait optimization methods used in the literature are the genetic algorithm (GA) [23] and artificial neural networks (ANNs) [24].

As stated above, ANNs and the GA are often used to control robot gaits. The works in [25–28] utilize the GA to evolve gaits for quadrupedal robots that are controlled by ANNs. In [25,28] a comparative study is performed between different types of ANNs to determine which one will produce a gait that is better than the rest of the ANNs. It was evident from their result that the gaits that resulted in symmetric movement often performed the best when their utility function favored speed; however, when faced with uneven terrains, the symmetric gaits performed badly and the asymmetric gaits performed well. Similarly, in [26], gaits were evolved for a quadrupedal robot, and the entire process was performed on a hardware setup. The robot was configured in such a way that its joints were symmetric about the center of the robot. The evolved gait performed very well and the robot successfully traversed even surfaces. However, evolving gaits on a physical robot is a slow process and it may take hundreds of hours to reach an optimal gait. The work in [27] is an extension of [26], and it still employs the same joint symmetry; however, the first few generations of the robot are simulated and, when the robot learns to walk properly, the controllers are transferred to a hardware robot. This saves the time it takes to evolve gaits for a complete hardware robot setup, and the loss of electronic components is also avoided.

The studies mentioned above demonstrate how joint symmetry and asymmetry aid in robot locomotion in different environments and on different terrains; however, no critical analysis has been carried out that relates the effects of joint symmetry to a robot's ability to traverse its environment. Therefore, in this research, we contribute to the field by experimenting with different joint symmetries and studying their effect on the ability of a robot to traverse a flat terrain. We will be using the GA to optimize an ANN controller for our robot. We have also designed a custom fitness function which allows us to evaluate the robot controllers. The results of this study will help in determining whether joint symmetry aids in robot movement or not; moreover, this study also aims to create a benchmark that can help future researchers in deciding which joint symmetries to choose for their robots.

The rest of the paper is arranged in the following order: Section 2 will discuss our methodology; our simulation will be detailed in Section 2.1; our controller will be discussed in Section 2.2; our optimization algorithm will be discussed in Section 2.3; and in Section 2.4 our selection criteria will be overviewed. Finally, the results will be discussed in Section 3.

2. Methodology

The simulations were performed on the PyroSim [29] software platform, a physics engine built on top of an Open Dynamics Engine [30]. To test the suggested algorithm, robot morphology was created in the simulator. Ten individual robots were created in the simulator: the so-called test-population. Every individual in the population was controlled by a fully connected neural network. The designed neural network-based controller could evolve itself using the GA. The robot morphology remained the same for each joint symmetry that was experimented on; however, the controller (ANN) was changed slightly each time to form different joint symmetries. A detailed explanation for all the steps involved in evolving the controllers can be found in Figure 1.

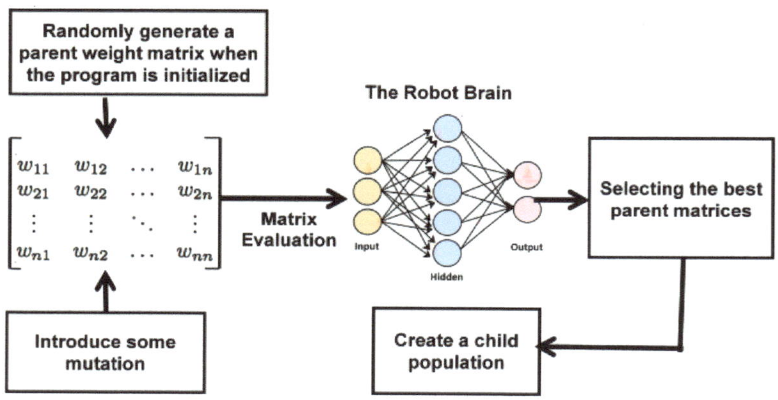

Figure 1. The block diagram of the controller evolution.

2.1. Simulation

As mentioned above, the simulations were performed in PyroSim. The robot body was created using geometric shapes, and a cuboid was used to make the main body while the limbs were made from cylindrical shapes. The robot body was completed by joining all the shapes through hinge joints, hence making a degree of freedom. Each robot leg had two joints, one for the hip and the second for the knee. A total of 4 limbs and 8 joints were created, as shown in Figure 2. The simulation environment was set specifically to suit the environment where the physical robot would be used, and the parameters are given in Table 1.

Table 1. Simulation parameters used in the simulator.

Parameter	Symbol	Value
Robot body length	l	0.2 m
Robot body width	w	0.2 m
Robot body height	h	0.05 m
Length of cylinder	cL	0.2 m
Radius of cylinder	cR	0.02 m
Mass of each robot body part in the simulation	m	1 kg

Table 1. *Cont.*

Parameter	Symbol	Value
Gravity	g	$-9.8\ \mathrm{ms}^{-2}$
Number of joints	J	8
Number of motors	M	8
Motor impulse	τ	0.15
Simulation world step time	dt	0.05
Total number of timesteps for the simulation	T	1000
ANN recall interval timesteps	Rc	60
ANN inputs	I	9
ANN outputs	O	8
Number of individuals in the population	P	10
Number of generations	G	200

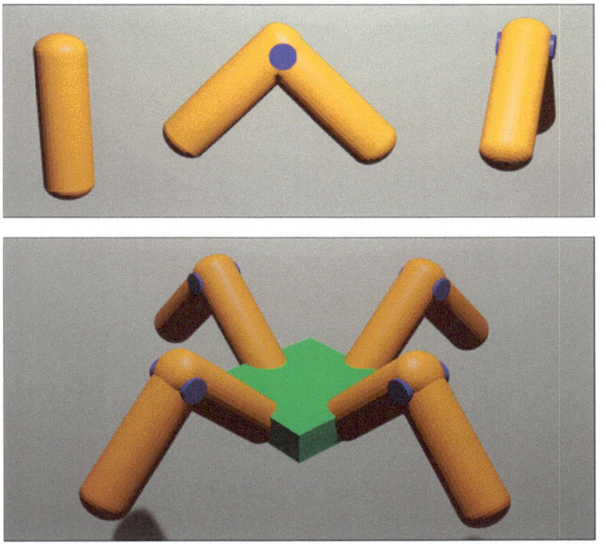

Figure 2. The figure shows the single components that form the robot. The single segment is represented by a capsule shape and the joining two of these capsules forms a joint. The figure at the bottom shows these jointed segments further joining to a cube-shaped body.

2.2. Controller

As stated earlier, the controller used in this study was an ANN. ANNs are mathematical models of the biological brain, and they are used to imitate the nervous system of biological beings. The ANN controls the robot using a feedforward neural network with a hyperbolic tangent activation function. The neural network has 9 inputs; the number of outputs is determined by the joint symmetry used and will be further discussed in this section. The first eight inputs of the neural network are the proprioceptive sensor values that return the current joint positions of the robot, and the ninth input is a bias neuron [31]. The data are passed through the neural network, the values of the output neurons are generated in the form of new joint positions, and the robot joints are actuated according to the joint positions. We did not include any hidden layers to keep the solution space small. The neural network was called once every 60 timesteps to avoid any jerky moment.

The equation below gives the output of the neural network:

$$O_j = \tanh \sum_{i=0}^{l} w_{ij} J_j \quad (1)$$

where O_j is the output value for joint j, w_{ij} represents the weight that connects input neuron i with output neuron j, and J_j is the current joint position of joint j.

To create different joint symmetries, the robot morphology is not altered; however, the controller was altered slightly to form the following symmetries.

2.2.1. Diagonal Joint Symmetry

The diagonal robot legs shared the same joint angles in this symmetry. The ANN controller, in this case, had 4 outputs, relating to 4 joint values. The ANN output angles were assigned to the two front legs of the robot and the same angle values were used for the legs that were diagonal to the front legs, i.e., the joint values of the front left leg were duplicated to the joint values of the hind right legs, as seen in Figure 3b.

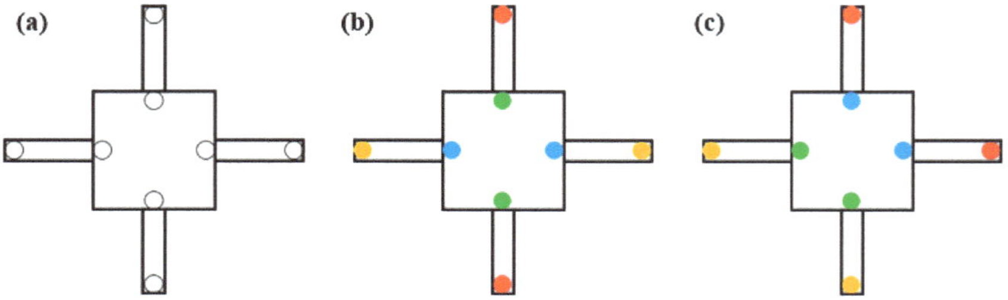

Figure 3. Top view of the quadrupedal robot. (**a**) Joint asymmetry in the quadrupedal robot. (**b**) Diagonal joint symmetry is shown in the quadrupedal robot. (**c**) Adjacent joint symmetry is shown in the robot.

2.2.2. Adjacent Joint Symmetry

For this joint symmetry, the ANN architecture still had 4 outputs; however, the leg symmetry was changed. The ANN output joint values were applied to the left front and left hind leg and the symmetry was formed between the front two legs and the hind two legs. Therefore, the joint values of the front left leg were replicated to the front right leg and the joint values of the hind left leg were replicated to the hind right leg, as seen in Figure 3c.

2.2.3. Diagonal Joint Reverse Symmetry

This joint symmetry was similar to the diagonal joint symmetry; the controller architecture remained the same (4 outputs), the only difference, in this case, was that when the joint values were replicated they were multiplied by -1, which made the diagonal legs move 180° out of phase.

2.2.4. Adjacent Joint Reverse Symmetry

The symmetric joints and the ANN controller remained the same as in the adjacent joint symmetry; however, in this case, the joint values were multiplied by -1 when they were replicated, shifting the phase of the symmetric legs by 180°.

2.2.5. Joint Asymmetry or Random Joint Movement

The joints did not form any symmetry: every joint moved in an arbitrary direction. This is shown in Figure 3a.

2.3. *Algorithm*

As mentioned before, we used the GA to optimize the neural network. The main steps of our proposed algorithm are shown in Figure 4.

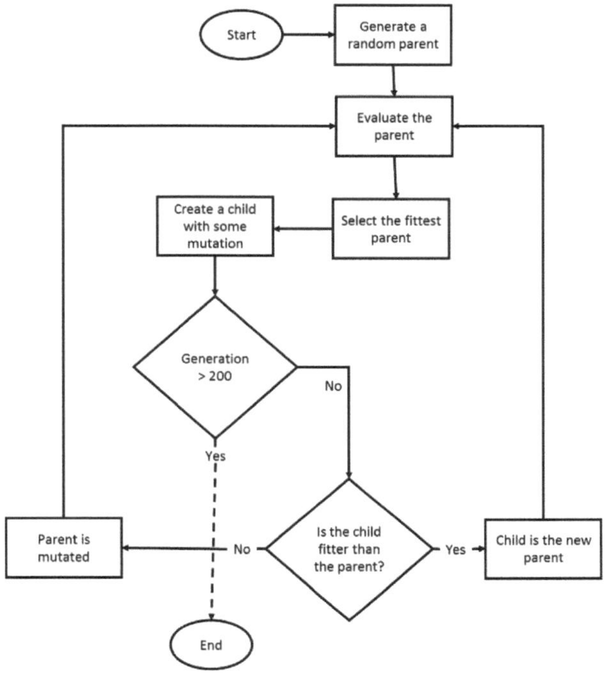

Figure 4. Flowchart of the algorithm used in this paper.

The simulator and the GA were coded in python. The simulation was run for 10 individual robots at a time where each robot had a random weight matrix that allowed the neural network to control the robot, and each simulation lasted for 1000 timesteps. When the simulation ended, the GA selected the best-performing individual based on the selection criteria, as mentioned in the next section. This individual was allowed to make copies of its weight matrix and the weight matrix of each copy was slightly changed to introduce some variation that would allow the child to perform slightly different from the parent. The different behavior could result in a child performing better than the parent; in such a case, the child becomes the new parent and the process continues until a set number of generations is completed, which in this case was 200.

2.4. Selection Criteria

After one simulation cycle was terminated, the robots from all the simulations were evaluated and the best performing controllers were selected using a mathematical function that was user-defined. This function was called the fitness function and is provided below:

$$F = (\sqrt{x^2 + y^2})(\textstyle\sum_{t=1}^{T} \sum_{j=0}^{J} J_{tj} - J_{(t-1)j})(1 - TS) \quad (2)$$

where x and y are the distances traveled by the robot in the x and y directions, respectively. J_{tj} and $J_{(t-1)}$ represent the positions of joint j at time t and $t-1$, respectively. TS is the value of a touch sensor attached to the back of the robot's body, such that TS remains one '1' when the robot flips itself over.

3. Results and Discussion

Once the simulations ended, the results from all the joint symmetries were analyzed. This section will deal with displaying and discussing the results obtained.

3.1. Adjacent Joint Symmetry and Reverse Symmetry

The result of the adjacent joint symmetry is shown in Figure 5. For two trials, the robot did not score a high fitness value; however, for the rest of the trials the robot scored a high value, showing that the robot could travel a notable amount of distance. This symmetry made the robot very stable, and the robot was able to gallop. This can be used to develop galloping gaits for robots.

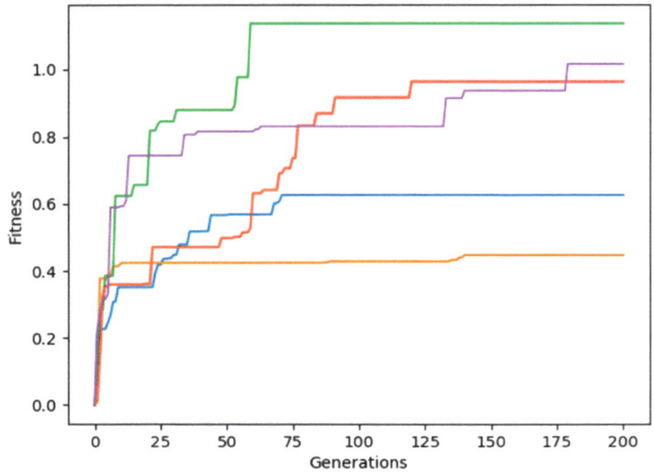

Figure 5. Adjacent joint symmetry fitness for five simulation runs, showing that the majority of the robots reached a high fitness value.

The fitness values for five evolution trials of adjacent joint reverse symmetry are shown in Figure 6. Note that most of the robots achieved high fitness values. This gait exhibited a jerky robot movement and sent the robot in arbitrary directions occasionally. However, this joint symmetry achieved the highest fitness values on average.

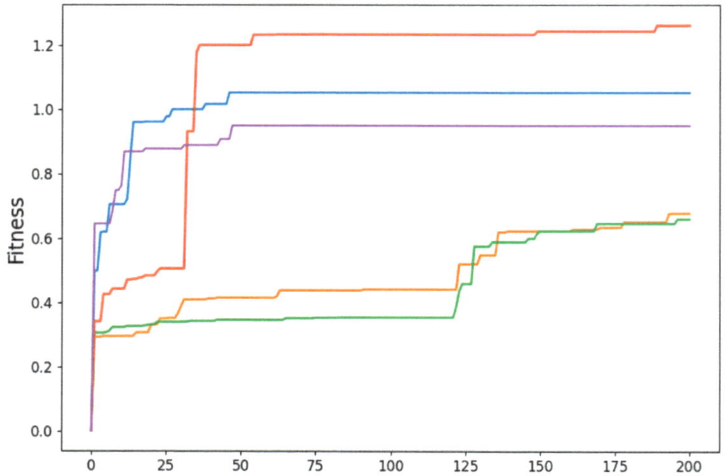

Figure 6. Adjacent joint reverse symmetry fitness for 5 simulation runs, showing that the majority of the robots reached a high fitness value.

3.2. Diagonal Joint Symmetry and Reverse Symmetry

The result for diagonal joint symmetry is shown in Figure 7. This joint symmetry had the least fitness score per evolution run, as well as on average. This is because when the diagonal joints were symmetrical, the forces applied by the robot's leg had the same magnitude but the direction was opposite, causing the forces to cancel each other out. Consequently, the robot stayed in the same spot.

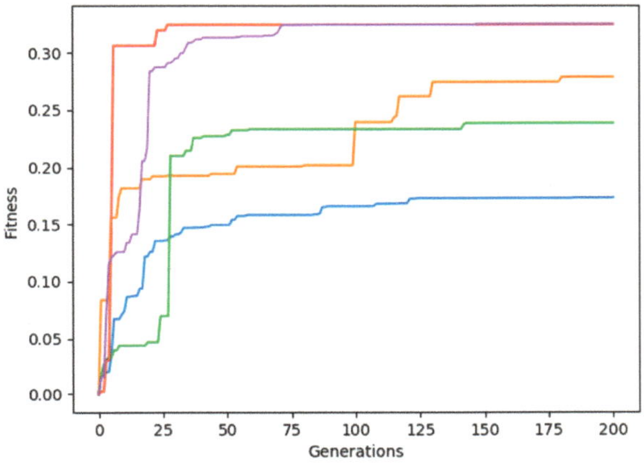

Figure 7. Diagonal joint symmetry fitness for 5 simulation runs.

The fitness graph for diagonal reverse symmetry is shown in Figure 8. The maximum fitness score and the average fitness score of this joint symmetry was the second highest. This high fitness value is attributed to the robot's joint movements that allowed it to walk stably by taking turns in using diagonal leg pairs to locomote. The gaits produced with this method resembled the creep gait, which is very common in nature.

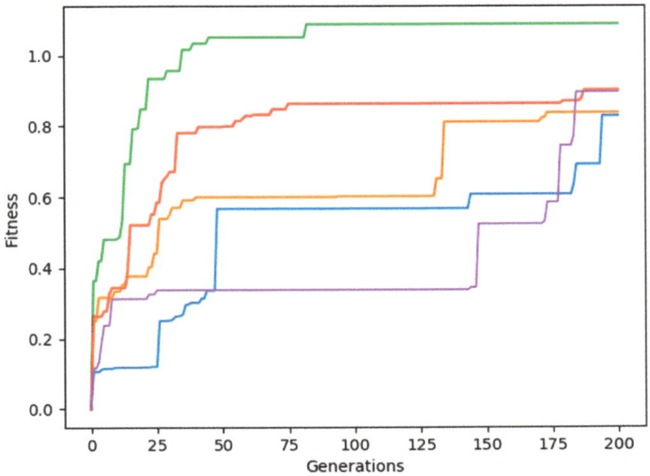

Figure 8. Diagonal joint reverse symmetry fitness for 5 simulation runs.

3.3. Random Joint Movement

The fitness for random joint movement is shown in Figure 9. The fitness values achieved by the robot were all caused by jerky movements that threw the robot around, causing the robot to gain a high fitness score. This robot gait was the least stable and produced gaits that were not optimal for a flat terrain. However, this random joint movement could be helpful over uneven terrains.

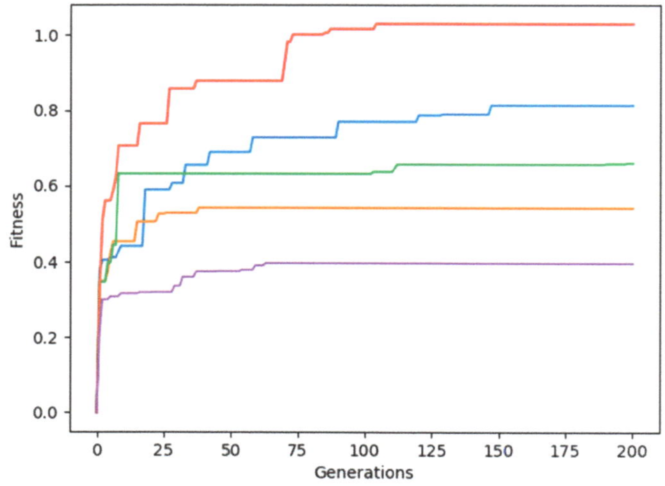

Figure 9. Random joint movement or joint asymmetry for 5 different simulation runs.

Figure 10 shows the average fitness scores of all the joint symmetries. As discussed above, the joints with adjacent reverse symmetry and diagonal reverse symmetry achieved the highest fitness scores, on average. The adjacent joint symmetry produced stable gaits that resembled a galloping motion, which could be useful in developing high-speed gaits. The gaits produced with adjacent joint reverse symmetry were able to generate the highest fitness score; however, these gaits were unstable and they would send the robot in arbitrary directions, which led to a high fitness score and no usable gaits. The diagonal joint symmetry produced the lowest fitness scores because the robot would stay at the origin, and the gaits produced were not useful at all in propelling the robot. The diagonal joint reverse symmetry produced useful and stable movements that resembled the creep gait. The robot was able to traverse the flat terrain with ease with this joint symmetry. Moreover, the gait speed was slow, which further made the robot more stable. The robot with no joint symmetry was able to achieve high fitness values and the robot mostly produced a jerky movement that would move the robot in arbitrary directions. Although the gait was useless for traversing a flat terrain, it could prove useful in moving over uneven terrain or an environment with obstacles.

Figure 11 shows a sequence of frames that show the robot movement in the simulator for all the joint symmetries used. Figure 11a shows the adjacent reverse symmetry; the robot travelled far, however, the robot body was very jerky and threw itself around. Figure 11b shows that the gaits produced by adjacent symmetry could jump high and were very stable. Figure 11c shows the diagonal joint reverse symmetry. The gaits produced mimicked nature, and they were the most stable gaits produced. Figure 11d shows the diagonal joint symmetry; in this case, the robot stayed near the origin. Finally, Figure 11e shows the robot with no joint symmetry. The robot was thrown around to an arbitrary position.

The importance of leg symmetry in gait generation has been highlighted in our research and our results agree with the literature that we reviewed. The gaits produced by [17–19] were symmetric and produced the best results for the authors. The gait that they generated was the trotting gait. Similarly, the authors in [20–22,32] produced galloping gaits that were also symmetric, and our experiments proved that the galloping gaits could be very fast and stable when we developed these gaits with adjacent joint symmetry. Finally, our results showed that gaits with no symmetry are not optimal for flat terrains due to the high variability in the joint movements; therefore, we also conclude that random joint movements are more suitable for uneven terrains, which agrees with [15,16,33].

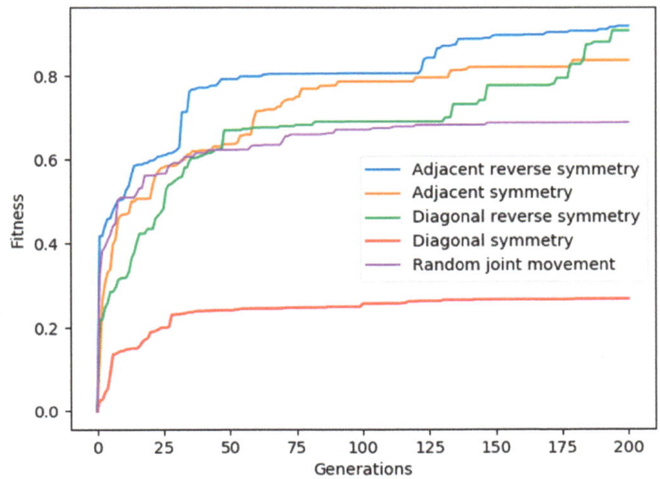

Figure 10. Average fitness values of all the joint symmetries and asymmetries used.

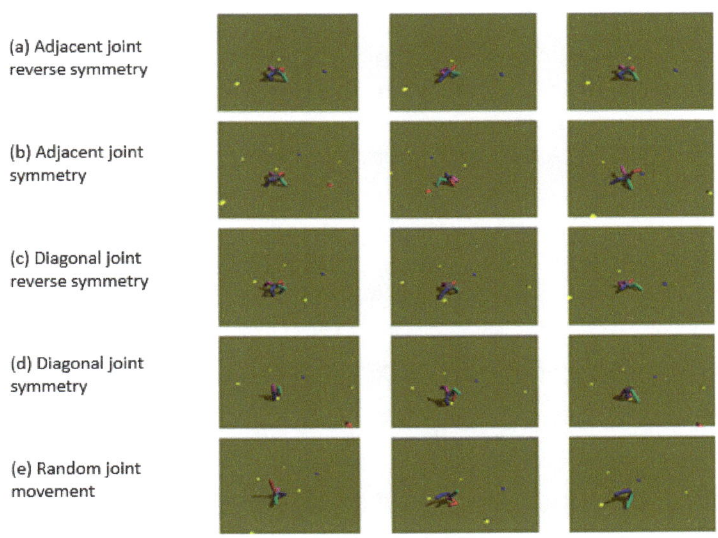

Figure 11. Robot movements in the simulator are based on the symmetry used. (**a**) shows the adjacent reverse symmetry. (**b**) shows the adjacent symmetry. (**c**) shows the diagonal reverse symmetry. (**d**) shows diagonal symmetry, and (**e**) shows random joint movement.

Our results give an idea about the joint symmetries that can produce the best results for a quadrupedal robot. The results showed that the wrong joint symmetry can result in unstable gaits that can damage a robot in the real world; moreover, the wrong joint symmetry can produce gaits that are suboptimal for a given terrain. In contrast to this, the right joint symmetry can produce agile and stable gaits that are optimal for a given terrain.

4. Conclusions

This paper performed a novel study to determine whether joint symmetry in quadrupedal robots could improve their locomotion gaits. The results showed that joint symmetry can help robots to evolve stable gaits. Our results showed that adjacent joint symmetry produces the fastest gaits, while the diagonal joint reverse symmetry produced gaits that were the slowest; however, both gaits exhibited qualities of gaits found in nature and were stable. The rest of the joint symmetries produced gaits that were not optimal and unstable. Moreover, symmetric gaits can travel greater distances. Therefore, choosing the right type of joint symmetry is important when generating robot gaits. This study can be further expanded to examine the effect of joint symmetry on a physical robot. It will also be interesting to experiment with evolving symmetric and asymmetric gaits for other robot morphologies.

Author Contributions: Conceptualization, Z.K., F.N. and Y.K.; methodology, Y.K., M.B. and M.A.B.; software, Z.K. and F.N.; validation, Y.K., M.B. and M.A.B.; formal analysis, Y.K. and M.B.; investigation, Z.K., Y.K., F.N. and M.B.; resources, Y.K., M.B. and M.A.B.; data curation, Y.K. and M.A.B.; writing—original draft preparation, Z.K., Y.K., F.N. and M.B.; writing—review and editing, Y.K., M.B. and M.A.B.; visualization, Y.K. and M.A.B.; supervision, Y.K, M.B. and M.A.B.; project administration, Y.K. and M.A.B.; funding acquisition, Y.K. and M.A.B. All authors have read and agreed to the published version of the manuscript.

Funding: The authors would like to thank the Higher Education Commission of Pakistan for research grant No. 21-2129 under Startup Research Grant Program (SRGP) which made this work possible. Moreover, the authors are also grateful to Ignite National Grassroots ICT Research Initiative (NGIRI) for granting final year project funding for this project.

Institutional Review Board Statement: Not appliable.

Informed Consent Statement: Not appliable.

Data Availability Statement: Not appliable.

Acknowledgments: The authors are thankful to Embedded Systems research group at BUITEMS Quetta for their valuable support to the project.

Conflicts of Interest: The authors declare no conflict of interests.

References

1. Pfeifer, R.; Bongard, J. *How the Body Shapes the Way We Think: A New View of Intelligence*; MIT Press: Cambridge, MA, USA, 2006.
2. Bongard, J.C. Evolutionary robotics. *Commun. ACM* **2013**, *56*, 74–83. [CrossRef]
3. Eckert, P.; Ijspeert, A.J. Benchmarking agility for multilegged terrestrial robots. *IEEE Trans. Robot.* **2019**, *35*, 529–535. [CrossRef]
4. Carpentier, J.; Wieber, P.B. Recent progress in legged robots locomotion control. *Curr. Robot. Rep.* **2021**, *2*, 231–238. [CrossRef]
5. Zhao, A.; Xu, J.; Konaković-Luković, M.; Hughes, J.; Spielberg, A.; Rus, D.; Matusik, W. Robogrammar: Graph grammar for terrain-optimized robot design. *ACM Trans. Graphics (TOG)* **2020**, *39*, 1–6. [CrossRef]
6. Hauert, S.; Zufferey, J.C.; Floreano, D. Reverse-engineering of artificially evolved controllers for swarms of robots. In Proceedings of the 2009 IEEE Congress on Evolutionary Computation, Trondheim, Norway, 18–21 May 2009; pp. 55–61.
7. De Santos, P.G.; Garcia, E.; Estremera, J. *Quadrupedal locomotion: An Introduction to the Control of Four-Legged Robots*; Springer: London, UK, 2006.
8. Goswami, A.; Thuilot, B.; Espiau, B. A study of the passive gait of a compass-like biped robot: Symmetry and chaos. *Int. J. Robot. Res.* **1998**, *17*, 1282–1301. [CrossRef]
9. Zanotto, D.; Stegall, P.; Agrawal, S.K. Adaptive assist-as-needed controller to improve gait symmetry in robot-assisted gait training. In Proceedings of the 2014 IEEE international conference on robotics and automation (ICRA), Hong Kong, China, 31 May–5 June 2014; pp. 724–729.
10. RunBin, C.; YangZhen, C.; WenQi, H.; Jiang, W.; HongXu, M. Trotting gait of a quadruped robot based on the time-pose control method. *Int. J. Adv. Robot. Syst.* **2013**, *10*, 148. [CrossRef]
11. Wong, L.H.; Sivanesan, S.; Faisol, M.F.; Othman, W.A.; Wahab, A.A.; Alhady, S.S. Development of quadruped walking robot with passive compliance legs using XL4005 buck converter. *J. Phys. Conf. Ser.* **2021**, *1969*, 012003. [CrossRef]
12. He, D. An Optimal Initial Foot Position for Quadruped Robots in Trot Gait. *J. Phys. Conf. Ser.* **2020**, *1624*, 052015. [CrossRef]
13. Kamimura, T.; Aoi, S.; Higurashi, Y.; Wada, N.; Tsuchiya, K.; Matsuno, F. Dynamical determinants enabling two different types of flight in cheetah gallop to enhance speed through spine movement. *Sci. Rep.* **2021**, *11*, 9631. [CrossRef] [PubMed]

14. Poulakakis, I.; Smith, J.A.; Buehler, M. Modeling and experiments of untethered quadrupedal running with a bounding gait: The Scout II robot. *Int. J. Robot. Res.* **2005**, *24*, 239–256. [CrossRef]
15. Pongas, D.; Mistry, M.; Schaal, S. A robust quadruped walking gait for traversing rough terrain. In Proceedings of the 2007 IEEE International Conference on Robotics and Automation, Roma, Italy, 10–14 April 2007; pp. 1474–1479.
16. Chen, J.P.; San, H.J.; Wu, X.; Xiong, B.Z. Structural design and gait research of a new bionic quadruped robot. *Proc. Inst. Mech. Eng. Part B J. Eng. Manuf.* **2021**, 0954405421995663. [CrossRef]
17. Havoutis, I.; Semini, C.; Buchli, J.; Caldwell, D.G. Quadrupedal trotting with active compliance. In Proceedings of the 2013 IEEE International Conference on Mechatronics (ICM), Vicenza, Italy, 27–28 February 2013; pp. 610–616.
18. Zhai, K.; Li, C.A.; Rosendo, A. Scaffolded Learning of In-place Trotting Gait for a Quadruped Robot with Bayesian Optimization. In Proceedings of the International Conference on Intelligent Autonomous Systems, Singapore, 22–25 June 2021; Springer: Cham, Switzerland; pp. 365–373.
19. Dini, N.; Majd, V.J. An MPC-based two-dimensional push recovery of a quadruped robot in trotting gait using its reduced virtual model. *Mech. Mach. Theory* **2020**, *146*, 103737. [CrossRef]
20. Wang, X.; Li, M.; Wang, P.; Sun, L. Running and turning control of a quadruped robot with compliant legs in bounding gait. In Proceedings of the 2011 IEEE International Conference on Robotics and Automation, Shanghai, China, 9–13 May 2011; pp. 511–518.
21. Owaki, D.; Ishiguro, A. A quadruped robot exhibiting spontaneous gait transitions from walking to trotting to galloping. *Sci. Rep.* **2017**, *7*, 277. [CrossRef] [PubMed]
22. Billard, A.; Ijspeert, A.J. Biologically inspired neural controllers for motor control in a quadruped robot. In Proceedings of the IEEE-INNS-ENNS International Joint Conference on Neural Networks. IJCNN 2000. Neural Computing: New Challenges and Perspectives for the New Millennium, Como, Italy, 27 July 2000; Volume 6, pp. 637–641.
23. Katoch, S.; Chauhan, S.S.; Kumar, V. A review on genetic algorithm: Past, present, and future. *Multimed. Tools Appl.* **2021**, *80*, 8091–8126. [CrossRef] [PubMed]
24. Lopez-Garcia, T.B.; Coronado-Mendoza, A.; Domínguez-Navarro, J.A. Artificial neural networks in microgrids: A review. *Eng. Appl. Artif. Intell.* **2020**, *95*, 103894. [CrossRef]
25. McHale, G.; Husbands, P. Quadrupedal locomotion: GasNets, CTRNNs and hybrid CTRNN/PNNs compared. In Proceedings of the 9th International Conference on the Simulation and Synthesis of Living Systems (ALIFE IX), Boston, MA, USA, 12–15 September 2004; MIT Press: Cambridge, MA, USA; pp. 106–112.
26. Yosinski, J.; Clune, J.; Hidalgo, D.; Nguyen, S.; Zagal, J.C.; Lipson, H. Evolving robot gaits in hardware: The HyperNEAT generative encoding vs. parameter optimization. In Proceedings of the ECAL, Paris, France, 8–12 August 2011; pp. 890–897.
27. Glette, K.; Klaus, G.; Zagal, J.C.; Torresen, J. Evolution of locomotion in a simulated quadruped robot and transferral to reality. In Proceedings of the Seventeenth International Symposium on Artificial Life and Robotics, Beppu, Japan, 19–21 January 2012; pp. 1–4.
28. Kim, J.; Ba, D.X.; Yeom, H.; Bae, J. Gait optimization of a quadruped robot using evolutionary computation. *J. Bionic Eng.* **2021**, *18*, 306–318. [CrossRef]
29. Bongard, J. GitHub-Jbongard/Pyrosim: A Python Robot Simulator. Available online: https://github.com/jbongard/pyrosim (accessed on 14 February 2022).
30. Smith, R. Open Dynamics Engine. 2005, p. 84. Available online: http://ode.org/ (accessed on 25 April 2022).
31. Phillips, A.; du Plessis, M. Towards the incorporation of proprioception in evolutionary robotics controllers. In Proceedings of the 2019 Third IEEE International Conference on Robotic Computing (IRC), Naples, Italy, 25–27 February 2019; pp. 226–229.
32. Bucolo, M.; Buscarino, A.; Famoso, C.; Fortuna, L.; Gagliano, S. Imperfections in Integrated Devices Allow the Emergence of Unexpected Strange Attractors in Electronic Circuits. *IEEE Access* **2021**, *9*, 29573–29583. [CrossRef]
33. Buscarino, A.; Fortuna, L.; Frasca, M.; Rizzo, A. Dynamical network interactions in distributed control of robots. *Chaos* **2006**, *16*, 015116. [CrossRef] [PubMed]

technologies

Article

Efficient Supervised Machine Learning Network for Non-Intrusive Load Monitoring

Muhammad Usman Hadi [1,*], Nik Hazmi Nik Suhaimi [1] and Abdul Basit [2]

1 School of Engineering, Ulster University, Newtownabbey BT37 0QB, UK; nik_suhaimi-nh@ulster.ac.uk
2 Department of Computer Engineering, Khawaja Farid University of Engineering and Information Technologies (KFUEIT), Rahim Yar Khan 64200, Pakistan; abdulbasit@kfueit.edu.pk
* Correspondence: usmanhadi@ieee.org

Abstract: From a single meter that measures the entire home's electrical demand, energy disaggregation calculates appliance-by-appliance electricity consumption. Non-intrusive load monitoring (NILM), also known as energy disaggregation, tries to decompose aggregated energy consumption data and estimate each appliance's contribution. Recently, methodologies based on Artificial Intelligence (AI) have been proposed commonly used in these models, which can be expensive to run on a server or prohibitive when the target device has limited capabilities. AI-based models are typically computationally expensive and require a lot of storage. It is not easy to reduce the computing cost and size of a neural network without sacrificing performance. This study proposed an efficient non-parametric supervised machine learning network (ENSML) architecture with a smaller size, and a quick inference time without sacrificing performance. The proposed architecture can maximise energy disaggregation performance and predict new observations based on past ones. The results showed that employing the ENSML model considerably increased the accuracy of energy prediction in 99 percent of cases.

Keywords: NILM; energy disaggregation; ENSML model

check for updates

Citation: Hadi, M.U.; Suhaimi, N.H.N.; Basit, A. Efficient Supervised Machine Learning Network for Non-Intrusive Load Monitoring. *Technologies* **2022**, *10*, 85. https://doi.org/10.3390/technologies10040085

Academic Editors: Manoj Gupta, Eugene Wong and Gwanggil Jeon

Received: 19 April 2022
Accepted: 14 July 2022
Published: 16 July 2022

1. Introduction

Artificial Intelligence (AI) has grown fast in recent decades, and it is no longer confined to science fiction literature and films. By 2030, AI is quite likely to exceed humans in the majority of cognitive skills. According to the World Economic Forum's latest study on the future of jobs, AI will create 58 million new jobs by 2022. Home automation is now used largely to provide a quick and efficient manner of integrating and connecting household equipment. AI may be used in a variety of ways, such as monitoring our daily utilisation of current or voltage in each device in a building. As an example, in a recent article, AI-generated simulations were demonstrated using MATLAB/SIMULINK.

AI is the greatest option for handling big data flows and storage in Internet of Things (IoT) networks [1]. Energy Efficiencies (EE) can provide a slew of benefits to energy customers and providers as a result of IoT demand. In 2011, homes utilised 21.54 percent of total energy consumption in the United States [2]. This solution is meant to minimise energy usage by utilising powerful optimisation algorithms to establish a better resource management system and flatten consumption peaks for each home.

Energy management systems to regulate peak energy demand [3] are examples of new technologies that have been developed to enhance EE. With a population of 67.22 million (2020), increasing the overall efficiency of the electricity grid by boosting EE in residential areas may be crucial [3]. Furthermore, giving precise information on the energy use of consumer appliances will enhance the EE. When considering the disaggregation of load consumption and the increased energy awareness of particular equipment, users can change their consumption behaviour, replace equipment, or install energy management systems to save energy or money [3,4].

The development of new buildings in cities throughout the world has transformed dwelling arrangements and boosted the demand for end-use appliances with energy conservation and control [5]. Furthermore, the move was accompanied by Smart Meter (SM), which enabled the computation of individual appliance power usage based on the building's aggregate measurements. The placement of current and voltage sensors at the SM is used to monitor energy usage and identify loads in a load disaggregation system [6]. This framework is far more proficient than the old intensive monitoring systems because it can reduce installation costs. By analysing the energy usage of each major appliance, inefficiencies in energy consumption of large appliances may be identified and eliminated [7]. These apps will provide useful information on the appliances that are being utilised. Figure 1 shows a categorical hierarchy of load disaggregation classes.

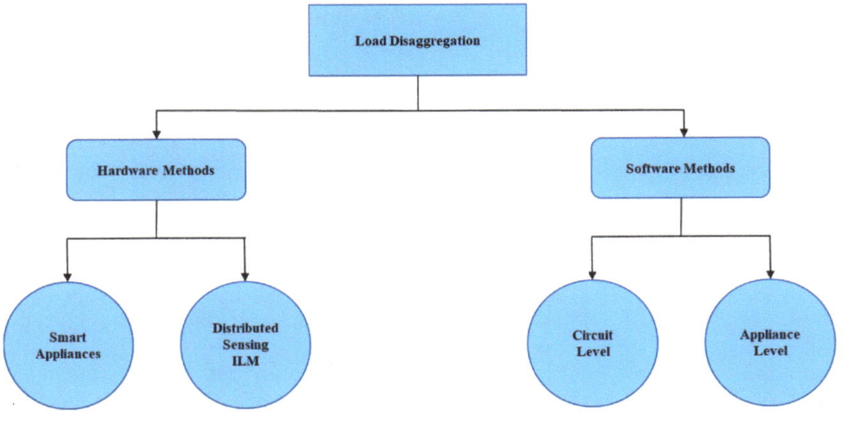

Figure 1. Load disaggregation hierarchy.

Load disaggregation is divided into two categories: hardware techniques and software approaches. Hardware-based techniques are simple to create, but they are limited by a number of factors, including implementation cost, reliability, and scalability. As a result, because it uses non-intrusive load monitoring, the software-based solution is preferred. Using a single primary metering point to aggregate load usage and dissect it into individual appliance use has grown common in recent years [8]. The benefits of adopting SM include (i) accurate billing; (ii) detecting defective appliances; and (iii) receiving detailed information on current appliance consumption.

Housing arrangements across the world have changed as a result of increased urbanisation, necessitating the development of high-rise buildings. Changes in dwelling patterns have also resulted in a system for breaking down the building's aggregate energy use at the appliance level. It is now feasible to estimate an appliance's energy use based on a building's overall energy data using smart meters [5].

Hart established a system in the 1950s that disaggregated electrical measurements such that the power consumption of each device could be discovered sequentially by reviewing load data gathered over time [9]. The suggested approach was deemed non-intrusive because no equipment had to be put on the customer's premises. The aggregated energy usage statistics may be gathered from the building or residence's main electrical panel. The separation of the total home construction data into its key energy components is a broad objective of this approach. Appliance monitoring may be conducted in two ways: intrusive appliance monitoring (ILM) or non-intrusive appliance monitoring (NIAM). NIAM, also known as Non-Intrusive Load Monitoring (NILM), is a technique for calculating energy disaggregation that may calculate device-specific energy disaggregation based on aggregate measurements gathered at a single site [5]. ILM necessitates the installation of hardware on each appliance, such as sensors and processors, in order to monitor each item independently. Meanwhile, NILM works on software algorithms that examine the resident's interior appliance functioning state using power data from the service panel.

Non-intrusive load monitoring is a technique for identifying and estimating the energy usage of each electrical item in a facility. It allows a homeowner to break down their home's energy use into specific appliances, allowing them to be recognised and conserved [10]. The operation of NILM is depicted in Figure 2.

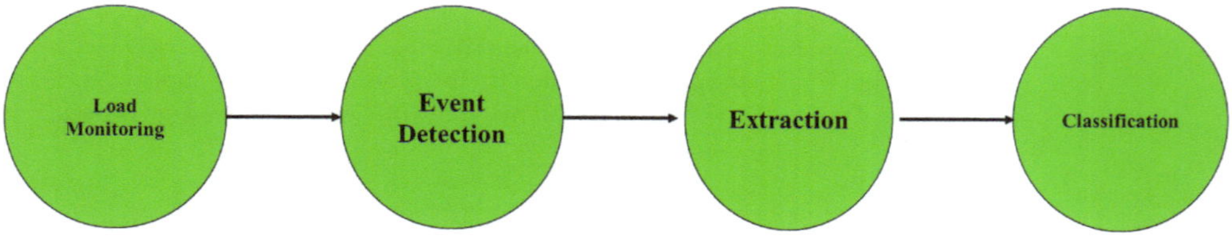

Figure 2. How NILM works.

Measurement kinds, sample rates, and sensing types are all key aspects to consider while designing NILM algorithm steps. The power on the line(s) of interest is measured first by NILM algorithms. While it is critical to keep an eye on the load on a home's main bus, it is also crucial to determine whether an incident has happened. Finding an event is difficult since a home's main bus is made up of several sorts of equipment. The NILM algorithm gathers data from several power signal monitoring systems to evaluate whether an incident has happened. When the proper characteristics are retrieved and matched to the labelled data, the classification process will be able to accurately identify appliances that have generated events.

Because SM is so commonly utilised, the NILM algorithm will be more beneficial to the customer. NILM is a device that analyses variations in the voltage and current entering a home. This approach may be used to detect appliances that are not performing well. The purpose of NILM is to minimise energy use by increasing user awareness. When one sensor is installed in an SM, it reads all of the appliances' energy use and sends it to the cloud. This installation differs from Intrusive Load Monitoring (ILM), in which each device requires its own sensor. The difference in installation between ILM and NILM is seen in Figure 3.

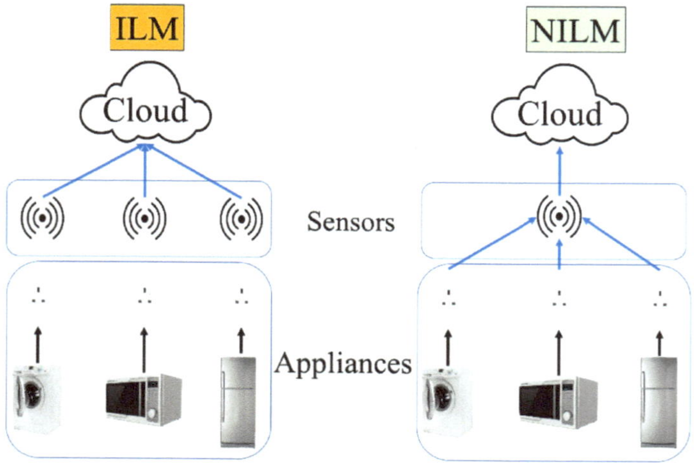

Figure 3. Installation of sensor(s) between ILM and NILM.

This article primarily focused on the appliance event detection and appliance usage prediction. The following list summarises the most important contributions:

1. An efficient non-parametric supervised machine learning network (ENSML) was proposed with fast inference speed and low storage requirements. The proposed method was used to create a realistic and adaptable NILM formulation model, with the parameter values following a supervised learning strategy.
2. The proposed ENSML has a lowered learning parameter; therefore, it takes up less space while performing as well as other state-of-the-art NILM systems.
3. The suggested ENSML methodology with the NILM system could recognise newly installed appliances, filling a critical research need.
4. A public dataset was used to validate the provided model and approach. All of the hypothesised potentials have been shown to be genuine, in addition to the high precision of load disaggregation.

The remainder of the paper is divided into the following sections. Section 2 presents the literature review while Section 3 highlights the visualisation of dataset and its preparation. Section 4 presents the ENSML model methodology while Section 5 presents the simulation method followed by results in Section 6. Finally, article is concluded in Section 7.

2. Literature Review

Table 1 in this section contains all of the data from prior NILM research. This will help us to have a better understanding of the algorithm. Machine learning (ML) is a forward-thinking method for predicting customer behaviour based on appliance usage.

Table 1. Visualisation of the Reference Energy Disaggregation Data (REDD) dataset.

No	House Number	Channels	Appliances
1	House 1	20	4 kitchen outlets, 3 lightings, 3 washer dryer, 2 mains, 2 ovens, 1 refrigerator, 1 dishwasher, 1 microwave, 1 electric heat, 1 stove, 1-bathroom.
2	House 2	11	2 mains, 2 kitchen outlets, 1 lighting, 1 stove, 1 microwave, 1 washer dryer, 1 refrigerator, 1 dishwasher, 1 disposal.
3	House 3	22	5 lightings, 3 unknown outlets, 2 mains, 2 washer dryer, 2 kitchen outlets, 1 electronic, 1 refrigerator, 1 dishwasher, 1 disposal, 1 microwave, 1 furnace, 1 smoke alarm, 1-bathroom.
4	House 4	20	4 lightings, 3 air-conditioner, 2 mains, 2-bathroom, 2 kitchen outlets, 1 unknown outlet, 1 washer dryer, 1 stove, 1 smoke alarm, 1 dishwasher, 1 miscellaneous, 1 furnace.
5	House 5	26	5 lightings, 4 unknown outlets, 2 mains, 2 washer dryers, 2 subpanel, 2 electric heat, 2 kitchen outlets, 1 microwave, 1 furnace, 1-bathroom, 1 dishwasher, 1 disposal, 1 electronics, 1 refrigerator.
6	House 6	17	3 air-conditioner, 2 mains, 2 kitchen outlets, 2 unknown outlets, 1 washer dryer, 1 stove, 1 electronics, 1 electrical heat, 1-bathroom, 1 refrigerator, 1 dishwasher, 1 lighting.

Deep learning in non-intrusive appliance monitoring learning techniques is now classified into three categories: supervised, unsupervised, and semi-supervised learning. Supervised algorithms can either learn from training data or build a model and then guess a new instance based on it. It offers the advantages of being simple to use, quick to calculate, compact to store, and yielding accurate analytical findings. There are, however, some issues. The performance of logistic regression is bad when the geographical features are considerable, for example. There are certain drawbacks, such as under- or over-fitting, and a lack of self-learning capacity.

The unsupervised algorithm is a data-processing approach that classifies samples without using category information by analysing data from multiple samples of the study item. It has a great ability to self-learn, and fresh data may be immediately added to the data set without retraining, but it also has the drawback of low analytical result accuracy. Semi-supervised learning is the most promising learning algorithm branch because it employs a huge quantity of unlabelled data while also using labelled data for pattern recognition. However, there is a scarcity of research on semi-supervised regression issues.

To identify variations in the electrical consumption signal owing to appliance on/off events, early NILM approaches analysed the electricity mains measurement and applied statistical techniques. The active and reactive power signatures were then matched to the right appliance using a best likelihood method, and similar "steady-state" elements of the power signal were grouped together. Certain two-state (on/off) appliances have been identified with good accuracy using such clustering approaches [10,11]. These methods, on the other hand, have major trouble detecting more complicated appliances with numerous states (e.g., washing machines) and have a tendency to fail in situations when multiple appliances are operating and switching at the same time [12]. Clustering approaches have also been used to uncover household features and trends in electricity use data [13,14].

Graph signal processing is another contemporary technique to NILM in the literature. Refs. [15,16] present a low-complexity unsupervised NILM technique based on entropy index limitations competitive agglomeration, a fuzzy clustering algorithm. This approach yielded encouraging results for NILM implementation in practice.

Ref. [17] described a spectrum-smoothing-based load disaggregation strategy for dealing well with many appliances turning on and off at the same time. There have also been proposals for NILM algorithms based on integer programming [18] and mixed-integer linear programming [19].

Since it may provide a considerably less intrusive and lower-cost solution than sub-metering, NILM has been included in a substantial number of mass-market home energy management products and services. Sense [20] employs NILM to discover trends in home energy usage in order to provide users advice on how to make their homes more energy efficient. Smappee [21] focuses on how to use NILM to provide precise feedback and advice on reducing energy and carbon footprints. A NILM device for commercial buildings has been developed by SmartB [22]. A variety of mass-market NILM gadgets are used to identify possible safety hazards when home appliances, such as the oven or iron, are left switched on and/or unattended [23]. Several commercial vendors claim to incorporate machine learning or artificial intelligence in their algorithms in their goods and services [24,25]. Bidgely et al. [24] has a number of patents in the field of machine learning-based NILM methods. Verv et al. [25] is a home energy management solution that uses high-resolution mains electricity measurements and artificial intelligence methodologies to perform NILM, with the output from the NILM classifier being used to offer advice and suggestions to consumers.

The use of deep learning techniques from other domains, such as image processing, to solve the NILM problem was presented in ref. [12], where preliminary findings revealed that deep learning approaches outperformed other approaches in the literature on unseen residential smart meter data sets. In several fields, such as image classification, automated speech recognition, and machine language translation, deep learning is currently the standard technique [26–31]. Deep learning approaches are expected to increase NILM performance, as one of the main challenges in NILM is selecting the most discriminative features to extract from a given household data set. Deep learning approaches can learn which characteristics to extract from a data set automatically and generalise to new and unknown data sets. This enables the creation of an unsupervised solution to the NILM issue, with the least amount of user involvement necessary to set up and train the system. Table 2 highlights the most important previous studies in the area.

Early NILM methods relied on statistical approaches to detect variations in the energy usage caused by both on/off appliances and electrical main readings. Based on identical steady-state components of the data, an algorithm matches the real and reactive power signatures of the data with the suitable appliance. The use of such ensemble methods for identifying certain two-state appliances has been found to be extremely accurate [32,33].

However, the technique has significant problems detecting devices with more intricate state-dependent behavior and in scenarios when many appliances are operating at the same time [34].

Table 2. Previous studies on NILM.

No	Author	Method	Advantage	Disadvantage
1	Kelly, Jack, Knottenbelt, Willian [27]	LSTM	Work best for two state appliances	Does not perform well when it comes to multi-state appliances such as washing machine and dish washer
2	Somchai, Boonyang [28]	ANN	With incomplete information, the data may still produce output.	Provides a probing solution, but it does not specify the why or how.
3	Barsim, Karim Said; Bin Yang [29]	SSL	It estimates the structure of the unlabelled data from its own predictions rather than relying on additional clustering components for this purpose.	Error propagation occurs when misclassified observations are chosen for an iteration, causing the prediction function to be increasingly skewed in subsequent iterations
4	Faustine, Anthony; Pereira, Lucas; Bousbiat, Hafsa and Kulkarni, Shridhar [30]	DNN	To be able to estimate the prediction's uncertainty by combining appliance states and power consumption values.	Single target regression, which ignores any correlations between targets, yielding a single model for each.
5	Jiang, Jie; Kong, Qiuqiang; Plumbley, Mark D; Gilbert, Nigel; Hoogendoorn, Mark and Roijers, Diederik M [31]	WaveNet	A reduction in filter sizes is achieved by reducing the size of the convolution filters as compared to conventional CNN(s).	Must minimise the loss with an optimizer with a learning rate of 0.001

Graph signal processing, which was described in ref. [35], is another new technique to NILM. The NILM method [36] presents a flexible and low-complexity entropy index constraint competitive agglomeration technique. The findings of this technique seemed promising for NILM application in the real world. A Cepstrum-based strategy for disaggregation load is described in ref. [37] to manage simultaneous on or off of several appliances. NILM has also been proposed using mixed-integer linear programming [38] and integer programming algorithms [39].

Hidden Markov Models (HMMs) are used to solve energy disaggregation problems [40,41]. Markovian models have hidden and visible states, with the hidden state being the appliance state. However, such approaches may be suitable for applications involving relatively continuous durations of time between states, such as speech recognition. As a result, energy disaggregation is impeded by the notion that run times might differ dramatically from one run to the next (and hence state durations). In addition, the HMM should include any appliances in the house that are either undesired or practical.

The implementation of machine learning by NILM to forecast what the SM will do based on data acquired from it is discussed in this section. Figure 4 depicts the entire operation of the system. The primary power distribution board has an acquisition circuit that gathers continuous data on current and voltage at the board (number one). The obtained data demonstrates a change in power at stage (2), when the appliance is turned on. The current and voltage behavior at the main distribution board can be used to determine this shift.

Aggregate data is collected whenever an appliance electrical signature is selected. Electrical signatures are the most important component of a NILM system at stage 3. Because the first form of signature requires a high sampling frequency, most home NILM systems use the steady-state type. The initial step in determining steady states is to recognise stable value sequences in the signal. This paper describes an approach for detecting Steady-State signatures using rectangular regions formed by successive data.

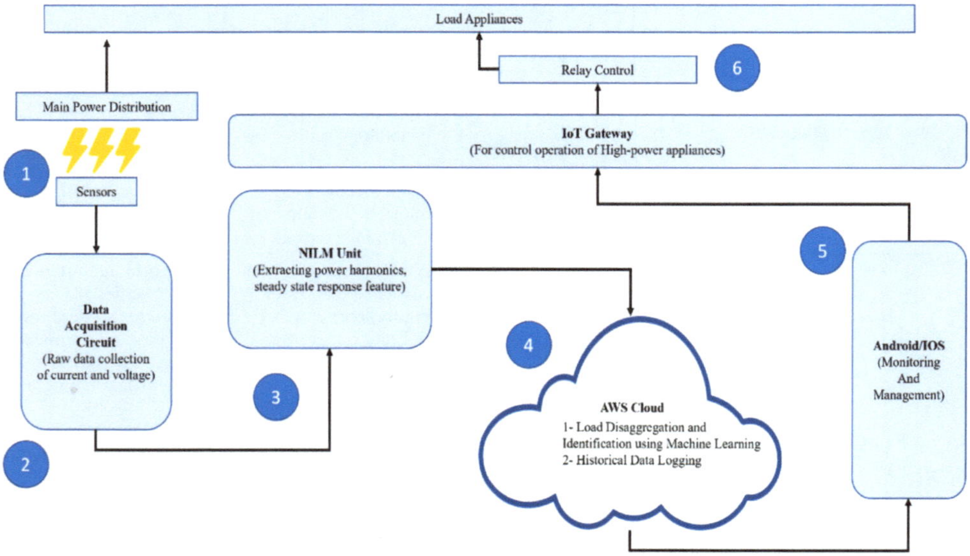

Figure 4. Overall system.

This approach enables the identification of a complete steady state from start to finish. The smart energy monitor can identify all monitored equipment using data from the NILM. A time-stamped active aggregate load is provided into the disaggregation process as well as the efficient non-parametric supervised machine learning network model for the household. During the disaggregation time, this method generates a comprehensive report for each appliance or event. This project can anticipate additional houses using only the same data set as the NILM data.

Stage (4) is where all of the data are kept. This is where the appliances from the energy disaggregation are labelled. This will also provide historical data logging, allowing the user to review the appliance's history when it is turned on or off. When the user is gone at stage (5), this is critical. This is where the appliance utilisation is monitored and managed. While the user is away from home, the user may keep an eye on what is going on at home. This is where the Internet of Things comes into play in the last stage (6). This is where all of the IoT-enabled appliances are installed. The gadgets can be used even while the user is not at home.

3. Visualisation of Dataset

REDD is acronym that stands for Reference Energy Disaggregation Data, created and managed by Massachusetts Institute of Technology (MIT). This aggregated data collection [5] contains extensive information on energy usage from a number of homes. REDD was used to monitor around 40 residences in Massachusetts and California. Monitoring devices were put in 30 residences around the state in 18 months. Each circuit breaker in each residence was obtained for two to four weeks. Having access to historical data helps to examine how the energy in a home has changed over time. Devices may be identified using the whole-home signal, which is made up of machine-readable waveforms, while devices with specialised data can offer information on behavior inside the house. These data were collected in six family households in the United States during a short period of time. This dataset is commonly used to assess NILM algorithms [5]. The process of creating REDD dataset for the project is shown in Figure 5.

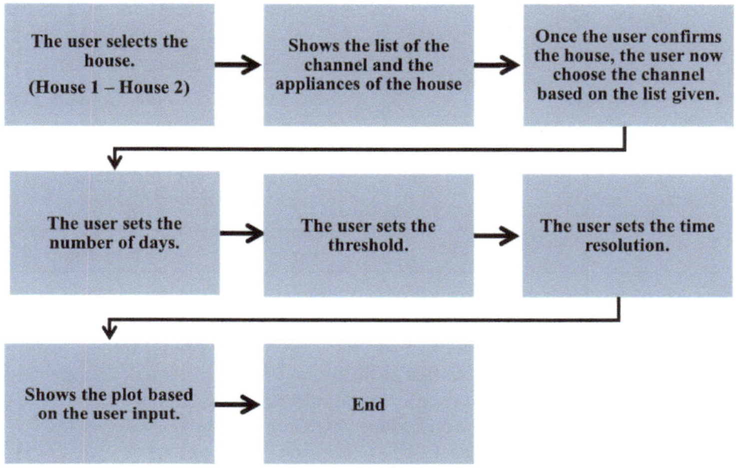

Figure 5. Process of the project using REDD dataset (Phase 1).

4. Efficient Non-Parametric Supervised Machine Learning (ENSML) Network as a Predictive Agent

One of the most widely used branches of networks comes from supervised learning. This paper proposed an efficient non-parametric supervised machine learning network (ENSML) having decision tree algorithm as the basic block that can be used to tackle both regression and classification problems. ENSML predictive models are created by combining a set of binary rules to calculate an objective value. Figure 6 shows the diagram of ENSML network while Table 3 summarises the explanation.

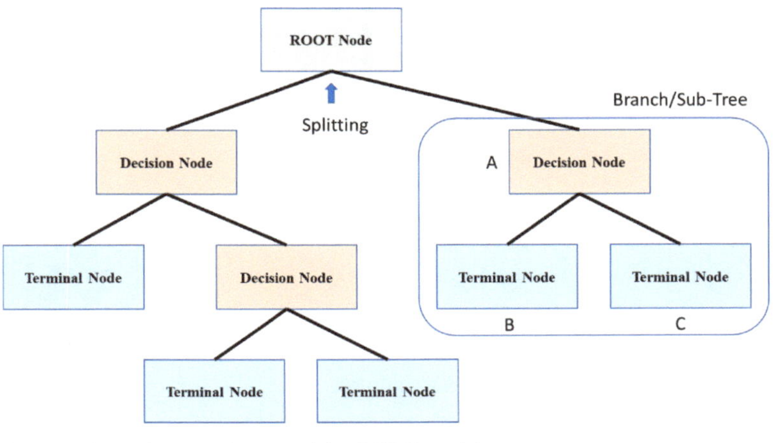

Figure 6. Block schematic diagram of the ENSML model.

Table 3. The explanation of terminology used in ENSML model.

No	Terminology	Explanation
1	Root Node	This is a sample of an entire population that is divided into two or more homogeneous groups.
2	Splitting	A process in which a node is divided into two or more sub-nodes
3	Decision Node	In a decision network, each subnode splits into further subnodes
4	Leaf/Terminal Node	Nodes that do not split are known as Leaf or Terminal nodes.
5	Prunning	Pruning is opposite to splitting. It is removing sub-nodes of a decision node
6	Branch/Sub-Tree	An individual branch or sub-tree is a part of an entire tree.
7	Parent and Child Node	Usually, the parent node of subnodes is referred to as the parent node, whereas subnodes are its children.

The hierarchical classifier has three types of nodes. The Root Node is the graph's initial node, and it symbolises the whole sample. It can be subdivided further into nodes. The inner nodes indicate the properties of a data collection, while the branches represent the decision criteria. Finally, the Leaf Nodes indicate the outcome. This strategy is quite useful for addressing challenges with decision-making.

The pseudocode and the process chart (Figure 7) explaining the ENSML Algorithm procedure is given below (Algorithm 1).

Algorithm 1 ENSML Algorithm

```
Initialise the control parameters
Input of the Model
for: prediction
do
for House 1
    | load file low_freq in House 1 to Spyder
    | convert dataset from House 1 to data frame
end
Training, validation, and test data for House 1.
for data frame House1
    | train data set to 11 days
    | validation data set to 5 days
    | test data set to 6 days
end
Tree regression Model
for train, validation and testing
    | x_axis as mains1, mains2
    | y_axis appliances
    | define mse_loss and mae_loss
end
plot the diagram
} return prediction
```

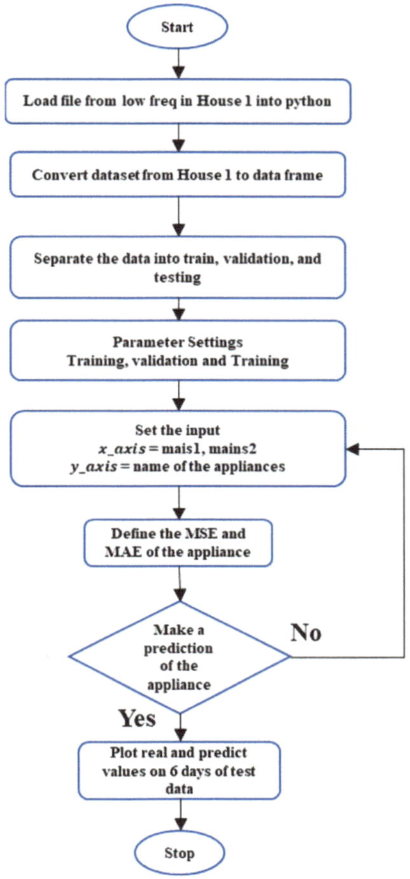

Figure 7. Process Flow of the ENSML method.

Advantages and Disadvantages

The ENSML methods are supervised learning algorithms that are mostly used for classification problems (because they have a predefined target variable). These models are used only in regression problems if and only if the target variable falls within the range of values seen in train data. Table 4 shows the advantages and disadvantages of using the ENSML method.

Table 4. Advantages and disadvantages of the ENSML method.

No	Benefits	Shortcomings
1	The model can be applied to both classification and regression.	Prone to overfitting.
2	Understanding, interpreting and visualising are easy.	No way to extrapolate.
3	There is no constraint on data type.	Regression can be unstable.

5. Setup Experiment

The power under 1 Hz for total signals and 0.2–0.3 Hz for individual appliances makes up the REDD low-frequency dataset. Individual appliance data were augmented to 1 Hz to ensure consistency. The REDD dataset utilising the NILM method deaggregated the lower and higher frequencies in the lower and higher frequencies. To examine the proposed study, with a reasonably basic appliance operation condition, the Dataset for House 1 was used. House 1 contains 23 days. To make the prediction successful, House 1 should have enough data. Figure 8a shows the power consumption (W) for House 1 while Figure 8b–e represents different appliances usage recorded at the simulator. The considered day was Monday, which is a working day.

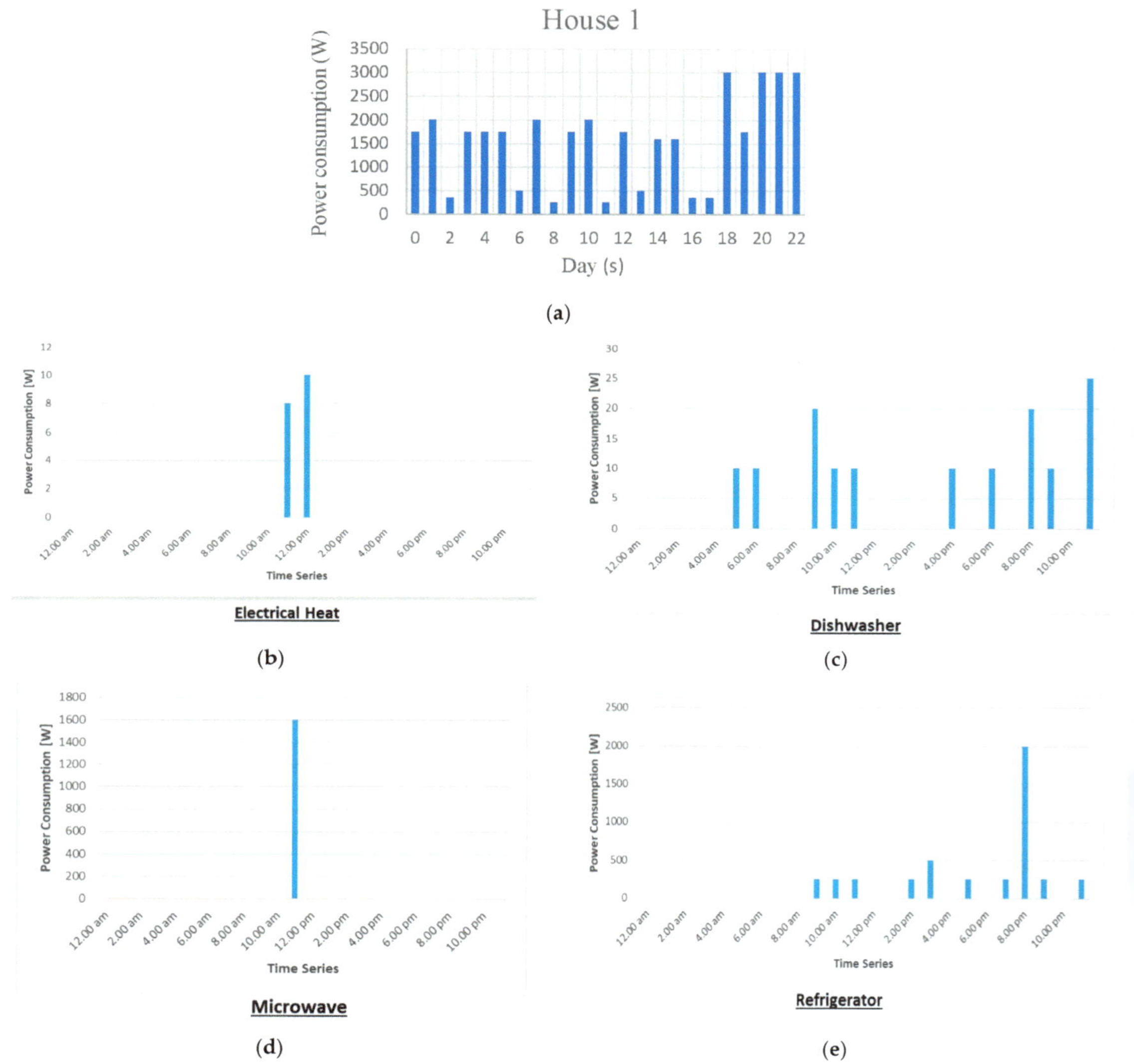

Figure 8. (**a**) represents Wattage (W) for House 1 while (**b**–**e**) represents power consumption usage for different appliances recorded on the simulator.

This shows that House 1 had enough data to train, validate, and test for this experiment as it contains 23 days; therefore, House 1 had appropriate data for training, validation, and testing. Continued with the setup, the input for training data, validation data, and test data were 1–10 days, 11–16 days, and 17 days onwards respectively. This shows that it had 10 days for training, 6 days for validation, and 7 days for testing.

5.1. ENSML Regression Model for Prediction

From these 20 appliances in House 1, this experiment used some of the appliances to make a prediction using the ENSML algorithm as explained in Section 4. Refrigerator, microwave, and electric heat are the appliances that were used for this experiment. The process flow diagram shown in Figure 9 summarises the process followed for the prediction of the power usage utilised by each appliance respectively.

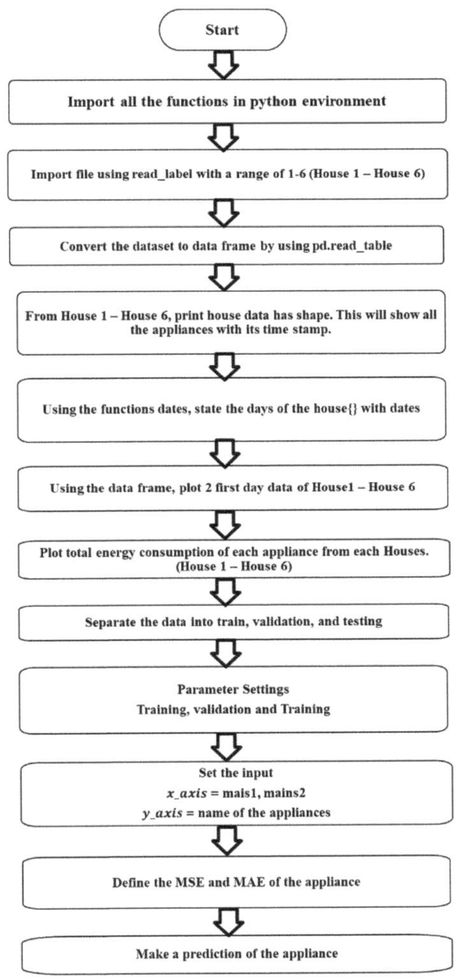

Figure 9. Process hierarchy summarising the steps simulator goes through from start to the end.

5.2. Performance Metrics

The proposed system's performance metrics are defined below. Here, precision, recall rate, $F1$ score, and absolute errors were used as the evaluation indicators.

$$Precision = \frac{TP}{TP + FP} \tag{1}$$

$$Recall\ Rate = \frac{TP}{TP + FN} \tag{2}$$

$$Accuracy = \frac{TP + TN}{TN + FP + TP + FN} \tag{3}$$

$$F1\text{-}score = \frac{2 * (Precision \times Recall\ Rate)}{2 * (Precision + Recall\ Rate)} \tag{4}$$

$$MAE = \frac{1}{T_1 - T_0} \sum_{t=T_0}^{T_1} \left| \left(\tilde{y_t} - \frac{y_t}{y_t} \right) \right| \tag{5}$$

where TP means the number of True Positives; FP stands for the number of False Positives; and FN means the number of False Negatives. TP represents the total number of sequence points for which the electrical appliance is truly operating and for which the disaggregation result is likewise working. The number of sequence points when the electrical appliance is truly operating but the outcome is non-functional is represented by FP. The number FN denotes the total number of sequence points, indicating that the electrical appliance is not in use but that the model decomposition result is. At time t, y_t reflects the real power of the electrical equipment. MAE is the average absolute error of the power disaggregation in the time period from T_0 to T_1. The disaggregated power y_t at time t, and MAE represents the average absolute error of the power disaggregation in the time period from T_0 to T_1. The *Precision, Recall Rate, Accuracy, F1-score*, and MAE are the fundamental indications of non-intrusive load disaggregation and can represent the model's accuracy in evaluating if the electrical appliance is in a functional state. The precision of the disaggregated power value at each time period can be reflected by MAE. The better the precision of the power decomposed value, the lower the value.

6. Results

In this section, results for refrigerator, microwave, and electrical heat showed some promising values on six test days. However, others showed spikes for predicting values caused by overfitting.

The performance characteristics for a refrigerator, microwave, and electrical heat are shown in Table 5. With excellent accuracy, recall rate, precision, and $F1$-score, the true category was predicted. The MAE prediction for the appliances was less than 1%. The results shown in Figure 10a,b presents the day 1 and day 6 usage by the refrigerator. This clarifies that the utilised ENSML model prediction and true value of the power usage by the refrigerator is accurate. Looking at Figure 11, at epoch 125, the acquired training and validation accuracy were 99.2 percent and 98.1 percent, respectively. Similarly, the training and validation losses were 0.04 and 0.05, respectively.

Table 5. Performance of the architecture.

Class	*Accuracy* (%)	*Recall Rate* (%)	*Precision* (%)	*F1-Score* (%)	*MAE* (%)
Refrigerator	99.556	99.667	99.336	99.501	0.64
Microwave	98.752	99.54	99.145	99.46	0.98
Electrical Heat	99.454	99.14	99.556	99.75	0.35

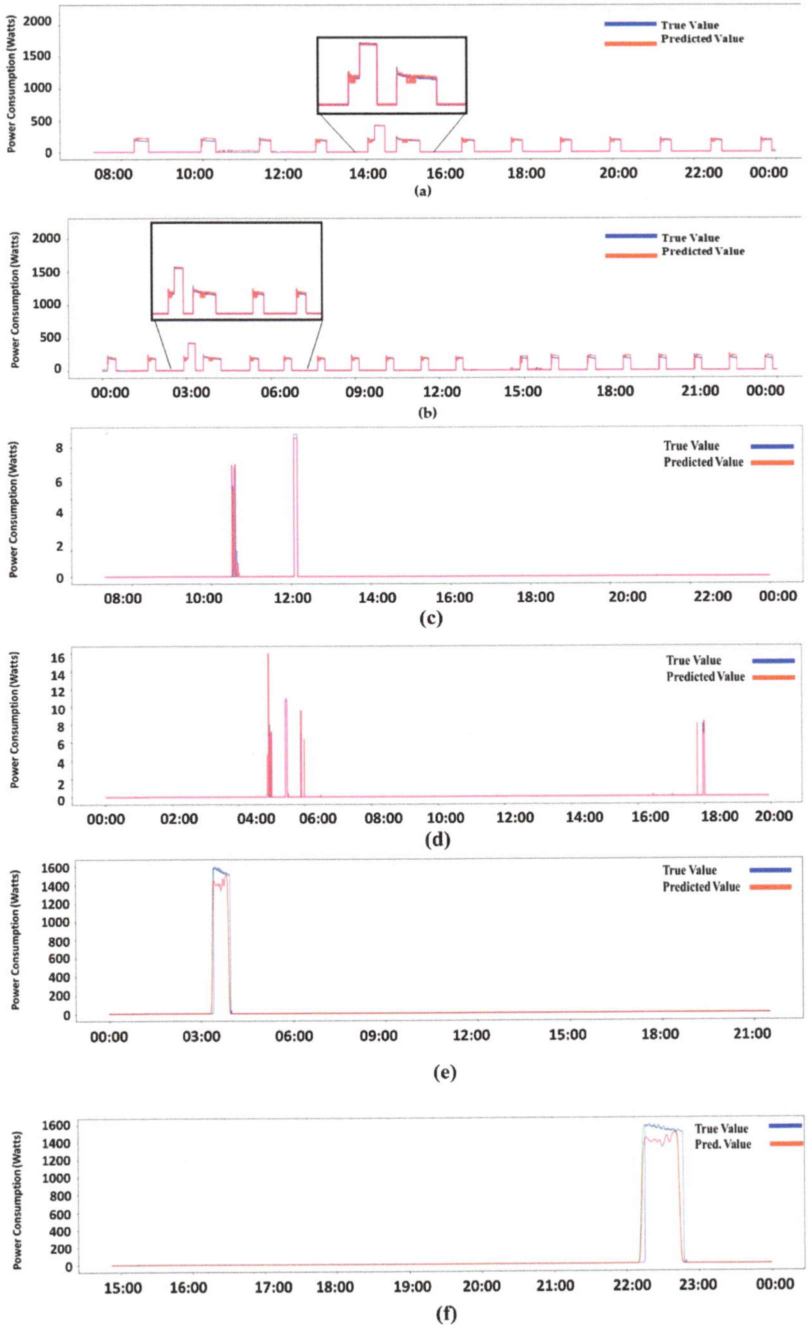

Figure 10. True vs. predicted result with day 1 and day 6 for (**a**,**b**) Refrigerator, (**c**,**d**) Electrical heat, and (**e**,**f**) Microwave.

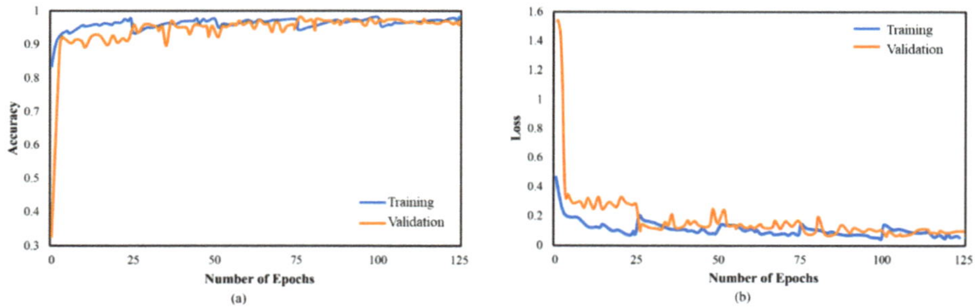

Figure 11. Evaluation of architecture in terms of (**a**) accuracy and (**b**) aoss.

The results presented showed that prediction of the house is very near to the true values of the aggregated powers. The reason for choosing refrigerator, microwave, and electrical heat as the appliances is that the pattern of data and aggregated powers is different. If we consider the refrigerator, the pattern was repetitive and was in the form of a square wave. The prediction of the data was accurately detected. However, if we consider the microwave, the usage was around 3500 W and is not repetitive; still, the prediction was in good proportions. A little dip in the shape of the predicted value was seen due to underfitting as these instances are quite fewer and data are scarce in this situation.

The performance of proposed method is compared with some recent methodologies that have been proposed in the recent past. The Table 6 summarises the contributions with respect to precision obtained by the proposed technique. The settings, appliance under test and dataset are similar. Table 6 shows that the proposed method has the best performance as compared to other methodologies.

Table 6. Comparison of the performance of other methodologies with proposed method.

Contribution	Methods/Techniques	Number of Appliances	Precision [%]
[42]	Factorial hidden Markov models	REDD	82
[43]	Deep Learning Approach	REDD	76
[44]	Back propagation neural network	REDD	45
[45]	*K*-means clustering algorithm	REDD	62
[46]	Unsupervised Linear Discrimination Method	REDD	81
[47]	CNN binary classifier	private	97
[48]	Deep CNN and a KNN classifier	private	93.8
Present Work	Efficient Non-parametric Supervised Machine Learning Network	REDD	99.55

Hardware cost is an important factor that needs to be taken into consideration. This is a cost that is usually proportional to the data resolution, i.e., higher resolutions mean higher costs [47,48]. In addition to the hardware expenditures, there are also training and operating costs to consider. For the pretrained model, a generic modelling technique will be adopted through training. The model for each appliance type utilised in all setups is an important factor; if we can find a similar model that predicts for all appliances, it reduces the complexities a lot. In any event, reliable models require data, patience, and, more often than not, many resources. We commonly refer to the cloud environment in terms of functioning because most of these services are hosted on the cloud. The hardware and software trainings and implementation with machine learning is an important factor that will be thoroughly investigated in future work.

7. Conclusions

The development of new technology to better regulate energy use has become a requirement in recent years. Using home smart meter data, this research proposed an appliance recognition and prediction system utilising the NILM technique based on a simple and low-complexity ENSML network, which can identify common household electrical equipment from a typical household smart meter reading. In this research, we offered an intelligent method to detect smart home loads without being obtrusive. The research focused on the prediction of different appliances which have different amounts and patterns of power usage. The evaluation results revealed that the proposed method has 99.9% accuracy. While utilising the proposed method, it is expected that future work will utilise other methodologies that can beat the performance as well as predict optimally for other appliances as well, and, finally, create a prototype of the model that can store the data in cloud. This data then can be stored in the internet operating system platform to continue monitoring for future use. Further work will also investigate the benefits of applying the NNs to convert smart meter data.

Author Contributions: Conceptualisation, M.U.H. and A.B.; methodology, M.U.H.; software, M.U.H., A.B. and N.H.N.S.; validation, M.U.H. and N.H.N.S.; formal analysis, A.B. and N.H.N.S.; investigation, M.U.H., A.B. and N.H.N.S.; resources, M.U.H.; data curation, N.H.N.S., M.U.H. and A.B.; writing—original draft preparation, M.U.H. and N.H.N.S.; supervision, M.U.H. All authors have read and agreed to the published version of the manuscript.

Funding: This research received no external funding.

Data Availability Statement: The data presented in this study are available on request.

Conflicts of Interest: The authors declare no conflict of interest.

References

1. Osuwa, A.A.; Ekhoragbon, E.B.; Fat, L.T. Application of Artificial Intelligence in Internet of Things. In Proceedings of the 2017 9th International Conference on Computational Intelligence and Communication Networks, Girne, Cyprus, 16–17 September 2017.
2. Subbulakshmi, V.; Aiswarya, D.; Arulselvi, A. Monitoring and controlling energy consumption using IOT and blockchain. *Int. J. Adv. Netw. Appl.* **2016**, *1*, 317–321.
3. Garcia, F.D.; Souza, W.A.; Diniz, I.S.; Marafão, F.P. NILM-based approach for energy efficiency assessment of household appliances. *Energy Inform.* **2020**, *3*, 10. [CrossRef]
4. Chang, H.; Wiratha, P.W.; Chen, N. A Non-intrusive Load Monitoring System Using an Embedded System for Applications to Unbalanced Residential Distribution Systems. *Energy Procedia* **2014**, *61*, 146–150. [CrossRef]
5. Nalmpantis, C.; Vrakas, D. On time series representations for multi-label NILM. *Neural Comput. Appl.* **2020**, *32*, 17275–17290. [CrossRef]
6. Sankar, L.; Rajagopalan, S.R.; Mohajer, S.; Poor, H.V. Smart Meter Privacy: A Theoretical Framework. *IEEE Trans. Smart Grid* **2013**, *4*, 837–846. [CrossRef]
7. Carrie Armel, K.; Gupta, A.; Shrimali, G.; Albert, A. Is disaggregation the holy grail of energy efficiency? The case of electricity. *Energy Policy* **2013**, *52*, 213–234. [CrossRef]
8. Devlin, M.A.; Hayes, B.P. Non-Intrusive Load Monitoring and Classification of Activities of Daily Living Using Residential Smart Meter Data. *IEEE Trans. Consum. Electron.* **2019**, *65*, 339–348. [CrossRef]
9. Hart, G.W. Nonintrusive appliance load monitoring. *Proc. IEEE* **1992**, *80*, 1870–1891. [CrossRef]
10. Klemenjak, C.; Makonin, S.; Elmenreich, W. Towards comparability in non-intrusive load monitoring: On data and performance evaluation. In Proceedings of the 2020 IEEE Power & Energy Society Innovative Smart Grid Technologies Conference (ISGT), Washington, DC, USA, 17–20 February 2020.
11. Virtsionis-Gkalinikis, N.; Nalmpantis, C.; Vrakas, D. SAED: Self-attentive energy disaggregation. *Mach. Learn.* **2021**, 1–20. [CrossRef]
12. Kelly, D.G. *Disaggregation of Domestic Smart Meter Energy Data*; London University: London, UK, 2016.

13. Batra, N.; Kukunuri, R.; Pandey, A.; Malakar, R.; Kumar, R.; Krystalakos, O.; Zhong, M.; Meira, P.; Parson, O. Towards reproducible state-of-the-art energy disaggregation. In Proceedings of the 6th ACM International Conference on Systems for Energy-Efficient Buildings, Cities, and Transportation, New York, NY, USA, 13–14 November 2019; pp. 193–202.
14. Jiang, J.; Kong, Q.; Plumbley, M.D.; Gilbert, N.; Hoogendoorn, M.; Roijers, D.M. Deep Learning-Based Energy Disaggregation and On/Off Detection of Household Appliances. *ACM Trans. Knowl. Discov. Data* **2021**, *15*, 1–21. [CrossRef]
15. He, K.; Stankovic, L.; Liao, J.; Stankovic, V. Non-intrusive load disaggregation using graph signal processing. *IEEE Trans. Smart Grid* **2018**, *9*, 1739–1747. [CrossRef]
16. Liu, Q.; Kamoto, K.M.; Liu, X.; Sun, M.; Linge, N. Low-complexity non-intrusive load monitoring using unsupervised learning and generalized appliance models. *IEEE Trans. Consum. Electron.* **2019**, *65*, 28–37. [CrossRef]
17. Kong, S.; Kim, Y.; Ko, R.; Joo, S. Home appliance load disaggregation using cepstrum-smoothing-based method. *IEEE Trans. Consum. Electron.* **2015**, *61*, 24–30. [CrossRef]
18. Bhotto, M.Z.A.; Makonin, S.; Bajic, I.V. Load disaggregation based on aided linear integer programming. *IEEE Trans. Circuits Syst. II Express Briefs* **2017**, *64*, 792–796. [CrossRef]
19. Wittmann, F.M.; Lopez, J.C.; Rider, M.J. Nonintrusive load monitoring algorithm using mixed-integer linear programming. *IEEE Trans. Consum. Electron.* **2018**, *64*, 180–187. [CrossRef]
20. Net2grid. 2019. Available online: http://www.net2grid.com (accessed on 19 March 2022).
21. Smappee. 2019. Available online: https://www.smappee.com/been/homepage (accessed on 19 March 2022).
22. SmartB Energy Management. 2019. Available online: http://www.smartb.de/ (accessed on 19 March 2022).
23. Watty. 2019. Available online: https://watty.io/ (accessed on 21 March 2022).
24. Bidgely. 2019. Available online: http://www.bidgely.com (accessed on 19 March 2022).
25. Verv. 2019. Available online: https://verv.energy/ (accessed on 19 March 2022).
26. Ronneberger, O.; Fischer, P.; Brox, T. U-Net: Convolutional Networks for Biomedical Image Segmentation. In *Medical Image Computing and Computer-Assisted Intervention–MICCAI 2015*; Navab, N., Hornegger, J., Wells, W.M., Frangi, A.F., Eds.; Springer International Publishing: Cham, Switzerland, 2015; Volume 9351, pp. 234–241.
27. Kelly, J.; Knottenbelt, W. Neural NILM. In: ACM, 55–64. (4 November 2015). Available online: http://dl.acm.org/citation.cfm?id=2821672 (accessed on 19 March 2022).
28. Barsim, K.S.; Yang, B. Toward a semi-supervised non-intrusive load monitoring system for event-based energy disaggregation. In Proceedings of the 2015 IEEE Global Conference on Signal and Information Processing (GlobalSIP), Orlando, FL, USA, 14–16 December 2015; pp. 58–62. Available online: https://ieeexplore.ieee.org/document/7418156 (accessed on 19 March 2022).
29. Biansoongnern, S.; Plangklang, B. Nonintrusive load monitoring (NILM) using an Artificial Neural Network in embedded system with low sampling rate. In Proceedings of the 2016 13th International Conference on Electrical Engineering/Electronics, Computer, Telecommunications and Information Technology (ECTI-CON), Chiang Mai, Thailand, 28 June–1 July 2016; pp. 1–4.
30. Faustine, A.; Pereira, L.; Bousbiat, H.; Kulkarni, S. UNet-NILM: A Deep Neural Network for Multi-Tasks Appliances State Detection and Power Estimation in NILM. In Proceedings of the 5th International Workshop on Non-Intrusive Load Monitoring, NILM'20, Virtual Event, Japan, 18 November 2020; Association for Computing Machinery: New York, NY, USA, 2020; pp. 84–88.
31. Zeifman, M.; Roth, K. Nonintrusive appliance load monitoring: Review and outlook. *IEEE Trans. Consum. Electron.* **2011**, *57*, 76–84. [CrossRef]
32. Wang, Z.; Zheng, G. Residential Appliances Identification and Monitoring by a Nonintrusive Method. *IEEE Trans. Smart Grid* **2012**, *3*, 80–92.
33. Weiss, M.; Helfenstein, A.; Mattern, F.; Staake, T. Leveraging smart meter data to recognize home appliances. In Proceedings of the 2012 IEEE International Conference on Pervasive Computing and Communications, Lugano, Switzerland, 19–23 March 2012; pp. 190–197.
34. Makriyiannis, M.; Lung, T.; Craven, R.; Toni, F.; Kelly, J. Smarter electricity and argumentation theory. In *Proceedings of the Smart Innovation, Systems and Technologies*; Springer Science and Business Media Deutschland GmbH: Berlin, Germany, 2016; Volume 46, pp. 79–95.
35. Hadi, M.U.; Murtaza, G. Enhancing distributed feedback-standard single mode fiber-radio over fiber links performance by neural network digital predistortion. *Microw. Opt. Technol. Lett.* **2021**, *63*, 1558–1565. [CrossRef]
36. Chang, H.H.; Lin, L.S.; Chen, N.; Lee, W.J. Particle-swarm-optimization-based nonintrusive demand monitoring and load identification in smart meters. *IEEE Trans. Ind. Appl.* **2013**, *49*, 2229–2236.
37. Kolter, J.Z.; Jaakkola, T. Approximate inference in additive factorial hmms with application to energy disaggregation. In Proceedings of the Machine Learning Research, Volume 22: Artificial Intelligence and Statistics, La Palma, Canary Islands, Spain, 21–23 April 2012; pp. 1472–1482.
38. Hadi, M.U.; Awais, M.; Raza, M.; Khurshid, K.; Jung, H. Neural Network DPD for Aggrandizing SM-VCSEL-SSMF-Based Radio over Fiber Link Performance. *Photonics* **2021**, *8*, 19. [CrossRef]
39. Hadi, M.U. Practical Demonstration of 5G NR Transport Over-Fiber System with Convolutional Neural Network. *Telecom* **2022**, *3*, 103–117. [CrossRef]
40. Egarter, D.; Bhuvana, V.P.; Elmenreich, W. PALDi: Online Load Disaggregation via Particle Filtering. *IEEE Trans. Instrum. Meas.* **2015**, *64*, 467–477. [CrossRef]

41. Ferrández-Pastor, F.J.; Mora-Mora, H.; Sánchez-Romero, J.L.; Nieto-Hidalgo, M.; García-Chamizo, J.M. Interpreting human activity from electrical consumption data using reconfigurable hardware and hidden Markov models. *J. Ambient. Intell. Humaniz. Comput.* **2016**, *8*, 469–483. [CrossRef]
42. Khurshid, K.; Khan, A.A.; Siddiqui, H.; Rashid, I.; Hadi, M.U. Big Data Assisted CRAN Enabled 5G SON Architecture. *J. ICT Res. Appl.* **2019**, *13*, 93–106. [CrossRef]
43. Hadi, M.U.; Awais, M.; Raza, M.; Ashraf, M.I.; Song, J. Experimental Demonstration and Performance Enhancement of 5G NR Multiband Radio over Fiber System Using Optimized Digital Predistortion. *Appl. Sci.* **2021**, *11*, 11624. [CrossRef]
44. Liu, Y.; Wang, J.; Deng, J.; Sheng, W.; Tan, P. Non-Intrusive Load Monitoring Based on Unsupervised Optimization Enhanced Neural Network Deep Learning. *Front. Energy Res.* **2021**, *9*, 718916. [CrossRef]
45. Wang, K.; Zhong, H.; Yu, N.; Xia, Q. Nonintrusive Load Monitoring Based on Sequence-To-Sequence Model with Attention Mechanism. *Proc. CSEE* **2019**, *39*, 75–83.
46. Liu, L.; Ding, J.; Zhong, J.; Fu, X.; Lv, Y. An unsupervised model for classification and recognition of household appliances. *J. Comput. Inf. Syst.* **2014**, *10*, 403–410.
47. Athanasiadis, C.; Doukas, D.; Papadopoulos, T.; Chrysopoulos, A. A Scalable Real-Time Non-Intrusive Load Monitoring System for the Estimation of Household Appliance Power Consumption. *Energies* **2021**, *14*, 767. [CrossRef]
48. Athanasiadis, C.L.; Papadopoulos, T.A.; Doukas, D.I. Real-time non-intrusive load monitoring: A light-weight and scalable approach. *Energy Build.* **2021**, *253*, 111523. [CrossRef]

technologies

MDPI

Article

Patterns Simulations Using Gibbs/MRF Auto-Poisson Models

Stelios Zimeras

Department of Statistics and Actuarial-Financial Mathematics, University of the Aegean, 83200 Karlovassi, Greece; zimste@aegean.gr

Abstract: Pattern analysis is the process where characteristics of big data can be recognized using specific methods. Recognition of the data, especially images, can be achieved by applying spatial models, explaining the neighborhood structure of the patterns. These models can be introduced by Markov random field (MRF) models where conditional distribution of the pixels may be defined by a specific distribution. Various spatial models could be introduced, explaining the real patterns of the data; one class of these models is based on the Poisson distribution, called auto-Poisson models. The main advantage of these models is the consideration of the local characteristics of the image. Based on the local analysis, various patterns can be introduced and models that better explain the real data can be estimated, using advanced statistical techniques like Monte Carlo Markov Chains methods. These methods are based on simulations where the proposed distribution must converge to the original (final) one. In this work, an analysis of a MRF model under Poisson distribution would be defined and simulations would be illustrated based on Monte Carlo Markov Chains (MCMC) process like Gibbs sampler. Results would be illustrated using simulated and real patterns data.

Keywords: patterns simulation; MRF; auto-Poisson; MCMC; Gibbs sampler

Citation: Zimeras, S. Patterns Simulations Using Gibbs/MRF Auto-Poisson Models. *Technologies* **2022**, *10*, 69. https://doi.org/10.3390/technologies10030069

Academic Editors: Manoj Gupta, Eugene Wong and Gwanggil Jeon

Received: 14 April 2022
Accepted: 2 June 2022
Published: 6 June 2022

1. Introduction

Pattern analysis is a class of methods that are used to recognized regular patterns in big data, like images. Considering these methods, models must be introduced, and analysis of the modeling process must be performed. Especially for the images, the spatial structure of the pixels is an important measure to explain the spatiality of the image.

Spatial structure of images can be analyzed based on specific models, where neighborhood structures are taking into consideration. Interactions between regions at different scales are characterized by their local dynamics, and the emergent spatial patterns are the outcome of different processes. Local dynamics could be explained by the local characteristics of the image with the main result being the construction of the images under homogeneous regions. The homogeneity of the image could be defined by the spatial pattern of the image, which is explained with way better presentation of real data. Under these local characteristics, special models considering the neighborhood structure of the image could established specific conditional models defined as Markov Random fields (MRF) models. These models are defined by Markov Random fields (MRF) models, explaining the spatial structure of the images by a conditional distribution of the pixels, defined as auto-models. These models under appropriate structure patterns have the ability to explain real patterns. Under the conditional distribution, various models could be introduced. In our work, the specific class of data that can be analyzed as best as it can is based on the Poisson distribution, defined as auto-Poisson models.

Referring to conditional probabilities, due to the largeness of the configuration space, it is impractical to sample from it by direct computation of the probabilities. Markov chains Monte Carlo (MCMC) methods have been investigated by various researchers as an alternative to exact probability computation [1,2]. The general method is to simulate a Markov chain with the required probability distribution as its equilibrium distribution. If the chain is aperiodic and irreducible, the convergence is guaranteed.

The main goal is to introduce a method where a model can be defined and investigated. Based on this investigation, characteristics for this model (auto-Poisson) can be held and generalized for similar image patterns. Simulations of the process are achieved by using MCMC method, like Gibbs sampler, in simulated and real data.

2. Materials and Methods

Markov Random Fields Modeling

Spatial data under investigation can be defined as regular or irregular areas (Figure 1). Under these areas, spatial patterns cam be illustrated based on the structure of the images. The main goal is to model the spatial patterns in a way to represent as best as they can the real data starting from a simple model to a more complicated one (Figure 2).

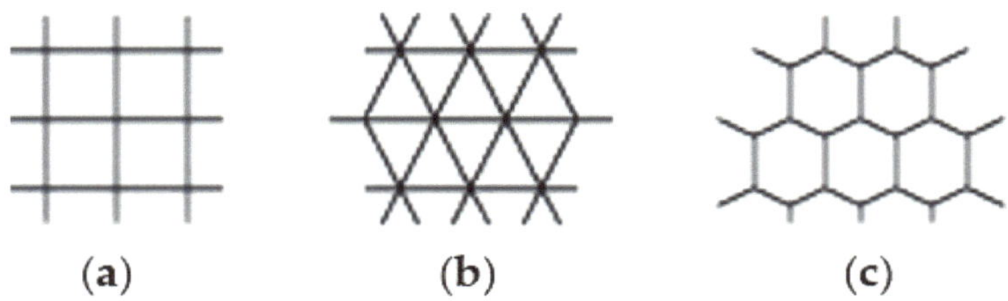

Figure 1. Different type of areas: (**a**) lattice; (**b**) hexagonal; and (**c**) hive.

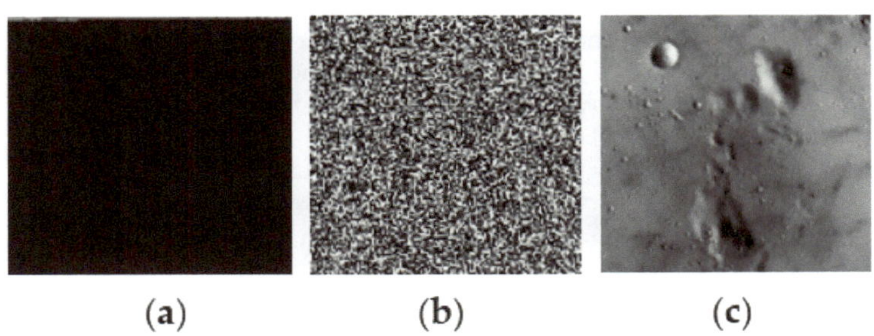

Figure 2. Various types of image models: (**a**) simple; (**b**) complicated; (**c**) very complicated.

Considering a 2-D space, which has been partitioned into n-pixels, labeled by the integers 1, 2, … n (rectangular space); x_{ij} or x_i denote the color for pixel (i) or (i,j); p(.. | ..) defined as the conditional probability distribution (Figure 3) ([3–6]).

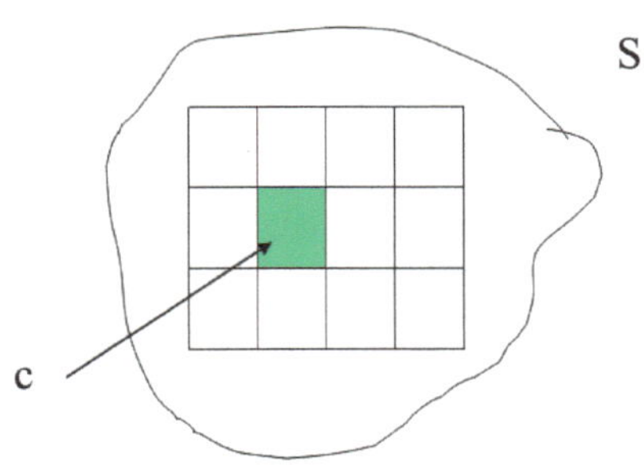

Figure 3. Definition of the 2-D space for image presentation.

If D is a finite lattice D = {(i, j), $1 \leq i \leq N, 1 \leq j \leq M$}, then each pixel of the finite lattice D can be colored from the set {0, 1, . . . ,c − 1}. Site j (^{1} i) is said to be a neighbor of site i iff the functional form of p (x_i | $x_1, \ldots, x_{i-1}, x_{i+1}, \ldots, x_n$) is defined upon the variables x_j as $p(x_i|x_1, \ldots, x_{i-1}, x_{i+1}, \ldots, x_n) = p(x_i|x_{\partial i})$, where ∂i is the set of pixels that are neighbous of pixel I, and $x_{\partial i}$ is the set of values of pixels that are neighbors of pixel i.

A neighborhood structure **N = {N$_i$, ∀ i∈ S}** is defined as a collection of subsets of S. The symmetry property is based on the following conditions: (i) $i \notin N_i$ (a site is not part of its neighborhood); (ii) $j \in N_i \Leftrightarrow i \in N_j$ (i is in the neighborhood of j if and only if j is in the neighborhood of i). Define a nearest-neighborhood set as the set of sites with the property that p (x_{ij} | all other values) depends only upon the neighbors $x_{i-1,j}, x_{i+1,j}, x_{i,j-1}, x_{i,j+1}$ for each internal site (i,j) (Figure 4).

	$(x_{i-1,j})$	
$(x_{i,j-1})$	$(x_{i,j})$	$(x_{i,j+1})$
	$(x_{i+1,j})$	

Figure 4. Nearest-neighborhood system.

The first order model (four neighbors) is denoted by N = {($x_{i,j}$, $x_{i-1,j}$), ($x_{i,j}$, $x_{i+1,j}$), ($x_{i,j}$, $x_{i,j-1}$), ($x_{i,j}$, $x_{i,j+1}$)} and second order (eight neighbors) is denoted by N = {($x_{i,j}$, $x_{i-1,j}$), ($x_{i,j}$, $x_{i+1,j}$), ($x_{i,j}$, $x_{i,j-1}$), ($x_{i,j}$, $x_{i,j+1}$), ($x_{i,j}$, $x_{i-1,j-1}$), ($x_{i,j}$, $x_{i-1,j+1}$), ($x_{i,j}$, $x_{i+1,j-1}$), ($x_{i,j}$, $x_{i+1,j+1}$)} (Figure 5).

	$(x_{i-1,j})$	
$(x_{i,j-1})$	$(x_{i,j})$	$(x_{i,j+1})$
	$(x_{i+1,j})$	

(a)

$(x_{i-1,j-1})$	$(x_{i-1,j})$	$(x_{i-1,j+1})$
$(x_{i,j-1})$	$(x_{i,j})$	$(x_{i,j+1})$
$(x_{i+1,j-1})$	$(x_{i+1,j})$	$(x_{i+1,j+1})$

(b)

Figure 5. Neighborhood structure. (**a**) first order and (**b**) second order.

A clique is defined as a set that consists either of a single pixel or a collection of pixels that are neighbors of each other. Given a 2-D space of a subset S and neighbor structure ∂i, a clique is any set of pixels $c \subset S$, where for all $i, j \in c, j \in \partial i$. C can be defined as the set of all cliques. The concept of clique is directly combined with the calculation of the energy of the image, which can be considered as a statistical measure of the weight of the correlation between the pixels.

For the first order, the neighborhood structure is given by: {(i,j)}, {(i − 1,j), (i,j)}, {(i,j − 1), (i,j)}; and for the second order the neighborhood structure is given by: {(i,j)}, {(i − 1,j), (i,j)}, {(i,j − 1), (i,j)},{(i − 1,j − 1), (i,j)}, {(i − 1,j + 1), (i,j)}, {(i,j + 1), (i − 1,j + 1), (i,j)}, {(i,j + 1), (i − 1,j), (i,j)}, {(i − 1,j − 1), (i − 1,j), (i,j)}, {(i − 1,j + 1), (i − 1,j), (i,j)}, {(i − 1,j − 1), (i − 1,j + 1), (i + 1,j + 1), (i + 1,j − 1),(i,j)}.

A Markov random field (MRF) is a joint probability density on the set of all possible coloring values **X** on a finite lattice D, following the conditions ([3,4,7]): (i) p (**x**) > 0 for all $\mathbf{x} \in S$. (Positivity). (ii) p (x_{ij} | all points (i,j)) = p (x_{ij} | neighbors). (Markovianity). (iii) p (x_{ij} | neighbors (i,j)) depends only on the configuration of the neighbors (Homogeneity). The probabilities on the condition (2) are called local characteristics. Condition (2) can be expressed as $p(x_i|x_j, i \neq j) = p(x_i|x_{\partial i})$, where it is clear that we need a representative distribution (Gibbs). If p (x) is a jpdf of $X_i \in S$ under any neighborhood model ∂i, the Gibbs distribution can be expressed based on the following form:

$$p(\mathbf{x}) = \frac{1}{Z}\exp\left\{-\frac{1}{T}[U(\mathbf{x})]\right\} = \frac{1}{Z}\exp\left\{-\frac{1}{T}\left[\sum_{c\in C} V_c(x_c)\right]\right\} \tag{1}$$

where C are all the potential cliques, Z is the normalized constant/partitions function, T is is the temperature with T = 1, and U (**x**) is the energy function with $U(\mathbf{x}) = \sum_{c\in C} V_c(x_c)$ [8]. Under the Hammersley–Clifford theorem ([9,10]), if X is a discrete or continuous variable assigned to a pixel x representing a random field with neighborhood structure ∂i and jpdf p (x), then X is a MRF. iff p (x) can be expressed as a Gibbs distribution. The general form for the energy function can be defined by

$$U(\mathbf{x}) = \sum_{i\in S} V_1(x_i) + \sum_{i\in S}\sum_{j\in S} V_2(x_i, x_j) = \sum_{1\le i\le S} x_i G_i(x_i) + \sum_{1\le i<}\sum_{j\le S} x_i x_j G_{ij}(x_i, x_j) + \ldots . \tag{2}$$

where G (.) is any arbitrary function ([3,4]). For example, for the second order neighborhood structure, the energy function is given by

$$U(\mathbf{x}) = \sum_{i\in S} V_1(x_i) + \sum_{i\in S}\sum_{j\in S} V_2(x_i, x_j) = \sum_{1\le i\le S} x_i G_i(x_i) + \sum_{1\le i<}\sum_{j\le S} x_i x_j G_{ij}(x_i, x_j) \tag{3}$$

defined as pairwise interaction MRF models.

Specific spatial patterns can be described by particular models based on their neighbors, defined as auto-models. Assumptions for these models are: (1) The probability structure only depends on contributions of sites taken either as singular or in pairs. (2) The conditional probability distribution is a member of the regular exponential family of distributions $p(x_i|x_{\partial i}) = \exp\{A_i(\theta_i)B_i(x_i) + C_i(x_i) + D_i(\theta_i)\}$ where θ_i is a model parameter associated with site i and is a function of the values at sites neighboring site i. A_i (θ_i) can be defined as a potential interaction between pixels. Assuming the conditional probabilities and the pairwise only dependence between sites, the A_i (θ_i) must satisfy ([2,5,6]):

$$A_i(\theta_i) = \alpha_i + \sum \beta_{ij} B_j(x_j) \tag{4}$$

where $\beta_{ij} = \beta_{ji}$ if i is neighbors with j and $\beta_{ij} = 0$ otherwise. As a final restriction, it is assumed that the function B_j (x_j) is linear in x_i with form $A_i(\theta_i) = \alpha_i + \sum \beta_{ij} x_j$. If $\beta_{ij} = \beta_{ji} = \beta$, it is an isotropic model, otherwise it is anisotropic. For the first-order system the models are: isotropic: $A_i(\alpha,\beta) = \alpha + \beta\,(x_{i-1,j} + x_{i+1,j} + x_{i,j-1} + x_{i,j+1})$ and anisotropic: $A_i(\alpha, \beta_1, \beta_2) = \alpha + \beta_1(x_{i-1,j} + x_{i+1,j}) + \beta_2(x_{i,j-1} + x_{i,j+1})$ (Figure 6a). For the second order, the models are: isotropic: $A_i(\alpha, \beta, \gamma) = \alpha + \beta\,(x_{i-1,j} + x_{i+1,j} + x_{i,j-1} + x_{i,j+1}) + \gamma\,(x_{i-1,j-1} + x_{i-1,j+1} + x_{i+1,j-1} + x_{i-1,j+1})$ and for anisotropic: $A_i(\alpha, \beta_1, \beta_2, \gamma_1, \gamma_2) = \alpha + \beta_1(x_{i-1,j} + x_{i+1,j}) + \beta_2(x_{i,j-1} + x_{i,j+1}) + \gamma_1(x_{i-1,j-1} + x_{i-1,j+1}) + \gamma_2(x_{i+1,j-1} + x_{i-1,j+1})$ (Figure 6b) ([8]).

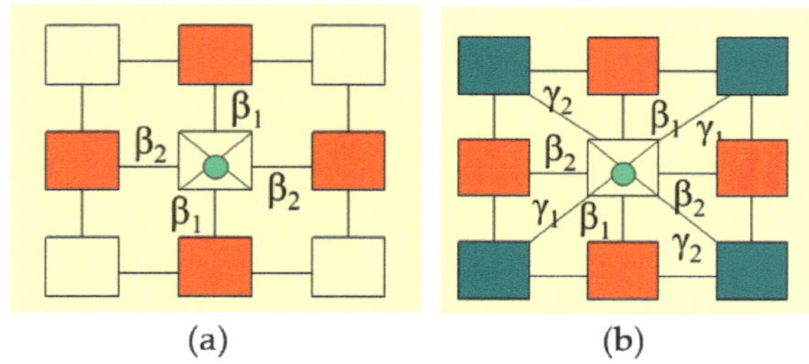

Figure 6. (**a**) First-order models and (**b**) second-order models.

Considering the general notation of the auto-models, the potential interaction between pixels for the auto-Poisson model is given by

$$A_i(\lambda_i) = \log(\lambda_i) \Rightarrow \lambda_i = \exp\left(a_i + \sum \beta_{ij} x_j\right) \quad (5)$$

under the assumption that the conditional distribution of the pixels x_i given their neighbors has a Poisson distribution mean λ_i with form

$$p(x_i|x_{\partial i}) = \frac{\lambda_i^{x_i} \exp(-\lambda_i)}{x!} = \exp[\log(\lambda_i)x_i - \lambda_i - \log(x_i!)] \quad (6)$$

The limitation of the model is that $|b| > \frac{1}{\lambda_i}$([2,5,6]), where b = mβc, c denotes number of colors, and m denotes the number of neighbors.

3. Results

Simulation Process Using MCMC Method

The investigation of the spatial patterns can be illustrated by the simulation procedures. A particular simulation process, where realizations from pseudo-samples can finally define the desired distribution, is the Monte Carlo Markov Chain (MCMC) ([11]). The goal of the process is to simulate the distribution p (x) using a particular realization $X^1, X^2, \ldots, X^N$ on the Markov chain with transition probability. Under the process, asymptotic results can be archived where:

$$X^t \underset{t\to\infty}{\overset{d}{\rightarrow}} X \sim p(x)\ ;\ \tfrac{1}{t}\sum_{i=1}^{t} f(x^i) \underset{t\to\infty}{\rightarrow} E_p\{f(x)\} \quad (7)$$

with the expectation E_p {f (x)} under estimation. The corresponding empirical average will be given by

$$\overline{f}_N = \frac{1}{N}\sum_{t+1}^{N} f\left(x^{(t)}\right) \quad (8)$$

Gibbs sampler is a special case of MCMC methods. Consider $X_i \in S$ a discrete or continuous value for a random field in a rectangular lattice system S with neighborhood structure ∂i. If X is defined as a pdf p (x) based on Gibbs distribution, then the conditional probability of a value given its neighbors can be defined as

$$p(x_s|x_j, s \neq j) = \frac{\exp\{-U(x_s|x_j, s \neq j)\}}{\sum_{x_s^* \in S} \exp\{-U(x_s^*|x_j, s \neq j)\}} \quad (9)$$

Each value could be replaced by the conditional probability

$$p(x_s\middle|x_j^k, j \neq s) = p(x_s\middle|x_{\partial s}) \quad (10)$$

The algorithm stopped when all the pixels were replaced. The Gibbs sampler algorithm is given by the following pseudo-algorithm ([12–15]):

Step 1: Chose randomly pixel x_i

Step 2: For t = 0 until ∞

Replace the value $\mathbf{x}^{(t)}$ with $\mathbf{x}^{(t+1)}$ based on

$$X^t \underset{t\to\infty}{\overset{d}{\to}} X \sim p(x)\,;\ \frac{1}{t}\sum_{i=1}^{t} f(x^i) \underset{t\to\infty}{\to} E_p\{f(x)\} \tag{11}$$

Step 3. Continue the process until all the pixels have been replaced

Realizations for the auto-Poisson model are given in Figure 7 with parameter a. First order isotropic with $\alpha = 0.75$ and $\beta = -0.25$; b. Second order isotropic with a = -0.93, $\beta = -0.33$, and $\gamma = -0.37$. Finally a comparison between biological images from microscopes with realizations from the simulated auto-Poisson model considering the first-order isotropic model ($\alpha = -1$, $\beta = -2$) is given in Figure 8, where it is clear that both images have similar patterns.

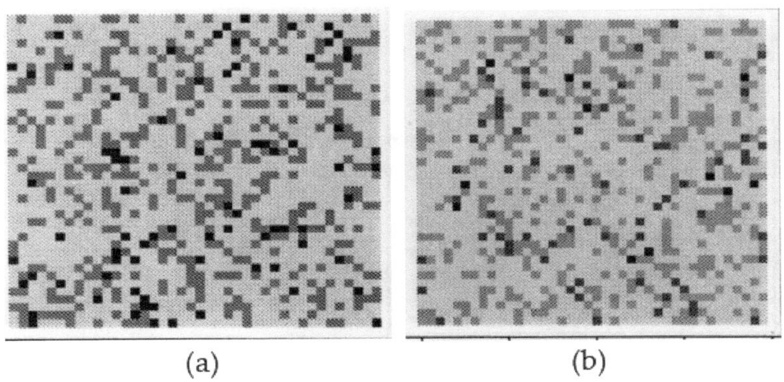

Figure 7. Realizations from auto-Poisson models: (**a**) first-order isotropic; (**b**) second-order isotropic.

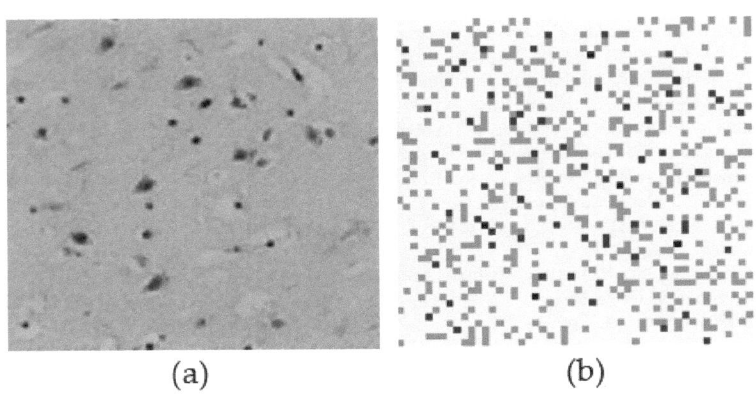

Figure 8. Patterns comparison between real image and simulated image under first order. Isotropic auto-Poisson model: (**a**) Real image and (**b**) Simulated image.

4. Conclusions

Pattern analysis is a process where regular synthesis or patterns can be recognized. Most of the time, these patterns can be analyzed using texture models based on the spatial structure of the images. The spatiality of the images can be defined by considering the homogeneity of the regions. These regions might be represented by considering identity patterns under the neighborhood structure of the image. Models for the estimation of these spatial patterns could be introduced considering the local characteristics of the image, leading us to Markov random fields (MRF) models. These patterns could be used to simulate real phenomena like biology, ecology, or medicine. In this work, a presentation of Markov random field models (MRF) considering specific conditional distributions like auto-models was analyzed. A specific auto-model under Poisson distribution (auto-Poisson) were defined and simulations using MCMC methods as Gibbs samples were illustrated using simulated and real images patterns. Modification of auto-Poisson models was successfully performed in many case like social networks [16], spreading disease modeling [17,18], ecological models [19–22]. In these works, extensions of Markov random fields (MRF) models were applied using network analysis considering the neighborhood structure of the pixels. Under this extension, homogeneous regions could be illustrated considering the connections between similar pixels. Another important application is based on the mapping analysis where simulations of auto-Poisson models might be used to reconstruct estimated maps under various regions, analyzing phenomena like cancer diseases [23]. Last but not least, the proposed models could be used for firm location analysis [24] to estimate the optimal positions for firm's establishment.

Funding: This research received no external funding.

Institutional Review Board Statement: Not applicable.

Informed Consent Statement: Not applicable.

Data Availability Statement: Not applicable.

Conflicts of Interest: The author declare no conflict of interest.

References

1. Zimeras, S.; Matsinos, Y. Modeling Uncertainty based on spatial models in spreading diseases: Spatial Uncertainty in Spreading Diseases. *Int. J. Reliab. Qual. E-Healthc.* **2019**, *8*, 55–66.
2. Aykroyd, R.G.; Zimeras, S. Inhomogeneous prior models for image reconstruction. *J. Am. Stat. Assoc.* **1999**, *94*, 934–946.
3. Besag, J. Spatial interaction and the statistical analysis of lattice systems. *J. R. Stat. Soc.* **1974**, *36*, 192–236.
4. Besag, J. On the statistical analysis of dirty pictures. *J. R. Stat. Soc.* **1986**, *48*, 259–302.
5. Zimeras, S. Statistical Models in Medical Image Processing. Ph.D. Thesis, Leeds University, Leeds, UK, 1997.
6. Zimeras, S.; Georgiakodis, F. Bayesian models for medical image biology using Monte Carlo Markov Chain techniques. *Math. Comput. Modeling* **2005**, *42*, 759–768.
7. Cross, G.R.; Jain, A.K. Markov Random Field Texture Models. *IEEE Trans. Pattern Anal. Mach. Intell.* **1983**, *5*, 25–39. [CrossRef]
8. Zimeras, S. Spreading Stochastic Models Under Ising/Potts Random Fields: Spreading Diseases. In *Quality of Healthcare in the Aftermath of the COVID-19 Pandemic*; IGI Global: Hershey, PA, USA, 2022; pp. 65–78.
9. Hamersley, J.A.; Clifford, P. Markov fields on finite graphs and lattices. 1971, *unpublished work*.
10. Kindermann, R.; Snell, J.L. *Markov Random Fields and Their Applications*; American Mathematical Society: Providence, RI, USA, 1980.
11. Hastings, W.K. Monte Carlo simulation methods using Markov chains, and their applications. *Biometrika* **1970**, *57*, 97–109.
12. Geman, S.; Geman, D. Stochastic relaxation, Gibbs distributions, and Bayesian restoration of images. *IEEE Trans. Pattern Anal. Mach. Intell.* **1984**, *6*, 721–741.
13. Green, P.J.; Han, X.L. Metropolis Methods, Gaussian Proposals and Antithetic Variables. In *Stochastic Models, Statistical Methods, and Algorithms in Image Analysis. Lecture Notes in Statistics*; Barone, P., Frigessi, A., Piccioni, M., Eds.; Springer: New York, NY, USA, 1992; Volume 74. [CrossRef]
14. Metropolis, N.; Rosenbluth, A.; Rosenbluth, M.; Teller, A.; Teller, E. Equations of state calculations by fast computing machines. *J. Chem. Physics* **1953**, *21*, 1087–1091.
15. Smith, A.F.M.; Robert, G.O. Bayesian computation via the Gibbs sampler and related Markov chain Monte Carlo methods. *J. R. Stat. Soc. B* **1993**, *55*, 3–23.
16. Agaskar, A.; Lu, Y.M. Alarm: A logistic auto-regressive model for binary processes on networks. In Proceedings of the IEEE Global Conference on Signal and Information Processing, Austin, TX, USA, 3–5 December 2013; pp. 305–308.
17. Kaiser, M.S.; Pazdernik, K.T.; Lock, A.B.; Nutter, F.W. Modeling the spread of plant disease using a sequence of binary random fields with absorbing states. *Spat. Stat.* **2014**, *9*, 38–50. [CrossRef]

18. Shin, Y.E.; Sang, H.; Liu, D.; Ferguson, T.A.; Song, P.X.K. Autologistic network model on binary data for disease progression study. *Biometrics* **2019**, *75*, 1310–1320. [CrossRef]
19. Zimeras, S.; Matsinos, Y. Spatial Uncertainty. In *Recent Researches in Geography, Geology, Energy, Environment and Biomedicine*; WSEAS Press: Kerkira, Greece, 2011; pp. 203–208.
20. Zimeras, S.; Matsinos, Y. Modelling Spatial Medical Data. In *Effective Methods for Modern Healthcare Service Quality and Evaluation*; IGI Global: Hershey, PA, USA, 2016; pp. 75–89.
21. Zimeras, S.; Matsinos, Y. Bayesian Spatial Uncertainty Analysis. In *Energy and Environment; Recent Researches in Environmental and Geological Sciences, Proceedings of the 7th International WSEAS International Conference on Energy & Environment, Kos Island, Greece, 14–17 July 2012*; WSEAS Press: Kerkira, Greece, 2012; pp. 377–385.
22. Aykroyd, R.; Haigh, J.; Zimeras, S. Unexpected Spatial Patterns in Exponential Family Auto Models. *Graph. Model. Image Process.* **1996**, *58*, 452–463. [CrossRef]
23. Morales-Otero, M.; Núñez-Antón, V. Comparing Bayesian Spatial Conditional Overdispersion and the Besag–York–Mollié Models: Application to Infant Mortality Rates. *Mathematics* **2021**, *9*, 282.
24. Brown, J.P.; Lambert, D.M. Extending a smooth parameter model to firm location analyses: The case of natural gas establishments in the United States. *J. Reg. Sci.* **2016**, *56*, 848–867. [CrossRef]

technologies

MDPI

Article

Application of 3D Virtual Prototyping Technology to the Integration of Wearable Antennas into Fashion Garments

Evridiki Papachristou [1,*] and Hristos T. Anastassiu [2]

1 Creative Design & Fashion Department, International Hellenic University, 61100 Kilkis, Greece
2 Department of Informatics, Computer and Communications Engineering, International Hellenic University, 62124 Serres, Greece; hristosa@ict.ihu.gr
* Correspondence: evridikipapa@ihu.gr; Tel.: +30-23410-29876

Abstract: A very large number of scientific papers have been published in the literature on wearable antennas of several types, structure and functionality. The main focus is always antenna efficiency from an engineering point of view. However, antenna integration into actual, realistic garments is seldom addressed. In this paper, 2D pattern and 3D virtual prototyping technology is utilized to develop regular clothing, available in the market, in which wearable antennas are incorporated in an automated manner, reducing the chances of compromising the garment elegance or comfort. The functionality of various commercial software modules is described, and particular design examples are implemented, proving the efficiency of the procedure and leading the way for more complex configurations.

Keywords: wearable antennas; textennas; garments; pattern software; fashion design software; integration

Citation: Papachristou, E.; Anastassiu, H.T. Application of 3D Virtual Prototyping Technology to the Integration of Wearable Antennas into Fashion Garments. *Technologies* **2022**, *10*, 62. https://doi.org/10.3390/technologies10030062

Academic Editors: Manoj Gupta, Eugene Wong and Gwanggil Jeon

Received: 6 April 2022
Accepted: 11 May 2022
Published: 17 May 2022

1. Introduction

Wearable antennas have been a topic of interest for more than a decade, with applications of a very broad scope, including security, health, sports, communications, glamor, etc. Hundreds of articles have been published in the literature, and a comprehensive review discussing all aspects of wearable antenna applicability is presented in [1]. Actual products based on this particular technology range from biometric insoles and interactive belts to connected T-shirts and Bluetooth jewelry. On the basis of their functionality and structure, wearable antennas may be categorized into two generic types: rigid and flexible. Typical rigid structures are used for off-body radio links, such as in smart watches [2] and life jackets [3], or even on-body configurations, firmly attached to the garment, such as in a military badge [4]. A wider variety of rigid antenna designs may be found in [5]. On the other hand, flexible antennas can easily be worn by a human body, which naturally moves, causing clothing deformation; however, such radiating structures are much more difficult to incorporate to a garment. This particular problem concerning textile antenna (*textenna*) manufacturing is the main focus of [6].

According to [6], textenna manufacturing may be implemented through (1) thin conductive layers attached to dielectric textiles; (2) woven or knitted conductive textile yarns attached or stitched onto the non-conductive textile substrate; (3) conductive textile yarns embroidered on the non-conductive textile substrate; and (4) inkjet and screen printing on non-conductive textile materials. Advantages and disadvantages of all four methods are discussed in [6]. Specifically, the third method (embroidery) is found to be preferable to the rest because computerized embroidery machines already exist in industry; so, it is easier to apply this technique for the mass production of garments with integrated embroidered textennas, allowing repeatable geometries to be made.

The aforementioned discussion raises the question of automated production of a large number of garment copies, where identical flexible antenna configurations are incorporated.

Obviously, suitable design-oriented software is necessary for this task. Moreover, an important factor should be taken into account, which is often neglected by engineers driven by a solely practical mentality: fashion and aesthetic design. Wearable antennas are not only expected to function well, but they should be comfortable and not awkward looking, like that shown in Figure 1. No matter how good a design is, from an electromagnetic point of view, very few people would be willing to carry such a device on their clothes.

Figure 1. A clumsy, wearable patch antenna attached to a shirt with a plastic, transparent sheet; computer simulation based on [6].

In this paper, realistic clothing design procedures are presented, incorporating antenna models to actual garments readily available in the market for sale. To this end, pattern and fashion design software tools are utilized, which are not well known to the engineering community, although much more widespread among design scientists. It is shown how an engineering approach may blend with a fashion designer's insight to produce garments combining elegance and efficiency. In Section 2, a selection of textennas amenable to such a procedure is presented. In Section 3, the operability of specific pattern and fashion design software tools is explained. In Section 4, examples of basic textenna patterns are integrated into actual garments, and the results are demonstrated. Section 5 discusses available options, whereas Section 6 summarizes the article and draws useful conclusions.

2. Challenges and Restrictions of Characteristic Textenna Types to Be Considered for Computer-Aided Garment Design

Several textenna configurations using woven, knitted or sewed conductive sheets/threads have been proposed in the literature. Reviewing all of them lies outside the scope of this paper; the interested reader may refer to [6]. In this section, only representative, simple designs are selected to show the fundamental concept of the applicability of fashion design software to antenna integration into garments.

The first ever compact fabric antenna design for commercial smart clothing was presented in [7]. It is a typical microstrip patch antenna intended for WLAN (wireless local area network) applications at a frequency equal to $f = 2.45$ GHz. The dimensions of the conductive patch shown in Figure 2 are $L = 56$ mm, $W = 51$ mm, whereas the ground plane dimensions are 76 mm and 71 mm, respectively. Conductive parts are made of knitted copper fabric, while the substrate is regular fleece with a relative permittivity equal to $\varepsilon_r = 1.04$, measured at $f = 2.45$ GHz.

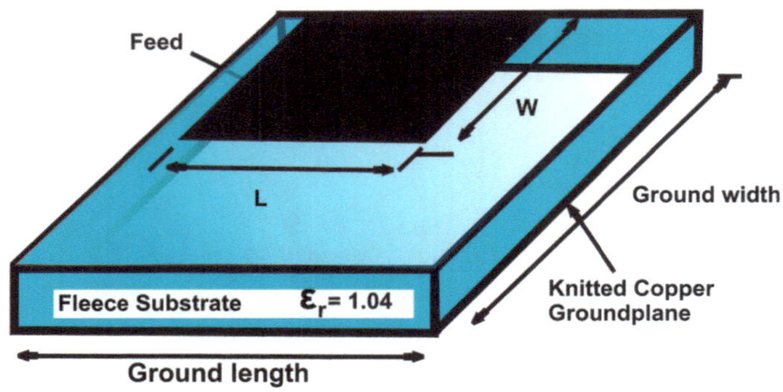

Figure 2. Geometry of the WLAN fabric antenna proposed in [7].

Apart from the antenna itself, additional instrumentation is necessary to facilitate actual radiation. For instance, a miniature feeding network for a patch antenna at f = 2.45 GHz is proposed in [8]. The main purpose is to reduce the length of the coupling aperture and the length of the stub to render the entire antenna structure more flexible and easier to handle. In order to compensate for performance deterioration due to miniaturization, periodic open conducting fingers are loaded to the coupling aperture, and a T-shaped structure is used at the end of the feed line (Figure 3). The substrate is made of nonconductive felt textile with relative permittivity and loss tangent equal to 1.3 and 0.044, respectively. The patch and the ground consist of ShieldIt superconductive textile with an estimated conductivity of 1.18×10^5 S/m, whereas FR-4 with permittivity 4.3 and loss tangent 0.025 is used as the feeding substrate.

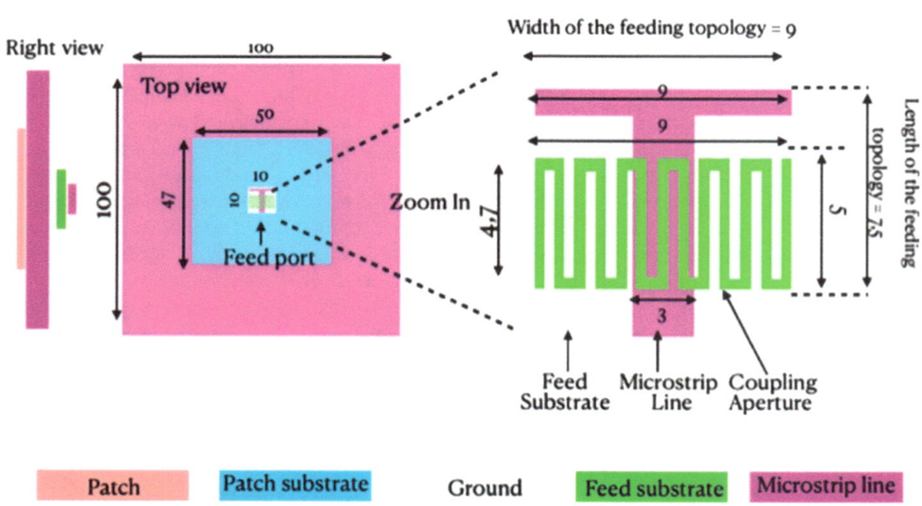

Figure 3. Feeding network of a patch antenna; design based on [8]. (**Left**): complete antenna. (**Right**): miniature feeding network. Dimensions are in mm.

More complicated geometries are presented in [9–11], where large, flexible wearable antennas are proposed for FM receivers (87–108 MHz) as an alternative to tunable small internal antennas. Shown in Figure 4, we chose a flexible third-order Minkowski fractal antenna that was designed to operate with land mobile radio systems at 136 MHz [11]. The work in [12] proposed a dual-band and dual-polarized button antenna for wearable body-centric communication (Figure 5). Button antenna can be easily integrated using copper as its conductive material, which is likely to outperform other more lossy conductive flexible materials in wearables [1].

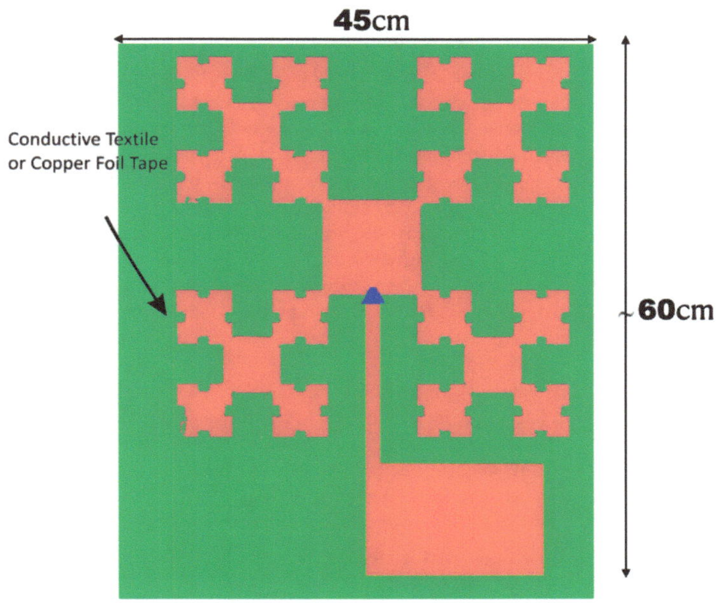

Figure 4. Large, flexible wearable antennas FM receiver (87–108 MHz); design based on [9].

Figure 5. Button antenna; design based on [12].

Although the aforementioned designs are easy to implement from a geometrical point of view, various restrictions should be taken into account prior to computer aided design (CAD) and manufacturing procedures. Corchia et al. [13] discuss several challenges of wearable antennas; among them, major ones are technology invisibility to the user and similar wearability guarantee of conventional clothes. Additionally, according to [14], operations such as washing and ironing are key aspects of antenna robustness. Additional factors were also considered herein that ensure stable antenna characteristics, such as (a) the antenna, combined with the supporting structure must be sufficiently flexible, but not drainable, (b) layouts are to be protected against stress [15] and crumbling [16]. Moreover, the literature suggests that (c) embroidered patch antennas, due to low-cost automatic embroidery machines that operate with conductive yarns, are highly popular [6]. Finally, the selection of the fabric material is also of great importance. A textile fabric may be described as a mixture of fibers, air, and water molecules [17,18]. Hertleer et al. [19] recommend hydrophobic fabrics, with low moisture regain (MR < 3%) as a rule of thumb.

The design process of integrating antennas into clothing will have to cope with all the above. Three-dimensional visualization tools for clothing design can assist in confronting some challenges when designing wearable antennas. According to [13], the robustness of the antenna performance to the operating scenario is one of these. More specifically, close proximity or direct contact with the human body can strongly affect the antenna performance. Thus, as proposed in [13], by adopting an appropriate design approach, this problem can be solved. The use of 3D software visualization has the potential to aid significantly in the development process [20,21]. Three-dimensional software visualization may transform the way that knowledge-gathering activities take place during software engineering phases [22,23].

By using digital prototyping tools, the design of such wearable items can be developed in a very fast and efficient manner, selecting different materials either from the software's fabric library or importing the digital representation of a specific material. This process ensures the design options for the decision-making process are numerous, resources are not wasted since everything is developed in a digital and virtual environment, and the visualization of the digital wearable is very close to the actual physical item when fabricated.

3. Description of Commercial Fashion Design Software Functionality

Below are presented the best-known commercial 3D solutions available, according to [24]. The list is believed to be fully up to date since it is based on recent developments, such as research by [25], the 5th edition of WhichPLM [26] focusing on 3D, a recent Ph.D. thesis on the effective integration of 3D prototype [21], the latest relevant Texprocess exhibition [27], and the 1st 3D Fashion Summit in Greece (2021) [28]. The following 3D systems are described below (in alphabetical order) based on the previous research:

Accumark3D (Gerber) [29]
CLO3D [30]
Modaris 3D (Lectra) [31]
Optitex3D [32]
Style3D [33]
V-Stitcher (Browzwear) [34]
Tukatech [35]

Accumark3D

Accumark3D is a fully integrated 3D tool to Accumark2D CAD system and Ynique-PLM. Like V-Stitcher, Accumark3D uses the powerful opensource simulation engine, Blender, a widely used technology tool in animation, movie, video game and simulation industries. The 3D tool for virtual sampling aims at assisting apparel companies to reduce time and cost of development and sample making. It is also possible to generate 3D renderings, incorporating parametric, fully customizable materials chosen from the 6000+ strong Substance Source library [36].

CLO3D

CLO3D is a commercial solution allowing the creation of the virtual fit process by inputting 2D patterns and virtually sewing them on a 3D digital human model (avatar). According to [37], users can visualize the fit of the garment in 3D at the time of sketching. This platform also encompasses a rich library of more than 900 "digital twins" of physical fabrics to eliminate excess waste [38].

Lectra Modaris 3D

Modaris classic and 3D are used for all stages of pattern development, from initial digitization to 3D virtual prototyping. According to [39], Modaris Pattern Cutting software is Lectra's leading solution which integrates on the same platform with the Modaris 3D Virtual Try-on of 2D garments' patterns. Like the rest of the 3D software tools mentioned in this section, this one also contains industry specific data libraries with more than 300 fabric samples [40].

Optitex3D

Pattern Design Software (PDS) 3D is the the name of Optitex's 3D virtual sample generator, fully integrated with the PDS 2D digital pattern solution The user can develop a new pattern, edit an existing one from the database or import a pattern file from another system in .dxf, .asthma and .aama formats. Similarly, the system provides an ingrate digital fabric library, but the user can measure and simulate new fabric in 3D based on its physical and visual properties.

Style3D

Style3D is a 3D software tool created by Linctex. Fabric management is performed through the Style 3D fabric solution, where existing materials can be scanned to produce photorealistic digital swatches that can be utilized for the same purpose as digitally designed fabrics.

V-Stitcher (Browzwear)

V-Stitcher by Browzwear is the digital tool for 3D fashion development aimed at pattern makers, cutters and technical designers. Lotta solution is suitable for designers, V-Stitcher for pattern makers and manufacturers, and Stylezone is a cloud platform for showcasing 3D designs on web and mobile. Browzwear's Fabric Analyzer (FAB) is part of the expanding digital ecosystem that gives users the ability to determine all physical properties of any fabric, from its thickness to the stretch and bend.

Tuka3D (Tukatech)

Tuka3D is the virtual prototyping making software system from Tukatech. It provides customized virtual fit models and builds life-like virtual clothing samples [41]. What is new in the 2022 version is that it offers an open system that allows designers, brands, retailers and their factories to work efficiently within a virtual process. This means that users of other 3D systems can start their workflow with 266 actual replica avatars from Tukatech's extensive library of over 700 models [42].

4. Application of Commercial Fashion Design Software to Textenna Integration into Garments

After investigating all the aforementioned 3D software solutions for prototype fashion modeling, the authors used CLO3D and V-Stitcher (Browzwear) to design a small collection of fashion garments that not only satisfy the needs of the wearer, but, most importantly, integrate popular textennas into their initial conceptual design. Moreover, the authors ensured that the integration of the chosen antennas was performed with special care for the characteristics of supporting structure, followed principles of construction and sewing assembly, selected appropriate fabric material (hydrophobic), took into consideration wearability and technology invisibility to the wearer, and lastly applied the selected design textennas to contemporary garment designs that can already be seen in several clothing brands worldwide.

Briefly, the process of the proposed antenna-equipped collection to be developed consisted of the following steps: (a) importing .dxf 2D CAD pattern files, (b) selecting an avatar in the appropriate size of the 2D pattern, (c) drawing/designing the selected antennas' outline in a 2D digital design software tool, (d) importing and tracing the .dxf outline of the antenna in the 3D prototyping tool, (e) virtually stitching the 2D patterns along with the designed antenna, (f) selecting fabrics from the preset library or scanning fabric/material with special scanning tools, (g) placing around the avatar, and finally (h) simulating and correcting the fit. The output file obviously contains implicitly all information related to the fabric mechanical and electromagnetic parameters of the materials involved in the design. Furthermore, it can be extracted in various formats, including .obj, meaning that it may be imported as input to widely used electromagnetic simulation software tools, such as CST EM Studio.

Shown in Figure 6, the patch antenna proposed in [8] was integrated into a dress, complete with its feeding network, including a T-shaped structure at the end of the feed line as shown in Figure 3 Due to the fact that this antenna consists of five layers and the actual miniature feeding network is of very small size, it was decided to be added as an interlining to the actual fabric of the garment. Therefore, it was integrated into the lower part of the proposed short-sleeved A-line dress. The size of the antenna was not adjusted to the pattern size, but the pattern was developed according to the size of the antenna instead. Figure 6A shows the virtual fitted dress without the embedded antenna, Figure 6B shows the dress with one patch antenna sewn internally at the lower part of the dress's hem and Figure 6C shows the possibility of this patch antenna to be placed in an antenna array (for instance, a 2X2 one). Figure 6D–F depicts a proposed style with digital textile design as an all-over print. When a textile all-over print is chosen, the somehow "bulky" look of the embedded antenna almost disappears. Figure 6G shows a closer look of the embedded antenna array; Figure 6H is a view of the inside garment by applying transparency to the back patterns; Figure 6I shows the placement of all the pattern pieces around the avatar with the virtual stitches applied (even on the antenna's five layers); and finally Figure 6J shows the actual construction and virtual sewing of these five layers of the antenna.

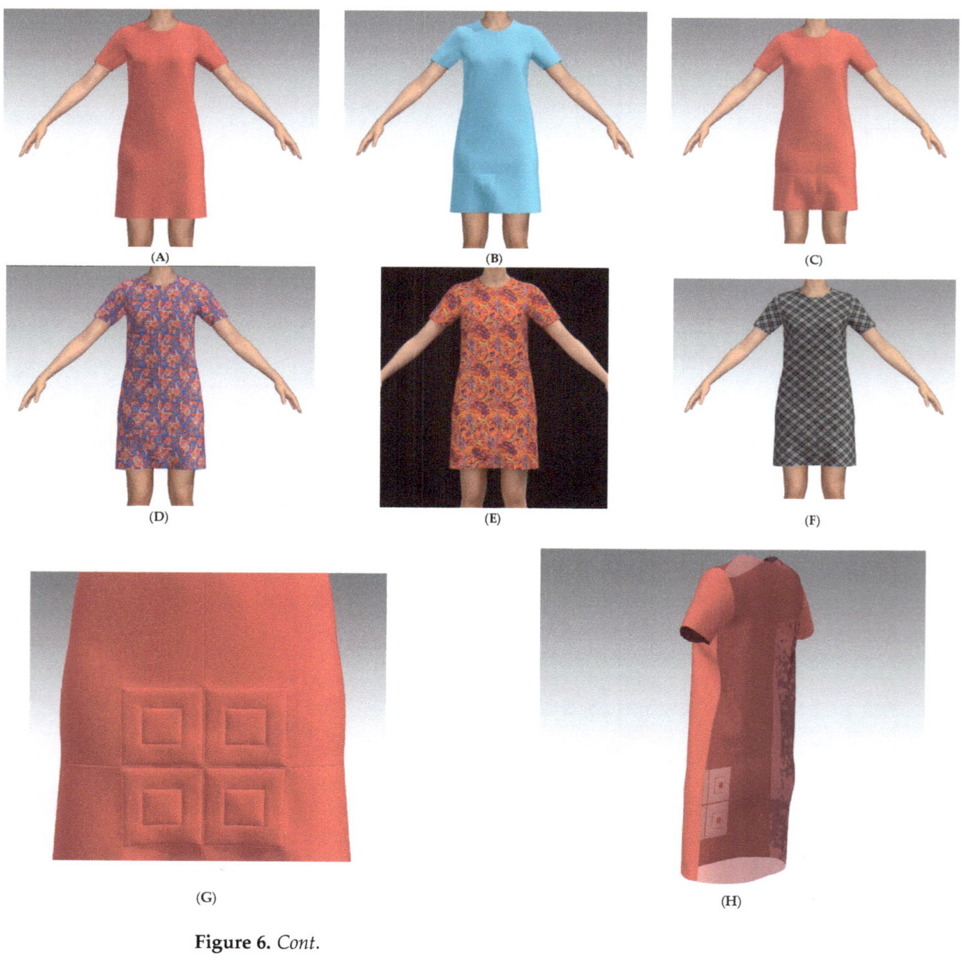

Figure 6. *Cont.*

(I)

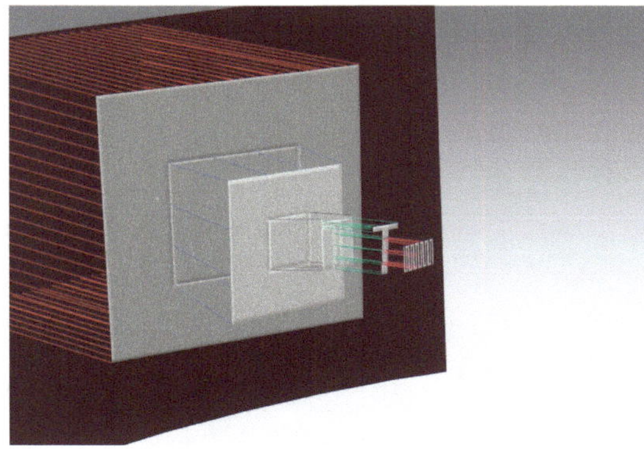

(J)

Figure 6. The patch antenna shown in Figure 3 [8] integrated in the lower part of the proposed short-sleeved A-line dress. Subfigure content is explained in the text.

Of course, placement of the antennas at that particular position of the dress is only indicative of the capabilities of the prototyping technology utilized. Depending on the actual radiation requirements, the antennas may be equally incorporated to any location, provided that there is enough space, for example, on the chest or the back, where deformation due to crumpling would be minimal.

The second garment is a women's bomber jacket. In this style, we integrated the flexible third-order Minkowski fractal antenna shown in Figure 4. Due to its large size, an appropriate pattern piece that accommodates the layout is the back of a garment. Figure 7A shows an antenna design integrated as an embroidered piece on top of the already virtually sewn jacket. Figure 7B shows the rendered design from five different angles, all in 3D virtual visualization software. In Figure 7C, a textile all-over print is applied, drawn from the 3D's software material library. The pattern piece of the antenna was colored with one of the textile's print pigments.

The third garment is a men's sleeveless zip jumper. Again, the integration involves the antenna shown in Figure 4. The back pattern piece of a menswear's jacket is large enough to host it. In Figure 8, the worn jacket is shown from five different angles and is followed by a similarly colored tracksuit trouser pant in a design that creates a smooth aesthetic combination of top and bottom garments.

The fourth garment is a denim-styled jacket with pockets positioned on the bust and metallic buttons as front fastening. In Figure 9, the patterns of the jacket in 2D are presented on the left, and the same pattern pieces positioned around the 3D avatar on the right. The lines/threads that connect the pattern pieces together are the virtual visualization of seams. After this preparation, the user "dresses" the avatar in the chosen body position. In this fashion garment, the antenna is hidden behind the metallic button; therefore, it ensures the invisibility to the wearer of the garment (as proposed in the work of Zhang et al. [12]). In V-Stitcher's interface, as can be seen in Figure 10, buttons can be easily imported as digital images with maps included and the user can insert/change several physical parameters (i.e., transparency, sheen, metallic hue, and diffusion) in order to create a 3D virtual vision of actual trims and other materials.

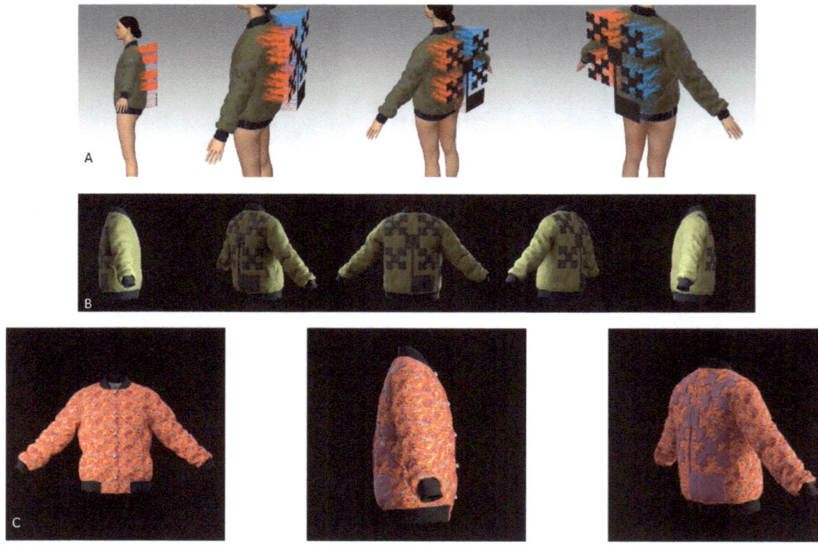

Figure 7. A third-order Minkowski fractal antenna integrated into a women's bomber jacket.

Figure 8. A third-order Minkowski fractal antenna integrated in a men's sleeveless zip jumper.

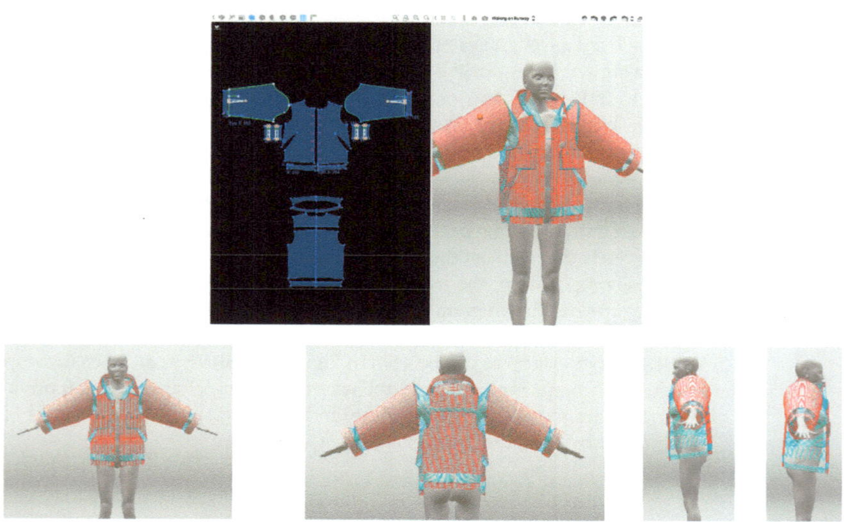

Figure 9. A denim-styled jacket with pockets positioned on the bust and metallic buttons as front fastening.

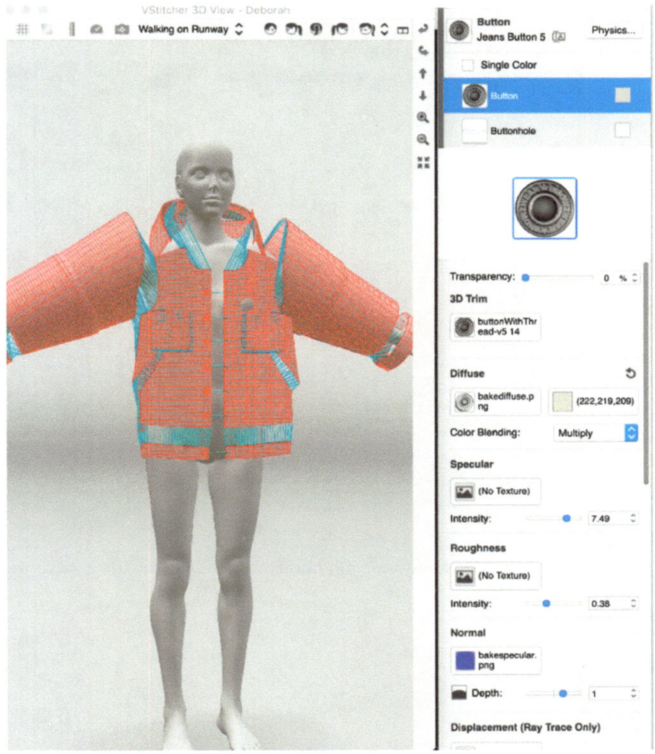

Figure 10. Buttons imported in the garment of Figure 10 as digital images.

5. Discussion

As already mentioned, the literature emphasizes the great importance of fabric material selection when integrating textennas into garments. Three-dimensional virtual prototyping technology tools, such as the ones used for this study, offer a full library of fabric materials with all their physical characteristics being fully parametric. The fabrics, as shown on the lefthand side of Figure 11, are categorized based on their fiber origin (i.e., natural fibers, polymer fibers, and synthetic). Virtual prototyping tools for typical garment visualization do not include woven fabrics with inserted metallic yarns or with plasma coating in their default libraries. Understandably enough, electromagnetic properties of the various fabrics are not explicitly shown in the menu since prototyping software was not originally intended for electromagnetic simulations. However, the user can create new libraries, especially for highly conductive materials, or for dielectrics with specific permittivity, after importing them via specialized hardware tools involving material scanning technologies or applications with embedded AI technologies. Examples of these devices are xTex by Vizoo [43] for scanning physically based samples of up to A4 paper size, and Scanatic™ Nuno Fabric Scanner [44] as a more affordable solution for a speedy material scanning. For the jacket in Figure 9, featuring an antenna integrated in the metallic buttons, authors used hydrophobic fabric (100% polyester). The result of the final visualization (render) of the denim-style womenswear jacket with embedded antenna behind the metallic buttons can be seen in various posture positions in Figure 12.

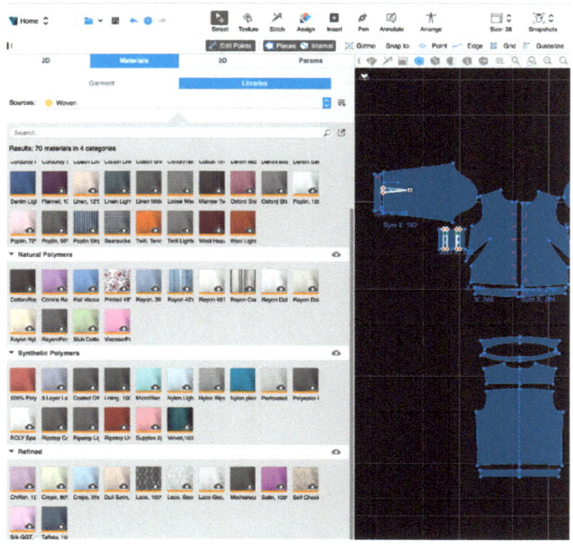

Figure 11. Fabric selection library, expandable by the user according to antenna design needs.

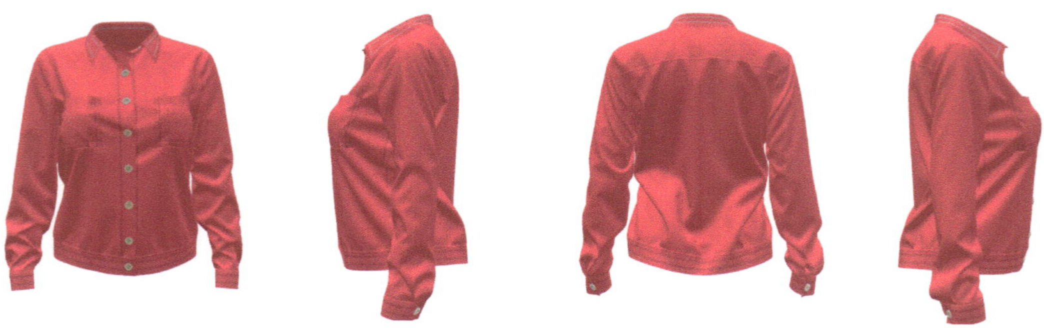

Figure 12. Final visualization (render) of the denim-style womenswear jacket with embedded antenna behind the metallic buttons (Figures 10 and 11) seen in various posture positions.

6. Conclusions

In this paper, the integration of wearable antennas into actual, comfortable and handsome clothes was discussed. Pure engineering design naturally focuses on electromagnetic efficiency of the antenna but often neglects the aesthetic dimension of a garment. A selection of commercially available software modules was presented, whose main functionality is fashion and pattern design. It was demonstrated how textennas are possible to incorporate into the fabric of various types of garments, by utilizing these software modules, without altering the antenna parameters, yet maintaining the elegance and reproducibility of the garment. Several antenna types already proposed and tested in the literature were selected and incorporated into actual clothes worn by human-like avatars. Even miniaturized feeding networks were possible to include in the design. The output files may be imported to commercial electromagnetic simulation packages for overall antenna performance evaluation. The prototyping procedure is important since it embeds often unattractive antenna structures into pleasant-looking clothes and offers an opportunity for antenna simulation in the presence of real-life garments. Finally, this procedure offers unlimited design options and minimizes resources waste since the entire development is completely digital, taking place in a virtual environment.

Author Contributions: Conceptualization, H.T.A.; methodology, E.P. and H.T.A.; software, E.P.; validation, E.P. and H.T.A.; formal analysis, E.P. and H.T.A.; investigation, E.P. and H.T.A.; resources, E.P. and H.T.A.; data curation, E.P. and H.T.A.; writing—original draft preparation, E.P. and H.T.A.; writing—review and editing, E.P. and H.T.A.; visualization, E.P. and H.T.A.; supervision, H.T.A.; project administration, H.T.A.; funding acquisition, E.P. and H.T.A. All authors have read and agreed to the published version of the manuscript.

Funding: This research has been co-financed by the European Regional Development Fund of the European Union and Greek national funds through the Operational Program Competitiveness, Entrepreneurship and Innovation, under the call RESEARCH—CREATE—INNOVATE (project code: T1EDK-03464).

Institutional Review Board Statement: Not applicable.

Informed Consent Statement: Not applicable.

Data Availability Statement: Not applicable.

Conflicts of Interest: The authors declare no conflict of interest.

References

1. Paracha, K.N.; Abdul Rahim, S.K.; Soh, P.J.; Khalily, M. Wearable Antennas: A Review of Materials, Structures, and Innovative Features for Autonomous Communication and Sensing. *IEEE Access* **2019**, *7*, 56694–56712. [CrossRef]
2. Su, S.W.; Lee, C.T. Metal-Frame GPS Antenna for Smartwatch Applications. *Prog. Electromagn. Res. Lett.* **2016**, *62*, 41–47. [CrossRef]
3. Serra, A.A.; Nepa, P.; Manara, G. A Wearable Two-Antenna System on a Life Jacket for Cospas-Sarsat Personal Locator Beacons. *IEEE Trans. Antennas Propag.* **2012**, *60*, 1035–1042. [CrossRef]
4. Joler, M.; Boljkovac, M. A Sleeve-Badge Circularly Polarized Textile Antenna. *IEEE Trans. Antennas Propag.* **2018**, *66*, 1576–1579. [CrossRef]
5. Pettitt, G.; Matthews, J.C.G.; Tyler, A.J.; Pirollo, B.P. Wide-Band Body Wearable Antennas. *BAE Syst. Def. Sci. Technol. Lab. IET* **2008**, *2008*, 111–127.
6. Tsolis, A.; Whittow, W.G.; Alexandridis, A.A.; Vardaxoglou, J. Embroidery and Related Manufacturing Techniques for Wearable Antennas: Challenges and Opportunities. *Electronics* **2014**, *3*, 314–338. [CrossRef]
7. Salonen, P.; Hurme, H. A Novel Fabric WLAN Antenna for Wearable Applications. In Proceedings of the IEEE Antennas and Propagation Society International Symposium, Columbus, OH, USA, 22–27 June 2003; Volume 2, pp. 700–703.
8. Zhang, J.; Yan, S.; Vandenbosch, G.A.E. A Miniature Feeding Network for Aperture-Coupled Wearable Antennas. *IEEE Trans. Antennas Propag.* **2017**, *65*, 2650–2654. [CrossRef]
9. Nepa, P.; Rogier, H. Wearable Antennas for Off-Body Radio Links at VHF and UHF Bands: Challenges, the State of the Art, and Future Trends below 1 GHz. *IEEE Antennas Propag. Mag.* **2015**, *57*, 30–52. [CrossRef]
10. Roh, J.-S.; Chi, Y.S.; Lee, J.H.; Tak, Y.; Nam, S.; Kang, T.J. Embroidered wearable multi resonant folded dipole antenna for FM reception. *IEEE Antennas Wirel. Propag. Lett.* **2010**, *9*, 803–806. [CrossRef]
11. Lee, E.C.; Soh, P.J.; Hashim, N.B.M.; Vandenbosh, G.A.E.; Volski, V.; Adam, I.; Mirza, H.; Aziz, M.Z.A.A. Design and fabrication of a flexible Minkowski fractal antenna for VHF applications. In Proceedings of the European Conference Antennas Propagation, Barcelona, Spain, 12–16 April 2010; pp. 521–524.
12. Zhang, X.Y.; Wong, H.; Mo, T.; Cao, Y.F. Dual-Band Dual-Mode Button Antenna for On-Body and Off-Body Communications. *IEEE Trans. Biomed Circuits Syst.* **2017**, *11*, 933–941. [CrossRef] [PubMed]
13. Corchia, L.; Monti, G.; Tarricone, L. Wearable Antennas: Nontextile versus Fully Textile Solutions. *IEEE Antennas Propag. Mag.* **2019**, *61*, 71–83. [CrossRef]
14. Khaleel, H. *Innovation in Wearable and Flexible Antennas*; WIT Press: Southampton, UK, 2015.
15. Sanjari, H.; Merati, A.; Varkiani, S.; Tavakoli, A. A study on the effect of compressive strain on the resonance frequency of rectangular textile patch antenna: Elastic and isotropic model. *J. Text. Inst.* **2014**, *105*, 156–162. [CrossRef]
16. Bai, Q.; Langley, R. Crumpling of PIFA textile antenna. *IEEE Trans. Antennas Propag.* **2012**, *60*, 63–70. [CrossRef]
17. Declercq, F.; Couckuyt, I.; Rogier, H.; Dhaene, T. Environmental high frequency characterization of fabrics based on a novel surrogate modeling antenna technique. *IEEE Trans. Antennas Propag.* **2010**, *61*, 5200–5213. [CrossRef]
18. Bal, K.; Kothari, V.K. Permittivity of woven fabrics: A comparison of dielectric formulas for air-fiber mixture. *IEEE Trans. Dielect. Electr. Insul.* **2010**, *17*, 881–889. [CrossRef]
19. Hertleer, C.; van Laere, C.; Rogier, H.; van Langenhove, L. Influence of relative humidity on textile antenna performance. *Text. Res. J.* **2010**, *80*, 177–183. [CrossRef]

20. Teyseyre, A.R.; Campo, M.R. An Overview of 3D Software Visualization. *IEEE Trans. Vis. Comput. Graph.* **2009**, *15*, 87–105. [CrossRef] [PubMed]
21. Papachristou, E. Effective Integration of 3D Virtual Prototype in Product Development of Textile and Clothing Industry. Ph.D. Thesis, School of Production Engineering & Management, Technical University of Crete, Chania, Greece, 2016.
22. Knight, C. System and Software Visualization. In *Handbook of Software Engineering and Knowledge Engineering;* World Scientific: Singapore, 2000.
23. Young, P.; Munro, M. Visualizing Software in Virtual Reality. In Proceedings of the Sixth Int'l Workshop Program Comprehension (IWPC '98), Ischia, Italy, 24–26 June 1998; p. 19.
24. Sayem, A.S. Virtual Prototyping for Fashion 4.0. In *Industry 4.0. Shaping the Future of the Digital World,* 1st ed.; da Silva Bartolo, P.J., da Silva, F.M., Jaradat, S., Bartolo, H., Eds.; Taylor & Francis Group: Manchester, UK, 2020; pp. 193–196.
25. Sayem, A.S.M.; Kennon, R.; Clarke, N. 3D CAD systems for the clothing industry. *Int. J. Fash. Des. Technol. Educ.* **2009**, *3*, 45–53. [CrossRef]
26. WhichPLM. *The WhichPLM Report, The 3D Issue,* 5th ed.; WhichPLM Limited: Lancashire, UK, 2015.
27. TexProcess. Available online: https://texprocess.messefrankfurt.com/frankfurt/en.html (accessed on 20 February 2022).
28. 3D Fashion Summit. Organised by International Hellenic University (Department of Creative Design & Clothing) & SEPEE (Hellenic Clothing Association), Chaired by Dr. Evridiki Papachristou (Assistant Professor IHU). Available online: https://www.ihu.gr/event/3d-fashion-summit. (accessed on 20 May 2021).
29. Modaris 3D. Available online: https://www.gerbertechnology.com/fashion-apparel/design/accumark-3d/ (accessed on 20 February 2022).
30. Clo3D. Available online: https://www.clo3d.com/ (accessed on 20 February 2022).
31. Lectra. Available online: https://www.lectra.com/en/products/modaris-expert (accessed on 20 February 2022).
32. Optitex. Available online: https://optitex.com/products/2d-and-3d-cad-software/ (accessed on 20 February 2022).
33. Style 3D. Available online: https://www.linctex.com/ (accessed on 20 February 2022).
34. V-Stitcher. Available online: https://browzwear.com/ (accessed on 20 February 2022).
35. Tuka 3D. Available online: https://tukatech.com/ (accessed on 20 February 2022).
36. The Interline, From Render to Real: Delivering on the Promise of Digital Design to On-Demand Production. The Interline (7 June 2021). Available online: https://www.theinterline.com/06/2021/from-render-to-real-delivering-on-the-promise-of-digital-design-to-on-demand-production/ (accessed on 20 February 2022).
37. Gupta, D. New directions in the field of anthropometry, sizing and clothing fit. In *Anthropometry, Apparel Sizing and Design;* Norsasdah, Z., Gupta, D., Eds.; The Textile Institute, Woodhead Publishing: Duxforth, UK, 2020.
38. Taylor, G. How this New Digital Fabric Library Cuts the Headaches of Textile Sampling. Sourcing Journal. 19 March 2021. Available online: https://sourcingjournal.com/topics/technology/swatchon-clo-digital-fabrics-sampling-visual-search-3d-design-textile-269153/ (accessed on 20 February 2022).
39. Grice, P. *Digital Pattern Cutting for Fashion with Lectra Modaris: From 2D Pattern Modification to 3D Prototyping;* Bloomsbury: Bloomsbury, UK, 2019.
40. Papachristou, E.; Bilalis, N. Should the fashion industry confront the sustainability challenge with 3D prototyping technology? *Int. J. Sustain. Eng.* **2017**, *10*, 207–214. [CrossRef]
41. Jhanji, Y. *Computer-Aided Design—Garment Designing and Patternmaking in Automation in Garment Manufacturing;* Elsevier, Woodhead Publishing: Duxforth, UK, 2018; pp. 253–290.
42. Apparel Resources, Tukatech Launches "Tuka3D 2022" That Eliminates the Need for Making FIT Samples. Apparel resources News Desk. 19 November 2021. Available online: https://vn.apparelresources.com/technology-news/manufacturing-tech/tukatech-launches-tuka3d-2022-eliminates-need-making-fit-samples/ (accessed on 20 February 2022).
43. xTex. Available online: https://www.vizoo3d.com/xtex-software (accessed on 20 February 2022).
44. Scanatic Nuno Fabric. Available online: https://www.tg3ds.com/3d-digital-fabric-scanner (accessed on 20 February 2022).

technologies

MDPI

Article

A Comparative Analysis on Suicidal Ideation Detection Using NLP, Machine, and Deep Learning

Rezaul Haque [1], Naimul Islam [1], Maidul Islam [2] and Md Manjurul Ahsan [3,*]

[1] Department of Computer Science and Engineering, East West University, Dhaka 1212, Bangladesh; rezaulh603@gmail.com (R.H.); naimul.islam.pulak@gmail.com (N.I.)
[2] Department of Aerospace Engineering, RMIT University, Melbourne, VIC 3000, Australia; s3801461@student.rmit.edu.au
[3] School of Industrial & Systems Engineering, University of Oklahoma, Norman, OK 73019, USA
* Correspondence: ahsan@ou.edu

Abstract: Social networks are essential resources to obtain information about people's opinions and feelings towards various issues as they share their views with their friends and family. Suicidal ideation detection via online social network analysis has emerged as an essential research topic with significant difficulties in the fields of NLP and psychology in recent years. With the proper exploitation of the information in social media, the complicated early symptoms of suicidal ideations can be discovered and hence, it can save many lives. This study offers a comparative analysis of multiple machine learning and deep learning models to identify suicidal thoughts from the social media platform Twitter. The principal purpose of our research is to achieve better model performance than prior research works to recognize early indications with high accuracy and avoid suicide attempts. We applied text pre-processing and feature extraction approaches such as CountVectorizer and word embedding, and trained several machine learning and deep learning models for such a goal. Experiments were conducted on a dataset of 49,178 instances retrieved from live tweets by 18 suicidal and non-suicidal keywords using Python Tweepy API. Our experimental findings reveal that the RF model can achieve the highest classification score among machine learning algorithms, with an accuracy of 93% and an F1 score of 0.92. However, training the deep learning classifiers with word embedding increases the performance of ML models, where the BiLSTM model reaches an accuracy of 93.6% and a 0.93 F1 score.

Keywords: suicide ideation; text classification; machine learning; NLP; text pre-processing; deep learning

Citation: Haque, R.; Islam, N.; Islam, M.; Ahsan, M.M. A Comparative Analysis on Suicidal Ideation Detection Using NLP, Machine, and Deep Learning. *Technologies* **2022**, *10*, 57. https://doi.org/10.3390/technologies10030057

Academic Editors: Manoj Gupta, Eugene Wong and Gwanggil Jeon

Received: 30 March 2022
Accepted: 27 April 2022
Published: 29 April 2022

1. Introduction

Suicide is regarded as one of today's most serious public health issues. Around 0.7 million individuals die every year, and many more, particularly the young and middle-aged, attempt suicide [1]. For persons aged 10 to 34, it is the second-largest cause of death [2]. Suicidal ideation affects individuals of all ages all over the globe due to shock, rage, guilt, and other symptoms of melancholy or anxiety. Suicidal ideation, often known as suicidal thoughts, refers to the conceptualizing or meanderings about terminating one's life. Although most people who have suicidal thoughts do not attempt suicide, long-term depression may lead to suicide if depressed persons do not get effective counseling [3]. Their suicidal ideation can be cured with the help of healthcare experts and drugs, but most of them shun medical treatments owing to societal stigma. Instead, individuals prefer to convey their suicidal intentions via social media. However, since mental illness can be detected and addressed, early detection of indications or risk factors may be the most effective strategy to avoid suicidal ideation.

Over the years, it has been discovered that online social media data, particularly tweets, include predictive information for various mental health disorders, including depression and suicide. The information provided on Twitter can be helpful in analyzing people's suicidal thoughts. However, with the widespread usage of internet and technology, the number of tweets has been increasing explosively. It will be very challenging and time consuming for us to go through these tweets and identify people with suicidal ideations. Early detection of suicidal ideations from tweets will help medical experts identify the suicidal intentions of an individual and provide appropriate treatment. In general, an automated suicidal ideation detection system would allow medical professionals to save many lives by detecting the early symptoms of depression from online tweets.

Sentiment Analysis (SA) is a growing field that automatically captures user sentiment [4,5]. We can recognize early suicidal thoughts and avert the majority of suicide attempts by properly using a combination of information from social media and SA. As a result, Machine Learning (ML) and Natural Language Processing (NLP) have arisen as techniques for predicting suicidal intent from social media data. In addition, Deep Learning (DL) architectures provide significant advantages for detecting suicidal thoughts since they perform at very high accuracy with lower-level engineering and processing. Prior research papers that have identified suicidal ideations from tweets using ML algorithms have conducted their studies on a limited dataset. The research of [6] utilized several ML models to conduct depression detection on a collection of 15,000 tweets. Due to the small amount of data, their ML models suffered from poor accuracy. A similar work can be shown by [7], where the author improved the performance of ML classifiers on a dataset of 50,000 tweets. They collected the tweets from news articles and websites using several keywords and manually labeled them to perform binary classification. In [8], the authors proposed an automatic depression detection system using ML models where the dataset was created from a Russian social networking site Vkontakte. All of these papers were unable to achieve a good accuracy score as their focus was on training ML algorithms on a small collection of tweets. With the use of appropriate annotation rules on a huge number of tweets and training DL models, it is possible to improve the classification score of ML models. Previously, DL classifiers have been used to identify the suicidal intentions of social media users with great accuracy [9]. Most of their research was based on humanly annotated datasets, collected from different suicide forums [10] and subreddits [11]. However, tweets are different in nature compared to Reddit or suicide forum posts, as users only get 280 characters to interact with others. The authors in [12] proposed a novel attention-based relational network that can identify mental disorders from 4800 tweets labeled into four classes with an accuracy score of 83%. However, the results can be further improved by collecting more tweets and using appropriate preprocessing and feature extraction techniques.

As discussed above, for the task of suicidal ideation detection from Twitter, most of the studies have focused on implementing only ML algorithms, which results in poor accuracy scores. In addition, DL algorithms were used in research on datasets of Reddit, suicide forums, or a small number of tweets. There is no work on a dataset of around 50,000 tweets, where DL classifiers attain great accuracy. Furthermore, there is still a lack of comparative analysis between the performances of ML and DL classifiers for the task of identifying suicidal ideation in live Tweets. The main contribution of this research is tackling these gaps by performing an experimental study to evaluate the performance of five ML and four DL classifiers on a dataset of around 50,000 tweets labeled as 'suicide' and 'non-suicide'. The primary purpose of our study is to use effective NLP and feature extraction techniques to train several ML and DL models to identify suicidal thoughts on Twitter and provide a comparative analysis between the performance of the classifiers. To our knowledge, this is the first research where around 50,000 tweets have been used to conduct experiments on ML and DL classifiers for the task of suicidal ideation detection. Our study represents that a DL model can outperform typical ML methods for the task of suicidal ideation detection if text pre-processing is appropriately executed. Our contributions are mentioned below:

- Using the Tweepy API [13], we built a dataset of 49,178 tweets gathered live from Twitter with 18 keywords associated with suicidal ideation. Furthermore, we annotated the tweets using VADER and TextBlob, labeling them as non-suicidal or suicidal;

- Tweets include informal language; thus, we used NLTK packages to clean the noisy text data to make the user's content seem more evident and enhance the text analysis. Feature extraction techniques such as CountVectorizer and word embedding were used for ML and DL, respectively, which helped lead to more accurate suicidal detection;
- Several DL classifiers such as Long-Short Term Memory (LSTM), Bi-directional LSTM (BiLSTM), Gated Recurrent Unit (GRU), Bi-directional GRU (BiGRU), and combined model of CNN and LSTM (C-LSTM) were trained using Keras, a high-level API of TensorFlow. Furthermore, their performance was evaluated based on accuracy, AUC, precision, recall, and F1-score;
- The performance of DL was compared with traditional ML approaches such as Random Forrest (RF), Support Vector classifier (SVC), Stochastic Gradient Descent classifier (SGD), Logistic Regression (LR), and Multinomial Naive Bayes (MNB) classifier.

The work is organized into five more sections, in addition to this introduction. The second section lays out the related research works on suicidal ideation detection. Research methodology is discussed in the third section. Experimental findings and analysis are presented in the fourth section. The fifth section interprets the discussion and lastly, in the sixth and final section, the concluding remarks and future works are summarized.

2. Related Works

Sentiment analysis is regarded as one of the fastest-growing study subjects in the field of computer science. According to [14], sentiment analysis can be traced back to the surveys on public sentiment research at the end of the 20th century. In addition, test analysis was started by a group known as computational linguistics around the 1990s. Computer-based sentiment analysis has emerged more prominently with the availability of texts on the web. Apart from that, various fields in terms of identifying the underlying emotions from any text or voice message were improved massively because of the impact of text-availability on the web. Several works on sentiment analysis using several approaches can be found in the literature.

In recent years, much research has been completed to investigate the link between mental health and linguistic use to get novel insights into identifying suicidal thoughts. Previous works on detecting suicidal thoughts made use of language elements from the psychiatric literature, such as LIWC [15], emotion features [16], and suicide notes [17]. However, the fundamental disadvantage of this analysis approach is that it utilizes language-specific strategies that evaluate individual posts in isolation and will not perform well enough when dealing with diverse or enormous quantities of data.

The use of social media in combination with NLP for mental health research is becoming more popular among researchers. With its mental health-related forums, online social media data have been a growing research field in sentiment analysis, such as Michael M. Tadesse [18], who developed a combined model of LDA, LIWCA, bigram, and MLP to reach an accuracy of 90%. In [6–8], the authors used a similar strategy to collect data from Twitter and train multiple ML approaches to classify suicidal thoughts.

DL approaches as LSTM and CNN have already made significant progress in the area of NLP, thanks to the rising popularity of word embedding. Since ML methods have certain limitations, such as dimension explosion, data sparsity, and time consumption, they are unsuitable for all applications. DL considerably improves traditional ML techniques by extracting more abstract characteristics from input data by increasing the number of layers in the model, making the model's final classification information more consistent and accurate. The ability of DL models compared to other ML classifiers was shown in [19,20], where they obtained greater prediction accuracy using DL models for suicidal ideation detection. In [21], Tadesse et al. employed a CNN-LSTM combination model with word2vec to predict suicidal ideation with a 93.8% accuracy, owing to the model's ability to extract long-term global dependencies as well as local semantic information. However, their research was conducted on a small dataset of suicidal ideation content.

Despite its solid foundation, the existing research in depression detection lacks certain critical key factors. Research is scarce on a comparative study of suicidal ideation in which all traditional ML classifiers are compared to DL models such as BiLSTM, LSTM, GRU, BiGRU, and CLSTM that use word embedding for feature extraction with high accuracy performance. Furthermore, most of them conducted their research on limited datasets gathered from social media, which must be carefully pre-processed to attain better results.

3. Experimental Methods

3.1. Data Collection and Labeling

3.1.1. Data Collection

Data collection is the initial step in the analysis process since we need data to train our classifiers. However, the absence of a public dataset is one of the most significant obstacles in the field of suicidal ideation detection. Conventionally, it has been complex extracting data that are related to mental illnesses or suicidal ideation because of social stigma. However, a growing number of people are surfing the Internet to vent their frustration, to seek help, and to discuss mental health issues. We chose Twitter as our primary source of data since it has been shown to be effective in assessing mental conditions, such as suicidal ideation [22–24]. Here, the main idea is to collect different types of posts that are connected to suicide other than those that more directly express suicidal ideation. To maintain the privacy of the individuals in the dataset, we do not present direct quotes from any data or any identifying information. Furthermore, a unique ID is generated as a replacement of their personal information to preserve the users' privacy.

Tweepy, a tweet extraction API, allows us to query through historical tweets with tokenized terms. We required a list of suicide-related search phrases that could be used as a query to acquire raw Twitter data. Therefore, we came up with suicide-related phrases in two stages. Firstly, we looked at several suicide-related tweets. We became acquainted with expressions expressing suicidal thoughts by reading tweets, such as "want to die", "kill myself", and so on. We attempted to collect the phrases used frequently to indicate suicidal thoughts or suicidal ideation. Secondly, we looked through some suicide-related research papers. These publications provided us with useful information regarding suicidal expressions. We extracted a significant list of terms from Twitter based on the previous two processes, including keywords connected to suicide or self-harm. We collected real-time tweets using the Tweepy API using the final keywords list. We manually went through the gathered tweets and modified the terms' list by inserting, removing, and changing terms. We terminated the operation after discovering that the majority of the tweets retrieved were about suicides. It took about 2 to 3 weeks to compile the list of suicide-related key phrases. From 20 February 2021 to 13 May 2021, a total of 65,516 tweets with these phrases were extracted. The following are some of the comprehensive lists of suicide-related terms:

Anxiety disorder, depression, help me out, suffering, trapped, kill myself, suffering, sleep forever, my sad life, suicide, struggle, depressed me, stressed out, crying, want to die, emotionally weak, hate myself, burden.

3.1.2. Data Annotation

The sentiment of the tweets was not known when they were gathered. For example, while collecting tweets with a term, it was unknown whether the tweet was intended for suicide awareness and prevention, the individual was discussing suicide ideas such as ways to kill himself, the tweet reflected a third person's suicide, or the tweet utilized suicide as a narrative. While several of the gathered tweets included suicidal-related phrases, they may have been discussing a suicide film or campaign that did not convey suicidal thoughts. We annotated the gathered tweets in two stages. First, we annotated the tweets with VADER and TextBlob, a Python program that extracts sentiment polarity (positive, neutral, or negative) from the text. Then, using the annotation rules in Table 1, we manually reviewed and corrected the labeled tweets. In total, there were 65,516 tweets in our Twitter dataset, with 24,458 tweets (around 38%) containing suicidal thoughts. Due to the imbalanced nature of this dataset, the training dataset had a maldistribution of classes, resulting in poor predictive performance, particularly for the minority (suicidal) category. We avoided the imbalance problem by deleting 16,338 non-suicidal tweets. Finally, our dataset contained 49,178 tweets with suicidal and non-suicidal class accounting for 50.3% and 49.7%, respectively, shown in Figure 1.

Table 1. Data Annotation Rules.

Label	Rule	Examples
Suicidal	Expressing suicidal thoughts	'I've got nothing left to live for', 'I hate my life sometimes'
	Potential suicidal thoughts	'I want myself dead I hate men and I hate this planet', 'I want to shoot myself'
Non-suicidal	Discussing suicide	'He wanted to die his partner wanted to live seriously you must watch', 'This dudes life is worthless'
	Irrelevant to suicide	'Can this back pain just end please I'm tired', 'My boyfriend's phone is dying a slow death this would be great for him'

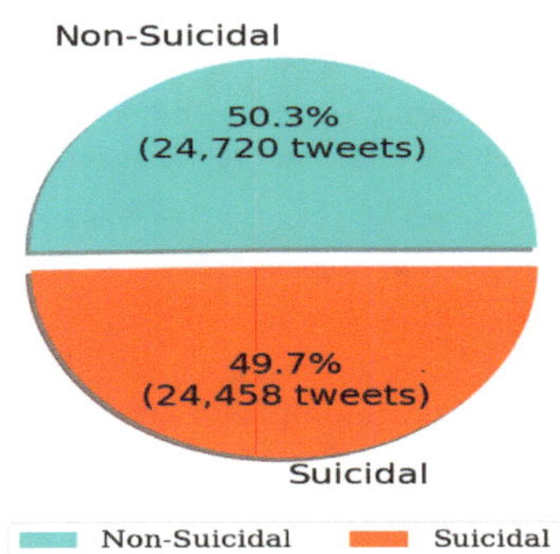

Figure 1. Class distribution of experimental dataset.

3.2. Pre-Processing

Preparing data entails converting raw data into a more usable format that can be fed into a classifier for improved performance. We correctly cleaned the textual data before executing the suicidal ideation detection task since most tweets included a significant noise. Figure 2 shows an example of a tweet under the suicidal class after performing all the pre-processing steps.

Figure 2. Example of an actual and pre-processed suicidal tweet.

3.2.1. Word Transformation

The majority of the tweets in our sample are made up of short conversational phrases and contractions. We used word segmentation to tokenize the text and replaced it with its complete form to turn them into meaningful words. For example, every tweet containing the phrase "AFAIK" was replaced with "As Far As I Know".

3.2.2. Removing Irrelevant Characters

ML models cannot comprehend nonsensical characters. Their presence in the text causes it to become noisy; thus, they must be deleted from tweets. Emojis, URLs, punctuation, whitespace, numerals, and user references were stripped from the text using regular expressions.

3.2.3. Stemming and Lemmatization

Stemming is a word-shortening approach that seeks to reduce a term to its root. Lemmatization is identical to stemming but combines vocabulary and morphological analysis to restore words to their dictionary form. We applied NLTK's Porter Stemmer and Wordnet Lemmatizer to perform stemming and lemmatization, which improved text categorization accuracy.

3.2.4. Stop Words Removal

A list of unimportant, frequently occurring words with little or no grammatical responsibility for text classification is known as a stop words list. We used NLTK's stop words corpus to eliminate them, to decrease the low-level information in our text and concentrate more on the relevant information. We also took out less frequently used words from the tweets.

3.3. Feature Extraction and Training

One dimensionality reduction approach used in ML is feature extraction, which maps higher dimensional data into a collection of low dimensional feature sets. Extracting valuable and crucial characteristics improves the performance of ML models while reducing computing complexity [25]. So, we will use feature extraction algorithms to turn text into a matrix (or vector) of possibilities. Word CountVectorizer and word embedding are two of the most successful feature extraction methods frequently utilized in ML and DL for text classification among all feature extraction approaches.

3.3.1. Count Vectorizer with ML Training

Text classification requires converting source documents into a vector representation. We implemented a word CountVectorizer using the unigram technique to vectorize our tweets by transforming the source texts into vector representations with the same length as the tweets and an integer count of the number of times a word occurred in each tweet. We obtained a lexicon of 36,121 unique words present in all tweets, which we fed into our vectorizer so that the ML models could conduct classification. To train our model, we split the data into two sets: training and testing, which were 80% and 20%, respectively. Various ML methods for performing suicidal ideation categorization are detailed in the related work. We used five ML algorithms, such as LR, SVC, RF, MNB, and SGD, to determine the existence of suicidal ideation among the users.

Logistic Regression employs a sigmoid function to convert the result to a probability. This minimizes the cost function to attain the best probability. The classifier was penalized using the 'l2' norm, stopping criteria with a tolerance of 0.0001, and the 'liblinear' optimizer for a maximum of 200 iterations. With a random state 42, the default value of the inverse of regularization strength is utilized. SVC estimates a hyperplane based on a feature set to categorize data points. We trained the SVC model with kernel type 'rbf', 0.0001 stopping criterion tolerance, and random state 42. One of the two fundamental Naive Bayes variations used in text classification is the MNB method, which implements the Naive

Bayes algorithm for multinomial distributed data. It presupposes that, given the class, the predictive qualities are conditionally independent, and it indicates that there are no hidden or underlying features that may impact the classification process. To train the MNB model, we employed Laplace smoothing and class prior probabilities based on the data. The RF algorithm is an ensemble-based approach that uses bagging techniques to train various decision trees. It has a low bias and a decent variance in prediction when it comes to categorization. Consequently, the algorithm can keep track of characteristics and predictors, and it is feasible to get excellent accuracy by utilizing it in text classification tasks. We used 200 trees for building a forest whose split quality was determined by 'entropy'.

3.3.2. Word Embedding with DL Training

Word embedding techniques, represented by DL, have recently received much attention and are now commonly employed in text classification. The Word2Vec word embedding generator seeks to discover the interpretation and semantic relationships between words by examining the co-occurrence of terms in texts within a specified corpus. This technique uses ML and statistics to model the proper context of words and generate a vector representation for each word in the dataset. With a post-padding of 60 vectors, we turned the encoded words of the tweets into a padded sequence. A matrix containing these padding sequences as input will be inserted into the embedding layer of 128 dimensions of the deployed DL models to translate the encoded textual comments to the correct word embeddings. We divided our dataset into training, validation, and testing sets of 80%, 10%, and 10%, respectively, where the model is trained on the training and validation sets.

RNN's implementation of the tree structure approach to extract the semantics of a phrase has been used to complete text classification, with good results in the previous research studies. However, developing a textual tree structure takes a long time for lengthy phrases of tweets, since the created model suffers from gradient vanishing and exploding. Two forms of RNN-based design approach, LSTM and GRU, were created, both of which include a gating mechanism to address the limitations of RNN [26]. It includes a 'forget' gate that allows the network to encapsulate longer-term relationships without encountering the vanishing gradient issue. GRUs are less complicated than LSTMs since they employ fewer parameters and do not need a memory unit [27]. The LSTM and GRU models both include bidirectional and directional approaches. To identify the best performing model for the subsequent trials, we used a few Deep Neural Network(DNN) architectures–BILSTM, LSTM, BIGRU, and CLSTM (a mix of LSTM and CNN). These models were trained using a 128 batch size, a memory unit of 128, Adam optimizer, a 0.0001 initial learning rate, and the Relu activation function. However, owing to the enormous set of parameters to be learned, DNNs are prone to overfitting. Due to noise, the learning accuracy of DNN models stops increasing or even worsens beyond a certain point. In order to minimize overfitting, we adjusted our network architecture and regularization parameters to fit the training data. In addition, we included ReduceLROnPlateau, an early stopping technique, which lowers the learning rate when the model stops improving. The proposed architecture of the CLSTM network is shown in Figure 3.

3.4. *Evaluation Matrices*

We opted to use the standard classification metrics, such as accuracy, precision, recall, f1 score, AUC, and confusion matrix, to evaluate the models. For binary classification tasks, such metrics are simple and can be obtained using Equations (1)–(4):

$$accuracy = \frac{TP + TN}{\mathrm{TP} + \mathrm{FP} + \mathrm{FN} + \mathrm{FP}} \quad (1)$$

$$precision = \frac{TP}{\mathrm{TP} + \mathrm{FP}} \quad (2)$$

$$recall = \frac{TP}{\mathrm{TP} + \mathrm{FN}} \quad (3)$$

$$f1 - score = \frac{2 * (\,precision * recall)}{\text{precision} + \text{recall}} \tag{4}$$

where, TP (True Positive) stands for the number of accurate positive predictions, FP (False Positive) for wrong positive predictions, FN (False Negative) for incorrect negative predictions, and TN (True Negative) for correct negative predictions.

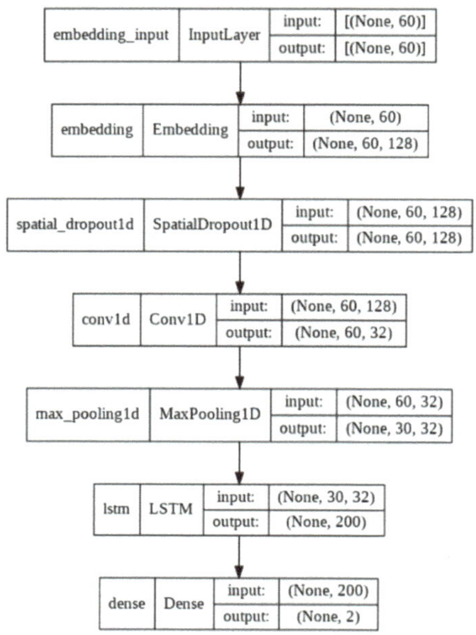

Figure 3. CLSTM model architecture.

4. Experimental Results and Analysis

4.1. Data Analysis Results

We examined the whole pre-processed textual dataset to evaluate the occurrence of suicidal thoughts to assess dissimilarities in the lexicon. We computed the frequencies of all unigrams in both suicidal and non-suicidal tweets. The top 200 unigrams from each category were chosen using Python's WordCloud visualization package to investigate their nature and relationship with suicidal thoughts. Figure 4 illustrates a WordCloud representation of the top 200 unigrams derived from the dataset, divided into two categories: suicidal and non-suicidal tweets.

Figure 4. WordCloud of suicidal and non-suicidal tweets.

The WordCloud of suicidal class shows that suicidal intent tweets include terms such as "fuck", "shit", "hate", "pain", "I'm tired," and "worthlessness", as well as negation phrases such as "don't want", "never", and "nothing". We then discovered that terms with death implications also represent the user's suicidal intentions ('death', 'want die', 'kill'). In contrast to the suicidal postings, the unigrams evaluated in the non-suicidal posts primarily include words expressing happy moments, positive attitudes, and emotions ("I'm happy", "want fun", "laugh loud", "beautiful feel"). Moreover, users are more likely to seek to maintain a positive perspective ("get better") or engage in social activities ("job", "work").

4.2. Classifiers Performance Analysis

Following the n-grams frequency analysis, we examined the experimental technique for detecting suicide thoughts using ML and DL models. We employed CountVectorizer feature extraction on ML models such as MNB, LR, SGD, RF, and SVC. Furthermore, we used word embedding on DL models such as LSTM, BiLSTM, BiGRU, and CLSTM. Then we used evaluation matrices to compare the performance of both the ML and DL models after training. Finally, we performed a comparative analysis of the performance of the classification model. Table 2 shows the classification results for all classifiers based on the evaluation matrices.

Table 2. Performance Table for Classifiers.

Method		Evaluation Metrics				
		Accuracy	Precision	Recall	F1-Score	AUC
DL	BiLSTM	93.6%	0.93	0.93	0.93	0.93
	LSTM	93.5%	0.93	0.93	0.93	0.93
	BiGRU	93.4%	0.93	0.93	0.93	0.93
	CLSTM	93.2%	0.91	0.93	0.93	0.93
ML	RF	93.0%	0.92	0.92	0.92	0.92
	SVC	91.9%	0.91	0.91	0.91	0.91
	SGD	91.7%	0.91	0.91	0.91	0.91
	LR	91.2%	0.91	0.91	0.91	0.91
	MNB	84.6%	0.84	0.84	0.84	0.84

Despite its extensive applicability, accuracy is not always the best performance statistic to use, particularly when the target variable classes in the dataset are imbalanced. Consequently, we employed the F1 score, which takes precision and recall into account when calculating an algorithm's efficiency. The classification scores for each ML and DL algorithm are shown in Figure 5. When comparing the performance of ML models, we found that RF exceeds other traditional ML approaches with an accuracy score of 93% and a 0.92 F1-score. The SVC, SGD, and LR classifiers' performance was slightly lower than RF's, where the obtained accuracy score was between the range of 91.2% and 91.9%, with an F1 score of 0.91. Though previous research [28,29] has shown that MNB performs well for the task of text classification, it fared the lowest in our study, with an accuracy of 84.6%. Comparing the performance of the DL classifiers, it can be seen that all of the models performed equally well in our experiment, with an accuracy score of 93.2%. With a performance gain of 93.6% accuracy and a 0.93 F1-score, the BiLSTM model outperformed other DL models. In our experiment, the LSTM and BiGRU models attained similar accuracy and an F1 score of 93.4% and 0.93, respectively. Lastly, the LSTM model performed the worst among the DL classifiers, with an accuracy of 93.2%. By observing the performance of both the ML and DL classifiers, it can be seen that the DL classifiers provide better results for the task of identifying suicidal tweets.

The ROC Curve, also known as the ROC-AUC, is used for binary classification, which shows the trade-off between sensitivity and specificity. To create the ROC curve, we must first compute the True Positive Rate (TPR) and False Positive Rate (FPR) for various thresholds. The FPR and TPR values are plotted in the x-axis and y-axis, respectively, for each threshold. As a starting point, a random classifier is supposed to provide points along the diagonal where FPR is equal to TPR. The further the curve gets to the ROC space's 45-degree diagonal, the more precise the test is. Figure 6 displays the area under the AUC-ROC curve for all ML models, indicating that the RF model has an AUC close to 1. The AUC of RF suggests it is better than other ML models in distinguishing between suicidal and non-suicidal classes.

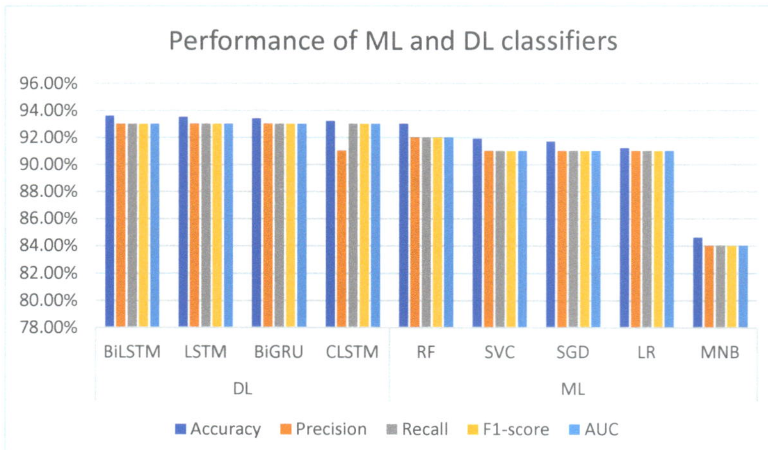

Figure 5. Accuracy score for ML models.

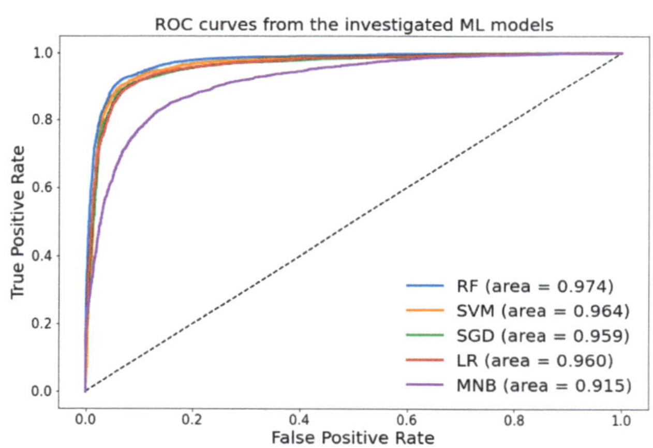

Figure 6. AUC ROC curve for each ML model.

One of the most informative and straightforward methods for evaluating the accuracy and completeness of an ML algorithm is the confusion matrix. Its principal use is in classification tasks where the output might comprise two or more forms of classes. From Figure 7a, we can see on the testing data of 9836 tweets, the confusion matrix of RF demonstrates that the model can predict non-suicidal tweets better than suicidal tweets, as TP (94.0%) is greater than TN (92.5%). Furthermore, the FN is 7.5%, indicating that the model incorrectly forecasts suicidal tweets as non-suicidal, making it challenging to identify the suicidal class. The confusion matrix of BiLSTM, shown in Figure 7b, reveals that the model can predict suicidal tweets better than non-suicidal tweets on a test dataset of 4918 tweets as TN (94.7%) seems to be higher than TP. The model's FN (5.3%) is lower than the FP (8.3%), indicating that it is less likely to misclassify suicidal tweets as non-suicidal.

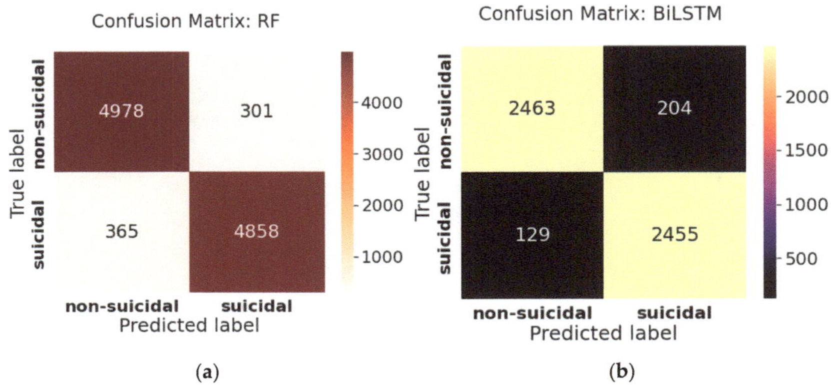

Figure 7. Confusion Matrix of best performed model of ML and DL (**a**) Random Forest; (**b**) BiLSTM.

4.3. Model Validation

To better understand the learning mechanism, we used k-fold cross-validation to determine the mean accuracy in our model. Cross-validation is one of the most extensively used data resampling strategies for assessing the generalization capabilities of predictive models and estimating the actual estimation error of models. The learning set is divided into k disjoint subgroups of roughly equal length in k-fold cross-validation. The number of subgroups produced is referred to as "fold." This partition is accomplished by randomly picking examples from the learning set without replacing them. Our ML models were trained using k = 10 subsets representing the training set as a whole. The model is then applied to the remaining subset, known as the validation set, and its performance is evaluated. This approach is continued until all k subsets have served as validation sets. Figure 8 demonstrates the accuracy of ML models for each fold.

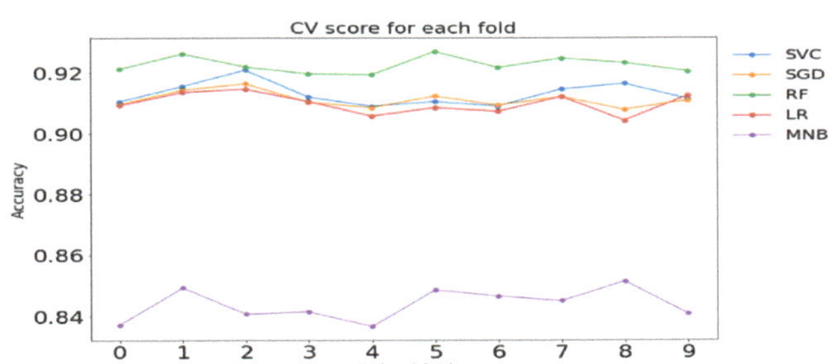

Figure 8. 10-fold Cross-Validation for each ML model.

The validation accuracy and loss for each epoch of the BiLSTM model are shown in Figure 9a,b, respectively. It shows that when the number of epochs increases, our BiLSTM models' validation accuracy and loss tend to increase and decrease, respectively. The 11th epoch model is kept for future testing on the test dataset since we employed a model checkpoint to monitor validation accuracy and saved the best model. We also used ReduceLROnPlateau to lower the learning rate when the model starts overfitting. The validation accuracy and loss appear to be constant after the 12th epoch.

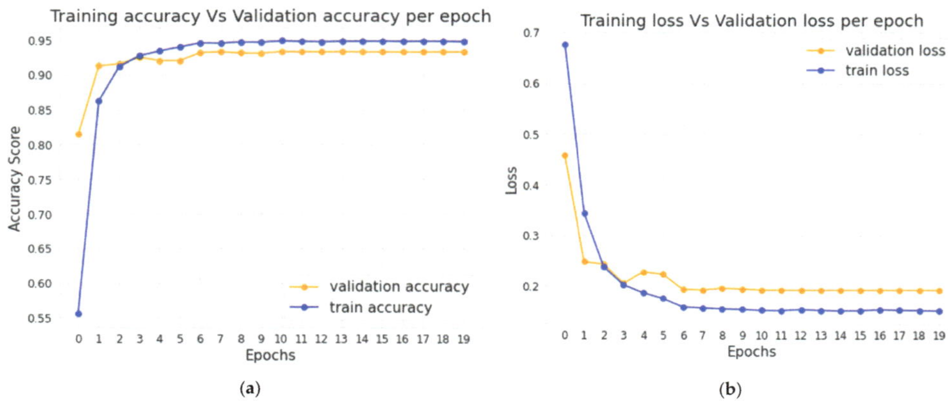

Figure 9. BiLSTM models training vs. validation performance comparison. (**a**) Training and validation accuracy per epoch; (**b**) Training and validation loss per epoch.

5. Discussion

It is generally accepted that for a better learning text classifier it needs to have a growing amount of contextual information. The BiLSTM processing chain replicates the LSTM processing chain, allowing inputs to be processed in both forward and backward time sequences. BiLSTM extends the unidirectional LSTM by allowing hidden-to-hidden interconnections to propagate in the opposite temporal sequence by adding a second hidden layer. Therefore, the model can exploit information from both the past and the future. It is advantageous for a model to have knowledge of both the past and future contexts for sentiment classification problems. The approach allows BiLSTM to consider the future context. At the same time, its layer learns bidirectional long-term dependency between time steps in time series or sequence data without maintaining duplicate context information. These dependencies are crucial when we want the network to learn from the entire time series at each time step while also having access to contextual information. Therefore, it demonstrated an excellent performance for our research. The BiLSTM model does, however, have the disadvantage of requiring more training data and effort than the other classifiers.

It is worth noting that most studies only provide unclear information on pre-processing procedures, which are an essential part of text classification. One of the main reasons for our high accuracy score was by using several NLP techniques to pre-process the tweets effectively. In addition, both the ML and DL models were trained with appropriate parameters to minimize overfitting. Models often appear to perform poorly in the unbalanced class due to the problem of class imbalance. We did not encounter the class imbalance issue since we conducted our experiment with two evenly split classes; as a result, all of the evaluation matrices performed equally well. Even though our experimental results indicate that evaluated matrices work relatively well, cross-validation was not completed on the DL classifiers, which would have resulted in an extremely time and resource-consuming setup. In addition, we ran the experiment on a single dataset of 49,178 tweets. A Tweet's length restriction is 280 characters, often insufficient to comprehend a person's suicidal intentions. If we had gathered more postings from other social media platforms, we could have detected a difference in classifier performance. Furthermore, the psychology of suicide attempts is complicated. Our models can extract statistical indications from suicidal tweets, but they cannot reason about risk variables by adding suicide psychology.

6. Conclusions

Early detection of suicidal thoughts is a crucial and effective method of preventing suicide. The majority of work on this topic has been completed by psychologists using statistical analysis, while computer scientists have used feature engineering-based ML- and DL-based representation learning. Detection of early suicidal intentions on microblogging sites such as Twitter will help medical experts identify and save many lives. The DL and ML approaches can offer new opportunities for improving suicidal ideation detection and early suicide prevention.

In this study, we attempted to compare and analyze several ML and DL models for detecting the presence of suicidal ideation signs in user tweets. The main purpose of the study was to find out the best performing model that can identify suicidal ideations of Twitter users with great accuracy. There are some publicly available datasets on several subreddits or suicide forums for the task of suicidal ideation, but there is no ground truth dataset for analyzing online tweets. As a result, we created the experimental dataset from live tweets by users using suicide-indicative and non-suicidal keywords, then pre-processed the text using different NLP approaches to train on ML and DL algorithms. Five ML algorithms were trained using CountVectorizer feature extractor and four DL models were trained with word embedding technique. Our experimental results show that the BiLSTM model performed best in training, validation, and testing. The model surpasses the other ML and DL models of our experiment with an accuracy of 93.6%. The reason behind the superior performance of BiLSTM is because it can extract relevant information from lengthy tweets more effectively by dealing with forward–backward dependencies from feature sequences resolving gradient disappearance and long-term dependence.

It is important to mention that the ML and DL classifiers were trained with CountVectorizer and word embeddings. Utilization of other feature extraction techniques such as Word2Vec, GloVe, Bag-of-words, and TF-IDF could have resulted in better classification scores. The experiments were carried out on a balanced dataset which implies that all the evaluation metrices performed equally well. On an imbalanced collection of datasets, the model would struggle to perform well on minority class. The problem can be solved with the use of a hierarchical ensemble model. Moreover, due to the lack of ground truth datasets, we have only focused on binary classification on a single dataset in our experiment. However, binary classification is not enough to comprehend the actual sentiments of the users. In the future, we will provide a better comparative analysis between the performances of DL and transfer learning classifiers on different word embedding techniques such as Word2Vec, GloVe, and FastText on several multi-class suicide related datasets. As the transfer learning algorithms can outperform DL classifiers for the task of text classification, the results can be further improved by hyperparameter fine tuning. Future work should reapply the same ML and DL approaches in other disease diagnoses such as heart disease [30–32], and COVID-19 [33–36] with explainable AI, which would also incorporate numerical, categorical, and text data.

It is important to develop a real-world web application that incorporates the classifiers that mental health professionals can utilize to identify online texts having suicidal thoughts in the hopes of preventing suicide. In the future, we also plan to improve the model's performance and produce a practical online application for clinical psychologists and healthcare practitioners.

Author Contributions: Conceptualization, R.H. and N.I.; methodology, R.H. and N.I.; Software, R.H. and N.I.; validation, M.I. and M.M.A.; writing—original draft preparation, R.H. and N.I.; writing—review and editing, R.H., N.I., M.I. and M.M.A.; formal analysis, M.M.A. and M.I. All authors have read and agreed to the published version of the manuscript.

Funding: This research received no external funding.

Institutional Review Board Statement: Not applicable.

Informed Consent Statement: Not applicable.

Data Availability Statement: Not applicable.

Conflicts of Interest: The authors declare no conflict of interest.

References

1. Suicide. Available online: https://www.who.int/news-room/fact-sheets/detail/suicide (accessed on 27 September 2021).
2. Ivey-Stephenson, A.Z.; Demissie, Z.; Crosby, A.E.; Stone, D.M.; Gaylor, E.; Wilkins, N.; Lowry, R.; Brown, M. Suicidal Ideation and Behaviors Among High School Students—Youth Risk Behavior Survey, United States, 2019. *MMWR Suppl.* **2020**, *69*, 47–55. [CrossRef] [PubMed]
3. Gliatto, M.F.; Rai, A.K. Evaluation and Treatment of Patients with Suicidal Ideation. *Am. Fam. Physician* **1999**, *59*, 1500. [PubMed]
4. Giachanou, A.; Crestani, F. Like it or not: A survey of Twitter sentiment analysis methods. *ACM Comput. Surv.* **2016**, *49*, 1–41. [CrossRef]
5. Oussous, A.; Benjelloun, F.-Z.; Lahcen, A.A.; Belfkih, S. ASA: A framework for Arabic sentiment analysis. *J. Inf. Sci.* **2019**, *46*, 544–559. [CrossRef]
6. Pachouly, S.J.; Raut, G.; Bute, K.; Tambe, R.; Bhavsar, S.; Students, U. Depression Detection on Social Media Network (Twitter) using Sentiment Analysis. *Int. Res. J. Eng. Technol.* **2021**, *8*, 1834–1839. Available online: www.irjet.net (accessed on 23 April 2022).
7. Machine Classification for Suicide Ideation Detection on Twitter. *Int. J. Innov. Technol. Explor. Eng.* **2019**, *8*, 4154–4160. [CrossRef]
8. Stankevich, M.; Latyshev, A.; Kuminskaya, E.; Smirnov, I.; Grigoriev, O. Depression detection from social media texts. *CEUR Workshop Proc.* **2019**, *6*, 2523.
9. Abdulsalam, A.; Alhothali, A. Suicidal Ideation Detection on Social Media: A Review of Machine Learning Methods. 2022. Available online: http://arxiv.org/abs/2201.10515 (accessed on 23 April 2022).
10. Aladag, A.E.; Muderrisoglu, S.; Akbas, N.B.; Zahmacioglu, O.; Bingol, H.O. Detecting suicidal ideation on forums: Proof-of-concept study. *J. Med. Internet Res.* **2018**, *20*, e215. [CrossRef]
11. Shah, F.M.; Haque, F.; Un Nur, R.; Al Jahan, S.; Mamud, Z. A Hybridized Feature Extraction Approach to Suicidal Ideation Detection from Social Media Post. In Proceedings of the 2020 IEEE Region 10 Symposium (TENSYMP), Dhaka, Bangladesh, 5–7 June 2020; pp. 985–988. [CrossRef]
12. Ji, S.; Li, X.; Huang, Z.; Cambria, E. Suicidal ideation and mental disorder detection with attentive relation networks. *arXiv* **2020**, arXiv:2004.07601. [CrossRef]
13. Tweepy. Available online: https://www.tweepy.org/ (accessed on 9 December 2021).
14. Mäntylä, M.V.; Graziotin, D.; Kuutila, M. The evolution of sentiment analysis—A review of research topics, venues, and top cited papers. *Comput. Sci. Rev.* **2018**, *27*, 16–32. [CrossRef]
15. Lumontod, R.Z., III. Seeing the invisible: Extracting signs of depression and suicidal ideation from college students' writing using LIWC a computerized text analysis. *Int. J. Res. Stud. Educ.* **2020**, *9*, 31–44. [CrossRef]
16. Masuda, N.; Kurahashi, I.; Onari, H. Suicide Ideation of Individuals in Online Social Networks. *PLoS ONE* **2013**, *8*, e62262. [CrossRef]
17. Pestian, J.; Nasrallah, H.; Matykiewicz, P.; Bennett, A.; Leenaars, A. Suicide Note Classification Using Natural Language Processing: A Content Analysis. *Biomed. Inform. Insights* **2010**, *3*, BII.S4706. [CrossRef] [PubMed]
18. Tadesse, M.M.; Lin, H.; Xu, B.; Yang, L. Detection of Depression-Related Posts in Reddit Social Media Forum. *IEEE Access* **2019**, *7*, 44883–44893. [CrossRef]
19. Sawhney, R.; Manchanda, P.; Mathur, P.; Shah, R.; Singh, R. Exploring and Learning Suicidal Ideation Connotations on Social Media with Deep Learning. In Proceedings of the 9th Workshop on Computational Approaches to Subjectivity, Sentiment and Social Media Analysis, Brussels, Belgium, 31 October 2018; pp. 167–175. [CrossRef]
20. Ji, S.; Yu, C.P.; Fung, S.-F.; Pan, S.; Long, G. Supervised Learning for Suicidal Ideation Detection in Online User Content. *Complexity* **2018**, *2018*, 6157249. [CrossRef]
21. Tadesse, M.M.; Lin, H.; Xu, B.; Yang, L. Detection of suicide ideation in social media forums using deep learning. *Algorithms* **2020**, *13*, 7. [CrossRef]
22. Abboute, A.; Boudjeriou, Y.; Entringer, G.; Azé, J.; Bringay, S.; Poncelet, P. Mining Twitter for suicide prevention. In *International Conference on Applications of Natural Language to Data Bases/Information Systems*; Lecture Notes in Computer Science (including Subseries Lecture Notes in Artificial Intelligence and Lecture Notes in Bioinformatics); Springer: Cham, Switzerland, 2014; Volume 8455, pp. 250–253. [CrossRef]
23. Colombo, G.B.; Burnap, P.; Hodorog, A.; Scourfield, J. Analysing the connectivity and communication of suicidal users on twitter. *Comput. Commun.* **2016**, *73*, 291–300. [CrossRef]
24. Hswen, Y.; Naslund, J.A.; Brownstein, J.S.; Hawkins, J.B. Monitoring Online Discussions About Suicide Among Twitter Users With Schizophrenia: Exploratory Study. *JMIR Mental Health* **2018**, *5*, e11483. [CrossRef]
25. Dara, S.; Tumma, P. Feature Extraction by Using Deep Learning: A Survey. In Proceedings of the 2nd International Conference on Electronics, Communication and Aerospace Technology, ICECA 2018, Coimbatore, India, 29–31 March 2018; pp. 1795–1801. [CrossRef]
26. Lu, Y.; Salem, F.M. Simplified gating in long short-term memory (LSTM) recurrent neural networks. *Midwest Symp. Circuits Syst.* **2017**, 1601–1604. [CrossRef]
27. ArunKumar, K.E.; Kalaga, D.V.; Kumar, C.M.S.; Kawaji, M.; Brenza, T.M. Comparative analysis of Gated Recurrent Units (GRU), long Short-Term memory (LSTM) cells, autoregressive Integrated moving average (ARIMA), seasonal autoregressive Integrated moving average (SARIMA) for forecasting COVID-19 trends. *Alex. Eng. J.* **2022**, *61*, 7585–7603. [CrossRef]

28. Singh, G.; Kumar, B.; Gaur, L.; Tyagi, A. Comparison between Multinomial and Bernoulli Naïve Bayes for Text Classification. In Proceedings of the 2019 International Conference on Automation, Computational and Technology Management, ICACTM 2019, London, UK, 24–26 April 2019; pp. 593–596. [CrossRef]
29. Zhao, L.; Huang, M.; Yao, Z.; Su, R.; Jiang, Y.; Zhu, X. Semi-supervised multinomial naive bayes for text classification by leveraging word-level statistical constraint. *Proc. AAAI Conf. Artif. Intell.* **2016**, *30*, 2877–2884.
30. Ahsan, M.M.; Mahmud, M.A.; Saha, P.K.; Gupta, K.D.; Siddique, Z. Effect of data scaling methods on machine learning algorithms and model performance. *Technologies* **2021**, *9*, 52. [CrossRef]
31. Ahsan, M.M.; Siddique, Z. Machine learning-based heart disease diagnosis: A systematic literature review. *Artif. Intell. Med.* **2022**, *128*, 102289. [CrossRef]
32. Ahsan, M.M.; Luna, S.A.; Siddique, Z. Machine-Learning-Based Disease Diagnosis: A Comprehensive Review. *Healthcare* **2022**, *10*, 541. [CrossRef] [PubMed]
33. Ahsan, M.M.; Gupta, K.D.; Islam, M.M.; Sen, S.; Rahman, M.; Shakhawat Hossain, M. COVID-19 symptoms detection based on nasnetmobile with explainable ai using various imaging modalities. *Mach. Learn. Knowl. Extr.* **2020**, *2*, 490–504. [CrossRef]
34. Ahsan, M.M.; EAlam, T.; Trafalis, T.; Huebner, P. Deep MLP-CNN model using mixed-data to distinguish between COVID-19 and Non-COVID-19 patients. *Symmetry* **2020**, *12*, 1526. [CrossRef]
35. Ahsan, M.M.; Ahad, M.T.; Soma, F.A.; Paul, S.; Chowdhury, A.; Luna, S.A.; Yazdan, M.M.S.; Rahman, A.; Siddique, Z.; Huebner, P. Detecting SARS-CoV-2 from chest X-Ray using artificial intelligence. *IEEE Access* **2021**, *9*, 35501–35513. [CrossRef]
36. Ahsan, M.; Nazim, R.; Siddique, Z.; Huebner, P. Detection of COVID-19 Patients from CT Scan and Chest X-ray Data Using Modified *MobileNetV2* and *LIME*. *Healthcare* **2021**, *9*, 1099. [CrossRef]

technologies

MDPI

Article

An Optimized Enhanced Phase Locked Loop Controller for a Hybrid System

Amritha Kodakkal [1], Rajagopal Veramalla [2], Narasimha Raju Kuthuri [1] and Surender Reddy Salkuti [3,*]

[1] Department of Electrical and Electronics Engineering, Koneru Lakshmaiah Education Foundation, Vaddeswaram 522502, India; amritha.k@bvrithyderabad.edu.in (A.K.); narasimharaju_eee@kluniversity.in (N.R.K.)
[2] Department of Electrical and Electronics Engineering, Kakatiya Institute of Technology and Science, Warangal 506015, India; vrg.eee@kitsw.ac.in
[3] Department of Railroad and Electrical Engineering, Woosong University, Daejeon 34606, Korea
* Correspondence: surender@wsu.ac.kr

Abstract: The use of renewable energy sources is the need of the hour, but the highly intermittent nature of the wind and solar energies demands an efficient controller be connected with the system. This paper proposes an adept control algorithm for an isolated system connected with renewable energy sources. The system under consideration is a hybrid power system with a wind power harnessing unit associated with a solar energy module. A controller that works with enhanced phase locked loop (EPLL) algorithm is provided to maintain the quality of power at the load side and ensure that the source current is not affected during the load fluctuations. EPLL is very simple, precise, stable, and highly efficient in maintaining power quality. The double-frequency error which is the drawback of standard phase locked loop is eliminated in EPLL. Optimization techniques are used here to tune the values of the PI controller gains in the controlling algorithm. Tuning of the controller is an important process, as the gains of the controllers decide the quality of the output. The system is designed using MATLAB/SIMULINK. Codes are written in MATLAB for the optimization. Out of the three different optimization techniques applied, the salp swarm algorithm is found to give the most suitable gain values for the proposed system. Solar power generation is made more efficient by implementing maximum power point tracking. Perturb and observe is the method adopted for MPPT.

Keywords: wind power generating unit; induction generator; enhanced phase locked loop; particle swarm optimization; selective particle swarm optimization; salp swarm optimization; voltage and frequency control; battery energy storage system

Citation: Kodakkal, A.; Veramalla, R.; Kuthuri, N.R.; Salkuti, S.R. An Optimized Enhanced Phase Locked Loop Controller for a Hybrid System. *Technologies* **2022**, *10*, 40. https://doi.org/10.3390/technologies10020040

Academic Editors: Manoj Gupta, Eugene Wong and Gwanggil Jeon

Received: 29 January 2022
Accepted: 7 March 2022
Published: 11 March 2022

1. Introduction

The fossil-fueled power plants emit nitrogen oxides, sulfur oxides, and other harmful particles. The rate at which the carbon dioxide in the atmosphere increases is quite alarming. It is found that the carbon dioxide content is increasing in the wintertime, and during summer, when photosynthesis is active, the CO_2 content is less. According to National Oceanic and Atmospheric Administration (NOAA), as stated in its Global Climate Summary 2021, the global land and ocean temperature is increasing at the rate of 0.070 °C per decade, and the average global surface temperature was the highest for July 2021 since 1880. Accepting wind and solar energies as the primary energy sources will relieve our power sector from contributing to global pollution.

National Wind Energy Mission has announced a target of 60 GW wind power generation, whereas the target for solar power is 100 GW by 2022. Floating wind farms are seen as the future of the global offshore wind sector. There are floating wind turbine structures that are installed in water depths where a fixed structure is not feasible. Siemens has built a huge floating wind farm in Scotland [1], which is the only commercial floating farm in the whole world. The vertical profile of the mean wind speed is given by

$$V(H) = \frac{v_0}{q} (\ln \frac{H}{Y_0} - \psi) \quad (1)$$

where $V(H)$ is the speed of the wind, H is the height, Y_0 is the roughness length, v_0 is the friction velocity, q is the Von Kaman constant, and ψ is the atmospheric stability function [2]. ψ value is greater than zero for the daytime and is less than zero in the night. The stability function is zero at neutral conditions. The wind speed varies a lot near the surface during the day and night, being higher during the daytime, but wind speed changes less with height during the day. At around 150 m from the ground level, there is not much difference between the day and night wind speeds.

Solar energy output from the panel depends upon different factors such as solar intensity, shading, relative humidity, and also the building up of heat in the module. Orientation of the solar panel and the cleanliness also influence the efficiency of the panel. The highly intermittent nature of convertible wind and solar energies makes them a little less reliable, but the presence of an energy storage device solves the problem. The varying wind speed produces oscillating torque, which in turn produces power fluctuations. The wind energy, which is converted to electrical energy, should be suitably controlled to satisfy various standards before being integrated with a grid or while independently handling loads in a standalone system. The solar gives a DC voltage which is to be converted to a suitable form of AC before applying it to the normal AC loads. Researchers adapt various control strategies to make sure that the electrical power is free from harmonics. The frequency and the magnitude of the supply voltage also should be maintained constant. The control technique used here is enhanced phase locked loop (EPLL).

Reference [3] proposes a method to estimate the damping factor along with phase angle, magnitude, and frequency. References [4–8] explain the efficiency of EPLL-based control algorithm in reactive power compensation and in maintaining the voltage and frequency constant in renewable energy-based power conversion units. The DC components present in the input cause lower frequency oscillations which are difficult to filter. Therefore, an improved linear time-invariant EPLL is modeled [9], which uses the linear transfer function approach for its design and is independent of the magnitude of the input signal. Reference [10] proposes a moving average filter (MAF) EPLL which removes the even order ripple and DC offset while deriving the reference quantities, while [11] applies the MAF on a grid-connected system. Here, a DC integrator loop is introduced for this purpose. References [12,13] focus on the applications of MDSC filters in the improvement of the dynamic performances of single-phase, two-phase, and three-phase PLL. Advancements in single-phase and three-phase PLLs are critically reviewed in references [14,15]. The disadvantages of each method, which resulted in advancement, are clearly explained here. An improved two-phase stationary frame EPLL, i.e., an αβ–PLL, is introduced to filter out DC offsets and the harmonics during the unbalance of a system [16] to avoid the complexity of calculations while implementing a three-phase PLL. Two integrators are used additionally here, and the moving average filter is used in the improved version. Reference [17] explains the application of a power-based PLL with MAF for single-phase systems. Five methods to tackle the issue of DC offset voltages with their design criteria, merits, and demerits, and the suggestions to overcome the disadvantages are proposed in Reference [18]. It suggests using the dq or αβ frame delayed signal cancellation operator and the notch filter in the in-loop as filtering agents. The use of cross-feedback networks and complex coefficient filters are also discussed. Reference [19] compares four phase locked loop (PLL) techniques in photovoltaic applications and proves the ability of EPLL in removing the harmonics. A three-phase PLL algorithm that works on the reforming of the signals effectively even for highly distorted grid conditions is suggested in [20]. The reforming process is carried out at every zero crossing. Designing aspects of an EPLL in a shunt active filter is discussed in [21]. Reference [22] explains the enhanced version of complex coefficient filter-based PLL. Here, the PI controller integrator output is given as the feedback signal to a complex coefficient filter, instead of the VCO output. The voltage normalization method which is additionally provided to PLL to improve the transient stability is explained in Reference [23]. Reference [24] recommends locating grid-forming converters to improve the small-signal stability of power systems when connected with large-scale PLL-based converters. The problems associated with the generation of the fictitious orthogonal signal in the implementation of single-phase PLL and the methods for the improvement and the design aspects associated with it are explained in [25]. The dynamic stability of a hybrid system by integrating an induction generator into the system is analyzed in Reference [26]. Reference [27] introduces repetitive learning-based PLL to enhance the power quality of a grid-connected DC microgrid. Reference [28] suggests a linear active disturbance rejection control-based nonlinear PLL which improves the filtering capacity and the dynamic response during synchronization. A stability analysis based on Lyapunov's approach was conducted in Reference [29] and it was concluded that SRF-EPLL is asymptotically stable when the gain values are positive. A new synchronization method called decoupling network $\alpha\beta$ frame PLL design is suggested in Reference [30]. The structures of the PLLs and the droop controllers are compared and the resemblance between them is stated in [31]. Reference [32] proposes to include an active disturbance rejection control with an EPLL for better frequency control and to mitigate the voltage drop in a complex grid system. An improved P&O technique with the confined search space is proposed in [33]. A battery charge controller based on a microcontroller with a maximum power point tracker (MPPT) is designed in [34]. Reference [35] proposes an H-bridge voltage converter for the predictive control of the wind energy system. A PI controller which ensures maximum coordination between the single-phase grid and the photovoltaic unit is discussed in [36]. A review on the suitability of renewable energy sources as reliable power resources performed based on Ethiopia is presented in [37]. The design of a microgrid consisting of a combination of wind and solar energy generating units and a battery is explained [38–40]. Various energy storage and control strategies for a hybrid system are explained in [41]. Analysis of power quality enhancement using a STATCOM is explained in [42]. The design of a hybrid system to power an irrigation system based on the geographic and climatic conditions of Sudan is explained in [43]. A hybrid system design is achieved using HOMER software in [44]. A review of the economic, legal, and regulatory aspects of the hybrid systems is presented in [45]. The sustainability of a renewable hybrid system for rural areas is discussed in [46]. The architecture of a hybrid system for a campus that covers the power, data handling, and application is proposed in [47]. A survey on different techniques used in the control of active power filters in the power quality improvement is carried out in [48]. A standalone hybrid system design based on the requirements of specific load in Kasuga city is proposed in [49]. A modified EPLL structure with improved stability margin is proposed in [50]. An overview of the different optimization techniques used in the design of electric machines is presented in [51].

In this paper, a hybrid system consisting of a wind energy conversion unit and a solar energy conversion system is studied. The wind generator output is directly connected to the load, with a controller in shunt, to regulate the quality of power delivered. An energy storage unit (ESU) is provided. The solar output charges the battery unit. MPPT technique is applied in the solar unit so that the maximum energy will be extracted from the panel.

Enhanced phase locked loop (EPLL) is used as the control strategy in this wind controller, considering its simplicity and effectiveness in maintaining the power quality. EPLL is an improved version of the standard PLL. The drawback of the presence of the double-frequency error in the basic PLL is removed here, by providing an inner loop for eliminating this frequency difference. PI controllers minimize the error between the standard values and the measured values of system parameters. The gain of these controllers decides the quality of the output. These gains are optimized by using optimization techniques. Three different swarm-based algorithms, namely particle swarm algorithm, selective particle swarm algorithm, and salp swarm algorithm are separately applied on the controller. The results are substituted for the PI controller gains and the resultant waveforms are compared. It is found that the outputs obtained from the salp swarm algorithm gave the best results. The controllers and optimization algorithms used are explained in Sections 3 and 4.

2. System Design

An isolated system, consisting of a combinational module of a wind energy conversion and a solar energy conversion unit is discussed here. The power output of the wind generator is regulated by a controller which is connected to the point of common coupling through a star-delta transformer. The wind energy unit employs an asynchronous generator. The generator is of 7.5 kW, 415 V, 50 Hz rating. The magnetic energy for the generator to produce the required voltage depends upon the excitation capacitor. A capacitor having a power rating of 8 kVAR is used here. The wind energy unit is directly connected to the load. A breaker is provided in one of the phases to disconnect the load for a small duration so that the efficiency of the controller during the unbalancing of the load can be studied. The measuring units are connected to sense and display the source and load quantities separately.

A star-delta transformer connects the wind power unit with the controller. The presence of the transformer helps the system in many ways. It enables the system to have a lesser rating for the converters and the battery. This reduces the losses in these units, and the system becomes more economical and more efficient. The connection between the transformer neutral and the load neutral gives a closed path for the neutral current in case of unbalance, which helps the source neutral current to be maintained at zero. A battery energy storage system is provided in the DC link. The presence of a constant voltage source at the DC link makes the system more stable. The rating of the battery depends upon the required DC link voltage and is given by the equation

$$V_{bs} = \frac{2\sqrt{2}\ V_{LL}}{m\ \sqrt{3}} \tag{2}$$

where m is the modulation index, V_{bs} is the DC link voltage, and V_{LL} is the line-to-line voltage.

The voltage source converter uses insulated-gate bipolar transistors (IGBTs) as the switching devices. The switching action of IGBTs is based on the triggering pulses applied to its gating circuit from the control circuit. Inductors are connected to the output of the inverter circuit. This arrangement reduces the ripple in the current. An appropriate selection of the inductor plays a very important role in maintaining a distortionless current waveform at the output terminals. The value of the inductor L_f is given by the equation

$$L_f = \frac{\sqrt{3}\ m\ V_{bs}}{12\ x\ f_s\ i_{pp}} \tag{3}$$

The peak-to-peak value of permissible ripple current is given by i_{pp} and x is the overloading factor and f_s is the switching frequency.

A solar photovoltaic (PV) panel is connected such that the output of the panel charges the battery. The panel consists of 84 cells connected in series. The open-circuit voltage is 64.2 volts, and the short-circuit current of the panel is 7.8 A. Two modules of the PV unit are connected in series and six modules are connected in parallel. A DC-to-DC converter is used to boost the generated voltage. The efficiency of the PV module is usually improved by employing different maximum power tracking techniques (MPPT). The perturb and observe method is used here. The schematic diagram of the hybrid unit under study is shown in Figure 1. The various system parameters are mentioned in Figure 1. V_{as}, V_{bs}, and V_{cs} are the source voltages; I_{as}, I_{bs}, and I_{cs} are the source currents in phase a, phase b, and phase c, respectively. I_{Las}, I_{Lbs}, and I_{Lcs} are the load currents in phase a, phase b, and phase c, respectively. I_c is the current supplied by the excitation capacitor and I_b is the battery current. V_b is the battery voltage and is selected as 400 V. R_{in} is the internal resistance of the battery and is taken as 0.1 Ω, C_b is the battery capacitance and is 50,000 F; Rb is the resistance of the capacitor and is 10 kΩ, L_f is the filtering inductor and is chosen as 1 mH. Load values in each phase are 30 Ω in series with 0.6 H.

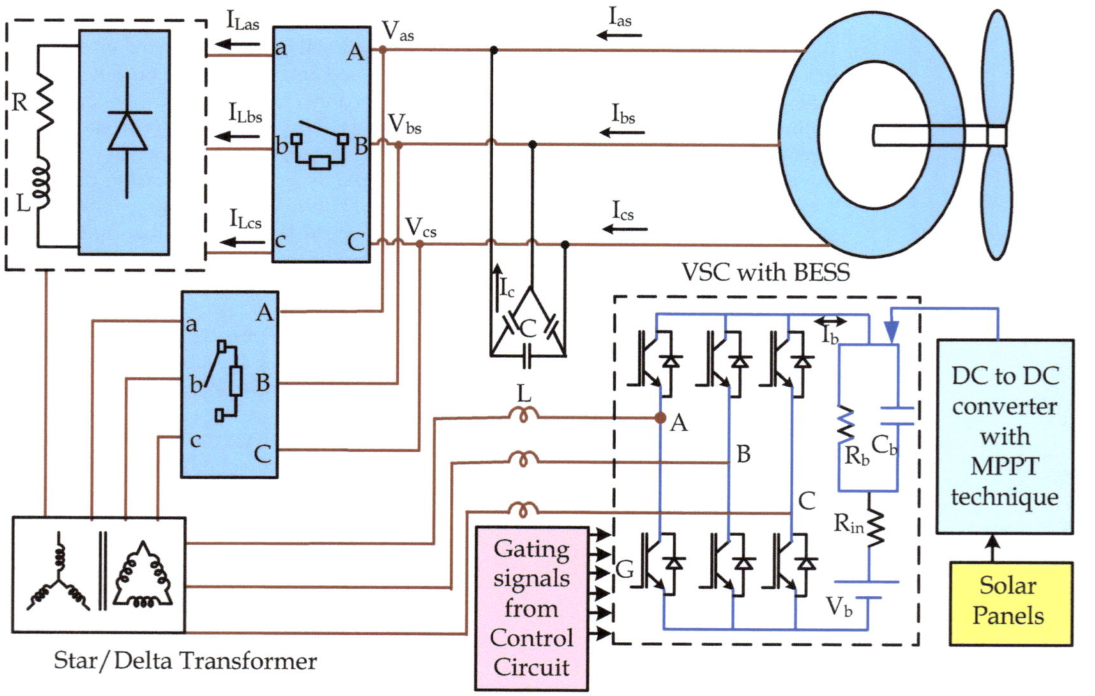

Figure 1. Layout of the hybrid system under study.

3. Control Algorithm

The enhanced phase locked loop (EPLL) is used as the control algorithm for maintaining the power quality at the wind generating unit and load side. Perturb and observe (P&O) is used as the maximum power point tracking algorithm.

3.1. Enhanced Phase Locked Loop (EPLL)

The enhanced phase locked loop (EPLL) is employed in the controller to effectively manage the reactive power and thus improve the quality of the delivered power. EPLL is a very simple and efficient method for power quality improvement. The drawback of double-frequency ripples in the basic phase-locked loop circuit is eliminated here, by incorporating an inner loop. For a three-phase balanced system, double-frequency ripples are not present, since they cancel each other. However, when the system becomes unbalanced, these ripples will be present. An EPLL is capable of handling these unbalanced conditions in a three-phase system.

The active and reactive phasors of the reference currents are formulated by combining the error signal with the average value of the in-phase and quadrature currents. The frequency error value is used for the calculation of the in-phase quantity. Error in the voltage value at the point of common coupling (PCC), when compared with the reference value of PCC voltage, is used to calculate quadrature components.

3.1.1. Computing In-Phase Reference Currents

The average magnitude of the voltage at the point of common coupling (PCC) is computed from the wind energy converted voltages (V_{as}, V_{bs}, and V_{cs}) as

$$V_{ts} = \sqrt{\frac{2}{3}}\left(V_{as}^2 + V_{bs}^2 + V_{cs}^2\right) \tag{4}$$

U_{as} ,U_{bs} , and U_{cs} are the unit components in phase with the phase voltages V_{as}, V_{bs}, and V_{cs}. These are derived as

$$U_{as} = \frac{V_{as}}{V_{ts}} \; ; \; U_{bs} = \frac{V_{as}}{V_{ts}} \; ; \; U_{cs} = \frac{V_{cs}}{V_{ts}} \tag{5}$$

The unit quadrature components of voltages w_{as} , w_{bs} , and w_{cs} are derived from in-phase unit voltages U_{as}, U_{bs}, and U_{cs} as

$$\begin{aligned} w_{as} &= \frac{-U_{bs}}{\sqrt{3}} + \frac{U_{cs}}{\sqrt{3}} \\ w_{bs} &= \frac{\sqrt{3}\,U_{as}}{2} + \frac{U_{bs}}{2\sqrt{3}} - \frac{U_{cs}}{2\sqrt{3}} \\ w_{cs} &= \frac{-\sqrt{3}\,U_{as}}{2} + \frac{U_{bs}}{2\sqrt{3}} - \frac{U_{cs}}{2\sqrt{3}} \end{aligned} \tag{6}$$

These unit vectors are combined with the load currents to produce the fundamental active and reactive components. These are again processed by combining with frequency error and voltage error appropriately to produce the components of the reference currents. In the enhanced phase-locked loop algorithm, the measured value of load current is compared with the fundamental component of the load current and is connected in a closed-loop. These two signals are continuously compared, and the error signal is produced. This error signal undergoes the process described by Equation (7) to derive the fundamental component of the load current. The control algorithm, along with the enhanced loop description, is given in Figure 2. The fundamental component of current is derived in the inner loop of EPLL. The equations governing this can be given as below:

$$\begin{aligned} \Delta\dot{\omega} &= K_2\, e\ Cos\ \delta_0 \\ \dot{\delta} &= \omega_n + \Delta\omega + \textstyle\int K_3\, e\ Cos\ \delta_0 .\, d\delta_0 \\ i_{LF} &= \left[\textstyle\int K_1\, e\ Sin\ \delta_0 .\, d\delta_0\right] \sin\ \delta_0 \end{aligned} \tag{7}$$

where $i_{L\,f\,a}$ is the fundamental value of the current in phase a, and e is the error between the actual load current and the fundamental value of the load current. This gives an estimate of the distortion in the signal. δ_0 is the angle between the fundamental component of current and the actual current. K_1,K_2, and K_3 are the constants that decide the transient and the steady-state behavior of EPLL. The distortion between the fundamental component and the actual load current appears as the error signal. The PI controller works on the error signal to maintain a constant phase angle, which means that the frequency should be the same. The system stability depends upon the value of the gains K_1, K_2, and K_3. The values chosen here are 20, 15, and 1.

The enhanced phase lock loop algorithm uses a zero-crossing detector (ZCD) and a sample and hold (S/H) circuit to produce the fundamental components. The unit vector which is in phase with the applied voltage U_{as} is made to pass through the ZCD. The zero detector gives the count of the number of times the waveform passes through the zero value. The other input signal to this block is the fundamental component of the load current. By properly combining these two signals, the output of the S/H block gives the phase component of current which goes in phase with the voltage. The average of three-phase components is taken out as the load active power component.

$$I_{LpA} = \frac{1}{3}\left[I_{Lpa} + I_{Lpb} + I_{Lpc}\right] \tag{8}$$

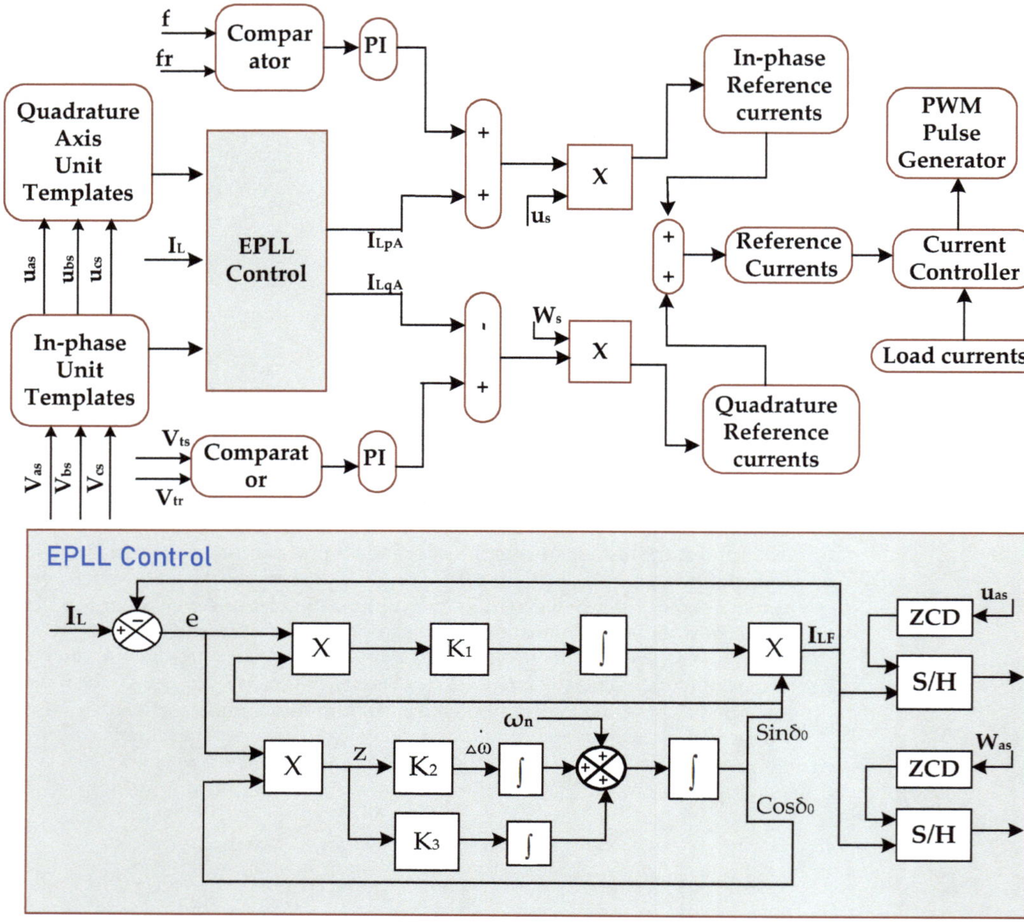

Figure 2. EPLL control technique.

3.1.2. Computing Quadrature-Phase Reference Currents

The quadrature component is also derived in a similar fashion to that of the active components. Here, the quadrature component is made to pass through the ZCD, the output of which is combined with S/H to give a reactive component of current in a particular phase. The average of three-phase components is taken out as the reactive power component of the load current.

$$I_{LqA} = \frac{1}{3}\left[I_{Lqa} + I_{Lqb} + I_{Lqc}\right] \tag{9}$$

The frequency error is produced by comparing the frequency of the measured value of the applied voltage with the reference value. The obtained error signal is modified using the PI gains. The output signal of a sample at the *t*th instant is given by

$$f_{e(t)} = f_r - f_{(t)} \tag{10}$$

The frequency PI controller output at the *t*th sampling instant is expressed as

$$P_{d(t)} = P_{d(t-1)} + K_{pd}\{f_{e(t)} - f_{e(t-1)}\} + K_{id} f_{e(t)} \tag{11}$$

where $P_{d(t)}$ is the active source power at the tth instant. K_{pd} and K_{id}, respectively, are the proportional and the integral gains of the PI controller. The frequency error is added with the active component of the current and is again multiplied by the unit vectors U_{as} , U_{bs} , and U_{cs} to generate the active component of the reference current. The voltage error is produced by comparing the terminal voltage with the reference value and amending this error with the controller gains. The voltage error at the tth instant is given by

$$V_{ts\ e(t)} = V_{wr} - V_{ts} \tag{12}$$

$V_{ts\ e(t)}$ is the error voltage, V_{wr} is the reference value of terminal voltage, and V_{ts} is the measured voltage. The PI controller output at the tth instant is

$$Q_{Vts(t)} = Q_{Vts(t-1)} + K_p \{ V_{tse(t)} - V_{tse(t-1)} \} + K_i V_{tse(t)} \tag{13}$$

This is the reactive power required to maintain a constant terminal voltage at the PCC. The voltage error is added with the reactive component of the current and is again multiplied by the unit vectors w_{as} , w_{bs} , and w_{cs} to produce the reactive component of the reference current. The total reference current is the sum of active and reactive components.

$$\begin{aligned} i_{as}^* &= i_{asp} + i_{asq} \\ i_{bs}^* &= i_{bsp} + i_{bsq} \\ i_{cs}^* &= i_{csp} + i_{csq} \end{aligned} \tag{14}$$

The measured values of the generated currents i_{as}, i_{bs}, and i_{cs} are matched with the reference currents i_{as}^* , i_{bs}^*, and i_{cs}^* . The result is applied to a hysteresis controller to produce the firing pulses for the insulated gate bipolar transistors (IGBT) inside an inverter. These IGBTs are switched on and off according to these gate signals and the compensating current flows to compensate for the distortions in the current.

3.2. Perturb and Observe (P&O)

The solar irradiance and the temperature are two important factors that decide the energy output of a solar cell. The I–V characteristics of a solar cell, given in Figure 3, provide information on the current–voltage relationship of a typical solar cell at a particular irradiance and temperature.

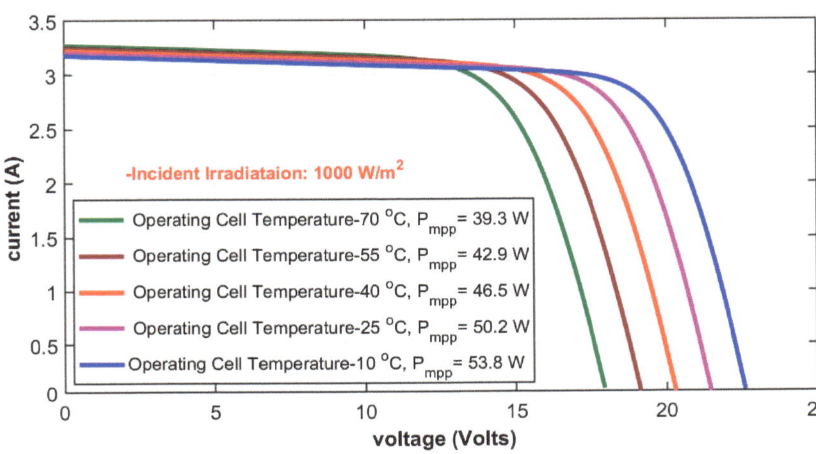

Figure 3. Current–voltage characteristics for different temperatures.

As the solar irradiation increases, the produced current increases, but when the temperature increases, the developed voltage decreases. The power generated from a solar panel is at its maximum when the current and voltage are at their maximum values. The current–voltage characteristics help formulate the methods for the solar cell to operate near its maximum power point.

The energy conversion using a solar panel is not very efficient. A normal PV panel is able to convert 11% to 15% of the solar input energy to useable electrical energy. The efficiency of the solar energy conversion is improved by MPPT techniques. These techniques help the PV panel to always operate at its maximum power point by adjusting its duty cycle.

Based on the power–voltage curve shown in Figure 4, the flowchart of the P&O algorithm is explained in Figure 5. In the P&O method, the duty cycle is adjusted in such a way that the power developed in the panel is at its maximum. Power is maximum when the voltage output of the cell and the current flowing are maximum. From the I–V characteristics of the solar cell, any change in voltage from the maximum value causes a reduction in the power extracted. The operating point is adjusted by adjusting the duty cycle of the converter. The relation between the voltage and the power is observed. Suppose an increase in the developed voltage results in an increase in the power, which indicates that the operating point is on the left side of the maximum point in the I–V characteristics, and the duty cycle should be adjusted in such a way that it moves more towards the right side and thus nearer to the maximum point. On the other hand, if an increase in the voltage causes a decrease in the power, that is an indication that the operating point is on the right side of the maximum point and directed towards the left.

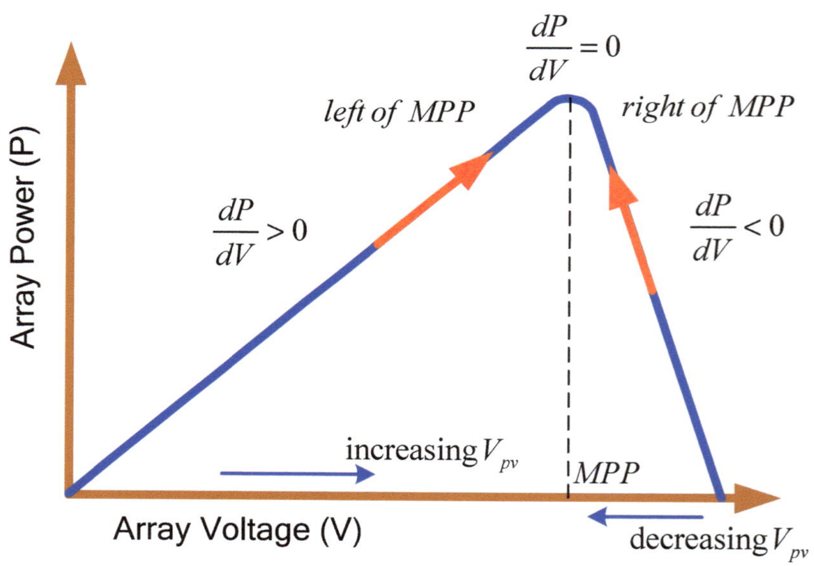

Figure 4. Power–voltage curve in MPPT.

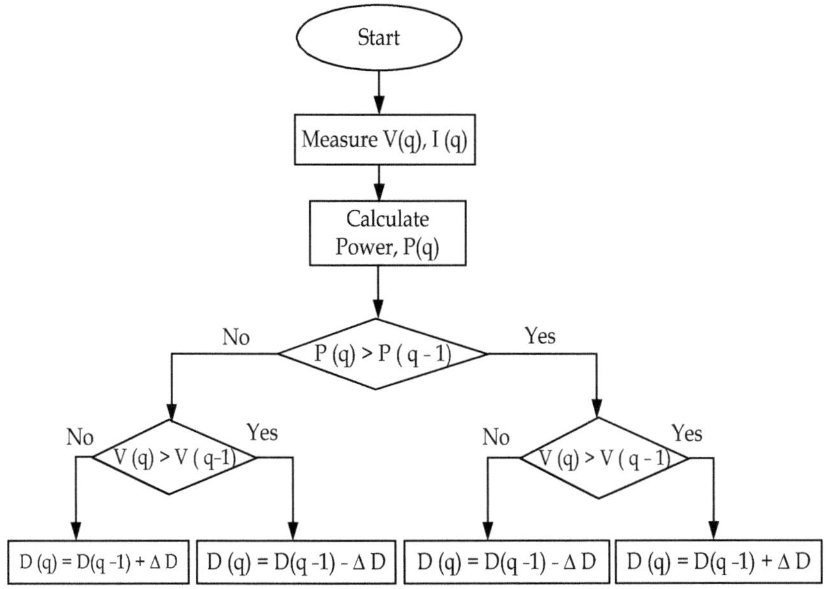

Figure 5. Perturb and observe algorithm for MPPT.

4. Optimization of PI Gains

The control circuit provided in the system generates reference signals based on which the triggering pulses are produced. These signals are applied to the voltage source converter to perform the switching operations. Due to these switching operations, the control circuit can effectively compensate for the reactive power, thus improving the power quality. The measured values are compared with the reference values to keep the frequency and terminal voltage constant. The error gets processed through the PI controller. The controller gains are adjusted such that the error value goes to zero.

An optimization problem finds its solution by determining a variable that can minimize or maximize the objective function. After going through several iterations of the fitness calculations and updations in the variables, the algorithm gives an optimal value. The final value of the objective function or fitness function proposes a solution close to the best solution possible. The proposed solution and the rate of convergence help the researcher decide how effective the optimization technique is, to solve a particular problem.

The PI controller gains are usually adjusted by trial and error, which is a lengthy and time-consuming process. Here, various optimization techniques are employed to find out the best values for the PI controller gains. The different techniques used are particle swarm optimization (PSO), selective particle swarm optimization (SPSO), and salp swarm algorithm (SSA). These techniques are applied individually, and results are noted down. The gain values obtained by these techniques are applied in the PI controller. The resultant output waveforms are compared.

All the techniques which are employed here are based on the characteristics of the swarms. Particle swarm optimization (PSO) is a popular technique to find the optimized solution for a problem. A new research area based on swarm intelligence was evolved based on this. PSO mainly refers to the intelligent behavior of a swarm of birds, while searching for food. The location of food is the optimized solution for the bird's search. In PSO, each particle refers to a potential solution to the problem. By performing continuous updates in the position and the velocity of the particle, the most probable solution is found.

PSO has a very simple concept, and it is easy to implement. It is faster and less complex compared to many other optimization techniques, but it suffers from the disadvantages of having less search precision and falling into local optima while solving complex problems. Many variations of the classical version are proposed in recent times to improve the quality of the solution, speed of convergence and to expand the applicability of the algorithm. Selective particle swarm optimization (SPSO) was suggested in 1997. Here, the search space is selected to give the best possible solution. In SPSO, similar to in a PSO, the velocity and positions of each particle are updated in each iteration. Since the search is confined to the selected search space, the rate of convergence is faster. Salp swarm algorithm (SSA) is a technique inspired by the swarming nature of the salps. Salps, with their barrel− shaped body, resemble jellyfishes. These form chain-like swarms in deep oceans.

The mathematical model of SSA divides the total population into two parts: the leader and the followers. The first salp in the chain takes the role of the leader. All other salps fall into the category of followers. The swarm is guided by the leader. The followers follow each other and, in that process, ultimately follow the leader. The purpose of the salp chain movement is to chase the source of food called F, which is placed in the search space. To start with, in the algorithm, random positions are assigned to the salps. It computes the fitness of each salp. The positions of the leader and the followers are updated by the respective equations. The algorithm assigns the position of the salp with the best fitness to the position of the food source. The best possible solution or position is taken as the global optimum. This process, except the initialization, is iterative and it continues until the end criterion is met. The search space is limited and is maintained within the boundaries by defining the boundary conditions. An n−dimensional search space with n number of variables is defined. A two−dimensional matrix X stores the positions of salps. In SSA, only the position of the leader is updated with respect to the food source and is given by the following equation:

$$X_k^1 = \left[\begin{array}{ll} F_k + m_1\left((ub_k - lb_k)\, m_2 + lb_k\right) & for\ m_3 \geq 0 \\ F_k - m_1\left((ub_k - lb_k)\, m_2 + lb_k\right) & for\ m_3 < 0 \end{array} \right. \tag{15}$$

Here, X_k^1 is the position of the leader in the kth dimension, F_k is the position of the food source in the kth dimension, ub_k and lb_k are the upper and lower boundaries of the kth dimension. m_1, m_2, and m_3 are random numbers. The coefficient m_1 is said to balance the exploration and exploitation and plays an important role in the optimization process. The value of m_1 is found out by

$$m_1 = 2\,e^{-\left(\frac{4\,(itr)}{(itr_{\max})}\right)^2} \tag{16}$$

where itr is the current iteration and $itr_{\max}$ is the maximum iteration. m_2 and m_3 are random numbers in the range [0,1]. These define if the position of the leader is towards the negative infinity or positive infinity. The follower position is updated by Newton's second law of motion and is given by

$$X_j^i = \frac{1}{2}\,a\,t^2 + v_0\,t \tag{17}$$

where a is the acceleration of the particle and v_0 is the initial velocity. In SSA, the food source position is updated according to the leader position. The follower positions are updated with respect to each other, which ensures the gradual movement towards the leader, which in turn helps to avoid the stagnation of local optima. Figure 6 shows the flowchart of SSA algorithm.

The cost function is selected such that the steady-state errors of the PI controllers used in the frequency comparator circuit and the PCC voltage comparator circuit are minimized. The cost function is given by

$$O = w1 * ITSE_1 + w2 * ITSE_2 \tag{18}$$

where $ITSE_1$ and $ITSE_2$ stand for integrated squared error and are the inputs to frequency and AC PI controllers.

$$\begin{aligned} ITSE_1 &= \int t\, f_{e(t)}^{\;2}\, dt \\ ITSE_2 &= \int t\, V_{ts(t)}^{\;2}\, dt \end{aligned} \quad (19)$$

$w1$ and $w2$ are the weights of $ITSE_1$ and $ITSE_2$, and are taken as 0.5. The Simulink model workspace data of $ITSE_1$ and $ITSE_2$ are extracted and then given to the algorithms for optimizing *PI* gains.

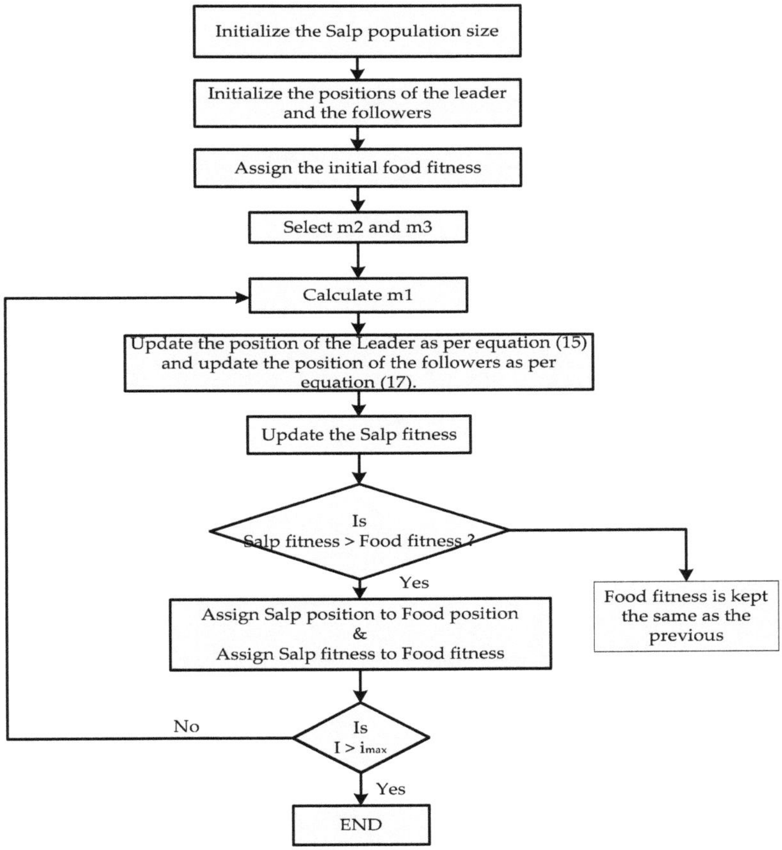

Figure 6. Flowchart of SSA.

The constraints incorporated for the values of *Kp* and *Ki* are given as

$$0 < Kp1 < 5;\ 0 < Ki1 < 5 \quad (20)$$

$$0 < Kp2 < 4;\ 0 < Ki2 < 4 \quad (21)$$

where *Kp*1 and *Ki*1 are the gains of the frequency PI controllers and *Kp*2 and *Ki*2 are the gains of the voltage *PI* controllers.

The results of the optimization techniques are displayed in Figures 7–9.

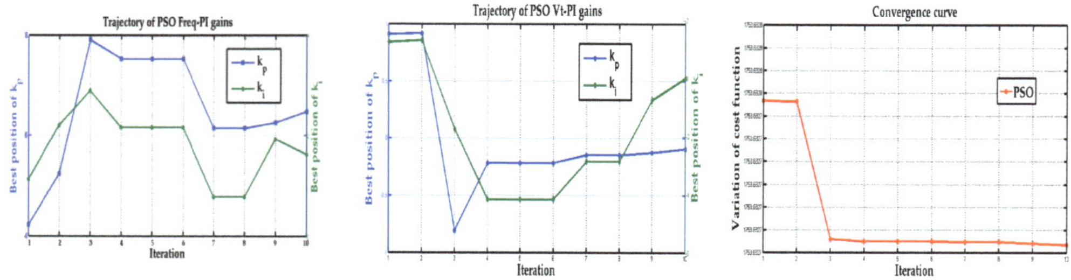

(a) (b) (c)

Figure 7. (**a**) Frequency gain; (**b**) Vt gain; (**c**) convergence curve for PSO algorithm.

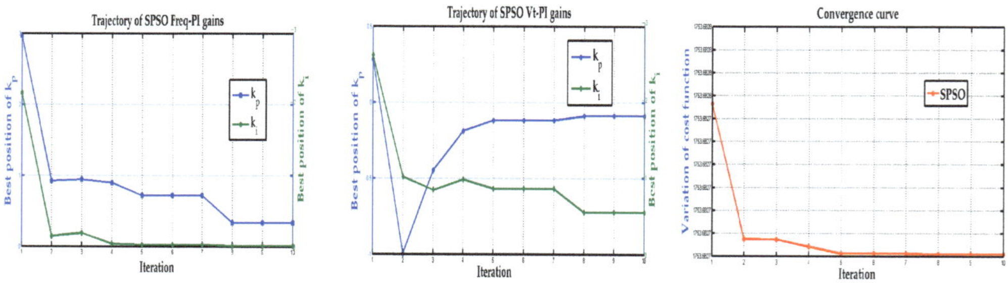

(a) (b) (c)

Figure 8. (**a**) Frequency gain; (**b**) Vt gain; (**c**) convergence curve of the SPSO algorithm.

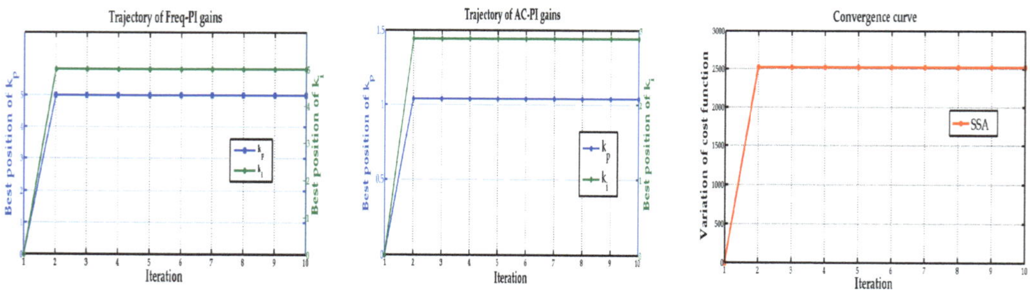

(a) (b) (c)

Figure 9. (**a**) Frequency gain; (**b**) Vt gain; (**c**) convergence curve for SSA algorithm.

For an easy comparison, the gain values while different optimization techniques are applied to the simulated circuit are tabulated in Table 1.

The frequency and Vt gains obtained from the optimization techniques are substituted in the PI controller blocks, and the waveforms of the simulated circuit of the wind generating unit are observed and analyzed. The gain values obtained from SSA gave exemplary waveforms for the output parameters, whereas the values from PSO and SPSO could not produce quality output. The convergence curves refer to the rate at which the algorithm proceeds towards the global optimum. A gradual convergence is preferred to obtain the best optimum solution. The convergence curve of SSA shows that the solution is moderately converged into its best.

Table 1. Gain values obtained from different optimization techniques.

Algorithm	Frequency PI Gains		AC PI Gains		Suitability of the Optimization Algorithm
	Kp	*Ki*	*Kp*	*Ki*	
PSO	6.5	1.7	−0.9	0	Not suitable
SPSO	0.3	0	1.8	0.6	Not suitable
SSA	5	5	1.1	2.8	Suitable

5. Simulation Results

The system under consideration is a wind power generating unit with a 7.5 KW generator. The generator starts developing the power at 0.35 s. The load is linked to the wind generator at 0.35 s. The wind energy conversion system is connected to a star−delta transformer. The delta side of the transformer is connected to the controller. The controller is included in the circuit at 0.5 s. The developed voltage, source current, load current, terminal voltage, frequency, and neutral current are observed during the simulation. The above parameters are monitored for different load conditions, such as linear load and nonlinear load, during the disconnection of a load and the load's reconnection. Different parameters of the wind energy system under disturbed load conditions are plotted in Figure 10.

The voltage at the generator output terminals, current at the same points, current drawn by the load, the current required to compensate the reactive power requirement, and the terminal voltage and frequency are plotted for the nonlinear load at constant wind speed. The battery current, load neutral current, and source neutral current are also plotted. The load in phase "a" is disconnected at 2.5 s. The controller acts in such a way that the imbalance in the load does not affect the source current. The source current maintains its sinusoidal nature when the load is disconnected and even when it is reconnected. Frequency and terminal voltage are also maintained constant during the load disturbance. When the load is disconnected, the battery current is found positive, which indicates that the battery is charging. This shows the role of the energy storage system in maintaining the system parameters to the reference values. When the load requirement is less than the rated value, the battery charges, accepting the additional power. When the load is restored, the battery current returns to the initial value.

The imbalance in one phase of the load sets up a current in the load neutral. Since the load neutral is connected with the transformer neutral, it becomes circulated between these two, leaving the source neutral current as zero. This helps the source current not to get distorted in case of a disturbance in the load side.

Figure 11 shows the variation in the current, voltage, and power produced in the solar panel, for different values of solar irradiation. The voltage developed in the solar cell is unaffected by the change in the irradiation, but the current and, thus, the power varies proportionally to the irradiation.

The FFT analysis of the terminal voltage and current waveforms is shown in Figure 12a–c. The displayed total harmonic distortion (THD) values of the generated voltage, generated current, and current at the load side, respectively, when a nonlinear load is connected, are shown. The THD values of the voltage at the generator terminals-generated current and load current was found to be 1.29%, 2.96%, and 39.41%, respectively. As per IEEE 519 standards, all these values are well within their allowable limits. It can be seen that the distortions in the load current are not allowed to pollute the source current, because of the action of the controller.

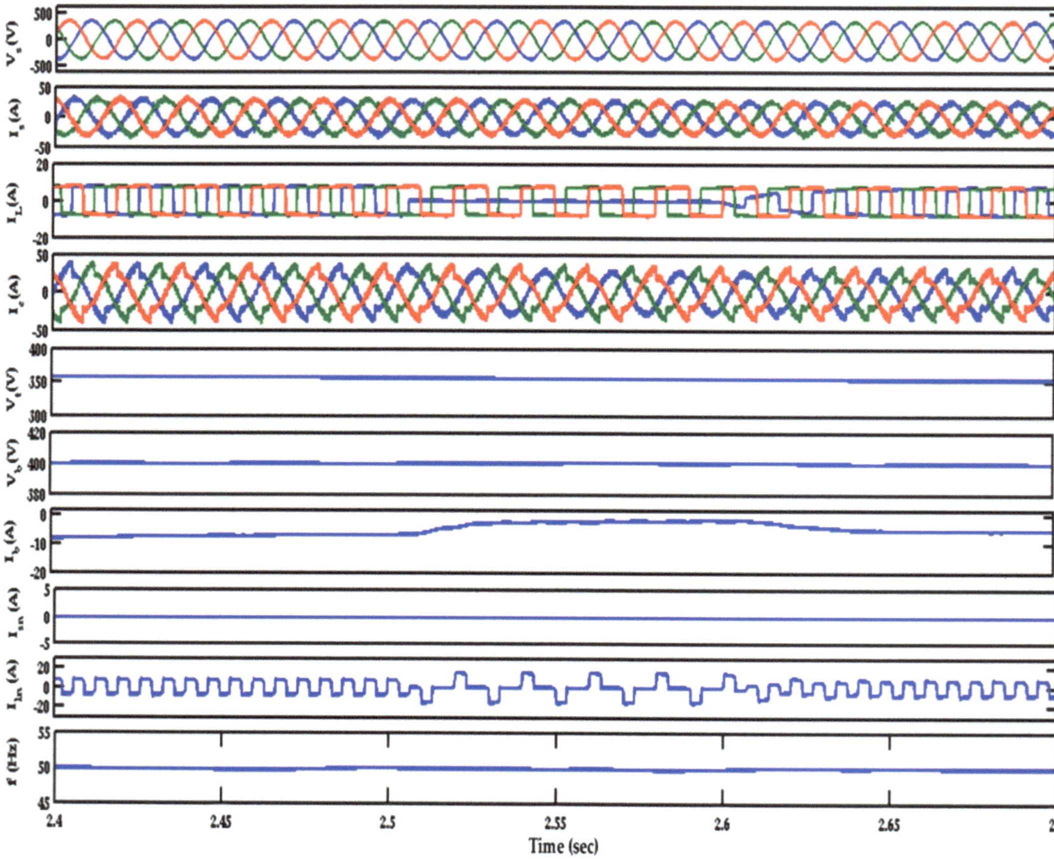

Figure 10. Performance characteristics of a wind energy unit with a nonlinear load.

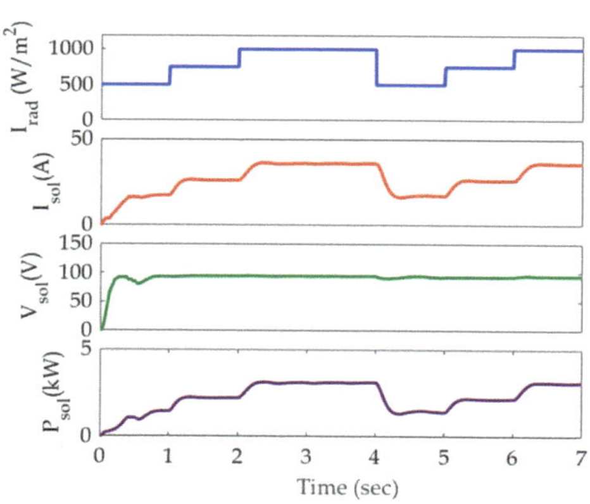

Figure 11. Solar voltage, current, and power at different irradiations.

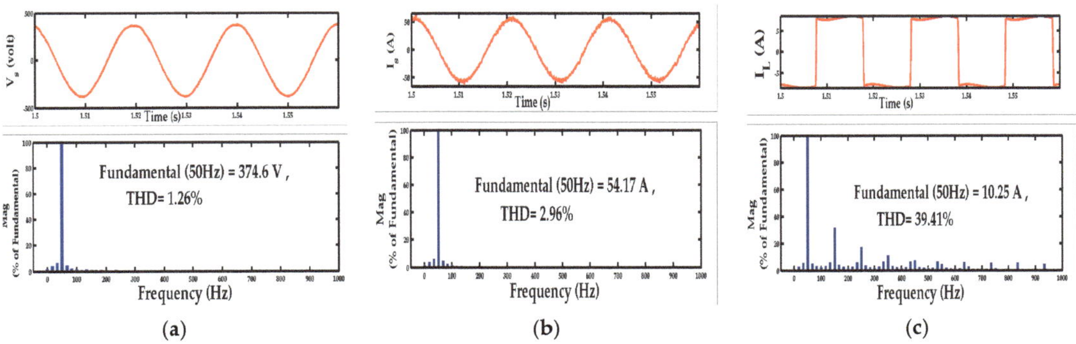

Figure 12. THD in (**a**) source voltage; (**b**) source current; (**c**) load current.

6. Conclusions

This paper deals with a controller which works very efficiently in maintaining the stability of a system during normal, as well as disturbed, load conditions. The system consists of wind and solar power generating units connected with an energy storage system. The generated power from the wind generator is directly connected to the load. The photovoltaic cell is connected as the alternate source of power, and it charges the battery. The controller works with enhanced phase-locked loop, which controls the power system quality parameters and maintains them at par with the reference values. The EPLL controller algorithm works very efficiently to compensate for the reactive power and thus reduce the harmonics under normal and disturbed load conditions. The double-frequency error, which is the drawback of standard PLL, is eliminated in EPLL by providing an inner loop, thus eliminating the frequency deviations. The battery stores energy when the generated power from the wind unit exceeds the load power. By absorbing and releasing the power according to the load conditions, the battery helps to improve the controller's efficiency. The perturb and observe method, which is used as the MPPT technique in the solar unit, is a proven technology in improving the efficiency of the solar panel. By properly implementing the control algorithms, the system is made to work very effectively to maintain quality power at the load and the source side. The total harmonic distortion in source voltage and source current are maintained below the limit specified by IEEE 519 standards. The distortions in the load current are not allowed to pollute the source current, because of the action of the controller. The optimization techniques are used to derive the gains of the PI controllers, which helps to fine-tune the system performance. The implementation of the optimization techniques avoids the difficulty of trial and error, thus making the tuning easy. The convergence curve and the trajectory of the gain values show that the tuning with the salp swarm algorithm suits the system better than the other algorithms used, and simulation results validate this conclusion.

In future, the system can be expanded by including a larger number of renewable energy sources.

Author Contributions: Conceptualization, A.K. and R.V.; methodology, R.V. and S.R.S.; software, A.K. and N.R.K.; validation, A.K., R.V., N.R.K. and S.R.S.; formal analysis, A.K.; investigation, R.V.; resources, N.R.K. and S.R.S.; data curation, A.K.; writing—original draft preparation, A.K. and N.R.K.; writing—review and editing, R.V. and S.R.S.; visualization, A.K.; supervision, R.V. and N.R.K.; project administration, R.V. and S.R.S.; funding acquisition, R.V. and S.R.S. All authors have read and agreed to the published version of the manuscript.

Funding: This research work was funded by Woosong University's Academic Research Funding-2022.

Institutional Review Board Statement: Not applicable.

Informed Consent Statement: Not applicable.

Data Availability Statement: Not applicable.

Conflicts of Interest: The authors declare no conflict of interest.

References

1. Giant Leap Forward in Floating Wind: Siemens Gamesa Lands the World's Largest Project, the First to Power Oil and Gas Offshore Platforms. Available online: https://www.siemensgamesa.com/-/media/siemensgamesa/downloads/en/newsroom/2019/10/pressrelease-siemens-gamesa-hywind-tampen_en.pdf (accessed on 20 January 2022).
2. Boro, D.; Donnou, H.E.V.; Kossi, I.; Bado, N.; Kieno, F.P.; Bathiebo, J. Vertical Profile of Wind Speed in the Atmospheric Boundary Layer and Assessment of Wind Resource on the Bobo Dioulasso Site in Burkina Faso. *Smart Grid Renew. Energy* **2019**, *10*, 257–278. [CrossRef]
3. Zamani, H.; Karimi-Ghartemani, M.; Mojiri, M. Analysis of Power System Oscillations from PMU Data Using an EPLL-Based Approach. *IEEE Trans. Instrum. Meas.* **2017**, *67*, 307–316. [CrossRef]
4. Singh, B.; Arya, S.R. Implementation of Single-Phase Enhanced Phase-Locked Loop-Based Control Algorithm for Three-Phase DSTATCOM. *IEEE Trans. Power Deliv.* **2013**, *28*, 1516–1524. [CrossRef]
5. Philip, J.; Singh, B.; Mishra, S. Performance Evaluation of an Isolated System Using PMSG Based DG Set, SPV Array and BESS, In Proceedings of the 2014 IEEE International Conference on Power Electronics, Drives and Energy Systems (PEDES), Mumbai, India, 16–19 December 2014.
6. Pathak, G.; Singh, B.; Panigrahi, B.K. Isolated Microgrid Employing PMBLDCG for Wind Power Generation and Synchronous Reluctance Generator for DG System. In Proceedings of the 2014 IEEE 6th India International Conference on Power Electronics (IICPE), Kurukshetra, India, 8–10 December 2014.
7. Kumar, S.; Verma, A.K. Performance of Grid Interfaced Solar PV System under Variable Solar Intensity. In Proceedings of the 2014 IEEE 6th India International Conference on Power Electronics (IICPE), Kurukshetra, India, 8–10 December 2014.
8. Verma, A.K.; Singh, B.; Shahani, D. Modified EPLL Based Control to Eliminate DC Component in a Grid Interfaced Solar PV System. In Proceedings of the 2014 6th IEEE Power India International Conference (PIICON), Delhi, India, 5–7 December 2014.
9. Singh, Y.; Hussain, I.; Singh, B. Power Quality Improvement in Single Phase Grid Tied Solar PV-APF Based System using Improved LTI-EPLL Based Control Algorithm. In Proceedings of the 2017 7th International Conference on Power Systems (ICPS), Pune, India, 21–23 December 2017.
10. Chandran, V.P.; Murshid, S. Power Quality Improvement for PMSG Based Isolated Small Hydro System Feeding Three-Phase 4-Wire Unbalanced Nonlinear Loads. In Proceedings of the 2019 IEEE Transportation Electrification Conference and Expo (ITEC), Detroit, MI, USA, 8 August 2019.
11. Liu, C.; Jiang, J.; Jiang, J.; Zhou, Z. Enhanced Grid-Connected Phase-Locked Loop Based on a Moving Average Filter. *IEEE Access* **2019**, *8*, 5308–5315. [CrossRef]
12. Gude, S.; Chu, C. Dynamic Performance Enhancement of Single-Phase and Two-Phase Enhanced Phase-Locked Loops by Using In-Loop Multiple Delayed Signal Cancellation Filters. *IEEE Trans. Ind. Appl.* **2020**, *56*, 740–751. [CrossRef]
13. Gude, S.; Chu, C.-C. Dynamic Performance Improvement of Multiple Delayed Signal Cancelation Filters Based Three-Phase Enhanced-PLL. *IEEE Trans. Ind. Appl.* **2018**, *54*, 5293–5305. [CrossRef]
14. Golestan, S.; Guerrero, J.; Vasquez, J.C. Single-Phase PLLs: A Review of Recent Advances. *IEEE Trans. Power Electron.* **2017**, *32*, 9013–9030. [CrossRef]
15. Golestan, S.; Guerrero, J.; Vasquez, J.C. Three-Phase PLLs: A Review of Recent Advances. *IEEE Trans. Power Electron.* **2016**, *32*, 1894–1907. [CrossRef]
16. Luo, S.; Wu, F. Improved Two-Phase Stationary Frame EPLL to Eliminate the Effect of Input Harmonics, Unbalance, and DC Offsets. *IEEE Trans. Ind. Inform.* **2017**, *13*, 2855–2863. [CrossRef]
17. Xie, M.; Zhu, C.Y.; Shi, B.W.; Yang, Y. Power Based Phase-Locked Loop Under Adverse Conditions with Moving Average Filter for Single-Phase System. *J. Electr. Syst.* **2017**, *13*, 332–347.
18. Golestan, S.; Guerrero, J.; Gharehpetian, G.B. Five Approaches to Deal with Problem of DC Offset in Phase-Locked Loop Algorithms: Design Considerations and Performance Evaluations. *IEEE Trans. Power Electron.* **2015**, *31*, 648–661. [CrossRef]
19. De Carvalho, M.M.; Medeiros, R.L.P.; Bessa, I.V.; Junior, F.A.C.; Lucas, K.E.; Vaca, D.A. Comparison of the PLL Control techniques applied in Photovoltaic System. In Proceedings of the 2019 IEEE 15th Brazilian Power Electronics Conference and 5th IEEE Southern Power Electronics Conference (COBEP/SPEC), Santos, Brazil, 1–4 December 2019.
20. Liu, B.; Zhuo, F.; Zhu, Y.; Yi, H.; Wang, F. A Three-Phase PLL Algorithm Based on Signal Reforming Under Distorted Grid Conditions. *IEEE Trans. Power Electron.* **2014**, *30*, 5272–5283. [CrossRef]
21. Agrawal, S.; Nagar, Y.K.; Palwalia, D.K. Analysis and implementation of shunt active power filter based on synchronizing enhanced PLL. In Proceedings of the 2017 International Conference on Information, Communication, Instrumentation and Control (ICICIC), Indore, India, 17–19 August 2017.
22. Ramezani, M.; Golestan, S.; Li, S.; Guerrero, J.M. A Simple Approach to Enhance the Performance of Complex-Coefficient Filter-Based PLL in Grid-Connected Applications. *IEEE Trans. Ind. Electron.* **2017**, *65*, 5081–5085. [CrossRef]
23. Wu, C.; Xiong, X.; Taul, M.G.; Blaabjerg, F. Enhancing Transient Stability of PLL-Synchronized Converters by Introducing Voltage Normalization Control. *IEEE J. Emerg. Sel. Top. Circuits Syst.* **2020**, *11*, 69–78. [CrossRef]

24. Yang, C.; Huang, L.; Xin, H.; Ju, P. Placing Grid-Forming Converters to Enhance Small Signal Stability of PLL-Integrated Power Systems. *IEEE Trans. Power Syst.* **2020**, *36*, 3563–3573. [CrossRef]
25. Golestan, S.; Guerrero, J.M.; Vidal, A.; Yepes, A.G.; Gandoy, J.D.; Freijedo, F.D. Small-Signal Modeling, Stability Analysis and Design Optimization of Single-Phase Delay-Based PLLs. *IEEE Trans. Power Electron.* **2015**, *31*, 3517–3527. [CrossRef]
26. Touti, E.; Zayed, H.; Pusca, R.; Romary, R. Dynamic Stability Enhancement of a Hybrid Renewable Energy System in Stand-Alone Applications. *Computation* **2021**, *9*, 14. [CrossRef]
27. Sahoo, S.; Prakash, S.; Mishra, S. Power Quality Improvement of Grid-Connected DC Microgrids Using Repetitive Learning-Based PLL Under Abnormal Grid Conditions. *IEEE Trans. Ind. Appl.* **2017**, *54*, 82–90. [CrossRef]
28. Sun, G.; Li, Y.; Jin, W.; Bu, L. A Nonlinear Three-Phase Phase-Locked Loop Based on Linear Active Disturbance Rejection Controller. *IEEE Access* **2017**, *5*, 21548–21556. [CrossRef]
29. Escobar, G.; Ibarra, L.; Valdez-Resendiz, J.E.; Mayo-Maldonado, J.C.; Guillen, D. Nonlinear Stability Analysis of the Conventional SRF-PLL and Enhanced SRF-EPLL. *IEEE Access* **2021**, *9*, 59446–59455. [CrossRef]
30. Hadjidemetriou, L.; Kyriakides, E.; Blaabjerg, F. A Robust Synchronization to Enhance the Power Quality of Renewable Energy Systems. *IEEE Trans. Ind. Electron.* **2015**, *62*, 4858–4868. [CrossRef]
31. Zhong, Q.-C.; Boroyevich, D. Structural Resemblance Between Droop Controllers and Phase-Locked Loops. *IEEE Access* **2016**, *4*, 5733–5741. [CrossRef]
32. Sun, D.; Long, H.; Zhou, K.; Wu, F.; Sun, L. An Improved $\alpha\beta$ -EPLL Based on Active Disturbance Rejection Control for Complicated Power Grid Conditions. *IEEE Access* **2019**, *7*, 139276–139293. [CrossRef]
33. Kamran, M.; Mudassar, M.; Fazal, M.R.; Asghar, M.U.; Bilal, M.; Asghar, R. Implementation of improved Perturb & Observe MPPT technique with confined search space for standalone photovoltaic system. *J. King Saud Univ.–Eng. Sci.* **2018**, *32*, 432–441.
34. Salman, S.; Ai, X.; Wu, Z. Design of a P-&-O algorithm based MPPT charge controller for a stand-alone 200W PV system. *Prot. Control Mod. Power Syst.* **2018**, *3*, 25.
35. Bodha, V.R.; Srujana, A.; Kuthuri, N.R. Predictive back-to-back SCHVC for renewable wind power system for scrutinizing quality and reliability. *Energy Sources Part A Recovery Util. Environ. Eff.* **2019**, *41*, 3058–3075. [CrossRef]
36. Sattenapalli, S.; Manohar, V.J. Performance analysis of reference current generation methods with pi controller for single-phase grid connected PV inverter system. *J. Green Eng.* **2019**, *9*, 658–672.
37. Tiruye, G.A.; Besha, A.T.; Mekonnen, Y.S.; Benti, N.E.; Gebreslase, G.A.; Tufa, R.A. Opportunities and Challenges of Renewable Energy Production in Ethiopia. *Sustainability* **2021**, *13*, 10381. [CrossRef]
38. Juma, M.I.; Mwinyiwiwa, B.M.M.; Msigwa, C.J.; Mushi, A.T. Design of a Hybrid Energy System with Energy Storage for Standalone DC Microgrid Application. *Energies* **2021**, *14*, 5994. [CrossRef]
39. Aguilar, R.S.; Michaelides, E.E. Microgrid for a Cluster of Grid Independent Buildings Powered by Solar and Wind Energy. *Appl. Sci.* **2021**, *11*, 9214. [CrossRef]
40. Al-Quraan, A.; Al-Qaisi, M. Modelling, Design and Control of a Standalone Hybrid PV-Wind Micro-Grid System. *Energies* **2021**, *14*, 4849. [CrossRef]
41. Farooq, Z.; Rahman, A.; Hussain, S.M.S.; Ustun, T.S. Power Generation Control of Renewable Energy Based Hybrid Deregulated Power System. *Energies* **2022**, *15*, 517. [CrossRef]
42. Tariq, M.; Zaheer, H.; Mahmood, T. Modeling and Analysis of STATCOM for Renewable Energy Farm to Improve Power Quality and Reactive Power Compensation. *Eng. Proc.* **2021**, *12*, 44. [CrossRef]
43. Khan, Z.A.; Imran, M.; Altamimi, A.; Diemuodeke, O.E.; Abdelatif, A.O. Assessment of Wind and Solar Hybrid Energy for Agricultural Applications in Sudan. *Energies* **2022**, *15*, 5. [CrossRef]
44. Tran, Q.T.; Davies, K.; Sepasi, S. Isolation Microgrid Design for Remote Areas with the Integration of Renewable Energy: A Case Study of Con Dao Island in Vietnam. *Clean Technol.* **2021**, *3*, 804–820. [CrossRef]
45. De Doile, G.N.D.; Rotella Junior, P.; Rocha, L.C.S.; Bolis, I.; Janda, K.; Coelho Junior, L.M. Hybrid Wind and Solar Photovoltaic Generation with Energy Storage Systems: A Systematic Literature Review and Contributions to Technical and Economic Regulations. *Energies* **2021**, *14*, 6521. [CrossRef]
46. Montisci, A.; Caredda, M. A Static Hybrid Renewable Energy System for Off-Grid Supply. *Sustainability* **2021**, *13*, 9744. [CrossRef]
47. Eltamaly, A.M.; Alotaibi, M.A.; Alolah, A.I.; Ahmed, M.A. IoT-Based Hybrid Renewable Energy System for Smart Campus. *Sustainability* **2021**, *13*, 8555. [CrossRef]
48. Das, S.R.; Ray, P.K.; Sahoo, A.K.; Ramasubbareddy, S.; Babu, T.S.; Kumar, N.M.; Elavarasan, R.M.; Mihet-Popa, L. A Comprehensive Survey on Different Control Strategies and Applications of Active Power Filters for Power Quality Improvement. *Energies* **2021**, *14*, 4589. [CrossRef]
49. Yoshida, Y.; Farzaneh, H. Optimal Design of a Stand-Alone Residential Hybrid Microgrid System for Enhancing Renewable Energy Deployment in Japan. *Energies* **2020**, *13*, 1737. [CrossRef]
50. Golestan, S.; Matas, J.; Abusorrah, A.M.; Guerrero, J.M. More-stable EPLL. *IEEE Trans. Power Electron.* **2022**, *37*, 1003–1011. [CrossRef]
51. Orosz, T.; Rassõlkin, A.; Kallaste, A.; Arsénio, P.; Pánek, D.; Kaska, J.; Karban, P. Robust Design Optimization and Emerging Technologies for Electrical Machines: Challenges and Open Problems. *Appl. Sci.* **2020**, *10*, 6653. [CrossRef]

technologies

Review

Aging Mechanism and Models of Supercapacitors: A Review

Ning Ma [1], Dongfang Yang [2], Saleem Riaz [3], Licheng Wang [4] and Kai Wang [1,*]

1 School of Electrical Engineering, Weihai Innovation Research Institute, Qingdao University, Qingdao 266000, China
2 Xi'an Traffic Engineering Institute, Xi'an 710300, China
3 School of Automation, Northwestern Polytechnical University, Xi'an 710072, China
4 School of Information Engineering, Zhejiang University of Technology, Hangzhou 310014, China
* Correspondence: wkwj888@163.com

Abstract: Electrochemical supercapacitors are a promising type of energy storage device with broad application prospects. Developing an accurate model to reflect their actual working characteristics is of great research significance for rational utilization, performance optimization, and system simulation of supercapacitors. This paper presents the fundamental working principle and applications of supercapacitors, analyzes their aging mechanism, summarizes existing supercapacitor models, and evaluates the characteristics and application scope of each model. By examining the current state and limitations of supercapacitor modeling research, this paper identifies future development trends and research focuses in this area.

Keywords: supercapacitors; models; aging mechanism; applications

check for updates

Citation: Ma, N.; Yang, D.; Riaz, S.; Wang, L.; Wang, K. Aging Mechanism and Models of Supercapacitors: A Review. *Technologies* **2023**, *11*, 38. https://doi.org/10.3390/technologies11020038

Academic Editors: Manoj Gupta, Eugene Wong and Gwanggil Jeon

Received: 29 January 2023
Revised: 27 February 2023
Accepted: 1 March 2023
Published: 3 March 2023

1. Introduction

As a new type of energy storage element, a supercapacitor has great potential in the energy field due to its high power density [1,2]. It has the advantages of high discharge power, long cycle life, wide operating temperature range, and environmental protection. It is the core device in the energy storage system [3,4]. Due to the pure electrostatic energy storage mechanism, compared with other energy storage systems based on electrochemical conversion (such as batteries), supercapacitors also have the characteristics of low internal series resistance, low-cost consumption, and fast charging and discharging speed.

A supercapacitor is a special capacitor between a traditional capacitor and rechargeable battery, which combines the high-current fast charging and discharging characteristics of an ordinary capacitor and the energy storage characteristics of a battery, filling the gap between an ordinary capacitor and battery [5,6]. According to different working principles, supercapacitors are mainly divided into two categories: electric double-layer supercapacitors and pseudo capacitance supercapacitors. The supercapacitor that has been described and mentioned in this paper is a double-layer capacitor.

The aging of supercapacitors can be divided into calendar aging and cycle aging [7]. The phenomenon of continuous aging of supercapacitors under actual working conditions is called calendar aging. The aging phenomenon of a supercapacitor in charge–discharge cycles is called cycle aging. The aging factors of a supercapacitor include external stress, self-acceleration, and manufacturer's production factors. The external stress includes voltage, temperature, charging and discharging power, etc.

The model of a supercapacitor has important theoretical value for analyzing its electrode structure and energy storage mechanism. Developing a model that accurately represents the operational characteristics of supercapacitors is essential for analyzing their electrochemical behavior. This is crucial for simulating and modeling supercapacitors, which can enable state monitoring and life prediction, leading to stable and efficient operation of energy storage systems. Such modeling can provide valuable insights into the internal mechanisms and phenomena of supercapacitors, enabling optimization of their design and performance. Accurate modeling can also help to identify and address potential failure modes and improve the safety and reliability of the supercapacitor system. Therefore, accurate modeling and simulation are of great significance in the development and application of supercapacitors. This paper introduces the working principle and applications of supercapacitors, analyzes the aging mechanism, summarizes various supercapacitor models, points out the characteristics of existing models, and looks forward to the development trend of supercapacitor modeling research.

The major key contributions of our study are summarized as follows:

- We have analyzed the aging mechanism and influence factors of supercapacitors in detail and have debated regarding recent studies.
- The various models of supercapacitors have been schematically summarized and their working principles are also debated.
- We have elaborated the advantages and disadvantages in detail for each category, as well as summarized the application of these models.

The rest of the paper has been divided into the following major sections. The basics and existing literature are explained in Section 1 (Introduction). The working principle of various types of electrochemical supercapacitors is given in Section 2. The aging mechanism and its key factors are discussed in Section 3. The various models according to their characteristics are briefly explained in Section 4. Finally, Section 5 includes the concluding remarks and future work recommendations of the study.

2. Working Principle and Applications

2.1. Working Principle

The principle of electric double-layer capacitance is electrostatic energy storage. The energy storage process is a physical process, without chemical reaction, and the process is completely reversible, which is different from the electrochemical energy storage of batteries. Since positive and negative ions are adsorbed on the surface between the solid electrode and the electrolyte, respectively, the potential difference between the two solid electrodes is caused, thereby realizing energy storage. During charging, under the action of the charge attraction on the solid electrode, the positive and negative ions in the electrolyte collect on the surfaces of the two solid electrodes, respectively. Meanwhile, during discharge, the cation and anion leave the surface of the solid electrode and return to the electrolyte body.

Technologies **2023**, *11*, 38 3 of 16

Simultaneously, the stored charge is released through the external circuit to supply power to the load [8]. This process is shown in Figure 1.

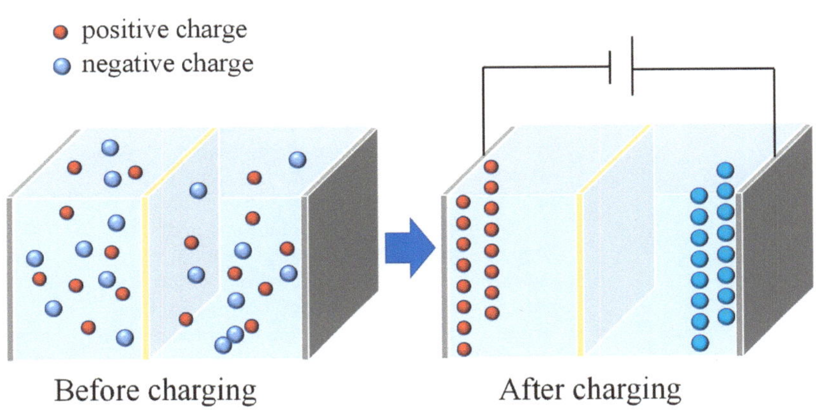

Figure 1. Operating principle of supercapacitors. Positive and negative charges are stored on the positive and negative plates, respectively, when the electrodes are connected to the external circuit.

Figure 1. Operating principle of supercapacitors. Positive and negative charges are stored on the positive and negative plates, respectively, when the electrodes are connected to the external circuit.

2.2. Applications

In today's society, there is a growing demand for superior standards of energy and power supply in terms of quality, safety, and reliability. In response to this need, a novel

2.2. Applications

In today's society, there is a growing demand for superior standards of energy and power supply in terms of quality, safety, and reliability. In response to this need, a novel power grid known as microgrid has emerged, which seamlessly integrates distributed power generation. In this context, supercapacitors have emerged as a new and innovative energy storage technology, capable of providing short-term power supply and energy buffering functions, ultimately enhancing the overall power quality of microgrids. As a result, supercapacitors have become one of the preferred energy storage devices for microgrids [9,10]. Supercapacitors are used as power sources in electric vehicles or hybrid electric vehicles to improve the service life of batteries. In addition to that, these are often used in wind power generation systems, photovoltaic power generation systems, distributed power generation systems, and large-scale power storage systems [11–14]. The application fields of supercapacitors are shown in Figure 2.

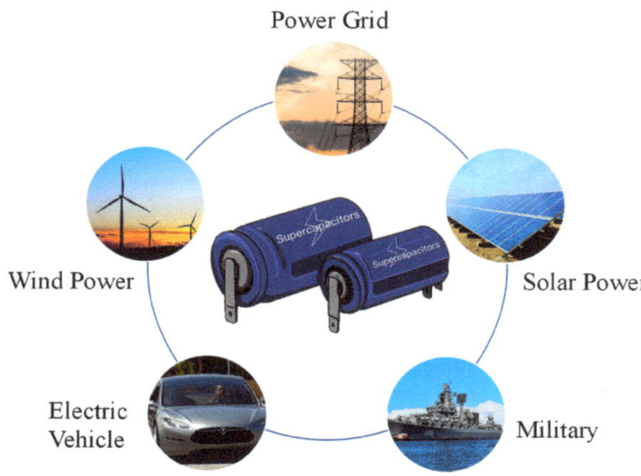

Figure 2. Application domain of supercapacitors.

3. Aging Mechanism

3.1. Overview

Activated carbon is the electrode material of supercapacitors widely used in industry at present, which mainly uses biomass resources to obtain porous activated carbon materials through physical activation or chemical activation treatment with an activator [15]. Because of its simple preparation process and low price, it is widely used in industrial manufacturing of supercapacitors. In the process of chemical activation treatment, corrosive activated materials will be used to make the porous electrode materials. After the activation treatment is completed, the electrode material will be cleaned to remove the activated substances remaining on the surface of the material. While cleaning, some residues will still be adsorbed on the electrode surface. The impurities left in the electrode during electrode manufacturing are the main source of aging and failure.

During the normal use of the supercapacitor, the residues on the electrode surface will react reversibly with the electrolyte to form solid and gaseous products [16], as shown in Figure 3. The gradual deposition of solid products on the electrode surface (as illustrated in area 1) can obstruct the porous structure and lead to a reduction in the contact area between the electrode and electrolyte. This phenomenon is commonly referred to as electrode fouling. Gaseous products may diffuse to multiple areas inside the supercapacitor. For example, when the gaseous products reach the free zone (shown in area 2), the air pressure inside the capacitor will rise; when gaseous products are adsorbed on the electrode surface (as shown in area 3), the contact area between the electrode and electrolyte will be reduced. Gaseous products may also be adsorbed on the membrane, thus blocking the path of ionic charge (shown in area 4). These products lead to an increase in the internal air pressure of the supercapacitor, which may cause electrode cracking (as shown in area 5) or shell deformation to damage the collector (as shown in area 6). The above series of reactions will lead to aging of the supercapacitor.

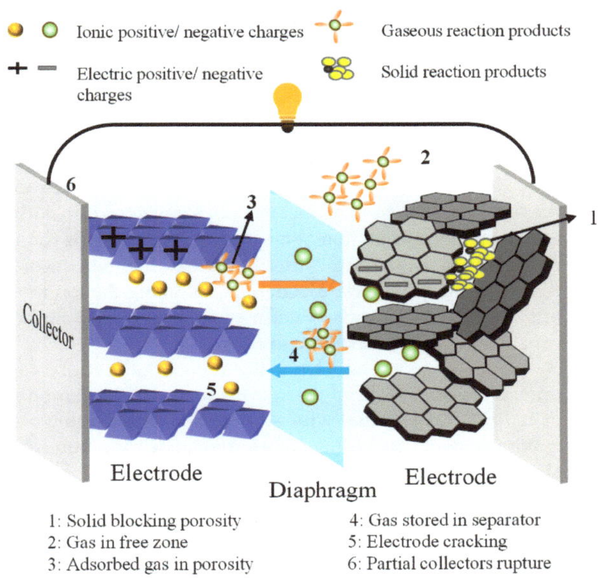

Figure 3. Supercapacitor aging principle [17]. Copyright 2022 Elsevier B.V.

Electrolyte decomposition, electrode deterioration, and shell damage are aging characteristics of supercapacitors. Capacitance loss, equivalent series internal resistance (ESR) increase, and deformation are typical effects of supercapacitor aging. Root causes, failure mechanisms, failure modes, and failure effects of EDLCs are shown in Figure 4.

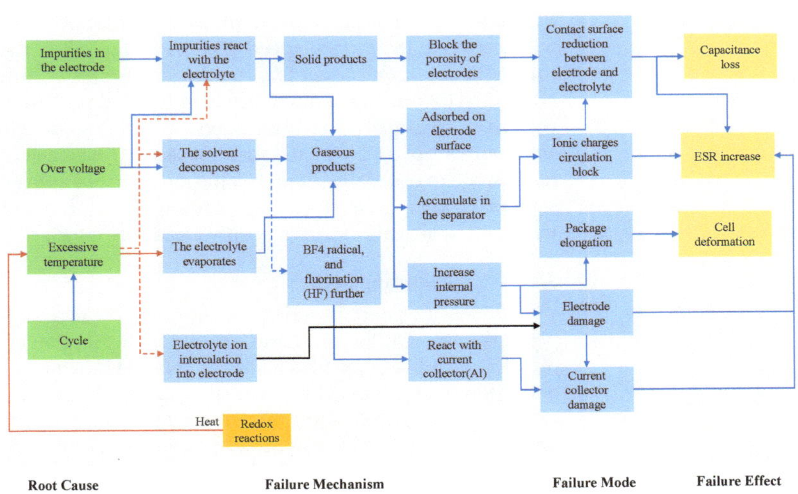

Figure 4. Root causes, failure mechanisms, failure modes, and failure effects of supercapacitors [18].

Therefore, the electrolyte of supercapacitors should meet these requirements: the conductivity should be high to minimize the internal resistance of the supercapacitor; the electrolyte should have higher electrochemical and chemical stability; the operating temperature range should be wide to meet the working environment of the supercapacitor; and the size of the ions in the electrolyte should match the aperture of the electrode material.

3.2. Aging Factor

3.2.1. External Stress

Taking the influence of temperature as an example, in reference [19], the supercapacitors were cyclically tested in 40–50 °C, 40–60 °C, and 50–60 °C, and the results showed that the degradation rate of the supercapacitor was significantly accelerated with the increase in temperature. The high temperature stimulates the chemical activity of each component of the supercapacitor and accelerates the aging speed. In addition, the high temperature accelerates the decomposition of electrolyte, which leads to the decrease in ion concentration. Simultaneously, the impurities generated from its decomposition block the pores of the diaphragm and electrode materials, reducing the ion mobility, reducing the accessibility of the ion to the porous structure on the electrode surface, thus causing the decrease in capacitance and the increase in ESR.

Voltage can also accelerate the attenuation rate of a supercapacitor. The authors of [20] draw a conclusion through experiments that 10 K temperature rise and a voltage increase of 100 mV have virtually the same effect on the aging behavior of a supercapacitor. The maximum working voltage of a supercapacitor is subject to the decomposition voltage of electrolyte solution; in contrast, the working voltage affects parameters such as current density and temperature, which are closely related to the stability of electrolyte.

In technical terms, the cycle life of a supercapacitor is impacted by its charge and discharge power. Higher charge and discharge power levels result in a more rapid decay of the supercapacitor's lifespan. This is due to the fact that increased power levels generate more Joule heat, which, in turn, accelerates the degradation of the supercapacitor's internal materials, increases its internal resistance, and accelerates the aging process [21].

3.2.2. Self-Acceleration of Aging

The aging process of supercapacitors is accompanied by self-acceleration, which is specifically shown as follows: (1) when supercapacitors are used in modules, due to the complexity and diversity of the application environment and uneven temperature distribution, individuals close to the heat source have a higher initial temperature, which will accelerate their aging speed and cause the ESR to rise faster, and the rise of ESR, in turn, will cause their own temperature to rise faster, thus forming a positive feedback effect [22]; (2) due to the difference in individual parameters of supercapacitors, there is a voltage imbalance between each individual in the module during charging. Specifically, the single supercapacitor with the smallest capacity has the highest charging voltage, which is most prone to overvoltage and has the most serious aging occurring during use. The more serious the aging is, the further the capacity is reduced and the charging voltage is further increased, which also forms a positive feedback effect.

3.2.3. Manufacturing Factors

Different materials and manufacturing processes also affect the service life of supercapacitors. On the one hand, the polymer that plays a bonding role in the electrode preparation process contains a large number of functional groups. During the preparation of the porous electrode, physical or chemical activation is carried out and water residues are inevitably introduced. During the normal use of the supercapacitor, the surface functional groups are decomposed by redox reaction; on the other hand, the impurity atoms on the surface of the carbon electrode that cause electrochemical phenomena also appear during the preparation of the electrode. In addition, different capacitor packaging methods will lead to significantly different life.

4. Models

4.1. Equivalent Circuit Models

Among the many models of supercapacitors, the most widely used is the equivalent circuit model. The equivalent circuit model, according to the electrical characteristics of the supercapacitor in the working process, uses various components in the circuit to characterize its internal deterioration mechanism. According to the circuit configuration and the number of components, different circuit models have different accuracy. Increasing the complexity of the circuit is helpful to improve the accuracy of the model.

4.1.1. Simple Series RC Models

The simplest equivalent circuit model is shown in Figure 5a [23]. RC series circuit only reflects the instantaneous dynamic response and can reflect the external electrical characteristics of the supercapacitor surface. Its advantages are simple parameter fitting process and high accuracy in charge and discharge calculation [24]. In order to significantly improve the accuracy of the simple RC circuit model, some online fitting methods have been proposed in the literature [25]. However, the equivalent circuit model is too simple, which also brings some disadvantages. In fact, supercapacitors are affected by various factors, resulting in performance degradation. However, this model does not consider the changes in the performance of supercapacitors in the working state and cannot deeply simulate the working principle of supercapacitors. Therefore, it is limited in complex energy storage systems. Therefore, it is very important to study an accurate degradation performance model.

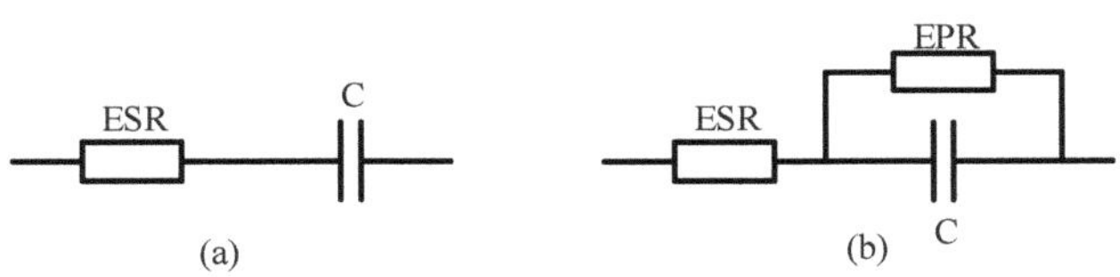

Figure 5. Simple series RC model: (**a**) simple series RC model; (**b**) the simple model of series-parallel connection.

Spyker and Nelms add a parallel resistance to account for the leakage current effect [26,27]. Compared with the simple RC series circuit, the model refines the resistance *R*. As shown in Figure 5b, the improved series RC model is composed of equivalent capacitance *C*, equivalent series internal resistance *ESR*, and equivalent parallel internal resistance *EPR*. Where, C is approximately equal to the nominal value of the capacitance, reflecting the aging speed of the supercapacitor, the magnitude of *ESR* is related to the energy loss of *R* during the aging process of supercapacitors, the capacitance tends to decrease, and the *ESR* increases; *EPR* represents the leakage current effect of the supercapacitor. However, the model can only fully represent the supercapacitor dynamics in a few seconds, which greatly limits its practical applicability.

4.1.2. Transmission Line Models

Pean, C. et al. introduced the transmission line model to simulate the distributed capacitance and electrolyte resistance determined by the porous electrode, as shown in Figure 6 [28,29]. The distributed parameter characteristics of supercapacitors are simulated by RC network and the model parameters are determined by impedance spectrum analysis. The model can have high fitting accuracy in a relatively wide frequency range. Its essence is to carry out high-order fitting of the charging and discharging curves of supercapacitors. The order of fitting can be determined according to the accuracy requirements of the model. Some of the literature has proposed the transmission lines with a variable number of branches and these range from 5 [30] to 15 branches [31]. The higher the order, the higher the accuracy of the model but the more the corresponding model parameters and the parameter identification will be very complex. In addition, the model cannot fully reflect the influence of leakage current of supercapacitors.

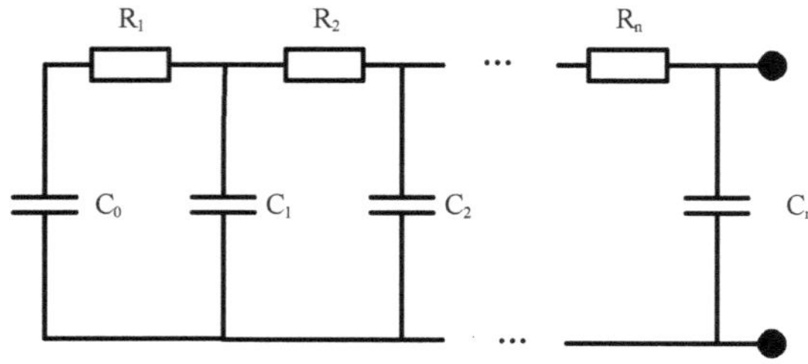

Figure 6. Transmission line model.

Saha P. and Dey S. et al. [32] used a combination system consisting of frequency and time techniques to describe the transmission line, taking the leakage effect into account.

4.1.3. Multi-Branch RC Network Models

The form of the multi-branch RC model is similar to that of the transmission line model. What is different from the transmission line model is that each branch of the model has a different time constant, and each branch acts independently in different time periods of the charging and discharging process. A ladder circuit composed of multiple RC branches with different time constants can be used to capture the distribution characteristics of capacitance and resistance of supercapacitors.

Most of the authors have used the three-branch model, which is proposed by Zubieta and Boner [33], as shown in Figure 7. It has included three RC branches: immediate branch, delay branch, and long-term branch, for which each branch has captured the characteristics of supercapacitors on different time scales. The immediate branch reflects the performance of the supercapacitor in the transient charging and discharging process, which is usually limited to a few seconds. The delay branch reflects the performance of the supercapacitor during charging and discharging in a few minutes. The long-term branch reflects the performance of the supercapacitor in the charging and discharging process within tens of minutes. In addition, the resistance R_L reflects the leakage current effect of the supercapacitor and the influence of a long time on the energy storage process. The nonlinear capacitor is connected in parallel with the constant capacitor as a voltage-dependent capacitor and is incorporated into the direct branch. Then, the parameters of the three branches are extracted by observing the terminal voltage evolution in the constant current charging process. For this model, Rajani et al. [34] presented a novel average point method to extract the model parameters. Analogous model representations were devised by other researchers with different characterization methods [31,35–37].

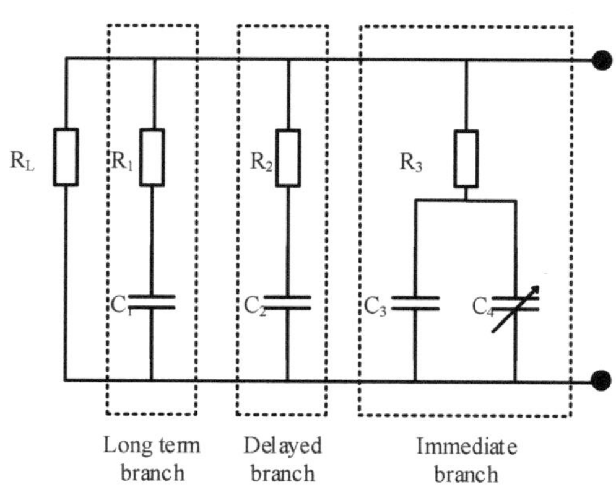

Figure 7. Third-order RC model.

4.2. Electrochemical Models

Helmholtz [38] first discovered the capacitive characteristics at the interface between solid conductor and liquid ionic conductor in 1853 and proposed the double-layer model in 1874. Helmholtz thought that the charge is evenly distributed at both ends of the electrode and electrolyte interface; he modeled this phenomenon as a conventional capacitor with distance for charge separation H, as shown in Figure 8a, where ψs is the local electric potential. Because the conductivity of the electrolyte is poor, the charge on the electrolyte side cannot be evenly distributed. The capacitance calculated according to the model is too large. However, this model shows the energy storage principle of supercapacitor in an intuitive and simple way and is a classic physical model of a supercapacitor. Gouy [39] put forward the model of side charge dispersion distribution in solution in 1910, and Chapman [40] made a detailed mathematical analysis of the model in 1913, as shown in Figure 8b. This model takes into account the spatial distribution of the charge on the electrolyte side, which is also called the diffusion layer. The calculated capacitance value based on the model is still larger than the actual value, because the model assumes that the ions are point charges, that is, they can be infinitely close to the electrode electrolyte interface. Stern proposed an improved model based on Gouy and Chapman's double-layer model. Stern believed that the double electric layer at the interface between the electrode and solution is composed of a compact layer and a diffusion layer. Under the action of electrostatic and thermal movement of particles, part of the ionic charges in the solution is adsorbed on the electrode surface to form a compact double electric layer, that is, the double electric layer capacitance can be seen as a series connection of the compact layer capacitance and the diffusion layer capacitance [41], as shown in Figure 8c. Later, Graham further established the metal solution interface model. He subdivided the compact layer into two layers: the inner Helmholtz layer and the outer Helmholtz layer. Generally, electrochemical models have high accuracy but low calculation efficiency.

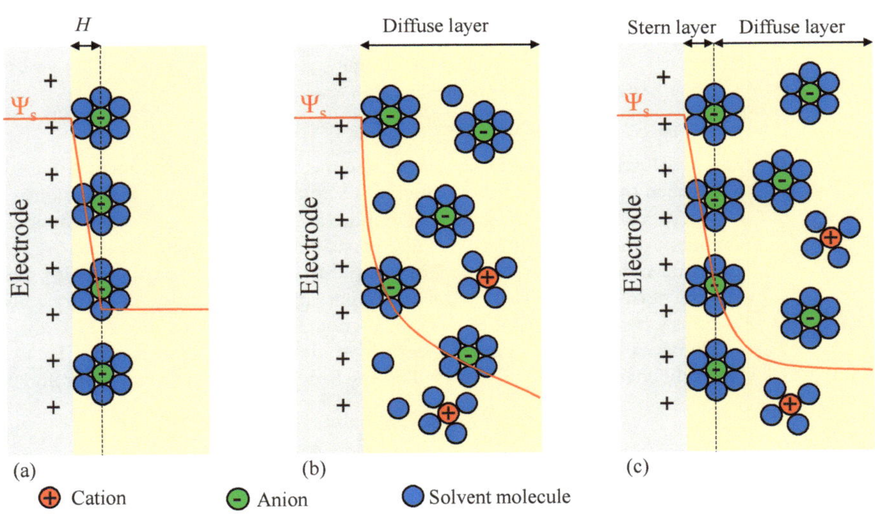

Figure 8. Schematics of the electric double−layer structure showing the arrangement of solvated anions and cations near the electrode/electrolyte interface in the Stern layer and the diffuse layer. Schematic of three basic electrochemical models of the supercapacitor: (**a**) Helmholtz model, (**b**) Chapman model, (**c**) combined mode. Reprinted with permission from Ref. [26].

4.3. Intelligent Models

This kind of model can be regarded as a black box. Without considering the internal mechanism of the supercapacitor, the relationship between input and output can be obtained by training a large amount of charging and discharging historical data. Sadiq Eziani et al. [42] took the voltage and current of supercapacitor as the input of ANN to estimate the SOC of the supercapacitor used for braking energy recovery of a railway system and achieved good results. Liu et al. [43] presented a stacked bidirectional long short-term memory recurrent neural network; the simulation results show that, when the number of hidden layers is two, the network has excellent performance and the predicted RMSE and MAE are 0.0275 and 0.0241, respectively.

The utilization of this model provides an advantageous capability to approximate the nonlinear properties of a given system, with infinite precision in theory. However, it is subject to the lack of a well-defined physical interpretation, and the parameters involved in its expression are highly intricate. Additionally, the model necessitates an extensive amount of training data, requiring a considerable amount of training time. It is noteworthy that the neural network learning algorithm, in its current state, has not fully addressed the issues of underfitting and overfitting, leading to suboptimal performance. Consequently, the applicability of this model may be limited. The neural network model is shown in Figure 9.

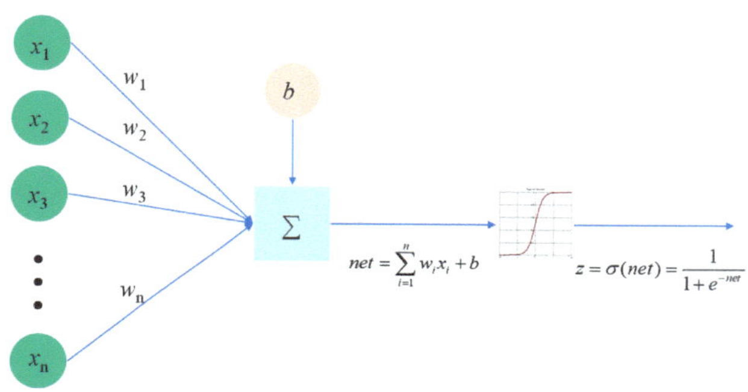

Figure 9. Neural network model. In the figure, x_1~x_n are input signals, w_1~w_n is the weight, b is the offset, and z is the activation function.

4.4. Fractional-Order Models

The resistance and capacitance parameters of supercapacitors are not constant and are affected by factors such as frequency. It is found that the fractional equivalent circuit model can more accurately describe the nonlinear characteristics of supercapacitors. It is found that the current and voltage of supercapacitors are fractional calculus [44], so the equivalent circuit model of a supercapacitor using fractional order components is proposed [45]. The typical fractional order model of a supercapacitor is shown in Figure 10, which consists of a series and a parallel resistor, a constant phase element (CPE), and Walburg-type elements [46]. Riu D and Retiere N et al. proposed a half-order FOM. Freeborn [47] established a simple FOM based on the series connection of a resistor and a CPE.

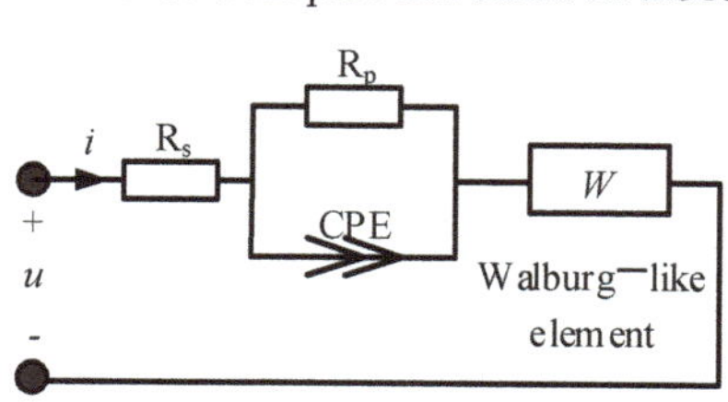

Figure 10. Fractional order model structure.

The fractional order model of supercapacitor uses fractional order components to describe the dynamic behavior of a supercapacitor. Compared with the integer order model, fractional order components can bring additional degrees of freedom to the model in order, which can not only improve the accuracy of the model, but also reduce the complexity of the model [48,49].

The model types for SC electrical behavior simulation are summarized in Table 1.

Table 1. Summary of model types for SC electrical behavior simulation.

Models	Advantages	Disadvantages
Equivalent circuit models	Simple and intuitive; convenient for analysis, calculation, and simulation; moderate accuracy	Susceptible to aging process
Electrochemical models	Description of inside physical–chemical reactions; high possible accuracy	Cannot reflect the dynamic process of charging and discharging; heavy computation; immeasurability of some parameters
Intelligent models	Can approximate the nonlinear characteristics of the system; good modeling capability	Absence of physical meanings; sensitive to training data quality and quantity; poor robustness
Fractional-order models	Better capability to fitting experimental data; few model parameters	Heavy computation

4.5. Self-Discharge Models

Among all kinds of electrochemical energy storage devices, supercapacitors had faced the most serious problem of self-discharge [50]. Smith and Tran et al. [50] compared the self-discharge behavior of commercial supercapacitors (2000F, Maxwell) and lithium-ion batteries (2.4 Ah, E-one moli energy corporation) in the charged state. The results show that the energy loss of the supercapacitor is as high as 22%, which is seven times more than the energy loss of the lithium-ion battery (3%) within 72 h of open-circuit storage time. The self-discharge problem can be seen. Conway and Pell et al. [51] have studied the self-discharge phenomenon of supercapacitors in the 20th century and proposed a mathematical model to describe the self-discharge process. Yang et al. [52] presented a self-discharge model in consideration of variable leakage resistance, as shown in Figure 11. The first branch contains R_1 and C_1 ($C_0 + K_V \times V$), which provides the instant behavior of the supercapacitor in response to the charging process. The second branch is the delayed branch, which represents the charge redistribution in the medium and long term, containing R_2 and C_2. The variable resistor R_3 in the third branch represents the leakage resistor corresponding to the self-discharge rate of the supercapacitor. Ricketts and Ton-That [53] pointed out that SC self-discharge is caused by two different mechanisms, i.e., ion diffusion and leakage current. Tete Tevi [53] proposed to cover the electrode with a thin insulating layer made of polyphenylene oxide (PPO) material, which can reduce the leakage current in the electric double-layer capacitor, thus slowing down the self-discharge of the supercapacitor. Increasing the thickness of the diaphragm (the thickness of the supercapacitor) can increase the diffusion distance of the reactant, thereby reducing the driving force of the concentration gradient of the self-discharge. Furthermore, it is worth noting that the pore structure of electrode materials is also an important factor [54].

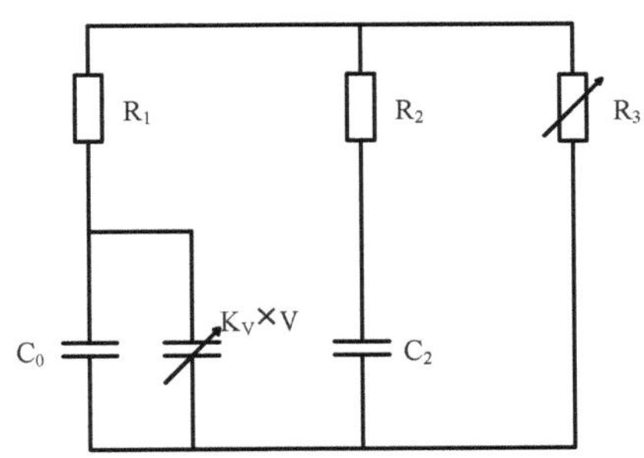

Figure 11. Self-discharge models.

4.6. Thermal Models

The previously proposed models are unable to predict the internal temperature of the supercapacitor. Thermal behavior is a very important aspect in the application of supercapacitors. Working in a bad thermal environment will reduce the performance parameters of supercapacitors, and the uneven distribution of temperature field in supercapacitors will cause the imbalance of individual performance. At present, most of the research on thermal behavior focuses on lithium-ion batteries, and the analysis of thermal behavior of electric double-layer supercapacitors is relatively lower. The development of a thermal model for a supercapacitor involves establishing a correlation between temperature variations and the corresponding changes in the supercapacitor's performance. This model is intended to serve as a theoretical framework for managing the thermal behavior of the supercapacitor.

Gualous H. [55] states that increment of the temperature increases the supercapacitor capacitance but reduces the ESR. Some studies have shown that higher temperatures will speed up the self-discharge process of the supercapacitor.

The thermal models of supercapacitors can be roughly divided into two categories:

- Heat generation: this kind of model describes the influence of its own heating on its temperature field [20,56–58]. The modeling purpose of this kind of model is to analyze the temperature change characteristics and temperature field distribution characteristics of supercapacitors when they work, which is mainly applied to the thermal management analysis of supercapacitor energy storage systems.
- Heat transmission: this kind of model describes the relationship between temperature and the change in model parameters [55,59]. Its modeling method is usually based on the equivalent circuit model to carry out a large number of experiments, determine the curve of model parameters with temperature, and then establish mathematical expressions through corresponding data processing. This kind of model is of great significance for studying the dynamic characteristics of supercapacitors under different ambient temperatures. The thermal model presented in Ref. [60] is shown in Figure 12. In the model, the heat generation is modeled as a current source, which is a function of the supercapacitor current; C_{th} represents the thermal capacity of the supercapacitor, R_{th} denotes the equivalent thermal resistance of the supercapacitor, and T_a denotes the surrounding air temperature.

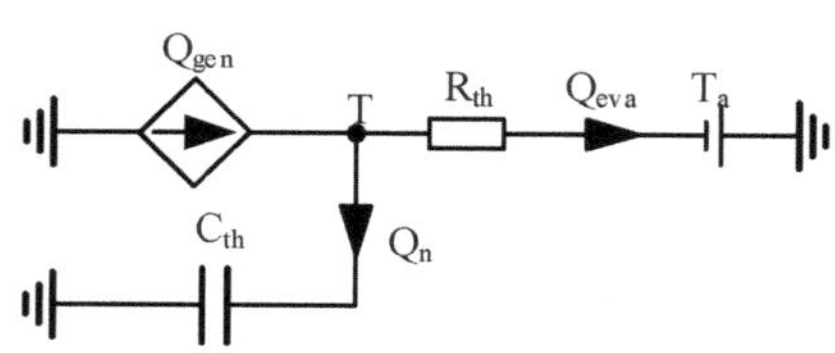

Figure 12. Supercapacitor thermal models.

4.7. Porous Electrode Models

The basic unit of the electric double-layer supercapacitor is composed of a pair of porous material electrodes, a collecting plate, an electrolyte, and an isolating film. Among them, the porosity of electrodes is the most important parameter to characterize the internal characteristics of supercapacitors, and it is also the main reason why supercapacitors differ from conventional electrolytic capacitors. The porous equivalent circuit model of an electric double-layer capacitor is an impedance ladder circuit model derived from the corresponding control equation and one-dimensional hole model [61]; its corresponding trapezoidal equivalent model is shown in Figure 13a. z_x and z_y are the impedances of unit length, and their directions are parallel to the x-axis of hole depth and perpendicular to the y-axis of hole depth, respectively.

The hole impedance model can be equivalent to the series connection of an ideal resistance and an ideal capacitor. Because the current is uniformly distributed on the corresponding resistance and capacitance, the equivalent circuit with several hole impedance models connected in series can approximate the porous characteristics. The equivalent circuit is composed of stray inductance L_S, equivalent series resistance R_s (sum of contact resistance, electrolyte solution resistance, and isolation film resistance), and porous electrode impedance Z_{pore} in series. The porous equivalent circuit model of a double-layer supercapacitor is shown in Figure 13b.

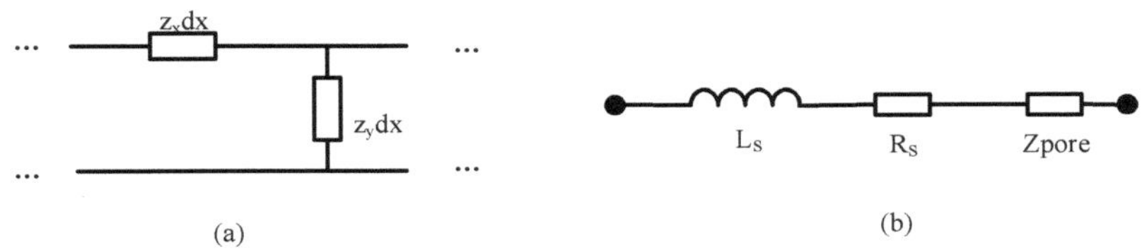

Figure 13. Porous electrode model: (**a**) trapezoidal equivalent model; (**b**) porous equivalent circuit model of double-layer supercapacitor.

4.8. Dynamic Models of Electrochemical Impedance Spectroscopy

The model consists of two RC networks in parallel and a series resistor. The dynamic model is used to replace the direct branch of the three-branch model, and a shunt leakage resistor is introduced to form a combined supercapacitor model. Among them, the series resistance and capacitance compose an equivalent circuit model with temperature-related parameters, which can estimate voltage or temperature through electrochemical impedance spectroscopy [62]. This model is shown in Figure 14.

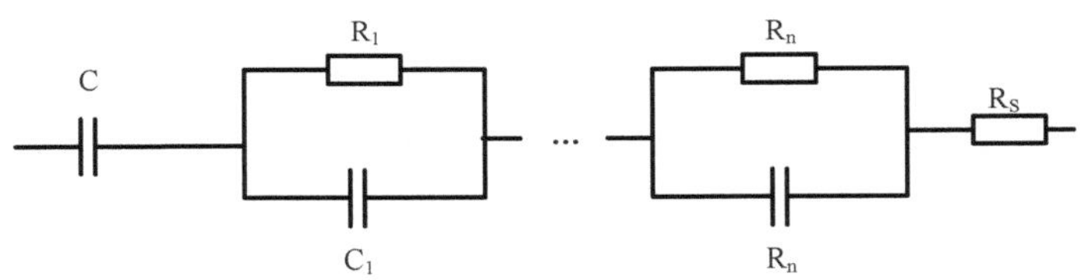

Figure 14. Dynamic model of electrochemical impedance spectroscopy.

5. Summary and Prospect

The modeling of supercapacitors is a key step to achieve different goals. This paper summarizes the aging mechanism and various models of supercapacitors. Through the above analysis of the current research status of supercapacitor modeling, it can be seen that different models can be established from different angles, and each model has its own scope of application. There is a contradiction between accuracy and complexity in the establishment of the model. Finding a compromise solution between the two is the key to the establishment of the actual system model.

The main problems of existing supercapacitor models are:

1. Some models have complex structures, such as the transmission line model, which is composed of many RC networks in series, parallel, and nested structures. The RC network parameters of each circuit are related to the internal structure and working state and are different from each other [63–65].
2. Parameter identification is difficult. At present, AC impedance analysis and circuit analysis are mainly used for supercapacitor model parameters. AC impedance analysis uses a lot of equipment, selects a lot of data when calculating parameters, and the calculation process is complex. The circuit analysis method uses the curve of voltage versus time to obtain the corresponding parameters. This method requires less equipment and is simple and convenient, but the structure of the model itself should not be too complex [66–68].

Although many achievements have been made in supercapacitor modeling, each model has its own advantages and disadvantages in limited applications. In the process of addressing the practical engineering problem, it is important to conduct an exhaustive evaluation of the benefits and drawbacks of diverse models in order to identify the most appropriate model. Given that there is no universally accepted model that can entirely and precisely capture the physical attributes of supercapacitors, the study of modeling for supercapacitors remains a critical area of interest in subsequent research [69–74].

Author Contributions: N.M.; D.Y.; S.R.; L.W.; K.W. have substantially contributed to conducting the underlying research and drafting this manuscript. All authors have read and agreed to the published version of the manuscript.

Funding: This work was supported by the Youth Fund of Shandong Province Natural Science Foundation (No. ZR2020QE212), Key Projects of Shandong Province Natural Science Foundation (No. ZR2020KF020), the Guangdong Provincial Key Lab of Green Chemical Product Technology (GC202111), Zhejiang Province Natural Science Foundation (No. LY22E070007) and National Natural Science Foundation of China (No. 52007170).

Data Availability Statement: The data and materials used to support the findings of this study are available from the corresponding author upon request.

Acknowledgments: This work was supported by the Youth Fund of Shandong Province Natural Science Foundation (No. ZR2020QE212), Key Projects of Shandong Province Natural Science Foundation (No. ZR2020KF020), the Guangdong Provincial Key Lab of Green Chemical Product Technology (GC202111), Zhejiang Province Natural Science Foundation (No. LY22E070007) and National Natural Science Foundation of China (No. 52007170).

Conflicts of Interest: The authors declare that they have no known competing financial interests or personal relationships that could have appeared to influence the work reported in this paper.

References

1. Yang, Y.; Han, Y.; Jiang, W.; Zhang, Y.; Xu, Y.; Ahmed, A.M. Application of the Supercapacitor for Energy Storage in China: Role and Strategy. *Appl. Sci.* **2022**, *12*, 19. [CrossRef]
2. Iqbal, M.Z.; Aziz, U. Supercapattery: Merging of battery-supercapacitor electrodes for hybrid energy storage devices. *J. Energy Storage* **2022**, *46*, 29. [CrossRef]
3. Li, Y.; Kong, Y. Energy storage devices based on supercapacitors. *Chin. J. Power Sources* **2011**, *35*, 409–411.
4. Sahin, M.E.; Blaabjerg, F.; Sangwongwanich, A. A Comprehensive Review on Supercapacitor Applications and Developments. *Energies* **2022**, *15*, 674. [CrossRef]
5. Ma, Y.; Xie, X.; Yang, W.; Yu, Z.; Sun, X.; Zhang, Y.; Yang, X.; Kimura, H.; Hou, C.; Guo, Z.; et al. Recent advances in transition metal oxides with different dimensions as electrodes for high-performance supercapacitors. *Adv. Compos. Hybrid Mater.* **2021**, *4*, 906–924. [CrossRef]
6. Chatterjee, D.P.; Nandi, A.K. A review on the recent advances in hybrid supercapacitors. *J. Mater. Chem. A* **2021**, *9*, 15880–15918. [CrossRef]
7. Chen, Y.; He, Y.G.; Li, Z.; Chen, L.P. A Combined Multiple Factor Degradation Model and Online Verification for Electric Vehicle Batteries. *Energies* **2019**, *12*, 12. [CrossRef]
8. Laadjal, K.; Cardoso AJ, M. A review of supercapacitors modeling, SoH, and SoE estimation methods: Issues and challenges. *Int. J. Energy Res.* **2021**, *45*, 18424–18440. [CrossRef]
9. Yang, H. A review of supercapacitor-based energy storage systems for microgrid applications. In Proceedings of the 2018 IEEE Power & Energy Society General Meeting (PESGM), Portland, OR, USA, 5–9 August 2018; pp. 1–5.
10. Zhang, L.; Hu, X.S.; Wang, Z.P.; Ruan, J.G.; Ma, C.B.; Song, Z.Y.; Dorrell, D.G.; Pecht, M.G. Hybrid electrochemical energy storage systems: An overview for smart grid and electrified vehicle applications. *Renew. Sustain. Energy Rev.* **2021**, *13*, 1105819. [CrossRef]
11. Wang, K.; Ren, B.S.; Li, L.W.; Li, Y.H.; Zhang, H.W.; Sui, Z.Q. A review of Modeling Research on Supercapacitor. In Proceedings of the Chinese Automation Congress (CAC), Jinan, China, 20–22 October 2017; pp. 5998–6001.
12. Huang, X.; Zhang, X.; Wei, T.; Qi, Z.; Ma, Y. Development and applications status of supercapacitors. *Adv. Technol. Electr. Eng. Energy* **2017**, *36*, 63–70.
13. Wu, J.; Zhou, Z.; Zha, F.; He, T.; Feng, B. Supercapacitor and their applications in power grids. *Chin. J. Power Sources* **2016**, *40*, 2095–2097.
14. Zhai, C.; Luo, F.; Liu, Y. Cooperative Power Split Optimization for a Group of Intelligent Electric Vehicles Travelling on a Highway with Varying Slopes. *IEEE Trans. Intell. Transp. Syst.* **2022**, *23*, 4993–5005. [CrossRef]
15. Wang, Y.F.; Zhang, L.; Hou, H.Q.; Xu, W.H.; Duan, G.G.; He, S.J.; Liu, K.M.; Jiang, S.H. Recent progress in carbon-based materials for supercapacitor electrodes: A review. *J. Mater. Sci.* **2021**, *56*, 173–200. [CrossRef]

16. Azais, P.; Duclaux, L.; Florian, P.; Massiot, D.; Lillo-Rodenas, M.A.; Linares-Solano, A.; Peres, J.P.; Jehoulet, C.; Beguin, F. Causes of supercapacitors ageing in organic electrolyte. *J. Power Sources* **2007**, *171*, 1046–1053. [CrossRef]
17. Li, D.; Li, S.; Zhang, S.; Sun, J.; Wang, L.; Wang, K. Aging state prediction for supercapacitors based on heuristic kalman filter optimization extreme learning machine. *Energy* **2022**, *250*, 123773. [CrossRef]
18. Liu, S.; Wei, L.; Wang, H. Review on reliability of supercapacitors in energy storage applications. *Appl. Energy* **2020**, *278*, 13. [CrossRef]
19. Ayadi, M.; Briat, B.; Lallemand, R.; Eddahech, A.; German, R.; Coquery, G.; Vinassa, J.M. Description of supercapacitor performance degradation rate during thermal cycling under constant voltage ageing test. *Microelectron. Reliab.* **2014**, *54*, 1944–1948. [CrossRef]
20. Bohlen, O.; Kowal, J.; Sauer, D.U. Ageing behaviour of electrochemical double layer capacitors—Part II. Lifetime simulation model for dynamic applications. *J. Power Sources* **2007**, *173*, 626–632. [CrossRef]
21. Zheng, F.H.; Li, Y.X.; Wang, X.S. Study on effects of applied current and voltage on the ageing of supercapacitors. *Electrochim. Acta* **2018**, *276*, 343–351. [CrossRef]
22. Sedlakova, V.; Sikula, J.; Majzner, J.; Sedlak, P.; Kuparowitz, T.; Buergler, B.; Vasina, P. Supercapacitor degradation assesment by power cycling and calendar life tests. *Metrol. Meas. Syst.* **2016**, *23*, 345–358. [CrossRef]
23. Zhang, L.; Hu, X.S.; Wang, Z.P.; Sun, F.C.; Dorrell, D.G. A review of supercapacitor modeling, estimation, and applications: A control/management perspective. *Renew. Sustain. Energy Rev.* **2018**, *81*, 1868–1878. [CrossRef]
24. Kim, S.-H.; Choi, W.; Lee, K.-B.; Choi, S. Advanced Dynamic Simulation of Supercapacitors Considering Parameter Variation and Self-Discharge. *IEEE Trans. Power Electron.* **2011**, *26*, 3377–3385.
25. Eddahech, A.; Ayadi, M.; Briat, O.; Vinassa, J.-M. Online parameter identification for real-time supercapacitor performance estimation in automotive applications. *Int. J. Electr. Power Energy Syst.* **2013**, *51*, 162–167. [CrossRef]
26. Berrueta, A.; Ursua, A.; San Martin, I.; Eftekhari, A.; Sanchis, P. Supercapacitors: Electrical Characteristics, Modeling, Applications, and Future Trends. *IEEE Access* **2019**, *7*, 50869–50896. [CrossRef]
27. Spyker, R.L.; Nelms, R.M. Classical equivalent circuit parameters for a double-layer capacitor. *IEEE Trans. Aerosp. Electron. Syst.* **2000**, *36*, 829–836. [CrossRef]
28. Pean, C.; Rotenberg, B.; Simon, P.; Salanne, M. Multi-scale modelling of supercapacitors: From molecular simulations to a transmission line model. *J. Power Sources* **2016**, *326*, 680–685. [CrossRef]
29. Torregrossa, D.; Bahramipanah, M.; Namor, E.; Cherkaoui, R.; Paolone, M. Improvement of Dynamic Modeling of Supercapacitor by Residual Charge Effect Estimation. *IEEE Trans. Ind. Electron.* **2014**, *61*, 1345–1354. [CrossRef]
30. Moayedi, S.; Cingoz, F.; Davoudi, A. Accelerated Simulation of High-Fidelity Models of Supercapacitors Using Waveform Relaxation Techniques. *IEEE Trans. Power Electron.* **2013**, *28*, 4903–4909. [CrossRef]
31. Logerais, P.O.; Camara, M.A.; Riou, O.; Djellad, A.; Omeiri, A.; Delaleux, F.; Durastanti, J.F. Modeling of a supercapacitor with a multibranch circuit. *Int. J. Hydrogen Energy* **2015**, *40*, 13725–13736. [CrossRef]
32. Saha, P.; Dey, S.; Khanra, M. Modeling and State-of-Charge Estimation of Supercapacitor Considering Leakage Effect. *IEEE Trans. Ind. Electron.* **2020**, *67*, 350–357. [CrossRef]
33. Zubieta, L.; Bonert, R. Characterization of double-layer capacitors for power electronics applications. *IEEE Trans. Ind. Appl.* **2000**, *36*, 199–205. [CrossRef]
34. Rajani, S.V.; Pandya, V.J.; Shah, V.A. Experimental validation of the ultracapacitor parameters using the method of averaging for photovoltaic applications. *J. Energy Storage* **2016**, *5*, 120–126. [CrossRef]
35. Faranda, R. A new parameters identification procedure for simplified double layer capacitor two-branch model. *Electr. Power Syst. Res.* **2010**, *80*, 363–371. [CrossRef]
36. Chai, R.Z.; Zhang, Y. A Practical Supercapacitor Model for Power Management in Wireless Sensor Nodes. *IEEE Trans. Power Electron.* **2015**, *30*, 6720–6730. [CrossRef]
37. Weddell, A.S.; Merrett, G.V.; Kazmierski, T.J.; Al-Hashimi, B.M. Accurate Supercapacitor Modeling for Energy Harvesting Wireless Sensor Nodes. *IEEE Trans. Circuits Syst. Ii-Express Briefs* **2011**, *58*, 911–915. [CrossRef]
38. Helmholtz, H.V. Studien über electrische Grenzschichten. *Ann. Phys.* **1879**, *243*, 337–382. [CrossRef]
39. Guoy, G. Constitution of the electric charge at the surface of an electrolyte. *J. Phys.* **1910**, *9*, 457–467.
40. Chapman DL, L.I. A contribution to the theory of electrocapillarity. *Lond. Edinb. Dublin Philos. Mag. J. Sci.* **1913**, *25*, 475–481. [CrossRef]
41. Yu, A.; Chabot, V.; Zhang, J. *Electrochemical Supercapacitors for Energy Storage and Delivery: Fundamentals and Applications*; Taylor & Francis: Abingdon, UK, 2013.
42. Eziani, S.; Ouassaid, M. State of Charge Estimation of Supercapacitor Using Artificial Neural Network for Onboard Railway Applications. In Proceedings of the 6th International Renewable and Sustainable Energy Conference (IRSEC), Rabat, Morocco, 5–8 December 2018; pp. 1076–1081.
43. Liu, C.L.; Zhang, Y.; Sun, J.R.; Cui, Z.H.; Wang, K. Stacked bidirectional LSTM RNN to evaluate the remaining useful life of supercapacitor. *Int. J. Energy Res.* **2022**, *46*, 3034–3043. [CrossRef]
44. Westerlund, S.; Ekstam, L. Capacitor theory. *IEEE Trans. Dielectr. Electr. Insul.* **1994**, *1*, 826–839. [CrossRef]
45. Zhang, L.; Hu, X.S.; Wang, Z.P.; Sun, F.C.; Dorrell, D.G. Fractional-order modeling and State-of-Charge estimation for ultracapacitors. *J. Power Sources* **2016**, *314*, 28–34. [CrossRef]

46. Zou, C.; Zhang, L.; Hu, X.; Wang, Z.; Wik, T.; Pecht, M. A review of fractional-order techniques applied to lithium-ion batteries, lead-acid batteries, and supercapacitors. *J. Power Sources* **2018**, *390*, 286–296. [CrossRef]
47. Freeborn, T.J. Estimating supercapacitor performance for embedded applications using fractional-order models. *Electron. Lett.* **2016**, *52*, 1478–1479. [CrossRef]
48. Dzielinski, A.; Sierociuk, D. Ultracapacitor modelling and control using discrete fractional order state-space model. *Acta Montan. Slovaca* **2008**, *13*, 136–145.
49. Smith, P.H.; Tran, T.N.; Jiang, T.L.; Chung, J. Lithium-ion capacitors: Electrochemical performance and thermal behavior. *J. Power Sources* **2013**, *243*, 982–992. [CrossRef]
50. Conway, B.E.; Pell, W.; Liu, T. Diagnostic analyses for mechanisms of self-discharge of electrochemical capacitors and batteries. *J. Power Sources* **1997**, *65*, 53–59. [CrossRef]
51. Yang, H.Z.; Zhang, Y. Self-discharge analysis and characterization of supercapacitors for environmentally powered wireless sensor network applications. *J. Power Sources* **2011**, *196*, 8866–8873. [CrossRef]
52. Ricketts, B.W.; Ton-That, C. Self-discharge of carbon-based supercapacitors with organic electrolytes. *J. Power Sources* **2000**, *89*, 64–69. [CrossRef]
53. Oickle, A.M.; Andreas, H.A. Examination of water electrolysis and oxygen reduction as self-discharge mechanisms for carbon-based, aqueous electrolyte electrochemical capacitors. *J. Phys. Chem. C* **2011**, *115*, 4283–4288. [CrossRef]
54. Gualous, H.; Bouquain, D.; Berthon, A.; Kauffmann, J.M. Experimental study of supercapacitor serial resistance and capacitance variations with temperature. *J. Power Sources* **2003**, *123*, 86–93. [CrossRef]
55. Guillemet, P.; Scudeller, Y.; Brousse, T. Multi-level reduced-order thermal modeling of electrochemical capacitors. *J. Power Sources* **2006**, *157*, 630–640. [CrossRef]
56. Lee, D.H.; Kim, U.S.; Shin, C.B.; Lee, B.H.; Kim, B.W.; Kim, Y.-H. Modelling of the thermal behaviour of an ultracapacitor for a 42-V automotive electrical system. *J. Power Sources* **2008**, *175*, 664–668. [CrossRef]
57. Guillemet, P.; Pascot, C.; Scudeller, Y. Compact Thermal Modeling of Electric Double-Layer-Capacitors. In Proceedings of the 14th International Workshop on Thermal Investigations of ICs and Systems, Rome, Italy, 24–26 September 2008; pp. 118–122.
58. Kötz, R.; Hahn, M.; Gallay, R. Temperature behavior and impedance fundamentals of supercapacitors. *J. Power Sources* **2006**, *154*, 550–555. [CrossRef]
59. Hijazi, A.; Kreczanik, P.; Bideaux, E.; Venet, P.; Clerc, G.; Di Loreto, M. Thermal Network Model of Supercapacitors Stack. *IEEE Trans. Ind. Electron.* **2012**, *59*, 979–987. [CrossRef]
60. Wang, K.; Zhang, L.; Ji, B.; Yuan, J. The thermal analysis on the stackable supercapacitor. *Energy* **2013**, *59*, 440–444. [CrossRef]
61. Buller, S.; Karden, E.; Kok, D.; De Doncker, R.W. Modeling the dynamic behavior of supercapacitors using impedance spectroscopy. *IEEE Trans. Ind. Appl.* **2002**, *38*, 1622–1626. [CrossRef]
62. Huang, S.F.; Zhu, X.L.; Sarkar, S.; Zhao, Y.F. Challenges and opportunities for supercapacitors. *APL Mater.* **2019**, *7*, 9. [CrossRef]
63. Wang, R.; Yao, M.J.; Niu, Z.Q. Smart supercapacitors from materials to devices. *Infomat* **2020**, *2*, 113–125. [CrossRef]
64. Lokhande, P.E.; Chavan, U.S.; Pandey, A. Materials and Fabrication Methods for Electrochemical Supercapacitors: Overview. *Electrochem. Energy Rev.* **2020**, *3*, 155–186. [CrossRef]
65. Wang, F.X.; Wu, X.W.; Yuan, X.H.; Liu, Z.C.; Zhang, Y.; Fu, L.J.; Zhu, Y.S.; Zhou, Q.M.; Wu, Y.P.; Huang, W. Latest advances in supercapacitors: From new electrode materials to novel device designs. *Chem. Soc. Rev.* **2017**, *46*, 6816–6854. [CrossRef]
66. Chen, X.; Paul, R.; Dai, L. Carbon-based supercapacitors for efficient energy storage. *Natl. Sci. Rev.* **2017**, *4*, 453–489. [CrossRef]
67. Meng, Q.; Cai, K.; Chen, Y.; Chen, L. Research progress on conducting polymer based supercapacitor electrode materials. *Nano Energy* **2017**, *36*, 268–285. [CrossRef]
68. Yedluri, A.K.; Kim, H.-J. Wearable super-high specific performance supercapacitors using a honeycomb with folded silk-like composite of NiCo 2 O 4 nanoplates decorated with NiMoO 4 honeycombs on nickel foam. *Dalton Trans.* **2018**, *47*, 15545–15554. [CrossRef] [PubMed]
69. Kulurumotlakatla, D.K.; Yedluri, A.K.; Kim, H.-J. Hierarchical NiCo2S4 nanostructure as highly efficient electrode material for high-performance supercapacitor applications. *J. Energy Storage* **2020**, *31*, 101619. [CrossRef]
70. Kumar, Y.A.; Kim, H.-J. Preparation and electrochemical performance of $NiCo_2$ O_4@ $NiCo_2$ O_4 composite nanoplates for high performance supercapacitor applications. *New J. Chem.* **2018**, *42*, 19971–19978. [CrossRef]
71. Guo, Y.; Yang, D.; Zhang, Y.; Wang, L.; Wang, K. Online estimation of SOH for lithium-ion battery based on SSA-Elman neural network. *Prot. Control. Mod. Power Syst.* **2022**, *7*, 40. [CrossRef]
72. Zhang, M.; Liu, Y.; Li, D.; Cui, X.; Wang, L.; Li, L.; Wang, K. Electrochemical Impedance Spectroscopy: A New Chapter in the Fast and Accurate Estimation of the State of Health for Lithium-Ion Batteries. *Energies* **2023**, *16*, 1599. [CrossRef]
73. Wang, L.; Xie, L.; Yang, Y.; Zhang, Y.; Wang, K.; Cheng, S.-j. Distributed Online Voltage Control with Fast PV Power Fluctuations and Imperfect Communication. *IEEE Trans. Smart Grid* **2023**. [CrossRef]
74. Zhang, M.; Wang, W.; Xia, G.; Wang, L.; Wang, K. Self-Powered Electronic Skin for Remote Human–Machine Synchronization. *ACS Appl. Electron. Mater.* **2023**, *5*, 498–508. [CrossRef]

technologies

Review

Risk Assessment of Heterogeneous IoMT Devices: A Review

Pritika [1], Bharanidharan Shanmugam [1,*] and Sami Azam [2]

1 Energy and Resources Institute, Faculty of Science and Technology, Charles Darwin University, Darwin, NT 0810, Australia
2 Faculty of Science and Technology, Charles Darwin University, Darwin, NT 0810, Australia
* Correspondence: bharanidharan.shanmugam@cdu.edu.au

Abstract: The adaptation of the Internet of Medical Things (IoMT) has provided efficient and timely services and has transformed the healthcare industry to a great extent. Monitoring patients remotely and managing hospital records and data have become effortless with the advent of IoMT. However, security and privacy have become a significant concern with the growing number of threats in the cyber world, primarily for personal and sensitive user data. In terms of IoMT devices, risks appearing from them cannot easily fit into an existing risk assessment framework, and while research has been done on this topic, little attention has been paid to the methodologies used for the risk assessment of heterogeneous IoMT devices. This paper elucidates IoT, its applications with reference to in-demand sectors, and risks in terms of their types. By the same token, IoMT and its application area and architecture are explained. We have also discussed the common attacks on IoMT. Existing papers on IoT, IoMT, risk assessment, and frameworks are reviewed. Finally, the paper analyzes the available risk assessment frameworks such as NIST, ISO 27001, TARA, and the IEEE213-2019 (P2413) standard and highlights the need for new approaches to address the heterogeneity of the risks. In our study, we have decided to follow the functions of the NIST and ISO 270001 frameworks. The complete framework is anticipated to deliver a risk-free approach for the risk assessment of heterogeneous IoMT devices benefiting its users.

Keywords: Internet of Things; Internet of Medical Things; framework; risk assessment; privacy risk; security risk

check for updates

Citation: Pritika; Shanmugam, B.; Azam, S. Risk Assessment of Heterogeneous IoMT Devices: A Review. *Technologies* **2023**, *11*, 31. https://doi.org/10.3390/technologies11010031

Academic Editor: Manoj Gupta

Received: 4 January 2023
Revised: 2 February 2023
Accepted: 7 February 2023
Published: 14 February 2023

1. Introduction

Kevin Ashton, a British technology pioneer, first introduced the term Internet of Things at Proctor & Gamble in 1999 in supply chain management. However, the definition has become more comprehensive in the past two decades, transforming various domains of our lives through agriculture, healthcare, transport, and the environment (smart buildings, energy-efficient cities, and infrastructure) [1,2]. In this section, the terms "Internet of Things" and "Internet of Medical Things" are defined along with a brief background. Statistics are provided to demonstrate their prevalence and level of integration in our lives. Furthermore, objectives, motivation, and contribution are outlined, and a quick overview of the structure of this paper is provided.

1.1. IoT (Internet of Things)

In general, the term refers to Internet-enabled objects like electronic devices and sensors interacting without human-to-human or human-to-computer interaction. These not only facilitate our life, but are also an integral part of it, providing services everywhere around the world [3].

Gartner's global government IoT revenue for electronics and communication will reach USD 21.3 billion in 2022, an increase of 22% over the previous year [4], and USD 58 billion by 2025 [5]. There are currently 31 billion "things" connected, which is estimated to balloon to 75 billion by 2025 [6]. A report published in 2018 by PWC [7] for the Australian

Computer Society (ACS) states that IoT has the potential to bring about an annual benefit of AUD 194–308 billion in Australia alone over the period of eight to eighteen years. Out of the top five industries that account for 25% of Australia's gross domestic product (GDP), the health industry alone contributes significantly. The number of IoT-connected devices globally from 2019 to 2030 is depicted in Figure 1.

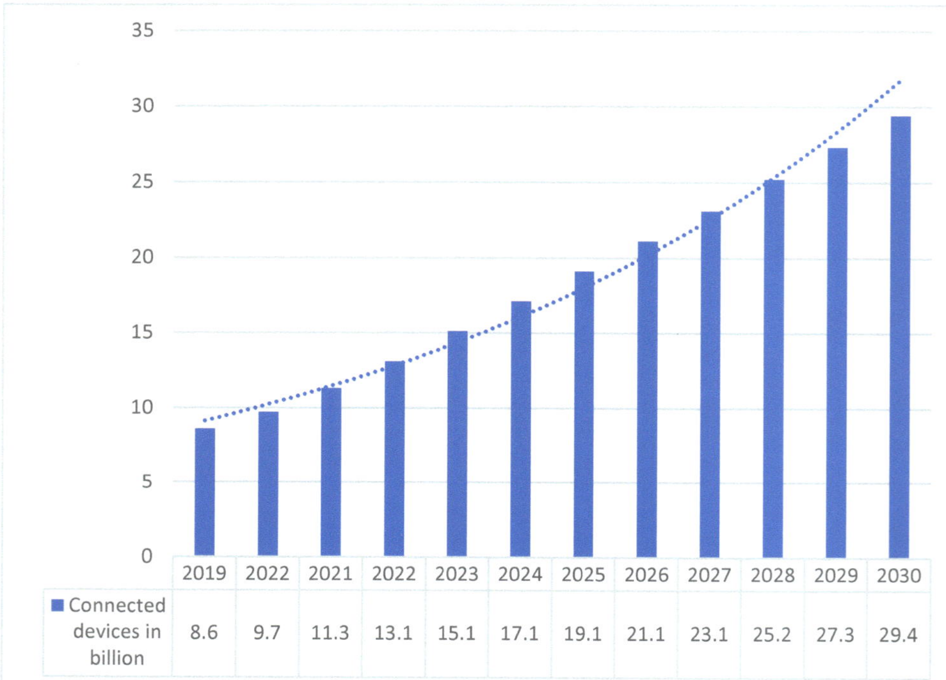

Figure 1. Number of globally connected IoT devices [8].

However, despite several benefits and economic potential, it is widely acknowledged that security and privacy have become key concerns that will affect the future development of IoT [9]. A compound annual growth rate of 33.7 percent is anticipated for the worldwide IoT security market from 2018 to 2023 due to the proliferation of IoT device cyberattacks, growing IoT security mandates, and rising security concerns [10]. This rise uncovers many new and emerging threats; the Kronos and Colonial Pipeline ransomware attacks of 2021 are high-profile examples [11].

Based on our literature search, it has been found that security and privacy issues have received massive research attention but the focus on risk assessment has not been adequately explored [12]. This fact emphasizes the need for a risk assessment methodology, since creating a generic, universal approach for all the devices would be challenging.

1.2. IoMT (Internet of Medical Things)

IoMT is a cloud-connected network of medical devices used to transmit data [13]. IoMT has gained popularity by incorporating connected medical devices, computing, and clinical systems due to its efficiency and quality of services. It is considered as a breakthrough in the medical world having billions of Internet-connected medical devices. Heterogeneous IoMT devices refer to the diversity of these medical devices which are used to connect to the Internet. These devices are interconnected and are able to share and collect data which include a wide range of standards and technologies [14]. The global

IoMT market is projected to increase from USD 72.5 billion in 2020 to USD 188.2 billion by 2025, and the highest compound annual growth rate (CAGR) expected during the forecast period is APAC (Asia Pacific) due to its advancement over the previous decade, globalization-inspired government policies, and expansion of digitalization [15]. IoMT can potentially be a 'Game-changer Technology' if the concepts are applied tactfully [16].

It is believed to be a one-stop solution to the absence of medical resources and has helped minimize unnecessary hospital visits [17]. However, IoMT devices are more susceptible to cyberattacks than any other sector, as they are positioned in networks without considering risks. Medical device security has been a weak spot for healthcare firms. An article published by Cynerio in January 2022 reports that 53 percent of IoT and IoMT devices in hospitals are vulnerable to cyberattacks [18].

There are many reasons behind these risks, and to help the healthcare industry protect its patients, the National Institute of Standards and Technology (NIST) recently updated its cybersecurity guidance for the medical sector on July 21, 2022. The healthcare services will benefit from this update to preserve the confidentiality, integrity, and availability of electronically protected health information [19]. Asimily, a leading risk management platform for IoMT devices that provides safe and trusted care, has prioritized understanding the risks of IoT devices.

Because risk assessment and threats are not static targets [20], we constantly need to monitor the devices, detect irregular behavior, and alert the handlers to remediate any identified anomalies. Additionally, there is a constant need for a risk assessment model structured to address the security and privacy risks of IoMT devices. To address these needs, this paper presents a risk assessment framework that will identify potential risks and recommends specific risk assessment attributes, which are discussed in detail in Section 3.

1.3. Research Questions

In this section, research questions have been formed to better understand the available risk assessment approaches and frameworks. They will help us derive various IoMT research and implementation gaps.

Research Question 1. What are the approaches used in the existing literature for the risk assessment of IoMT devices? (Please refer to Section 2.5 for the approaches)

Research Question 2. Which of the available frameworks and standards can be applied for the risk assessment of IoMT devices? (Please refer to Section 2.7 for the approaches)

1.4. Research Objectives

Based on the above research questions, the following objectives have been formed.

Objective 1. The objective here is to provide an overview of the available risk assessment frameworks and standards employed for IoMT devices.

Objective 2. The objective here is to derive the standard criteria and limitations of these frameworks, and based on these criteria, how they can be modified or merged to provide a risk assessment for the IoMT devices.

1.5. Motivation

This research is motivated by the understanding that the assessment frameworks intended for use in various IoT scenarios may not directly address the need for IoMT-based devices. It aims to investigate the potential that existing assessment frameworks offer in addressing security and privacy concerns of IoMT-based devices by identifying their key functions. This research will share new knowledge with other researchers in the respective field and help those using the methodology for the risk assessment of IoMT-based devices.

1.6. Contribution

The contribution of this paper is highlighted by comparing several aspects of other papers such as techniques for security and privacy risks, risk assessment frameworks developed for various IoT- and IoMT-based scenarios, application areas, and architecture.

Between 2014 and 2022, several publications have been reviewed, but none of them have addressed all the necessary aspects including applications, architecture, risks, and common attacks for the risk assessment of IoMT devices, and only a small number of papers have discussed the need for a framework, which encourages us to further our research on the subject. Our primary contribution is the risk assessment of smartwatches, portable wireless vital monitors, and lung monitors. The heterogeneous properties of these devices have led us to propose a framework for risk assessment.

This framework classifies the methodology process into five steps so that it can be effectively understood and can also be applied by other researchers across the heterogeneous network.

1.7. How the Paper Is Organized

To discuss in detail, the paper is organized as follows: Section 2 presents the IoT applications and related risks. Furthermore, the focus is shifted towards the most prominent area of IoT, i.e., IoMT, also known as Healthcare IoT, and explains frequently used applications of IoMT. The architecture of IoMT devices and common attacks are also covered. Several papers on IoT, IoMT, risk assessment, and frameworks are also discussed in this section. Additionally, risk assessment and currently used frameworks are described. Section 3 anticipates the proposed methodology and the steps expected to be performed in the risk assessment of IoMT devices. Finally, a conclusion has been provided for this research paper, along with a recommendation for future research.

2. Literature Review

The overall literature review process is structured by first describing various IoT applications in detail and their associated risks. Furthermore, we limit our research to IoMT devices, their applications, architecture, and some of most common cyberattacks. The literature is then thoroughly evaluated, and the Preferred Reporting Items for Systematic Reviews and Meta-Analysis (PRISMA) model is employed to illustrate the various stages of the review. It displays the number of identified, added, and removed records, and based on these identified records, related works are reviewed. Lastly, the study of some of the available frameworks and directions for using them in our research are provided.

2.1. Applications of IoT and Its Associated Risks

Given the unique nature of cyber risks and vulnerabilities of the IoMT devices, coming out with a new risk assessment framework requires understanding both IoT applications and their associated risks, since IoMT is a subset of IoT and the risks can be similar to IoT risks as well.

2.1.1. Applications

IoT's cosmic evolution has significantly contributed to advancing technology and assisting humanity in many ways. The potential application domains for IoT are depicted in Figure 2. In the agriculture domain, IoT helps farmers to earn profit, reduce labor costs, and increase their agricultural output, as well as improve the quality of products. Some IoT applications involved in agriculture are crop growth environment monitoring, water-saving irrigation, intelligent agricultural machinery, etc. In addition to these applications, China is using IoT in farmland planting, aquaculture, animal husbandry, and product safety traceability [21]. Limited energy has been a substantial issue for IoT devices, as they are expected to have a superior battery life to run smoothly for an extended time. Some of the energy management research perspectives in IoT are energy-efficient cognitive radio IoT, 5G IoT, and energy-efficient social network software IoT. With IoT, residential and commercial buildings are undergoing a drastic change. Automation using IoT plays a vital role in smart buildings providing efficient, comfortable, and secure environments. The research in [22] presents all the potential applications of smart buildings. In the automotive industry, IoT is considered a blessing and is envisioned as shaping its future [23]. One of the core technologies is self-driving cars, which are being tested in some countries and will soon be available. There is a rapid growth in the next application of IoT, i.e., security and surveillance, to protect private organizations. Supply chain and logistics are the most common examples of industrial applications.

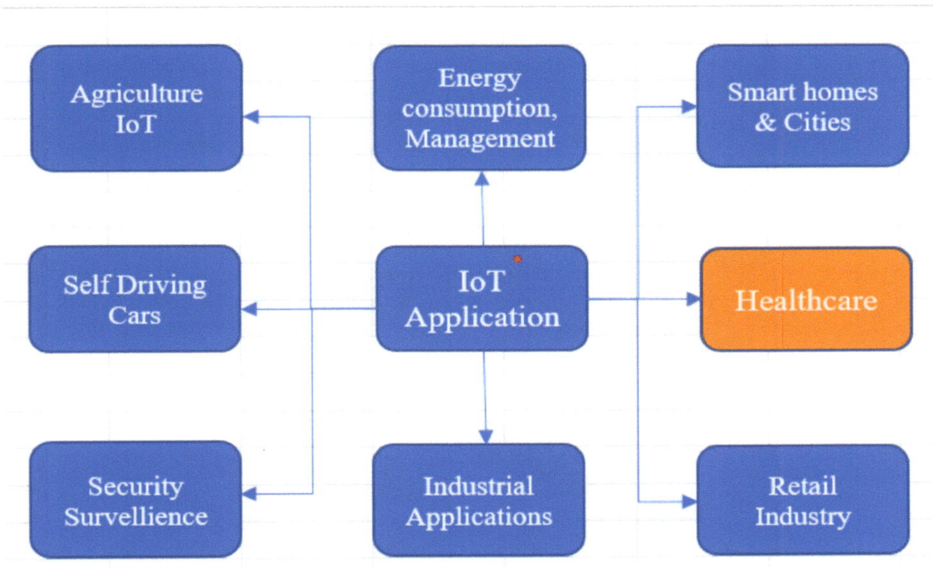

Figure 2. Applications of IoT.

While IoT focuses on multiple domains, in our research, we have concentrated on one of its subsets, IoMT. It incorporates small intelligent equipment and devices to support the healthcare system. These devices may access various health issues, fitness levels, and number of calories burnt in the fitness center. They are also used to monitor the critical health conditions of the patients in hospitals and trauma centers. Hence, IoMT has completely changed the structure of the medical domain by facilitating it with high technology and smart devices. Moreover, IoT developers are actively involved in elevating the lifestyle of the disabled and senior citizens and in making it accessible to the masses [24]. Despite IoMT's growth, there are still challenges with its implementation which need to be addressed. Hence, it is regarded as a critical area, because even the slightest error can be fatal [25].

All possible IoT application areas are discussed above, including agriculture, retail, security, the automotive industry, energy consumption, smart homes, smart cities, industries, and healthcare. As IoT continues to gain popularity in agriculture, we have included a few applications that have been widely used across several countries, including China. With the world moving towards less energy consumption, the energy sector also plays a significant role in our everyday lives. Self-driving cars are an example of IoT increasingly becoming prominent in the automotive industry. In a similar manner, we have discussed IoT's growth in security and supply chains. Due to IoT's broader scope, it is important to know what other applications are available before moving on to IoMT, our primary focus.

2.1.2. Risks

Contrary to the dominance of IoT devices in our lives, its downsides cannot be overlooked. As the applications of IoT continue to escalate, it brings in significant security, privacy, and ethical challenges which introduce the need for a comprehensive overview covering all these challenges [26].

- Privacy risks—With the advancement of IoT and the diffusion of technology, privacy has become a prominent issue. IoT devices collect, analyze, and transmit a massive amount of confidential data that must be protected from adversaries [27]. The reason behind this privacy concern is the ubiquitous connectivity of IoT devices and the universal distribution of information [28]. For users of IoT medical devices, the concern grows wider due to the fear of sharing personal data such as dietary habits, exercise regimens, sleep patterns, and running routes. Hence, safeguarding them becomes more challenging when using medical devices such as smart monitors, smart test kits, and smart assistive technologies at home [29]. In October 2016, malware Mirai generated tens of millions of IP addresses on Dyn, resulting in parts of the Internet going down, including Twitter, Netflix, Cable News Network (CNN), Reddit, etc. [30].
- Security risks—A system is considered secure if it satisfies three primary objectives: confidentiality, integrity, and availability. It is commonly called the CIA Triad. Confidentiality signifies that private information is not accessed by unauthorized users. Integrity is keeping the message intact between the sender and receiver, meaning that IoT devices are not utilized or modified by unauthorized services, and availability is the continuation of computing resources, information, and services against disruption attacks. The security of IoT is essential, as most of the data collected in IoT devices are personal and need security. These sensitive data in IoT could be an open invitation to attackers to take and consume them in many ways [31].

  In the context of security risks to IoMT devices, confidentiality pertains to safeguarding the medical information that a patient shares with doctors. This information must be protected from intrusion, eavesdropping, and organizations that could cause harm to the patient or use the patient's medical information against him. Integrity safeguards against unauthorized users alerting or destroying patient data, primarily ensuring that they reach their destination intact during wireless transmission. Availability is the efficiency of servers and medical devices to provide for users when required. The system needs to be modified to provide a suspect data storage or transmission channel in the event of a DoS attack [32].

- Ethical risks—In general terms, ethics means what is morally good or bad and ethically right or wrong. Ethical risks in the context of IoT devices are actions that are outside of a professional standard. Any new technology designed for the convenience of people will also have adverse effects on individuals and society. Thus, it is essential to define ethical rules and legal regulations to protect them. Since personal data will be in the system owner's hand, it may not be possible to control each data flow; thus, ethical manners and observing user rights are highly significant [33]. For example, Volkswagen, a vehicle manufacturing company, developed and installed software to elude diesel emissions tests. This action violated the USA's Clean Air Act, compromised organization and industry standards, and resulted in massive reputational and financial losses [34].

  Given that the healthcare ecosystem is highly interconnected and generates a significant amount of data containing personal health information, some of the ethical risks associated with IoT, such as difficulty in identification, unpredictable behavior, life threats, and difficulty in controlling the data, may also exist in the IoMT environment.

An object needs to be identified to be connected, but data collected by these objects make it difficult to precisely identify the owner of the object. Therefore, collecting these data without the consent of the user makes it a significant issue, as most of the data collected on IoMT devices are personal. Similarly, a data breach in the IoMT network of connected devices can harm a patient's life directly, as collective information about their health is being shared [26].

Being a relatively new technology and an extension of IoT, all the risks associated with IoT are also associated with IoMT [34]. We are keen to investigate and see if these risks can be mitigated. In the above section, it is noted that security, privacy, and ethical risks are common to IoT devices and need to be addressed to protect them. The ubiquitous connectivity of IoT devices makes these risks omnipresent. We also discussed the triad that risk-free devices should meet: confidentiality, integrity, and availability. Furthermore, we have discussed how these risks apply to IoMT devices as well.

The tabular representation of the different types of risks is shown in Table 1 below.

Table 1. Types of IoT risks [26,31].

Type of Risks	Concerns	Concern for IoMT Devices
Privacy Risk	Ubiquitous connectivity Universal distribution of information	Sharing of personal data (dietary habits, sleep pattern, running routes, etc.)
Security Risk	Data breaching Data compromised during the wireless transmission	Attackers may gain access to and modify patient data
Ethical Risk	Ethical rules may be obstructed	Adverse effect on individual using the IoMT device

2.2. How IoMT Works

The healthcare industry has improvised from the time it was more doctor-centric, to patient-centric, now to technology-centric using IoT, cloud computing, fog computing, and tele healthcare technologies for sharing data [35]. The transformation of the healthcare sector has largely been improvised through the adaptation of IoMT by providing efficient and accurate services in a timely manner [36]. IoT is not a panacea, but if implemented wisely and strategically, it can change the healthcare industry for the better [37]. IoMT has the power to connect various devices, users, sensors, databases, etc., and is designed to facilitate medical services in a unified manner. Many of the medical tasks such as disease diagnosis and chronic disease monitoring can be performed remotely with more efficient and less costly healthcare services. IoMT is considered as networked communication between doctors and their patients through the sensors connected to the patient's body which can be used for monitoring, diagnosis, and further treatment [35]. The demand for IoMT devices has been soaring over the past few years, and in [38], it was forecasted that IoMT is ready to claim the most significant share of the IoT market after analyzing the trend and developments in the global industry. The changing face of the old healthcare system into a smart system is attributed to the arrival of newly developed devices revolving around IoMT technology. The sudden outbreak of COVID-19 in the past two years has forced people to take precautionary measures and prioritize their health [25]. To save and improve quality of life, IoMT has opened new opportunities in the healthcare sector and changed the way of doing things. Areas such as clinical decision making, patient record management, and data acquisition have become effortless [39]. A few of the familiar IoMT applications have been described below and are represented in Figure 3 along with the systems to which they are connected, such as hospitals, diagnostic centers, and ambulances.

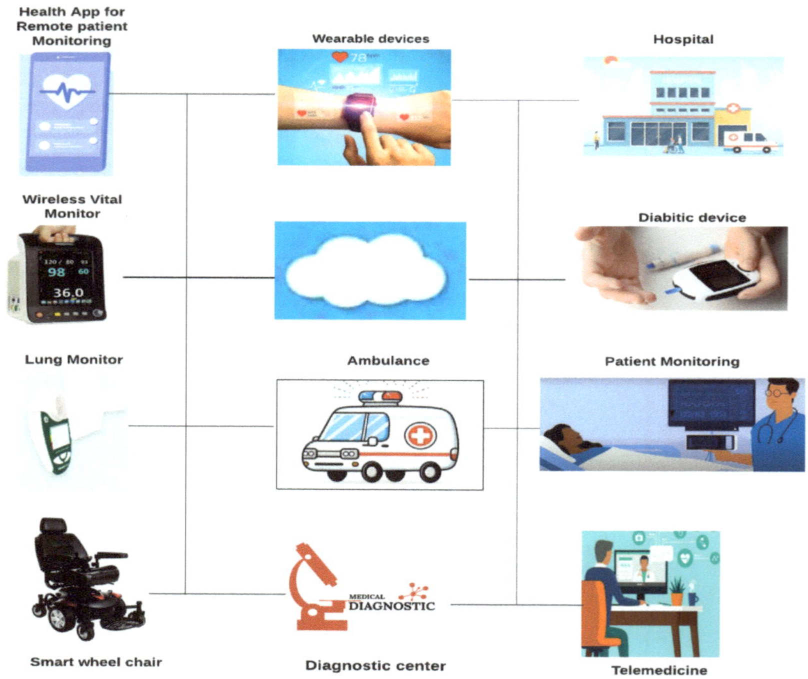

Figure 3. Applications of IoMT.

2.2.1. Applications of IoMT

- Built-in sensors and wearable devices help to improve the healthcare industry and provide real-time health information, reducing the load on the medical staff. Assisting in the early detection of disease and infection symptoms enhances the efficiency of health monitoring. A common example is a glucose monitor linked to an insulin pump with an automatic suspension of insulin infusion, which is continuously monitored using these technologies [40]. Wearable smart devices are easy to use and are capable of monitoring heart rate, ECG (electrocardiogram patterns), blood glucose level, and cardiac pacemaker's activity in real-time and transmitting the data to the doctor [41]. It has greatly benefited patients, doctors, and healthcare professionals. The smart devices are connected to the user's smartphone and to the remote system to transmit data in a faster way.
- Telemedicine, commonly known as e-medicine or telehealth, is a new concept referring to the remote delivery of healthcare services, like consultations and tests [42]. Without physically seeing the patient, healthcare professionals can examine and treat patients. Similarly, patients can communicate with their doctors from the comfort of their homes by utilizing personal technologies. Blood sugar level, blood pressure, temperature, and other vital measures can be captured by the patients and can be provided to the doctors. Telemedicine systems are based on futuristic developing technologies and are used for efficient infection prevention [43].
- Remote patient monitoring helps in monitoring glucose levels and heart activities of the patients. Doctors can receive real-time updates if anything goes wrong [40]. It proved appropriate during the COVID-19 pandemic, as doctors were able to monitor patients remotely with fingertip medical data like blood pressure level, glucose level, ECG, temperature, pulse rate, heart rate, etc. [44]. Patients could monitor the status of

their disease and receive required medical needs on their phone without visiting the doctor [45].
- Diabetic devices are very commonly used, and most diabetic patients keep a glucose monitor and keep track of their glucose level, thus saving their time. IoMT devices also help insurers to view users' data more quickly, making the health insurance claim process faster.
- Smart wheelchair—The world today makes a massive difference in the life of people with restricted mobility. A smart wheelchair works depending upon the mood of the disabled person and helps effectively in different weather conditions, improving quality of life.
- A wireless vital monitor can be used both in hospitals to ease the load on a nurse and at home after the patient is discharged. It allows continuous recording of vital signs. It can measure heart rate, temperature, respiratory flow, ECG, etc., and these data are sent directly to an interactive monitoring device via Bluetooth for the doctor to check regularly [43,44].
- Lung monitors are mainly used by discharged patients to measure their vitals. They provide accurate and effective monitoring of lung function for respiratory conditions, including COPD, cystic fibrosis, and post-transplant patients.

There are a number of application areas for IoMT that we have discussed. Wearable devices with built-in sensors are becoming more popular, easing the workload of medical staff as they can also be used remotely. Another useful application is telemedicine, which allows patients to monitor their disease status and receive medication recommendations. Patients with diabetes are widespread users of the IoMT application. In addition, smart wheelchairs, lung monitors, and wireless vital monitors are all prevalent application fields [46].

However, the interconnectivity between numerous devices makes them vulnerable to security breaches in a way similar to how other networked computing systems are vulnerable, but the consequences can be pernicious, as they can be dangerous to users' lives [47]. It is a necessity to perform a risk assessment for these IoMT devices. In our study, we plan to perform a risk assessment for the wireless vital monitor, the smartwatch, and a lung monitor.

Different researchers have explained IoT, IoMT, and their components differently with respect to their own interests and aspects. In the next section, the criteria used to perform the literature review are explained. Firstly, the data sources are enumerated, and then the search and selection process is explained.

2.2.2. IoMT Architecture

A discussion on IoMT application and technology is not complete without reference to the architecture model, and not all applications use the same IoMT architecture. Most of the present IoMT systems are divided into three layers: the perception layer, the communication network layer, and the application layer.

- Sensor layer or perception layer—This is a foundational layer and deals with the collection of data from the source, providing the necessary viewpoint from the gathered data [48]. This layer ensures the precise sensing of the parameters related to health issues [32] and consists of hardware such as sensors, controllers, and actuators. The hardware presently in use includes the radio frequency identification (RFID) reader/tag, GPRS, facial recognition camera, fitness smartwatch, health-monitoring sensors, insulin pumps, and infrared temperature sensors. Wearable sensor devices, implanted sensor devices, and ambient sensor devices are three categories for the sensors [32]. As depicted in Figure 4, the perception layer comprises two sublayers, the data access sublayer and the data acquisition sublayer. The primary task carried out by the data acquisition sublayer is perception of the gathered data, for which it makes use of various medical perception equipment and signal acquisition equipment. Some of the major signal acquisition methods can be GPRS, RFID, graphic code, etc. [49]. Short-range data transfer technologies like Bluetooth, Wi-Fi, Zigbee, 4G, 5G [50,51], etc., are then used to transfer this obtained data to the network layer via the data access sublayer.

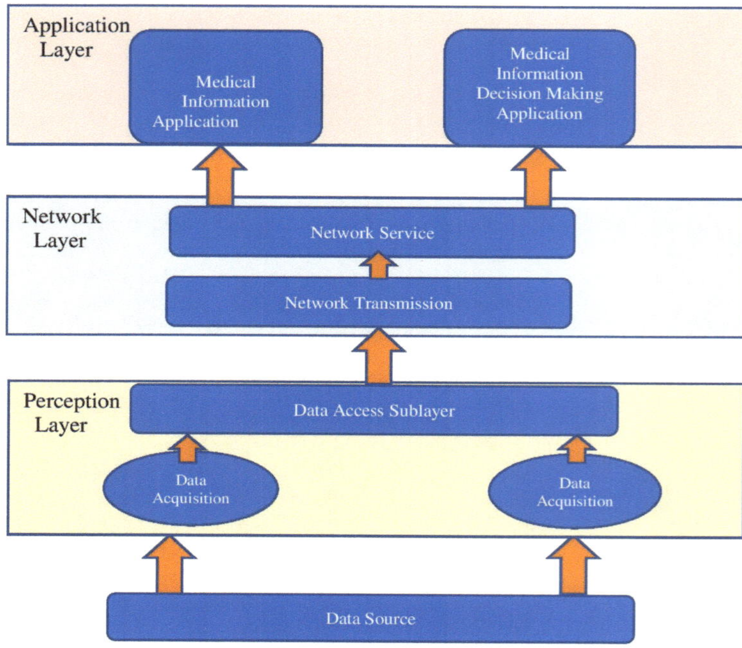

Figure 4. IoMT architecture [49].

- Network layer—This is the subsequent layer, and it offers a wide range of platforms, interface-related services, and data transmission methods. This layer consists of two sublayers, which are the network transmission layer and the network service layer. The network transmission sublayer transmits the data it receives from the perception layer in real time and with accuracy using the Internet, mobile communication networks, wireless sensor networks, etc. The integration of various networks, information description formats, and data warehouses is accomplished via the service layer, which also offers a variety of platform-related services and open interface services for these integrations [49].
- Application layer—This is the topmost layer which utilizes the information taken from the network layer to manage medical records by means of various applications [32]. Like the previous two layers, this layer is also composed of two sublayers: the medical information decision-making application layer and the medical information application layer. The medical information application layer incorporates various healthcare equipment and other materials related to information for maintaining patient information, such as inpatient, outpatient, medical treatment, tracking system, fitness/ health system, remote diagnostic system, telemedicine, medical e-record, etc. [52]. On the other hand, the medical information decision-making application layer deals with the analysis of various pieces of information, such as patients, disease, medication, diagnosis, treatment, etc.

In [53], the researcher has adopted a three-tier architecture consisting of the sensor level, personal servers, and medical servers. As the name suggests, the sensor level contains sensors and medical devices, which form a local network knows as the Body Sensor Network (BSN). For wireless communications at the sensors and personal server level, low-power wireless technology protocols including RFID, Bluetooth Low Energy (BLE), and Near Field Communication (NFC) are employed. Data collected by the medical devices will be sent to personal servers, which could be either on-body devices or off-body devices. Prior to being transferred to the centralized medical servers, this layer will locally process and store patient data. It is needed when a network connection is lost or when a user needs access to the patient's data remotely. The last layer is the medical server layer which consists of an algorithm or program for early diagnosis, rehabilitation progress assessment, or continuous patient monitoring like MobiCare [54] or BSN-Care [55]. This architecture prioritizes usability and power consumption, but it does not cater to any security or privacy risks, leaving these considerations to future work.

In [56], the researcher here proposes an end-to-end architecture called the mHealth System, which is able to connect the IoT smart sensors directly with the Smart Healthcare System (SHS). This architecture consists of three layers: the data processing layer, the data collection layer, and the data storage layer. The data collection layer, which is the bottom layer, consists of IoT devices that can sense and collect medical parameters. The next layer, which is the data storage layer, stores medical data on wide-scale and high-speed storage racks. The topmost layer, the data processing layer, involves various techniques to analyze collected sensor data.

A foundational layer, which is called the perception layer, deals with data collection and making interpretations about gathered information. It facilitates the sensing of health parameters. The data are collected and transferred to the network layer where they are transmitted in real-time. The top layer is the application layer, where medical records are managed and information is gathered from the previous layer. Comparing the architectures, we can conclude that the bottom layer has sensors with direct contact with the human body. The middle layer is used for storage and processing of data, and the last layer is used for providing services to the users.

2.2.3. Most Common Cyberattacks on IoMT

- Denial-of-service attack—This type of attack occurs when an IoT system is prevented from uploading patients' health information onto the respective cloud-based services or medical database or when the medical professional is unable to retrieve patient information through the IoMT system. Frequent data backups would be essential for recovering historical data, but real-time services would be disrupted. Time stamping and strong authentication on IoMT devices may be taken into consideration to minimize these types of attacks [52].
- Injection attack—Data integrity is essential to ensure that the data received have not been altered or distorted in any way during communication channels. False data injection attacks, which cause false data to be transmitted to a hospital data center, are one example of such attacks. Another frequent attack is an SQL injection, which provides back doors for cyber criminals to access medical databases.
- Data leakage and privacy—Compilation and storing of an individual's health and movement records should conform to legal and ethical laws on privacy. Owing to the transparent and accessible nature of wireless messages, IoMT systems are also more likely to suffer from data leakage through sniffing attacks, and these include eavesdropping, traffic analysis, and brute force attacks (trial and error to guess login info) [52].

As IoMT devices increase in popularity, attacks and hacking opportunities also increase [57]. Insecure devices can put patients at risk and damage a healthcare organization's entire infrastructure. Above, we have discussed some common cyberattacks on IoMT devices.

Some of the most common attacks are presented in Figure 5 below.

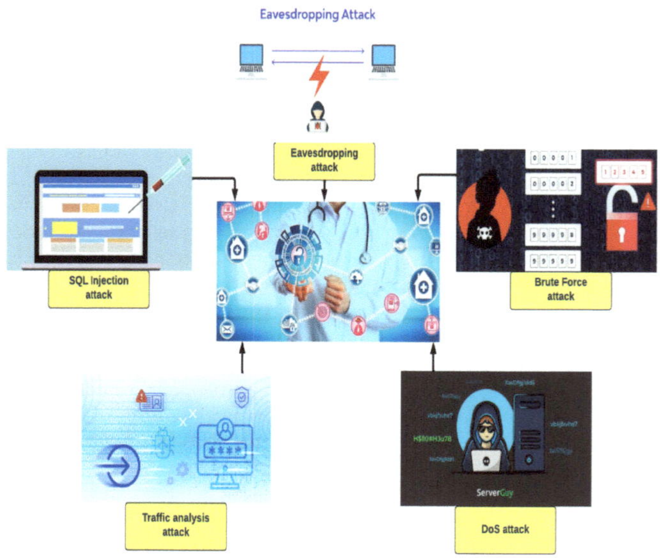

Figure 5. Common attacks of IoMT.

In the next section, the criteria used to perform the literature review are explained. Firstly, the data sources are enumerated, and then the search and selection process is explained.

2.3. Data Collection

To address the research objective, a systematic literature search was carried out on major indexing databases following PRISMA guidelines and is represented in Figure 6.

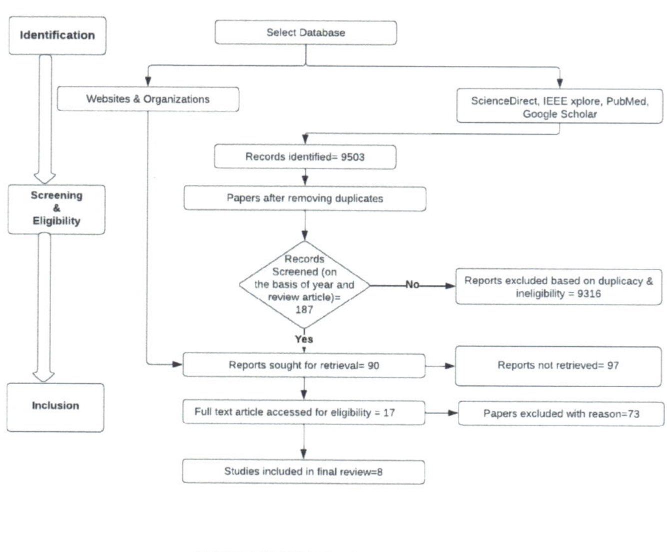

Figure 6. PRISMA.

The electronic search uses IEEE Xplore, ScienceDirect, MDPI, PubMed, Google Web browser, and Springer using the terminologies such as "Internet of Things", "Risk Assessment in IoMT", "Internet of Medical Things", and "IoMT Frameworks", as depicted in the Table 2. Academic review papers published between 2014 and 2022 were searched and were further studied based on their title/abstract. PubMed was particularly useful in gaining additional information about IoMT. In addition, some results have been removed to ensure that the paper contains only data from journals, top-quality review papers, and conferences. This step was performed by selecting studies published in journals and conference papers with competitive acceptance rates. Some studies were eliminated due to quality restrictions such as a slight increase from previous studies, technical issues, full-text unavailability, and were ruled out. It is worth noting that a keyword search of "IoT" returned maximum results on all the indexing databases, and the least number of papers were found on IoMT risk assessment and their framework.

Table 2. Keyword search of all indexing databases.

Source of Database	IoT	IoMT	Risk Assessment in IoMT	IoMT Framework
ScienceDirect	2599	79	45	63
IEEE Xplore	3551	82	3	18
PubMed	373	21	3	2
Springer	1923	55	13	42
MDPI	617	13	0	1
Total	9063	250	64	126

In the identification phase (phase 1), 9503 records were identified in the original and umbrella search, taken from the five aforementioned databases, websites, and organizations. Based on duplication and ineligibility, 9316 records from these studies were eliminated and 187 records remained. Phase 2 involved screening a total of 187 papers based on the year and review article. Out of those, 90 records were sought for retrieval so that they could be further examined based on the title and abstract, while 97 papers were not retrieved since they did not fit into the objective of this work. Out of the 90 full-text papers that were retrieved in the third phase, 73 papers were excluded because they met the exclusion criteria, which included papers that only discussed IoT devices but did not contain much information about IoMT-based devices or did not discuss risk assessment methodologies. The remaining 17 papers were sought for full-text review, out of which 8 papers were finally included in this systematic review, as they adhered to the aims/objectives of this study and met the inclusion criteria.

The pictorial representation of the selection of keywords from various databases is shown in Figure 7 below.

Figure 8 presents the search keyword "IoT" in all five databases, which brought about a total number of 9063 open-access review articles between 2014 and 2022.

Figure 9 presents the search keyword "IoMT" in all five databases, which brought about a total number of 250 review articles between 2014 and 2022.

Figure 10 presents the search keyword "Risk assessment in IoMT" in all five databases, which brought about a total of 64 open-access review articles between 2014 and 2022.

Figure 11 presents the search keyword "IoMT Framework" in all the databases, which brought about a total of 126 open-access review articles between 2014 and 2022.

Applicable studies from the aforementioned data sources were carried out in three rounds.

- Round 1—An electronic search was conducted to identify and categorize the literature review related to primary studies. The title, abstract, and introduction were read to narrow down the selection of relevant papers, thereby removing the irrelevant studies.
- Round 2—The relevant papers selected in round 1 were carefully examined, and those found irrelevant were removed.

- Round 3—A snowball search using the reference list of papers from round 2 was applied to distinguish relevant papers and include them. If found applicable, they were read carefully and included.

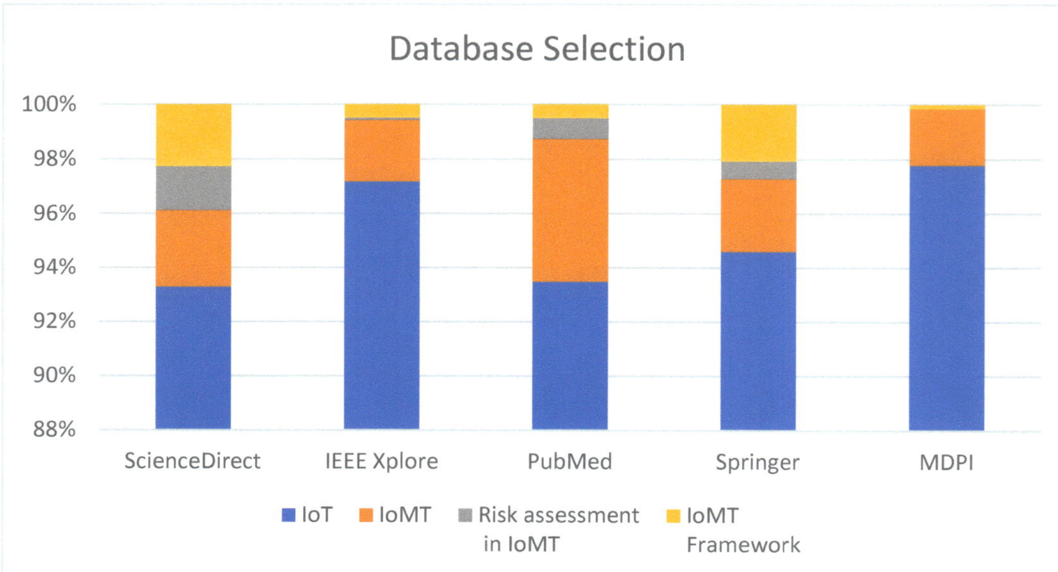

Figure 7. Database search selection.

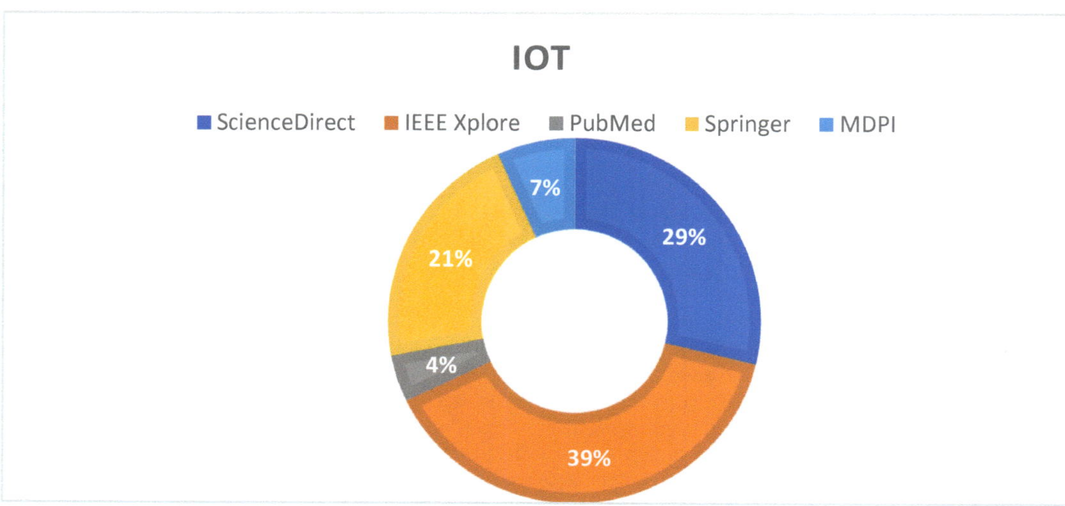

Figure 8. Search result for term IoT.

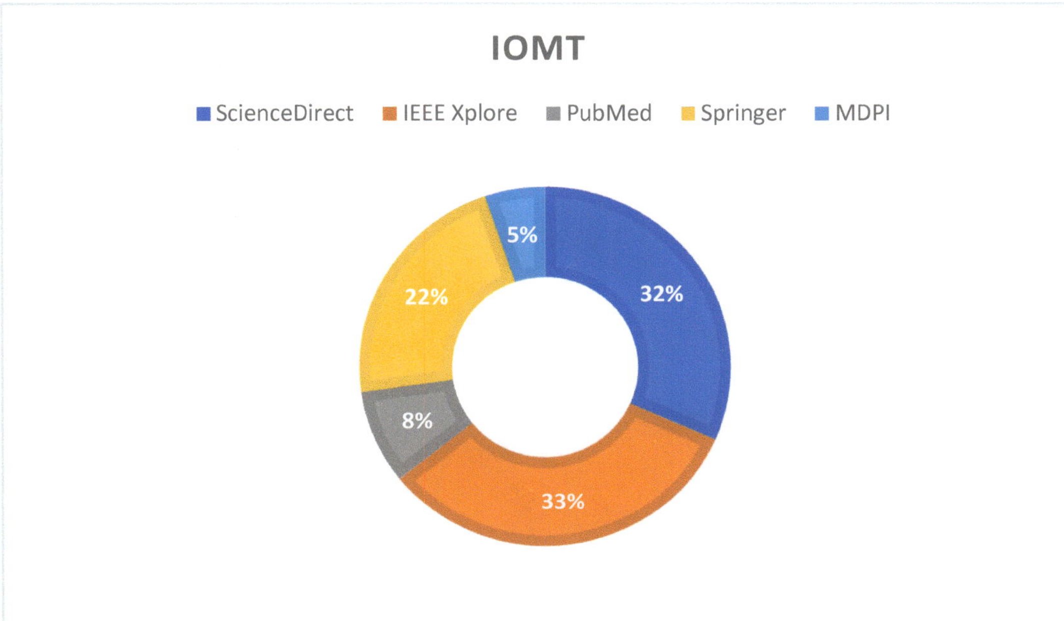

Figure 9. Search result for term IoMT.

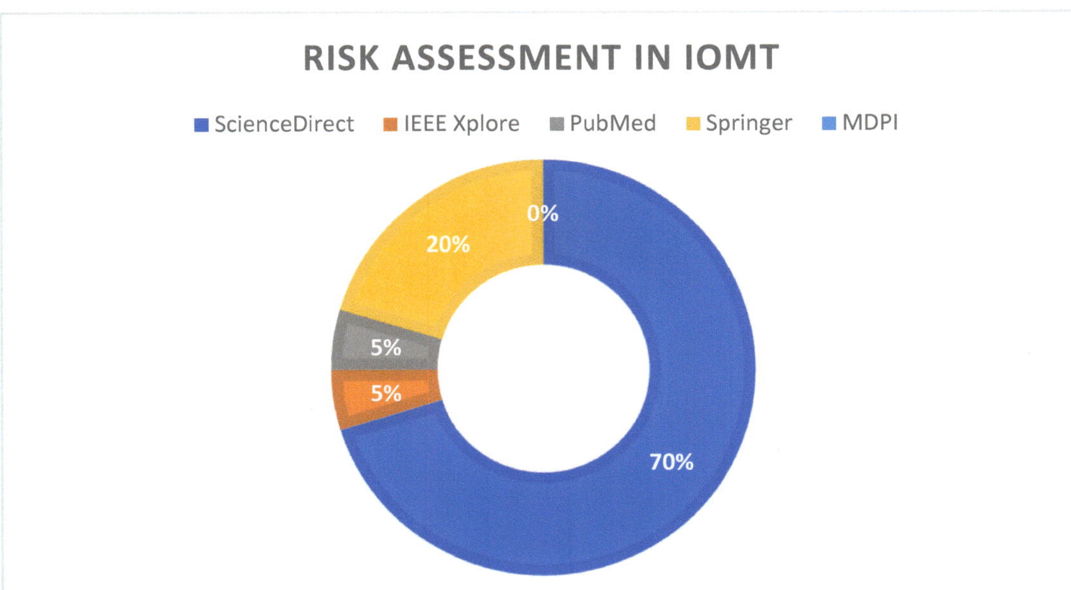

Figure 10. Search result for term Risk Assessment of IoMT devices.

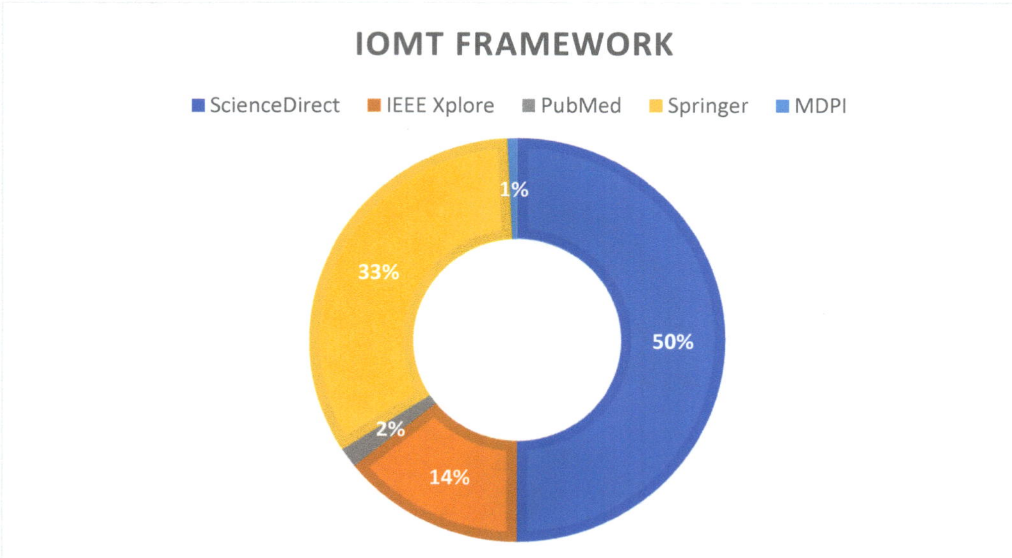

Figure 11. Search result for term IoMT Framework.

2.4. Quality of the Selected Papers

Different inclusion and exclusion approaches were applied to the remaining series of studies generated for the subsequent second and third rounds. In order to narrow down our search, we applied some exclusion criteria to the number of papers retrieved. In the selection process, an English-language criterion was first applied, while duplicates were removed based on keyword searches. All the papers were then reviewed for relevance based on their titles.

Our next step was to access the abstracts and introductions of the retrieved papers, which helped us decide whether or not to add a paper to our database for further research, following which an in-depth analysis of papers related to IoT, IoMT risk assessment, their frameworks, and countermeasures was conducted. Additionally, some papers were excluded during this step and were sorted based on the reason for exclusion. Retention was used only for the purpose of analyzing the literature review and answering the stated research questions.

2.5. Review of the Existing Literature

In this section, papers based on the aforementioned keywords are explained along with their contribution. Various papers relevant to IoT applications, security, and architecture have gained considerable attention. Papers related to IoT frameworks and risk assessment have also been discussed. After the COVID-19 pandemic, special attention has been paid to IoMT applications and their security issues, but only a few papers discussed the risk assessment for IoMT devices. Below are the reviewed papers related to IoT, IoMT, risk assessment, and frameworks.

The author in [39] introduces the status of the healthcare sector, including applications and their research and development plans. The existing IoMT applications are classified into body-centric and object-centric applications. Data acquisition, communication gateways, and servers of IoMT architecture are discussed. Furthermore, the paper discusses the gaps, challenges, and open research issues. The main objective of this review paper is to offer a new research perception for the development and advancement of the IoMT ecosystem.

Even though both software and hardware aspects of IoMT are explained in detail, this paper does not come up with any assessment framework as a solution.

The author reviews the existing proposed security standards and assessment frameworks as well as those under development for assessing the security of IoT-based smart environments in [58]. They present the conclusion that most of the assessment frameworks and security standards do not directly address current needs but have the potential to be adapted to IoT-based smart environments. A taxonomy of challenges is proposed to address the current and future IoT security issues. A further study is necessary to enhance the quality of the research conducted and spark discussion about the development of new security standards and assessment frameworks for IoT-based smart environments.

The work in [59] performed a holistic analysis of the available technologies, system architecture, optimization factors, and challenges that emerge by incorporating IoT in a hospital environment. This article has come out as a bridge between business applications and sensors in the unified network, which will be successful in creating step-by-step interoperable smart hospital design, but the research work does not cover any privacy and security risks which may arise in creating the hospital design.

In [60], the researchers have reviewed the security requirements, security techniques, architecture, and various new attacks. Since the attacks are unique and none of the proposed frameworks can satisfy the systems, a framework is proposed covering all data and device security stages such as data collection, data storage, and data sharing. The aforementioned framework is only limited to fourteen mentioned attacks in the paper and faces certain challenges. Thus, there is a need to create a system that can sustain a remotely secure primary setup and an alternate access method. Moreover, the years covered by the selected papers are not specified. The proposed framework is presented in Figure 12.

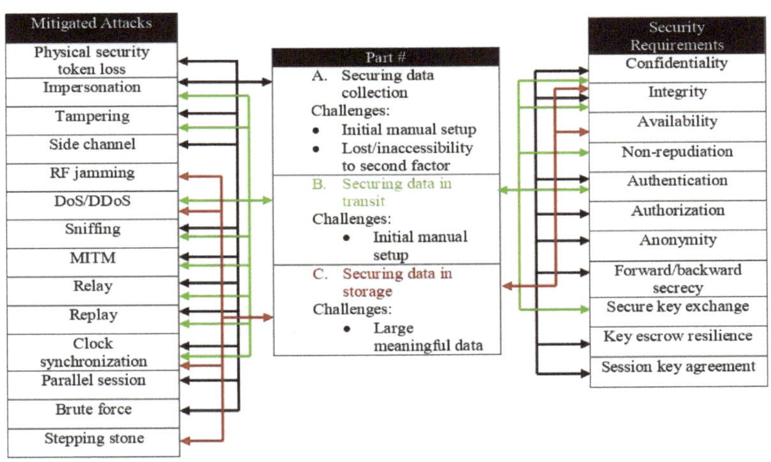

Figure 12. Framework by Ghubais with security features [60].

The review paper [61] reflects an overview of IoT used in the healthcare industry along with the challenges faced by IoMT applications. It surveys the literature on the Internet of Things in healthcare. It suggests that even though there exist a plethora of studies, they are lacking in conceptual and theoretical approaches. In their examination, six major categories of healthcare are taken into consideration, and light is shed on the gap existing in the articles related to the field.

A novel key management framework is presented in [36], which provides point-to-point secure communication channels between devices of the IoMT platform. It is designed for continuous patient monitoring and general medical applications and claims that the framework will give the patients full control of their personal data.

The paper in [62] presents a comprehensive history of the growth of IoMT applications and the related machine-learning-based frameworks from 2010 to 2019, focusing primarily on monitoring health through mobile applications, controlling rural health, detecting stress in drivers, identifying e-health applications, recognizing other health-related human movements, etc. The paper also presents previous challenges which are still unresolved and discusses how the deployment of the discussed approaches has challenges ranging from leakage of patients' personal information to the unaffordable price range. The article is useful for the deployment of future healthcare units, but it does not provide much information about the mentioned frameworks and techniques.

In [63], an IoMT risk assessment framework is designed to indicate security and protection features in IoMT devices and other IoMT platforms. IoMT Security Assessment Framework (IoMT-SAF) enables users to make security decisions based on a quantitative assessment method that uses recommended scenario-based security assessment criteria. A case study has been considered to understand the potential security issues based on consumption scenarios. The paper presents a framework comprising two modules: the recommendation module and the assessment module. The recommendation module identifies IoMT security threats and recommends security measures needed to respond to these threats. In the assessment module, these threats are ranked based on their degree of security, and this hierarchy is used as assessment criteria. Furthermore, based on these criteria and additional user requirements, the solutions (device, service, and platform) are assessed. Finally, a detailed ranking result is generated to allow IoMT end users to choose a secure solution.

2.6. Risk Assessment

The security vulnerabilities of modern IoT systems are unique, mainly due to the complexity and heterogeneity of technology and data. From a security and trust management perspective, organizations need to invest effectively in IoT cybersecurity. However, the challenge for IoT is its existing risk assessment methods, which were established before its development and used in many locally deployed organizations. These methods may not be effective when trying to manage the complexity and pervasive nature of these automated systems. Extending the existing risk assessment methods to the IoT could lead us to overlook new risks in the ecosystem [64].

Risk assessment is a process of identifying and assessing the risks associated with an organization's assets. This process includes estimating the risks and ranking them based on their importance. Risk assessment is a necessary part of the risk management process as it constitutes an essential step towards addressing risks. The likelihood and impact of an attack are some of the features considered in the risk assessment process. Risk treatment includes (a) accepting the risk when it is under a harmless level, (b) mitigating the risk by applying security measures, (c) transferring the risk, or (d) avoiding the risk by removing the affected asset itself. Some of the core concepts in risk assessment include assets, vulnerabilities, threats, attacks, and their impact [65].

Assets are defined as the value of any enterprise, whether tangible or intangible. Vulnerabilities are the points of weakness in an asset that can be exploited by others. Threat is explained as a possible action that could exploit these vulnerabilities. These actions can be deliberately done or happen accidentally, therefore resulting in the likelihood of attack and harm to the assets [65].

2.7. Risk Assessment Framework

Although the required process for risk assessment is defined, we still need various methods, guides, and tools for undertaking a risk assessment. Therefore, there is a persistent need to implement an effective cybersecurity framework due to the heterogeneity of IoMT devices. Examples of the most popular and well-regarded approaches include NIST SP800-30, ISO/IEC 27001, the Operationally Critical Threat, Asset, and Vulnerability Evaluation (OCTAVE), the CCTA Risk Analysis and Management Method (CRAMM), and the Expression of Needs and Identification of Security Objectives (EBIOS); their origins range from standard-setting bodies (such as NIST and ISO/IEC) to governments (CRAMM from the UK and EBIOS from France).

These approaches were by and large designed with specific application circumstances in mind; hence, they cannot be applied in the same way for the IoMT environment requirements because threats tend to be unique. As there is no standard framework available, these existing frameworks can be slightly modified for the risk assessment of IoMT devices in our research. Inferring from the aforementioned summarized research work, most of it does not provide risk assessment for IoMT devices and is only limited to a specific area; thus, it does not completely cater to the needs of IoMT-based devices. This section reviews the assessment frameworks suggested from the review literature and their limitations and forms the basis of the existing research.

2.7.1. NIST (National Institute of Standard and Technology) Framework

NIST's framework was created based on a set of organization standards to help them manage their cybersecurity requirements [58]. The framework's design aims to secure critical infrastructure but is used by private organizations to secure themselves from cyber threats [66]. It is suitable for organizations that are more technology-oriented and need to create a strong baseline strategy. NIST delivers regulatory and legal advantages that extend well for the organization which adopts it early [67]. It is not a one-size-fits-all approach to manage the threats to critical infrastructure because organizations will continue to have unique threats [68]. It has a structured and planned format, making it easier to execute at the enterprise level. The NIST framework is broken down into five functions: Identify, Protect, Detect, Respond, and Recover (as represented in Figure 13). These functions provide a systematic way to classify security risks, making it easier to implement controls [58].

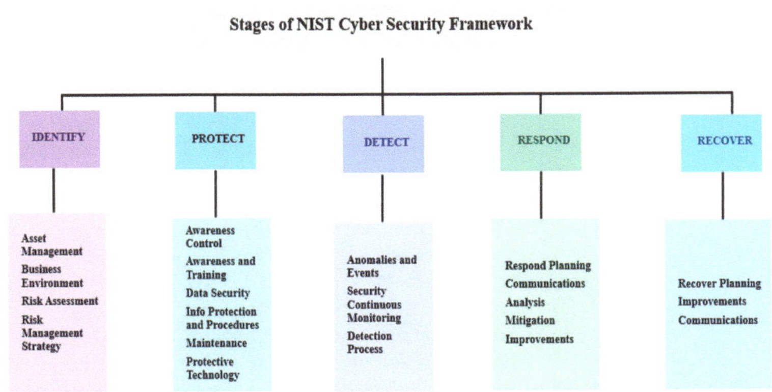

Figure 13. NIST Cybersecurity Framework (based on NIST model) [68].

- Identify—Helps organizations to develop an understanding to manage cybersecurity risks to people, systems, data, assets, and capabilities. It also incorporates asset management, business environment, risk assessment, and governance [68].
- Protect—Assists organizations in developing and implementing adequate safeguards to ensure the delivery of critical services. This phase includes developing security controls to protect data and information systems, such as access control, data security, information protection procedures, and maintaining protective technologies [58].
- Detect—Supports organizations to develop and implement appropriate activities to identify the presence of a cybersecurity event. It also offers guidelines for detecting anomalies in security, monitoring systems, and networks to uncover security

incidences. It also incorporates access control, communication processes, detection processes, anomalies, and events [58].

- Respond—Once a cybersecurity incident is detected, it helps organizations to develop and implement appropriate activities to act. This includes planning response, security resilience, mitigation, and communication during a response [58]
- Recovery—Develops and implement steps needed to maintain plans for resilience and restore capabilities and services compromised during a cybersecurity incident [58].

Most of the categories and sub-categories of the NIST framework use reference to other frameworks such as ISO 27001, combining significant features of these frameworks. Below are the limitations of the NIST framework:

- Because of the voluntary nature of the NIST framework, it does not provide proper risk management. Therefore, it cannot be used as a long-term replacement for information security management frameworks.
- The NIST framework is not a one-size-fits-all approach to handle the breaches and threats, as the organizations are complex and threats are unique [66].

NIST is making a unique contribution to meet the interoperation capability. It is uniquely qualified to undertake this task because of its technical capability, industry knowledge, standards and testing expertise, and international influence. Ensuring interoperability requires the integration of technical expertise in numerous disciplines. NIST brings an understanding of various industries through its research in supporting technology and testing; expertise in advanced networking technology; expertise in controls and their interfaces; and expertise in technology, computer, and network security. It has a long track record of working closely with industry and standards development organizations to develop consensus standards for industry use and, where needed, for regulatory agencies. NIST has extensive experience establishing testing and certification programs in critical areas, including cybersecurity. Finally, it has a strong presence and leadership in key international standards organizations. Moreover, NIST Special Publication 800-53 provides the foundation for security controls and a method for tailoring security controls to an organization.

2.7.2. ISO 27001 Cybersecurity Framework

ISO 27001 is a globally recognized standard developed in 2005 by the International Organization for Standardization (ISO). It takes a broader approach, and its methodology is based on Plan-Do-Check-Act (PDCA) cycle, which means that it builds the management system that not only plans and implements cybersecurity but also maintains and improves the complete system. This framework provides a series of requirements for an information security management system (ISMS) that an organization must follow to secure their data and is best suited for commercial companies. One of the most significant advantages of ISO 27001 is that companies can become certified against it and gain client confidence in providing a safe and effective risk management framework. One more advantage of ISO 27001 is that its documentation, such as incident management, change management, BYOD policy, password policy, etc., is structured and streamlined [66]. Below are the limitations of the ISO 27001 cybersecurity framework:

- It does not provide any specific risk management method.
- Organizations are expected to define their own method for risk management depending on their own requirements [10].

The ISO 27001 standard defines the requirements for establishing, implementing, maintaining, and improving an ISMS. Through risk management, ISMS assures confidentiality, integrity, and availability of information and provides confidence to interested parties. ISO/IEC 27001 (2013) specifies a total of 114 security controls across the following areas: A.5 Security policy, A.6 Organization of information security, A.7 Asset management, A.8 Human resources security, A.9 Physical and environmental security, A.10 Communications and operations management, A.11 Access control, A.12 Information systems acquisition, development and maintenance, A.13 Information security incident management, A.14

Business continuity management, A.15 Compliance. It also provides guidelines for organizations to address common cybersecurity risks such as social engineering attacks, hacking, malicious software, spyware, or other potentially unwanted software. Moreover, it provides a framework for sharing information, coordinating efforts, and controlling incidents [69]. Every device has its own risks; therefore, even though the standards might not apply to all of them, the number of controls used to address these risks will depend on each device's risks. While interoperability is a concern, the controls can be chosen according to the risks [70–72].

2.7.3. TARA Cybersecurity Framework

TARA (Threat Analysis and Risk Remediation) is a predictive framework initially developed within Intel to address complicated security risks. It is a qualitative approach to risk assessment that lists the expected attacks, communicates risks to the organizations, reduces the effort of risk analysis, and produces a better decision. TARA is mostly used alongside the NIST framework, applying IoT considerations of the NIST framework [34].

2.7.4. IEEE 2413-2019 (P2413) Standard

It is a standard that defines an architectural framework for IoT and conforms to the international standard ISO/IEC/IEEE 42010:2011. This framework is motivated by the concerns shared by stakeholders across several domains such as home, health, energy, transport, etc., and identifies intersection points between various domains. It does not define a specific standard for the IoMT platform but briefly outlines a domain of interest focused on health. The architectural framework in Figure 14 identifies sections like information, kind of model and viewpoints, architecture development, the rationale for key decisions, stakeholders' concerns, and viewpoint catalogue, where the last section serves as a reference for the adaptation of the standard to IoMT systems [72]. The standard focuses on two objectives: a) to deliver an interoperable and secure IoT systems framework for diverse application disciplines; b) to present a framework for the assessment and comparison between available IoT systems that will help in accelerating operations, design, and deployment of IoT systems [52]. Limitations of the framework include:

- It does not provide a specific standard for the IoMT platform.

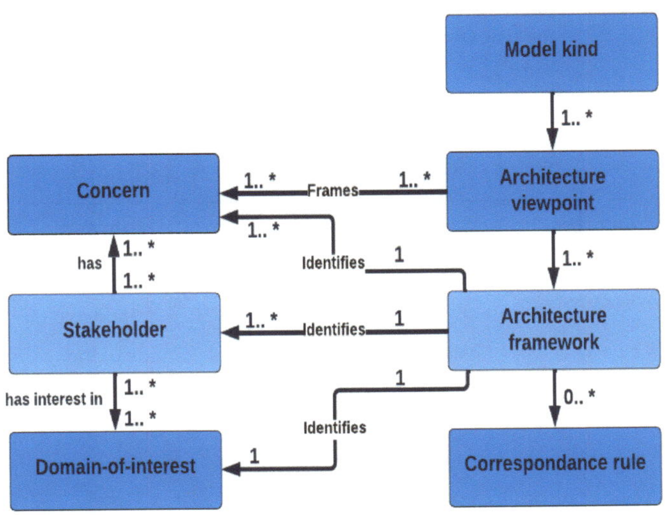

Figure 14. Conceptual model of an architectural framework of P2413 standard (based on ISO/IEC/IEEE 42010:2011) [72].

Table 3 summarizes the focus area, strengths, limitations, and application area of these frameworks. Due to the heterogeneity of devices, none of the frameworks are universally accepted; therefore, they do not entirely address IoMT cybersecurity and its consequences.

Table 3. Summary of existing frameworks.

Name of the Framework	Owner	Focus Area	Strength	Limitation	Application Area
NIST Framework	National Institute of Standard and Technology	Standards, Technology, Publications, government adoption, market intelligence	Structured and planned format, easy to execute, good for disaster and recovery planning	Not suitable for long-term approach, need more work with other standards to address compliance	Healthcare, manufacturing, government and private firms, insurance, financial organizations
TARA	Intel	Threat analysis and risk remediation	Predictive for crucial threats, provides definition of a list of attacks	Risk impact quantification is not available	Manufacturing and healthcare, financial organizations
ISO 27001	International Standard Organization	Global standardization of risk assessment	Suitable for crucial risk, international experience	Expects organizations to develop their own method	Small business, private and government firms
IEEE 2413-2019 (P2413) Standard	IEEE	Cross-domain interaction, system interoperability, functional compatibility	Provides methodology for privacy and security	Does not provide standard for IoMT design	Energy, health, home, transport

By analyzing available frameworks and the applications of IoT and IoMT devices, we now have the understanding to use the methodology in the proposed framework. The papers which have been reviewed demonstrate that only a small number of studies discuss the risk assessment of internet of medical devices. In our research, for the risk assessment of IoMT devices, we will adopt the methodologies followed in NIST and ISO frameworks, which is covered in more detail in the following section. In Table 4 we have presented the statistical analysis of the papers reviewed.

Table 4. Statistical analysis of the papers [49].

No	Ref	Authors	Year	Type	Citation	Publisher	Journal Name	Impact Factor
1.	[1]	Vashi et al.	2017	Conference	245	IEEE	IEEEXplore	Q1
2.	[2]	Gulzar and Abbas	2019	Journal	30	IEEE	IEEEXplore	Q1
3.	[3]	Van Kranenburg and Bassi	2012	Journal	154	Springer	Communications in Mobile Computing	Q2
4.	[6]	Schiller et al.	2022	Journal	18	ScienceDirect	Computer Science Review	Q1
5.	[9]	Wang, Zhang and Taleb	2018	Journal	96	Springer	World wide web	Q1
6.	[10]	Lee	2020	Journal	74	MDPI	Future Internet	Q2
7.	[12]	Aven	2016	Journal	1313	ScienceDirect	European Journal of Operational Research	Q1
8.	[13]	Wang et al.	2020	Journal	58	IEEE	IEEE Access	Q2
9.	[14]	Rubi and Gondim	2020	Journal	37	SAGE	Distributed Sensor Networks	Q2

Table 4. *Cont.*

No	Ref	Authors	Year	Type	Citation	Publisher	Journal Name	Impact Factor
10.	[16]	Pratap Singh et al.	2020	Journal	165	ScienceDirect	Journal of Clinical Orthopedics and Trauma	Q3
11.	[17]	Li et al.	2020	Journal	51	ScienceDirect	Computer Communications	Q1
12.	[21]	Xu, Gu, and Tian	2022	Journal	33	ScienceDirect	Artificial Intelligence in Agriculture	Q1
13.	[22]	Lawal and Rafsanjani	2022	Journal	43	ScienceDirect	Energy and Built Environment	Q1
14.	[23]	Rahim et al.	2021	Journal	69	ScienceDirect	Vehicular Communications	Q1
15.	[24]	Kumar, Tiwari and Zymbler	2019	Journal	432	Springer	Journal of Big Data	Q1
16.	[25]	Dwivedi, Mehrotra and Chandra	2022	Journal	40	ScienceDirect	Journal of Oral Biology and Craniofacial Research	Q2
17.	[26]	Karale	2021	Journal	45	ScienceDirect	Internet of Things	Q1
18.	[27]	Ogonji, Okeyo, and Wafula	2020	Journal	74	ScienceDirect	Computer Science Review	Q1
19.	[28]	Tawalbeh et al.	2020	Journal	286	MDPI	Applied Sciences	Q2
20.	[30]	Bertino and Islam	2017	Journal	639	IEEE	IEEEXplore	Q1
21.	[31]	Hameed	2019	Conference	59	IEEE	IEEEXplore	Q1
22.	[32]	Hireche, Mansouri and Pathan	2022	Journal	6	MDPI	Journal of Cybersecurity and Privacy	Q1
23.	[33]	Mercan et al.	2020	Conference	5	IEEE	IEEEXplore	Q1
24.	[34]	Kandasamy et al.	2020	Journal	65	Springer	EURASIP Journal on Information Security	Q2
25.	[35]	Kakhi et al.	2022	Journal	10	ScienceDirect	Biocybernetics and Biomedical Engineering	Q2
26.	[36]	Ree et al.	2021	Conference	-	IEEE	IEEEXplore	Q1
27.	[37]	Furtado et al.	2022	Journal	1	ScienceDirect	Digital Communications and Networks	Q1
28.	[39]	Al-Turjman, Hasan Nawaz and Deniz Ulusar	2020	Journal	175	ScienceDirect	Computer Communications	Q1
29.	[40]	Haleem et al.	2022	Journal	8	ScienceDirect	Internet of Things and Cyber-Physical Systems	Q2
30.	[42]	Chau and Hu	2002	Journal	1558	ScienceDirect	Information & Management	Q1
31.	[43]	Moazzami et al.	2020	Journal	391	ScienceDirect	Journal of Clinical Virology	Q1
32.	[44]	Swayamsiddha and Mohanty	2020	Journal	168	ScienceDirect	Diabetes & Metabolic Syndrome: Clinical Research & Reviews	Q1
33.	[45]	Yang et al.	2020	Journal	100	MDPI	Diagnostics	Q2
34.	[49]	Srivastava et al.	2022	Journal	5	Hindawi	Computational Intelligence and Neuroscience	Q2
35.	[51]	Sengupta, Ruj and Das Bit	2020	Journal	477	ScienceDirect	Journal of Network and Computer Applications	Q1
36.	[52]	Mohd Aman et al.	2021	Journal	120	ScienceDirect	Journal of Network and Computer Application	Q1
37.	[53]	Sun, Lo and Lo	2019	Journal	122	IEEE	IEEEXplore	Q1
38.	[56]	Algarni	2019	Journal	50	IEEE	IEEEXplore	Q1
39.	[58]	Karie et al.	2021	Journal	30	IEEE	IEEEXplore	Q1
40.	[59]	Çalı¸s, Uslu and Dursun	2020	Journal	76	Springer	Journal of Cloud Computing	Q1
41.	[60]	Ghubais et al.	2020	Journal	80	IEEE	IEEEXplore	Q1

Table 4. *Cont.*

No	Ref	Authors	Year	Type	Citation	Publisher	Journal Name	Impact Factor
42.	[61]	Lederman, Ben-Assuli and Vo	2021	Journal	-	ScienceDirect	Health Policy and Technology	Q1
43.	[62]	Din et al.	2019	Journal	72	IEEE	IEEEXplore	Q1
44.	[63]	Alsubaei et al.	2019	Journal	104	ScienceDirect	Internet of Things	Q1
45.	[64]	Radoglou Grammatikis, Sarigiannidis and Moscholios	2019	Journal	217	ScienceDirect	Internet of Things	Q1
46.	[65]	Nurse, Creese and De Roure	2017	Journal	182	IEEE	IEEEXplore	Q1
47.	[66]	Roy	2020	Conference	21	IEEE	IEEEXplore	Q1
48.	[72]	Talaminos-Barroso, Reina-Tosina and Roa	2022	Journal	-	ScienceDirect	Measurement: Sensors	Q3
49.	[73]	Kheirkhahan et al.	2019	Journal	67	ScienceDirect	Journal of Biomedical Informatics	Q1

A list of the publications is presented in Table 4. The information contains reference, author's name, journal and publisher names, type of article, number of citations, and year of publication. The papers have all been published in peer-reviewed journals or at conferences. Overall, the research community is showing an increase in interest year after year. The worldwide pandemic probably contributed to a dip in research in 2021. Journal and peer-reviewed articles have been the focus of this review, followed by conference publications. There were 48 publications, five of which were conference proceedings, and the rest were journals. The referenced papers are compared according to their publication dates in Figure 15. It highlights the distribution of referenced papers based on the type of journal. Out of the total referenced papers, 45 originate from reputed journals, while 4 come from conferences. Figure 16 represents the frequency of papers concerning IoT and IoMT. Based on papers cited from 2016 to 2022, we found that numbers have increased yearly except for the decline in 2021 due to the global pandemic, and the PDF version of a few papers from 2018 cannot be found. We have reviewed only 2 papers that are published in or before the year 2016.

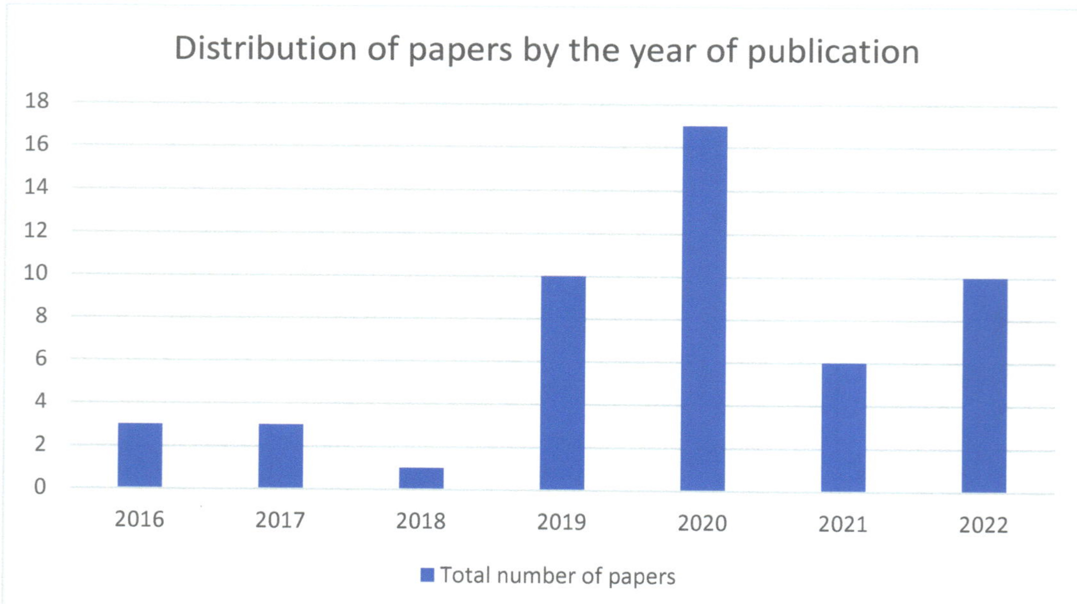

Figure 15. Distribution of papers by the publication year.

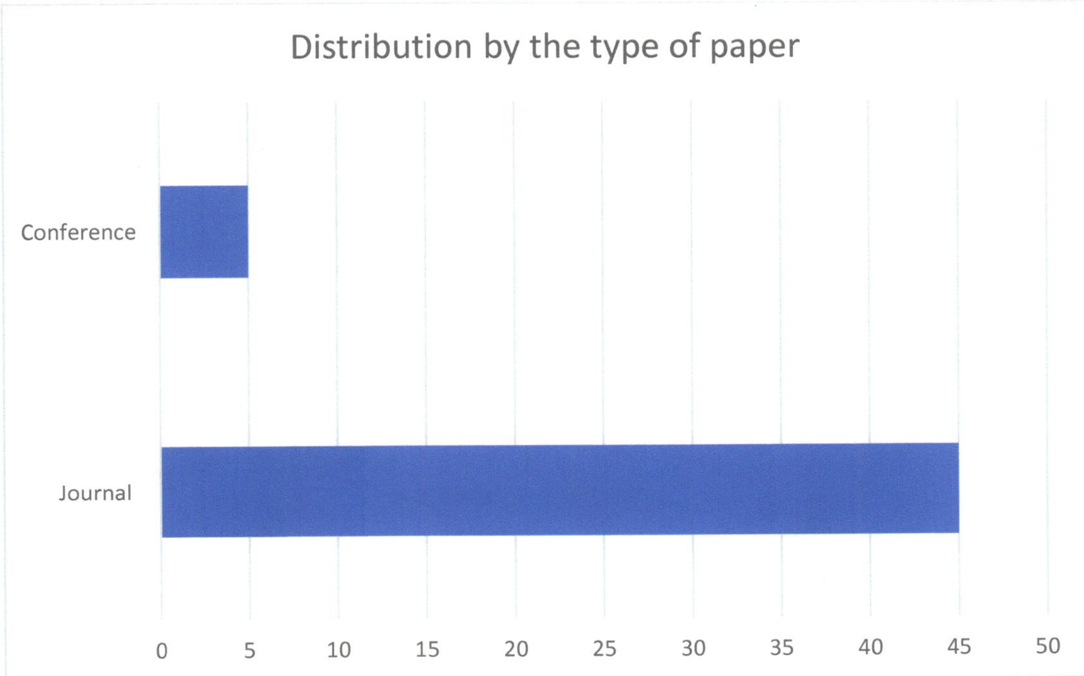

Figure 16. Distribution by the type of paper.

3. Methodology

The objective is to identify and predict a framework for evaluating the risk associated with IoMT devices because of their immunity to security measures and a large number of devices on the market that communicate private and sensitive information. Below is a flowchart in Figure 17 to represent the steps performed in the risk assessment.

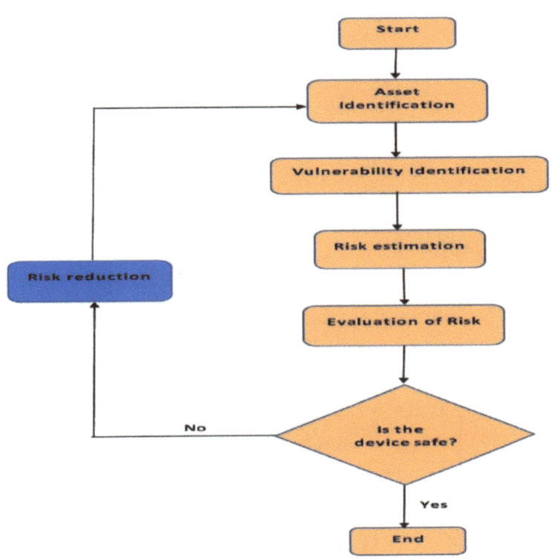

Figure 17. Risk assessment flowchart.

We will test the portable wireless vital monitor, smartwatches, and lung monitor. As the primary research methodology, this study will adopt a NIST- and ISO-based framework for the risk assessment. Since the NIST framework is threat-oriented, it can be tailored based on the requirement and will be appropriate to address the present threat landscape. The limitation of one size not fitting with all the approaches gives the flexibility to establish a strong baseline and augment compliance with new regulations.

ISO 27001, on the other hand, takes a more comprehensive approach. It bases its methodology on the PDCA cycle and creates a management system to plan, implement, maintain, and enhance the entire cybersecurity system. This framework provides a series of requirements that an organization must follow to secure its data. Its documentation is structured and streamlined. Based on both frameworks, we have proposed a framework, as shown in Figure 18, where we will create a checklist to perform the risk assessment. The methodology process entails the following steps: Asset identification, Identify vulnerabilities, Risk estimation, Evaluation of risk, and Risk reduction, which are described below.

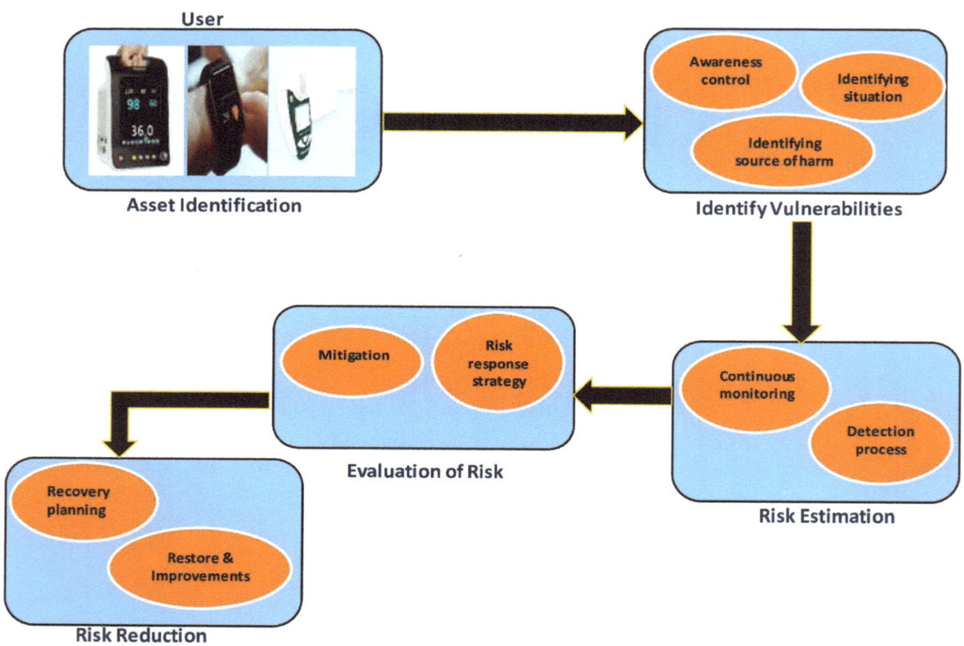

Figure 18. Risk assessment flowchart.

3.1. Asset Identification

The first step is the identification of the asset, which focuses on monitoring and baselining, where an asset will be a device using the IoMT application. As explained in Section 2, there is a range of applications, and we anticipate testing a few, such as a wireless vital monitor, lung monitor, and wearable device, like a smartwatch. The first two are critical devices, but they are unapparent when it comes to performing risk assessment, and the third is chosen since it is a device with a high rate of user acceptance. This is why we selected them as assets. We classify these assets into two categories, where the value depends on the sensitivity of the data and their potential impact on the CIA.

- High-Value Asset: The wireless vital monitor and lung monitor will fall under this category, as the level of concern given to the asset will be high, because they need more security implementation. A wireless vital monitor is a portable device capable of

monitoring vital signs, including heart rate, electrocardiogram (ECG), blood pressure, temperature, and other vitals. It transmits the readings wirelessly through Bluetooth to an interactive monitoring device [44]. It is mainly used by patients who have been discharged but still need their vitals to be measured. For patients with respiratory issues, such as Chronic Obstructive Pulmonary Disease (COPD) or cystic fibrosis, and those who have undergone a lung transplant, a lung monitor provides accurate and effective monitoring of lung function [47]. Despite having a low asset value, they have high usability. Both the devices are used by patients with critical conditions, making them highly important.
- Low-Value Asset: Smartwatches will be considered low-value assets. Thus, concern will be low. They are convenient to wear and are equipped with several sensors suitable for gathering physical activity throughout the day [73].

Since the risk can come from both the use and misuse of the devices, in this step, we will set the limit of use, where a use statement will be created to get an exact idea of the precise data to be taken from these devices. This step will also help to identify the scenarios of predictable misuse. Next, the time limit will be determined. It is essential to describe estimates for how long each device component should endure because it may eventually wear out. In addition, we shall identify the safety characteristics of the device.

3.2. *Identify Vulnerabilities*

While baselining is the primary focus of asset identification, this is the step where the framework starts to take action and become proactive. Adequate safeguards are implemented to ensure device safety, and security controls are developed to protect sensitive data and information. In this step, we will identify all the potential risks and the harmful situations that may arise. It is necessary to describe all the dangerous situations in this step, since failing to do so increases the likelihood that we may overlook them in the following steps, where the risk must be eliminated or reduced. For identifying, we will go through each step required to operate the device and note the potential source of damage along the way. It is necessary since the threat is not only limited to one user, but also to everyone using the device. Based on this identification, we can generate an awareness control (gathering, understanding, and anticipating information) for the user.

3.3. *Risk Estimation*

The next step is to estimate the risk after it has been identified. As the risk assessment is an iterative process, risks can also be found in this step and may become apparent when previously found risks are estimated. The primary objective of the risk estimation process is to analyze the risk and determine the severity and probability of risk occurrence. A qualitative risk assessment will be employed in our study to understand this likelihood and severity. The amount of risk will be broken down into high, medium, and low categories to help assess whether the current safeguards and controls are adequate or if more needs to be done to recover from the impact. Various methods can be used to estimate risk, such as a risk matrix or a risk graph.

Estimation is a critical step because the faster a risk is estimated, the faster the repercussions can be mitigated in the next step. As the IoMT devices may contain personal information, we anticipate using the best among the methods to lower the risk.

3.4. *Evaluation of Risk*

As the name suggests, in this step, we determine the actions that need to be taken to reduce the identified risks. We will consider two objectives in evaluating the risk:

1. Determining whether a hazardous situation requires further risk reduction.
2. Determining whether risk reduction has introduced any new risk or has increased the level of other risks.

Based on these two objectives, we will determine the action that needs to be taken to reduce the risk while making sure that these actions do not introduce any new risks. If there is a high risk involved, this process will be performed repeatedly until the above two objectives are met. This will be our response strategy to the risks to ensure the device is in a state of continuous improvement. After the risk estimation, it is necessary to evaluate risk, and if risks are found, we will need to go back and repeat the estimation for those risks. Although this is the last step of risk assessment, we will highlight some of the relevant risk reduction information, as it is connected to the risk assessment process.

3.5. Risk Reduction

Risk reduction entails reducing the risks to an acceptable level, putting resilience strategies into practice, and regaining access to the skills and services that were lost during a cybersecurity event. To minimize the impact of a cybersecurity event, this function will support prompt recovery. This step is closely connected to the risk assessment process, as every time risk reduction is not achieved, we will go back and perform the complete risk assessment process. The recovery function is required to ensure that, if a breach does occur, the employed device can stay on the right path to achieve the appropriate goals and objectives.

3.6. Summary

Throughout the risk assessment process, the goal is to understand potential risks before attempting to prevent them. By performing all the steps mentioned above, we will be able to safeguard patients from potential risks, recognize and evaluate the magnitude of these risks, and implement and monitor efficient control measures to reduce and eliminate them. The complete risk assessment procedure has been divided into four parts, starting with identifying the device that needs to be tested and concluding with evaluating the risk. We have also emphasized the significance of risk mitigation as the fifth phase. For every risk, there are possible scenarios that can unfold at any step, so given that it has to do with human life, we must be very cautious.

While IoT has been a dominant field of study for more than a decade and has received many accolades, only recently has the Internet of medical devices been receiving significant attention. Our literature review in Table 5 is based on papers published between 2019 and 2022 covering both IoT and IoMT. Our review is based on their findings, risks, and whether they propose a framework for risk assessment as a solution to these risks. These papers discuss IoMT application areas, challenges, architecture, risks associated with devices, and risk assessment frameworks. Despite the fact that privacy and security risks are significant issues with IoMT devices, more than half of the existing literature has not taken them into account. A few papers that included risks and challenges in their survey failed to offer an assessment framework for addressing them. There is a paper that discusses framework and security risks, but it only covers fourteen attacks, which is inadequate, as there will be new attacks for which we need to be prepared.

Table 5. Summary of literature review.

References	Year	Proposed Framework	Findings	Limitations	Privacy Risk	Security Risk
[10]	2020	Yes	IoT architecture; qualitative and quantitative approach for risk management; four-layer IoT cyber risk management framework; risk identification	Framework proposed for IoT systems but it may not work with all the security requirements for IoMT applications	×	×

Table 5. *Cont.*

References	Year	Proposed Framework	Findings	Limitations	Privacy Risk	Security Risk
[16]	2020	No	IoMT solutions and treatments for health issues related to orthopedic patients; challenges faced during COVID-19; digital connectivity of IoMT devices to the hospital; expected applications in the future	Challenges and applications mentioned are only limited to orthopedic patients	×	×
[25]	2022	No	Role of IoMT applications for the improvement of healthcare industry; challenges faced by IoMT in developing smart healthcare system	There are no frameworks designed for challenges faced	×	×
[36]	2021	No	Presents ad hoc, point-to-point secure channels between devices and IoMT system	Provides complete key management solution for IoMT patient monitoring system but does not present a framework	×	✓
[39]	2020	No	Surveys existing IoMT technologies, sensors, and communication protocols; provides new research perception	The paper does not present any assessment framework for the challenges mentioned	✓	✓
[58]	2021	No	Reviews security standards and frameworks for IoT-based environments, potential solutions for identified challenges	Taxonomy of challenges based on various categories are mentioned but further study is required to enhance the quality of work conducted	✓	✓
[59]	2020	No	Analyzes different factors affecting IoT-based smart hospitals based on various architectural layers	Provides an architecture for interoperable smart hospital design but this architecture needs further research and experimentation	×	×
[60]	2021	Yes	Reviews security requirements, architecture, techniques, and new attacks and presents a framework covering all device and data security stages	Framework is limited to only fourteen attacks and faces challenges	×	✓

Table 5. *Cont.*

References	Year	Proposed Framework	Findings	Limitations	Privacy Risk	Security Risk
[62]	2019	No	Describes a comprehensive view of IoMT-based applications developed and deployed over the last decade	Paper presents the limitations and challenges of IoMT applications but does not provide a way to overcome the difficulties	✓	✓
[63]	2019	Yes	Recommends detailed list of assessment attributes covering security measures	Missing on some security features needed for IoMT device users	×	✓
Our Work	2023	Yes	Discusses recent advances, probable risks, IoMT application areas, and risk assessment frameworks	This paper provides IoT and IoMT application areas, probable risks, architecture, and frameworks for the risk assessment	✓	✓

Traditional risk assessment methodologies cannot always cater to the new risks generated by the integration of IoT in a critical sector like healthcare. Contrary to the existing papers, we provide a comprehensive approach towards a risk-free IoMT device, starting with the application, the architecture, and plausible risks, followed by a framework for risk assessment.

4. Conclusions

IoMT is evolving rapidly, and it can potentially change the healthcare industry cost-effectively, focusing on treatment, early diagnosis, and prevention of spread. It is becoming more diverse, prevalent, and highly successful at identifying, predicting, and monitoring recently emerging infectious diseases. However, it is still in its early stages of growth, and heterogeneity and associated risk are still significant concerns. Due to the rapid advancement and breakthroughs, security measures must be considered; if these risks are disregarded, there will be more cyber breaches.

This study has initially focused on the broader IoT domain and narrowed it down to IoMT and its risk assessment. To fully comprehend the concept, the paper reviewed previous research publications, and it was discovered that the current risk assessment approaches do not always cater to the new threat landscape generated by the integration of IoT in the healthcare sector. Despite the recent surge in interest in the IoMT sector, a detailed review of risk assessment methodology and the security precautions for IoMT devices is still in its infancy. Therefore, the paper examines the currently available risk assessment frameworks, standards, and their limitations to provide a comparative analysis. Lastly, a framework is proposed for the risk assessment of the selected devices.

5. Future Work

This study will give readers a thorough understanding of the subject and aid future researchers in creating new IoMT risk assessment methodologies or enhancing those that already exist. The suggested methodology will be put to the test and implemented.

As a future direction, we plan to test heterogeneous devices, including a lung monitor, smartwatch, and wireless vital monitor. We intend to apply the proposed methodology to these devices, and a risk assessment will be conducted, which will address every aspect of data and device security, from data collection to storage and sharing. Based on the risks mentioned in our paper, we will assess the efficiency and efficacy of the performance, and we anticipate having risk-free heterogeneous IoMT devices. Given the security and privacy risks, we will also study how the current taxonomy can be adapted and integrated into different IoMT-based systems.

This finding has the potential to spark additional IoMT research and advance society's ability to function effectively. Additionally, it will benefit the stakeholders and policymakers in the healthcare industry. Although IoMT has gained attention in the past few years, the research is still fragmented, with increased heterogeneity in approaches and devices. However, we strongly believe that IoMT risk assessment is an ongoing hot research topic, and we expect a significant amount of related literature to be produced in the near future.

Author Contributions: Conceptualization, P. and B.S.; Methodology, P., B.S. and S.A.; Formal analysis, P. and B.S.; Data curation, P., B.S. and S.A.; Writing—original draft preparation, P.; Writing—review and editing, B.S. and S.A.; Visualization, P., B.S. and S.A.; Supervision, B.S. and S.A. All authors have read and agreed to the published version of the manuscript.

Funding: This research received no external funding.

Institutional Review Board Statement: Not applicable.

Informed Consent Statement: Not applicable.

Data Availability Statement: Not applicable.

Conflicts of Interest: The authors declare no conflict of interest.

References

1. Vashi, S.; Ram, J.; Modi, J.; Verma, S.; Prakash, C. Internet of Things (IoT) A Vision, Architectural Elements, and Security Issues. In Proceedings of the 2017 International Conference on I-SMAC (IoT in Social, Mobile, Analytics and Cloud) (I-SMAC), Palladam, India, 10–11 February 2017.
2. Gulzar, M.; Abbas, G. Internet of Things Security: A Survey and Taxonomy; In Proceedings of the 2019 International Conference on Engineering and Emerging Technologies (ICEET), Lahore, Pakistan, 21–22 February 2019.
3. Van Kranenburg, R.; Bassi, A. IoT Challenges. *Commun. Mob. Comput.* **2012**, *1*, 9. [CrossRef]
4. Global Government IoT Revenue for Endpoint Electronics and Communications to Total $21 Billion in 2022. Available online: https://www.gartner.com/en/newsroom/press-releases/2021-06-30-gartner-global-government-iot-revenue-for-endpoint-electronics-and-communications-to-total-us-dollars-21-billion-in-2022 (accessed on 11 July 2022).
5. Forecast: IT Services for IoT, Worldwide, 2019–2025. Available online: https://www.gartner.com/en/documents/4004741 (accessed on 11 July 2022).
6. Schiller, E.; Aidoo, A.; Fuhrer, J.; Stahl, J.; Ziörjen, M.; Stiller, B. Landscape of IoT security. *Comput. Sci. Rev.* **2022**, *44*, 100467. [CrossRef]
7. Australia's IoT Opportunity-Driving Future Growth. Available online: https://www.acs.org.au/insightsandpublications/reports-publications/iot-opportunity.html (accessed on 2 February 2023).
8. • IoT Total Revenue Worldwide 2019–2030 | Statista. Available online: https://www.statista.com/statistics/1194709/iot-revenue-worldwide/ (accessed on 17 July 2022).
9. Wang, H.; Zhang, Z.; Taleb, T. Special Issue on Security and Privacy of IoT. *World Wide Web* **2017**, *21*, 1–6. [CrossRef]
10. Lee, I. Internet of Things (IoT) Cybersecurity: Literature Review and IoT Cyber Risk Management. *Future Internet* **2020**, *12*, 157. [CrossRef]
11. IoT Security in 2022: Defending Data during the Rise of Ransomware. Available online: https://www.perle.com/articles/iot-security-in-2022-defending-data-during-the-rise-of-ransomware-40193618.shtml (accessed on 24 July 2022).
12. Aven, T. Risk assessment and risk management: Review of recent advances on their foundation. *Eur. J. Oper. Res.* **2016**, *253*, 1–13. [CrossRef]
13. Wang, L.; Ali, Y.; Nazir, S.; Niazi, M. Special Section on Lightweight Security and Provenance for Internet of Health Things ISA Evaluation Framework for Security of Internet of Health Things System Using AHP-TOPSIS Methods. *IEEE Access* **2020**, *8*, 152316–152332. [CrossRef]
14. Rubí, J.N.S.; Gondim, P.R.D.L. Interoperable Internet of Medical Things platform for e-Health applications. *Int. J. Distrib. Sens. Netw.* **2020**, *16*, 1550147719889591. [CrossRef]
15. 2025 Forecast: Global IoT Healthcare Market Looks Good—A $188.2 Billion Opportunity | TechRepublic. Available online: https://www.techrepublic.com/article/2025-forecast-global-iot-looks-good-a-188-2-billion-opportunity/ (accessed on 11 July 2022).

16. Pratap Singh, R.; Javaid, M.; Haleem, A.; Vaishya, R.; Ali, S. Internet of Medical Things (IoMT) for orthopaedic in COVID-19 pandemic: Roles, challenges, and applications. *J. Clin. Orthop. Trauma* **2020**, *11*, 713–717. [CrossRef]
17. Li, X.; Dai, H.-N.; Wang, Q.; Imran, M.; Li, D.; Imran, M.A. Securing Internet of Medical Things with Friendly-jamming schemes. *Comput. Commun.* **2020**, *160*, 431–442. [CrossRef]
18. 53% of Connected Medical Devices Contain Critical Vulnerabilities. Available online: https://healthitsecurity.com/news/53-of-connected-medical-devices-contain-critical-vulnerabilities (accessed on 26 July 2022).
19. Marron, J.A. *Implementing the Health Insurance Portability and Accountability Act (HIPAA) Security Rule*; NIST Special Publication: Gaithersburg, MD, USA, 2022. [CrossRef]
20. Asimily: Healthcare & Medical Device Security (IoMT). Available online: https://www.asimily.com/ (accessed on 24 July 2022).
21. Xu, J.; Gu, B.; Tian, G. Review of agricultural IoT technology. *Artif. Intell. Agric.* **2022**, *6*, 10–22. [CrossRef]
22. Lawal, K.; Rafsanjani, H.N. Trends, benefits, risks, and challenges of IoT implementation in residential and commercial buildings. *Energy Built Environ.* **2022**, *3*, 251–266. [CrossRef]
23. Rahim, M.A.; Rahman, M.A.; Rahman, M.M.; Asyhari, A.T.; Bhuiyan, M.Z.A.; Ramasamy, D. Evolution of IoT-enabled connectivity and applications in automotive industry: A review. *Veh. Commun.* **2021**, *27*, 100285. [CrossRef]
24. Kumar, S.; Tiwari, P.; Zymbler, M. Internet of Things is a revolutionary approach for future technology enhancement: A review. *J. Big Data* **2019**, *6*, 1–21. [CrossRef]
25. Dwivedi, R.; Mehrotra, D.; Chandra, S. Potential of Internet of Medical Things (IoMT) applications in building a smart healthcare system: A systematic review. *J. Oral Biol. Craniofac. Res.* **2021**, *12*, 302–318. [CrossRef]
26. Karale, A. The Challenges of IoT Addressing Security, Ethics, Privacy, and Laws. *Internet Things* **2021**, *15*, 100420. [CrossRef]
27. Ogonji, M.M.; Okeyo, G.; Wafula, J.M. A survey on privacy and security of Internet of Things. *Comput. Sci. Rev.* **2020**, *38*, 100312. [CrossRef]
28. Tawalbeh, A.I.; Muheidat, F.; Tawalbeh, M.; Quwaider, M. IoT Privacy and Security: Challenges and Solutions. *Appl. Sci.* **2020**, *10*, 4102. [CrossRef]
29. Kathryn Cormican, S.M.; Dhanapathi, C. Analysis of critical success factors to mitigate privacy risks in IoT Devices. *Procedia Comput. Sci.* **2022**, *196*, 191–198. [CrossRef]
30. Bertino, E.; Islam, N. Botnets and Internet of Things Security. *Computer* **2017**, *50*, 76–79. [CrossRef]
31. Hameed, A.; Alomary, A. Security Issues in IoT: A Survey. In Proceedings of the 2019 International Conference on Innovation and Intelligence for Informatics, Computing, and Technologies (3ICT), Sakhier, Bahrain, 22–23 September 2019.
32. Hireche, R.; Mansouri, H.; Pathan, A.-S.K. Security and Privacy Management in Internet of Medical Things (IoMT): A Synthesis. *J. Cybersecur. Priv.* **2022**, *2*, 640–661. [CrossRef]
33. Mercan, S.; Akkaya, K.; Cain, L.; Thomas, J. Security, Privacy and Ethical Concerns of IoT Implementations in Hospitality Domain. In Proceedings of the 2020 International Conferences on Internet of Things (iThings) and IEEE Green Computing and Communications (GreenCom) and IEEE Cyber, Physical and Social Computing (CPSCom) and IEEE Smart Data (SmartData) and IEEE Congress on Cybermatics (Cybermatics), Rhodes, Greece, 2–6 November 2020; pp. 198–203.
34. Kandasamy, K.; Srinivas, S.; Achuthan, K.; Rangan, V.P. IoT cyber risk: A holistic analysis of cyber risk assessment frameworks, risk vectors, and risk ranking process. *EURASIP J. Inf. Secur.* **2020**, *2020*, 8. [CrossRef]
35. Kakhi, K.; Alizadehsani, R.; Kabir, H.M.D.; Khosravi, A.; Nahavandi, S.; Acharya, U.R. The internet of medical things and artificial intelligence: Trends, challenges, and opportunities. *Biocybern. Biomed. Eng.* **2022**, *42*, 749–771. [CrossRef]
36. De Ree, M.; Vizár, D.; Mantas, G.; Bastos, J.; Kassapoglou-Faist, C.; Rodriguez, J. A Key Management Framework to Secure IoMT-enabled Healthcare Systems. In Proceedings of the 2021 IEEE 26th International Workshop on Computer Aided Modeling and Design of Communication Links and Networks (CAMAD), Porto, Portugal, 25–27 October 2021; pp. 1–6.
37. Furtado, D.; Gygax, A.F.; Chan, C.A.; Bush, A.I. Time to forge ahead: The Internet of Things for healthcare. *Digit. Commun. Netw.* 2022. [CrossRef]
38. Internet of Medical Things (IoMT) Market: Global Industry Analysis, Trends, Market Size, and Forecasts up to 2026. Available online: https://www.researchandmarkets.com/reports/5338262/internet-of-medical-things-iomt-market-global (accessed on 20 June 2022).
39. Al-Turjman, F.; Hassan Nawaz, M.; Deniz Ulusar, U. Intelligence in the Internet of Medical Things era: A systematic review of current and future trends. *Comput. Commun.* **2020**, *150*, 644–660. [CrossRef]
40. Haleem, A.; Javaid, M.; Pratap Singh, R.; Suman, R. Medical 4.0 technologies for healthcare: Features, capabilities, and applications. *Internet Things Cyber-Phys. Syst.* **2022**, *2*, 12–30. [CrossRef]
41. Lu, L.; Zhang, J.; Xie, Y.; Gao, F.; Xu, S.; Wu, X.; Ye, Z. Wearable Health Devices in Health Care: Narrative Systematic Review. *JMIR mHealth uHealth* **2020**, *8*, e18907. [CrossRef]
42. Chau, P.Y.K.; Hu, P.J.H. Investigating healthcare professionals' decisions to accept telemedicine technology: An empirical test of competing theories. *Inf. Manag.* **2002**, *39*, 297–311. [CrossRef]
43. Moazzami, B.; Razavi-Khorasani, N.; Dooghaie Moghadam, A.; Farokhi, E.; Rezaei, N. COVID-19 and telemedicine: Immediate action required for maintaining healthcare providers well-being. *J. Clin. Virol.* **2020**, *126*, 104345. [CrossRef]
44. Swayamsiddha, S.; Mohanty, C. Application of cognitive Internet of Medical Things for COVID-19 pandemic. *Diabetes Metab. Syndr. Clin. Res. Rev.* **2020**, *14*, 911–915. [CrossRef]

45. Yang, T.; Gentile, M.; Shen, C.-F.; Cheng, C.-M. Diagnostics Combining Point-of-Care Diagnostics and Internet of Medical Things (IoMT) to Combat the COVID-19 Pandemic. *Diagnostics* **2020**, *10*, 224. [CrossRef]
46. Kaputa, D.; Price, D.; Enderle, J.D. A portable, inexpensive, wireless vital signs monitoring system. *Biomed Instrum Technol.* **2010**, *44*, 350–353. [CrossRef] [PubMed]
47. Lung Monitor | Healthcare | Vitalograph. Available online: https://vitalograph.com/intl/product/lung-monitor/ (accessed on 31 July 2022).
48. Williams, P.A.H.; Woodward, A.J. Cybersecurity vulnerabilities in medical devices: A complex environment and multifaceted problem. *Med. Devices (Auckl.)* **2015**, *8*, 305. [CrossRef]
49. Srivastava, J.; Routray, S.; Ahmad, S.; Waris, M.M.; Asghar, M.Z. Internet of Medical Things (IoMT)-Based Smart Healthcare System: Trends and Progress. *Comput. Intell. Neurosci.* **2022**, *2022*, 7218113. [CrossRef]
50. Ahad, A.; Tahir, M.; Kok-Lim, A.; Yau, A. 5G-Based Smart Healthcare Network: Architecture, Taxonomy, Challenges and Future Research Directions. *IEEE Access* **2019**, *7*, 100747–100762. [CrossRef]
51. Sengupta, J.; Ruj, S.; Das Bit, S. A Comprehensive Survey on Attacks, Security Issues and Blockchain Solutions for IoT and IIoT. *J. Netw. Comput. Appl.* **2020**, *149*, 102481. [CrossRef]
52. Mohd Aman, A.H.; Hassan, W.H.; Sameen, S.; Attarbashi, Z.S.; Alizadeh, M.; Latiff, L.A. IoMT amid COVID-19 pandemic: Application, architecture, technology, and security. *J. Netw. Comput. Appl.* **2021**, *174*, 102886. [CrossRef] [PubMed]
53. Sun, Y.; Lo, P.-W.; Lo, B. Security and Privacy for the Internet of Medical Things Enabled Healthcare Systems: A Survey. *IEEE Access* **2019**, *7*, 183339–183355. [CrossRef]
54. Chakravorty, R. A Programmable Service Architecture for Mobile Medical Care. In Proceedings of the Fourth Annual IEEE International Conference on Pervasive Computing and Communications Workshops (PERCOMW'06), Pisa, Italy, 13–17 March 2006. [CrossRef]
55. Yeh, K.H. A Secure IoT-Based Modern Healthcare System with Body Sensor Networks. *IEEE Access* **2022**, *4*, 10288–10299. [CrossRef]
56. Algarni, A. A Survey and Classification of Security and Privacy Research in Smart Healthcare Systems. *IEEE Access* **2019**, *7*, 101879–101894. [CrossRef]
57. Increase in Health-Care Security Breach by Proliferation of IoMT Devices—dynamicCISO. Available online: https://dynamicciso.com/increase-in-health-care-security-breach-by-proliferation-of-iomt-devices/ (accessed on 2 February 2023).
58. Karie, N.M.; Sahri, N.M.; Yang, W.; Valli, C.; Kebande, V.R. A Review of Security Standards and Frameworks for IoT-Based Smart Environments. *IEEE Access* **2021**, *9*, 121975–121995. [CrossRef]
59. Çalı¸s, B.; Uslu, Ç.; Dursun, E. Analysis of factors affecting IoT-based smart hospital design. *J. Cloud Comput.* **2020**, *9*, 67. [CrossRef]
60. Ghubaish, A.; Salman, T.; Zolanvari, M.; Al-Ali, A.; Jain, R. Recent Advances in the Internet-of-Medical-Things (IoMT) Systems Security; Recent Advances in the Internet-of-Medical-Things (IoMT) Systems Security. *IEEE Internet Things J.* **2021**, *8*, 8707–8718. [CrossRef]
61. Lederman, R.; Ben-Assuli, O.; Vo, T.H. The role of the Internet of Things in Healthcare in supporting clinicians and patients: A narrative review. *Health Policy Technol.* **2021**, *10*, 100552. [CrossRef]
62. Din, I.U.; Member, S.; Almogren, A.; Guizani, M.; Zuair, M. Special Section on Data Mining for Internet of Things A Decade of Internet of Things: Analysis in the Light of Healthcare Applications. *Ieee Access* **2019**, *7*, 89967–89979. [CrossRef]
63. Alsubaei, F.; Abuhussein, A.; Shandilya, V.; Shiva, S. IoMT-SAF: Internet of Medical Things Security Assessment Framework. *Internet Things* **2019**, *8*, 100123. [CrossRef]
64. Radoglou Grammatikis, P.I.; Sarigiannidis, P.G.; Moscholios, I.D. Securing the Internet of Things: Challenges, threats and solutions. *Internet Things* **2019**, *5*, 41–70. [CrossRef]
65. Nurse, J.R.C.; Creese, S.; De Roure, D. Trusting the Internet of Things Security Risk Assessment in Internet of Things Systems. *IT Prof.* **2017**, *19*, 20–26. [CrossRef]
66. Roy, P.P. A High-Level Comparison between the NIST Cyber Security Framework and the ISO 27001 Information Security Standard. *2020 Natl. Conf. Emerg. Trends Sustain. Technol. Eng. Appl.* **2020**. [CrossRef]
67. Institute of Standards, N. *Framework for Improving Critical Infrastructure Cybersecurity, Version 1.1*; National Institute of Standards and Technology: Gaithersburg, MD, USA, 2014. [CrossRef]
68. Strengthen Security of Your Data Center with the NIST Cybersecurity Framework | Dell Technologies United States. Available online: https://www.dell.com/en-us/blog/strengthen-security-of-your-data-center-with-the-nist-cybersecurity-framework/ (accessed on 12 July 2022).
69. Lechner, N.H. An Overview of Cybersecurity Regulations and Standards for Medical Device Software. *Cent. Eur. Conf. Inf. Intell. Syst.* **2017**, 237–249. Available online: https://cve.mitre.org (accessed on 15 July 2022).
70. Moreira, A.; Guimarães, T.; Duarte, R.; Salazar, M.M.; Santos, M. Interoperability and Security Issues on Multichannel Interaction In Healthcare Services. *Procedia Comput. Sci.* **2022**, *201*, 714–719. [CrossRef]
71. Barata, J.; Cardoso, A.; Haenisch, J.; Chaure, M. Interoperability standards for circular manufacturing in cyber-physical ecosystems: A survey. *Procedia Comput. Sci.* **2022**, *207*, 3320–3329. [CrossRef]

72. Talaminos-Barroso, A.; Reina-Tosina, J.; Roa, L.M. Adaptation and application of the IEEE 2413-2019 standard security mechanisms to IoMT systems. *Meas. Sensors* **2022**, *22*, 100375. [CrossRef]
73. Kheirkhahan, M.; Nair, S.; Davoudi, A.; Rashidi, P.; Wanigatunga, A.A.; Corbett, D.B.; Mendoza, T.; Manini, T.M.; Ranka, S. A smartwatch-based framework for real-time and online assessment and mobility monitoring. *J. Biomed. Inform.* **2019**, *89*, 29–40. [CrossRef] [PubMed]

technologies

Review

Production Technologies, Regulatory Parameters, and Quality Control of Vaccine Vectors for Veterinary Use

Raquel de M. Barbosa [1,2,*], Amélia M. Silva [3,4], Classius F. da Silva [5], Juliana C. Cardoso [6,7], Patricia Severino [6,7], Lyghia M. A. Meirelles [1], Arnobio A. da Silva-Junior [1], César Viseras [2], Joel Fonseca [8] and Eliana B. Souto [8,9,*]

1 Department of Pharmacy, Federal University of Rio Grande do Norte, R. Gen. Gustavo Cordeiro de Faria, S/N—Petrópolis, Natal 59012-570, Brazil
2 Department of Pharmacy and Pharmaceutical Technology, School of Pharmacy, University of Granada, Campus of Cartuja s/n, 18071 Granada, Spain
3 Department of Biology and Environment, School of Life Sciences and Environment, University of Trás-os-Montes and Alto Douro (UTAD), 5001-801 Vila Real, Portugal
4 Centre for Research and Technology of Agro-Environmental and Biological Sciences (CITAB), University of Trás-os-Montes and Alto Douro UTAD, 5001-801 Vila Real, Portugal
5 Departamento de Ciências Exatas e da Terra, Universidade Federal de São Paulo, Rua Arthur Riedel, 275 Diadema, São Paulo 09972-270, Brazil
6 Laboratory of Nanotechnology and Nanomedicine (LNMED), Institute of Technology and Research (ITP), Av. Murilo Dantas, 300, Aracaju 49010-390, Brazil
7 Industrial Biotechnology Program, University of Tiradentes (UNIT), Av. Murilo Dantas 300, Aracaju 49032-490, Brazil
8 Department of Pharmaceutical Technology, Faculty of Pharmacy, University of Porto, Rua de Jorge Viterbo Ferreira, No. 228, 4050-313 Porto, Portugal
9 REQUIMTE/UCIBIO, Faculty of Pharmacy, University of Porto, Rua de Jorge Viterbo Ferreira, No. 228, 4050-313 Porto, Portugal
* Correspondence: rbarbosa@ugr.es (R.d.M.B.); ebsouto@ff.up.pt (E.B.S.)

Citation: Barbosa, R.d.M.; Silva, A.M.; Silva, C.F.d.; Cardoso, J.C.; Severino, P.; Meirelles, L.M.A.; Silva-Junior, A.A.d.; Viseras, C.; Fonseca, J.; Souto, E.B. Production Technologies, Regulatory Parameters, and Quality Control of Vaccine Vectors for Veterinary Use. *Technologies* **2022**, *10*, 109. https://doi.org/10.3390/technologies10050109

Academic Editor: Manoj Gupta

Received: 19 August 2022
Accepted: 20 October 2022
Published: 21 October 2022

Abstract: This paper presents a comprehensive review of the main types of vaccines approaching production technology, regulatory parameters, and the quality control of vaccines. Bioinformatic tools and computational strategies have been used in the research and development of new pharmaceutical products, reducing the time between supposed pharmaceutical product candidates (R&D steps) and final products (to be marketed). In fact, in the reverse vaccinology field, in silico studies can be very useful in identifying possible vaccine targets from databases. In addition, in some cases (subunit or RNA/ DNA vaccines), the in silico approach permits: (I) the evaluation of protein immunogenicity through the prediction of epitopes, (II) the potential adverse effects of antigens through the projection of similarity to host proteins, (III) toxicity and (IV) allergenicity, contributing to obtaining safe, effective, stable, and economical vaccines for existing and emerging infectious pathogens. Additionally, the rapid growth of emerging infectious diseases in recent years should be considered a driving force for developing and implementing new vaccines and reassessing vaccine schedules in companion animals, food animals, and wildlife disease control. Comprehensive and well-planned vaccination schedules are effective strategies to prevent and treat infectious diseases.

Keywords: vaccines; veterinary application; bacteria; toxins; antigenic residues

1. Introduction

In the last century, the relationship between humans and pets has grown considerably in different societies, although it is not culturally universal. Only in the US do pet owners spend thousands of dollars a year to maintain care of their dog or cat. In addition, many works have shown how pets might play an essentially positive role in animal-assisted therapy in several conditions such as post-traumatic stress disorders or autism, for example [1].

In parallel with these benefits, pets can become harmful transmitters of various diseases such as brucellosis, roundworm, skin mites, *E. coli*, salmonella, giardia, ringworms, and cat-scratch fever [2,3]. No less important and also necessary is the vaccination of poultry, cattle, horses, sheep, goats and pigs. Vaccine use promotes animal health, safety for humans and financial protection for farmers. The animal vaccination process is fundamental, preventing and eradicating the spread of multiple diseases [4].

Vaccines are biological agents exploiting the humoral immune system's capacity and/or cell-mediated immunity safely to induce an immune response by inducing the production of immunological memory against a specific antigen derived from an infectious disease-causing pathogen [5,6].

Most of the vaccines currently available for animals have protein or polysaccharide antigens in their composition [7]. It is generally classified as liquid or lyophilized preparations of live (attenuated) or non-live (inactivated or killed) microorganisms [6,8]. In the last few years, viral vectors and RNA/DNA vaccines have contributed significantly to developing new immunizing products for animals use [8,9].

Bacterial vaccines and toxoids are produced from cell cultures in vitro, or in embryonated eggs using appropriate and validated methods. The cases described in this review do not apply to bacterial vaccines prepared from cell cultures or live animals. The bacterial strain employed may be genetically engineered and the identification, the antigenic power, and the purity of each bacterial culture used must be carefully controlled. Bacterial toxoids or anatoxins are prepared from toxins by reducing their toxicity to an undetectable level or by complete toxicity neutralization using physical or chemical methods; therefore, toxoids induce the production of neutralizing antibodies [10].

There are cases of bacterial toxins that are weakened until no toxicity exhibition but with enough strength to induce the formation of antibodies and specific disease immunity caused by the toxin. Toxins are derived from selected strains of specific microorganisms cultured in suitable media, or they may also be obtained by other appropriate methods (e.g., chemical synthesis) Toxins are derived from selected strains of specific microorganisms cultured in suitable media. They may also be obtained by other appropriate methods, for example, chemical synthesis. However, bacterial toxins should be weakened to be used as the bioactive compounds in vaccine products, and they present low or any toxicity as a fundamental requirement [11].

Toxoids can be purified by adsorption using adjuvants such as aluminum phosphate, aluminum hydroxide, calcium phosphate, and others. Bacterial toxoids may be in the form of a clear, transparent, or slightly opalescent liquid. Adsorbed toxoids are presented in the form of suspensions or emulsions, and some may be lyophilized. Unless otherwise indicated, provisions and requirements specified for bacterial vaccines also apply to vaccines based on bacterial toxoids and products containing a mixture of bacterial cells and toxoids [12]. Although alum-precipitated tetanus and diphtheria toxoids had been used for human immunization for many years, their use has declined considerably because of the variability in the production of alum precipitated toxoids.

Viral vaccines are prepared from viruses grown in suitable cell cultures, tissues, microorganisms, or embryonic eggs. If there is no other possibility, viral vaccines may also be produced in live animals. The used virus strains can be genetically engineered. Liquid or lyophilized preparations are composed of one or more virus or viral subunits or peptides. Live viral vaccines are prepared from viruses with attenuated virulence or low virulence for the native target species. Inactivated vaccines are subjected to a validated method of virus inactivation and can be purified and concentrated [13,14].

In this way, vaccines based on vectors are liquids or lyophilized preparations of one or more non-pathogenic or low pathogenic live microorganisms (bacteria or virus), in which one or more antigen-expressing genes, which elicit a protective immune response against other microorganisms, are inserted [15]. The preparation methods, which vary depending on the type of vaccine, should ensure the integrity and the immunogenic power of the antigen and the prevention of contamination by foreign agents. The origin of the animal products used in the production of vaccines for veterinary use shall meet the regulatory requirements [16].

Substances from other sources must meet the requirements of Regulatory Agencies and should be prepared to prevent any contamination of the vaccine by living microorganisms or toxins. Cell cultures used in the preparation of vaccines for veterinary use should also satisfy the requirements. It may be necessary to demonstrate the effectiveness of the inactivation method against specific potential contaminants. The use of embryonic eggs from specific pathogens flocks is required for the production of the primary seed batch in every passage of a microorganism to the working seed batch [17]. If there is no alternative to the use of animals or animal tissues in the production of veterinary vaccines, these must be free from specific pathogens, and their nature will depend on the species of origin and the target vaccination species [18].

Vaccination is the main approach to achieve the best cost-effective relationship to prevent economic losses and to increase the quality of life of animals. Figure 1 shows several vaccine technologies available for animals. In veterinary medicine, many immunogens are still produced using conventional technologies, such as attenuated vaccines. However, with the development of biotechnological tools, these are being used in vaccine development. These "modern" vaccine technologies are not just used to control infectious diseases but also to increase their productivity and the control of ectoparasites.

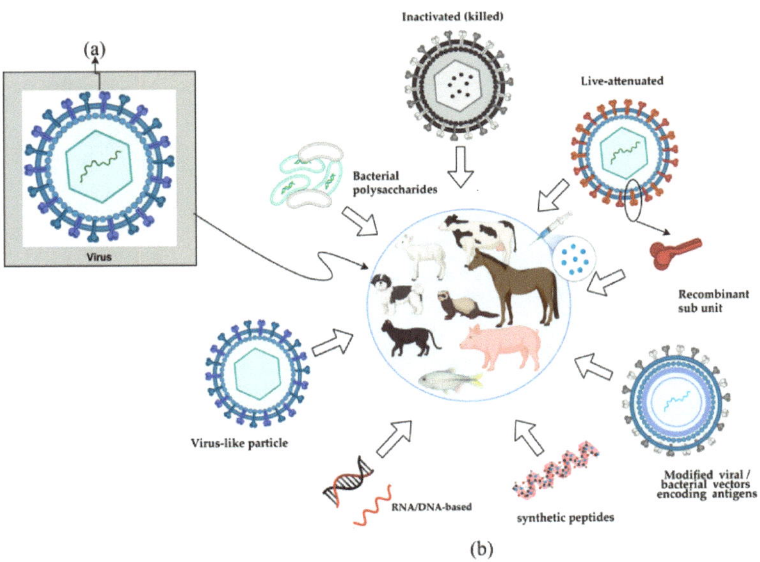

Figure 1. Schematic representation of common virus (**a**) and the main platforms adopted in the development of vaccines against pathogens (**b**).

There are currently several approaches to obtain a vaccine capable of promoting acquired immunity, from the most traditional ones, based on the intact pathogen (attenuated or inactive), or even based on the use of subunits, such as isolated proteins or self-assembled structural molecules, which are called virus-like particles, nucleic acids, or viral vectors. Table 1 shows the main characteristics of the mentioned vaccine platforms available for animals.

Among them, bacterial polysaccharide vaccines consist of inactivated or subunits that are characterized by structures that are part of the bacterial cell. It also constitutes purified molecules such as capsular polysaccharides, native or even recombinant proteins. With the advent of reverse vaccinology, several proteins identified the important targets in bacterial infections being expressed in different vectors, purified, and tested as potential vaccine targets.

Another technology concerns the use of synthetic peptides, which are designed from studies by computational prediction, defining the possible sequences that contained immunogenic determinants [19]. The synthetic vaccine against *Rhipicephalus microplus* called SBm7462® was developed by the Laboratory of Biology and Control of Hematozoa and Vectors using this technique [20].

Table 1. Description of the most common types of vaccines for animal use.

Platform	Characteristics	Restrictions	Refs
Whole virus			
Attenuated	Entire virus passed on in successive cultivations to lose infective capacity.	Production requires cell culture of the virus and exhaustive safety tests as they are more immunogenic.	[21]
Inactivated	The intact virus is inactivated by chemical or physical methods.	Vaccines based on the inactivated virus require an initial high amount of virus.	
Subunit			
Proteins	Proteins or their fragments are injected directly into the animal.	Generally, require adjuvants or multiple doses to achieve the desired immune response.	[21]
Virus-like particle	Self-assembled viral structural proteins that resemble the virus, however, lack genetic material.	The biggest challenge of this platform is to ensure that the epitopes are in an adequate conformation after translation and that the expressed proteins are not allergenic.	[22]
Nucleic acid			
DNA	Insertion of the DNA that encodes the viral antigen into a plasmid.	They are platforms under experimentation for animal purposes.	[23]
RNA	Messenger RNA encapsulated in a lipid membrane.		
Vector encoding antigen			
Replicating and non-replicating	Non-infective pathogens are genetically modified with the insertion of one or more genes that express antigenic particles, which may or may not multiply in the animal organism.	Requires level 2 biosafety labs for production; it has reduced efficacy due to pre-existing immunity to selected vectors.	[24]

DNA vaccine is developed from a plasmid, and its expression contains genes encoding one or more immunogenic antigens of interest. Once these recombinant plasmids are inserted inside the host cell, the target gene will be transcribed. Recombinant RNA vaccines consist of fragments of the sequence of the genetic material of messenger RNA (mRNA), which can be designed to encode any viral, bacterial, or parasitic protein. When virus mRNA is inside host cells, they are translated into proteins, which induce an immune response to the host's body. In addition, customized RNA/DNA sequences allow researchers to create vaccines that produce virtually any protein desired [4,25–27].

Vaccines are essential to prevent and control zoonotic infectious diseases in humans and animals (domestic and wild). The use of vaccines in animals impacts positively the production and their quality of life. Some examples of veterinary vaccines are described in Tables 2 and 3.

Zoonoses have always been a great concern for the scientific community, with a strong worldwide public health impact, as recently happened with the COVID-19 pandemic. Data released by the WHO in 28 June of 2022 [28] confirmed 542,188,789 cases of COVID-19, including 6,329,275 deaths worldwide. In addition, the International Monetary Fund predicted an estimated global cost to the economy of US$12.5 trillion by 2024 due to the new coronavirus pandemic. However, other zoonoses that also deserve mention are avian influenza and MERS, both with a high risk of becoming a new pandemic; and, as the regional transboundary epizootics, we can mention yellow fever, Venezuelan equine encephalitis, and Rift Valley fever [29].

Table 2. Examples veterinary vaccines using different strategies.

Vaccine Strategy	Disease	Animal	Consequences	Refs
Bacterial ghost construction	Avian Colibacillosis	Avian	Mortality of poultry bacterial infections—it causes a variety of disease manifestations in poultry including yolk sac infection, omphalitis, respiratory tract infection, swollen head syndrome, septicemia, polyserositis, coligranuloma, enteritis, cellulitis and salpingitis	[30,31]
Avirulent suspension of *Salmonella typhimurium* AWC 591	Salmonellosis	Commercial poultry	Economic losses and risks to public health such as diarrhea, fever, and stomach cramps	[32]
Modified live vaccine (MLV) infectious bovine rhinotracheitis	Rhinotracheitis	Cattle	Respiratory disease complex	[33]
The gene for protein 2 (VP2) of infectious bursal disease virus was cloned into a *Pichia pastoris* expression system	Infectious bursal disease (also known as Gumboro disease)	Avian	Immunosuppressive viral disease due to widespread destruction of lymphocytes	[34]
Replacement of the capsid-encoding gene (P1) from the vaccine strain O1 Manisa	Foot-and-mouth disease virus	Cattle, pigs, sheep, and many wildlife species	Economically devastating disease; reduced animal productivity and the restrictions on international trade in animal products	[35,36]
Recombinant vaccines based on *Brucella* Outer Membrane Protein (OMP) antigens	Brucellosis	Calves, sheep, cattle, goats, pigs, and dogs, among others	High economic losses due to restrictions on international trade in animal products; the signs and symptoms include fever, joint pain (arthritis, spondylitis, sacroiliitis), endocarditis and fatigue.	[37]
Recombinant vaccines based on their major toxins and their genetic origins (iota (ia), alpha (cpa), beta (cpb), and epsilon (etx), and toxoid vaccines, bacterin-toxoid vaccine	*Clostridial* diseases	Cattle, sheep, and goats	botulism, tetanus, enterotoxaemia, gas gangrene, necrotic enteritis, pseudomembranous colitis, blackleg, and black disease causing severe economic losses in livestock and poultry industries	[38]

Different zoonotic diseases are annually responsible for the death and economic loss due to the substantial reductions in livestock production. In general, the large-scale slaughter of herds negatively impacts the livestock sector, but this practice is essential to prevent human infections. In addition, wild animals can be mortally affected by other diseases such as West Nile disease (birds), yellow fever (neotropical monkeys), plague (black-footed ferrets), and Ebola (great apes) [29].

Table 3. Domestic animals' vaccination (cats, dogs, and rabbit) schedule examples. Recommendation from the National Office of Animal Health, representing the UK animal health industry.

Disease	Example (Supplier)/Vaccine Strategy	Recommended Vaccination Schedule
Feline Panleukopenia/ Infectious Enteritis (Parvovirus)	Fevaxyn® Pentofel (Zoetis Belgium SA)/Fevaxyn Pentofel contains the following inactivated viruses: feline panleukopenia virus, feline rhinotracheitis virus, feline calicivirus, feline leukemia virus, and the inactivated bacterium feline *Chlamydophila felis*.	Cats of 9 weeks or older. Two doses at an interval of 3 to 4 weeks.
Feline Calicivirus	Purevax RC (Boehringer Ingelheim, Ingelheim am Rhein, Germany)/Attenuated feline rhinotracheitis herpesvirus (FVH F2 strain) and inactivated *feline calicivirus* antigens (FCV 431 and G1 strains)	Only cats of 8 weeks or older receive the first injection; the second injection is 3 to 4 weeks later. Revaccination: the first revaccination should be carried out one year after the primary vaccination, and subsequent revaccinations: at intervals of up to three years.
	Feligen RCP (Virbac)/a modified live vaccine providing immunization of healthy cats against *feline rhinotracheitis* virus, *feline calicivirus* and *feline panleucopaenia* virus.	Cats from minimum 9 weeks of age. Two doses at an interval of 3 to 4 weeks. Annual boosters are recommended after that
Feline Leukaemia Virus	Purevax FeLV (Boehringer Ingelheim, Ingelheim am Rhein, Germany)/virus canaripox recombinante FeLV (vCP97). The vaccine strain is a recombinant canarypox virus that expresses the FeLV-A env and gag genes. Under natural conditions, only subgroup A is infectious and immunization against subgroup A induces total protection against subgroups A, B, and C. After inoculation, the virus expresses the protective proteins but does not replicate in the cat. Thus, the vaccine induces an immune state against the *feline leukemia* virus.	Cats of 8 weeks of age or older. Primary vaccination: first injection: from the age of 8 weeks. Second injection: 3 to 4 weeks later. Revaccination: annual
Feline Rhinotracheitis (Herpesvirus)	Purevax RC (Boehringer Ingelheim, Ingelheim am Rhein, Germany)/Attenuated *feline rhinotracheitis* herpesvirus (FHV F2 strain) and inactivated *feline calicivirus* (FCV 431 and G1 strains) antigens	Cats of 8 weeks of age or older. Against feline viral rhinotracheitis, for the reduction in clinical signs and against calicivirus infection for the reduction in clinical signs. Primary vaccination: first injection: from 8 weeks. Second injection: 3 to 4 weeks later. Revaccination: the first revaccination should be carried out one year after the primary vaccination, subsequent revaccinations at intervals of up to three years.
Feline Rabies	Purevax Rabies (Boehringer Ingelheim, Ingelheim am Rhein, Germany)/Contains rabies recombinant canarypox virus (vCP65); Rabisin (Boehringer Ingelheim, Ingelheim am Rhein, Germany)/inactivated rabies antigen (viral glycoproteins)	Cats 12 weeks of age and older. The cats should be revaccinated every year
Canine Rabies	Rabvac 1 (Boehringer Ingelheim Ingelheim am Rhein, Germany)/a inactivated virus vaccine; Defensor (Zoetis, Belgium SA)/ Rabico virus strain PV-Paris (*Pasteur*) replicated in a stable cell line, chemically inactivated; Rabisin (Boehringer Ingelheim, Ingelheim am Rhein, Germany)/inactivated rabies antigen (viral glycoproteins).	Rabvac 1:3 months of age or older. Revaccinate one year later and annually thereafter. Defensor: heath dogs and cats: a single dose at 3 months of age or older. Annual revaccination with a single dose is recommended. Rabisin: inactivated rabies antigen (viral glycoproteins)

Table 3. *Cont.*

Disease	Example (Supplier)/Vaccine Strategy	Recommended Vaccination Schedule
Canine distemper virus, *Canine Adenovirus* Type 2, infectious hepatitis, *Canine Parvovirus* (modified live viruses), Coronavirose canina, and Leptospira Canicola-Icterohaemorrhagiae (*L. canicola* and *L. icterohaemorrhagiae*)	V8 Nobivac® Canine (MSD, NJ, USA)/vaccine combination—modified live virus vaccine and a live attenuated vaccine	Puppies from 45 days of age, there are 3 or 4 doses in a row with intervals of 21 to 30 days between them
Canine distemper, infectious hepatitis, parainfluenza, parvovirus, coronavirus, and leptospirosis (Canicola and *Icterohaemorrhagiae serovars*), leptospirosis (Grippotyphosa and Pomona)	V10 Vanguard Plus (Zoetis, Belgium SA)/live attenuated vaccine	After V8 applications, the adult dog must be vaccinated with V10 from 6 weeks of age or older.

Nowadays, the target species focuses on vaccination schedules on species that are "almost" always directly affected; unfortunately, there is still a lack of strategies that indirectly prevent human diseases through the immunization of domestic animals and sources of infection. On the other hand, the vaccination of wild animals aimed at preventing diseases in humans or domestic animals is even more challenging and scarce. Furthermore, the primary sources of funding for research on human and animal diseases tend to be channeled to different government agencies, stifling cross-cutting approaches.

Some examples of vaccines are already available on the market and were developed to protect humans and economically valuable animals, such as Japanese encephalitis. Vaccinating horses and pigs is available, especially in countries where the disease is endemic, but, unfortunately, the costs often outweigh the benefits [39]. Other vaccines target domestic animals and aim to reduce the infection between animals and humans (as presented in Table 3).

There are a few examples of vaccines for wild animals; in this case, the objective is disease eradication and/or transmission from wild animals to humans and domesticated animals. For instance, in the State of Texas, USA, the oral rabies vaccination program led to the eradication of rabies among dog–coyote by distributing baits containing the vaccine with the aid of aircraft [40].

Some factors may suggest additional care concerning the vaccine schedules, given that there is no single ideal vaccine schedule solution for all species and regions (or countries). Instead, there are instructions, government regulations, scientific standards, professional organization guidelines, and veterinarian recommendations for vaccination programs. Any decision to adopt the vaccination schedule needs to be made on a case-by-case basis, considering the vaccination history of the animal in association with the epidemiological context of the analyzed region.

Another excellent example to illustrate the concerns transmission disease from animal to human is brucellosis. It is caused by *Brucella* spp., which are Gram-negative bacteria that have been found primarily in mammals such as goats, sheep, cattle, dogs, pigs, dolphin, porpoise, and whale, among others. Symptoms begin as an acute febrile illness with little or no localized signs and may progress to a chronic phase characterized by relapses of fever, weakness, sweating, and vague pain [41].

In cattle, the infection of *Brucella* spp. can be identified by clinical signs such as the births of weak calves, retained placenta, vaginal discharge, inflammation of the joints, and inflammation of the testicles. The most widely used vaccine for the prevention of brucellosis in cattle is the *B. abortus* S19 vaccine, but there are important differences in the dose in dependence of age and sex of cattle. The females aged 3–8 months must be vaccinated (limited to sexually immature female animals) as a single subcutaneous dose of 5–8 $\times$ 10^{10}; however, a reduced dose of viable organisms is necessary (from 3 $\times$ 10^8 to 3 $\times$ 10^9) to vaccinate adult cattle by the same administration route. Alternatively, it can be administered to cattle of any age as either one or two doses of 5 $\times$ 10^9 viable organisms, given via the conjunctival route. It is worth mentioning that specialized veterinarians must perform the procedure due to the susceptibility of infection for those who handle it (vaccine produced by a live bacterium). To ensure the correct application of the immunizer, the veterinarian provides the vaccination certificate to the producer, which is a governmental mandatory requirement in the most of countries. Another important strategy to control and eradicate the disease is the running of brucellosis tests at least once a year, which is crucial to carry out quarantine and new exams to incorporate new the animals into the herd [41].

Factors that may influence the effectiveness of animal vaccination may be related to the vaccine (platform used in the development, interval required for application of the booster dose, addition of adjuvants in the formulation), to the host (maternal antibodies, immune system functionality, concurrent diseases, different races), to humans (storage condition, preparation, administration), and the environment (endemicity of the region and contact with strains of wild animals) [42].

In the case of bovine tuberculosis, the dose administered by the parenteral route is one hundred times lower than the dose required to ensure the effectiveness of protection by the oral route. Revaccination of cattle against tuberculosis is contraindicated, as it induces the strongest antigen-specific IFN-g responses [43].

Therefore, following the practices and protocols described in the literature and regulatory parameters is imperative. As mentioned above, each disease has a peculiarity concerning the active pharmaceutical ingredient and period to be applied in the animal's life. In addition, each country has its legislation that must be strictly followed to avoid animal and human health problems [41,44].

2. Production of Vaccines

Several methods of vaccine production have been described in the literature. The methodologies are divided into two groups denominating inactivated (killed) or live attenuated (weakened) microorganisms technologies. These techniques have been successfully used to control many diseases in the veterinary application. Each technique shows advantages and disadvantages as well as the ability to influence protective efficacy, affecting the economy of production [45].

In addition to choosing the correct strains, the qualitative composition of media used in the preparation and the production of seed cultures must be specified, namely by referring to the quality of each ingredient, and an adequate description should be registered of them. In the case of ingredients from animal origin, the species and the country of their source should be indicated and should meet the regulatory requirements. The methods used for media preparation must also be documented, including the inactivation process. The addition of antibiotics during production should usually be limited to cell cultures, inoculums injected into the eggs, and the material collected from the skin or other tissues [46].

2.1. Bacteria Seed

Bacteria used in the production of vaccines are characterized by genus and species. Whenever possible, bacteria used in production must be grown according to a seed batch system. For each primary seed batch, the origin, the date of isolation, the history of passages (including purification and characterization methods), and conservation conditions should be kept on record. Each primary seed batch should be assigned a specific identification code, whereas the minimum and the maximum number of subcultures made in each primary seed batch before the production stage should also be specified [47].

In addition, the methods used for preparing the seed crops and seed suspensions, the techniques for seed inoculation, the title and the concentration of the inoculum, and the used means should be documented. It should be demonstrated that subcultures do not modify seed characteristics (e.g., dissociation or antigenic power). Storage conditions of each seed batch also should be documented. It must be demonstrated that each primary seed batch consists solely of bacteria of the species or the indicated strain [45].

Briefly, the method used to identify the biochemical, serological, and morphological characteristics of each strain should be registered to distinguish the strains as much as possible. Furthermore, the method applied to determine purity should also be properly registered for easy tracking if needed. If the primary seed batch contains any live microorganism other than the bacteria of the species or the indicated strain, the batch cannot be used in the production of vaccines [48].

2.2. *Virus Seed*

Viruses used in the production of vaccines are cultured according to a seed batch system. In this case, also for each primary seed batch, a record of the origin, date of isolation, history of passages (including the methods of purification and characterization), and storage conditions should be kept on storage and labeled with a specific code. Typically, in the production of a vaccine, the used virus must not be subjected to more than five passages from the primary seed batch. Unless otherwise indicated, the tests carried out on each primary seed batch are the ones briefly described here. It should normally not relate to microorganisms with a greater number of passages than five from the primary seed batch at the beginning of the tests [49].

The tests described below must be conducted with an appropriate volume of virus from the lysis of primary cell bank cells when the primary seed batch consists of a primary cell bank chronically infected with a virus. Appropriate tests have already been carried out in lysed cells for primary cells database validation, so it is not necessary to repeat the tests [50].

The multiplication of the primary seed batch virus and all subsequent passages must be carried out in cell culture in embryonic eggs or suitable animals to produce vaccines. Materials of animal origin must satisfy their specific requirements. A suitable method must be used to identify the vaccine strain and, as much as possible, to distinguish it from closely related strains. The primary seed batch must meet the sterility and the mycoplasmas tests. For inactivation of the complement, serum batches must be kept at 56 °C for 30 min. It must be proved that batches of serum are free of antibodies to potential contaminants of the seed virus, and they have no nonspecific inhibitory effects able to prevent infection or virus multiplication in cells (or eggs, as appropriate). If there is no possibility to use a serum with these characteristics, other methods should be used to counteract or specifically eliminate seed virus [49].

The sample of the primary seed batch should be treated with the lowest possible amount of monoclonal or polyclonal antibodies so that the virus can be neutralized as much as possible or removed [16]. The final serum–virus mixture will contain (where appropriate) a quantity of virus at least equivalent to 10 doses of vaccine per 0.1 mL or 1.0 mL, in the case of poultry vaccine or the other, respectively [51].

Next, as indicated below, the presence of foreign agents in the mixture should be investigated. For the remaining vaccines, the inoculated mixture should be at least 70 cm^2 of appropriate cell culture. Cells can be seeded in any growth phase at a lower confluence that corresponds to 70%. At least one cell of each type should be kept. Cultures should be observed daily for a week. At the end of this period, cultures are frozen and thawed three times; then, they are centrifuged to remove cell debris and re-inoculated in the same type of previous crops twice [13].

The number of cells obtained in the last passage, in suitable containers, should be sufficient to achieve the following tests [52]. Techniques, such as immunofluorescence, can be used for the detection of specific contaminants in cell cultures [53]. The primary seed batch must be inoculated in primary cells of the species origin of the virus, susceptible cells to viral pathogens for the target species of the vaccine, and sensitive cells to pestiviruses. If the primary seed batch contains any living microorganisms other than the virus species and the indicated strain, or viral or foreign antigens, the batch cannot be used in the production of vaccines.

2.3. Computational Based Vaccine

In the last three decades, significant advances have been made in the genetic sequencing field with the use of innovative technologies such as Next-Generation Sequencing (NGS). The associated progress in the NGS area associated with the precise analysis of the sequences, the in-depth study of structural and molecular modeling and machine learning have allowed the growing interest of researchers of different areas providing the emergence of a team with interdisciplinary training, which has provided promising results in vaccinology [4] or reverses vaccinology (RV). RV is based in the rational design and development of vaccines using computational tools which identify and examine immunogenic antigens without the need for cell culture [54].

In recent years, the exponential growth of datasets with genomes of bacteria, viruses, archaebacteria, and eukaryotes has been observed, which are all freely available in databases on the web. Thus, computational techniques, bioinformatics, and immunoinformatic approaches have become essential for the better prediction and analysis of high-throughput data, aiming to identify, design, and develop new drugs or vaccines for human or veterinary and human use [55]. The main aim is the routine use of in silico techniques to favor the reduction in the time and cost of laboratory experimentation and production that generally lasts from 5 to 15 years which can also provide faster, convergent, and cost-effective discoveries of drugs [56] or vaccines [57] against new and emerging diseases.

Corynebacterium pseudotuberculosis mainly affects small ruminants such as goats and sheep. However, it can also infect horses, cattle, llamas, alpacas, and buffaloes, causing lymphadenitis clinically presented in its cutaneous, mastitis, or visceral form, which causes a significant loss in agribusiness worldwide [58]. To solve part of the problem, Soares and collaborators (2013) identified the genomic sequence of *C. pseudotuberculosis biovar equi* strain 258 to select antigenic targets and used them in reverse vaccinology to develop new vaccines for the hosts [59]. In addition, Araujo et al. (2019) also studied strains of *C. pseudotuberculosis* with the aim of in silico prospecting the development of new targets [60].

Works focused on trypanosomiasis, also known as Chagas disease, which can be caused by a protozoan of the species *Trypanosoma cruzi*. The transmission occurs through the feces that the "barber" deposits on the skin, while sucking the blood. It is endemic in South America and affects mainly humans; however, rats, dogs and cats can be a reservoir host. Ruminants are not affected. Despite efforts by different research groups, there are still no vaccines against *Plasmodium vivax*, which is one of several etiologic agents of malaria. *P. vivax* protozoan affects chimpanzees and gorillas (wild animals). In 2011, Bueno and co-authors [61] presented a selected list of antigenic and immunogenic epitopes within the Apical membrane antigen 1, which was considered the leading candidate antigens for developing a malaria vaccine. In 2020, Michel-Todó et al. published preliminary data on a rationally optimized vaccine development based on multiple epitopes of multiple antigens to neutralize the biological complexity of parasites with the aid of computational techniques for the analysis and prediction of biological data [62].

Other works have been published with a focus on the production of vaccines against brucellosis [63] and toxoplasmosis, both of which have been extensively studied with significant prevalence in humans and several animal species globally for human and veterinary use [64], having been optimally planned from reverse vaccinology with massive use of bioinformatics and computational tools.

2.4. Challenges in Vaccine Production

Viruses, parasites, bacteria, fungi, and prions are agents that cause zoonoses, all of which have extraordinarily varied life cycles and modes of transmission, providing complex epidemiological patterns. In this context, deep knowledge of the genomic and antigenic diversity of each microorganism involved in the target disease and their epidemiological profiles are mandatory information for effective vaccine development.

Despite developing new vaccines for emerging diseases, researchers are currently addressing vaccines that can bypass inhibitory maternal antibodies, reduce dependence on the cold chain, or even adapt to husbandry management or animal owner lifestyles.

Drug delivery systems can be used to enhance the vaccine's performance, either by the slow delivery of the antigens or even by targeting specific sites. Slow delivery systems can reduce the number of doses, e.g., a vaccine that would be taken every year could be taken every two years because such systems behave like a reservoir that delivers the antigens slowly. Liposomes, lipid nanoparticles, and polymeric nanoparticles are among the delivery systems studied in veterinary vaccine development. All those delivery nanosystems have already been widely described in review papers about their application in veterinary vaccines [65–68]. Such nanoparticles can encapsulate the antigens, protect them from the body's chemical and enzymatic attacks, and even enhance the antigen's internalization into the specific body cells, improving efficiency.

Another obstacle is that most vaccines currently available must be refrigerated at 2–8 °C, and they must be protected from high temperatures as well as freezing to ensure their effectiveness. Such sensitivity is linked to the antigen used in the preparation of the vaccine, which may consist of attenuated organisms or a protein subunit, sensitive to moderate heating, or even consist of inactive organisms that are more affected by low temperatures. Such a scenario proves to be more complex when vaccines must serve herds in regions far from large urban centers, lacking the support of an adequate cold chain [69].

In addition, many of the countries endemic for diseases whose control can already be achieved using vaccines are developing, limiting investments to ensure the adequate storage and distribution of inputs in rural areas [70]. The number of vaccines for veterinary use commercially available with thermostability is still limited, such as the vaccine against Conventional Newcastle disease for chickens [71] or against rabies for dogs [72].

Although vaccination is an efficient approach to disease prevention and control, it is known that exposure of the pathogen to vaccinated animals can result in the emergence of resistant variants of the vaccine in question, with the evolution of the pathogenicity of the strain. This situation manifests itself more commonly among RNA viruses due to the high mutation rate during replication. Thus, the genome that best adapts to a given environment will prevail [73].

To ensure the effectiveness of vaccines, monitoring strategies must therefore be implemented. Adjustments in vaccination schedules or cases of resistance can thus be detected, avoiding unnecessary expenses. The immune response of the vaccinated population must be evaluated based on different indicators. In the case of foot and mouth disease, the number of outbreaks and the levels of virus circulation are determined by means of serosurveys, measuring the proportion of vaccinated animals that did not have the disease during an outbreak, compared to unvaccinated animals. It is worth noting that serological control is not sufficient to monitor the success of a vaccination program, as they are influenced by the type of vaccine and the test used to determine the antibody titer [74].

2.5. Production Methods

The vaccine production process comprises four phases: product profile, pre-development, development, registration commercial (Figure 2). The production of vaccine concentrate is characterized by the origin of the vaccine. The viral vaccines process consists of cell replication from a reference strain. The classic methodology of viral vaccine production consists in the technology of viral cultivation directly in embryonated chicken eggs free of pathogenic organisms, such as yellow fever, for example. Otherwise, the bacterial vaccines are produced by a process of fermentation of inputs and conjugation of active principles. The concentrated vaccine can only be made available for final processing after completion of the qualitative analysis, as this involves a sequence of physical, chemical, biological, and microbiological tests that take place simultaneously. The concentrated produced vaccine is stored in cold at a suitable temperature to maintain the product's characteristics [75].

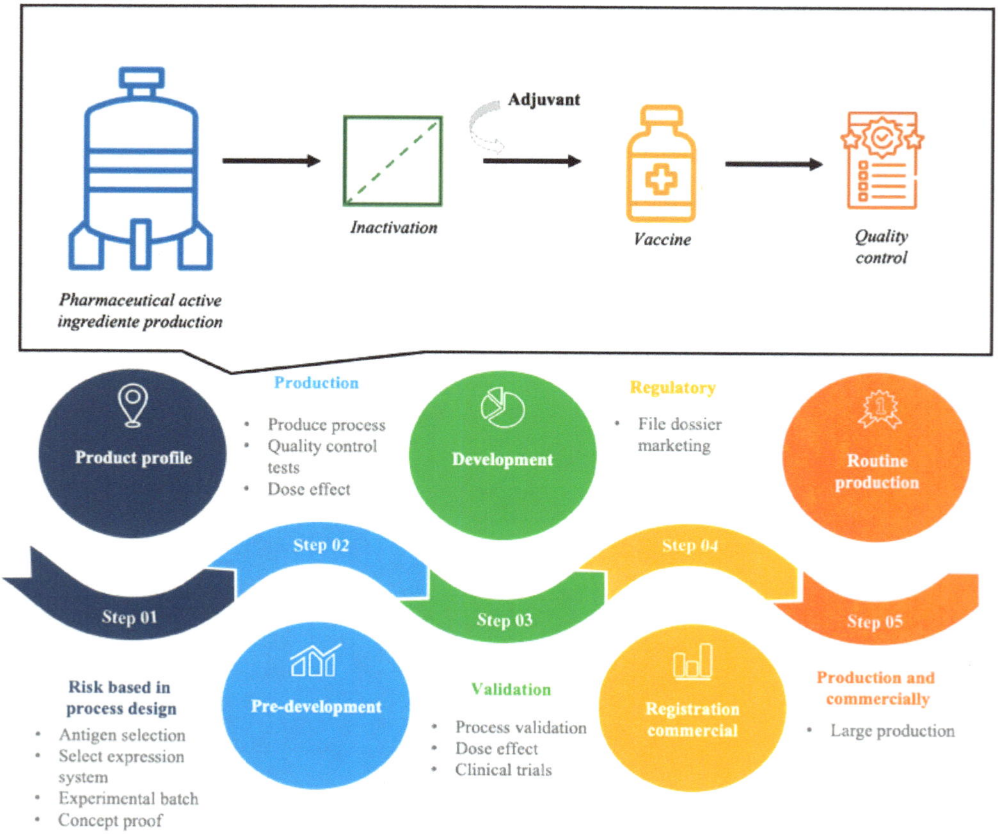

Figure 2. Steps of development vaccine from conception to production.

The active pharmaceutical ingredient (API) is the main component of the vaccine. However, other components are added to stabilize the formulation and diluting the API to the ideal fraction for veterinary application. The key adjuvants focus on improving immune response (aluminum salt), preservatives (thimerosal), stabilizers to protect against adverse conditions such as freeze and thaw (gelatin and monosodium glutamate), antibiotics to prevent contamination (neomycin, streptomycin, and polymyxin B), and microorganism suspension fluid (egg and yeast protein). As a result, you have the vaccine in bulk [76].

The final step in the process is divided into three stages: filling, lyophilization, and labeling and packaging. In the bottling, the bulk vaccine is transferred from the stainless-steel tanks to the glass bottles. The filling machine starts an in-line process of washing and sterilizing the bottles. After the vials receive the vaccine, they are closed with a butyl rubber stopper. For liquid vaccines, this closure is total, and the vials are directed via a conveyor to an aluminum cap fixing machine. The lyophilized vaccines are partially closed, and the vials are transported via trays to equipment called a lyophilizer [77].

After the freeze-drying cycle, the vials are completely closed with the stoppers they received in the filling process. When removed from the lyophilizer, the vials immediately go to a machine for applying an aluminum seal that seals each vial individually. These are stored in a cold room separated by batches, which is followed by labeling and packaging. The completion of final processing is to package the vaccine. The vials containing the lyophilized vaccine, the liquid vaccine, or the diluent for the lyophilized vaccine are labeled with the product identification, batch number, manufacturing date, and product expiration date, among other information. Cartridges are packed in a box and then transferred to the finished products warehouse but remain in a segregated area for quarantined products until the completion of quality control and issuance of the product release certificate [78] (Figure 2).

3. Inactivation

Chemical or physical agents can carry out inactivation of virus. Among the most common inactivating agents are formaldehyde [79–83] and β-propiolactone [79,80,83,84]. Other chemical agents have also been explored, such as binary ethylenimine [80] and even natural compounds such as catechins obtained from green tea extract [85]. Green tea extract could be the first non-toxic natural compound to prepare inactivated viral vaccines with improved efficacy, productivity, safety, and public acceptance. In terms of antibody titer, cross-reactivity to heterosubtypic of viruses, and avidity to viral antigens, the quality of antibody responses to the green tea-inactivated virus was superior to that of the formaldehyde-inactivated virus [85].

Hydrogen peroxide was also used as an inactivating agent for the rabies virus. The results showed that hydrogen peroxide could replace β-propiolactone to reduce the time and cost of the inactivation process [86]. Ascorbic acid was also tested as an inactivating agent for rabies virus, but further studies are required to evaluate its effect on the cell-associated virus, probable therapeutic potential, and feasibility of replacing β-propiolactone in the production of inactivated rabies vaccine [87].

Concerning the physical inactivating agents, heat inactivation [82,88] and UV light [79] were also found in the literature. The etiological agent for Hydropericardium Syndrome (HPS) in broiler birds was inactivated by heat treatment at 56 °C for one hour and 80 °C for 10 min followed by formalin inactivation. They verified that the autogenous vaccination was extremely successful in both preventing and lowering illness in affected flocks [82]. The immunogenicity of the virus was unaffected by dual inactivation of the virus by heat and formalin treatment. Gupta et al. 1987 evaluated five inactivating methods for the diphtheria–pertussis–tetanus (DPT) vaccine [88]. Heat-inactivated pertussis (HIP) preparation was less potent than thimerosal-inactivated pertussis preparation, but the HIP was more potent than acetone-inactivated pertussis. However, HIP was similar to formaldehyde-inactivated pertussis (FIP) and glutaraldehyde-inactivated pertussis (GIP) preparations. They also checked that the inactivating agents did not affect the stability of the vaccine. On the other hand, Egorova et al. (2020) compared UV light at 253.7 nm to formaldehyde and β-propiolactone for viral inactivation during the development of a whole-virion vaccine against hemorrhagic fever with renal syndrome (HFRS). Although UV light was able to inactivate the virus, the β-propiolactone was the most promising of the tested inactivators [79].

Inactivated vaccines should be subjected to a validated process of inactivation. The described assay below for inactivation kinetics is performed only once for a given production. The other described tests are carried out in each production cycle. When the inactivation test is performed, it should have an eye out for the possibility of certain conditions of manufacture. Microorganisms can be physically protected from the inactivating agent [89].

Kinetics of inactivation must be proved if the inactivating agent and the method effectively ensure the inactivation of microorganisms in the vaccine manufacturing conditions. Data on the inactivation kinetics must be obtained. The time typically required for inactivation should not be higher than 67% of the duration of the inactivation process. If the formaldehyde is used as an inactivating agent, the test must be carried out free of formaldehyde [90]. To neutralize the residue of preparations of aziridine, sodium thiosulfate is added to promote the hydrolysis of this inactivating agent [91].

When using other inactivation methods, the assays must be carried out to show that the inactivating agent was eliminated or reduced to an acceptable concentration. The inactivation assay must be realized immediately after the inactivation process, or, depending on the case, after the neutralization or the disposal of the inactivating agent. If the vaccine contains an adjuvant impossible to achieve the inactivation test in the final blend inactivation, one test should be conducted during the mixture of the bulk antigen, immediately before the addition of adjuvants instead of being administered on the final batch [92].

3.1. Bacterial Vaccines

The test must be appropriate for the used bacteria and should comprise at least two passages in the culture medium used in production or, if the production is carried out in a solid medium, a suitable liquid medium or a semi-prescribed liquid in the specific monograph. The product meets the specifications if no living microorganisms are detected [93].

3.2. Bacterial Toxoids

The detoxification tests should be performed immediately after the preparation of the anatoxin and, as appropriate, after the neutralization or the elimination of the inactivating agent. The selected test should be adapted to the toxin or toxins involved, especially when in the case of sensitive assays. If there is any risk of reversion of the toxicity, one supplementary test should be performed in the earlier stage of the manufacturing process [94].

3.3. Viral Vaccines

To develop and manufacture a viral vaccine, the selection of a cell substrate is an important factor as it relies several parameters, such as cell susceptibility and permissiveness to the viral pathogen, performance in terms of viral antigens quality and production yield, primary versus continuous cells, ethical point of view, tumorigenicity status, anchorage-dependent versus suspension culture, culture medium, manufacturing cost, free of adventitious agents, and so on. Another step that has also to be considered is the format of the vaccines, as they influence the cell substrate selection, (e.g., inactivated versus live-attenuated viral vaccines; administration routes; preventive or therapeutic vaccines). The last factors to take into account are the safety and industrial considerations that deeply impact the choice of the suitable/optimal cell substrate [95].

Based on regulatory considerations, it is important make sure that all parameters are studied. These parameters included: (i) evolution of regulatory requirements for vaccine safety [96]; (ii) characterization of cell substrates used for the manufacturing of viral vaccines [97], related to, e.g., source of the cell substrate [98], history of the cell substrate [99], characteristics of the cell substrate and detection of adventitious agents, assessment of tumorigenic and oncogenic potency [100].

4. Choice of Composition and Strain of Vaccines

Among the several important aspects to be considered when choosing the composition and the vaccine strain are the safety, efficacy, and stability. Requirements to assess the safety and effectiveness have also been previously described. These requirements can be explained or supplemented by the requirements of the specific monographs. The validity must be justified by the stability studies. These comprise the titration of viruses, bacteria count and the determination of the activity. This determination is carried out at regular intervals until three months beyond the expiration date on, at least, employing three successive representative lots of vaccines stored under recommended conditions. If appropriate, the determination of moisture is also performed in lyophilized vaccines [101,102].

5. Final Bulk and Final Batch

The final bulk is formed by mixing one or more batches of the antigen, which should meet all the specified requirements, including adjuvants, such as stabilizers, antimicrobial preservatives, and solvents. The antimicrobial preservatives are used to prevent tampering or adverse-side effects caused by the vaccine microbial contamination during use. Antimicrobial preservatives cannot be incorporated in the lyophilized product. However, their use can be justified taking into account the recommended maximum duration of use of the vaccine after reconstitution, and they should be incorporated in the diluent of the lyophilized products for multiple dose [103].

Usually, the incorporation of antimicrobial preservatives in liquid preparations is not acceptable for single dose, but it may be acceptable when the same product is distributed in single-dose containers and in multiple-dose ones. In the case of multi-dose liquid preparations, the need for the use of antimicrobial preservatives must be evaluated considering the possibility of contamination during the use of the vaccine and the maximum recommended usage time after opening the container. When an antimicrobial preservative is incorporated, its efficacy must be demonstrated throughout the period of validity [104].

For inactivated vaccines, if the auxiliary substances interfere with the inactivation test, the test must be carried out for the preparation of the final bulk. This should be performed after mixing the different antigen batch but before the addition of the auxiliary substances; in case of dismissing inactivation, this should be tested at the bulk batch. Among these tests, the determination of the antimicrobial preservative free and formaldehyde, the safety test and the determination of the activity of inactivated vaccines are included [104].

As otherwise indicated in the monograph, the final bulk should be distributed aseptically into sterile containers with tamper-proof closure and sealed to prevent contamination. For the physical tests, vaccines with oil adjuvants must be submitted to the viscosity test by an appropriate method. The viscosity should be between the accepted limits for the product, and it must demonstrate the stability of the emulsion.

The chemical tests shall demonstrate, through adequate assays, that the concentrations of certain substances, such as antimicrobial preservatives and aluminum derivates, are within the set limits for the product, namely: (i) to determine the pH of liquids and diluents and demonstrate that those values lie within the limits set for the product; (ii) in certain cases, the lyophilization process is verified by determining the water content, which must comply with the approved limits for the product.

The compliance of each of the requirements prescribed in "Identification", "Test", or "Activity", and also described in the individual monographs, allows the product delivery [102,105].

6. Vaccines Assays and Quality Control

The quality of human vaccines can be evidenced by validated tests defined by regulatory agencies (WHO, FDA, EDQM, ANVISA) described in their guidelines [106–109], which defines the minimal requirements to the product. These requirements assure the products are safe and have a high quality level. However, for veterinary vaccines production, the international standard of production and quality control is described by The World Organization for Animal Health (OIE) guidelines [110]. The guidelines are discussed and prepared by VICH (International Cooperation on Harmonization of Technical Requirements for Registration of Veterinary Medical Products), a trilateral program aimed at harmonizing technical requirements for veterinary product registration between the European Union, Japan, and the USA since 1996 [111].

For biological products, such as the vaccines, VICH presented guidelines to check the quality (impurities, stability, specifications) and the safety (batch safety testing and target animal safety) (Table 4).

Table 4. VICH quality guidelines for biological products.

Issue		Test	Guideline	Refs
Quality	Impurities	Test for the detection of *Mycoplasma* contamination	VICH GL34	[112]
		Test of residual moisture	VICH GL26	[113]
		Test of residual formaldehyde	VICH GL25	[114]
	Stability	Stability testing of new biotechnological/biological veterinary medicinal products	VICH GL17	[115]
	Specification	Test procedures and acceptance criteria for new biotechnological/biological veterinary medicinal products	VICH GL40	[116]
Safety	Target animal batch safety	Harmonization of criteria to waive target animal batch safety testing for inactivated vaccines for veterinary use	VICH GL50 (R)	[117]
		Harmonization of criteria to waive target animal batch safety testing for live vaccines for veterinary use	VICH GL55	[118]
		Harmonization of criteria to waive laboratory animal batch safety testing for vaccines for veterinary use	VICH GL 59	[119]
	Target animal safety	Examination of live veterinary vaccines in target animals for absence of reversion to virulence	VICH GL41	[120]
		Target animal safety for veterinary live and inactivated vaccines	VICH GL44	[121]

The impurities tests include (i) test for the detection of *Mycoplasma* contamination [112], (ii) test of residual moisture [113], and (iii) test of residual formaldehyde [114].

i *Test for the detection of* Mycoplasma *contamination*

Mycoplasmas are contaminants of the biological products and can be inserted by the cell culture (master seeds, stock, starting materials of animal origin). Since they can cause several disturbances, such as polyserositis, pneumonia, arthritis, otitis media and reproductive syndromes, they must be absent in vaccines [122]. The test for the detection of *Mycoplasma* contamination is apply to vaccines produced in embryonated eggs from a qualified farm. The supplier farm is responsible for the quality control of the hens, which are submitted to tests of serology for viral, avian, *Mycoplasma* and bacterial agents.

In addition, the same tests are performed by the industry quality control department [123]. In the industry, the verification of the quality of the eggshell is also performed, considering the porosity and integrity of the same in each batch of eggs supplied. The *Mycoplasma* contamination test is based on the Japan and European Pharmacopeias methods [124,125]. The vaccine formulation must be free of contaminant with *Mycoplasma* to guarantee the consistency and safety of the product. The test must be performed in working seeds and harvest seeds, starting materials (master seed, master cell seed and ingredients of animal origin) and final product. Three tests are recommended: (i) expansion in broth culture and detection by colony formation on nutrient agar plates; (ii) expansion in cell culture and characteristic fluorescent staining of DNA; (iii) nucleic acid amplification. The last one is currently approved or under consideration by regulatory authorities for more rapid detection confirmation and strain identification. This technique must be validated for inclusion in the guideline [112].

ii *Test of residual moisture (RM)*

Freeze-dried vaccines generally have RM that can impact in their shelf-life. Therefore, RM assay is applied to freeze-dried vaccines formulations. The effectivity of the freeze-dried step process is controlled by the amount of RM. The high amount of RM can interfere with the shelf life of the product; therefore, it must be limited concerning the specifications. For the determination of RM, the guideline recommends a titrimetric method (Karl Fischer), azeotropic method or gravimetric method [113].

iii *Test of residual formaldehyde*

The inactivation of botulinum neurotoxin for toxoid vaccine production occurs by formaldehyde treatment [47]. The presence of this chemical is common in inactivated vaccines. Bacterin-based vaccine (suspension of killed or attenuated bacteria) containing residual levels of formaldehyde must be analyzed by the residual formaldehyde test. The determination of the quantity of this compound refers to the vaccine safety, assuring the formaldehyde is active, it has no impact on the vaccine shelf life, and any clostridial toxoids will be antigenic and safe. The methods for the determination of residual free formaldehyde in inactivated vaccines are acetyl acetone titration, ferric chloride titration and the basic fuchsin test [114].

For new biotechnological/biological veterinary medicinal products, it is necessary to follow the stability guideline presented by VICH GL17 [115]. For this study, the selection of batches that involve drug substance (bulk material), intermediates, and drug products (finished product) is necessary for a minimum of six months after production to test their potency and purity and enable molecular characterization.

The potency tests for live and attenuated vaccine material are performed determining the number of live particles in each batch, counting or by titration. In vivo tests are required when a new seed strain is used. However, for each batch of inactivated vaccines, an in vivo potency test is required. To evaluate the purity and molecular characterization, the guideline indicated the followed methodologies: electrophoresis (SDS-polyacrylamide gel electrophoresis, immunoelectrophoresis, Western blot, isoelectrofocusing), high-resolution chromatography (e.g., reversed-phase chromatography, gel filtration, ion exchange, affinity chromatography), and peptide mapping [115]. In this step, storage conditions are also defined and controlled. The performance of the product in different temperature and humidity conditions (normal and stress conditions) is tested. The photo sensibility test may be necessary [126,127].

The specifications of procedures and acceptance criteria for new biological veterinary products to prove the adequate quality control are declared in VICH GL40 [116]. This guideline explains principles to characterize a biotechnological or biological product (determination of physicochemical properties, biological activity, immunochemical properties, purity, and impurities).

Regarding the target animal batch safety, the organization makes available three documents involving issues of Good Laboratory Practices (GLP), Good Manufacturing Practices (GMP), Pharmacovigilance and standards for the production batch and seed batch system [117–119]. Concerning the target animal safety, the documents for live veterinary vaccines for the absence of reversion to virulence [120] and veterinary live and inactivated vaccines [121] are available.

Several methods are used to carry out the quality control of vaccines. The quality control was based on the uniqueness of each batch of vaccine. Consistency in vaccine production means that each batch of product is of the same quality and within the specifications of the batch described and effective in testing. Therefore, the development and validation of methods are crucial before the vaccine becomes a product to be marketed [128].

7. Vaccines' Labeling

The label must indicate the following: indication of the vaccine for veterinary use, the total volume and the number of doses contained in the container, the route of administration, the type or types of used bacteria or virus—in case of live vaccines, and the minimum number of live bacteria or the minimum title viruses. In the case of inactivated vaccines, the label information should comprise the minimum activity (in international units), and, if necessary, the name and the amount of any antimicrobial preservative or any other substance added to the vaccine. The presence of any substance likely to cause adverse side reactions should also be described. For lyophilized vaccines, the name, composition, and the volume of the liquid used to reconstitute the vaccine, and the time period during which the vaccine may be used after reconstitution must be present. In the case of vaccines containing an oily adjuvant, the need for emergency medical treatment should be noted in the case of accidental injection in humans [129]. The species of animals for which the vaccine is intended should be included, in addition to the indication of the vaccine, the instructions for use as well as recommended doses for the different species.

8. Conclusions

The development and production of safe, effective, stable, and economically viable vaccines is a challenge. Over many years, the entire process has been very costly and required extensive research. Currently, the use of bioinformatic and pharmaceutical technology encompassing interdisciplinary teams that change information all over the world has reduced the time of production and development. In addition, the USA, South America, Europe and Asia have shown a large evolution in regulatory parameters connecting the product to animals through multinational industries. Veterinary vaccines are instrumental not only on animal welfare, health, and reproduction but also to human health. The COVID-19 pandemic showed that under emergency, many parties will come together to ensure that vaccines are being developed at an unprecedented speed, in addition to addressing the worldwide commercial challenges.

Author Contributions: E.B.S., A.M.S., L.M.A.M., R.d.M.B., J.F., J.C.C., A.A.d.S.-J. and C.F.d.S. contributed to the conceptualization, methodology, validation, formal analysis, and investigation, and writing—original draft, preparation. P.S., C.V., E.B.S. and R.d.M.B. contributed to the methodology, supervision, writing—review and editing, project administration, resources, and funding acquisition. All authors have made a substantial contribution to the work. All authors have read and agreed to the published version of the manuscript.

Funding: This research was funded by the Portuguese Foundation for Science and Technology for the project UIDB/04033/2020 (CITAB), receiving financial support from FCT/MEC through national funds, and co-financed by FEDER, under the Partnership Agreement PT2020; also by the Coordenação Aperfeiçoamento de Pessoal de Nivel Superior (CAPES), Fundação de Ámparo à Pesquisa do Estado de Sergipe (FAPITEC) (PROCESSO: 88887.159533/2017-00 extração, encapsulação e caracterização de bioativos para o interesse biotecnologico) and Conselho Nacional de Desenvolvimento Científico e Tecnológico (CNPq 301964/2019-0 Chamada 06/2019, and Chamada CNPq n 01/2019). This research also was financially supported by Junta de Andalucía, under the project reference PT18 RT 3786.

Institutional Review Board Statement: Not applicable.

Informed Consent Statement: Not applicable.

Data Availability Statement: Not applicable.

Conflicts of Interest: The authors declare no conflict of interest.

References

1. Tedeschi, P.; Fine, A.H.; Helgeson, J.I. Assistance Animals: Their Evolving Role in Psychiatric Service Applications. In *Handbook on Animal-Assisted Therapy: Theoretical Foundations and Guidelines for Practice*; Fine, A.H., Ed.; Academic Press: Cambridge, UK, 2010; pp. 421–438. [CrossRef]
2. Flegr, J.; Preiss, M. Friends with Malefit. The Effects of Keeping Dogs and Cats, Sustaining Animal-Related Injuries and Toxoplasma Infection on Health and Quality of Life. *PLoS ONE* **2019**, *14*, 1–30. [CrossRef]
3. Maaten, T.S.D.; Turner, D.; Tilburg, J.V. Benefits and Risks for People and Livestock of Keeping Companion Animals: Searching for a Healthy Balance. *J. Comp. Pathol.* **2016**, *155*, S8–S17. [CrossRef]
4. Jorge, S.; Dellagostin, O.A. The Development of Veterinary Vaccines: A Review of Traditional Methods and Modern Biotechnology Approaches. *Biotechnol. Res. Innov.* **2017**, *1*, 6–13. [CrossRef]
5. Meeusen, E.N.T.; Walker, J.; Peters, A.; Pastoret, P.; Jungersen, G. Current Status of Veterinary Vaccines. *Clin. Microbiol. Rev.* **2007**, *20*, 489–510. [CrossRef]
6. Aida, V.; Pliasas, V.C.; Neasham, P.J.; North, J.F.; Mcwhorter, K.L.; Glover, S.R.; Kyriakis, C.S. Novel Vaccine Technologies in Veterinary Medicine: A Herald to Human Medicine Vaccines. *Front. Vet. Sci.* **2021**, *8*, 654289. [CrossRef]
7. Pollard, A.J. A Guide to Vaccinology: From Basic Principles to New Developments. *Nat. Rev. Immunol.* **2021**, *21*, 83–100. [CrossRef]
8. Besnard, L.; Fabre, V.; Fettig, M.; Gousseinov, E.; Kawakami, Y.; Laroudie, N.; Scanlan, C.; Pattnaik, P. Clarification of Vaccines: An Overview of Filter Based Technology Trends and Best Practices. *Biotechnol. Adv.* **2016**, *34*, 1–13. [CrossRef]

9. Rauch, S.; Jasny, E.; Schmidt, K.E.; Petsch, B. New Vaccine Technologies to Combat Outbreak Situations. *Front. Immunol.* **2018**, *9*, 1963. [CrossRef]
10. Ferreira, M.R.A.; Moreira, G.M.S.G.; Eduardo, C.; Cunha, P.; Mendonça, M.; Salvarani, F.M.; Moreira, Â.N.; Conceição, F.R. Clostridium Perfringens: Production Strategies and Applications as Veterinary Vaccines. *Toxins* **2016**, *2*, 340. [CrossRef]
11. Fabbri, A.; Travaglione, S.; Falzano, L.; Fiorentini, C. Bacterial Protein Toxins: Current and Potential Clinical Use. *Curr. Med. Chem.* **2008**, *15*, 1116–1125. [CrossRef]
12. Ghimire, T.R. The Mechanisms of Action of Vaccines Containing Aluminum Adjuvants: An in Vitro vs in Vivo Paradigm. *Springerplus* **2015**, *4*, 181. [CrossRef]
13. Van Gelder, P.; Makoschey, B. Production of Viral Vaccines for Veterinary Use. *Berl. Munch. Tierarztl. Wochenschr.* **2012**, *125*, 103–109.
14. Hansen, L.J.J.; Daoussi, R.; Vervaet, C.; Remon, J.; Beer, T.R.M.D. Freeze-Drying of Live Virus Vaccines: A Review. *Vaccine* **2015**, *33*, 5507–5519. [CrossRef]
15. Choi, Y.; Chang, J. Viral Vectors for Vaccine Applications. *Clin. Exp. Vaccine Res.* **2013**, *2*, 97–105. [CrossRef]
16. Barrett, P.N.; Mundt, W.; Kistner, O.; Howard, M.K.; Barrett, P.N.; Mundt, W.; Kistner, O.; Vero, M.K.H.; Barrett, P.N.; Mundt, W.; et al. Towards Cell Culture-Based Viral Vaccines Vero Cell Platform in Vaccine Production: Moving towards Cell Culture-Based Viral Vaccines. *Expert Rev. Vaccines* **2014**, *0584*, 607–618. [CrossRef]
17. Xia, J.; Cui, J.; He, X.; Liu, Y.; Yao, K.; Cao, S.; Han, X. Genetic and Antigenic Evolution of H9N2 Subtype Avian Influenza Virus in Domestic Chickens in Southwestern China, 2013–2016. *PLoS ONE* **2017**, *12*, 2013–2016. [CrossRef]
18. Ewer, K.J.; Lambe, T.; Rollier, C.S.; Spencer, A.J.; Hill, A.V.S.; Dorrell, L. Viral Vectors as Vaccine Platforms: From Immunogenicity to Impact. *Curr. Opin. Immunol.* **2016**, *41*, 47–54. [CrossRef]
19. Rabie, N.S.; Girh, Z.M.S.A. Bacterial Vaccines in Poultry. *Bull. Natl. Res. Cent.* **2020**, *44*, 15. [CrossRef]
20. Guzman, F.; Vargas, M.I.; Sossai, S.; Patarroyo, J.H.; Patarroyo, A.M.V.; Gonza, C.Z.L. Use of Biodegradable PLGA Microspheres as a Slow Release Delivery System for the Boophilus Microplus Synthetic Vaccine SBm7462. *Vet. Immunol. Immunopathol.* **2005**, *107*, 281–290. [CrossRef]
21. Callaway, E. The Race for Coronavirus Vaccines: A Graphical Guide. *Nature* **2020**, *580*, 576–577. [CrossRef]
22. Maslow, J.N. Vaccine Development for Emerging Virulent Infectious Diseases. *Vaccine* **2017**, *35*, 5437–5443. [CrossRef]
23. Khuroo, M.S.; Khuroo, M.; Khuroo, M.S.; Sofi, A.A.; Khuroo, N.S. COVID-19 Vaccines: A Race Against Time in the Middle of Death and Devastation! *J. Clin. Exp. Hepatol.* **2020**, *10*, 610–621. [CrossRef]
24. Koirala, A.; Jin Joo, Y.; Khatami, A.; Chiu, C.; Britton, P.N. Vaccines for COVID-19: The Current State of Play. *Paediatr. Respir. Rev.* **2020**, *35*, 43–49. [CrossRef]
25. Heppell, J.; Davis, H.L. Application of DNA Vaccine Technology to Aquaculture. *Adv. Drug Deliv. Rev.* **2000**, *43*, 29–43. [CrossRef]
26. Fomsgaard, A.; Liu, M.A. The Key Role of Nucleic Acid Vaccines for One Health. *Viruses* **2021**, *13*, 258. [CrossRef]
27. Dhama, K.; Mahendran, M.; Gupta, P.K.; Rai, A. DNA Vaccines and Their Applications in Veterinary Practice: Current Perspectives. *Vet. Res. Commun.* **2008**, *32*, 341–356. [CrossRef]
28. WHO. WHO Coronavirus (COVID-19). Available online: https://covid19.who.int/ (accessed on 30 July 2022).
29. Monath, T.P. Vaccines against Diseases Transmitted from Animals to Humans: A One Health Paradigm. *Vaccine* **2013**, *31*, 5321–5338. [CrossRef]
30. Ebrahimi-nik, H.; Bassami, M.R.; Mohri, M.; Rad, M.; Khan, M.I. Bacterial Ghost of Avian Pathogenic E. Coli (APEC) Serotype O78: K80 as a Homologous Vaccine against Avian Colibacillosis. *PLoS ONE* **2018**, *13*, e0194888. [CrossRef]
31. Kabir, S.M.L. Avian Colibacillosis and Salmonellosis: A Closer Look at Epidemiology, Pathogenesis, Diagnosis, Control and Public Health Concerns. *Int. J. Environ. Res. Public Health* **2010**, *7*, 89. [CrossRef]
32. Muniz, E.C.; Verdi, R.; Leão, J.A.; Back, A.; Pinheiro, V.; Correa, E.; Verdi, R.; Leão, J.A.; Back, A. Evaluation of the Effectiveness and Safety of a Genetically Modified Live Vaccine in Broilers Challenged with Salmonella Heidelberg. *Avian Pathol.* **2017**, *46*, 676–682. [CrossRef]
33. Lark, C.; William, W.; Kayla, M.G.; Jillian, J.A. Influence of a Bovine Respiratory Disease Vaccine with a Temperature—Sensitive Modified Live or Killed Infectious Bovine Rhinotracheitis Component on Oestrous Cycle Parameters and Anti—Müllerian Hormone Concentration in Nulliparous Heifers. *Reprod. Domest. Anim.* **2019**, *54*, 1470–1476. [CrossRef]
34. Pitcovski, J.; Gutter, B.; Gallili, G.; Goldway, M.; Perelman, B.; Gross, G.; Krispel, S.; Barbakov, M.; Michael, A. Development and Large-Scale Use of Recombinant VP2 Vaccine for the Prevention of Infectious Bursal Disease of Chickens. *Vaccine* **2003**, *21*, 4736–4743. [CrossRef]
35. You, S.; Jo, H.; Choi, J.; Ko, M.; Shin, S.H.; Lee, M.J.; Kim, S.; Kim, B.; Park, J. Evaluation of Novel Inactivated Vaccine for Type C Foot-and-Mouth Disease in Cattle and Pigs. *Vet. Microbiol.* **2019**, *234*, 44–50. [CrossRef]
36. Jamal, S.M.; Belsham, G.J. Foot-and-Mouth Disease: Past, Present and Future. *Vet. Res.* **2013**, *44*, 116. [CrossRef]
37. Rezaei, M.; Rabbani-khorasgani, M.; Zarkesh-Esfahani, S.H.; Emamzadeh, R.; Abtahi, H. Prediction of the Omp16 Epitopes for the Development of an Epitope-Based Vaccine Against Brucellosis. *Infect. Disord. Drug Targets* **2019**, *19*, 36–45. [CrossRef]
38. Abdolmohammadi Khiav, L.; Zahmatkesh, A. Vaccination against Pathogenic Clostridia in Animals: A Review. *Trop. Anim. Health Prod.* **2021**, *53*, 284. [CrossRef]
39. Mans, K.L.; Hernández-triana, L.M.; Banyard, A.C.; Fooks, A.R.; Johnson, N. Japanese Encephalitis Virus Infection, Diagnosis and Control in Domestic Animals. *Vet. Microbiol.* **2017**, *201*, 85–92. [CrossRef]

40. Maki, J.; Guiot, A.L.; Aubert, M.; Brochier, B.; Cliquet, F.; Hanlon, C.A.; King, R.; Oertli, E.H.; Rupprecht, C.E.; Schumacher, C.; et al. Oral Vaccination of Wildlife Using a Vaccinia—Rabies-Glycoprotein Recombinant Virus Vaccine (RABORAL V-RG®): A Global Review. *Vet. Res.* **2017**, *48*, 57. [CrossRef]
41. Corbel, M.J. *Brucellosis in Humans and Animals Brucellosis in Humans and Animals*; World Health Organization: Geneva, Switzerland, 2006.
42. Rashid, A.; Rasheed, K.; Akhtar, M. Factors Influencing Vaccine Efficacy—A General Review. *J. Anim. Plant Sci.* **2009**, *19*, 22–25.
43. Buddle, B.M.; Vordermeier, H.M.; Chambers, M.A. Efficacy and Safety of BCG Vaccine for Control of Tuberculosis in Domestic Livestock and Wildlife. *Front. Vet. Sci.* **2018**, *5*, 259. [CrossRef]
44. Pascual, D.W.; Yang, X.; Wang, H.; Goodwin, Z.; Hoffman, C.; Clapp, B. Alternative Strategies for Vaccination to Brucellosis. *Microbes Infect.* **2019**, *20*, 599–605. [CrossRef]
45. Francis, M.J. Recent Advances in Vaccine Technologies. *Vet. Clin. Small Anim. Pract.* **2018**, *48*, 231–241. [CrossRef]
46. Demain, A.L. Microbial Biotechnology. *Trends Biotechnol.* **2000**, *18*, 26–31. [CrossRef]
47. Zaragoza, N.E.; Orellana, C.A.; Moonen, G.A.; Moutafis, G.; Marcellin, E. Vaccine Production to Protect Animals Against Pathogenic Clostridia. *Toxins* **2019**, *11*, 525. [CrossRef]
48. Campos, I.B.; Cardoso, C.P.; Fernando, J.; Muriel, F.; Leite, L.C.C.; Moffitt, K.L.; Jie, Y.; Richard, L.; Gonçalves, V.M. Process Intensification for Production of Streptococcus Pneumoniae Whole—Cell Vaccine. *Biotechnol. Bioeng.* **2020**, *117*, 1661–1672. [CrossRef]
49. Fang, C.; Guilbault, C.; Li, X.; Elahi, S.M.; Ansorge, S.; Kamen, A.; Gilbert, R. Development of Suspension Adapted Vero Cell Culture Process Technology for Production of Viral Vaccines. *Vaccine* **2019**, *37*, 6996–7002. [CrossRef]
50. Ottiger, H.P. Monitoring Veterinary Vaccines for Contaminating Viruses. *Dev. Biol. Stand.* **2006**, *126*, 309–319.
51. Yang, D.; Nakagawa, K.; Ito, N.; Kim, H.; Hyun, B.; Nah, J.; Sugiyama, M.; Song, J. A Single Immunization with Recombinant Rabies Virus (ERAG3G) Confers Complete Protection against Rabies in Mice. *Clin. Exp. Vaccine* **2014**, *3*, 176–184. [CrossRef]
52. Reciilard, P. Cell-Culture Vaccines for Veterinary Use. In *Laboratory Techniques in Rabies*; Meslin, F.X., Ed.; World Health Organization: Geneva, Switzerland, 1958; pp. 314–324.
53. Prkno, A.; Ho, D.; Kaiser, M.; Goerigk, D.; Pfe, M.; Winter, K.; Vahlenkamp, T.W.; Beer, M.; Starke, A. Field Trial Vaccination against Cowpox in Two. *Viruses* **2020**, *12*, 234. [CrossRef]
54. Dalsass, M.; Brozzi, A.; Medini, D.; Rappuoli, R. Comparison of Open-Source Reverse Vaccinology Programs for Bacterial Vaccine Antigen Discovery. *Front. Immunol.* **2019**, *10*, 113. [CrossRef]
55. Soleymani, S.; Tavassoli, A.; Reza, M. An Overview of Progress from Empirical to Rational Design in Modern Vaccine Development, with an Emphasis on Computational Tools and Immunoinformatics Approaches. *Comput. Biol. Med.* **2022**, *140*, 105057. [CrossRef]
56. Souza, J.G.D.; Fernandes, M.A.C. A Novel Deep Neural Network Technique for Drug—Target Interaction. *Pharmaceutics* **2021**, *14*, 625. [CrossRef]
57. Awasthi, A.; Sharma, G.; Agrawal, P. Chapter 20-Computational Approaches for Vaccine Designing. In *Bioinformatics—Methods and Applications*; Academic Press: Cambridge, MA, USA, 2022; pp. 317–335. [CrossRef]
58. Orellaa, F.A.D.; Gustavo, L.; Achecoa, C.P.; Liveirab, S.C.O.; Iyoshia, A.M.; Zevedoa, V.A. Corynebacterium Pseudotuberculosis: Microbiology, Biochemical Properties, Pathogenesis and Molecular Studies of Virulence. *Vet. Res.* **2006**, *37*, 201–218. [CrossRef]
59. Soares, S.C.; Trost, E.; Ramos, R.T.J.; Carneiro, A.R.; Santos, A.R.; Pinto, A.C.; Barbosa, E.; Aburjaile, F.; Ali, A.; Diniz, C.A.A.; et al. Genome Sequence of Corynebacterium Pseudotuberculosis Biovar Equi Strain 258 and Prediction of Antigenic Targets to Improve Biotechnological Vaccine Production. *J. Biotechnol.* **2013**, *167*, 135–141. [CrossRef]
60. Leonardo, C.; Alves, J.; Nogueira, W.; César, L.; Cybelle, A.; Ramos, R.; Azevedo, V.; Silva, A.; Folador, A. Prediction of New Vaccine Targets in the Core Genome of Corynebacterium Pseudotuberculosis through Omics Approaches and Reverse Vaccinology. *Gene* **2019**, *702*, 36–45. [CrossRef]
61. Bueno, L.L.; Lobo, F.P.; Guimara, C.; Soares, I.S.; Fontes, C.J.; Viru, M.; Bartholomeu, D.C.; Fujiwara, R.T. Identification of a Highly Antigenic Linear B Cell Epitope within Plasmodium Vivax Apical Membrane Antigen 1. *PLoS ONE* **2011**, *6*, e21289. [CrossRef]
62. Michel-tod, L.; Bigey, P.; Reche, P.A.; Pinazo, M.; Gasc, J.; Alonso-padilla, J. Design of an Epitope-Based Vaccine Ensemble for Animal Trypanosomiasis by Computational Methods. *Vaccines* **2020**, *8*, 130. [CrossRef]
63. Shan, R.; Lina, G.; Yao, D.; Chaoli, W.; Li, F.; Yongen, X. Design and Evaluation of a Multi-Epitope Assembly Peptide Vaccine against Acinetobacter Baumannii Infection in Mice. *Swiss Med. Wkly.* **2019**, *149*, 2324. [CrossRef]
64. Majidiani, H.; Dalimi, A.; Ghaffarifar, F.; Pirestani, M.; Ghaffari, A.D. Microbial Pathogenesis Computational Probing of Toxoplasma Gondii Major Surface Antigen 1 (SAG1) for Enhanced Vaccine Design against Toxoplasmosis. *Microb. Pathog.* **2020**, *147*, 104386. [CrossRef]
65. Roopngam, P.E. Liposome and Polymer-Based Nanomaterials for Vaccine Applications. *Nanomedicine* **2019**, *6*, 1–10. [CrossRef]
66. Sadozai, H.; Saeidi, D. Recent Developments in Liposome-Based Veterinary Therapeutics. *Hindawi* **2013**, *2013*, 16752. [CrossRef]
67. Schwendener, R.A. Liposomes as Vaccine Delivery Systems: A Review of the Recent Advances. *Ther. Adv. Vaccines* **2014**, *2*, 159–182. [CrossRef]
68. Celis-giraldo, C.T.; Julio, L.; Muro, A.; Patarroyo, M.A. Nanovaccines against Animal Pathogens: The Latest Findings. *Vaccines* **2021**, *9*, 988. [CrossRef]

69. Williams, P.D.; Paixão, G. On-Farm Storage of Livestock Vaccines May Be a Risk to Vaccine Efficacy: A Study of the Performance of on-Farm Refrigerators to Maintain the Correct Storage Temperature. *BMC Vet. Res.* **2018**, *14*, 136. [CrossRef]
70. Fanelli, A.; Mantegazza, L.; Hendrickx, S.; Capua, I. Thermostable Vaccines in Veterinary Medicine: State of the Art and Opportunities to Be Seized. *Vaccines* **2022**, *10*, 245. [CrossRef]
71. Abdi, R.D.; Amsalu, K.; Merera, O.; Asfaw, Y.; Gelaye, E.; Yami, M.; Sori, T. Serological Response and Protection Level Evaluation in Chickens Exposed to Grains Coated with I2 Newcastle Disease Virus for Effective Oral Vaccination of Village Chickens. *BMC Vet. Res.* **2016**, *12*, 150. [CrossRef]
72. Lankester, F.J.; Wouters, P.A.W.M.; Czupryna, A.; Palmer, G.H.; Mzimbiri, I.; Cleaveland, S.; Francis, M.J.; Sutton, D.J.; Sonnemans, D.G.P. Thermotolerance of an Inactivated Rabies Vaccine for Dogs. *Vaccine* **2016**, *34*, 5504–5511. [CrossRef]
73. Schat, K.; Baranowski, E. Animal Vaccination and the Evolution of Viral Pathogens. *Rev. Sci. Tech.* **2007**, *26*, 327–338. [CrossRef]
74. Ferrari, G.; Paton, D.; Duffy, S.; Bartels, C.; Knight-jones, T. *Foot and Mouth Disease Vaccination and Post-Vaccination Monitoring (Guidelines)*; Metwally, S., Münstermann, S., Eds.; Food and Agriculture Organization of the United Nations (FAO): Rome, Italy, 2016.
75. Cunningham, A.L.; Garçon, N.; Leo, O.; Friedland, L.R.; Strugnell, R.; Laupèze, B.; Doherty, M.; Stern, P. Vaccine Development: From Concept to Early Clinical Testing. *Vaccine* **2016**, *34*, 6655–6664. [CrossRef]
76. Shah, R.R.; Hassett, K.J.; Brito, L.A. Overview of Vaccine Adjuvants: Introduction, History, and Current Status. In *Vaccine Adjuvants: Methods and Protocols, Methods in Molecular Biology*; Fox, C.B., Ed.; Springer Science + Business Media: New York, NY, USA, 2017; Volume 1494, pp. 1–13. [CrossRef]
77. Gomez, P.L.; Robinson, J.M. Section 1: General Aspects of Vaccination. In *Plotkin's Vaccines*; Plotkin, S.A., Orenstein, W.A., Offit, P.A., Edwards, K.M., Eds.; Elsevier: Philadelphia, PA, USA, 2018; pp. 51–60. [CrossRef]
78. Kamminga, T.; Slagman, S.; Martins, V.A.P.; Bijlsma, J.J.E.; Schaap, P.J. Risk-Based Bioengineering Strategies for Reliable Bacterial Vaccine Production. *Trends Biotechnol.* **2019**, *37*, 805–816. [CrossRef]
79. Egorova, M.S.; Kurashova, S.S.; Dzagurova, T.K.; Balovneva, M.V.; Ishmukhametov, A.A.; Tkachenko, E.A. Effect of Virus-Inactivating Agents on the Immunogenicity of Hantavirus Vaccines against Hemorrhagic Fever with Renal Syndrome. *Appl. Biochem. Microbiol.* **2020**, *56*, 940–947. [CrossRef]
80. Tang, L.; Kang, H.; Duan, K.; Guo, M.; Lian, G. Effects of Three Types of Inactivation Agents on the Antibody Response and Immune Protection of Inactivated IHNV Vaccine in Rainbow Trout. *Viral Immunol.* **2016**, *29*, 430–435. [CrossRef]
81. Jagt, H.J.M.; Bekkers, M.L.E.; Bommel, S.A.J.T.V.; Marel, P.V.D.; Schrier, C.C. The Influence of the Inactivating Agent on the Antigen Content of Inactivated Newcastle Disease Vaccines Assessed by the in Vitro Potency Test. *Biologicals* **2010**, *38*, 128–134. [CrossRef]
82. Kuma, R.; Chandra, R.; Shukla, S.K.; Agrawal, D.K.; Kumar, M. Hydropericardium Syndrome (HPS) in India: A Preliminary Study on the Causative Agent and Control of the Disease by Inactivated Autogenous Vaccine. *Trop. Anim. Health Prod.* **1997**, *29*, 158–164. [CrossRef]
83. Herrera-rodriguez, J.; Signorazzi, A.; Holtrop, M.; Vries-idema, J.D.; Huckriede, A. Inactivated or Damaged? Comparing the Effect of Inactivation Methods on Influenza Virions to Optimize Vaccine Production. *Vaccine* **2020**, *37*, 1630–1637. [CrossRef]
84. Fan, C.; Ye, X.; Ku, Z.; Kong, L.; Liu, Q. Beta-Propiolactone Inactivation of Coxsackievirus A16 Induces Structural Alteration and Surface Modification of Viral Capsids. *J. Virol.* **2017**, *91*, e00038-17. [CrossRef]
85. Lee, Y.H.; Jang, Y.H.; Byun, Y.H.; Cheong, Y.; Kim, P.; Lee, Y.J.; Lee, Y.J.; Sung, J.M.; Son, A.; Lee, H.M.; et al. Green Tea Catechin-Inactivated Viral Vaccine Platform. *Front. Microbiol.* **2017**, *8*, 2469. [CrossRef]
86. Abd-elghaffar, A.A.; Ali, A.E.; Boseila, A.A.; Amin, M.A. Inactivation of Rabies Virus by Hydrogen Peroxide. *Vaccine* **2016**, *34*, 98–802. [CrossRef]
87. Narayan, S.; Shamsundar, R.; Seetharaman, S. In Vitro Inactivation of the Rabies Virus by Ascorbic Acid. *Int. J. Infect. Dis.* **2004**, *8*, 21–25. [CrossRef]
88. Gupta, R.K.; Sharma, R.; Ahuja, S.; Saxena, S.N. The Effects of Different Inactivating Agents on the Potency, Toxicity and Stability of Pertussis Vaccine. *J. Biol. Stand.* **1987**, *15*, 87–98. [CrossRef]
89. Osterholm, M.T.; Kelley, N.S.; Sommer, A.; Belongia, E.A. Efficacy and Effectiveness of Influenza Vaccines: A Systematic Review and Meta-Analysis. *Lancet* **2012**, *12*, 36–44. [CrossRef]
90. Sanders, B.; Koldijk, M.; Schuitemaker, H. Inactivated Viral Vaccines. In *Vaccine Analysis: Strategies, Principles, and Control*; Nunnally, B.K., Turula, V.E., D.Sitrin, R., Eds.; Springer: Berlin/Heidelberg, Germany, 2015; pp. 45–80. [CrossRef]
91. Mondal, S.K.; Neelima, M.; Seetha Rama Reddy, K.; Ananda Rao, K.; Srinivasan, V.A. Validation of the Inactivant Binary Ethylenimine for Inactivating Rabies Virus for Veterinary Rabies Vaccine Production. *Biologicals* **2005**, *33*, 185–189. [CrossRef] [PubMed]
92. Hiatt, C.W. Kinetics of the Inactivation of Viruses. *Bact. Rev.* **1964**, *28*, 150–163. [CrossRef]
93. Moghadam, A.T.; Afsharpad, K. Application of Fermentor Technology in Production of Diphtheria Toxin. *Jundishapur J. Microbiol.* **2017**, *1*, 24–27.
94. Yuen, C.-T.; Asokanathan, C.; Cook, S.; Lin, N.; Xing, D. Effect of Different Detoxification Procedures on the Residual Pertussis Toxin Activities in Vaccines. *Vaccine* **2016**, *34*, 2129–2134. [CrossRef] [PubMed]
95. Perugi, F.; Léon, A.; Guéhenneux, F.; Champion-arnaud, P.; Lahmar, M.; Schwamborn, K. Cell Substrates for the Production of Viral Vaccines. *Vaccine* **2015**, *33*, 6–13. [CrossRef]

96. Knezevic, I.; Stacey, G.; Petricciani, J. WHO Study Group on Cell Substrates for Production of Biologicals Geneva, Switzerland, 11–12 June 2007. *Biologicals* **2008**, *36*, 203–211. [CrossRef]
97. Coecke, S.; Balls, M.; Bowe, G.; Davis, J.; Hartung, T.; Hay, R.; Price, A.; Stokes, W.; Schechtman, L.; Stacey, G. Guidance on Good Cell Culture Practice. a Report of the Second ECVAM Task Force on Good Cell Culture Practice. *Altern. Lab. Anim.* **2005**, *33*, 261–287. [CrossRef]
98. Lucey, B.P.; Nelson-rees, W.A.; Hutchins, G.M. Henrietta Lacks, HeLa Cells, and Cell Culture Contamination. *Arch. Pathol. Lab. Med.* **1951**, *133*, 1463–1467. [CrossRef]
99. Marcus-sekura, C.; Richardson, J.C.; Harston, R.K.; Sane, N.; Sheets, R.L. Biologicals Evaluation of the Human Host Range of Bovine and Porcine Viruses That May Contaminate Bovine Serum and Porcine Trypsin Used in the Manufacture of Biological Products. *Biologicals* **2011**, *39*, 359–369. [CrossRef]
100. Lee, H.; Tang, H. Next-Generation Sequencing Technologies and Fragment Assembly Algorithms. In *Evolutionary Genomics: Statistical and Computational Methods*; Anisimova, M., Ed.; Springer Science + Business Media, LLC: New York, NY, USA, 2012; Volume 855, pp. 155–174. [CrossRef]
101. Metwally, S.; Viljoen, G.; El Idrissi, A. (Eds.) *Veterinary Vaccines: Principles and Applications*, 1st ed.; The Food and Agriculture Organization of the United Nations: Rome, Italy; John Wiley & Sons Limited: Oxford, UK, 2021. [CrossRef]
102. European Pharmacopoeia. *Technical Guide for the Elaboration and Use of Monographs for Vaccines and Immunological Veterinary Medicinal Products*, 1st ed.; The European Pharmacopoeia: Strasbourg, France, 2016.
103. Dodet, B. An Important Date in Rabies History. *Vaccine* **2020**, *25*, 8647–8650. [CrossRef] [PubMed]
104. Meyer, B.K.; Ni, A.; Hu, B.; Shi, L.I. Antimicrobial Preservative Use in Parenteral Products: Past and Present. *J. Pharm. Sci.* **2007**, *96*, 3155–3167. [CrossRef] [PubMed]
105. European Pharmacopeia. *European Directtorate for the Quality and Healthcare*; Updated pa; The European Pharmacopoeia: Strasbourg, France, 2020.
106. United States Pharmacopeial Convention. *The United States Pharmacopeia 2018: USP 41*; The National Formulary: NF 36; United States Pharmacopeial Convention, Inc.: Rockville, MD, USA, 2018.
107. Brazil. *Brazilian Pharmacopeia, 6a edição.*; Agência Nacional de Vigilância Sanitária: Brasilia, Brazil, 2019.
108. European Directorate for the Quality of Medicines & HealthCare. *European Pharmacopoeia*, 10th ed.; Deutscher Apotheker Verlag: Strasbourg, France, 2019.
109. WHO—World Health Organization. *The International Pharmacopoeia*, 8th ed.; World Health Organization: Wellington, New Zeland, 2018.
110. OIE—World Organisation for Animal Health. *Minimum Requirements for the Production and Quality Control of Vaccines*; OIE: Paris, France, 2018.
111. VICH. *Organisational Charter of Vich*; VICH: Bruxelles, Belgium, 2019.
112. EMEA. *VICH GL34—Biologicals: Testing for the Detection of Mycoplasma Contamination*; EMEA: London, UK, 2014.
113. EMEA. *VICH Topic GL26—Biologicals: Testing of Residual Moisture*; EMEA: London, UK, 2003.
114. EMEA. *VICH Topic GL25—Biologicals: Testing of Residual Formaldehyde*; EMEA: London, UK, 2003.
115. EMEA. *VICH GL17—Stability Testing of Veterinary Medicine Products*; EMEA: London, UK, 2001.
116. EMEA. *VICH Topic GL40—Guideline on Test Procedures and Acceptance Criteria for New Biotechnological/Biological Veterinary Medicinal Products*; EMEA: London, UK, 2006.
117. EMA. *VICH GL50: Harmonisation of Criteria to Waive Target Animal Batch Safety Testing for Inactivated Vaccines for Veterinary Use*; EMA: London, UK, 2018.
118. EMA. *VICH GL55—Harmonisation of Criteria to Waive Target Animal Batch Safety Testing for Live Vaccines for Veterinary Use*; EMA: London, UK, 2018.
119. EMA. *VICH GL59—Harmonisation of Criteria to Waive Laboratory Animal Batch Safety Testing for Vaccines for Veterinary Use Testing for Vaccines for Veterinary Use*; EMA: Amsterdam, The Netherlands, 2021.
120. EMEA. *VICH Topic GL41: Guideline on Target Animal Safety: Examination of Live Veterinary Vaccines in Target Animals for Absence of Reversion to Virulence*; EMEA: London, UK, 2008.
121. EMEA. *VICH Topic GL44—GGuideline on Target Animal Safety for Veterinary Live and Inactied Vaccines*; EMEA: London, UK, 2009.
122. Mbelo, S.; Gay, V.; Blanchard, S.; Abachin, E.; Falque, S.; Lechenet, J.; Poulet, H.; Saint-Vis, B.D. Biologicals Development of a Highly Sensitive PCR / DNA Chip Method to Detect Mycoplasmas in a Veterinary Modi Fi Ed Live Vaccine. *Biologicals* **2018**, *54*, 22–27. [CrossRef] [PubMed]
123. Fiorentin, L.; Mores, M.; Trevisol, I.; Antunes, S.; Costa, J.; Soncini, R.; Vieira, N. Test Profiles of Broiler Breeder Flocks Housed in Farms with Endemic Mycoplasma Synoviae Infection. *Braz. J. Poult. Sci.* **2003**, *3*, 37–43. [CrossRef]
124. European Directorate for the Quality of Medicines & HealthCare. 2.6.7. Mycoplasmas. In *European Pharmacopoeia*; European Directorate for the Quality of Medicines & HealthCare: Strasbourg, France, 2020.
125. Mycoplasma Testing for Cell Substrates Used for the Production of Biotechnological/Biological Products. In *Japanese Pharmacopoeia*; Stationery Office: London, UK, 2016; pp. 2460–2464.
126. Harrak, M.E.; Belkourati, I.; Boumart, Z.; Fakri, F.; Hamdi, J. The Manufacture of Veterinary Vaccines: Quality Control of the Manufacturing Process. In *Veterinary Vaccines: Principles and Applications*; Metwally, S., Viljoen, G., Idrissi, A.E., Eds.; John Wiley & Sons Ltd.: Hoboken, NJ, USA, 2021; pp. 147–159.

127. Francis, M.J. A Veterinary Vaccine Development Process Map to Assist in the Development of New Vaccines. *Vaccine* **2020**, *38*, 4512–4515. [CrossRef]
128. Metz, A.B.; Dobbelsteen, G.V.D.; Els, C.V.; Gun, J.V.D.; Levels, L.; Pol, L.V.D. Quality-Control Issues and Approaches in Vaccine Development. *Expert Rev. Vaccines* **2009**, *8*, 227–238. [CrossRef]
129. FDA. *Guidance for Industry: FDA Review of Vaccine Labeling Requirements for Warnings, Use Instructions, and Precautionary Information*; FDA: Rockville, MD, USA, 2004.

technologies

MDPI

Review

Thermal Inkjet Printing: Prospects and Applications in the Development of Medicine

Md Jasim Uddin [1,2,3], Jasmin Hassan [3] and Dennis Douroumis [1,2,*]

[1] Faculty of Engineering and Science, University of Greenwich at Medway, Chatham Maritime, Chatham, Kent ME4 4TB, UK
[2] Center for Innovation, Process Engineering & Research, University of Greenwich at Medway, Chatham Maritime, Chatham, Kent ME4 4TB, UK
[3] Drug Delivery & Therapeutics Lab, Dhaka 1212, Bangladesh
* Correspondence: d.douroumis@greenwich.ac.uk

Abstract: Over the last 10 years, inkjet printing technologies have advanced significantly and found several applications in the pharmaceutical and biomedical sector. Thermal inkjet printing is one of the most widely used techniques due to its versatility in the development of bioinks for cell printing or biosensors and the potential to fabricate personalized medications of various forms such as films and tablets. In this review, we provide a comprehensive discussion of the principles of inkjet printing technologies highlighting their advantages and limitations. Furthermore, the review covers a wide range of case studies and applications for precision medicine.

Keywords: thermal inkjet printing; TIJ; bubble jet printing; personalized treatment; precision medicine

Citation: Uddin, M.J.; Hassan, J.; Douroumis, D. Thermal Inkjet Printing: Prospects and Applications in the Development of Medicine. *Technologies* **2022**, *10*, 108. https://doi.org/10.3390/technologies10050108

Academic Editors: Manoj Gupta, Eugene Wong and Gwanggil Jeon

Received: 2 August 2022
Accepted: 17 October 2022
Published: 21 October 2022

1. Introduction

Over the last 20 years, we have encountered a transformation in manufacturing technologies in the area of medicinal products [1–3]. Traditionally marketed medicines are manufactured at fixed doses (one size fits all) targeting a large number of patients in order to reduce the production costs and time to the market. However, the widely varied responses to a particular therapeutic dose in patient populations especially for medicines with narrow therapeutic windows points out the limitations of generalized mass manufacturing [1,4]. Moreover, there is a growing number of patients worldwide with chronic diseases who have to take multiple doses of medicines per day, called polypharmacy, which increases the risk for side effects and drug–disease interactions [5]. Currently, swift advances in gene sequencing technology along with increased knowledge of genomics and better understanding of diseases on molecular level coupled with the use of toxicogenomic markers have opened a door for personalized medicine that will possibly bring a revolution in the conventional treatment approaches as well as in pharmaceutical industry [6–9]. For the materialization of these advances in personalized medicines, a wide range of 2D and 3D printing technologies have been introduced as appropriate for manufacturing print-on-demand medicinal products. Inkjet printing (IJP) technology is considered an ideal approach as it is cost effective [1,10–14] with high precision, repeatability, robustness, and high-throughput (Figure 1). Due to its wide applicability, inkjet printing has been extensively used for pharmaceutical applications and tissue engineering [15–24].

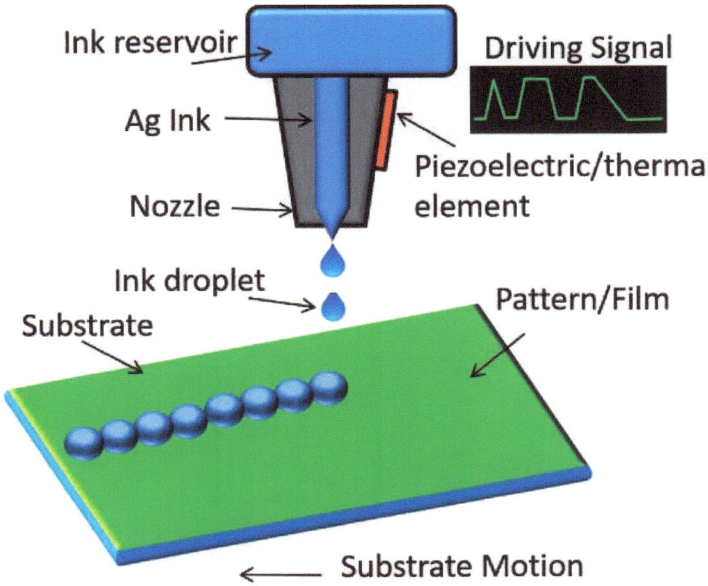

Figure 1. Schematic illustration of inkjet printing technology [25].

1.1. Inkjet Printing Technology

Inkjet printing is a reprographic method that provides for the controlled deposition of a small drop of ink (e.g., biological, synthetic and any form of therapeutic or nontherapeutic substances) on a substrate [26,27].

Today, this widely known digital printing technology that was originally developed to transfer electronic data on paper is present in almost every office and household as a common technique for printing text or graphics [26–29]. Being a noncontact deposition and direct-patterning technique, it provides minimal contamination and waste of therapeutic sample, respectively [11,30–32]. It has also caught the attention of researchers worldwide because of its drop placement accuracy in a precisely fixed amount (typically in volumes of picolitres, pL) of material that can be dispensed without any prior pattern [11,30,33,34].

This material-conserving patterning technique is usually used for the deposition of liquid phase materials that are technically termed ink and that contain the solute as dissolved or dispersed in a solvent. A piezoelectric inkjet printer uses a piezo-ceramic plate to apply ink droplets in order to regulate the ejection. To avoid unwanted interactions between the inks and the plate, a tiny diaphragm is connected to the piezo-ceramic plate. An electric impulse causes a piezo-ceramic plate to distort, and subsequently, the droplet is ejected from the nozzle as a result of the pressure wave this creates. The piezo-ceramic plate returns to its original shape after the electric pulse is removed, and the ink is replaced. These ejected droplets gravitate towards and settled on the surface of the substrate using the momentum obtained during the motion. Subsequently, droplets dry (see detailed schematic representation in Figure 2) via the evaporation of the solvent [35].

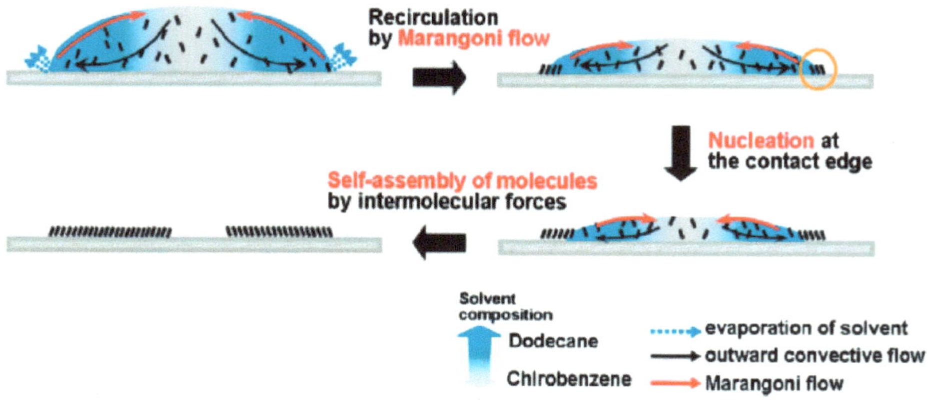

Figure 2. Schematic representation of the potential drying process and crystal formation of an organic semiconductor, 6,13-bis((triisopropylsilylethynyl) pentacene (form 25% dodecane) after deposition from a drop-on-demand piezoelectric print head [36].

Before the drying process starts, the droplet first reaches an equilibrium temperature due to the continuous heat loss and evaporation of the solvent into a warmer environment of surroundings at a certain pressure and temperature [37,38]. In the first stage of droplet drying, the drying rate (which is commonly expressed in kg m^{-2} s^{-1}) is limited and determined via the energy essential to evaporate the solvent, which leads the heat transport towards the droplet's surface [38,39]. After that, the drying process starts from the surface of the droplets, and the molecules of solvent keep drifting towards the surface from the center, which can be mediated via diffusion allied to the solute, convection of liquid within the droplet or capillary fluid flow [37,38,40]. If the temperature of the surroundings is constant, then the drying rate remains unchanged and determined only by the temperature transfer towards the droplet surface [38,41]. For this reason, the first stage of droplet drying is known as the constant-rate drying stage [38].

The second stage of droplet drying is elucidated by the materials present in droplets. Since the evaporation of liquid occurs in the surface of a droplet, the material concentration increases at the surface. This growing concentration gradient results in diffusional material flux far from the surface and towards the center of the droplet, which is a complex phenomenon [38,39]. Consequently, the diffusional motion of the material towards the droplet's center becomes less than the reduction rate of the droplet diameter because of the constant rate for solvent loss. At this point, crust forms because the higher concentration of materials at the droplet surface leads to a decrease in the drying rate [38,42]. This point is called the locking point or critical point. In the beginning of the second drying stage, a porous solid crust with an internally wet core might be observed in the drying droplet, and drying rate here is now determined by the diffusion or capillary flow rate of the liquid from the wet core via the porous crust. A slowed liquid evaporation still causes the shrinkage of wet core and a considerable increase in the crust towards the droplet's focal point [42,43]. The condensed crust will influence a growing resistance to mass transfer, and therefore a decrease in the drying rate can be observed. Because of that, this second stage is called the falling-rate stage in the droplet-drying process [37]. It infers the presence of lowest possible amount of residual liquid in a single droplet, which can be either an equilibrium amount or residual solvent that cannot be eliminated by drying [38,44]. Hence, the droplet drying rate in the course of time at the falling-rate period might take on different shapes depending on the mechanism and factors of the drying [43,45]. Figure 3 shows a simplified illustration of the two stages of droplet drying.

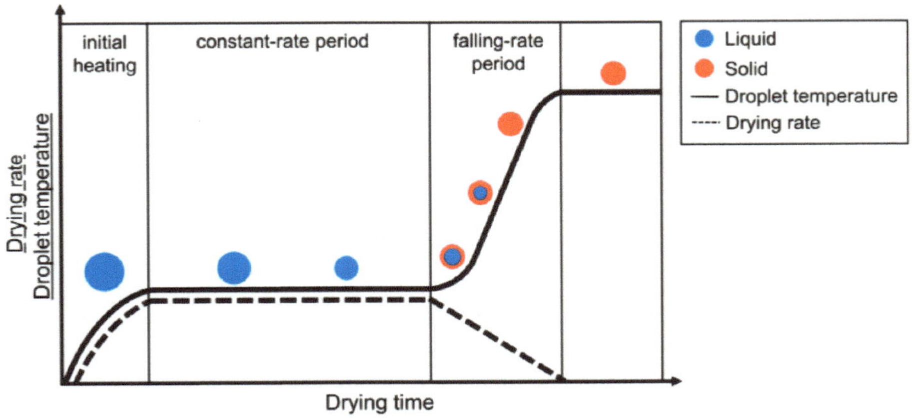

Figure 3. Simplified schematic depiction of the different stages a droplet goes through while drying. Drying rate is represented by a dotted line and temperature evolution by a solid line. The solid fraction is shown in red and the liquid fraction in blue [43,45].

In short, the mechanism of inkjet printing comprises three main steps: (1) ejection of ink and droplet formation, (2) liquid–solid interaction after the placement of droplets on the substrate's surface and (3) drying of ink droplet and subsequent solidifying of the printed features to generate a solid deposit [46,47].

Moreover, inkjet printing does not require sophisticated infrastructure such as clean rooms and large-scale facilities [11]. Merely tiny drops are enough to produce superior quality images with higher resolution [48]. Overall, inkjet printing is a better patterning technique in comparison with some of the other available technologies in the market (Table 1) in terms of cost, efficiency, resolution, compatibility with polymer, process, mode of action, flexibility, requirement of environment and material consumption [49,50].

Table 1. Comparison of some typical patterning technologies [50].

SN	Properties	Photolithography	Micro-Contact Printing	Shadow Mask	Inkjet Printing
1.	Cost	Extremely high	Medium	Low	Low
2.	Efficiency	Low	High	High	High
3.	Resolution	Extremely high	High	Low	High
4.	Compatibility with polymer	Bad	Bad	Good	Excellent
5.	Process	Multi step	Multi step	Multi step	All in one
6.	Mode of action	Noncontact	Contact	Contact	Noncontact
7.	Flexibility	Bad	Bad	Bad	Good, digital lithography
8.	Requirement of environment	Clean rooms, vibration isolation	Medium	Low	Low
9.	Material consumption	High	Low	Medium	Low

Inkjet printers present some drawbacks. Mainly, they are expensive because of their two basic requirements: (a) printheads must be well suited to different kinds of inks, for example polymeric or metal-based inks and (b) printing cycles must be executed repeatedly.

Low-cost inkjet printers that are used generally in household and office might be considered for use as simple devices, but they usually cannot perform due to the incompatibility with the bioinks or inks used for laboratorial research purposes. This is especially the case with metal inks due to their viscosity and nozzle occlusion problems. In addition, the printers' multilayer patterned structure makes it quite difficult to print with them [51–54]. An additional issue is the printing of polymers on material surfaces, which leads to adsorbed patterns that are poorly adhesive [55]. Another disadvantage is the coffee ring effect, which causes inkjet-printed insulators to generate a wave-shaped profile where other methods produce perfectly smooth profiles, such as spin-coating [56].

To overcome these issues, researchers have come up with different strategies for both maximizing the benefits that inkjet printing provides and minimizing the disadvantages as well. For example, screen printing provides high-speed printing that is adaptive to commonly available materials and complex multilayer devices. In addition, low-cost inkjet piezoelectric printers provide good spatial resolution (for example, 5760 × 1440 drops per inch), low-cost printing and production and good repeatability (range ~300 µm). Professional inkjet printers provide high spatial resolution and low production cost, are compatible properties with several materials and show repeatability ranges from 5 µm to 25 µm. Finally, mixed-screen printing and low-cost inkjet have demonstrated adaptability to numerous materials, good spatial resolution and repeatability (~300 µm) [57].

There are a few reported studies regarding the development of multilayered structures based on the development of ink properties such as viscosity, surface tension and pH combined with printing parameters (voltage and duration) [58]. Control over the evaporation rate is a key parameter for increasing accuracy and resolution, and the rate can be adjusted by the addition of cosolvents (e.g., alcohol)

1.2. Inks for Inkjet Printing

The ink used for material deposition and its physical properties is considered to be the most crucial part of inkjet printing technology [28]. Inkjet printing involves various processing steps such as droplet generation, motion, interaction with substrate, and drying that are influenced by the quality and properties of the ink for successful printing [11]. The resolution, uniformity and quality of the patterns significantly depend on the viscosity and surface tension of the ink [19,46]. These two parameters can determine the three main steps of inkjet printing process [46]. The speed and accuracy of droplet ejection will decrease if there is an increase in viscosity. At high viscosity and drop rate, the printing process will fail as the ink might not move towards the ink cartridge swiftly to refill it in-between the jetting [19]. Glycols (e.g., glycerol, propylene glycol and polyethylene glycol) are the typically used excipients for viscosity adjustment [59–61]. Contrastingly, surface tension influences the propensity of the ink to draw off of the nozzle to produce a droplet, and it is usually adjusted by adding surfactants [19]. Fromm et al. introduced an equation to determine an apt balance among the physical properties in one of his published articles in 1984, which is as follows:

$$Z = \frac{\sqrt{\gamma \rho l}}{\eta} = \frac{\sqrt{We}}{Re} = \frac{1}{Oh}$$

In this equation, the surface tension, density, viscosity and printing mesh aperture diameter are denoted by γ, ρ, η and l, respectively [27,62]. This equation also explains the dimensionless numbers from fluid physics. i.e., Reynolds number, Weber number and Ohnesorge number designated by Re, We and Oh, respectively. It was suggested by Fromm that proper inkjet printing would be possible if Z is greater than 4 ($Z > 4$). Further investigations led to the introduction of a limiting range for inkjet printing where $1 < Z < 10$ [27,63]. This range explains the viscous dissipation of a droplet formation if Z is greater than 1 ($Z > 1$), and if Z is less than 10 ($Z < 10$), then the formation of satellite droplets occurs (also termed as secondary droplets) [27]. These satellite droplets have an effect on primary droplets and influence their positioning on the substrate. In fact, the droplet deposition should be accurate, uniform and precise to facilitate successful inkjet printing [1]. Further experimental studies revealed a new range for Z, that is $1 < Z < 14$ [62,64,65]. Figure 4 shows the formation of satellite droplets.

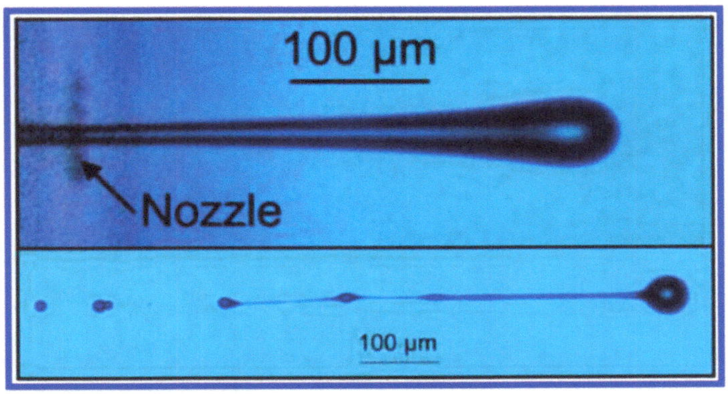

Figure 4. The upper image is a high-speed photograph of a droplet coming out of a nozzle, and the bottom one is a high-speed photograph of a satellite droplet formation [27].

The modulation of droplet size is a major challenge in inkjet printing. To date, a total of eight mechanisms have been recorded that are capable of changing the droplet volume utilizing same ink and printhead.

Usually, optical techniques are used to measure the droplet coming out of the nozzle [66]. To illustrate, a 6 ns short illumination with a laser induced fluorescent stroboscopic recording using iLIF (illumination by laser induced fluorescence) or ultra-high-speed cameras (up to 25 Mfps) are usually used to measure the drop formation with pL sized droplets at approx. 100 kHz repetition rate [67].

One of the inkjet printing limitations is the use of inks of low viscosities. Table 2 lists the compositions of some typically used printing inks including their viscosity and surface tension.

Table 2. Some of the recently used inks in inkjet printing.

SN	Ink	Ink Viscosity (mPa·s)	Ink Surface Tension (m Nm^{-1})	Z Value	Ref.
1.	Ethylene glycol	15.8	45.5	2.08	[68]
	Ethylene Glycol: Water (5/95)	1.16	69.5	33.2	
	Ethylene Glycol: Water (10/90)	1.47	68.9	26.1	
	Ethylene Glycol: Water (15/85)	2.32	67.7	16.5	
	Ethylene Glycol: Water (25/75)	2.72	67.0	14.1	
	Ethylene Glycol: Water (50/50)	5.05	46.7		
	Ethylene Glycol: Water (50/50)	4.39	60.3	8.40	
	Ethylene Glycol: Water (75/25)	7.81	52.7	4.47	
	Ethylene Glycol: Water (85/15)	10.5	50.2	3.28	

Table 2. *Cont.*

SN	Ink	Ink Viscosity (mPa·s)	Ink Surface Tension (m Nm^{-1})	Z Value	Ref.
2.	De-ionized water	1.07	72.7	36.8	[68]
3.	Gallium-indium (75/25)	1.7	624		[46]
4.	Glycerol-Water	1–22.5	66.4–7.6		[69]
5.	CuNO4- Water	~4.45	88		[70]
6.	Dowanol	10.17	15.55		[71]
7.	Ethyl acetate	0.452	2.367		[13]
8.	5 Fe_3O_4-95 (nanoparticles + UV Curable matrix resin)	18.03	23.91	1.72	[72]
9.	10 Fe_3O_4-90 (nanoparticles + UV Curable matrix resin)	18.08	20.91	1.57	[72]
10.	Hydroxypropyl cellulose:Water (6/94)	45	44.5		[73]
11.	Commercial AgNp	6.8 ± 0.7	30 ± 1		[74]
12.	Diethylene glycol	27.1	42.7	1.17	[68]
13.	Glycerol	934.0	76.2	0.05	[68]
14.	$MnCo_2O_4$	10		6.17	[75]
15.	$MnCo_{1.8}Fe_{0.2}O_4$	>15		4.77	[75]
16.	PVDF: $BaTiO_3$ (40/8)	13.6	30.2	1.17	[76]
	PVDF: $BaTiO_3$ (32/6.4)	9.7	31.7	1.72	[76]
	PVDF: $BaTiO_3$ (24/4.8)	6.0	32.4	2.79	[76]
	PVDF: $BaTiO_3$ (16/3.2)	3.7	33.5	4.59	[76]
	PVDF: $BaTiO_3$ (8/1.6)	2.1	34.8	8.23	[76]
	PVDF: $BaTiO_3$ (1/0.2)	1.3	36.0	13.56	[76]
17.	DNTF: Hexogen (13.86/0)	1.2	23.33	36.94	[77]
	DNTF: Hexogen (12.47/1.39)	1.0	23.09	44.56	[77]
	DNTF: Hexogen (11.09/2.7)	0.8	23.77	58.01	[77]
	DNTF: Hexogen (9.70/4.16)	0.6	24.15	75.51	[77]
	DNTF: Hexogen (8.32/5.54)	0.8	24.52	58.2	[77]
	DNTF: Hexogen (6.93/6.93)	1.3	23.66	35.44	[77]
18.	8 mol% Y_2O_3-stabilized ZrO_2 (8YSZ)	1.5	18.8 ± 0.3	7.6	[78]

Note: concentration of Ethylene Glycol: Water ratios are in v/v; concentration of PVDF: $BaTiO_3$ ratios are in mg mL^{-1}; concentration of DNTF: Hexogen ratios are in wt%; PVDF = Polyvinylidene difluoride; DNTF = 3,4-dinitrofurazanofuroxan.

1.3. Overview of Different Types of Inkjet Printing Technology

Based on the physical process of droplet generation, this automated, high-throughput technology is predominantly classified into two categories: (a) continuous inkjet printing (CIJ) and (b) drop-on-demand printing (DOD) (Figure 5) [16,23,27,29,79]. Continuous inkjet printers generate droplets as a continuous stream of ink discharged on the target, while in drop-on-demand printers, the droplets are ejected in a discontinuous manner only when they are needed [80]. Droplets deposited by continuous inkjet method are usually twice the size of the orifice [27].

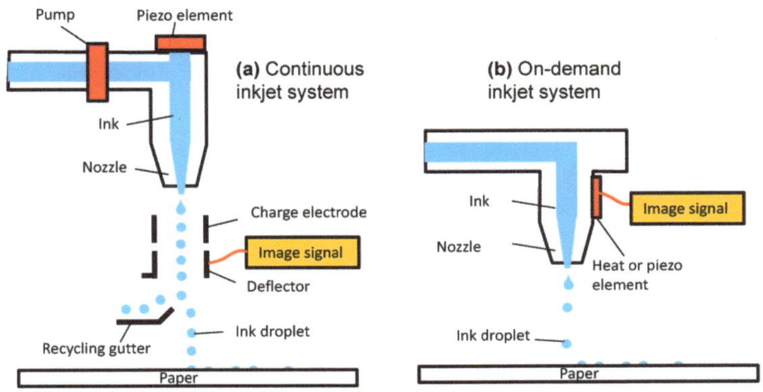

Figure 5. Simplified representation of two different categories of inkjet printing mechanism: (**a**) continuous inkjet printing (CIJ) and (**b**) drop-on-demand printing (DOD) [81].

In CIJ printing, a high-pressure pump under an electric field allows the continuous flow of liquid material that is ejected via the orifice, diameter 50–80 μm, which then disintegrate into a stream of droplets under the surface tension forces due to Reayleigh instability [13,15,79]. Continuous inkjet printing is predominantly used in textile printing, labelling and other high-speed graphical works [28]. Depending on actuation technique, IJP can be classified into another two categories which are: (a) piezoelectric and (b) thermal [82–85]. Both of the techniques reserve the material to be printed in a chamber and emit the droplets through the printhead via a nozzle, but they differ in the process of droplet formation [86–88].

In DoD inkjet printing, the liquid droplet is emitted through a nozzle only when it is required. Typically, a DOD printhead consists of multiple nozzles (usually 100–1000, aside from specialized printheads, which might have only one nozzle). The formation of droplets occurs swiftly after the deformation of the piezoelectric wall, which compresses the ink due to the applied wave. The ink material comes out of reservoir as a form of jet through the printhead, gravitates down afterwards and gets ejected via the nozzle under the surface tension forces to generate one or more droplets [15].

In contrast to CIJ, droplets produced by DOD inkjet printing are comparable with the diameter of the orifice, usually ranging from 10 to 50 μm, in accordance with drop volumes, which vary from 1 to 70 pL [15,27]. Due to the capability of smaller droplet formation, it has become a method of choice for several studies [28].

Biological ink materials can be affected by the electrostatic inkjet process due to the shear pressure (sonication with the frequency of 15–25 kHz), and they can clog easily since the diameter of the nozzle is not only fixed but also small [89].

The use of higher amount of solid material in the ink solution can increase the printing efficiency and decrease the cost notably which is one of the advantages of thermal inkjet printing [79].In addition, inks comprising aqueous solvents are usually more feasible for jetting with thermal inkjet printing. In contrast, organic solvents are generally more suitable for piezoelectric inkjet processing. Furthermore, thermal inkjet printers are usually inexpensive compared with the piezoelectric devices [1,90].

There are several other inkjet printing processes such as electrostatic, electrohydrodynamic and acoustic, but they are not frequently used because of their major drawbacks [82]. Some disadvantages of electrostatic inkjet printing are that it requires high voltage (sometimes over 2 kV) to operate, utilizes conductive metal pipe, requires placing one electrode externally to the device and requires placing a substrate between the nozzle and the electrode [91].

One of the drawbacks of electrohydrodynamic inkjet printing is that it cannot deposit single droplets at a time. The droplet generation occurs using an electric field and not by shrinkage of the ink with thermal energy or chamber deformation [89]. The low throughput (low production speed) of electrohydrodynamic inkjet printing is considered to be the most severe drawback that has retarded its widespread application [92,93]. Figure 6 illustrates an overview of the available inkjet printing methods.

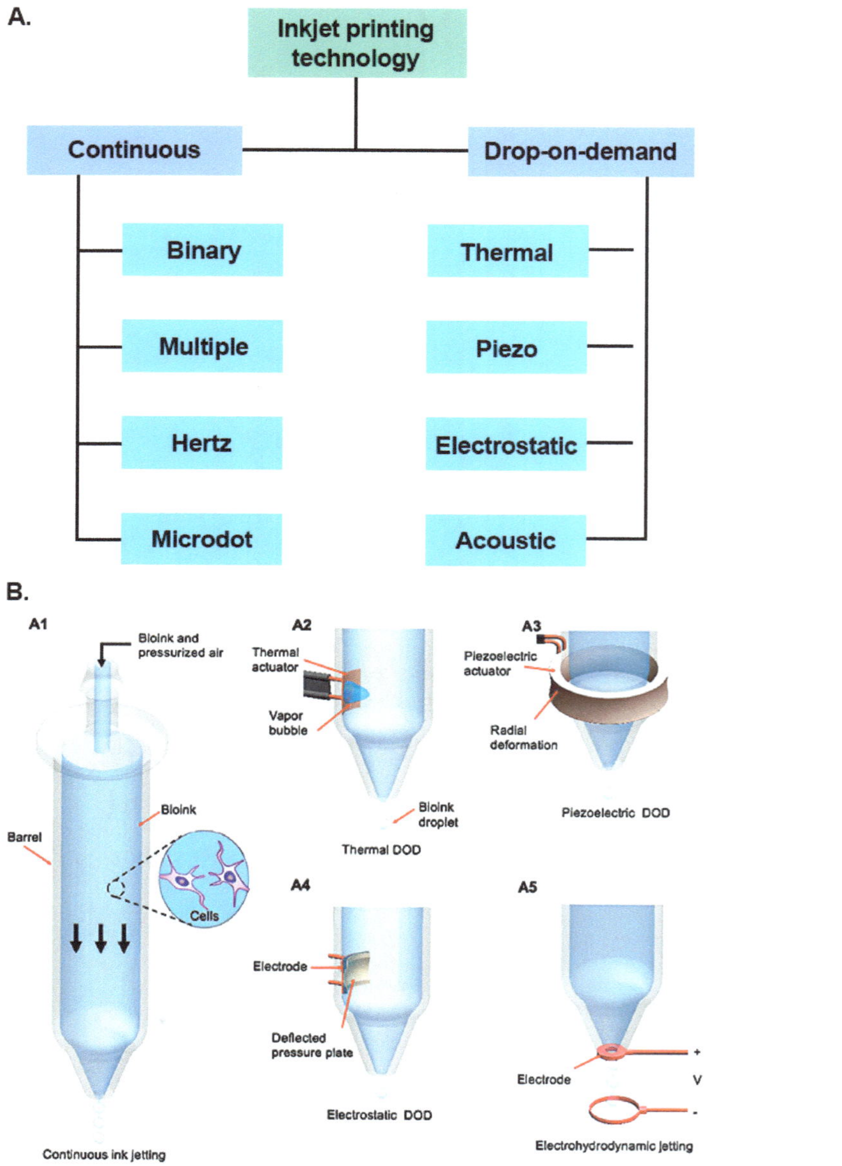

Figure 6. (**A**) Classification of IJP technology (reproduced from [94]). (**B**) Working principles of some of the different IJP techniques: (**A1**) continuous, (**A2**) thermal, (**A3**) piezoelectric, (**A4**) electrostatic, (**A5**) electrohydrodynamic (modified from [95]).

2. Thermal Inkjet Printing

Thermal inkjet printing (TIJP) is a noncontact DOD printing system that was basically developed for printing digital data on media [90,96,97]. It is also known as bubble inkjet printing since the droplet ejection occurs via bubble nucleation [28,87,98]. TIJ printers can eject droplets in a range of 2–180 pL of volume [99,100].

A TIJ printer consists of an ink (desired fluid material to be printed), the cartridge and a printhead. The printhead comprises several nozzles (column like small channels) filled with the fluid material from the ink chamber, and a transducer (which is a thin film resistor for thermal inkjet printing) is attached to each nozzle [79,100].

Due to its reproducibility, low cost and high throughput, this printing technique dominates the market over other printing technologies [47,48,101–104].

In TIJP, a thin-film resistive heater that creates a frequency ranging from 1 kHz to 5 kHz with an approximate rectangular wave of 3–6 μs pulse width is attached to the printhead, which instantaneously heats the ink in the cartridge. A small vapor bubble forms and puffs up using the heat, which generates a pressure pulse essential for droplet emission through the nozzle. Once a droplet emission is completed, the current is withdrawn, which facilitates a prompt reduction in the vapor pressure and temperature. Consequently, the bubble collapses inside the printhead, which somewhat creates a vacuum (negative pressure) that pulls the liquid ink to refill the chamber [4,47,80,98,105–111]. Thermal gradient, viscosity of ink material and electric pulse frequency determine the size of the droplets to be generated [22,47,109,111].

In TIJP, the thermal resistor can momentarily (approximately 3 to 10 μs) produce up to 300 °C temperature, and merely around 0.5% of the ink encounters a thermal rise in the nucleation of a vapor bubble [89,105].

Types of Thermal Inkjet Printer

Depending on the droplet emission principle, there are three types of TIJ printer available: (a) side shooter, (b) roof shooter and (c) back shooter [13,48]. For the side shooter printer, the droplet is ejected tangential to the surface of heater. On the other hand, the droplet ejects straight (at 90° angle to the heater surface) in a roof shooter. In the back shooter printer, the droplet ejects straight, but the vapor bubble nucleation occurs in the opposite direction of droplet emission [48]. Simplified versions of the working mechanisms behind the side shooter, roof shooter and back shooter are shown in Figure 7.

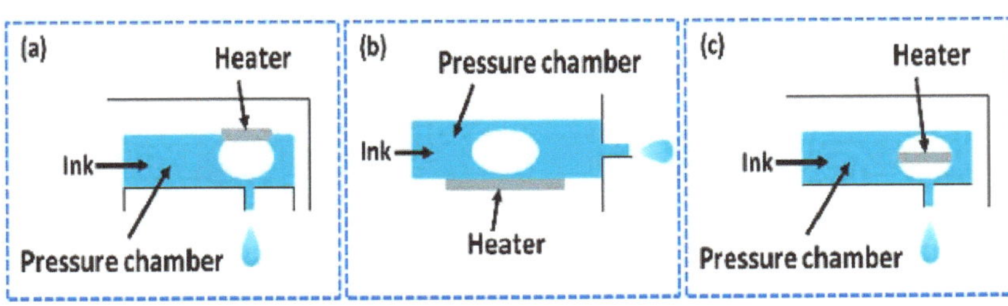

Figure 7. (**a**) Roof shooter, (**b**) side shooter, and (**c**) back shooter system [13].

3. Application of Thermal Inkjet Printing

TIJP has found several applications for printing of medicines in various forms by controlling the printing pattern and the deposited amount, most importantly for the printing of a wide range of drug formulations. In addition, it has found a strong ground in bioprinting applications for cell-laden bioinks in tissue engineering and regenerative medicine.

3.1. Bioprinting

An engineered tissue can be used as a physiological replica for better understanding of basic biology. Moreover, the problem with finding suitable organ donors for repairing or replacing damaged or injured organs, regenerative medicine, cell transplantation and tissue engineering may be resolved by using bioprinting technology (see Figure 8) [84,112–115]. Thermal inkjet printing-based bioprinting is one of the promising approaches in the field of tissue engineering [111]. There are some other bioprinting strategies apart from TIJP, for instance, extrusion bioprinting (mechanistically similar to traditional fused deposition modeling (FDM) 3D printing, which extrudes droplets via pneumatic pressure or mechanical forces through a nozzle onto a previously fixed location called the fabrication platform) [116] and vat polymerization-based bioprinting (mostly used for tissue scaffold fabrication utilizing conventional cell-seeding approach) [117].

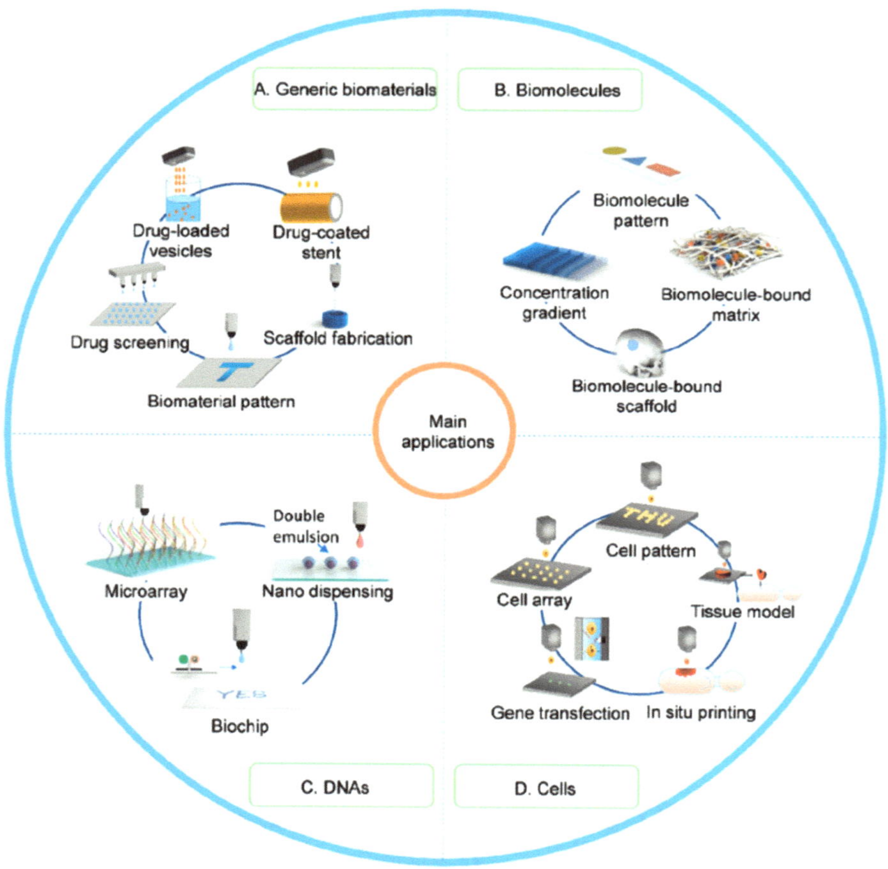

Figure 8. Schematic illustration of major applications of inkjet bioprinting. (**A**) Generic biomaterials were printed for pharmaceutical applications including drug screening, drug loading and drug coating and for biomaterial patterning and scaffold fabrication. (**B**) Biomolecules were printed for concentration gradient, biomolecule pattern, biomolecule-bound matrix and biomolecule-bound scaffold. (**C**) DNAs were printed for microarray, nano-dispensing and biochip. (**D**) Cells were printed for cell array, cell pattern, tissue model, in situ printing and gene transfection [89].

Printers used in bioprinting fabricate the biological elements in a layer-by-layer manner often described as a bottom-up process since the deposition of the first layer is followed by the buildup of more layers. [82,106]. In bioprinting, the conventional ink material is replaced with bioink, which is a liquid and contains water [112] along with biological substances, e.g., enzymes [118], proteins [65], saline or other media with suspended cells [119]. Some studies observed a 10–15% decrease in the enzyme activity [111,120], and hence, an increase in temperature for bubble nucleation was suspected to affect the biological substances in the bioink chamber [47,106,111]. To overcome this issue, heat is transmitted for merely 2 μs, which causes a 4–10 °C increase in temperature that ensures a count of 90% viable cell, enough to conduct bioprinting efficiently [89,106,111,112].

Despite the significant number of studies, there is still a lack of understanding of how the viability of human primary cells is affected during the inkjet printing of sub-nanolitre amounts. In a recent study, Ng et al. used TIJP to dispense cell-laden inks to investigate cell viability and proliferation [121]. It was observed that increased cell concentrations had a minimal impact on the droplet velocities but led to better cell viability. By regulating the droplet volumes at 20 nL, it was possible to eliminate the evaporation-induced damage to the cells, which also resulted in high viability.

Park et al. developed a strategy for the formation of self-organized 3D collagen microstructures [122]. By applying DoD inkjet printing with predefined patterns, it was revealed that cell-to-extracellular matrix interactions facilitate the self-organization of microstructures on hydrogels comprising collagen, while actin polymerization inhibits the formation of the microstructures. Further manipulation of the print patterns and cell densities assisted with the formation of a human skin model with papillary microstructures.

Suntornnond et al. introduced significant advances in TIJP by expanding the use of printable bioinks [123]. The authors applied saponification in gelatin methacrylate (GelMA), a very common bioink, to study the printability in a thermal inkjet printer (HP Inc. D300e Digital Dispenser). The two-step process, which comprised saponification and heat treatment, led to the formation of excellent bioinks with good cell viability and proliferation. Saponification is an exothermic reaction where the hydrolysis of triglycerides with alkali produces salts of fatty acid and glycerol [124,125]. Sun et al. presented a detailed description of the saponification process that was investigated via isothermal titration calorimetry, attenuated total reflection infrared and small-angle X-ray scattering spectroscopies [124]. Another group of researchers, Tan et al., published a review in 2012 on glycerol (which is a valuable byproduct of the saponification process) in which they discussed different methods of glycerol production as a byproduct and its purification process [125]. In bioprinting, both saponification and heat treatment are usually used to enable better jetting behavior of the ink.

Yoon et al. employed the saponification of gelatine methacryloyl to stabilize the jet formation and to reduce the viscoelasticity of the inks [126]. The use of inkjet printing allowed for the large-scale production of multilayers and ensured high shape fidelity. The use of alginate, cellulose nanofiber, fibrinogen blended with human dermal fibroblasts facilitated the generation of structures that mimic native tissue functions. Freeman et al. (Figure 9) mentioned that heat treatment of their material (gelatin) induced favorable rheological properties that increased the printability of ink; however, it also affected the cell viability, which they optimized by increasing cell density and tissue volume, and that ultimately increased the collagen deposition and mechanical strength of printed material [127].

(A) 1 x 10⁶ cells per mL

Heat treatment time (h); rows: Gelatin concentration in bioink %(w/v)

	1	3	5	7	9
5	87.80	81.92	85.78	90.76	91.72
6.25	86.95	81.94	81.33	89.45	91.70
7.5	83.76	83.24	80.72	89.66	90.35
8.75	77.49	80.71	80.04	89.49	89.61
10	69.34	77.38	84.48	90.51	89.85

Not printable / Printable

(B) 3 x 10⁶ cells per mL

Heat treatment time (h); rows: Gelatin concentration in bioink %(w/v)

	1	3	5	7	9
5	93.44	91.66	91.82	93.70	94.46
6.25	93.17	91.49	91.06	93.59	93.51
7.5	91.38	91.75	92.01	92.97	92.93
8.75	89.43	88.67	91.34	92.45	93.24
10	84.96	87.40	91.83	92.00	92.82

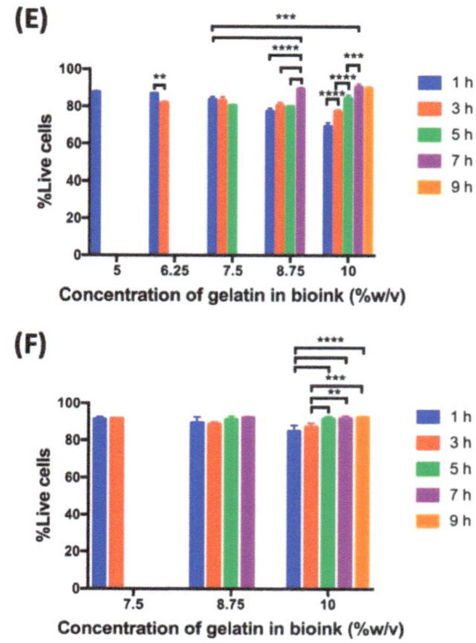

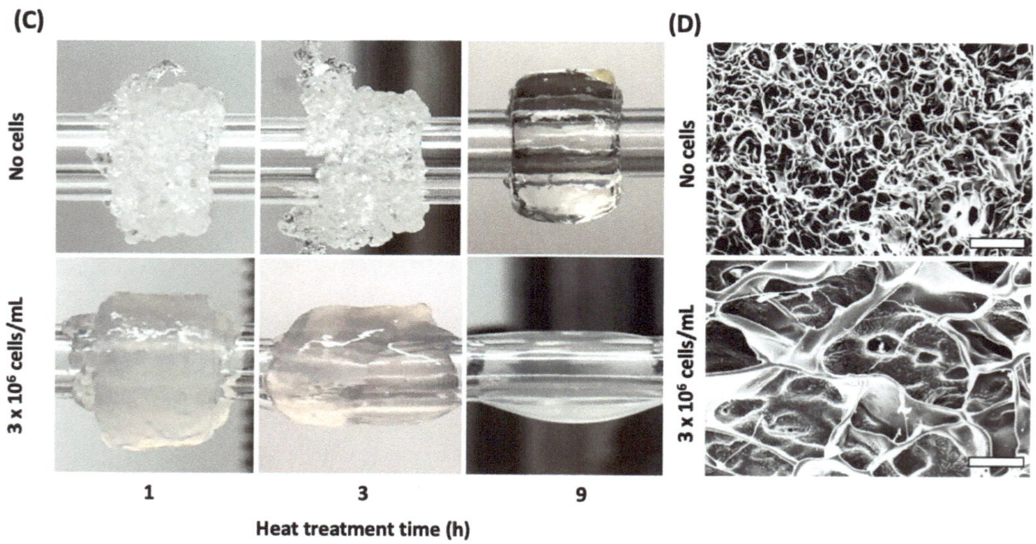

Figure 9. Effect of heat treatment of gelatin on cell viability during 3D rotary printing. (**A**) 1×10^6 or (**B**) 3×10^6 cells/mL neonatal human dermal fibroblasts (HDF-n) was mixed with the gelatin–fibrinogen bioink for vascular 3D rotary printing. The gelatin was heat treated at 90 C before use in preparing the bioinks. The red regions indicated poor printability of the cell-laden bioinks, while the green regions indicated conditions that held. Average percentage of live cells detected by ethidium homodimer staining are shown in the table. (**C**) The appearance of tubular tissue constructs printed using cell-laden bioinks prepared using 10 mg/mL of fibrinogen and 7.5% (*w*/*v*) heat-treated gelatin. (**D**) SEM micrographs of the 5% (*w*/*v*) 1 h heat-treated gelatin + 10 mg/mL fibrinogen. Scale bar: 10 mm. Percentage of live cells in gelatin–fibrinogen bioinks containing (**E**) 1×10^6 cells/mL or (**F**) 3×10^6 cells/mL. Data are presented as mean ± S.D. **, $p < 0.0021$; ***, $p < 0.0002$; and ****, $p < 0.0001$ [127].

Solis et al. employed TIJP to develop human microvascular endothelial cells for the formation of microvasculature to print implantable graft–host anastomoses [128]. Flow cytometry revealed 75% apoptosis during printing, but the viability improved after 3D incubation. Further investigations showed the overexpression of various cytokines such as HSP70, IL-1α, VEGF-A, IL-8 and FGF-1 for the printed cells. The activation of the HSP-NF-κB pathway led to the production of VEGF and consequently to the immense formation of capillary blood vessels after implantation.

Figure 10 shows another example of how TIJP was used for tissue engineering and regeneration by Gao et al. [129]. The authors used a TIJP with continuous photopolymerization to develop a bioprinting platform for 3D cartilage tissue engineering. The created cartilage showed native zonal arrangement with excellent extracellular matrix architecture, and the required material performance. The vitality of the printed cells with concurrent photopolymerization was noticeably higher than with the control tissue creation method, which necessitates prolonged UV exposure. Substantial glycosaminoglycan (GAG) and collagen type II production was seen in printed neocartilage that was compatible with the gene expression pattern.

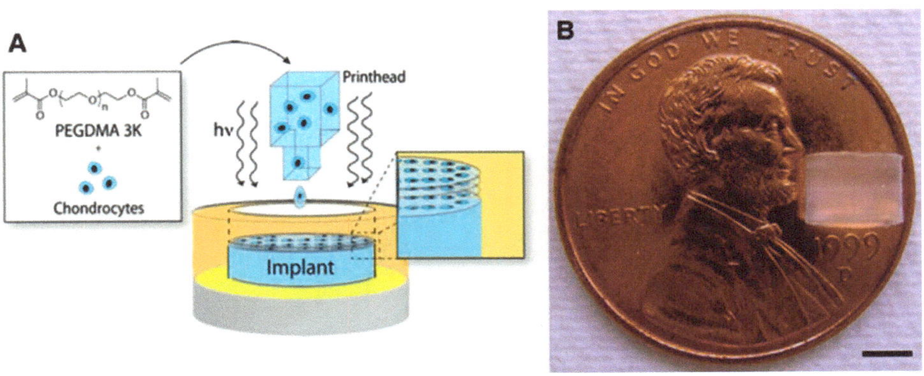

Figure 10. Bioprinted neocartilage tissue using TIJP with continuous photopolymerization. (**A**) Schematic of cartilage bioprinting with simultaneous photopolymerization and layer-by-layer assembly. (**B**) A printed neocartilage tissue of 4 mm in diameter and 4 mm in height. Scale bar, 2 mm [130].

In another study, Kador et al. combined the formed electrospun cell transplantation scaffolds with retinal ganglion cells and positioned them accurately on the scaffolds using thermal inkjet printing. This procedure preserved the printed cells' functioning electrophysiological capabilities, cell growth, and neurite expansion [131].

Furthermore, Gao et al. used TIJP to determine the effectiveness of bioactive ceramic nanoparticles in promoting osteogenesis in human bone marrow mesenchymal stem cells that were printed on poly(ethylene glycol) dimethacrylate scaffolds [132]. The printing of the stem cells suspended in poly(ethylene glycol) dimethacrylate scaffolds with nanoparticles of bioactive glass and hydroxyapatite during simultaneous polymerization enabled the deposition of the printed substrates with extremely precise placement in 3D locations. Further analysis revealed that the printed scaffolds produced far more collagen and had the most alkaline phosphatase activity, comparable with the gene expression found by quantitative PCR. The study was an example of how inkjet printing can be employed for the engineering of both soft and hard tissues with biomimetic assemblies.

As summary of various studies that have used thermal inkjet printing is presented in Table 3, illustrating the different applications and the main positive or negative outcomes reported for each work.

Table 3. List of thermal inkjet-based bioprinters with their applications & outcomes in biopharmaceutical research area.

SN	Printer	Bioink	Area of Application	Outcome (Positive, Negative or Both)
1.	HP Deskjet 500 printer (modified) [Hewlett-Packard, Inc., Palo Alto, CA]	Rat tail collagen type I	Cell printing	Around 89% cell viability was reported [111].
2.	HP DeskJet 550C printer (modified)	hAFSCs cell line	Stem cell printing	Data revealed that printed hAFSCs are capable of forming a firm bony tissue that can withstand high compressive force [107].
3.	Prototype of thermal inkjet printer combined with amperometric GOD electrode [developed by Lesepidado srl (Bologna, Italy) & supplied by Olivetti Tecnost (Ivrea, Italy)]	Glucose oxidase (GOD) from *Aspergillus niger* and poly(3,4-ethylene di-oxy thiophene/ polystyrene sulfonic acid)	Biosensor	Approximately 15% decrease in the efficiency of enzyme was noted [120].
4.	Canon inkjet printer (Pixma ip4500) (modified)	Fluoroscein isothiocyanate-conjugated bovine albumin and horseradish peroxidase	Microfluidic patterned paper	Bioactivity was retained by patterned paper. However, the percentage was not measured [118].
5.	Hewlett-Packard (HP) Deskjet 560 (Modified)	Herring sperm DNA in pure water, surfactant, alcohol, or a water-soluble polymer	Microarray	Was reported as a dependable printing option [133]
6.	Bubble Jet (BJC-2100, Canon, Tokyo, Japan)	Rat tail collagen solution	Cell patterning	Spatial resolution of around 350 mm was obtained, and adherence of neuronal and smooth muscle cells to the printed area was reported [134].
7.	BJ F850 (Canon, Tokyo Japan)	Insulin related growth factors	Cell patterning and analysis	Intensified proliferation of cells on patterned area was observed [133].
8.	HP Deskjet 500 inkjet printer (modified) [Hewlett-Packard, Inc., Palo Alto, CA]	Chinese hamster ovary (CHO) cells	Cell patterning	Cellular viability count of 80% was reported that improved after changing the carrier fluid. Transient membrane damage of cells was observed after printing [111].

Table 3. *Cont.*

SN	Printer	Bioink	Area of Application	Outcome (Positive, Negative or Both)
9.	HP DeskJet 692C and 55uC	CHO cells and porcine aortic endothelial	Gene transfection	Transfection rate of 10% and cellular viability of 90% were reported [135].
10.	HP Desktop printer (HP 55uC) (modified)	Mouse myoblast	Biosensor	Myotube generation alongside printed substrate was demonstrated [136].
11.	Hewlett Packard (HP) Deskjet 500	Mammalian cells	Cell printing	Cellular viability varied 85–95% [112].
12.	HP-2225C Think Jet ink jet printer 7470A graphics plotter	Fibronectin	Cell patterning	Stickiness of cells to patterned fibronectin was noticed [137].
13.	BJC-600 (Canon, Tokyo) and BJC-700J printer	5′-terminal-thiolated oligonucleotides	DNA microarrays	No trouble was encountered by researchers while ejecting DNA solution using bubble jet printer rather than heat generation, which was stated as an added advantage as it provided efficient reaction energy [138].
14.	Prototype model of TIJ printer from Olivetti Tecnost developed by Lesepidado srl	β-Galactosidase (GAL) from *Aspergillus oryzae*	Biosensor	Aside from approximately 15% reduction in enzyme activity, TIJP was determined to be a promising option for enzyme or other biological material micro-deposition [110].
15.	HP60 inkjet printer	Unmentioned cell	Cell printing	Successful concurrent simulation of thermal transfer, interaction between cell and fluid, transition of phase and increased cell viability was reported [47].
16.	Hewlett Packard Deskjet 500 thermal inkjet printer (modified) [Hewlett–Packard Company (Palo Alto, CA, USA)]	Human microvascular endothelial cells (HMVEC) and fibrin	Cell printing	Printed HMVEC proliferated and the formation of microvascular endothelial cells along with fibrin scaffolding was observed [139].

Table 3. *Cont.*

SN	Printer	Bioink	Area of Application	Outcome (Positive, Negative or Both)
17.	Canon inkjet printer (Pixma ip4500) (modified)	Horseradish peroxidase (HRP) and alkaline phosphatase (ALP, from bovine intestinal mucosa)	Enzymatic paper	Bioactivity was retained by patterned enzymatic paper, but the percentage was not mentioned [140].
18.	Hewlett Packard (HP) 550C printer (modified)	Suspensions were made using embryonic motoneurons of rat and Chinese Hamster Ovary (CHO)	Cell printing	Successful printing of embryonic motoneurons and CHO cells with >90% viability was reported [104].
19.	Hewlett-Packard (HP) Deskjet thermal inkjet printer	Bone-marrow derived hMSCs	Cell printing	Viability of the printed cells was significantly higher [129].
20.	HP TIPS print head (Hewlett-Packard Packard, Corvallis	Retinal ganglion cells	Cell printing	Comparatively better cell survival, neurite outgrowth and functional electrophysiological properties of the printed cells were observed [131].

3.2. Oral Dosage Form

Today, most of the market oral drug products are in the form of tablets or capsules where more than 40% of the APIs (active pharmaceutical ingredients) are water insoluble [41]. The norm for the increase of drug solubility and hence the bioavailability [103,104] involves a range of processing technologies such as particle size reduction, salt formation, cocrystals, granulation, or solid dispersions [141,142].

As previously mentioned, those technologies are designed for mass production and are not suitable for the design and administration of personalized dosage forms that address the specific needs of individual patients such as children or elders. Over the last 10 years, thermal inkjet printing has been adopted as an ideal technology (Figure 11) for the printing of unique dosage forms with various features. [105,143]. A well-known technology named binder jetting (a nonfusion powder additive manufacturing technology) has been implemented for the printing of pharmaceutical drug dosage form [144]. Aside from the FDA approval for the Spritam (a binder jet 3D printed dosage form) [145], there have been experimental approaches using binder jetting technology. For example, Chang et al. used binder jet 3D printing (BJ-3DP) to print solid dosage forms. They used feedstock materials of pharmaceutical grade to make printed tablets and also developed a molding method for the selection of suitable powder and binder materials [146]. Rahman et al. discussed that the advantages of using binder jet 3D printing are that no polymers with special properties are needed, and any available FDA-approved excipients can be used for dosage form preparation [147]. Hong et al. used BJ-3DP to develop multicompartmental structure-dispersible tablets of levetiracetam-pyridoxine hydrochloride (LEV-PN), and they also managed to trounce the coffee ring effect by modifying the drying method in their study [148]. Kozakiewicz-Latała et al. developed a fast-dissolving tablet using BJ-3DP. Here, they used two easily available FDA-approved model APIs, (a) quinapril hydrochloride (hydrophilic, logP = 1.4) and (b) clotrimazole (hydrophobic, logP = 5.4) [149].

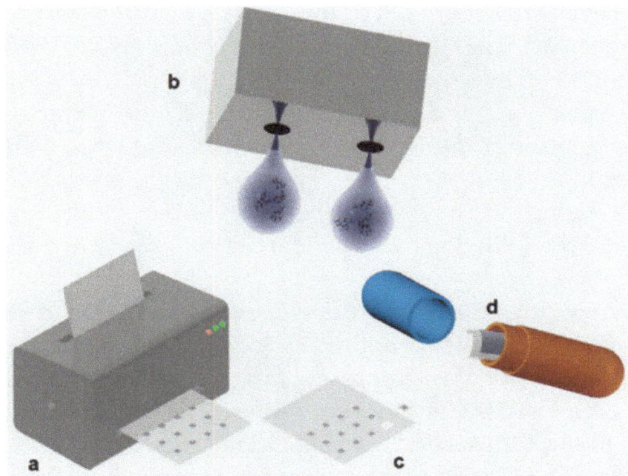

Figure 11. Schematic demonstration of the printing concept for pharmaceutical oral dosage form: (**a**) inkjet printer, (**b**) therapeutic material deposition, (**c**) unit doses on a paper substrate and (**d**) doses inserted to be into capsules or directly fabricated into oral dosage forms [61].

TIJP has demonstrated excellent capability of producing drug crystals rapidly and generating fine particles by dispensing volumes of drug solutions ranging from 5 to 15 pL. A Hewlett-Packard HP460 Deskjet thermal inkjet printer was used for the formulation of orally administrable co-crystal dosage forms. Various solutions of carbamazepine, nicotinamide, benzoic acid, isonicotinamide, theophylline and saccharin were printed and evaluated for their capacity to produce co-crystals using water/ethanol inks [150].

In another study, TIJP was employed for the printing of APIs on a substrate followed by polymer coating. As ink material, riboflavin sodium phosphate or paracetamol were dissolved in a glycerol and purified water solution following printing at two different dose intensities [151]. Wilts et al. used a combination of a thermal inkjet printhead HP_{11} and ZCorp Spectrum Z510 printer to formulate acetaminophen tablets through the comparison of 4-arm star and linear poly(vinyl pyrrolidone) as binder materials [152]. The molecular weight and polymer concentration were found to affect the ink jetability, tablet porosity, hardness and drug loading on the tablet.

TIJP has been successfully used for the printing of prednisolone, a poorly water-soluble drug, in polymorphic forms. Using a mixture of glycerol, water and ethanol at a ratio of 3:17:80, the drug droplets were deposited on substrates comprising fiberglass films [86]. Raman analysis showed that the selection of the ink solvents was the main reason for the formation of the prednisolone polymorphs on the substrate, and this could be important for the design of oral solid forms and the printing of the most stable polymorphs with the fastest dissolution rates.

Montenegro-Nicolini and coworkers introduced inkjet printing for the deposition of biological molecules and the formation of lysozyme [153]. Polymeric films of hydroxypropyl methylcellulose and chitosan were further developed by applying polycaprolactone fibers using electrospinning prior to the molecule deposition. The printed drug amounts did not affect the muco-adhesiveness or mechanical properties of the films. The same group used a Hewlett Packard Deskjet 1000 to print buccal dosage forms comprising lysozyme and ribonuclease-A [154]. Printing proteins and peptides is challenging, but the use of thermal inkjet printing was proved a promising approach.

Buanz and coworkers used a modified cartridge of a HP Deskjet D1660 TIJ printer to print oral thin films of salbutamol sulphate. Potato starch was used as substrate for the dispensing of salbutamol sulphate dissolved in distilled water and glycerin in a series of different concentration ratios [100]. The surface tension had no effect on the droplet deposition, while the calibration of the printing process allowed the design of personalized dosage forms. The same group used a modified Hewlett-Packard HP 5940 Deskjet to formulate orodispersible vitamin films (ODF) of warfarin at dose strengths varying from 1.25–2.5 mg [99]. The film substrates comprised HPMC/glycerol blends and showed rapid disintegration. The study demonstrated the capabilities of inkjet printing for the development of personalized dosage forms for API with narrow therapeutic index. Wickström et al. used an unmodified TIJ Canon Pixma desktop printer to print ODFs for pediatric administration [33]. A multicomponent formulation comprising B, B1, B2, B3, and B6 vitamins was printing on rice paper (Easybake® edible rice paper) that was used as a film substrate. The technology was validated for the accuracy and reproducibility of the printed doses.

More recently, Tam et al. developed paracetamol ODFs for point-of-care applications using HPMC as ink component [155]. The technology used for this study differed from typical thermal inkjet printing as it incorporates a piezoelectric micro-dispensing system to overcome viscosity limitations. Another advantage of the printed ODFs is that there is no need to use film substrates as they are formed during the dispensing. Even at high ink viscosities (32–818 mPa·s), it was possible to develop transparent films with homogenous distribution of materials.

Inkjet printing has been used for water-based inks and the development of pharmaceutical dosage forms [156]. A Fujifilm Dimatix printer was employed to develop polyvinylpyrrolidone and thiamine hydrochloride inks for the printing of tablet forms. Interestingly, the printing process optimization prevented the recrystallization of theophylline in the PVP matrix, and the dissolution rates were fast and were not related to the number of printed layers (Figure 12).

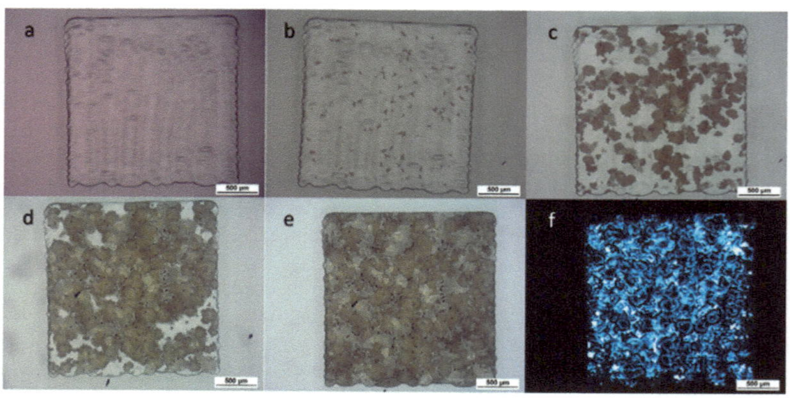

Figure 12. Optical reflection images of printed films on glass slide: (**a**) 30 min, (**b**) 1 day, (**c**) 4 days, (**d**) 8 days, (**e**) 12 days, (**f**) 12 days after orienting. All confirm the formation of a crystalline phase (cross-polarized transmission OM). Scale bar = 500 μm [156].

By developing macrogol inks on edible substrates, Thabet at al. prepared ODFs of hydrochlorothiazide and enalapril maleate [157]. The dynamic viscosities varied from 7 to 17 mPa.s through the addition of small ethanol amounts. More importantly, the careful selection of formulation composition prevented the drug's recrystallization on the edible substrates without affecting the mechanical properties as well.

Aerosol drug delivery has also been explored by scientists using TIJP technology. A combination of thermal inkjet printing and spray freeze drying (TIZ-SFD) was applied to formulate inhalable powders of terbutaline sulphate and compared with the marketed Bricanyl product [158]. By modifying a Hewlett-Packard thermal printer, it was feasible to atomize terbutaline sulphate (excipient-free) solution incorporated in liquid nitrogen followed by freeze drying of the produced droplets afterwards. The process resulted in spherical and porous particles with a volume median diameter of 14.1 ± 0.8 μm and mass median aerodynamic diameter of 4.0 μm, respectively. The measured fines in the TIZ-SFD process were found to be 22.9%, in contrast with the 25.7 μm of the Bricanyl Turbohaler.

Similarly, it was demonstrated that TIZ-SFD is capable of processing up to 15% *w/v* salbutamol sulphate (SS) solution and producing droplets of around 35 μm diameter. The samples analyzed with a twin-stage impinger showed 24.0 ± 1.2% and 26.4 ± 2.2% fine particle fraction. The result is scalable, and TIZ-SFD showed better outcome for inhalable particle formulation in comparison to standard Cyclocap® [159].

One of the great advantages of TIJP is the feasibility of processing a range of materials and hence producing multimaterial objects with complex geometries. Lion et al. investigated the development of multilayer dosage forms with tailored drug doses and release profiles using solvent-free thermal inkjet printing [160]. Using Compritol HD5 ATO as the core material, they printed multilayer structures of complex geometry (Figure 13) that could tune the release of Fenofibrate (loadings varied from 5–30%) and provide both sustained or immediate release rates. The printer features allow the production of droplets of 30 pL in volume. The printing consistency facilitated the fabrication of honeycomb internal integrity and various channel diameters.

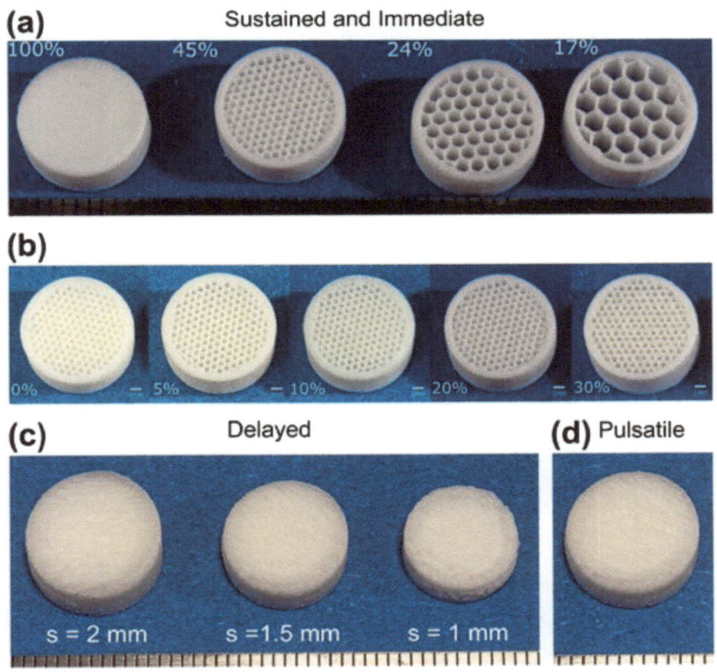

Figure 13. Images of sustained- and immediate-release tablets (**a**) with constant drug content (10% drug loading) and varying infill and (**b**) with varying drug content (5–30% drug loading) and constant infill (45%). (**c**) Delayed-release tablet with 1, 1.5 and 2 mm shell thickness (10% drug loaded core), and (**d**) pulsatile-release tablets. Ruler unit: 1 mm [160].

Table 4 summarizes typical examples of TIJP that have been investigated for printing oral solid dosage forms using thermal inkjet printing.

Table 4. TIJ printing examples used in printing of oral solid dosage forms.

Sl No.	Printer	Dosage Form	Ink Material	Ref.
1.	Hewlett-Packard HP460 Deskjet	Cocrystal	carbamazepine, nicotinamide, benzoic acid, isonicotinamide, theophylline and saccharin	[150]
2.	HP Photosmart B010	Cocrystal	riboflavin sodium phosphate and paracetamol	[151]
3.	Combination of thermal inkjet printhead HP_{11} and ZCorp Spectrum Z510	Tablets	acetaminophen	[152]
4.	Hewlett-Packard 970 Cxi DeskJet	Tablets	prednisolone	[86]
5.	Hewlett Packard Deskjet 1000	Buccal film	lysozyme	[153]
6.	Hewlett Packard Deskjet 1000	Buccal film	lysozyme and ribonuclease-A	[154]
7.	HP Deskjet D1660	Oral film	salbutamol sulphate	[100]
8.	HP 5940 Deskjet	Orodispersible films	warfarin	[99]
9.	TIJ Canon Pixma (unmodified)	Orodispersible films	vitamin B B1, B2, B3, and B6	[33]
10.	Nanojet Piezo Valve NJ-K-4020	Orodispersible films	paracetamol	[155]
11.	Fujifilm Dimatix DMP-2850 Series	Tablet	polyvinylpyrrolidone and thiamine hydrochloride	[156]
12.	PIXDRO JS 20	Orodispersible films	hydrochlorothiazide and enalapril maleate	[157]
13.	TIZ-SFD	Powder particle	terbutaline sulphate	[158]
14.	TIZ-SFD	Powder particle	salbutamol sulphate	[159]

3.3. Antimicrobial Resistance Control

Antimicrobial therapies are not limited to bacterial infections [161,162] but are also used in the treatment of cancer [163–165], Alzheimer's disease [166] and other neurological disorders [167,168]. In this situation of the rapidly growing use of antimicrobial therapies, antimicrobial resistance is also increasing at a similar pace. Sadly, the dramatically increasing rate of antimicrobial resistance has become a global concern [169–172].

Therefore, to overcome the issues associated with the inappropriate administration of antimicrobial drugs, MIC (minimum inhibitory concentration) is being assessed these days [173–176]. MIC assessment as a quantitative analysis determines the lowest concentration of an antimicrobial therapeutic by which a specific microorganism's growth can be inhibited [175,177]. However, the conventional techniques available for MIC assessment such as agar and broth dilution and broth microdilution have drawbacks, i.e., trouble attaining different therapeutic concentrations in extensive scale and room for potential errors [97,176]. Hence, automated technologies like thermal inkjet printing are being explored in this area. Since any technique for MIC assessment should have the properties of accuracy, controlled deposition and high throughput, TIJP is the best match for performing MIC. Moreover, TIJP requires only tiny volume of sample which is complimentary in this case [101,133,178].

A group of researchers used a Hewlett Packard (HP) 5940 Deskjet thermal inkjet printer and broth microdilution for their studies to evaluate the MICs of a few antibiotics, amoxicillin, ampicillin, doxycycline and tetracycline (all of them were 92.5–100.5% pure) against Lactobacillus acidophilus. Data obtained from their experiment (see Table 5) show that the MICs for the tested antibiotics were within the acceptable range for TIJP, in contrast with broth microdilution [97].

Table 5. Calculated MICs of antibiotics against Lactobacillus acidophilus determined via thermal inkjet printing and broth microdilution.

SN	Antibiotic	Thermal Inkjet Printed MIC(μg/ mL)	Broth Microdilution MIC(μg/mL)
1.	Amoxicillin	0.20	0.5
		0.23	0.5
		0.15	0.5
		0.19	0.5
2.	Ampicillin	0.12	0.25
		0.12	0.25
		0.15	0.25
3.	Doxycycline	0.29	1
		0.31	1
		0.29	1
		0.35	1
4.	Tetracycline	0.59	2
		0.55	2

4. Conclusions

Thermal inkjet printing has found several applications in the fields of tissue engineering and pharmaceutics as it is a versatile technology. There are several remarkable studies in which TIJP has been used for the development of cell-laden bioinks with excellent viability and tissue regeneration. This is not always the case, and therefore, print process optimization is a prerequisite, including understanding the effect on the cell viability and proliferation. Nevertheless, TIJP can be used to investigate new materials or combinations thereof with unique properties such as biocompatibility and high print resolution. A major advantage of the technology is the precise and accurate printing of cells in comparison with cell seeding.

Furthermore, it is a very promising technology that can lead to the commercialization of various pharmaceutical products, especially for personalized dosage forms at the point of care. However, there are several considerations that need to be taken into account. TIJP is not very easy to operate and requires significant expertise by the operator including a good understanding of troubleshooting. Specific attention should be given in the development of printable inks and the selection of the drug carriers. For the first, the choice of liquid ink is important for ensuring complete drug solubilization and fast drying times, which quite often limit the applicability of the technology: Large doses with long dry times will increase production times and thus limit the manufacturability of the technology. The selection of the polymers is equally important as they should ensure the stability of the embedded drug in the matrix, prevent recrystallization and definitely improve or tune the dissolution rates of the drug substances. Additionally, it is clear that the regulatory environment is not yet developed for entirely flexible dosage and patient-adaptable multidrug drugs, and this means that these components of printed drug delivery devices will require careful attention and experience difficulties. Quality control is a crucial component that has not been addressed yet; there are no studies in this direction. In our opinion, TIJP could be applicable for the printing of potent APIs in small doses and particularly for the printing of QR codes. Nevertheless, there is still much work to go before we witness the full exploitation of TIJP at commercial scale or at the point of care such as in clinical pharmacies.

Author Contributions: Conceptualization, M.J.U. and D.D.; resources, M.J.U. and D.D.; writing—original draft preparation, M.J.U., J.H. and D.D.; writing—review and editing, M.J.U., J.H. and D.D.; supervision, M.J.U. and D.D.; project administration, M.J.U. and D.D.; funding acquisition, M.J.U. and D.D. All authors have read and agreed to the published version of the manuscript.

Funding: This research received no external funding.

Institutional Review Board Statement: Not applicable.

Informed Consent Statement: Not applicable.

Data Availability Statement: There are no additional raw data for this paper. The paper only uses secondary data from published papers, and all credits for these data have been made via citations and copyright permissions.

Conflicts of Interest: The authors declare no conflict of interest.

References

1. Alomari, M.; Mohamed, F.H.; Basit, A.W.; Gaisford, S. Personalised Dosing: Printing a Dose of One's Own Medicine. *Int. J. Pharm.* **2015**, *494*, 568–577. [CrossRef] [PubMed]
2. Douroumis, D. *Hot-Melt Extrusion: Pharmaceutical Applications*; Wiley: Hoboken, NJ, USA, 2012.
3. Douroumis, D.; Fahr, A. *Drug Delivery Strategies for Poorly Water-Soluble Drugs*; John Wiley & Sons, Ltd.: Hoboken, NJ, USA, 2013. [CrossRef]
4. Vaz, V.M.; Kumar, L. 3D Printing as a Promising Tool in Personalized Medicine. *AAPS PharmSciTech* **2021**, *22*, 49. [CrossRef] [PubMed]
5. Scoutaris, N.; Alexander, M.R.; Gellert, P.R.; Roberts, C.J. Inkjet Printing as a Novel Medicine Formulation Technique. *J. Control. Release* **2011**, *156*, 179–185. [CrossRef]
6. Ginsburg, G.S.; McCarthy, J.J. Personalized Medicine: Revolutionizing Drug Discovery and Patient Care. *Trends Biotechnol.* **2001**, *19*, 491–496. [CrossRef]
7. Brittain, H.K.; Scott, R.; Thomas, E. The Rise of the Genome and Personalised Medicine. *Clin. Med.* **2017**, *17*, 545. [CrossRef] [PubMed]
8. Mathur, S.; Sutton, J. Personalized Medicine Could Transform Healthcare. *Biomed. Rep.* **2017**, *7*, 3. [CrossRef]
9. Goetz, L.H.; Schork, N.J. Personalized Medicine: Motivation, Challenges and Progress. *Fertil. Steril.* **2018**, *109*, 952. [CrossRef] [PubMed]
10. Azizi Machekposhti, S.; Mohaved, S.; Narayan, R.J. Inkjet Dispensing Technologies: Recent Advances for Novel Drug Discovery. *Expert Opin. Drug Discov.* **2019**, *14*, 101–113. [CrossRef]
11. Fang, M.; Li, T.; Zhang, S.; Rao, K.V.; Belova, L. Design and Tailoring of Inks for Inkjet Patterning of Metal Oxides. *R. Soc. Open Sci.* **2020**, *7*, 200242. [CrossRef]
12. Gao, Y.; Shi, W.; Wang, W.; Leng, Y.; Zhao, Y. Inkjet Printing Patterns of Highly Conductive Pristine Graphene on Flexible Substrates. *Ind. Eng. Chem. Res.* **2014**, *53*, 16777–16784. [CrossRef]
13. Shah, M.A.; Lee, D.G.; Lee, B.Y.; Hur, S. Classifications and Applications of Inkjet Printing Technology: A Review. *IEEE Access* **2021**, *9*, 140079–140102. [CrossRef]
14. Evans, S.E.; Harrington, T.; Rodriguez Rivero, M.C.; Rognin, E.; Tuladhar, T.; Daly, R. 2D and 3D Inkjet Printing of Biopharmaceuticals—A Review of Trends and Future Perspectives in Research and Manufacturing. *Int. J. Pharm.* **2021**, *599*, 120443. [CrossRef]
15. Daly, R.; Harrington, T.S.; Martin, G.D.; Hutchings, I.M. Inkjet Printing for Pharmaceutics—A Review of Research and Manufacturing. *Int. J. Pharm.* **2015**, *494*, 554–567. [CrossRef]
16. Zhu, X.; Zheng, Q.; Yang, H.; Cai, J.; Huang, L.; Duan, Y.; Xu, Z.; Cen, P. Recent Advances in Inkjet Dispensing Technologies: Applications in Drug Discovery. *Expert Opin. Drug Discov.* **2012**, *7*, 761–770. [CrossRef]
17. Uddin, M.J.; Scoutaris, N.; Klepetsanis, P.; Chowdhry, B.; Prausnitz, M.R.; Douroumis, D. Inkjet Printing of Transdermal Microneedles for the Delivery of Anticancer Agents. *Int. J. Pharm.* **2015**, *494*, 593–602. [CrossRef]
18. Ross, S.; Scoutaris, N.; Lamprou, D.; Mallinson, D.; Douroumis, D. Inkjet Printing of Insulin Microneedles for Transdermal Delivery. *Drug Deliv. Transl. Res.* **2015**, *5*, 451–461. [CrossRef]
19. Ihalainen, P.; Määttänen, A.; Sandler, N. Printing Technologies for Biomolecule and Cell-Based Applications. *Int. J. Pharm.* **2015**, *494*, 585–592. [CrossRef]

20. Zheng, Q.; Lu, J.; Chen, H.; Huang, L.; Cai, J.; Xu, Z. Application of Inkjet Printing Technique for Biological Material Delivery and Antimicrobial Assays. *Anal. Biochem.* **2011**, *410*, 171–176. [CrossRef]
21. Preis, M.; Breitkreutz, J.; Sandler, N. Perspective: Concepts of Printing Technologies for Oral Film Formulations. *Int. J. Pharm.* **2015**, *494*, 578–584. [CrossRef]
22. Scoutaris, N.; Ross, S.; Douroumis, D. Current Trends on Medical and Pharmaceutical Applications of Inkjet Printing Technology. *Pharm. Res.* **2016**, *33*, 1799–1816. [CrossRef]
23. Thabet, Y.; Sibanc, R.; Breitkreutz, J. Printing Pharmaceuticals by Inkjet Technology: Proof of Concept for Stand-Alone and Continuous in-Line Printing on Orodispersible Films. *J. Manuf. Process.* **2018**, *35*, 205–215. [CrossRef]
24. Visser, J.C.; Wibier, L.; Kiefer, O.; Orlu, M.; Breitkreutz, J.; Woerdenbag, H.J.; Taxis, K. A Pediatrics Utilization Study in The Netherlands to Identify Active Pharmaceutical Ingredients Suitable for Inkjet Printing on Orodispersible Films. *Pharmaceutics* **2020**, *12*, 164. [CrossRef]
25. Khan, S.; Ali, S.; Bermak, A. Smart Manufacturing Technologies for Printed Electronics. *Hybrid Nanomater. Flex. Electron. Mater.* **2019**. [CrossRef]
26. Boland, T.; Xu, T.; Damon, B.; Cui, X. Application of Inkjet Printing to Tissue Engineering. *Biotechnol. J.* **2006**, *1*, 910–917. [CrossRef]
27. Saunders, R.E.; Derby, B. Inkjet Printing Biomaterials for Tissue Engineering: Bioprinting. *Int. Mater. Rev.* **2014**, *59*, 430–448. [CrossRef]
28. De Gans, B.J.; Duineveld, P.C.; Schubert, U.S. Inkjet Printing of Polymers: State of the Art and Future Developments. *Adv. Mater.* **2004**, *16*, 203–213. [CrossRef]
29. Komuro, N.; Takaki, S.; Suzuki, K.; Citterio, D. Inkjet Printed (Bio)Chemical Sensing Devices. *Anal. Bioanal. Chem.* **2013**, *405*, 5785–5805. [CrossRef]
30. Delaney, J.T.; Smith, P.J.; Schubert, U.S. Inkjet Printing of Proteins. *Soft Matter* **2009**, *5*, 4866–4877. [CrossRef]
31. Tekin, E.; Smith, P.J.; Schubert, U.S. Inkjet Printing as a Deposition and Patterning Tool for Polymers and Inorganic Particles. *Soft Matter* **2008**, *4*, 703–713. [CrossRef]
32. Moya, A.; Gabriel, G.; Villa, R.; Javier del Campo, F. Inkjet-Printed Electrochemical Sensors. *Curr. Opin. Electrochem.* **2017**, *3*, 29–39. [CrossRef]
33. Wickström, H.; Nyman, J.O.; Indola, M.; Sundelin, H.; Kronberg, L.; Preis, M.; Rantanen, J.; Sandler, N. Colorimetry as Quality Control Tool for Individual Inkjet-Printed Pediatric Formulations. *AAPS PharmSciTech* **2017**, *18*, 293–302. [CrossRef] [PubMed]
34. Vakili, H.; Wickström, H.; Desai, D.; Preis, M.; Sandler, N. Application of a Handheld NIR Spectrometer in Prediction of Drug Content in Inkjet Printed Orodispersible Formulations Containing Prednisolone and Levothyroxine. *Int. J. Pharm.* **2017**, *524*, 414–423. [CrossRef]
35. Singh, M.; Haverinen, H.M.; Dhagat, P.; Jabbour, G.E. Inkjet Printing—Process and Its Applications. *Adv. Mater.* **2010**, *22*, 673–685. [CrossRef] [PubMed]
36. Lim, J.A.; Lee, W.H.; Lee, H.S.; Lee, J.H.; Park, Y.D.; Cho, K. Self-Organization of Ink-Jet-Printed Triisopropylsilylethynyl Pentacene via Evaporation-Induced Flows in a Drying Droplet. *Adv. Funct. Mater.* **2008**, *18*, 229–234. [CrossRef]
37. Farid, M. A New Approach to Modelling of Single Droplet Drying. *Chem. Eng. Sci.* **2003**, *58*, 2985–2993. [CrossRef]
38. Boel, E.; Koekoekx, R.; Dedroog, S.; Babkin, I.; Vetrano, M.R.; Clasen, C.; Van den Mooter, G. Unraveling Particle Formation: From Single Droplet Drying to Spray Drying and Electrospraying. *Pharmaceutics* **2020**, *12*, 625. [CrossRef] [PubMed]
39. Vehring, R. Pharmaceutical Particle Engineering via Spray Drying. *Pharm. Res.* **2008**, *25*, 999–1022. [CrossRef]
40. Singh, A.; Van den Mooter, G. Spray Drying Formulation of Amorphous Solid Dispersions. *Adv. Drug Deliv. Rev.* **2016**, *100*, 27–50. [CrossRef]
41. Mezhericher, M.; Levy, A.; Borde, I. Modelling the Morphological Evolution of Nanosuspension Droplet in Constant-Rate Drying Stage. *Chem. Eng. Sci.* **2011**, *66*, 884–896. [CrossRef]
42. de Souza Lima, R.; Ré, M.I.; Arlabosse, P. Drying Droplet as a Template for Solid Formation: A Review. *Powder Technol.* **2020**, *359*, 161–171. [CrossRef]
43. Mezhericher, M.; Levy, A.; Borde, I. Modelling of Particle Breakage during Drying. *Chem. Eng. Process. Process Intensif.* **2008**, *47*, 1404–1411. [CrossRef]
44. Perdana, J.; Bereschenko, L.; Fox, M.B.; Kuperus, J.H.; Kleerebezem, M.; Boom, R.M.; Schutyser, M.A.I. Dehydration and Thermal Inactivation of Lactobacillus Plantarum WCFS1: Comparing Single Droplet Drying to Spray and Freeze Drying. *Food Res. Int.* **2013**, *54*, 1351–1359. [CrossRef]
45. Mezhericher, M.; Levy, A.; Borde, I. Spray Drying Modelling Based on Advanced Droplet Drying Kinetics. *Chem. Eng. Process. Process Intensif.* **2010**, *49*, 1205–1213. [CrossRef]
46. Machekposhti, S.A.; Movahed, S.; Narayan, R.J. Physicochemical Parameters That Underlie Inkjet Printing for Medical Applications. *Biophys. Rev.* **2020**, *1*, 011301. [CrossRef]
47. Sohrabi, S.; Liu, Y. Modeling Thermal Inkjet and Cell Printing Process Using Modified Pseudopotential and Thermal Lattice Boltzmann Methods. *Phys. Rev. E* **2018**, *97*, 033105. [CrossRef]
48. Lee, S.W.; Kim, H.C.; Kuk, K.; Oh, Y.S. A Monolithic Inkjet Print Head: DomeJet. *Sens. Actuators A Phys.* **2002**, *95*, 114–119. [CrossRef]
49. Kholghi Eshkalak, S.; Chinnappan, A.; Jayathilaka, W.A.D.M.; Khatibzadeh, M.; Kowsari, E.; Ramakrishna, S. A Review on Inkjet Printing of CNT Composites for Smart Applications. *Appl. Mater. Today* **2017**, *9*, 372–386. [CrossRef]

50. Yin, Z.P.; Huang, Y.A.; Bu, N.B.; Wang, X.M.; Xiong, Y.L. Inkjet Printing for Flexible Electronics: Materials, Processes and Equipments. *Chin. Sci. Bull.* **2010**, *55*, 3383–3407. [CrossRef]
51. Andò, B.; Baglio, S.; Bulsara, A.R.; Emery, T.; Marletta, V.; Pistorio, A. Low-Cost Inkjet Printing Technology for the Rapid Prototyping of Transducers. *Sensors* **2017**, *17*, 748. [CrossRef]
52. Matsuda, Y.; Shibayama, S.; Uete, K.; Yamaguchi, H.; Niimi, T. Electric Conductive Pattern Element Fabricated Using Commercial Inkjet Printer for Paper-Based Analytical Devices. *Anal. Chem.* **2015**, *87*, 5762–5765. [CrossRef]
53. Huang, Q.; Shen, W.; Xu, Q.; Tan, R.; Song, W. Properties of Polyacrylic Acid-Coated Silver Nanoparticle Ink for Inkjet Printing Conductive Tracks on Paper with High Conductivity. *Mater. Chem. Phys.* **2014**, *147*, 550–556. [CrossRef]
54. Huang, Q.; Shen, W.; Xu, Q.; Tan, R.; Song, W. Room-Temperature Sintering of Conductive Ag Films on Paper. *Mater. Lett.* **2014**, *123*, 124–127. [CrossRef]
55. Garcia, A.; Hanifi, N.; Jousselme, B.; Jégou, P.; Palacin, S.; Viel, P.; Berthelot, T. Polymer Grafting by Inkjet Printing: A Direct Chemical Writing Toolset. *Adv. Funct. Mater.* **2013**, *23*, 3668–3674. [CrossRef]
56. Moon, S.J.; Robin, M.; Wenlin, K.; Yann, M.; Bae, B.S.; Mohammed-Brahim, T.; Jacques, E.; Harnois, M. Morphological Impact of Insulator on Inkjet-Printed Transistor. *Flex. Print. Electron.* **2017**, *2*, 035008. [CrossRef]
57. Ando, B.; Baglio, S. All-Inkjet Printed Strain Sensors. *IEEE Sens. J.* **2013**, *13*, 4874–4879. [CrossRef]
58. Salaoru, I.; Zhou, Z.; Morris, P.; Gibbons, G.J. Inkjet-Printed Polyvinyl Alcohol Multilayers. *J. Vis. Exp.* **2017**, *2017*, 55093. [CrossRef]
59. Genina, N.; Fors, D.; Vakili, H.; Ihalainen, P.; Pohjala, L.; Ehlers, H.; Kassamakov, I.; Haeggström, E.; Vuorela, P.; Peltonen, J.; et al. Tailoring Controlled-Release Oral Dosage Forms by Combining Inkjet and Flexographic Printing Techniques. *Eur. J. Pharm. Sci.* **2012**, *47*, 615–623. [CrossRef]
60. Genina, N.; Janßen, E.M.; Breitenbach, A.; Breitkreutz, J.; Sandler, N. Evaluation of Different Substrates for Inkjet Printing of Rasagiline Mesylate. *Eur. J. Pharm. Biopharm.* **2013**, *85*, 1075–1083. [CrossRef]
61. Sandler, N.; Määttänen, A.; Ihalainen, P.; Kronberg, L.; Meierjohann, A.; Viitala, T.; Peltonen, J. Inkjet Printing of Drug Substances and Use of Porous Substrates-towards Individualized Dosing. *J. Pharm. Sci.* **2011**, *100*, 3386–3395. [CrossRef]
62. McManus, D.; Vranic, S.; Withers, F.; Sanchez-Romaguera, V.; Macucci, M.; Yang, H.; Sorrentino, R.; Parvez, K.; Son, S.K.; Iannaccone, G.; et al. Water-Based and Biocompatible 2D Crystal Inks for All-Inkjet-Printed Heterostructures. *Nat. Nanotechnol.* **2017**, *12*, 343–350. [CrossRef]
63. Reis, N.; Ainsley, C.; Derby, B. Ink-Jet Delivery of Particle Suspensions by Piezoelectric Droplet Ejectors. *J. Appl. Phys.* **2005**, *97*, 094903. [CrossRef]
64. Derby, B. Inkjet Printing Ceramics: From Drops to Solid. *J. Eur. Ceram. Soc.* **2011**, *31*, 2543–2550. [CrossRef]
65. Jang, D.; Kim, D.; Moon, J. Influence of Fluid Physical Properties on Ink-Jet Printability. *Langmuir* **2009**, *25*, 2629–2635. [CrossRef]
66. Wijshoff, H. Drop Dynamics in the Inkjet Printing Process. *Curr. Opin. Colloid Interface Sci.* **2018**, *36*, 20–27. [CrossRef]
67. Van Der Bos, A.; Van Der Meulen, M.J.; Driessen, T.; Van Den Berg, M.; Reinten, H.; Wijshoff, H.; Versluis, M.; Lohse, D. Velocity Profile inside Piezoacoustic Inkjet Droplets in Flight: Comparison between Experiment and Numerical Simulation. *Phys. Rev. Appl.* **2014**, *1*, 014004. [CrossRef]
68. Liu, Y.; Derby, B. Experimental Study of the Parameters for Stable Drop-on-Demand Inkjet Performance. *Phys. Fluids* **2019**, *31*, 032004. [CrossRef]
69. Choi, H.W.; Zhou, T.; Singh, M.; Jabbour, G.E. Recent Developments and Directions in Printed Nanomaterials. *Nanoscale* **2015**, *7*, 3338–3355. [CrossRef]
70. Padilla-Martinez, J.P.; Ramirez-San-Juan, J.C.; Berrospe-Rodriguez, C.; Korneev, N.; Aguilar, G.; Zaca-Moran, P.; Ramos-Garcia, R. Controllable Direction of Liquid Jets Generated by Thermocavitation within a Droplet. *Appl. Opt.* **2017**, *56*, 7167. [CrossRef]
71. Oktavianty, O.; Haruyama, S.; Ishii, Y. Enhancing Droplet Quality of Edible Ink in Single and Multi-Drop Methods by Optimization the Waveform Design of DoD Inkjet Printer. *Processes* **2022**, *10*, 91. [CrossRef]
72. Saleh, E.; Woolliams, P.; Clarke, B.; Gregory, A.; Greedy, S.; Smartt, C.; Wildman, R.; Ashcroft, I.; Hague, R.; Dickens, P.; et al. 3D Inkjet-Printed UV-Curable Inks for Multi-Functional Electromagnetic Applications. *Addit. Manuf.* **2017**, *13*, 143–148. [CrossRef]
73. Barlow, N.E.; Kusumaatmaja, H.; Salehi-Reyhani, A.; Brooks, N.; Barter, L.M.C.; Flemming, A.J.; Ces, O. Measuring Bilayer Surface Energy and Curvature in Asymmetric Droplet Interface Bilayers. *J. R. Soc. Interface* **2018**, *15*, 20180610. [CrossRef]
74. Mypati, S.; Dhanushkodi, S.R.; McLaren, M.; Docoslis, A.; Peppley, B.A.; Barz, D.P.J. Optimized Inkjet-Printed Silver Nanoparticle Films: Theoretical and Experimental Investigations. *RSC Adv.* **2018**, *8*, 19679–19689. [CrossRef]
75. Pandiyan, S.; El-Kharouf, A.; Steinberger-Wilckens, R. Formulation of Spinel Based Inkjet Inks for Protective Layer Coatings in SOFC Interconnects. *J. Colloid Interface Sci.* **2020**, *579*, 82–95. [CrossRef]
76. Abdolmaleki, H.; Agarwala, S. PVDF-$BaTiO_3$ Nanocomposite Inkjet Inks with Enhanced β-Phase Crystallinity for Printed Electronics. *Polymers* **2020**, *12*, 2430. [CrossRef]
77. Xu, C.; An, C.; Long, Y.; Li, Q.; Guo, H.; Wang, S.; Wang, J. Inkjet Printing of Energetic Composites with High Density. *RSC Adv.* **2018**, *8*, 35863–35869. [CrossRef]
78. Zhu, Z.; Gong, Z.; Qu, P.; Li, Z.; Rasaki, S.A.; Liu, Z.; Wang, P.; Liu, C.; Lao, C.; Chen, Z. Additive Manufacturing of Thin Electrolyte Layers via Inkjet Printing of Highly-Stable Ceramic Inks. *J. Adv. Ceram.* **2021**, *10*, 279–290. [CrossRef]
79. Li, C.; Shi, H.; Ran, R.; Su, C.; Shao, Z. Thermal Inkjet Printing of Thin-Film Electrolytes and Buffering Layers for Solid Oxide Fuel Cells with Improved Performance. *Int. J. Hydrogen Energy* **2013**, *38*, 9310–9319. [CrossRef]

80. Kolakovic, R.; Viitala, T.; Ihalainen, P.; Genina, N.; Peltonen, J.; Sandler, N. Printing Technologies in Fabrication of Drug Delivery Systems. *Expert Opin. Drug Deliv.* **2013**, *10*, 1711–1723. [CrossRef]
81. Lau, G.K.; Shrestha, M. Ink-Jet Printing of Micro-Electro-Mechanical Systems (MEMS). *Micromachines* **2017**, *8*, 194. [CrossRef]
82. Kumar, P.; Ebbens, S.; Zhao, X. Inkjet Printing of Mammalian Cells—Theory and Applications. *Bioprinting* **2021**, *23*, e00157. [CrossRef]
83. Alamán, J.; Alicante, R.; Peña, J.I.; Sánchez-Somolinos, C. Inkjet Printing of Functional Materials for Optical and Photonic Applications. *Materials* **2016**, *9*, 910. [CrossRef]
84. Murphy, S.V.; Atala, A. 3D Bioprinting of Tissues and Organs. *Nat. Biotechnol.* **2014**, *32*, 773–785. [CrossRef]
85. Calvert, P. Materials Science. Printing Cells. *Science* **2007**, *318*, 208–209. [CrossRef]
86. Gaisford, S. 3D Printed Pharmaceutical Products. *3D Print. Med.* **2017**, 155–166. [CrossRef]
87. Özkol, E.; Ebert, J.; Uibel, K.; Wätjen, A.M.; Telle, R. Development of High Solid Content Aqueous 3Y-TZP Suspensions for Direct Inkjet Printing Using a Thermal Inkjet Printer. *J. Eur. Ceram. Soc.* **2009**, *29*, 403–409. [CrossRef]
88. Mannerbro, R.; Ranlöf, M.; Robinson, N.; Forchheimer, R. Inkjet Printed Electrochemical Organic Electronics. *Synth. Met.* **2008**, *158*, 556–560. [CrossRef]
89. Li, X.; Liu, B.; Pei, B.; Chen, J.; Zhou, D.; Peng, J.; Zhang, X.; Jia, W.; Xu, T. Inkjet Bioprinting of Biomaterials. *Chem. Rev.* **2020**, *120*, 10793–10833. [CrossRef]
90. Uzun, S.; Schelling, M.; Hantanasirisakul, K.; Mathis, T.S.; Askeland, R.; Dion, G.; Gogotsi, Y. Additive-Free Aqueous MXene Inks for Thermal Inkjet Printing on Textiles. *Small* **2021**, *17*, 2006376. [CrossRef]
91. Lee, S.; Byun, D.; Jung, D.; Choi, J.; Kim, Y.; Yang, J.H.; Son, S.U.; Tran, S.B.Q.; Ko, H.S. Pole-Type Ground Electrode in Nozzle for Electrostatic Field Induced Drop-on-Demand Inkjet Head. *Sensors Actuators A Phys.* **2008**, *141*, 506–514. [CrossRef]
92. Khan, A.; Rahman, K.; Hyun, M.-T.; Kim, D.-S.; Choi, K.-H.; Khan, A.; Rahman, K.; Hyun, M.; Choi, K.; Kim, D. Multi-Nozzle Electrohydrodynamic Inkjet Printing of Silver Colloidal Solution for the Fabrication of Electrically Functional Microstructures. *Appl. Phys. A* **2011**, *104*, 1113–1120. [CrossRef]
93. Khan, A.; Rahman, K.; Kim, D.S.; Choi, K.H. Direct Printing of Copper Conductive Micro-Tracks by Multi-Nozzle Electrohydrodynamic Inkjet Printing Process. *J. Mater. Process. Technol.* **2012**, *212*, 700–706. [CrossRef]
94. Wijshoff, H. The Dynamics of the Piezo Inkjet Printhead Operation. *Phys. Rep.* **2010**, *491*, 77–177. [CrossRef]
95. Gudapati, H.; Dey, M.; Ozbolat, I. A Comprehensive Review on Droplet-Based Bioprinting: Past, Present and Future. *Biomaterials* **2016**, *102*, 20–42. [CrossRef] [PubMed]
96. Ferris, C.J.; Gilmore, K.G.; Wallace, G.G.; In Het Panhuis, M. Biofabrication: An Overview of the Approaches Used for Printing of Living Cells. *Appl. Microbiol. Biotechnol.* **2013**, *97*, 4243–4258. [CrossRef]
97. Dodoo, C.C.; Alomari, M.; Basit, A.W.; Stapleton, P.; Gaisford, S. A Thermal Ink-Jet Printing Approach for Evaluating Susceptibility of Bacteria to Antibiotics. *J. Microbiol. Methods* **2019**, *164*, 105660. [CrossRef]
98. Prasad, L.K.; Smyth, H. 3D Printing Technologies for Drug Delivery: A Review. *Drug Dev. Ind. Pharm.* **2016**, *42*, 1019–1031. [CrossRef]
99. Vuddanda, P.R.; Alomari, M.; Dodoo, C.C.; Trenfield, S.J.; Velaga, S.; Basit, A.W.; Gaisford, S. Personalisation of Warfarin Therapy Using Thermal Ink-Jet Printing. *Eur. J. Pharm. Sci.* **2018**, *117*, 80–87. [CrossRef]
100. Buanz, A.B.M.; Saunders, M.H.; Basit, A.W.; Gaisford, S. Preparation of Personalized-Dose Salbutamol Sulphate Oral Films with Thermal Ink-Jet Printing. *Pharm. Res.* **2011**, *28*, 2386–2392. [CrossRef]
101. Alper, J. Biology and the Inkjets. *Science* **2004**, *305*, 1895. [CrossRef]
102. Wilson, W.C.; Boland, T. Cell and Organ Printing 1: Protein and Cell Printers. *Anat. Rec. Part A Discov. Mol. Cell. Evol. Biol.* **2003**, *272*, 491–496. [CrossRef]
103. Lemmo, A.V.; Rose, D.J.; Tisone, T.C. Inkjet Dispensing Technology: Applications in Drug Discovery. *Curr. Opin. Biotechnol.* **1998**, *9*, 615–617. [CrossRef]
104. Xu, T.; Jin, J.; Gregory, C.; Hickman, J.J.; Boland, T. Inkjet Printing of Viable Mammalian Cells. *Biomaterials* **2005**, *26*, 93–99. [CrossRef] [PubMed]
105. Meléndez, P.A.; Kane, K.M.; Ashvar, C.S.; Albrecht, M.; Smith, P.A. Thermal Inkjet Application in the Preparation of Oral Dosage Forms: Dispensing of Prednisolone Solutions and Polymorphic Characterization by Solid-State Spectroscopic Techniques. *J. Pharm. Sci.* **2008**, *97*, 2619–2636. [CrossRef] [PubMed]
106. Patra, S.; Young, V. A Review of 3D Printing Techniques and the Future in Biofabrication of Bioprinted Tissue. *Cell Biochem. Biophys.* **2016**, *74*, 93–98. [CrossRef]
107. Xu, T.; Zhao, W.; Zhu, J.M.; Albanna, M.Z.; Yoo, J.J.; Atala, A. Complex Heterogeneous Tissue Constructs Containing Multiple Cell Types Prepared by Inkjet Printing Technology. *Biomaterials* **2013**, *34*, 130–139. [CrossRef]
108. Gao, G.; Cui, X. Three-Dimensional Bioprinting in Tissue Engineering and Regenerative Medicine. *Biotechnol. Lett.* **2016**, *38*, 203–211. [CrossRef]
109. Zhou, H.; Gué, A.M. Simulation Model and Droplet Ejection Performance of a Thermal-Bubble Microejector. *Sens. Actuators B Chem.* **2010**, *145*, 311–319. [CrossRef]
110. Setti, L.; Piana, C.; Bonazzi, S.; Ballarin, B.; Frascaro, D.; Fraleoni-Morgera, A.; Giuliani, S. Thermal Inkjet Technology for the Microdeposition of Biological Molecules as a Viable Route for the Realization of Biosensors. *Anal. Lett.* **2007**, *37*, 1559–1570. [CrossRef]

111. Cui, X.; Dean, D.; Ruggeri, Z.M.; Boland, T. Cell Damage Evaluation of Thermal Inkjet Printed Chinese Hamster Ovary Cells. *Biotechnol. Bioeng.* **2010**, *106*, 963–969. [CrossRef]
112. Cui, X.; Boland, T.; D'Lima, D.D.; Lotz, M.K. Thermal Inkjet Printing in Tissue Engineering and Regenerative Medicine. *Recent Pat. Drug Deliv. Formul.* **2012**, *6*, 149. [CrossRef]
113. Agarwal, S.; Saha, S.; Balla, V.K.; Pal, A.; Barui, A.; Bodhak, S. Current Developments in 3D Bioprinting for Tissue and Organ Regeneration–A Review. *Front. Mech. Eng.* **2020**, *6*, 589171. [CrossRef]
114. Dey, M.; Ozbolat, I.T. 3D Bioprinting of Cells, Tissues and Organs. *Sci. Rep.* **2020**, *10*, 14023. [CrossRef] [PubMed]
115. Khanna, A.; Ayan, B.; Undieh, A.A.; Yang, Y.P.; Huang, N.F. Advances in Three-Dimensional Bioprinted Stem Cell-Based Tissue Engineering for Cardiovascular Regeneration. *J. Mol. Cell. Cardiol.* **2022**, *169*, 13–27. [CrossRef] [PubMed]
116. Jiang, T.; Munguia-Lopez, J.G.; Flores-Torres, S.; Kort-Mascort, J.; Kinsella, J.M. Extrusion Bioprinting of Soft Materials: An Emerging Technique for Biological Model Fabrication. *Appl. Phys. Rev.* **2019**, *6*, 011310. [CrossRef]
117. Ng, W.L.; Lee, J.M.; Zhou, M.; Chen, Y.W.; Lee, K.X.A.; Yeong, W.Y.; Shen, Y.F. Vat Polymerization-Based Bioprinting—Process, Materials, Applications and Regulatory Challenges. *Biofabrication* **2020**, *12*, 022001. [CrossRef] [PubMed]
118. Khan, M.S.; Fon, D.; Li, X.; Tian, J.; Forsythe, J.; Garnier, G.; Shen, W. Biosurface Engineering through Ink Jet Printing. *Colloids Surf. B. Biointerfaces* **2010**, *75*, 441–447. [CrossRef]
119. Christensen, K.; Xu, C.; Chai, W.; Zhang, Z.; Fu, J.; Huang, Y. Freeform Inkjet Printing of Cellular Structures with Bifurcations. *Biotechnol. Bioeng.* **2015**, *112*, 1047–1055. [CrossRef]
120. Setti, L.; Fraleoni-Morgera, A.; Ballarin, B.; Filippini, A.; Frascaro, D.; Piana, C. An Amperometric Glucose Biosensor Prototype Fabricated by Thermal Inkjet Printing. *Biosens. Bioelectron.* **2005**, *20*, 2019–2026. [CrossRef]
121. Ng, W.L.; Huang, X.; Shkolnikov, V.; Goh, G.L.; Suntornnond, R.; Yeong, W.Y. Controlling Droplet Impact Velocity and Droplet Volume: Key Factors to Achieving High Cell Viability in Sub-Nanoliter Droplet-Based Bioprinting. *Int. J. Bioprinting* **2021**, *8*, 424. [CrossRef]
122. Park, J.A.; Lee, H.R.; Park, S.Y.; Jung, S. Self-Organization of Fibroblast-Laden 3D Collagen Microstructures from Inkjet-Printed Cell Patterns. *Adv. Biosyst.* **2020**, *4*, 1900280. [CrossRef] [PubMed]
123. Suntornnond, R.; Ng, W.L.; Huang, X.; Yeow, C.H.E.; Yeong, W.Y. Improving Printability of Hydrogel-Based Bio-Inks for Thermal Inkjet Bioprinting Applications via Saponification and Heat Treatment Processes. *J. Mater. Chem. B* **2022**, *10*, 5989–6000. [CrossRef] [PubMed]
124. Sun, M.; Liu, S.; Zhang, Y.; Liu, M.; Yi, X.; Hu, J. Insights into the Saponification Process of Di(2-Ethylhexyl) Phosphoric Acid Extractant: Thermodynamics and Structural Aspects. *J. Mol. Liq.* **2019**, *280*, 252–258. [CrossRef]
125. Tan, H.W.; Abdul Aziz, A.R.; Aroua, M.K. Glycerol Production and Its Applications as a Raw Material: A Review. *Renew. Sustain. Energy Rev.* **2013**, *27*, 118–127. [CrossRef]
126. Yoon, S.; Park, J.A.; Lee, H.R.; Yoon, W.H.; Hwang, D.S.; Jung, S. Inkjet–Spray Hybrid Printing for 3D Freeform Fabrication of Multilayered Hydrogel Structures. *Adv. Healthc. Mater.* **2018**, *7*, 1800050. [CrossRef]
127. Freeman, S.; Ramos, R.; Alexis Chando, P.; Zhou, L.; Reeser, K.; Jin, S.; Soman, P.; Ye, K. A Bioink Blend for Rotary 3D Bioprinting Tissue Engineered Small-Diameter Vascular Constructs. *Acta Biomater.* **2019**, *95*, 152–164. [CrossRef] [PubMed]
128. Solis, L.H.; Ayala, Y.; Portillo, S.; Varela-Ramirez, A.; Aguilera, R.; Boland, T. Thermal Inkjet Bioprinting Triggers the Activation of the VEGF Pathway in Human Microvascular Endothelial Cells in Vitro. *Biofabrication* **2019**, *11*, 045005. [CrossRef]
129. Gao, G.; Hubbell, K.; Schilling, A.F.; Dai, G.; Cui, X. Bioprinting Cartilage Tissue from Mesenchymal Stem Cells and PEG Hydrogel. *Methods Mol. Biol.* **2017**, *1612*, 391–398. [CrossRef]
130. Cui, X.; Breitenkamp, K.; Lotz, M.; D'Lima, D. Synergistic Action of Fibroblast Growth Factor-2 and Transforming Growth Factor-Beta1 Enhances Bioprinted Human Neocartilage Formation. *Biotechnol. Bioeng.* **2012**, *109*, 2357–2368. [CrossRef]
131. Kador, K.E.; Grogan, S.P.; Dorthé, E.W.; Venugopalan, P.; Malek, M.F.; Goldberg, J.L.; D'Lima, D.D. Control of Retinal Ganglion Cell Positioning and Neurite Growth: Combining 3D Printing with Radial Electrospun Scaffolds. *Tissue Eng. Part A* **2016**, *22*, 286–294. [CrossRef]
132. Gao, G.; Schilling, A.F.; Yonezawa, T.; Wang, J.; Dai, G.; Cui, X. Bioactive Nanoparticles Stimulate Bone Tissue Formation in Bioprinted Three-Dimensional Scaffold and Human Mesenchymal Stem Cells. *Biotechnol. J.* **2014**, *9*, 1304–1311. [CrossRef]
133. Allain, L.R.; Stratis-Cullum, D.N.; Vo-Dinh, T. Investigation of Microfabrication of Biological Sample Arrays Using Piezoelectric and Bubble-Jet Printing Technologies. *Anal. Chim. Acta* **2004**, *518*, 77–85. [CrossRef]
134. Roth, E.A.; Xu, T.; Das, M.; Gregory, C.; Hickman, J.J.; Boland, T. Inkjet Printing for High-Throughput Cell Patterning. *Biomaterials* **2004**, *25*, 3707–3715. [CrossRef]
135. Xu, T.; Rohozinski, J.; Zhao, W.; Moorefield, E.C.; Atala, A.; Yoo, J.J. Inkjet-Mediated Gene Transfection into Living Cells Combined with Targeted Delivery. *Tissue Eng. Part A* **2008**, *15*, 95–101. [CrossRef]
136. Cui, X.; Gao, G.; Qiu, Y. Accelerated Myotube Formation Using Bioprinting Technology for Biosensor Applications. *Biotechnol. Lett.* **2012**, *35*, 315–321. [CrossRef]
137. Klebe, R.J. Cytoscribing: A Method for Micropositioning Cells and the Construction of Two- and Three-Dimensional Synthetic Tissues. *Exp. Cell Res.* **1988**, *179*, 362–373. [CrossRef]
138. Okamoto, T.; Suzuki, T.; Yamamoto, N. Microarray Fabrication with Covalent Attachment of DNA Using Bubble Jet Technology. *Nat. Biotechnol.* **2000**, *18*, 438–441. [CrossRef]

139. Cui, X.; Boland, T. Human Microvasculature Fabrication Using Thermal Inkjet Printing Technology. *Biomaterials* **2009**, *30*, 6221–6227. [CrossRef]
140. Khan, M.S.; Li, X.; Shen, W.; Garnier, G. Thermal Stability of Bioactive Enzymatic Papers. *Colloids Surf. B Biointerfaces* **2010**, *75*, 239–246. [CrossRef]
141. Nandi, U.; Trivedi, V.; Ross, S.A.; Douroumis, D. Advances in Twin-Screw Granulation Processing. *Pharmaceutics* **2021**, *13*, 624. [CrossRef]
142. Douroumis, D.; Ross, S.A.; Nokhodchi, A. Advanced Methodologies for Cocrystal Synthesis. *Adv. Drug Deliv. Rev.* **2017**, *117*, 178–195. [CrossRef]
143. Preis, M.; Pein, M.; Breitkreutz, J. Development of a Taste-Masked Orodispersible Film Containing Dimenhydrinate. *Pharmaceutics* **2012**, *4*, 551. [CrossRef] [PubMed]
144. Mostafaei, A.; Elliott, A.M.; Barnes, J.E.; Li, F.; Tan, W.; Cramer, C.L.; Nandwana, P.; Chmielus, M. Binder Jet 3D Printing—Process Parameters, Materials, Properties, Modeling, and Challenges. *Prog. Mater. Sci.* **2021**, *119*, 100707. [CrossRef]
145. Sen, K.; Mehta, T.; Sansare, S.; Sharifi, L.; Ma, A.W.K.; Chaudhuri, B. Pharmaceutical Applications of Powder-Based Binder Jet 3D Printing Process—A Review. *Adv. Drug Deliv. Rev.* **2021**, *177*, 113943. [CrossRef]
146. Chang, S.Y.; Li, S.W.; Kowsari, K.; Shetty, A.; Sorrells, L.; Sen, K.; Nagapudi, K.; Chaudhuri, B.; Ma, A.W.K. Binder-Jet 3D Printing of Indomethacin-Laden Pharmaceutical Dosage Forms. *J. Pharm. Sci.* **2020**, *109*, 3054–3063. [CrossRef] [PubMed]
147. Rahman, Z.; Charoo, N.A.; Kuttolamadom, M.; Asadi, A.; Khan, M.A. Printing of Personalized Medication Using Binder Jetting 3D Printer. *Precis. Med. Investig. Pract. Provid.* **2020**, 473–481. [CrossRef]
148. Hong, X.; Han, X.; Li, X.; Li, J.; Wang, Z.; Zheng, A. Binder Jet 3D Printing of Compound LEV-PN Dispersible Tablets: An Innovative Approach for Fabricating Drug Systems with Multicompartmental Structures. *Pharmaceutics* **2021**, *13*, 1780. [CrossRef]
149. Kozakiewicz-Latała, M.; Nartowski, K.P.; Dominik, A.; Malec, K.; Gołkowska, A.M.; Złocińska, A.; Rusińska, M.; Szymczyk-Ziółkowska, P.; Ziółkowski, G.; Górniak, A.; et al. Binder Jetting 3D Printing of Challenging Medicines: From Low Dose Tablets to Hydrophobic Molecules. *Eur. J. Pharm. Biopharm.* **2022**, *170*, 144–159. [CrossRef] [PubMed]
150. Buanz, A.B.M.; Telford, R.; Scowen, I.J.; Gaisford, S. Rapid Preparation of Pharmaceutical Co-Crystals with Thermal Ink-Jet Printing. *CrystEngComm* **2013**, *15*, 1031–1035. [CrossRef]
151. Takala, M.; Helkiö, H.; Sundholm, J.; Genina, N.; Kivikuoma, P.; Widmaier, T.; Sandler, N.; Kuosmanen, P. Ink-Jet Printing of Pharmaceuticals. In Proceedings of the 8th International DAAAM Baltic Conference "INDUSTRIAL ENGINEERING", Tallinn, Estonia, 19–21 April 2012; Volume 6.
152. Wilts, E.M.; Ma, D.; Bai, Y.; Williams, C.B.; Long, T.E. Comparison of Linear and 4-Arm Star Poly(Vinyl Pyrrolidone) for Aqueous Binder Jetting Additive Manufacturing of Personalized Dosage Tablets. *ACS Appl. Mater. Interfaces* **2019**, *11*, 23938–23947. [CrossRef]
153. Montenegro-Nicolini, M.; Reyes, P.E.; Jara, M.O.; Vuddanda, P.R.; Neira-Carrillo, A.; Butto, N.; Velaga, S.; Morales, J.O. The Effect of Inkjet Printing over Polymeric Films as Potential Buccal Biologics Delivery Systems. *AAPS PharmSciTech* **2018**, *19*, 3376–3387. [CrossRef]
154. Montenegro-Nicolini, M.; Miranda, V.; Morales, J.O. Inkjet Printing of Proteins: An Experimental Approach. *AAPS J.* **2016**, *19*, 234–243. [CrossRef]
155. Tam, C.H.; Alexander, M.; Belton, P.; Qi, S. Drop-on-Demand Printing of Personalised Orodispersible Films Fabricated by Precision Micro-Dispensing. *Int. J. Pharm.* **2021**, *610*, 121279. [CrossRef] [PubMed]
156. Cader, H.K.; Rance, G.A.; Alexander, M.R.; Gonçalves, A.D.; Roberts, C.J.; Tuck, C.J.; Wildman, R.D. Water-Based 3D Inkjet Printing of an Oral Pharmaceutical Dosage Form. *Int. J. Pharm.* **2019**, *564*, 359–368. [CrossRef]
157. Thabet, Y.; Lunter, D.; Breitkreutz, J. Continuous Inkjet Printing of Enalapril Maleate onto Orodispersible Film Formulations. *Int. J. Pharm.* **2018**, *546*, 180–187. [CrossRef] [PubMed]
158. Sharma, G.; Mueannoom, W.; Buanz, A.B.M.; Taylor, K.M.G.; Gaisford, S. In Vitro Characterisation of Terbutaline Sulphate Particles Prepared by Thermal Ink-Jet Spray Freeze Drying. *Int. J. Pharm.* **2013**, *447*, 165–170. [CrossRef]
159. Mueannoom, W.; Srisongphan, A.; Taylor, K.M.G.; Hauschild, S.; Gaisford, S. Thermal Ink-Jet Spray Freeze-Drying for Preparation of Excipient-Free Salbutamol Sulphate for Inhalation. *Eur. J. Pharm. Biopharm.* **2012**, *80*, 149–155. [CrossRef]
160. Lion, A.; Wildman, R.D.; Alexander, M.R.; Roberts, C.J. Customisable Tablet Printing: The Development of Multimaterial Hot Melt Inkjet 3D Printing to Produce Complex and Personalised Dosage Forms. *Pharmaceutic* **2021**, *13*, 1679. [CrossRef]
161. Makabenta, J.M.V.; Nabawy, A.; Li, C.H.; Schmidt-Malan, S.; Patel, R.; Rotello, V.M. Nanomaterial-Based Therapeutics for Antibiotic-Resistant Bacterial Infections. *Nat. Rev. Microbiol.* **2020**, *19*, 23–36. [CrossRef] [PubMed]
162. Coates, A.R.M.; Hu, Y. Novel Approaches to Developing New Antibiotics for Bacterial Infections. *Br. J. Pharmacol.* **2007**, *152*, 1147–1154. [CrossRef]
163. Mohamed, A.; Menon, H.; Chulkina, M.; Yee, N.S.; Pinchuk, I.V. Drug–Microbiota Interaction in Colon Cancer Therapy: Impact of Antibiotics. *Biomedicines* **2021**, *9*, 259. [CrossRef]
164. Helmink, B.A.; Khan, M.A.W.; Hermann, A.; Gopalakrishnan, V.; Wargo, J.A. The Microbiome, Cancer, and Cancer Therapy. *Nat. Med.* **2019**, *25*, 377–388. [CrossRef]
165. Roy, S.; Trinchieri, G. Microbiota: A Key Orchestrator of Cancer Therapy. *Nat. Rev. Cancer* **2017**, *17*, 271–285. [CrossRef]
166. Kumari, S.; Deshmukh, R. β-Lactam Antibiotics to Tame down Molecular Pathways of Alzheimer's Disease. *Eur. J. Pharmacol.* **2021**, *895*, 173877. [CrossRef]

167. Cryan, J.F.; O'Riordan, K.J.; Sandhu, K.; Peterson, V.; Dinan, T.G. The Gut Microbiome in Neurological Disorders. *Lancet Neurol.* **2020**, *19*, 179–194. [CrossRef]
168. Yimer, E.M.; Hishe, H.Z.; Tuem, K.B. Repurposing of the β-Lactam Antibiotic, Ceftriaxone for Neurological Disorders: A Review. *Front. Neurosci.* **2019**, *13*, 2260–2271. [CrossRef]
169. Spellberg, B.; Gilbert, D.N. The Future of Antibiotics and Resistance: A Tribute to a Career of Leadership by John Bartlett. *Clin. Infect. Dis.* **2014**, *59*, S71–S75. [CrossRef] [PubMed]
170. Frieri, M.; Kumar, K.; Boutin, A. Antibiotic Resistance. *J. Infect. Public Health* **2017**, *10*, 369–378. [CrossRef]
171. Cars, O.; Chandy, S.J.; Mpundu, M.; Peralta, A.Q.; Zorzet, A.; So, A.D. Resetting the Agenda for Antibiotic Resistance through a Health Systems Perspective. *Lancet Glob. Health* **2021**, *9*, e1022–e1027. [CrossRef]
172. Aslam, B.; Wang, W.; Arshad, M.I.; Khurshid, M.; Muzammil, S.; Rasool, M.H.; Nisar, M.A.; Alvi, R.F.; Aslam, M.A.; Qamar, M.U.; et al. Antibiotic Resistance: A Rundown of a Global Crisis. *Infect. Drug Resist.* **2018**, *11*, 1645. [CrossRef]
173. Nguyen, M.; Brettin, T.; Long, S.W.; Musser, J.M.; Olsen, R.J.; Olson, R.; Shukla, M.; Stevens, R.L.; Xia, F.; Yoo, H.; et al. Developing an in Silico Minimum Inhibitory Concentration Panel Test for Klebsiella Pneumoniae. *Sci. Rep.* **2018**, *8*, 421. [CrossRef] [PubMed]
174. Lamy, B.; Carret, G.; Flandrois, J.P.; Delignette-Muller, M.L. How Does Susceptibility Prevalence Impact on the Performance of Disk Diffusion Susceptibility Testing? *Diagn. Microbiol. Infect. Dis.* **2004**, *49*, 131–139. [CrossRef] [PubMed]
175. Kowalska-Krochmal, B.; Dudek-Wicher, R. The Minimum Inhibitory Concentration of Antibiotics: Methods, Interpretation, Clinical Relevance. *Pathogens* **2021**, *10*, 165. [CrossRef] [PubMed]
176. Andrews, J.M. Determination of Minimum Inhibitory Concentrations. *J. Antimicrob. Chemother.* **2001**, *48*, 5–16. [CrossRef] [PubMed]
177. Michael, A.; Kelman, T.; Pitesky, M. Overview of Quantitative Methodologies to Understand Antimicrobial Resistance via Minimum Inhibitory Concentration. *Animals* **2020**, *10*, 1405. [CrossRef]
178. Derby, B. Bioprinting: Inkjet Printing Proteins and Hybrid Cell-Containing Materials and Structures. *J. Mater. Chem.* **2008**, *18*, 5717–5721. [CrossRef]

technologies

MDPI

Review

Synthetic Micro/Nanomotors for Drug Delivery

Eduardo Guzmán [1,2,*] **and Armando Maestro** [3,4]

1 Departamento de Química Física, Facultad de Ciencias Químicas, Universidad Complutense de Madrid, Ciudad Universitaria s/n, 28040 Madrid, Spain
2 Instituto Pluridisciplinar, Universidad Complutense de Madrid, Paseo Juan XXIII 1, 28040 Madrid, Spain
3 Centro de Física de Materiales (CSIC, UPV/EHU), Paseo Manuel de Lardizabal 5, 20018 San Sebastián, Spain
4 IKERBASQUE—Basque Foundation for Science, Plaza Euskadi 5, 48009 Bilbao, Spain
* Correspondence: eduardogs@quim.ucm.es; Tel.: +34-9-1394-4107

Abstract: Synthetic micro/nanomotors (MNMs) are human-made machines characterized by their capacity for undergoing self-propelled motion as a result of the consumption of chemical energy obtained from specific chemical or biochemical reactions, or as a response to an external actuation driven by a physical stimulus. This has fostered the exploitation of MNMs for facing different biomedical challenges, including drug delivery. In fact, MNMs are superior systems for an efficient delivery of drugs, offering several advantages in relation to conventional carriers. For instance, the self-propulsion ability of micro/nanomotors makes possible an easier transport of drugs to specific targets in comparison to the conventional distribution by passive carriers circulating within the blood, which enhances the drug bioavailability in tissues. Despite the promising avenues opened by the use of synthetic micro/nanomotors in drug delivery applications, the development of systems for in vivo uses requires further studies to ensure a suitable biocompatibility and biodegradability of the fabricated engines. This is essential for guaranteeing the safety of synthetic MNMs and patient convenience. This review provides an updated perspective to the potential applications of synthetic micro/nanomotors in drug delivery. Moreover, the most fundamental aspects related to the performance of synthetic MNMs and their biosafety are also discussed.

Keywords: biomedicine; drug delivery; machines; micro/nanomotors; nanomedicine; propulsion

check for updates

Citation: Guzmán, E.; Maestro, A. Synthetic Micro/Nanomotors for Drug Delivery. *Technologies* **2022**, *10*, 96. https://doi.org/10.3390/technologies10040096

Academic Editors: Manoj Gupta, Eugene Wong and Gwanggil Jeon

Received: 5 July 2022
Accepted: 12 August 2022
Published: 17 August 2022

1. Introduction

Molecular biological motors have inspired research on the seeking of self-propelled supramolecular engines, so-called synthetic micro/nanomotors (MNMs). These offer a broad range of applications in different fields, including drug delivery, environmental remediation, biosensing, and precision surgery at the micro-/nanoscale. In fact, synthetic micro/nanomotors are structures characterized by reduced dimensionality and have the ability to convert energy obtained from diverse sources into kinetic energy, i.e., into motion. Therefore, it is of a paramount importance to deepen the understanding of the mechanisms driving the energy conversion in synthetic MNMs to ensure an accurate control over the motion of the manufactured devices [1]. This is essential for the fabrication of suitable micro/nanomotors for biomedical applications.

The potential biomedical applications of synthetic micro/nanomotors have opened new avenues for overcoming some of the current challenges of precision medicine, e.g., low permeation of the biological barriers, towards a more efficient and personalized treatment of different diseases [2]. Furthermore, synthetic micro/nanomotors add to the characteristics of traditional drug delivery systems, e.g., drug protection, selectivity and biocompatibility, and the capability of swimming and penetration through cellular barriers [1]. Therefore, it may be expected that synthetic MNMs present the capability for encapsulation, transport, and supply of drugs directly to the disease sites, which in turn contributes to enhancing the therapeutic efficiency and decreasing the systemic side effects associated with the administration of toxic drugs [3].

There are available a broad range of fabrication methods enabling the fabrication of synthetic micro/nanomotors with different chemical composition and geometries, e.g., nanowires, microtubular microrockets, Janus microspheres, and supramolecular-based motors [4]. Nevertheless, the most common micro/nanomotors are miniaturized human-made engines characterized by an asymmetric structure and/or chemical heterogeneity. This enables the rupture of the thermodynamic balance, which must be considered an essential aspect for driving the conversion of the energy from chemical reactions or external sources to mechanical motion [5–7]. This motion results from an intricate balance between the drag force and the propulsion, operating over the micro/nanomotors, which guides the motion displacement under low Reynolds number conditions [8,9]. This is possible by using different propulsion mechanisms, based on both endogenous (i.e., chemotaxis) or exogenous (e.g., ultrasound, magnetic fields, light) stimuli, which can allow simultaneously the transport of drugs to specific targets and a triggered release at the right time [10,11]. Moreover, MNMs can be also functionalized using different strategies, enabling their use for in vivo imaging [12]. Therefore, nanotechnology and nanomaterials can contribute to finding solutions to several challenges in the precise therapy for different diseases, and the design of smart synthetic micro/nanomotors has opened up exciting opportunities that can contribute to solving complex problems that are not always easy to address with conventional approaches [13].

The last decade has been very fruitful in the fabrication of synthetic micro/nanomotors with high biocompatibility, multifunctionality, and efficient propulsion in biological fluids, which has provided the bases for closing the gap between lab-scale studies and the potential in vivo biomedical application of synthetic motors [14]. Nevertheless, the optimal application in the biomedical field and practical clinical translation of synthetic micro/nanomotors requires understanding of the interactions of these untethered tiny machines with the immune system [15,16]. It may be expected that the entrance of the micro/nanomotors into the bloodstream can lead to undesirable interactions with the immune cells, which in turn may hinder the capacity of the engines to reach their targets and fulfill their task [17]. Therefore, it is necessary to evaluate the biocompatibility of the materials constituting the micro/nanomotors and the energy sources [18]. Furthermore, the physico-chemical properties of the synthetic MNMs should be modulated in such a way that can preclude their clearance from the hosts [19].

This review tries to guide to the reader along the applications of synthetic micro/nanomotors to address some of the current challenges in the drug delivery field, providing an updated perspective on the potential interest in these human-made engines as tools for improving the efficiency of the treatment of different diseases. This requires a careful analysis of the characteristics and properties of this type of system. For this purpose, the review starts with two general sections devoted to the main physico-chemical aspects influencing the fabrication and characteristics of the synthetic MNMs, including their chemical and morphological characteristics, and the effect of the environmental conditions (viscosity of the medium and temperature) on their motion. Then, a detailed discussion of the main mechanisms exploited for guiding the motion of MNMs is included. The last part of the work presents a discussion of some important aspects related to the biosafety of MNMs followed by the introduction of some examples of MNMs exploited for specific drug delivery applications.

2. Designing Micro/Nanomotors

The geometry of MNMs is essential for controlling the flow field and pressure distribution during motor propulsion, and hence the optimization of the design of MNMs has driven extensive research activity trying to find the most suitable conditions for the dynamic performance of this type of engine [20]. For instance, the velocity of the motor motion depends on an intricate balance involving the drag forces and the driving forces acting on the motor, with the increase of the former leading to an increase in the motor speed. On the other side, if both forces reach an equilibrium point, the velocity of the motor motion reaches its maximum speed. Therefore, it is essential to design MNMs with an optimal geometry for optimizing the motion pathway and maximum velocity of the manufactured engines.

2.1. Tubular and Rod Motors

Two different geometrical aspects are of a paramount importance in the performance of asymmetric motors: (i) semi-cone angle, and (ii) aspect ratio (ratio between the length and the larger radius) [20]. The former parameter presents importance only in tubular motors, affecting the drag coefficient. For instance, conical shaped motors can reach higher speeds than cylindrical ones with similar aspect ratios [8]. This can be understood as the aspect ratio, which depends on the geometry and chemical properties, of concave surfaces being smaller than that corresponding to convex ones [21]. The importance of the geometrical characteristics of the motor was proven by Wang et al. [22], who showed that the velocity of the engines was significantly increased as the semi-cone angle increases. This is the result of a most favored detachment of the bubbles from concave surfaces in comparison to convex ones. Moreover, the semi-cone angle modifies the aspect ratio, which influences the size of the produced bubbles and their production frequency and enlarges the contact area, favoring the progress of catalytic reactions. This is important because a recent theory suggests that the speed of synthetic motors can be approximately defined by the product of the bubble radius and the generation frequency. Therefore, it is essential to optimize the aspect ratio for controlling the motor speed [23]. According to Li et al. [8], two different regimes may be expected for the dependence of the drag coefficient on the aspect ratio. Thus, when the aspect ratio remains below three, a decrease of the drag coefficient can occur with the increase of the aspect ratio, whereas the drag coefficient increases with the aspect ratio for values of the latter above six. This indicates that the aspect ratio affects the characteristics of the motor motion. In fact, conical motors undergo a faster motion than motors with any other geometry. This can be explained considering the decrease of the drag force acting on the motor as the aspect ratio decreases. Therefore, the propulsion efficiency can be ensured by increasing the semi-cone angle and the aspect ratio in such a way that the drag force operating over the motors can be minimized.

2.2. Janus Motors

Janus motors can be prepared with a broad range of structures, including bimetallic structures, shells, and capsules. In the case of bimetallic Janus motors, their motion is commonly the result of a catalytic reaction between the motor surface and the environment. This is the result of the generation of bubbles as a consequence of the catalytic reaction, and their subsequent ejection, which push the motion forward in the motors. The motors can be pulled backward as a result of the instantaneous depression occurring when the bubbles burst [24,25].

The speed of Janus motors can be tuned by modifying their shape. For instance, the use of multilayered Janus hollow capsules can result in a new type of self-propelled engine that can undergo a motion equivalent to 125 times their main dimension per second (about 1 mm/s). Another alternative to increase the velocity of the motor motion is the use of nanoshell motors [21].

2.3. Roughness

The origin of the high speed associated with the motion of Janus motors can be found in their inherent surface roughness, which increases the specific area involved in the catalytic process. The work by Orozco et al. [26] reported that the increase of the surface roughness plays an important role in the motion of synthetic motors, controlling the speed of Janus motors, in agreement with the findings of Jurado-Sánchez et al. [27]. They reported that the increase of the roughness of the motors increases the dimension of the catalytic layer and contributes to an efficient bubble generation and propulsion. The control of the surface roughness as a strategy for modulating the motor speed was also applied to tubular motors. However, the role of the roughness in the motion of the latter is less intuitive because the speed depends on an intricate balance between two forces. The first is the driving force, which contributes to the decomposition of the fuel in the inner region of the motor, and the second is the friction force operating within the outer surface of the motor [28]. This latter contribution is increased with the surface roughness and tends to reduce the speed of the motor motion.

In summary, for Janus motors, the surface roughness can contribute to an effective increase of the speed of the motion. However, the roughness can also introduce undesired friction contributions to the motion, which in turn reduces the speed of the motion.

3. Environmental Factors Affecting the Motor Motion

3.1. Viscosity

The viscous properties of the fluid surrounding the MNMs play an essential role on their propulsion and speed. For instance, the increase of the environmental viscosity increases the strength of the drag forces operating over the motors, which is opposed to the motion and reduces the motor velocity. Wang et al. [29] reported on the existence of a linear relationship between the velocity of microrockets and the viscosity of the solution. Thus, the increase of the viscosity of the medium reduces the diameter of the produced bubbles and their generation rate, which in turn results in a reduction of the speed of the motion.

The Reynolds number can also affect the motor motion as a result of its dependence on the viscosity. At the highest viscosities (low Reynolds number conditions), the motor motion is commonly linear, whereas it becomes circular as the viscosity decreases and the Reynolds number increases. Moreover, the increase of the viscosity reduces the velocity of the motion due to the increase of the drag force operating over the motor [30].

3.2. Temperature

In general, the velocity of the motor increases with temperature, which allows modulating the efficiency of the motor motion. This is, in part, related to the decrease of the solution viscosity as demonstrated by the linearity of the motion at low temperature (higher viscosity) and its circular character at high temperature (lower viscosity) [31].

4. Powering the Motion of Micro/Nanomotors

4.1. Endogenous Powered Micro/Nanomotors

Endogenous powered self-propelled micro/nanomachines exploit the energy obtained from specific chemical or biochemical reactions for driving their motion [32]. This requires combining two components: a catalytic material (generally a metal or enzyme) and an inert one. Thus, it is possible to fabricate an asymmetric supramolecular architecture that is coated with a specific catalytic material to ensure a continuous energy input from the environment. The asymmetry is essential for the performance of chemically/biochemically propelled MNMs because it ensures the existence of an asymmetry field across the supramolecular system that contributes to power its motion by the generation of local gradients of electrical potential and/or concentration and gas bubbles as the result of surface reactions [33]. Therefore, it is possible to define up to three different mechanisms of propulsion based on chemical reactions. This depends on the mechanism used for the conversion of chemical energy into kinetic energy: self-electrophoresis, self-diffusiophoresis, and bubble propulsion [34]. Self-electrophoresis is associated with an asymmetric production and consumption of ions surrounding the motors. This results in a local electric field guiding the motion. In the case of the self-diffusiophoresis, it is the asymmetry of the chemical species around the motors that is the driving force of the motion. Finally, the diffusion mediated by the generation of gas bubbles is associated with the recoil force of the chemical reactions resulting from the production and growth of bubbles that are separated from the motors, pushing their motion.

A popular alternative is exploiting redox reactions for driving the motion of micro/nanorobots, with the decomposition reaction of hydrogen peroxide being a very frequently approach. This takes advantage of the instability of the hydrogen peroxide, which eases its decomposition into water and oxygen with the participation of several types of catalysts, including metals, enzymes, or alkaline environments. Furthermore, the decomposition of hydrogen peroxide has been widely used as fuel for motors based on bimetallic nanorods, Janus particles, and polymer vesicles [35–37]. These are typical examples of self-electrophoresis–powered motors where the chemical gradient originated as a result of the reaction induces a local electric field, which guides the motor motion. Thus, considering the motion of bimetallic (gold-platinum) Janus nanorods in hydrogen peroxide, a redox reaction of the hydrogen peroxide can be expected, resulting in a production of protons at the platinum end (anode) and their consumption at the gold end (cathode), producing protons. This results in an asymmetric proton distribution within the Janus particle, which induces a local electrical field from the platinum end to the gold one, driving the motion of such electric fields [38].

Solute concentration gradients can be also exploited as an effective method for guiding the micro/nanomotor motion (self-diffusiophoresis). This depends on the nature of the specific solute, which can lead to two different situations: electrolyte diffusiophoresis and non-electrolyte diffusiophoresis [39,40]. Electrolyte diffusiophoresis occurs when the surface reactions produce different anions and cations, which diffuse very differently in the solution. This induces a local electric field that pushes the motion of the motors. In contrast, the non-electrolyte diffusiophoresis relies on the release of non-ionic products in the solution [41].

The propulsion due to the formation of gas bubbles is very common when micro/nanomotors having large catalytic surfaces are considered [20,42]. Thus, it is possible to push the directional motion of the motors as a result of a surface catalytic reaction, which drives the decomposition of the hydrogen peroxide into O_2 or H_2. In fact, when the concentration of gas accumulated within the motor overcomes its solubility limit, it will overflow as bubbles, pushing the motor motion [26]. It should be noted that the motion speed depends on the concentration of hydrogen peroxide, and hence the higher the hydrogen peroxide concentration, the higher the motion speed. This was demonstrated by Gao et al. [43], who fabricated microtubular engines consisting of a platinum cylinder covered by a poly(aniline) layer 8 μm of length, which can be self-propelled at high speed (around 350 times the microtubule length per second). Furthermore, this type of motor can move even at low concentrations of hydrogen peroxide (0.2% v/v). Indeed, the hydrogen peroxide concentration allows modulating the motor speed by tuning the radius of the bubbles and their production frequency. For instance, the increase of the bubble sizes coupled to the decrease of their production frequency leads to a decrease of the speed of the motor motion. On the other hand, the bubble-continuous medium interfacial tension also influences the velocity of the microtubular motors. In fact, the smaller the surface tension, the smaller the bubble size and the higher the production frequency, which in turn enhances the mobility of the motors.

Self-propelled Layer-by-Layer (LbL) Janus capsules obtained by covering with Au and Ni layers the hemispheres of silicon dioxide particles coated with five bilayers formed by the alternate deposition of poly(4-styrenesulfonate of sodium) (PSS) and poly(allylamine hydrochloride) (PAH) have been used for drug delivery applications. For this purpose, catalase was conjugated to the Au layer for driving the catalytic breakdown of hydrogen peroxide to produce gas bubbles that can push the motor motion. The combination of this motion with the magnetic field guidance allows guiding the obtained particles to the target cells, where it is possible to release the encapsulated drugs upon Near Infrared (NIR) irradiation [36]. Figure 1 shows a sketch of the triggered transport and release of doxorubicin (DOX) using LbL Janus capsule motors.

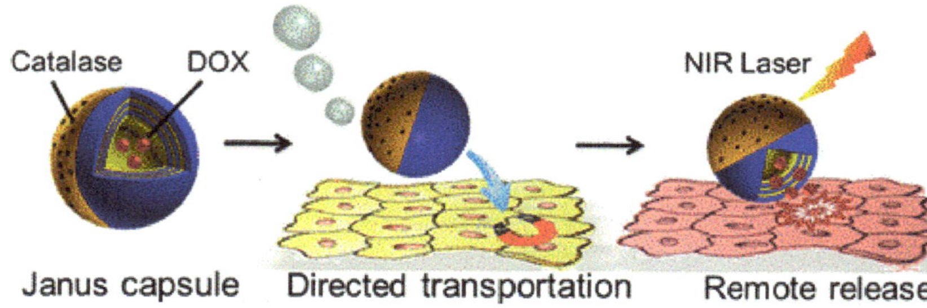

Figure 1. Sketch of the triggered transport and release of DOX using LbL Janus capsule-based motors. Reprinted from Wu et al. [36], with permission from the American Chemical Society. Copyright (2014).

It should be stressed that the use of high concentrations of hydrogen peroxide or acid as fuel in chemically driven micro/nanomotors is not suitable when biomedical applications are considered due to the well-known oxidative toxicity of such chemicals [44]. Therefore, it is required to seek more convenient fuels to propel motors when biological media are involved. This can be partially solved by using magnesium-based motors, which present a high biocompatibility and can react with water to produce gas bubbles [6]. This was exploited by Mou et al. [45] for fabricating hemocompatible Janus motors with one hemisphere coated with Pt and a second one with Mg. This type of motor can be propelled by the hydrogen bubbles generated as a result of the reaction between the water and the magnesium, which can lead to the mechanism as summarized in Figure 2.

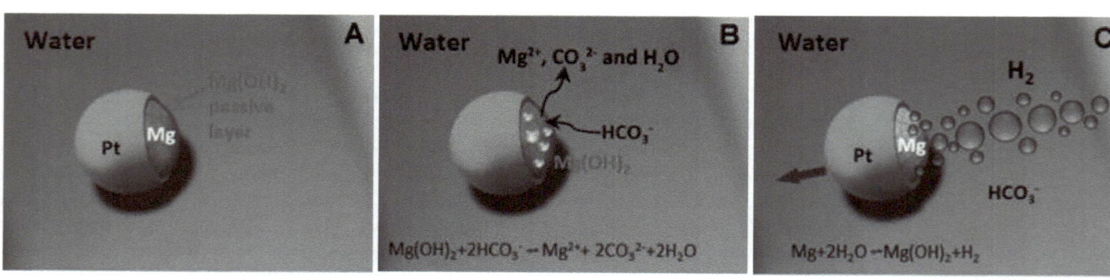

Figure 2. Schematic representation of the mechanism driving the motion of Janus motors of Mg and Pt: (**A**) Formation of a passivation layer formed by $Mg(OH)_2$ as a result of the magnesium–water reaction. (**B**) Removal of the $Mg(OH)_2$ passivation layer upon reaction in aqueous media with $NaHCO_3$. (**C**) Release of H_2 bubbles from the Mg surface triggered by $NaHCO_3$ to drive the propulsion of the Janus motor. Reprinted from Mou et al. [45], with permission from John Wiley and Sons, Co., Ltd, Hoboken, NJ, USA, Copyright (2013).

The previous example exploits the reaction of water and metals for propelling the motion of micro/nanomotors. This approach, together with photocatalytic water-splitting reactions, can be considered an excellent alternative for pushing the motion of micro/nanomotors by the generation of hydrogen or oxygen [46,47]. This results from active metals, which can react with water smoothly, e.g., magnesium and aluminum. This type of metal can generally form a passivation layer on the surface, reducing the violence of the reaction with water [48]. The seminal work on motors using water as fuel was performed by Gao et al. [49]. They fabricated a Janus microparticle composed of a hemisphere of Ti and a second one having an alloy of Al and Ga. This latter hemisphere in the presence of water reacts, producing hydrogen bubbles that can push the motion of the Janus motor with a remarkable speed of about 3 mm/s (around 150 times the particle diameter per second), exerting a force higher than 500 pN. The high speed of this type of water-driven Janus motor can be in part ascribed to the large diameter of the generated bubbles (around 10 μm) and the large size of the fabricated motor (average diameter of 20 μm), which ensure a large catalytic surface area and the formation of large bubbles. The propulsion behavior and lifetime of the motors can be tuned by changing the ionic strength or pH of the medium. On the other side, it may be expected that the control of the alloy reactivity by changing the composition or microstructure can contribute to improve the locomotion of the Janus motors. Figure 3 represents the displacement of Janus motors composed of a hemisphere of Ti and a second one of an Al-Ga alloy in water. Wu et al. [50] designed red blood cells coated on one of their sides with an Mg layer that allows an asymmetric generation of hydrogen bubbles, driving the propulsion of the motors without any external fuel and reaching an average velocity of about 172 μm/s.

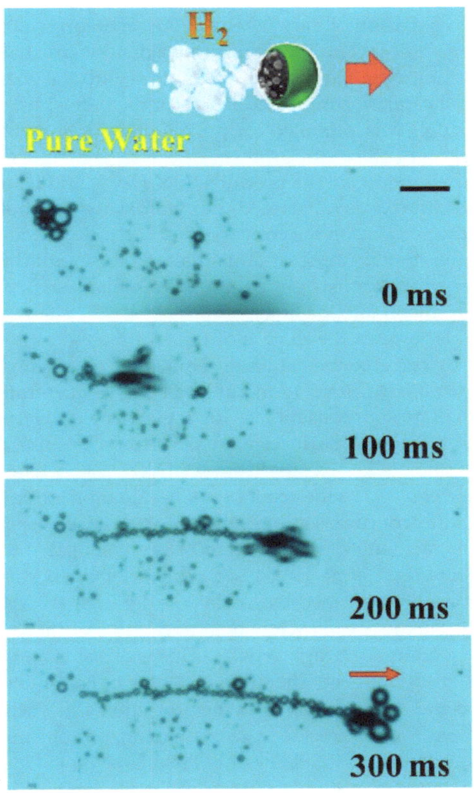

Figure 3. Displacement of Janus motors composed of a hemisphere of Ti and a second one of an Al-Ga alloy in water at different times. Reprinted from Gao et al. [49], with permission from the American Chemical Society. Copyright (2012).

It should be stressed that endogenous powered micro/nanomotors are helpful because this type of motor does not require the use of any external stimulus to control the motor during the entire operational time. Unfortunately, they need, in most cases, an external trigger ensuring that the motor is driven to the specific target. On the other hand, the motion of chemically driven motors is not always easy to control and can be easily disturbed, especially in ionic mediums. Furthermore, the use of chemical reactions for powering micro/nanomotors can result in a significant reduction of the power as the reaction approaches the end [16].

Enzyme-Actuated Micro/Nanomotors

Some micro/nanomachines take advantage of specific enzyme-triggered bio-catalytic reactions, which allows the conversion of chemical energy into a mechanical force [51]. This propulsion mechanism based on enzymatic reactions presents as the main advantage the biofriendly character of the used fuels, increasing the value of enzyme-actuated micro/nanomotors (EMNMs) for biomedical applications [52]. Hortelao et al. [53] proposed the design of mesoporous silica-based core-shell motors for the efficient delivery of doxorubicin to cells. For this purpose, they conjugated urease on the surface of the particles. This urea acts as catalyst to stimulate the decomposition of the urea contained in the medium into carbon dioxide and ammonia, which pushes the motion of the motors. This type of propulsion increases significantly (four times) the efficiency of the drug release in comparison with passive systems, resulting in an enhanced anticancer efficiency towards HeLa cells as a result of the synergistic interactions resulting from the combination of the drug release process and the ammonia production due to the catalytic reaction.

Schattling et al. [54] used Janus motors with the enzymatic pair formed by glucose oxidase and catalase conjugated to one of the hemispheres. The catalytic activity of this enzymatic pair was exploited for pushing the motion of the manufactured motor using glucose as fuel. The main problem of this approach is that the enzymatic degradation of the glucose results in the production of H_2O_2, and hence it is necessary to ensure that its concentration does not exceed the cytotoxicity threshold. This can be achieved by the action of the catalase on the conversion of the H_2O_2 into oxygen and water. Ma et al. [51] reported the first example of Janus nanomotors with dimensions below 100 nm. This type of nanomotor consists of a mesoporous silica nanoparticle coated on one of its sides with a thin silicon dioxide layer, while on the other hemisphere, catalase is bound. Thus, it is possible to push the motion of the motor as a result of the decomposition of H_2O_2 triggered by the enzyme catalase. This leads to an enhanced diffusion in comparison to the Brownian diffusion found at low H_2O_2 concentrations, opening new avenues for the design of mesoporous motors for active drug delivery. Further studies on the design of mesoporous motor particles were performed by Simmchen et al. [55]. They designed a motor particle where a single-strand DNA was conjugated to one of their faces, and the enzyme catalase to the other one. These motors offer the opportunity to capture and transport different cargos as a result of the presence of specific oligonucleotide sequences that can interact with the single-strand DNA conjugated to the particles.

Abdelmohsen et al. [56] designed bowl-shaped stomatocytes functionalized with the enzymatic pair formed by catalase and glucose oxidase (see Figure 4). This type of nanomotor can move even at low fuel concentrations (hydrogen peroxide or glucose), while the enzyme maintain a high activity. This is possible by the confinement of the enzymes in the designed platform. Therefore, this type of nanomotor combines a high control over the motion and directionality with the protection of the active molecules, offering different opportunities for application, e.g., biosensing, protein and DNA isolation and detection, and immunoassays. It should be noted that enzyme-loaded stomatocytes can be propelled three times faster than stomatocytes based in the reaction of Pt and hydrogen peroxide [57].

It should be noted that the locomotion efficiency of EMNMs can be enhanced by controlling their geometries. In fact, tubular motors present better performance than spherical ones [42].

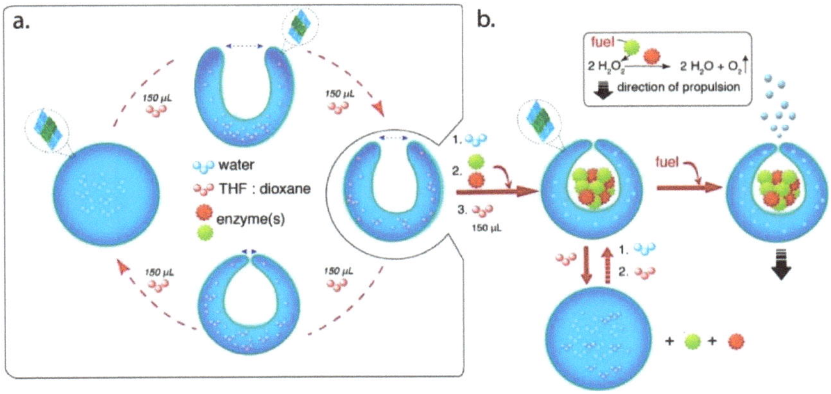

Figure 4. (**a**) Schematic representation of the assembly process of active stomatocytes. (**b**) Sketch of the propulsion mechanism of active stomatocytes. Reprinted from Abdelmohsen et al. [56], with permission from the American Chemical Society. Copyright (2016).

4.2. Externally Actuated Micro/Nanomotors

The requirement of autonomous motion in the micro/nanoscale in drug delivery applications using synthetic motors or robots may require, in certain cases, driving their motion using an external power input. A broad range of power sources are currently available that can be used as external triggers of the motion of micro/nanomotors in drug delivery applications, e.g., magnetic and electrical fields, light irradiation, acoustic waves, and heat. These can be used independently or in combination to provide multifunctionality to the drug delivery platforms, which allows the exploration of new avenues in the treatment of several diseases [16].

4.2.1. Magnetically Guided Micro/Nanomotors

Magnetic actuation is probably the most promising alternative in the design of self-propelling micro/nanomotors, offering a broad range of swimming strategies, e.g., helical swimmers, flexible swimmers, and surface walkers [58]. Furthermore, the magnetic fields present several advantages in relation to other actuation methodologies, e.g., high penetration, non-invasiveness, and strong controllability. On the other side, magnetic fields can penetrate freely within biological tissues, becoming a very simple operational strategy for driving motor motion [34].

The preparation of magnetically actuated micro/nanomotors commonly requires the presence of ferromagnetic elements (Fe, Co, Ni, etc.), which can be magnetized under the application of an external magnetic field [59]. Thus, it is possible to obtain several types of propulsion depending on the nature of the applied magnetic field: rotating or oscillating magnetic fields. Rotating magnetic fields are characterized by a magnetic induction vector that rotates at a fixed frequency, generating a torque. This drives a forward rotation of the motor, allowing the modification of the motion direction by changing the direction of the applied magnetic field [60]. Oscillating magnetic fields are the result of the combination of a uniform static magnetic field and a rotating one. Thus, it is possible to push a back-and-forth motion of the motor along the direction of the resultant magnetic field, offering long-range driving and navigation capabilities to the manufactured motors [61].

Zhang et al. [60] fabricated artificial bacterial flagella actuated by weak magnetic fields. This type of motor (helical swimmer) consists of a helical tail (InGaAs/GaAs or InGaAs/GaAs/Cr) similar to natural flagellum and a thin magnetic head (Cr/Ni/Au) on one end, allowing a swimming locomotion that can be precisely controlled by three orthogonal electromagnetic coil pairs. In fact, the artificial bacterial flagella work as helical propellers, converting rotary motion to linear motion. The polarity of this motion can be switched by reversing the rotation direction of the magnetic field. Further studies on the design of motors based on artificial bacterial flagella were performed by Schamel et al. [62]. They designed nano-sized artificial bacterial flagella (around 70 nm), which offer advantages for the control of their motion even through viscoelastic fluids.

A very interesting approach, especially for drug delivery purposes, is the combination of artificial bacterial flagella and drug-loaded liposomes, as demonstrated Qiu et al. [63]. They showed that the application of low-strength rotating magnetic fields to titanium -coated artificial bacterial flagella is an excellent tool for a precise 3D navigation in fluids, allowing the transport of calcein-loaded liposomes deposited on their external surface. This leads to a targeted release of about 73% of the loaded model drug. Following a similar concept, artificial Ni-based bacterial flagella were decorated with lipoplexes containing pDNA, allowing an effective gene delivery to human embryonic kidney cells under in vitro conditions. This is possible by taking advantage of the motion of the magnetic engine under low-strength magnetic fields. Unfortunately, the use of Zn can result in chronic toxicity under in vivo conditions, and it is necessary to carefully analyze the potential applications of this type of system [64].

Medina-Sánchez et al. [65] designed metal-coated polymer microhelices for the capture, transport, and release of immotile sperm cells under conditions mimicking those occurring during the fertilization process. Thus, it is possible to deliver single sperm cells on the oocyte wall under the action of magnetic fields. The use of magnetically driven micromotors was later extended to other aspects related to the reproductive medicine, such as embryo implantation. For this purpose, Schwarz et al. [66] tested two types of magnetic micro-propellers, helixes and spirals, for capturing and transporting bovine and murine zygotes, taking advantage of the propulsions induced by the application of a rotating magnetic field. This approach allows propulsion and cargo transportation within high-viscosity mediums and confined microfluidic channels. Moreover, it is possible to transfer cell-loaded motors between different environments, offering new opportunities for in vivo application in embryo transfer.

Flexible swimmers such as Au-Ag-Ni nanowires based on an undulatory locomotion mechanisms are also interesting alternatives in the fabrication of magnetically driven synthetic motors. The motion of this type of engine under the application of a rotating magnetic field occurs by the rotation of the Au and Ni segments at different amplitudes. This is possible because the torque produced by the rotation of the Ni segment as a response to the magnetic stimulus is transmitted along the Ag flexible segment to the gold head forcing its rotation. Therefore, it may be assumed that the motion of this type of swimmer is triggered by the breaking of the system symmetry. The modification of the lengths of the Au and Ni segments and the modulation of the applied magnetic field allow designing engines with forward (pushing) and backward (pulling) magnetically powered motion, and a precise switch "on/off" of the motion, respectively, providing a promising strategy for transport of biological media such as urine [67]. Figure 5 represents a comparison of the mechanism of forward and backward motion.

Jang et al. [68] fabricated magnetically composite multilink nanowire-based chains (diameter 200 nm), which undergo an undulatory motion under the application of a planar-oscillating magnetic field. This type of swimmer can be considered as eukaryote-like systems constituted by a polypyrrole tail and a series of rigid magnetic nickel links that are connected through flexible polymer bilayer hinges.

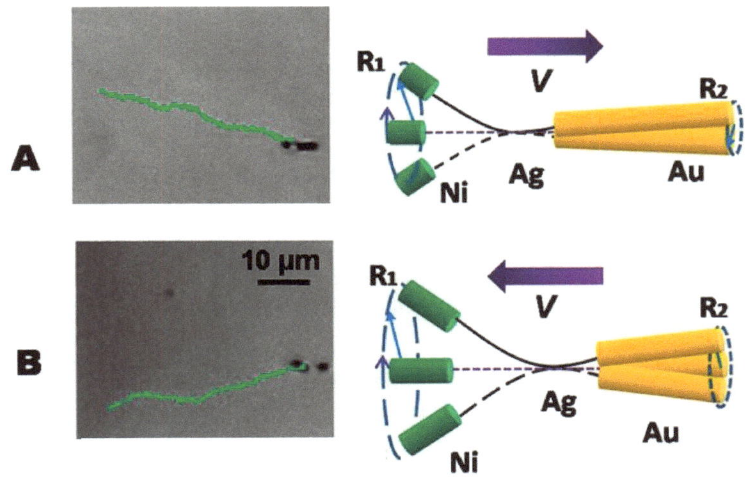

Figure 5. Comparison of the forward (**A**) and backward (**B**) motions of Ni-Ag-Au flexible microswimmers. The panel on the left represents the motion of the rods under the application of the magnetic field, and the panel on the right represents the corresponding strategies leading to the microswimmer motion. Adapted from Gao et al. [67], with permission from the American Chemical Society. Copyright (2010).

4.2.2. Electric Propulsion of Micro/Nanomotors

The fabrication of MNMs using conductive materials allows for the exploitation of the conversion of the electric energy into motion through two types of mechanisms: (i) electro-osmotic flow propulsion and (ii) electric current dynamic flow propulsion. Thus, it is possible to tune the motion direction as well as the speed by regulating the surface charge of the motor or the electrochemical reactions occurring at the motor/fluid interface [34]. The propulsion guided by an electroosmotic flow occurs in polarized particles when they accumulate opposite charges in their electrical double layer as a result of the application of an AC electric field. This generates a DC local electric field and the electro-osmotic flow, pushing the motion of the motors as a result of the constant electric field emerging between the electrode and the local electro-osmotic flow [69]. On the other hand, electric current propulsion occurs upon the application of a high-intensity electric field in a medium with low conductivity. Thus, it is possible to generate a current body dynamics that propel the liquid motion, pushing the active translational and rotational motion of the motors [70].

One of the seminal works on electrically driven MNMs was authored by Calvo-Marzal et al. [71]. They proved that the motion of Pt/Au nanowire motors within a hydrogen peroxide medium may be triggered under the application of an external electrical field. This strategy allows a cyclic on/off activation of the motor motion as well as a fine control over the speed. For instance, the decrease of the applied voltage from 1 to 0.4 V leads to an increase of the motor speed from 4 μm/s to 20 μm/s, whereas in the absence of any applied electric field, the motion occurs at 9 μm/s. Fan et al. [72] designed synthetic motors based on the use of gold nanowires decorated with cytokines. These motors can be moved under the application of constant and alternating currents to generate electrophoretic and dielectrophoretic forces. Moreover, the application of the electrical field can contribute to the cell stimulation once the motors stick on their surfaces. It is worth mentioning that the combination of constant and alternating current allows simultaneous control of the motion and positioning directions, allowing a displacement more than two times faster than those obtained with the motors designed by Calvo-Marzal et al. [71], i.e., reaching velocities of up to 50 μm/s.

Rahman et al. [73] demonstrated that carbon nanotubes in aqueous medium can rotate upon the application of a rotating electric field of a specific magnitude and angular speed by taking advantage of the orientations of the water dipole. Thus, a rotational motion at ultrahigh speed (higher than 10^{11} rpm) is possible, allowing the rotational transport of attached loads. Guo et al. [74] designed Pt-based bimetallic nanorods with versatility for the transport of cargos to specific targets and the ability to integrate into chemically powered nanomechanical actuators. The combination of AC and DC electric fields allows the alignment and control of the motor, respectively. Moreover, the use of a set of 3D orthogonal microelectrodes for turning on and off the motor motion also allows the motion of the motors within the vertical direction.

It should be stressed that even though synthetic MNMs actuated by external fields are very promising, their applications in drug delivery are limited by the unfriendly character of some of their components in terms of human health [75].

4.2.3. Light-Actuated Micro/Nanomotors

Light can be considered among the most ubiquitous energy sources, hence their use for triggering the motion of MNMs emerges as a promising strategy, especially because light can be transmitted wirelessly and remotely. Moreover, the properties of light, e.g., intensity, frequency, polarization, and propagation direction, can be temporally and spatially adjusted, which creates interesting opportunities to control the motion of MNMs. The basic concept for light-driven motion relies on the breaking of the pressure distribution symmetry by the generation of a light-induced asymmetric field within photoactive particles, pushing their motion. This is possible through different strategies [34].

The most common method of light-induced propulsion relies on the combination of two asymmetric gradient fields, which drive the motion following self-electrophoresis, self-diffusion electrophoresis, or self-thermophoresis mechanisms. The first type of gradient presents a chemical origin and is derived from the photocatalytic reactions occurring upon irradiation, resulting in an asymmetric field that propels the motion [76]. The second type of gradient results from the change of the temperature associated with the irradiation, pushing the motion by autothermal migration [77]. The motor motion can be also triggered by using the bubble recoil principle. This relies on the asymmetric distribution of the bubbles produced as a result of photochemical reactions, which induce a bubble concentration gradient that drives the motor motion [78].

Light can also generate surface tension gradients as a result of photochromic reactions that modify the interfacial properties. This forces the fluid to flow from low interfacial tension regions to high interfacial tension ones, pushing the motion of the motors [79–81].

The last, but not the least, application of light for driving the motion of synthetic motors relies on the ability to induce deformations in specific materials, such as crystal elastomers with high elasticity and liquid crystal order. The irradiation of this type of material with the light of a specific wavelength allows inducing their dilational deformation, changing the fluidity of the materials and powering their motion [82,83].

Zhan et al. [84] designed light-driven motors formed by core-shell Sb_2Se_3/ZnO nanomotors, which exploit the ability of the Sb_2Se_3 to adsorb light polarized parallel to the nanowires. This leads to a strong dichroic swimming behavior, which presents higher velocity when the incident light is parallelly polarized than when it is perpendicularly polarized. On the other hand, the combination of two cross-aligned dichroic motors can drive the behavior of polarotactic artificial microswimmers, which can move by controlling the direction of polarization of the incident light.

Wang et al. [85] designed micromotors based in glucose-fueled composites formed by cuprous oxide and N-doped carbon nanotubes activated under the action of environmentally friendly visible light. Thus, it is possible to move the manufactured motors with a velocity of up to 18.71 μm/s, which is similar to that obtained for Pt-based catalytic Janus micromotors fueled in H_2O_2 medium. Moreover, the velocity of the new type of motor can be regulated by tuning the glucose concentration or the light intensity. On the other hand, the glucose-fueled composite motors formed by cuprous oxide and N-doped carbon nanotubes can undergo a highly controllable negative phototaxis behavior. Figure 6 represents a sketch of the locomotion mechanisms of motors formed by cuprous oxide and N-doped carbon nanotubes as well as their trajectories in the absence and presence of glucose fuel.

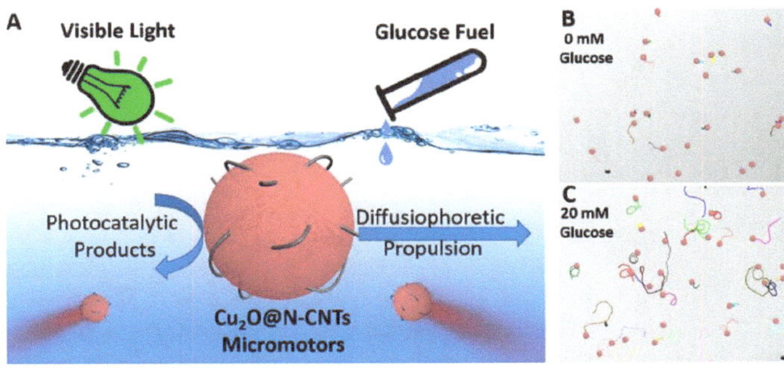

Figure 6. (**A**) Sketch of the locomotion mechanism of motors formed by cuprous oxide and N-doped carbon nanotubes. (**B**) Trajectories of motors formed by cuprous oxide and N-doped carbon nanotubes in the absence of glucose fuel. (**C**) Trajectories of motors formed by cuprous oxide and N-doped carbon nanotubes in the presence of glucose fuel. Reprinted from Wang et al. [85], with permission from the American Chemical Society. Copyright (2019).

Xing et al. [86] designed Janus tubular motors constituted of hollow mesoporous carbon particles, with one of the hemispheres coated by a Pt layer. Thus, it was possible to combine a dual actuation mechanism based on the catalytic degradation of the hydrogen peroxide and the irradiation with near infrared light (NIR). The latter provides a directional motion of the fabricated motors as a result of the thermal gradient generated during the irradiation process, favoring adhesion to carcinogenic cells. Similarly, Xuan et al. [87] designed Janus motors based on silica nanoparticles, with one hemisphere coated by a gold layer. This type of system presents a fuel-free motion that is externally actuated by NIR irradiation. This induces a photothermal effect on the Au-coated hemisphere, which leads to thermal gradients across the motor, driving a self-thermophoresis phenomenon. Thus, it is possible to push an ultrafast motion (up to 950 body lengths/s for motors of 50 nm). The use of a remote NIR laser provides a powerful strategy for a reversible "on/off" motion, simultaneously controlling the motion directionality and offering an excellent tool for improving the maneuverability of fuel-free motors. He et al. [88] exploited Janus motor particles guided by the external actuation of magnetic fields or NIR irradiation for inducing a photothermal tissue welding, with results comparable to those expected for common medical sutures. It should be noted that the use of MNMs actuated by NIR irradiation offers several advantages for in vivo applications due to the ability of this radiation to focus on regions of small specific area and to penetrate deeply into the tissues without significant damage [89]. The drawbacks associated with the use of other types of radiation do not preclude the design of motors actuated for them. For instance, the irradiation of AgCl particles with UV light induces a self-diffusiophoretic response due to the asymmetric photodecomposition of the particles, which drives the particle motion in aqueous medium [41].

4.2.4. Ultrasound-Actuated Micro/Nanomotors

The use of ultrasound radiation as an external trigger of the motion of synthetic MNMs relies on forcing the particle motion upon the application of acoustic radiation. This can be understood considering that the application of an ultrasonic field to fluids with suspended particles leads to particle–fluid interactions that can induce different motion states [90]. There are two different mechanisms of ultrasonic propulsion, depending on whether the radiation acts directly on the motor or not. The former is ultrasonic wave propulsion, which requires asymmetric motors and uneven sound pressure distribution, making possible the motion as a result of the pressure gradient [91]. The second mechanism is the so-called acoustic droplet vaporization, which results from the evaporation of liquid droplets. This increases the enthalpy and momentum, generating a powerful pulse to promote the motor motion [92].

The seminal work in ultrasound-powered synthetic motors proved that the application of ultrasonic standing waves in the MHz range can drive the levitation, propulsion, rotation, and assembly in metallic microrods (2 μm long and 330 nm diameter) in aqueous medium [91]. This occurs through an acoustophoretic mechanism resulting from the microrod asymmetry, which drives its axial propulsion. The asymmetry associated with the convex and concave regions leads to an anisotropic contribution of the ultrasound pressure, which results in a pressure gradient at the microrod surface. Thus, the motors are unidirectionally propelled. On the other hand, the metallic rods can align and self-assemble to form spinning chains, which in the case of Janus microrods occurs following a head-to-tail alternating structure. Moreover, the chains can form ring or streak patterns in the levitation plane, with such patterns having a characteristic distance that is about a half of the wavelength of the ultrasonic excitation radiation. Figure 7 shows the different possible motion pathways of ultrasonic-powered Au/Ru microrods. It should be noted that the asymmetry of the microrods plays a central role in the control of their motion. For instance, the higher the asymmetry, the higher the motion speed, as was proven by Garcia-Gradilla et al. [93], who designed multifunctional rod-like nanomotors formed by three different segments (Au-Ni-Au). This type of motor can be propelled by ultrasound radiation and magnetic fields. The interaction of the latter with the Ni segments can be exploited for providing a predefined and controlled motion to the manufactured motors. Moreover, the Ni segments can be exploited for picking up and transporting magnetic materials.

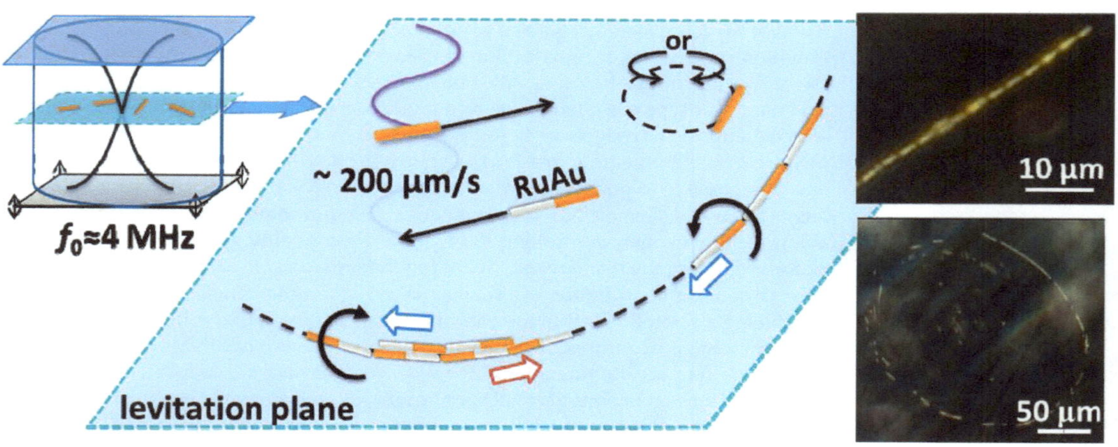

Figure 7. Different motion pathways of Au/Ru microrods powered by ultrasonic standing waves. Reprinted from Wang et al. [56], with permission from the American Chemical Society. Copyright (2012).

Tubular motors also offer interesting opportunities for ultrasound-triggered motion, as was demonstrated by Kagan et al. [92]. They showed that ultrasound-triggered vaporization of electrostatically bonded perfluorocarbon droplets contained inside the motor can propel the tubular engines with speeds of up to 6 m/s (two orders of magnitude faster than previous systems). It should be noted that the mechanisms driving the motion of tubular motors present higher efficiency than those occurring in rod-shape ones [34].

Xu et al. [94] showed that the application of an ultrasound field can help in the control of the motion of tubular catalytic motors propelled by hydrogen peroxide. This is possible because ultrasound radiation can disrupt the normal evolution and ejection of the generated bubbles, which is essential to the propulsion of the manufactured motors. Therefore, ultrasound radiation can be used for a precise control of the velocity of the motors by increasing and decreasing sharply the speed of the engine at low and high powers, respectively. For instance, the application of an ultrasound field of up to 10 V reduces the velocity of the manufactured motors from 231 μm/s to 6 μm/s in less than 0.1 s, undergoing a fast recovery upon removal of the applied field. Therefore, this strategy allows extremely fast changes in the motion speed and reproducible "on/off" reversible activation of the motor, improving the efficiency of the energy conversion. Figure 8 shows the ultrasound-modulated motion of chemically powered motors.

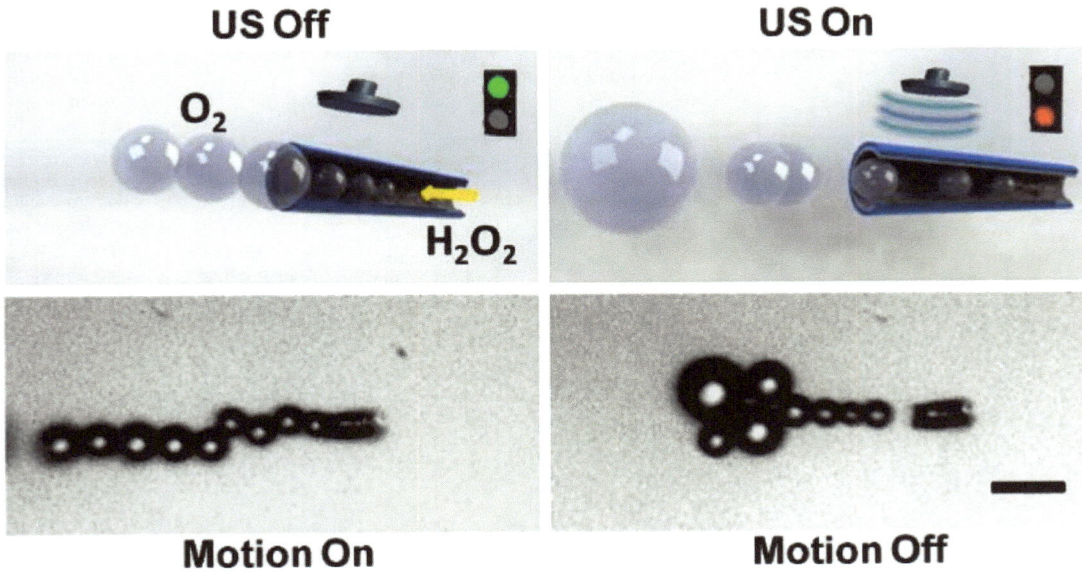

Figure 8. Set images (schemes, **top**, and true micrograph, **bottom**) showing the ultrasound-modulated motion of chemically powered motors. Reprinted from Xu et al. [94], with permission from the American Chemical Society. Copyright (2014).

Table 1 presents a comparison between the characteristics of chemically powered and externally actuated motors.

Table 1. Comparison of the main characteristics of chemically powered and externally actuated motors. Reprinted from Hu et al. [16], with permission under open access CC BY 4.0 license, https://creativecommons.org/licenses/by/4.0/ (accessed on 4 July 2022).

Type	Energy	Penetration	Motion Ability	Persistence	Safety
Endogenous powered motors	Chemical	Not applicable	Requires external force for positioning	Not as good, chemical energy can be depleted when it decreases gradually, limiting the motion of the engines	Depends on the fuel: hydrogen peroxide is a toxic fuel, whereas glucose and urea are safe
Exogenous powered motors (Externally triggered)	Magnetic	Good, weak magnetic fields can be enough	Precise 3D navigation in fluids under the action of rotating magnetic fields	Good, engines can keep moving under the guidance of the external field	Used magnetic fields are generally safe, metallic components can present toxicity upon long-term exposure
	Electric	Weak, strong electric fields are needed	Requires the combination of electric fields and additional fields for ensuring the directional motion		Strong electric field can affect human body, metallic components can present toxicity upon long-term exposure
	Light	Depends on the type of light, different penetration	Normally exploited for triggering other reactions. However, it can provide directional motion		Depend on the type of light, ultraviolet light may be harmful, whereas other lights are commonly safe
	Ultrasound	Good	Commonly combined with magnetic fields, provides directional motion		Ultrasound irradiation can cause oxidative stress in cells, metallic components can present toxicity upon long-term exposure

5. Towards the Biocompatibility of Micro/Nanomotors

One of the main challenges associated with the use of micro/nanomotors in biomedicine is the introduction of biocompatible and biodegradable components for the fabrication of miniaturized devices. This is important when biomedical applications of MNMs are considered because of the use of the manufactured motors in biological environments, including cells and tissue [95]. Therefore, the contact between MNMs and the human environment requires analysis of a broad range of factors, including protein adhesion, stimulation of the immune response, biodistribution, toxicity, degradation, and elimination profiles [18]. These factors may influence the activity of MNMs and reduce their effectiveness.

Synthetic MNMs can be manufactured almost at will to provide a suitable response to a specific demand. However, they present different concerns associated with their biological safety under in vivo conditions. This has driven important research activity towards the fabrication of biological motors using natural cells. For instance, different self-driving biological motors have been fabricated using motile bacteria, neutrophils, sperm cells, and cardiomyocytes [96–99]. These MNMs are characterized by a good compatibility without causing any adverse immune response. However, the number of cells that can be used as substrates for the fabrication of MNMs is limited, and the size of these motors is limited to the micrometric scale. On the other hand, the effectiveness of biological motors in the treatment of different diseases should be carefully examined. The situation changes when truly synthetic MNMs are considered. These can be manufactured with a broad range of shapes and sizes (from a few nanometers to several millimeters). On the other hand, there is a growing interest on the fabrication of synthetic motors following a bionics approach, i.e., by combining biological and synthetic components. This allows combining the biocompatibility of biological blocks and the modularity of synthetic ones, which can create promising opportunities for specific applications [50]. An example of this type of approach is the engineering modification of natural cells by physical or chemical methods, which provides the basis for the introduction of new functionalities to the cells, respecting their biocompatibility. For instance, to ensure the locomotion of this new motor, it is common to include magnetic materials, e.g., iron oxides or magnesium, or enzymes that catalyze specific chemical reactions in aqueous medium or urea [95].

Chemical and physical MNMs can be designed using degradable or self-destructive materials, which can be destructed once they have completed a specific task. However, there are many concerns about the final fate of biological motors after they complete their life cycle. In some cases, they can follow their own metabolic cycles without any safety concerns [100]. Unfortunately, motors having self-proliferation capabilities, e.g., bacteria, should be analyzed more carefully. In particular, it is necessary to design them in such a way that they cannot move to undesired sites where they can proliferate. This requires specific strategies, as proposed by Stanton et al. [101], who used a the local NH_3 concentration resulting from urea hydrolysis to stop the motion of bacterial motors on demand by killing the biological activity of the bacteria.

The application of MNMs also requires ensuring the biosafety of the power sources [95]. This is of a paramount importance because a broad number of works have dealt with the use of catalytic motors based on noble metals or metal oxides characterized by a motion ability triggered by the degradation of the hydrogen peroxide, which is toxic to the human body [26,43]. A common alternative for reducing the potential harmful effects of the toxic fuels is the reduction of the concentration of specific molecules, e.g., hydrogen peroxide or mineral acids, in the fuel used for the motor motion. Unfortunately, this reduces the efficiency and speed of the motor motion [95]. The drawbacks associated with the use of hydrogen peroxide as fuel have driven an important piece of research towards the use of enzyme-catalyzed reactions for powering motors. Some examples include urease, which can convert urea to NH_3 and CO_2; catalase, which converts the hydrogen peroxide in water and oxygen; and glucose oxidase, which catalyzes the conversion of glucose in gluconic acid and hydrogen peroxide [21,102]. This type of system is important because it allows extending the number of driving systems available for MNMs, reducing the possible toxicity. However, they require in most cases higher fuel concentration than that corresponding to the physiological conditions.

In recent years, MNMs using water as fuel have been designed, taking advantage of the reaction of water with metals, e.g., magnesium, aluminum, or other reactive metals, to produce hydrogen, which is theoretically possible. However, the true situation is far from satisfactory due to the formation of passivation layers on the metal surface that reduce the efficiency of the reaction for powering the motor motion, requiring the additional chemicals, e.g., sodium bicarbonate, to the environment and limiting the use of this type of motor to non-basic media, e.g., the acidic environment of the human stomach [95].

In the case of physical motors, it was previously stated that their motion is driven by different external fields without the addition of any fuel [103], and hence the biosafety of this type of system depends on whether the intensity and time of the applied stimulus can cause any damage to normal tissues.

Biological motors that present the capability of autonomous motion, e.g., sperm cells, can move within the human body without adding any fuel or using additional devices. Therefore, their biosafety is almost ensured [104]. However, the biosafety of composite motors containing biological and synthetic pieces depends on the specific characteristics of the driving system [95].

6. Micro-/Nano-Motors in Drug Delivery

The exploitation of human-made micro/nanomotors to deliver therapeutic drugs to specific targets represents a novel approach for the treatment of different diseases [5]. This is possible because micro/nanomotors offer different advantages in comparison to more classical drug vectors. These advantages include a rapid drug transport, high tissue penetration, and controllable motion [6]. Indeed, the autonomous motion of micro/nanomotors provides the bases for the controlled transport of drugs to reach tissues that are difficult to access [16]. The fabrication of micro/nanomotors for drug delivery is commonly based on the combination of an internal payload with an external shell, which contributes to the active transport of the device to reach specific targets [105]. Many drugs, including small molecules, small interfering ribonucleic acid (siRNA), DNA, peptides, antibodies, and proteins can be delivered from MNMs [106].

The design of MNMs for drug delivery applications requires consideration of the material used for the fabrication, the cargo, and the mechanism used for guiding the motion. This is important because they affect the different characteristics of the final products, including their size, shape, charge, and the fate of MNMs as defined by their tissue accumulation, intracellular transport, biodegradability, or biocompatibility [107].

The effectiveness of MNMs for drug delivery within the gastrointestinal tract was proven by Esteban-Fernández de Ávila et al. [108], who used Mg-based motors loaded with clarithromycin to treat mice infected with *Helicobacter pylori*. The fabricated motors were administrated orally, evidencing a good ability to be propelled within the gastric fluid. Moreover, the administration of these engines contributed to a reduction of almost two orders of magnitude of the population of *Helicobacter pylori*, without significant toxicity for the mice. A similar approach was followed by Gao et al. [109], who fabricated Zn-based motors. These motors undergo an acid-driven propulsion in the stomach, presenting an effective binding and retention in the stomach as well as good cargo payloads on the stomach walls. Moreover, the motors can be progressively dissolved in the gastric acid, which leads to an autonomous release of the encapsulated drugs without any evidence of toxicity.

Baylis et al. [110] used gas-generating particles formed by the combination of carbonate and tranexamic acid for halting hemorrhage through blood vessels in mice and pigs. This is possible by the locomotion of the particles loaded with thrombin at velocities of up to 1.5 cm/s through aqueous medium, mimicking the main characteristics of the blood. Thus, the combination of different mechanisms, including lateral propulsion, buoyant rise, and convection, push the motors through the fluid, allowing the use of this type of system as an effective hemostatic agent, making possible the halting of hemorrhages in several animal models of intraoperative and traumatic bleeding.

Kim et al. [111] designed magnetically actuated hydrogel engines for the transport and subsequent release of (poly-D,L-lactic-co-glycolic acid particles) loaded with doxorubicin into the eyes. The particularity of this type of engine is its ability to remove the magnetic components from the eyes after drug delivery, which minimizes the possible side effects. Thus, the application of alternating magnetic fields at the target point leads to the dissolution of the therapeutic layers, resulting in the release of the drug; then, the magnetic components are retrieved again under the application of a new magnetic field. Moreover, ex vivo and in vitro studies showed the ability of this type of engine to migrate towards the vitreous, which enables a significant therapeutic effect against retinoblastoma cancer cells.

Cancer induces strong oxidative stress in cells, which leads to the production of high amounts of H_2O_2. This can be exploited as an energy source for driving the motion of drug carriers [112]. Villa et al. [113] designed superparamagnetic/catalytic robots consisting of Janus micromotors formed by an iron oxide particle decorated with tosyl groups, with one of its hemispheres coated by a platinum layer (see Figure 9). This provides a multifunctional character to the motor: (i) the tosyl group layer provides the capacity for binding molecules and biological materials; (ii) the Pt layer contributes to the catalytic decomposition of hydrogen peroxide, helping in the propulsion of the motor; and (iii) the magnetic particle allows manipulation under the application of magnetic fields. In fact, this latter part makes it possible for the motor to work as a single unit or assembly of chains for performing collective actions (e.g., capture and transport of cancer cells). These motors can be exploited for the release of anticancer drugs, e.g., doxorubicin, showing a significant reduction in the proliferation of carcinogenic cells.

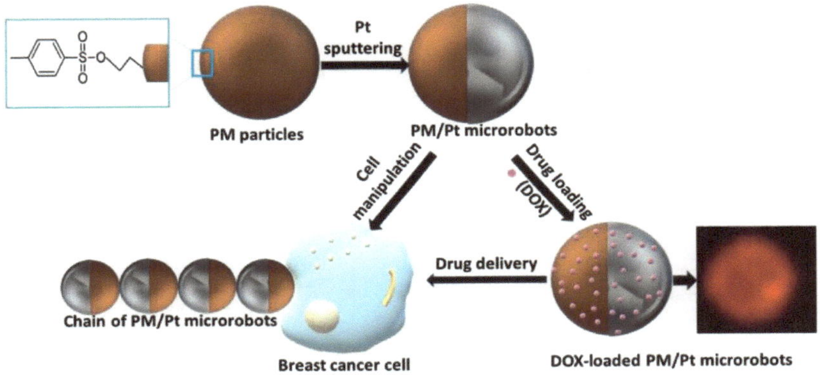

Figure 9. Sketch of the fabrication process and action mechanism of catalytic Janus motors loaded with the anticancer drug doxorubicin (DOX). Reprinted from Villa et al. [113], with permission from John Wiley and Sons, Co., Ltd, Hoboken, NJ, USA, Copyright (2018).

Kagan et al. [35] designed very complex Ni/(Au_{50}/Ag_{50})/Ni/Pt) nanowire motors with the ability of picking up, transporting, and releasing doxorubicin encapsulated in biodegradable particles (poly-D,L-lactic-co-glycolic acid particles) and liposomes doped with Fe_3O_4 (sizes in the range 100 nm–3 μm) to specific targets. These catalytic motors combine two actuation mechanisms for ensuring a fast propulsion and a directional motion. In fact, they are propelled by the decomposition of hydrogen peroxide, whereas the application of a magnetic field ensures their directional guiding, with a velocity three times faster than that expected for passive motors. Gao et al. [114] designed fuel-free motors based on magnetically actuated flexible nickel-silver swimmers (5–6 μm in length and 200 nm in diameter). This type of engine allows the transport of doxorubicin to HeLa cells at high speeds (more than 10 μm/s, which is equivalent to more than 0.2 body lengths per revolution in dimensionless speed), following a process including several steps: capture drug-loaded magnetic polymeric particles, transport the particles through the channel, approach and stick onto the HeLa cells, and release the doxorubicin. However, this type of motor can be affected by a poor adhesion between the particles and the motors, which can lead to the failure of the motor system before reaching its target. This can be avoided by coating the drug-loaded particles with a polymer layer, which can contribute to the drug–cell membrane binding process.

Manganese oxide–based (PEDOT/MnO_2) catalytic tubular micromotors were evaluated as a tool for delivery of the chemotherapeutic drug camptothecin. This is possible by the catalytic decomposition of hydrogen peroxide, which propels an effective autonomous motion in biological media with high speed (318.80 μm/s). One of the main advantages of this type of motor is related to its capability of operating at low fuel concentrations (below 0.4% *w*/*w*) [115]. Wu et al. [116] fabricated Pt nanorockets with a biocompatible coating formed by a Layer-by-Layer film of chitosan and alginate for the transport of doxorubicin. This type of engine allows a targeted transport of the drug following a propulsion mechanism driven by the catalytic degradation of hydrogen peroxide, and a controlled release of the drug. In fact, the catalytic degradation of the hydrogen peroxide by the internal core of the nanorockets releases a tail of oxygen bubbles, propelling the engines at a 74 μm/s. The polydispersity of the fabricated engines and the distribution of the Pt nanoparticles within the nanorockets results in different motion pathways, including straight, circular, curved, and self-rotating motions. On the other hand, the application of an ultrasound field was used for triggering the rupture of the capsules, ensuring the release of the drug.

Xuan et al. [117] fabricated a self-propelled Janus nanomotor (diameter about 75 nm) for the transport and controlled release of doxorubicin encapsulated within liposomes on cells. These motors were based on mesoporous silica nanoparticles with caps of chromium and platinum, using the bubble generated as a result of the catalytic decomposition of hydrogen peroxide as the driving force of the motor motion, which can occur at speeds of up to 20.2 μm/s. The study of the in vitro intracellular localization evidences that the fabricated motors can enter into the cells, and the release of the encapsulated drug occurs by the decomposition of the liposomes within the intracellular region. Figure 10 displays the fabrication process and performance mechanism of the Janus motors.

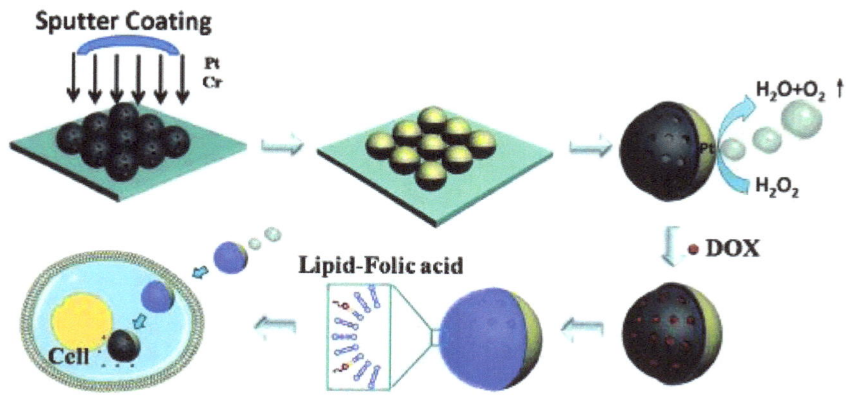

Figure 10. Sketch of the fabrication process and action mechanism of catalytic Janus motors on the transport of liposomes loaded with the anticancer drug doxorubicin. Reprinted from Xuan et al. [117], with permission from John Wiley and Sons, Co., Ltd, Hoboken, NJ, USA, Copyright (2014).

Hortelao et al. [53] fabricated urease-powered mesoporous silica-based core-shell motors for the loading, transport, and efficient release of doxorubicin to cells. This is possible due to their ability to undergo self-propulsion in ionic media. This allows a release of the encapsulated drug four times faster than that resulting from passive systems. Thus, the efficiency of the anticancer drug towards HeLa cells is enhanced due a synergistic action of the drug release and the ammonia produced by the degradation of the urea in the medium. The high efficiency of this type of motor may create new opportunities for biomedical applications.

Kehzri et al. [118] fabricated high-speed tubular electrically conductive engines by combining reduced graphene oxide as a platform for an effective drug delivery and platinum as a catalytic core. This type of machine can be loaded with doxorubicin, which can be expelled during motion due to the application of an electron current, resulting in a high therapeutic efficiency against cancer cells, with a significant reduction of the side effects towards healthy tissues. Therefore, this type of motor provides an important step forward in the fabrication of advanced drug delivery systems.

Table 2 summarizes some examples of synthetic MNMs exploited for drug delivery purposes and their power source.

Table 2. Examples of synthetic MNMs used for drug delivery purposes.

Type of MNM	Power Source	Disease	Drug	Reference
Mg-based motors	Chemical (catalytic motors powered by gastric acids)	gastrointestinal bacteria	clarithromycin	Esteban-Fernández de Ávila et al. [108]
Zn-based motors				Gao et al. [109]
Carbonate and tranexamic acid particles	Chemical	hemorrhages	thrombin	Baylis et al. [110]
Hydrogel/magnetic particle hybrid	Magnetic fields	eye diseases	doxorubicin	Kim et al. [111]
Iron oxide particles with one hemisphere coated by Pt and decorated with tosylated groups	Chemical (catalytic motors powered by decomposition of hydrogen peroxide)	cancer	doxorubicin	Villa et al. [113]
Ni/(Au_{50}/Ag_{50})/Ni/Pt) nanowires and Fe_3O_4 particles	Chemical (catalytic motors powered by decomposition of hydrogen peroxide) combined with magnetic field (directionality control)			Kagan et al. [35]
Pt nanorockets coated by a Layer-by-Layer film of chitosan and alginate (tubular motors)	Chemical (catalytic motors powered by decomposition of hydrogen peroxide)			Wu et al. [116]
Janus nanomotors with caps of chromium and platinum				Xuan et al. [117]
Silica-based nanoparticles decorated with urease	Chemical (enzymatic degradation of urea by urease)			Hortelao et al. [53]
Flexible nickel-silver swimmers	Magnetic field			Gao et al. [114]
Carbon-platinum tubular Janus motors	Chemical (catalytic motors powered by decomposition of hydrogen peroxide) and light (near infrared radiation)			Xing et al. [86]
Tubular motors of platinum and reduced graphene oxide	Electric field			Kehzri et al. [118]
Tubular (PEDOT/MnO_2) micromotors	Chemical (catalytic motors powered by decomposition of hydrogen peroxide)		camptothecin	Feng et al. [115]

7. Concluding Remarks

Synthetic micro/nanomotors (MNMs) are at the forefront of the nanomedical tools designed for improving the diagnosis and treatment of a broad range of diseases. However, while important progress has been made for ensuring the efficient motion of these types of engine, the true application of these small-scale devices is still in its infancy, presenting different problems and challenges. For instance, the biosafety of power sources and fuels as well as that of the materials used for engine fabrication must be taken into consideration to reduce the risks and hazards associated with their use in the human body. Moreover, the capability of targeting the motors also is of paramount importance in their design. However, in most cases there is a poor understanding of the true framework involving the performance of this type of engine, which is in part the result of the poor understanding of their behavior under in vivo conditions. The understanding of the in vivo performance of MNMs is the only way to verify the effect of their interactions with complex body fluid environments as well as the effects associated with the side effects of long-term persistence of motors within the human body. Therefore, the fabrication of ideal MNMs requires consideration of their capabilities of precise targeted motion and autonomous drug delivery, without compromising their biosafety. The field is open to research and innovation for building safe and efficient engines for drug delivery. In fact, the incorporation of synthetic MNMs as a true therapeutic option for the treatment and prevention of different diseases is far from reality, and additional research on materials and fuels as well as efficiency tests under in vivo relevant conditions are required.

Author Contributions: E.G., conducted the review and wrote the draft; E.G. and A.M. revised the manuscript and contributed to substantial enhancement of the manuscript. All authors have read and agreed to the published version of the manuscript.

Funding: This work was funded in part by MICINN under Grant PID2019-106557GB-C21 and by E.U. on the framework of the European Innovative Training Network—Marie Sklodowska-Curie Action Nano Paint (Grant Agreement 955612).

Institutional Review Board Statement: Not applicable.

Informed Consent Statement: Not applicable.

Data Availability Statement: Not applicable.

Conflicts of Interest: The authors declare no conflict of interest. The funders had no role in the design of the study; in the collection, analyses, or interpretation of data; in the writing of the manuscript, or in the decision to publish the results.

References

1. Sánchez, S.; Soler, L.; Katuri, J. Chemically Powered Micro-and Nanomotors. *Angew. Chem. Int. Ed.* **2015**, *54*, 1414–1444. [CrossRef] [PubMed]
2. Zhang, J.; Chen, Z.; Kankala, R.K.; Wang, S.-B.; Chen, A.-Z. Self-propelling micro-/nano-motors: Mechanisms, applications, and challenges in drug delivery. *Int. J. Pharm.* **2021**, *596*, 120275. [CrossRef] [PubMed]
3. Suhail, M.; Khan, A.; Rahim, M.A.; Naeem, A.; Fahad, M.; Badshah, S.F.; Jabar, A.; Janakiraman, A.K. Micro and nanorobot-based drug delivery: An overview. *J. Drug Target.* **2022**, *30*, 349–358. [CrossRef] [PubMed]
4. Parmar, J.; Ma, X.; Katuri, J.; Simmchen, J.; Stanton, M.M.; Trichet-Paredes, C.; Soler, L.; Sanchez, S. Nano and micro architectures for self-propelled motors. *Sci. Technol. Adv. Mater.* **2015**, *16*, 014802. [CrossRef]
5. Gao, W.; Wang, J. Synthetic micro/nanomotors in drug delivery. *Nanoscale* **2014**, *6*, 10486–10494. [CrossRef]
6. Medina-Sánchez, M.; Xu, H.; Schmidt, O.G. Micro-and nano-motors: The new generation of drug carriers. *Ther. Deliv.* **2018**, *1*, 303–316. [CrossRef]
7. Abdelmohsen, L.K.E.A.; Peng, F.; Tu, Y.; Wilson, D.A. Micro-and nano-motors for biomedical applications. *J. Mater. Chem. B* **2014**, *2*, 2395–2408. [CrossRef]
8. Li, L.; Wang, J.; Li, T.; Song, W.; Zhang, G. A unified model of drag force for bubble-propelled catalytic micro/nano-motors with different geometries in low Reynolds number flows. *J. Appl. Phys.* **2015**, *117*, 104308. [CrossRef]
9. Li, J.; Ávila, B.E.-F.d.; Gao, W.; Zhang, L.; Wang, J. Micro/nanorobots for biomedicine: Delivery, surgery, sensing, and detoxification. *Sci. Robot.* **2017**, *2*, eaam6431. [CrossRef]
10. Solovev, A.A.; Xi, W.; Gracias, D.H.; Harazim, S.M.; Deneke, C.; Sanchez, S.; SchmidT, O.G. Self-Propelled Nanotools. *ACS Nano* **2012**, *6*, 1751–1756. [CrossRef]
11. Campuzano, S.; Ávila, B.E.-F.d.; Yáñez-Sedeño, P.; Pingarrón, J.M.; Wang, J. Nano/micro-vehicles for efficient delivery and (bio)sensing at cellular level. *Chem. Sci.* **2017**, *8*, 6750–6763. [CrossRef] [PubMed]
12. Wang, J.; Dong, R.; Wu, H.; Cai, Y.; Ren, B. A Review on Artificial Micro/Nanomotors for Cancer-Targeted Delivery, Diagnosis, and Therapy. *Nano-Micro Lett.* **2020**, *12*, 11. [CrossRef] [PubMed]
13. Zhao, Q.; Cui, H.; Wang, Y.; Du, X. Microfluidic platforms toward rational material fabrication for biomedical applications. *Small* **2019**, *16*, 1903798. [CrossRef] [PubMed]

14. Karshalev, E.; Ávila, B.E.-F.D.; Beltrán-Gastélum, M.; Angsantikul, P.; Tang, S.; Mundaca-Uribe, R.; Zhang, F.; Zhao, J.; Wang, L.Z. Micromotor Pills as a Dynamic Oral Delivery Platform. *ACS Nano* **2018**, *12*, 8397–8405. [CrossRef]
15. Fu, D.; Wang, Z.; Tu, Y.; Peng, F. Interactions between Biomedical Micro-/Nano-Motors and the Immune Molecules, Immune Cells, and the Immune System: Challenges and Opportunities. *Adv. Healthc. Mater.* **2021**, *7*, 2001788. [CrossRef]
16. Hu, M.; Ge, X.; Chen, X.; Mao, W.; Qian, X.; Yuan, W.-E. Micro/Nanorobot: A Promising Targeted Drug Delivery System. *Pharmaceutics* **2020**, *12*, 665. [CrossRef]
17. Hato, T.; Dagher, P.C. How the Innate Immune System Senses Trouble and Causes Trouble. *Clin. J. Am. Soc. Nephrol.* **2015**, *10*, 1459–1469. [CrossRef]
18. Ou, J.; Liu, K.; Jiang, J.; Wilson, D.A.; Liu, L.; Wang, F.; Wang, S.; Tu, Y.; Peng, F. Micro-/Nanomotors toward Biomedical Applications: The Recent Progress in Biocompatibility. *Small* **2020**, *16*, 1906184. [CrossRef]
19. Mitragotri, S.; Lahann, J. Physical approaches to biomaterial design. *Nat. Mater.* **2009**, *8*, 15–23. [CrossRef]
20. Liu, L.; Bai, T.; Chi, Q.; Wang, Z.; Xu, S.; Liu, Q.; Wang, Q. How to Make a Fast, Efficient Bubble-Driven Micromotor: A Mechanical View. *Micromachines* **2017**, *8*, 267. [CrossRef]
21. Huang, W.; Manjare, M.; Zhao, Y. Catalytic Nanoshell Micromotors. *J. Phys. Chem. C* **2013**, *117*, 21590–21596. [CrossRef]
22. Wang, Z.; Chi, Q.; Liu, L.; Liu, Q.; Bai, T.; Wang, Q. A Viscosity-Based Model for Bubble-Propelled Catalytic Micromotors. *Micromachines* **2017**, *8*, 198. [CrossRef] [PubMed]
23. Solovev, A.A.; Mei, Y.; Ureña, E.B.; Huang, G.; Schmidt, O.G. Catalytic Microtubular Jet Engines Self-Propelled by Accumulated Gas Bubbles. *Small* **2009**, *5*, 1688–1692. [CrossRef] [PubMed]
24. Shklyaev, S. Janus droplet as a catalytic micromotor. *Eur. Phys. Lett.* **2015**, *110*, 54002. [CrossRef]
25. Manjare, M.; Yang, B.; Zhao, Y.-P. Bubble Driven Quasioscillatory Translational Motion of Catalytic Micromotors. *Phys. Rev. Lett.* **2012**, *109*, 128305. [CrossRef]
26. Orozco, J.; Mercante, L.A.; Pol, R.; Merkoçi, A. Graphene-based Janus micromotors for the dynamic removal of pollutants. *J. Mater. Chem. A* **2016**, *4*, 3371–3378. [CrossRef]
27. Jurado-Sánchez, B.; Sattayasamitsathit, S.; Gao, W.; Santos, L.; Fedorak, Y.; Singh, V.V.; Orozco, J.; Galarnyk, M.; Wang, J. Self-Propelled Activated Carbon Janus Micromotors for Efficient Water Purification. *Small* **2015**, *11*, 499–506. [CrossRef]
28. Maria-Hormigos, R.; Jurado-Sanchez, B.; Vazquez, L.; Escarpa, A. Carbon Allotrope Nanomaterials Based Catalytic Micromotors. *Chem. Mater.* **2016**, *28*, 8962–8970. [CrossRef]
29. Wang, L.; Li, T.; Li, L.; Wang, J.; Song, W.; Zhang, G. Microrocket Based Viscometer. *ECS J. Solid State Sci. Technol.* **2015**, *4*, S3020–S3023. [CrossRef]
30. Zhao, G.; Nguyen, N.-T.; Pumera, M. Reynolds numbers influence the directionality of self-propelled microjet engines in the 10−4 regime. *Nanoscale* **2013**, *5*, 7277–7283. [CrossRef]
31. Sanchez, S.; Ananth, A.N.; Fomin, V.M.; Marlitt, V.; Schmidt, O.G. Superfast Motion of Catalytic Microjet Engines at Physiological Temperature. *J. Am. Chem. Soc.* **2011**, *133*, 14860–14863. [CrossRef] [PubMed]
32. Sokolov, I.L.; Cherkasov, V.R.; Tregubov, A.A.; Buiucli, S.R.; Nikitin, M.P. Smart materials on the way to theranostic nanorobots: Molecular machines and nanomotors, advanced biosensors, and intelligent vehicles for drug delivery. *Biochim. Biophys. Acta (BBA) Gen. Subj.* **2017**, *1861*, 1530–1544. [CrossRef] [PubMed]
33. Luo, M.; Feng, Y.; Wang, T.; Guan, J. Micro-/Nanorobots at Work in Active Drug Delivery. *Adv. Funct. Mater.* **2018**, *28*, 1706100. [CrossRef]
34. Wang, Y.; Tu, Y.; Peng, F. The Energy Conversion behind Micro-and Nanomotors. *Micromachines* **2021**, *12*, 222. [CrossRef]
35. Kagan, D.; Laocharoensuk, R.; Zimmerman, M.; Clawson, C.; Balasubramanian, S.; Kang, D.; Bishop, D.; Sattayasamitsathit, S.; Zhang, L.; Wang, J. Rapid Delivery of Drug Carriers Propelled and Navigated by Catalytic Nanoshuttles. *Small* **2010**, *6*, 2741–2747. [CrossRef]
36. Wu, Y.; Lin, X.; Wu, Z.; Möhwald, H.; He, Q. Self-Propelled Polymer Multilayer Janus Capsules for Effective Drug Delivery and Light-Triggered Release. *ACS Appl. Mater. Interfaces* **2014**, *6*, 10476–10481. [CrossRef]
37. Tu, Y.; Peng, F.; André, A.A.M.; Men, Y.; Srinivas, M.; Wilson, D.A. Biodegradable Hybrid Stomatocyte Nanomotors for Drug Delivery. *ACS Nano* **2017**, *11*, 1957–1963. [CrossRef]
38. Jang, B.; Wang, W.; Wiget, S.; Petruska, A.J.; Chen, X.; Hu, C.; Hong, A.; Folio, D.; Ferreira, A.; Pané, S.; et al. Catalytic Locomotion of Core–Shell Nanowire Motors. *ACS Nano* **2016**, *10*, 9983–9991. [CrossRef]
39. Popescu, M.; Uspal, W.; Dietrich, S. Self-diffusiophoresis of chemically active colloids. *Eur. Phys. J. Spec. Top.* **2016**, *225*, 2189–2206. [CrossRef]
40. Velegol, D.; Garg, A.; Guh, R.; Kar, A.; Kumara, M. Origins of concentration gradients for diffusiophoresis. *Soft Matter* **2016**, *12*, 4686–4703. [CrossRef]
41. Ibele, M.; Mallouk, T.E.; Sen, A. Schooling Behavior of Light-Powered Autonomous Micromotors in Water. *Angew. Chem. Int. Ed.* **2009**, *48*, 3308–3312. [CrossRef] [PubMed]
42. Chi, Q.; Wang, Z.; Tian, F.; You, J.A.; Xu, S. A Review of Fast Bubble-Driven Micromotors Powered by Biocompatible Fuel: Low-Concentration Fuel, Bioactive Fluid and Enzyme. *Micromachines* **2018**, *9*, 537. [CrossRef]
43. Gao, W.; Sattayasamitsathit, S.; Orozco, J.; Wang, J. Highly Efficient Catalytic Microengines: Template Electrosynthesis of Polyaniline/Platinum Microtubes. *J. Am. Chem. Soc.* **2011**, *133*, 11862–11864. [CrossRef] [PubMed]

44. Sanchez, S.; Solovev, A.A.; Mei, Y.; Schmidt, O.G. Dynamics of Biocatalytic Microengines Mediated by Variable Friction Control. *J. Am. Chem. Soc.* **2010**, *32*, 13144–13145. [CrossRef] [PubMed]
45. Mou, F.; Chen, C.; Ma, H.; Yin, Y.; Wu, Q.; Guan, J. Self-Propelled Micromotors Driven by the Magnesium-Water Reaction and Their Hemolytic Properties. *Angew. Chem. Int. Ed.* **2013**, *52*, 7208–7212. [CrossRef]
46. Wang, J.; Cui, W.; Liu, Q.; Xing, Z.; Asiri, A.M.; Sun, X. Recent progress in cobalt-based heterogeneous catalysts for electrochemical water splitting. *Adv. Mater.* **2016**, *28*, 215–230. [CrossRef]
47. Ran, J.; Zhang, J.; Yu, J.; Jaroniec, M.; Qiao, S.Z. Earth-abundant cocatalysts for semiconductor-based photocatalytic water splitting. *Chem. Soc. Rev.* **2015**, *46*, 7787–7812. [CrossRef]
48. Ouyang, L.; Ma, M.; Huang, M.; Duan, R.; Wang, H.; Sun, L.; Zhu, M. Enhanced hydrogen generation properties of MgH2-based hydrides by breaking the magnesium hydroxide passivation layer. *Energies* **2015**, *8*, 4237–4252. [CrossRef]
49. Gao, W.; Pei, A.; Wang, J. Water-Driven Micromotors. *ACS Nano* **2012**, *6*, 8432–8438. [CrossRef]
50. Wu, Z.; Li, J.; de Ávila, B.E.F.; Li, T.; Gao, W.; He, Q.; Zhang, L.; Wang, J. Water-Powered Cell-Mimicking Janus Micromotor. *Adv. Funct. Mater.* **2015**, *15*, 7497–7501. [CrossRef]
51. Ma, X.; Sánchez, S. Bio-catalytic mesoporous Janus nano-motors powered by catalase enzyme. *Tetrahedron* **2017**, *73*, 4883–4886. [CrossRef]
52. Yuan, H.; Liu, X.; Wang, L.; Ma, X. Fundamentals and applications of enzyme powered micro/nano-motors. *Bioact. Mater.* **2021**, *6*, 1727–1749. [CrossRef] [PubMed]
53. Hortelão, A.C.; Patiño, T.; Perez-Jiménez, A.; Blanco, À.; Sánchez, S. Enzyme-Powered Nanobots Enhance Anticancer Drug Delivery. *Adv. Funct. Mater.* **2018**, *28*, 1705086. [CrossRef]
54. Schattling, P.; Thingholm, B.; Städler, B. Enhanced Diffusion of Glucose-Fueled Janus Particles. *Chem. Mater.* **2015**, *27*, 7412–7418. [CrossRef]
55. Simmchen, J.; Baeza, A.; Ruiz, D.; Esplandiu, M.J.; Vallet-Regí, M. Asymmetric Hybrid Silica Nanomotors for Capture and Cargo Transport: Towards a Novel Motion-Based DNA Sensor. *Small* **2012**, *8*, 2053–2059. [CrossRef]
56. Abdelmohsen, L.K.E.A.; Nijemeisland, M.; Pawar, G.M.; Janssen, G.-J.A.; Nolte, R.J.M.; Hest, J.C.M.v.; Wilson, D.A. Dynamic Loading and Unloading of Proteins in Polymeric Stomatocytes: Formation of an Enzyme-Loaded Supramolecular Nanomotor. *ACS Nano* **2016**, *10*, 2652–2660. [CrossRef]
57. Wilson, D.A.; Nolte, R.J.M.; Hest, J.C.M.v. Autonomous movement of platinum-loaded stomatocytes. *Nat. Chem.* **2012**, *4*, 268–274. [CrossRef]
58. Chen, X.-Z.; Hoop, M.; Mushtaq, F.; Siringil, E.; Hu, C.; Nelson, B.J.; Pané, S. Recent developments in magnetically driven micro-and nanorobots. *Appl. Mater. Today* **2017**, *9*, 37–48. [CrossRef]
59. Wang, H.; Pumera, M. Fabrication of Micro/Nanoscale Motors. *Chem. Rev.* **2015**, *115*, 8704–8735. [CrossRef]
60. Zhang, L.; Abbott, J.J.; Dong, L.; Kratochvil, B.E.; Bell, D.; Nelson, B.J. Artificial bacterial flagella: Fabrication and magnetic control. *Appl. Phys. Lett.* **2009**, *94*, 064107. [CrossRef]
61. Dreyfus, R.; Baudry, J.; Roper, M.L.; Fermigier, M.; Stone, H.A.; Bibette, J. Microscopic artificial swimmers. *Nature* **2005**, *437*, 862–865. [CrossRef] [PubMed]
62. Schamel, D.; Mark, A.G.; Gibbs, J.G.; Miksch, C.; Morozov, K.I.; Leshansky, A.M.; Fischer, P. Nanopropellers and Their Actuation in Complex Viscoelastic Media. *ACS Nano* **2014**, *8*, 8794–8801. [CrossRef] [PubMed]
63. Qiu, F.; Mhanna, R.; Zhang, L.; Ding, Y.; Fujita, S.; Nelson, B.J. Artificial bacterial flagella functionalized with temperature-sensitive liposomes for controlled release. *Sens. Actuators B Chem.* **2014**, *195*, 676–681. [CrossRef]
64. Qiu, F.; Fujita, S.; Mhanna, R.; Zhang, L.; Simona, B.R.; Nelson, B.J. Magnetic Helical Microswimmers Functionalized with Lipoplexes for Targeted Gene Delivery. *Adv. Funct. Mater.* **2015**, *25*, 1666–1671. [CrossRef]
65. Medina-Sánchez, M.; Schwarz, L.; Meyer, A.K.; Hebenstreit, F.; Schmidt, O.G. Cellular Cargo Delivery: Toward Assisted Fertilization by Sperm-Carrying Micromotors. *Nano Lett.* **2016**, *16*, 555–561. [CrossRef]
66. Schwarz, L.; Karnaushenko, D.D.; Hebenstreit, F.; Naumann, R.; Schmidt, O.G.; Medina-Sánchez, M. A Rotating Spiral Micromotor for Noninvasive Zygote Transfer. *Sci. Adv.* **2020**, *7*, 2000843. [CrossRef]
67. Gao, W.; Sattayasamitsathit, S.; Manesh, K.M.; Weihs, D.; Wang, J. Magnetically Powered Flexible Metal Nanowire Motors. *J. Am. Chem. Soc.* **2010**, *132*, 14403–14405. [CrossRef]
68. Jang, B.; Gutman, E.; Stucki, N.; Seitz, B.F.; Wendel-García, P.D.; Newton, T.; Pokki, J.; Ergeneman, O.; Pané, S.; Or, Y.; et al. Undulatory Locomotion of Magnetic Multilink Nanoswimmers. *Nano Lett.* **2015**, *15*, 4829–4833. [CrossRef]
69. Chang, S.T.; Paunov, V.N.; Petsev, D.N.; Velev, O.D. Remotely powered self-propelling particles and micropumps based on miniature diodes. *Nat. Mater* **2007**, *6*, 235–240. [CrossRef]
70. Ni, S.; Marini, E.; Buttinoni, I.; Wolf, H.; Isa, L. Hybrid colloidal microswimmers through sequential capillary assembly. *Soft Matter* **2017**, *13*, 4252–4259. [CrossRef]
71. Calvo-Marzal, P.; Manesh, K.M.; Kagan, D.; Balasubramanian, S.; Cardona, M.; Flechsig, G.-U.; Posner, J.; Wang, J. Electrochemically-triggered motion of catalytic nanomotors. *Chem. Commun.* **2009**, *2009*, 4509–4511. [CrossRef] [PubMed]
72. Fan, D.; Yin, Z.; Cheong, R.; Zhu, F.Q.; Cammarata, R.C.; Chien, C.L.; Levchenko, A. Subcellular-resolution delivery of a cytokine through precisely manipulated nanowires. *Nat. Nanotechnol.* **2010**, *5*, 545–551. [CrossRef] [PubMed]
73. Rahman, M.; Chowdhury, M.M.; Alam, K. Rotating-Electric-Field-Induced Carbon-Nanotube-Based Nanomotor in Water: A Molecular Dynamics Study. *Small* **2017**, *13*, 1603978. [CrossRef] [PubMed]

74. Guo, J.; Gallegos, J.J.; Tom, A.R.; Fa, D. Electric-Field-Guided Precision Manipulation of Catalytic Nanomotors for Cargo Delivery and Powering Nanoelectromechanical Devices. *ACS Nano* **2018**, *12*, 1179–1187. [CrossRef] [PubMed]
75. Xu, Y.; Bian, Q.; Wang, R.; Gao, J. Micro/nanorobots for precise drug delivery via targeted transport and triggered release: A review. *Int. J. Pharm.* **2022**, *616*, 121551. [CrossRef] [PubMed]
76. Singh, D.P.; Choudhury, U.; Fischer, P.; Mark, A.G. Non-Equilibrium Assembly of Light-Activated Colloidal Mixtures. *Adv. Mater.* **2017**, *29*, 1701328. [CrossRef] [PubMed]
77. Govorov, A.O.; Richardson, H.H. Generating heat with metal nanoparticles. *Nano Today* **2007**, *2*, 30–38. [CrossRef]
78. Tang, X.; Tang, S.-Y.; Sivan, V.; Zhang, W.; Mitchell, A.; Kalantar-zadeha, K.; Khoshmanesha, K. Photochemically induced motion of liquid metal marbles. *Appl. Phys. Lett.* **2013**, *103*, 174104. [CrossRef]
79. Li, W.; Wu, X.; Qin, H.; Zhao, Z.; Liu, H. Light-Driven and Light-Guided Microswimmers. *Adv. Funct. Mater.* **2016**, *26*, 3164–3171. [CrossRef]
80. Ryazantsev, Y.S.; Velarde, M.G.; Rubio, R.G.; Guzmán, E.; Ortega, F.; López, P. Thermo-and soluto-capillarity: Passive and active drops. *Adv. Colloid Interface Sci.* **2017**, *247*, 52–80. [CrossRef]
81. Ryazantsev, Y.S.; Velarde, M.G.; Guzmán, E.; Rubio, R.G.; Ortega, F.; Montoya, J.J. On the Autonomous Motion of Active Drops or Bubbles. *J. Colloid Interface Sci.* **2018**, *527*, 180–186. [CrossRef] [PubMed]
82. Jiang, H.; Lia, C.; Huang, X. Actuators based on liquid crystalline elastomer materials. *Nanoscale* **2013**, *5*, 5225–5240. [CrossRef] [PubMed]
83. Ikeda, T.; Mamiya, J.-i.; Yu, Y. Photomechanics of liquid-crystalline elastomers and other polymers. *Angew. Chem. Int. Ed.* **2007**, *46*, 506–528. [CrossRef] [PubMed]
84. Zhan, X.; Zheng, J.; Zhao, Y.; Zhu, B.; Cheng, R.; Wang, J.; Liu, J.; Tang, J.; Tang, J. From Strong Dichroic Nanomotor to Polarotactic Microswimmer. *Adv. Funct. Mater.* **2019**, *31*, 1903329. [CrossRef] [PubMed]
85. Wang, Q.; Dong, R.; Wang, C.; Xu, S.; Chen, D.; Liang, Y.; Ren, B.; Gao, W.; Cai, Y. Glucose-Fueled Micromotors with Highly Efficient Visible-Light Photocatalytic Propulsion. *ACS Appl. Mater. Interfaces* **2019**, *11*, 6201–6207. [CrossRef] [PubMed]
86. Xing, Y.; Zhou, M.; Du, X.; Li, X.; Li, J.; Xu, T.; Zhang, X. Hollow mesoporous carbon@Pt Janus nanomotors with dual response of H_2O_2 and near-infrared light for active cargo delivery. *Appl. Mater. Today* **2019**, *17*, 85–91. [CrossRef]
87. Xuan, M.; Wu, Z.; Shao, J.; Dai, L.; Si, T.; He, Q. Near Infrared Light-Powered Janus Mesoporous Silica Nanoparticle Motors. *J. Am. Chem. Soc.* **2016**, *138*, 6492–6497. [CrossRef]
88. He, W.; Frueh, J.; Hu, N.; Liu, L.; Gai, M.; He, Q. Guidable Thermophoretic Janus Micromotors Containing Gold Nanocolorifiers for Infrared Laser Assisted Tissue Welding. *Adv. Sci.* **2016**, *3*, 1600206. [CrossRef]
89. Srivastava, S.K.; Clergeaud, G.; Andresen, T.L.; Boisen, A. Micromotors for drug delivery in vivo: The road ahead. *Adv. Drug Deliv. Rev.* **2019**, *138*, 41–55. [CrossRef]
90. Doinikov, A.A. Acoustic radiation forces: Classical theory and recent advances. In *Recent Research Developments in Acoustics*; Pandalai, S.G., Ed.; Transworld Research Network: Trivandrum, India, 2003; Volume 1, pp. 39–67.
91. Wang, W.; Castro, L.A.; Hoyos, M.; Mallouk, T.E. Autonomous Motion of Metallic Microrods Propelled by Ultrasound. *ACS Nano* **2012**, *6*, 6122–6132. [CrossRef]
92. Kagan, D.; Benchimol, M.J.; Claussen, J.C.; Chuluun-Erdene, E.; Esener, S.; Wang, J. Acoustic Droplet Vaporization and Propulsion of Perfluorocarbon-Loaded Microbullets for Targeted Tissue Penetration and Deformation. *Angew. Chem. Int. Ed.* **2012**, *51*, 7519–7522. [CrossRef] [PubMed]
93. Garcia-Gradilla, V.; Orozco, J.; Sattayasamitsathit, S.; Soto, F.; Kuralay, F.; Pourazary, A.; Katzenberg, A.; Gao, W.; Shen, Y.; Wang, J. Functionalized Ultrasound-Propelled Magnetically Guided Nanomotors: Toward Practical Biomedical Applications. *ACS Nano* **2013**, *7*, 9232–9240. [CrossRef] [PubMed]
94. Xu, T.; Soto, F.; Gao, W.; Garcia-Gradilla, V.; Li, J.; Zhang, X.; Wang, J. Ultrasound-Modulated Bubble Propulsion of Chemically Powered Microengines. *J. Am. Chem. Soc.* **2014**, *136*, 8552–8555. [CrossRef]
95. Wan, M.; Li, T.; Chen, H.; Mao, C.; Shen, J. Biosafety, Functionalities, and Applications of Biomedical Micro/nanomotors. *Angew. Chem. Int. Ed.* **2021**, *60*, 13158–13176. [CrossRef] [PubMed]
96. Shao, J.; Xuan, M.; Zhang, H.; Lin, X.; Wu, Z.; He, Q. Chemotaxis-Guided Hybrid Neutrophil Micromotors for Targeted Drug Transport. *Angew. Chem. Int. Ed.* **2017**, *56*, 12935–12939. [CrossRef]
97. Xu, H.; Medina-Sánchez, M.; Magdanz, V.; Schwarz, L.; Hebenstreit, F.; Schmidt, O.G. Sperm-Hybrid Micromotor for Targeted Drug Delivery. *ACS Nano* **2018**, *12*, 327–337. [CrossRef]
98. Alapan, Y.; Yasa, O.; Schauer, O.; Giltinan, J.; Tabak, A.F.; Sourjik, V.; Sitti, M. Soft erythrocyte-based bacterial microswimmers for cargo delivery. *Sci. Robot.* **2018**, *3*, eaar4423. [CrossRef]
99. Williams, B.J.; Anand, S.V.; Rajagopalan, J.; Saif, M.T. A self-propelled biohybrid swimmer at low Reynolds number. *Nat. Commun.* **2014**, *5*, 3081. [CrossRef]
100. Tang, S.; Zhang, F.; Gong, H.; Wei, F.; Zhuang, J.; Karshalev, E.; Ávila, B.E.F.d.; Huang, C.; Zhou, Z.; Li, Z.; et al. Enzyme-powered Janus platelet cell robots for active and targeted drug delivery. *Sci. Robot.* **2020**, *5*, eaba6137. [CrossRef]
101. Stanton, M.M.; Park, B.W.; Miguel-Lopez, A.; Ma, X.; Sitti, M.; Sánchez, S. Biohybrid Microtube Swimmers Driven by Single Captured Bacteria. *Small* **2017**, *13*, 1603679. [CrossRef]
102. Ma, X.; Jannasch, A.; Albrecht, U.R.; Hahn, K.; Miguel-Lopez, A.; Schaffer, E.; Sánchez, S. Enzyme-Powered Hollow Mesoporous Janus Nanomotors. *Nano Lett.* **2015**, *15*, 7043–7050. [CrossRef] [PubMed]

103. Wang, J.; Xiong, Z.; Zheng, J.; Zhan, X.; Tang, J. Light-Driven Micro/Nanomotor for Promising Biomedical Tools: Principle, Challenge, and Prospect. *Acc. Chem. Res.* **2018**, *51*, 1957–1965. [CrossRef] [PubMed]
104. Nijemeisland, M.; Abdelmohsen, L.; Huck, W.; Wilson, D.A.; Van Hest, J.C. A Compartmentalized Out-of-Equilibrium Enzymatic Reaction Network for Sustained Autonomous Movement. *ACS Cent. Sci.* **2016**, *2*, 843–849. [CrossRef] [PubMed]
105. Grifantini, K. The State of Nanorobotics in Medicine. *IEEE Pulsa* **2019**, *10*, 13–17. [CrossRef] [PubMed]
106. Tezel, G.; Timur, S.S.; Kuralay, F.; Gürsoy, R.N.; Ulubayram, K.; Öner, L.; Eroğlu, H. Current Status of Micro/Nanomotors in Drug Delivery. *J. Drug Target.* **2021**, *29*, 29–45. [CrossRef] [PubMed]
107. Reinisova, L.; Hermanova, S.; Pumera, M. Micro/nanomachines: What is needed for them to become a real force in cancer therapy? *Nanoscale* **2019**, *11*, 6519–6532. [CrossRef]
108. Esteban-Fernández de Ávila, B.; Angsantikul, P.; Li, J.; Lopez-Ramirez, M.A.; Ramírez-Herrera, D.E.; Thamphiwatana, S.; Chen, C.; Delezuk, J.; Samakapiruk, R.; Ramez, V.; et al. Micromotor-enabled active drug delivery for in vivo treatment of stomach infection. *Nat. Commun.* **2017**, *8*, 272. [CrossRef]
109. Gao, W.; Dong, R.; Thamphiwatana, S.; Li, J.; Gao, W.; Zhang, L.; Wang, J. Artificial Micromotors in the Mouse's Stomach: A Step toward in Vivo Use of Synthetic Motors. *ACS Nano* **2015**, *9*, 117–123. [CrossRef]
110. Baylis, J.R.; Yeon, J.H.; Thomson, M.H.; Kazerooni, A.; Wang, X.; John, A.E.S.; Lim, E.B.; Chien, D.; Lee, A.; Zhang, J.Q.; et al. Self-propelled particles that transport cargo through flowing blood and halt hemorrhage. *Sci. Adv.* **2015**, *1*, e1500379. [CrossRef]
111. Kim, D.-I.; Lee, H.; Kwon, S.-H.; Sung, Y.J.; Song, W.K.; Park, S. Bilayer Hydrogel Sheet-Type Intraocular Microrobot for Drug Delivery and Magnetic Nanoparticles Retrieval. *Adv. Healthc. Mater.* **2020**, *9*, 2000118. [CrossRef]
112. Szatrowski, T.P.; Nathan, C.F. Production of large amounts of hydrogen peroxide by human tumor cells. *Cancer Res.* **1991**, *51*, 794–798. [CrossRef]
113. Villa, K.; Krejčová, L.; Novotný, F.; Heger, Z.; Sofer, Z.; Pumera, M. Cooperative Multifunctional Self-Propelled Paramagnetic Microrobots with Chemical Handles for Cell Manipulation and Drug Delivery. *Adv. Funct. Mater.* **2018**, *28*, 1804343. [CrossRef]
114. Gao, W.; Kagan, D.; Pak, O.S.; Clawson, C.; Campuzano, S.; Chuluun-Erdene, E.; Shipton, E.; Fullerton, E.E.; Zhang, L.; Lauga, E.; et al. Cargo-Towing Fuel-Free Magnetic Nanoswimmers for Targeted Drug Delivery. *Small* **2011**, *8*, 460–467. [CrossRef]
115. Feng, X.; Wang, l.; Chen, J.; Zeng, W.; Liu, R.; Lin, X.-J.; Ma, Y.; Jiahui, W. Self-propelled Manganese Oxide-Based Catalytic Micromotors for Drug Delivery. *RSC Adv.* **2016**, *6*, 65624–65630. [CrossRef]
116. Wu, Z.; Wu, Y.; He, W.; Lin, X.; Sun, J.; He, Q. Self-Propelled Polymer-Based Multilayer Nanorockets for Transportation and Drug Release. *Angew. Chem. Int. Ed.* **2013**, *52*, 7000–7003. [CrossRef] [PubMed]
117. Xuan, M.; Shao, J.; Lin, X.; Dai, L.; He, Q. Self-Propelled Janus Mesoporous Silica Nanomotors with Sub-100 nm Diameters for Drug Encapsulation and Delivery. *ChemPhysChem* **2014**, *15*, 2255–2260. [CrossRef]
118. Khezri, B.; Mohsen, S.; Mousavi, B.; Krejčová, L.; Heger, Z.; Sofer, Z.; Pumera, M. Ultrafast Electrochemical Trigger Drug Delivery Mechanism for Nanographene Micromachines. *Adv. Funct. Mater.* **2019**, *29*, 1806696. [CrossRef]

technologies

Review

Multimodal Semantic Segmentation in Autonomous Driving: A Review of Current Approaches and Future Perspectives

Giulia Rizzoli , **Francesco Barbato** and **Pietro Zanuttigh** *

Department of Information Engineering, University of Padova, Via Gradenigo 6/A, 35131 Padova, Italy; giulia.rizzoli@dei.unipd.it (G.R.); francesco.barbato@dei.unipd.it (F.B.)
* Correspondence: zanuttigh@dei.unipd.it

Abstract: The perception of the surrounding environment is a key requirement for autonomous driving systems, yet the computation of an accurate semantic representation of the scene starting from RGB information alone is very challenging. In particular, the lack of geometric information and the strong dependence on weather and illumination conditions introduce critical challenges for approaches tackling this task. For this reason, most autonomous cars exploit a variety of sensors, including color, depth or thermal cameras, LiDARs, and RADARs. How to efficiently combine all these sources of information to compute an accurate semantic description of the scene is still an unsolved task, leading to an active research field. In this survey, we start by presenting the most commonly employed acquisition setups and datasets. Then we review several different deep learning architectures for multimodal semantic segmentation. We will discuss the various techniques to combine color, depth, LiDAR, and other modalities of data at different stages of the learning architectures, and we will show how smart fusion strategies allow us to improve performances with respect to the exploitation of a single source of information.

Keywords: semantic segmentation; autonomous driving; multimodal; LiDAR; depth; modality fusion; deep learning

Citation: Rizzoli, G.; Barbato, F.; Zanuttigh, P. Multimodal Semantic Segmentation in Autonomous Driving: A Review of Current Approaches and Future Perspectives. *Technologies* **2022**, *10*, 90. https://doi.org/10.3390/technologies10040090

Academic Editors: Manoj Gupta, Eugene Wong and Gwanggil Jeon

Received: 22 June 2022
Accepted: 19 July 2022
Published: 25 July 2022

1. Introduction

In recent years, the autonomous driving field has experienced an impressive development, gaining a huge interest and expanding into many sub-fields that cover all aspects of the self-driving vehicle [1,2]. Examples are vehicle-to-vehicle communications [3], energy-storage devices, sensors [4], safety devices [5], and more. Among them, a fundamental field is scene understanding, a challenging Computer Vision (CV) task which deals with the processing of raw environmental data to construct a representation of the scene in front of the car that allows for the subsequent interaction with the environment (e.g., route planning, safety breaks engagement, packet transmission optimizations, etc.).

Scene understanding is the process of perceiving, analysing, and elaborating on an interpretation of an observed scene through a network of sensors [6]. It involves several complex tasks, from image classification, to more advanced ones like object detection and Semantic Segmentation (SS). The first task deals with the assignment of a global label to an input image; however, it is of limited use in the autonomous driving scenario, given the need for localizing the various elements in the environment [1]. The second task provides a more detailed description, localizing all identified objects and providing classification information for them [7]. The third task is the most challenging one, requiring the assignment of a class to each pixel of an input image. Due to the accurate semantic description this problem provides, it requires complex machine learning architectures and can be identified as the basic goal for a scene understanding pre-processor. It will be the subject of this work.

Most approaches for semantic segmentation were originally developed by using as input a single RGB camera (see Section 1.1 for a brief review of the task). However, the development of self-driving vehicles provided with many onboard sensors requires the generalization toward different modalities. The joint employment of the various data streams coming from the sensors (RGB, LiDAR, RADAR, stereo setups, etc.) allows a much more in-depth understanding of the environment.

The importance of multimodal data for autonomous driving applications came under the spotlight for the first time in the DARPA's Grand Challenge in 2007. All three teams on the podium underlined the necessity of such an approach, especially focusing on LiDAR perception systems.

In later years, LiDARs found many applications in the development of large-scale datasets for the training of deep architectures, e.g., the well-known KITTI [8] benchmark. Although such sensors provide very high accuracy, they come with a couple of major downsides, namely the high cost, the presence of delicate moving parts, and the fact that the depth map produced is sparse, rather than dense as the images from standard cameras. To tackle the first problem, more cost-effective, consumer-grade technologies have been used, such as stereo cameras, matricial Time-of-Flight or structured-light sensors [9,10]. On the other hand, these technologies are less accurate and suffer the effect of sunlight, claiming for approaches accounting for the unreliability of their data.

The investigation of approaches able to leverage multiple heterogeneous datastreams (like those produced by the aforementioned sensors) is the focus of this survey, wherein we investigate the various proposed approaches for multi-modal semantic segmentation in autonomous driving. In particular, we will focus on 2.5D scenes (RGB and depth, including stereo vision setups), 2D + 3D fusion (RGB and LiDAR), and also report some additional, specific setups (e.g., by also using thermal data).

1.1. Semantic Segmentation with Deep Learning

In this section, we will report the main approaches for semantic segmentation from a single data source, overviewing the task history and highlighting the landmarks of its evolution. A graphic example of a possible deployment of the task in autonomous driving scenarios is reported in Figure 1.

Figure 1. The car screen shows an example of semantic segmentation of the scene in front of the car.

Early approaches to semantic segmentation were based on the use of classifiers on small image patches [11–13], until the introduction of deep learning, which has enabled great improvements in this field as well.

The first approach to showcase the deep learning potential on this task is found in [14], which introduced an end-to-end convolutional model, the so-called Fully Convolutional Network (FCN) model, which is made of an encoder (or contraction segment) and a decoder (or expansion segment). The former maps the input into a low-resolution feature representation, which is then upsampled in the expansion block. The encoder (also called backbone) is typically a pretrained image classification network used as a feature extractor. Among these networks, popular choices are VGG [15], ResNet [16], or the more lightweight MobileNet [17].

Other remarkable architectures that followed FCN are ParseNet (Liu et al. [18]), which models global context directly rather than only relying on a larger receptive field, and DeconvNet (Noh et al. [19]) which proposes an architecture that contains overlapping deconvolution and unpooling layers to perform nonlinear upsampling, resulting in improving the performance at the cost of increasing the complexity of the training procedure.

A slightly different approach is proposed in the Feature Pyramid Network (FPN), developed by Lin et al. [20], where a bottom-up pathway, a top-down pathway, and lateral connections are used to join low-resolution and high-resolution features and to better propagate the low-level information into the network. Inspired by the FPN model, Chen et al. [21,22] proposes the DeepLab architecture, which adopts pyramid pooling modules wherein the feature maps are implicitly downsampled through the use of dilated convolutions of different rates. According to the authors, dilated convolutions allow for an exponential increase in the receptive field without a decrease in resolution or increase in parameters, as may happen in the traditional pooling or stride-based approaches. Chen et al. [22] further extended the work by employing depth-wise separable convolutions.

Nowadays the current objective in semantic segmentation consists of improving the multiscale feature learning while making a trade-off between keeping the inference time low and increasing the receptive field/upsampling capability.

One recent strategy is feature merging through attention-based methods. Recently, such techniques gained a lot of traction in Computer Vision, following its success in Natural Language Processing (NLP) tasks. The most famous approach of this class is the transformer architecture [23], introduced by Vaswani et al. in 2017 in an effort to reduce the dependence of NLP architectures on recurrent blocks, which have difficulty in handling long-time relationships between input data. This architecture has been adapted to the image understanding field in the Vision Tranformers (ViT) [24,25] work, which presents a convolution-free, transformer-based vision approach able to surpass previous state-of-the-art techniques in image classification (at the cost of much higher memory and training data requirements). Transformers have been used as well in semantic segmentation in numerous works [26–28].

Although semantic segmentation was originally tackled by RGB data, recently many researchers started investigating its application for LiDAR data [29–34]. The development of such approaches is supported by an ever-increasing number of datasets that provide labeled training samples, e.g., Semantic KITTI [35]. More in detail, PointNet [29,30] was one of the first general-purpose 3D pointcloud segmentation architectures, but although it achieved state-of-the-art results on indoor scenes, the sparse nature of LiDAR data led to a significant performance decrease in outdoor settings, limiting its applicability in autonomous driving scenarios. An evolution of this technique is developed in RandLANet [31], where an additional grid-based downsampling step is added as preprocessing, together with a feature aggregation based on random-centered KD-trees, to better handle the sparse nature of LiDAR samples. Other approaches are SqueezeSeg [33] and RangeNet [36], wherein the segmentation is performed through a CNN architecture. In particular, the LiDAR data is converted to a spherical coordinate representation allowing one to exploit 2D semantic segmentation techniques developed for images. The most recent and better-performing architecture is Cylinder3D [34], which exploits the prior knowledge of LiDAR topologies—in particular their cylindrical aspect—to better represent the data fed into the architecture. The underlying idea is that the density of points in each voxel is inversely dependent on the distance from the sensor; therefore the architecture samples the data according to a cylindrical grid, rather than a cuboid one, leading to a more uniform point density.

Given the recent growth in the availability of heterogeneous data, the exploitation of deep multimodal methods attracted great research interest (in Section 4, a detailed overview is reported). RGB data carries a wealth of visual and textual information, which in many cases has successfully been used to enable semantic segmentation. Nevertheless, depth measurements provide useful geometric cues, which help significantly in the discrimination of visual ambiguities, e.g., to distinguish between two objects with a similar appearance. Moreover, RGB cameras are sensitive to light and weather conditions which can lead to failures in outdoor environments [37]. Thermal cameras give temperature-based characteristics of the objects, which can better enhance the recognition of some objects, thereby improving the resilience of semantic scene understanding in challenging lighting conditions [38].

1.2. Outline

In this paper, we focus on analyzing and discussing deep learning based fusion methods in multimodal semantic segmentation. The survey is organized as follows: Section 2 describes the most common sensors and their arrangements in autonomous driving setups; in Section 3 the main datasets for this application are listed, pointing out their features with particular attention to data diversity; finally, Section 4 reports several methods to address data fusion. As a conclusion, in Section 5 the open challenges and future outlooks are remarked upon.

2. Multimodal Data Acquisition and Preprocessing

One of the key aspects of an autonomous driving system is the choice of the acquisition devices and the infrastructure which allows them to exchange information among themselves and to the central perception system. Over the years many setups have been proposed, introducing different cameras, LiDARs, RADAR sensors, GPS systems, and IMU units. In this section, we will report an overview of the most commonly employed sensors, their placement, and the post-processing steps needed to convert the provided data into a machine-friendly format. In Figure 2 we report an example of sensor setup. The vehicle shown was used during the generation of [39]. In the work, the authors remark how it was chosen to be close to real autonomous vehicles (such as TESLA https://www.tesla.com/autopilot (accessed on 21 July 2022), Waymo https://waymo.com/ (accessed on 21 July 2022), and Argo https://www.argoverse.org/ (accessed on 21 July 2022), ...).

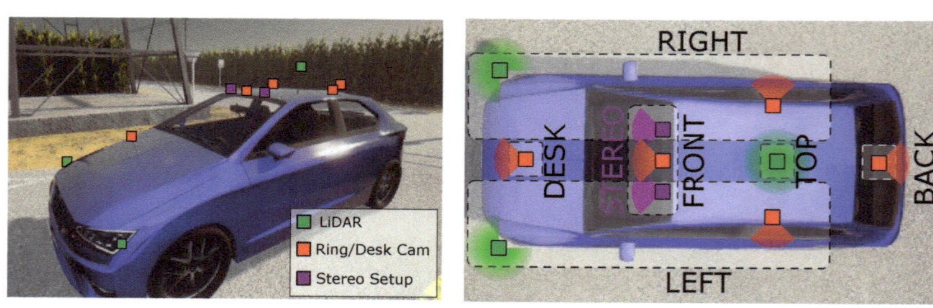

Figure 2. Figure (derived from the one in [39]) showcasing the multi-sensor setup used in the data collection.

2.1. RGB Cameras

Standard color cameras are employed in almost all setups (as underlined by the datasets reported in Section 3). Due to their limited cost, many systems rely on the combination of multiple cameras looking in different directions, both to improve the scene understanding and to allow a 360° Field-of-View (which may be helpful in the identification of obstacles/dangers coming from directions different than the heading one, but incurs additional processing costs related to the stitching and understanding of the bigger scene). Even if standard cameras provide an extremely useful representation of the scene, the data they provide suffers from some key limitations. First of all, they do not provide distance information, making it impossible to access precise information about the positions and sizes of the objects. Secondly, they are strongly affected by the illumination and weather conditions. Dark environments, direct sunlight, rain, or fog can strongly reduce the usefulness of the data provided by these devices [37]. This suggests that the combination of color cameras with other devices is a goal worth investigating, particularly with sensors resilient to the weaknesses of the cameras themselves.

2.2. Thermal Cameras

A thermal (or thermographic) camera is a special type of camera, which rather than acquiring information from the visible light spectrum (380∼750 nm) captures information in the near-infrared range (1∼14 µm) [38]. These wavelengths have the particular property of being the vector of irradiation heat, allowing to capture the heat sources in the scene (e.g., the heat produced by the vehicles).

This implies that they are able to work even in dark (in the usual sense) conditions because each object can be considered as a light source. Due to this property, these cameras can be very useful in night-time autonomous driving scenarios. A thermal camera output, in general, has two forms: the raw heatmap of the scenes (computed from the wavelength emitted by each object in the scene) or a color-coded post-processing. The second format is usually more meaningful than the first because the encoding uses special perceptive functions to map differences in temperature to differences in color [40].

2.3. Depth Cameras

Another approach to solving the problems affecting color cameras is to use depth sensors. As in the case of thermal cameras, the idea is to change the captured quantity from visible light to something more resilient to illumination/environmental changes. In the case of depth cameras, the acquired quantity is the distance-from-the-camera information for each pixel. Depth information cannot be directly inferred from a single standard image, and this has led to the development of multiple, complementary, active and passive techniques to acquire the depth information, e.g., stereo setups [9], matricial Time-of-Flight [10], RADARs [41], and LiDARs [42]. The last three actually belong to the same macro-class of techniques, which is ToF and differ in the way the time delay is computed (directly or indirectly) and on the medium used to extract the information (radio waves or light). In Table 1 we summarize in a qualitative manner the various sensors, classifying them depending on:

- the resilience to environmental conditions;
- the working range;
- the sparsity of the output depthmap; and
- the cost.

Table 1. Qualitative comparison between depth sensors. More details reported at [10,41,43].

Sensor	Range	Sparsity	Robustness	Direct Sun Perf.	Night Perf.	Cost
Passive Stereo	Far	Dense	Low	Medium	Low	Very Low
Active Stereo	Medium	Dense	Medium	Medium	Good	Low
Matricial ToF	Medium	Dense	High	Low	Good	Medium
LiDAR	Far	Sparse	High	Good	Good	High
RADAR	Far	Very Sparse	Medium	Good	Good	Low

2.3.1. Stereo Camera

Passive Stereo camera systems are one of the most common and cost-effective approaches for depth estimation. They employ two or more color cameras positioned at a known distance with respect to each other (commonly referred to as "baseline") to reconstruct a dense depthmap of the scene. The estimation procedure follows two main steps. The first is pixel matching between the two images (i.e., pixels representing the same location in the scene are found and coupled with each other). The second is the actual depth computation, wherein the distance between coupled pixels (disparity) is converted into the depthmap applying the well-known relation $d = {}^{bf}/_{p}$ pixel-wise, where d is the distance, b is the baseline, f is the camera focal length and p is the disparity. Clearly, the challenging part in the depth computation lies in the first step, the stereo matching, and many efficient algorithms were proposed to tackle the problem (from traditional computer vision algorithms like SGM [44] to recent deep learning-based strategies [45–47]).

In a similar fashion as for thermal data, alternative encodings for depthmaps exist. One example is HHA [48], which encodes in the three channels the horizontal disparity, the height above ground, and the angle that the pixel's local surface normal makes with the inferred gravity direction.

Active Stereo camera systems aid the stereo matching by adding a light projector to the stereo setup. This allows one to artificially increase the texture contrast, reducing the number of wrongly matched pixels. These systems, however, suffer in strong sunlight conditions, because the sunlight can overshadow or add noise to the projected light, thus strongly limiting the performance of the approach, that can instead be quite useful at night or in low light conditions.

2.3.2. Time-of-Flight

A matricial Time-of-Flight camera is a device able to calculate the distance between each scene point and the device [10]. This is done by measuring the round-trip time of the light traveling from the light transmitter, which illuminates the target to the photodetector. ToF sensors are categorized into indirect (iToF) and direct (dToF) sensors. In iToF the distance is measured by calculating the shift in phase of the original emitted light signal, which is continuously modulated, and the received light signal. iToF sensors have demonstrated good spatial resolution with a greater ability to detect multiple objects over a wide (but still limited by the camera optics) field of view (FoV) [49]. However, such sensors come with a significant drawback, that being that their light source modulation frequency is directly proportional to the maximum range, but inversely proportional to the precision attainable, thereby constraining them to a short range of typically less than 30 m. This limitation makes them less suited for autonomous driving applications. In dToF, the depth information is collected by measuring the time the light pulse takes to hit the target and return to the sensor, which requires the pulsing laser and the camera acquisition to be synchronized. dToF are typically employed also in LiDARs due to their longer range and reliability.

2.3.3. LiDAR

A LiDAR is a long-range, omnidirectional depth sensor, which comes with high robustness in geometry acquisition at the expense of a higher cost [39]. It employs one or multiple focused laser beams whose ToF is measured to generate a 3D representation of the environment in the form of a point cloud. Generally speaking, a point cloud consists of the 3D location and the intensity of the incident light collected at every frame. LiDARs have different operating principles [50]. In the scanning type, a collimated laser beam illuminates a single point at a time, and the beam is raster-scanned to illuminate the field of view point-by-point. In the flash type, a wide diverging laser beam illuminates the whole field of view in a single pulse. In the latter approach, the acquired frames do not need to be patched together, and the device is not sensitive to platform motion, which allows for more precise imaging. Motion can produce "jitter" in scanning LiDAR due to the delay in time as the laser rasters over the area.

Due to the sparsity and uneven distribution of point clouds, LiDAR-only perception tasks are challenging [50]. Whereas images are dense tensors, 3D point clouds can be represented in a variety of ways, resulting in several families of preprocessing algorithms. Besides directly representing the 3D coordinates of the acquired points, projection methods are the most intuitive approaches to having a direct correspondence with RGB images. Common choices for multi-modal applications consist of:

- spherical projection;
- perspective projection; and
- bird's-eye view.

In the first case, each 3D point is projected onto a sphere by using azimuth and zenith angles to create a spherical map. The result is a dense representation; however, it can differ in terms of size from the camera image. This does not happen in perspective projection where the 3D points are projected into the camera coordinate system; hence the depthmap has the same size. The main drawback of this method is that it leaves many pixels empty, and upsampling techniques are required to reconstruct the image. The latter approach, as the name suggests, directly provides the objects' positions on the ground plane. Although it preserves the objects' length and width, it loses height information and, as a result, some physical characteristics.

Point-based approaches utilize a raw pointcloud as input and provide point-by-point labeling as output. These algorithms can handle any unstructured pointcloud. As a direct consequence, the key challenge in processing raw pointclouds is extracting local contextual information. Several approaches were used to create an ordered feature sequence from unordered 3D LiDAR data, which was subsequently translated to 3D LiDAR data by by using convolutional deep networks [51].

- **Voxel-based** : convert 3D LiDAR data to voxels in order to represent structured data. These algorithms typically accept voxels as input and predict one semantic label for each voxel [32,34].
- **Graph-based**: create a graph by using 3D LiDAR data. A vertex generally represents a single point or a set of points, whereas edges indicate vertexes' adjacency connections [52,53].
- **Point Convolution**: establish a similarity between points e.g., by sorting the K-nearest points according to their spatial distance from the centers [29–31].
- **Lattice Convolution**: provide a transformation between pointclouds and sparse permutohedral lattices so that convolutions can be performed efficiently [54,55].

Despite their high cost and moving components (in spindle-type lidars, whereas other technologies like solid-state lidars do not have this issue), LiDARs are being used as part of the vision systems of several high-level autonomous vehicles.

2.3.4. RADAR

RADAR (Radio Detection and Ranging) sensors can also give distance information; however, depth information coupled with RGB data is rarely produced by them. RADARs send out radio waves to be reflected by an obstacle, measure the signal runtime, and use the Doppler effect to estimate the object's radial motion. They can withstand a variety of lighting and weather situations; however, due to their low resolution, semantic understanding with RADARs is difficult. Their application in driving is usually restricted to directional proximity sensors, usually to aid in cruise control or assistive parking. Nevertheless, some works [56,57] propose strategies that allow their use in semantic segmentation setups. An interesting approach to automatic RADAR samples labelling is presented in [58], wherein the authors exploit both image- and LiDAR-labeled samples to infer the correct RADAR-point classification.

2.4. Position and Navigation Systems

Many devices allow the absolute position and orientation of the vehicle to be established. Global Positioning System (GPS) receivers and Inertial Measurement Unit (IMU) are common examples of such devices. Global Navigation Satellite Systems (GNSS) were first utilized in cars as navigation tools in driver assistance features [59], but they are now also used in conjunction with HD Maps for autonomous vehicle path planning and autonomous vehicle self-localization. Internal vehicle information (i.e., "proprioceptive sensor") is provided by IMUs and odometers. IMUs measure the acceleration and rotational rates of cars and are currently employed in autonomous driving for accurate localization. These sensors can be leveraged to aid camera segmentation architectures in the creation of lane-level HD Maps [60]. On the other hand, it is possible to improve coarse GPS measurements through camera-vision systems [61].

3. Datasets

One of the biggest challenges involved in the use of deep learning-based architectures is the need for large amounts of labeled data, fundamental for their optimization [62]. This is reflected in a very active and diverse field [63,64] that deals with the generation (in case of synthetic datasets) or collection (in case of real-world datasets) and subsequent labeling of data suitable for training deep learning models. A fundamental task for autonomous driving that suffers greatly from the data availability problem is semantic segmentation. In this task, the action of producing a label coincides with assigning to each pixel in an image (or to each point in a pointcloud) a semantic class. The complexity of this task is the main reason for the huge time and cost involved in the collection of datasets for semantic segmentation. In Table 2 a high-level summary is reported for each of the datasets used in the methods described in Section 4 differentiating them by the type of scene content (e.g., indoor or outdoor).

In the following, we will focus on semantic segmentation datasets, with special attention to the current problems and challenges of the available datasets. For a comprehensive list of general datasets for autonomous driving applications one may refer to [63], and to [64] for RGB-D tasks. Very few large-scale (more than 25k labeled samples) semantic segmentation datasets are available for autonomous driving settings, and even fewer take care of the multimodal aspect of the sensors present in vehicles.

In Section 3, we will go over the most commonly used driving datasets that support this task, reporting their characteristics and classifying them according to the following criteria in Table 2:

- modalities provided (i.e., type of available sensors);
- tasks supported (i.e., provided labeling information);
- data variability offered (i.e., daytime, weather, season, location, etc.); and
- acquisition domain (i.e., real or synthetic).

For the dataset description, we will follow the order reported in Table 2, which summarizes the discussed datasets. Some of the dataset names were compressed into acronyms, the expanded name can be found at the end of the document in the abbreviations listing.

Table 2. Comparison between multi-modal datasets. Shorthand notation used: *Type* Real/Synthetic; *Cameras* Grayscale/Color/FishEye/Thermal/Polarization/Event/MultiSpectral/Depth; *Daytime* Morning/Day/Sunset/Night; *Location* City/Indoor/Outdoor/Region/Traffic (left/right-handed), † indicates that the cities/regions considered belong to the same state, + indicates that single views of the 3D scene are labeled, ∗ indicates variability with no control or categorization. The table is color-coded to indicate the scenarios present in each dataset: Driving, Exterior, In/Out, Interior.

Metadata					Sensors					Diversity					Labels			Size	
Name	Created	Update	Type	Cameras	LiDARs	Stereo	GT Depths	RADARs	IMU	Daytime	Seasons	Location	Weather	Env. Control	Sem Seg	Bboxes	Opt. Flow	Sequences	Labeled Sampl.
KITTI [8,65–67]	2012	2015	R	2G/2C	1	2	-	-	+	-	-	-	-	-	-	+	-	1(6h)	200
Cityscapes [68]	2016	2016	R	2C	-	1	-	-	+	-	-	27C †	-	-	+	-	-	-	5000
Lost and found [69]	2016	2016	R	2C	-	1	-	-	-	-	-	-	-	-	+	-	-	112	2104
Synthia [70–72]	2016	2019	S	1C	-	-	1	-	-	DS	+	-	2	-	+	-	-	-	9400
Virtual KITTI [73,74]	2016	2020	S	2C	-	1	1	-	+	MDS	-	-	4	+	+	+	+	35	17k
MSSSD/MF [75]	2017	2017	R	1C/1T	-	-	-	-	-	DN	-	-	-	-	+	-	-	-	1569
RoadScene-Seg [76]	2018	2018	R	1C/1T	-	-	-	-	-	DN	-	-	-	-	-	-	-	-	221
AtUlm [77]	2019	2019	R	1G	4	-	-	-	-	-	-	-	-	-	+	-	-	-	1446
nuScenes [78]	2019	2020	R	6C	1	1	-	5	+	-	-	T	-	-	+	+	-	-	40k
SemanticKITTI [35]	2019	2021	R	-	1	-	-	-	-	-	-	-	-	-	+	-	-	22	43,552
ZJU [79]	2019	2019	R	2C/1FE/1P	-	1	-	-	-	DN	-	-	-	-	-	-	-	-	3400
A2D2 [80]	2020	2020	R	6C	5	-	-	-	+	-	-	-	-	-	+	+	-	-	41,280
ApolloScape [81]	2020	2020	R	6C	2	1	-	-	+	*	-	4R †	*	-	+	+	-	-	140k
DDAD [82]	2020	2020	R	6C	4	-	-	-	-	-	-	2R	-	-	-	-	-	-	16,600
KITTI 360 [83]	2021	2021	R	2C/2FE	1	1	-	-	+	-	-	-	-	-	+	+	-	-	78k
WoodScape [84]	2021	2021	R	4FE	1	-	-	-	+	-	-	10C	-	-	+	+	-	-	10k
EventScape [85]	2021	2021	S	1C/1E	-	-	1	-	+	-	-	4C	-	-	+	+	-	743(2 h)	-
SELMA [39]	2022	2022	S	8C	3	3	7	-	-	DSN	-	7C	9	+	+	+	-	-	31k×27
Freiburg Forest [86]	2016	2016	R	2C/1MS	-	1	-	-	-	-	-	-	-	-	+	-	-	-	336
POLABOT [87]	2019	2019	R	2C/1P/1MS	-	1	-	-	-	-	-	-	-	-	+	-	-	-	175
SRM [88]	2021	2021	R	1C/1T	-	-	-	-	-	-	-	-	2	-	+	-	-	-	2458
SSW [88]	2021	2021	R	1C/1T	-	-	-	-	-	-	-	-	2	-	+	-	-	-	1571
MVSEC [89]	2018	2018	R	2G/2E	1	1	-	-	+	DN	-	IO	-	-	-	-	-	14(1h)	-
PST900 [90]	2019	2019	R	2C/1T	-	1	-	-	-	-	-	IO	-	-	+	-	-	-	4316
NYU-depth-v2 [91]	2012	2012	R	1C + 1D	-	-	1	-	-	-	-	-	-	-	+	-	-	-	1449 $^+$
SUN-RGBD [92]	2015	2015	R	1C + 1D	-	-	1	-	-	-	-	-	-	-	+	-	-	-	10k $^+$
2D-3D-S [93]	2017	2017	R	1C + 1D	-	-	1	-	-	-	-	-	-	-	+	-	-	-	270
ScanNet [94]	2017	2018	R	1C + 1D	-	-	1	-	-	-	-	-	-	-	+	-	-	-	1513
Taskonomy [95]	2018	2018	R	1C + 1D	-	-	1	-	-	-	-	-	-	-	∼	-	-	-	4 m $^+$

Summary

KITTI [8,65–67] was the first large-scale dataset to tackle the important issue of multimodal data in autonomous vehicles. The KITTI vision benchmark was introduced in 2012 and contains a real-world 6-h-long sequence recorded using a LiDAR, an IMU, and two stereo setups (with one grayscale and one color camera each). Although the complete suite is very extensive (especially for depth estimation and object detection), the authors did not focus much on the semantic labeling process, opting to label only 200 training samples for semantic (and instance) segmentation and for optical flow.

Cityscapes [68] became one of the most common semantic segmentation datasets for autonomous driving benchmarks. It is a real-world dataset containing 5000 finely labeled, high-definition (2048 × 1024) images captured in multiple German cities. Additionally, the authors provide 25,000 coarsely labeled samples—polygons rather than object borders, with many unlabeled areas (see Figure 3)—to improve deep architectures' performance through data variability. The data was captured with a calibrated and rectified stereo setup in high-visibility conditions, allowing the authors to provide high-quality estimated depthmaps for each of the 30,000 samples. Given its popularity in semantic segmentation settings, this dataset is also one of the most used for monocular depth estimation or 2.5D segmentation tasks.

Lost and Found [69] is an interesting road-scene dataset that tackles lost cargo scenarios, it includes pixel-level segmentation of the road and of the extraneous objects present on the surface. It was introduced in 2016 and includes around 2000 samples. The dataset comprises 112 stereo video sequences with 2104 annotated frames in a real-world scenario.

Synthia [70–72] is one of the oldest multimodal synthetic datasets providing labeled semantic segmentation samples. First introduced in 2016, it provides color, depth, and semantic information generated from the homonym simulator. The authors tackled data diversity by simulating the four seasons, by rendering the dataset samples from multiple PoVs, not only from road-view, but also from building height, and by considering day/night times. The dataset provides multiple versions, but only one supports (partially) the Cityscapes dataset label-set; it contains 9400 total samples.

Virtual KITTI [73,74] is an extension of the KITTI dataset. It is a synthetic dataset produced in Unity (https://unity.com/ (accessed on 21 July 2022)) which contains scenes modeled after the ones present in the original KITTI dataset. The synthetic nature of the dataset allowed the authors to produce a much greater number of labeled samples than those present in KITTI, while also maintaining a higher precision (due to the automatic labeling process). Unfortunately, the dataset does not provide labels for the LiDAR pointclouds.

MSSSD/MF [75] is a real-world dataset and one of the few that provides multispectral (thermal + color) information. It is of relatively small size, with only 1.5k samples, recorded in day and night scenes. Regardless, it represents an important benchmark for real-world applications, because thermal cameras are one of the few dense sensors resilient to low-visibility conditions such as fog or rain for which consumer-grade options exist.

RoadScene-Seg [76] is real-world dataset that provides 200 unlabeled road-scene images captured with an aligned color + infrared setup. Given the absence of labels, the only validation metric supported for architectures in this dataset is a qualitative evaluation by humans.

AtUlm [77] is a non-publicly available real-world dataset developed by Ulm University in 2019. It has been acquired with a grayscale camera and 4 LiDARs. In total the dataset contains 1446 finely annotated samples (grayscale images).

nuScenes [78] is a real-world dataset and one of the very few providing RADAR information. It is the standard for architectures aiming to use such sensor modality. The number of sensors provided is very impressive, as the dataset contains samples recorded from six top ring-cameras (two of which form a stereo setup), one top-central LiDAR, five ring RADARs placed at headlight level, and an IMU. The labeled samples are keyframes extracted with a frequency of 2 Hz from the recorded sequences, totaling 40k samples. The environmental variability lies in the recording location. The cities of Boston and Singapore were chosen as they offer different traffic handedness (Boston right-handed, Singapore left-handed).

Semantic KITTI [35] is an extension to the KITTI dataset. Here the authors took on the challenge of labeling in a point-wise manner all the LiDAR sequences recorded in the original set. It has rapidly become one of the most common benchmarks for LiDAR semantic segmentation, especially thanks to the significant number of samples made available.

ZJU [79] is a real-world dataset and the only among the one listed supporting the light polarization modality. It was introduced in 2019 and features 3400 labeled samples provided with color, (stereo) depth, light polarization, and an additional fish-eye camera view to cover the whole scene.

A2D2 [80] is another real-world dataset which focuses highly on the multimodal aspect of the data provided. It was recorded by a research team from the AUDI car manufacturer and provides five ring LiDARs, six ring cameras (two of which form a stereo setup) and an IMU. The semantic segmentation labels refer to both 2D images and LiDAR pointclouds, for a total of 41k samples. The daytime variability is very limited, offering only high-visibility day samples. The weather variability is slightly better, as it was changing throughout the recorded sequences, but no control over the conditions is offered.

ApolloScape [81] is a large-scale real-world dataset that supports a multitude of different tasks (semantic segmentation, lane segmentation, trajectory estimation, depth estimation, and more). As usual, we focus on the semantic segmentation task, for which ApolloScape provides ~150k labeled RGB images. Together with the color samples, the dataset also provides depth information. Unfortunately the depth maps contain only static objects, and all information about vehicles or other road occupants is missing. This precludes the possibility of directly exploiting the dataset in multimodal settings because a deployed agent wouldn't have access to such static maps.

DDAD [82] is a real-world dataset developed by the Toyota Research Institute, whose main focus is on monocular depth estimation. The sensors provided include six ring cameras and four ring LiDARs. The data was recorded in seven cities across two states: San Francisco, the Bay Area, Cambridge, Detroit, and Ann Arbor in the USA, and Tokyo and Odaiba in Japan. The dataset provides semantic segmentation labels only for the validation and test (non-public) sets, significantly restricting its use-case.

KITTI 360 [83] is a real-world dataset first released in 2020, which provides many different modalities (Stereo Color, LiDAR, Spherical, and IMU) and labeled segmentation samples for them. The labeling is performed in the 3D space, and the 2D labels are extracted by re-projection. In total, the dataset contains 78K labeled samples. Like KITTI, the dataset is organized in temporal sequences, recorded from a synchronized sensor setup mounted on a vehicle. As such, it offers very limited environmental variability.

WoodScape [84] is another real-world dataset providing color and LiDAR information. As opposed to its competitors, its 2D information is extracted only by using fish-eye cameras. In particular, the dataset provides information coming from four fish-eye ring cameras and a single top-LiDAR (360° coverage), recorded from more than ten cities in five different states. In total, the dataset contains 10k 2D semantic segmentation samples.

EventScape [85] is a very recent (2021) synthetic dataset developed by using the CARLA simulator [96], providing color, (ground truth) depth, event camera, semantic segmentation, bounding boxes, and IMU information for 743 sequences for a total of 2 h of video across four cities.

SELMA [39] is a very recent (2022) synthetic dataset developed in a modified CARLA simulator [96] whose goal is to provide multimodal data in a multitude of environmental conditions, while also allowing a researcher to control such conditions. It is heavily focused on semantic segmentation, providing labels for all of the sensors offered (seven co-placed RGB/depth cameras, and three LiDARs). The environmental variability takes the form of three daytimes (day, sunset, night), nine weather conditions (clear, cloudy, wet road, wet road and cloudy, soft/mid-level/heavy rain, mid-level/heavy fog), and 8 synthetic towns. The dataset contains 31k unique scenes recorded in all 27 environmental conditions, resulting in 800k samples for each sensor.

Figure 3. Example of finely (**left**) and coarsely (**right**) labeled Cityscapes [68] samples.

4. Multimodal Segmentation Techniques in Autonomous Driving

This section is the core of this work, wherein we present a detailed review of recent and well-performing approaches for multi-modal semantic segmentation.
We will start with a brief overview of the field and of the most common design choices, before moving to an in-depth description of the works, starting with RGB and depth data fusion in Section 4.1 (the most common choice). Then, we will discuss approaches combining RGB with LiDAR data in Section 4.2. Finally, approaches exploiting less conventional data sources (e.g., RADAR, event or thermal cameras) will be discussed in Section 4.4. Table 3 shows a summarized version of the methods discussed in the following sections, comparing them according to

- modalities used for the fusion;
- datasets used for training and validation;
- approach to feature fusion (e.g., sum, concatenation, attention, etc.); and
- fusion network location (e.g., encoder, decoder, specific modality branch, etc.).

On the other hand, in Table 4, we report the numerical score (mIoU) attained by the methods in three benchmark datasets, respectively: Cityscapes [68] for 2.5D SS in Table 4a, KITTI [8] for 2D + 3D SS in Table 4b and MSSSD/MF [75] for RGB + Thermal SS in Table 4c.

Table 3. Summary of recent multimodal semantic segmentation architectures. Modality shorthand: Dm, raw depth map; Dh, depth HHA; De, depth estimated internally; E, event camera; T, thermal; Lp, light polarization; Li, LiDAR; Ls, LiDAR spherical; F, optical flow. Location: D, decoder; E, encoder. Direction: D, decoder; C, color; B, bi-directional; M, other modality.

Metadata				Fusion Approach						Fusion Architecture				
Name	Year	Dataset(s)	Modality(ies)	+	×	⊙	Ad-Hoc Block	Ad-Hoc Loss	Multi-Task	Location	Direction	Parallel Branches	Skip Connections	Multi-Level Fusion
LWM [97]	2021	[68,91,92]	DmDe	+	-	+	-	+	+	D	D/C	2	+	+
SSMA [98]	2019	[68,70,86,92,94]	DmDhT	-	+	+	+	+	-	E	D	2	+	+
CMX [99]	2022	[68,75,79,85,91–94]	EDhLpT	+	+	-	+	-	-	E	D/B	2	+	+
AsymFusion [100]	2021	[68,91,95]	Dm	+	-	-	+	-	-	E	B	2	-	+
SA-Gate [101]	2020	[68,91]	Dh	+	-	+	+	-	-	E	B	2	+	+
ESANet [102]	2021	[68,91,92]	Dm	+	-	-	-	-	-	E	C	2	+	+
DA-Gate [103]	2018	[68,91–93]	DmDe	-	-	-	-	+	-	N/A	N/A	1	-	-
RFBNet [104]	2019	[68,94]	Dh	+	+	+	+	-	-	E	B	2	-	+
MMSFB-snow [88]	2021	[68,70,88]	DmT	-	-	+	+	-	-	E	D	2	+	+
AdapNet [105]	2017	[68,70,86]	DmT	+	+	-	+	-	-	D	D	2	-	-
RFNet [106]	2020	[68,69]	Dm	+	-	-	+	-	-	E	C	2	+	+
RSSAWC [77]	2019	[68,77]	DmLi	+	-	+	-	-	-	E	D	2	-	-
PMF [107]	2021	[35,78]	Li	+	+	+	+	+	-	E	M	2	+	+
MDASS [108]	2019	[68,73]	DmF	+	-	-	-	-	-	E	D	2/3	+	+
CMFnet [109]	2021	[68,87]	DmLp	-	+	+	-	-	-	E	D/B	3+	-	+
CCAFFMNet [110]	2021	[75,76]	T	-	-	+	+	-	-	E	C	2	+	+
DooDLeNet [111]	2022	[75]	T	-	+	+	-	-	-	E	D	2	+	+
GMNet [112]	2021	[75,90]	T	+	+	-	+	-	+	E	D	2	+	+
FEANet [113]	2021	[75]	T	+	+	-	-	-	-	E	C	2	-	+
EGFNet [114]	2021	[75,90]	T	+	+	+	+	-	-	E	D	2	-	+
ABMDRNet [115]	2021	[75]	T	+	+	+	+	+	+	E	D	2	-	+
AFNet [116]	2021	[75]	T	+	+	-	+	-	-	E	D	2	-	-
FuseSeg-Thermal [117]	2021	[75]	T	+	-	+	-	-	-	E	C	2	+	+
RTFNet [106]	2019	[75]	T	+	-	-	-	-	-	E	C	2	-	+
FuseSeg-LiDAR [118]	2020	[8]	LsLi	-	-	+	-	-	-	E	M	2	+	+
RaLF3D [119]	2019	[8]	LsLi	+	-	+	-	-	-	E	D	2	+	+
DACNN [120]	2018	[91–93]	DmDh	+	-	-	-	-	-	E	D	2	-	-
xMUDA [121]	2020	[35,78,80]	Li	-	-	+	-	+	+	D	D	2	-	+

Table 4. Architectures Performance Comparison.

Name	Backbone	mIoU
	(a) Cityscapes dataset (2.5D SS).	
LWM [97]	ResNet101 [16]	83.4
SSMA [98]	ResNet50 [16]	83.29
CMX [99]	MiT-B4 [27]	82.6
AsymFusion [100]	Xception65 [122]	82.1
SA-Gate [101]	ResNet101 [16]	81.7
ESANet [102]	ResNet34 [16]	80.09
DA-Gate [103]	ResNet101 [16]	75.3
RFBNet [104]	ResNet50 [16]	74.8
MMSFB-snow [88]	ResNet50 [16]	73.8
AdapNet [105]	AdapNet [105]	71.72
RFNet [106]	ResNet18 [16]	69.37
RSSAWC [77]	ICNet [123]	65.09
MDASS [108]	VGG16 [15]	63.13
CMFnet [109]	VGG16 [15]	58.97
	(b) KITTI dataset (2D + 3D SS).	
PMF [107]	ResNet34 [16]	63.9
FuseSeg-LiDAR [118]	SqueezeNet [124]	52.1
RaLF3D [119]	SqueezeSeg [33]	37.8
xMUDA [121]	SparseConvNet3D [125] ResNet34 [16]	49.1
	(c) MSSSD/MF dataset (RGB + Thermal SS).	
CMX [99]	MiT-B4 [27]	59.7
CCAFFMNet [110]	ResNeXt50 [126]	58.2
DooDLeNet [111]	ResNet101 [16]	57.3
GMNet [112]	ResNet50 [16]	57.3
FEANet [113]	ResNet101 [16]	55.3
EGFNet [114]	ResNet152 [16]	54.8
ABMDRNet [115]	ResNet50 [16]	54.8
AFNet [116]	ResNet50 [16]	54.6
FuseSeg-Thermal [117]	DenseNet161 [127]	54.5
RTFNet [106]	ResNet152 [16]	53.2

Early attempts of multimodal semantic segmentation approaches combine RGB data and other modalities into multi-channel representations that were then fed into classical semantic segmentation networks based on the encoder–decoder framework [128,129]. This simple early fusion combination strategy is not too effective because it struggles to capture the different type of information carried by the different modalities (e.g., RGB images contain color and texture, whereas the other modalities typically better represent the spatial relations among objects). Within this reasoning, feature-level and late-fusion approaches have been developed. Fusion strategies have typically been categorized in early, feature and late-fusion strategies, depending on the fact that the fusion happens at the input level, in some intermediate stage or at the end of the understanding process. However, most recent approaches try to get the best of the three modalities by performing multiple fusion operations at different stages of the deep network [98,115,118].

A very common architectural choice is to adopt a multi-stream architecture for the encoder with a network branch processing each modality (e.g., a two-stream architecture for RGB and depth) and additional network modules connecting the different branches that combine modality-specific features into fused ones and/or carry information across the branches [98,99,101]. This hierarchical fusion strategy leverages multilevel features via progressive feature merging and generate a refined feature map. It entails fusing features at various levels rather than at early or late stages.

The feature fusion can take place through simple operations e.g., concatenation, element-wise addition, multiplication, etc., or a mixture of these, which is typically addressed as a fusion block, attention, or gate module. In this fashion, multi-level features can be fed from one modality to another, e.g., in [102] where depth cues are fed to the RGB branch, or mutually between modalities. The fused content can either reach the next layer or the decoder directly through skip connections [98].

The segmentation map is typically computed by a decoder taking in input the fused features and/or the output of some of the branches. Multiple decoders can also be used but it is a less common choice [121]. We also remark that both symmetrical approaches (by using the same architecture for all modalities) and asymmetrical ones (setting a main modality from which the output is computed and by using the others as side information) have been proposed. Finally, the loss function can be just the cross-entropy, or any other loss for semantic segmentation on the output maps. Furthermore multi-task strategies employing different losses on the estimate of some of the modalities from others have also been proposed as further described in the following sub-sections [97,103].

4.1. Semantic Segmentation from RGB and Depth Data

Wang et al. [100] claim that typical methods relying on fusing the multimodal features into one branch in a hierarchical manner are still lacking rich feature interactions. They design a bidirectional fusion scheme (AsymFusion) wherein they maintain the two branches with shared weights and promote the propagation of informative features at later fusion layers by making use of an asymmetric fusion block (see Figure 4). In their architecture, the encoders of the two modalities are sharing convolutional parameters (except for the batch normalization layers which are modality-specific) and at each layer a mutual fusion is performed introducing two operations: channel shuffle and pixel shift. The authors hold that features fused by symmetrical fusion methods at both branches tend to learn similar representations, therefore asymmetric operations might be significant. To avoid bringing redundant information at both the encoder branches, channel shuffle fuses two features by exchanging features corresponding to a portion of channels, whereas pixel shift constantly shifts one pixel on a feature map introducing zero padding.

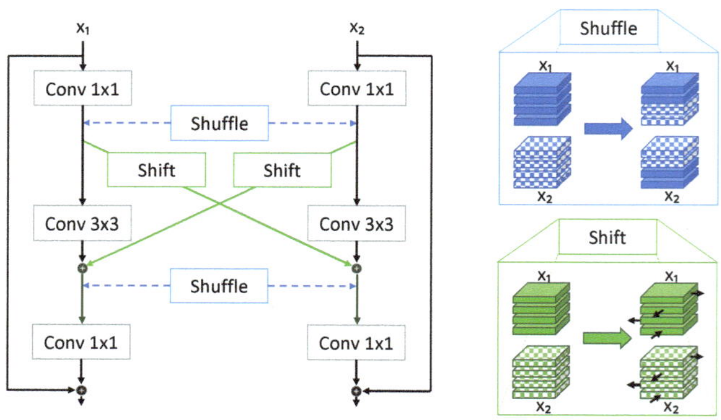

Figure 4. Asymmetric fusion block of [100].

Chen et al. [101] propose a unified and efficient cross-modality guided encoder whose architecture is depicted in Figure 5. It not only effectively re-calibrates RGB feature responses, but also takes into account the noise of the depth and accurately distills its information via multiple stages, alternately aggregating the two re-calibrated representations. The separation-and-aggregation gate (SA-Gate) is designed with two operations to ensure informative feature propagation between modalities. Formerly, feature re-calibration is performed for each individual modality. It is then followed by feature aggregation across modality boundaries. The operations are classified as feature separation and feature combination. The first consists of a global average pooling along the channel-wise dimensions of two modalities, which is followed by concatenation and a MLP operation to obtain an attention vector. This operation finds its motivation in filtering out exceptional depth activations that may overshadow confident RGB responses, reducing the probability of misleading information propagation. The same principle is implemented as a re-calibration step in a symmetric and bi-directional manner. Feature combination generates spatial-wise gates for both modalities to control information flow of each modality feature map with a soft attention mechanism. At each layer, the normalized output of the SA-Gate is added to each modality branch; thus the refined result will be passed on to the encoder's next layer, resulting in more precise and efficient encoding of the two modalities.

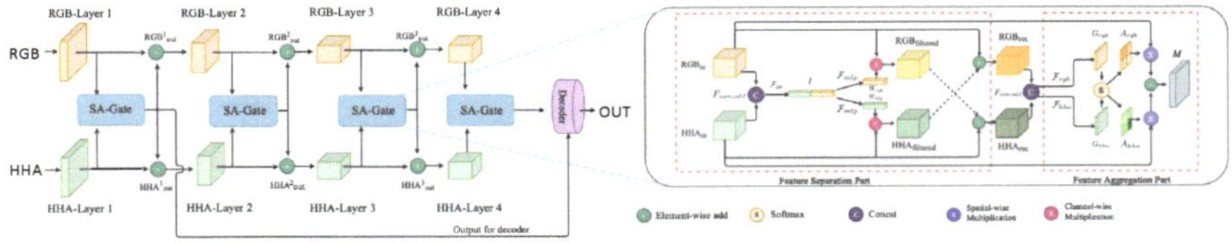

Figure 5. Figure from [101] showing its cross-modality feature propagation scheme. Adapted with authors' permission from [101]. Copyright 2020, Springer Nature Switzerland AG.

Valada et al. [98] present a multimodal fusion framework that incorporates an attention mechanism for effectively correlating multimodal features at mid- and high levels, and for better object boundary refinement (see Figure 6). Each modality is individually fed into a computationally efficient unimodal semantic segmentation architecture, AdapNet++ [105], that includes a strong encoder with skip refinement phases, as well as an efficient atrous spatial pyramid module and a decoder with multiscale residual units. By using the proposed Self-Supervised Model Adaptation (SSMA) block, the encoder uses a late fusion approach to join feature maps from modality-specific streams. In the SSMA block, the features are concatenated and re-weighted through a bottleneck which is used for dimensionality reduction and to improve the representational capacity.

Vachmanus et al. [88] adapt the SSMA architecture with the addition of another parallel bottleneck, with the aim of better capturing the temperature feature in snowy environments. To this end, they introduced two thermal datasets, SRM and SSW (see Table 2), while still testing their network on depth data.

A similar approach is presented in the work by **Zhang et al. [109]**, wherein the modalities are mixed together in a central branch through cross-attention mechanisms. Differently from SSMA, the weighting is performed in each branch separately and the features mixed correspond to the re-weighted outputs. Moreover, the final prediction is performed exploiting a statistics-aware module, able to extract more meaningful information from the concatenated multi-resolution features.

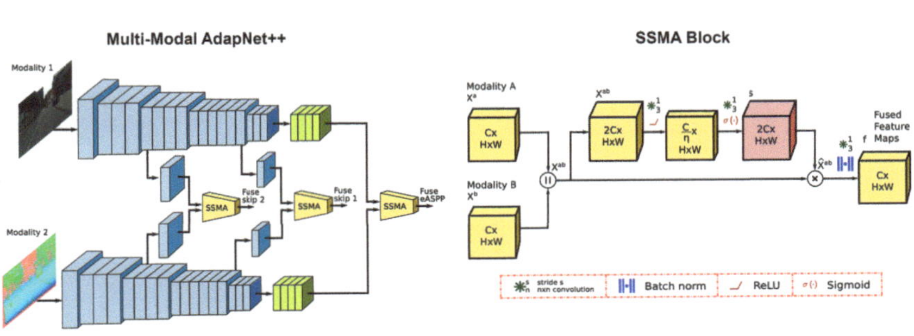

Figure 6. Figure from [98] that explains the work's multimodal semantic segmentation scheme. Reprinted with permission from the authors of [98]. Copyright 2019, Springer Nature Switzerland AG.

Deng et al. [104] adapt the SSMA model and propose an interactive fusion structure to compute the inter-dependencies between the two modality-specific streams and to propagate them through the network. Their residual fusion block (RFB) is composed of two residual units and a gating function unit which adaptively aggregates the features and generates complementary ones. These are fed to the residual units as well as the next layer. In this way, the gating unit exploit the complementary relationship in a soft-attention manner (see Figure 7).

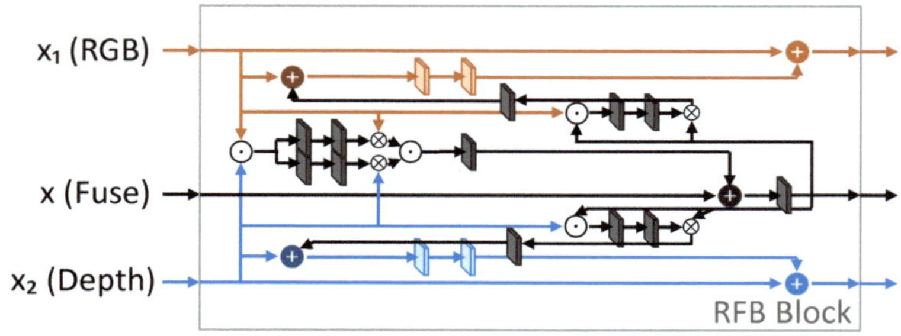

Figure 7. Architecture of the modified version of SSMA proposed by [104].

Seichter et al.'s [102] contribution, although mainly intended for indoor scenes, achieves good segmentation performance in outdoor settings as well. They target an efficient segmentation for embedded hardware, rather than by using high-end GPUs, meaning that their two branches encoder (depicted in Figure 8) is optimized to enable much faster inference than a single deep unimodal encoder. The depth encoder provides geometric information to the RGB one at several stages by using an attention mechanism. The latter aims for understanding which modality to focus on and which to suppress. It consists in an addition between the features reweighted through a squeeze-and-excitation (SA) module [130].

A similar approach is presented by Sun et al. [106], wherein the SA blocks and concatenation are used to merge the features into the RGB branch at multiple levels.

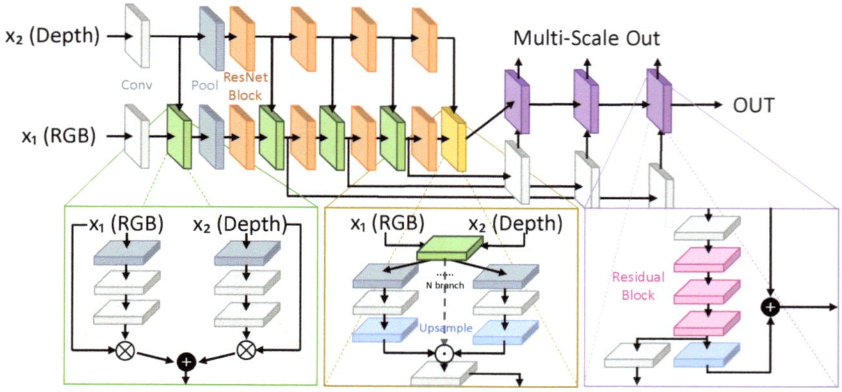

Figure 8. Two branches encoder architecture proposed in [102].

Kong et al. [103], differently from the common multi-scale approaches, exploit the benefit of processing the input image at a single fixed scale, but performing pooling at multiple convolutional dilate rates. Semantic segmentation is carried out by combining a CNN, used as a feature extractor, and a recurrent convolutional neural network, that includes a depth-aware gate. The gating module selects the size over which the features must be pooled, following the idea that larger depth values should have a smaller pooling field to precisely segment small objects. The module works with either estimated depth ("raw" measurements) or directly from monocular cues. A graphic representation of the fusion architecture may be found in Figure 9.

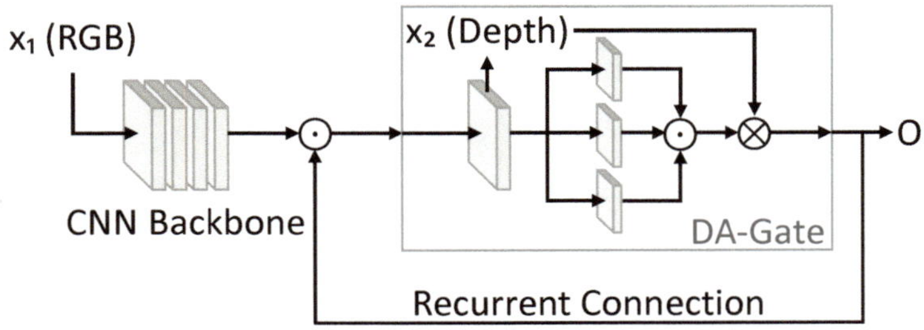

Figure 9. Fusion architecture proposed in [103].

Gu et al. [97] take a similar approach in the self-estimation of depth, noting how such information is not always available in real case scenarios. Therefore in their network (LWM) they establish a depth-privileged paradigm in which depth is provided only during the training process (Figure 10). They pay special attention to hard pixels, which are defined as pixels with a high probability of being misclassified. For this reason, they employ at different multi-scale outputs a loss weight module whose aim is to generate a loss weight map by additively fusing two metrics: depth prediction error and depth-aware segmentation error. The latter have the objective of measuring the "hardness" of a pixel. In the first case, for example, when the depth of two adjacent objects with a considerable distance gap is mispredicted, the delineation of the depth boundary between them may fail, resulting in the segmentation error. In the other, a local region of similar depth becomes a hard region when the categories of distinct subregions are confused due to similar visual appearance. Their network is based on a multi-task learning framework, which has one shared encoder branch and two distinct decoder branches for the segmentation and depth prediction branches. The final output, as well as four side outputs of the segmentation decoder branch, are feeded to the loss weight module.

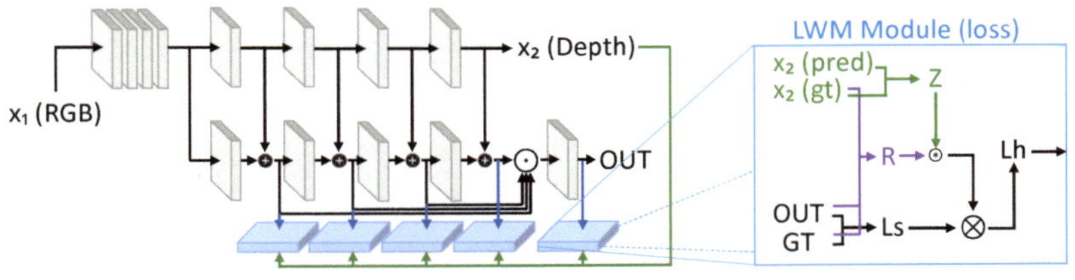

Figure 10. Architecture of [97], exploiting the LWM module.

Rashed et al. [108] focus on sensor fusion for an autonomous driving scenario wherein the dense depth map and the optical flow are considered. They establish a mid-fusion network (MDASS) that performs feature extraction for each modality separately and combines the modality cues at feature-level by using skip connections. In their experiments, they try to fuse at different stages by using a combination of two or three modalities. In addition, they analyzed the effect of using the ground truth measurement or a monocular depth estimate. A graphic representation of the architecture is available in Figure 11.

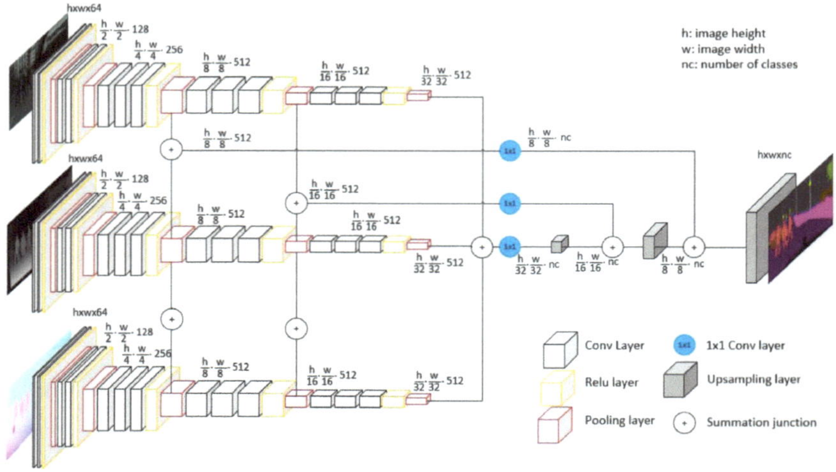

Figure 11. Architecture of [108] where the parallel, multimodal architecture is reported. Reprinted with permission from [108]. Copyright 2019, IEEE.

Liu et al. [99] propose an architecture (CMX, Figure 12) whose fundamental goal is to achieve enough flexibility to generalize across various multi-modal combinations (their approach is not limited to the fusion of RGB and depth data). They do so by exploiting a two-stream network (RGB and X-modality) with two ad-hoc modules for feature interaction and fusion: the cross-modal feature rectification module leverages the spatial and channel correlations to filter noise and calibrate the modalities, and the fusion module merges the rectified features by using a cross-attention mechanism. The latter finds its motivation behind the success of vision transformers and it is modeled into two stages. In the first stage, a cross-modal global reasoning is performed via a symmetric dual-path structure, and in the second stage a mixed channel embedding is applied to produce enhanced output features. The authors achieved remarkable results not just in fusing depth with RGB color, but also in fusing thermal data with color information.

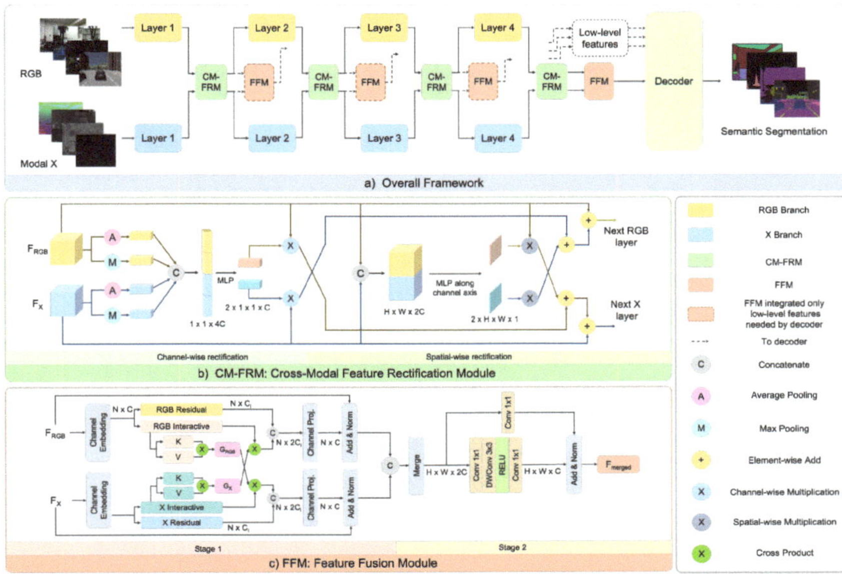

Figure 12. Figure from [99] where the CMX architecture and its modules are shown. Reprinted with permission from the authors of [99]. Copyright 2022, H. Liu.

4.2. Semantic Segmentation from RGB and LiDAR Data

LiDAR acquisitions offer an accurate spatial representation of the physical world. However, the pointclouds from these sensors are relatively sparse and lack color information, which results in a significant classification error in fine-grained segmentation [42]. Due of the sparsity and irregular structure of LiDAR data, the combination with standard camera data for multimodal sensor fusion remains a challenging problem. A possible workaround is to obtain a dense pointcloud by merging multiple LiDAR sensors as in the work by Pfeuffer et al. [77] (unfortunately the employed dataset is not public). However, most of the existing approaches use a projection of the original pointcloud over the color frame and try to find an alignment that can be exploited for the fusion between the cross-modality features. Pointcloud data processing has been tackled in Section 2, whereas the main fusion strategies for LiDAR data are now described.

Zhuang et al. [107] present an approach (PMF) whereby RGB data and LiDAR's projected data (using a perspective projection model) are fed to a two-stream architecture with residual-based fusion modules toward the LiDAR branch (see Figure 13). The modules are designed to learn the complementary features of color and LiDAR data (i.e., the appearance information from color data and the spatial information from pointclouds). The output of the network are two distinct semantic predictions that are used for the optimization through several losses. Among them, a perception-aware loss, based on the predictions and on the perceptual confidence, is introduced to be able to measure the difference between the two modalities. A similar approach is proposed by Madawi et al. [119], wherein RGB images and LiDAR data are converted to a polar-grid mapping representation to be fed into an hybrid early and mid-level fusion architecture. The first is achieved by establishing a mapping between the LiDAR scan points and the RGB pixels. The network is composed of two branches. The first uses the LiDAR measurements, whereas in the second the RGB images are concatenated with the depth and intensity map from LiDAR. The features from the two streams are then fused additively at different levels of the upsampling by using skip connections.

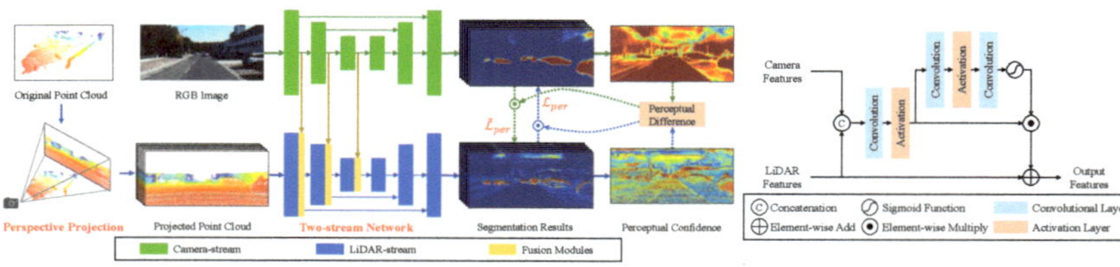

Figure 13. Figure from [107] showing the perception-aware multi-sensor fusion (PMF) architecture and fusion module. Reprinted with permission of the authors from [107]. Copyright 2021, IEEE.

Krispel et al. [118], in the architecture we refer to as FuseSeg-LiDAR, adopt a multi-layer concatenation of the features from the color information in a network for LiDAR data segmentation as depicted in Figure 14. The LiDAR data is spherically projected; hence alignment is required to enable a RGBD representation. Each RGB feature is the bilinear interpolation from the pixels adjacent to a non-discrete position computed as the alignment to the LiDAR range image by using a first-order polyharmonic spline interpolation.

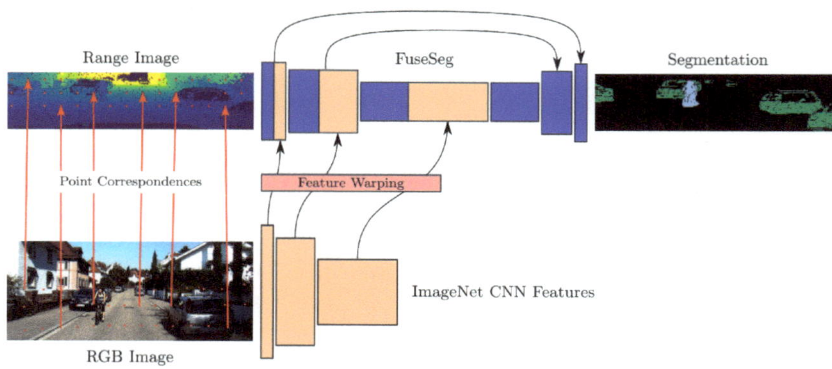

Figure 14. Figure from [118] that explains the FuseSeg-LiDAR architecture. Reprinted with permission from the authors of [118]. Copyright 2020, IEEE.

4.3. Pointcloud Semantic Segmentation from RGB and LiDAR Data

An alternative to the computation of a semantic map in the image space is to produce a semantically labeled pointcloud of the surrounding environment [51]. This approach is particularly well suited for LiDAR data, which typically have this structure.

Early works following this strategy aimed at 3D classification problems, where 3D representations were obtained by applying CNNs to 2D rendering pictures and combining multi-view features [131]. Then the attention moved to 3D semantic segmentation for indoor scenarios. Cheng et al. [132] proposed a method in which they back-project 2D image features into 3D coordinates. Then the network learns both 2D textural appearance and 3D structural features in a unified framework. The work of Jaritz et al. [133] instead aggregates 2D multi-view image features into 3D pointclouds, and then uses a point-based network to fuse the features in 3D canonical space to predict 3D semantic labels.

Jaritz et al. [121] provide a complex pipeline (xMUDA, see Figure 15) that can exchange 2D and 3D information to achieve an unsupervised domain adaptation for 3D semantic segmentation, leveraging the fact that LiDAR is robust to day-to-night domain shifts, and RGB camera images are deeply impacted by it. The architecture consists of a 2D and 3D network inspired by the U-Net model [134] that produces a feature vector of length equal to the number of points in the pointcloud. To obtain such a representation for the RGB image, the 3D points are projected to sample the 2D features at the corresponding pixel location. Each vector is fed to two classifiers to produce the segmentation prediction of the modality and the complementary one, obtaining four distinct segmentation outputs. With the aim of establishing a link between the 2D and 3D, they introduce a "mimicry" loss between the output probabilities. Each modality should be able to predict the output of the other. The final prediction is computed on the concatenated feature vectors of the two modalities.

Similarly to the previous approach, **Liu et al. [135]** adopt a 2D and 3D network called AUDA. Nevertheless, they believe that instead of sampling sparse 2D points in the source domain, the domain adaptation may benefit from using the entire 2D picture. The semantic prediction for the RGB image is achieved directly in this manner, and the calculated loss is used as supervision for the 3D prediction. They also offer an adaptive threshold-moving post-processing phase for boosting the recall rate for uncommon classes, as well as a cost-sensitive loss function to mitigate class imbalance.

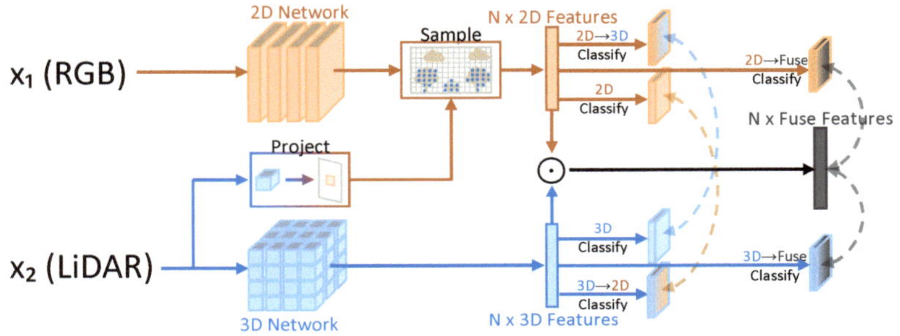

Figure 15. xMUDA [121] 2D/3D architecture.

4.4. *Semantic Segmentation from Other Modalities*

Even if color and 3D data are the two key sources of information for semantic understanding, other imaging techniques have also been exploited in combination with them. Some recent works combine color and 3D data with thermal imaging, radar acquisitions, and other sources of information.

Zhang et al. [115] employs a bi-directional image-to-image translation to reduce modality differences between RGB and thermal features (ABMDRNet, depicted in Figure 16). The RGB image is first fed to a feature extractor, then is upsampled and fed to a translation network, which is an encoder–decoder architecture, to obtain the corresponding thermal image. The same is done for the thermal image. The difference between the real and the pseudoimages is used as supervision to another decoder which takes as input the cross-modality features at multiple layers and fuses them. In their fusion strategy, the complementary information is exploited by re-weighting the importance of the single-modality features in a channel-dependent way, rather than in a spatial position-dependent way. Additionally, two modules are designed to exploit the multi-scale contextual information of the fused features.

Deng et al. [113] also addresses the fusion of RGB and thermal images by designing an encoder with a two-stream architecture, wherein each convolutional layer is followed by an attention module to re-weight the features. The idea is to enhance the difference between modalities, given that an object at night may be invisible in RGB maps but clearly visible in thermal maps. The information from the thermal branch is additively fused at each layer in the RGB one.

In **Zhou et al.'s GMNet [112]** the multi-layer RGB and thermal features are integrated by using two different fusion modules accounting the fact that deep-layer features provide richer contextual information. For the latter case, they design a densely connected structure to transmit global contextual inception data and a residual module to preserve original information. As opposed to other similar strategies, their decoder has multiple streams wherein different level features are joined. The semantic prediction is decoupled in the foreground, background, and boundary maps which all contribute to the optimization of the model.

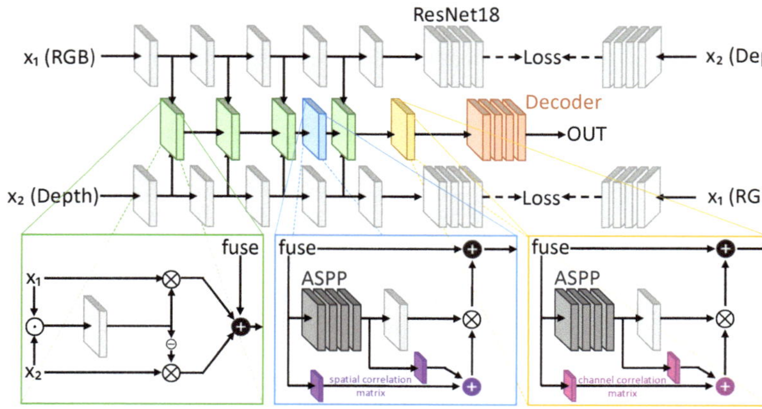

Figure 16. Architecture of the approach of Zhang et al. [115].

Sun et al. [106] propose RTFNet, whereby the encoder and the decoder are asymmetrically designed. The features are extracted through a large encoder for each modality whereas the upsampling is made by a small decoder. The modalities are combined into the RGB branch at multiple levels of the encoder.

Sun et al. [117] propose a two-branch architecture, FuseSeg-Thermal (Figure 17), in which the thermal feature maps are hierarchically added to the RGB feature maps in the RGB encoder in the first step of a two-stage fusion. The fused feature maps, except for the bottom one, are then fused again in the second stage with the matching feature maps in the decoder by tensor concatenation, which is inspired by the U-Net design [134].

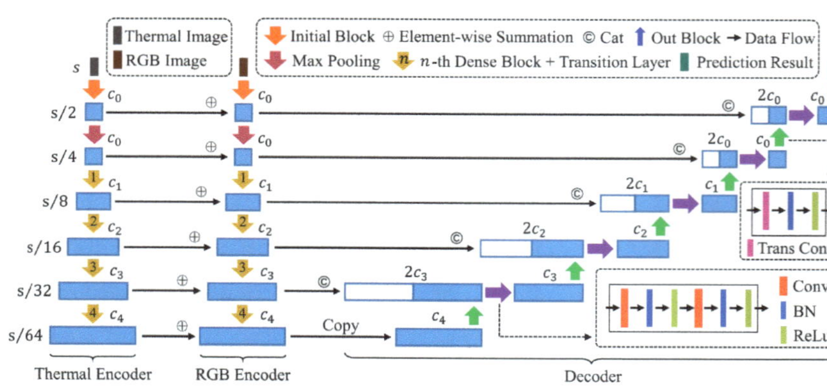

Figure 17. Figure from [117] showcasing the U-net-like architecture presented in the work. Reprinted with permission from the authors [117]. Copyright 2021, IEEE.

Another similar approach, which exploits the coarse-to-fine U-Net architecture, is the one presented by **Yi et al. [110]**, wherein thermal and color modalities are mixed through weights computed from multi-level attention blocks.

Similarly to previous approaches, in **Xu et al. [116]** a fusion module is used on the features extracted from a two-stream encoder to feed a single decoder. The modalities are scaled via cosine similarity, obtaining a channel-wise normalized product, and then the attention map is multiplied with the features that are then summed.

5. Conclusions and Outlooks

In this work, we overviewed the current approaches for multimodal road-scenes segmentation, with particular attention to the imaging modalities and datasets used. Several different approaches have been discussed and compared, showing how the combination of multiple inputs allows for improving the performance with respect to each modality when used alone. Even if there is a variety of different solutions, it is possible to notice a quite common design strategy based on having one network branch for each modality and some additional modules moving the information across them or merging the extracted features.

During our investigation, we were able to recognize some important issues that may be worth tackling by the research community. First of all, as is common when employing deep learning, data availability (and in particular labeled samples for supervised training) is a big bottleneck. This is particularly critical for semantic segmentation wherein labeling is extremely costly and the task itself is notably data-hungry. Therefore many—real and synthetic—datasets are required for optimization. Many of them have been introduced, but they are still far from being able to represent all the situations that can appear in a real-world driving scenario. In particular, the shortage is more critical for thermal data, where no "standard" large-scale dataset is currently available, precluding thorough training and evaluation, and leaving open the question of whether the availability of more data could make the exploitation of these sensors more effective (both alone or combined with standard cameras). On the other hand, a field where data is abundant but that is still mostly unexplored (due to the significant modality difference) is RGB+LiDAR fusion, especially when exploiting the LiDAR samples as raw pointclouds and not after projection. In fact, working in a fully three-dimensional environment can bring some additional understanding capabilities with respect to the 2D projection given by images. Also, the fusion of radar data with other approaches is still quite unexplored.

For the time being, there is no indication that one fusion scheme is preferable to the others. The search for an optimal fusion architecture is often driven by empirical results. In turn, current metrics compare the networks' accuracy on the semantic prediction directly rather than considering multi-modal resilience. The formulation of a metric for assessing multi-modal network robustness could help future improvements.

Author Contributions: Conceptualization, G.R., F.B. and P.Z.; investigation, G.R. and F.B.; resources, G.R., F.B. and P.Z.; writing—original draft preparation, G.R. and F.B.; writing—review and editing, G.R., F.B. and P.Z.; visualization, G.R. and F.B.; supervision, P.Z.; project administration, P.Z.; funding acquisition, P.Z. All authors have read and agreed to the published version of the manuscript.

Funding: This research was partly funded by the SID project "Semantic Segmentation in the Wild".

Institutional Review Board Statement: Not applicable.

Informed Consent Statement: Not applicable.

Data Availability Statement: All datasets used in the works analyzed in this survey can be found in Section 3 "Datasets", together with a reference to their presentation paper.

Conflicts of Interest: The authors declare no conflict of interest.

Abbreviations

The following abbreviations are used in this manuscript:

A2D2	Audi Autonomous Driving Dataset.
CV	Computer Vision.
DARPA	Defense Advanced Research Projects Agency.
DDAD	Dense Depth for Autonomous Driving.
dToF	Direct Time-of-Flight.
FCN	Fully Convolutional Network.
FoV	Field-of-View.
FPN	Feature Pyramid Network.
GNSS	Global Navigation Satellite Systems.
GPS	Global Positioning System.
IMU	Inertial Measurement Unit.
iToF	Indirect Time-of-Flight.
LiDAR	Light Detection and Ranging.
mIoU	mean Intersection over Union.
MLP	Multi-Layer Perceptron.
MSSSD	Multi-Spectral Semantic Segmentation Dataset.
MVSEC	MultiVehicle Stereo Event Camera.
NLP	Natural Language Processing.
PoV	Point of View.
RADAR	Radio Detection and Ranging.
SELMA	SEmantic Large-scale Multimodal Acquisitions.
SGM	Semi-Global Matching.
SRM	Snow Removal Machine.
SS	Semantic Segmentation.
SSMA	Self-Supervised Model Adaptation.
SSW	Snowy SideWalk.
ToF	Time-of-Flight.
VGG	Visual Geometry Group.
ViT	Vision Tranformers.

References

1. Yurtsever, E.; Lambert, J.; Carballo, A.; Takeda, K. A survey of autonomous driving: Common practices and emerging technologies. *IEEE Access* **2020**, *8*, 58443–58469. [CrossRef]
2. Liu, L.; Lu, S.; Zhong, R.; Wu, B.; Yao, Y.; Zhang, Q.; Shi, W. Computing Systems for Autonomous Driving: State of the Art and Challenges. *IEEE Internet Things J.* **2021**, *8*, 6469–6486. [CrossRef]
3. Wang, J.; Liu, J.; Kato, N. Networking and Communications in Autonomous Driving: A Survey. *IEEE Commun. Surv. Tutor.* **2019**, *21*, 1243–1274. [CrossRef]
4. Broggi, A.; Buzzoni, M.; Debattisti, S.; Grisleri, P.; Laghi, M.C.; Medici, P.; Versari, P. Extensive Tests of Autonomous Driving Technologies. *IEEE Trans. Intell. Transp. Syst.* **2013**, *14*, 1403–1415. [CrossRef]
5. Okuda, R.; Kajiwara, Y.; Terashima, K. A survey of technical trend of ADAS and autonomous driving. In Proceedings of the Technical Papers of 2014 International Symposium on VLSI Design, Automation and Test, Hsinchu, Taiwan, 28–30 April 2014; pp. 1–4. [CrossRef]
6. Bremond, F. Scene Understanding: Perception, Multi-Sensor Fusion, Spatio-Temporal Reasoning and Activity Recognition. Ph.D. Thesis, Université Nice Sophia Antipolis, Nice, France, 2007.
7. Gu, Y.; Wang, Y.; Li, Y. A survey on deep learning-driven remote sensing image scene understanding: Scene classification, scene retrieval and scene-guided object detection. *Appl. Sci.* **2019**, *9*, 2110. [CrossRef]
8. Geiger, A.; Lenz, P.; Urtasun, R. Are we ready for Autonomous Driving? The KITTI Vision Benchmark Suite. In Proceedings of the Conference on Computer Vision and Pattern Recognition (CVPR), Providence, RI, USA, 16–21 June 2012.
9. Fan, R.; Wang, L.; Bocus, M.J.; Pitas, I. Computer stereo vision for autonomous driving. *arXiv* **2020**, arXiv:2012.03194.
10. Zanuttigh, P.; Marin, G.; Dal Mutto, C.; Dominio, F.; Minto, L.; Cortelazzo, G.M. *Time-of-Flight and Structured Light Depth Cameras*; Springer International Publishing: Cham, Switzerland, 2016.
11. Brostow, G.J.; Shotton, J.; Fauqueur, J.; Cipolla, R. Segmentation and recognition using structure from motion point clouds. In Proceedings of the European Conference on Computer Vision, Marseille, France, 12–18 October 2008; pp. 44–57.
12. Sturgess, P.; Alahari, K.; Ladicky, L.; Torr, P.H. Combining appearance and structure from motion features for road scene understanding. In Proceedings of the BMVC-British Machine Vision Conference, London, UK, 7–10 September 2009.

13. Zhang, C.; Wang, L.; Yang, R. Semantic segmentation of urban scenes using dense depth maps. In Proceedings of the European Conference on Computer Vision, Heraklion, Greece, 5–11 September 2010; pp. 708–721.
14. Long, J.; Shelhamer, E.; Darrell, T. Fully convolutional networks for semantic segmentation. In Proceedings of the IEEE Conference on Computer Vision and Pattern Recognition, Boston, MA, USA, 7–12 June 2015; pp. 3431–3440.
15. Simonyan, K.; Zisserman, A. Very deep convolutional networks for large-scale image recognition. *arXiv* **2014**, arXiv:1409.1556.
16. He, K.; Zhang, X.; Ren, S.; Sun, J. Deep residual learning for image recognition. In Proceedings of the IEEE Conference on Computer Vision and Pattern Recognition, Honolulu, HI, USA, 21–26 July 2016; pp. 770–778.
17. Howard, A.G.; Zhu, M.; Chen, B.; Kalenichenko, D.; Wang, W.; Weyand, T.; Andreetto, M.; Adam, H. Mobilenets: Efficient convolutional neural networks for mobile vision applications. *arXiv* **2017**, arXiv:1704.04861.
18. Liu, W.; Rabinovich, A.; Berg, A.C. Parsenet: Looking wider to see better. *arXiv* **2015**, arXiv:1506.04579.
19. Noh, H.; Hong, S.; Han, B. Learning deconvolution network for semantic segmentation. In Proceedings of the IEEE International Conference on Computer Vision, Santiago, Chile, 11–18 December 2015; pp. 1520–1528.
20. Lin, T.Y.; Dollár, P.; Girshick, R.; He, K.; Hariharan, B.; Belongie, S. Feature pyramid networks for object detection. In Proceedings of the IEEE Conference on computer Vision and Pattern Recognition, Honolulu, HI, USA, 21–26 July 2017; pp. 2117–2125.
21. Chen, L.C.; Papandreou, G.; Kokkinos, I.; Murphy, K.; Yuille, A.L. Deeplab: Semantic image segmentation with deep convolutional nets, atrous convolution, and fully connected crfs. *IEEE Trans. Pattern Anal. Mach. Intell.* **2017**, *40*, 834–848. [CrossRef]
22. Chen, L.C.; Zhu, Y.; Papandreou, G.; Schroff, F.; Adam, H. Encoder-decoder with atrous separable convolution for semantic image segmentation. In Proceedings of the European Conference on Computer Vision (ECCV), Munich, Germany, 8–14 September 2018; pp. 801–818.
23. Vaswani, A.; Shazeer, N.; Parmar, N.; Uszkoreit, J.; Jones, L.; Gomez, A.N.; Kaiser, Ł.; Polosukhin, I. Attention is all you need. In Proceedings of the 31st Conference on Neural Information Processing Systems (NIPS 2017), Long Beach, CA, USA, 4–9 December 2017.
24. Dosovitskiy, A.; Beyer, L.; Kolesnikov, A.; Weissenborn, D.; Zhai, X.; Unterthiner, T.; Dehghani, M.; Minderer, M.; Heigold, G.; Gelly, S.; et al. An image is worth 16x16 words: Transformers for image recognition at scale. *arXiv* **2020**, arXiv:2010.11929.
25. Kolesnikov, A.; Dosovitskiy, A.; Weissenborn, D.; Heigold, G.; Uszkoreit, J.; Beyer, L.; Minderer, M.; Dehghani, M.; Houlsby, N.; Gelly, S.; et al. An image is worth 16x16 words: Transformers for image recognition at scale. In Proceedings of the International Conference on Learning Representations (ICLR), Virtual Event, Austria, 3–7 May 2021.
26. Zheng, S.; Lu, J.; Zhao, H.; Zhu, X.; Luo, Z.; Wang, Y.; Fu, Y.; Feng, J.; Xiang, T.; Torr, P.H.; et al. Rethinking semantic segmentation from a sequence-to-sequence perspective with transformers. In Proceedings of the IEEE/CVF Conference on Computer Vision and Pattern Recognition, Nashville, TN, USA, 20–25 June 2021; pp. 6881–6890.
27. Xie, E.; Wang, W.; Yu, Z.; Anandkumar, A.; Alvarez, J.M.; Luo, P. SegFormer: Simple and efficient design for semantic segmentation with transformers. *Adv. Neural Inf. Process. Syst.* **2021**, *34*, 12077–12090.
28. Strudel, R.; Garcia, R.; Laptev, I.; Schmid, C. Segmenter: Transformer for semantic segmentation. In Proceedings of the IEEE/CVF International Conference on Computer Vision, Montreal, QC, Canada, 10–17 October 2021; pp. 7262–7272.
29. Qi, C.R.; Su, H.; Mo, K.; Guibas, L.J. Pointnet: Deep learning on point sets for 3d classification and segmentation. In Proceedings of the IEEE Conference on Computer Vision and Pattern Recognition, Honolulu, HI, USA, 21–26 July 2017; pp. 652–660.
30. Qi, C.R.; Yi, L.; Su, H.; Guibas, L.J. Pointnet++: Deep hierarchical feature learning on point sets in a metric space. In Proceedings of the 31st Conference on Neural Information Processing Systems (NIPS 2017), Long Beach, CA, USA, 4–9 December 2017.
31. Hu, Q.; Yang, B.; Xie, L.; Rosa, S.; Guo, Y.; Wang, Z.; Trigoni, N.; Markham, A. Randla-net: Efficient semantic segmentation of large-scale point clouds. In Proceedings of the IEEE/CVF Conference on Computer Vision and Pattern Recognition, Seattle, WA, USA, 13–19 June 2020; pp. 11108–11117.
32. Tchapmi, L.; Choy, C.; Armeni, I.; Gwak, J.; Savarese, S. SEGCloud: Semantic Segmentation of 3D Point Clouds. In Proceedings of the 2017 International Conference on 3D Vision (3DV), Qingdao, China, 10–12 October 2017; pp. 537–547. [CrossRef]
33. Wu, B.; Wan, A.; Yue, X.; Keutzer, K. Squeezeseg: Convolutional neural nets with recurrent crf for real-time road-object segmentation from 3d lidar point cloud. In Proceedings of the 2018 IEEE International Conference on Robotics and Automation (ICRA), Brisbane, Australia, 21–25 May 2018; pp. 1887–1893.
34. Zhu, X.; Zhou, H.; Wang, T.; Hong, F.; Ma, Y.; Li, W.; Li, H.; Lin, D. Cylindrical and asymmetrical 3d convolution networks for lidar segmentation. In Proceedings of the IEEE/CVF Conference on Computer Vision and Pattern Recognition, Nashville, TN, USA, 20–25 June 2021; pp. 9939–9948.
35. Behley, J.; Garbade, M.; Milioto, A.; Quenzel, J.; Behnke, S.; Stachniss, C.; Gall, J. Semantickitti: A dataset for semantic scene understanding of lidar sequences. In Proceedings of the IEEE/CVF International Conference on Computer Vision, Seoul, Korea, 27 October–2 November 2019; pp. 9297–9307.
36. Milioto, A.; Vizzo, I.; Behley, J.; Stachniss, C. Rangenet++: Fast and accurate lidar semantic segmentation. In Proceedings of the 2019 IEEE/RSJ International Conference on Intelligent Robots and Systems (IROS), Macao, China, 3–8 November 2019; pp. 4213–4220.
37. Secci, F.; Ceccarelli, A. On failures of RGB cameras and their effects in autonomous driving applications. In Proceedings of the IEEE 31st International Symposium on Software Reliability Engineering (ISSRE), Coimbra, Portugal, 12–15 October 2020.
38. Gade, R.; Moeslund, T.B. Thermal cameras and applications: A survey. *Mach. Vis. Appl.* **2014**, *25*, 245–262. [CrossRef]

39. Testolina, P.; Barbato, F.; Michieli, U.; Giordani, M.; Zanuttigh, P.; Zorzi, M. SELMA: SEmantic Large-scale Multimodal Acquisitions in Variable Weather, Daytime and Viewpoints. *arXiv* **2022**, arXiv:2204.09788.
40. Moreland, K. Why we use bad color maps and what you can do about it. *Electron. Imaging* **2016**, *2016*, 1–6. [CrossRef]
41. Zhou, Y.; Liu, L.; Zhao, H.; López-Benítez, M.; Yu, L.; Yue, Y. Towards Deep Radar Perception for Autonomous Driving: Datasets, Methods, and Challenges. *Sensors* **2022**, *22*, 4208. [CrossRef]
42. Gao, B.; Pan, Y.; Li, C.; Geng, S.; Zhao, H. Are We Hungry for 3D LiDAR Data for Semantic Segmentation? A Survey of Datasets and Methods. *IEEE Trans. Intell. Transp. Syst.* **2022**, *23*, 6063–6081. [CrossRef]
43. Jang, M.; Yoon, H.; Lee, S.; Kang, J.; Lee, S. A Comparison and Evaluation of Stereo Matching on Active Stereo Images. *Sensors* **2022**, *22*, 3332. [CrossRef]
44. Hirschmuller, H. Accurate and efficient stereo processing by semi-global matching and mutual information. In Proceedings of the 2005 IEEE Computer Society Conference on Computer Vision and Pattern Recognition (CVPR'05), San Diego, CA, USA, 20–25 June 2005; Volume 2, pp. 807–814.
45. Zhou, K.; Meng, X.; Cheng, B. Review of stereo matching algorithms based on deep learning. *Comput. Intell. Neurosci.* **2020**, *2020*, 8562323. [CrossRef]
46. Li, J.; Wang, P.; Xiong, P.; Cai, T.; Yan, Z.; Yang, L.; Liu, J.; Fan, H.; Liu, S. Practical Stereo Matching via Cascaded Recurrent Network with Adaptive Correlation. In Proceedings of the 2022 IEEE Computer Society Conference on Computer Vision and Pattern Recognition (CVPR), New Orleans, LO, USA, 19–24 June 2022.
47. Tonioni, A.; Tosi, F.; Poggi, M.; Mattoccia, S.; Stefano, L.D. Real-time self-adaptive deep stereo. In Proceedings of the IEEE/CVF Conference on Computer Vision and Pattern Recognition, Long Beach, CA, USA, 15–20 June 2019; pp. 195–204.
48. Gupta, S.; Girshick, R.; Arbeláez, P.; Malik, J. Learning rich features from RGB-D images for object detection and segmentation. In Proceedings of the European Conference on Computer Vision, Zurich, Switzerland, 6–12 September 2014; pp. 345–360.
49. Padmanabhan, P.; Zhang, C.; Charbon, E. Modeling and analysis of a direct time-of-flight sensor architecture for LiDAR applications. *Sensors* **2019**, *19*, 5464. [CrossRef]
50. Li, Y.; Ibanez-Guzman, J. Lidar for autonomous driving: The principles, challenges, and trends for automotive lidar and perception systems. *IEEE Signal Process. Mag.* **2020**, *37*, 50–61. [CrossRef]
51. Camuffo, E.; Mari, D.; Milani, S. Recent Advancements in Learning Algorithms for Point Clouds: An Updated Overview. *Sensors* **2022**, *22*, 1357. [CrossRef] [PubMed]
52. Landrieu, L.; Simonovsky, M. Large-scale point cloud semantic segmentation with superpoint graphs. In Proceedings of the IEEE Conference on Computer Vision and Pattern Recognition, Salt Lake City, UT, USA, 18–23 June 2018; pp. 4558–4567.
53. Wang, Y.; Sun, Y.; Liu, Z.; Sarma, S.E.; Bronstein, M.M.; Solomon, J.M. Dynamic graph cnn for learning on point clouds. *ACM Trans. Graph. (TOG)* **2019**, *38*, 1–12. [CrossRef]
54. Su, H.; Jampani, V.; Sun, D.; Maji, S.; Kalogerakis, E.; Yang, M.H.; Kautz, J. Splatnet: Sparse lattice networks for point cloud processing. In Proceedings of the IEEE Conference on Computer Vision and Pattern Recognition, Salt Lake City, UT, USA, 18–23 June 2018; pp. 2530–2539.
55. Rosu, R.A.; Schütt, P.; Quenzel, J.; Behnke, S. Latticenet: Fast point cloud segmentation using permutohedral lattices. *arXiv* **2019**, arXiv:1912.05905.
56. Prophet, R.; Deligiannis, A.; Fuentes-Michel, J.C.; Weber, I.; Vossiek, M. Semantic segmentation on 3D occupancy grids for automotive radar. *IEEE Access* **2020**, *8*, 197917–197930. [CrossRef]
57. Ouaknine, A.; Newson, A.; Pérez, P.; Tupin, F.; Rebut, J. Multi-view radar semantic segmentation. In Proceedings of the IEEE/CVF International Conference on Computer Vision, Montreal, QC, Canada, 10–17 October 2021; pp. 15671–15680.
58. Kaul, P.; De Martini, D.; Gadd, M.; Newman, P. RSS-Net: Weakly-supervised multi-class semantic segmentation with FMCW radar. In Proceedings of the 2020 IEEE Intelligent Vehicles Symposium (IV), Las Vegas, NV, USA, 19 October–13 November 2020; pp. 431–436.
59. Bengler, K.; Dietmayer, K.; Farber, B.; Maurer, M.; Stiller, C.; Winner, H. Three decades of driver assistance systems: Review and future perspectives. *IEEE Intell. Transp. Syst. Mag.* **2014**, *6*, 6–22. [CrossRef]
60. Zhou, Y.; Takeda, Y.; Tomizuka, M.; Zhan, W. Automatic Construction of Lane-level HD Maps for Urban Scenes. In Proceedings of the 2021 IEEE/RSJ International Conference on Intelligent Robots and Systems (IROS), Prague, Czech Republic, 27 September–1 October 2021; pp. 6649–6656. [CrossRef]
61. Guo, C.; Lin, M.; Guo, H.; Liang, P.; Cheng, E. Coarse-to-fine Semantic Localization with HD Map for Autonomous Driving in Structural Scenes. In Proceedings of the 2021 IEEE/RSJ International Conference on Intelligent Robots and Systems (IROS), Prague, Czech Republic, 27 September–1 October 2021; pp. 1146–1153. [CrossRef]
62. Aggarwal, C.C. *Neural Networks and Deep Learning*; Springer: Berlin/Heidelberg, Germany, 2018; Volume 10, p. 978.
63. Yin, H.; Berger, C. When to use what data set for your self-driving car algorithm: An overview of publicly available driving datasets. In Proceedings of the 2017 IEEE 20th International Conference on Intelligent Transportation Systems (ITSC), Yokohama, Japan, 16–19 October 2017; pp. 1–8.
64. Lopes, A.; Souza, R.; Pedrini, H. A Survey on RGB-D Datasets. *arXiv* **2022**, arXiv:2201.05761.
65. Geiger, A.; Lenz, P.; Stiller, C.; Urtasun, R. Vision meets Robotics: The KITTI Dataset. *Int. J. Robot. Res. (IJRR)* **2013**, *32*, 1231–1237. [CrossRef]

66. Fritsch, J.; Kuehnl, T.; Geiger, A. A New Performance Measure and Evaluation Benchmark for Road Detection Algorithms. In Proceedings of the International Conference on Intelligent Transportation Systems (ITSC), The Hague, The Netherlands, 6–9 October 2013.
67. Menze, M.; Geiger, A. Object Scene Flow for Autonomous Vehicles. In Proceedings of the Conference on Computer Vision and Pattern Recognition (CVPR), Boston, MA, USA, 7–12 June 2015.
68. Cordts, M.; Omran, M.; Ramos, S.; Rehfeld, T.; Enzweiler, M.; Benenson, R.; Franke, U.; Roth, S.; Schiele, B. The cityscapes dataset for semantic urban scene understanding. In Proceedings of the IEEE Conference on Computer Vision and Pattern Recognition, Honolulu, HI, USA, 21–26 July 2016; pp. 3213–3223.
69. Pinggera, P.; Ramos, S.; Gehrig, S.; Franke, U.; Rother, C.; Mester, R. Lost and found: Detecting small road hazards for self-driving vehicles. In Proceedings of the 2016 IEEE/RSJ International Conference on Intelligent Robots and Systems (IROS), Daejeon, Korea, 9–14 October 2016; pp. 1099–1106.
70. Ros, G.; Sellart, L.; Materzynska, J.; Vazquez, D.; Lopez, A.M. The SYNTHIA Dataset: A Large Collection of Synthetic Images for Semantic Segmentation of Urban Scenes. In Proceedings of the IEEE Conference on Computer Vision and Pattern Recognition (CVPR), Las Vegas, NV, USA, 27–30 June 2016.
71. Hernandez-Juarez, D.; Schneider, L.; Espinosa, A.; Vazquez, D.; Lopez, A.M.; Franke, U.; Pollefeys, M.; Moure, J.C. Slanted Stixels: Representing San Francisco's Steepest Streets. In Proceedings of the British Machine Vision Conference (BMVC), London, UK, 4–7 September 2017.
72. Zolfaghari Bengar, J.; Gonzalez-Garcia, A.; Villalonga, G.; Raducanu, B.; Aghdam, H.H.; Mozerov, M.; Lopez, A.M.; van de Weijer, J. Temporal Coherence for Active Learning in Videos. In Proceedings of the IEEE International Conference in Computer Vision, Workshops (ICCV Workshops), Seoul, Korea, 27 Octorber–2 November 2019.
73. Gaidon, A.; Wang, Q.; Cabon, Y.; Vig, E. Virtual Worlds as Proxy for Multi-Object Tracking Analysis. In Proceedings of the IEEE Conference on Computer Vision and Pattern Recognition (CVPR), Honolulu, HI, USA, 21–26 July 2016.
74. Cabon, Y.; Murray, N.; Humenberger, M. Virtual kitti 2. *arXiv* **2020**, arXiv:2001.10773.
75. Ha, Q.; Watanabe, K.; Karasawa, T.; Ushiku, Y.; Harada, T. MFNet: Towards real-time semantic segmentation for autonomous vehicles with multi-spectral scenes. In Proceedings of the 2017 IEEE/RSJ International Conference on Intelligent Robots and Systems (IROS), Vancouver, BC, Canada, 24–28 September 2017; pp. 5108–5115. [CrossRef]
76. Xu, H.; Ma, J.; Le, Z.; Jiang, J.; Guo, X. FusionDN: A Unified Densely Connected Network for Image Fusion. In Proceedings of the Thirty-Fourth AAAI Conference on Artificial Intelligence, New York, NY, USA, 7–12 February 2020.
77. Pfeuffer, A.; Dietmayer, K. Robust Semantic Segmentation in Adverse Weather Conditions by means of Sensor Data Fusion. In Proceedings of the 22th International Conference on Information Fusion (FUSION), Ottawa, ON, Canada, 2–5 July 2019; pp. 1–8.
78. Caesar, H.; Bankiti, V.; Lang, A.H.; Vora, S.; Liong, V.E.; Xu, Q.; Krishnan, A.; Pan, Y.; Baldan, G.; Beijbom, O. nuscenes: A multimodal dataset for autonomous driving. In Proceedings of the IEEE/CVF Conference on Computer Vision and Pattern Recognition, Seattle, WA, USA, 13–19 June 2020; pp. 11621–11631.
79. Xiang, K.; Yang, K.; Wang, K. Polarization-driven semantic segmentation via efficient attention-bridged fusion. *Opt. Express* **2021**, *29*, 4802–4820. [CrossRef]
80. Geyer, J.; Kassahun, Y.; Mahmudi, M.; Ricou, X.; Durgesh, R.; Chung, A.S.; Hauswald, L.; Pham, V.H.; Mühlegg, M.; Dorn, S.; et al. A2d2: Audi autonomous driving dataset. *arXiv* **2020**, arXiv:2004.06320.
81. Wang, P.; Huang, X.; Cheng, X.; Zhou, D.; Geng, Q.; Yang, R. The apolloscape open dataset for autonomous driving and its application. *IEEE Trans. Pattern Anal. Mach. Intell.* **2019**, *42*, 2702–2719.
82. Guizilini, V.; Ambrus, R.; Pillai, S.; Raventos, A.; Gaidon, A. 3D Packing for Self-Supervised Monocular Depth Estimation. In Proceedings of the IEEE/CVF Conference on Computer Vision and Pattern Recognition (CVPR), Seattle, WA, USA, 14–19 June 2020.
83. Liao, Y.; Xie, J.; Geiger, A. KITTI-360: A Novel Dataset and Benchmarks for Urban Scene Understanding in 2D and 3D. *arXiv* **2021**, arXiv:2109.13410.
84. Yogamani, S.; Hughes, C.; Horgan, J.; Sistu, G.; Varley, P.; O'Dea, D.; Uricár, M.; Milz, S.; Simon, M.; Amende, K.; et al. Woodscape: A multi-task, multi-camera fisheye dataset for autonomous driving. In Proceedings of the IEEE/CVF International Conference on Computer Vision, Seoul, Korea, 27 Octorber–2 November 2019; pp. 9308–9318.
85. Gehrig, D.; Rüegg, M.; Gehrig, M.; Hidalgo-Carrió, J.; Scaramuzza, D. Combining events and frames using recurrent asynchronous multimodal networks for monocular depth prediction. *IEEE Robot. Autom. Lett.* **2021**, *6*, 2822–2829. [CrossRef]
86. Valada, A.; Oliveira, G.L.; Brox, T.; Burgard, W. Deep multispectral semantic scene understanding of forested environments using multimodal fusion. In Proceedings of the International Symposium on Experimental Robotics, Nagasaki, Japan, 3–8 October 2016; pp. 465–477.
87. Zhang, Y.; Morel, O.; Blanchon, M.; Seulin, R.; Rastgoo, M.; Sidibé, D. Exploration of Deep Learning-based Multimodal Fusion for Semantic Road Scene Segmentation. In Proceedings of the VISIGRAPP (5: VISAPP), Prague, Czech Republic, 25–27 February 2019; pp. 336–343.
88. Vachmanus, S.; Ravankar, A.A.; Emaru, T.; Kobayashi, Y. Multi-Modal Sensor Fusion-Based Semantic Segmentation for Snow Driving Scenarios. *IEEE Sens. J.* **2021**, *21*, 16839–16851. [CrossRef]
89. Zhu, A.Z.; Thakur, D.; Özaslan, T.; Pfrommer, B.; Kumar, V.; Daniilidis, K. The multivehicle stereo event camera dataset: An event camera dataset for 3D perception. *IEEE Robot. Autom. Lett.* **2018**, *3*, 2032–2039. [CrossRef]

90. Shivakumar, S.S.; Rodrigues, N.; Zhou, A.; Miller, I.D.; Kumar, V.; Taylor, C.J. PST900: RGB-Thermal Calibration, Dataset and Segmentation Network. In Proceedings of the 2020 IEEE International Conference on Robotics and Automation (ICRA), 31 May–31 August 2020; pp. 9441–9447.
91. Silberman, N.; Hoiem, D.; Kohli, P.; Fergus, R. Indoor segmentation and support inference from rgbd images. In Proceedings of the European Conference on Computer Vision, Florence, Italy, 7–13 October 2012; pp. 746–760.
92. Song, S.; Lichtenberg, S.P.; Xiao, J. SUN RGB-D: A RGB-D Scene Understanding Benchmark Suite. In Proceedings of the IEEE Conference on Computer Vision and Pattern Recognition (CVPR), Boston, MA, USA, 7–12 June 2015.
93. Armeni, I.; Sax, A.; Zamir, A.R.; Savarese, S. Joint 2D-3D-Semantic Data for Indoor Scene Understanding. *arXiv* **2017**, arXiv:cs.CV/1702.01105.
94. Dai, A.; Chang, A.X.; Savva, M.; Halber, M.; Funkhouser, T.; Niessner, M. ScanNet: Richly-Annotated 3D Reconstructions of Indoor Scenes. In Proceedings of the IEEE Conference on Computer Vision and Pattern Recognition (CVPR), Honolulu, HI, USA, 21–26 July 2017.
95. Zamir, A.R.; Sax, A.; Shen, W.; Guibas, L.J.; Malik, J.; Savarese, S. Taskonomy: Disentangling task transfer learning. In Proceedings of the IEEE Conference on Computer Vision and Pattern Recognition, Salt Lake City, UT, USA, 18–23 June 2018; pp. 3712–3722.
96. Dosovitskiy, A.; Ros, G.; Codevilla, F.; Lopez, A.; Koltun, V. CARLA: An Open Urban Driving Simulator. In Proceedings of the 1st Annual Conference on Robot Learning, Mountain View, CA, USA, 13–15 November 2017; pp. 1–16.
97. Gu, Z.; Niu, L.; Zhao, H.; Zhang, L. Hard pixel mining for depth privileged semantic segmentation. *IEEE Trans. Multimed.* **2020**, *23*, 3738–3751. [CrossRef]
98. Valada, A.; Mohan, R.; Burgard, W. Self-supervised model adaptation for multimodal semantic segmentation. *Int. J. Comput. Vis.* **2020**, *128*, 1239–1285. [CrossRef]
99. Liu, H.; Zhang, J.; Yang, K.; Hu, X.; Stiefelhagen, R. CMX: Cross-Modal Fusion for RGB-X Semantic Segmentation with Transformers. *arXiv* **2022**, arXiv:2203.04838.
100. Wang, Y.; Sun, F.; Lu, M.; Yao, A. Learning deep multimodal feature representation with asymmetric multi-layer fusion. In Proceedings of the 28th ACM International Conference on Multimedia, Seattle, WA, USA, 12–16 October 2020; pp. 3902–3910.
101. Chen, X.; Lin, K.Y.; Wang, J.; Wu, W.; Qian, C.; Li, H.; Zeng, G. Bi-directional cross-modality feature propagation with separation-and-aggregation gate for RGB-D semantic segmentation. In Proceedings of the European Conference on Computer Vision, Glasgow, UK, 23–28 August 2020; pp. 561–577.
102. Seichter, D.; Köhler, M.; Lewandowski, B.; Wengefeld, T.; Gross, H.M. Efficient rgb-d semantic segmentation for indoor scene analysis. In Proceedings of the 2021 IEEE International Conference on Robotics and Automation (ICRA), Xi'an, China, 30 May–5 June 2021; pp. 13525–13531.
103. Kong, S.; Fowlkes, C.C. Recurrent scene parsing with perspective understanding in the loop. In Proceedings of the IEEE Conference on Computer Vision and Pattern Recognition, Salt Lake City, UT, USA, 18–23 June 2018; pp. 956–965.
104. Deng, L.; Yang, M.; Li, T.; He, Y.; Wang, C. RFBNet: Deep multimodal networks with residual fusion blocks for RGB-D semantic segmentation. *arXiv* **2019**, arXiv:1907.00135.
105. Valada, A.; Vertens, J.; Dhall, A.; Burgard, W. Adapnet: Adaptive semantic segmentation in adverse environmental conditions. In Proceedings of the 2017 IEEE International Conference on Robotics and Automation (ICRA), Singapore, 29 May–3 June 2017; pp. 4644–4651.
106. Sun, Y.; Zuo, W.; Liu, M. Rtfnet: Rgb-thermal fusion network for semantic segmentation of urban scenes. *IEEE Robot. Autom. Lett.* **2019**, *4*, 2576–2583. [CrossRef]
107. Zhuang, Z.; Li, R.; Jia, K.; Wang, Q.; Li, Y.; Tan, M. Perception-aware Multi-sensor Fusion for 3D LiDAR Semantic Segmentation. In Proceedings of the IEEE/CVF International Conference on Computer Vision, Montreal, QC, Canada, 10–17 October 2021; pp. 16280–16290.
108. Rashed, H.; El Sallab, A.; Yogamani, S.; ElHelw, M. Motion and depth augmented semantic segmentation for autonomous navigation. In Proceedings of the IEEE/CVF Conference on Computer Vision and Pattern Recognition Workshops, Long Beach, CA, USA, 15–20 June 2019.
109. Zhang, Y.; Morel, O.; Seulin, R.; Mériaudeau, F.; Sidibé, D. A central multimodal fusion framework for outdoor scene image segmentation. *Multimed. Tools Appl.* **2022**, *81*, 12047–12060. [CrossRef]
110. Yi, S.; Li, J.; Liu, X.; Yuan, X. CCAFFMNet: Dual-spectral semantic segmentation network with channel-coordinate attention feature fusion module. *Neurocomputing* **2022**, *482*, 236–251. [CrossRef]
111. Frigo, O.; Martin-Gaffé, L.; Wacongne, C. DooDLeNet: Double DeepLab Enhanced Feature Fusion for Thermal-color Semantic Segmentation. In Proceedings of the 2022 IEEE Computer Society Conference on Computer Vision and Pattern Recognition (CVPR), New Orleans, LO, USA, 19–24 June 2022; pp. 3021–3029
112. Zhou, W.; Liu, J.; Lei, J.; Yu, L.; Hwang, J.N. GMNet: Graded-Feature Multilabel-Learning Network for RGB-Thermal Urban Scene Semantic Segmentation. *IEEE Trans. Image Process.* **2021**, *30*, 7790–7802. [CrossRef]
113. Deng, F.; Feng, H.; Liang, M.; Wang, H.; Yang, Y.; Gao, Y.; Chen, J.; Hu, J.; Guo, X.; Lam, T.L. FEANet: Feature-Enhanced Attention Network for RGB-Thermal Real-time Semantic Segmentation. In Proceedings of the 2021 IEEE/RSJ International Conference on Intelligent Robots and Systems (IROS), Prague, Czech Republic, 27 September–1 October 2021; pp. 4467–4473. [CrossRef]
114. Zhou, W.; Dong, S.; Xu, C.; Qian, Y. Edge-aware Guidance Fusion Network for RGB Thermal Scene Parsing. In Proceedings of the Thirty-Sixth AAAI Conference on Artificial Intelligence, Washington, DC, USA, 7–14 February 2022; pp. 3571–3579.

115. Zhang, Q.; Zhao, S.; Luo, Y.; Zhang, D.; Huang, N.; Han, J. ABMDRNet: Adaptive-weighted Bi-directional Modality Difference Reduction Network for RGB-T Semantic Segmentation. In Proceedings of the IEEE/CVF Conference on Computer Vision and Pattern Recognition, Nashville, TN, USA, 20–25 June 2021; pp. 2633–2642.
116. Xu, J.; Lu, K.; Wang, H. Attention fusion network for multi-spectral semantic segmentation. *Pattern Recognit. Lett.* **2021**, *146*, 179–184. [CrossRef]
117. Sun, Y.; Zuo, W.; Yun, P.; Wang, H.; Liu, M. FuseSeg: Semantic segmentation of urban scenes based on RGB and thermal data fusion. *IEEE Trans. Autom. Sci. Eng.* **2020**, *18*, 1000–1011. [CrossRef]
118. Krispel, G.; Opitz, M.; Waltner, G.; Possegger, H.; Bischof, H. Fuseseg: Lidar point cloud segmentation fusing multi-modal data. *arXiv* **2020**, arXiv:1912.08487. [CrossRef]
119. El Madawi, K.; Rashed, H.; El Sallab, A.; Nasr, O.; Kamel, H.; Yogamani, S. Rgb and lidar fusion based 3d semantic segmentation for autonomous driving. In Proceedings of the 2019 IEEE Intelligent Transportation Systems Conference (ITSC), Auckland, New Zealand, 27–30 October 2019; pp. 7–12.
120. Wang, W.; Neumann, U. Depth-aware CNN for RGB-D Segmentation. In Proceedings of the European Conference on Computer Vision (ECCV), Munich, Germany, 8–14 September 2018; pp. 135–150.
121. Jaritz, M.; Vu, T.H.; Charette, R.d.; Wirbel, E.; Pérez, P. xmuda: Cross-modal unsupervised domain adaptation for 3d semantic segmentation. In Proceedings of the IEEE/CVF Conference on Computer Vision and Pattern Recognition, Seattle, WA, USA, 13–19 June 2020; pp. 12605–12614.
122. Chollet, F. Xception: Deep Learning with Depthwise Separable Convolutions. In Proceedings of the 2018 IEEE/CVF Conference on Computer Vision and Pattern Recognition, Honolulu, HI, USA, 21–26 July 2017; pp. 1251–1258. [CrossRef]
123. Zhao, H.; Qi, X.; Shen, X.; Shi, J.; Jia, J. ICNet for Real-Time Semantic Segmentation on High-Resolution Images. In Proceedings of the ECCV, Munich, Germany, 8–14 September 2018.
124. Iandola, F.N.; Moskewicz, M.W.; Ashraf, K.; Han, S.; Dally, W.J.; Keutzer, K. SqueezeNet: AlexNet-level accuracy with 50x fewer parameters and <1 MB model size. *arXiv* **2016**, arXiv:1602.07360.
125. Graham, B.; Engelcke, M.; Maaten, L.V.D. 3D Semantic Segmentation with Submanifold Sparse Convolutional Networks. In Proceedings of the 2018 IEEE/CVF Conference on Computer Vision and Pattern Recognition, Salt Lake City, UT, USA, 8–23 June 2018; pp. 9224–9232. [CrossRef]
126. Xie, S.; Girshick, R.; Dollár, P.; Tu, Z.; He, K. Aggregated Residual Transformations for Deep Neural Networks. In Proceedings of the 2018 IEEE/CVF Conference on Computer Vision and Pattern Recognition, Honolulu, HI, USA, 21–26 July 2017; pp. 1492–1500. [CrossRef]
127. Huang, G.; Liu, Z.; Van Der Maaten, L.; Weinberger, K.Q. Densely connected convolutional networks. In Proceedings of the 2018 IEEE/CVF Conference on Computer Vision and Pattern Recognition, Honolulu, HI, USA, 21–26 July 2017; pp. 4700–4708. [CrossRef]
128. Couprie, C.; Farabet, C.; Najman, L.; LeCun, Y. Indoor semantic segmentation using depth information. *arXiv* **2013**, arXiv:1301.3572.
129. Pagnutti, G.; Minto, L.; Zanuttigh, P. Segmentation and semantic labelling of RGBD data with convolutional neural networks and surface fitting. *IET Comput. Vis.* **2017**, *11*, 633–642. [CrossRef]
130. Hu, J.; Shen, L.; Sun, G. Squeeze-and-Excitation Networks. In Proceedings of the 2018 IEEE/CVF Conference on Computer Vision and Pattern Recognition, Salt Lake City, UT, USA, 8–23 June 2018; pp. 7132–7141. [CrossRef]
131. Su, H.; Maji, S.; Kalogerakis, E.; Learned-Miller, E. Multi-view convolutional neural networks for 3d shape recognition. In Proceedings of the IEEE International Conference on Computer Vision, Santiago, Chile, 11–18 December 2015; pp. 945–953.
132. Chiang, H.Y.; Lin, Y.L.; Liu, Y.C.; Hsu, W.H. A unified point-based framework for 3d segmentation. In Proceedings of the 2019 International Conference on 3D Vision (3DV), Québec City, QC, Canada, 16–19 September 2019; pp. 155–163.
133. Jaritz, M.; Gu, J.; Su, H. Multi-view pointnet for 3d scene understanding. In Proceedings of the IEEE/CVF International Conference on Computer Vision Workshops, Seoul, Korea, 27–28 October 2019.
134. Ronneberger, O.; Fischer, P.; Brox, T. U-net: Convolutional networks for biomedical image segmentation. In Proceedings of the International Conference on Medical Image Computing and Computer-Assisted Intervention, Munich, Germany, 5–9 October 2015; pp. 234–241.
135. Liu, W.; Luo, Z.; Cai, Y.; Yu, Y.; Ke, Y.; Junior, J.M.; Gonçalves, W.N.; Li, J. Adversarial unsupervised domain adaptation for 3D semantic segmentation with multi-modal learning. *ISPRS J. Photogramm. Remote Sens.* **2021**, *176*, 211–221. [CrossRef]

technologies

MDPI

Review

Explainable AI (XAI) Applied in Machine Learning for Pain Modeling: A Review

Ravichandra Madanu [1], Maysam F. Abbod [2,*], Fu-Jung Hsiao [3], Wei-Ta Chen [3,4,5] and Jiann-Shing Shieh [1,*]

1 Department of Mechanical Engineering, Yuan Ze University, Taoyuan 32003, Taiwan; madanuravichandra@gmail.com
2 Department of Electronic and Computer Engineering, Brunel University London, Uxbridge UB8 3PH, UK
3 Brain Research Center, National Yang-Ming University, Taipei 112, Taiwan; fujunghsiao@gmail.com (F.-J.H.); wtchen71@gmail.com (W.-T.C.)
4 School of Medicine, National Yang-Ming University, Taipei 112, Taiwan
5 Neurological Institute, Taipei Veterans General Hospital, Taipei 112, Taiwan
* Correspondence: maysam.abbod@brunel.ac.uk (M.F.A.); jsshieh@saturn.yzu.edu.tw (J.-S.S.)

Abstract: Pain is a complex term that describes various sensations that create discomfort in various ways or types inside the human body. Generally, pain has consequences that range from mild to severe in different organs of the body and will depend on the way it is caused, which could be an injury, illness or medical procedures including testing, surgeries or therapies, etc. With recent advances in artificial-intelligence (AI) systems associated in biomedical and healthcare settings, the contiguity of physician, clinician and patient has shortened. AI, however, has more scope to interpret the pain associated in patients with various conditions by using any physiological or behavioral changes. Facial expressions are considered to give much information that relates with emotions and pain, so clinicians consider these changes with high importance for assessing pain. This has been achieved in recent times with different machine-learning and deep-learning models. To accentuate the future scope and importance of AI in medical field, this study reviews the explainable AI (XAI) as increased attention is given to an automatic assessment of pain. This review discusses how these approaches are applied for different pain types.

Keywords: pain; healthcare; neural networks; artificial intelligence; explainable AI

check for updates

Citation: Madanu, R.; Abbod, M.F.; Hsiao, F.-J.; Chen, W.-T.; Shieh, J.-S. Explainable AI (XAI) Applied in Machine Learning for Pain Modeling: A Review. *Technologies* **2022**, *10*, 74. https://doi.org/10.3390/technologies10030074

Academic Editors: Manoj Gupta, Eugene Wong and Gwanggil Jeon

Received: 19 May 2022
Accepted: 10 June 2022
Published: 14 June 2022

1. Introduction

Artificial intelligence (AI) has been a great opportunity for the progress of the economy, with its ability for solving problems that cannot be solved precisely in a short time using human intelligence. In recent years, the utilization of computer-assisted approaches in every domain, especially in healthcare, has increased with advancements in AI incorporated by reducing and optimizing the cost, time, workforce required for assessing, testing and completing the tasks performed by humans and increasing the quality of healthcare. However, there will be some challenges to overcome with the availability and development of clinical facilities using AI systems, equipment and trained professionals, etc. [1,2]. The availability of AI for use in healthcare and in different domains has already achieved great heights and has much influence on the present generation of mankind. Investing in AI has increased tremendously, from over 37.5 billion USD in the year 2019 to nearly 97.9 billion USD by 2023 [1–3].

For improving health-associated records of individual patients' Electronic Health Records (EHR) [4], National Health Insurance (NHI) [5–8] for developing the Health Information Technology (HIT) [8], along with the deployment of some assistive tools, has made AI more reachable to people for easy access and more convenient healthcare [3–8]. However, the treatment and diagnosis of patients is a challenging task with machine-learning or AI models, as they are not necessarily sufficient in and of themselves, which is further endorsed by medical staff [9]. The artificial-intelligence approach for storing the health status of individuals in an Electronic Medical Record (EMR) gives the possibility for physicians to know the patient's disease history, diagnosis, planned or unplanned treatments, severity and reoccurrence, medications used, test results of laboratory, etc. This information in one click makes the physician's analysis about the patient more accurate and faster [10–12]. Rather than chart reviews, these data are well-organized and easy to access. The chip card, which has an electromagnetic chip inside, will render the data useful for the physician within no time once it is scanned, and give a scope to analyze the condition of the patient for diagnosis and treatment [12]. The drawback when progress is made in this is that the medical and health data are noisy, which when sampled irregularly makes it difficult to combine the data from different sources [10].

The inclusion of these advancements in technology in medicine and healthcare improves digitization and informatization [13]. Even with the boom in machine learning (ML) and AI, failure of the automated system dysfunction leads to losses of human lives. Diagnosis and treatment of patients at an early stage is key for technology utilization and minimizing the risk of the disease advancing. Some diseases require long treatments such as cancer, chronic pain, diabetes, etc. There should be accountability and transparency in medical data. Thus, the questions to be answered are: (1) who can be accountable if something goes wrong? (2) Can we explain why things are going wrong? (3) How do we leverage them if they go well? Many studies have suggested different methods and ideas that focused on interpretability, and furthermore, in explainable artificial intelligence (XAI) [14].

The explanations of AI models are more practically applied to global AI processes but should be careful while with individual decisions. There should be thorough validation before applying explainable artificial intelligence [15]. According to [15], the explanations of ML decisions have been categorized as inherent explainability and post hoc explainability. Inherent explainability means the clear, understandable and limited complex data by which the simple input and output can quantify their relation. The very simple way to understand is a calculation of a car's fuel efficiency versus the weight of the car, using regression. It is understandable by explaining how a kilogram increase in weight changes the fuel efficiency on an average. On the other hand, post hoc explainability is where the data and models are difficult in complexity and high-dimensional to understand. This can be seen in medical-image analysis. Papers [15,16] describe examples of heat-map images for diagnosing pneumonia. The data, which contain useful and non-useful information after localizing the region, do not reveal exactly what in that area that the model considers useful. It is hard for clinicians to know if the model appropriately established that the presence of an airspace opacity was important in the decision, if the shapes of the heart border or left pulmonary artery were the deciding factor, or if the model had relied on an inhuman feature, such as a particular pixel value or texture that might have more to do with the image-acquisition process than the underlying disease [15]. XAI applied in diagnosis of some diseases is explained, such as Chronic, Ophthalmic, etc.

Interpretability is significant for AI models, by which the user knows the reason for the decision of the model as optimistic compared with the others [17]. If the data is high-dimensional [15], it leads to a lack of good explainability, and also may not create trustworthiness and transparency in usage [17]. Paper [18] explains how explainability techniques are used on a heart-disease dataset. The model created by [18] was used for the detection of explaining 13 attributes from the Heart Disease Cleveland UC Irvine dataset. Some of the feature-based techniques include Local Interpretable Model-Agnostic Explanations (LIME) and SHapley Additive exPlanations (SHAP), in which LIME is capable for local explanations and SHAP can explain globally as well as locally. Local explanations are limited to individual predictions, whereas global explanations are for the whole model, although global explanations can be used for individual predictions but are less accurate [3,18].

Retinal diseases such as glaucoma turn severe when they are untreated. In some cases, they cause irreversible vison loss. Even though there is an emergence of many deep-learning methods for use in diagnosing retinal diseases, their practical implementation is limited, with drawbacks with trust in the models in providing optimal and accurate decisions. The work reported [19] includes the quantitative analysis of the attribution methods using multiple measures, including robustness, runtime and sensitivity [19]. As explained before, the decisions are more transparent and explainable for users. These are the ethical and legal challenges for the model to be used [19] before taking decisions.

AI is also applied to the diagnosing of many other diseases that will end up with pain that lasts for a shorter period or even more after treatment ends. Pain is subjective in nature and cannot be same with two persons and for the same illness [20]. As explained in [20], pain cannot be concluded with one experiment or analysis for an individual. It needs to be observed, or experiments need to be conducted, numerous times. In olden times there was no alternative to mitigate pain rather than to accept it, but with the invention of first anesthetic drug named ether, the scenario was altered. Now, unpleasant pain is mitigated by injecting anesthetics. This was invented to be used for certain situations where tolerance of pain is not acceptable, and the patient needs to take it to reducing its severity [20]. This review focuses on pain types and XAI approaches applied in the pain model, and how helpful it can be in pain detection.

2. Scope of Review

This review is solely related to pain-scaling approaches and machine learning or deep learning, by further extending the machine-learning model's decision of classifying by explainability. Pain is experienced due to different factors, for different ages, for different genders, differing from one another with no common standard that can be relied on. A person with different conditions of medical illness will experience pain that cannot be treated with same medication dosage [20]. For instance, patients under surgery are given anesthesia drugs, while people with headache pain are given a dosage of medicine, and for some others, pain opioids are used, which when misused or addicted to lead to other chronic complications. To allow the AI model to predict the associated pain, using XAI is discussed. Furthermore, the difficulties involved while assessing pain by clinicians and AI models and the bridge between these two is discussed.

3. Pain Measurement and Variation

The measurement of pain catches eyes as it is not reliable even today, after many advancements in science and technology. It received attention when different researchers approached it in different ways with ML and AI models, despite the outcome not being trustworthy.

3.1. Pain Measurement

The treatment of pain after the invention of anesthetic drugs (ether) became more trouble-free, and the mortality rate during surgeries reduced significantly. The treatment of postoperative pain remained unsophisticated and largely opioid-based, receiving scant attention in the literature. The invention of anesthesia in 1846 and the advancements in the anesthetic agent's research for hundred years after it led to the foundation for the measurement of pain in 1940. There was a group of students from Connell University named James Hardy, Harold Wolff and Helen Goodell, who began working on a method to measure the intensity of pain. The study group, first assuming pain as the end of any overstimulation of recognized sensory mechanisms, found that pain had its own neurological pathways and was most likely to have its peripheral receptors and cerebral centers. They later devised a dolorimeter, a device that focused light on a blackened area of skin, and exhibited a painful stimulus at 45 °C or 113 °F [20–22]. This led them to devise a scoring system to record the intensity of pain experienced: "Twenty-one discriminable intensities of pain were observed between the threshold pain and the ceiling pain, a scale of pain intensity is proposed, the unit of which is called a 'dol'." In year 1951, the dolorimeter was used for the first time on patients in an attempt to assess the effectiveness of analgesia during labor. [21,22]. Two years later, a group of anesthetists evaluated the dolorimeter as an instrument for assessing pain, principally in pain-clinic patients. They conducted over a hundred hours of testing on themselves and reported. Their conclusion was that the dolorimeter might have an application as a tool for evaluating analgesic drugs, but felt it had little application as a measuring tool in patients [22].

The main critic of the dolorimeter is Henry Beecher, who is a Professor and Chair of the Department of Anesthesia at Massachusetts General Hospital. He insisted that pain research could have been carried out only by studying real pain in patients, taking into account all the subjective, emotional overlays that accompanied the origins of the pain. Thus, his work on the measuring of pain became extensive and it was one of the many areas where he tried to quantify subjective responses. His randomized, blinded trials involved the use of placebos—then a very new concept—and a crossover design where the patients served as their own controls, receiving two or more analgesics during a given painful episode. He measured a single response, the presence of a 50% reduction in pain. Beecher's meticulous methodology became the foundation for future research into clinical pain management and analgesic efficacy [21,22]. Hence, pain is subjective, individually centered and usually measured by the self-report taken from the suffering person. There are a number of tools for measuring pain, which include the Visual Analogue scale (VAS), Verbal Rating scale (VRS) and some multidimensional tools. These scalings were developed for pain assessment after many experiments and clinical trials, which found them to be cost effective and robust. However, the patient's self-report, which is the quantification of the pain experienced, is reported when talking with the clinician [20,23,24]. It can be interpreted by a clinician, although it can be taken as a standard for treating the patient even today. There is no skill involved for assessing pain [24]. Among those, the automatic detection of facial expression in particular is of high importance due to its applications in many fields such as in biometrics, forensics, medical diagnosis, monitoring, defense and surveillance, etc. It is not a continuous experience; pain varies and intensifies with time and cannot be predicted.

3.2. Pain Variation

Pain in some surgeries and injuries last long, thus making it difficult for the patient to conclude the cause. For example, pain could have been caused by other factors that are related to the musculoskeletal system. On the other hand, surgical pain is severe as it involves a loss of blood [20]. Hence, it received much attention from all the researchers to assess the pain using some AI models. There has been a lot of work carried out to automatically detect the pain from ML to deep learning (DL) [25–29].

In [20], the mechanism of pain is discussed elaborately; the nerve fibers A-delta and C fibers are sensitive to sharp and dull pain sensations, respectively. As pain is sensitive to the environment, distress and emotional conditions, people can experience it with no such remarks observed on the face. The influence of social, economic and cultural factors may also include people observing pain differently from one another [20,26–35].

In anesthetics, pain is the key parameter to deal with and regulate smoothly during the process of surgical operations on the patient. For this, preoperative and postoperative monitoring of pain is of high importance on different types of surgeries, including eyes, heart, brain or other organs [28]. Additionally, pain from patients sometimes cannot be verbally communicated, as they are under anesthesia for a long time, children, dementia patients, noncooperative, etc. In such cases, the above-mentioned measurement of pain using self-reports is not useful. The variation of pain with such patients cannot be dealt with easily [29–34].

4. Explainability in Pain Models

In [36], the concept of human–computer Interaction that has roots in cognitive science, particularly on the intelligence of humans and knowledge discovery/data mining in computer science with AI, is discussed. The definition of intelligence is with human intelligence in cognitive science and AI in computer science. In these two cases, the intelligence should be usable, and the factors that are needed are prior data, knowledge, generalization of data, dimensionality and its explanatory factors. In the medical sector, with pain, the raw data are available, and the problem lies with the generalization of data to different pain types and a lack of explanation factors. To date, the misuse of the interpretability and exploitability in many contexts leads to the model being perceived as ineffective by humans. The sense of a model to a human end observer is called interpretability; in other words, 'transparency' [37,38]. Depending on the transparency, the models can be simulated, decomposable models and algorithmically transparent models, wherein if the model is understandable to humans, it is called an explainable model [38]. Table 1 gives some explainable features related to different pain types. Chest pain in most cases is related with the heart and is also caused sometimes by problems with the lungs, esophagus, muscles, ribs or nerves. In most chest-pain cases, the doctor needs to evaluate the electrocardiogram (ECG), vital signs, past medical history of the patient (PMHx), the patient's symptoms (Sx) and heart-rate variability (HRV), which is the peak-to-peak variation of the time interval. The explainable features mentioned in Table 1 are of high variable importance, as those are the features that the researchers are using for the detection of pain using machine-learning or deep-learning methods.

Table 1. Comparison of various pain for explainability of the features.

Pain Types	Pain-Affected Organs	AI/ML Techniques Used	Explainable Features in the Pain
Chest pain	Heart	Random forest (RF), support vector machine (SVM), artificial neural network (ANN), linear regression (LR), gradient boosting.	ECG, Vitals, HEART Score, Troponin, Labs, Exam, PMHx, Sx, HRV
Back pain	Back bone	K-nearest neighbor (K-NN), principal component analysis (PCA), random forest (RF), ANN, SVM, multilayer perceptron (MLP), LR, stochastic gradient boosting (SGM), naïve Bayes (NB).	EMG, HRV, pelvic incidence, pelvic tilt, lumbar lordosis angle, sacral slope, direct tilt, pelvic radius, degree spondylolisthesis, pelvic slope, thoracic slope, cervical tilt, sacrum angle and scoliosis slope, gait features, data from pressure sensors to assess sitting posture, erector spine muscle activity.
Shoulder pain	Shoulder joint/muscle	SVM, ResNet	Facial images, landmarks.
Headache pain	Brain	Random oracle model (ROM), linear neural network (LNN), SVM, K-NN, ANN	Age, visual analog scale rating, duration of pain, facial images, landmarks.
Surgical/post-operative pain	Body cells	AlexNet, VGGNet, CifarNet, ResNet, DenseNet	Electroencephalogram (EEG)

Back pain has many features, such that research implemented AI to model features such as electromyography (EMG), HRV, pelvic incidence, tilt, slope or direct tilt, etc., which is explained in Section 5.2. Shoulder pain and headache pain have common features that are used for detection using AI, which are facial images and facial landmarks. However, for headache pain, some other features such as age and visual analog scale (VAS) readings are taken into consideration to determine the cause. Surgical pain is a critical pain for the patient in operation theatre. The electroencephlogram (EEG) signal gives more relevant data for the anesthesiologists in surgical/postoperative types of pain.

4.1. Chest Pain

Physiological signs that are used for modeling pain are the only data that are trustworthy; the data from the behavioral signs will have low levels of reliability. In chest pain, there is a risk of evaluation involved of vital signs without involving any new variable [39].

Patients with chest pain at the emergency department (ED) constitute a greater logistic challenge as the majority have noncardiac-related symptoms and often benign disorders that do not need emergency treatment or hospitalization. Acute chest pain, which comes under primary cardiovascular disease, ranges from severe to no pain, from acute coronary syndromes to harmless conditions. Chest pain constitutes the most emergency department cases and is diagnosed using the HEART (history, EEG, age, risk factors, troponin) score in Table 2. Acute coronary syndrome-related mortality is more common in patients presented to the emergency department. Hence, the need of the emergency physician to assess the diagnosis of myocardial ischemia, including unstable angina, non-ST elevation myocardial infarction (NSTEMI) and ST elevation myocardial infarction (STEMI) [39–43]. The patient history, i.e., past medical history (PMHx) including smoking, drugs, medication, etc., and examination of the patient physically for vital-sign changes may not be reliable for the determination of treatment in emergency care by physicians [41]. Therefore, the data will be an input to the computer algorithm. The risk due to the disease can be predicted using an AI model only after the time when the patient's health condition is stable and transferred to the general ward from the emergency department. This means the risk is less with the condition that the pain may be a chronic cause of the criticality. Chest pain records in the patient's historical information are critical in determining the underlying cause. Treatment and clinical assessment of the patient is determined by many factors, including data such as age, history of medication and surgeries, therapies, physiological signs, etc. [42–45].

Table 2. Predicting chances of hospitalization using HEART score.

HEART Score Points	MACE Occurrence	Hospitalization
0–3	2.5%	Not Necessary
4–6	20.3%	Necessary
≥7	72.7%	Immediate

4.2. Back Pain

Lower back pain remains the most possible musculoskeletal disorder in the whole world, with the population at risk from different conditions. A common back pain can transform into more stressed chronic back pain when not diagnosed for a long time. This type is more common in occupational workers. Even physicians cannot determine the cause of chronic back pain from MR (magnetic resonance) images. Due to numerous advances in medical image processing and AI, physicians are now able to optimize the time for their diagnosis and treatment. Around 60% to 80% of the population in the UK may have experienced back pain at any time in their life [46], and among chronic diseases worldwide, one-fourth of the population suffers back pain. Advances in AI have reduced the risk by having a fast diagnosis system, with early prevention of acute back pain becoming chronic back pain. Electromyography (EMG), heart-rate variability (HRV), pelvic incidence, pelvic tilt (Figure 1), lumbar lordosis angle, sacral slope, pelvic radius, degree spondylolisthesis, pelvic slope, direct tilt, thoracic slope, cervical tilt, sacrum angle and scoliosis slope, gait features, data from pressure sensors to assess sitting posture and erector spinae muscle activity are some explainable features for back pain diagnosis using AI [46–51].

Figure 1 indicates the pelvic tilt that is a feature to describe lower back pain (LBP), where PT represents pelvic tilt, PI represents pelvic incidence and SS represents sacral slope. PT is the angle between the vertical line from the midpoint of the two hip-joint centers and a line connecting the midpoint of the two hip-joint centers with the midpoint of the sacral end plate. PI is the angle that connects the midpoints of the two hip-joint centers with the midpoint of the sacral end plate with the line perpendicular to the center of the sacral end plate. SS is the angle between the horizontal line and the sacral end plate.

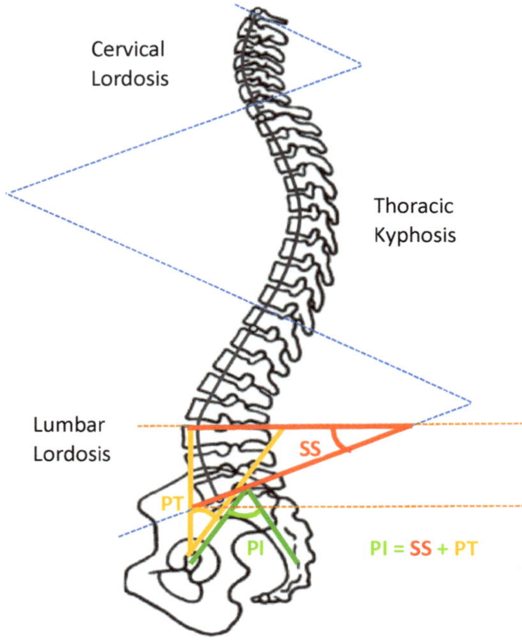

Figure 1. Pelvic tilt "Adapted from Ref. [52].

4.3. Shoulder Pain

The shoulder joint has bones packed in contact with the muscles, tendons and ligaments. Usually, pain is caused due to the rotator-cuff tendons that are packed under the bony area of the shoulder. For most of the shoulder pain research, the labeled faces were taken for assessing the level of the pain as one of three classes, namely no pain, medium pain and high pain, or even more. Patients with shoulder pain underwent some physical tests on the abnormal shoulder, and appropriate levels of pain were considered from relative expressions of the face [53]. Different motion tests were conducted on patients to know the level of pain by the physiotherapist. These included passive and active motion tests that are performed when the patient is resting in a supine position on the bed and in standing position, respectively. The passive test is achieved when the patient's limb is rotated by a physiotherapist until the maximum range was achieved or the patient feels pain and asks to stop. The active test was performed before the passive test clinically [54]. The UNBC shoulder pain database has the publicly accessible data for pain research as of today, where most research related to pain is carried out. In [55], the UNBC database is tested for the ensemble deep-learning model (EDLM) and achieved good accuracy. In [56], the facial-muscle-based action units (AU) are used to assess the pain from UNBC shoulder-pain-archive facial images. Figure 2 shows the tendon torn or tearing as an example of having shoulder pain.

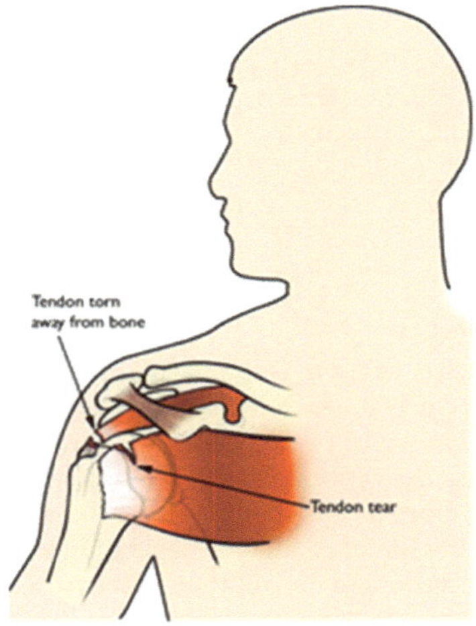

Figure 2. Tendon torn/tear.

4.4. Headache Pain

Headache pain refers to pain that arises due to a sensation occurring in the nerve fibers of the brain. Headache is a neurological disorder that can be due to different stimulus caused within the head. The dynamics of headache pain differs with the severity. Types of headache-pain perception varies with the person, as explained in [20,57,58]. Pain caused due to damage of tissues or organs inside the human body will also cause the sensation of headache pain. The source of pain, severity, time period, etc., gives different types of headaches [57–60]. The most common type of headache classification is primary and secondary headaches, and is explained in the International Classification of Headache Disorders (ICHD), 3rd edition [61].

4.4.1. Primary Headaches

Primary headaches are caused with no medical illness or condition involved, meaning they have no known serious cause for stimulus of pain. These include cluster headaches, migraines, tension-type headaches and new daily persistent headaches (NDPH) [60], as explained in Table 3. Cluster headaches are triggered by nitroglycerin, histamine and alcohol consumption. This is usually accompanied by eye watering, nasal congestion and swelling around the eye on the affected side, while symptoms last from 15 min to 3 h. The attacks of clusters last for weeks or months [57,58]. Migraine headache pain is on one side of the head. Migraine with aura and without aura are the two subtypes of migraine. Aura is the sensation perceived by a patient that leads to a condition affecting the brain. Migraine with aura has been associated with an increased risk of ischemic stroke, and not much risk is associated with migraine without aura. The diagnosis of migraine with or without aura is greater than or equal to 5 to 60 min and 4 to 72 h, respectively. Tension-type headache is more common, with a lifetime prevalence in the general population, and impacts the socioeconomic life of an individual. NDPH lasts for 24 h and is persistent in some days from the day it starts [57–62].

Table 3. Primary headache types.

Primary Headache Type	Duration of Symptoms	Occurrence
Cluster-type	15 min to 3 h	Frequent
Migraine with Aura	≥5 min to 60 min	Frequent
Migraine without Aura	4 to 72 h	Rare
NDPH	24 h	Rare

4.4.2. Secondary Headaches

Secondary headaches occur in response to other conditions that cause headache. It is classified in ICHD, which states that secondary headache occurring in close temporal relation to a disorder known to cause headache should be considered secondary unless proved otherwise. Headache worsening or improvement depends on the causative disorder that has worsened or improved. Sometimes it can be mistreated as a primary headache. Essentially, headaches last for 3 to 6 months depending on the severity of the cause [63–65]. As explained in [20], according to the duration of the pain and acuteness, chronic pain, it is associated with each headache pain.

4.5. Surgical/Postoperative Pain

Surgical pain has many risks that may lead to death. The diagnosis of pain is highly important when undergoing surgery or after surgical operation. To reduce the pain as explained in [20–23,66–69], general anesthesia is a safe and fundamental component for performing surgeries. Pain is monitored by anesthesiologists and a proper dosage of anesthesia is recommended according to the patients' health, age and type of surgery. Hence, it becomes a difficult task for anesthesiologists to maintain the levels of anesthesia. The pain relates to the brain dynamics, and thus provides potential to trace differences in the brain's activity under different anesthetics. In [66–69], work was carried out on how to access the depth of anesthesia (DoA) levels using different signals from the brain, and how to relate it with the bispectral index (BIS) value to make a more convenient and easy way for the surgical operation to continue, as the patient will experience pain if they become awake.

5. Explainable AI Models

Medical AI is used in performing clinical diagnosis and treatment suggestions. The application of DL in many biomedical fields from genomic applications such as gene expression, public medical health management and epidemic prevention have much importance. Explainable artificial intelligence (XAI) models are needed to relate the context-based explanations with the decisions made by machines in clinical decision making. Depending on the application domain, the decision made by the machines is to be explained. AI models with their broad application in all fields of science or in technology are usually skill-based decisions made from the datasets that the model is trained in or tested on, which means a clear understanding of the AI model-made decision is acceptable. In medicine, there are two distinct types of areas: one is the science of medicine, and the other is clinical medicine. Clinical medicine normally focuses on the patient at bedside. The physician's communication with the patient is common, and the physician's medical advice to the patient and selecting the therapy required for the diagnosing is made by explainable AI models. The need for explanations in these decisions to the patients is to be communicated effectively in an understandable way, which is possible with XAI [37,70–82].

XAI models have an increased interest as they gain high importance, which gives an explainable output of the machine-learning algorithm. Reinforcement learning (RL) is one such type of algorithm in machine learning that uses a goal-directed learning. RL has an agent that is used to learn by interacting with the environment for achievement of the goal. Rewards are to be returned by the environment. The reinforcement learning in healthcare domain is well-explained in [83]. Pain is a critical area in medical diagnostics, especially with the headache pain that is almost not explainable. This paper aimed at giving the view of developing an XAI model to further achieve a true, unbiased result with diagnosis and treatment. To date, there are few deep-learning models that are explainable for pain diagnosis or treatment, as it is a subjective experience and varies with time and treatment. We can say that every treatment for illness will lead somehow to pain. The patient feels pain with the given dosage of medicine or with the mentioned cause of illness. This is one of the first papers to review the features of the different pain types with machine-learning and deep-learning models that are to be explainable to make the decisions of the algorithm to be understood by the end user [70,83,84]. In future, a detailed survey can be addressed in upcoming review papers in this field. This review paper turns the interest of the researchers to focus on the pain, and gives an idea of how advancements are making the patient's pain an understandable feature to be implemented using AI that is explainable. The variable importance of the explainable features is mentioned in Table 4. The variable importance is based on the reviewed papers that used the variable to determine the pain.

Table 4. Variable importance of explainable features.

Pain Type	High Variable Importance Features	Less Variable Importance Features
Chest pain	ECG, Vitals, HEART Score, PMHx, Sx, HRV	Troponin, Labs, Exam.
Back pain	Pelvic incidence, pelvic tilt, lumbar lordosis angle, sacral slope, direct tilt, pelvic radius, degree spondylolisthesis, pelvic slope, thoracic slope, cervical tilt, sacrum angle and scoliosis slope, gait features.	EMG, HRV, data from pressure sensors to assess sitting posture, Erector spine muscle activity.
Shoulder pain	Facial images, landmarks.	-
Headache pain	Facial images, landmarks.	Age, visual analog scale rating, duration of pain.
Surgical/postoperative pain	Electroencephalogram (EEG)	-

5.1. AlexNet

AlexNet is a convolutional neural network (CNN) architecture that works with large datasets. The architecture consists of eight convolutions and three fully connected layers. As there are many parameters, there is a problem of overfitting, which is solved by data-augmentation and dropout methods [71]. In [72], the AlexNet model is used in the application of detecting the brain's computer tomography (CT) hemorrhage from normal brain CT images. The long-term problem of classifying the normal and hemorrhage CT images of the brain is solved with the AlexNet model, which extracts more features with trained filters.

5.2. VGGNet

After AlexNet, CNN gained popularity via VGG, with 16 and 19 layers being used. This also presents the best performance on ImageNet, and reduces computing speed and accuracy. The depth of the network is increased by adding convolutional layers [73]. As we know, COVID diagnosis and treatment in early stages may help patients to survive with few complications. In [74], the authors concluded that the ultrasound images provide much more reliable data and superior detection accuracy compared with X-ray at 86% and CT scan with 84%. The ultrasound images are almost 100% accurate, as it is easy to assess for the patient at bedside.

5.3. ResNet

ResNet is a deeper neural network and is difficult to train the model. This was also trained on ImageNet data and achieved higher performance than VGG as this is 8x deeper than VGG. This also worked on COCO detection and segmentation. As can be seen in [75], 50-, 101- and 152-layer-deep ResNets are used, depending on the data. The color fundus images in diabetic retinopathy are classified in [76] using inception ResNet V2 as an application of AI in the biomedical field, with high accuracies over 80%.

5.4. DenseNet

Although CNN is introduced over two decades, improvements in hardware and network structure have recently allowed the training of truly deep CNNs. After the above discussed deep CNNs, which surpassed 100 layers, creating a problem of gradient when it passes through many layers may vanish and wash out by the time it reaches the end. DenseNet layers are very narrow (i.e., 12 filters per layer) [77]. In [78], the author proposed that the DenseNet-121 model reported a more accurate patient-recognition rate (PRR), and image-recognition rate (IRR) metrics improved by 2–8% and 2–9% with the VGG 16 and ResNet50 convolutional neural network models.

6. Discussion

Deep-learning methods are widely used in many applications related to healthcare. The utilization of AI in healthcare has reduced the burden on the system exclusively. The datasets available to train models have increased in time and have achieved good results with a small amount of medical data. The investment of AI in healthcare has increased tremendously in recent years, with surgical operations also being assisted by AI systems that detect and assess the health of patients instantly with the electroencephalogram (EEG), electromyogram (EMG), pulse rate and electrocardiogram (ECG).

Pain detection using these DL methods is a difficult task, as it cannot be accurate enough for diagnosing. For research purposes, datasets that are used for pain detection are developed, but the recorded pain cannot be constant and dynamic. It is a sensation caused due to some medical disorder; hence, it is highly difficult to predict as it can be with other medical conditions. Pain varies with time and is not evaluated using one's facial image. Different research has been conducted with some publicly available datasets and has proved that headache pain is the most difficult to predict, among others. There was little research that went on to discuss headache pain. However, we achieved certain progress in detecting other types of pain as discussed above in this paper. Incorporating XAI models may increase the features that are explainable to a certain illness and what causes it, and feature the importance in diagnosis.

7. Conclusions

The different AI-based approaches to pain are reviewed, and the importance of explainable AI in health and medicine is explained. This review gives an overview of all the pains that are automatically diagnosed using facial emotions and expressions, vital signs, and other important signs for detecting pain from the available data. There is a gap between the engineering systems in real-time diagnosis of patients with pain; this should be filled using some AI approaches. Pain from different sources such as injury, illness and tissue damage are not the same sensations as each other. Pain is so persistent that it leads to stress [20]. The meaning of intelligence in explaining the features of pain that are studied is discussed. The pain scale, which is verbal or rated by the physician, is time-consuming and not reliable, with more stress on the health system. The application of AI models for healthcare-system improvement by diagnosing without any invasive methods gives much scope nowadays to every other disease prior to diagnosis. The features explainable for the diagnosis of pain, as described for each pain in Table 1, should be concentrated on as a solution.

Author Contributions: Conceptualization, M.F.A. and J.-S.S.; methodology, R.M., M.F.A., F.-J.H., W.-T.C. and J.-S.S.; formal analysis, J.-S.S. and M.F.A.; investigation, R.M. and J.-S.S.; resources, R.M. and J.-S.S.; data curation, R.M. and J.-S.S.; writing—original draft preparation, R.M.; writing—review and editing, J.-S.S. and M.F.A.; visualization, J.-S.S. and M.F.A.; supervision, M.F.A., F.-J.H., W.-T.C. and J.-S.S.; project administration, J.-S.S.; funding acquisition, J.-S.S. All authors have read and agreed to the published version of the manuscript.

Funding: This research was funded by Ministry of Science and Technology (MOST) of Taiwan, grant number: MOST 110-2221-E-155-004-MY2.

Institutional Review Board Statement: Not applicable.

Informed Consent Statement: Not applicable.

Data Availability Statement: Not applicable.

Acknowledgments: The authors would like to thank the support in part by the Ministry of Science and Technology, Taiwan.

Conflicts of Interest: The authors declare no conflict of interest.

References

1. Matheny, M.; Sonoo, T.I.; Mahnoor, A.; Danielle, W. (Eds.) *Artificial Intelligence in Health Care: The Hope, the Hype, the Promise, the Peril;* NAM Special Publication; National Academy of Medicine: Washington, DC, USA, 2019.
2. Bohr, A.; Memarzadeh, K. The rise of artificial intelligence in healthcare applications. *Artif. Intell. Healthc.* **2020**, 25–60. Available online: https://www.sciencedirect.com/science/article/pii/B9780128184387000022 (accessed on 1 April 2022).
3. Aniek, F.M.; Jan, A.K.; Peter, R.R. The role of explainability in creating trustworthy artificial intelligence for health care: A comprehensive survey of the terminology, design choices, and evaluation strategies. *J. Biomed. Inform.* **2021**, *113*, 103655.
4. Rajkomar, A.; Oren, E.; Chen, K.; Dai, A.M.; Hajaj, N.; Hardt, M.; Liu, P.J.; Liu, X.; Marcus, J.; Sun, M.; et al. Scalable and accurate deep learning with electronic health records. *NPJ Digit. Med.* **2018**, *1*, 18. [CrossRef] [PubMed]
5. Wu, T.Y.; Majeed, A.; Kuo, K.N. An overview of the healthcare system in Taiwan. *Lond. J. Prim. Care* **2010**, *3*, 115–119. [CrossRef]
6. Lee, S.Y.; Chun, C.B.; Lee, Y.G.; Seo, N.K. The National Health Insurance system as one type of new typology: The case of South Korea and Taiwan. *Health Policy* **2008**, *85*, 105–113. [CrossRef]
7. Victor, B.K.; Yang, C.T. The equality of resource allocation in health care under the National Health Insurance System in Taiwan. *Health Policy* **2011**, *100*, 203–210.
8. Chi, C.; Lee, J.L.; Schoon, R. Assessing Health Information Technology in a National Health Care System—An Example from Taiwan. *Adv. Health Care Manag.* **2012**, *12*, 75–109.
9. Tonekaboni, S.; Joshi, S.; McCradden, M.D.; Goldenberg, A. What Clinicians Want: Contextualizing Explainable Machine Learning for Clinical End Use. In Proceedings of the 4th Machine Learning for Healthcare Conference, Ann Arbor, MI, USA, 9–10 August 2019; Volume 106, pp. 359–380.
10. Qinghan, X.; Mooi, C.C. Explainable deep learning based medical diagnostic system. *Smart Health* **2019**, *13*, 100068.
11. Bonnie, B.D.; Jessica, L.; Jaime, L.N.; Qiana, B.; Daniel, A.; Robert, J.N. Use of Electronic Medical Records for Health Outcomes Research: A Literature Review. *Med. Care Res. Rev.* **2009**, *66*, 611–638.
12. Lau, E.C.; Mowat, F.S.; Kelsh, M.A.; Legg, J.C.; Engel-Nitz, N.M.; Watson, H.N.; Collins, H.L.; Nordyke, R.J.; Whyte, J.L. Use of electronic medical records (EMR) for oncology outcomes research: Assessing the comparability of EMR information to patient registry and health claims data. *Clin. Epidemiol.* **2011**, *3*, 259–272. [CrossRef]
13. Shuo, T.; Wenbo, Y.; Jehane, M.L.G.; Peng, W.; Wei, H.; Zhewei, Y. Smart healthcare: Making medical care more intelligent. *Glob. Health J.* **2019**, *3*, 62–65.
14. Tjoa, E.; Guan, C. A Survey on Explainable Artificial Intelligence (XAI): Toward Medical XAI. *IEEE Trans. Neural Netw. Learn. Syst.* **2021**, *32*, 4793–4813. [CrossRef] [PubMed]
15. Marzyeh, G.; Luke, O.R.; Andrew, L.B. The false hope of current approaches to explainable artificial intelligence in health care. *Lancet Digit. Health* **2021**, *3*, 745–750.
16. Rajpurkar, P.; Irvin, J.; Zhu, K.; Yang, B.; Mehta, H.; Duan, T.; Ding, D.; Bagul, A.; Langlotz, C.; Shpanskaya, K.; et al. CheXNet: Radiologist-Level Pneumonia Detection on Chest X-rays with Deep Learning. *arXiv* **2017**, arXiv:1711.05225.
17. Han, H.; Liu, X. The challenges of explainable AI in biomedical data science. *BMC Bioinform.* **2022**, *22*, 443–445. [CrossRef]
18. Dave, D.; Het, N.; Smiti, S.; Pankesh, P. Explainable AI meets Healthcare: A Study on Heart Disease Dataset. *arXiv* **2020**, arXiv:2011.03195.
19. Singh, A.; Sengupta, S.; Mohammed, A.R.; Faruq, I.; Jayakumar, V.; Zelek, J.; Lakshminarayanan, V. What is the Optimal Attribution Method for Explainable Ophthalmic Disease Classification. In *Ophthalmic Medical Image Analysis;* Springer: Cham, Switzerland, 2020; Volume 12069, pp. 21–31.
20. Chen, J.; Abbod, M.; Shieh, J.-S. Pain and Stress Detection Using Wearable Sensors and Devices—A Review. *Sensors* **2021**, *21*, 1030. [CrossRef]

21. Myles, P.S.; Christelis, N. Measuring pain and analgesic response. *Eur. J. Anaesthesiol.* **2011**, *28*, 399–400. [CrossRef]
22. Noble, B.; Clark, D.; Meldrum, M.; Ten Have, H.; Seymour, J.; Winslow, M.; Paz, S. The measurement of pain, 1945–2000. *J. Pain Symptom Manag.* **2005**, *29*, 14–21. [CrossRef]
23. Virrey, R.A.; Liyanage, C.D.S.; Petra, M.I.B.P.H.; Abas, P.E. Visual data of facial expressions for automatic pain detection. *J. Vis. Commun. Image Represent.* **2019**, *61*, 209–217. [CrossRef]
24. Yang, R.; Tong, S.; Bordallo, M.; Boutellaa, E.; Peng, J.; Feng, X.; Hadid, A. On pain assessment from facial videos using spatio-temporal local descriptors. In Proceedings of the 2016 Sixth International Conference on Image Processing Theory, Tools and Applications (IPTA), Oulu, Finland, 12–15 December 2016; pp. 1–6.
25. Sourav, D.R.; Mrinal, K.B.; Priya, S.; Anjan, K.G. An Approach for Automatic Pain Detection through Facial Expression. *Procedia Comput. Sci.* **2016**, *84*, 99–106.
26. Ashraf, A.B.; Lucey, S.; Cohn, J.F.; Chen, T.; Ambadar, Z.; Prkachin, K.M.; Solomon, P.E. The painful face—Pain expression recognition using active appearance models. *Image Vis. Comput.* **2009**, *27*, 1788–1796. [CrossRef] [PubMed]
27. Ilyas, C.; Haque, M.; Rehm, M.; Nasrollahi, K.; Moeslund, T. Facial Expression Recognition for Traumatic Brain Injured Patients. In Proceedings of the 13th International Joint Conference on Computer Vision, Imaging and Computer Graphics Theory and Applications (VISIGRAPP 2018), Funchal, Portugal, 27–29 January 2018; Volume 4, pp. 522–530.
28. McGrath, H.; Flanagan, C.; Zeng, L.; Lei, Y. Future of Artificial Intelligence in Anesthetics and Pain Management. *J. Biosci. Med.* **2019**, *7*, 111–118. [CrossRef]
29. Garcia-Chimeno, Y.; Garcia-Zapirain, B.; Gomez-Beldarrain, M.; Fernandez-Ruanova, B.; Garcia-Monco, J.C. Automatic migraine classification via feature selection committee and machine learning techniques over imaging and questionnaire data. *BMC Med. Inf. Decis Mak.* **2017**, *17*, 38. [CrossRef] [PubMed]
30. Liu, D.; Cheng, D.; Houle, T.T.; Chen, L.; Zhang, W.; Deng, H. Machine learning methods for automatic pain assessment using facial expression information: Protocol for a systematic review and meta-analysis. *J. Med.* **2018**, *97*, e13421. [CrossRef] [PubMed]
31. Pranti, D.; Nachamai, M. Facial Pain Expression Recognition in Real-Time Videos. *J. Healthc. Eng.* **2018**, *2018*, 7961427.
32. Lucey, P.; Cohn, J.F.; Matthews, I.; Lucey, S.; Sridharan, S.; Howlett, J.; Prkachin, K.M. Automatically Detecting Pain in Video Through Facial Action Units. *IEEE Trans. Syst. Man Cybern. Part B* **2011**, *41*, 664–674. [CrossRef]
33. Jörn, L.; Alfred, U. Machine learning in pain research. *Pain* **2018**, *159*, 623–630.
34. Keight, R.; Aljaaf, A.J.; Al-Jumeily, D.; Hussain, A.J.; Özge, A.; Mallucci, C. An Intelligent Systems Approach to Primary Headache Diagnosis. In *Intelligent Computing Theories and Application*; Springer: Cham, Switzerland, 2017; Volume 10362, pp. 61–72.
35. Evan, C.; Angkoon, P.; Erik, S. Feature Extraction and Selection for Pain Recognition Using Peripheral Physiological Signals. *Front. Neurosci.* **2019**, *13*, 437.
36. Rasha, M.A.-E.; Hend, A.-K.; AbdulMalik, A.-S. Deep-Learning-Based Models for Pain Recognition: A Systematic Review. *Appl. Sci.* **2020**, *10*, 5984.
37. Holzinger, A. From Machine Learning to Explainable AI. In Proceedings of the 2018 World Symposium on Digital Intelligence for Systems and Machines (DISA), Košice, Slovakia, 23–25 August 2018; pp. 55–66.
38. Arrieta, A.B.; Díaz-Rodríguez, N.; Del Ser, J.; Bennetot, A.; Tabik, S.; Barbado, A.; García, S.; Gil-López, S.; Molina, D.; Benjamins, R.; et al. Explainable Artificial Intelligence (XAI): Concepts, taxonomies, opportunities and challenges toward responsible AI. *Inf. Fusion* **2020**, *58*, 82–115. [CrossRef]
39. Liu, N.; Koh, Z.X.; Goh, J.; Lin, Z.; Haaland, B.; Ting, B.P.; Ong, M.E.H. Prediction of adverse cardiac events in emergency department patients with chest pain using machine learning for variable selection. *BMC Med. Inf. Decis. Mak.* **2014**, *14*, 75. [CrossRef] [PubMed]
40. Six, A.J.; Backus, B.E.; Kelder, J.C. Chest pain in the emergency room: Value of the HEART score. *Neth. Heart J.* **2008**, *16*, 191–196. [CrossRef] [PubMed]
41. Stewart, J.; Lu, J.; Goudie, A.; Bennamoun, M.; Sprivulis, P.; Sanfillipo, F.; Dwivedi, G. Applications of machine learning to undifferentiated chest pain in the emergency department: A systematic review. *PLoS ONE* **2021**, *16*, e0252612. [CrossRef]
42. Stepinska, J.; Lettino, M.; Ahrens, I.; Bueno, H.; Garcia-Castrillo, L.; Khoury, A.; Lancellotti, P.; Mueller, C.; Muenzel, T.; Oleksiak, A.; et al. Diagnosis and risk stratification of chest pain patients in the emergency department: Focus on acute coronary syndromes. A position paper of the Acute Cardiovascular Care Association. *Eur. Heart J.* **2020**, *9*, 76–89. [CrossRef]
43. Amsterdam, E.A.; Kirk, J.D.; Bluemke, D.A.; Diercks, D.; Farkouh, M.E.; Garvey, J.L.; Kontos, M.C.; McCord, J.; Miller, T.D.; Morise, A.; et al. Testing of Low-Risk Patients Presenting to the Emergency Department with Chest Pain: A scientific statement from the American Heart Association. *Circulation* **2010**, *17*, 1756–1776. [CrossRef]
44. Backus, B.E.; Six, A.J.; Kelder, J.C.; Bosschaert, M.A.R.; Mast, E.G.; Mosterd, A.; Veldkamp, R.F.; Wardeh, A.J.; Tio, R.; Braam, R.; et al. A prospective validation of the HEART score for chest pain patients at the emergency department. *Int. J. Cardiol.* **2013**, *168*, 2153–2158. [CrossRef]
45. Zhang, P.I.; Hsu, C.C.; Kao, Y.; Chen, C.J.; Kuo, Y.W.; Hsu, S.L.; Liu, T.L.; Lin, H.J.; Wang, J.J.; Liu, C.F.; et al. Real-time AI prediction for major adverse cardiac events in emergency department patients with chest pain. *Scand. J. Trauma Resusc. Emerg. Med.* **2020**, *28*, 93. [CrossRef]
46. Al Kafri, A.S.; Sudirman, S.; Hussain, A.J.; Fergus, P.; Al-Jumeily, D.; Al-Jumaily, M.; Al-Askar, H. A Framework on a Computer Assisted and Systematic Methodology for Detection of Chronic Lower Back Pain Using Artificial Intelligence and Computer Graphics Technologies. *Intell. Comput. Theor. Appl.* **2016**, *9771*, 843–854.

47. Tagliaferri, S.D.; Angelova, M.; Zhao, X.; Owen, P.J.; Miller, C.T.; Wilkin, T.; Belavy, D.L. Artificial intelligence to improve back pain outcomes and lessons learnt from clinical classification approaches: Three systematic reviews. *NPJ Digit. Med.* **2020**, *3*, 93. [CrossRef]
48. Chen, D.; Zhang, H.; Kavitha, P.T.; Loy, F.L.; Ng, S.H.; Wang, C.; Phua, K.S.; Tjan, S.Y.; Yang, S.Y.; Guan, C. Scalp EEG-Based Pain Detection Using Convolutional Neural Network. *IEEE Trans. Neural Syst. Rehabil. Eng.* **2022**, *30*, 274–285. [CrossRef] [PubMed]
49. Azimi, P.; Yazdanian, T.; Benzel, E.C.; Aghaei, H.N.; Azhari, S.; Sadeghi, S.; Montazeri, A. A Review on the Use of Artificial Intelligence in Spinal Diseases. *Asian Spine J.* **2020**, *14*, 543–571. [CrossRef] [PubMed]
50. Goldstein, P.; Ashar, Y.; Tesarz, J.; Kazgan, M.; Cetin, B.; Wager, T.D. Emerging Clinical Technology: Application of Machine Learning to Chronic Pain Assessments Based on Emotional Body Maps. *Neurotherapeutics* **2020**, *17*, 774–783. [CrossRef] [PubMed]
51. Nitish, A. Prediction of low back pain using artificial intelligence modeling. *J. Med. Artif. Intell.* **2021**, *4*, 1–9.
52. Abelin-Genevois, K. Sagittal Balance of the Spine. *Orthop. Traumatol. Surg. Res.* **2021**, *107*, 102769. [CrossRef]
53. Pikulkaew, K.; Boonchieng, E.; Boonchieng, W.; Chouvatut, V. Pain Detection Using Deep Learning with Evaluation System. Proceedings of Fifth International Congress on Information and Communication Technology. *Adv. Intell. Syst. Comput.* **2020**, *1184*, 426–435.
54. Lucey, P.; Cohn, J.F.; Prkachin, K.M.; Solomon, P.E.; Chew, S.; Matthews, I. Painful monitoring: Automatic pain monitoring using the UNBC-McMaster shoulder pain expression archive database. *Image Vis. Comput.* **2012**, *30*, 197–205. [CrossRef]
55. Ghazal, B.; Xujuan, Z.; Ravinesh, C.D.; Jeffrey, S.; Frank, W.; Hua, W. Ensemble neural network approach detecting pain intensity from facial expressions. *Artif. Intell. Med.* **2020**, *109*, 101954.
56. Guglielmo, M.; Zhanli, C.; Diana, J.W.; Rashid, A.; Yasemin, Y.; Çetin, A.E. Pain Detection from Facial Videos Using Two-Stage Deep Learning. In Proceedings of the 2019 IEEE Global Conference on Signal and Information Processing (GlobalSIP), Ottawa, ON, Canada, 11–14 November 2019; pp. 1–5.
57. Straube, A.; Andreou, A. Primary headaches during lifespan. *J. Headac. Pain* **2019**, *20*, 35. [CrossRef]
58. Sharma, T.L. Common Primary and Secondary Causes of Headache in the Elderly. *Headache* **2018**, *58*, 479–484. [CrossRef]
59. Paul, R.; William, J.M. Headache. *Am. J. Med.* **2018**, *131*, 17–24.
60. Yamani, N.; Olesen, J. New daily persistent headache: A systematic review on an enigmatic disorder. *J. Headac. Pain* **2019**, *20*, 80. [CrossRef] [PubMed]
61. HIS Classification ICHD-3. Available online: https://ichd-3.org/classification-outline/ (accessed on 18 January 2022).
62. Hansen, J.M.; Charles, A. Differences in treatment response between migraine with aura and migraine without aura: Lessons from clinical practice and RCTs. *J. Headac. Pain* **2019**, *20*, 96. [CrossRef] [PubMed]
63. Vij, B.; Tepper, S.J. Secondary Headaches. In *Fundamentals of Pain Medicine*; Springer: Berlin/Heidelberg, Germany, 2018; pp. 291–300.
64. Keight, R.; Al-Jumeily, D.; Hussain, A.J.; Al-Jumeily, M.; Mallucci, C. Towards the discrimination of primary and secondary headache: An Intelligent Systems Approach. In Proceedings of the 2017 International Joint Conference on Neural Networks (IJCNN), Anchorage, AK, USA, 14–19 May 2017; pp. 2768–2775.
65. Sanchez-Sanchez, P.A.; García-González, J.R.; Rúa Ascar, J.M. Automatic migraine classification using artificial neural networks. *F1000Research* **2020**, *9*, 618. [CrossRef] [PubMed]
66. Liu, Q.; Cai, J.; Fan, S.Z.; Abbod, M.F.; Shieh, J.S.; Kung, Y.; Lin, L. Spectrum Analysis of EEG Signals Using CNN to Model Patient's Consciousness Level Based on Anesthesiologists' Experience. *IEEE Access* **2019**, *7*, 53731–53742. [CrossRef]
67. Liu, Q.; Ma, L.; Fan, S.Z.; Abbod, M.F.; Ai, Q.; Chen, K.; Shieh, J.S. Frontal EEG Temporal and Spectral Dynamics Similarity Analysis between Propofol and Desflurane Induced Anesthesia Using Hilbert-Huang Transform. *BioMed Res. Int.* **2018**, *2018*, 4939480. [CrossRef]
68. Zi–Xiao, W.; Faiyaz, D.; Yan-Xin, L.; Shou-Zen, F.; Jiann-Shing, S. An Optimized Type-2 Self-Organizing Fuzzy Logic Controller Applied in Anesthesia for Propofol Dosing to Regulate BIS. *IEEE Trans. Fuzzy Syst.* **2020**, *28*, 1062–1072.
69. Yi-Feng, C.; Shou-Zen, F.; Maysam, F.A.; Jiann-Shing, S.; Mingming, Z. Electroencephalogram variability analysis for monitoring depth of anesthesia. *J. Neural Eng.* **2021**, *18*, 066015.
70. Lötsch, J.; Kringel, D.; Ultsch, A. Explainable Artificial Intelligence (XAI) in Biomedicine: Making AI Decisions Trustworthy for Physicians and Patients. *BioMedInformatics* **2022**, *2*, 1–17. [CrossRef]
71. Alex, K.; Ilya, S.; Geoffrey, E.H. ImageNet classification with deep convolutional neural networks. *Community* **2017**, *60*, 84–90.
72. Awwal, M.D.; Kamil, Y.; Huseyin, O. Application of Deep Learning in Neuroradiology: Brain Haemorrhage Classification Using Transfer Learning. *Comput. Intell. Neurosci.* **2019**, *2019*, 4629859.
73. Simonyan, K.; Zisserman, A. Very Deep Convolutional Networks for Large-Scale Image Recognition. In Proceedings of the 3rd International Conference on Learning Representations, ICLR 2015, San Diego, CA, USA, 7–9 May 2015; pp. 1–14.
74. Horry, M.J.; Chakraborty, S.; Paul, M.; Ulhaq, A.; Pradhan, B.; Saha, M.; Shukla, N. COVID-19 Detection Through Transfer Learning Using Multimodal Imaging Data. *IEEE Access* **2020**, *8*, 149808–149824. [CrossRef] [PubMed]
75. Kaiming, H.; Xiangyu, Z.; Shaoqing, R.; Jian, S. Deep Residual Learning for Image Recognition. In Proceedings of the IEEE Conference on Computer Vision and Pattern Recognition (CVPR), Las Vegas, NV, USA, 27–30 June 2016; pp. 770–778.
76. Weijun, H.; Yan, Z.; Lijie, L. Study of the Application of Deep Convolutional Neural Networks (CNNs) in Processing Sensor Data and Biomedical Images. *Sensors* **2019**, *19*, 3584.

77. Huang, G.; Liu, Z.; van der Maaten, L.; Weinberger, K.Q. Densely Connected Convolutional Networks. *IEEE Conf. Comput. Vis. Pattern Recognit.* **2017**, *2017*, 2261–2269.
78. Li, X.; Shen, X.; Zhou, Y.; Wang, X.; Li, T.-Q. Classification of breast cancer histopathological images using interleaved DenseNet with SENet (IDSNet). *PLoS ONE* **2020**, *15*, e0232127. [CrossRef] [PubMed]
79. Chan, Y.K.; Chen, Y.F.; Pham, T.; Chang, W.; Hsieh, M.Y. Artificial Intelligence in Medical Applications. *J. Healthc. Eng.* **2018**, *2018*, 4827875. [CrossRef]
80. Zemouri, R.; Zerhouni, N.; Racoceanu, D. Deep Learning in the Biomedical Applications: Recent and Future Status. *Appl. Sci.* **2019**, *9*, 1526. [CrossRef]
81. Moraes, J.L.; Rocha, M.X.; Vasconcelos, G.G.; Vasconcelos Filho, J.E.; De Albuquerque, V.H.C.; Alexandria, A.R. Advances in Photopletysmography Signal Analysis for Biomedical Applications. *Sensors* **2018**, *18*, 1894. [CrossRef]
82. Johnson, K.W.; Torres Soto, J.; Glicksberg, B.S.; Shameer, K.; Miotto, R.; Ali, M.; Ashley, E.; Dudley, J.T. Artificial Intelligence in Cardiology. *J. Am. Coll. Cardiol.* **2018**, *71*, 2668–2679. [CrossRef]
83. Coronato, A.; Naeem, M.; De Pietro, G.; Paragliola, G. Reinforcement learning for intelligent healthcare applications: A survey. *Artif. Intell. Med.* **2020**, *109*, 101964. [CrossRef]
84. Wells, L.; Bednarz, T. Explainable AI and Reinforcement Learning—A Systematic Review of Current Approaches and Trends. *Front. Artif. Intell.* **2021**, *4*, 550030. [CrossRef] [PubMed]

technologies

MDPI

Review

Advanced Security Framework for Internet of Things (IoT)

Abid Ali [1], Abdul Mateen [1], Abdul Hanan [2] and Farhan Amin [3,*]

1 Department of Computer Science, Federal Urdu University of Arts, Science & Technology, Islamabad 44000, Pakistan; abid.khawaja11@gmail.com (A.A.); abdul.mateen.cs@gmail.com (A.M.)
2 Department of Computer Science, CECOS University, Peshawar 25000, Pakistan; hanan@cecos.edu.pk
3 Department of Information and Communication Engineering, Yeungnam University, Gyeongsan 38541, Korea
* Correspondence: farhanamin10@hotmail.com

Abstract: The stimulus to carry out this research was to identify and propose a secure framework for the Internet of Things (IoT). Due to the massive accessibility and interconnection of IoT devices, systems are at risk of being exploited by hackers. Therefore, there is a need to find an advanced security framework that covers data security, data confidentiality, and data integrity issues. The study uses a systematic literature review (SLR) technique and complete substantive literature is reviewed to find out the constructs and themes in the existing literature. We performed it in four steps, which were inclusion, eligibility, screening, and identification. We reviewed around 568 articles from well-reputable journals, and after exclusion, 260 articles and 54 reports were analyzed. We performed an analysis using MAXQDA in which the nodes and themes were first identified. After the classification, a qualitative model was generated using MAXQDA. The proposed model is supported by the literature so it will be useful for the IT managers, developers, and the users of IoT.

Keywords: Internet of Things; data availability; data security; data confidentiality; data integrity

Citation: Ali, A.; Mateen, A.; Hanan, A.; Amin, F. Advanced Security Framework for Internet of Things (IoT). *Technologies* **2022**, *10*, 60. https://doi.org/10.3390/technologies10030060

Academic Editors: Manoj Gupta, Eugene Wong and Gwanggil Jeon

Received: 9 April 2022
Accepted: 10 May 2022
Published: 12 May 2022

1. Introduction

The twenty-first century is known as the era of interconnectivity and wireless communication where the world has witnessed some major technological revolutions in computer networking. The term Internet of Things (IoT) was coined by Kevin Ashton in 1999 [1]. The IoT provides a way of connectivity of things to things. The "thing" refers to all the things around us that are connected to the network. For example, the household appliances at home that are connected to the internet. IoT technology is used to share information and generate useful information between "things". It can operate without human intervention. The IoT concept is illustrated in Figure 1.

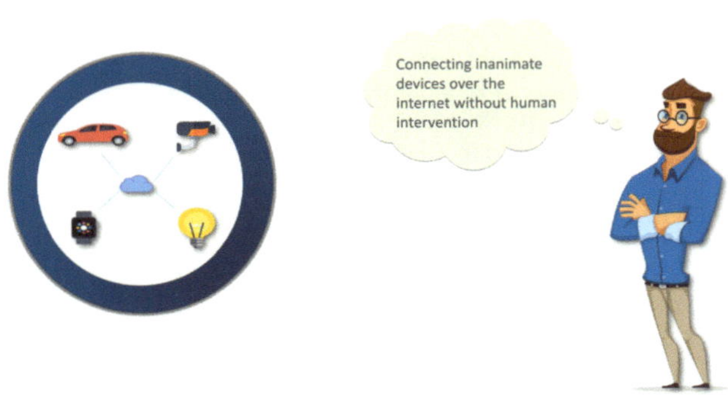

Figure 1. The concept of the Internet of Things.

In this Figure, the things are connected without human intervention. The traditional role of human command has been overpowered by the analytical capability of the IoT. Mobile phones, actuators, transceivers, protocol stacks, and microcontrollers have been developed to provide a firm connection and communication through the IoT. The data are collected and transmitted back to these devices with certain commands. The automated actions are made based on these suggested commands. The concepts of the IoT have been updated to improve the current Internet infrastructure to advanced network infrastructure, and have brought a technological revolution to the IT industry. The concept of the IoT suggests some interconnection between devices that include the facility of device autonomy, contextual awareness, sensing capability, and so on. To implement the IoT platform, many technologies and sensors, such as radiofrequency identifiers and networks of wireless sensors, are being used nowadays. However, in a conventional Internet protocol (IP), the security mechanisms need to be extended and modified to support IoT applications. The current IoT architecture is usually divided into three layers: the perception layer, the network layer, and the application layer. Figure 2 illustrates this architecture. The other forms are four-layer, five-layer, and seven-layer architecture, etc. However, we will use the three-layer architecture for illustration. The interaction of the sensors, actuators, and edge devices is the key part of this layer. The perception layer is used to identify the objects, perceive objects, collect information, and automatic control. This layer contains different types of control modules and collecting devices, such as the sound sensors, the temperature and pressure sensors, vibration sensors, etc., as shown in Figure 2. The perception layer is further divided into two parts: the perception node (controllers and sensors, etc.) and the perception network (transportation communication network) [2]. The use of the perception layer is to control data and data acquisition, while the perception network sends control instructions to the controller. The perception layers include implantable medical devices (IMDs), Global Positioning Systems (GPS), implantable medical devices (IMDs), Radio Frequency Identification (RFI), etc. The identification of abnormal sensor nodes is the one of security issues. It occurs when the node is attacked physically (e.g., destroyed or disabled). In general, these nodes are also known as faulty nodes. To ensure the standards of service, it is necessary to detect the fault codes and overcome the causes of lower standard services [3,4]. Another security concern of the perception layer is the key management mechanism and cryptography algorithms. For node authentication, public keys have been considered convenient. It is better to secure the entire network without any management protocol of complicated keys and to have large scalability [5]. According to [6], the most promising candidates for wireless sensor networks are three low-power public key encryption algorithms, namely, Rabin's Scheme, Ntru Encrypt, and the Elliptic Curve Cryptography. The network layer mainly realizes the transmission of information, routing (deciding the way of information transmission), and control (how to control the transmission of information). It is divided into two parts; one part is the communication technology and the other is the communication protocol of the Internet of Things. Communication technology is responsible for physically linking things with things to enable them to communicate. The communication protocol is responsible for establishing communication rules [7]. The application layer provides users with professional services and functional data processing and storage [8]. It has the support of the cloud and servers for the storage of data in the network. Our study is more focused on the aspect of data security in the IoT. The key data security aspects are given below:

1.1. Data Security in IoT

Currently, data security and privacy protection should be adopted equally to offer robust data security. Accessing and securing data by a static approach has become unacceptable because it fails to address the scalable data security IoT [5]. The security support is not always maintained. Consumer knowledge of IoT security is weak: security incidents can be difficult to detect or to resolve for usage [9].

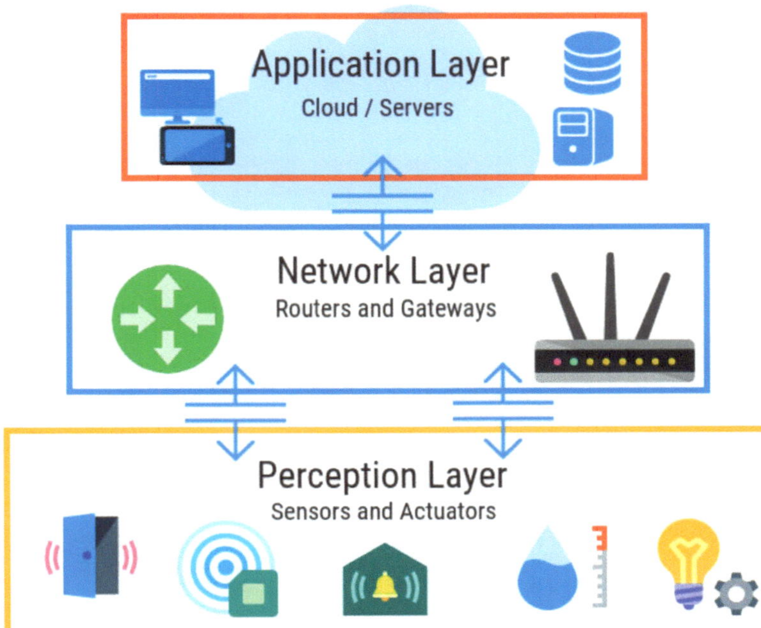

Figure 2. A three-tier IoT.

1.2. Data Integrity in the IoT

Data integrity is necessary for up-to-date and accurate data. It is very important to store data by any person or organization for integrity [10]. It is significant that data integrity in the IoT is measured, as data need to be secure and every transaction of data needs to be secure. Defining the integrity of data is easy but it is hard to ensure.

1.3. Data Confidentiality in the IoT

To keep data private in the public domain is called 'data privacy'. Data privacy terms can be applied to any organization or a person. Data are always limited and related to any person's life and existence [11]. He or she can keep the data private or public. An organization can also keep its data private, such as for financial statement reports or business plans. If there is no framework available for establishing personal privacy, then the privacy of any individual is very limited [12,13]. Data security and data privacy are used in many situations in the same context, but there is a distinct difference; data security is broadly thought to be about protection and saving your data from other unknown persons, whereas data privacy is to control where your data are collected, shared, and used for which, and for what, purpose.

1.4. Data Validity in the IoT

Data validity ensures that IoT services are practically available. If these services are unavailable, total progress can be decreased; it will also facilitate and provide help to hackers and attackers who are working in different smart industries, smart cities, and smart home etc. [6]. With the development of connected objects, users entrust part of their privacy to improve their environment and make their living environment more efficient and safer. There are risks to the person and his data; for example, a hacked surveillance camera lets you know if the owner is away or not from their home; a smart electricity meter: the meter can quickly become a spy if you are not careful [14].

1.5. Current IoT Security Framework

1. It consists of sensors, actuators, and other embedded systems [15].
2. Fog set of connections: A class of exchange ideas, technologies, and protocols by several IoT policies with the prerequisite to expand and enforce an entire confidence policy [16].
3. Core Complex: It provides a set of connection center platforms and IoT devices. The issues at this time are individuals confronted with conventional fundamental networks [17]. The measureless number of endpoints act together and get by to create a considerable precautions burden. Thus, based on the suggestions made in previous research papers, the current study proposes a security framework for the IoT in terms of data confidentiality, availability, and integrity.

The study has used the Systematic Literature Review (SLR) approach to find out the best security framework, which covers and identifies any problems. This study has provided a detailed analysis of prior published literature on the topic and compared the strengths and weaknesses of at least 20 security frameworks to evaluate and find out the best security framework for the IoT. This research mainly focuses on the three major security requirements, namely, data confidentiality, data availability, and data integrity. Therefore, the IoT has built a strong impact on commercial to domestic spheres of life, but besides this positive side, the IoT has introduced another darker side to the security and privacy of the person. The accessibility and interconnectivity of IoT devices have put the system at risk of being exploited by hackers [5,9].

1.6. Motivation of This Study

To the best of our knowledge, the literature still lacks research on extracting useful studies from a large pool based on the security aspects of data such as integrity, etc. Therefore, the stimulus to carry out this research was to identify and propose a secure framework for the emerging technologies.

1.7. Contribution of This Study

This study proposes a security model. In this model, the literature is reviewed from a large pool and, hence, suitable literature was extracted. We herein defined an article's inclusion and disillusion criteria and applied them to a large dataset. The model can select or discard the most relevant literature. It can easily be applied to emerging technologies such as the Internet of things (IoT). Briefly, we highlighted the different aspects and security concerns of the IoT. We also discussed recent solutions, along with comparisons and contrast. Our model is useful for IT managers, developers, and users, in extracting the most relevant literature from the databases. The rest of our study is organized as follows. In Section 2, we performed a literature review and discussed our proposed model. In Section 3, we have discussed inclusion criteria and explained that how we generate the results. In Section 4, we discuss the proposed model based on the achieved results. Finally, Section 5 offers conclusions from this study and suggests future work.

2. Literature Review and the Selection Criteria

In this study, we proposed a systematic literature review (SLR). We adopted an advanced method by Brinner and Denver. The detailed steps of the adopted methodology include the following steps. First, we performed systematic identification of the need for a systematic literature review and finalized the review protocol. On this site, we performed the election of studies and assessed their quality, and took notes to extract the relevant data. Finally, we reported and discussed the results. The details of our proposed methodology are given below.

2.1. Selection of Relevant Studies

To address the primary objective of this study, we performed a systematic literature review. This review was conducted in May 2020 without time restrictions, and the result was updated in June 2020. In this review study, we extracted relevant literature from esteemed journals, such as Scopus, the Web of Science, Google Scholar, ScienceDirect, etc. The relevant grey literature, such as government publications and unpublished material, was searched systematically [18]. To locate the grey literature, the first 150 hits from Google Scholar were evaluated. The keyword search and alternate key work searching were used to locate the relevant studies that aligned with our objectives. The hand search reference list further locates various other sources of grey literature, particularly, committee and research documents and policy briefs from both public and private sector organizations. Accordingly, the flow chart of the strategy for locating the studies is shown in Figure 3. Furthermore, various refinement features of the Web of Science, Google Scholar, and Scopus were applied to find the most relevant studies. The articles with missing abstracts were retrieved and scrutinize for relevance. All the articles accessed through different journals were retrieved in full text.

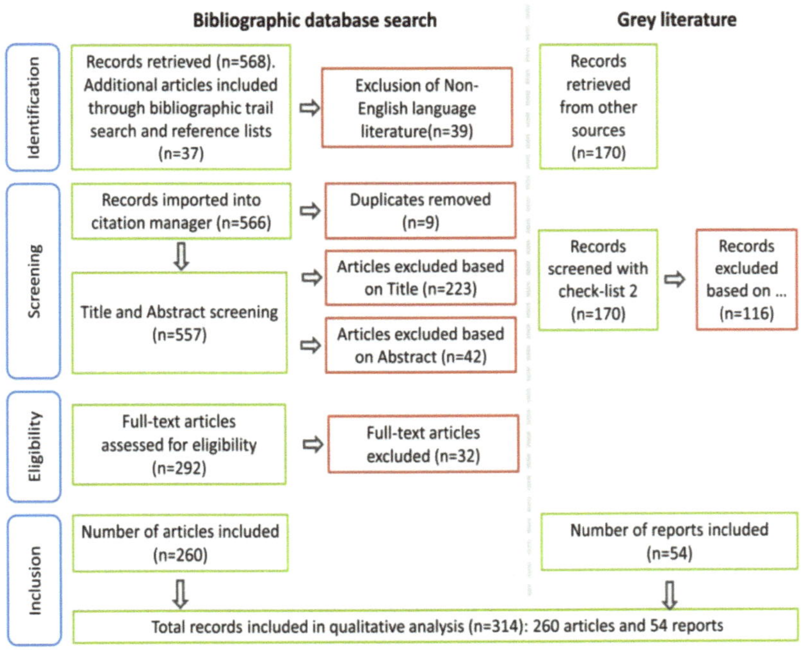

Figure 3. Flow chart of the strategy for locating the studies.

2.2. Evaluation

In this step, the selection and evaluation were performed using a systemic literature review. The eligibility of the accessed articles was examined independently based on pre-defined criteria that were outlined for inclusion and exclusion [19]. The exclusion criteria were applied especially when the search was performed in selected databases, such as custom range in terms of year, language, and subject. At first, the abstract of the paper was evaluated to determine its relevance. The studies that met the exclusion criteria were excluded and sorted by cause of exclusion. After carefully evaluating and scrutinizing the abstract of the retrieved articles, a full-text review was made and additional articles were discarded by using the exclusion criteria. The discrepancy concerning the relevance of the articles was resolved through the specific criteria for inclusion of the articles. The articles that remained out of the scope were excluded, and a refined list of articles was finalized. Articles from the Web of Science and Google Scholar that did not fit the inclusion criteria were discarded to avoid ambiguity.

2.3. Analysis and Synthesis

The retrieved and evaluated articles were finalized based on the inclusion criteria and processed through qualitative analysis software (MAXQDA11). The processing of the data results was performed in major themes. The thematic content analysis was made to determine the major theme that emerged in the selected articles. Thematic analysis is one of the commonly used qualitative research techniques; it analyses and interprets various patterns of qualitative data. In our context, the qualitative data were extracted from the selected papers [19]. Thematic analysis is a widely employed technique in contrast to most other qualitative analytic approaches, such as narrative analysis and discourse analysis, which are also widely used in a systematic literature review (SLR). The thematic analysis in a systematic literature review enables the detection of major trends and patterns in the collected papers. The significant themes remain the ones that predominate and remain prominent; after completing the thematic analysis, the coding is complete. The coding is a systematic process of indexing the text to develop a framework of major themes. The coding enables the entire process to be effective and robust. Aligned with past studies expounded in past literature, categorically, there are two types of coding that were identified; one is data-driven, and second is the concept-driven coding [20,21]. We aligned our study with past studies that used data-driven coding, and the data extracted from the selected papers were coded accordingly.

3. Reporting and Discussion of the Results

The retrieved data and articles were finalized based on the inclusion criteria. These data were processed through qualitative analysis software, i.e., MAXQDA20, which processed the data results in terms identifying different themes. The core objective of thematic content analysis is to determine various major themes. Thematic analysis is one of the commonly used qualitative research techniques. It is used to perform analyses and interprets various patterns of qualitative data. In our context, the qualitative data were extracted from the selected papers. Figure 4 illustrates how the article files were imported to the MAXQDA 20 for inferring the results. The first step in MAXQDA 20 was to conduct a quantitative analysis, whereby the file is imported and proceeded to further analysis. Once the required file has been imported, the next step is to run the auto coding. The auto code results are significant to determine which variables were reputedly used in past studies. Figure 4 illustrates the auto-coding results and confirms how many times the given variables that remain significant to the IoT remain significant. The details of the rest of the steps are given below.

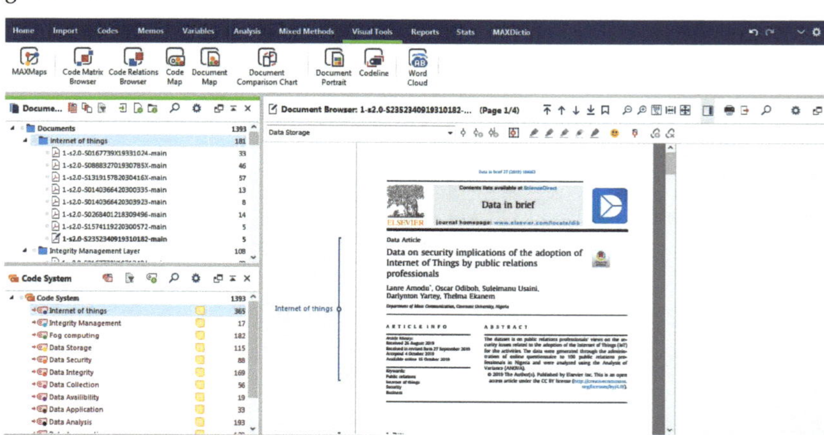

Figure 4. Data File in MAXQDA20.

3.1. Auto Codes Results from SLR

Figure 5 shows the auto-coding result. In this Figure, the auto-coding results of the selected articles reflect that articles were selected for analysis 365 times. Integrity management was used seventeen times while fog computing was used 182 times. Accordingly, data storage also remains one of the most important features of IoT as it was extensively examined and discussed 115 times in the past literature. Data security and data integrity were used 88 and 169 times, respectively. The data collection and data availability were used 56 and 19 times in articles that were selected for analysis. Data application and data analysis were used 33 and 193 times in the articles that were used for the analysis. Data aggregation and data confidentiality reflect that they were used 128 times and 28 times, respectively. Based on the auto coding, it was inferred that data analysis and data integrity along with fog computing remain the main determinants of IoT. Our efficient model contains the features of data analysis and data integrity along with fog computing to develop and implement the most robust digital system for an organization.

	Parent code	Code	Code alias	Cod. seg. (all ...
		Internet of things		365
		Integrity Manag...		17
		Fog computing		182
		Data Storage		115
		Data Security		88
		Data Integrity		169
		Data Collection		56
		Data Availibility		19
		Data Application		33
		Data Analysis		193
		Data Aggregation		128
		Data Confidentia...		28

Figure 5. Auto Codes Results from SLR.

3.2. The Codes Cloud

Based on auto coding, the codes cloud was generated. The codes cloud and auto coding are integrated. The codes cloud remains more convenient to interpret and is widely used in information technology (IT) research to make robust analyses. Figure 6 presents the phenomenon that the main codes cloud is generated based on auto coding. The codes cloud shown in this Figure state that the IoT remains one of the most significant themes appearing in the articles examined. A quantitative analysis was performed through software that enabled us to detect the major themes used in studies expounded in past literature. The codes cloud reflect that besides IoT, fog computing also remains the second major theme used in the studies analyzed. Data aggregation was also outlined as the third major theme that remains critical for effective IoT. Besides these four major themes, fog computing, and data aggregation, other minor themes were discovered through the codes of the cloud. The minor themes mainly include data collection, integrity management, data confidentiality, and data application. These minor themes all remain the key determinants of the IoT. Based on codes cloud, it remains essential to infer that collectively the IoT contains various major and minor determinants that should be considered when implementing a framework relevant to the IoT. The organization of major themes such as fog computing and data aggregation is important considering their significance, along with the other elements of IoT, such as data collection, integrity management, data confidentiality, and data application. These should be considered to be important to develop and implement effective IoT. The

IoT contains all the elements of fog computing, namely, data aggregation, data collection, integrity management, data confidentiality, and data application. These can be used to improve the effectiveness and efficiency of the system. The efficiency and effectiveness of the IoT is the main attribute that should be fulfilled to run the affairs of the organization effectively. Therefore, based on auto coding and code cloud, an analysis of the articles was selected for SLR through qualitative analysis software (MAXQDA20), it is asserted that IoT is an integrated and multifaceted phenomenon. The organizations that aim to develop and implement effective IoT should conduct internal and external analyses. The IoT remains standardized but should be aligned with organizational strategy to promote efficiency.

Figure 6. Codes Cloud are generated based on auto coding.

3.3. Word Frequencies

The IoT is a relatively new concept in communication studies as it is still developing with the evolution of the IoT and its dominations. Thus, the growing influence of the IoT on commercial and domestic spheres has raised concerns regarding the availability, confidentiality, and integrity of data. For auto coding and code cloud, an analysis of the articles was selected for SLR through qualitative analysis software (MAXQDA20). Besides auto coding and code cloud, the keywords in the literature were examined to determine which keywords remain significant and had been used widely in past studies. Figure 7 reflects the most significant and insignificant keywords used in the current literature. The keyword remains dominant and widely used in past literature. Data availability remains the first prerequisite while dealing with IoT. Data availability is very important, the other determinants of IoT remain useless, as one cannot ensure the computations and processing of the data without its availability. Data availability is based on an SLR keyword search and remains one of the primary features of the IoT. Data security after data availability remains vital to keep the privacy of the information. In a connected world, data security and privacy sensitivity, and in recent times, an increase in exponential data availability, have become a big challenge. However, with the increase in security sophistication information needed for a launch, any attack decreases. That is why the security measurement and privacy protection should be adopted equally to offer robust data security and end to end. For regulating access and securing data, a static approach is not acceptable because it fails to address the necessity that a mechanism of scalable data security IoT is conceivably generally involved and an immature part of net safety. The third keyword that is significant remains critical in ensuring the effectiveness of cloud computing. Cloud computing is popular due to advancements in information and cloud technology; it remains robust to ensure data security and effective backups so that the processing and accuracy of the data are achieved effectively. The next dominant and significant keyword search that is highlighted in the above figure is known as data integrity. Data integrity is defined as the reliability and validity of the data being used for analysis. It is the most vital feature of IoT as it is the primary concern of the entire stakeholder who uses such a system to assist their decision making through information. The information is accessed through the processing of data, which provides valuable information to the stakeholder to make a different decision. Therefore, if the data integrity remains minimal, the data reliability and validity will jeopardize the stakeholder's decision making. The information extracted, based on data that have integrity pitfalls, remains misleading and will result in economic losses. Therefore, one of the most important things that needs to be ensured during the process of the IoT is data integrity. The studies that were expounded in past literature confirmed the significance of data integrity, and it is a repeated keyword that was found to be a significant keyword search. However, it was found in the above figure that in our keyword search of the literature, confidentiality of data was found as being the least popular keyword search. These security issues have received the attention of academics, policymakers, and security experts toward ensuring the confidentiality and security of IoT devices and consumers' privacy. Hundreds of surveys were published to address these security challenges, but very limited efforts were made to design a framework that can resolve these security challenges.

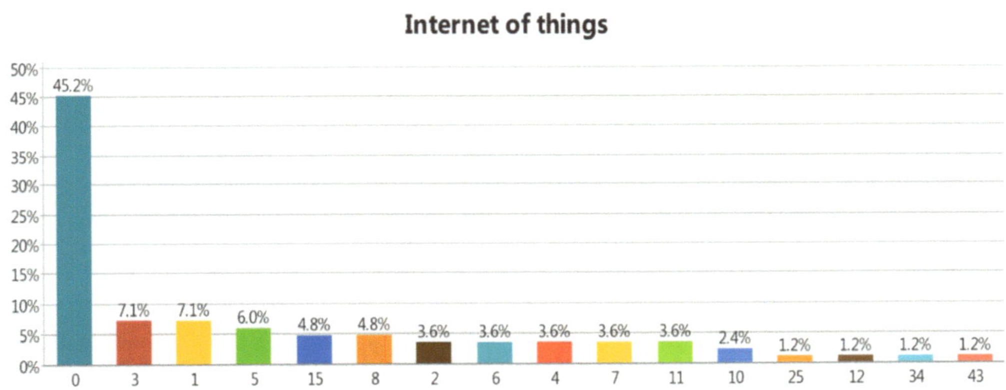

Figure 7. Significant and insignificant keywords being used in the literature.

3.3.1. Data Confidentiality

The growing influence of the IoT on commercial and domestic spheres has raised concerns regarding the availability, confidentiality, and integrity of data. By auto coding and cloud code, an analysis of the articles was selected for SLR through qualitative analysis software (MAXQDA11). Besides auto coding and code cloud, the keywords in the literature were identified. The significance of each feature of IoT was examined to determine which keywords remain significant and had been widely used by past studies expounded in literature. Figure 8 reflects the data confidentiality used in past studies. Data confidentiality remains one of the most important features of the IoT. This connection between the physical and visual world with the help of software and sensors has opened up possibilities to connect the required data or information at any time. However, these possibilities have also added certain threats to human security and confidentiality in the world of interconnected devices, where sensitive private information of users can be manipulated or leaked by hackers. As per past studies, our results also confirm the significance of data confidentiality. The studies expounded in past literature proclaim that 25% of the studies remain concerned with data confidentiality. The IoT exposes an organization to various types of risk. The information that remains private and confidential may be used by an unauthorized user to adversely affect the reputation of the business. Trust remains one of the most important elements in the IoT. Therefore, the breaching of security and privacy has introduced a whole new degree of online privacy concerns for consumers because these devices not only can collect personal information such as users' names and telephone numbers, but can also monitor users' activities. Due to the utmost significance of data confidentiality, most organizations have separately established a cyber security system that ensures data confidentiality and prevents the data's unauthorized use. The number of studies and applied research have surged and studies have devoted their attention to developing frameworks and models that robustly contain the feature of data confidentiality. The IoT without data confidentiality remains ineffective in meeting stakeholder and organizational needs effectively.

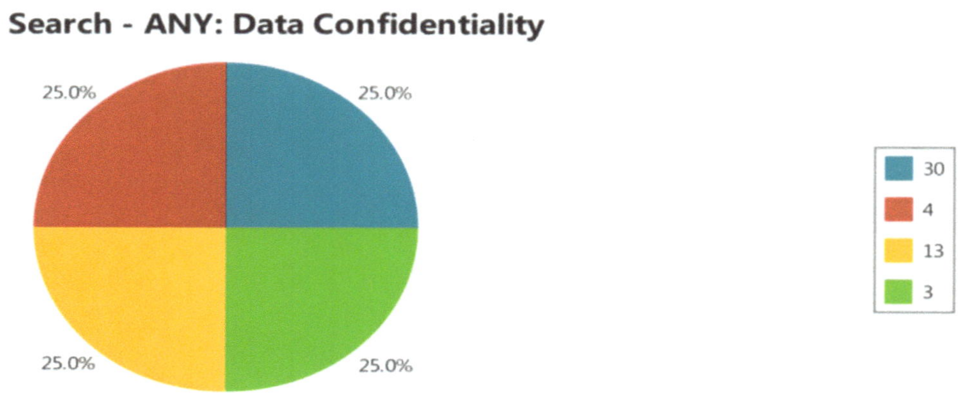

Figure 8. Data confidentiality of past studies.

3.3.2. Internet of Things

This is the second major theme that remains dominant in the literature. The SLR was conducted based on selected articles and analyzed through software to predict the most significant themes being discussed in the literature. The past few decades have witnessed an increased devotion to empirical studies toward examining the role of IoT and its determinants. The analysis of the selected articles states that the IoT has been discussed most frequently in past studies and has been examined by various methods. The objectives of these studies that remain concerned with the IoT are similar. The underlying objective of these studies remains concerned with methods and frameworks that remain robust for the effectiveness and efficiency of the IoT. The IoT is relatively a new concept in communication studies as it is still developing with the evolution of the IoT and its dominations. Thus, the growing influence of the IoT on commercial and domestic spheres has raised concerns regarding the availability, confidentiality, and integrity of data. Therefore, the underlying objective of this study was to analyze the significance of SLR. The results of our study suggest that most of the literature remains devoted to the IoT. The underlying theories and framework being postulated in past studies remain robust to improve the reliability of the IoT and improve organizational capabilities. Researchers have tried to find the role of the IoT in human life along with proposed challenges to data availability, confidentiality, and integrity, but very limited data have been published on security mechanisms that can address these challenges. The adoption of the IoT appears to occur regardless of the type of organization. Figure 9 reflects the significance of IoT and its appearance in earlier studies. The studies remain devoted to understanding the phenomenon that is the IoT.

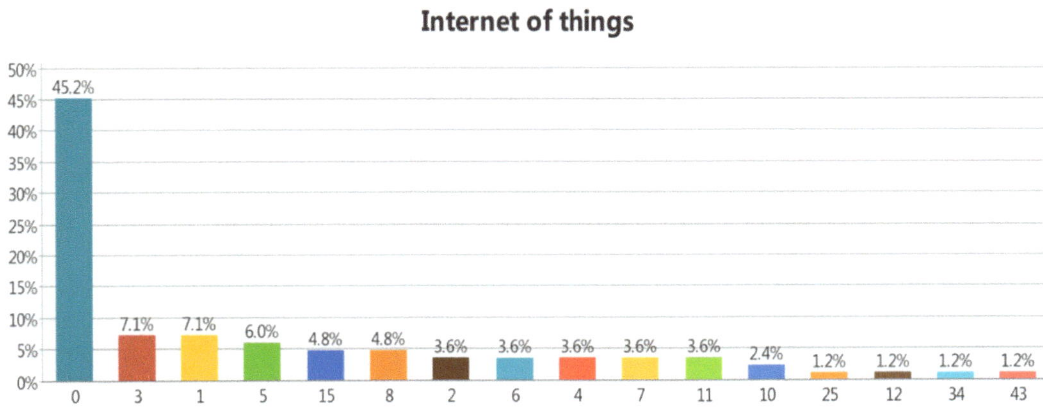

Figure 9. Significance of the IoT and its appearance in past studies.

3.3.3. Integrity Management Layer

The third major theme that appeared significant in past studies is known as the integrity management layer. This layer has become one of the most robust determinants of the IoT. The integrity management layer ensures the reliability and validity of the data. Thus, the growing influence of the IoT on commercial and domestic spheres has raised concerns regarding the availability, confidentiality, and integrity of data. Therefore, these limitations and loopholes in the security framework of the IoT have been considered during the implementation of safety mechanisms, and it is expected that this proposed research will bring new insights regarding the current security practices of the IoT and provide a solution to address any problems. Our study suggests that integrity management is considered the most pivotal and robust element of the IoT as it is essential to determine the effectiveness and efficiency as shown in Figure 10. The selected papers were chosen based on a specified threshold and reflect the past studies that remain devoted to integrity management. The integrity ensures that data input and processing and its output remain reliable and valid to ensure the effectiveness and efficiency of the IoT.

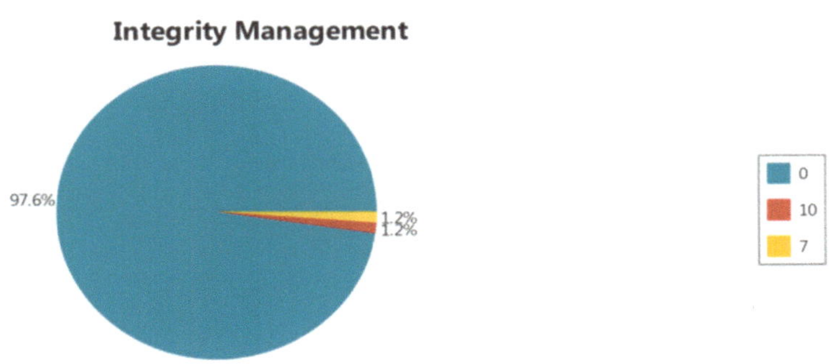

Figure 10. Effectiveness and efficiency of the IoT.

3.3.4. Fog Computing Layer

Fog computing is nowadays considered the most vital element of the IoT as it is presumed that it makes data storage and access more reliable and safer.

3.3.5. Data Storage Layer

The IoT should contain enough capacity to store the collected and processed data with an element of high privacy. The data storage should be robust so that its access can be granted only to authorized users. The use of data storage, due to advancements in technology, has risen, and it has become convenient for companies to manage their data storage effectively. The SLR technique shown in Figure 11 reflects that 59% of the studies are being investigated. It shows that data storage is one of the most potent determinants of IoT. The traditional cryptography solutions cannot work anymore on IoT systems since these devices have limited and less space for storage. It cannot manage the heavyweight and advanced cryptography algorithm storage requirements. Therefore, alternative storage frameworks and models were developed through empirical examination to uplift the data storage, which ensures the dynamic needs of the organization.

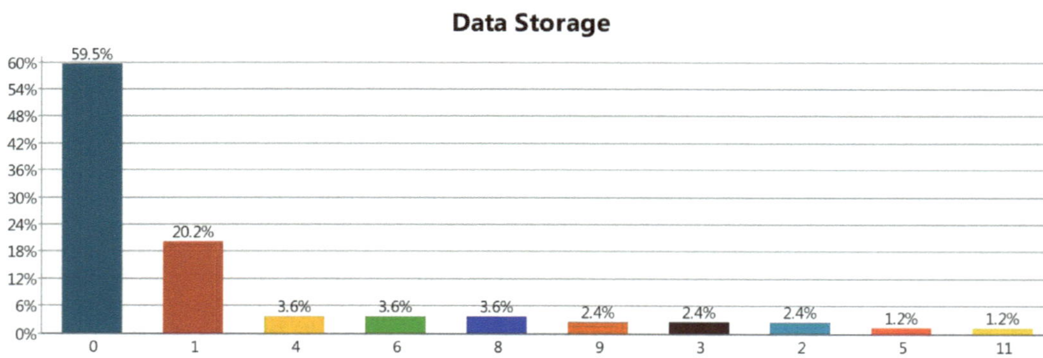

Figure 11. Potential determinants of IoT—Data storage.

3.3.6. Data Security Layer

Data security also remains very critical to ensure the effectiveness of the IoT. Data security ensures the privacy of information and safeguards it from unauthorized usage. The information system that contains loopholes in terms of data security does not meet the needs of the standard organization. Therefore, organizations have established separate arrangements to ensure data security. Figure 12 reflects the keyword search based on auto coding and it suggests that data security remains the most important determinant of the IoT. This is why security measurements and privacy protection should be adopted equally to offer robust end-to-end data security. For regulating access and securing data, a static approach is not acceptable because it fails to address the necessity that a mechanism of a scalable data security IoT is a conceivably and generally involved immature part of a net safety.

Figure 12. Data Security in IoT.

3.3.7. Data Collection Layer

The results shown in Figure 13 reflect that data collection has been widely discussed in past studies expounded in the literature. Data collection remains critical as the initial input to the IoT; the processing remains highly dependent on the data collection phase. Unless and until the data collection has been made effective, the other elements of data storage remain useless.

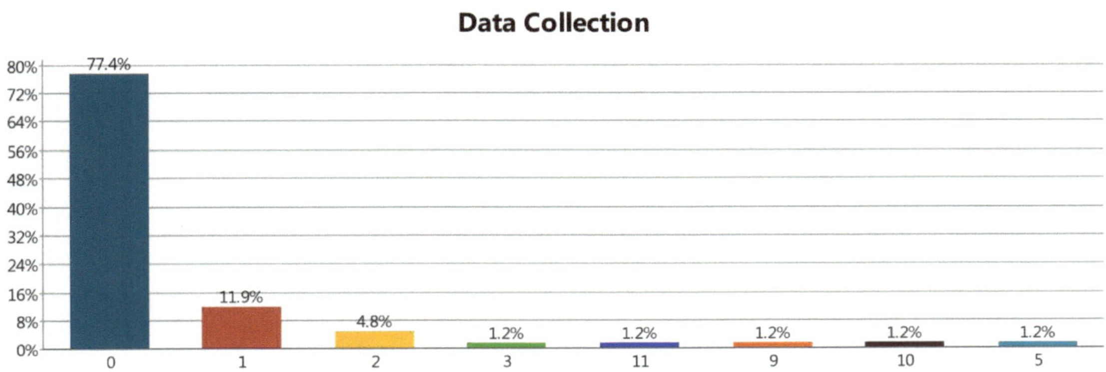

Figure 13. Data collection layer in the IoT.

3.3.8. Data Availability Layer

The data availability and the data collection are critical issues. They are used to ensure the effectiveness and efficiency of the IoT. The results shown in our study suggest that studies expounded in past literature remain concerned with data availability. It is impossible to ensure the efficient working of the information system without data availability. Therefore, besides the data collection, data availability also remains essential for the IoT to work effectively.

3.3.9. Data Application Layer

Data application is also considered as being a very important thing, which has been widely acknowledged in past studies, as shown in Figure 14. The application layer, particularly applications from the processed data, accord to the demands or requirements of the user. Therefore, the application layer should be user-friendly so that the IoT can be used with ease and convenience.

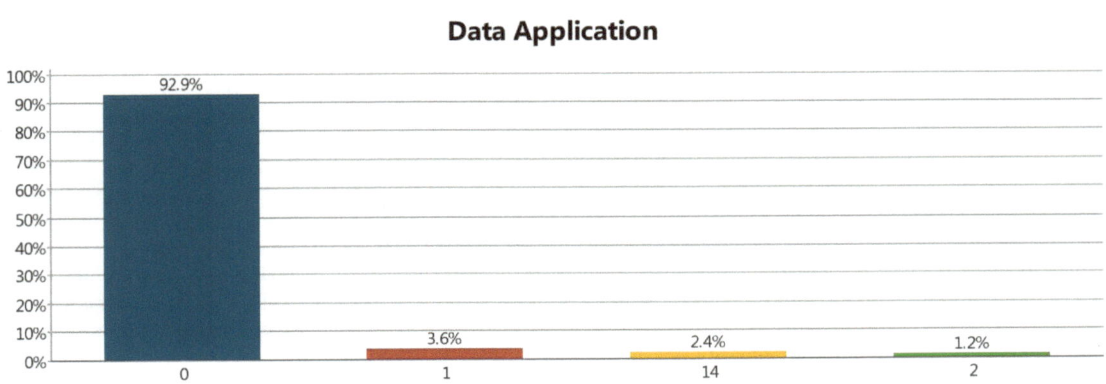

Figure 14. Data application layer.

3.3.10. Data Analysis Layer

Data analysis and the processing of the data are very important. Data analysis remains a critical challenge as it provides valuable insight to the stakeholder to issue information and decisions. Figure 15 reflects the significance of data analysis based on SLR. The SLR of the selected papers reflects that the data analysis layer gives importance to collecting the data for the development and the experimentation of smart decisions.

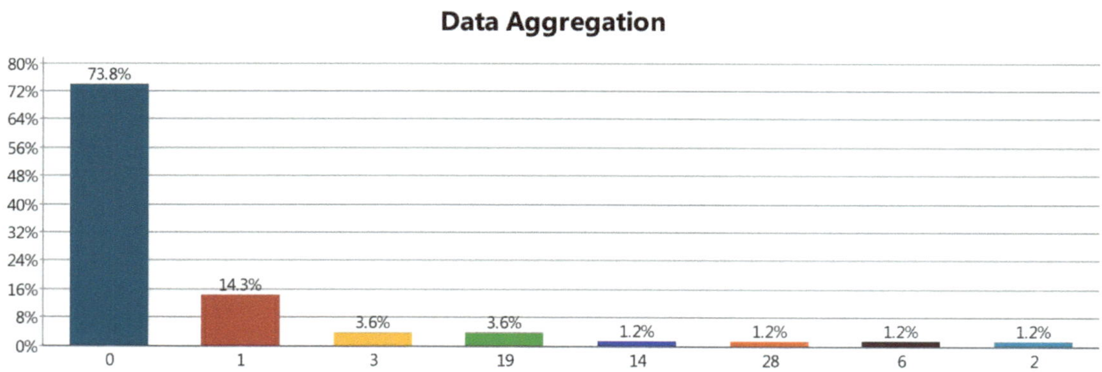

Figure 15. Data Aggregation.

3.3.11. Data Aggregation Layer

Data aggregation is one of the dominant themes that has been pointed out in past studies. Figure 15 reflects the significance of data aggregation. The storage, data supply, and the reduction size for improvement in storage and transmission of data are the major challenges of this layer. That is why the layer of these data is concentrated on merging and summarizing data. The modules that are the key to this layer are the heterogeneity, aggregation, filtering, interoperability, and transformation manager. Data that are received from the integrity management layer are more redundant, raw, and very large, as shown in Figure 15.

4. Qualitative Model

Based on the substantive SLR and the results, the following model was deduced through a deductive approach, which provides the essential elements that should be robust to build an effective and efficient IoT. The qualitative models that were deduced based on past studies are aligned with our proposed framework. The proposed model and qualitative model categorically contain nine layers, namely, computing, fog, management, integrity, security, data analysis, data aggregation, and data storage layer. Every layer of the framework contributes to the management process of the next layer. These nine layers are also considered the most robust determinants of the IoT. Each layer ensures the effectiveness and efficiency of the system to meet the individual's and organization's needs effectively. Our study aimed to design an advanced security framework for the IoT that can be used to analyze possible threats or challenges. This process began with elaborating the concept of the IoT, its characteristics, and layers of IoTs, all possible threats or challenges to the different layers of the IoT, and then moved on to find the best security framework to address these threats. The core objective of this study was to propose a security framework in terms of data confidentiality, availability, and integrity. The proposed model is shown in Figure 16. It comprises security frameworks in terms of computing layer, fog, management layer, integrity, security layer, data analysis, and data aggregation, where the data storage layer remains robust. The proposed and deduced security framework for the IoT remains to be aligned, which reflects the notion of the major determinants or features that should be an inclusive part of a security model for the IoT.

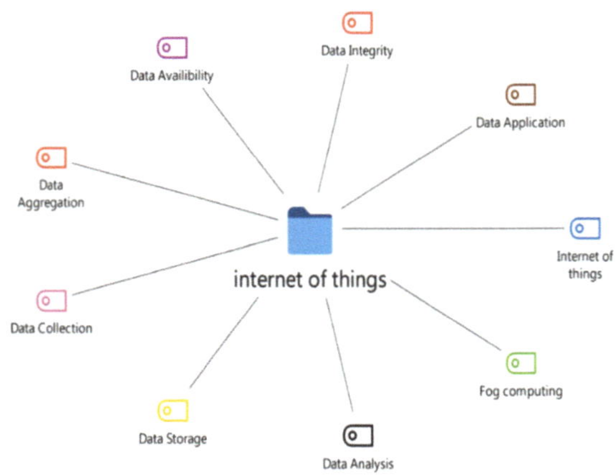

Figure 16. Security Model for the IoT.

5. Findings and Implications of This Study

In this study, we performed a thematic analysis and built a qualitative model. In this model, we extracted relevant literature from various databases using MAXQDA. The proposed security framework is completely based on different layers. These layers are data availability, data integrity, data application, IoT, fog computing, data analysis, data storage, data collection, data aggregation, etc. The research study in [22] concludes that data confidentiality in the IoT is a primary constraint that guarantees access and modification to certified entities via an access control mechanism and object authentication practice with a related identity supervision system. Our study concludes that data confidentiality is an important characteristic that needs to be included in the security framework. Similarly, the findings of [10] indicate that data integrity [23] is necessary for accurate data. Integrity is very important for storing data by any person or any organization. Data integrity is an important characteristic that needs to be included in the security framework. Hence, the key findings of our study are consistent. The results from [24] indicate that the aspects of data management should be kept in mind. The proposed model was made with a fog computing layer. This layer facilitates the devices to analyze, process, and partially store data on the node's edge. Our study also concludes that data application is an important characteristic that needs to be included in the security framework. The key findings of our study are consistent with the other findings. The studies conducted by [25–30] indicate that fog computing should also be considered as being important to develop and implement an effective IoT. One more finding concludes that data application is an important characteristic that needs to be included in the advanced security framework [31].

6. Conclusions

This research study proposed a security framework based on the available literature by using the SLR technique, using the the current literature we first identified. The coding was performed using an extensive literature review. Thematic analysis was conducted and a qualitative model was developed. During communication, data can be altered by cybercriminals. These methods are used to ensure the accuracy and originality of the data, including methods such as Checksum and Cyclic Redundancy Check (CRC). Moreover, the continuous syncing of data for backup requires the use of features such as version control, etc. These are used to keep a record of the file changes in the system to restore the file in case of an accidental deletion of data; this can also ensure the integrity of data such that the data on IoT-based devices are in their original form when accessed by permitted users. In the future, we will extend this security framework by employing advanced CRC for errors.

Author Contributions: Conceptualization, A.A.; methodology, A.A.; software, A.H.; validation, A.H.; formal analysis, A.H.; investigation, A.H.; resources, F.A.; data curation, F.A.; writing—original draft preparation, A.A.; writing—review and editing, F.A.; visualization, A.M.; supervision, A.M.; project administration, F.A. and A.M.; funding acquisition, F.A. All authors have read and agreed to the published version of the manuscript.

Funding: This research received no external funding.

Institutional Review Board Statement: Not applicable.

Informed Consent Statement: Not applicable.

Data Availability Statement: Not applicable.

Acknowledgments: We thank our families and colleagues who provided us with moral support.

Conflicts of Interest: The authors declare they have no conflicts of interest regarding the present study.

References

1. Amin, F.; Abbasi, R.; Rehman, A.; Choi, G.S. An Advanced Algorithm for Higher Network Navigation in Social Internet of Things Using Small-World Networks. *Sensors* **2019**, *19*, 2007. [CrossRef] [PubMed]
2. Qian, Y.; Jiang, Y.; Chen, J.; Zhang, Y.; Song, J.; Zhou, M.; Pustišek, M. Towards decentralized IoT security enhancement: A blockchain approach. *Comput. Electr. Eng.* **2018**, *72*, 266–273. [CrossRef]
3. Khattak, H.A.; Shah, M.A.; Khan, S.; Ali, I.; Imran, M. Perception layer security in Internet of Things. *Futur. Gener. Comput. Syst.* **2019**, *100*, 144–164. [CrossRef]
4. Amin, F.; Choi, G.S. Hotspots Analysis Using Cyber-Physical-Social System for a Smart City. *IEEE Access* **2020**, *8*, 122197–122209. [CrossRef]
5. Kumar, N.M.; Mallick, P.K. The Internet of Things: Insights into the building blocks, component interactions, and architecture layers. *Procedia Comput. Sci.* **2018**, *132*, 117–119. [CrossRef]
6. Sun, W.; Cai, Z.; Li, Y.; Liu, F.; Fang, S.; Wang, G. Security and Privacy in the Medical Internet of Things: A Review. *Secur. Commun. Netw.* **2018**, *2018*, 5978636. [CrossRef]
7. Jose, D.V.; Vijyalakshmi, A. An Overview of Security in Internet of Things. *Procedia Comput. Sci.* **2018**, *143*, 744–748. [CrossRef]
8. Yang, A.; Li, Y.; Kong, F.; Wang, G.; Chen, E. Security Control Redundancy Allocation Technology and Security Keys Based on Internet of Things. *IEEE Access* **2018**, *6*, 50187–50196. [CrossRef]
9. Chen, K.; Zhang, S.; Li, Z.; Zhang, Y.; Deng, Q. Internet-of-Things security and vulnerabilities: Taxonomy, challenges, and practice. *J. Hardw. Syst. Secur.* **2018**, *2*, 97–110. [CrossRef]
10. Javaid, U.; Aman, M.N.; Sikdar, B. Blockpro: Blockchain based data provenance and integrity for secure IoT environments. In Proceedings of the ACM Blocksys 2018, New York, NY, USA, 4 November 2018; pp. 13–18.
11. El-Hajj, M.; Chamoun, M.; Fadlallah, A.; Serhrouchni, A. Analysis of authentication techniques in Internet of Things (IoT). In Proceedings of the 2017 1st Cyber Security in Networking Conference (CSNet), Rio de Janeiro, Brazil, 18–20 October 2017; pp. 1–13.
12. Huang, Q.; Yang, Y.; Wang, L. Secure Data Access Control with Ciphertext Update and Computation Outsourcing in Fog Computing for Internet of Things. *IEEE Access* **2017**, *5*, 12941–12950. [CrossRef]
13. Sahmim, S.; Gharsellaoui, H. Privacy and security in internet-based computing: Cloud computing, internet of things, cloud of things: A review. *Procedia Comput. Sci.* **2017**, *112*, 1516–1522. [CrossRef]
14. Meddeb, M.; Dhraief, A.; Belghith, A.; Monteil, T.; Drira, K.; Alahmadi, S. Cache Freshness in Named Data Networking for the Internet of Things. *Comput. J.* **2018**, *61*, 1496–1511. [CrossRef]
15. Angin, P.; Mert, M.B.; Mete, O.; Ramazanli, A.; Sarica, K.; Gungoren, B. A Blockchain-Based Decentralized Security Architecture for IoT. In Proceedings of the International Conference on Internet of Things, Seattle, WA, USA, 25–30 June 2018. [CrossRef]
16. Aman, M.N.; Sikdar, B.; Chua, K.C.; Ali, A. Low Power Data Integrity in IoT Systems. *IEEE Internet Things J.* **2018**, *5*, 3102–3113. [CrossRef]
17. Amin, F.; Choi, G.S. Advanced Service Search Model for Higher Network Navigation Using Small World Networks. *IEEE Access* **2021**, *9*, 70584–70595. [CrossRef]
18. Colicchia, C.; Strozzi, F. Supply chain risk management: A new methodology for a systematic literature review. *Supply Chain. Manag. Int. J.* **2012**, *17*, 403–418. [CrossRef]
19. Dziopa, F.; Ahern, K. A systematic literature review of the applications of Q-technique and its methodology. *Eur. J. Res. Methods Behav. Soc. Sci.* **2011**, *7*, 39–55. [CrossRef]
20. Si, K.; Wolfson, C.; Fi, B. A multidisciplinary systematic literature review on frailty: Overview of the methodology used by the Canadian Initiative on Frailty and Aging. *BMC Med. Res. Methodol.* **2009**, *9*, 68–72.
21. Liu, C.; Yang, C.; Zhang, X.; Chen, J. External integrity verification for outsourced big data in cloud and IoT: A big picture. *Future Gener. Comput. Syst.* **2015**, *49*, 58–67. [CrossRef]

22. Farooq, M.U.; Waseem, M.; Khairi, A.; Mazhar, P.S. A critical analysis on the security concerns of internet of things (IoT). *Int. J. Comput. Appl.* **2015**, *111*, 1–6.
23. Amin, F.; Lee, W.-K.; Mateen, A.; Hwang, S.O. Integration of Network science approaches and Data Science tools in the Internet of Things based Technologies. In Proceedings of the 2021 IEEE Region 10 Symposium (TENSYMP), Jeju, Korea, 23–25 August 2021; pp. 1–6. [CrossRef]
24. Atlam, H.F.; Walters, R.J.; Wills, G.B. Fog computing and the internet of things: A review. *Big Data Cogn. Comput.* **2018**, *2*, 10. [CrossRef]
25. Hameed, K.; Khan, A.; Ahmed, M.; Reddy, A.G.; Rathore, M.M. Towards a formally verified zero watermarking scheme for data integrity in the Internet of Things based-wireless sensor networks. *Futur. Gener. Comput. Syst.* **2018**, *82*, 274–289. [CrossRef]
26. Bin Qaim, W.; Ometov, A.; Molinaro, A.; Lener, I.; Campolo, C.; Lohan, E.S.; Nurmi, J. Towards Energy Efficiency in the Internet of Wearable Things: A Systematic Review. *IEEE Access* **2020**, *8*, 175412–175435. [CrossRef]
27. Navas, R.E.; Cuppens, F.; Cuppens, N.B.; Toutain, L.; Papadopoulos, G.Z. MTD, Where Art Thou? A Systematic Review of Moving Target Defense Techniques for IoT. *IEEE Internet Things J.* **2020**, *8*, 7818–7832. [CrossRef]
28. Valadares, D.C.G.; Will, N.C.; Caminha, J.; Perkusich, M.B.; Perkusich, A.; Gorgonio, K.C. Systematic Literature Review on the Use of Trusted Execution Environments to Protect Cloud/Fog-Based Internet of Things Applications. *IEEE Access* **2021**, *9*, 80953–80969. [CrossRef]
29. Amjad, A.; Azam, F.; Anwar, M.W.; Butt, W.H. A Systematic Review on the Data Interoperability of Application Layer Protocols in Industrial IoT. *IEEE Access* **2021**, *9*, 96528–96545. [CrossRef]
30. Reilly, E.; Maloney, M.; Siegel, M.; Falco, G. A smart city IoT integrity-first communication protocol via an ethereum blockchain light client. In Proceedings of the SERP4IoT, Colocated with the 44th ACM/IEEE International Conference on Software Engineering ICSE 2022, Marrakech, Morocco, 19 May 2022; pp. 15–19.
31. Amin, F.; Ahmad, A.; Sang Choi, G.S. Towards Trust and Friendliness Approaches in the Social Internet of Things. *Appl. Sci.* **2019**, *9*, 166. [CrossRef]

technologies

MDPI

Review

Strategic Investment in Open Hardware for National Security

Joshua M. Pearce [1,2]

[1] Department of Electrical & Computer Engineering, Western University, London, ON N6A 5B9, Canada; joshua.pearce@uwo.ca

[2] Ivey Business School, Western University, London, ON N6G 0N1, Canada

Abstract: Free and open-source hardware (FOSH) development has been shown to increase innovation and reduce economic costs. This article reviews the opportunity to use FOSH as a sanction to undercut imports and exports from a target criminal country. A formal methodology is presented for selecting strategic national investments in FOSH development to improve both national security and global safety. In this methodology, first the target country that is threatening national security or safety is identified. Next, the top imports from the target country as well as potentially other importing countries (allies) are quantified. Hardware is identified that could undercut imports/exports from the target country. Finally, methods to support the FOSH development are enumerated to support production in a commons-based peer production strategy. To demonstrate how this theoretical method works in practice, it is applied as a case study to a current criminal military aggressor nation, who is also a fossil-fuel exporter. The results show that there are numerous existing FOSH and opportunities to develop new FOSH for energy conservation and renewable energy to reduce fossil-fuel-energy demand. Widespread deployment would reduce the concomitant pollution, human health impacts, and environmental desecration as well as cut financing of military operations.

Keywords: energy policy; energy conservation; climate change; global safety; open hardware; open source; photovoltaic; renewable energy; solar energy; national security

Citation: Pearce, J.M. Strategic Investment in Open Hardware for National Security. *Technologies* **2022**, *10*, 53. https://doi.org/10.3390/technologies10020053

Academic Editors: Manoj Gupta, Eugene Wong and Gwanggil Jeon

Received: 14 March 2022
Accepted: 13 April 2022
Published: 18 April 2022

1. Introduction

Free and open-source software (FOSS) is released under a license that allows anyone to use, copy, study, and change it, and the source code is openly shared so that people are encouraged to voluntarily improve the design in exchange for requiring adaptations to be re-shared with the same license [1]. This gift economy [2] results in rapid innovation [3,4] and using FOSS licenses has been widely [5] and repeatedly [6] successful [7]. FOSS has become a dominant form of technical development in the software industry and now 90% of cloud servers [8] run open-source operating systems (this includes most internet companies such as Facebook, Twitter, Yahoo, Google and Amazon) as do 90% of the Fortune Global 500 (e.g., including less-tech-focused companies such as Wal-Mart and McDonalds) [9]. Similarly, 100% of supercomputers [10], over 84% of the global smartphone market [11] and more than 80% of the internet of things (IOT) market [12] also use FOSS.

The same open-source development paradigm [13,14] has started to democratize [15] manufacturing of physical products [16]. This is known as free and open-source hardware (FOSH). The Open Source Hardware Association defines open-source hardware [17] as:

> *Hardware whose design is made publicly available so that anyone can study, modify, distribute, make, and sell the design or hardware based on that design. The hardware's source, the design from which it is made, is available in the preferred format for making modifications to it. Ideally, open source hardware uses readily-available components and materials, standard processes, open infrastructure, unrestricted content, and open-source design tools to maximize the ability of individuals to make and use hardware. Open source hardware gives people the freedom to control their technology while sharing knowledge and encouraging commerce through the open exchange of designs.*

Open hardware uses viral licenses (e.g., CERN [18]) that demand if users make modifications, they must share their improvements with the global community [19]. FOSH is demonstrating rapid innovation [20–22] and is approximately 15 years behind FOSS in terms of technical development [23]. Both technologies have followed an exponential rate of growth in the peer-reviewed literature [23].

One of the core strengths of FOSH is the ability to replicate the hardware from digital designs [24,25] that themselves can be customized [26] with FOSS [27]. Digital fabrication of open-source designs enables wealth growth [28–30] and helps even the poor access high-value products such as state-of-the-art equipment [31]. It is well known that open hardware can create opportunities for distributed manufacturing that radically undercut commercial products [32–36]. For scientific hardware, for example, researchers can expect to save approximately 87% compared to proprietary products [37]. The savings are strongest when a form of distributed manufacturing is used (e.g., the open-source self-replicating rapid prototyper (or RepRap) [38–41] dramatically reduces additive manufacturing costs [42] and increases the number of 3D printing designs exponentially [43] that now number in the millions). The literature shows that low-cost open-source 3D printers can even reduce costs for mass-manufactured consumer goods, on average by 90–99% [43,44].

These savings can be scaled to the national level by investing in the development of new open-source software [45] and hardware of strategic interest to a specific country [46]. In the analysis completed in Finland, one of the secondary advantages is that imported products could be offset by manufacturing products internally from open-source designs [46]. This advantage can be leveraged to act in the same way as a sanction if applied to undercut imports from a specific country and technology sector to increase national security and global safety. Although FOSH is becoming well known, the strategic development of it to meet national goals outside of scientific research has not been explored.

This article reviews this opportunity by formalizing a methodology for selecting strategic national investments in open hardware development to improve national security and global safety. In the methodology, first the target criminal country that is threatening national security is identified. Next, the top imports from the target criminal country as well as potentially other importing countries (allies) are quantified. Then, hardware is identified that could undercut those imports as well as potentially other strategic exports from the target criminal country. Finally, methods to support the FOSH development are enumerated. The FOSH are designed in a way that facilitates distributed manufacturing from digital designs using local materials and tools. Thus, in addition to sanctions or instead of sanctions, supporting FOSH development can undercut the export market for the target criminal countries. To demonstrate how this theoretical method works in practice, it is applied as a case study to a current military aggressor nation. It is classified as a fossil-fuel exporter. Thus, how the development of energy conservation and renewable energy-based open hardware could reduce fossil-fuel-energy demand and the concomitant pollution, human health impacts, environmental desecration as well as financing of military operations is discussed.

2. Methods

The method used to determine strategic investments in open hardware is summarized in Figure 1.

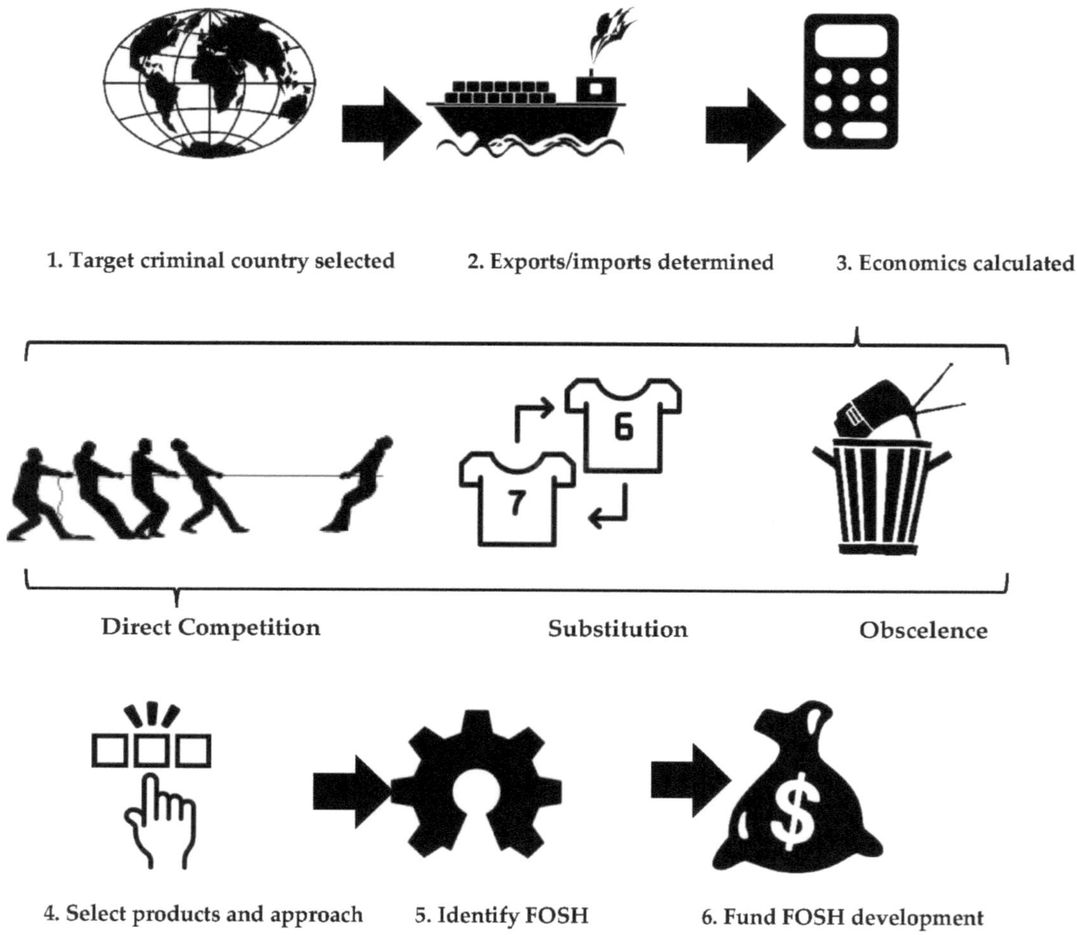

Figure 1. Flow diagram of method to determine strategic investments in open hardware.

As shown in Figure 1, first, the target criminal country is identified for disruption with strategic investment in FOSH. The target criminal country can be a political rival, an enemy or a country threatening the stability or safety of a region or the world. Next, both the major exports for the target criminal country are identified from public data as well as the imports to the country doing the analysis are identified. These imports and exports are quantified in economic terms for specific products. Then, the highest-value exports (and/or imports) are evaluated for the technical capacity to be disrupted by innovation. This can be determined using three approaches. First, direct competition is producing the same product at lower costs. For example, a USD $20,000 potentiostat/galvanostat for characterizing thin-film batteries can be replaced with a USD $100 open-source model [47]. With such opportunities, products can be directly open sourced to undercut the existing market for the product. Second, substitution is replacing a product with a different one that serves the same function. For example, a study of a single open hardware toy repository found that users were offsetting USD $60 million purchases per year by 3D printing 100 FOSH toy designs rather than purchasing similar proprietary toys [48]. For toys that were functionally equivalent, the most common savings fell by 40–90%, which represents savings expected for open-source design replications of low-value products. Third, elimination of the demand for an existing product by producing a new product that makes the need for the target product obsolete. For example, Wikipedia, an open-source digital encyclopedia running on open-source Wikimedia software, made the 244-year-old paper-based Encyclopedia Britannica obsolete and it is no longer produced [49]. Wikipedia did this by creating enormous laterally scaled value, as for example, the images on Wikimedia alone have been valued at more than USD $28.9 billion [50]. Wikipedia has essentially eliminated the market for hard-copy encyclopedias. Finally, after the target products and innovations are selected, specific open hardware is identified to reach the goal of reducing or eliminating the export of the products from the target criminal nation. Imports into one's own nation are likely the easiest to offset, but open hardware can be used to out innovate and thus reduce costs or increase functionality faster than the target criminal country's existing system (or companies within it) can innovate on its own. In addition to the imports into one's own nation, these same sources can determine other countries (allies) that are importing goods from the target criminal nation. These allies could form alliances to open hardware and leverage digital technologies [51] to use the do-it-together (DIT) methodology [52] to accomplish more than possible by going alone [53]. Since the industrial revolution, a conventional centralize manufacturing model has held sway (Figure 2a), where all intellectual property is held by an organization that manufactures products and ships it to consumers. DIT consists of participatory design and collaborative production (e.g., global design and home or local manufacturing), as shown in Figure 2b. As can be seen in Figure 2, rather than have the design ideas only come from large companies, communities and individuals can assist in developing open-source designs and then community organizations (e.g., fablabs, makerspaces as well as small- and medium-sized businesses) can manufacture the products locally. DIT encourages local production in a commons-based peer production strategy. Such social manufacturing generates positive externalities for all the involved stakeholders including customers, professionals and the local producers [54]. DIT is well suited for small-scale production on local sites, offering significant potential for new businesses and employment as DIT can provide competitive advantages while limiting costs and risks associated with innovation [55]. These imports, which can be converted to distributed local production, need to be quantified and hardware is identified that could undercut those imports. Thus, in addition to sanctions (or instead of sanctions), supporting FOSH development can undercut the market for the target countries by enabling distributed manufacturing of products that reduce or eliminate exported products from a targeted criminal nation.

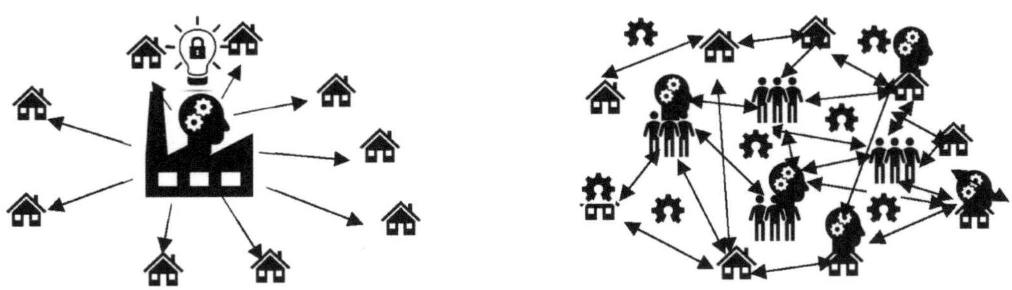

(a) (b)

Figure 2. (**a**) Conventional centralized and proprietary manufacturing model and (**b**) the decentralized open-source hardware model (design and share globally, while manufacturing locally).

3. Case Study

In order to illustrate this method, Russia was selected as an example country (step 1) for the case study. Russia, a country of 142 million citizens, has a gross domestic product per capita of $26,500 [56]. Russia also possesses thousands of nuclear weapons [57] and has continued to reproduce existing nuclear warhead designs even as it has reduced stockpiles [58]. Thus, Russia still controls enough nuclear weapons to be past the rational limit (e.g., where using them would hurt Russians even in the best-case scenario by aggravating food shortages) and worse, plunge the world into nuclear winter single handedly, which would result in mass starvation throughout the world [59]. Despite its inflated military, Russia is an otherwise developing country, where the average wage is only 51,100 Russian rubles per month or USD $7284/year [60] (e.g., the average Russia is below the poverty line in the U.S.). Note, these figures were taken prior to the 2022 sanctions. Russia, however, is described as an energy superpower due only to its fossil-fuel reserves [61]. Russia has the largest natural gas reserves in the world (followed by Iran) [62]. Still lamenting the loss of the USSR, Russian leadership has ambitions to expand its control over regions near it [63]. This was most clearly seen by several recent acts of aggression towards its neighboring countries that came in the form of invasions. First in 2008 the Russo–Georgian War, the European Court of Human Rights ruled that Russia maintained direct control over Abkhazia and South Ossetia [64] and was responsible for grave human rights abuses [65]. In 2014, Russia invaded and annexed Crimea [66,67]; the UN General Assembly condemned the occupation of the Autonomous Republic of Crimea and part of the territory of Ukraine [68]. Finally, most recently, Russia is the clear illegal aggressor in the 2022 full-scale invasion of Ukraine [69]. A full-scale war with Russia would be catastrophic even if nuclear war is prevented, so the U.S. and allies have retaliated with a long and growing list of sanctions [70,71]. The sanctions are meant to apply economic pressure on Russia to stop aggression, the destabilization of Europe and the rest of the world; but even as these sanctions are in place, they do not permanently disable the economic engine that makes Russia's threat to global safety a reality: exporting fossil fuels. Not only do fossil-fuel sales finance Russia's military [72], but fossil-fuel pollution is destabilizing the global climate, with severe impacts of climate change [73]. These risks include forcing 1/3 of global food production outside of a safe climate space [74] and creating human health risks [75]; and climate change also threatens the global economy [76].

4. Results

4.1. Case Study Risks from Exports

To complete steps 2 and 3, the top Russian exports are identified—(1) crude petroleum (USD $123B), (2) refined petroleum (USD $66.2B), (3) petroleum gas (USD $26.3B), and (4) coal briquettes (USD $17.6B) [77]. All of the top exports from Russia are fossil fuels, which is a serious threat to global safety if combusted, resulting in greenhouse gas (GHG) emissions and concomitant climate change [78,79]. Such human-caused global climate destabilization is established with a 95% confidence [80] as are the overwhelmingly detrimental repercussions on the environment as well as human social systems [81]. The impacts of burning Russia's fossil-fuel exports resulting in further climate change include

(1) Increased global temperatures and heat waves, which are already responsible for thousands of human deaths [82–84];
(2) Increased crop failures throughout the world [85,86], which aggravates global hunger and starvation [87–89];
(3) Increased electric grid failures and intermittent power outages [90,91];
(4) Increased droughts [92–94];
(5) Increased number and severity of forest fires [95–97];
(6) Increased sea level rise, which submerges low-lying coastal areas and increases shoreline erosion [98,99];
(7) Increased saltwater intrusion [99,100], which can threaten drinking water supplies [101];
(8) Increased storm damage to coast lines and increased flood risks [102–106].

4.2. Case Study FOSH Targets

4.2.1. FOSH for Electric Vehicles

For step 4, the profit centers of Russia's fossil-fuel industry will be targeted systematically with open hardware (step 5). As the first two profit centers for Russia focus on both crude and refined petroleum, targeted investments in open-source development revolve around those that antiquate the internal combustion engine that burns gasoline or diesel fuel. Any hardware that reduces the need for gas-based automobiles could be of some help (e.g., improved public transportation and equipment for telecommuting). Currently, however, electric vehicles (EV) offer the best potential for elimination of fossil-fuel-based land transport and are already gaining in market share [107], having more than doubled in 2021 [108]. Despite this growth, there are still several technical challenges to overcome [109] to reduce EV costs (and their batteries [110]) to accelerate the obsolesce of the use of oil for internal combustion engines altogether.

There is already some FOSH development revolving around EVs. Support for open-source development to support EV charging stations [111] exists. In addition, Tesla unlocked its EV patents [112]. Shortly after, Ford also announced opening its portfolio of previously patented EV technologies in an effort to accelerate industry-wide development [113]. Open-source battery management has been developed [114]. FOSH has also been developed for in situ monitoring of Li-ion cells [115], a maintenance tool for light-electric-vehicle batteries [116] and research has been performed on a line of open-source all-iron batteries [117,118].

Further open-source development of batteries and electronics could all be expected to reduce the capital and operating costs of EVs, helping to expand the diffusion of the technology and for eliminating oil for transportation. In addition, there is an opportunity for distributed manufacturing of EV components and potentially entire vehicles in high-population-density regions.

4.2.2. FOSH for Energy Conservation

The third profit center is petroleum gas or natural gas, which is primarily used to heat buildings in Europe although it is also burned to generate electric power. There are several areas that could benefit from open-source development that would reduce natural gas use for heating and some are already underway. Open hardware that is part of the internet of things (IOT) [119,120] is used for low-energy-consumption devices [121], monitoring power quality and energy savings [122]. There are numerous FOSH methods that have been developed for power monitoring [123–126], and smart meters [127–129] including those for institutional buildings (i.e., schools) [130]. FOSH is also used to improve energy efficiency and demand response [131] as well as smart converters [132] and microgrid communications [133]. There are also opportunities to develop open-source smart sockets, programmable thermostats, and high-efficiency LED lighting.

Related to conservation, development of open-source technologies could also take the form of those that help improve energy efficiency for physical testing such as an open-source blower door [134] to help identify air leaks in buildings. This technology could enable library-check out style of the device to help retrofit homes in an area. Other devices in this class would be thermal imaging cameras to detect improper insulation or faulty windows. Similarly, reducing heat loss can also be reduced with insulation. So open-source development could reduce the cost of insulation by providing the tools to turn local materials into insulating materials. The clearest opportunity would be the machinery at the fablab scale [135,136] to convert newspaper and fire retardant into cellulose insulation [137]. However, it should be pointed out that the costs of all of the types of insulation may benefit from open-source distributed manufacturing such as FOSH methods to make fibers [138].

4.2.3. FOSH for Heat Pumps

With the invasion of the Ukraine, the EU has already unveiled a strategy to eliminate their dependence on Russian natural gas [139]. Energy conservation, however, is not going to be enough to do it and the other approach is to electrify heating to not only cut down on natural gas, but to eliminate its use (and thus the demand for natural gas from Russia) all together. One promising upcoming technology to do this is heat pumps—both ground source and air source. Heat pump systems are often uneconomic unless coupled with solar power for offsetting natural gas or propane fuels [140,141]. Open-source development for a heat pump is already underway [142]. Future work could focus on FOSH for local fabrication of heat exchangers, pumps, motors, and piping. Ground source heat pumps would benefit from open hardware development of bore hole drilling units, such as an open-source ecology tractor [143] attachment.

4.2.4. FOSH for Renewable Energy

The fourth profit center for Russia is exporting coal, which is most commonly combusted for electricity. It should be pointed out here that even if EV scaling eliminates internal combustion engine-based vehicles and heat pumps offset the need for burning natural gas for heat, to completely eliminate the need for fossil fuels from Russia, electricity from other sources must be generated. According to the International Energy Agency, wind and solar are the fastest-growing sources of energy [144]. Choi et al. developed a renewable energy monitoring system [145], which is a start towards FOSH development in the renewable energy space. Although there are now FOSH-based hardware in the loop simulators for wind turbine testing [146] and there is some potential for open-source small wind turbines [147], there is less of an opportunity for small-scale distributed production with wind than solar.

Solar photovoltaic (PV) is particularly promising because it is more geographically diverse—available to most of humanity and growing rapidly [148,149]. PV is sustainable [150], a net energy producer [151] with an excellent ecological balance sheet [152,153]. The capital costs for PV are dropping rapidly (60% in the last decade) [154–157]. Because of this, already large-scale PV is generally the lowest-cost option for generating electricity [158]. Despite the lifetime economic benefits of PV, capital costs are the primary barrier to faster deployment, in the developing [159] and developed nations [160]. There are several approaches to reducing upfront PV costs including small-scale do it yourself (DIY) [161], which can cut more than half of the cost by eliminating most labor and soft costs.

The majority of PV system cost declines were caused by the $/W declines in the cost of PV modules, but racking, electronics, and wiring have seen far lower rates of cost decline [162]. PV costs can be more manageable if broken in small components that can be purchased over time, rather than all at once. For example, plug-and-play solar, where PV modules are connected with microinverters directly to the household circuits by consumers is technically possible [163–167], but regulations have stalled scaling of the technology in much of the world even though it allows the poor to enter the PV market [168]. There are currently no FOSH based microinverters, which is a large opportunity to reduce solar costs. Such microinverters would need to be developed for all of the world's standard voltages.

Secondly, racking is focused proprietary and costly designs, and racking costs dominate the cost of small PV systems [161]. There are, however, FOSH designs that substantially reduce racking costs for low-tilt-angle ground mounts, which result in 85% and 92% savings from proprietary alternatives [169], large-scale ground mounts with low-concentration reflectors [170], tensegrity structures (saved up to 77%) [171], and small-scale agrivoltaic cold frames [172]. For buildings, there are also FOSH designs for flat roofs that save over 80% [173], RV rooftops mounts [174], and post-market module retrofits for building integrated PV [175]. In addition to fixed mounts, FOSH solar trackers [176–178] and dual axis trackers [179] have been developed. Far more work is needed to develop the lowest-cost FOSH designs for PV racks based on the availability of materials locally for all parts of the world so that all communities can take advantage of these generally >75% savings.

Research into improving the PV industry can also benefit from open-source design [180]. Currently, no PV manufacturer is able to use all the known improvements to device performance and opportunities exist for opening up patents in the PV space similar to what Tesla and Ford have done for EVs. Open-source is mature, however, in part of the PV technical ecosystem: software. There is already extensive open-source software to assist in PV system design including PVLib [181,182] and the National Renewable Energy Lab's Systems Advisory Model (SAM) [183,184] as well as a module emulator [185] and FOSS for modeling advanced inverters [186]. FOSS has also been developed to determine the PV potential in urban areas [187] or over entire regions with open-source geographic information system software [188].

In addition, the Open Source Outdoors Testing Facility provides open access data on PV systems in northern environments [189] and could be replicated in other parts of the world. Any outdoor testing facility can reduce capital cost for monitoring using open-source systems for real time measurement of the sunlight incident angle [190], for UV–Vis-NIR radiation measurements [191] and radiation shields for environmental sensors [192]. Modules can be tested on site or in the lab with a solar simulator using an open-source IV curve tracer for solar [193]. There is also considerable FOSH developed for monitoring PV systems including data logging [194] and those for monitoring PV device performance [195], in situ monitoring of smart PV modules [196], system monitoring [197], monitoring PV plants [198], and remote monitoring [199,200]. FOSH has also been developed to monitor PV integrated into microgrid [201,202], agrivoltaic weather stations [203] and smart monitoring [204]. All of these solar-related FOSH can be further improved to be completely digitally manufactured so that communities could locally manufacture as much of the system as possible from local materials.

5. Discussion

5.1. Countries Positioned to Use the FOSH Model

Russia primarily exports to the EU (USD $188), China (USD $58.1B), the Netherlands (USD $41.7B), Belarus (USD $20.5B), Germany (USD $18.9B), and Italy (USD $16.7B) [205]. Thus, these countries are in the best position to spearhead the FOSH development recommended from the results of this study. Europe is particularly well endowed with fablabs and makerspaces as shown in Figure 3 [206]. Thus, they are well prepared to follow a distributed manufacturing (e.g., DIT) model to fabricate EVs, energy conservation equipment and renewable energy such as PV FOSH. In addition, China is already a major open-source technology proponent [207]. China, for example, developed the open-source Kylin operating system, and by 2019, a NeoKylin variant was compatible with more than 4000 software and hardware products and ships pre-installed on most computers sold in China [208]. Combined, Kylin and Neokylin dominate the domestic Chinese market with over 90% of the operating system market share in the government sector [209]. In addition, China is already the leading manufacturer of solar photovoltaics modules, and thus appears well positioned to benefit from FOSH development of peripheral technologies (e.g., racking and electronics) that would increase the size of their market throughout the world faster than it is already increasing.

5.2. Target Response

If a wave of FOSH was developed that made energy conservation, heat pumps, EVs and PV extremely inexpensive to manufacture locally, and countries that import Russia's goods adopted the 'design global—manufacture local' system, Russia's current fossil-fuel-export model would be made obsolete. If Russia attempted to maintain business as usual, it would be economically devastating. As this would be a distributed method of resistance and any retaliation would be against customers, such retaliation would be futile. Instead of maintaining the status quo as an aggressor nation and fossil-fuel exporter, Russia has the opportunity to lift its own citizens out of poverty [210] by leveraging the FOSH funded by external countries to manufacture fossil-fuel-conserving products to meet their own domestic demand and help transition them to a sustainable more diversified economy. This would not only help improve climate stability, but it would also directly improve domestic economic security and thus the perceived need for militarization and aggression.

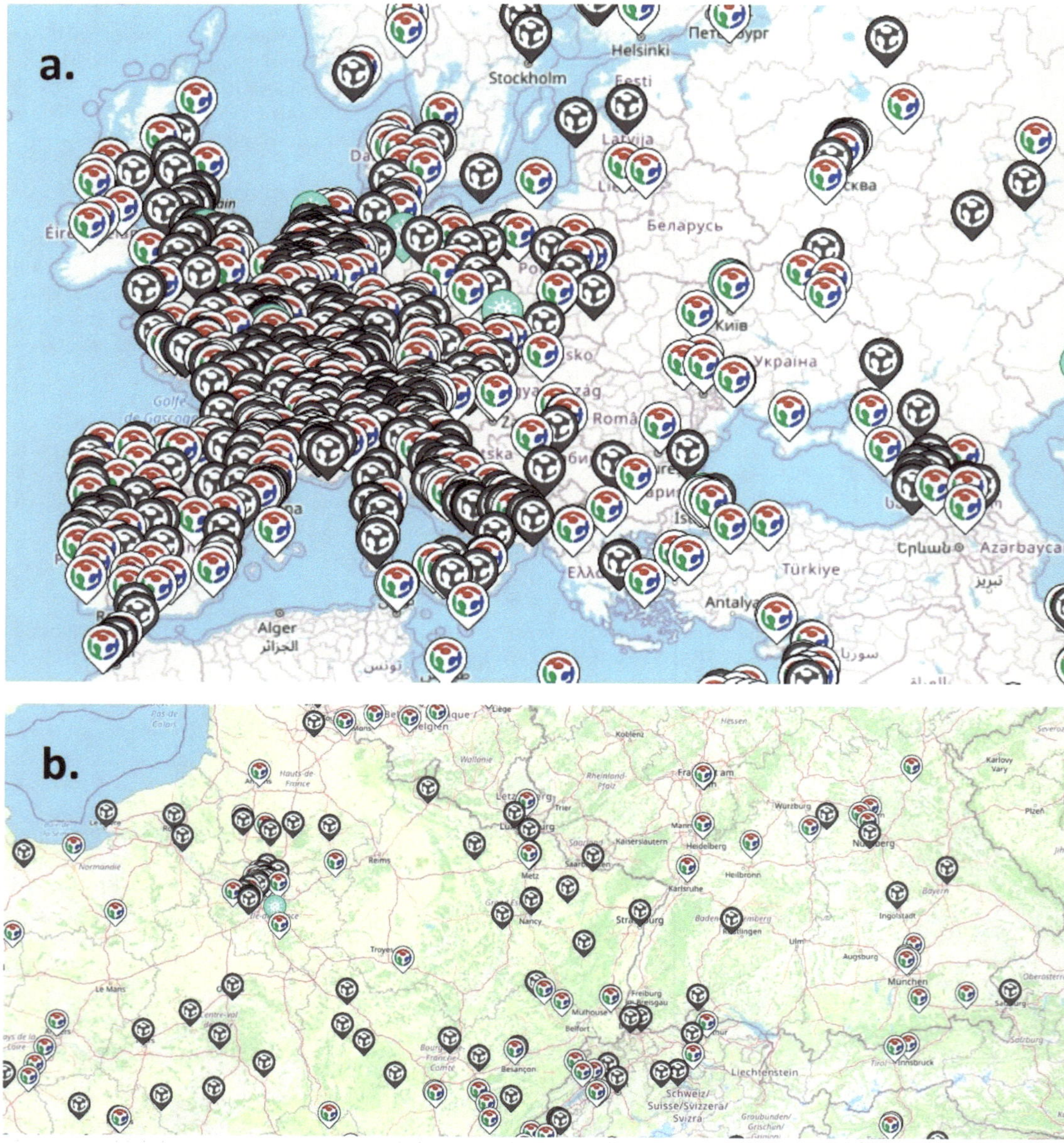

Figure 3. Fablab locations (**a**) throughout Europe and (**b**) zoomed-in view showing clustering of fablabs in population centers [205].

5.3. Funding National Strategic FOSH Development

There are several ways the open hardware development identified in Section 4 could be funded. First, federal governments can use standard calls for proposals (CFPs) specifically requiring open-source licensing of the FOSH technologies listed. Already, for example, the U.S. National Science Foundation (NSF) is investing USD $20 million in the Pathways to Enable Open-Source Ecosystems (POSE) program [211]. The NSF aims to "harness the power of open-source development for the creation of new technology solutions to problems of national and societal importance" [211]. Prior work has shown that open-source investment should result in an extremely high return on investment (ROI) in FOSH [29]. The funding would work as normal university or industry grants/contracts, with the exception that rather than fund researchers and allow them to gain a monopoly on the intellectual property, instead there would be an open-source license agreement mandate. In this way, the researchers are still funded, but the benefits of the research accrue to society more directly. Surveys indicate that the vast majority of faculty would be amenable to open-source development as they would accept an open-source-endowed chair requiring them to open source all of their work [212]. Additionally, national and international funding agencies may wish to sponsor challenges or contests such as the XPRIZE to promote development of FOSH toward specific technical goals by offering "bounties", scholarships, tax breaks, national park passes, lottery entries, awards or even citizenship. The latter rewarding of innovators of citizenship could be a particularly strong incentive to innovate given the current demand in some countries.

In addition to funding and incentivizing FOSH development, governments can also use their purchasing power to accelerate the adoption of FOSH developed in the national interest. This can be achieved by having purchasing policy preferences for open-source technologies. This would include prioritizing funding for open-source technologies over purchasing proprietary commercial products. The government could also make bulk purchases of materials or provide tax breaks for those manufacturing or purchasing FOSH that supports the national interests. Lastly, national governments have the opportunity of creating a free online database of tested, vetted, and validated FOSH to further national interests. It could act as an equivalent to a digital twin model being used in industry [213]. The database would include the bill of materials (BOMs), digital designs files (e.g., CAD), instructions for assembly and operation, and raw source code for all software and firmware. In order to vet designs, governments could provide funding to universities, companies, and/or utilize technical staff at government labs. Already, the U.S. National Institute of Health maintains an open design database called the 3D Print Exchange [214] and the United Nations is evaluating starting an open hardware database for appropriate technology to meet its sustainable development goals [215].

6. Conclusions

This article has reviewed the opportunity to take advantage of the innovation acceleration and economic savings provided by open-source design coupled to distributed manufacturing by formalizing a methodology for selecting strategic national investments in open hardware development to improve national security and global safety. The method was explained in detail, along with a summary of ways to support the FOSH development. For this method to work and scale, FOSH are designed in a way that facilitates distributed manufacturing from digital designs and then made using local materials and tools. Thus, in addition to sanctions or instead of sanctions, nations now have the option of supporting FOSH development to undercut the export market for target criminal countries.

Some of the largest current threats to humanity come from both acts of military aggression and climate destabilization. As the case study results in this review show, by investing in FOSH development of technologies for energy conservation, EVs, heat pumps and renewable energy technologies, nations of good will have the opportunity to radically reduce costs and thus accelerate the diffusion of these technologies that can move humanity towards a sustainable state. With freely available open-source designs of these technologies able to be manufactured locally, as for example in makerspaces and fablabs, the adoption of these technologies can play a strategic role in economically limiting criminal fossil-fuel states, while also helping to limit the negative impacts of global warming.

Funding: This research was supported by the Thompson Endowment.

Data Availability Statement: Data are available upon request.

Conflicts of Interest: The author declares no conflict of interest.

References

1. Lakhani, K.R.; von Hippel, E. How Open Source Software Works: "Free" User-to-User Assistance. In *Produktentwicklung Mit Virtuellen Communities: Kundenwünsche Erfahren und Innovationen Realisieren*; Herstatt, C., Sander, J.G., Eds.; Gabler Verlag: Wiesbaden, Germany, 2004; pp. 303–339, ISBN 978-3-322-84540-5.
2. Zeitlyn, D. Gift Economies in the Development of Open Source Software: Anthropological Reflections. *Res. Policy* **2003**, *32*, 1287–1291. [CrossRef]
3. Raymond, E. The Cathedral and the Bazaar. *Know Technol. Pol.* **1999**, *12*, 23–49. [CrossRef]
4. Herstatt, C.; Ehls, D. *Open Source Innovation: The Phenomenon, Participant's Behaviour, Business Implications*; Routledge: London, UK, 2015; ISBN 978-1-317-62425-7.
5. Comino, S.; Manenti, F.M.; Parisi, M.L. From Planning to Mature: On the Success of Open Source Projects. *Res. Policy* **2007**, *36*, 1575–1586. [CrossRef]
6. Lee, S.-Y.T.; Kim, H.-W.; Gupta, S. Measuring Open Source Software Success. *Omega* **2009**, *37*, 426–438. [CrossRef]
7. Weber, S. *The Success of Open Source*; Harvard University Press: Cambridge, MA, USA, 2004; ISBN 978-0-674-01292-9.
8. Hiteshdawda. Realising the Value of Cloud Computing with Linux. Available online: https://www.rackspace.com/en-gb/blog/realising-the-value-of-cloud-computing-with-linux (accessed on 24 February 2022).
9. Parloff, R. How Linux Conquered the Fortune 500 | Fortune. Available online: https://fortune.com/2013/05/06/how-linux-conquered-the-fortune-500/ (accessed on 24 February 2022).
10. Vaughan-Nichols, S. Supercomputers: All Linux, All the Time. Available online: https://www.zdnet.com/article/supercomputers-all-linux-all-the-time/ (accessed on 24 February 2022).
11. IDC—Smartphone Market Share. Available online: https://www.idc.com/promo/smartphone-market-share (accessed on 24 February 2022).
12. Eclipse. IoT Developer Survey 2019 Results. Available online: https://iot.eclipse.org/community/resources/iot-surveys/assets/iot-developer-survey-2019.pdf (accessed on 24 February 2022).
13. Gal, M.S. Viral Open Source: Competition vs. Synergy. *J. Compet. Law Econ.* **2012**, *8*, 469–506. [CrossRef]
14. Hausberg, J.P.; Spaeth, S. Why Makers Make What They Make: Motivations to Contribute to Open Source Hardware Development. *RD Manag.* **2020**, *50*, 75–95. [CrossRef]
15. Powell, A. Democratizing production through open source knowledge: From open software to open hardware. *Media Cult. Soc.* **2012**, *34*, 691–708. [CrossRef]
16. Spaeth, S.; Hausberg, P. Can Open Source Hardware Disrupt Manufacturing Industries? The Role of Platforms and Trust in the Rise of 3D Printing. In *The Decentralized and Networked Future of Value Creation: 3D Printing and Its Implications for Society, Industry, and Sustainable Development*; Ferdinand, J.-P., Petschow, U., Dickel, S., Eds.; Progress in IS; Springer International Publishing: Cham, Switzerland, 2016; pp. 59–73, ISBN 978-3-319-31686-4.
17. Open Hardware Definition (English). Available online: https://www.oshwa.org/definition/ (accessed on 20 September 2021).
18. Cern Ohl Version 2 Wiki Projects/CERN Open Hardware Licence. Available online: https://ohwr.org/project/cernohl/wikis/Documents/CERN-OHL-version-2 (accessed on 24 February 2022).
19. Gibb, A. *Building Open Source Hardware: DIY Manufacturing for Hackers and Makers*; Pearson Education: London, UK, 2014.
20. Yip, M.C.; Forsslund, J. Spurring Innovation in Spatial Haptics: How Open-Source Hardware Can Turn Creativity Loose. *IEEE Robot. Autom. Mag.* **2017**, *24*, 65–76. [CrossRef]
21. Dosemagen, S.; Liboiron, M.; Molloy, J. Gathering for Open Science Hardware 2016. *J. Open Hardw.* **2017**, *1*, 4. [CrossRef]
22. Hsing, P.-Y. Sustainable Innovation for Open Hardware and Open Science -Lessons from The Hardware Hacker. *J. Open Hardw.* **2018**, *2*, 4. [CrossRef]
23. Pearce, J.M. Sponsored Libre Research Agreements to Create Free and Open Source Software and Hardware. *Inventions* **2018**, *3*, 44. [CrossRef]
24. Fernando, P. Tools for Public Participation in Science: Design and Dissemination of Open-Science Hardware. In Proceedings of the 2019 on Creativity and Cognition, San Diego, CA, USA, 13 June 2019; Association for Computing Machinery: New York, NY, USA; pp. 697–701.
25. Pearce, J.M. Quantifying the Value of Open Source Hard Ware Development. *Mod. Econ.* **2015**, *6*, 1–11. [CrossRef]
26. Daniel, K.F.; Peter, J.G. Open-Source Hardware Is a Low-Cost Alternative for Scientific Instrumentation and Research. *Mod. Instrum.* **2012**, *2012*, 18950. [CrossRef]
27. Oberloier, S.; Pearce, J.M. Open Source Low-Cost Power Monitoring System. *HardwareX* **2018**, *4*, e00044. [CrossRef]
28. Thompson, C. Build it. Share it. Profit. Can open source hardware work. *Wired Magazine*, 20 October 2011.
29. Pearce, J.M. Return on Investment for Open Source Scientific Hardware Development. *Sci. Public Policy* **2016**, *43*, 192–195. [CrossRef]
30. Pearce, J.M. Impacts of Open Source Hardware in Science and Engineering. *The Bridge* **2017**, *47*, 24–31.
31. Harnett, C. Open Source Hardware for Instrumentation and Measurement. *IEEE Instrum. Meas. Mag.* **2011**, *14*, 34–38. [CrossRef]
32. Pearce, J.M. Building Research Equipment with Free, Open-Source Hardware. *Science* **2012**, *337*, 1303–1304. [CrossRef]
33. Pearce, J.M. *Open-Source Lab: How to Build Your Own Hardware and Reduce Research Costs*; Elsevier: Amsterdam, The Netherlands, 2014.

34. Chagas, A.M. Haves and have nots must find a better way: The case for open scientific hardware. *PLoS Biol.* **2018**, *16*, e3000014. [CrossRef]
35. Gibney, E. 'Open-Hardware' Pioneers Push for Low-Cost Lab Kit. *Nature* **2016**, *531*, 147–148. [CrossRef] [PubMed]
36. Pearce, J.M. Cut Costs with Open-Source Hardware. *Nature* **2014**, *505*, 618. [CrossRef] [PubMed]
37. Pearce, J.M. Economic Savings for Scientific Free and Open Source Technology: A Review. *HardwareX* **2020**, *8*, e00139. [CrossRef] [PubMed]
38. Sells, E.; Bailard, S.; Smith, Z.; Bowyer, A.; Olliver, V. RepRap: The Replicating Rapid Prototyper: Maximizing Customizability by Breeding the Means of Production. In *Handbook of Research in Mass Customization and Personalization*; World Scientific Publishing Company: Singapore, 2009; pp. 568–580, ISBN 978-981-4280-25-9.
39. Jones, R.; Haufe, P.; Sells, E.; Iravani, P.; Olliver, V.; Palmer, C.; Bowyer, A. RepRap- The replicating rapid prototyper. *Robotica* **2011**, *29*, 177–191. [CrossRef]
40. Kentzer, J.; Koch, B.; Thiim, M.; Jones, R.W.; Villumsen, E. An open source hardware-based mechatronics project: The replicating rapid 3-D printer. In Proceedings of the 2011 4th International Conference on Mechatronics (ICOM), Kuala Lumpur, Malaysia, 17–19 May 2011; pp. 1–8.
41. Bowyer, A. 3D Printing and Humanity's First Imperfect Replicator. *3D Print. Addit. Manuf.* **2014**, *1*, 4–5. [CrossRef]
42. Rundle, G. *A Revolution in the Making*; Simon and Schuster: New York, NY, USA, 2014; ISBN 978-1-922213-48-8.
43. Wittbrodt, B.T.; Glover, A.G.; Laureto, J.; Anzalone, G.C.; Oppliger, D.; Irwin, J.L.; Pearce, J.M. Life-cycle economic analysis of distributed manufacturing with open-source 3-D printers. *Mechatronics* **2013**, *23*, 713–726. [CrossRef]
44. Petersen, E.E.; Pearce, J. Emergence of Home Manufacturing in the Developed World: Return on Investment for Open-Source 3-D Printers. *Technologies* **2017**, *5*, 7. [CrossRef]
45. Lewis, J.A. Government Open Source Policies. p. 66. Available online: https://openforumeurope.org/wp-content/uploads/2015/06/100416_Open_Source_Policies-1.pdf (accessed on 24 February 2022).
46. Heikkinen, I.T.S.; Savin, H.; Partanen, J.; Seppälä, J.; Pearce, J.M. Towards National Policy for Open Source Hardware Research: The Case of Finland. *Technol. Forecast. Soc. Chang.* **2020**, *155*, 119986. [CrossRef]
47. Dobbelaere, T.; Vereecken, P.M.; Detavernier, C. A USB-Controlled Potentiostat/Galvanostat for Thin-Film Battery Characterization. *HardwareX* **2017**, *2*, 34–49. [CrossRef]
48. Petersen, E.E.; Kidd, R.W.; Pearce, J.M. Impact of DIY Home Manufacturing with 3D Printing on the Toy and Game Market. *Technologies* **2017**, *5*, 45. [CrossRef]
49. Olanoff, D. The Internet Has Just Finally Killed The Encyclopedia. Available online: https://thenextweb.com/news/wikipedia-and-the-internet-just-killed-244-year-old-encyclopaedia-britannica (accessed on 24 February 2022).
50. Erickson, K.; Perez, F.R.; Perez, J.R. What Is the Commons Worth? Estimating the Value of Wikimedia Imagery by Observing Downstream Use. In Proceedings of the 14th International Symposium on Open Collaboration, Paris, France, 22–24 August 2018; Association for Computing Machinery: New York, NY, USA, 2018; pp. 1–6.
51. Fauchart, E.; Bacache-Beauvallet, M.; Bourreau, M.; Moreau, F. Do-It-Yourself or Do-It-Together: How Digital Technologies Affect Creating Alone or with Others? *Technovation* **2021**, *112*, 102412. [CrossRef]
52. Dupont, L.; Kasmi, F.; Pearce, J.M.; Ortt, R.J. *"Do-It-Together": Towards the Factories of the Future. Cosmolocal Reader*; José, M.R., Michel, B., Sharon, E., James, G.W., Eds.; Cosmo-local Reader; 2021; pp. 52–59. Available online: https://clreader.net/ (accessed on 24 February 2022).
53. Mahajan, S.; Luo, C.-H.; Wu, D.-Y.; Chen, L.-J. From Do-It-Yourself (DIY) to Do-It-Together (DIT): Reflections on Designing a Citizen-Driven Air Quality Monitoring Framework in Taiwan. *Sustain. Cities Soc.* **2021**, *66*, 102628. [CrossRef]
54. Hirscher, A.-L.; Niinimäki, K.; Joyner Armstrong, C.M. Social Manufacturing in the Fashion Sector: New Value Creation through Alternative Design Strategies? *J. Clean. Prod.* **2018**, *172*, 4544–4554. [CrossRef]
55. Cullmann, S.; Guittard, C.; Schenk, E. Participative Creativity Serving Product Design in SMEs: A case study. *J. Innov. Econ. Manag.* **2015**, *18*, 79–98. [CrossRef]
56. Russia. *The World Factbook*; CIA: Washington, DC, USA, 2022. Available online: https://www.cia.gov/the-world-factbook/countries/russia (accessed on 24 February 2022).
57. Kristensen, H.M.; Norris, R.S. Russian Nuclear Forces. *Bull. At. Sci.* **2016**, *72*, 125–134. [CrossRef]
58. Norris, R.S.; Kristensen, H.M. Global Nuclear Weapons Inventories, 1945–2010. *Bull. At. Sci.* **2010**, *66*, 77–83. [CrossRef]
59. Pearce, J.M.; Denkenberger, D.C. A National Pragmatic Safety Limit for Nuclear Weapon Quantities. *Safety* **2018**, *4*, 25. [CrossRef]
60. Russia: Average Nominal Wage per Month 2020. Available online: https://www.statista.com/statistics/1010660/russia-average-monthly-nominal-wage/ (accessed on 25 February 2022).
61. Rutland, P. Russia as an Energy Superpower. *New Political Econ.* **2008**, *13*, 203–210. [CrossRef]
62. Global Natural Gas Reserves by Country 2020. Available online: https://www.statista.com/statistics/265329/countries-with-the-largest-natural-gas-reserves/ (accessed on 25 February 2022).
63. Putin Says He Moonlighted as Taxi Driver after Fall of Soviet Union. Available online: https://www.nbcnews.com/news/world/russia-s-putin-laments-soviet-collapse-says-he-moonlighted-taxi-n1285807 (accessed on 25 February 2022).
64. Makszimov, V. Strasbourg Court Rules Russia Has "Direct Control" over Abkhazia, South Ossetia. Available online: https://www.euractiv.com/section/europe-s-east/news/strasbourg-court-rules-russia-has-direct-control-over-abkhazia-south-ossetia/ (accessed on 25 February 2022).

65. European Court Finds Russia Guilty of Georgia Violations in 2008. Available online: https://www.euronews.com/2021/01/26/russia-guilty-of-violations-during-2008-war-with-georgia-says-europe-s-top-court (accessed on 25 February 2022).
66. Mankoff, J. Russia's Latest Land Grab: How Putin Won Crimea and Lost Ukraine. *Foreign Aff.* **2014**, *93*, 60.
67. Treisman, D. Why Putin Took Crimea: The Gambler in the Kremlin. *Foreign Aff.* **2016**, *95*, 47.
68. Resolutions Calling on Withdrawal of Forces from Crimea, Establishing Epidemic Preparedness International Day among Texts Adopted by General Assembly | Meetings Coverage and Press Releases. Available online: https://www.un.org/press/en/2020/ga12295.doc.htm (accessed on 25 February 2022).
69. Putin Shatters Peace in Europe as Russia Invades Ukraine. Available online: https://www.aljazeera.com/news/2022/2/24/russia-putin-shatters-peace-europe-ukraine-invasion (accessed on 25 February 2022).
70. Toh, M.; Ogura, J.; Humayun, H.; McGee, C.; Yee, I.; Cheung, E.; Fossum, S.; Kennedy, N. CNN The List of Global Sanctions on Russia for the War in Ukraine. Available online: https://www.cnn.com/2022/02/25/business/list-global-sanctions-russia-ukraine-war-intl-hnk/index.html (accessed on 25 February 2022).
71. CNN, K.L. Biden Imposes Additional Sanctions on Russia: "Putin Chose This War". Available online: https://www.cnn.com/2022/02/24/politics/joe-biden-ukraine-russia-sanctions/index.html (accessed on 25 February 2022).
72. Henderson, J.; Mitrova, T. Implications of the Global Energy Transition on Russia. In *The Geopolitics of the Global Energy Transition*; Hafner, M., Tagliapietra, S., Eds.; Lecture Notes in Energy; Springer International Publishing: Cham, Switzerland, 2020; pp. 93–114, ISBN 978-3-030-39066-2.
73. Jamet, S.; Corfee-Morlot, J. *Assessing the Impacts of Climate Change: A Literature Review*; OECD: Paris, France, 2009.
74. Kummu, M.; Heino, M.; Taka, M.; Varis, O.; Viviroli, D. Climate Change Risks Pushing One-Third of Global Food Production Outside the Safe Climatic Space. *One Earth* **2021**, *4*, 720–729. [CrossRef] [PubMed]
75. Tong, S.; Ebi, K. Preventing and Mitigating Health Risks of Climate Change. *Environ. Res.* **2019**, *174*, 9–13. [CrossRef] [PubMed]
76. Stern, N. *Stern Review: The Economics of Climate Change*; Government of the United Kingdom: London, UK, 2006.
77. Russia (RUS) Exports, Imports, and Trade Partners | OEC -The Observatory of Economic Complexity. Available online: https://oec.world/en/profile/country/rus/ (accessed on 25 February 2022).
78. Hansen, J.; Kharecha, P.; Sato, M.; Masson-Delmotte, V.; Ackerman, F.; Beerling, D.J.; Hearty, P.J.; Hoegh-Guldberg, O.; Hsu, S.-L.; Parmesan, C.; et al. Assessing "Dangerous Climate Change": Required Reduction of Carbon Emissions to Protect Young People, Future Generations and Nature. *PLoS ONE* **2013**, *8*, e81648. [CrossRef] [PubMed]
79. Ripple, W.J.; Wolf, C.; Newsome, T.M.; Galetti, M.; Alamgir, M.; Crist, E.; Mahmoud, M.I.; Laurance, W.F. World Scientists' Warning to Humanity: A Second Notice. *BioScience* **2017**, *67*, 1026–1028. [CrossRef]
80. Pachauri, R.K.; Allen, M.R.; Barros, V.R.; Broome, J.; Cramer, W.; Christ, R.; Church, J.A.; Clarke, L.; Dahe, Q.; Dasgupta, P.; et al. *Climate Change 2014: Synthesis Report. Contribution of Working Groups I, II and III to the Fifth Assessment Report of the Intergovernmental Panel on Climate Change*; Pachauri, R., Meyer, L., Eds.; IPCC: Geneva, Switzerland, 2014; 151p, ISBN 978-92-9169-143-2.
81. Moss, R.H.; Edmonds, J.A.; Hibbard, K.A.; Manning, M.R.; Rose, S.K.; Van Vuuren, D.P.; Carter, T.R.; Emori, S.; Kainuma, M.; Kram, T.; et al. The next generation of scenarios for climate change research and assessment. *Nature* **2010**, *463*, 747–756. [CrossRef]
82. Dhainaut, J.F.; Claessens, Y.E.; Ginsburg, C.; Riou, B. Unprecedented heat-related deaths during the 2003 heat wave in Paris: Consequences on emergency departments. *Crit. Care* **2003**, *8*, 1. [CrossRef]
83. Poumadère, M.; Mays, C.; Le Mer, S.; Blong, R. The 2003 Heat Wave in France: Dangerous Climate Change Here and Now: The 2003 Heat Wave in France. *Risk Anal.* **2005**, *25*, 1483–1494. [CrossRef]
84. Fouillet, A.; Rey, G.; Laurent, F.; Pavillon, G.; Bellec, S.; Guihenneuc-Jouyaux, C.; Clavel, J.; Jougla, E.; Hémon, D. Excess mortality related to the August 2003 heat wave in France. *Int. Arch. Occ. Env. Health* **2006**, *80*, 16–24. [CrossRef]
85. D'Amato, G.; Cecchi, L. Effects of climate change on environmental factors in respiratory allergic diseases. *Clin. Exp. Allergy* **2008**, *38*, 1264–1274. [CrossRef]
86. Gislason, A.; Gorsky, G. (Eds.) *Proceedings of the Joint ICES/CIESM Workshop to Compare Zooplankton Ecology and Methodologies between the Mediterranean and the North Atlantic (WKZEM)*; ICES, International Council for the Exploration of the Sea: Copenhagen, Denmark, 2010.
87. Parry, M.L.; Rosenzweig, C.; Iglesias, A.; Livermore, M.; Fischer, G. Effects of climate change on global food production under SRES emissions and socio-economic scenarios. *Glob. Environ. Chang.* **2004**, *14*, 53–67. [CrossRef]
88. Parry, M.; Rosenzweig, C.; Livermore, M. Climate change, global food supply and risk of hunger. Philosophical Transactions of the Royal Society. *Bio. Sci.* **2005**, *360*, 2125–2138. [CrossRef] [PubMed]
89. Schmidhuber, J.; Tubiello, F.N. Global food security under climate change. *Proc. Natl. Acad. Sci. USA* **2007**, *104*, 19703–19708. [CrossRef] [PubMed]
90. Vine, E. Adaptation of California's electricity sector to climate change. *Clim. Chang.* **2012**, *111*, 75–99. [CrossRef]
91. Val, D.V.; Yurchenko, D.; Nogal, M.; O'Connor, A. Chapter Seven-Climate Change-Related Risks and Adaptation of Interdependent Infrastructure Systems. In *Climate Adaptation Engineering*; Bastidas-Arteaga, E., Stewar, M.G., Eds.; Butterworth-Heinemann: Oxford, UK, 2019; pp. 207–242, ISBN 978-0-12-816782-3.
92. Dai, A. Drought under global warming: A review. *WIREs Clim. Chang.* **2011**, *2*, 45–65. [CrossRef]
93. Diffenbaugh, N.S.; Swain, D.L.; Touma, D. Anthropogenic warming has increased drought risk in California. *Proc. Natl. Acad. Sci. USA* **2015**, *112*, 3931–3936. [CrossRef]

94. Mann, M.E.; Gleick, P.H. Climate change and California drought in the 21st century. *Proc. Natl. Acad. Sci. USA* **2015**, *112*, 3858–3859. [CrossRef]
95. Dale, V.H.; Joyce, L.A.; Mcnulty, S.; Neilson, R.P.; Ayres, M.P.; Flannigan, M.D.; Hanson, P.J.; Irland, L.C.; Lugo, A.E.; Peterson, C.J.; et al. Climate Change and Forest Disturbances. *BioScience* **2001**, *51*, 723. [CrossRef]
96. Amiro, B.D.; Stocks, B.J.; Alexander, M.E.; Flannigan, M.D.; Wotton, B.M. Fire, climate change, carbon and fuel management in the Canadian boreal forest. *Int. J. Wildland Fire* **2001**, *10*, 405–413. [CrossRef]
97. Flannigan, M.; Stocks, B.; Turetsky, M.; Wotton, M. Impacts of climate change on fire activity and fire management in the circumboreal forest. *Glob. Chang. Biol.* **2009**, *15*, 549–560. [CrossRef]
98. Moorhead, K.K.; Brinson, M.M. Response of Wetlands to Rising Sea Level in the Lower Coastal Plain of North Carolina. *Ecol. App.* **1995**, *5*, 261. [CrossRef]
99. Frihy, O.E. The Nile delta-Alexandria coast: Vulnerability to sea-level rise, consequences and adaptation. *Mitig. Adapt. Strateg. Glob. Chang.* **2003**, *8*, 115–138. [CrossRef]
100. Bobba, A.G. Numerical modelling of salt-water intrusion due to human activities and sea-level change in the Godavari Delta, India. *Hydro. Sci. J.* **2002**, *47*, S67–S80. [CrossRef]
101. Post, V.E.A. Fresh and Saline Groundwater Interaction in Coastal Aquifers: Is Our Technology Ready for the Problems Ahead? *Hydrogeol. J.* **2005**, *13*, 120–123. [CrossRef]
102. Nicholls, R.J.; Hoozemans, F.M.; Marchand, M. Increasing flood risk and wetland losses due to global sea-level rise: Regional and global analyses. *Glob. Environ. Chang.* **1999**, *9*, S69–S87. [CrossRef]
103. Desantis, L.R.G.; Bhotika, S.; Williams, K.; Putz, F.E. Sea-level rise and drought interactions accelerate forest decline on the Gulf Coast of Florida, USA. *Glob. Chang. Biol.* **2007**, *13*, 2349–2360. [CrossRef]
104. Cox, D.; Hunt, J.; Mason, P.; Wheater, H.; Wolf, P.; Poff, N.L. Ecological Response to and Management of Increased Flooding Caused by Climate Change. *Philos. Trans. R. Soc. Lond. Ser. A Math. Phys. Eng. Sci.* **2002**, *360*, 1497–1510. [CrossRef]
105. Knox, J.C. Sensitivity of Modern and Holocene Floods to Climate Change. *Quat. Sci. Rev.* **2000**, *19*, 439–457. [CrossRef]
106. Carnicer, J.; Coll, M.; Ninyerola, M.; Pons, X.; Sánchez, G.; Peñuelas, J. Widespread crown condition decline, food web disruption, and amplified tree mortality with increased climate change-type drought. *Proc. Natl. Acad. Sci. USA* **2011**, *108*, 1474–1478. [CrossRef]
107. Zhang, Q.; Ou, X.; Yan, X.; Zhang, X. Electric Vehicle Market Penetration and Impacts on Energy Consumption and CO2 Emission in the Future: Beijing Case. *Energies* **2017**, *10*, 228. [CrossRef]
108. Lambert, F. Global Market Share of Electric Cars More than Doubled in 2021 as the EV Revolution Gains Steam. *Electrek* **2022**. Available online: https://electrek.co/2022/02/02/global-market-share-of-electric-cars-more-than-doubled-2021/ (accessed on 24 February 2022).
109. Sanguesa, J.A.; Torres-Sanz, V.; Garrido, P.; Martinez, F.J.; Marquez-Barja, J.M. A Review on Electric Vehicles: Technologies and Challenges. *Smart Cities* **2021**, *4*, 372–404. [CrossRef]
110. Deng, J.; Bae, C.; Denlinger, A.; Miller, T. Electric Vehicles Batteries: Requirements and Challenges. *Joule* **2020**, *4*, 511–515. [CrossRef]
111. Vaughan-Nichols, S. EVerest: The Open Source Software Stack for EV Charging Infrastructure. Available online: https://www.zdnet.com/article/everest-the-open-source-software-stack-for-electric-vehicle-charging-infrastructure/ (accessed on 25 February 2022).
112. All Our Patent Are Belong to You. Available online: https://www.tesla.com/blog/all-our-patent-are-belong-you (accessed on 25 February 2022).
113. Ford Motor Company Announces Open Source Portfolio of EV Patents. Available online: http://greenlivingguy.com/2015/06/ford-motor-company-announces-open-source-portfolio-of-ev-patents/ (accessed on 27 February 2022).
114. Sylvestrin, G.R.; Scherer, H.F.; Hideo Ando Junior, O. Hardware and Software Development of an Open Source Battery Management System. *IEEE Lat. Am. Trans.* **2021**, *19*, 1153–1163. [CrossRef]
115. Fleming, J.; Amietszajew, T.; McTurk, E.; Towers, D.P.; Greenwood, D.; Bhagat, R. Development and Evaluation of In-Situ Instrumentation for Cylindrical Li-Ion Cells Using Fibre Optic Sensors. *HardwareX* **2018**, *3*, 100–109. [CrossRef]
116. Carloni, A.; Baronti, F.; Di Rienzo, R.; Roncella, R.; Saletti, R. An Open-Hardware and Low-Cost Maintenance Tool for Light-Electric-Vehicle Batteries. *Energies* **2021**, *14*, 4962. [CrossRef]
117. Yensen, N.; Allen, P.B. Open Source All-Iron Battery for Renewable Energy Storage. *HardwareX* **2019**, *6*, e00072. [CrossRef]
118. Koirala, D.; Yensen, N.; Allen, P.B. Open Source All-Iron Battery 2.0. *HardwareX* **2021**, *9*, e00171. [CrossRef]
119. Loukatos, D.; Dimitriou, N.; Manolopoulos, I.; Kontovasilis, K.; Arvanitis, K.G. Revealing Characteristic IoT Behaviors by Performing Simple Energy Measurements via Open Hardware/Software Components. In Proceedings of the Sixth International Congress on Information and Communication Technology, London, UK, 25–26 February 2021; Yang, X.-S., Sherratt, S., Dey, N., Joshi, A., Eds.; Springer: Singapore, 2022; pp. 1045–1053.
120. Raval, M.; Bhardwaj, S.; Aravelli, A.; Dofe, J.; Gohel, H. Smart Energy Optimization for Massive IoT Using Artificial Intelligence. *Internet Things* **2021**, *13*, 100354. [CrossRef]
121. Lopez, L.J.R.; Aponte, G.P.; Garcia, A.R. Internet of Things Applied in Healthcare Based on Open Hardware with Low-Energy Consumption. *Healthc. Inform. Res.* **2019**, *25*, 230–235. [CrossRef]

122. Viciana, E.; Alcayde, A.; Montoya, F.G.; Baños, R.; Arrabal-Campos, F.M.; Manzano-Agugliaro, F. An Open Hardware Design for Internet of Things Power Quality and Energy Saving Solutions. *Sensors* **2019**, *19*, 627. [CrossRef]
123. Makonin, S.; Popowich, F.; Moon, T.; Gill, B. Inspiring Energy Conservation through Open Source Power Monitoring and In-Home Display. In Proceedings of the 2013 IEEE Power Energy Society General Meeting, Vancouver, BC, Canada, 21–25 July 2013; pp. 1–5.
124. Makonin, S.; Sung, W.; Dela Cruz, R.; Yarrow, B.; Gill, B.; Popowich, F.; Bajić, I.V. Inspiring Energy Conservation through Open Source Metering Hardware and Embedded Real-Time Load Disaggregation. In Proceedings of the 2013 IEEE PES Asia-Pacific Power and Energy Engineering Conference (APPEEC), Kowloon, Hong Kong, 8–11 December 2013; pp. 1–6.
125. Adamo, F.; Cavone, G.; Di Nisio, A.; Lanzolla, A.M.L.; Spadavecchia, M. A Proposal for an Open Source Energy Meter. In Proceedings of the 2013 IEEE International Instrumentation and Measurement Technology Conference (I2MTC), Minneapolis, MN, USA, 6–9 May 2013; pp. 488–492.
126. Ferry, C.; Connolly, J. Open Source Power Quality Meter with Cloud Monitoring. In Proceedings of the 2020 31st Irish Signals and Systems Conference (ISSC), Letterkenny, Ireland, 11–12 June 2020; pp. 1–6.
127. Klemenjak, C.; Egarter, D.; Elmenreich, W. YoMo: The Arduino-Based Smart Metering Board. *Comput. Sci. Res. Dev.* **2016**, *31*, 97–103. [CrossRef]
128. Klemenjak, C.; Jost, S.; Elmenreich, W. YoMoPie: A User-Oriented Energy Monitor to Enhance Energy Efficiency in Households. In Proceedings of the 2018 IEEE Conference on Technologies for Sustainability (SusTech), Long Beach, CA, USA, 11–13 November 2018; pp. 1–7.
129. Jameel, H.; Farhan, H.K. Low-Cost Energy-Efficient Smart Monitoring System Using Open-Source Microcontrollers. *IREACO* **2016**, *9*, 423. [CrossRef]
130. Pocero, L.; Amaxilatis, D.; Mylonas, G.; Chatzigiannakis, I. Open Source IoT Meter Devices for Smart and Energy-Efficient School Buildings. *HardwareX* **2017**, *1*, 54–67. [CrossRef]
131. Andreou, G.T.; Chatzigeorgiou, I.M. Open Source Hardware and Software to Support Energy Efficiency and Demand Response in LV Installations. In Proceedings of the 2015 IEEE Eindhoven PowerTech, Eindhoven, The Netherlands, 29 June 2015; pp. 1–5.
132. Merenda, M.; Iero, D.; Pangallo, G.; Falduto, P.; Adinolfi, G.; Merola, A.; Graditi, G.; Della Corte, F.G. Open-Source Hardware Platforms for Smart Converters with Cloud Connectivity. *Electronics* **2019**, *8*, 367. [CrossRef]
133. Mlakić, D.; Baghaee, H.R.; Nikolovski, S.; Vukobratović, M.; Balkić, Z. Conceptual Design of IoT-Based AMR Systems Based on IEC 61850 Microgrid Communication Configuration Using Open-Source Hardware/Software IED. *Energies* **2019**, *12*, 4281. [CrossRef]
134. A DIY Blower Door—Easy to Build—Easy to Use—Cheap. Available online: https://www.builditsolar.com/Projects/Conservation/BlowerDoor/BlowerDoor.htm (accessed on 25 February 2022).
135. Walter-Herrmann, J.; Büching, C. *FabLab: Of Machines, Makers and Inventors*; Transcript Verlag: Bielefeld, Germany, 2014; ISBN 978-3-8394-2382-0.
136. Redlich, T.; Buxbaum-Conradi, S.; Basmer-Birkenfeld, S.-V.; Moritz, M.; Krenz, P.; Osunyomi, B.D.; Wulfsberg, J.P.; Heubischl, S. OpenLabs—Open Source Microfactories Enhancing the FabLab Idea. In Proceedings of the 2016 49th Hawaii International Conference on System Sciences (HICSS), Kauai, HI, USA, 5–8 January 2016; pp. 707–715.
137. Open Source Cellulose Insulation Manufacturing. Available online: https://www.appropedia.org/Open_source_cellulose_insulation_manufacturing (accessed on 25 February 2022).
138. Domínguez, J.E.; Olivos, E.; Vázquez, C.; Rivera, J.M.; Hernández-Cortes, R.; González-Benito, J. Automated Low-Cost Device to Produce Sub-Micrometric Polymer Fibers Based on Blow Spun Method. *HardwareX* **2021**, *10*, e00218. [CrossRef]
139. E.U. Will Unveil a Strategy to Break Free from Russian Gas, after Decades of Dependence. *Washington Post*. Available online: https://www.washingtonpost.com/climate-environment/2022/02/23/russia-ukraine-eu-nordstream-strategy-energy/ (accessed on 24 February 2022).
140. Pearce, J.M.; Sommerfeldt, N. Economics of Grid-Tied Solar Photovoltaic Systems Coupled to Heat Pumps: The Case of Northern Climates of the U.S. and Canada. *Energies* **2021**, *14*, 834. [CrossRef]
141. Padovani, F.; Sommerfeldt, N.; Longobardi, F.; Pearce, J.M. Decarbonizing Rural Residential Buildings in Cold Climates: A Techno-Economic Analysis of Heating Electrification. *Energy Build.* **2021**, *250*, 111284. [CrossRef]
142. Rowntree, D. Arduino Powered Heat Pump Controller Helps Warm Your Toes. Hackaday 2021. Available online: https://hackaday.com/2021/09/08/arduino-powered-heat-pump-controller-helps-warm-your-toes/ (accessed on 24 February 2022).
143. Thomson, C.C.; Jakubowski, M. Toward an Open Source Civilization: Innovations Case Narrative: Open Source Ecology. *Innov. Technol. Gov. Glob.* **2012**, *7*, 53–70. [CrossRef]
144. Renewable Electricity Growth Is Accelerating Faster than Ever Worldwide, Supporting the Emergence of the New Global Energy Economy—News. Available online: https://www.iea.org/news/renewable-electricity-growth-is-accelerating-faster-than-ever-worldwide-supporting-the-emergence-of-the-new-global-energy-economy (accessed on 25 February 2022).
145. Choi, C.-S.; Jeong, J.-D.; Lee, I.-W.; Park, W.-K. LoRa Based Renewable Energy Monitoring System with Open IoT Platform. In Proceedings of the 2018 International Conference on Electronics, Information, and Communication (ICEIC), Honolulu, HI, USA, 24–27 January 2018; pp. 1–2.

146. Vidal, Y.; Acho, L.; Luo, N.; Tutiven, C. Hardware in the Loop Wind Turbine Simulator for Control System Testing. In *Wind Turbine Control and Monitoring*; Luo, N., Vidal, Y., Acho, L., Eds.; Advances in Industrial Control; Springer International Publishing: Cham, Switzerland, 2014; pp. 449–466, ISBN 978-3-319-08413-8.
147. Reinauer, T.; Hansen, U.E. Determinants of Adoption in Open-Source Hardware: A Review of Small Wind Turbines. *Technovation* **2021**, *106*, 102289. [CrossRef]
148. Solar Industry Research Data. Available online: https://www.seia.org/solar-industry-research-data (accessed on 13 April 2020).
149. Vaughan, A. Time to shine: Solar power is fastest-growing source of new energy. *Guardian*, 4 October 2017.
150. Pearce, J.M. Photovoltaics—A Path to Sustainable Futures. *Futures* **2002**, *34*, 663–674. [CrossRef]
151. Pearce, J.; Lau, A. Net Energy Analysis For Sustainable Energy Production From Silicon Based Solar Cells. In Proceedings of the American Society of Mechanical Engineers Solar 2002: Sunrise on the Reliable Energy Economy, Reno, NV, USA, 15–20 June 2002.
152. Fthenakis, V.M.; Moskowitz, P.D. Photovoltaics: Environmental, health and safety issues and perspectives. *Prog. Photovolt. Res. Appl.* **2000**, *8*, 27–38. [CrossRef]
153. Fthenakis, V.; Alsema, E. Photovoltaics energy payback times, greenhouse gas emissions and external costs: 2004–early 2005 status. *Prog. Photovolt. Res. Appl.* **2006**, *14*, 275–280. [CrossRef]
154. Feldman, D.; Barbose, G.; Margolis, R.; Bolinger, M.; Chung, D.; Fu, R.; Seel, J.; Davidson, C.; Darghouth, N.; Wiser, R. *Photovoltaic System Pricing Trends: Historical, Recent, and Near-Term Projections 2015 Edition*; NREL: Golden, CO, USA, 2015.
155. Barbose, G.L.; Darghouth, N.R.; LaCommare, K.H.; Millstein, D.; Rand, J. *Tracking the Sun: Installed Price Trends for Distributed Photovoltaic Systems in the United States-2018 Edition*; LBL: Berkeley, CA, USA, 2018.
156. Barron, A.R. Cost reduction in the solar industry. *Mater. Today* **2015**, *18*, 2–3. [CrossRef]
157. Matasci, S. Solar Panel Cost: Avg. Solar Panel Prices by State in 2019: EnergySage. Solar News, Energy Sage. 5 June 2019. Available online: news.energysage.com/how-much-does-the-average-solar-panel-installation-cost-in-the-u-s/ (accessed on 24 February 2022).
158. Dudley, D. Renewable Energy Will Be Consistently Cheaper Than Fossil Fuels by 2020, Report Claims [WWW Document]. Forbes. 2019. Available online: https://www.forbes.com/sites/dominicdudley/2018/01/13/renewable-energy-cost-effective-fossil-fuels-2020/ (accessed on 13 April 2020).
159. Minigrids in the Money. Available online: https://rmi.org/insight/minigrids-money/ (accessed on 25 February 2022).
160. Alafita, T.; Pearce, J.M. Securitization of residential solar photovoltaic assets: Costs, risks and uncertainty. *Energy Policy* **2014**, *67*, 488–498. [CrossRef]
161. Grafman, L.; Pearce, J.M. *To Catch the Sun*; Humboldt University Press: Arcata, CA, USA, 2021.
162. Feldman, D.G.; Barbose, R.; Margolis, R.; Wiser, N.D.; Goodrich, A. *Photovoltaic (PV) Pricing Trends: Historical, Recent, and Near-Term Projections, Sunshot*; NREL: Golden, CO, USA, 2019.
163. Renewables International. Photovoltaics after Grid Parity Plug-and-Play PV: The Controversy 2013. Renewables. 2013. Available online: http://www.renewablesinternational.net/plug-and-play-pv-the-controversy/150/452/72715/ (accessed on 18 December 2015).
164. Mundada, A.S.; Nilsiam, Y.; Pearce, J.M. A review of technical requirements for plug-and-play solar photovoltaic microinverter systems in the United States. *Sol. Energy* **2016**, *135*, 455–470. [CrossRef]
165. Khan, M.T.A.; Norris, G.; Chattopadhyay, R.; Husain, I.; Bhattacharya, S. Autoinspection and Permitting with a PV Utility Interface (PUI) for Residential Plug-and-Play Solar Photovoltaic Unit. *IEEE Trans. Ind. Appl.* **2017**, *53*, 1337–1346. [CrossRef]
166. Khan, M.T.A.; Husain, I.; Lubkeman, D. Power electronic components and system installation for plug-and-play residential solar PV. In Proceedings of the 2014 IEEE Energy Conversion Congress and Exposition (ECCE), Pittsburgh, PA, USA, 14–18 September 2014; pp. 3272–3278.
167. Lundstrom, B.R. *Plug and Play Solar Power: Simplifying the Integration of Solar Energy in Hybrid Applications; Cooperative Research and Development Final Report, CRADA Number CRD-13-523*; National Renewable Energy Lab. (NREL): Golden, CO, USA, 2017.
168. Mundada, A.S.; Prehoda, E.W.; Pearce, J.M. US market for solar photovoltaic plug-and-play systems. *Renew. Energy* **2017**, *103*, 255–264. [CrossRef]
169. Wittbrodt, B.; Pearce, J.M. 3-D Printing Solar Photovoltaic Racking in Developing World. *Energy Sustain. Dev.* **2017**, *36*, 1–5. [CrossRef]
170. Hollman, M.R.; Pearce, J.M. Geographic Potential of Shotcrete Photovoltaic Racking: Direct and Low-Concentration Cases. *Sol. Energy* **2021**, *216*, 386–395. [CrossRef]
171. Arefeen, S.; Dallas, T. Low-Cost Racking for Solar Photovoltaic Systems with Renewable Tensegrity Structures. *Sol. Energy* **2021**, *224*, 798–807. [CrossRef]
172. Pearce, J.M. Parametric Open Source Cold-Frame Agrivoltaic Systems. *Inventions* **2021**, *6*, 71. [CrossRef]
173. Wittbrodt, B.T.; Pearce, J.M. Total U.S. Cost Evaluation of Low-Weight Tension-Based Photovoltaic Flat-Roof Mounted Racking. *Sol. Energy* **2015**, *117*, 89–98. [CrossRef]
174. Wittbrodt, B.; Laureto, J.; Tymrak, B.; Pearce, J.M. Distributed Manufacturing with 3-D Printing: A Case Study of Recreational Vehicle Solar Photovoltaic Mounting Systems. *J. Frugal Innov.* **2015**, *1*, 1–7. [CrossRef]
175. Pearce, J.M.; Meldrum, J.; Osborne, N. Design of Post-Consumer Modification of Standard Solar Modules to Form Large-Area Building-Integrated Photovoltaic Roof Slates. *Designs* **2017**, *1*, 9. [CrossRef]

176. Motahhir, S.; EL Hammoumi, A.; EL Ghzizal, A.; Derouich, A. Open Hardware/Software Test Bench for Solar Tracker with Virtual Instrumentation. *Sustain. Energy Technol. Assess.* **2019**, *31*, 9–16. [CrossRef]
177. Carballo, J.A.; Bonilla, J.; Roca, L.; Berenguel, M. New Low-Cost Solar Tracking System Based on Open Source Hardware for Educational Purposes. *Sol. Energy* **2018**, *174*, 826–836. [CrossRef]
178. Carballo, J.A.; Bonilla, J.; Berenguel, M.; Fernández-Reche, J.; García, G. New Approach for Solar Tracking Systems Based on Computer Vision, Low Cost Hardware and Deep Learning. *Renew. Energy* **2019**, *133*, 1158–1166. [CrossRef]
179. Gómez-Uceda, F.J.; Ramirez-Faz, J.; Varo-Martinez, M.; Fernández-Ahumada, L.M. New Omnidirectional Sensor Based on Open-Source Software and Hardware for Tracking and Backtracking of Dual-Axis Solar Trackers in Photovoltaic Plants. *Sensors* **2021**, *21*, 726. [CrossRef]
180. Buitenhuis, A.J.; Pearce, J.M. Open-Source Development of Solar Photovoltaic Technology. *Energy Sustain. Dev.* **2012**, *16*, 379–388. [CrossRef]
181. Stein, J.S.; Holmgren, W.F.; Forbess, J.; Hansen, C.W. PVLIB: Open Source Photovoltaic Performance Modeling Functions for Matlab and Python. In Proceedings of the 2016 IEEE 43rd Photovoltaic Specialists Conference, Portland, OR, USA, 5–10 June 2016; pp. 3425–3430.
182. Andrews, R.W.; Stein, J.S.; Hansen, C.; Riley, D. Introduction to the Open Source PV LIB for Python Photovoltaic System Modelling Package. In Proceedings of the 2014 IEEE 40th Photovoltaic Specialist Conference, Denver, CO, USA, 8–13 June 2014; pp. 170–174.
183. Freeman, J.M.; DiOrio, N.A.; Blair, N.J.; Neises, T.W.; Wagner, M.J.; Gilman, P.; Janzou, S. *System Advisor Model (SAM) General Description (Version 2017.9.5)*; National Renewable Energy Lab. (NREL): Golden, CO, USA, 2018.
184. SAM Open Source—System Advisor Model (SAM). Available online: https://sam.nrel.gov/about-sam/sam-open-source.html (accessed on 26 February 2022).
185. Merenda, M.; Iero, D.; Carotenuto, R.; Della Corte, F.G. Simple and Low-Cost Photovoltaic Module Emulator. *Electronics* **2019**, *8*, 1445. [CrossRef]
186. Sunderman, W.; Dugan, R.C.; Smith, J. Open Source Modeling of Advanced Inverter Functions for Solar Photovoltaic Installations. In Proceedings of the 2014 IEEE PES T D Conference and Exposition, Chicago, IL, USA, 14 April 2014; pp. 1–5.
187. Hofierka, J.; Kaňuk, J. Assessment of Photovoltaic Potential in Urban Areas Using Open-Source Solar Radiation Tools. *Renew. Energy* **2009**, *34*, 2206–2214. [CrossRef]
188. Nguyen, H.T.; Pearce, J.M. Estimating Potential Photovoltaic Yield with r.Sun and the Open Source Geographical Resources Analysis Support System. *Sol. Energy* **2010**, *84*, 831–843. [CrossRef]
189. Pearce, J.; Babasola, A.; Andrews, R. Open Solar Photovoltaic Systems Optimization. In Proceedings of the Open 2012: NCIIA 16th Annual Conference, San Francisco, CA, USA, 21–24 May 2012.
190. Botero-Valencia, J.S.; Valencia-Aguirre, J.; Gonzalez-Montoya, D.; Ramos-Paja, C.A. A Low-Cost System for Real-Time Measuring of the Sunlight Incident Angle Using IoT. *HardwareX* **2022**, *11*, e00272. [CrossRef]
191. Botero-Valencia, J.S.; Mejia-Herrera, M. Modular System for UV–Vis-NIR Radiation Measurement with Wireless Communication. *HardwareX* **2021**, *10*, e00236. [CrossRef]
192. Botero-Valencia, J.S.; Mejia-Herrera, M.; Pearce, J.M. Design and Implementation of 3-D Printed Radiation Shields for Environmental Sensors. *HardwareX* **2022**, *11*, e00267. [CrossRef]
193. González, I.; Portalo, J.M.; Calderón, A.J. Configurable IoT Open-Source Hardware and Software I-V Curve Tracer for Photovoltaic Generators. *Sensors* **2021**, *21*, 7650. [CrossRef]
194. Singh, T.; Thakur, R. Design and Development of PV Solar Panel Data Logger. *IJCSE* **2019**, *7*, 364–369. [CrossRef]
195. Montes-Romero, J.; Piliougine, M.; Muñoz, J.V.; Fernández, E.F.; De la Casa, J. Photovoltaic Device Performance Evaluation Using an Open-Hardware System and Standard Calibrated Laboratory Instruments. *Energies* **2017**, *10*, 1869. [CrossRef]
196. Papageorgas, P.; Piromalis, D.; Antonakoglou, K.; Vokas, G.; Tseles, D.; Arvanitis, K.G. Smart Solar Panels: In-Situ Monitoring of Photovoltaic Panels Based on Wired and Wireless Sensor Networks. *Energy Procedia* **2013**, *36*, 535–545. [CrossRef]
197. Charaabi, L. Open Monitoring System for Photovoltaic Solar Installations. In Proceedings of the 2020 6th IEEE International Energy Conference (ENERGYCon), Gammarth, Tunisia, 1 October 2020; pp. 1068–1071.
198. De Arquer Fernández, P.; Fernández Fernández, M.Á.; Carús Candás, J.L.; Arboleya Arboleya, P. An IoT Open Source Platform for Photovoltaic Plants Supervision. *Int. J. Electr. Power Energy Syst.* **2021**, *125*, 106540. [CrossRef]
199. López-Vargas, A.; Fuentes, M.; García, M.V.; Muñoz-Rodríguez, F.J. Low-Cost Datalogger Intended for Remote Monitoring of Solar Photovoltaic Standalone Systems Based on Arduino™. *IEEE Sens. J.* **2019**, *19*, 4308–4320. [CrossRef]
200. López-Vargas, A.; Fuentes, M.; Vivar, M. On the Application of IoT for Real-Time Monitoring of Small Stand-Alone PV Systems: Results from a New Smart Datalogger. In Proceedings of the 2018 IEEE 7th World Conference on Photovoltaic Energy Conversion (WCPEC) (A Joint Conference of 45th IEEE PVSC, 28th PVSEC 34th EU PVSEC), Waikoloa, HI, USA, 10–15 June 2018; pp. 605–607.
201. Portalo, J.M.; González, I.; Calderón, A.J. Monitoring System for Tracking a PV Generator in an Experimental Smart Microgrid: An Open-Source Solution. *Sustainability* **2021**, *13*, 8182. [CrossRef]
202. González Pérez, I.; Calderón Godoy, A.J.; Portalo Calero, J.M.; Calderón Godoy, M. *Monitoring Interfaces for Photovoltaic Systems and DC Microgrids: Brief Survey and Application Case*; Universidade da Coruña, Servizo de Publicacións: A Coruña, Spain, 2021; pp. 183–189. [CrossRef]
203. Botero-Valencia, J.S.; Mejia-Herrera, M.; Pearce, J.M. Low Cost Climate Station for Smart Agriculture Applications with Photovoltaic Energy and Wireless Communication. *HardwareX* **2022**, *11*, e00296. [CrossRef]

204. Kadhim Abed, J. Smart Monitoring System of DC to DC Converter for Photovoltaic Application. *IJPEDS* **2018**, *9*, 722. [CrossRef]
205. Russia Exports—January 2022 Data—1994-2021 Historical—February Forecast. Available online: https://tradingeconomics.com/russia/exports (accessed on 25 February 2022).
206. Labs Map | FabLabs. Available online: https://www.fablabs.io/labs/map (accessed on 26 February 2022).
207. China Bets on Open-Source Technologies to Boost Domestic Innovation. Available online: https://merics.org/en/short-analysis/china-bets-open-source-technologies-boost-domestic-innovation (accessed on 26 February 2022).
208. Cimpanu, C. Two of China's Largest Tech Firms Are Uniting to Create a New 'Domestic OS' | ZDNet. 2019. Available online: https://www.zdnet.com/google-amp/article/two-of-chinas-largest-tech-firms-are-uniting-to-create-a-new-domestic-os (accessed on 24 February 2022).
209. Winning Bid Software + Tianjin Kirin = the New Flagship of China's Domestic Operating System-China Electronics. Available online: https://www.cec.com.cn/jtxw/2019/1209/8ac085cc6e112a0f016ee947c8ac00b5.html (accessed on 24 February 2022).
210. One-Fifth Of Russians Live In Poverty, 36 Percent In "Risk Zone", Study Finds. Radio Free Europe/Radio Liberty 14:19:18Z. Available online: https://www.rferl.org/a/study-22-percent-of-russians-live-in-poverty-36-percent-in-risk-zone-/29613059.html (accessed on 24 February 2022).
211. Pathways to Enable Open-Source Ecosystems (POSE). Available online: https://beta.nsf.gov/funding/opportunities/pathways-enable-open-source-ecosystems-pose (accessed on 26 February 2022).
212. Pearce, J.; Pascaris, A.S.; Schelly, C. Professors Want to Share: Preliminary Survey Results on Establishing Open Source Endowed Professorships. *Res. Sq.* **2022**. [CrossRef]
213. Sierla, S.; Sorsamäki, L.; Azangoo, M.; Villberg, A.; Hytönen, E.; Vyatkin, V. Towards Semi-Automatic Generation of a Steady State Digital Twin of a Brownfield Process Plant. *Appl. Sci.* **2020**, *10*, 6959. [CrossRef]
214. Coakley, M.F.; Hurt, D.E.; Weber, N.; Mtingwa, M.; Fincher, E.C.; Alekseyev, V.; Chen, D.T.; Yun, A.; Gizaw, M.; Swan, J.; et al. The NIH 3D Print Exchange: A Public Resource for Bioscientific and Biomedical 3D Prints. *3D Print. Addit. Manuf.* **2014**, *1*, 137–140. [CrossRef]
215. UNCTAD. *Note on a Proposed United Nations Centralised Database of Open-Source Appropriate Technologies*; UNCTAD: Geneva, Switzerland, 2021. Available online: https://unctad.org/webflyer/note-proposed-united-nations-centralised-database-open-source-appropriate-technologies (accessed on 24 February 2022).

technologies

Communication

Exciting of Strong Electrostatic Fields and Electromagnetic Resonators at the Plasma Boundary by a Power Electromagnetic Beam

O. M. Gradov

Kurnakov Institute of General and Inorganic Chemistry, Russian Academy of Sciences, Leninsky pr. 31, 119991 Moscow, Russia; lutt.plm@igic.ras.ru

Abstract: The interaction of an electromagnetic beam with a sharp boundary of a dense cold semi-limited plasma was considered in the case of a normal wave incidence on the plasma surface. The possibility of the appearance of an electrostatic field outside the plasma was revealed, the intensity of which decreased according to the power law with a distance from the plasma and the center of the beam. It was possible to form cavities with a reduced electron density, being each electromagnetic resonators, which probed deeply into the dense plasma and couldexist in a stable state for a long period.

Keywords: nonlinear properties; electrostatic field; resonator; electromagnetic beam; irradiation; surface charge

Citation: Gradov, O.M. Exciting of Strong Electrostatic Fields and Electromagnetic Resonators at the Plasma Boundary by a Power Electromagnetic Beam. *Technologies* **2022**, *10*, 78. https://doi.org/10.3390/technologies10040078

Academic Editors: Manoj Gupta, Eugene Wong and Gwanggil Jeon

Received: 20 May 2022
Accepted: 27 June 2022
Published: 29 June 2022

1. Introduction

The origination of many phenomena taking place during the interaction of electromagnetic radiation with a dense plasma occurs on the interface of media where the possibility of the appearance of certain effects is determined depending on the conditions and the ratio of parameters. Therefore, identifying such conditions and characterizing interactions in simple modeling cases appearsto be an important primary step toward detecting and predicting many interesting phenomena. It was within the framework of the simplest models of cold plasma with a sharp boundary that the effects of nonlinear transparency [1–3], complete absorption [4,5], and anomalous radiation [6] of electromagnetic radiation were investigated. The construction of such a model impliedan accurate representation of the physical essence of the phenomenon under study and those basic features of the interaction of radiation with plasma that ensured its existence.

In this work, the possibility of the formation of globe-shaped resonators, being cavities with a rarefied electron density created at the plasma boundary under the influence of a beam of powerful electromagnetic radiation was considered. The main features of this phenomenon couldbe best studied in a simple model of a semi-infinite plasma with a sharp boundary and stationary ions for the case when electromagnetic radiation normally reached it in the form of a beam with an exponential intensity distribution in the frontal plane. The possibility of forming a cavity with a low electron density followedfrom the physical essence of the interaction of radiation with a plasma. This wasdue to the fact that, on the one hand, a powerful electromagnetic flux wasable to remove electrons from a certain volume and to hold the boundary in the equilibrium against forces of the thermal pressure and the charge separation field [7]. However, on the other hand, such cavities in the plasma couldacquire, under certain conditions, the properties of an electromagnetic resonator [8]. This occurredwhen the size of the cavity, the amplitude, the frequency and spatial structure of the electromagnetic field, the thermal pressure of electrons, and other characteristics reached certain resonant values, for which the stable state of the cavity couldbe maintained for a long periodin the absence of dissipation. The formation of the surface of the cavity and the spatial structure of the electromagnetic field inside the resonator wereinterrelated processes, the parameters of which maintained equilibrium by mutual correction of their values. At the same time, depending on the ratio of the characteristics of the task, the shape of the cavity couldbe either spherical, ellipsoidal, or cylindrical. In the latter case, a situation is possible when such a cylinder crosses the entire thickness of the plasma layer so that the radiation can pass through this layer of dense plasma into a region where it could not penetrate at low values of its intensity. At the same time, for small amplitudes of the electromagnetic signal, when the plasma boundary remains flat, a nonlinear surface charge couldbe formed on it under certain conditions [9], which createdan electrostatic field outside the plasma with a large localization region, when its amplitude decreasedwith a distance from the boundary and the center of the beam according to the power law, in contrast to the strength of the electromagnetic wave field. The possibilities of such a field may arouse interest, both from the point of view of the practical application (for example, for particle acceleration) and from the standpoint of the probable need to prevent undesirable effects.

2. Basic Equations

Consider a semi-infinite plasma consisting of electrons with mass m, density n_e, charge -e, and immobile ions with density n_i ($x \geq 0$), forming a sharp boundary, to which a beam of plane-polarized electromagnetic radiation with a frequency ω and a wave number k propagates along itsnormal on the axis $0X$ (Figure 1). In the region ($x \leq 0$) surrounding the plasma, the following expression can be written for the intensity of $\mathbf{E}_0 = \{0, E_0, 0\}$ of the electric field having auniform spatial distribution in azimuth in the front plane $0YZ$.

$$\mathbf{E}(\mathbf{r},t) = -\nabla\varphi_e(\mathbf{r}) + \hat{\mathbf{y}}E_0\sin(\omega t - kx) + \hat{\mathbf{y}}E_{0r}\sin(\omega t + kx)\,, \quad (1)$$

where $\varphi_e(\mathbf{r})$ is the electrostatic potential of the surface charge formed at the plasma boundary [9–11], and E_{0r} is the amplitude of the reflected electromagnetic signal.

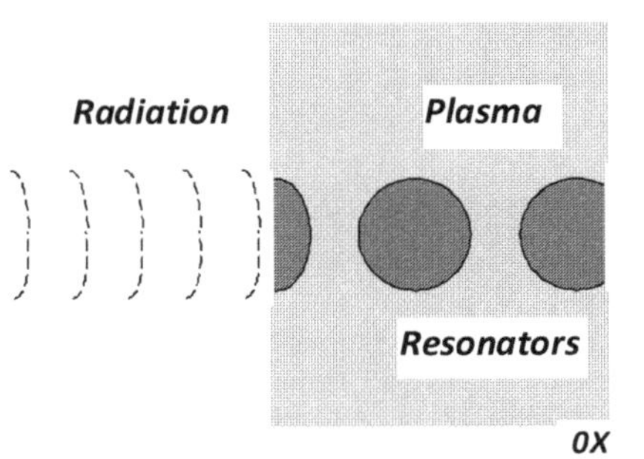

Figure 1. Scheme of interaction of the electromagnetic beam with the surface of the plasma.

The motion of electrons with the velocity $\mathbf{v}$ is described by the equation:

$$\partial_t\mathbf{v} + (\mathbf{v}\cdot\nabla)\mathbf{v} = \frac{e}{m}[\mathbf{E} + \mathbf{v}\times\mathbf{B}] - \frac{v_T^2}{n_0}\nabla n_e \quad (2)$$

where $\mathbf{B}$ is the strength of the magnetic field of external radiation, $v_T{}^2 = T_e/m$, T_e is the temperature of electrons, thermal pressure is taken into account in (2) only to estimate the parameters of the equilibrium state, and n_0 is the equilibrium density of plasma particles in the stationary state ($n_e = n_i \equiv n_0$).

The field strengths in (1), (2) satisfy Maxwell's equations:

$$\partial_t\mathbf{B} = -\nabla\times\mathbf{E}, \quad (3)$$

$$\nabla \times \mathbf{B} = \frac{1}{c^2}\partial_t \mathbf{E} + \mu_0 e n_e \mathbf{v}, \tag{4}$$

$$\nabla \cdot \mathbf{E} = \mathrm{e}(n_e - n_i)/\varepsilon_0. \tag{5}$$

Here, ε_0 is the dielectric density of a vacuum, and μ_0 is its magnetic permeability.

Due to the azimuthhomogeneity of the electromagnetic beam, it waspossible to use a cylindrical coordinate system with an axis of *0X* and coordinates ρ,χ in the frontal plane ($z = \rho\cos\chi$, $y = \rho\sin\chi$). In this case, the amplitude of $E_0(y, z)$ woulddepend only on the coordinate ρ, and it waspossible to consider different intensity distributions in the plane of the wave front, for example, an exponential one.

$$E_0(\rho) = E_a \exp\{-\rho/\rho_0\}, \rho_0 = const, \; k\rho_0 >> 1. \tag{6}$$

For harmonic analysis, the velocity **v** should be divided into a fast-variable $\mathbf{v}_\mathrm{E}$ component and a static $\delta\mathbf{v}(\mathbf{r})$ part ($\mathbf{v}_\mathrm{E} = \mathrm{e}\,\mathbf{E}_0/m\omega$). As a result, the following expression can be derived from Equation (2)

$$(\delta\mathbf{v} \cdot \nabla)\delta\mathbf{v} - \frac{1}{2}\nabla \mathrm{v}_\mathrm{E}^2 - \frac{e}{m}\nabla\varphi - \frac{v_\mathrm{T}^2}{n_0}\nabla n_e = 0 \tag{7}$$

Therefore, the function $F(x,\rho)$ defined by the formula.

$$F(x,\rho) = \frac{1}{2}\delta v^2 - \frac{1}{2}v_\mathrm{E}^2 - \frac{e}{m}\varphi - \frac{v_\mathrm{T}^2}{n_0}n_e \tag{8}$$

This is a continuous quantity both along the polar coordinate ρ and normally to the surface of the plasma (axis *0X*) in the case where the velocity $\delta\mathbf{v}$ is determined by the potential ψ ($\delta\mathbf{v} = \nabla\psi$). With its help, it waspossible to estimate the change in individual physical quantities, as compared to their values at selected points.

3. Analytical and Numerical Results

For high-power radiation, the continuity of the function $F(x,\rho)$ wasreduced to the balance of electromagnetic and thermal energy:

$$\frac{1}{2}v_\mathrm{E}^2(x,\rho) + \frac{v_\mathrm{T}^2}{n_0}n_e(x,\rho) \cong const. \tag{9}$$

From the equilibrium ratio (9), it followedthat the total pressure (the sum of radiation and heat) of electrons wasa continuous quantity, and this balance wasobserved everywhere, including along the *0X* axis and along the ρ axis. It also enabled us to understand how many electrons wereforced out of the cavern formed by the electromagnetic beam incident on the plasma. Along the polar radius ρ, the amplitude of E_0changedsmoothly, and when the density of $n_e$reachedthe critical value of n_c, that is $\omega = \omega_\mathrm{p}(\omega_\mathrm{p}{}^2 = e^2 n_e/(m\cdot\varepsilon_0))$, the plasma becameopaque to this wave field, as a result of which it droppedexponentially rapidly into the dense plasma, the density of which, in turn, increasedas rapidly, according to (9). At this increase in density on the surface ($x = x_\mathrm{b}$, $\rho = \rho_\mathrm{b}$), thermal pressure $v_\mathrm{T}{}^2$ dominated one side, and on the other, electromagnetic, characterized by $v_\mathrm{E}{}^2$, which in a stationary state wouldbalance each other. Therefore, an approximate condition

$$v_\mathrm{E}^2 \approx v_\mathrm{T}^2 \tag{10}$$

wouldbe executed at this boundary to determinethe threshold value of the amplitude $E_0(x_\mathrm{b},\rho_\mathrm{b})$ for the formation of the cavern.

3.1. Conditions for the Formation of Globe-Shaped Resonators of the Electromagnetic Field

The dynamics of the development of the cavity were represented as follows. First, a small space formednear the surface of the plasma and close to the center of the beam with a boundary separating the bulk of the electrons and having a surface shape similar to function (6). As it moveddeeper into the plasma, this cavity wasformed in accordance with the values of the plasma and radiation parameters acting at each time. Since the ions remained stationary, a bulk electric charge formed in the cavern, creating an electric field φ_e, which attemptedto return the displaced electrons back to their positions. The shape of the cavern varied depending on the ratio of the characteristics of the task. However, for example, in the case of a spherical cavity, its radius R in the equilibrium state was determined bycondition (8), in which the potential φ_e that dependedonly on the radial coordinate $r$had tobe substituted from the solution of Equation (5) for the sphere, within which $n_e \approx 0$. In this case, one couldobtain from (5):

$$\varphi_e = -\frac{e}{6\varepsilon_0} n_i r^2 \tag{11}$$

By substituting (11) into condition (8) taken at the boundary $r = R$, it was possible to obtain an estimate of the magnitude of the cavity radius:

$$R = \frac{\sqrt{6}}{\omega_p}\sqrt{v_E^2 - v_T^2} \sim \frac{\sqrt{6}}{\omega_p} v_E \tag{12}$$

When a spherical resonator formed simultaneously with the electronic surface of the cavity, structural changes in the spatial distribution of electric (and magnetic) fields occurred, which beganto reflect from the curved boundary and, according to (3) and (4), were described by the equation:

$$\Delta \mathbf{E} + \frac{\omega^2}{c^2}\varepsilon(\omega)\mathbf{E} = 0,\ \varepsilon(\omega) = 1 - \frac{\omega_p^2}{\omega^2}. \tag{13}$$

The general solution of this equation wasgiven in [8] for a spherical coordinate system (r, ϑ, χ), beginning in the center of the cavity. It has a cumbersome appearance, but for a spherically symmetric case, it was written in a simple form for the radial intensity E_r of the electric field

$$E_r(r \le R) = E_a \frac{\sin kr}{kr}, k = \frac{\omega}{c}\sqrt{\varepsilon_1}, \varepsilon_1 = \varepsilon(\omega, r \le R). \tag{14}$$

$$E_r(r \ge R) = E_a \frac{1}{\kappa r} e^{-\kappa r}, \kappa = \frac{\omega}{c}\sqrt{\varepsilon_2}, \varepsilon_2 = -\varepsilon(\omega, r \ge R). \tag{14a}$$

The oscillations described by formulas (14) and (14a) did not have a wave structure along the surface of the sphere and hada frequency of ω_m (m = 1,2,3, ...), as defined from the following dispersion equation:

$$\varepsilon_2 k e^{-\kappa R} = \varepsilon_1 \kappa \sin(kR). \tag{15}$$

For large values of the parameter $a_p = \omega_p R/c$, the approximate value of the frequency of natural oscillations was in the form $\omega_m = a_m c/R$, where the constant a_m is determined from the solution of the following transcendental equation:

$$a_p e^{-a_p} = a_m \sin a_m. \tag{16}$$

The expression (16) together with (12) allowed us to derive the value of the amplitude of the electric field and frequency, at which it was possible to form a spherical resonator with the parameters presented herein in the form of estimates. We determined that at the value of the velocity δv, the resonator moved deeply into the plasma. To do this, using expression (8), it was necessary to take the parameter values near the surface of the cavity close to the center of the beam where the velocity δv wasentirely directed along the 0X axis. The result wasthe following approximation:

$$\delta v^2 \sim V_E^2 - V_T^2 \tag{17}$$

It should be noted that for other resonant combinations between the parameters of plasma and external radiation, an ellipsoidal form of the resonator couldbe realized. In addition, when the thickness of the plasma along the $0X$ axis wasnarrow, it could-have the appearance of a cylinder through which radiation wasable to penetrate through dense plasma.

3.2. Generation of Electrostatic Fields of Surface Charge near Plasma Space by a Beam of Electromagnetic Radiation

In the case when the force effect of the electromagnetic beam wassmall, as compared to the pressure of electrons, the surface of the plasma remainedflat when interacting with the radiation. However, as shown in [9–11], it formeda nonlinear surface charge associated with the electrostatic field $\varphi_e(\mathbf{r})$, which hada large localization region near the plasma boundary and couldaccelerate charged particles [12–14]. The description of this surface charge, performed in [9–11], was based on the theory of the potential [8,15], which could express the value of the potential $\varphi_e(\mathbf{r})$ throughout space viaits value on the surface of the plasma:

$$\varphi_e(\mathbf{r}) = \frac{1}{2\pi}\int_{-\infty}^{\infty} dy' \int_{-\infty}^{\infty} dz' \frac{|x|\Phi(y',z')}{\left[(y-y')^2+(z-z')^2+x^2\right]^{3/2}}, \; \Phi(y,z) = \varphi_e(x=0,y,z). \quad (18)$$

The integral in (18) should be interpreted as the principal value (p.v.). It indicatedthat in the limit $x \to \pm 0$, when the peculiar point appearedin (18), the path of the integration must have had the form of a small sphere that surrounded this point.

In polar coordinates (χ,ρ), Equation (18), after the integration at the azimuthal angle χ for $x \leq 0$ values not close to the plasma boundary, could be written as follows:

$$\varphi_e(x,\rho) = -\int_0^{\infty} \frac{\Phi(\rho')x\rho' d\rho'}{[\rho^2+\rho'^2+x^2]^{3/2}}. \quad (19)$$

The value of the function $\Phi(\rho)$ couldbe obtained from the equation of motion (2) at the boundary (for $x = 0$) under conditions when nonlinear corrections from the stationary velocity of movement of electrons in the surface charge zone couldbe neglected, and the representation (6) was valid:

$$\Phi(\rho) \cong \frac{eE_0^2(\rho)}{m\omega^2} = \Phi_0 \exp\{-\rho/\rho_0\} \quad (20)$$

In this case, the expression (19) could be expressed as follows:

$$\varphi_e(x,\rho) = \pi\Phi_0 \frac{x}{\rho_0}\left\{\beta \mathbf{H}_0(\beta) - \beta N_0(\beta) - \frac{2}{\pi}\right\}, \; \beta = \frac{\rho^2+x^2}{\rho_0^2}. \quad (21)$$

Here, $\mathbf{H}_0(x)$ and $N_0(x)$ are Struve and Neumann functions, respectively [16].

The asymptotic value of the potential in the region far from the boundary $|x| > \rho$ was described, as follows from (21), by the formula:

$$\varphi_e(x,\rho) \cong -2\Phi_0 \frac{x\rho_0}{\rho^2+x^2}. \quad (22)$$

Based on (22), the electrostatic field component along the plasma $E_\rho = -\partial_\rho \varphi_e$ decreased with increasing distance $|x|$ proportionally $1/|x|$ (component $E_x = -\partial_x \varphi_e$ fell $1/|x|^2$). As compared tothe amplitude of the electromagnetic beam, the magnitude of the electrostatic strength of the field decreasedat a distance from its center not according to the exponential but according to the power law, that is, the area of its localization was much larger.

As an example of the acceleration of charged particles in the electrostatic field of a surface charge (22), it was possible to consider the motion of a particle with a charge e_0 and a mass M from a point (x,ρ) and calculate the final velocity of its movement at infinity. From the equation of motion for the velocity $\mathbf{v}_\mathrm{p}$ of the particle:

$$\partial_t \mathbf{v}_\mathrm{p} + (\mathbf{v}_\mathrm{p} \cdot \nabla)\mathbf{v}_\mathrm{p} = -\frac{e_0}{M}\nabla\varphi_\mathrm{e} \qquad (23)$$

one can write

$$V_\mathrm{p} = \frac{E_a}{\omega}\sqrt{\frac{e_0 e}{Mm}} \qquad (24)$$

It followedfrom (24) that a particle with a mass M in the electrostatic field of the surface charge acquireda constant velocity, which in $(m/M)^{1/2}$ times wasless than the amplitude of electron oscillation at the center of the electromagnetic beam.

4. Summary and Conclusions

The electrostatic field of the surface charge that arosein the process of interaction of the electromagnetic radiation beam with the plasma appearedand affectedthe environment due to the specific movement of electrons [3,9–11] and the complex of conditions that supported its existence (e.g., sharp boundary, quasi-neutrality, absence of non-harmonic perturbations, etc.). The power law of the decrease inthis field in space for a distance from the boundary and from the axis of the electromagnetic beam determinedthe large size of the region of its localization. This circumstance couldbe useful for achieving practical application (e.g., for particle acceleration) or couldbe considered in cases where its effect is likely to have negative consequences. In its magnitude, the strength of this electrostatic field was comparable to the amplitude of electromagnetic oscillations, but it didnot have a spatial and temporal oscillatory structure.

For high intensities of the electromagnetic beam, when the rate of oscillation of electrons wascomparable to their thermal velocity in the plasma, the flat boundary of the electrons wascurved, which in the model of stationary ions ledto the appearance of a charge separation field. As a result of the self-consistent deformation of the surface of electrons and the spatial structure of the electromagnetic field, it waspossible, under certain conditions, to form a cavity, which wasan electromagnetic resonator where the shape of its surface and the structure of the field could exist together for a long period, unchanged. Such conditions werefound in the present work for a resonator in aspherical shape. However, under other conditions, ellipsoidal cavities and even cylindrical cavities can occur. The latter, in the case of a relatively narrowthickness of the plasma layer, wereable to ensure the passage of radiation through a non-transparent medium (in other words, burn through it). The movement of electromagnetic resonators of various shapes also contributedto the penetration of the electromagnetic radiation deeply into the dense plasma and couldbe used to create a number of special nonlinear interactions [3,17–20]. It should be noted that the appearance of resonators waspossible not only in plasma, but hasalso been actively investigated in plasmonic materials, such as hyperbolic metamaterials with giant enhancements [21], metamaterial cavities with broadband strong coupling, and metamaterials with large index sensitivities [22]. The results obtained in these and other similar works couldbe useful for continuing research in plasma with similar configurations.

Funding: This research received no external funding.

Institutional Review Board Statement: Not applicable.

Informed Consent Statement: Not applicable.

Data Availability Statement: Not applicable.

Acknowledgments: The publication was conducted within the State Assignment on Fundamental Research to the Kurnakov Institute of General and Inorganic Chemistry of the Russian Academy of Sciences.

Conflicts of Interest: The author declares no conflict of interest.

References

1. Rosenbluth, M.N.; Liu, C.S. Excitation of plasma waves by two laser beams. *Phys. Rev. Lett.* **1972**, *29*, 701. [CrossRef]
2. Gradov, O.M.; Stenflo, L. On the parametric transparency of a magnetized plasma slab. *Phys. Lett.* **1981**, *83*, 257. [CrossRef]
3. Gradov, O.M.; Ramazashvili, R.R.; Stenflo, L. Parametrlc transparency of a magnetized plasma. *Plasma Phys.* **1982**, *24*, 1101. [CrossRef]
4. Aliev, Y.M.; Gradov, O.M.; Kyrie, A.Y.; Čadež, V.M.; Vuković, S. Total absorption of electromagnetic radiation in a dense inhomogeneous plasma. *Phys. Rev. A* **1977**, *15*, 2120. [CrossRef]
5. Forslund, D.W.; Kindel, J.M.; Lee, K.; Lindman, E.L. Absorption of laser light on self-consistent plasma-density profiles. *Phys. Rev. Lett.* **1976**, *36*, 35. [CrossRef]
6. Gradov, O.M.; Larsson, J.; Lindgren, T.; Stenflo, L.; Tegeback, R.; Uddholm, P. Anomalous radiation from a nonstationary plasma. *Phys. Scr.* **1980**, *22*, 151. [CrossRef]
7. Berger, J.M.; Newcomb, W.A.; Dawson, J.M.; Frieman, E.A.; Kulsrud, R.M.; Lenard, A. Heating of a confined plasma by oscillating electromagnetic fields. *Phys. Fluids* **1958**, *1*, 301. [CrossRef]
8. Stratton, J.A. *Electromagnetic Theory*; McGraw-Hill: New York, NY, USA, 1941.
9. Gradov, O.M. Self-consistent plasma boundary distortions during the interaction of a normally incident electromagnetic beam and a nonlinear surface charge. *Chin. J. Phys.* **2021**, *72*, 360–365. [CrossRef]
10. Gradov, O.M. Three-dimensional surface charge nonlinear waves at a plasma boundary. *Phys. Scr.* **2019**, *94*, 125601. [CrossRef]
11. Gradov, O.M. Nonlinear behavior of a surface charge on the curved plasma boundary with a moving cavity. *Phys. Lett. A* **2020**, *384*, 126566. [CrossRef]
12. Yamagiva, M.; Koga, J. MeV ion generation by an ultra-intense short-pulse laser: Application to positron emitting radionuclide production. *J. Phys. D Appl. Phys.* **1999**, *32*, 2526. [CrossRef]
13. Kovalev, V.F.; Bychenkov, V.Y. Analytic theory of relativistic self-focusing for a Gaussian light beam entering a plasma: Renormalization-group approach. *Phys. Rev. E* **2019**, *99*, 043201. [CrossRef] [PubMed]
14. Boyer, C.N.; Destler, W.W.; Kim, H. Controlled collective field propagation for ion-acceleration using a slow-wave structure. *IEEE Trans. Nucl. Sci.* **1977**, *24*, 1625–1627. [CrossRef]
15. Jeffreys, H.; Swirles, B. *Methods of Mathematical Physics*; Cambridge University Press: Cambridge, UK, 1956.
16. Bateman, H.; Erdélyi, A. *Higher Transcendental Functions*; McGraw-Hill: New York, NY, USA, 1953; Volume I–II.
17. Ma, J.Z.G.; Hirose, A. Parallel propagation of ion solitons in magnetic flux tubes. *Phys. Scr.* **2009**, *79*, 045502. [CrossRef]
18. Brodin, G.; Stenflo, L. Large amplitude electron plasma oscillations. *Phys. Lett. A* **2014**, *378*, 1632. [CrossRef]
19. Vladimirov, S.V.; Yu, M.Y.; Tsytovich, V.N. Recent advances in the theory of nonlinear surface waves. *Phys. Rep.* **1994**, *241*, 1–63. [CrossRef]
20. Vladimirov, S.V.; Yu, M.Y.; Stenflo, L. Surface-wave solitons in an electronic medium. *Phys. Lett. A* **1993**, *174*, 313. [CrossRef]
21. Xu, H.; Zhu, Z.; Xue, J.; Zhan, Q.; Zhou, Z.; Wang, X. Giant enhancements of high-order upconversion luminescence enabled by multiresonant hyperbolic metamaterials. *Photonics Res.* **2021**, *9*, 395. [CrossRef]
22. Gu, P.; Chen, J.; Chen, S.; Yang, C.; Zhang, Z.; Du, W.; Chen, Z. Ultralarge Rabi splitting and broadband strong coupling in a spherical hyperbolic metamaterial cavity. *Photonics Res.* **2021**, *9*, 829. [CrossRef]

technologies

MDPI

Communication

An a Priori Discussion of the Fill Front Stability in Semisolid Casting

Anders E. W. Jarfors [1,*], Qing Zhang [1] and Stefan Jonsson [2]

[1] Materials and Manufacturing, School of Engineering, Jönköping University, P.O. Box 1026, 511 11 Jönköping, Sweden; qing.zhang@ju.se

[2] Materials Science and Engineering, School of Industrial Engineering and Management, KTH Royal Institute of Technology, Brinellvägen 23, 100 44 Stockholm, Sweden; jonsson@kth.se

* Correspondence: anders.jarfors@ju.se

Abstract: Metal casting is an industrially important manufacturing process offering a superior combination of design flexibility, productivity and cost-effectiveness, but has limitations due to filling related defects. Several semisolid casting processes are available capable of casting at a range of solid fractions to overcome this. The current communication aims to review the filling front behaviour and give a new perspective to the gate design in semisolid processing compared to conventional high-pressure die-casting. It is shown that solid fraction and gate widths are critical to avoid instability and spraying.

Keywords: high-pressure die-casting; semisolid; gate; filling; stability; solid fraction; speed

Citation: Jarfors, A.E.W.; Zhang, Q.; Jonsson, S. An a Priori Discussion of the Fill Front Stability in Semisolid Casting. *Technologies* **2022**, *10*, 67. https://doi.org/10.3390/technologies10030067

Academic Editors: Manoj Gupta, Eugene Wong and Gwanggil Jeon

Received: 13 May 2022
Accepted: 27 May 2022
Published: 30 May 2022

1. Introduction

Metal casting is an industrially important manufacturing process offering a superior combination of design flexibility, productivity and cost-effectiveness [1,2]. Aluminium is a vital material offering lightweight solutions for the transportation industry and cooling solutions for the electronics and telecom industries. Using semisolid casting processes, the ability to replace heavier materials and designs with more efficient solutions has significantly increased. Many examples exist from the electronics industry [3] and automotive and truck components [3,4]. Today, several processes are available with different capabilities and characteristics, ranging from low solid fractions, such as the GISS process, to high fraction solid processing, such as the SEED process [5].

The success of these processes is often referred to as a reduction in turbulence from an increase in viscosity [6–8]. Very little attention has been given to fill front stability that has been a focus for improvements in gravity die casting processes [1]. Fill front stability is characterised by the Weber number or similar, but rarely used in the discussion of filling [1,2]. A possible reason for this is that the filling process in high pressure die casting (HPDC) is very violent, with high gate speed very far from what would be required for a stable front [1,2,9]. In the GISS process, a measure with the ratio of gate speed, v (m/s), over solid fraction, f_s (-), was developed as a process index where values from 22 below gave a stable filling for a thin plate [10].

In reality, there are many mechanisms active in jet break up where break up can occur in many different modes, starting from laminar flow and growth of instabilities, to so-called Rayleigh break-up. In the current study, this type of break up is considered stable and does not occur within the lengths available in the die, as shown in Figure 1a [11]. It is essential to understand that spray formation in HPDC is not the same as atomisation and spraying that for a flat jet would appear as in Figure 1b, with surface tension-driven hole formation generating ligaments and droplet formation. The break-up is instead a consequence where a break-up takes place in the transition regime and turbulent regime, where the travelled distance is reduced before the gate or jet-speed reaches the actual spray regime under normal gate speed, with speed below 55 m/s [9,12,13]. Depending on the degree of filling of the cavity cross-section and cavity geometry, two scenarios are possible. For a cavity cross-section not fully filled, the break-up would have the possibility to occur in a similar fashion as the flat jet break up. For a filled cavity, there would be undulations on the surface, entraining gas and possible droplet formation similar to gravity casting. These latter two are illustrated in Figure 1c.

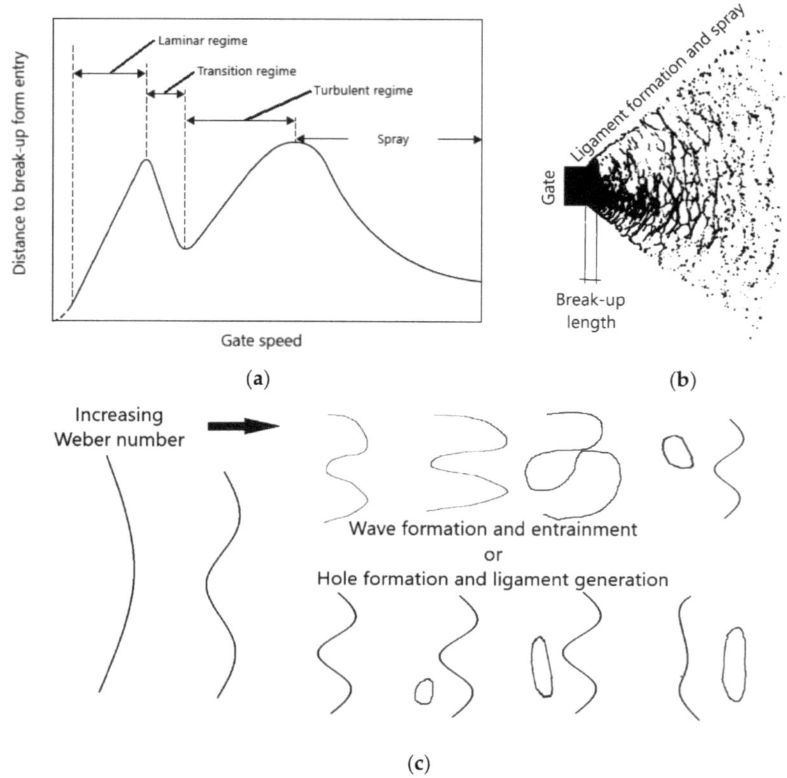

Figure 1. Break-up illustration with (**a**) Schematic illustration of the travelled distance before break-up and gate speed adapted from Lefebvre and McDonnel [11]. (**b**) Break-up distance of flat jet with ligament formation for partially filled die cavities and (**c**) different instabilities with surface wave formation and entrainment for a filled fie cavity and hole formation and ligament formation for a partially filled cavity.

Saeedipour et al. [14] analysed HPDC processes and break-up and concluded that the atomisation regime reached speeds as high as 70 m/s. The first and second wind break-up was the critical regime for all the other speeds. First wind break-up corresponds to break-up in the transition regime and would be more related to folding and ligament formation for a flat jet that would enter the die cavity. Second wind break-up is related to break up in the turbulent regime and corresponds to droplet formation with beads or cold shot formation.

The current communication aims to review the filling front behaviour and give a new perspective to the gate design in HPDC, and especially SSM processing using HPDC.

2. Methodology

This communication is an a priori analysis of the fill front behaviour, taking a literature foundation in developing a theoretical framework for the analysis of fill front stability. The example used is that of rheocasting an A356 alloy. The A356 alloy is a preferred type of alloy in rheocasting, due to its large solidification range [3,4].

3. Theoretical Framework

3.1. Turbulence, Surface Stability and Fill Front Break-Up

The problems related to filling in HPDC involve all types of behaviour, ranging from a stable front to a wavy fill front and a fully developed spray and atomisation flow state [12]. Turbulence is mainly characterised by the Reynolds number, with turbulence starting as low as 2500, shown in Equation (1). The Reynolds number is the ratio of inertial forces to viscous forces within a fluid volume subjected to motion at different fluid velocities.

$$Re = \frac{\rho v D_H}{\mu} \tag{1}$$

where ρ is density (kg/m^3), v is gate speed (m/s), μ is viscosity (Pa s) and D_H is the hydraulic diameter (m). The analysis takes its foundation in the fill front stability developed by Campbell [1] and by Miller [15], who worked mainly with gravity-driven processes. At low viscosity and low flow speeds, gravity matters and is characterized by the Froude number. The Froude number is a ratio of the flow inertia to the external field and is based on a speed–length ratio. The external field under the current conditions is gravity and that is only relevant for flow speed up to 0.25 m/s according to Miller [15]. At higher speeds, the surface tension phenomenon becomes important, and the stability can be assessed based on the Weber number instead, *We*, as in Equation (2) [11]. The Weber number is the ratio of drag forces/cohesion forces

$$We = \frac{\rho v^2 D_H}{\sigma} \tag{2}$$

where σ is surface tension (N/m). The absolute stability of a fill front is with $We < 0.8$, but a practical limit is given by $We < 2$ [2].

In atomisation and spray theory, there are several modes for the break-up with first and second wind break up. First wind break up is similar to the Weber number stability criterion as waves are formed, and with time in a flat jet, ligaments will form. Second wind break-up involves the formation of droplets similar to what is found as cold shots or beads. This can be analysed using the Ohnesorge number, *Oh*, Equation (3) [11]:

$$Oh = \frac{\sqrt{We}}{Re} \tag{3}$$

The boundaries for the first and second wind break-up are straight lines in a logarithmic plot of the Ohnesorge number versus the Reynolds adapted from Lefebvre and McDonnel [11] and Saeedipour et al. [14], as seen in Figure 2.

3.2. Surface Tension and Shear Strength Build-Up in the Mushy State

The surface tension of the aluminium melt is 0.8 Pa. The strength build-up in the two-phase region was studied by Pan et al. [16], which analysed two microstructure types based on the A356 alloy. The magnetohydrodynamically stirred material is the most similar to what is expected in SSM processing and also the slurry with the lower strength. The expression for the strengths was given by Equation (4):

$$\tau = \frac{1000}{\frac{0.43}{f_s^2} - 1.87} \tag{4}$$

Under the assumption that the internal resistance to motion in the slurry can be seen acting in the direction of the surface tension, it is possible to add this as a cohesive force of the slurry and, as such different from viscosity.

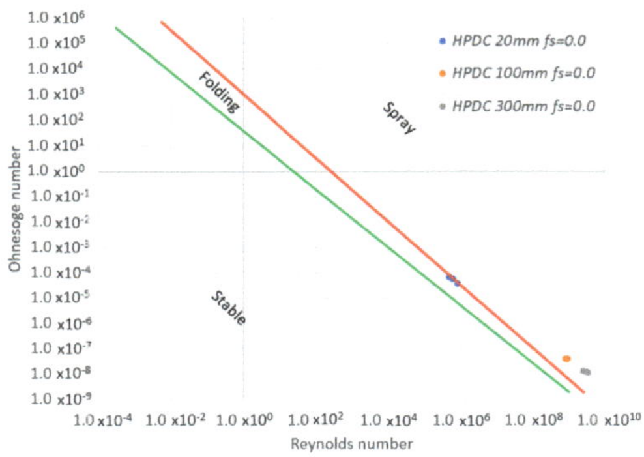

(**a**)

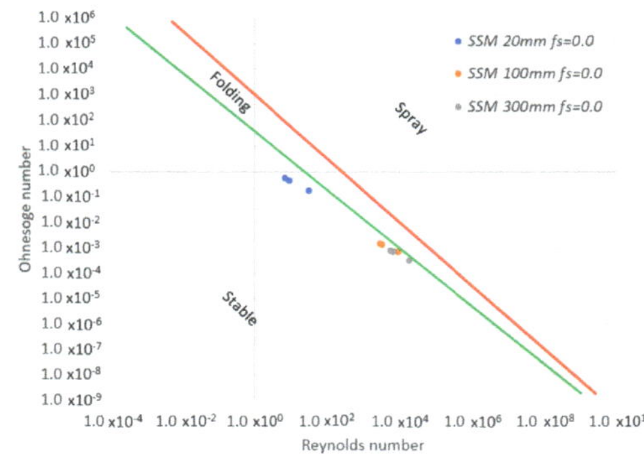

(**b**)

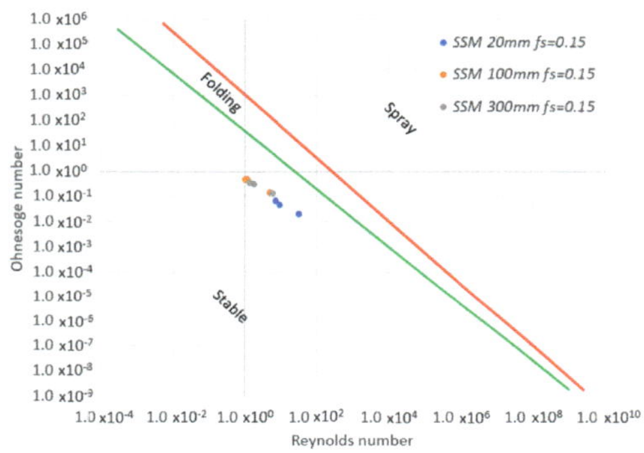

(**c**)

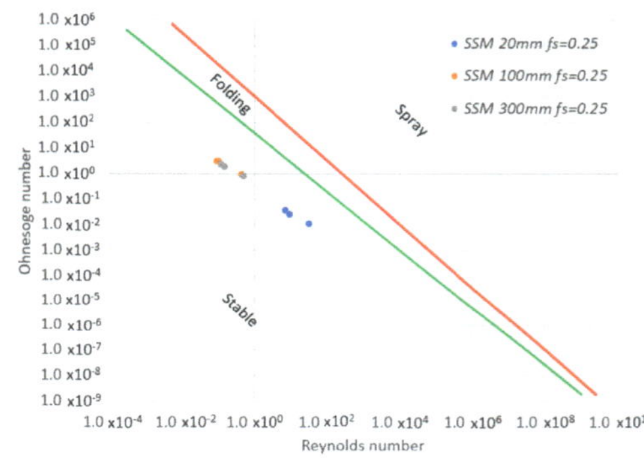

(**d**)

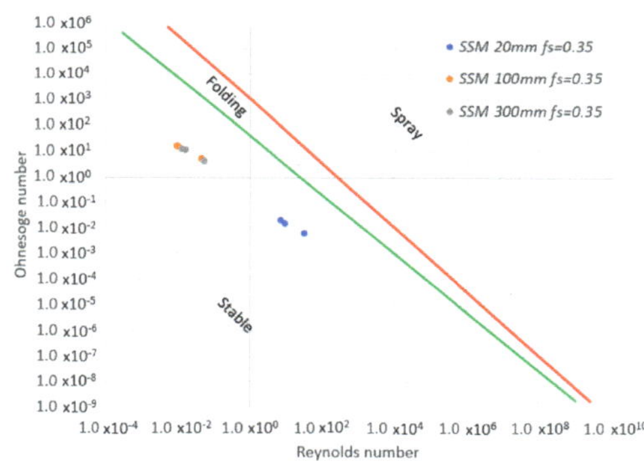

(**e**)

Figure 2. Break-up analysis with the Ohnesorges number plotted against the Reynolds number for (**a**) HPDC conditions without solid content; (**b**) SSM conditions without any solid fraction; (**c**) SSM conditions with f_s = 0.15 where practical stability based on We was found; (**d**) SSM conditions with f_s = 0.25 where absolute stability based on We was found; (**e**) SSM conditions with f_s = 0.35, a typical fraction used in high fraction SSM processing. The green line is first wind break-up, and the red line is second wind break-up. Blue markers are 20 mm gate width, orange markers are 100 mm gate width and grey is for 300 mm gate widths.

4. Discussion

4.1. Effect of a Solid Fraction Present on the Weber Number and Front Stability

Starting with the fill front stability, the Weber number is one measure [1,2], and a complementary criterion was developed by Janudom et al. [10] based on the gate speed and the solid fraction. In Table 1, typical data for casting A356, cast under different conditions, gate speed and thickness, and gate width and used in the analysis are collated. For HPDC and semisolid casting, a practical limit is to keep $We < 2$, but $We < 0.8$ results in absolute stability [1,2]. All conditions in Table 1 will result in folding or spraying for the typical HPDC conditions without a solid phase present. The conditions will be similar for the typical SSM conditions without a solid phase present as a hypothetical case as $We > 2$ for all cases. It should be noted that the liquid aluminium viscosity was approximated to 1 mPa s [17,18] and for the SSM, approximately to 2 Pa s [19,20].

Table 1. Weber number (We) for different processed and conditions.

Process	Gate Speed	v/f_s	Gate Thickness	Weber Numbers for the Gate Widths [3]		
	(m/s)	(m/s)	(mm)	(mm)		
				10	100	300
HPDC [1] $f_s = 0.0$	45	N/A	2	6242.98	6732.63	6821.80
	35	N/A	4	6923.80	7989.00	8199.24
	30	N/A	6	7043.37	8638.09	8976.84
SSM [2] $f_s = 0.0$	8	N/A	2	197.31	212.78	215.60
	4	N/A	4	90.43	104.35	107.09
	3.5	N/A	6	95.87	117.57	122.18
SSM [2] $f_s = 0.15$	8	53	2	2.68	2.89	2.93
	4	27	4	1.23	1.42	1.46
	3.5	23	6	1.30	1.60	1.66
SSM [2] $f_s = 0.25$	8	32	2	0.79	0.85	0.86
	4	16	4	0.36	0.42	0.43
	3.5	14	6	0.38	0.47	0.49
SSM [2] $f_s = 0.35$	8	23	2	0.26	0.28	0.28
	4	11	4	0.12	0.14	0.14
	3.5	10	6	0.13	0.15	0.16

[1] Viscosity approximated to 1 mP s [17,18]. [2] Viscosity approximated to 2 Pa s [19,20]. [3] Density 2700 kg/m^3 [2].

The gate speeds for RheoMetal processing (Bromma, Sweden) are significantly lower than those found in HPDC. Comparing the Weber numbers for the gate geometries analysed will give instability with folding and possibly spraying. Increasing the solid fraction gradually will reduce the Weber number since the shear strengths are added to the surface tension term. At a solid fraction $f_s = 0.15$, the practical stability limit, based on the Weber number, is reached within the conditions investigated. The ratio v/f_s had a practical maximum of approximately 22, resulting in a slightly more conservative measure than the Weber number.

Increasing the solid fraction further results in that the absolute limit, based on the Weber number, being reached from $f_s = 0.25$ or higher and at $f_s = 0.35$ all conditions result in absolute stability. The ratio v/f_s is, in general, more conservative but does not give a critical value at a constant Weber number. This is concluded by comparing the v/f_s is 23 for both $f_s = 0.15$ with a 6 mm gate and $f_s = 0.35$ and with a 2 mm gate but where the $We = 1.30$ and $We = 0.26$, respectively. The foundation of the v/f_s is more related to turbulence and the Reynolds number Re [10].

4.2. Effect of the Solid Fraction of the Spray Behaviour

In the analysis of the folding and spraying of the SSM processed material, additional gate thicknesses and gate speed recommendations were added with an 8 mm gate with a speed of 3 m/s, 10 mm gate with 2.5 m/s and a 12 mm gate with 2 m/s. The gate widths of 20 mm, 100 mm and 300 mm were kept.

Starting with HPDC conditions, Figure 2a shows similar results as Saeedipour et al. [13] in terms of Ohnesorges number but higher Reynolds number due to a geometric difference. The results indicate similar break-up behaviour where the 20 mm gate width is on the second win break-up boundary (red line), and the wider gates of 100 and 300 mm are well into the droplet formation range. Shifting the speeds to those recommended for SSM processing (RheoMetal^{TM} process) moves the conditions in the safe region where laminar flow Rayleigh break-up may occur. The Weber number, Table 1, for SSM with $f_s = 0$ does not fulfil the practical of 2, not the absolute stability of 0.8. The Weber number is thus a more conservative measure.

Adding the effect of solid fraction reduced the Webers number to the practical limit at $f_s = 0.15$ corresponding to Figure 2c. Absolute stability was reached at $f_s = 0.25$, corresponding to Figure 2d. The levels used on many higher solid fraction SSM processes is $f_s = 0.35$ provides absolute stability, is well inside the stable region and fulfils the v/f_s condition for all geometries.

5. Conclusions

In the current paper, the effect of the solid phase on the filling conditions was analysed a priori using three different tools, (1) the Weber number for fluid dynamics and used in gravity die casting, (2) the first and second wind break-up analysis utilising Ohnesoges number and the Reynolds number and (3) the criterion developed for the GISS process with the gate speed divided by the solid fraction.

The three different measures all showed similar results, with gate speed divided by the solid fraction being the most conservative. The Weber number and thus also the Ohnesorges number were corrected for the presence of a solid fraction, resulting in a practical fill front stability level being reached for a solid fraction of 0.15.

Absolute stability based on the Weber number was reached at a solid fraction of 0.25. Not even at this high fraction could a sufficiently low value of the gate speed divided with the solid fraction be reached for all geometries. The conclusion for the v/f_s ratio measure is that it will force the gate speed to very low values and likely hinder the users from choosing processing conditions of SSM processes to achieve the possible extended flow lengths possible, which is one of the benefits [5,8].

The break-up analysis utilising Ohnesorges and Reynolds numbers in the assessment of the first and second wind break-up was the most forgiving, and no solid fraction was required to obtain stability allowing for fill speed to be even higher than the recommended values for the RheoMetal process, suggesting that there is a significant margin to fill front instability. However, the diagram shows that the stable regime in the diagram is a laminar break-up regime and break-up is possible with extended flow lengths. A break-up is likely possible for large components and long flow lengths, and then the Webernumber criterion should be adhered to.

In all this analysis, it was assumed that for A356, the shear strength of the slurry was determined by Pan et al. [16], which is one uncertainty. The slurry quality, slurry structure and solid fraction are essential, and ideally, for the high solid fraction, this parameter is an essential metric for the gating design and choice of fill parameters for a good quality casting.

Author Contributions: Conceptualisation, A.E.W.J.; methodology, A.E.W.J.; formal analysis, A.E.W.J. and Q.Z.; investigation, A.E.W.J. and Q.Z.; writing—original draft preparation, A.E.W.J.; writing—review and editing, A.E.W.J., Q.Z. and S.J.; supervision, A.E.W.J. and S.J.; project administration, A.E.W.J. and S.J.; funding acquisition, A.E.W.J. All authors have read and agreed to the published version of the manuscript.

Funding: This research was funded by Vinnova under the ReCKA project (contract No. 2018-02831).

Institutional Review Board Statement: Not applicable.

Informed Consent Statement: Not applicable.

Data Availability Statement: All data used are contained within the paper.

Acknowledgments: The authors are indebted to Comptech AB for supporting with gate speed parameters and discussions leading to the idea behind this paper.

Conflicts of Interest: The authors declare no conflict of interest. The funders had no role in the study's design, in the collection, analyses, or interpretation of data, in the writing of the manuscript, or in the decision to publish the results.

References

1. Campbell, J. *Complete Casting Handbook Metal Casting Processes, Metallurgy, Techniques and Design*, 2nd ed; Butterworth-Heinemann: Oxford, UK, 2015; ISBN 978-0-444-63509-9.
2. Jarfors, A.E.W.; Seifeddine, S. *Metal Casting*; Springer: Berlin, Germany, 2015; ISBN 9781447146704/9781447146698.
3. Jarfors, A.E.W.; Zheng, J.C.; Chen, L.; Yang, J. Recent Advances in Commercial Application of the Rheometal Process in China and Europe. *Solid State Phenom.* **2019**, *285*, 405–410. [CrossRef]
4. Li, D.Q.; Zhang, F.; Midson, S.P.; Liang, X.K.; Yao, H. Recent Developments of Rheo-Diecast Components for Transportation Markets. *Solid State Phenom.* **2019**, *285*, 417–422. [CrossRef]
5. Jarfors, A.E.W. A Comparison between Semisolid Casting Methods for Aluminium Alloys. *Metals* **2020**, *10*, 1368. [CrossRef]
6. Atkinson, H.V. Semisolid Processing of Metallic Materials. *Mater. Sci. Technol.* **2010**, *26*, 1401–1413. [CrossRef]
7. Jarfors, A.E.W. Pressure Different Casting. *Encycl. Mater. Met. Alloys* **2022**, *4*, 117–128. [CrossRef]
8. Jarfors, A.E.W. *Semisolid Casting of Metallic Parts and Structures*; Elsevier Ltd.: Amsterdam, The Netherlands, 2022; Volume 4, ISBN 9780128197264.
9. Street, A.C. *The Diecasting Handbook*; Portcullis Press Ltd.: Redhill, UK, 1977.
10. Janudom, S.; Wannasin, J.; Basem, J.; Wisutmethangoon, S. Characterization of Flow Behavior of Semi-Solid Slurries Containing Low Solid Fractions in High-Pressure Die Casting. *Acta Mater.* **2013**, *61*, 6267–6275. [CrossRef]
11. Lefebvre, A.H.; Mcdonell, V.G. *Atomization and Sprays*, 2nd ed.; CRC Press: Boca Raton, FL, USA, 1989; ISBN 9781498736268.
12. Hao, S.; Hu, B.; Pehlke, R. Atomization in High Pressure Die Casting—A Problem and a Challenge. *Die Cast. Eng.* **1998**, *42*, 42.
13. Saeedipour, M.; Pirker, S.; Schneiderbauer, S. Numerical Study on Liquid Jet Breakup and Droplet-Wall Interaction in High Pressure Die Casting Process. In Proceedings of the ICLASS 2015—13th International Conference on Liquid Atomization and Spray Systems, Tainan, Taiwan, 23–27 August 2015.
14. Saeedipour, M.; Schneiderbauer, S.; Pirker, S.; Bozorgi, S. Prediction of Surface Porosity Defects in High Pressure Die Casting. *TMS Annu. Meet.* **2015**, 155–163. [CrossRef]
15. Miller, R.A. *Casting Solutions for Readiness, Thin Wall and High Strength Die Casting Alloys*; The Ohio State University: Columbus, OH, USA, 2017; Available online: https://files.core.ac.uk/pdf/23/84591373.pdf (accessed on 26 May 2022).
16. Pan, Q.Y.; Apelian, D.; Alexandrou, A.N. Yield Behavior of Commercial Al-Si Alloys in the Semisolid State. *Metall. Mater. Trans. B* **2004**, *35*, 1187–1202. [CrossRef]
17. Dinsdale, A.T.; Quested, P.N. The Viscosity of Aluminium and Its Alloys—A Review of Data and Models. *J. Mater. Sci.* **2004**, *39*, 7221–7228. [CrossRef]
18. Zhang, F.; Du, Y.; Liu, S.; Jie, W. Modeling of the Viscosity in the AL-Cu-Mg-Si System: Database Construction. *Calphad* **2015**, *49*, 79–86. [CrossRef]
19. Ma, Z.; Zhang, H.; Fu, H.; Fonseca, J.; Yang, Y.; Du, M.; Zhang, H. Modelling Flow-Induced Microstructural Segregation in Semi-Solid Metals. *Mater. Des.* **2022**, *213*, 110364. [CrossRef]
20. Das, P.; Samanta, S.K.; Dutta, P. Rheological Behavior of Al-7Si-0.3Mg Alloy at Mushy State. *Metall. Mater. Trans. B Process Metall. Mater. Process. Sci.* **2015**, *46*, 1302–1313. [CrossRef]

technologies

MDPI

Perspective

Developments and Applications of Artificial Intelligence in Music Education

Xiaofei Yu [1], Ning Ma [2], Lei Zheng [3], Licheng Wang [4] and Kai Wang [2,*]

1 Conservatory of Music, Qingdao University, Qingdao 266000, China
2 School of Electrical Engineering, Weihai Innovation Research Institute, Qingdao University, Qingdao 266000, China
3 Science and Technology Department, Qingdao University, Qingdao 266000, China
4 School of Information Engineering, Zhejiang University of Technology, Hangzhou 310014, China
* Correspondence: wangkai@qdu.edu.cn or wkwj888@163.com; Tel.: +86-15863060145; Fax: +86-532-85951980

Abstract: With the continuous developments of information technology, advanced computer technology and information technology have been promoted and used in the field of music. As one of the products of the rapid development of information technology, Artificial Intelligence (AI) involves many interdisciplinary subjects, adding new elements to music education. By analyzing the advantages of AI in music education, this paper systematically summarizes the application of AI in music education and discusses the development prospects of AI in music education. With the aid of AI, the combination of intelligent technology and on-site teaching solves the lack of individuation in the traditional mode and enhances students' interest in learning.

Keywords: artificial intelligence; music education; applications; developments

Citation: Yu, X.; Ma, N.; Zheng, L.; Wang, L.; Wang, K. Developments and Applications of Artificial Intelligence in Music Education. *Technologies* **2023**, *11*, 42. https://doi.org/10.3390/technologies11020042

Academic Editor: Carmelo J. A. Bastos-Filho

Received: 21 February 2023
Revised: 10 March 2023
Accepted: 14 March 2023
Published: 16 March 2023

1. Introduction

Artificial Intelligence (AI) is a new technical science to research and develop the theory, method, technology and application system for simulating and expanding human intelligence [1–5]. It is a branch of computer science, involving philosophy, cognitive science, mathematics, neurophysiology, psychology and a range of other disciplines. It is also a challenging subject [6–10]. Since AI was formally put forward in 1956, AI has experienced more than 50 years of development and become an interdisciplinary and frontier science.

The emergence of computers has promoted the development of modern electronic music technology. With the rapid development of computer multimedia technology and signal processing technology and its penetration into the field of music appreciation and creation, modern music technology represented by electronic music has developed rapidly, and the field of technological innovation is gradually expanding [11–13]. When it comes to the application of AI in the field of music, music technology has to be mentioned [14,15]. Music technology is an interdisciplinary subject, divided into art and technology parts [16–18]. The art part mainly studies the use of various audio softwares for music creation and production; the science and technology part mainly studies the use of computer technology to provide technical support for music production. Figure 1 shows the relationship between AI and music education. Due to the combination and development of music education, AI technology has become the future trend of music education, exerting a huge influence on traditional teaching concepts and methods and forming a diversified and multi-level development direction. In recent years, digital music has become a huge part of the music industry [19,20]. The combination of audio big data and AI generates Music Information Retrieval (MIR), which is based on music acoustics and extracts audio features based on audio signal processing. Various machine learning technologies in AI are widely used at the back end, which is the most important part of music technology. With MIR, we can use music as a kind of information to carry out information retrieval, so that we can classify the huge music library and conduct more detailed study on the elements of music, such as pitch and rhythm. In addition to MIR, current music technologies include AI composition [21], song synthesis technology [22] and digital audio watermarking technology [23]. Although these music technologies are not perfect and have certain limitations, they have played a particular role in promoting the development of the music industry and have their own theoretical and practical value, which will be widely used after continuous improvement in the future [24–26].

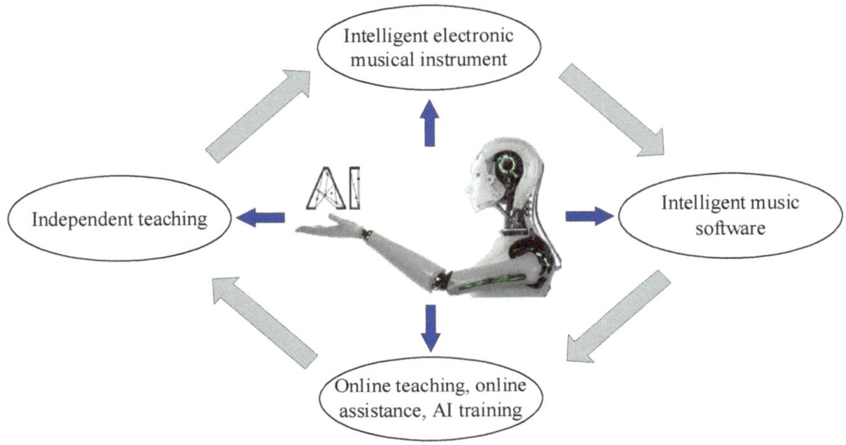

Figure 1. The Relationship between AI and music education.

The organic integration of AI technology and music education has enriched classroom teaching resources, expanded the functions of intelligent instruments and improved the technical means of music education. It supports personalized learning, analyzes the melody and rhythm of music, effectively evaluates the teaching effect, and inspires music teachers to use artificial intelligence technology to innovate music teaching. This paper systematically summarizes the application of AI in music education, including the application of AI in intelligent electronic instruments, intelligent music software, online teaching and autonomous teaching. In the rest of the paper, Section 2 discusses the application of combining AI and music education, Section 3 discusses the development, significance, and prospect of AI and music education.

2. The Application of Combining AI and Music Education

2.1. Application in Intelligent Electronic Musical Instruments

In recent years, the continuous development of AI technology has made electronic musical instruments more intelligent, humanized and specialized to bring forth the new [27]. The intelligent electronic instrument can not only store all kinds of musical instrument timbre, but also realize the effective combination of all kinds of timbre, so that all kinds of timbre can be performed according to different action instructions. This function is obviously difficult to achieve for traditional musical instruments. With these advantages, intelligent electronic instruments are gradually introduced into music teaching to guide students to learn new intelligent electronic instruments. It is because of the introduction of intelligent electronic musical instruments for music education, that a new mode of education is provided. More than ever, one person alone can play, and through various combinations of effective sounds, expand creative thinking. Music provides great convenience for the students of music practice, and further gain a higher quality of teaching [28–32].

Xu et al. [33] studied the collaboration between electronic music creation and online performance of music education and wireless networks under AI. Today, with the rapid development of science and technology, AI technology continues to progress, and the development of digital technology, electronic music online performance, and wireless network collaborative research is more important. Through the research of AI, an electronic music creation system was designed, which realizes the cooperation between electronic music online education and wireless networks. The concept and technology of computer sensor networks, intelligent algorithms and wireless networks are studied, and it becomes a new type of intelligent electronic musical instrument. Through the simulation experiment, the matching degree of AI electronic music course resources are verified, and the oscilloscope is used to transform the sound characteristics into the corresponding sound and image patterns, so as to achieve the purpose of online electronic music intelligent matching and to realize the function of online education.

Guo et al. [34] proposed a new method of piano teaching. Under the framework of an AI environment and wireless network optimization, they adopted a new piano teaching method of "people + equipment", and constantly improved two piano teaching modes: "complementary" piano teaching mode and "remote network" piano teaching mode, which conforms to the trend of the integration of piano performance form and current high-tech development. The function and role of AI is reflected in intelligent teaching, intelligent scoring, networked piano classrooms, and automatic playing functions. The combination of traditional piano teaching and modern AI technology innovation promotes the renewal of new piano education concepts, the continuous advancement of the piano education industry, the continuous improvement of the system's power, and gradually improves the standardization and specialization of the piano education industry. Liu et al. proposed the design of an intelligent piano performance system based on AI and studied the realization of the piano teaching system. The teaching system from the angle of the simulation of the teachers, based on the piano teaching evaluation system was put forward. For the system as a whole, the function of the piano, including signal extraction, play interface, etc., through the experiment verified the feasibility of the teaching system. The system also can simulate teachers guiding students to play the piano, which is of great significance. By using this kind of software, students can detect the music they play in the process of piano practice at home, and more intuitively, see the problems in their playing mistakes. For some common mistakes, they can solve them directly without bringing them to the classroom. At the same time, for some mistakes, specific and correct playing positions will be provided on these software detection pages, which can help students to quickly find them on the piano, greatly improving the efficiency of piano practice and the quality of the class. At present, intelligent electronic instruments have been favored by consumers in the market, such as the intelligent electronic piano. Compared with the complex and expensive traditional piano, the intelligent electronic piano is cheap and easy to use. At the same time, it is also equipped with a self-study software app, which is more suitable for self-study at home.

The comparison between traditional instruments and intelligent electronic instruments with AI is shown in Table 1.

Table 1. Comparison between traditional instruments and intelligent electronic instruments with AI.

Traditional Instruments	Intelligent Electronic Instrument with AI
Need relatively solid basic skills	Assist the performer to complete the music performance and reduce the difficulty of performance
One person only can play the instrument	One person can play multiple instruments
No such function	Realizes the cooperation between electronic music online education and the wireless network
No such function	Intelligent teaching, intelligent scoring

2.2. Application of Intelligent Music Software

The application of AI music software depends on the output of electronic equipment and whether the processing ability of music data has been restricted by conditions, however, the storage of music information is more stable. Users can edit, adjust, record freely, and process various music elements with AI. With the popularity of music teaching, AI music software provides an interactive platform for teachers and students to share learning resources, where teachers or students can find their own resources to improve. The traditional way of music teaching has undergone a huge change. The knowledge that the teacher teaches in the music teaching class and the content that the student is interested in expanding can be completed by AI music software. Advanced music software includes all kinds of music elements, which broadens students' music vision and deepens their music perception. At the same time as spreading the charm of music elements, it can provide a platform for teachers and students to communicate with each other, or leave feedback or play together, so that music teaching class is no longer limited to the interaction of imparting and absorbing, presenting a positive communication between teachers and students [35–37].

Zhao et al. [38] through the combination of AI and professional platform analysis, evaluated the changes in education and teaching with the coming of the AI era, and put forward alternative topics for music education ability and development. They first discuss the key link between the ability and development of music teachers in primary schools, and demonstrate a sound system and teaching environment of music education in primary schools in the era of AI. Summarizing the experience in practical education provides an important reference for promoting the development of students' personality and ability, and provides a powerful data reference and effective methods for the key abilities and professional development of primary music education. The research will also provide better experimental methods and research models for the key competencies and professional development of teachers in other disciplines.

In addition to normal students, music AI can also effectively support students with learning disabilities to participate in classroom teaching, overcome the defects of traditional education methods, and achieve inclusive teaching. Della et al. [39] explores the impact of using AI in music education on the learning process of students with learning disabilities. Through the auditory and motor systems involved in music, students can become more independent and achieve learning goals in this situation. Students with learning disabilities are not people who don't want to learn or aren't committed enough. Not all students have a disability, but each student has a specific condition that may involve different skills at different levels. Teachers can use different approaches to meet the learning needs of all students and can use compensation tools to support students, including any technology that enhances, maintains, or improves the abilities of students affected by any type of disability (and anyone indirectly affected). Artificial music intelligence systems provide an opportunity that should be more thoroughly integrated into pedagogy, including formal and informal learning environments, teachers and their methods, available resources and activities undertaken by students. Instead of entrusting problem solving to AI-based technology, teachers must monitor dyslexic students throughout the learning process. The development of technology requires teachers to have innovative training that keeps pace with the times and can lead students in the learning process. At present, some AI technologies have begun to be applied to specific disabled groups, bringing benefits to their learning. The School of Special Education of Beijing United University uses the iFLYTEK voice transcription system to teach students with hearing disabilities. A team from the Department of Computer Science at Oxford University has developed a new AI system called LipNet to help people with hearing disabilities read lips.

2.3. Application to Online Teaching, Online Assistance, AI Sparring

As an important driving force of the new round of scientific and industrial revolution, AI is profoundly changing the way people live, work, and learn in education. The application of AI technology in teaching will effectively improve the quality and efficiency of education, from classroom teaching to course guidance, from AI examinations to college entrance planning.

In the traditional teaching method, teachers mainly impart knowledge to students through dictation and PPT, which lacks interest and interaction between teachers and students, and the teaching process is boring. With the arrival of the 5G era, the technological standards have broken down the barriers to acquiring knowledge, and the quality of online education has also been improved, which can meet more personalized education needs [40].

Hua et al. [41] combine the multi-user detection algorithm of artificial intelligence to provide a good online design example for online music education, the conclusion analysis shows that the music online education system based on the SCMA system multiuser detection algorithm and artificial intelligence designed in this paper can significantly improve the audience's music learning efficiency and has obvious benefits to the student group. The system module involves basic information management, student music assignments, online courses, and other levels, providing an excellent educational system design example for music online education. The combination of AI and system has a positive impact on the future sustainable development of online music education. Through the application of online teaching of music, teachers and students get enough learning. With the aid of AI technology, the system analysis and design method are used to analyze and design a functional system of music teaching. As shown in Figure 2, when the user operating system enters data or selects options, it can be submitted by the system to the server. The physical structure of the system is connected to the physical network of the system and associated with the user terminal. Each user can connect to the mobile communication network through a mobile phone, query information through the desktop system or access the background. Students can complete music learning online through mobile phones. The system module involves basic information management, students' music homework, network courses and so on. It provides an excellent example of education system design for music network education.

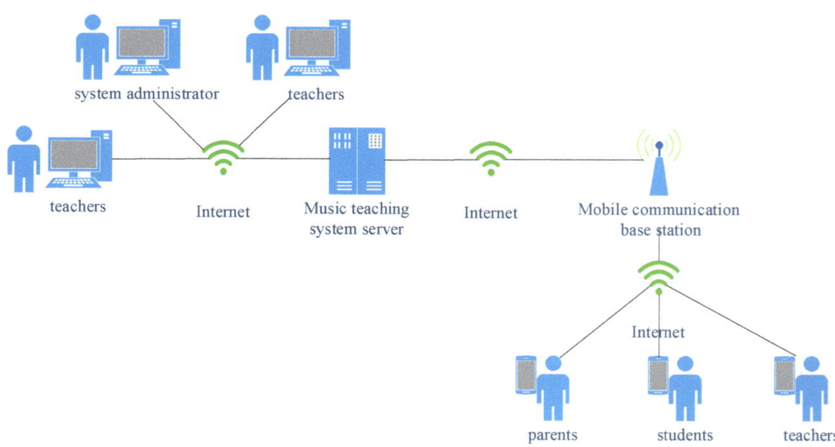

Figure 2. Schematic diagram of physical system structure.

Dai et al. [42] proposed and improved the "7 − 7" teaching mode based on artificial intelligence on the basis of the "4 + 3" mode of traditional classroom, which fully demonstrates the characteristics of the teaching mode under AI. The "4 + 3" model, that is, the four-operation links of teachers (lesson preparation, teaching, assignment, and evaluation) and the three learning links of students (preview, listening, and completing homework). The "7 + 7" model, that is, in smart teaching, teachers' "teaching" has become seven steps (resource release, goal setting, sensory introduction, task distribution, guidance and explanation, detection and evaluation, and extension and push), and students' "learning" has also become seven steps (independent preview, learning expectation, situational experience, cooperative learning, onstage explanation, consolidation of quiz, and breakthrough points), and the interaction between teachers and students is more vivid and rich. The model of "7(medium) + 2(excellent) + 1(low)" is introduced to analyze and judge the learning situation of students in a class in a certain region. As shown in Figure 3, simple classifications combined with big data can help teachers make basic judgments on students, adopt reasonable teaching strategies and carry out classroom teaching design for all students. The teaching mode based on AI is more student-centered and focuses on the interaction between education and learning. It does not consider the single element of education or learning in the teaching process, but the complete cycle mode based on pre-class, in-class and after-class. It uses big data, internet of things, mobile internet, AI and other new generations of information technology to build a set of scientific, intelligent music teaching design models. Wisdom teaching provides reference for the whole process before, during and after class, helps guide teachers to better carry out wisdom teaching, helps students to explore cooperative autonomous learning, and promotes the wisdom transformation of teaching methods and learning methods to a certain extent. Music classroom teaching becomes more targeted and effective.

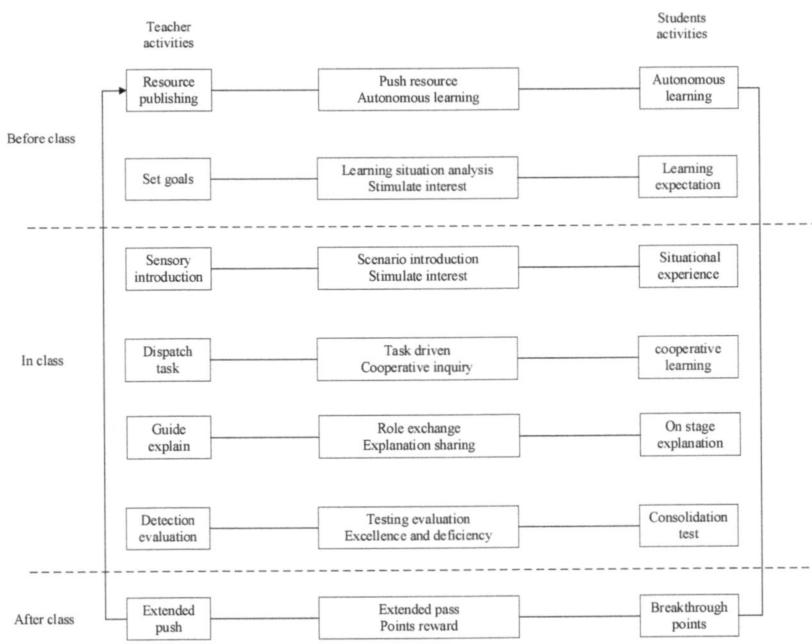

Figure 3. "7 + 7" mode of intelligent music teaching.

At present, some online assisted teaching software has been put on the market. In China, the Little Leaf Piano application can identify and mark the wrong tone and wrong rhythm in real-time during piano practice, through the millisecond level of artificial intelligence piano sound recognition technology and the billion level database, with high recognition accuracy. In India, a platform such as Artium Academy, provides an AI-enabled music learning experience. Learners can track and improve their learning based on AI's immediate feedback and can perform regularly online in the Artium community.

2.4. Application to Autonomous Teaching

For students majoring in music education, autonomous teaching using AI in the classroom is a huge challenge, because music is a field that requires students and teachers to constantly ask questions and discuss innovation. From original music education to innovative music education, teaching methods are combined with information technology to enhance the effect of education. In order for students to have a better educational experience, effective discussion and interaction between teachers and students will facilitate teachers to better understand students and help students make progress. Music teachers can continue to develop effective assessment methods, follow different methods, and evaluate students to continuously improve music knowledge.

Wei et al. [43] proposed a music education method based on AI, they thought that AI can make more optimized environments and professional music classes so that teachers and students can make the most of this and ensure smooth improvement in the network's teaching model. With the development of modern information technology, music education continues to improve. The use of AI in music education has broken the traditional mode of music teaching, greatly improving the level of music teaching and music education teaching mode. The online teaching platform for music majors based on AI technology can provide a more optimized environment and more professional music courses, so that teachers and students can make full use of this and ensure the smooth improvement of the online teaching mode. In addition, the evaluation method is described in detail in the system framework to support the development of music education. By choosing the AI system instead of other machine learning methods, both teachers and students can obtain sufficient benefits and ensure the effective improvement of the network teaching mode. Music teachers are increasingly using technology to develop new student engagement strategies. Acoustic tools and pencils used to be the primary training tools. Now students can use technology like the iPad and educational apps for creative music learning. The proposed music education and teaching based on AI techniques enhanced music education in music education management and proved a deep-level AI implementation could enable management services to be intelligent and informatized.

3. Development Significance and Prospects of AI and Music Education

With the advent of the era of AI, the form of education has improved with the age of network information, the singleness of school music teaching has improved, students' interest in music has been stimulated, learning efficiency has improved, and the music education model in the age of network information has been made more perfect and developed in deeper directions. On the basis of completing basic teaching, AI can also perfect each stage, so that the education concept can delve deeper into society [44–49].

In summary, with the continuous development of AI, AI has been widely popularized and penetrated the field of music education, realizing the integration and interaction between music and modern science and technology, and greatly promoting the development of the music education industry [50–52]. The combination of AI and music education is the general trend. In the future, no matter what kinds of intelligent equipment and virtual technologies, they will be more and more applied in education. The emergence of all kinds of intelligent tools will also promote the improvement of students' learning efficiency and quality. In order to assist teachers to complete the course arrangement more effectively and accurately, the prospect of introducing AI into the classroom is very broad. It can provide teachers with an auxiliary tool and pay more attention to teaching in accordance with their aptitude. Therefore, in the development of music education, we need to uphold innovative ideas, deepen the effective understanding of AI in the music education industry, strengthen the professional application of AI in music education, closely follow the development trend of AI, and promote the long-term healthy and sustainable development of the music education industry [53–55].

Author Contributions: Conceptualization, X.Y. and K.W.; methodology, N.M.; software, N.M.; formal analysis, L.W.; investigation, L.Z.; writing—original draft preparation, X.Y.; writing—review and editing, N.M.; visualization, L.Z.; supervision, K.W.; project administration, L.W.; funding acquisition, K.W. The named authors have substantially contributed to conducting the underlying research and drafting this manuscript. All authors have read and agreed to the published version of the manuscript.

Funding: This work was supported by the Shandong University Youth Innovation Team Development Plan (No. 2021RW012), National Social Science Foundation Art Program (20CD173), the Youth Fund of Shandong Province Natural Science Foundation (No. ZR2020QE212), Key Projects of Shandong Province Natural Science Foundation (No. ZR2020KF020), the Guangdong Provincial Key Lab of Green Chemical Product Technology (No. GC 202111), Zhejiang Province Natural Science Foundation (No. LY22E070007) and National Natural Science Foundation of China (No. 52007170).

Data Availability Statement: The data and materials used to support the findings of this study are available from the corresponding author upon request.

Conflicts of Interest: The authors declare that they have no known competing financial interest or personal relationship that could have appeared to influence the work reported in this paper.

References

1. Santos, O.C. Artificial Intelligence in Psychomotor Learning: Modeling Human Motion from Inertial Sensor Data. *Int. J. Artif. Intell. Tools* **2019**, *28*, 1940006. [CrossRef]
2. Graili, P.; Ieraci, L.; Hosseinkhah, N.; Argent-Katwala, M. Artificial intelligence in outcomes research: A systematic scoping review. *Expert Rev. Pharm. Outcomes Res.* **2021**, *21*, 601–623. [CrossRef] [PubMed]
3. Benetos, E.; Dixon, S.; Duan, Z.; Ewert, S. Automatic Music Transcription An overview. *IEEE Signal Process. Mag.* **2019**, *36*, 20–30. [CrossRef]
4. Byrd, D. Music notation software and intelligence. *Comput. Music J.* **1994**, *18*, 17–20. [CrossRef]
5. Chen, X. Research and Application of Interactive Teaching Music Intelligent System Based on Artificial Intelligence. In Proceedings of the International Conference on Artificial Intelligence, Virtual Reality, and Visualization (AIVRV), Sanya, China, 19–21 November 2021.
6. Hsieh, Y.-Z.; Lin, S.-S.; Luo, Y.-C.; Jeng, Y.-L.; Tan, S.-W.; Chen, C.-R.; Chiang, P.-Y. ARCS-Assisted Teaching Robots Based on Anticipatory Computing and Emotional Big Data for Improving Sustainable Learning Efficiency and Motivation. *Sustainability* **2020**, *12*, 5605. [CrossRef]
7. Fang, G.; Chan, P.W.K.; Kalogeropoulos, P. Secondary School Teachers' Professional Development in Australia and Shanghai: Needs, Support, and Barriers. *Sage Open* **2021**, *11*, 21582440211026951. [CrossRef]
8. Su, W.; Tai, K.h. Case Analysis and Characteristics of Popular Music Creative Activities Using Artificial Intelligence. *J. Humanit. Soc. Sci.* **2022**, *13*, 1937–1948.
9. SungHoon, L. Artificial Intelligence Applications to Music Composition. *J. Converg. Cult. Technol.* **2018**, *4*, 261–266. [CrossRef]
10. Tai, K.h.; Kim, S.y. Artificial intelligence(AI) Composition Technology Trends & Creation Platform. *Cult. Converg.* **2022**, *44*, 207–228.
11. Park, D. A Study on the production of Music Content Using Artificial Intelligence Composition Program. *Trans* **2022**, *13*, 35–58.
12. Park, J.-R. A Study on Technology and Artificial Intelligence Applied to Music Production. *J. Music Theory* **2019**, *33*, 108–143. [CrossRef]
13. Shin, W.; Cheol, K.M. Music artificial intelligence: A Case of Google Magenta. *J. Tour. Ind. Res.* **2020**, *40*, 21–28. [CrossRef]
14. Chen, J.; Ramanathan, L.; Alazab, M. Holistic big data integrated artificial intelligent modeling to improve privacy and security in data management of smart cities. *Microprocess. Microsyst.* **2021**, *81*, 103722. [CrossRef]
15. Lee, J.; Nazki, H.; Baek, J.; Hong, Y.; Lee, M. Artificial Intelligence Approach for Tomato Detection and Mass Estimation in Precision Agriculture. *Sustainability* **2020**, *12*, 9138. [CrossRef]
16. Kladder, J. Digital audio technology in music teaching and learning: A preliminary investigation. *J. Music Technol. Educ.* **2021**, *13*, 219–237. [CrossRef]
17. Zhang, Y.; Yi, D. A New Music Teaching Mode Based on Computer Automatic Matching Technology. *Int. J. Emerg. Technol. Learn.* **2021**, *16*, 117–130. [CrossRef]
18. Zhao, Y. Analysis of Music Teaching in Basic Education Integrating Scientific Computing Visualization and Computer Music Technology. *Math. Probl. Eng.* **2022**, *2022*, 3928889. [CrossRef]
19. Chu, H.; Moon, S.; Park, J.; Bak, S.; Ko, Y.; Youn, B.-Y. The Use of Artificial Intelligence in Complementary and Alternative Medicine: A Systematic Scoping Review. *Front. Pharmacol.* **2022**, *13*, 826044. [CrossRef] [PubMed]
20. Wang, X. Design of Vocal Music Teaching System Platform for Music Majors Based on Artificial Intelligence. *Wirel. Commun. Mob. Comput.* **2022**, *2022*, 5503834. [CrossRef]

21. Yang, T.; Nazir, S. A comprehensive overview of AI-enabled music classification and its influence in games. *Soft Comput.* **2022**, *26*, 7679–7693. [CrossRef]
22. Ma, M.; Sun, S.; Gao, Y. Data-Driven Computer Choreography Based on Kinect and 3D Technology. *Sci. Program.* **2022**, *2022*, 2352024. [CrossRef]
23. Xiang, Y.; Natgunanathan, I.; Rong, Y.; Guo, S. Spread Spectrum-Based High Embedding Capacity Watermarking Method for Audio Signals. *IEEE-ACM Trans. Audio Speech Lang. Process.* **2015**, *23*, 2228–2237. [CrossRef]
24. Moon, H.; Yunhee, S. A Study on the Understanding of Artificial Intelligence (AI) and the Examples and Applications of AI-based Music Tools. *J. Learn.-Cent. Curric. Instr.* **2022**, *22*, 341–358. [CrossRef]
25. Nicholls, S.; Cunningham, S.; Picking, R.; ACM. Collaborative Artificial Intelligence in Music Production. In Proceedings of the Conference on Interation with Sound (Audio Mostly): Sound in Immersion and Emotion (AM), Wrexham, UK, 12–14 September 2018.
26. Park, B. Analysis of Research Trends Related to Artificial Intelligence in Korean Music Field. *J. Next-Gener. Converg. Technol. Assoc.* **2022**, *6*, 570–578. [CrossRef]
27. Zhang, J.; Wan, J. A summary of the application of artificial intelligence in music education. In Proceedings of the International Conference on Education, Economics and Information Management (ICEEIM 2019), Wuhan, China, 21–22 December 2019; Atlantis Press: Zhengzhou, China, 2020; pp. 42–44.
28. Wei, J.; Marimuthu, K.; Prathik, A. College music education and teaching based on AI techniques. *Comput. Electr. Eng.* **2022**, *100*, 107851. [CrossRef]
29. Yan, H. Design of Online Music Education System Based on Artificial Intelligence and Multiuser Detection Algorithm. *Comput. Intell. Neurosci.* **2022**, *2022*, 9083436. [CrossRef]
30. Yang, Y. Piano Performance and Music Automatic Notation Algorithm Teaching System Based on Artificial Intelligence. *Mob. Inf. Syst.* **2021**, *2021*, 3552822. [CrossRef]
31. Yoo, H.-J. A Case Study on Artificial Intelligence's Music Creation: Focusing on. *J. Next-Gener. Converg. Technol. Assoc.* **2022**, *6*, 1737–1745. [CrossRef]
32. YoungGun, K. Study on Artificial Intelligence Technology Used in Popular Music Harmony Arrangement. *Korean J. Pop. Music* **2021**, *27*, 9–47.
33. Xu, N.; Zhao, Y. Online Education and Wireless Network Coordination of Electronic Music Creation and Performance under Artificial Intelligence. *Wirel. Commun. Mob. Comput.* **2021**, *2021*, 1–9. [CrossRef]
34. Guo, Y.; Yu, P.; Zhu, C.; Zhao, K.; Wang, L.; Wang, K. A state-of-health estimation method considering capacity recovery of lithium batteries. *Int. J. Energy Res.* **2022**, *46*, 23730–23745. [CrossRef]
35. YoungGun, K. Study on Music Arrangement Education Content Development Using Artificial Intelligence. *Cult. Converg.* **2021**, *43*, 275–296.
36. Yu, Z.; IEEE. Selection Method Of Linear Thinking Path Of Chinese Piano Music Based On Artificial Intelligence. In Proceedings of the 5th International Conference on Smart Grid and Electrical Automation (ICSGEA), Zhangjiajie, China, 13–14 June 2020; pp. 327–333.
37. Zeng, Y.-f.; Gao, J.-h. Application of Artificial Intelligence in Digital Music. In Proceedings of the International Conference on Applied Mechanics and Mechatronics Engineering (AMME), Bangkok, Thailand, 25–26 October 2015; pp. 479–481.
38. Zhao, X.; Guo, Z.; Liu, S.; Gupta, P. Exploring Key Competencies and Professional Development of Music Teachers in Primary Schools in the Era of Artificial Intelligence. *Sci. Program.* **2021**, *2021*, 5097003. [CrossRef]
39. Della Ventura, M. Exploring the Impact of Artificial Intelligence in Music Education to Enhance the Dyslexic Student's Skills. In *Learning Technology for Education Challenges: 8th International Workshop, LTEC 2019, Zamora, Spain, 15–18 July 2019, Proceedings 8*; Springer International Publishing: Berlin/Heidelberg, Germany, 2019; pp. 14–22. [CrossRef]
40. Jiang, Q. Application of Artificial Intelligence Technology in Music Education Supported by Wireless Network. *Math. Probl. Eng.* **2022**, *2022*, 2138059. [CrossRef]
41. Li, D.; Yang, D.; Li, L.; Wang, L.; Wang, K. Electrochemical Impedance Spectroscopy Based on the State of Health Estimation for Lithium-Ion Batteries. *Energies* **2022**, *15*, 6665. [CrossRef]
42. Dai, D.D.; Ding, B. Artificial Intelligence Technology Assisted Music Teaching Design. *Sci. Program.* **2021**, *2021*, 9141339. [CrossRef]
43. Sun, H.; Yang, D.; Wang, L.; Wang, K. A method for estimating the aging state of lithium-ion batteries based on a multi-linear integrated model. *Int. J. Energy Res.* **2022**, *46*, 24091–24104. [CrossRef]
44. Venugopal, K.; Madhusudan, P. Feasibility of Music Composition using Artificial Neural Networks. In Proceedings of the International Conference on Computing Methodologies and Communication (ICCMC), Surya Engn Coll, Erode, India, 18–19 July 2017; pp. 524–525.
45. Zheng, H.; Dai, D. Construction and Optimization of Artificial Intelligence-Assisted Interactive College Music Performance Teaching System. *Sci. Program.* **2022**, *2022*, 3199860. [CrossRef]
46. Zhang, M.; Wang, W.; Xia, G.; Wang, L.; Wang, K. Self-Powered Electronic Skin for Remote Human–Machine Synchronization. *ACS Appl. Electron. Mater.* **2023**, *5*, 498–508. [CrossRef]
47. Wang, W.; Yang, D.; Yan, X.; Wang, L.; Hu, H.; Wang, K. Triboelectric nanogenerators: The beginning of blue dream. *Front. Chem. Sci. Eng* **2023**. [CrossRef]

48. Wang, W.; Yang, D.; Huang, Z.; Hu, H.; Wang, L.; Wang, K. Electrodeless Nanogenerator for Dust Recover. *Energy Technol.* **2022**, *10*. [CrossRef]
49. Wang, W.; Pang, J.; Su, J.; Li, F.; Li, Q.; Wang, X.; Wang, J.; Ibarlucea, B.; Liu, X.; Li, Y. Applications of nanogenerators for biomedical engineering and healthcare systems. *InfoMat* **2022**, *4*, e12262. [CrossRef]
50. Guo, Y.; Yang, D.; Zhang, Y.; Wang, L.; Wang, K. Online estimation of SOH for lithium-ion battery based on SSA-Elman neural network. *Prot. Control Mod. Power Syst.* **2022**, *7*, 40. [CrossRef]
51. Cui, Z.; Kang, L.; Li, L.; Wang, L.; Wang, K. A combined state-of-charge estimation method for lithium-ion battery using an improved BGRU network and UKF. *Energy* **2022**, *259*, 124933. [CrossRef]
52. Zhang, M.; Wang, K.; Zhou, Y.-t. Online state of charge estimation of lithium-ion cells using particle filter-based hybrid filtering approach. *Complexity* **2020**, *2020*, 8231243. [CrossRef]
53. Zhang, M.; Liu, Y.; Li, D.; Cui, X.; Wang, L.; Li, L.; Wang, K. Electrochemical Impedance Spectroscopy: A New Chapter in the Fast and Accurate Estimation of the State of Health for Lithium-Ion Batteries. *Energies* **2023**, *16*, 1599. [CrossRef]
54. Wang, L.; Xie, L.; Yang, Y.; Zhang, Y.; Wang, K.; Cheng, S.-j. Distributed Online Voltage Control with Fast PV Power Fluctuations and Imperfect Communication. *IEEE Trans. Smart Grid* **2023**. [CrossRef]
55. Ma, N.; Yang, D.; Riaz, S.; Wang, L.; Wang, K. Aging Mechanism and Models of Supercapacitors: A Review. *Technologies* **2023**, *11*, 38. [CrossRef]

MDPI AG
Grosspeteranlage 5
4052 Basel
Switzerland
Tel.: +41 61 683 77 34

Technologies Editorial Office
E-mail: technologies@mdpi.com
www.mdpi.com/journal/technologies

www.ingramcontent.com/pod-product-compliance
Lightning Source LLC
Chambersburg PA
CBHW040853230926
51850CB00014BA/3425
* 9 7 8 3 7 2 5 8 3 0 8 4 8 *